Keys to Neotropical Hexapoda

Thorp and Covich's Freshwater Invertebrates

A Global Series of Books on the Identification,
Ecology, and General Biology of Inland Water Invertebrates
by Experts from Around the World

Fourth Edition
Series Editor: James H. Thorp

Volume I: Ecology and General Biology
Edited by James H. Thorp and D. Christopher Rogers
Published 2015

Volume II: Keys to Nearctic Fauna
Edited by James H. Thorp and D. Christopher Rogers
Published 2016

Volume III: Keys to Neotropical Hexapoda
Edited by Neusa Hamada, James H. Thorp, and D. Christopher Rogers
Published 2018

Volume IV: Keys to Palaearctic Fauna
Edited by D. Christopher Rogers and James H. Thorp
In Press: 2019

Volumes in Preparation and Under Contract
Keys to Neotropical and Antarctic Fauna
Keys to Australasian Fauna
Keys to Afrotropical Fauna

Possible Future Volumes of the Fourth Edition
Keys to Oriental and Oceanic Hexapoda
Keys to Oriental and Oceanic Fauna
Keys to Palaearctic Hexapoda
Keys to Nearctic Hexapoda

Related Publications
Field Guide to Freshwater Invertebrates of North America
by J.H. Thorp and D.C. Rogers

Keys to Neotropical Hexapoda

Thorp and Covich's Freshwater Invertebrates - Volume III

Fourth Edition

Edited by

Neusa Hamada
Coordenação de Biodiversidade, Instituto Nacional de Pesquisas da Amazônia
Manaus, Amazonas, Brazil

James H. Thorp
Kansas Biological Survey and Department of Ecology and Evolutionary Biology
University of Kansas, Lawrence, KS, USA

D. Christopher Rogers
Kansas Biological Survey and The Biodiversity Institute
University of Kansas, Lawrence, KS, USA

ELSEVIER

ACADEMIC PRESS
An imprint of Elsevier

Academic Press is an imprint of Elsevier
125 London Wall, London EC2Y 5AS, United Kingdom
525 B Street, Suite 1650, San Diego, CA 92101, United States
50 Hampshire Street, 5th Floor, Cambridge, MA 02139, United States
The Boulevard, Langford Lane, Kidlington, Oxford OX5 1GB, United Kingdom

Library of Congress Cataloging-in-Publication Data
A catalog record for this book is available from the Library of Congress

British Library Cataloguing-in-Publication Data
A catalogue record for this book is available from the British Library

ISBN: 978-0-12-804223-6

Last digit is the print number: 9 8 7 6 5 4 3

For information on all Academic Press publications visit our website at https://www.elsevier.com/books-and-journals

Cover Photos' credits

Odonata Picture
Fernando da Costa Pinheiro
Conselho Nacional de Desenvolvimento Científico e Tecnológico (CNPq), Brazil

Trichoptera Picture
Ana Maria Oliveira Pes
Instituto Nacional de Pesquisas da Amazônia (INPA), Brazil

Collembola Picture
Diego D. da Silva
Instituto Nacional de Pesquisas da Amazônia, Manaus (INPA), Brazil

Publisher: Candice Janco
Acquisitions Editor: Louisa Hutchins
Editorial Project Manager: Emily Thomson
Production Project Manager: Paul Prasad Chandramohan
Cover Designers: Greg Harris, with contributions from the academic editors

Typeset by TNQ Technologies

Dedications from the Editors

To all the authors for agreeing to face this journey,
making the construction of this book possible
Neusa Hamada

To where the wild things are and the children,
who grow up to study and protect them
James H. Thorp

To our authors who made this book possible
D. Christopher Rogers

Contents

Section II.3
Order Coleoptera

Section II.4
Order Diptera

Contributors to Volume III

Rosaly Ale-Rocha [Chapters 16.13, 16.14] Coordenação de Biodiversidade, Instituto Nacional de Pesquisas da Amazônia, Manaus, Amazonas, Brazil; email: alerocha@inpa.gov.br

André Pereira Amaral [Chapters 16.7, 16.8] Departamento de Ecologia e Zoologia, Universidade Federal de Santa Catarina, Florianópolis, Santa Catarina, Brazil; email: depereira@gmail.com

Trond Andersen [Chapter 16.2] Department of Natural History, University Museum of Bergen, Bergen, Norway; email: trond.andersen@uib.no

Danielle Anjos-Santos [Chapters 14.4, 14.6] Centro de Investigaciones Esquel de Montaña y Estepa Patagónicas (CIEMEP), Esquel, Chubut, Argentina; email: danielleanjos2@yahoo.com.br

Miguel Archangelsky [Chapters 15, 15.4] Laboratorio de Investigaciones en Ecología y Sistemática Animal (LIESA), Centro de Investigación Esquel de Montaña y Estepa Patagónica (CIEMEP, CONICET-UNPSJB), Esquel, Chubut, Argentina; email: hydrophilinae@gmail.com

Adrian Ardila-Camacho [Chapters 8, 9] Laboratorio de Entomología, Centro de Investigaciones en Microbiología y Parasitología Tropical – CIMPAT, Universidad de los Andes, Bogotá, Colombia; email: aardilac88@gmail.com

Fernanda Avelino-Capistrano [Chapter 4] Faculdades São José, and Laboratório de Entomologia, UFRJ, Rio de Janeiro, Rio de Janeiro, Brazil; email fernandaacsilva@yahoo.com.br

Leandro Silva Barbosa [Chapter 4] Departamento de Entomologia, Museu Nacional, UFRJ, Rio de Janeiro, Brazil; email: leanbarbosa@gmail.com

Patrik Barcelos-Silva [Chapter 10] Divisão do Curso de Pós-graduação em Entomologia, Instituto Nacional de Pesquisas da Amazônia, Manaus, Amazonas, Brazil; email: patrikbarcelos@gmail.com

Bruno C. Bellini [Chapter 2] Universidade Federal do Rio Grande do Norte, Natal, Rio Grande do Norte, Brazil; email: entobellini@gmail.com

Cesar J. Benetti [Chapters 15, 15.2, 15.3] Coordenação de Biodiversidade, Instituto Nacional de Pesquisas da Amazônia, Manaus, Amazonas, Brazil; email: cjbenetti@gmail.com

Rafael Boldrini [Chapter 3] Centro de Estudos da Biodiversidade, Universidade Federal de Roraima, Boa Vista, Roraima, Brazil; email: rafael.boldrini@ufrr.br

Carlos Elias de Souza Braga [Chapter 5] Department of Natural Sciences, State University of Pará, Belém, Pará, Brazil; email: bragaelias@yahoo.com.br

Rafael Benzi Braga [Chapter 15.3] Laboratório de Entomologia, Instituto de Biologia, Universidade Federal do Rio de Janeiro, Rio de Janeiro, Rio de Janeiro, Brazil; email: rafaelbenzi@gmail.com

Lucas M. Camargos [Chapter 10] Department of Entomology, University of Minnesota, St. Paul, MN, United States; email: lmcamargosbio@gmail.com

Alcimar do Lago Carvalho [Chapter 14.1] Departamento de Entomologia, Museu Nacional, UFRJ, Rio de Janeiro, Rio de Janeiro, Brazil; email: alagoc@acd.ufrj.br

Jeane Marcelle Cavalcante do Nascimento [Chapters 16.3, 16.5] Coordenação de Biodiversidade, Curso de Pós-Graduação em Entomologia, Instituto Nacional de Pesquisas da Amazônia, Manaus, Amazonas, Brazil; email: jeanemarcelle@gmail.com

Nikolas G. Cipola [Chapter 2] Coordenação de Biodiversidade, Curso de Pós-Graduação em Entomologia, Instituto Nacional de Pesquisas da Amazônia, Manaus, Amazonas, Brazil; email: nikolasgc@gmail.com

Bruno Clarkson [Chapter 15.4] Laboratório de Sistemática e Bioecologia de Coleoptera, Departamento de Zoologia, Universidade Federal do Paraná; Curitiba, Paraná, Brazil; email: brclarkson@gmail.com

Daniara Colpani [Chapter 15.2] Coordenação de Biodiversidade, Curso de Pós-Graduação em Entomologia, Instituto Nacional de Pesquisas da Amazônia, Manaus, Amazonas, Brazil; email: daniara.entomo@gmail.com

Atilano Contreras-Ramos [Chapter 8] Departamento de Zoología, Instituto de Biología, Universidad Nacional Autónoma de México, Ciudad Universitaria, Ciudad de México, México; email: acontreras@ib.unam.mx

Danilo P. Cordeiro [Chapter 16.10] Coordenação de Biodiversidade, Instituto Nacional de Pesquisas da Amazônia, Manaus, Amazonas, Brazil; email: d.pacheco.c@gmail.com

Caio C.D. Corrêa [Chapter 16.6] Departamento de Entomologia, Museu Nacional, Universidade Federal do Rio de Janeiro, Rio de Janeiro, Rio de Janeiro, Brazil; email: caio.entomologia@gmail.com

Galileu P.S. Dantas [Chapters 16, 16.2] Coordenação de Biodiversidade, Instituto Nacional de Pesquisas da Amazônia, Manaus, Amazonas, Brazil; email: galileu.psd@gmail.com

Diego D. da Silva [Chapter 2] Coordenação de Biodiversidade, Curso de Pós-Graduação em Entomologia, Instituto Nacional de Pesquisas da Amazônia, Manaus, Amazonas, Brazil; email: diegocollembola@gmail.com

Fabio Siqueira Pitaluga de Godoi [Chapter 16.11] Departamento de Biologia, Universidade Federal do Amazonas, Manaus, Amazonas, Brazil; Email: fgodoi@gmail.com

Albert Deler-Hernández [Chapter 15.5] Department of Zoology, Faculty of Science, Charles University, Prague, Czech Republic; email: adeler1982@gmail.com

Juan A. Delgado [Chapter 15.5] Departamento de Zoología y Antropología Física, Facultad de Biología, Universidad de Murcia, Murcia, Spain; email: jdelgado@um.es

Lucimar Gomes Dias [Chapter 3] Departamento de Ciencias Biológicas, Facultad de Ciencias Exactas y Naturales, Grupo de Investigación Bionat, Universidad de Caldas, Manizales, Caldas, Colômbia; email: lucimar.dias@ucaldas.edu.co

Florentina Díaz [Chapter 16.1] Centro de Estudios Parasitológicos y de Vectores (CEPAVE), Facultad de Ciencias Naturales y Museo (UNLP), Consejo Nacional de Investigaciones Científicas y Técnicas (CONICET), Buenos Aires, Argentina, UNLP, FCNYM, División Entomología, Museo de La Plata, Buenos Aires, Argentina; email: mfdiaz@fcnym.unlp.edu.ar

Eduardo Domínguez [Chapter 3] Instituto de Biodiversidad Neotropical (IBN). CONICET - U.N.T., & Facultad de Ciencias Naturales, Universidad Nacional de Tucumán, San Miguel de Tucumán, Argentina; email: eduardo.mayfly@gmail.com

Daniell R.R. Fernandes [Chapter 12] Coordenação de Biodiversidade, Instituto Nacional de Pesquisas da Amazônia, Manaus, Amazonas, Brazil; email: daniellrodrigo@hotmail.com

Ruth Leila Ferreira-Keppler [Chapters 16.1, 16.5, 16.13, 16.14] Coordenação de Biodiversidade, Instituto Nacional de Pesquisas da Amazônia, Manaus, Amazonas, Brazil; email: ruth@inpa.gov.br

Nelson Ferreira Jr. [Chapter 15.3] Laboratório de Entomologia, Instituto de Biologia, Universidade Federal do Rio de Janeiro, Rio de Janeiro, Rio de Janeiro, Brazil; email: nferrejr@gmail.com

Gunther Fleck [Chapter 14.3] Ver à soie Musée-Magnanerie, Lagorce, Ardèche, France; email: fleckgunther@gmail.com

Lívia Maria Fusari [Chapter 16] Departamento de Hidrobiologia, Universidade Federal de São Carlos, São Carlos, São Paulo, Brazil; email: liviafusari@gmail.com

Josefina Garrido [Chapter 15.5] Departamento de Ecología y Biología Animal, Facultad de Biología, Universidad de Vigo, Vigo, Spain; email: jgarrido@uvigo.es

Leonardo H. Gil-Azevedo [Chapters 16.6, 16.9] Departamento de Entomologia, Museu Nacional, Universidade Federal do Rio de Janeiro, Rio de Janeiro, Rio de Janeiro, Brazil; email: lhgazevedo@mn.ufrj.br

Enrique González-Soriano [Chapter 14.2] Instituto de Biología, Departamento de Zoología, Universidad Nacional Autónoma de México, Ciudad de México, México; email: esoriano@ib.unam.mx

Márlon Breno Graça [Chapter 11] Coordenação de Biodiversidade, Curso de Pós-Graduação em Entomologia, Instituto Nacional de Pesquisas da Amazônia, Manaus, Amazonas, Brazil; email: marlon_lgp@hotmail.com

Bruno A.C. Guimarães [Chapter 15.2] Laboratório de Entomologia, Departamento de Zoologia, Instituto de Biologia, Universidade Federal do Rio de Janeiro, Rio de Janeiro, Rio de Janeiro, Brazil; email: brunoaguilar0044@gmail.com

Neusa Hamada [Chapters 1, 12, 16.3, 16.5] Coordenação de Biodiversidade, Instituto Nacional de Pesquisas da Amazônia, Manaus, Amazonas, Brazil; email: nhamada@inpa.gov.br

Ralph W. Holzenthal [Chapter 10] Department of Entomology, University of Minnesota, St. Paul, MN, United States; email: holze001@umn.edu

Heron Huerta [Chapters 16.4, 16.7] Instituto de Diagnóstico y Referencia Epidemiológicos, Laboratorio de Entomología, Ciudad de México, México; email: cerato_2000@yahoo.com

Maria Inês S Passos [Chapter 15.6] Laboratório de Insetos Aquáticos, Departamento de Zoologia/ Instituto de Biociências, Universidade Federal do Estado do Rio de Janeiro (Unirio), Rio de Janeiro, Rio de Janeiro, Brazil; email: minespassos@gmail.com

Fabio Laurindo da Silva [Chapter 16.2] Department of Zoology, Institute of Biosciences, University of São Paulo, São Paulo, São Paulo, Brazil; email: fabiologia@gmail.com

María L. Libonatti [Chapter 15.7] Laboratory of Entomology, IBBEA, CONICET-UBA, DBBE-FCEN, University of Buenos Aires, Buenos Aires, Argentina; email: libonatti.marialaura@gmail.com

Sonia Maria Lopes [Chapter 6] Laboratório de Biodiversidade e Sistemática de Blattaria, Departamento de Entomologia, Museu Nacional, Universidade Federal do Rio de Janeiro, Rio de Janeiro, Rio de Janeiro, Brazil; email: sonialfraga@gmail.com

Federico Lozano [Chapter 14.6] Laboratorio de Biodiversidad y Genética Ambiental (BioGeA), Universidad Nacional de Avellaneda (UNDAV), Buenos Aires, Buenos Aires, Argentina; email: lozano@undav.edu.ar

Renato Jose Pires Machado [Chapter 13] Instituto de Biociências, Universidade Federal do Mato Grosso, Cuiabá, Mato Grosso, Brazil; email: rjpmachado@gmail.com

Oséias Martins Magalhães [Chapter 7] Laboratório de Biodiversidade Entomológica, Fundação Oswaldo Cruz, Rio de Janeiro, Rio de Janeiro, Brazil; email: oseiasmm@bol.com.br

Crystal A. Maier [Chapter 15.6] Gantz Family Collection Center, Field Museum of Natural History, Chicago, IL, United States; email: cmaier@fieldmuseum.org

Veronica Manzo [Chapter 15.6] Instituto de Biodiversidad Neotropical (IBN), CONICET - UNT. Horco Molle, Yerba Buena, Tucumán, Argentina; email: vmanzo@csnat.unt.edu.ar

Pablo Ignacio Marino [Chapter 16.1] UNLP, FCNYM, División Entomología, Museo de La Plata, La Plata, Buenos Aires, Argentina; email: pmarino@fcnym.unlp.edu.ar

Caleb C. Martins [Chapter 9] Laboratório de Morfologia e Evolução de Diptera, Departamento de Biologia, FFCLRP, Universidade de São Paulo, Ribeirão Preto, São Paulo, Brazil; email: calebcalifre@gmail.com

Yoandri S. Megna [Chapter 15.3] Departamento de Biología, Facultad de Ciencias Naturales y Exactas, Universidad de Oriente, Santiago de Cuba, Cuba; email: ysuarez1976@yahoo.es

Mariano C. Michat [Chapters 15, 15.3] Laboratory of Entomology, IBBEA, CONICET-UBA., DBBE-FCEN, University of Buenos Aires, Buenos Aires, Argentina; email: marianoide@gmail.com

Gil Felipe Gonçalves Miranda [Chapter 16.12] Gerência da Região Administrativa Calha Norte III, Diretoria de Gestão e Monitoramento de Unidades de Conservação, Instituto de Desenvolvimento Florestal e da Biodiversidade, Monte Alegre, Pará, Brazil; email: gilfgm@gmail.com

Carlos Molineri [Chapter 3] Instituto de Biodiversidad Neotropical (IBN). CONICET - U.N.T., & Facultad de Ciencias Naturales, Universidad Nacional de Tucumán, San Miguel de Tucumán, Tucumán, Argentina; email: carlosmolineri@gmail.com

Felipe Ferraz Figueiredo Moreira [Chapter 7] Laboratório de Biodiversidade Entomológica, Fundação Oswaldo Cruz, Rio de Janeiro, Rio de Janeiro, Brazil; email: felipe.moreira@ioc.fiocruz.br

Javier Muzón [Chapter 14, 14.5, 14.6] Laboratorio de Biodiversidad y Genética Ambiental (BioGeA), Universidad Nacional de Avellaneda (UNDAV), Buenos Aires, Buenos Aires, Argentina; email: jmuzon@gmail.com

Ulisses Gaspar Neiss [Chapters 14, 14.3, 14.4] Instituto de Criminalística, Departamento de Polícia Técnica-Científica, Manaus, Amazonas, Brazil; email: ulisses.neiss@gmail.com

Carolina Nieto [Chapter 3] Instituto de Biodiversidad Neotropical (IBN). CONICET - U.N.T., & Facultad de Ciencias Naturales, Universidad Nacional de Tucumán, San Miguel de Tucumán, Tucumán, Argentina; email: macarolina_nieto@yahoo.com.ar

Rodolfo Novelo-Gutiérrez [Chapter 14.2] Instituto de Ecología, A. C., Xalapa, Veracruz, México; email: rodolfo.novelo@inecol.mx

Ana Lúcia Nunes Gutjahr [Chapter 5] Department of Natural Sciences, State University of Pará, Belém, Pará, Brazil; email: melcam@uol.com.br

Ana Maria Pes [Chapter 10] Coordenação de Biodiversidade, Instituto Nacional de Pesquisas da Amazônia, Manaus, Amazonas, Brazil; email: anampes@gmail.com

Pablo Pessacq [Chapters 4, 14, 14.1, 14.3, 14.4, 14.6] Centro de Investigaciones Esquel de Montaña y Estepa Patagónicas (CIEMEP), Esquel, Chubut, Argentina; email: pablopessacq@yahoo.com.ar

Luiz Carlos Pinho [Chapters 16, 16.2, 16.4, 16.7, 16.8] Departamento de Ecologia e Zoologia, Universidade Federal de Santa Catarina, Florianópolis, Santa Catarina, Brazil; email: luiz.pinho@ufsc.br

José Roberto Pujol-Luz [Chapter 16.11] Departamento de Zoologia, Universidade de Brasília, Brasília, Distrito Federal, Brazil; email: jrpujolluz@gmail.com

Ranyse B. Querino [Chapter 12] Empresa Brasileira de Pesquisa Agropecuária, Embrapa Meio-Norte, Teresina, Piauí, Brazil; email: ranyse.silva@embrapa.br

Alonso Ramírez [Chapter 14.2] Department of Environmental Sciences, University of Puerto Rico, San Juan, Puerto Rico; email: aramirez@ramirezlab.net

Higor Daniel Duarte Rodrigues [Chapter 7] Museu de Zoologia, Universidade de São Paulo, São Paulo, São Paulo, Brazil; email: higorddr@gmail.com

D. Christopher Rogers [Chapter 1] Kansas Biological Survey and The Biodiversity Institute, University of Kansas, Lawrence, KS, United States; email: branchiopod@gmail.com

Maria Marcela Ronderos [Chapter 16.1] Centro de Estudios Parasitológicos y de Vectores (CEPAVE), Facultad de Ciencias Naturales y Museo (UNLP), Consejo Nacional de Investigaciones Científicas y Técnicas (CONICET), Buenos Aires, Argentina, UNLP, FCNYM, División Entomología, Museo de La Plata, La Plata, Buenos Aires, Argentina; email: ronderos@fcnym.unlp.edu.ar

Graham Rotheray [Chapter 16.12] Department of Natural Sciences, National Museums, Scotland, Edinburgh, Great Britain; email: grahamrotheray@gmail.com

Rafał Ruta [Chapter 15.7] Department of Biodiversity and Evolutionary Taxonomy, Faculty of Biological Sciences, Wroclaw University, Poland; email: rafal.ruta@uwr.edu.pl

Frederico Falcão Salles [Chapter 3] Departamento de Ciências Agrárias e Biológicas, Universidade Federal do Espírito Santo, São Mateus, Espírito Santo, Brazil; email: ffsalles@gmail.com

Brunno H.L. Sampaio [Chapter 15.1] Laboratório de Entomologia, Departamento de Zoologia, Instituto de Biologia, Universidade Federal do Rio de Janeiro, Rio de Janeiro, Rio de Janeiro, Brazil; email: brunnosampaio@ufrj.br

Allan P.M. Santos [Chapter 10] Departamento de Zoologia, Universidade Federal do Estado do Rio de Janeiro – UNIRIO, Rio de Janeiro, Rio de Janeiro, Brazil; email: allanpms@gmail.com

Julieta V. Sganga [Chapter 10] Departamento de Biodiversidad y Biología Experimental, Facultad de Ciencias Exactas y Naturales, Universidad de Buenos Aires, Buenos Aires, Buenos Aires, Argentina; email: jsganga@gmail.com

Andrew E.Z. Short [Chapters 15.1, 15.4] Department of Ecology & Evolutionary Biology, Division of Entomology, Biology Institute, University of Kansas, Lawrence, KS, United States; email: aezshort@ku.edu

Isabelle da Rocha Silva Cordeiro [Chapter 7] Laboratório de Biodiversidade Entomológica, Fundação Oswaldo Cruz, Rio de Janeiro, Rio de Janeiro, Brazil; email: rocha.belle@gmail.com

Augusto Siri [Chapter 16.2] Instituto de Limnología 'Dr. Raúl A. Ringuelet', (ILPLA, CONICET, UNLP), La Plata, Buenos Aires, Argentina; email: augustosiri@yahoo.com.ar

Robert W. Sites [Chapter 7] Enns Entomology Museum, Division of Plant Sciences, University of Missouri, Columbia, MO, United States; email: SitesR@missouri.edu

M. Alma Solis [Chapter 11] MAS, Systematic Entomology Laboratory, USDA, National Museum of Natural History, Washington, DC, United States; email: alma.solis@ars.usda.gov

Kenneth J. Tennessen [Chapter 14.3] Florida State Collection of Arthropods, Gainesville, Florida, United States; email: ktennessen@centurytel.net

James H. Thorp [Chapter 1] Kansas Biological Survey and Department of Ecology and Evolutionary Biology, University of Kansas, Lawrence, KS, United States; email: thorp@ku.edu

Mario Toledo [Chapters 15.2, 15.3] Museo di Storia Naturale, Parma, Italy; email: toledo.pinguicula.mario3@gmail.com

Patrícia L.M. Torres [Chapter 15.4] IBBEA, CONICET-UBA. Laboratory of Entomology, DBBE-FCEN, University of Buenos Aires, Buenos Aires, Buenos Aires, Argentina; email: patricialmtorres@yahoo.com.ar

Susana Trivinho-Strixino [Chapter 16.2] Department of Hydrobiology, Federal University of São Carlos, São Carlos, São Paulo, Brazil; email: strixino@ufscar.br

Luis F. Valladares [Chapter 15.5] Departamento de Biodiversidad y Gestión Ambiental, Facultad de Ciencias Biológicas y Ambientales, Universidad de León, León, Spain; email: lfvald@unileon.es

Rüdiger Wagner [Chapter 16.10] Schlitz, Germany; email: ruediger2.wagner@t-online.de

Sofia Wiedenbrug [Chapter 16.2] Zoologische Staatssammlung München, Munich, Germany; email: sofia.wig@gmail.com

About the Editors

Dr. Neusa Hamada has been working for more than 30 years at the National Institute for Research in Amazonia (INPA) in Manaus, Amazonas, Brazil, where she is a Research Professor. She received her Master's degree in Biological Sciences from INPA's Entomology graduate program and her Ph.D. in Entomology from Clemson University, Clemson, SC, United States. Neusa's research began with black flies (Diptera: Simuliidae) and subsequently widened to other aquatic taxonomic groups. Since 1999 she has advised students at both the Master's and Ph.D. levels in INPA's Entomology graduate program and collaborates with other graduate schools in the Amazon region, advising students on the taxonomy, biology, and ecology of aquatic insects. She has published more than 200 refereed journal articles, books, and book chapters on aquatic insects. She has also been active in outreach initiatives, including publishing children books and creating exhibits that use aquatic insects as model to popularize aquatic science among school-age children and the general public.

Dr. James H. Thorp has been Professor in the Department of Ecology and Evolutionary Biology at the University of Kansas (Lawrence, KS, United States) and a senior scientist in the Kansas Biological Survey since 2001. Prior to returning to his alma mater, Prof. Thorp was a Distinguished Professor and Dean at Clarkson University, Department Chair and Professor at the University of Louisville, Associate Professor and Director of the Calder Ecology Center of Fordham University, Visiting Associate Professor at Cornell, and Research Ecologist at the University of Georgia's Savannah River Ecology Laboratory. He received his Baccalaureate from the University of Kansas (KU) and both Master's and Ph.D. degrees from North Carolina State University. Those degrees focused on zoology, ecology, and marine biology, with an emphasis on the ecology of freshwater and marine invertebrates. Dr. Thorp has been on the editorial board of three freshwater journals and is a former President of the International Society for River Science. He teaches freshwater, marine, and invertebrate courses at KU, and his Master's and Doctoral graduate students work on various aspects of the ecology of communities through macrosystems in rivers, reservoirs, and wetlands. Prof. Thorp's research interests and background are highly diverse and span the gamut from organismal biology to community, ecosystem, and macrosystem ecology. He works on both fundamental and applied research topics using descriptive, experimental, and modeling approaches in the field and lab. While his research emphasizes aquatic invertebrates, he also studies fish ecology, especially as related to food webs. He has published more than 100 refereed journal articles, books, and chapters, including three single-volume editions of *Ecology and Classification of North American Freshwater Invertebrates* (edited by J.H. Thorp and A.P. Covich) and two volumes in the current fourth edition of *Thorp and Covich's Freshwater Invertebrates*.

Dr. D. Christopher Rogers is a research zoologist at the University of Kansas with the Kansas Biological Survey and is affiliated with the Biodiversity Institute. He received his Ph.D. degree from the University of New England in Armidale, NSW, Australia. Christopher specializes in freshwater crustaceans (particularly Branchiopoda and Malacostraca) and the invertebrate fauna of seasonally astatic wetlands on a global scale. He has numerous peer reviewed publications in crustacean taxonomy and invertebrate ecology, as well as published popular and scientific field guides and identification manuals to freshwater invertebrates. Christopher is an associate editor for the Journal of Crustacean Biology and a founding member of the Southwest Association of Freshwater Invertebrate Taxonomists. He has been involved in aquatic invertebrate conservation efforts all over the world.

Preface to the Fourth Edition

Those readers familiar with the first three editions of our invertebrate book (*Ecology and Classification of North American Freshwater Invertebrates*, edited by J.H. Thorp and A.P. Covich in 1991, 2001, and 2010) will note that the fourth edition has expanded from a North American focus to worldwide coverage of inland water invertebrates. We gave this revised book series the name *Thorp and Covich's Freshwater Invertebrates* to: (1) associate present with past related editions, unite current volumes, and link all to future editions; (2) establish a connection between the ecological and general biology coverage in Volume I with the taxonomic keys in the remaining volumes; and (3) give credit to Professor Alan Covich for his work on the first three editions. For the sake of brevity, we refer to the current edition as T&C IV. Whether the fifth edition of T&C will ever appear is certainly problematic, but who knows! At present we are considering producing 10–12 total volumes in the fourth edition.

While I am the sole editor of the overall book series, Christopher Rogers has been a major and highly valued partner in developing ideas for the fourth edition and has been a senior or coeditor on all our volumes. He will also play a major role in many of the remaining volumes because of his diverse and global knowledge of freshwater invertebrates, especially in the area of taxonomy. After publishing the first two volumes, we began contacting some potential coeditors and authors to develop volumes for other zoogeographic regions. Out of the seven volumes in this edition, three are now published and four are under contract. In addition, we have agreements from scientists willing to lead development of two volumes in another bioregion. However, we are still seeking experts in fields of invertebrate taxonomy for various zoogeographic regions to serve as highly dependable authors or even coeditors, especially those who both study fauna and live in the zoogeographic regions covered by the various future volumes.

Our concept for T&C IV included producing one book (Volume I, published in 2015) with six chapters on general environmental issues applicable to many invertebrates, followed by 35 chapters devoted to individual taxa at various levels (order to phylum, or even multiple phyla in the case of the protozoa). Volume I was designed both as an independent book on ecology and general biology of various invertebrate taxa and as a companion volume for users of the keys in the regional taxonomic volumes, thereby reducing the amount of information duplicated in the taxonomic volumes. The perhaps 10–11 taxonomic volumes to be published in the next decade or so will contain both keys for identifying invertebrates in specific zoogeographic regions and descriptions of detailed anatomical features needed to employ those keys. While the vast majority of authors in T&C editions I–III were from the United States or Canada, we attempted in T&C IV to attract as many editors and authors as possible from around the world but especially from the zoogeographic regions described in the keys. Our goals for T&C IV are to improve the global state of taxonomic and ecological knowledge of inland water invertebrates, to help protect our aquatic biodiversity, and to encourage more students to devote their careers to working with these fascinating organisms. These goals are especially important because the verified and probable percent losses of species in wetlands, ponds, lakes, creeks, and rivers around the globe exceed those in most terrestrial habitats.

James H. Thorp

Preface to Volume III

This is the third volume of the fourth edition of *Thorp and Covich's Freshwater Invertebrates* (T&C IV), and the first to focus on a bioregion outside the Nearctic. Information on the ecology and general biology of the groups can be found in Volume I (Ecology and General Biology, edited by Thorp & Rogers, 2015), the companion text for the current and all remaining books in this series. All taxonomic volumes are expected to consist of an introductory chapter followed by chapters on different taxonomic groups (phyla, class, orders, or smaller levels), depending on the nature of the volume. Some chapters in the current volume are very small and others are quite large, such as coverage of all Trichoptera in Chapter 10.

A typical chapter includes a short introduction, a brief discussion of limits to identification of taxa in that chapter, important information on terminology and morphology that is needed to use the keys, techniques for preparing and preserving material for identification, the taxonomic keys, and a few references. Three chapters serve as introductions to an order, which is then covered in greater detail in subsequent chapters; these consist of Chapters 14 (Odonata), 15 (Coleoptera), and 16 (Diptera).

The multilevel keys are formatted to enable users to work easily at the level of their taxonomic expertise and the needs of their project. For that reason, we separated keys by major taxonomic divisions. For example, a student in a college course might work through one or more of the initial Coleoptera keys to determine the family in which a freshwater beetle belongs. In contrast, someone working on an environmental monitoring project might need to identify a dragonfly or midge to genus, and thus would use the relevant, detailed keys that require more background experience. We also designed most keys, where possible, to proceed from a general to a specific character within a couplet.

We have asked authors to include only taxa that are recognized internationally by publication in reputable scientific journals which follow the International Code of Zoological Nomenclature. Thus, no taxa that have merely been proposed should have been included even if they have been identified by the world's expert on that group. "Common" species are not designated in the keys themselves because a common species in one area may not be common in another, and this designation can easily lead to false identifications. Authors have been encouraged to end the keys at the point where further identification without genetic analysis is not practical or when it is clear that too many of the extant fauna have yet to be described in scientific publications. Where the majority of taxa at a given taxonomic level (e.g., genus) have not been identified, we have requested that the author's key end at the taxonomic level above that (e.g., subfamily).

Users of these keys need to realize that taxonomy is a growing and vibrant field in which new taxa are being described and previously accepted relationships reevaluated. For some users, this volume may be sufficient for their needs, but for others a companion text listing known species in a smaller geographic region may also be helpful.

This edition is focused on species found in fresh surface waters, thereby reflecting the bias of existing scientific literature and the habitats of springtails and insects. The majority of taxa described in this volume spend their larval stage entirely in water, and some also occur there as adults. Most of the taxa, however, have an aerial adult life stage that rarely enters the water.

It is our hope that scientists and students from around the world will benefit from this volume, especially those studying organisms in the Neotropics. Suggestions for improving future volumes are welcome.

Editors
Neusa Hamada
James H. Thorp
D. Christopher Rogers

Acknowledgments for Volume III

Many people contributed to this volume in addition to the chapter authors and those acknowledged in individual chapters. We greatly appreciate all our colleagues who have contributed information, figures, or reviews for Volume III and also thank those who provided similar services for the earlier editions, upon which the present book partially relies. We are again grateful to the highly competent people at Academic Press/Elsevier who helped in many aspects of the book's production from the original concept to the final marketing. In particular, we appreciate our association with Elsevier editors Emily Thomson and Louisa Hutchins, and the production teams in Great Britain, the United States, and other parts of the globe. We also thank the hard workers in India led by Paul Prasad Chandramohan who efficiently produced the proofs for this volume.

Neusa Hamada
James H. Thorp
D. Christopher Rogers

Division I

Introduction and Subclass Entognatha

Chapter 1

Introduction[1]

James H. Thorp
Kansas Biological Survey and Department of Ecology and Evolutionary Biology, University of Kansas, Lawrence, KS, United States

D. Christopher Rogers
Kansas Biological Survey and The Biodiversity Institute, University of Kansas, Lawrence, KS, United States

Neusa Hamada
Coordenação de Biodiversidade, Instituto Nacional de Pesquisas da Amazônia, Manaus, Amazonas, Brazil

Chapter Outline

INTRODUCTION TO THIS SERIES, VOLUME, AND CHAPTER

The first volume of *Thorp and Covich's Freshwater Invertebrates* (Thorp and Rogers, 2015) provided global coverage of the ecology, general biology, phylogeny, and collection techniques for inland water invertebrates. All subsequent volumes have now or will have focused on taxonomic keys for individual bioregions of the world. The taxonomic keys provide the users with means to work at the level of their need and expertise. Our authors have tried to limit the specificity of the keys to those taxa where comprehensive knowledge is available for most of the taxa in a particular group and where users cannot identify all life stages.

We hope to have six volumes in print by the end of 2019: (1) an introductory volume on ecology and general biology of inland water invertebrates of the world (Thorp and Rogers, 2015); (2) *Keys to Nearctic Fauna* (Thorp and Rogers, 2016); (3) the present volume, *Keys to Neotropical Hexapoda* (Hamada et al., 2018); (4) *Keys to Palaearctic Fauna* (Rogers and Thorp, 2018); (5) *Keys to Neotropical and Antarctic Fauna* (Hexapoda only to order) (Damboreana et al., in prep.); and (6) *Keys to Australasian Fauna* (Ahyong et al., in prep). Future volumes will cover Oriental/Oceanic fauna and Afrotropical fauna (each with 1−2 volumes) as well as more detailed coverage of Hexapoda in the Nearctic and Palaearctic Bioregions. We consider this series of 10−12 volumes to be a globally expanded, fourth edition of the earlier *Ecology and Classification of North American Freshwater Invertebrates* (edited by Thorp & Covich in 1991, 2001, and 2010). While some labs may have multiple copies of the "Keys to Fauna" in their region, we also recommend that they have at least one copy of Volume I to obtain useful background information on each invertebrate group.

The current chapter is organized into an introduction, a description of the taxonomic focus of this particular volume, a section explaining the organization of most taxonomic chapters, and a key to larger taxonomic groups. This chapter's key is designed to help the reader locate the most pertinent chapter (a task probably useful only for students and relatively novice taxonomists) and begin identifying organisms in their samples. Readers will note that chapters within and among volumes vary in the specificity of their taxonomic keys. This reflects both the likely

[1] This chapter was written to be a useful starting point for the previous, current, and future taxonomic volumes in all zoogeographic regions. Consequently, there will be only minor to moderate differences in this chapter among various volumes.

Thorp and Covich's Freshwater Invertebrates, Fourth Edition. https://doi.org/10.1016/B978-0-12-804223-6.00001-9

percent of the fauna that has been named and how easily taxa can be separated by alpha taxonomic methods and associated keys. While there are occasional references to genetic separation of taxa, the keys are primarily based on microscope identifications.

NATURE AND FAUNA OF THE NEOTROPICS

As is clear from the title, this volume covers inland water Hexapoda of the Neotropics, which to some readers, especially those from temperate zones, may suggest a warm climate. It is important to recognize, however, that the Neotropical Bioregion stretches from northern Mexico (warm) through the Caribbean and equatorial countries (often hot) to the frigid areas of Tierra del Fuego (Argentina) and the islands of Cape Horn (Chile). It also includes both low- and high-altitude sites which vary greatly in their temperature range. Moreover, some regions are hot or cold deserts and others range up to very humid environments. The freshwater invertebrate fauna of these diverse regions will vary accordingly. Many of these diverse habitats are described in Chapter 2 of Volume I.

COMPONENTS OF TAXONOMIC CHAPTERS

This volume is an identification manual to the inland water hexapods (springtails and insects) of the Neotropics where we present information needed to diagnose and assign these organisms to various taxonomic levels. Other information concerning ecology, morphology, physiology, phylogeny, and both collecting and culturing techniques can be found in Volume I of this series. Each of the remaining 42 chapters in the current volume is limited to a single order or lower taxonomic level, except for Chapter 2's coverage of all orders of Collembola. We have attempted to include the following six sections in those chapters: (1) a brief overview of the higher-level taxon; (2) a description of identification limitations for each taxon; (3) details of pertinent terminology and morphology; (4) information on preparing and preserving specimens for identification; (5) taxonomic keys (separated by level of identification); and (6) limited list of references. Readers can find a much more extensive list of references to their group in Volume I (Chapters 3 and 7–41) along with more details on collecting, preparation, and preserving major taxa. Figures in each chapter are limited to those needed for effective use of the keys. For additional anatomical information, including figures, see the relevant chapter in Volume I.

HOW TO USE THIS VOLUME

There is an old maxim that says "keys are written by people who do not need them for people who cannot use them." Nonetheless, we have made every effort to make these keys as user friendly as publication limitations would permit.

Each chapter includes a basic introduction to the morphology and terminology used in diagnosing the taxa of that section. Limitations to the current state of taxonomic knowledge are also presented so that the reader may gage the reliability of the information presented. Only the established, peer-reviewed scientific literature is used to define the taxonomic categories and epithets included. All names, as far as we are aware, conform to the International Code of Zoological Nomenclature. All nomina and taxonomic arrangements used, as well as the rejection of old names are based on peer-reviewed scientific literature. Names from unpublished manuscripts, dissertations, "in house" designations, or records that have not been validated are not included. Provisional names and species designated "taxon 1" or "species 1" are not used unless they were previously recognized and accepted in the peer-reviewed scientific literature (Richards and Rogers, 2011). No new species descriptions or previously unpublished taxonomic arrangements are presented.

The keys are dichotomous (no triplets or quadruplets are used) and are hierarchical. Thus, for a given group, the first keys are to the highest taxonomic category. The second set of keys is to the next level, the third set to the level below that one, and so on, down to the lowest justifiable taxonomic level based on current knowledge of that group. This final level varies for different groups depending upon the state of resolution in the scientific literature. Organisms not identifiable beyond a particular taxonomic level are left at that level.

Properly prepared keys typically employ specific, primary, diagnostic characters. Older keys often use different characters than the more recent keys. This shift in primary characters results from systematists and taxonomists testing the importance of characters. The ultimate goal of the systematist is to have the taxonomic characters used in our keys converge as closely as possible with biological reality. To a nontaxonomist, the resulting process may erroneously seem merely to be "lumping and splitting" taxonomic groups rather than the result of employing the scientific method to reveal natural relationships.

Surprisingly, many users do not know how to interpret a dichotomous key, making the fundamental assumption that a correct identification answer is always present in the key. This assumption generally takes one of the following three forms:

1. *All species are identifiable using a given key.* Many new species have yet to be discovered, let alone described. Generalized geographic ranges are provided for most taxa presented herein, yet species ranges shrink, swell, and change elevation constantly, particularly as weather and climate patterns shift. Species disperse, colonize, and suffer stochastic local extinctions. In addition to these natural processes, some species are introduced intentionally or accidentally by humans, and sometimes their establishment allows other species to invade as well. Furthermore, some regions are better studied than others: many parts of the world have not been explored, nor the fauna described. There are still plenty of new taxa to encounter.
2. *All variation is accounted for in the key.* As stated, identification keys use specific, primary, diagnostic characters. Problems in identification are compounded by taxa that: (a) have different character states at different times; (b) only have diagnostic characters at certain life stages or in certain genders; and/or (c) have severely truncated morphology (often due to lack of sexual selection) and lack morphological characters to separate the species. Furthermore, new variation within taxa is continually developing, and thus one cannot assume that species are immutable or develop tools predicting those changes.
3. *The key is a sufficient identification tool in and of itself.* A key is just a tool. The fact that one has a bolt that needs removing and a wrench of the correct size does not mean that the bolt can be loosened. Similarly, identification keys are tools to aid in taxon identification. They are primarily tools to eliminate incorrect taxa from the range of possible choices, narrowing the field to the names that may be applicable. Keys are the process of elimination. The possibility that the specimen to be identified is new, a hybrid, anomalous, or a recent invasive colonist is always a possible answer. This is fundamental to using any identification key.

Once one arrives at a name or group of possible names for a specimen in hand, the specimen should then be compared against descriptions, distribution maps, and figures of that and other taxa in that group. The descriptions, figures, and maps are other tools to be used in identification. Direct comparison of the specimen at hand with identified museum material or using molecular comparisons is also sometimes necessary for a correct identification.

Species are not immutable, fixed in location and form. They change constantly and will continue to do so, confounding keys and any other identification method, such as trait tables, character matrices, or even genetic analyses. This is one reason why biology is far behind physics in the development of unified theories: biology is far more complex than physics, as it involves more mutable interacting parts and processes.

KEYS TO MAJOR TAXA OF HEXAPODA

A major change in the identification keys for our fourth edition has been to include multiple keys per chapter that generally start with a class-level key (order in this volume) and proceed to finer and finer divisions. These allow users to work at their levels of interest, need, and skill without having to wade through extraneous taxa not in the direct line to the taxon of interest.

The general characteristics of Hexapoda (insects and springtails) are: genitalia positioned at abdominal terminus; limbs present or absent, if present, not more than three pairs; pseudolegs may be present in larvae of some orders; one pair of antennae; wings generally present in adults.

The following keys will allow you to move to the next level of keys, which are in the individual chapters. Chapters 14 (Odonata), 15 (Coleoptera), and 16 (Diptera) are introductory chapters to their groups and lead to 6, 7, or 14 additional chapters, respectively, for those orders. The key to larval insects was derived in part from Chapter 1 in Volume I of the fourth edition.

Hexapoda: Classes

1 Mouthparts internal; abdomen forked distally, and bent ventrally to meet the collophore (tubular projection) depending from the first abdominal segment; wings absent (Fig. 1.1) .... **Collembola** [Chapter 2]
1' Mouthparts external; abdomen various, collophore absent; wings present or not .... **Insecta [p. 6]**

Hexapoda: Insecta: Orders

Based in part on DeWalt et al. (2010).

1 Thorax without jointed legs 2
1' Thorax with three pairs of jointed legs 3
2(1) Mummy-like pupa, with developing adult structures enveloped in a case which may contain vegetable or mineral matter bound by silk secretions; occur in holometabolous insects within inland waters (Coleoptera, Diptera, Hymenoptera, Lepidoptera, Mecoptera, Megaloptera, Neuroptera, and Trichoptera); generally not keyed below, but pupal keys present in some dipteran chapters .... insect pupae
2' Not completely enclosed in a case; mobile larvae, mostly with legs, prolegs, pseudopods, or creeping welts on one or more segments; larvae .... **Diptera** [Chapter 16]
3(1) With large functional wings; forewings may be rigid and chitinized, at least at the base 4
3' Wingless, or with developing wings (wingpads) or brachypterous 7
4(3) Not the combination of characteristics described below 5
4' Abdomen narrowly constricted at junction with thorax; basal half of forewings parallel to the leading edge; last leg pair not greatly enlarged for jumping; usually <3 mm; enter water to deposit eggs on or in aquatic insect hosts; adult Fig. 1.2 .... **Hymenoptera** [Chapter 12]
5(4) Mesothoracic wings rigid and chitinized (Figs. 15.7–9) or thickened in basal half (Figs. 7.29–30) 6
5' Meso- and metathoracic wings completely membranous with numerous distinct veins, forewings leathery; hind legs with enlarged femora suitable for jumping; order primarily terrestrial but with some semiaquatic adult species in the tropics; Fig. 1.3 .... **Orthoptera** [Chapter 5]
6(5) Chewing mouthparts; or if appearing as a tube, then antennae are situated half way down the tube; mesothoracic wings as elytra (e.g., rigid and chitinized) covering metathoracic wings, elytra may be abbreviated; adults; Fig. 1.4 .... **Coleoptera** [Chapter 15]
6' Sucking mouthparts formed into a broad or narrow tube; mesothoracic wings hardened in basal half; adults; Fig. 1.5 .... **Hemiptera** [Chapter 7]
7(3) Without extensile labium; abdomen not terminating in lamellae or points 8
7' Labium formed into elbowed geniculate, extensile grasping organ; abdomen terminating in three lamellae or five usually short, triangular points; larvae; Fig. 1.6 .... **Odonata** [Chapter 14]
8(7) Mouthparts for sucking, formed into a broad or narrow tube or a pair of long stylets 9

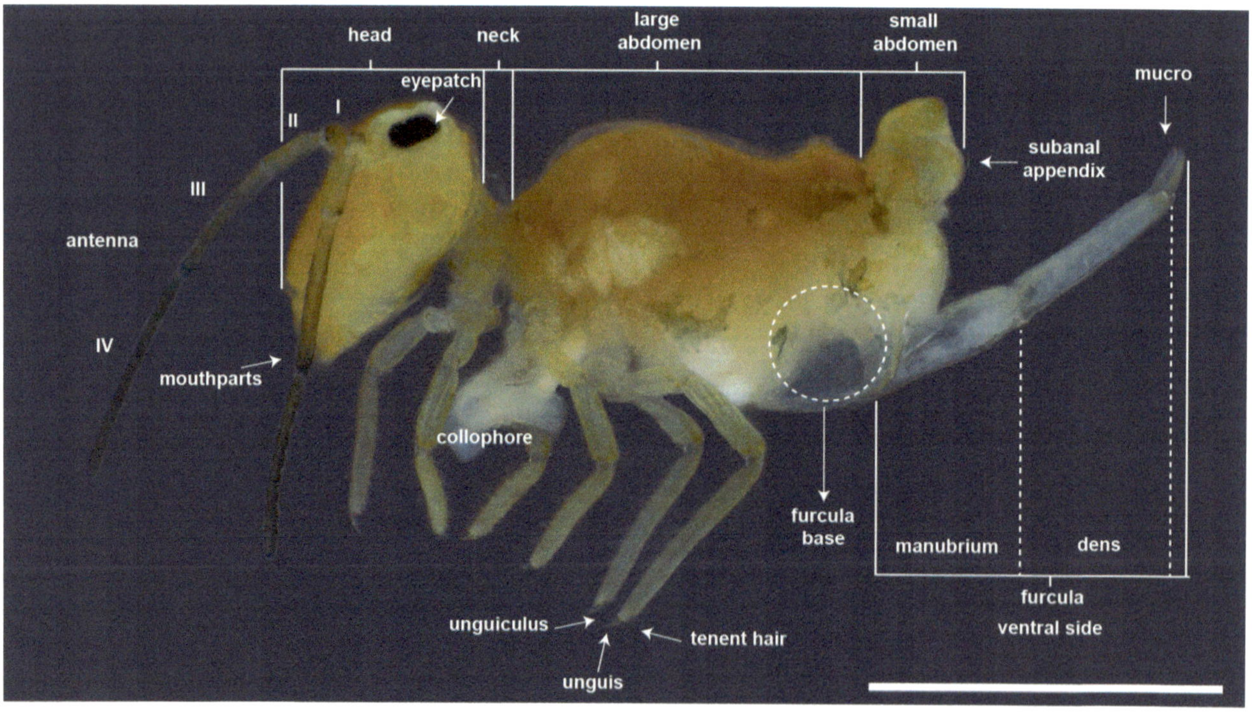

FIGURE 1.1 Lateral view of the springtail *Symphypleona* (*Arlesminthurus*, Collembola) showing its general morphology. *From Chapter 2. Bar = 0.5 mm.*

FIGURE 1.2 Lateral view of the head and part of the thorax of a semiaquatic aquatic wasp (Hymenoptera) Chalcididae. *From Chapter 12.*

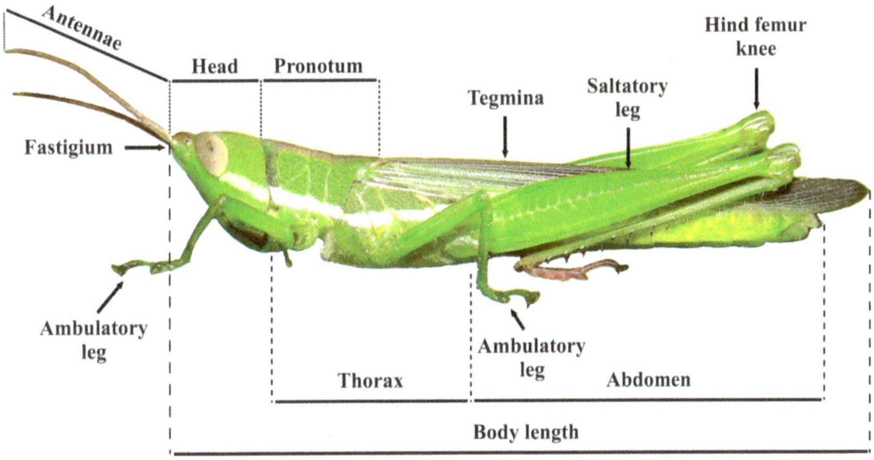

FIGURE 1.3 External structures of a semiaquatic grasshopper (Orthoptera, Acrididae). *From Chapter 5.*

8' Mouthparts not for sucking, not formed into a tube or pair of stylets........................ 10

9(8). Mouthparts a pair of long stylets, feeds on sponges; all tarsi with one claw; larvae; Fig. 1.7........................ **Neuroptera** [Chapter 9]

9' Mouthparts a broad or narrow tube; mesotarsi with two claws; adults and nymphs; Fig. 1.5 .... **Hemiptera** [Chapter 7]

10(9) Abdomen without ventral prolegs on abdominal segments 3–6 11

10' Ventral prolegs on abdominal segments 3–6, each with a ring of fine hooks; larvae; Fig. 1.8 .... **Lepidoptera** [Chapter 11]

11(10) Antennae elongate, with three or more antennomeres 12

11' Antennae extremely small, inconspicuous, with one antennomere; larvae; Fig. 1.9........................ **Trichoptera** [Chapter 10]

12(11) Terminal appendages absent, or composed of one or two short appendages uni or bi-articulated........................ 13

12' Terminal appendages composed of two or three long, multiarticulated filaments 15

13(12) Antennae with three or greater than five antennomeres 14

13' Antennae with four antennomeres; larvae; Fig. 1.10........................ **Megaloptera** [Chapter 8]

14 (13) Abdomen posterior end with a pair of hooks inserted dorsally, curved ventrally with the apices pointing anteriorly; larvae........................ **Mecoptera** (Nannochoristidae) [Chapter 13]

14' Abdomen posterior end not as described above larvae .... **Coleoptera** [Chapter 15]

15(12) Abdomen with plumose, forked, or lamellar gills; with two cerci and a median caudal filament, occasionally latter is reduced to stub; tarsus with one claw; larvae; Fig. 1.11........................ **Ephemeroptera** [Chapter 3]

15' Abdomen with or without gills; if present gills digitiform in shape on abdominal segments 1–2 or 3, and/or at the base of the cerci; two cerci, no median filament; tarsi with two claws; larvae; Fig. 1.12........................ **Plecoptera** [Chapter 4]

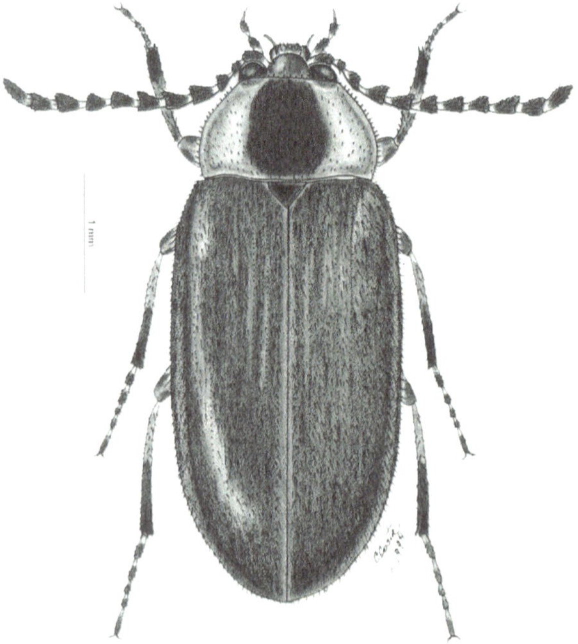

FIGURE 1.4 Example of an adult aquatic beetle (Coleoptera, Cneoglossidae). *From Chapter 15.*

FIGURE 1.5 Dorsal view of the adult, aquatic bug *Ambrysus* sp. (Hemiptera, Naucoridae). *From Chapter 7.*

FIGURE 1.6 Dorsal view of the dragonfly larva *Epigomphus subsimilis* Calvert (Odonata, Gomphidea) showing wing sheaths strongly divergent. *From Chapter 14.2.*

FIGURE 1.7 Lateral view of the larval stage of the Spongilla fly *Climacia* (Neuroptera, Sisyridae). Bar = 0.5 mm. *From Chapter 9.*

FIGURE 1.8 Lateral view of the moth larva *Petrophila* sp. (Lepidoptera, Crambidae). *From Chapter 11.*

FIGURE 1.9 Lateral view of the larva of the *Cheumatopsyche* (Trichoptera, Hydropsychidae). *From Chapter 10.*

FIGURE 1.11 Nymph of the mayfly *Cloeon smaeleni* Lestage (Ephemeroptera, Baetidae). *From Chapter 3.*

FIGURE 1.10 Dorsal view of a larval *Ilyobius* sp. (Megaloptera, Sialidae). *From Chapter 8.*

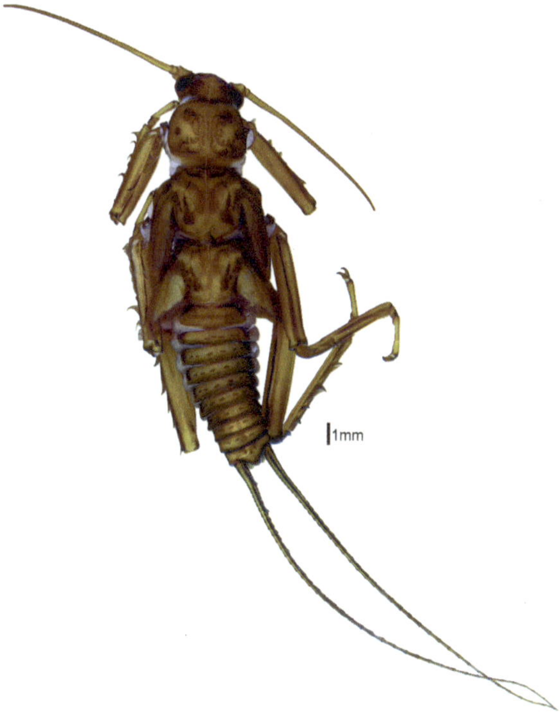

FIGURE 1.12 Stonefly nymph (Plecoptera, *Notoperla*). *From Chapter 4.*

REFERENCES

Fraulob, M., B. Wipfler, F. Hünefeld, H. Pohl & R.G. Beutel. 2012. The larval abdomen of the enigmatic Nannochoristidae (Mecoptera, Insecta). Arthropod Structure & Development 41: 187–198.

Richards, A.B. & D.C. Rogers. 2011. Southwest Association of Freshwater Invertebrate Taxonomists (SAFIT) list of freshwater macroinvertebrate taxa from California and adjacent states including standard taxonomic effort levels. 266 p.

Rogers, D.C. & J.H. Thorp (eds.). 2019. Keys to Palaearctic Fauna. Volume IV in: Thorp and Covich's Freshwater Invertebrates, Fourth Edition. Academic Press, Elsevier, Boston, MA. In Press.

Thorp, J.H. & D.C. Rogers (eds.). 2015. Ecology and General Biology. Volume I in: Thorp and Covich's Freshwater Invertebrates, Fourth Edition. Academic Press, Elsevier, Boston, MA. 1118 p.

Thorp, J.H. & D.C. Rogers (eds.). 2016. Keys to Nearctic Fauna. Volume II in: Thorp and Covich's Freshwater Invertebrates, Fourth Edition. Academic Press, Elsevier, Boston, MA. 740 p.

Chapter 2

Class Collembola

Nikolas G. Cipola
Coordenação de Biodiversidade, Curso de Pós-Graduação em Entomologia, Instituto Nacional de Pesquisas da Amazônia, Manaus, Amazonas, Brazil

Diego D. da Silva
Coordenação de Biodiversidade, Curso de Pós-Graduação em Entomologia, Instituto Nacional de Pesquisas da Amazônia, Manaus, Amazonas, Brazil

Bruno C. Bellini
Universidade Federal do Rio Grande do Norte, Natal, Rio Grande do Norte, Brazil

CLASS COLLEMBOLA

Chapter Outline

INTRODUCTION

Springtails (Hexapoda: Collembola) are small terrestrial arthropods similar to insects, possessing three tagmata, one pair of antennae, and three pairs of thoracic legs. They differ from all other hexapods by the presence of three abdominal appendages: a ventral tube (or collophore), the tenaculum (or retinaculum), and the furcula. These appendages are considered as synapomorphies of the taxon (Hopkin, 1997). The tenaculum and furcula are secondarily lost in some taxa. The taxonomic name "Collembola" was proposed by Lubbock in 1870 based on the presence of the collophore in species of the genus *Podura* Linnæus. Lubbock observed that the appendage was responsible for attaching specimens to a surface and used Latin/Greek terminology to systematically redefine *Podura*: **colla** (Latin) or **kolla** (Greek) which means glue, and **embolon** (Greek) which means "that which has been thrown into something," in this case, like a plug or tube (Bellinger et al., 2017).

Collembola is an independent class of arthropods with four orders (Poduromorpha, Entomobryomorpha, Symphypleona, and Neelipleona, Fig. 2.1) which currently has about 8700 described species in 708 genera and 29 families (Soto-Adames et al., 2008; Bellinger et al., 2017). However, this substantial species richness probably represents a small slice of the real diversity of the class, which may surpass 50,000 species (Hopkin, 1997; Deharveng et al., 2008; Cicconardi et al., 2013).

Thorp and Covich's Freshwater Invertebrates, Fourth Edition. https://doi.org/10.1016/B978-0-12-804223-6.00002-0

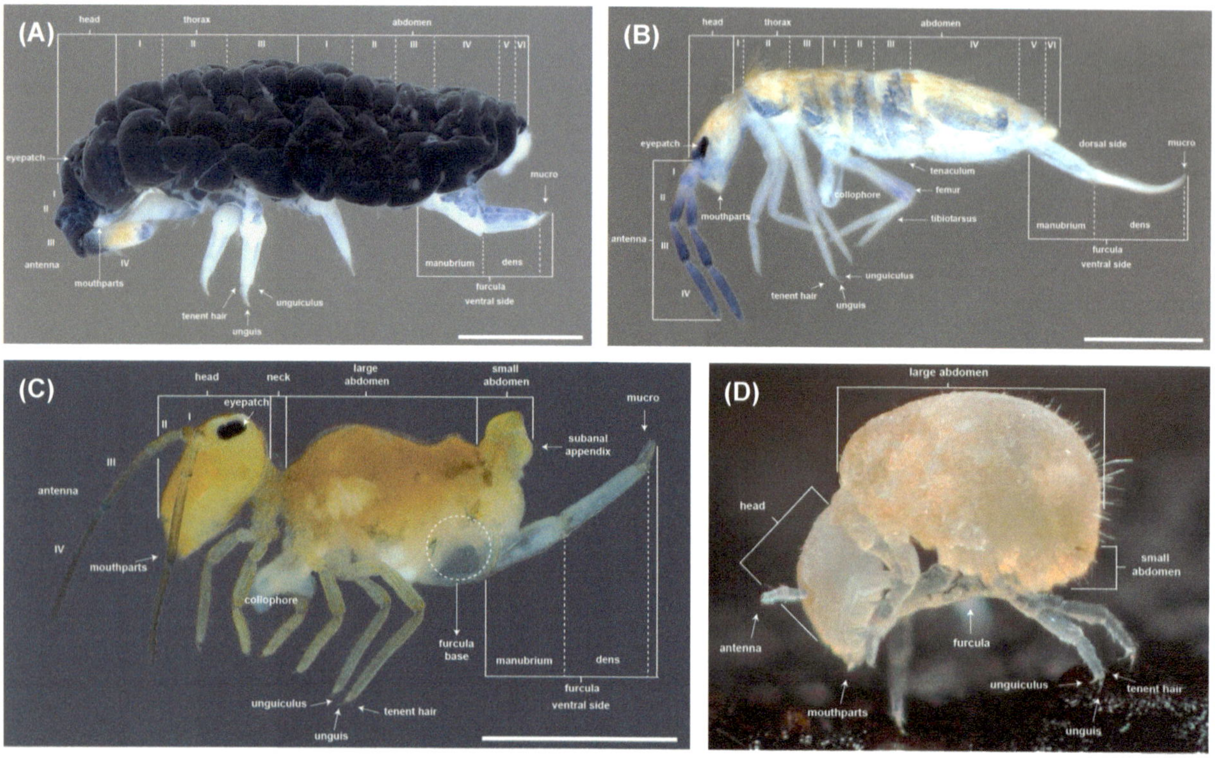

FIGURE 2.1 (A) Poduromorpha (*Neotropiella carli*) habitus, lateral view, general morphology. Bar = 1 mm. (B) Entomobryomorpha (*Seira dowlingi*) habitus, lateral view, general morphology. Bar = 0.5 mm. (C) Symphypleona (*Arlesminthurus*) habitus, lateral view, general morphology. Bar = 0.5 mm. (D) Neelipleona (*Neelus murinus*) habitus, lateral view, general morphology. *(D) Photo courtesy of Andy Murray. (https://www.chaosofdelight.org).*

The potentially high species richness of springtails is possibly related to small body size, restricted dispersal capacity, widespread distribution in terrestrial and aquatic surface ecosystems, and specialization to different microhabitats. Concerning aquatic habitats, springtails also are specialized for different microhabitats and can be ecologically categorized as primary (water-dependent) or secondary (epigean hydrophilous) aquatic-associated species (Waltz & McCafferty, 1979; Deharveng et al., 2008). The primary aquatic-associated springtails can be cryophilous species, which live permanently on the top of snow in temperate mountains or glaciers; epineustic species, found on the surface of freshwater, usually with adaptive morphological structures, such as in some species of *Sminthurides*, *Arlesminthurus* and *Pseudobourletiella*; and intertidal (or littoral) species, which are associated with beach soil, rocks, and water puddles and are capable of being submerged, such as *Spinactaletes* and *Isotogastrura*. The secondary aquatic associated springtails are usually hygrophilous epiedaphic species that accidentally fall on the surface of freshwater (Arlé, 1971; Waltz & McCafferty, 1979; Christiansen & Snider, 2008; Palacios-Vargas, 2013). Adaptation to aquatic or semiaquatic habitats is a derived condition that has evolved independently several times in springtails (D'Haese, 2002).

Although several species of Collembola can be hygrophilous and may be found both in land or directly associated with water habitats, these aquatic associations are infrequently recorded in taxonomic or ecological papers. At present, 525 water-dependent species of springtails are known worldwide, of which 465 are freshwater-dependent. In the Neotropical Region approximately 45 water-dependent species have been reported, of which 28 are freshwater-dependent (Deharveng et al., 2008).

Systematics of Collembola Orders

Phylogenetic relationships among Collembola orders are much discussed and numerous hypotheses have been proposed so far. Poduromorpha and Neelipleona are possibly monophyletic (D'Haese, 2002, 2003; Gao et al., 2008; Xiong et al., 2008; Schneider et al. 2011; Schneider & D'Haese, 2013). On the other hand, Entomobryomorpha and Symphypleona appear as either monophyletic or polyphyletic taxa in different approaches (D'Haese, 2002; Xiong et al., 2008; Schneider & D'Haese, 2013).

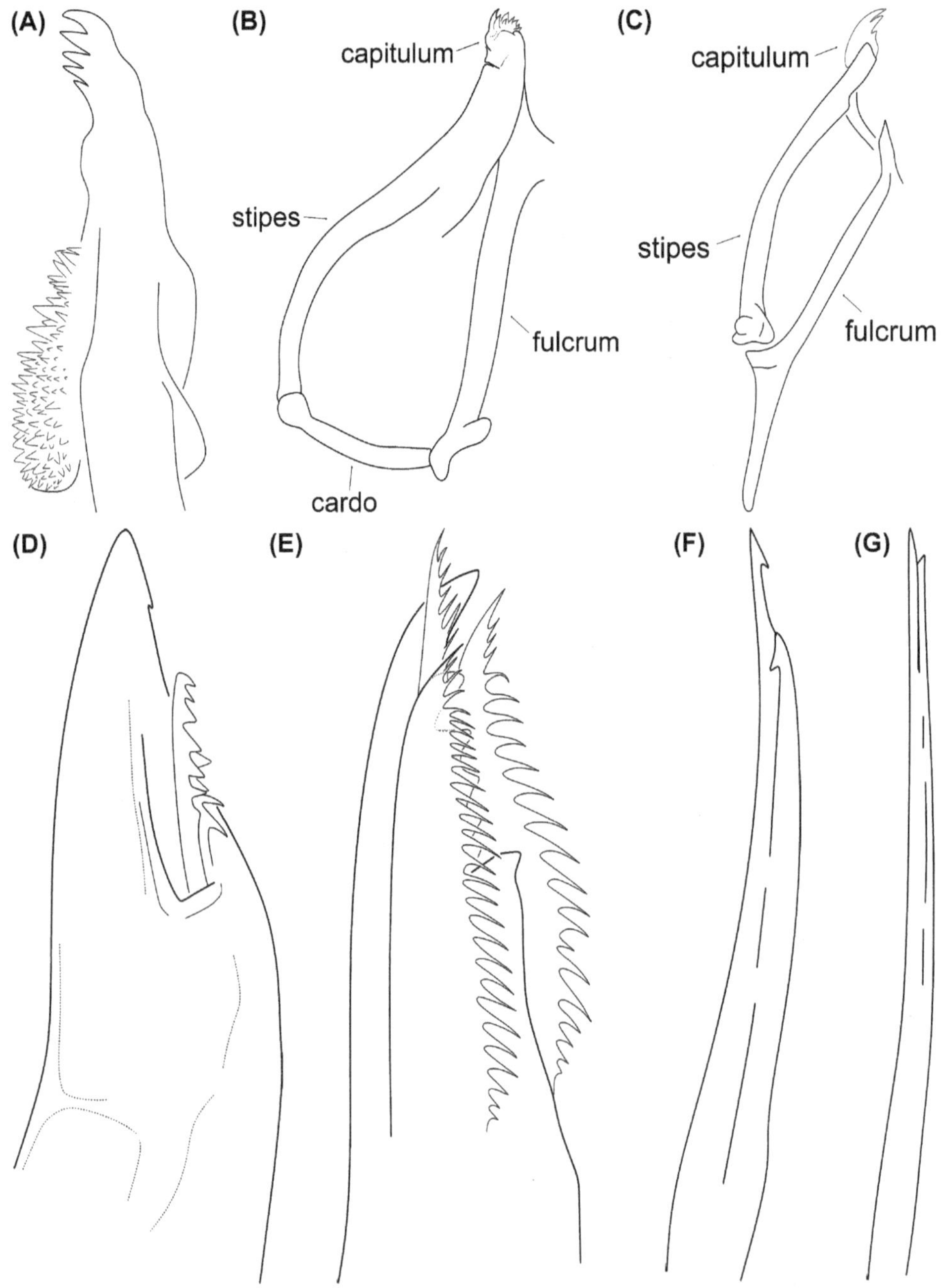

FIGURE 2.2 (A) *Ceratophysella* (Hypogastruridae), mandible apex. (B) *Brachystomella* (Brachystomellidae), maxilla. (C) *Superodontella nana* (Cassagnau, 1954) (Odontellidae), maxilla. (D) *Friesea jeanneli* Denis, 1947 (Neanuridae), maxilla capitulum. (E) *Anurida maritima* (Guérin-Méneville, 1836) (Neanuridae), maxilla capitulum. (F) *Pseudachorutes auricularius* Cassagnau & Rapoport, 1962 (Neanuridae), maxilla capitulum. (G) *Neotropiella murphyi* Massoud, 1965 (Neanuridae), maxilla capitulum. *(C) Modified from Jordana et al. (1997). (D–G) Modified from Massoud (1967).*

Due to body similarities (including elongated trunk with clear tergal segmentation), Entomobryomorpha and Poduromorpha were once grouped as the order Arthropleona. Since this group is not supported by phylogenetic analyses and an elongated segmented trunk is considered a plesiomorphic character among springtails, Arthropleona is no longer used (Xiong et al., 2008; Schneider et al., 2011; Schneider & D'Haese, 2013; Bellinger et al., 2017). D'Haese (2003) considered Poduromorpha as the sister group of the other orders, and Entomobryomorpha as the sister group of Symphypleona plus Neelipleona. In the same sense, the relationships between springtails with globular body (Symphypleona and Neelipleona) are also uncertain. Due to the globular body form Symphypleona

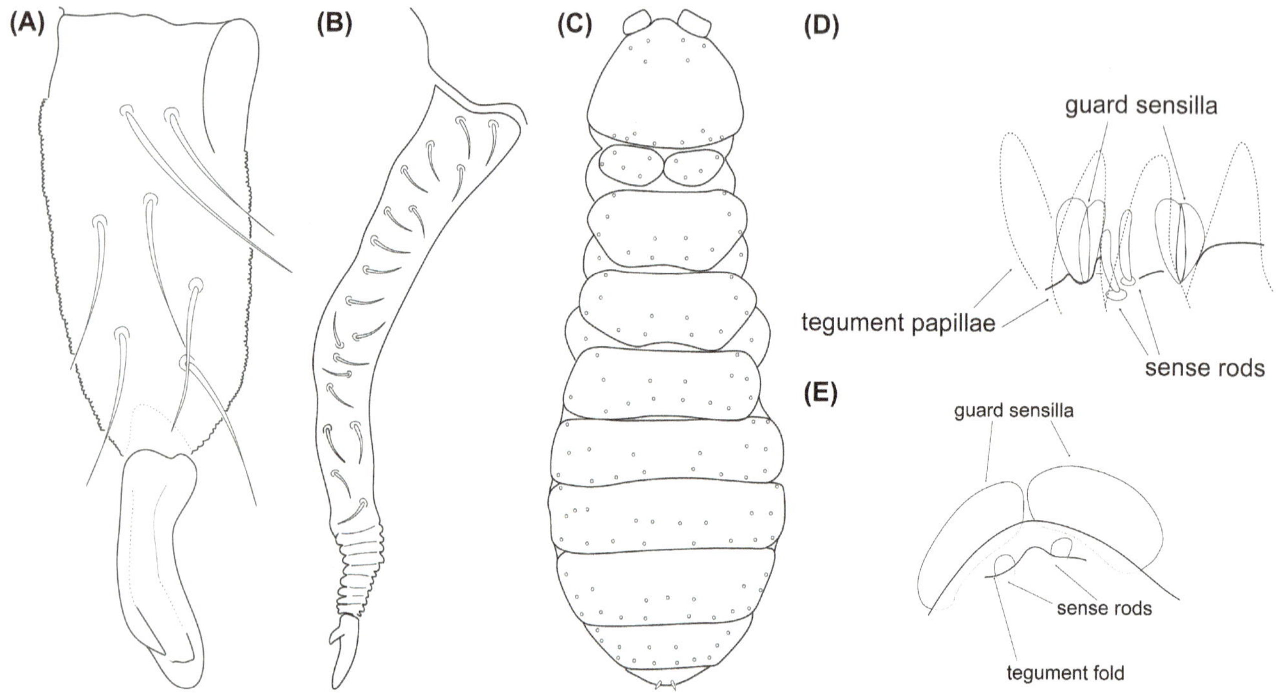

FIGURE 2.3 (A) *Ceratophysella* (Hypogastruridae), dens and mucro. (B) *Podura aquatica* Linnaeus, 1758 (Poduridae), dens and mucro. (C) *Onychiurus argus* Denis, 1925 (Onychiuridae), head and trunk dorsal pseudocelli (setae omitted). (D) *Onychiurus argaensis* Arbea & Jordana, 1994 (Onychiuridae), third antennal segment sense organ. (E) *Mesaphorura critica* Ellis, 1976 (Tullbergiidae), third antennal segment sense organ. *(B–E) Modified from Jordana et al. (1997).*

has been considered a sister group of Neelipleona (D'Haese, 2003; Xiong et al., 2008). Other analyses indicate that Symphypleona is the sister group of Poduromorpha plus Entomobryomorpha (Gao et al., 2008), or all other orders (Schneider & D'Haese, 2013).

Order Poduromorpha (*sensu* Janssens & Christiansen, 2011)

Poduromorpha (Fig. 2.1 A) specimens are characterized by a strongly granulated cuticle, distinct tergal segmentation and presence usually of setae on the protergite or rarely papillae (e.g., Isotogastruridae) (Palacios-Vargas et al., 2013). The order has about 3342 species in 333 genera; the Neotropics has at least 70 genera (Mari-Mutt & Bellinger, 1990; Bellinger et al., 2017). Poduromorph species are classified in 6 superfamilies and 11 families: Neanuroidea (Brachystomellidae, Neanuridae, Odontellidae), Poduroidea (Poduridae), Hypogastruroidea (Hypogastruridae, Pachytullbergiidae), Gulgastruroidea (Gulgastruridae), Isotogastruroidea (Isotogastruridae), and Onychiuroidea (Onychiuridae, Paleotullbergiidae, Tullbergiidae) (Bellinger et al., 2017).

Neanuridae is the most diverse family, with 167 genera and about 1400 species described worldwide. They are characterized by reduced (rarely totally absent) mandibles without a molar plate; modified maxillae with or without denticles and large fringed lamellae; and mucro, if present, with fewer than three lamellae (Figs. 2.2 D–G). Brachystomellidae has about 130 species in 18 genera; it is the sister group of Neanuridae, with absent mandibles and typical maxilla structure (Fig. 2.2 B), which serves as the main characteristics for differentiating the two families (Massoud, 1967; D'Haese, 2002; Najt et al., 2005; Bellinger et al., 2017). Odontellidae presents 135 species and differs from these other Neanuroids by maxilla lacking the cardo (Fig. 2.2 C) (Jordana et al., 1997; Bellinger et al., 2017). Genera such as *Aethiopella*, *Anurida*, *Friesea*, *Furculanurida*, *Halachorutes*, *Neotropiella*, *Pseudanurida*, and *Pseudachorutes* of Neanuridae; *Brachystomella*, *Brachystomellides*, and *Rapoportella* of Brachystomellidae; and *Odontella* and *Superodontella* of Odontellidae may have aquatic or semiaquatic species in the Neotropical Region, but most are known from the Nearctic Region (Waltz & McCafferty, 1979; Christiansen & Snider, 2008).

Poduridae has only one recognized species, *Podura aquatica*, with Holarctic distribution, and it is typically found on water. *Podura* is characterized by dentes more than three times as long as the manubrium and surpassing the ventral tube, with distal rings of granules and modified mucro (Fig. 2.3 B).

Hypogastruridae is the second largest family of Poduromorpha, with about 700 species in 40 genera, of which *Ceratophysella*, *Hypogastrura*, *Schoettella*, and

Xenylla are found in aquatic environments in the Neotropical Region. Hypogastruridae is defined on symplesiomorphies such as presence of the molar plate on mandibles (Fig. 2.2. A) and absence of pseudocelli (D'Haese, 2002; Deharveng, 2004). Pachytullbergiidae groups 13 species in three genera characterized by lateral projections on the abdomen. *Sensiphorura* (Pachytullbergiidae) apparently is a semiaquatic genus.

Isotogastruridae is represented by the monotypic genus *Isotogastrura* with 10 species found in beaches intertwined of tropical and subtropical regions (Bellinger et al., 2017). They are recognized by the prothorax devoid of setae (Palacios-Vargas et al., 2013).

Onychiuridae is the most diverse family of Onychiuroidea with approximately 650 species in 56 genera, while Tullbergiidae has 226 species and 32 genera. Onychiuridae is apparently polyphyletic, due to divergence in ancestry between the subfamilies Onychiurinae and Tetrodontophorinae (D'Haese, 2002, 2003). Onychiurids differ by third antennal article sense organ with five conical papillae and five setae protecting the microsensilla or rod-like sensilla, while tullbergids lack such structures and present two microsensilla and two or three dorsal sensilla (Jordana et al., 1997). Onychiuroids genera that have semiaquatic members are: *Lophognathella*, *Onychiurus*, *Thalassaphorura* and *Tullbergia* (Waltz & McCafferty, 1979; Christiansen & Snider, 2008). Paleotullbergiidae is a monotypic family from Africa without any known aquatic association (Bellinger et al., 2017).

Order Entomobryomorpha (*sensu* Soto-Adames et al., 2008)

Entomobryomorpha (Fig. 2.1 B) is the most diverse order of Collembola, with more than 4000 described species in 200 genera, distributed among 8 families and 4 superfamilies: Coenaletoidea, Entomobryoidea, Isotomoidea, and Tomoceroidea (Soto-Adames et al., 2008).

Entomobryoidea (Entomobryidae, Microfalculidae, and Paronellidae) is a monophyletic group, characterized by mostly multilaterally ciliate body setae, cylindrical or broadening macrosetae, presence of a postocular bothriotrichum, and second and third abdominal segments with 2 and 3 bothriotricha, respectively (Zhang et al., 2015). Entomobryidae is the largest family of Collembola, with 1839 species in 64 genera, widely distributed and found in almost all terrestrial and freshwater environments. Paronellidae is also distributed worldwide, with about 550 species in 38 genera. The two families are distinguished from each other by having crenulate dentes with small falcate or bidentate mucrones in Entomobryidae, while the Paronellidae have smooth dentes and square-to-elongate mucrones with up to 10 teeth (Soto-Adames et al., 2008). Microfalculidae is a monotypic family restricted to Magadascar.

Isotomoidea (Actaletidae and Isotomidae) and Coenaletoidea (Coenaletidae) are characterized by having the body without scales and setae usually smooth or unilaterally ciliate. The dorsal chaetotaxy is more homogenous when compared to Entomobryoidea and macrosetae, if present, are always acuminate. The phylogenetic relationships among these groups are still unclear, but Coenaletidae differs from Isotomoidea by having the third abdominal segment fused with the fourth. Actaletidae and Coenaletidae are families of marine Collembola with a few species recorded from Mexico, Central America, and Caribbean Islands, which are present in marine water the first, and associated with hermit crabs of the genus *Coenobita* (Crustacea: Decapoda) the second. Within Isotomoidea, Isotomidae differs from Actaletidae by tenent hairs never leaf-like and fourth to sixth abdominal segments distinctly separated; when fused, then segments are shorter than the length of mesothorax to third abdominal segment. Isotomidae is the second most diverse family of the order, with 112 genera and about 1,400 species, which are usually related to the soil environment (Soto-Adames et al., 2008; Bellinger et al., 2017). However, at least 30 genera of Isotomidae have been recorded in freshwater habitats in North America (Waltz & McCafferty, 1979; Christiansen & Snider, 2008).

Tomoceroidea (Oncopoduridae and Tomoceridae) is possibly a monophyletic group, but with unclear relationships among other groups of Entomobryomorpha. They are similar to Entomobryidae in having ciliate body setae, but differ by lacking bothriotricha on the head, second, and third abdominal segments. Tomoceridae species are distinguished by dens basally with spines, while in Oncopoduridae the dens lacks spines, or if present, on the distal half (Soto-Adames et al., 2008; Bellinger et al., 2017). Genera found in freshwater habitats are *Tomocerus* from North America and *Oncopodura* from the Brazilian Amazon and Mexico (Arlé, 1960; Waltz & McCafferty, 1979).

Order Symphypleona (*sensu* Bretfeld, 1994, 1999)

Symphypleona (Fig. 2.1 C) springtails have antennae longer than head and a globular body, formed by fusion of most thoracic and abdominal segments. They are widely distributed and quite diverse with about 1,300 species in 115 genera, 10 families, and 5 superfamilies: Dicyrtomoidea (Dicyrtomidae), Katiannoidea (Arrhopalitidae, Collophoridae, Katiannidae, Spinothecidae), Sminthuridoidea (Mackenziellidae, Sminthurididae), Sturmioidea (Sturmiidae), and Sminthuroidea (Bourletiellidae, Sminthuridae) (Bellinger et al., 2017).

Sminthuridoidea differs from other superfamilies in the presence of a clasping modification of the second and third antennal articles of males. Sminthurididae differs from Mackenziellidae by the presence of well-developed mandibles and a hypognathous head. Sminthurididae has 154 species in 12 genera, but in Neotropical Region species often found on water surfaces belong to *Denisiella*, *Sminthurides*, and *Sphaeridia* spp. (Waltz & McCafferty, 1979; Bretfeld & Gauer, 1994; Fjellberg, 2007; Christiansen & Snider, 2008).

The remaining superfamilies of Symphypleona form a monophyletic clade, the Appendiciphora, characterized by the presence of a pair of modified setae (subanal appendages) on the inferior anal lobes of the small abdomen of females. The subanal appendix may be directed toward the anus (up) in Dicyrtomoidea and Sminthuroidea, posteriorly in Sturmioidea, or toward the genital opening (down) in Katiannoidea (Bretfeld, 1999).

Sminthuridae is one the most diverse family of Symphypleona, with 260 species in 33 genera, followed by Bourletiellidae with 241 species in 35 genera and Dicyrtomidae 207 species in 8 genera. Sminthuridae differs from Bourletiellidae in having a pair of bothriotricha on fifth abdominal segment and pretarsus usually with a pair of setae. Dicyrtomidae resembles Sminthuridae but the fourth antennal article is much shorter than the third.

In Katiannoidea, Katiannidae is the most diverse family with about 210 species in 19 genera, with at least six genera recorded from the Neotropical Region. Katiannidae differs from other families of the group by the presence of 6–8 eyes and spatulate setae on tibiotarsus. Collophoridae has only 9 species all of genus Collophora, while Arrhopalitidae 150 species in 3 genera, both families are characterized by reduction of eyes (0–4). Collophoridae has a capitate bothriotrix on each side of the small abdomen; these setae are pointed in Katiannidae and Arrhopalitidae (Bretfeld, 1999; Bellinger et al., 2017).

Genera of Appendiciphora considered aquatic or semi-aquatic in North America are: *Arrhopalites* (Arrhopalitidae), *Collophora* (Collophoridae), *Sminthurinus* (Katiannidae), *Bourletiella*, *Deuterosminthurus*, *Heterosminthurus*, *Pseudobourletiella* (Bourletiellidae), *Neosminthurus*, *Sminthurus*, and *Sphyrotheca* (Sminthuridae) (Waltz & McCafferty, 1979; Christiansen & Snider, 2008).

Order Neelipleona, Family Neelidae

Neelidae (Fig. 2.1 D) is the only family in Neelipleona, and similarly to Symphypleona, species have globular bodies. They differ from Symphypleona by minute body size (some Symphypleona species may be also minute), antennae shorter than head, absence of eyes, sensory fields on the large abdomen, and reduced small abdomen hidden ventrally (Schneider & D'Haese, 2013).

Currently Neelidae comprises 44 species in 5 genera, of which *Megalothorax*, *Neelides*, *Neelus*, and *Spinaethorax* are recorded in the Neotropical Region. The species are usually associated with euedaphic environments and caves, but species such as *Megalothorax aquaticus* Stach, 1951 are found on freshwater in Europe, as well as *Neelus* spp. in North America (Christiansen & Snider, 2008; Bellinger et al., 2017).

LIMITATIONS

Identification of springtails is challenging because of chaetotaxy, use of immature specimens, polymorphisms, inattention to type specimens, and imprecise location of collected individuals. Improved descriptions should include chaetotaxy (distribution, morphology, size and presence or absence of some head, trunk or appendicular setae) as well as other morphological characters. The lack of details in descriptions can hide a range of similar species under a single name or make separation of two or more previously described species difficult. In Brazil, a clear example is seen with *Entomobrya* species, which in most cases were described mostly based in color patterns, a character which may be useful to depict species, but seen as highly plastic at least in some taxa (Heckman, 2001; Katz et al., 2015). Descriptions of immature forms lead to uncertain species identities since they differ from adults in size, color, morphometry, absence of genital plate, and pore chaetotaxy. For example, within populations of some Entomobryoidea species variations occur in color pattern, dorsal body setae, labial chaetotaxy, and number of teeth on unguis (Cipola et al., 2016, 2017; Soto-Adames, 2016). Juveniles of many Poduromorpha have reduced chaetotaxy and less sensilla on fourth antennal article, as well as less setae on dens and sometimes different type of anal spines. Few specimens can even present punctual asymmetries or polymorphisms, especially regarding chaetotaxic elements on the left and right sides.

The misuse of corrosive chemicals during clarification, improper choice of conservation medium or storage of slides under inadequate conditions can lead to material degradation. Another problem is the absence of important collecting data attached to the material, especially regarding type locality, which hinders recollection of species from their original collection sites. An example of this problem is *Dicranocentrus silvestrii* Absolon, with the type locality given as "South America" leading to a very unclear type locality (Mari Mutt, 1979). *Mastigoceras camponoti* Handschin, a monotypic endemic taxon from Brazil, was stated by the author to have been collected from "South of Minas Gerais state," which is one of the largest federal units of Brazil (Heckman, 2001). Another example is provided by Börner (1906) who described three species from

"São Franscisco, Brazil," which could be several locations in different regions of the country.

Neotropical springtails are not well known compared to other regions. Numerous ecosystems and even large phytogeographic domains in Central and South America are much understudied. Most of the species recorded from the New World have been described from the United States, Mexico, Puerto Rico, and Brazil, with few or no species from other countries (Mari-Mutt & Bellinger, 1990). At the same time, large countries with comparatively well-known faunas like Brazil are still undersampled in many regions and/or federal units (Zeppelini et al., 2017). Second, the Neotropical Region potentially has a high biodiversity of land arthropods, including Collembola, so it is expected that many new species will be found in unexplored habitats. Lack of knowledge combined with a large undescribed fauna limits the overall comprehension of the distribution and composition of Collembola fauna in Neotropical Region. For instance, regarding the aquatic fauna, only 45 species in 15 genera have been recorded so far, which we believe does not reflect the real species richness of aquatic springtails in the region since these numbers were gathered from isolated data, not from directed collections (Deharveng et al., 2008).

All of these factors lead to the lack of identification keys and specific diagnoses for freshwater springtails from the

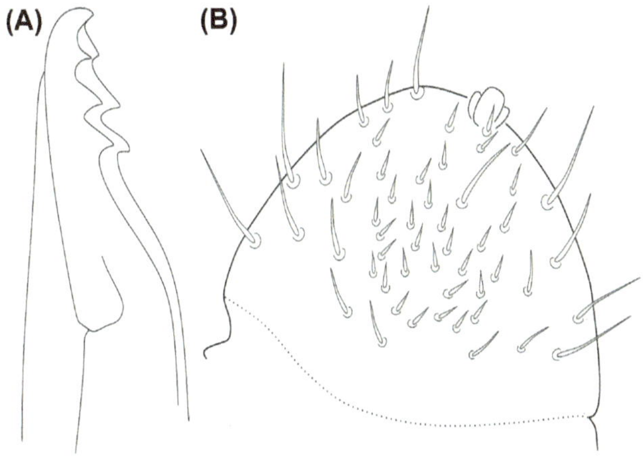

FIGURE 2.4 (A) *Rapoportella mucronata* Najt & Massoud, 1974 (Brachystomellidae), maxilla capitulum. (B) *Brachystomellides navarinensis* Weiner & Najt, 1997 (Brachystomellidae), fourth antennal segment ventral side. *(A, B) Modified from the original description.*

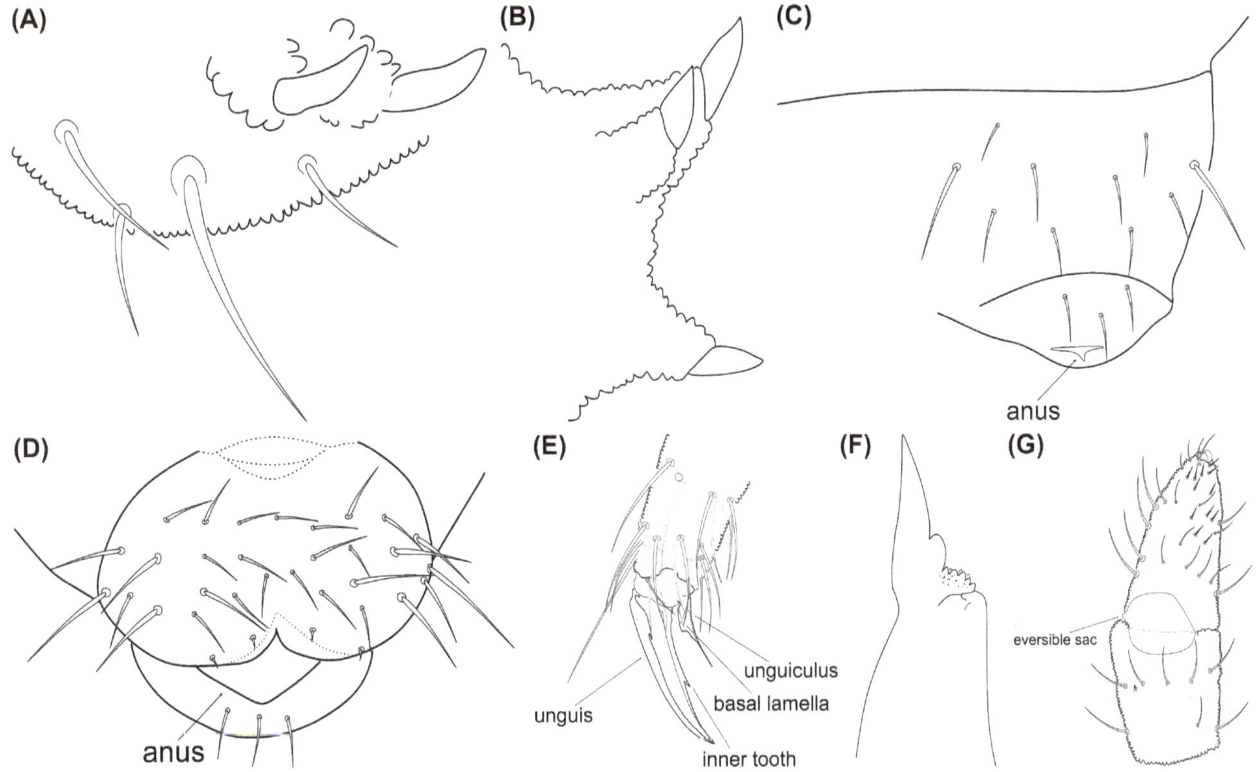

FIGURE 2.5 (A) *Ceratophysella* (Hypogastruridae), sixth abdominal segment spines. (B) *Triacanthella nivalis* Cassagnau & Deharveng, 1974 (Hypogastruridae), sixth abdominal segment spines. (C) *Paraxenylla affiniformis* Stach, 1929 (Hypogastruridae), fifth and sixth abdominal segments and anus position (dorsal view). (D) *Xenylla capixaba* Fernandes & Mendonça, 2010 (Hypogastruridae), sixth abdominal segment and anus position (ventral view). (E) *Ceratophysella* (Hypogastruridae), third tibiotarsus distal region, and empodical complex. (F) *Microgastrura sofiae* Vázquez & Palacios-Vargas, 1996 (Hypogastruridae), mandible apex. (G) *Ceratophysella* (Hypogastruridae), fourth and third antennal segments ventral faces. *(B, C) Modified from Thibaud et al. (2004). (D, F) Modified from the original description.*

Neotropics. The only specific key to freshwater springtails of the New World is for the Nearctic taxa (Waltz & McCafferty, 1979), since the key to Neotropical species provided by Heckman (2001) is not specific for aquatic taxa, therefore, is not directly relevant to Neotropical Collembola fauna.

TERMINOLOGY AND MORPHOLOGY

The general morphology of Collembola is based mostly on the following references: Hopkin (1997), Christiansen & Bellinger (1998), Fjellberg (1999, 2007), Heckman (2001), Deharveng (2004), and Bellinger et al. (2017). Specific references for each order follows: Massoud (1967), Jordana et al. (1997), Thibaud et al. (2004), and Najt et al. (2005) for Poduromorpha; Potapov (2001) and Soto-Adames et al. (2008) for Entomobryomorpha; Richards (1968), Betsch (1980), and Bretfeld (1999) for Neelipleona and Symphypleona.

Springtails, as well as other hexapods, have the body divided into three tagmata: head, thorax and abdomen (Figs. 2.1 A, B). The body is entirely covered by a thin granulous cuticle, which is variably hydrophobic. The granules of epicuticle are typically hexagonal patches formed by the connection of primary triangular microgranules. This pattern of cuticular organization is quite evident in scanning electron microscope photographs of Poduromorpha species. Other taxa may have reduction or fusion of such granules giving the epicuticle another microprofile. In some species of Symphypleona, the cuticular granules can also be square (formed by four linked granules), not hexagonal. The cuticle has various sizes of setae (usually smooth or ciliated), spines, sensilla and/or scales, which combined are mostly diagnostic for different taxonomical groups (Figs. 2.3, 4 B, 5 A–D, 6, 7, 9, 13, 16, 20–24, 34, 37, 42, 45–48, 50, 52, 53, 58–63, 65, 66). An example of this is the presence of scales in some Entomobryomorpha (Figs. 2.8 A–D, 9 B, 23).

The head is dorsally rounded, elliptical, or subtriangular and bears a pair of antennae, eyepatches and postantennal organs (which can be totally absent), and mouthparts (Figs. 2.1, 8, 18, 25, 32, 33, 38, 48, 55, 57, 66A). Springtails are prognathous (most Poduromorpha and Entomobryomorpha) or hypognathous (Neelipleona and most Symphypleona) (Fig. 2.1).

Antennae are inserted anterodorsally on head and have intrinsic muscles that allow independent movement of all four antennal articles (I–IV). Antenna size is variable within different taxa and they can be shorter than the head (e.g., Neelidae, Figs. 2.1 D, 66 A) to much longer than the whole body (e.g., *Campylothorax*, *Salina*, and *Temeritas*) (Figs. 2.32 B, C, 55 A). The antennal articles I–II are usually simple and undivided, but in some cases the first (e.g., *Mastigoceras*) and the second (e.g., *Dicranocentrus*) segments may be subdivided, creating the impression of V or VI segments (Figs. 2.8 D, E). The antennal article III is usually undivided and possesses an apical sense organ with distinctive setae or and associated sensilla (Figs. 2.3 D, E). In males of Sminthurididae (e.g., *Sminthurides*) and in Coenaletidae (*Coenaletes*) the antennal articles II–III are modified into a clasping organ for holding hold females during courtship (Fig. 2.46). The antennal article IV can be undivided (more common), annulated (e.g., *Dicranocentrus* and *Mastigoceras*) or pseudosegmented (e.g., *Sminthurides*, *Temeritas*) (Figs. 2.8 D–E, 55). This segment often has an apical bulb and numerous sensory modifications (sensilla) of taxonomic significance, especially in Poduromorpha (Figs. 2.4 B, 2.5 G, 6 F, 7 A, 21). In most Collembola taxa the fourth segment is the longest, but in Dicyrtomidae and most Tomoceridae it is much shorter than the third segment (Fig. 2.38).

The eyepatches are dorsal and bear a small set of eyes (ommatidia) with up to 8 lenses on each side (Figs. 2.1 A–C, 18, 25, 33 B, 57 A, 58 A). Eyepatches are usually pigmented (black, brown, or dark blue). Reduction in number of ommatidia is common and nearly all families have taxa with fewer than 8 + 8 eyes. In a few families the number is always reduced, as in Collophoridae (4 + 4 eyes), Arrhopalitidae (2 + 2 or less), and Neelidae and Cyphoderinae (Paronellidae) (without eyes) (Figs. 2.1 D, 32 A, 33 A, 66 A).

Between the eyes and antenna in some families a postantennal organ (PAO) may be present (e.g., Brachysto mellidae, Neanuridae, Isotomidae, and rarely in Entomobryidae), simple and oval (e.g., part of Isotomidae) to quite complex or formed by many circular vesicles (moruliform, e.g., part of *Anurida*, *Morulina*, and *Neotropiella*) (Figs. 2.6 G–H, 7 E, 18–19). The structure of the PAO is variable and of taxonomic importance for being a hydrophilic structure and it is possibly a chemo, thermo, or hygroreceptor.

The dorsal region of the mouth is covered by a variably developed clypeus usually carrying some setae and terminating in the prelabrum and an anterior labrum, both with a variable number of setae (Fig. 2.67). The labium is found on the ventral side of the head. In most Collembola it consists of a pair of palps often with five main papillae and a variable number of accessory (guard) setae. Each palp is associated with three labial plates, each with a few setae. The maxillary palps (or maxillary outer lobes) are basically conical projections seen laterally of labial papillae. Each palp usually has one apical and one basal setae, with rare exceptions of one or two additional setae in Isotomidae (e.g., *Desoria*). The sublobal plate may have up to four setae-like projections which can be totally absent. Finally, labial and postlabial regions present several typical setae of taxonomic importance.

Inside the cephalic capsule are the mouthparts, mostly formed by a pair of strong chewing mandibles and a pair of

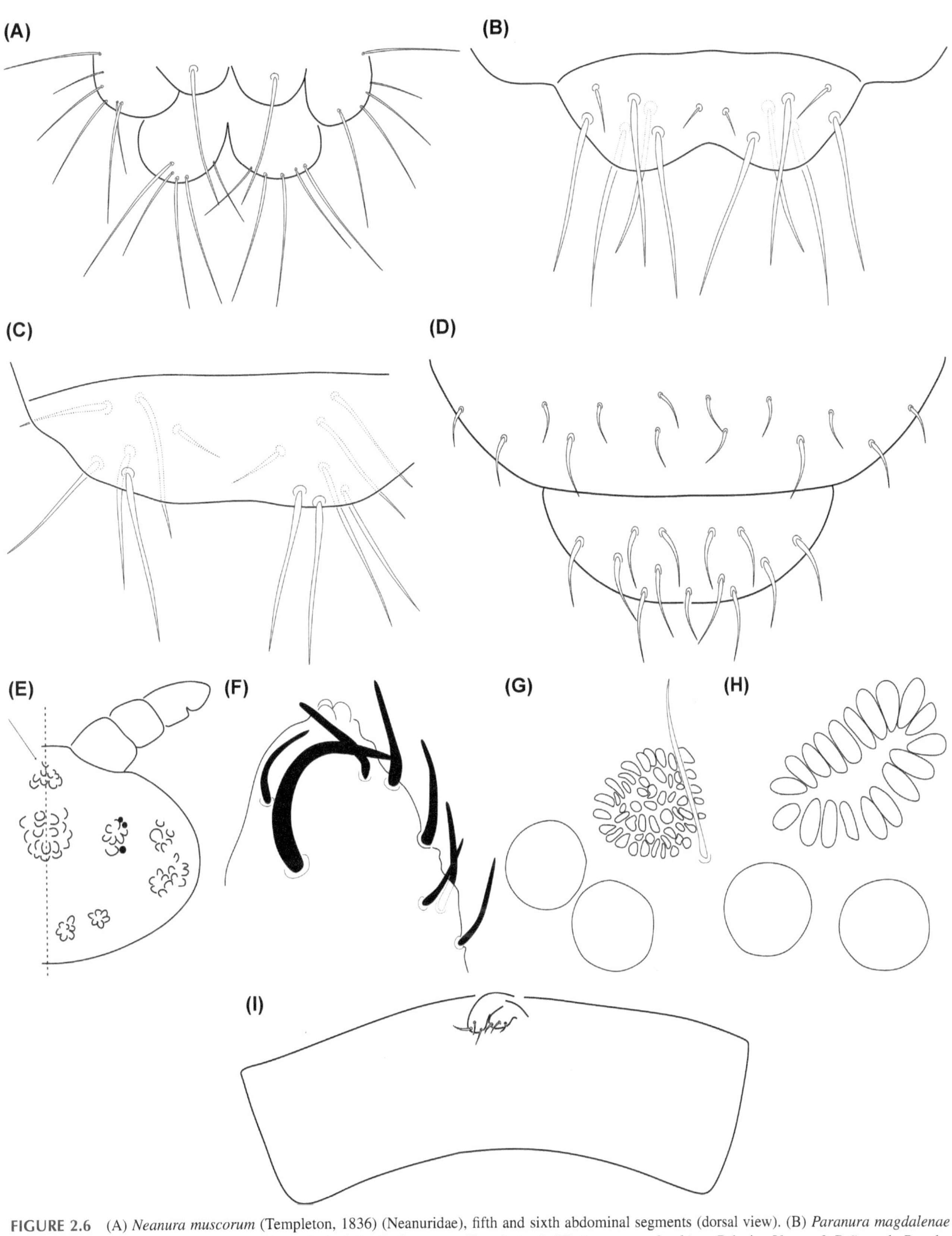

FIGURE 2.6 (A) *Neanura muscorum* (Templeton, 1836) (Neanuridae), fifth and sixth abdominal segments (dorsal view). (B) *Paranura magdalenae* Benito & Palacios-Vargas, 2008 (Neanuridae), sixth abdominal segment (dorsal view). (C) *Paranura colombiana* Palacios-Vargas & Peñaranda-Parada, 2005 (Neanuridae), sixth abdominal segment (dorsal view). (D) *Pseudachorutes parvulus* Börner, 1901 (Neanuridae), fifth and sixth abdominal segments (dorsal view). (E) *Neanura muscorum* (Templeton, 1835) (Neanuridae), head tubercles, setae omitted (dorsal view). *Arrow* points to unpaired antenno-frontal tubercule (clypeal). (F) *Sensillanura santizaragozai* Palacios-Vargas & Catalán, 2010 (Neanuridae), fourth antennal segment sensilla (setae omitted). (G) *Aethiopella condei* Massoud, 1963 (Neanuridae), frontal eyes and postantennal organ. (H) *Pseudachorutes subabdominalis* Steiner, 1958 (Neanuridae), frontal eyes and postantennal organ. (I) *Hylaeanura mendoncae* Zeppelini & Palacios-Vargas, 2013 (Neanuridae), fourth abdominal segment ventral side (setae omitted) and reduced furcula. *(A, G) Modified from Massoud (1967). (B, C, F, I) Modified from the original description. (D, E, H) Modified from Jordana et al. (1997).*

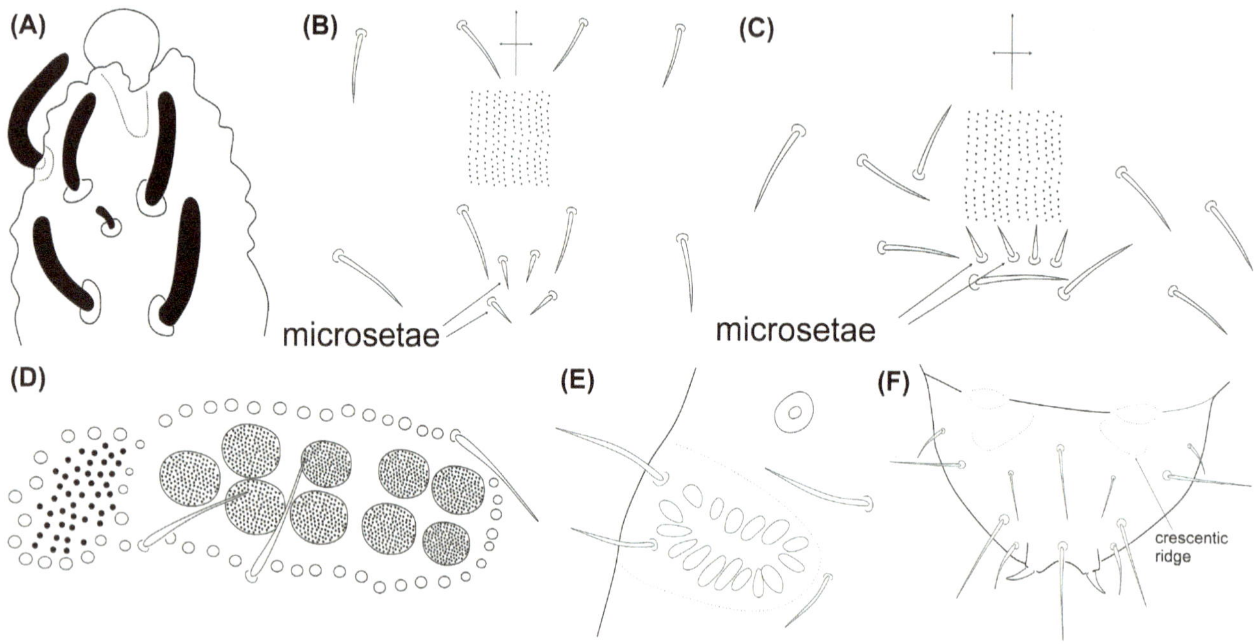

FIGURE 2.7 (A) *Odontellina nivalis* (Cassagnau, 1959) (Odontellidae), fourth antennal segment apex. (B) *Thalassaphorura brevisetosa* Sun, Gao & Potapov, 2013 (Onychiuridae), fourth abdominal segment ventral view (furcula area). (C) *Onychiurus gulinensis* Sun & Zhang, 2012 (Onychiuridae), fourth abdominal segment ventral view (furcula area). (D) *Agraphorura calvoi* Arbea, 2005 (Onychiuridae), postantennal organ. (E) *Thalassaphorura brevisetosa* Sun, Gao & Potapov, 2013 (Onychiuridae), postantennal organ. (F) *Mesaphorura matilei* Thibaud & Palacios-Vargas, 2000 (Tullbergiidae), sixth abdominal segment dorsal view. *(A) Modified from Massoud (1967). (B–F) Modified from the original description.*

lateral maxillae (Figs. 2.2, 4 A, 5 F, 64). The mandibles consist apically by an inner row of incisors teeth (4–6) followed by a basal molar plate with numerous conical denticles (Fig. 2.2 A). Mandibles are absent in Brachystomellidae (Fig. 2.2 B) and lacking molar plates in Neanuridae (Figs. 2.2 D–E). The maxillae are usually formed by a variable number of apical teeth plus some fringed associated lamellae (Figs. 2.2 B–E). Maxillae teeth and lamellae could also be reduced or modified, as seen in styliform maxillae of Pseudachorutinae (Neanuridae) (Figs. 2.2 F, G). The mandibles and maxillae can be elongated in *Rhynchocyrtus*, *Stenognathriopes*, and *Stenognathellus* (Figs. 2.63 B and 64).

The trunk is formed by a thorax with ambulacral legs and a legless abdomen. It can be cylindrical or elongated and have distinct segments (Entomobryomorpha and Poduromorpha), or the segments can be mostly fused to each other into a rounded or globose body (Symphypleona and Neelipleona) (Fig. 2.1).

The thorax has three segments, prothorax (I), mesothorax (II), and metathorax (III), each with a pair of legs. The prothorax is the shortest segment. Poduromorpha has a distinct plate-like pronotum with setae and papillae (rarely), while Entomobryomorpha lacks a distinct pronotum and dorsal prothoracic setae (Figs. 2.1 A, B). The mesothorax and metathorax are simple and similar to each other, although in few cases one or the other may be enlarged and strongly projected along the midline (e.g., *Lepidocyrtoides*, *Lepidocyrtinus*, and *Campylothorax*) (Figs. 2.8 A, B, 32 C).

Each leg is formed by epicoxa, subcoxa (epicoxa or both may be fused with body), coxa, trochanter, femur, tibiotarsus, pretarsus, and one claw (with or without an empodial appendage—unguiculus). Coxae usually have only a few setae. In some Symphypleona, such as Sminthuridae, the hind subcoxa possesses a robust spine-like seta (Fig. 2.40). The next segment, the trochanter, usually presents few setae, but in Entomobryoidea and Tomoceroidea the hind trochanter may have a field of small, smooth spine-like setae (trochanteral organ). The Symphypleonan families Katiannidae and Sminthuridae may also have trochanteral organ composed by a single robust spine (Fig. 2.40). In these families such organ can occur on both middle and hind trochanters, and may not be homologous to the trochanteral organ in Entomobryoidea.

The femur (plural, femora) is an undivided segment with setae and/or scales (Fig. 2.1). This segment usually has minor taxonomic value but a remarkably exception is the entomobryid *Tyranosseira*, in which males have broadened femora with distinct spines on the first pair of legs (Bellini & Zeppelini, 2011).

The tibiotarsus (plural, tibiotarsi) is presumed to be formed by the fusion of tibia and tarsomeres seen in other hexapods. This segment usually is long and covered by

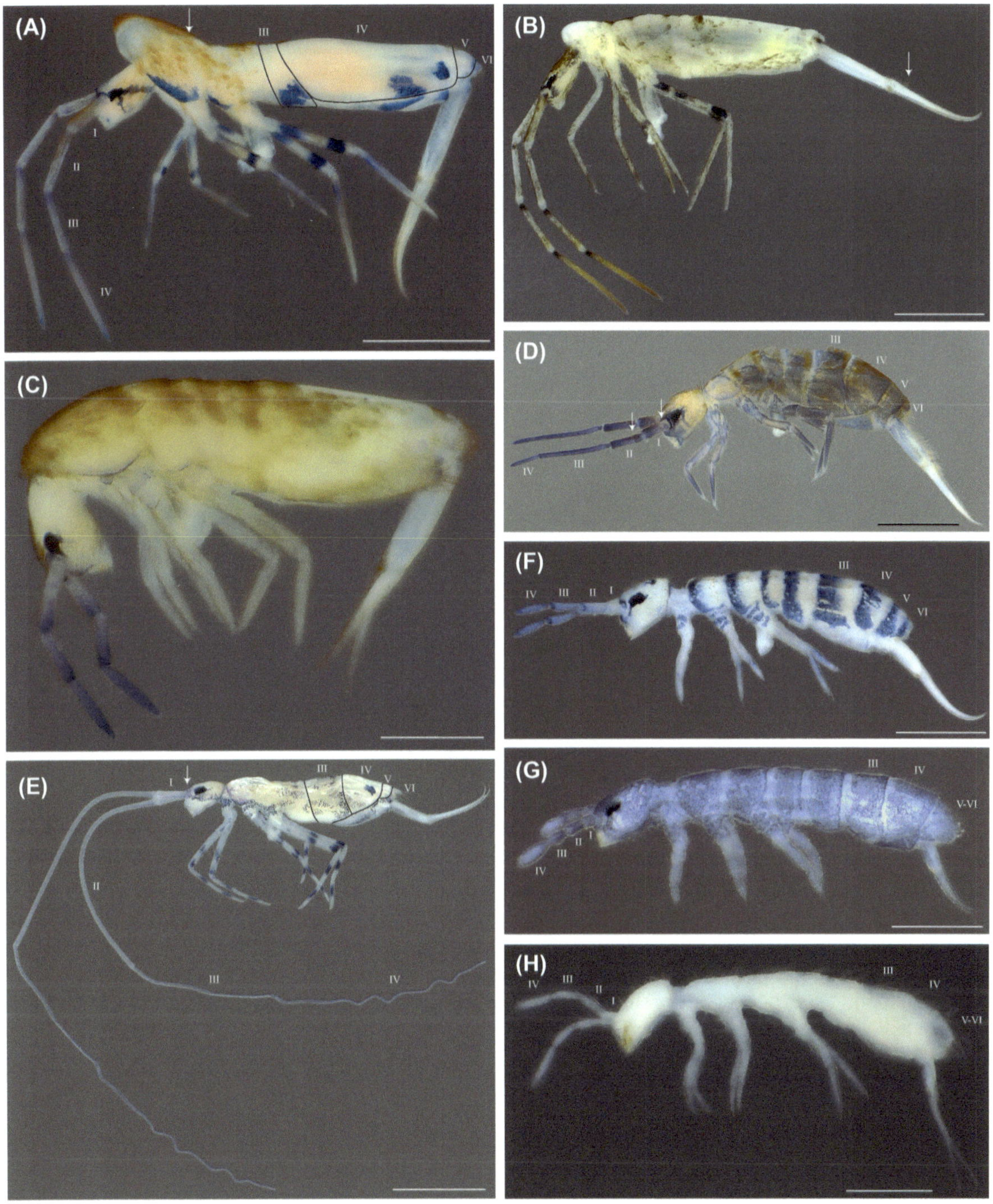

FIGURE 2.8 (A) *Lepidocyrtoides* (Entomobryidae), habitus, lateral view, *arrow* indicate scales on body. Bar = 1 mm. (B) *Lepidocyrtinus* (Entomobryidae), habitus, lateral view, *arrow* indicates long blunt setae on dorsal dens. Bar = 1 mm. (C) *Lepidocyrtus* (Entomobryidae), habitus, lateral view. Bar = 0.3 mm. (D) *Dicranocentrus* (Entomobryidae), habitus, lateral view, *arrows* indicate subdivision of the first and second antennal segment. Bar = 1 mm. (E) *Mastigoceras camponoti* (Entomobryidae), habitus, lateral view, *arrows* indicate subdivision of the first antennal segment. Bar = 1 mm. (F) *Isotomurus* gr. *palustris* (Isotomidae), habitus, lateral view. Bar = 0.5 mm. (G) *Hemisotoma thermophila* (Isotomidae), habitus, lateral view. Bar = 0.2 mm. (H) *Isotomiella amazonica* (Isotomidae), habitus, lateral view. Bar = 0.2 mm.

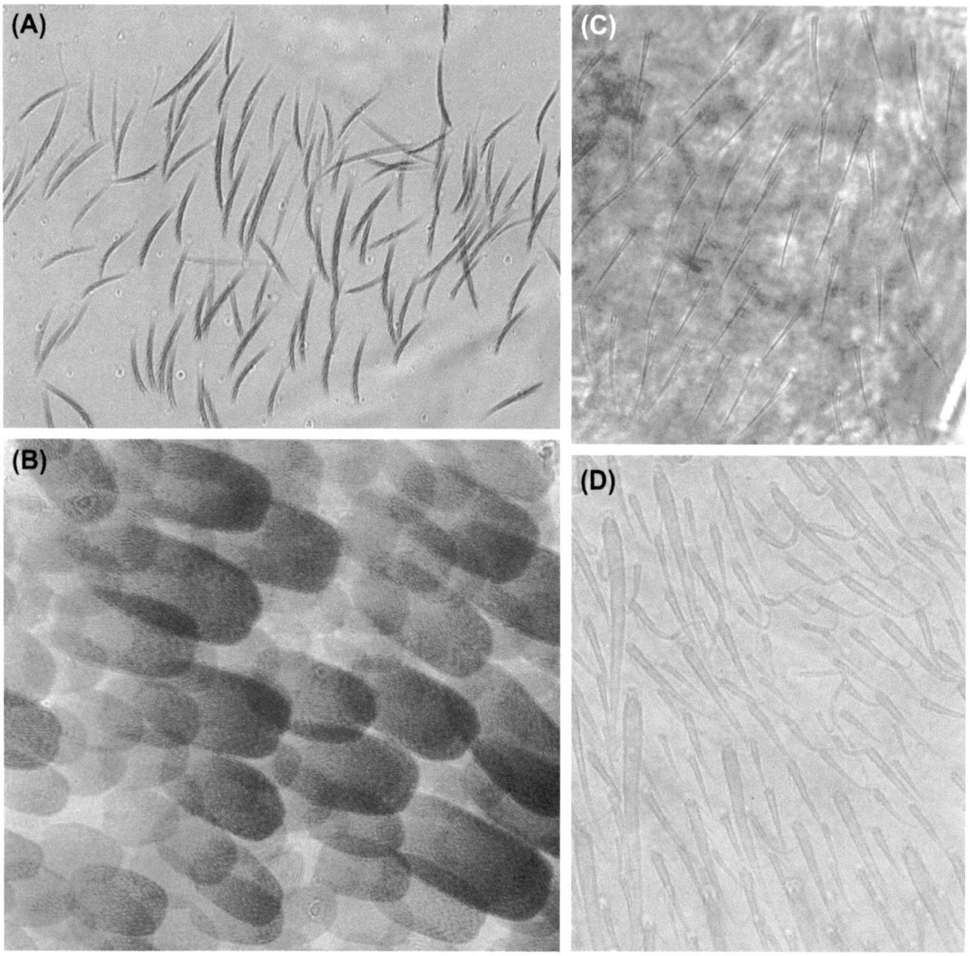

FIGURE 2.9 (A) *Orchesella* (Entomobryidae), tergal ciliate setae. (B) *Seira* (Entomobryidae), tergal scales. (C) *Hemisotoma* (Isotomidae), tergal smooth setae. (D) *Isotomurus* gr. *plumosus* (Isotomidae), multilaterally ciliate macrosetae.

numerous setae, which are almost all ciliate in Entomobryoidea and smooth to serrate in Poduromorpha, Isotomidae, Symphypleona, and Neelipleona. Depending on taxon, tibiotarsi can possess various shapes of setae, scales, or spines (Figs. 2.1, 5 E, 59, 60 C). The distal region (usually on external side) of tibiotarsi may bear one (e.g., Entomobryidae) or more (e.g., Katiannidae and Bourletiellidae) modified setae called tenent hairs, which can be pointed, clavate, spatulate, or capitate (Figs. 2.27, 36, 42, 60). In Symphypleona, the tibiotarsi may also have small, pit-like tibiotarsal organs, which aid in the diagnosis of some genera and species (Figs. 2.44, 47).

The pretarsus is a small membranous structure to which are attached the unguis and unguiculus. Together they all form the empodial or foot complex (Figs. 2.5 E, 27, 54). Pretarsus usually has two very small smooth setae, one on anterior and another on posterior face, but the latter is absent in Bourletiellidae and both are apparently absent in *Campylothorax*. The unguis (plural, ungues) is the single tibiotarsal claw of springtails (Figs. 2.1, 5 E, 27, 36, 60). It consists of an external convex transverse plate tapering distally joined on its inner concave side by a central lamella which may contain one or more teeth, paired or not. In some isotomids and symphypleonans (e.g., Katiannidae and Arrhopalitidae) the external ungual edge may have an inflated membrane (tunica), as well as complex toothed lamellae (pseudonychia) (Fig. 2.49). The unguiculus (plural, unguiculi), or empodial appendage, is a secondary smaller claw-like structure with an outer plate and an inner lamella. The external side is usually acuminate and can terminate as an apical or subapical filament in some Poduromorpha and Symphypleona (Fig. 2.60 C). The inner unguicular lamella is acuminate, truncate, or even concave. In several families of Poduromorpha, the unguiculi are reduced to a minute spine or entirely missing (e.g., Neanuridae) (Figs. 2.36, 60).

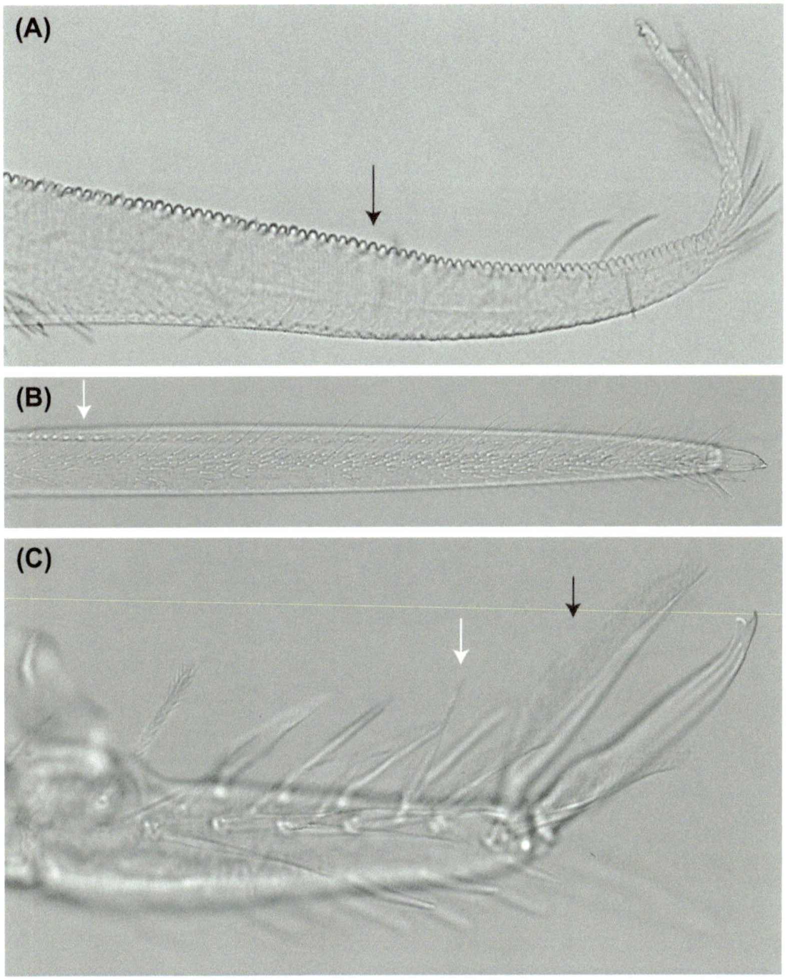

FIGURE 2.10 (A) *Orchesella* (Entomobryidae), dens, lateral view, *arrow* indicates dorsal crenulation. (B) *Campylothorax* (Paronellidae), dens, dorsal view, *arrow* indicates dental spines. (C) *Cyphoderus* (Paronellidae), dens, lateral view, *white arrow* indicates ciliate seta and *black arrow* indicates fringed scale-like seta.

The springtail abdomen has six segments (I–VI) (Figs. 2.1 A, B). In Arthropleona (Poduromorpha + Entomobryomorpha) they are separated with sutures, but in some Isotomidae genera the segments IV–VI or V–VI can be completely fused to each other (e.g., *Folsomina*, *Folsomia*, *Cryptopygus*, *Isotomiella*, and *Hemisotoma*) (Figs. 2.8 G, 17, 22 B). In Poduromorpha, Isotomidae, Tomoceridae, and some Entomobryidae (e.g., *Dicranocentrus*, *Mastigoceras*) abdominal segments III and IV have similar lengths along the midline (Figs. 2.1 A, 8 D–H), while in most Entomobryidae and in all Paronellidae the abdominal segment IV is at least twice the length of the segment III (Figs. 2.1 B, 8 A, 32). In Symphypleona, the thoracic and abdominal segments are largely fused into the "large" and "small" abdomen (Figs. 2.1 C, D, 38, 55). In most families the large abdomen is formed by the thoracic segments II–III and the abdominal segments I–IV, and the small abdomen consists of the abdominal segments V–VI (Arrhopalitidae, Bourletiellidae, Katiannidae, and Sminthuridae). In some Sminthuridae the prothorax is incorporated into the large abdomen. In Dicyrtomidae the large abdomen contains the abdominal segment V (Fig. 2.38). In Collophoridae and *Neosminthurus* (Sminthuridae) the large abdomen includes almost all trunk segments, excluding only the abdominal segment VI.

Features on the ventral side of the abdomen include the collophore, tenaculum, furcula, genital plate, and anus in the abdominal segments I, III, IV, V, and VI, respectively (Fig. 2.1).

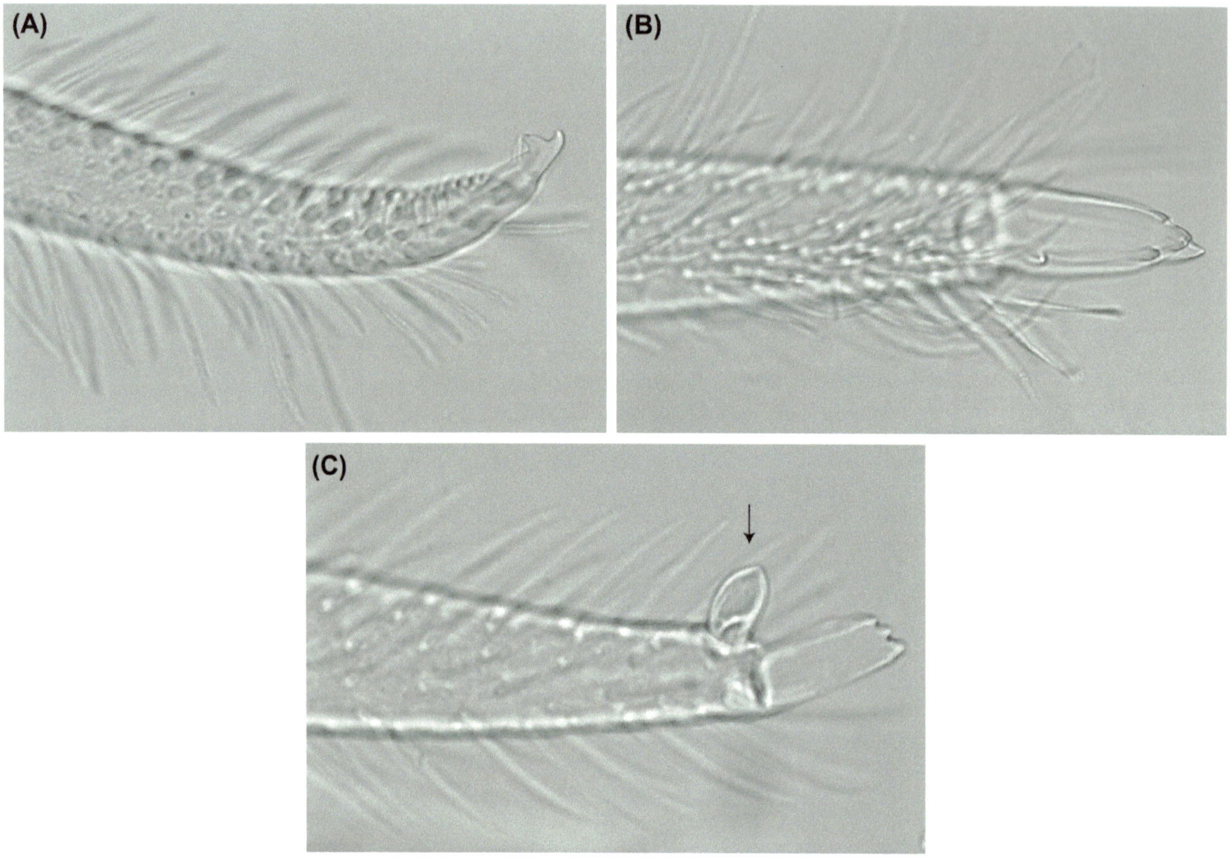

FIGURE 2.11 (A) *Entomobrya* (Entomobryidae), distal dens and mucro, lateral view. (B) *Campylothorax* (Paronellidae), distal dens and mucro, dorsal view. (C) *Salina* (Paronellidae), distal dens and mucro, dorsal view.

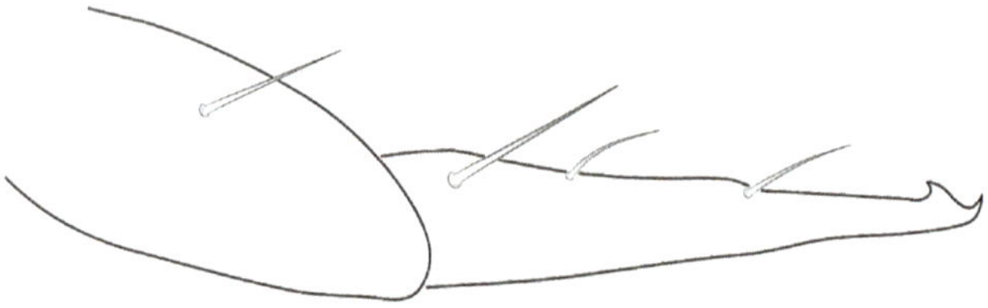

FIGURE 2.12 *Folsomides parvulus* Stach, 1922 (Isotomidae), furcula, lateral view. *Modified from Potapov (2001).*

The collophore (or ventral tube, Figs. 2.1 B, C) is formed by a cylindrical corpus attached to a basal plate, from which two sacs are distally everted. The corpus is anteriorly split by the ventral groove, and it is a unique piece (except in Sturmiidae in which is biparted). The anterior, laterodistal, and posterior faces of collophore corpus can bear setae of taxonomical importance. The eversible sacs (or filaments, or vesicles) can be short (Poduromorpha and Entomobryomorpha) or long (Symphypleona). The tenaculum (or retinaculum) consists of a basal corpus with or without setae, and a pair of lateral rami with 2–4 distal teeth. The tenaculum has the function of holding the furcula (or furca) in place before a leap (Fig. 2.1 B). When the furcula is absent or vestigial, the tenaculum is also usually absent (e.g., *Friesea*, *Americanura*, *Stachia*, and *Stachiomella*).

The furcula is the jumping appendage of springtails and can be long and well-developed (e.g., Entomobryoidea), to short, reduced, or even absent in Onychiuridae, Tullbergiidae, and some Brachystomellidae, Hypogastruridae, and Neanuridae (Figs. 2.1, 6 I, 12). This organ is actually the highly modified and partially fused appendages of the abdominal segment IV. Furcula consists of a proximal manubrium, a pair of dentes (singular dens), and a pair of distal mucrones (singular mucro). The manubrium is a single segment usually with setae. Some taxa may have ventral manubrial scales (e.g., some Entomobryoidea), spines (*Isotoma*), or even lack setae completely (*Archisotoma*, *Ballistura*, and *Folsomides*) (Fig. 2.13). Manubrium can also hold dorso-distal pseudopores and modified setae, in a small region called manubrial plate (Fig. 2.31). The dentes are paired cylindrical segments, generally longer than the manubrium and covered by setae, spines, and/or scales. Rarely the dens can present modifications such as

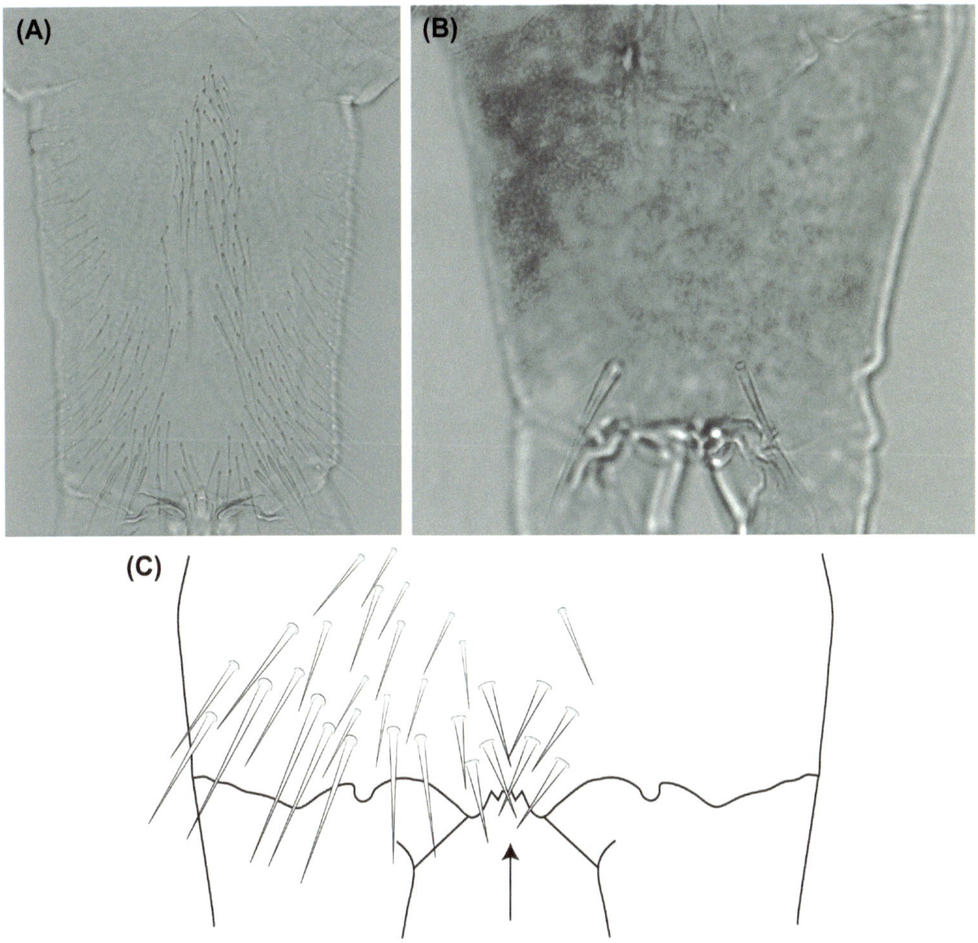

FIGURE 2.13 (A) *Isotomurus* sp. (Isotomidae), ventral manubrium. (B) *Hemisotoma thermophila* (Axelson, 1900) (Isotomidae), ventral manubrium. (C) *Isotoma viridis* Bourlet, 1839 (Isotomidae), ventro-distal manubrium, ventral view, *arrow* indicates spine-like setae.

tubercles (e.g., in *Granisotoma*, Fig. 2.14 B) and appendages in the anterior (e.g., neotropical *Lepidocyrtus*) or distal regions (e.g., *Salina*) (Fig. 2.11 C). In Entomobryidae, some genera of Isotomidae and rarely in *Arlesminthurus* (Bourletiellidae) the dorsal surface of dens is crenulated (Figs. 2.10 A, 14 A, 61). Another modification is the annulated form of dens in Poduridae (Fig. 2.3 B). Moreover, some Entomobryoidea, Isotomidae, and Tomoceridae may have dorsal spines on dens or large scale-like setae (Figs. 2.10 B, C, 24). At the tip of each dens there is a mucro, which shape vary widely among families and genera. To most Poduromorpha mucro can be spoon-like, relatively large when compared to the dens length, with smooth edges and with or without a small lateral tooth (Figs. 2.3 A, B). In Isotomidae mucro can bear one to four teeth and usually is very short (Figs. 2.12, 15). To Entomobryidae mucro is also short, with one or two teeth, with or without a single mucronal seta (Figs. 2.11 A, 26). In Paronellidae mucro is quite variable, and can be long with up to 10 teeth (as in Cyphoderinae), or short with up to 6 teeth (in Paronellinae) and in both cases lack any mucronal seta (Figs. 2.10 C, 11 B, C). In Symphypleona the mucro is large and flattened dorsoventrally forming concave spatula with smooth edges (Bourletiellidae), or a spoon-like structure with at least one serrated edge (most Sminthuridae) (Fig. 2.61). The mucro may have a basolateral seta, especially in some Isotomidae and Symphypleona (Figs. 2.14 A, 45).

On females the genital plate is ventral in abdominal segment V. It consists of a transverse opening with several surrounding setae, while males present a circular plate with a longitudinal short aperture surrounded by setae. In Entomobryomorpha, the male genital plate can be oligochaetotic (with few setae), multisetaceous (with many

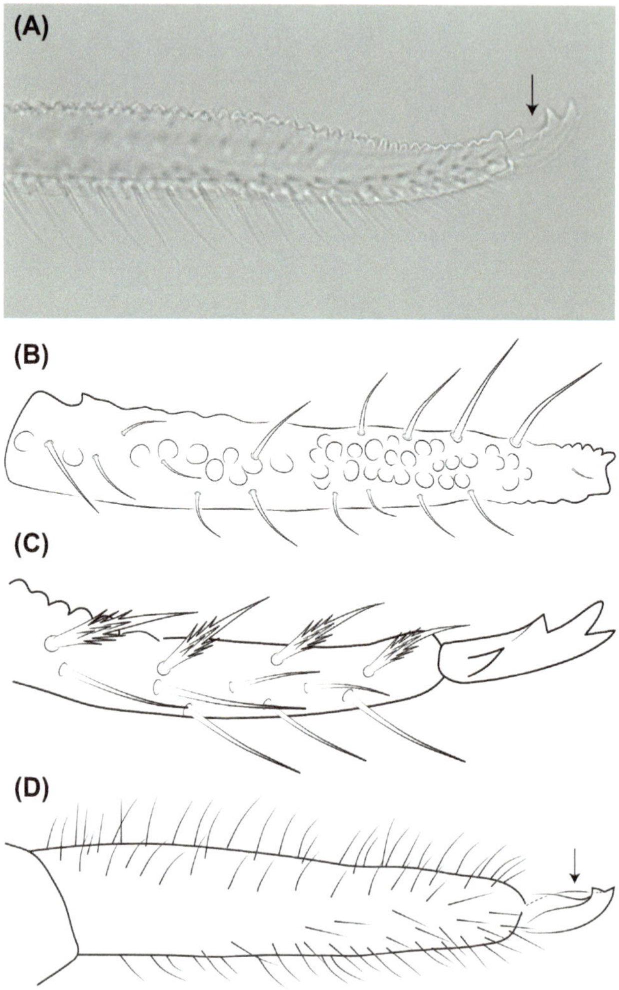

FIGURE 2.14 (A) *Isotomurus* sp. (Isotomidae), dens and mucro, lateral view, *arrow* indicates dorsal lamella. (B) *Granisotoma rainieri* (Folsom, 1937) (Isotomidae), dorsal dens. (C) *Paracerura* sp. (Isotomidae), distal dens and mucro, lateral view. (D) *Ballistura schoetti* (Dalla Torre, 1895) (Isotomidae), dens and mucro, lateral view, *arrow* indicates dorsal lamella. *(B, D) Modified from Potapov (2001). (C) Modified from Abrantes & Duarte (2013).*

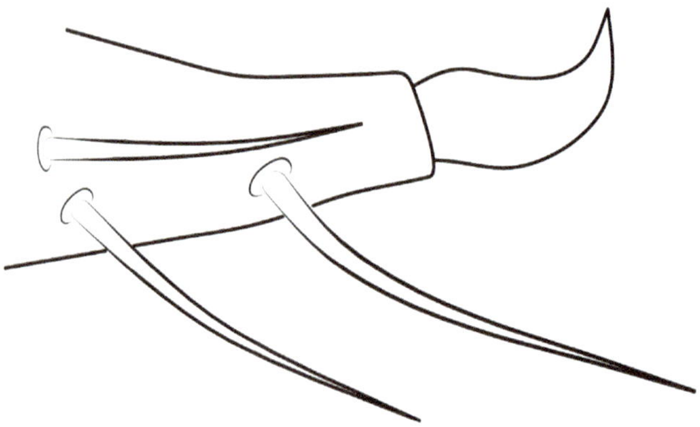

FIGURE 2.15 *Arlea arenicola* Abrantes & Mendonça (2005) (Isotomidae), distal dens and mucro, lateral view. *Modified from the original description.*

setae), circinate (almost smooth without setae), and papillate (with papillae). The male plate has been used taxonomically mainly for some Entomobryidae, but it also may be of value in some Tullbergiidae and Symphypleona.

The abdominal segment VI is divided into one dorsal and two subventral lobes that together surround the triradiate anus. The three anal lobes (or valves) bear setae of taxonomic value at least in Poduromorpha and Symphypleona. In most Symphypleona females have a pair of thick, spiniform subanal appendages (Fig. 2.37). The dorsal anal lobe carries one or more pairs of anal spines in some Hypogastruridae, Isotomidae, Neanuridae, Odontellidae, Onychiuridae, and Tullbergiidae (Figs. 2.5 A, B).

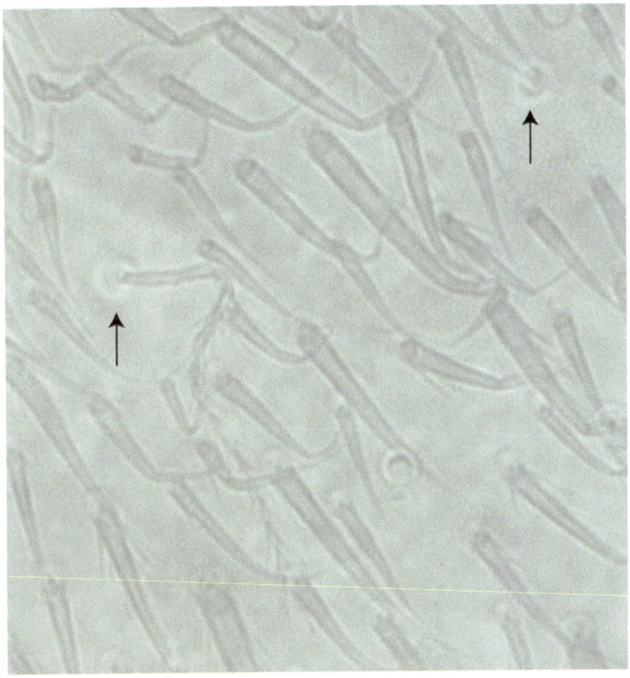

FIGURE 2.16 *Isotomurus* sp. (Isotomidae), fourth abdominal segment, *arrow* indicates bothriotrichum.

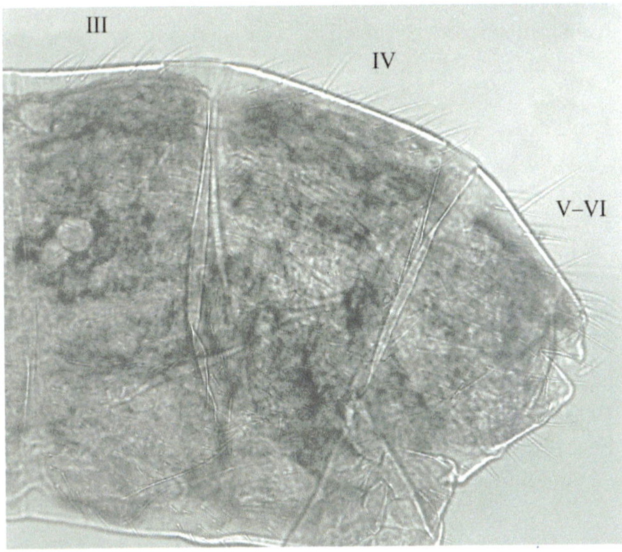

FIGURE 2.17 *Hemisotoma thermophila* (Isotomidae), third to sixth abdominal segment, lateral view.

Collembola Terminology

Antennal bulb: Apical lobe or vesicle on antennal article IV (Figs. 2.4 B, 5 G, 7 A).

Antennal sense organ: Two laterodistal rods of antennal article III usually with one guard sensillum on each side (Figs. 2.3 D, E).

Bothriotrichum (bothriotrix or trichobothrium; plural bothriotricha): Long, thin, flexible seta, smooth or finely ciliate (Figs. 2.16, 43, 52, 62).

Collophore (or ventral tube): Ventral appendage on the abdominal segment I, formed by basal cylindrical tube with two distal eversible vesicles (Figs. 2.1 B, C).

Circumanal setae: Setae arranged around the anal opening (Fig. 2.37 C).

Dens (plural, dentes): Paired segment of furcula appendage, located between manubrium and mucro (Figs. 2.1, 10, 14, 28, 29, 61).

Dental spines: Short smooth or ciliated spine-like setae arranged in rows or grouped on the dens (Fig. 2.24).

Empodial complex (= foot complex): Distal region of tibiotarsus formed by pretarsus, unguis, and unguiculus (Figs. 2.1, 5 E, 27, 36, 42, 44, 49, 54, 60).

Furcula (or furca): Jumping organ inserted ventrally on the abdominal segment IV, composed of manubrium, a pair of dentes, and a pair of distal mucrones (Figs. 2.1, 12).

Furcula base: Lateroventral region of large abdomen near the manubrium where it may have neosminthuroid setae in some Symphypleona (Figs. 2.1 C, 50, 57 A, 65).

Large abdomen: Globular region of trunk composed of fused thoracic and abdominal segments (in Symphypleona and Neelipleona) (Figs. 2.1 C, D, 38, 55, 57).

Macroseta (or macrochaeta): Robust, long seta present on head, trunk, and appendages (Figs. 2.30 A, 48 B).

Mandibles: Main chewing mouthparts, usually with apical teeth and basal molar plate (Figs. 2.2 A, 5 F).

Manubrial plate: Dorso-distal region of manubrium, which can bear pseudopores and modified setae (Fig. 2.31).

Manubrium: Unpaired proximal piece of furcula (Figs. 2.1, 12, 13).

Maxillae: Accessory chewing mouthparts, usually consisting of a capitulum (apical region usually with teeth and fringed lamellae) connected with 3 articulating arms: stipes, cardo, and fulcrum (Figs. 2.2 B–G, 64).

Mesothorax: Second segment of the thorax (Figs. 2.1 A, B).

Mesothoracic vesicles: Finger-like papillae present on each side of the dorsal mesothorax in some Sminthuridae (Fig. 2.51).

Metathorax: Third segment of the thorax (Figs. 2.1 A, B).

Microseta (or microchaeta): Small setae present on head, trunk, and appendages (Figs. 2.9 C, 30 B, 48 A).

Mucro: Furcula distal piece, with variable morphology (Figs. 2.11, 15, 26).

Nasal organ: Organ that occurs in the clypeal region of some Symphypleona and consists of a group of thick setae or cuticular structures in a circle with apexes directed toward the center (Fig. 2.56 A).

CLASS COLLEMBOLA

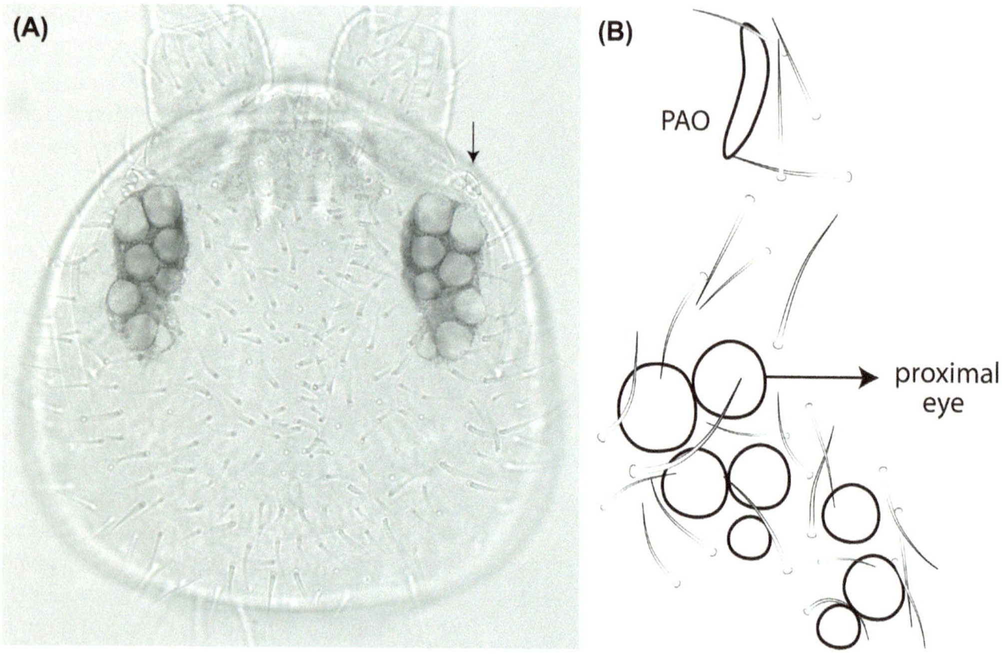

FIGURE 2.18 (A) *Isotomurus* sp. (Isotomidae), head, dorsal view, dorsal head, *arrow* indicates postantenal organ. (B) *Paracerura cristinae* Abrantes & Duarte (2013) (Isotomidae), eyes and postantenal organ right. *Modified from the original description.*

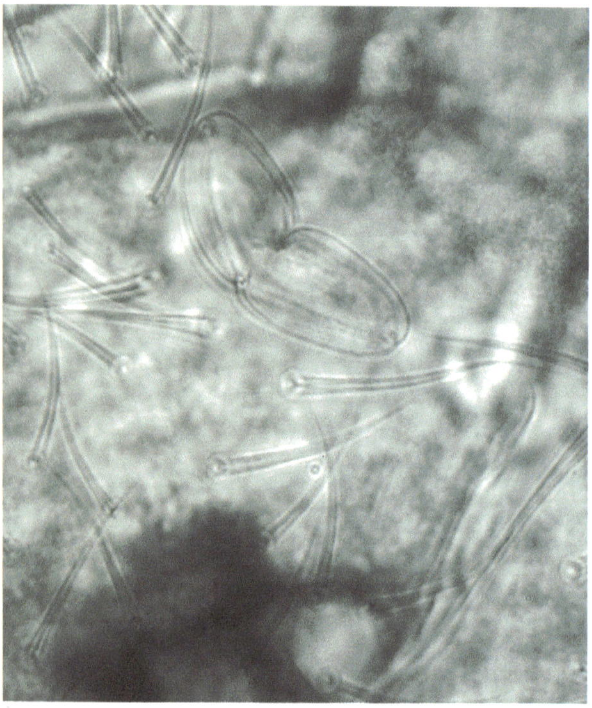

FIGURE 2.19 *Hemisotoma thermophila* (Isotomidae), postantenal organ left, dorsal view.

Neosminthuroid setae: Differentiated ciliate setae on furcular base, close to furcular base in some Symphypleona (Figs. 2.50, 57 A, 65).

Pretarsus: Distal part of the tibiotarsus, set off by suture, and bearing the unguis and unguiculus (Fig. 2.54).

Postantennal organ (PAO): Dorsal paired sensory organ on head between the anterior eyes and antennal base (Figs. 2.6 G–H, 7 E, 18–19, 25).

Prelabral setae: Transverse row of four or six setae inserted on anterior region of the head, between the labrum and the frontoclypeal field.

Prothorax: First segment of the thorax (Fig. 2.1 A).

Sensillum (plural sensilla): Swollen or thin-walled smooth seta with rounded apex (Figs. 2.3 D, 6 F, 7 A, 21, 22 A).

Seta (or chaeta): Slender, pointed hair connected to the body wall by a socket (Figs. 2.5 E, G, 6 A, B, D, 9 A, C, D, 13–19, 37, 46, 48, 59–63).

Sensory field: Rounded spot present on the head and body with sensilla (in Neelidae) (Fig. 2.66).

Small abdomen: Formed by terminal abdominal segments in Symphypleona and Neelipleona (Figs. 2.1 C, D, 37, 38, 62).

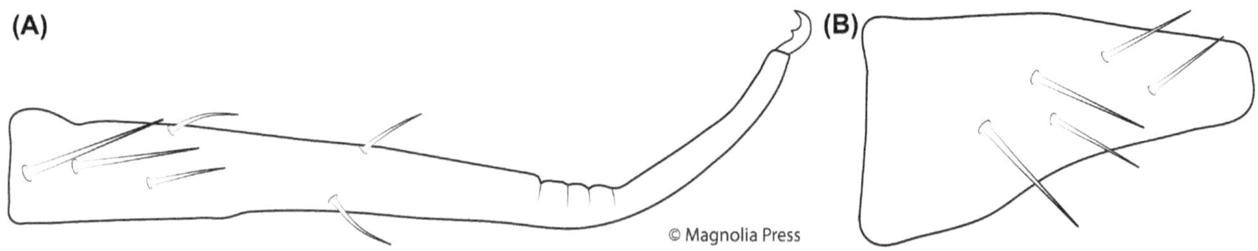

FIGURE 2.20 (A) *Cryptopygus tingus* Queiroz & Mendonça (2010) (Isotomidae), dens and mucro, ventral view. (B) *Dagamaea tenuis* (Folsom, 1934) (Isotomidae), dens, ventral view. *(A) Modified from the original description with permission from copyright holder. (B) Modified from the Christiansen & Bellinger (1998).*

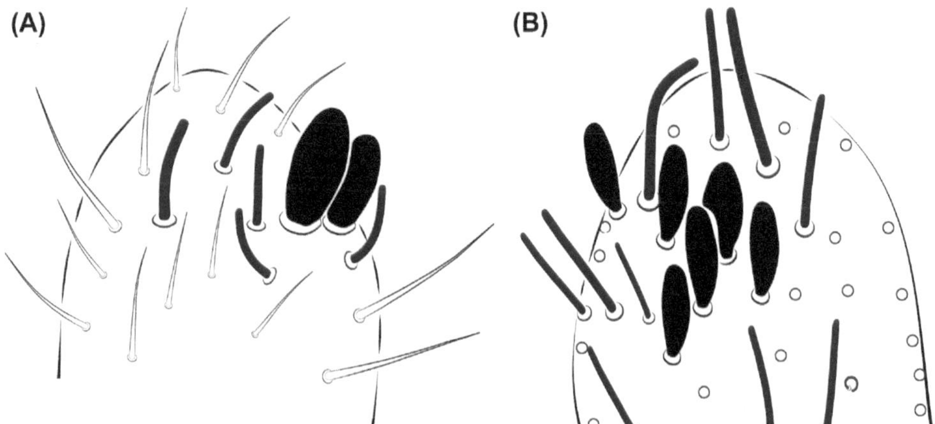

FIGURE 2.21 (A) *Folsomina onychiurina* Denis, 1931 (Isotomidae), apex of the fourth antennal segment. (B) *Isotomiella alulu* Christiansen & Bellinger, 1992 (Isotomidae), apex of the fourth antennal segment. *(A) Modified from Christiansen & Bellinger (1998).*

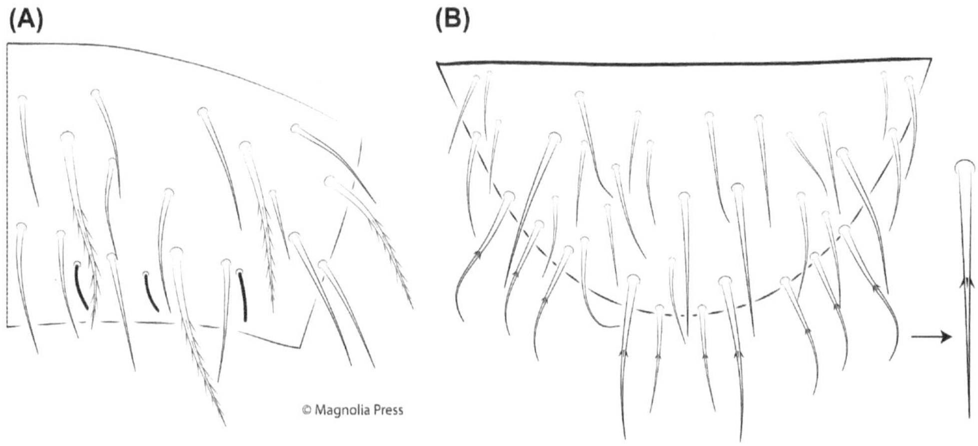

FIGURE 2.22 (A) *Isotomiella barrana* Mendonça & Abrantes, 2007 (Isotomidae), tergal chaetotaxy right of the fourth abdominal segment. (B) *Hemisotoma thermophila* (Isotomidae), tergal chaetotaxy of the sixth abdominal segment. *(A) Modified from the original description with permission from copyright holder.*

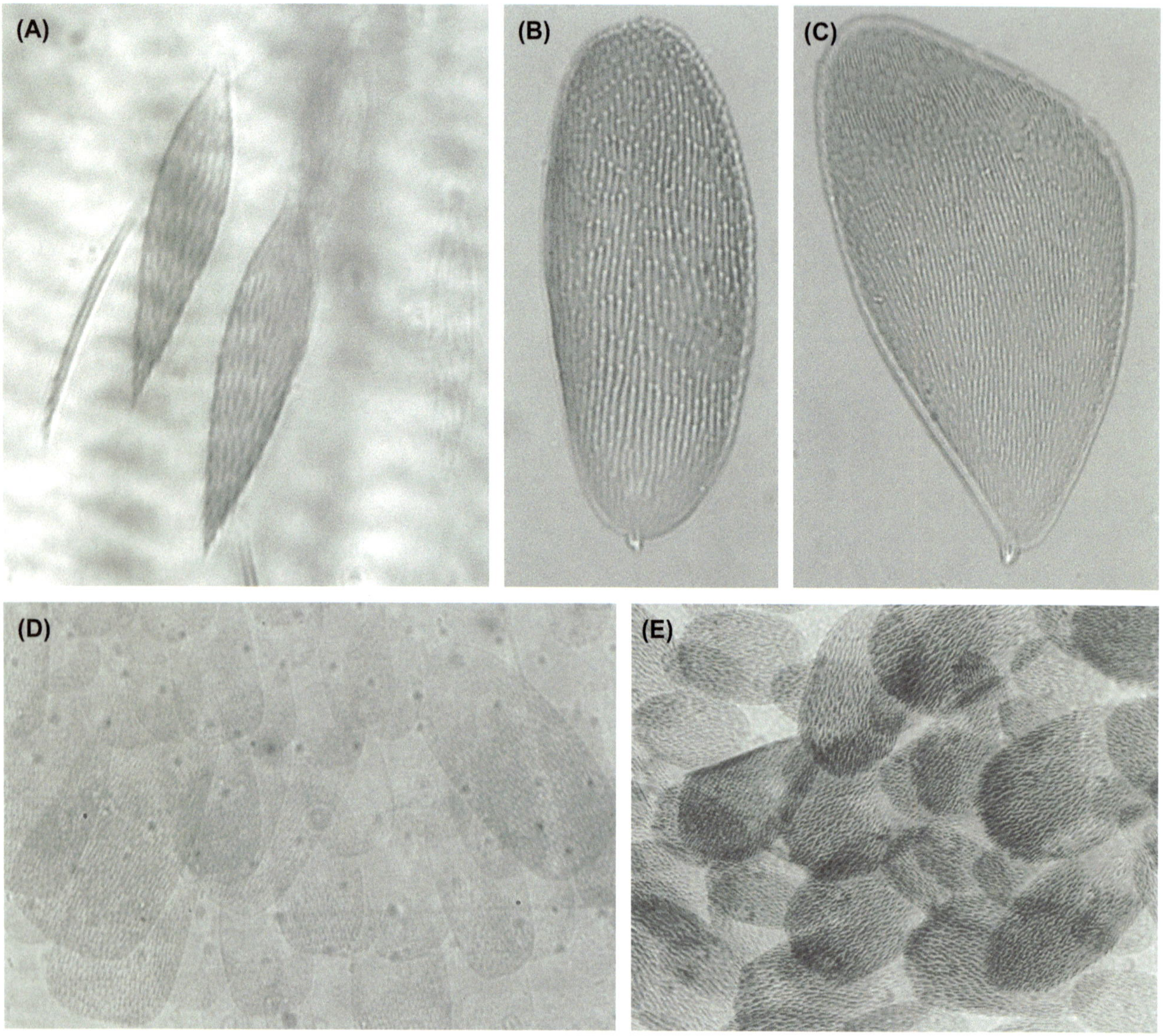

FIGURE 2.23 (A) *Mastigoceras camponoti* (Entomobryidae), pointed scales. (B) *Dicranocentrus* (Entomobryidae), rounded scales. (C) *Dicranocentrus* (Entomobryidae), truncate scales. (D) *Lepidocyrtus* (Entomobryidae), scales gently striate. (E) *Lepidocyrtoides* (Entomobryidae), scales heavily striate.

Subanal appendage: Pair of modified setae present on subventral anal lobes in most females of Symphypleona (Figs. 2.1 C, 37).

Tenaculum (or retinaculum): Ventral appendage of the abdominal segment III that holds the furcula at rest.

Tenent hair: Differentiate seta of the tibiotarsus apex. It can be pointed, knobbed, or clavate (Figs. 2.1, 27 B, 36 B, 42, 60 B, C).

Tibiotarsal organ: Depressed organ present on tibiotarsi of some Symphypleona (Figs. 2.44, 47).

Tibiotarsus (plural, tibiotarsi): Last segment of the legs, formed possibly by fusion of tibia and tarsomeres seen in other hexapods (Figs. 2.1, 47, 60 A–C).

Trochanteral organ: Group of smooth, spine-like setae seen on trochanter of hind legs.

Unguis (plural, ungues): Claw inserted on the pretarsus (Figs. 2.1, 5 E, 27, 36, 44, 49, 54, 60).

Unguiculus (plural, unguiculi): A pretarsal appendage similar to a smaller claw (Figs. 2.1, 5 E, 27, 36, 44, 49, 54, 60).

MATERIAL PREPARATION AND PRESERVATION

Collected specimens must be properly preserved in 70 percent ethanol or higher concentrations. For freshwater

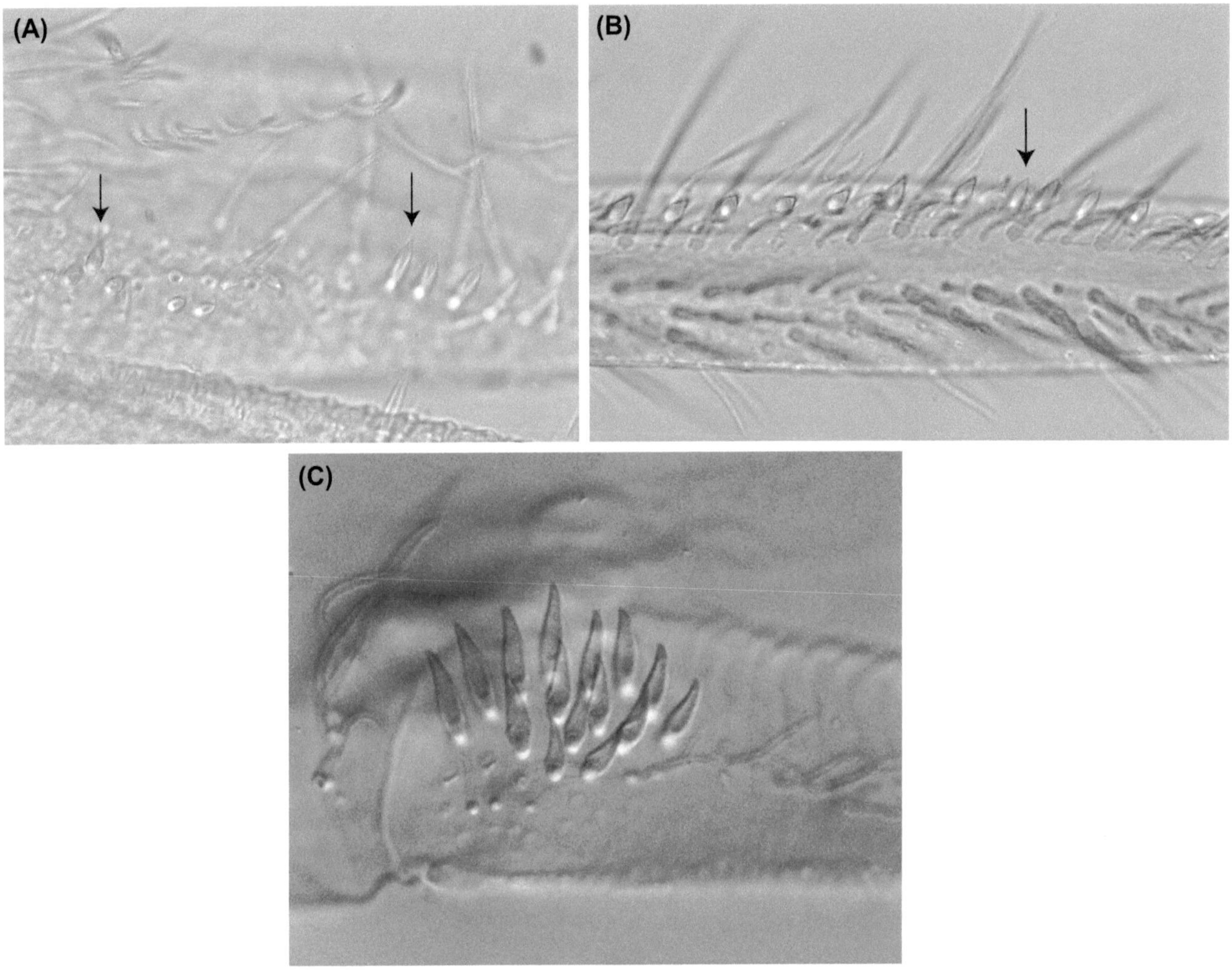

FIGURE 2.24 (A) *Dicranocentrus* (Entomobryidae), dorsal dens, *arrows* indicate spines. (B) *Amazhomidia* (Entomobryidae), dorsal dens, *arrows* indicate spines. (C) *Lepidocyrtus* (*Fractocyrtus*) (Entomobryidae), dorsal dens.

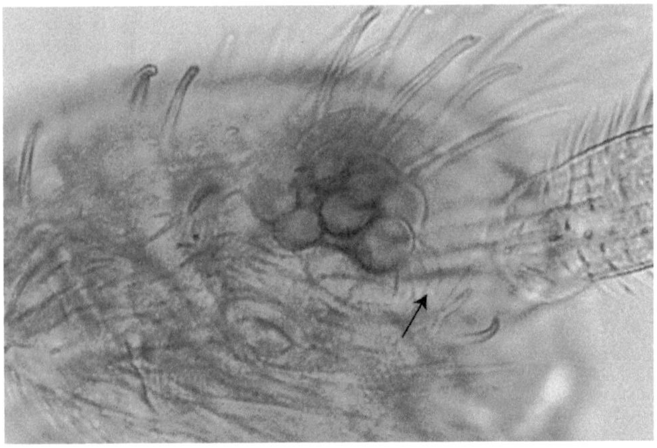

FIGURE 2.25 *Nothobrya* (Entomobryidae), lateral head, *arrow* indicates postantenal organ.

springtails we recommend 95 percent ethanol, which prevents loss of quality and leaves it in good condition for preparation and identification. Concerning molecular analysis, springtails should be kept in 100 percent ethanol, within a refrigerator at −20°C.

For observation of basic morphology and identification, most taxa need to be cleared and mounted under glass slides and observed with a phase-contrast microscope and/or differential interference contrast microscopes (capable of magnification of 400× or more) and a drawing tube.

Specimens need to be cleared, especially when they have strong pigmentation. Options include: (1) solution of potassium hydroxide 5 percent−10 percent (KOH), which converts the pigmentation of Collembola to a red color and serves to dissolve the cuticular oil, and must be used for a short time (5−10 min); (2) lactic acid ($C_3H_6O_3$) or

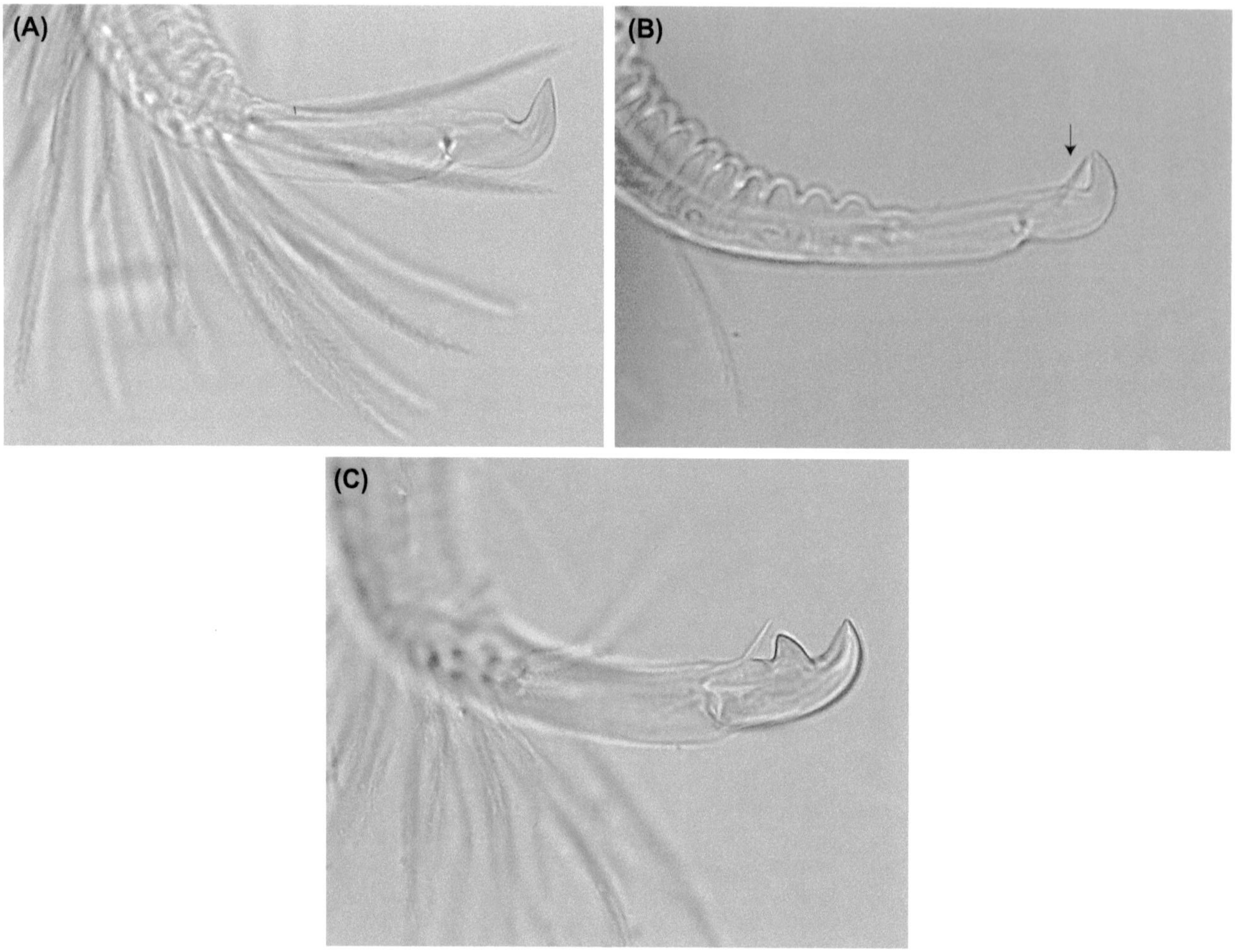

FIGURE 2.26 (A) *Lepidocyrtinus* (Entomobryidae), distal dens and mucro. (B) *Drepanura* (Entomobryidae), distal dens and mucro, *arrow* indicates basal mucronal seta. (C) *Lepidocyrtoides* (Entomobryidae), distal dens and mucro.

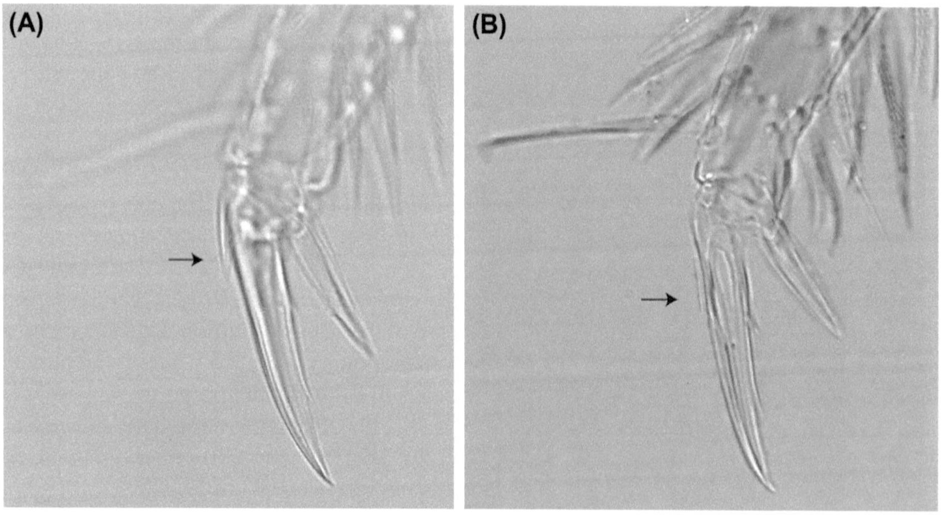

FIGURE 2.27 (A) *Lepidocyrtinus* (Entomobryidae), third empodial complex, posterior view, *arrow* indicates outer tooth of unguis. (B) *Seira* (Entomobryidae), third empodial complex, posterior view, *arrow* indicates outer tooth of unguis.

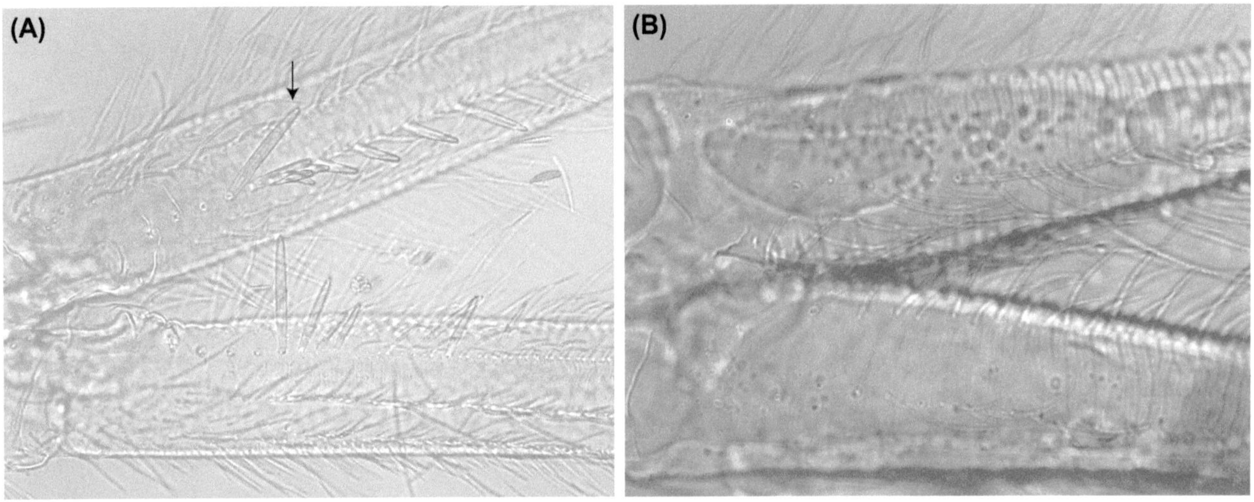

FIGURE 2.28 (A) *Lepidocyrtinus* (Entomobryidae), dorsal dens, *arrow* indicates long blunt setae. (B) *Seira* (Entomobryidae), dorsal dens.

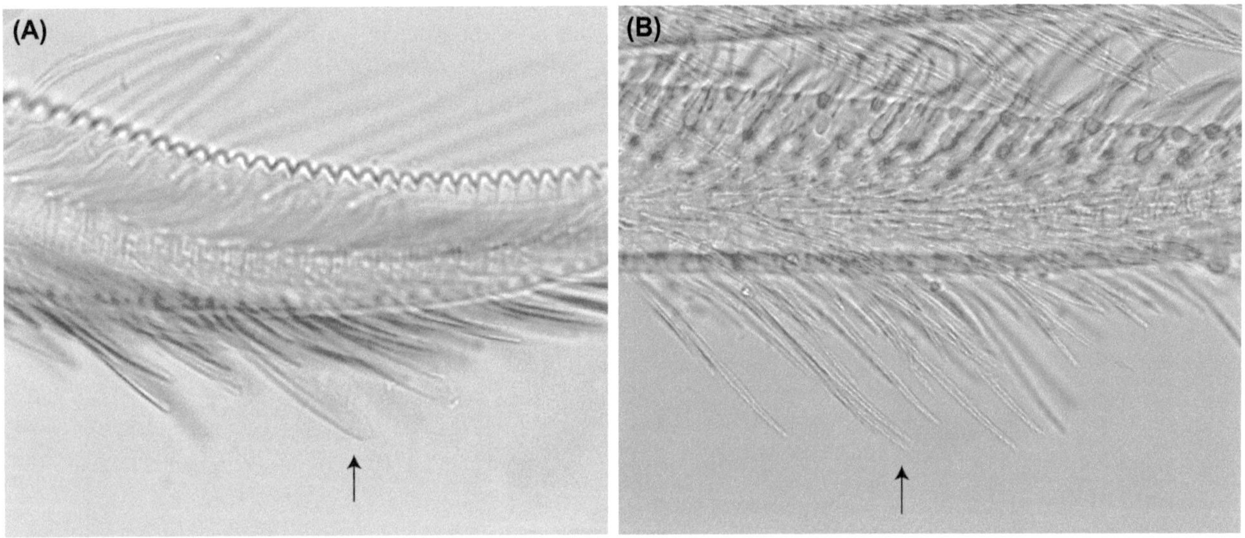

FIGURE 2.29 (A) *Lepidocyrtus* (*Setogaster*) (Entomobryidae), dens, lateral view, *arrow* indicates scales. (B) *Amazhomidia* (Entomobryidae), dens, lateral view, *arrow* indicates ciliate setae.

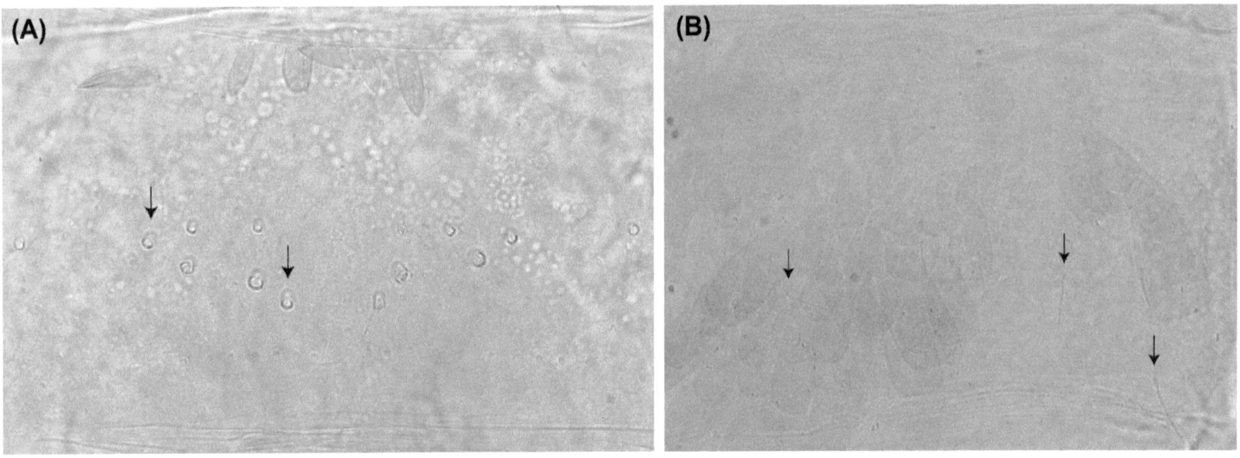

FIGURE 2.30 (A) *Amazhomidia* (Entomobryidae), metathorax, *arrows* indicate insertion of macrosetae. (B) *Lepidocyrtus* (*Setogaster*) (Entomobryidae), dorsal dens.

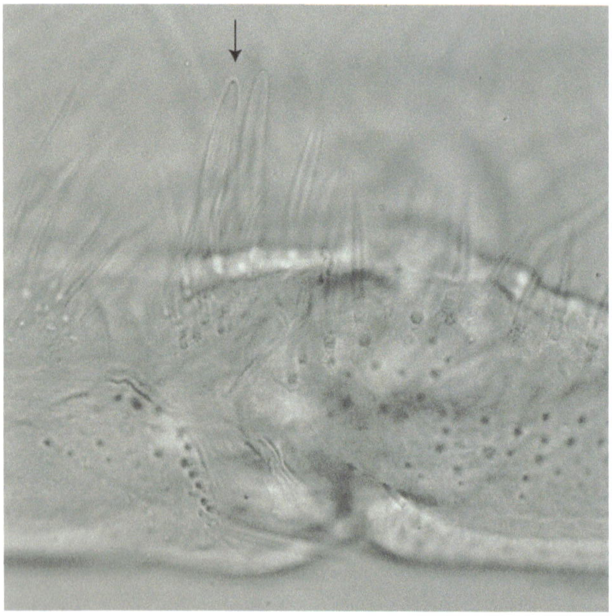

FIGURE 2.31 *Lepidocyrtoides* (Entomobryidae), distal manubrium and proximal dens, lateral view, *arrows* indicate insertion of macrosetae on manubrial plate.

lactophenol ($C_3H_6O_3 + C_6H_6O$), which serves to macerate the internal specimen tissues, and in which the specimen can be left for a long period of time (weeks), or be heated for several seconds; or (3) Nesbitt's solution ($C_2H_3Cl_3O_2 + H_2O + HCl$) is also useful for rapid clarification when the medium is heated on a hot plate (55°, 10–20 min.), and also suitable for specimens that are preserved in ethanol for a long time. These methods can be combined, modified, and/or used for different periods depending on the specimen and the results to be obtained. For dark specimens it is appropriate to wash the specimens first in potassium hydroxide solution, remove the excess of KOH, wash them in lactic acid or lactophenol, and drop the excess of chemical before mounting. Jordana et al. (1997) suggested a modification of Nesbitt's fluid, adding chloral-lactophenol to the solution ($C_2H_3Cl_3O_2 + C_3H_6O_3 + C_6H_6O$).

After clarification, a temporary mounting slide can be used when the specimen needs to be analyzed only for a short time. A temporary slide can be made on glass slides (with or without cavity) with lactic acid, lactophenol, or glycerine. The advantage of this method is the specimen

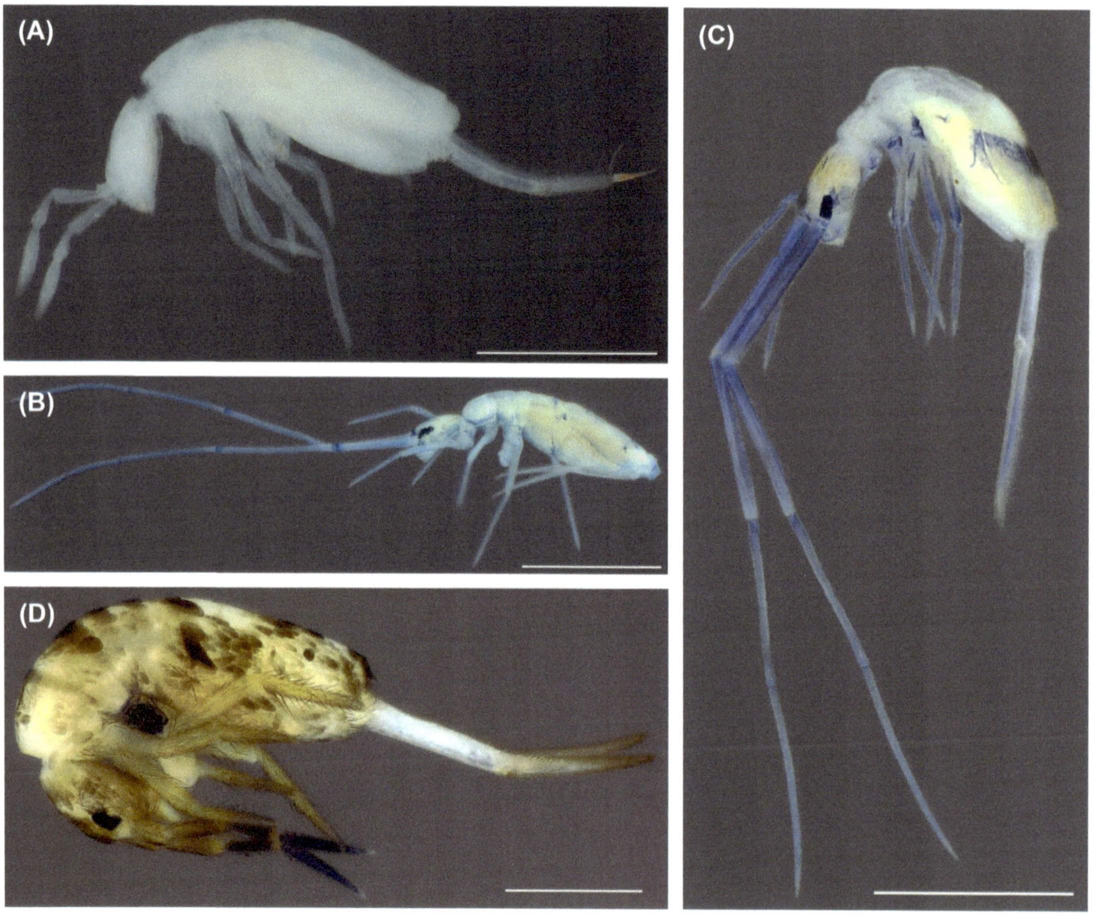

FIGURE 2.32 (A) *Cyphoderus* (Paronellidae), habitus, lateral view. Bar = 0.5 mm. (B) *Salina tristani* (Paronellidae), habitus, lateral view. Bar = 1 mm. (C) *Campylothorax* (Paronellidae), habitus, lateral view. Bar = 1 mm. (D) *Trogolaphysa* (Paronellidae), habitus, lateral view. 0.5 mm.

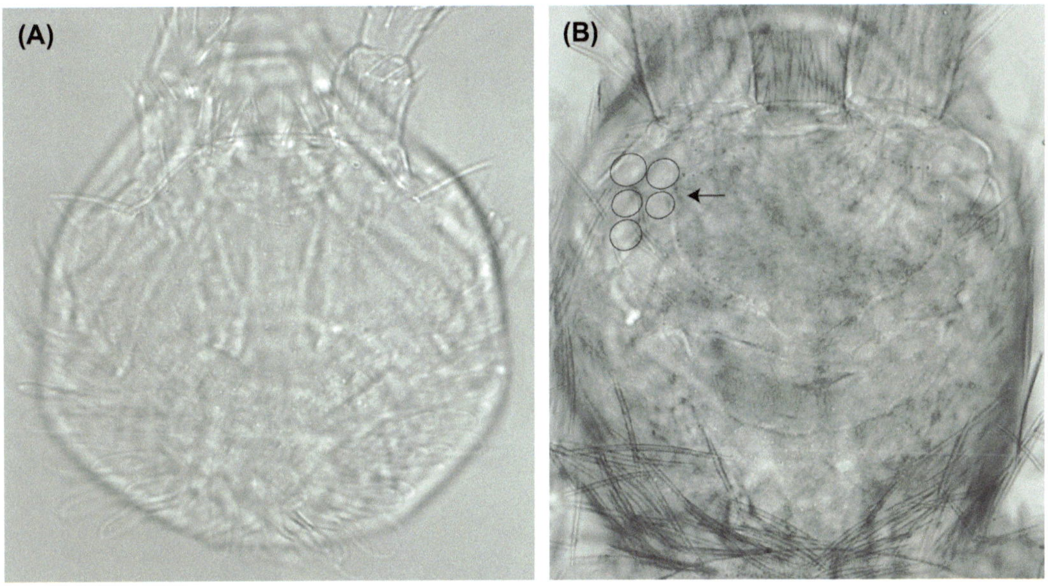

FIGURE 2.33 (A) *Cyphoderus* (Paronellidae), head, dorsal view. (B) *Trogolaphysa* (Paronellidae), head, dorsal view, *arrow* indicates eyepatches.

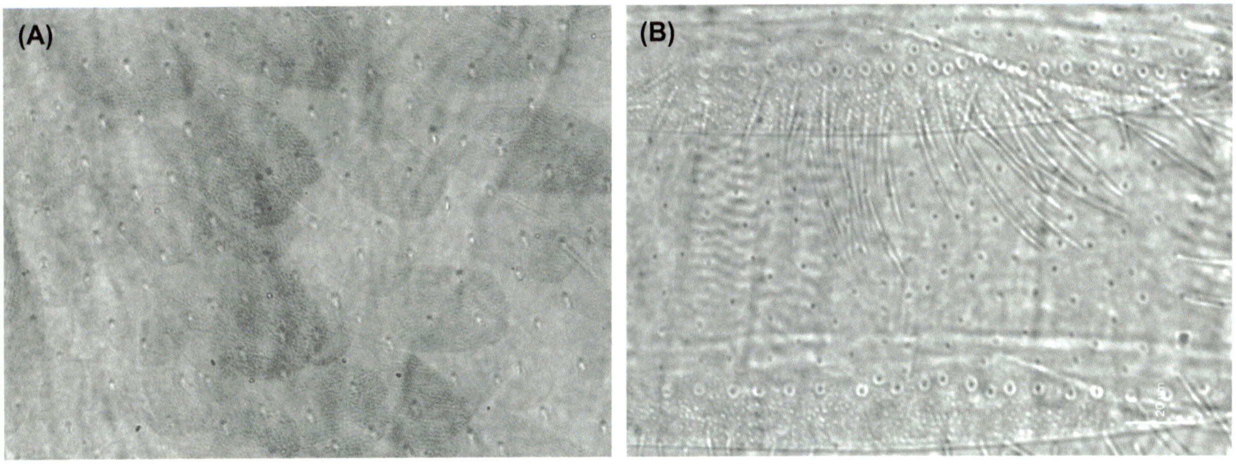

FIGURE 2.34 (A) *Trogolaphysa* (Paronellidae), tergal scales. (B) *Salina* (Entomobryidae), tergal ciliate setae.

may be recovered with a diluting medium and returned to alcohol. It is also an appropriate method when there are several specimens of the same taxon and there is no problem with damaging or losing some individuals. In some cases, if the specimen is very large, it is appropriate to use an excavated slide to dissect it, but not to mount.

Permanent mounting is the most common method to preserve specimens for identification. The mounted springtails can be preserved in glass slides during long periods (over 50 years) without the need of remounting. A disadvantage of this method is that each specimen will remain in a fixed position and generally become flattened, which can cause distortion and possibly obscure some

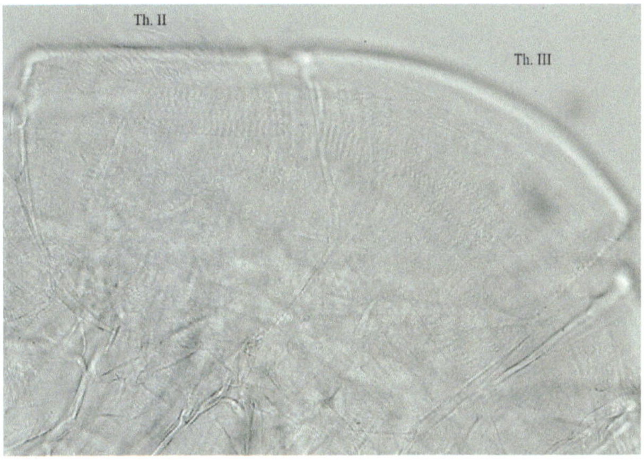

FIGURE 2.35 *Campylothorax* (Paronellidae), thorax, lateral view.

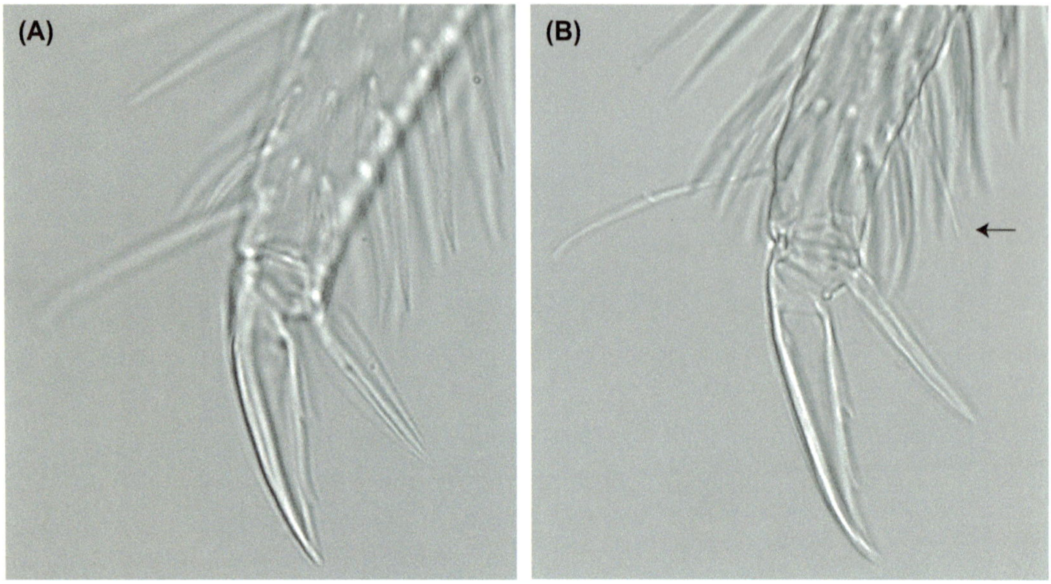

FIGURE 2.36 (A) *Campylothorax* (Paronellidae), third empodial complex, posterior view. (B) *Trogolaphysa* (Paronellidae), third empodial complex, posterior view, *arrow* indicates inner smooth seta.

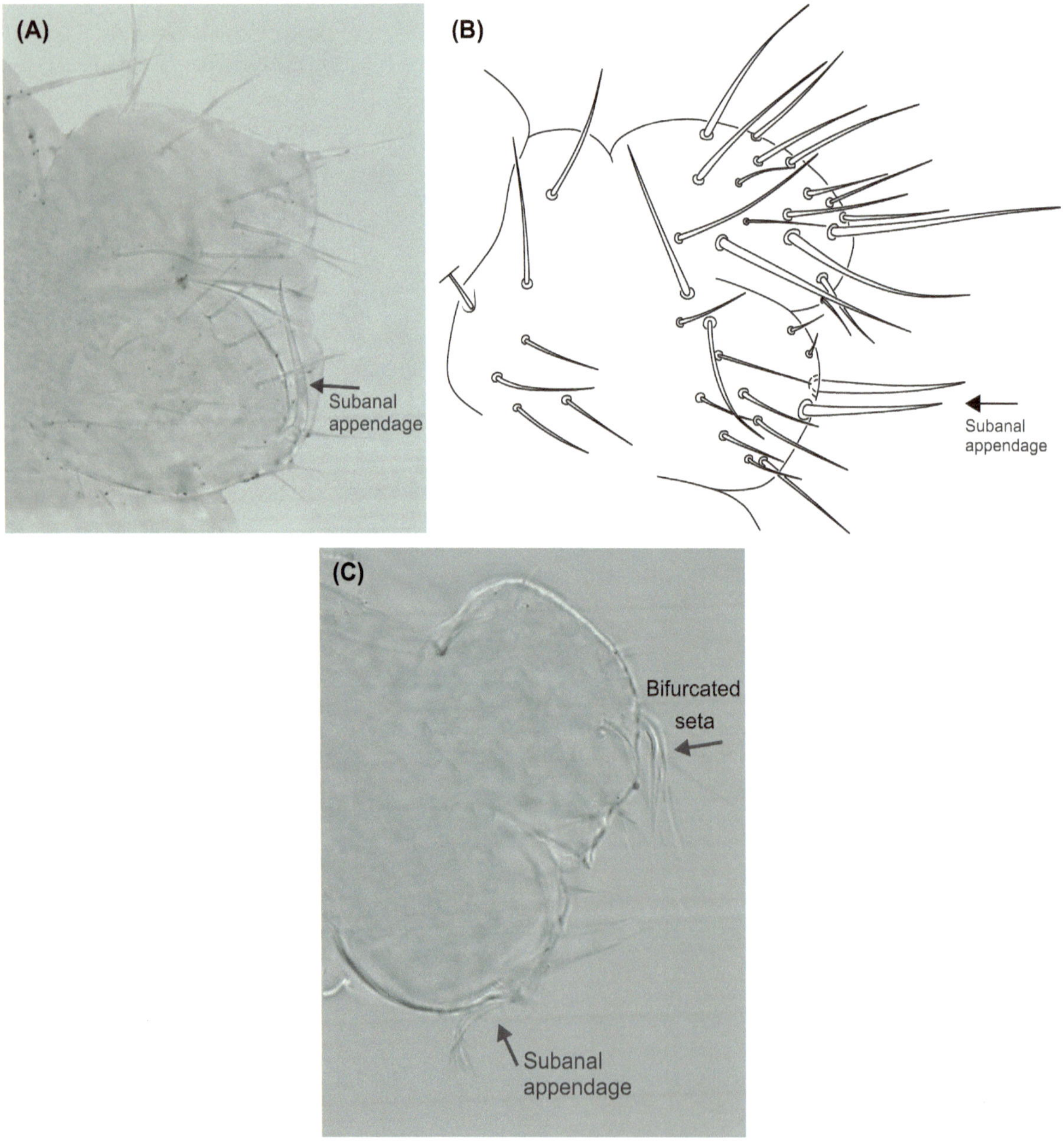

FIGURE 2.37 (A) *Pararrhopalites* (Sminthuridae), small abdomen, lateral view. (B) *Sturmius epiphytus* (Sturmiidae), small abdomen, lateral view. (C) *Sminthurinus* (Katiannidae), small abdomen, lateral view. *(B) Modified from Bretfeld (1994).*

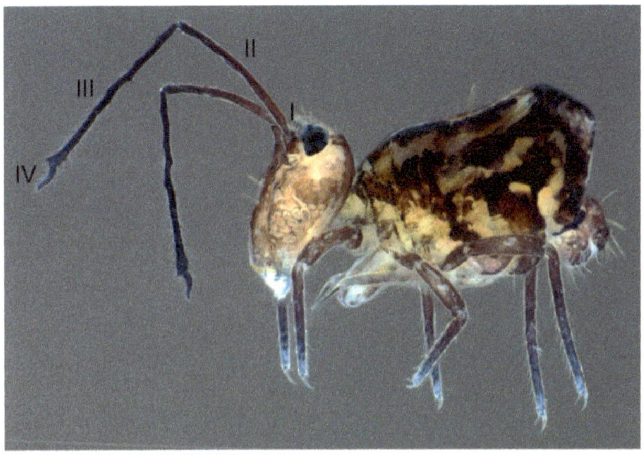

FIGURE 2.38 *Ptenothrix* (Dicyrtomidae), habitus, lateral view, numbers indicate antennal segments.

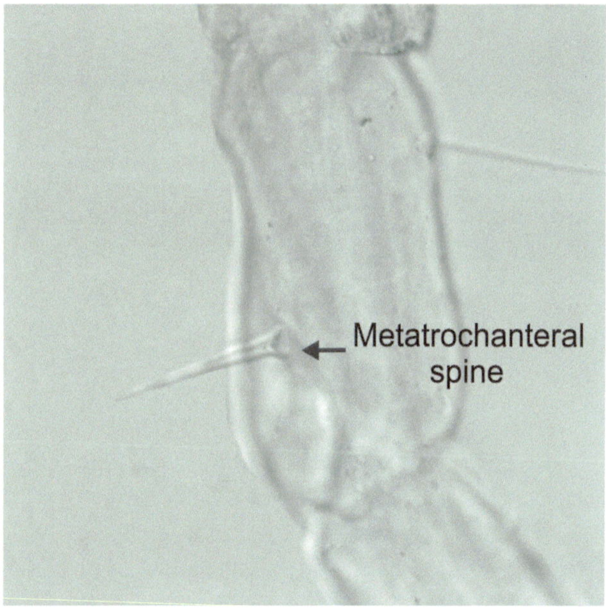

FIGURE 2.40 *Pararrhopalites* (Sminthuridae), third trochanter.

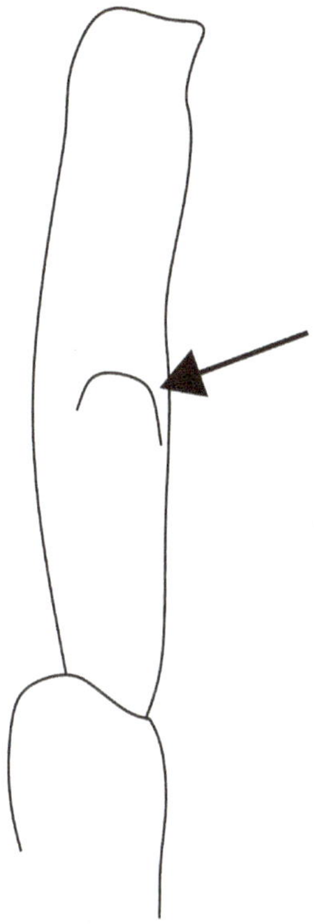

FIGURE 2.39 *Sturmius epiphytus* (Sturmiidae), third antennal segment with papilla. *Modified from Bretfeld (1994).*

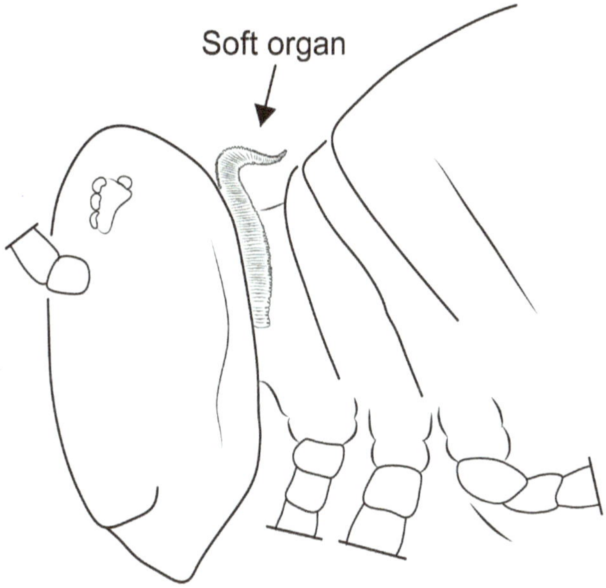

FIGURE 2.41 *Spinotheca patagonica* (Spinothecidae), head and anterior region of body. *Modified from Delamare Deboutteville (1961).*

structures. The most common medium used for semipermanent slides is Hoyer's solution (distilled H_2O + fragments of arabic gum + glycerin + chloral hydrate). As a semipermanent medium, Hoyer's solution is soluble in water, so it is possible to dissolve the medium and carefully remount specimens, if necessary.

In institutional collections, as well as other small arthropods, springtails can be kept in liquid vials, preserved in 92 percent ethanol or higher concentrations, or on semipermanent slides. In liquid, they are usually kept in small vials with ethanol (4 cm length × 0.5 in diameter), which can be closed with cotton and placed vertically in a large pot also containing ethanol. Permanent mounts can be stored in glass slides boxes in dry and fungi proof environments.

More information on preservation, clarification, and mounting techniques are presented in Jordana et al. (1997), Christiansen & Bellinger (1998), and Palacios-Vargas & Mejía-Recamier (2007).

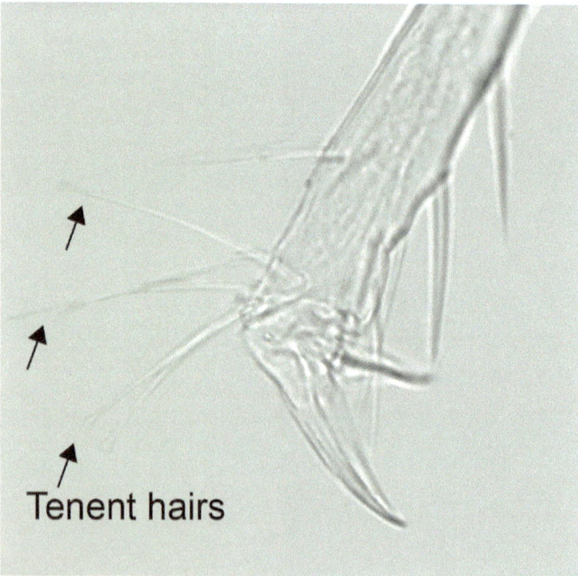

FIGURE 2.42 *Polykatianna* (Katiannidae), distal third tibiotarsus and empodial complex.

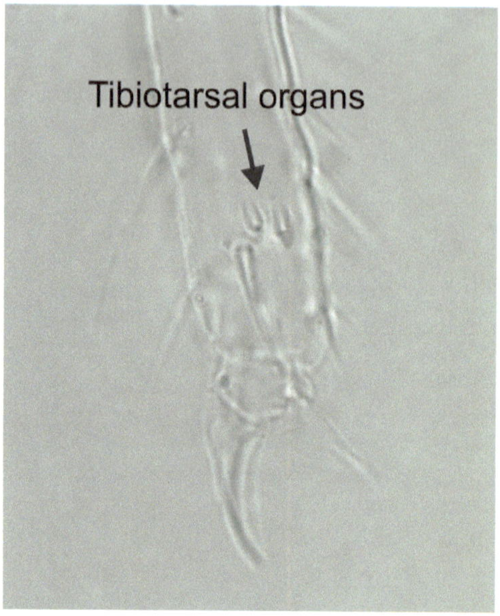

FIGURE 2.44 *Sminthurides* (Sminthurididae), distal third tibiotarsus and empodial complex.

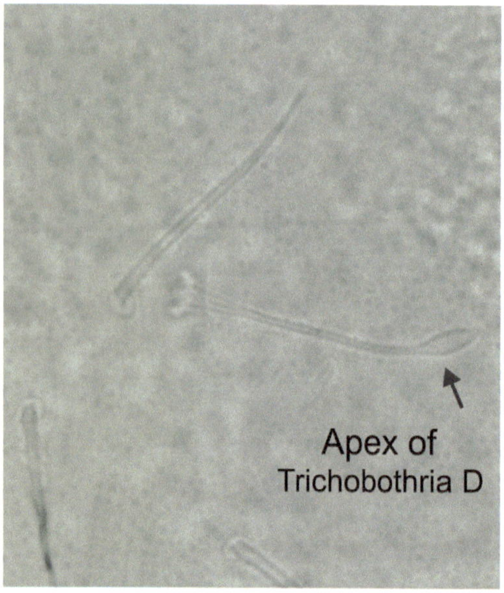

FIGURE 2.43 *Collophora* (Collophoridae), trichobothrium D region.

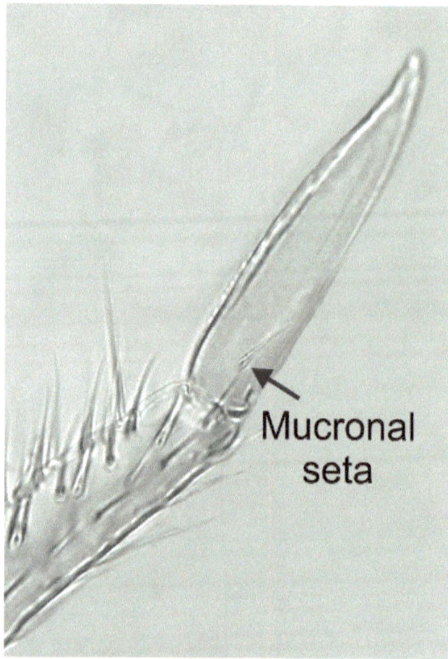

FIGURE 2.45 *Denisiella* (Sminthurididae), distal dens and mucro, lateral view.

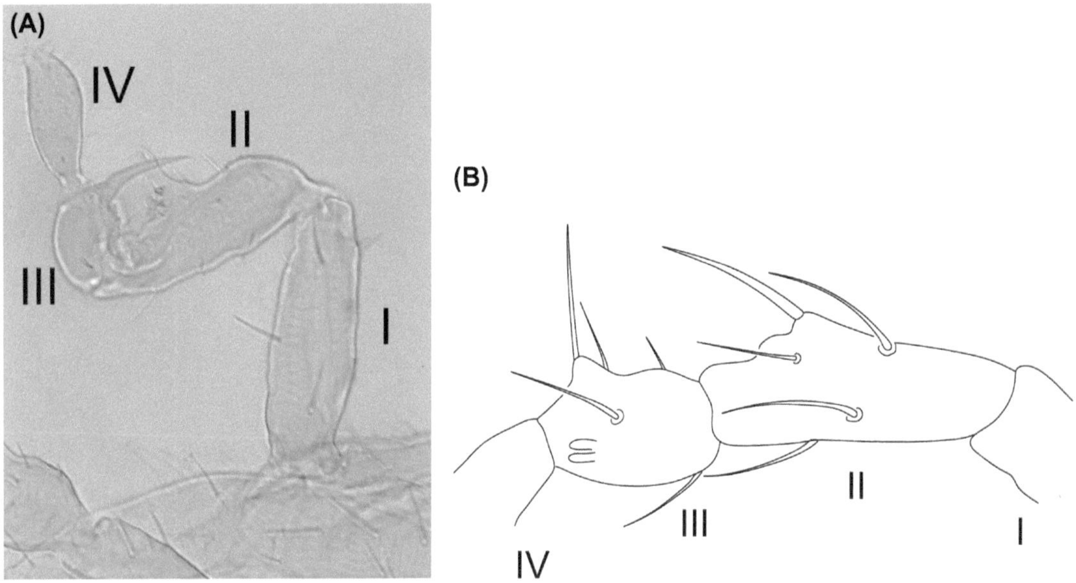

FIGURE 2.46 (A) *Denisiella* (Sminthurididae), male antenna, numbers indicate antennal segments. (B) *Sphaeridia paroara* (Sminthurididae), antennal segments. *(B) Modified from Arlé (1984).*

KEY TO NEOTROPICAL FRESHWATER COLLEMBOLA

Symbols: (†) often in freshwater; (*) sometimes in freshwater environments; (§) rarely in freshwater environments.

Collembola: Orders

1	Elongated body, with distinct tergal segmentation (Figs. 2.1 A, B)	2
1'	Globular body, formed by thoracic fusion and abdominal segments (Figs. 2.1 C, D)	3
2(1)	Prothorax distinct, with setae or papillae (Fig. 2.1 A)	**Poduromorpha [p. 39]**
2'	Prothorax reduced, without setae or papillae (Fig. 2.1 B)	**Entomobryomorpha [p. 49]**
3(1)	Eyes usually present; antennae longer than head; small abdomen distinct; specimens of coloration and sizes variable (Fig. 2.1 C)	**Symphypleona [p. 51]**
3'	Eyes absent; antennae shorter than head; small abdomen indistinct, difficult to see; very minute specimens; usually depigmented (Fig. 2.1 D)	**Neelipleona [p. 53]**

Collembola: Poduromorpha: Families

1	Mandibles absent or rudimentary, if present, without molar plates	2
1'	Mandibles present and well-developed, with molar plates (Fig. 2.2 A)	4
2(1)	Maxilla cardo present (Fig. 2.2 B)	3
2'	Maxilla cardo absent (Fig. 2.2 C)	**Odontellidae [p. 40]**
3(2)	Maxilla capitulum with apical, well-developed teeth (Fig. 2.2 B); mandibles usually absent	**Brachystomellidae [p. 43]**
3'	Maxilla capitulum otherwise, maxillary teeth usually reduced (Figs. 2.2 D–G); mandibles usually present	**Neanuridae [p. 45]**
4(1)	Dens, if present, short and straight, without distal rings (Fig. 2.3 A)	5
4'	Dens long and curved, distally crenulate (Fig. 2.3 B)	Poduridae, one species: *Podura aquatica* Linnaeus, 1758† [Mexico. Nearctic]
5(4)	Head and trunk dorsally with pseudocelli (Fig. 2.3 C)	6
5'	Head and trunk dorsally without pseudocelli	**Hypogastruridae [p. 48]**
6(5)	Antennal article III sense organ hidden by well-developed conical tegument papillae (Fig. 2.3 D)	**Onychiuridae [p. 48]**
6'	Antennal article III sense organ with none or one small inner papilla; sense rods partially hidden by weakly developed tegument fold (Fig. 2.3 E)	**Tullbergiidae [p. 49]**

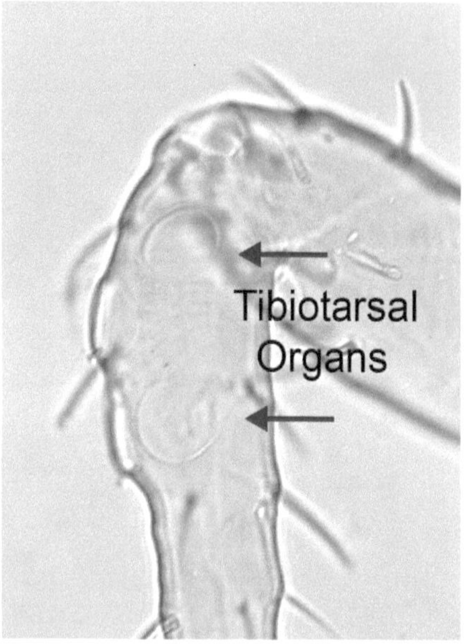

FIGURE 2.47 *Denisiella* (Sminthurididae), first tibiotarsus of male.

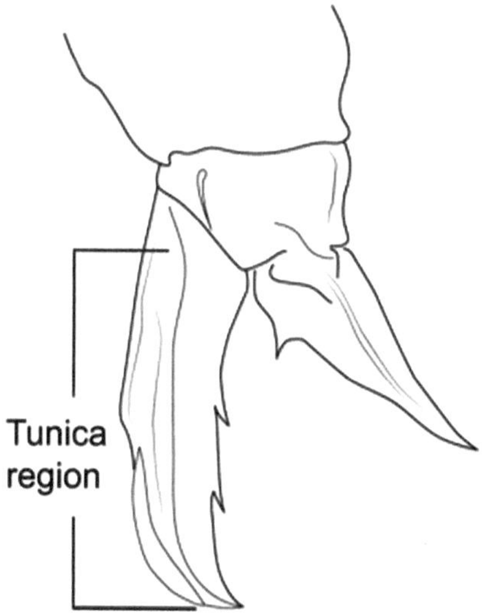

FIGURE 2.49 *Calvatomina* (Dicyrtomidae), empodial complex.

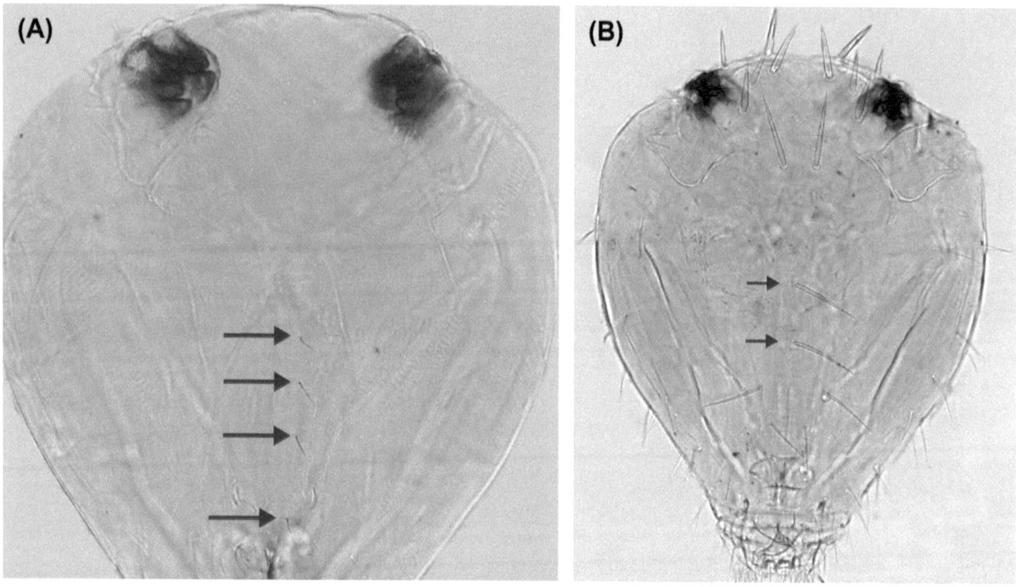

FIGURE 2.48 (A) *Calvatomina* (Dicyrtomidae), head, frontal view. (B) *Ptenothrix* (Dicyrtomidae), head, frontal view.

Collembola: Poduromorpha: Odontellidae: Genera

1 Antennal article IV with apical bulb (Fig. 2.7 A); antenna with eversible sac at junction of segments III and IV (as in Fig. 2.5 G) .... *Odontella*§

1' Antennal article IV without apical bulb; antenna lacking eversible sac at junction of segments III and IV .... *Superodontella**

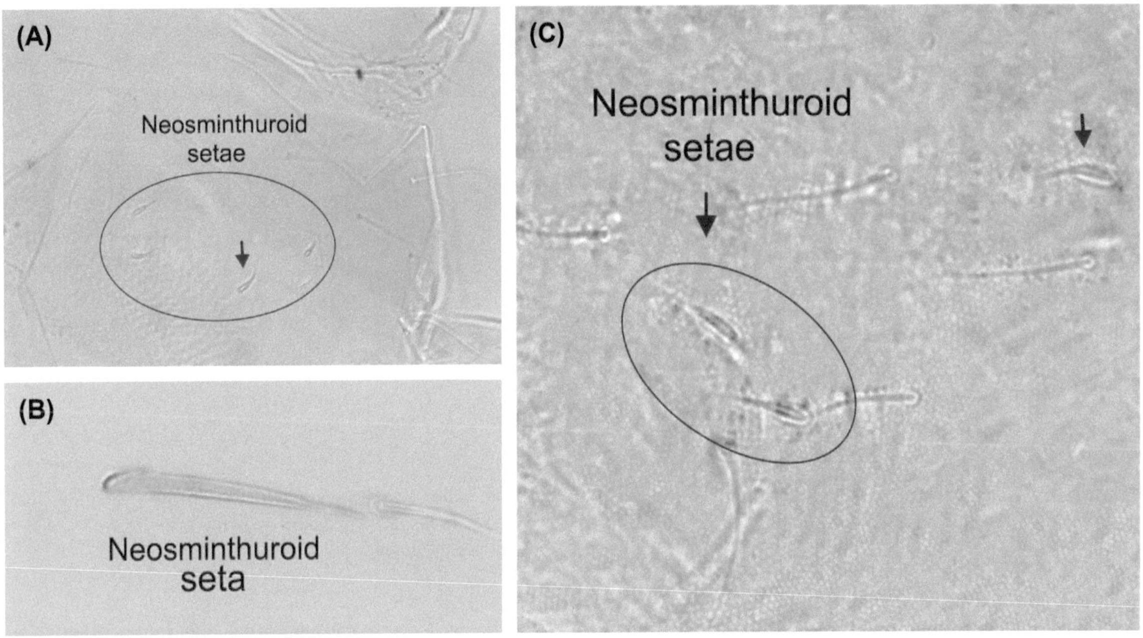

FIGURE 2.50 (A) *Calvatomina* (Dicyrtomidae), neosminthuroid seta on furcula base. (B) *Szeptyckitheca* (Sminthuridae), neosminthuroid seta on furcula base. (C) *Songhaica* (Sminthuridae), neosminthuroid seta on furcula base.

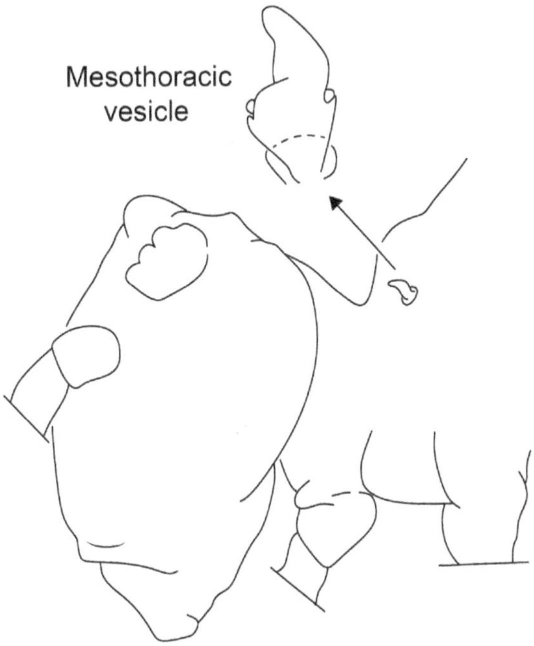

FIGURE 2.51 *Neosminthurus clavatus* (Sminthuridae), head and anterior body. *Modified from Betsch (1980).*

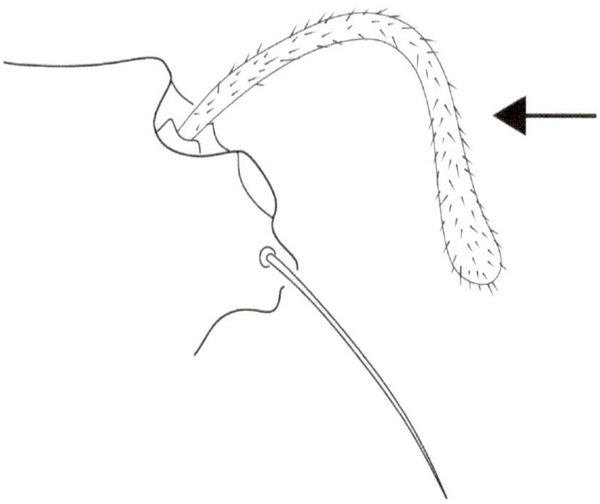

FIGURE 2.52 *Austrosminthurus mirabilis* (Sminthuridae), bothriotrichum D. *Modified from Delamare Deboutteville & Massoud (1963).*

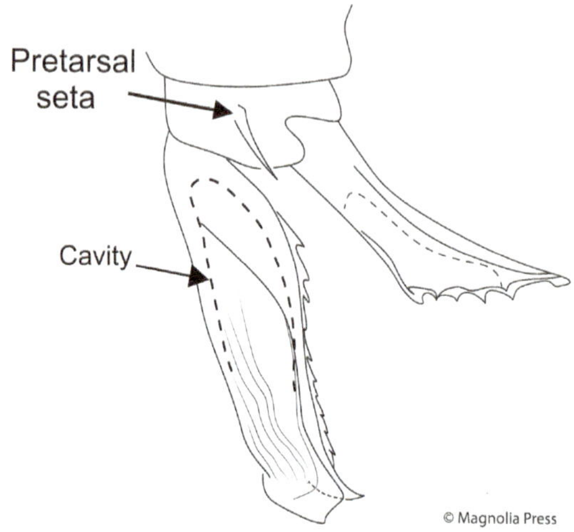

FIGURE 2.54 *Varelasminthurus potiguarus* (Sminthuridae), third empodial complex. *Modified from the original description with permission from copyright holder.*

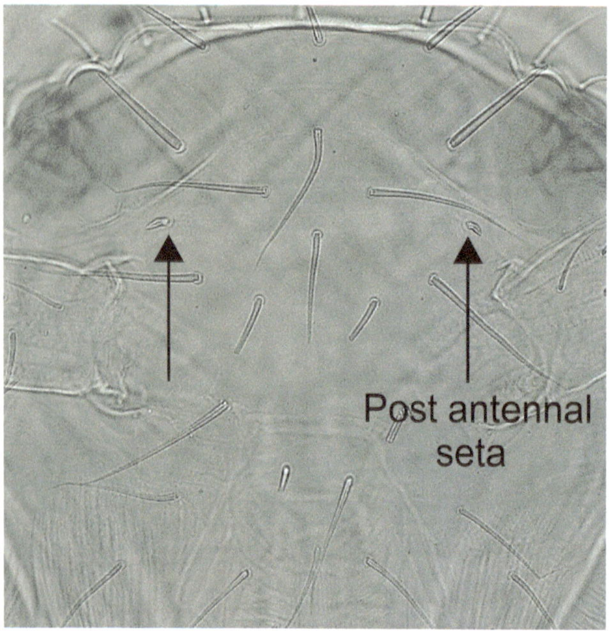

FIGURE 2.53 *Sminthurus viridis* (Sminthuridae), head frontal region, postantennal setae.

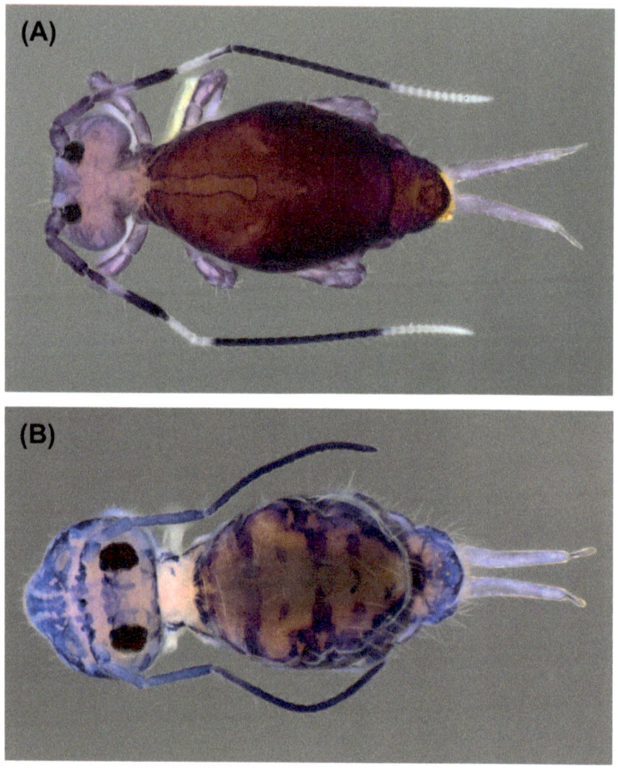

FIGURE 2.55 (A) *Temeritas* (Sminthuridae), dorsal view. (B) *Pararrhopalites* (Sminthuridae), dorsal view.

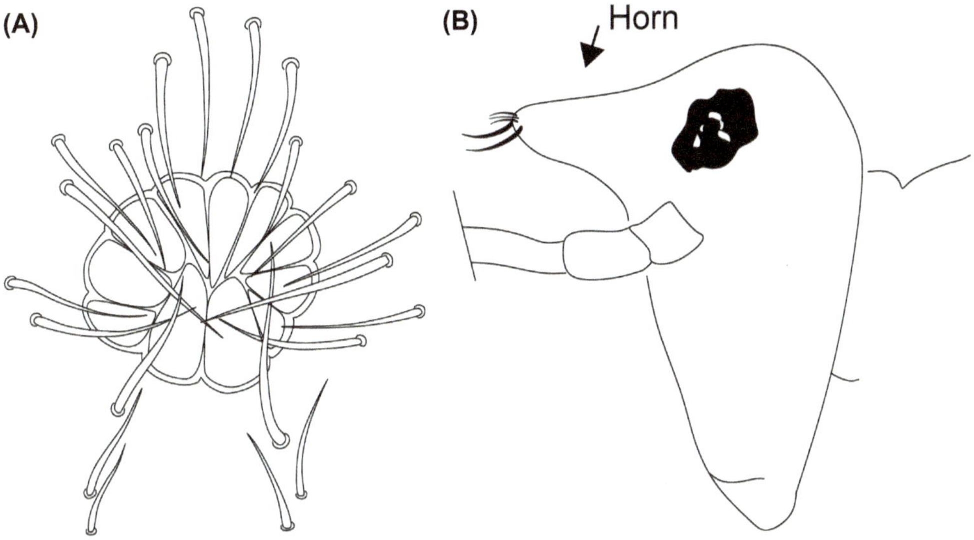

FIGURE 2.56 (A) *Pseudobourletiella spinata* (Bourletiellidae), nasal organ. (B) *Bovicornia* cf. *greensladei* (Bourletiellidae), lateral view of the head showing the interocular dorsal horn. *(A) Modified from Richards (1968). (B) Modified from Betsch (1974).*

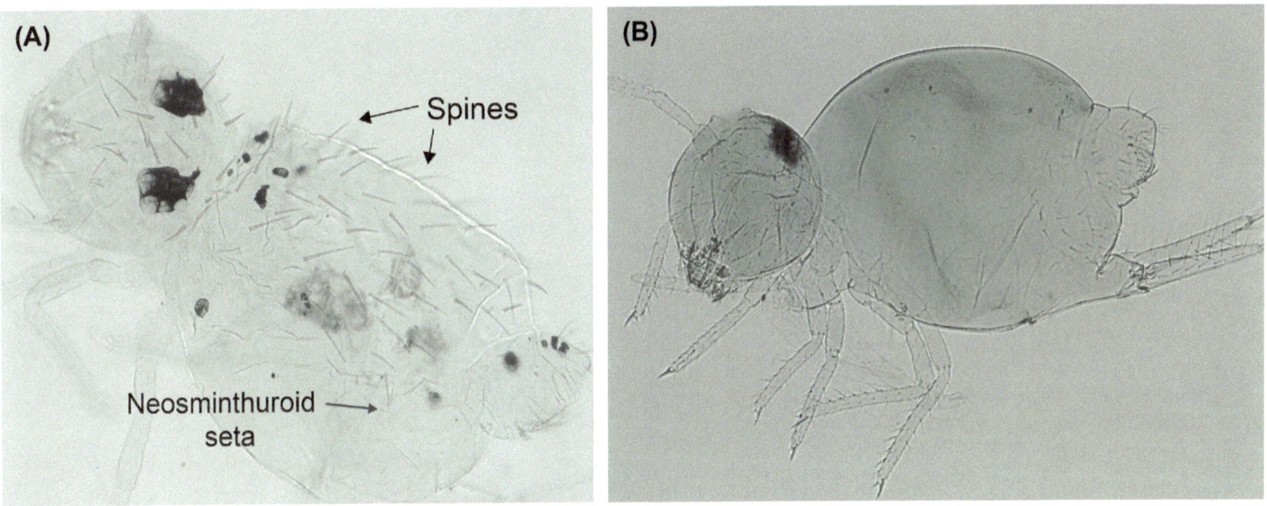

FIGURE 2.57 (A) *Adisianus* (Bourletiellidae), head and body, dorso-lateral view. (B) *Arlesminthurus* (Bourletiellidae), head and body, dorso-lateral view.

Collembola: Poduromorpha: Brachystomellidae: Genera

1	Eyes present (Figs. 2.1 A, 6 G, H)	2
1'	Eyes absent	*Folsomiella**
2	Maxilla globular with apical teeth (Fig. 2.2 B)	3
2'	Maxilla elongated with lateral teeth (Fig. 2.4 A)	*Rapoportella*†
3(2)	Antennal article IV ventrally with several modified short setae (sensory field) (Fig. 2.4 B)	*Brachystomellides*§
3'	Antennal article IV ventrally without modified short setae	*Brachystomella*†

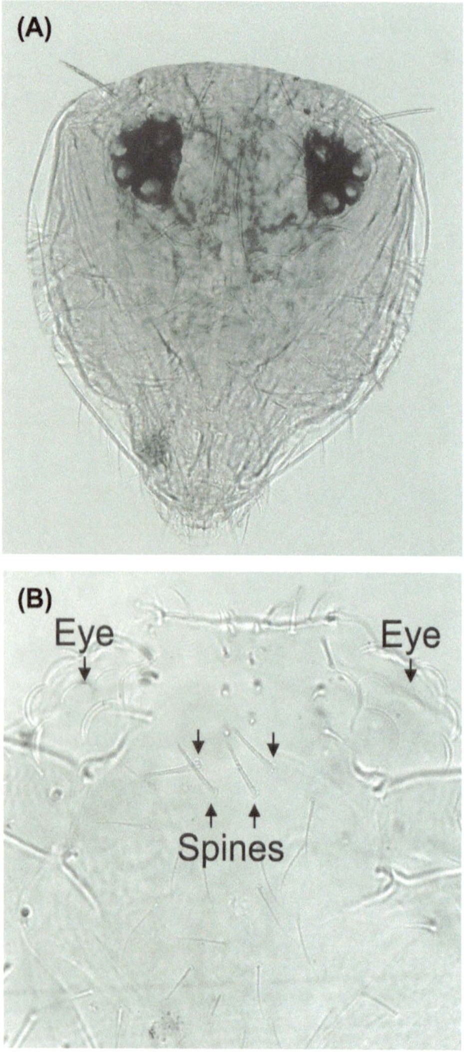

FIGURE 2.58 (A) *Stenognathriopes* (Bourletiellidae), head, frontal view. (B) *Arlesminthurus* (Bourletiellidae), head, frontal view.

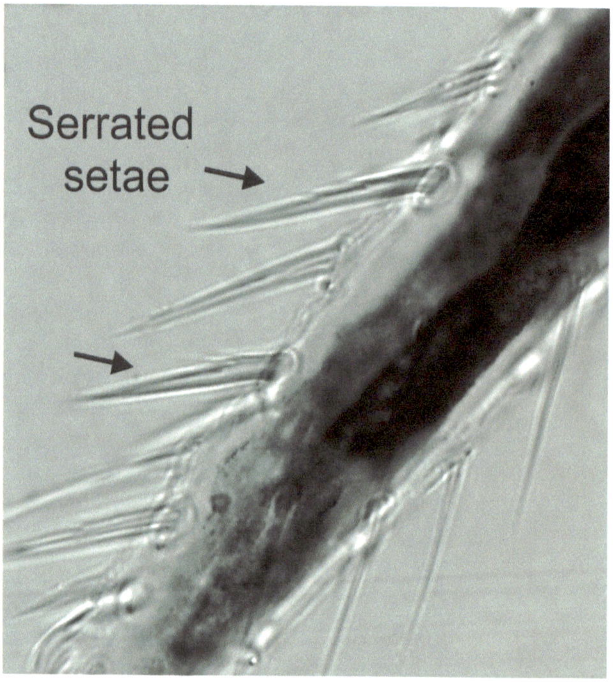

FIGURE 2.59 *Stenognathriopes* (Bourletiellidae), third tibiotarsus, lateral view.

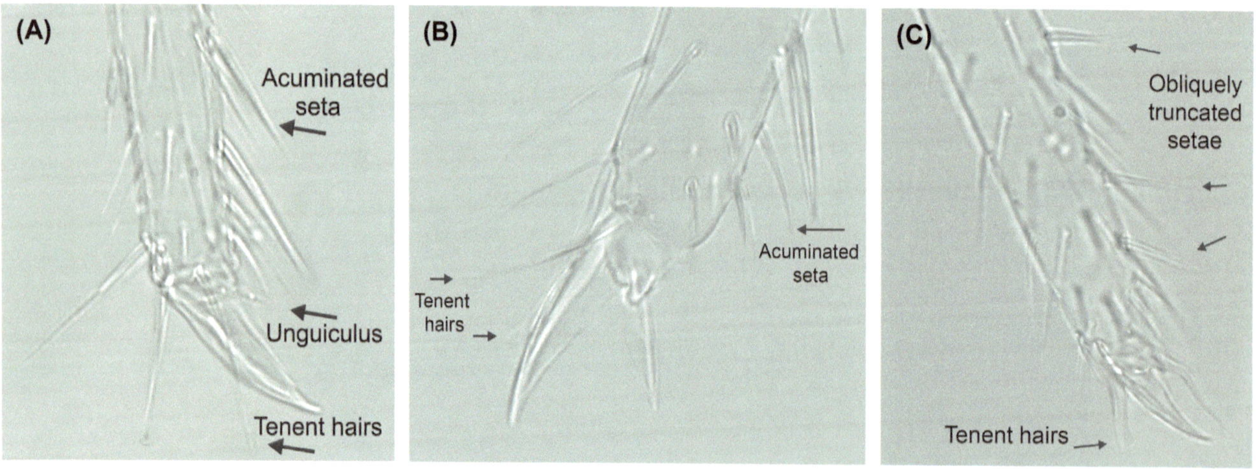

FIGURE 2.60 (A) *Adisianus* (Bourletiellidae), distal third tibiotarsus and empodial complex. (B) *Arlesminthurus* (Bourletiellidae), distal third tibiotarsus and empodial complex. (C) *Prorastriopes* (Bourletiellidae), distal third tibiotarsus and empodial complex.

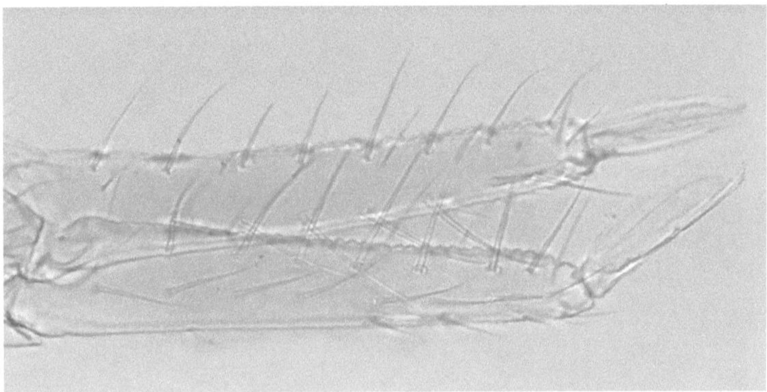

FIGURE 2.61 *Arlesminthurus* (Bourletiellidae), dens and mucro, lateral view.

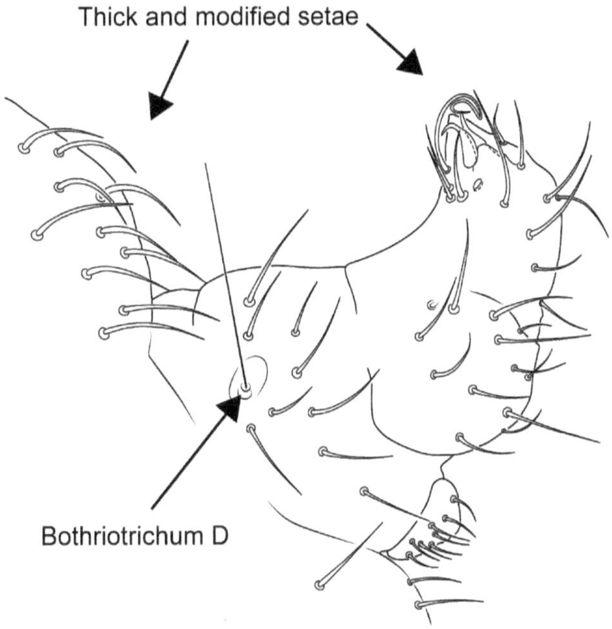

FIGURE 2.62 *Bourletiella hortensis* (Bourletiellidae), male posterior region of the body. *Modified from Betsch (1980).*

Collembola: Poduromorpha: Neanuridae: Genera

1 Abdominal segment VI dorsally bilobed (Figs 2.6 A, 6 B) or truncated (Fig. 2.6 C); furcula absent; postantennal organ absent............ 2
1' Abdominal segment VI dorsally simple, rounded (Fig. 2.6 D); furcula present or absent; postantennal organ present or absent 5
2(1) Abdominal segment VI dorsally clearly bilobed (Fig. 2.6 A); eyes, if present, usually lacking dark pigment 3
2' Abdominal segment VI dorsally slightly bilobed (Fig. 2.6 B) or truncated (Fig. 2.6 C); eyes with dark pigment .... *Paranura*§
3(2) Tubercle antenno-frontal (unpaired) absent; eyes 2 + 2 to 0 + 0; specimens not blue (yellowish, reddish, or lacking pigments)........... 4
3' Tubercule antenna-frontal (unpaired or clypeal tubercle *sensu* Jordana et al. 1997) present (Fig. 2.6 E); eyes 3 + 3; specimens mostly blue .... *Neanura**
4(3) Antennal article IV sensilla subequal .... *Paleonura**
4' Antennal article IV with one sensillum clearly thicker than others (Fig. 2.6 F)... *Sensillanura*§

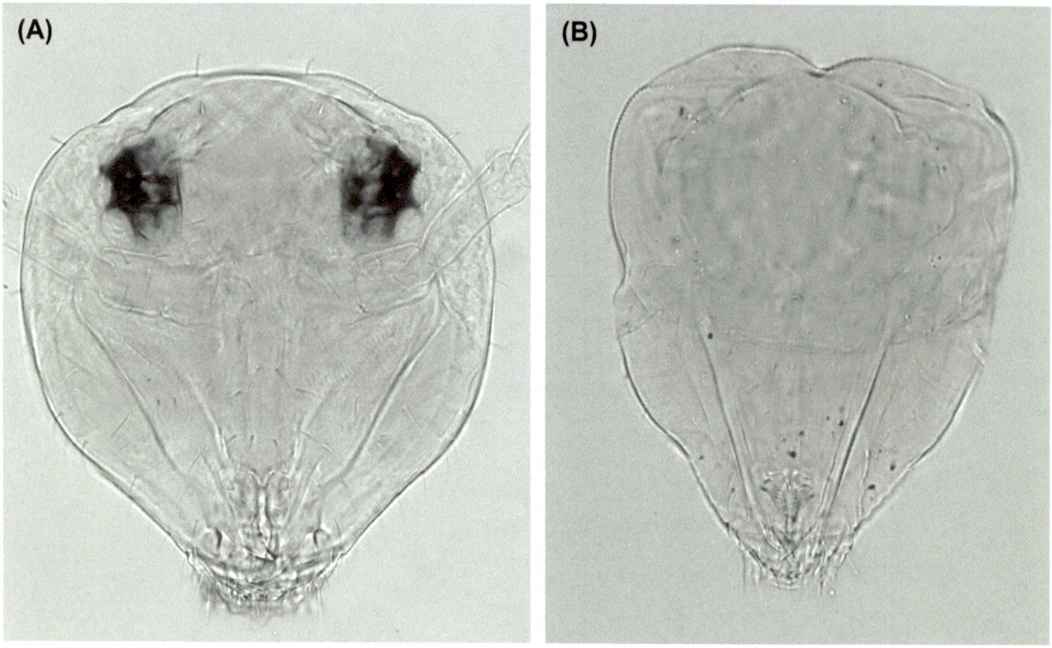

FIGURE 2.63 (A) *Polykatianna* (Katiannidae), head, frontal view. (B) *Stenognathellus* (Katiannidae), head, frontal view.

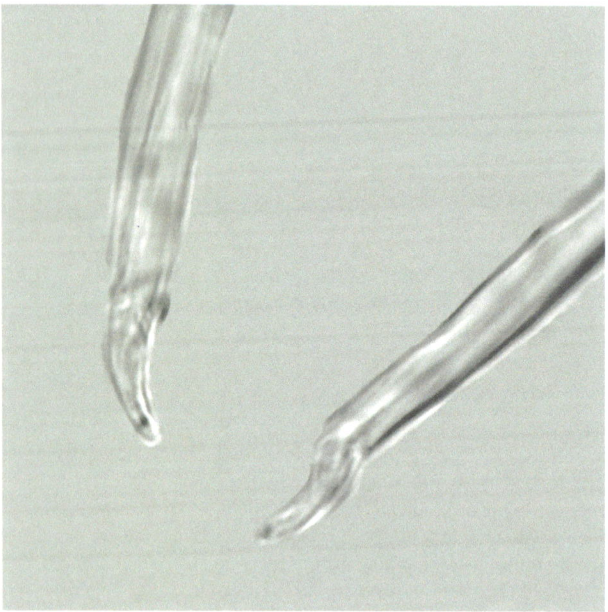

FIGURE 2.64 *Stenognathellus* (Katiannidae), apex of the maxillae.

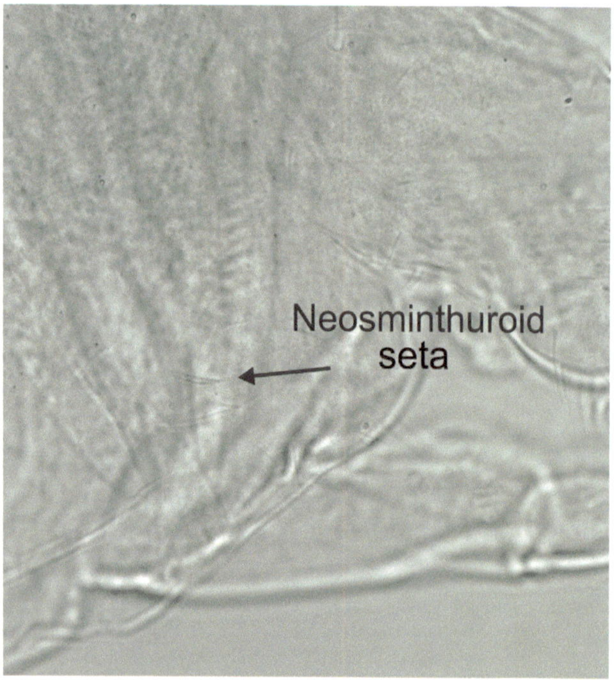

FIGURE 2.65 *Sminthurinus* (Katiannidae), posteroinferior region of large abdomen.

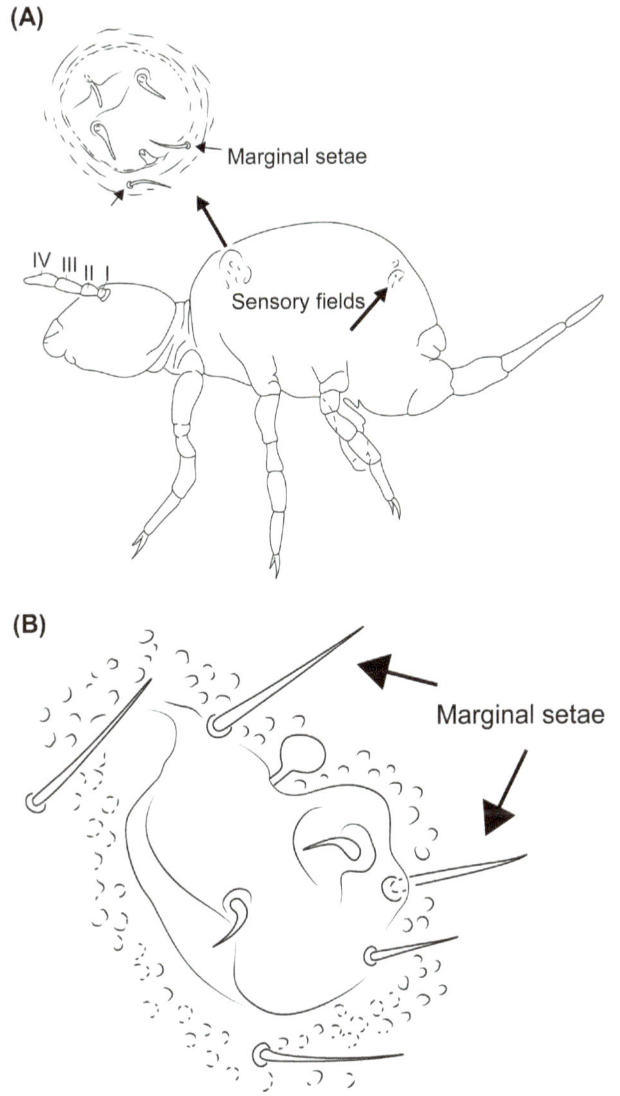

FIGURE 2.66 (A) *Neelus murinus* (Neelidae), habitus and sensory fields. (B) *Megalothorax minimus* (Neelidae), sensory field. *(A) Modified from Fjellberg (2007). (B) Modified from Fjellberg (2007).*

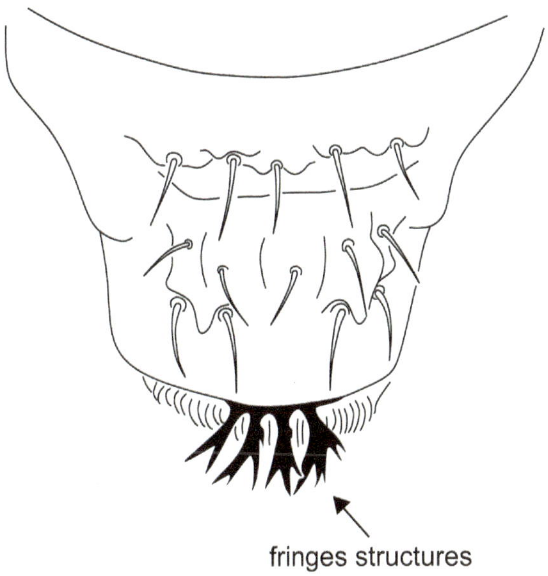

FIGURE 2.67 *Neelides minutus* (Neelidae), labral region. *Modified from Fjellberg (2007).*

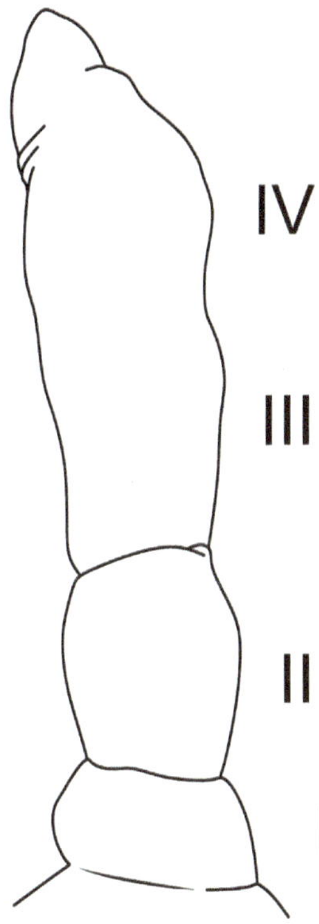

FIGURE 2.68 *Megalothorax minimus* (Neelidae), antenna, numbers indicate antennal segments. *Modified from Fjellberg (2007).*

[Mexico to Costa Rica]

5(1) Maxilla capitulum elongated, with or without teeth (Figs. 2.2 E–G) 6

5' Maxilla capitulum short and triangular; with two small toothed lamellae (Fig. 2.2 D) ..... *Friesea*[†]

6(5) Maxilla without fringed or toothed lamellae (Figs. 2.2 F, G) 7

6' Maxilla with fringed and/or toothed lamellae (Fig. 2.2 E) ..... *Anurida**

7(6) Eyes 8 + 8 8

7' Eyes 7 + 7 or less 10

8(7) Postantennal organ present; furcula short, at rest not reaching the ventral tube when near the trunk 9

8' Postantennal organ absent; furcula long, reaching past the ventral tube ..... *Pseudanurida**

9(8) Postantennal organ moruliform (Fig. 2.6 G) ..... *Aethiopella**

9' Postantennal organ with one ring of vesicles (Fig. 2.6 H) ..... *Pseudachorutes**

10(7) Postantennal organ present 11

10' Postantennal organ absent 13

11(10) Postantennal organ not moruliform 12

11' Postantennal organ moruliform (as in Fig. 2.6 G) ..... *Neotropiella**

12(11) Furcula present ..... *Furculanurida*[§]

12' Furcula absent or reduced to small ventral projections (as in Fig. 2.6 I) ..... *Micranurida**

13(10) Furcula present and well developed; antennal article IV sensilla subequal ..... *Arlesia**

13' Furcula reduced to small ventral projections (Fig. 2.6 I); antennal article IV with one sensillum clearly thicker than others ..... *Hylaeanura**

Collembola: Poduromorpha: Hypogastruridae: Genera

1 Abdominal segment VI with two or zero anal spines (Fig. 2.5 A) 2

1' Abdominal segment VI with three anal spines (Fig. 2.5 B) ..... *Triacanthella*[§]
[Chile]

2(1) Postantennal organ absent 3

2' Postantennal organ present 5

3(2) Pigmented, usually dark blue; 5 + 5 (rarely 4 + 4) eyes; furcula and tenaculum usually present 4

3' Lacking pigments; eyes, furcular, and tenaculum absent ..... *Acherontiella*[§]

4(3) Anal spines absent; terminal or subterminal position of anus (Fig. 2.5 C) ..... *Paraxenylla*[§]

4' Anal spines present (Fig. 2.5 A); ventral position of anus (Fig. 2.5 D) ..... *Xenylla**

5(2) Eyes and furcula absent 6

5' Eyes and furcula present 7

6(5) Unguiculus usually present; unguis without teeth ..... *Willemia*[§]

6' Unguiculus absent; unguis with inner tooth (Fig. 2.5 E) ..... *Tafallia*[§]
[Central America. Nearctic]

7(5) Unguiculus present, with enlarged basal lamella (Fig. 2.5 E) 8

7' Unguiculus absent, or if present, setiform ..... *Schoettella*[§]

8(7) Mandibles normal, with apical teeth and strong molar plate (Fig. 2.2 A) 9

8' Mandibles reduced, without apical teeth and with rudimentary molar plate (Fig. 2.5 F) ..... *Microgastrura*[§]
[Central America, invasive in Puerto Rico]

9(8) Antenna with eversible sac at junction of articles III and IV (Fig. 2.5 G); mucro spatulate (Fig. 2.3 A) ..... *Ceratophysella**

9' Antenna without eversible sac; mucro triangular ..... *Hypogastrura**

Collembola: Poduromorpha: Onychiuridae: Genera

1 Furcula reduced to a finely granulated area with 2 + 2 microsetae in two rows (Fig. 2.7 B); abdominal segment VI with or without anal spines 2

1' Furcula reduced to a finely granulated area with four microsetae in one transverse row (Fig. 2.7 C); abdominal segment VI with anal spines ..... *Onychiurus*[§]

2 Postantennal organ with compound vesicles (Fig. 2.7 D); tibiotarsus with seven setae in the distal whorl; abdominal segment VI devoid of anal spines ..... *Agraphorura**

2' Postantennal organ with simple vesicles (Fig. 2.7 E); tibiotarsus with nine or more setae in the distal whorl; abdominal segment VI with or without anal spines ..... *Thalassaphorura*[§]

Collembola: Poduromorpha: Tullbergiidae: Genera

1 Abdominal segment VI with crescentic ridges (Fig. 2.7 F); antennal article III sense organ with two enlarged sensory clubs (Fig. 2.3 E); prothorax without pseudocelli .... *Mesaphorura**

1' Abdominal segment VI without crescentic ridges; antennal article III sense organ with three (more common) or two enlarged sensory clubs; prothorax with or without one pair of pseudocelli .... *Tullbergia*§

Collembola: Entomobryomorpha: Families

1 Abdominal segment IV length >1.5× abdominal segment III along midline (Figs. 2.1 B, 8 A–C), if shorter, then antennal article I and/or II subdivided (Figs. 2.8 D, E); scales present or absent (Figs. 2.8 A, 9 A, B); body setae mostly ciliate, multilaterally ciliate macrosetae usually present, cylindrical, or with truncate or broadened apices (Fig. 2.25) 2

1' Abdominal segment IV length <1.5× abdominal segment III at midline (Fig. 2.8 F–H); scales absent; body setae usually smooth or unilaterally ciliate, multilaterally ciliate macrosetae can be present, but always acuminate (Figs. 2.9 C, D) .... **Isotomidae [p. 49]**

2(1) Dens dorsally crenulate; mucro short with one or two teeth always pointed dorsally (Figs. 2.10 A, 11 A) .... **Entomobryidae [p. 50]**

2' Dens not crenulate; mucro short with at least two teeth projected apically (Figs. 2.10 B, 11b–c), when inserted dorsally then mucro ≥1/3 the dens length (Fig. 2.10 C) .... **Paronellidae [p. 51]**

Collembola: Entomobryomorpha: Isotomidae: Genera

1 Furcula present, reduced or long (Figs. 2.8 F–H) 2

1' Furcula absent .... *Pseudanurophorus*§

2(1) Manubrium ventrally without distal setae (Fig. 2.12) 3

2' Manubrium ventrally with distal setae (Figs. 2.13 A, B) 6

3(2) Eyes present; abdominal segments separated (Figs. 2.8 F, 18) 4

3' Eyes absent; abdominal segments V–VI fused (Figs. 2.8 G, 17) .... *Isotomodes*§

4(3) Cuticle normal, discretely granulated (Figs. 2.9 C, D) 5

4' Cuticle strongly granulated (Poduromorpha-like) .... *Bonetrura*§
[Central America. Nearctic]

5(4) Collophore with four or more laterodistal setae per side; dens with more than seven posterior setae; mucro with two teeth (Fig. 2.14 D) .... *Ballistura*§

5' Collophore with three laterodistal setae per side; dens with less than six posterior setae; mucro (if present) with one to two teeth (Fig. 2.12) .... *Folsomides*§

6(2) Fourth to sixth Abdominal segments IV–VI individualized (Fig. 2.8 F) 7

6' Abdominal segments IV–VI or V–VI fused (Figs. 2.8 G, H, 17) 12

7(6) Eyes 6 + 6 to 8 + 8; manubrium ventrally with 11 + 11 or more setae (Fig. 2.13 A) 8

7' Eyes 0 + 0 to 8 + 8; manubrium ventrally with 6 + 6 or less setae (Fig. 2.13 B) .... *Proisotoma**

8(7) Manubrium ventrodistal region without modified setae (Figs. 2.13 A, B) 9

8' Manubrium ventrodistal region with thickened or spine-like setae (Fig. 2.13 C) .... *Isotoma*§
[Central America to coastal western South America]

9(8) Abdominal segments without bothriotricha; body with smooth setae (Fig. 2.9 C) 10

9' Abdominal segments III and/or IV with bothriotricha (Fig. 2.16); thorax, abdomen and tibiotarsi with multilaterally ciliate macrosetae (Fig. 2.9 D) .... *Isotomurus**

10(9) Dens distally narrowed and crenulated (Fig. 2.14 A); mucro without lamella (Fig. 2.14 C) 11

10' Dens distally tuberculate (Fig. 2.14 B); mucro with dorsal lamella (Figs. 2.14 A, D) .... *Granisotoma**
[Central America. Nearctic]

11(10) Distance between postantennal organ and the nearest eye <2× proximal eye diameter; dens with smooth setae (Figs. 2.14 A, 18 A) .... *Desoria**

11' Distance between postantennal organ and the nearest eye ≥2× proximal eye diameter; dorsal dens usually with foil setae (Figs. 2.14 C, 18 B) .... *Paracerura**
[Brazil]

12(6) Postantennal organ absent 13

12' Postantennal organ present (Figs. 2.18, 19) 14

13(12) Abdominal segments IV–VI fused; antennal article IV with two apical oval sensilla; tergal segments only with smooth setae (Figs. 2.9 C, 21 A) .... *Folsomina**

13' Abdominal segments V–VI fused; antennal article IV with six apical oval sensilla; tergal segments with some serrated setae (Figs. 2.17, 21 B, 22 A) .... *Isotomiella**
14(12) Mucro with 2 or more teeth (Figs. 2.12, 14 A, C) 15
14' Mucro falcate (Fig. 2.15) .... *Arlea*§
[South America]
15(14) Abdominal segments V–VI fused (Figs. 2.8 G, 17) 16
15' Abdominal segments IV–VI fused .... *Folsomia**
16(15) Abdominal segments V–VI without modified setae 17
16' Abdominal segments V and/or VI with foil setae (Fig. 2.22 B) .... *Hemisotoma**
17(16) Body generally pigmented; dens long with six or more ventral setae (Figs. 2.8 G, 20 A) .... *Cryptopygus**
17' Body depigmented; dens short with setae five or less ventral setae (Figs. 2.8 H, 20 B) .... *Dagamaea*§
[Central America. Nearctic]

Collembola: Entomobryomorpha: Entomobryidae: Genera

1 Antennae with five or six apparent articles (first and second subdivided); abdominal segment IV length <1.5× abdominal segment III along midline (Figs. 2.8 D, E) 2
1' Antennae with four articles; abdominal segment IV length >2× abdominal segment III along midline (Figs. 2.8 A–C) 5
2(1) Body with scales (Figs. 2.8 D, 9 B, 23 A–E) 3
2' Body without scales (Fig. 2.9 A) 4
3(2) Body dorsally with apically pointed scales, antennae and furcula without scales; antennae longer than body, with five apparent articles (first subdivided); dens without spines (Figs. 2.8 E, 23 A) .... *Mastigoceras**
[Brazil]
3' Body dorsally with apically rounded or truncate scales, antennae and furcula with scales; antennae shorter than body, with six apparent articles (first and second subdivided); dens proximal region usually with dorsal spines (Figs. 2.8 D, 23 B, C, 24 A) .... *Dicranocentrus**
4(2) Postantennal organ present; mucro with one tooth, without basal spine (Figs. 2.25, 26 A) .... *Nothobrya**
[Brazil]
4' Postantennal organ absent; mucro with two teeth and basal spine (Fig. 2.26 C) .... *Orchesella**
[Mexico]
5(1) Body covered by ciliate setae, scales absent (Fig. 2.9 A) 6
5' Body with scales (Figs. 2.8 A–C, 9 B, 23 A–E) 7
6(5) Mucro falcate (Fig. 2.26 B) .... *Drepanura**
[South America]
6' Mucro bidentate (Fig. 2.11 A) .... *Entomobrya**
7(5) Mucro falcate, without basal spine (Fig. 2.26 A) 8
7' Mucro bidentate, with basal spine (Fig. 2.26 C) 9
8(7) Antennae length ≥ body; mesothorax strongly projecting over head; unguis with large lateral tooth; dens dorsally with long, blunt setae (Figs. 2.8 B, 27 A, 28 B) .... *Lepidocyrtinus**
[South America]
8' Antennae length < body; mesothorax rounded, not projecting over head; unguis with small lateral tooth, dens dorsally without blunt setae (Figs. 2.1 B, 27 B, 28 B) .... *Seira* *
9(7) Dental spines present (Figs. 2.24 B, C) 10
9' Dental spines absent (Fig. 2.28 B) 11
10(9) Body scales heavily ciliated and apically pointed (Fig. 2.23 E); mesothorax, metathorax and abdominal segment I with macrosetae (Fig. 2.30 A); unguis with large lateral tooth (Fig. 2.27 A); dens ventrally without scales (Fig. 2.29 B); dens with dental spines in rows (Fig. 2.24 B) .... *Amazhomidia**
[Amazon]
10' Body scales lightly ciliated and apically truncate (Fig. 2.23 D); mesothorax, metathorax and abdominal segment I without macrosetae (Fig. 2.30 B); unguis with normal lateral tooth (Fig. 2.27 B); dens ventrally with scales (Fig. 2.29 A); dens with dental spines clustered near base (Fig. 2.24 C); .... *Lepidocyrtus* (in part)*
[South America]
11(9) Mesothorax usually projecting over head (Figs. 2.8 A, C); dens ventrally with scales (Fig. 2.29 A) 12
11' Mesothorax not projected over head (Fig. 2.1 B); ventral dens without scales (Fig. 2.29 B) .... *Willowsia*§
[Central America to coastal western South America]
12(11) Body scales weakly striate and apically truncate (Fig. 2.23 D); mesothorax, metathorax, and abdominal segment I without macrosetae (Fig. 2.30 B); unguis with normal lateral tooth (Fig. 2.27 B); manubrial plate without macrosetae .... *Lepidocyrtus* (s.l.)*

12'	Body scales heavily striate and apically pointed (Fig. 2.23 E); mesothorax, metathorax, and abdominal segment I with macrosetae (Fig. 2.30 A); unguis with large lateral tooth (Fig. 2.27 A); manubrial plate with at least two macrosetae per side (Fig. 2.31) [South America]	*Lepidocyrtoides**

Collembola: Entomobryomorpha: Paronellidae: Genera

1	Body pale, lacking pigment and eyes (Figs. 2.32 A, 33 A); dens dorsally with fringed scaliform setae; mucro long (Fig. 2.10 C)	2
1'	Body usually with pigments (Figs. 2.32 B–D); eyes 0 +0 to 8 + 8 (Figs. 33 A, B); dens dorsally without fringed scaliform setae; mucro short (Figs. 2.10 B, 11 B, C)	3
2(1)	Dens dorsally with ciliated setae between large scales (Fig. 2.10 C)	*Cyphoderus*§
2'	Dens dorsally without setae between large scales [South America. Oceana]	*Serroderus*§
3(1)	Body with scales (Figs. 2.32 D, 34 A); dens dorsally with spines (Fig. 2.10 B); mucro base without appendage (Fig. 2.11 B)	4
3'	Body without scales (Fig. 2.34 B); dens without spines; mucro base with enlarged spatulate appendage (Fig. 2.11 C)	*Salina*†
4(3)	Antennae length > body; metathorax enlarged and conspicuously humped, resulting in a bent body between metathorax and abdominal segment I (Figs. 2.32 C, 35); metatibiotarsus with only ciliated setae (Fig. 2.36 A)	*Campylothorax**
4'	Antennal length ≤ body (longer only in cave forms); metathorax normal (Fig. 2.32 D); metatibiotarsus with one distal smooth seta (Fig. 2.36 B)	*Trogolaphysa**

Collembola: Symphypleona: Families

1	Female with subanal appendages (Figs. 2.37 A–C); male antennal articles normal, without any modification	2
1'	Female without subanal appendage; males with antennal articles II-III modified as a prehensile organ (Figs. 2.46 A, B)	**Sminthurididae [p. 51]**
2(1)	Female subanal appendage directed toward the anus or posteriorly (Figs. 2.37 A, B)	3
2'	Female subanal appendage directed toward the genital opening (Fig. 2.37 C)	6
3(2)	Antennal article IV length > third (Figs. 2.1 C, 58 A, B)	4
3'	Antennal article IV length < third (Fig. 2.38)	**Dicyrtomidae [p. 52]**
4(3)	Antennal article III without papilla; female subanal appendage directed toward anus (Fig. 2.37 A)	5
4'	Antennal article III with a low papilla at dorsal and medial position (Fig. 2.39); female subanal appendage directed posteriorly (Fig. 2.37 B)	Sturmiidae, one genus: *Sturmius*§
5(4)	Metatrochanter often with a posterior spine (Fig. 2.40); tibiotarsi often without capitate tenent hairs	**Sminthuridae [p. 52]**
5'	Metatrochanter always without posterior spine; tibiotarsi always with capitate tenent hairs (Figs. 2.60 A–C)	**Bourletiellidae [p. 52]**
6(2)	Neck without peculiar organs (Fig. 2.1 C)	7
6'	Neck with one pair of lateral soft organs (Fig. 2.41)	**Spinothecidae [p. 53]**
7(6)	Eyes at most with four lenses per side; tibiotarsi without capitate tenent hairs (Fig. 2.49)	8
7'	Eyes with six to eight lenses per side; tibiotarsi with capitate tenent hairs (Fig. 2.42)	**Katiannidae [p. 53]**
8(7)	Abdominal segment V completely included in large abdomen; bothriotrichum D short and capitated (Fig. 2.43)	Collophoridae, one genus: *Collophora**
8'	Abdominal segment V included in large abdomen; bothriotrichum D long and with normal form	Arrhopalitidae, one genus: *Arrhopalites**

Collembola: Symphypleona: Sminthurididae: Genera

1	Metatibiotarsus without organs	2
1'	Metatibiotarsus with a distal pair of tibiotarsal organs (Fig. 2.44)	*Sminthurides*†
2(1)	Mucro with basal seta (Fig. 2.45); male with antennal articles II–III highly modified (Fig. 2.46 A); protibiotarsus with four proximal round organs (Fig. 2.47)	*Denisiella**
2'	Mucro without seta; male with antennal articles II–III weakly modified (Fig. 2.46 B); protibiotarsus without organs	*Sphaeridia**

Collembola: Symphypleona: Dicyrtomidae: Genera

1 Adults without bothriotrichum D; head clypeal region with four or more median setae (Fig. 2.48 A) 2
1' Adults with bothriotrichum D; head clypeal region with two median setae (Fig. 2.48 B) .... *Ptenothrix**
2(1) Unguis with tunica well developed (Fig. 2.49) 3
2' Unguis with tunica poorly developed or absent .... *Dicyrtoma**
[Central America. Nearctic]
3(2) Head and large abdomen anterior region with spiniform setae (Fig. 2.48 B); furcula base without neosminthuroid setae .... *Dicyrtomina**
[Argentina, Central America. Nearctic]
3' Head and large abdomen anterior region with normal setae (Fig. 2.48 A); furcula base with neosminthuroid setae (Fig. 2.50 A) .... *Calvatomina**

Collembola: Symphypleona: Sminthuridae: Genera

1 Furcula base with neosminthuroid setae (Figs. 2.50 B, C) 2
1' Furcula base without neosminthuroid setae 5
2(1) Furcula base with one neosminthuroid setae per side (Fig. 2.50 B) 3
2'(1) Furcula base with three neosminthuroid setae per side (Fig. 2.50 C) .... *Songhaica**
3(2) Antennal article IV clearly subdivided (Fig. 2.1 C); mesothoracic vesicles absent; large abdomen in adults with bothriotricha B and D 4
3' Antennal article IV weakly or not subdivided; mesothoracic vesicle one per side (Fig. 2.51); adult large abdomen without bothriotricha B and D .... *Neosminthurus**
[Central America. Nearctic]
4(3) Pro- and mesotrochanters without spine .... *Sphyrotheca**
4' Pro- and sometimes mesotrochanters with spine (Fig. 2.40) .... *Szeptyckitheca**
5(1) Small abdomen with bothriotrichum D smooth and acuminate (Fig. 2.62) 6
5' Small abdomen with bothriotrichum D clavated and ciliated (Fig. 2.52) .... *Austrosminthurus**
[Patagonia]
6(5) Metatrochanter with spine (Fig. 2.40); head without postantennal seta 7
6' Metatrochanter without spine; head with postantennal seta (Fig. 2.53) .... *Sminthurus**
7(6) Unguis normal (Figs. 2.60 A–C) 8
7' Unguis with an anterior cavity (Fig. 2.54) .... *Varelasminthurus**
[Brazil]
8(7) Antennal article IV larger than body and with 20–40 subarticles; large abdomen without spines (Fig. 2.55 A) .... *Temeritas**
8' Antennal article IV smaller than body and with fewer than 20 subarticles; large abdomen with dorsal spines (Fig. 2.55 B) .... *Pararrhopalites**

Collembola: Symphypleona: Bourletiellidae: Genera

1 Metatibiotarsus with three capitate tenent hairs 2
1' Metatibiotarsus with two capitate tenent hairs (Fig. 2.60 B) 3
2(1) Male frons with median nasal organ (Fig. 2.56 A); unguiculus length <0.5 unguis (as in Fig. 2.60 A); dens with long posterior internal and external setae, longer than space between them .... *Pseudobourletiella**
2' Male interocular region with a large dorsal salience (horn) with two or more long setae (Fig. 2.56 B); unguiculus long, reaching, or surpassing unguis apex; dens with normal setae .... *Bovicornia**
[Virgin Islands]
3(1) Large abdomen with dorsal spines (Fig. 2.57 A) 4
3' Large abdomen without spines (Fig. 2.57 B) 5
4(3) Head triangular (Fig. 2.58 A), mouthparts elongated; tibiotarsi inner side with thick, serrated setae (Fig. 2.59); unguiculus longer than unguis; furcula base without neosminthuroid setae .... *Stenognathriopes**
4' Head usually roundish (Fig. 2.57 A), mouthparts normal; tibiotarsi inner side with acuminate setae; unguiculus shorter than unguis (Fig. 2.60 A); furcula base with neosminthuroid setae (Fig. 2.57 A) .... *Adisianus**
5(3) Tibiotarsal inner side with acuminate setae (Fig. 2.60 B); males with modified setae between antennae and/or on small abdomen (Figs. 2.58 B, 62) 6
5' Tibiotarsal inner side with obliquely truncated or distally flattened setae (Fig. 2.60 C); males without any special sexual dimorphism .... *Prorastriopes**
6(5) Dens often crenulate with long internal and external setae (Fig. 2.61); male interantennal space with two pairs of spines (Fig. 2.58 B); large abdomen posterior region and small abdomen without thick or modified setae .... *Arlesminthurus**
6' Dens never crenulate with normal size setae; male interantennal space without particular spines; large abdomen posterior region and small abdomen with thick and modified setae (Fig. 2.62) .... *Bourletiella**

Collembola: Symphypleona: Spinothecidae: Genera

1 Eyes present .... *Spinotheca**
[Patagonia]
1' Eyes absent .... *Troglospinotheca**
[Patagonia]

Collembola: Symphypleona: Katiannidae: Genera

1 Head roundish and mouthparts normal (Fig. 2.63 A) 2
1' Head and mouthparts elongate (Figs. 2.63 B, 64) .... *Stenognathellus**
2(1) Antennal article IV subdivided (Fig. 2.1 C), furcula base without neosminthuroid seta 3
2' Antennal article IV not subdivided, furcula base with neosminthuroid seta (Fig. 2.65) 4
3(2) Head with frontal thick spines (Fig. 2.57 A) .... *Katianna**
3' Head with normal setae (Fig. 2.63 A) .... *Polykatianna**
4(2) Small abdomen with bifurcated median circumanal seta (Fig. 2.37 C) .... *Sminthurinus**
4' Small abdomen with normal median circumanal seta .... *Arborianna**
[Brazil, Amazon]

Collembola: Neelipleona: Neelidae: Genera

1 Head and large abdomen with sensory fields (Figs. 2.66 A, B); labrum without fringes or ciliate edges 2
1' Head and large abdomen without sensory fields; labrum with fringes and ciliate anterior edges (Fig. 2.67) .... *Neelides**
2(1) Antennal articles III–IV fused (Fig. 2.68); abdomen sensory field with more than two marginal setae (Fig. 2.66 B) 3
2' Antennal articles III–IV distinctly separate; abdomen sensory field with two marginal setae (Fig. 2.66 A) .... *Neelus**
3(2) Abdomen sensory field with five marginal setae (Fig. 2.66 A) .... *Megalothorax**
3' Abdomen sensory field with three marginal setae .... *Spinaethorax*§
[Mexico]

CLASS COLLEMBOLA

ACKNOWLEDGMENTS

We thank Drs. José G. Palacios-Vargas (Universidad Nacional Autónoma de México), Douglas Zeppelini (Universidade Estadual da Paraíba), Enrique Baquero (University of Navarra), Ernest C. Bernard (University of Tennessee), Gabriel C. Queiroz (Museu Nacional da Universidade do Rio de Janeiro), Mikhail Potapov (Moscow State Pedagogical University), Rafael Jordana (University of Navarra), and Wanda M. Weiner (Polish Academy of Sciences) for review and several additions to this chapter. We also thank all authors and journals for permission to use copyrighted figures in our chapter.

REFERENCES

Abrantes, E.A. & M.C.Mendonça. 2005. Uma nova espécie de *Arlea* Womersley do sudeste do Brasil (Collembola, Isotomidae). Revista Brasileira de Zoologia 22: 936–939.

Abrantes, E.A. & M.Duarte. 2013. New Species of *Paracerura* (Collembola: Isotomidae) from the State of São Paulo, Brazil. The Florida Entomologist 96: 1392–400.

Arbea, J.I. 2005. *Agraphorura calvoi* n. sp. from Venezuelan caves (Collembola: Onychiuridae). International Journal of Speleology 34: 19–24.

Arlé, R. 1960. Notas sôbre a família Oncopoduridae, com descrição de duas espécies novas do Brasil (Collembola). Arquivos do Museu Nacional Rio de Janeiro 50: 9–23.

Arlé, R. 1971. Collemboles dÁmazonie, III. Quelques Symphypléones du Bas-Amazone et complément à la description de *Deuterosminthurus aueti* Arlé, 1961. Boletim do Museu Paraense Emilio Goeldi 75: 1–11.

Arlé, R. 1984. O gênero *Sphaeridia* Linnaniemi, 1912 no Brasil com descrição de novas espécies (Collembola-Symphypleona). Boletim do Museu Paraense Emilio Goeldi, Zoología, 1: 229–255.

Bellinger, P.F., K.A. Christiansen & F. Janssens. 2017. Checklist of the Collembola of the World. Available from: http://www.collembola.org. accessed: 07 June 2017.

Bellini, B.C. & D. Zeppelini. 2011. New genus and species of Seirini (Collembola, Entomobryidae) from Caatinga Biome, Northeastern Brazil. Zoosystema 33: 545–555.

Benito, J.C.S. & J.G. Palacios-Vargas. 2008. Two new species of *Paranura* (Collembola: Neanuridae) from Southeastern Mexico. Annals of the Entomological Society of America 101: 683–688.

Betsch, J.-M. 1974. Contribution à l'étude de la reproduction chez les Bourletiellinae (Collembola Symphypléones). Pedobiologia 14: 179–181.

Betsch, J.M. 1980. Éléments pour une monographie des Collemboles Symplyplêones (Hexapodes, Aptérygotes). Mémoires du Muséum National d'Histoire Naturelle, Nouvelle Série, Série A, Zoologie 116: 1–227.

Börner, C. 1906. Das system der Collembolen nebst Beschreibung neuer Collembolen des Hamburger Naturhistorischen Museums. Mitteilungen aus dem Naturhistorischen Museum in Hamburg 23: 147–188.

Bretfeld, G. 1994. *Sturmius epiphytus* n. gen. n. spec. from Colombia, a taxon of the Symphypleona (Insecta, Colembola) with an unexpected character combination. Description and position in non-Linnean and Linnean classifications of the Symphypleona. Journal of Zoological Systematics and Evolutionary Research 32: 264–281.

Bretfeld, G. 1999. Synopses on Palaearctic Collembola, volume 2. Symphypleona. Abhandlungen und Berichte des Naturkundemuseums Görlitz, 71, 318 pp.

Bretfeld, G. & U. Gauer. 1994. Diagnostic description of the males of new *Sphaeridia* species (Insecta, Collembola) from South America. Andrias 13: 113–136.

Christiansen, K. & R.F. Snider. 2008. Aquatic Collembola. pp. 165–179 *in*: R.W. Merritt, K.W. Cummins and M.B. Berg (eds.), An introduction to the aquatic insects of North America, 4th edition. Kendall/Hunt Publishing Company, Dubuque, IA.

Christiansen, K. & P. Bellinger. 1992. Insects of Hawaii, volume 15. Collembola, Honolulu. 445 pp.

Christiansen, K. & P. Bellinger. 1998. The Collembola of North America North of the Rio Grande, A taxonomic analysis, Grinnell College, Grinnell, IA. 1520 pp.

Cicconardi, F., P.P. Fanciulli & B.C. Emerson. 2013. Collembola, the biological species concept and the underestimation of global species richness. Molecular Ecology 22: 5382–5396.

Cipola, N.G., J.W. Morais & B.C. Bellini. 2016. A new genus of Entomobryinae (Collembola, Entomobryidae) from Brazilian Amazon with body scales and dental spines. Zootaxa 4105: 261–273.

Cipola, N.G., J.W. Morais & B.C. Bellini. 2017. The discovery of *Lepidocyrtoides* Schött, 1917 (Collembola, Entomobryidae, Entomobryinae) from the New World, including three new species from Brazil and one from Australia. Zootaxa 4324: 201–248.

Deharveng, L. 2004. Recent advances in Collembola systematics. 6th International Seminar on Apterygota, Siena, Italy, 2002, Pedobiologia 48: 415–433.

Deharveng, L., C.D'Haese & A. Bedos. 2008. Global diversity of springtails (Collembola; Hexapoda) in freshwater. Hydrobiologia, 595: 329–338.

Delamare Deboutteville, C. 1961. Matériaux pour une révisíon des Collemboles Symphypléones. II Le genre *Spinotheca* Stach 1956 et la sous-famille des Spinothecinae, subfam. nov. Revue Francaise d'Entomologie 28: 101–111.

Delamare Deboutteville, C. & Z. Massoud. 1963. Collemboles Symphypléones., In: Delamare Deboutteville, C. & Rapoport, E. Biologie de l'Amérique Australe, volume II, Études sur la faune du sol, Éditions du centre National de la Recherche scientifique, Paris, pp.169–289.

D'Haese, C. 2002. Were the first springtails semi-aquatic? A phylogenetic approach by means of 28S rDNA and optimization alignment. Proceedings of the Royal Society of London 269: 1143–1151.

D'Haese, C.A. 2003. Homology and morphology in Poduromorpha (Hexapoda, Collembola). European Journal of Entomology 101: 385–407.

Fernandes, L.H. & M.C. Mendonça. 2010. A new species of *Xenylla* (Collembola: Poduromorpha: Hypogastruridae) with new records for the state of Espirito Santo, southeastern Brazil. Zoologia 27: 123–127.

Fjellberg, A. 1999. The labial palp in Collembola. Zoologischer Anzeiger 237: 309–330.

Fjellberg, A. 2007. The Collembola of Fennoscandia and Denmark, Part II: Entomobryomorpha and Symphypleona, Fauna Entomologica Scandinavica, volume 42. Brill Publishers, Boston, MA. 264 pp.

Gao, Y., Y.Bu & Y.-X. Luan. 2008. Phylogenetic relationships of basal hexapods reconstructed from nearly complete 18S and 28S rRNA gene sequences. Zoological Science 25: 1139–1145.

Heckman, C.W. 2001. Encyclopedia of South American Aquatic Insects: Collembola. Kluwer Academic Publishers, Berlin, Germany. 418 pp.

Hopkin, S.P. 1997. Biology of the Springtails (Insecta: Collembola). Oxford University Press, New York, NY. 322 pp.

Janssens, F. & K.A. Christiansen. 2011. Class Collembola Lubbock, 1870. pp. 192–194 *in*: Z.-Q. Zhang (ed.), Animal biodiversity: an outline of higher-level classification and survey of taxonomic richness. Zootaxa, 3148.

Jordana, R., J.I.Arbea, C. Simón & M.J. Luciáñez. 1997. Fauna Iberica, volume 8: Collembola, Poduromorpha. Museo Nacional de Ciencias Naturales, Consejo Superior de Investigaciones Científicas, Madrid. 807 pp.

Katz, A.D., R. Giordano, & F.N. Soto-Adames. 2015. Operational criteria for cryptic species delimitation when evidence is limited, as exemplified by North American *Entomobrya* (Collembola: Entomobryidae). Zoological Journal of the Linnean Society: 173: 818–840.

Mari-Mutt, J.A. & P.F. Bellinger. 1990. A Catalog of the Neotropical Collembola, Including Nearctic Areas of Mexico. Flora & Fauna Handbook, n.5. Sandhill Crane Press, Gainesville, Florida, 237 pp.

Massoud, Z. 1967. Monographie des Neanuridae, Collemboles Poduromorphes a pièces buccales modifiées, Biologie de l'Amérique Australe, volume III, Paris. 399 pp.

Mendonça, C. & E.A. Abrantes. 2007. A new Brazilian species of *Isotomiella* (Collembola: Isotomidae), with notes on *I. bidentata* Delamare Deboutteville, 1950 and *I. amazonica* Oliveira & Deharveng, 1990. Zootaxa 1652: 41–48.

Najt, J. & Z. Massoud. 1974. Contribution à l'étude des Brachystomellinae (Insects, Collemboles). I. — Nouvelles espèces récoltées em Argentine. Revue d'Ecologie et de Biologie du Sol 11: 367–372.

Najt, J., W.M. Weiner & P. Grandcolas. 2005. Phylogeny of the Brachystomellidae (Collembola) - were the mandibles ancestrally absent and did they re-appear in this family? Zoologica Scripta 34: 305–312.

Palacios-Vargas, J.G. 2013. Biodiversidad de Collembola (Hexapoda: Entognatha) en México. Revista Mexicana de Biodiversidad 84: 220–231.

Palacios-Vargas, J.G. & B.M. Mejía-Recamier. 2007. Técnicas de coleta montaje y preservatión de microartrópodos edáficos. Universidad Nacional Autónoma de México, Ciudad de México. 74 pp.

Palacios-Vargas, J.G. & E. Catalán. 2010. First Mexican species of *Sensillanura* (Collembola: Neanuridae). Dugesiana 17: 161–166.

Palacios-Vargas, J.G., E.C.A. Lima & D. Zeppelini. 2013. A new species of *Isotogastrura* (Collembola: Isotogastruridae) from Northeastern Brazil. Florida Entomologist 96: 1579–1587.

Palacios-Vargas, J.G. & M.R. Peñaranda-Parada. 2005. Two new species of *Paranura* (Collembola: Neanuridae) from Colombia and México. Journal of the Kansas Entomological Society 78: 308–314.

Potapov, M. 2001. Synopses on Palaearctic Collembola. Isotomidae. Abhandlungen und Berichte des Naturkundemuseums Görlitz, volume 3. 603 pp.

Queiroz, G.C. & M.C. Mendonça. 2010. Two new Isotomidae species (Collembola) from Espírito Santo State, Brazil. Zootaxa 2480: 37–44.

Richards, W.R. 1968. Generic classification, evolution, and biogeography of the Sminthuridae of the world (Collembola). Memoirs of the Entomological Society of Canada 53: 1–54.

Schneider, C. & C.A. D'Haese. 2013. Morphological and molecular insights on *Megalothorax*: the largest Neelipleona genus revisited (Collembola). Invertebrate Systematics 27: 317–364.

Schneider, C., C. Cruaud & C.A. D'Haese. 2011. Unexpected diversity in Neelipleona revealed by molecular phylogeny approach (Hexapoda, Collembola). Soil Organisms 83: 383–398.

Silva, D.D., J.G. Palacios-Vargas & B.C. Bellini. 2015. A new genus and a new species of Sminthuridae (Collembola: Symphypleona) from Atlantic Forest of Brazil. Zootaxa 3990: 410–418.

Soto-Adames, F.N., J.A. Barra, K. Christiansen & R. Jordana. 2008. Suprageneric Classification of the Entomobryomorph Collembola. Annals of the Entomological Society of America 101: 501–513.

Soto-Adames, F.N., R. Jordana & E. Baquero. 2014. Comparative Analysis of the Dorsal Chaetotaxy of *Troglopedetes*, *Trogolaphysa*, and *Campylothorax* Supports the Synonymization of Tribes Paronellini and Troglopedetini (Collembola: Paronellidae). Journal of Insect Science 14: 1–16.

Soto-Adames, F.N. 2016. Chaetotaxy of first-instar *Campylothorax sabanus* (Wray), and description of three new *Campylothorax* species from Hispaniola (Collembola, Paronellidae). Journal of Natural History 50: 1583–1612.

Sun, X., Y. Gao & M.B. Potapov. 2013. Review of Chinese littoral *Thalassaphorura* (Collembola: Onychiuridae), with the description of two new species. Journal of Natural History 1: 1–15.

Sun, X. & F. Zhang. 2012. Two new species of *Onychiurus* (Collembola: Onychiuridae) from Eastern China. Journal of Natural History 46: 1895–1904.

Thibaud, J.M. & J.G. Palacios-Vargas. 2000. Une nouvelle espéce de genre *Mesaphorura* des sables littoraux du Mexique. Revue française d' Entomologie 22(4): 146–148.

Thibaud, J.M., H.J. Schulz & M.M.G. Assalino. 2004. Synopses on Palaeartic Collembola volume 4: Hypogastruridae. Staaliches Museum für Naturkunde Görlitz, Görlitz. 287pp.

Vázquez, M.M. & J.G. Palacios-Vargas. 1996. Two new Mexican species of *Microgastrura* (Collembola: Hypogastruridae) associated with mushrooms. Folia Entomológica Mexicana 98: 59–65.

Waltz, R.D. & W.P. McCafferty. 1979. Freshwater Springtails (Hexapoda: Collembola) of North America. Purdue University Agricultural Experiment Station West Lafayette, Indiana, USA. 32 pp.

Weiner, W.M. & J. Najt. 1997. Collembola Poduromorpha from the Magallanes Province (Chile). Bonner Zoologische Beiträge 47: 99–110.

Xiong, Y., Y. Gao, W.-Y. Yina & Y.-X. Luan. 2008. Molecular phylogeny of Collembola inferred from ribosomal RNA genes. Molecular Phylogenetics and Evolution 49: 728–735.

Zeppelini, D. & J.G. Palacios-Vargas. 2013. A new Brazilian species of *Hylaeanura* (Collembola: Neanuridae). The Florida Entomologist 96: 1401–1405.

Zeppelini, D.F., G.C. Queiroz & B.C. Bellini. 2017. *In*: Collembola Lubbock, 1870. Catálogo Taxonômico da Fauna do Brasil. PNUD. Available online at: http://fauna.jbrj.gov.br. Access: 22 Jun 2016.

Zhang, F., D.-D. Sun, D.-Y. Yu & B.X. Wang. 2015. Molecular phylogeny supports S-chaetae as a key character better than jumping organs and body scales in classification of Entomobryoidea (Collembola). Scientific Reports 5 12471: 1–12.

Division II

Subclass Insecta

Orders Ephemeroptera, Plecoptera, Orthoptera, Blattodea, Hemiptera, Neuroptera, Megaloptera, Hymenoptera, Trichoptera, Lepidoptera, and Mecoptera

Chapter 3

Order Ephemeroptera

Frederico Falcão Salles
Departamento de Ciências Agrárias e Biológicas, Universidade Federal do Espírito Santo, São Mateus, Espírito Santo, Brazil

Eduardo Domínguez
Instituto de Biodiversidad Neotropical (IBN). CONICET - U.N.T., & Facultad de Ciencias Naturales, Universidad Nacional de Tucumán, San Miguel de Tucumán, Tucumán, Argentina

Carlos Molineri
Instituto de Biodiversidad Neotropical (IBN). CONICET - U.N.T., & Facultad de Ciencias Naturales, Universidad Nacional de Tucumán, San Miguel de Tucumán, Tucumán, Argentina

Rafael Boldrini
Centro de Estudos da Biodiversidade, Universidade Federal de Roraima, Boa Vista, Roraima, Brazil

Carolina Nieto
Instituto de Biodiversidad Neotropical (IBN). CONICET - U.N.T., & Facultad de Ciencias Naturales, Universidad Nacional de Tucumán, San Miguel de Tucumán, Tucumán, Argentina

Lucimar Gomes Dias
Departamento de Ciencias Biológicas, Facultad de Ciencias Exactas y Naturales, Grupo de Investigación Bionat, Universidad de Caldas, Manizales, Caldas, Colômbia

Chapter Outline

INTRODUCTION

Mayflies comprise a small order of insects, with approximately 3500 species, 450 genera, and 42 families (numbers updated based on Sartori & Brittain, 2015). The Neotropical realm, even with a relatively few families (16) is the most diverse in terms of genera and species (130 and 820, respectively). These numbers, however, increased exponentially in the last 10 years (Domínguez et al., 2014; Cardoso et al., 2015) and many species and probably some genera remain to be described. Cardoso et al. (2015), for example, predicted more than 11,000 species to occur Brazil alone, although their predictive model is possibly biased by the recent increase in species descriptions. Da-Silva & Salles (2012), on the other hand, estimated 600 undescribed species for the same country, while Sartori & Brittain (2015) estimate 1,000 undescribed species worldwide.

Thorp and Covich's Freshwater Invertebrates, Fourth Edition. https://doi.org/10.1016/B978-0-12-804223-6.00003-2

The systematics of Ephemeroptera, and consequently the classification of this order, need a major revision. Based on morphology (McCafferty & Edmunds, 1979; McCafferty, 1991; Kluge, 2004), molecular evidence (Ogden & Whiting, 2005), and combined data (Ogden et al., 2009), several efforts have been made attempting to uncover the relationship among mayfly families. Despite that, higher classification of mayflies is still a matter of debate (Sartori & Brittain, 2015). While some historical groups were recovered by Ogden et al. (2009) (e.g., Furcatergalia, Fossoriae, Pannota), others were not supported as monophyletic (e.g., Setisura and Pisciforma), as a consequence, we avoid the use of taxa above the family group.

LIMITATIONS

There are several existing keys to South and Central American Ephemeroptera, but none so far that can be used for the whole Neotropical region. The first reliable keys that could be used for the region were in the "Mayflies of North and Central America" by Edmunds et al. (1976). It was intended for North and Central America, but included several genera with a distribution going well into the Neotropical region. Valuable as it was, it is now very outdated. The first generic keys for nymphs and adults made especially for continental South America were published by Domínguez et al. (1992). Since then, several versions and improvements (with different authors were published (Domínguez et al., 1995, 2001, 2006, 2009). With regards to specific keys, Heckman published in 2002 a book on South American Ephemeroptera, which unfortunately was plagued with errors (see Peters, 2007). More recently, Domínguez et al. (2006) published the book *Ephemeroptera of South America*, which included keys for genera and species and descriptions of all the known species for the region. Since its publication, around 20 genera and 180 species were described, so its keys must be used with caution. With regards to Central America, the chapter Ephemeroptera by Flowers & De la Rosa in the book *Macroinvertebrados de Costa Rica*, published in 2010, included generic keys for the nymphs recorded in the region. Finally, although outdated, Peters published in 1971 generic and specific keys for the Ephemeroptera of the Caribbean. The nymph keys provided here allow for the first time the determination of all known Neotropical mayfly genera. Nevertheless, it must be kept in mind that there are still several undescribed genera and species.

TERMINOLOGY AND MORPHOLOGY

Head: Variable in shape, but in general aspect similar to the head of most mandibulate insects. Compound eyes are well developed and located above the gena; in flattened nymphs they are very close to the posterolateral corners. Three ocelli, forming a characteristic triangle, are present on the vertex. An ecdysial inverted Y line, or epicranial suture, is also present on the vertex. Lateral ocelli are inserted above or at the end of the frontal suture. The filiform antennae are variable in length, from short (shorter or as long as the head capsule) to very long (longer than the body). Mouthparts are oriented anteriorly (prognathous) or ventrally (hypognathous), they are extremely important for distinguishing genera and species and bear the typical elements of the chewing type. The clypeus-labrum, forming the upper lip (clypeus can be fused to the frons in some taxa, e.g. Leptophlebiidae). The heavily sclerotized mandibles, consisting of outer and inner incisors, prostheca, and a molar area. Mandibular tusks, variable in length and armature, are present in some taxa. The maxillae is composed by the basal cardo and stipes and by the apical galea-lacinia, an area formed by the fusion of these two lobes. Denticles (also referred to as canines) and usually two rows of setae, variable in quantity and length, are present at the inner apex of the galea-lacinia. Three- or two-articulate maxillary palpi are present at the outer apex of the stipes. The labium is composed by the submentum, connected to the head, and the mentum, which bears the glossae, paraglossae (sometimes fused to each other) and the three-articulate labial palpi. Between mandibles and maxillae lies a pair of superlinguae and a central hypopharynx. Besides eyes and appendages, tubercles, or projections may be present on the dorsal surface of the head.

Thorax: Pronotum generally small in the shape of a transversal rectangular plate, well defined. Mesonotum and metanotum fused, not well distinguishable, bearing the wing pads (sometimes the hind wing pads are very small to absent). Wing pads may be fused to each other in varying degrees, and in the last stages the venation of the adult wing may be recognized. At the final period of the last larval stage, the developing wing folds inside the wing pad and wing characters turn out unobservable. The thorax has three distinct segments: the pro-, meso-, and the metathorax. Pleurae formed by small plates and membranous areas with little use in taxonomy, sterna with varying structure, sometimes with strong setae, or membranous gills arising from membranous areas between coxae and other thoracic sclerites. The main characters used in systematic are in the legs. The legs always present include coxae, trochanter, femur, tibia, uniarticulate tarsus, and single tarsal claw. Coxae and trochanter may show dorsal projections or membranous gills. Femora are the more complex structures showing many useful characters: general shape, setation on dorsal, and ventral surfaces, and on leading and posterior edges, presence of tubercles or projections, long food filtering setae, etc. Tibia and tarsus are generally long cylindrical or flat, sometimes strongly flattened in burrowing nymphs, with spurs or projections on distal apex, with varying setation mainly arranged as longitudinal rows on inner and outer margins. Tarsus are uniarticulate (except in one Chinese genus), and bears an apical large claw, with variable shape and denticulation. Denticles on tarsal claw, when present, are generally arranged in a longitudinal row along inner margin (marginal denticles), sometimes are located on distal third and just at the side of the margin (submarginal subdistal denticles). In the legs, the thicker setae may be referred as "spines" or "chalaza." The structure and morphology of the legs are generally related to their habits: swimming, crawling, clinging, burrowing, filtering food, and gill protection.

Abdomen: Formed by 10 distinct segments, the long median filament (or terminal filament) is interpreted as derived from the 11th abdominal segment. The long distal cerci at each side of the terminal filament are the modified appendages of the 10th segment. All three caudal filaments are movable, annulated, and setated throughout their length. Abdominal segments are formed by two plates (the notum and the sternum); pleura has the form of a thin membrane separating them. Abdominal nota generally have spines, scales, and setae of systematic importance; the posterior margin may be denticulated, serrated, or smooth. Tubercles may be present on some nota, they may be median and single, or paired submedian. The lateral margins of abdominal nota form flanges in some groups, generally these flanges protect the gills from below, or more rarely are curved upwards forming a gill basket. Also the posterolateral corners of the segments are spiniform projecting in some groups forming posterolateral acute projections (or posterolateral spines). Segments I to VII generally bears the tracheal gills (one arising at each posterior corner), in some fossils gills even reached the IXth segment. Gills arise laterally or dorsolaterally in most groups but some have modified gills arising dorsally or even ventrally. Gills form the main surface to dissolved gas exchange, and thus are generally much more developed in groups inhabiting oxygen poor habitats. The abdominal tracheal gill is formed by a dorsal and a ventral thin cuticular layer with the trachea inside, showing different patterns of ramification and coloration. In most families each gill is formed by two lamellae, a dorsal and a ventral one, with many important characters including their shape, presence of filaments, number of tracheae, etc. Some groups show single laminate gills (formed by just one lamellae), for example the large family Baetidae, sometimes these gills are folded seeming composed by more than one lamella (e.g., *Callibaetis*). Gills are named after their respective segment, from I to VII, and sometimes in the keys (Leptohyphidae) the number of lamellae present in each gill is given as a "gill formula," for example 2/3/3/3/1, meaning that the first gill (in Leptohyphidae is the gill II) is formed by two lamellae, the following gill III is formed by three lamellae, and so on. Finally, paraprocts may be of importance for some groups (mainly Baetidae), these are paired small sclerited surrounding the annus and bearing different number of spines. The long cerci arise dorsally to the paraprocts, and bear setae of different length and thickness, in a similar form to the terminal filament.

MATERIAL PREPARATION AND PRESERVATION

The best way to preserve mayfly nymphs is to fix them in 80% ethanol, although molecular studies often require specimens preserved in 95% alcohol and stored in the refrigerator. In the field, and after that if necessary, ethanol must be replaced every time it becomes yellowish. They are fragile, especially antennae, legs, and gills, thus air bubbles must be removed from the vials during sample transportation from the field to the laboratory. For the identification of the nymphs at generic and specific levels, specimens must be dissected and mounted with body parts on slides. The slides can be mounted on permanent medium, with Euparal or Canada balsam, or on nonpermanent medium, with glycerin or alcohol gel. Then, the structures can be visualized in compound light microscope. For the preparation of slides, fine forceps and stylets are essential. Structures such as antennae, mouthparts, legs, abdominal segments, gills, paraprocts, and filaments are the most important and, therefore, are usually mounted on slides. These structures can be mounted on the same slide, but in distinct areas with their own cover glass, or in different slides when nymphs are too large.

KEYS TO THE ORDER EPHEMEROPTERA

Ephemeroptera: Families

1	Mandibular tusk present (Figs. 3.3, 4 A, 5 A); gills II–VII lanceolate, forked and with fringed margins (Fig. 3.3)	2
1'	Mandibular tusk absent (Figs. 3.6, 7 A, 10–20); gills II–VII variable (Figs. 3.10–20)	4
2(1)	Mandibular tusk usually shorter or as long as head, curved medially close to apex in dorsal view (Figs. 3.4 A, 5 A); fore tibia without distomedial spine (Figs. 3.4 B, 5 B); gills II to VII held dorsally	3
2'	Mandibular tusk distinctly longer than head, curved medially from base to apex in dorsal view (Fig. 3.3); fore tibia with distomedial spine (Fig. 3.23); gills II to VII held laterally (Fig. 3.3)	**Euthyplociidae [p. 66]**
3(2)	Tusks down curved to straight (Fig. 3.4 A); base of tusk with double row of filtering setae (Fig. 3.4 A); fore tibia and tarsus fused (Fig. 3.4 B), apex of hind tibia without an acute point (Fig. 3.4 C)	**Polymitarcyidae [p. 67]**
3'	Tusks upcurved (Fig. 3.5 A); base of tusk without double row of filtering setae (Fig. 3.5 A); fore tibia and tarsi not fused (Fig. 3.5 B), hind tibia end in an acute point (Fig. 3.5 C)	Ephemeridae, one genus: *Hexagenia*
4(1')	Fore leg, at least, with rows of filtering setae (Figs. 3.6, 7, 8 B)	5
4'	Fore leg without rows of filtering setae (Figs. 3.10–14, 16–20)	7

FIGURE 3.1 Schematic representation of a mayfly nymph in dorsal view. *Roman numbers* represent abdominal segments.

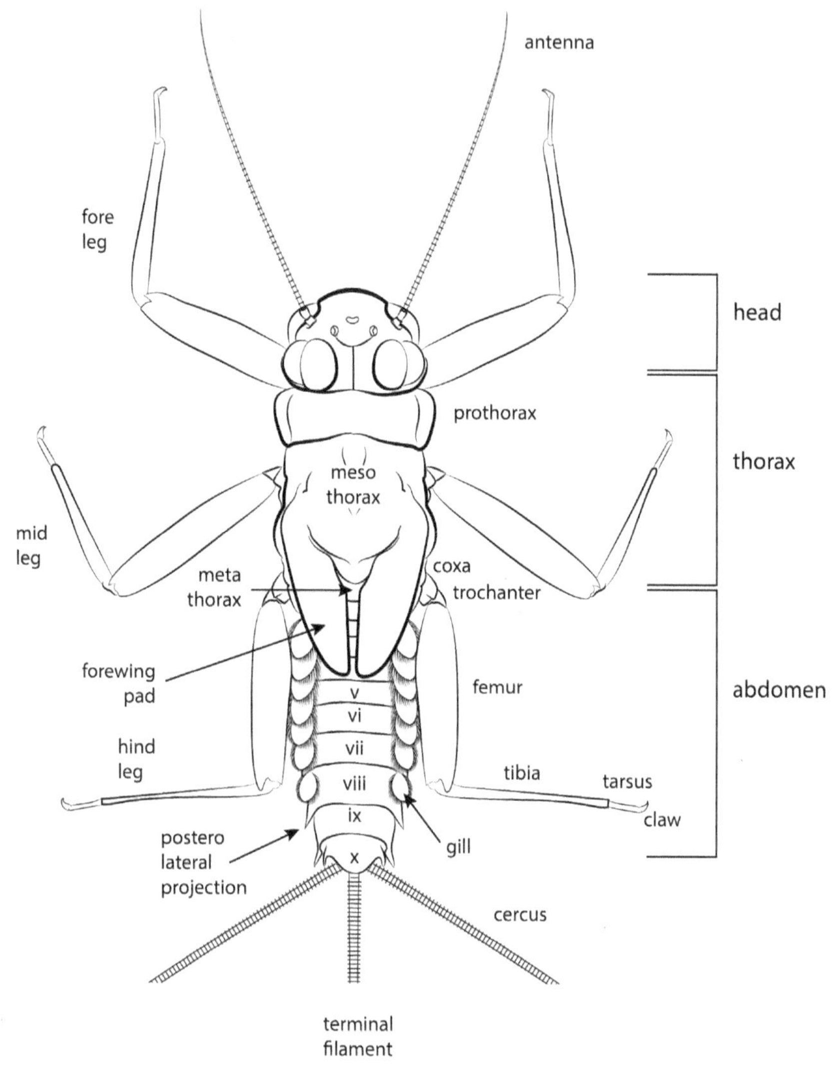

5(4) Gills present (Fig. 3.7); fore legs only with filtering setal rows (Figs. 3.7 B, 8 B) 6

5' Gills absent (Fig. 3.6); fore and mid legs with filtering setal rows (Fig. 3.6) .... Coloburiscidae, one species *Murphyella needhami* Lestage [Andean-Patagonic Region]

6(5) Gill I dorsal (Fig. 3.7 A); base of fore coxa with branched gills (Fig. 3.7 B) .... Isonychiidae, one species *Isonychia sicca* (Walsh) [Neotropics]

6' Gill I ventral (Fig. 3.8 A); base of fore coxa without gills (Fig. 3.8 B) .... **Oligoneuriidae [p. 68]**

7(4') Opercular gill present (Figs. 3.9–13) 8

7' Opercular gill absent (Figs. 3.14, 16–20) 12

8(7) Gill II opercular (gill I reduced or absent) (Figs. 3.10–13); terminal filaments without secondary swimming setae (primary setae sometimes present) 9

8' Gill I opercular (Fig. 3.9); terminal filaments with primary and secondary swimming setae (i.e., row of setae present on inner and outer margin) .... Onsicigrastridae, *Siphlonella ventilans* Needham & Murphy [Andean-Patagonic Region]

9(8) Gill I absent (Figs. 3.11–13); operculate gill without dorsal forked crest (Figs. 3.11–13) 10

9' Gill I present, filiform (Fig. 3.10); operculate gill with dorsal forked crest (Fig. 3.10) .... **Caenidae [p. 69]**

10(9) Head anterior margin never densely bordered with setae (Figs. 3.12, 13); femora of fore and mid legs usually not dilated (Figs. 3.12, 13); abdominal tubercles usually absent (if present usually single, except in some species of *Leptohyphes*) (Figs. 3.12, 13) 11

10' Head anterior margin densely bordered with setae (Fig. 3.11 A); femora of all legs dilated (Fig. 3.11); abdominal tubercles present, paired (Fig. 3.11) .... Melanemerellidae, one species: *Melanemerella brasiliana* Ulmer [Atlantic Forest]

11(10) Posterolateral projections of gill bearing segments bent dorsally, forming a gill chamber (Fig. 3.12); tubercles present on all tagma (Fig. 3.12) .... Coryphoridae, *Coryphorus aquilus* Peters [Mostly restricted to the Amazon]

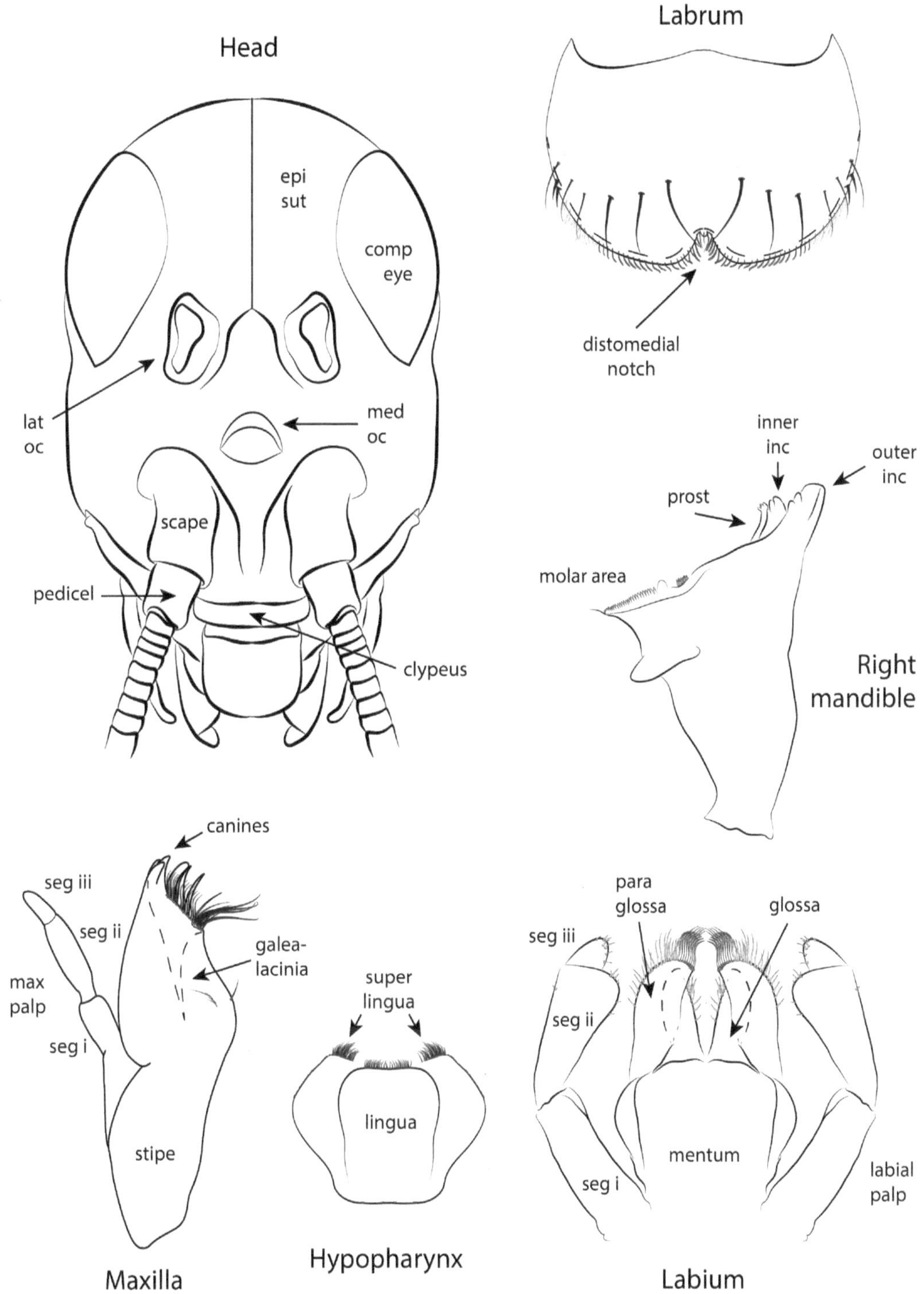

FIGURE 3.2 Schematic representation of a mayfly head (frontal view) and mouthparts. Abbreviations as follows: head (*comp eye*, compound eye; *epi sut*, epicranial suture; *lat oc*, lateral ocellus; *med oc*, median ocellus); right mandible (*outer inc*, outer incisors; *inner inc*, inner incisors; *prost*, prostheca); maxilla (*max palp*, maxillary palp; *seg i, ii, and iii*, maxillary palp segments i, ii, and iii); labium (*seg i, ii, and iii*, labial palp segments i, ii, and iii).

11' Posterolateral projections of gill bearing segments not upcurved to form gill chamber (Fig. 3.13); tubercles usually absent, when present restricted to one or two tagma (except *Loricyphes*) (Fig. 3.13) .... **Leptohyphidae [p. 79]**

12(7') Head never enlarged (may be wide, but flat in this case) (Figs. 3.15–21 A, 22 A); maxillary and labial palpi with two or three articles ... 13

12' Head enlarged (Fig. 3.14); maxillary and labial palpi multiarticulate .... **Ameletopsidae [p. 83]** [Andean-Patagonian region]

FIGURE 3.3 Dorsolateral view of a nymph of *Campylocia* (Euthyplociidae).

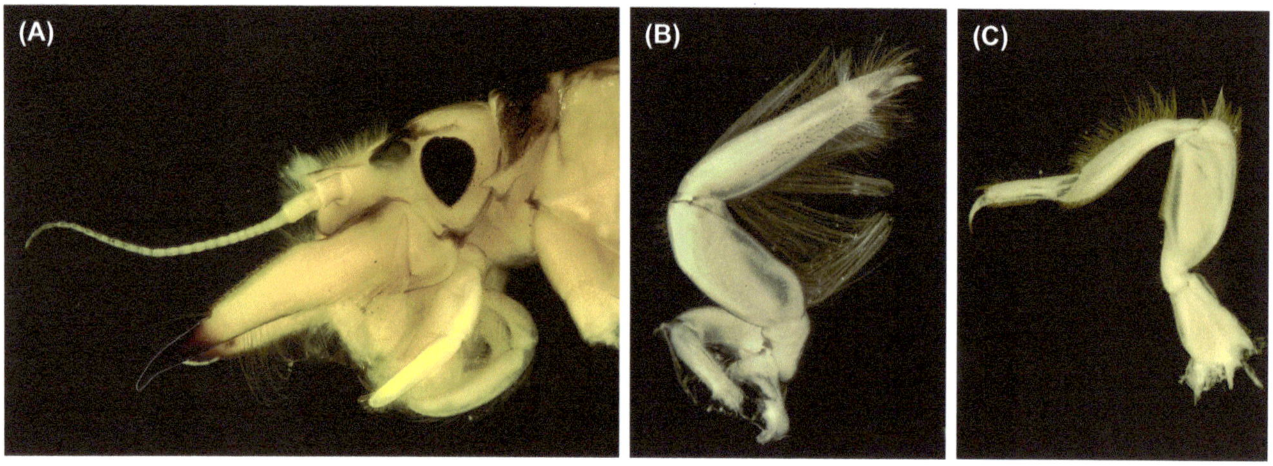

FIGURE 3.4 Nymphal structures of *Tortopsis* (Polymitarcyidae): (A) head (lateral view); (B) fore leg; and (C) hind leg.

13(12) Body usually depressed, head prognathous (body hydrodynamic in some Leptophlebiidae) (Figs. 3.16, 17); gills with dorsal and ventral lamella (Figs. 3.16, 17) 14

13' Body hydrodynamic, pisciform (Figs. 3.19, 20); gills with a single lamella (folds, however, may be present) (Figs. 3.19, 20)........ 15

14(13) Mandibles and labrum visible in dorsal view (Fig. 3.15) .... **Leptophlebiidae [p. 88]**

14' Mandibles and labrum concealed by head in dorsal view (Fig. 3.18)........ **Heptageniidae [p. 86]**

15(13') Epicranial suture with lateral branch located below lateral ocelli (Fig. 3.21 A); distal lobe of femora ventrally oriented (Fig. 3.21 B). .... **Baetidae [p. 107]**

15' Epicranial suture with lateral branch located on or above lateral ocelli (Fig. 3.22 A); femora without a distal lobe (Fig. 3.22 B) .... Nesameletidae, one species: *Metamonius anceps* (Eaton) [Andean-Patagonic Region]

Euthyplociidae: Genera

Modified from Domínguez et al. (2006). The most diversified Neotropical genus is *Campylocia*, with five species (Gonçalves et al., 2017). They are the largest living mayflies, with body lengths reaching up to 35 mm (Edmunds et al., 1976). Euthyplociidae nymph do not burrow, they are crawlers and live under rocks or among pebbles and leaf packs in areas with slow to intermediate current of streams.

1 Antenna distinctly longer than tusk; fore tibial projection length ~1/2–1/3 tarsus (Figs. 3.23 A, B) 2

1' Antenna length shorter or subequal to tusk; fore tibial projection length ~1/4 tarsus (Fig. 3.23 C) .... *Campylocia*

2(1) Fore tibial projection length ~1/3 tarsus (Fig. 3.23 B)........ *Mesoplocia*

2' Fore tibial projection length ~1/2 tarsus (Fig. 3.23 C)........ *Euthyplocia* [Neotropics. Nearctic]

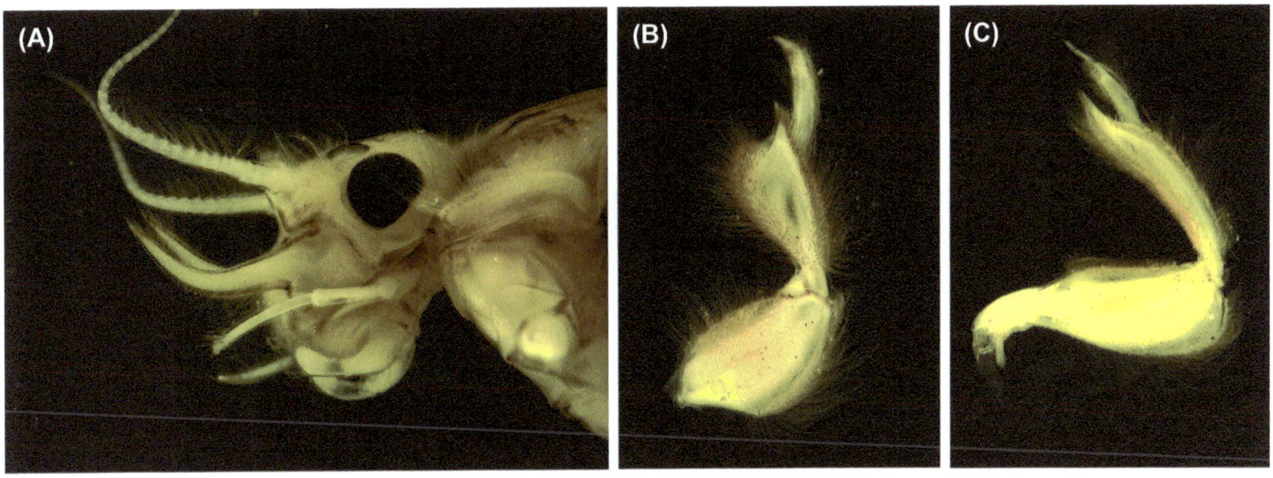

FIGURE 3.5 Nymphal structures of *Hexagenia* (Ephemeridae): (A) head (lateral view); (B) fore leg; and (C) hind leg.

FIGURE 3.6 Habitus of a living nymph of *Murphyella needhami* Lestage (Coloburiscidae).

Polymitarcyidae: Genera

Modified from Molineri et al. (2015). The nymphs make U-shaped burrows (Molineri, 2010). *Asthenopus*, at least in the Amazon, often build tunnels in wood debris and Asthenopodinae may be miners in plant tissue, such as *Eichhornia* and *Typha* (Domínguez et al., 2006). Some *Campsurus* species build silk cases on rocky substrates (Molineri et al., 2010).

1 Head dorsally mostly glabrous, without large tufts of tightly grouped short setae (Figs. 3.24 A, B); occiput roundly convex (Fig. 3.24 A); apex of left mandibular tusk with 3–4 pointed processes (Fig. 3.24 A) (**Asthenopodinae**) 2

1' Head dorsally with dense patches of short setae, mainly anteriorly to lateral ocelli (Figs. 3.25 A, B); occiput flat, subquadrate in dorsal view (Fig. 3.25 B); apex of left mandibular tusk with one pointed process (Fig. 3.25 A) (**Campsurinae**) 4

2(1) Mandibular tusks robust, stout, with one or more tubercles on medial margin (Figs. 3.27, 28 A, 29 A, 30) 3

2' Mandibular tusks relatively long and slender, without tubercles or large spines on medial margin (Fig. 3.26) .... *Hubbardipes*

3(2) Mandibular tusks without large subbasal tubercle on medial margin (Fig. 3.28 A); fore tarsal claw with double row of denticles (Fig. 3.28 B) .... *Asthenopodes*

3' Mandibular tusks with a large subbasal tubercle on medial margin (Fig. 3.29 A); foretarsal claw with single row of denticles (Fig. 3.29 B) .... *Asthenopus*

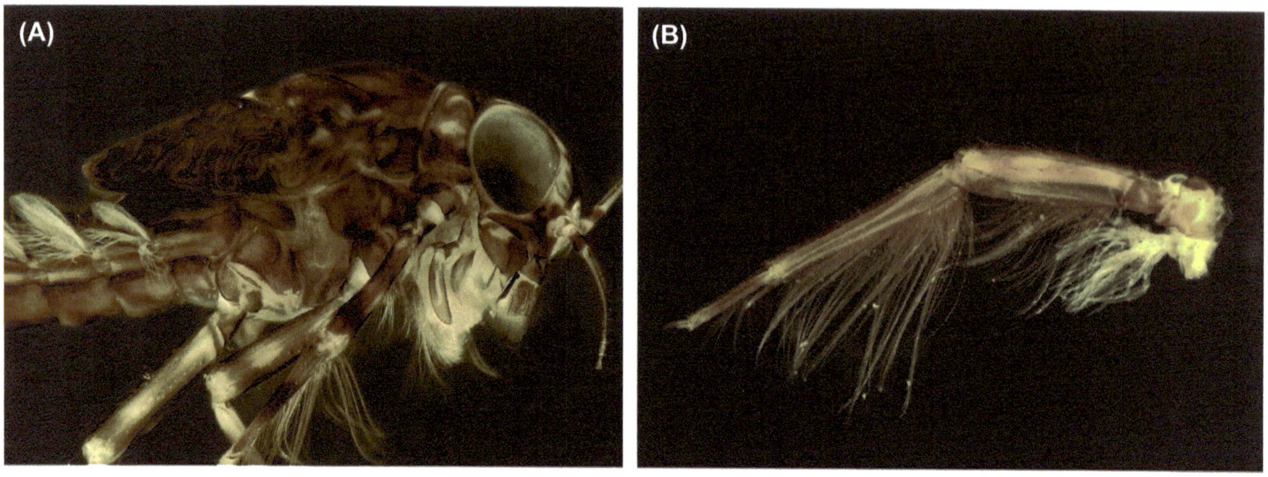

FIGURE 3.7 Nymphal structures of *Isonychia* (Isonychiidae): (A) head and thorax lateral view; and (B) fore leg.

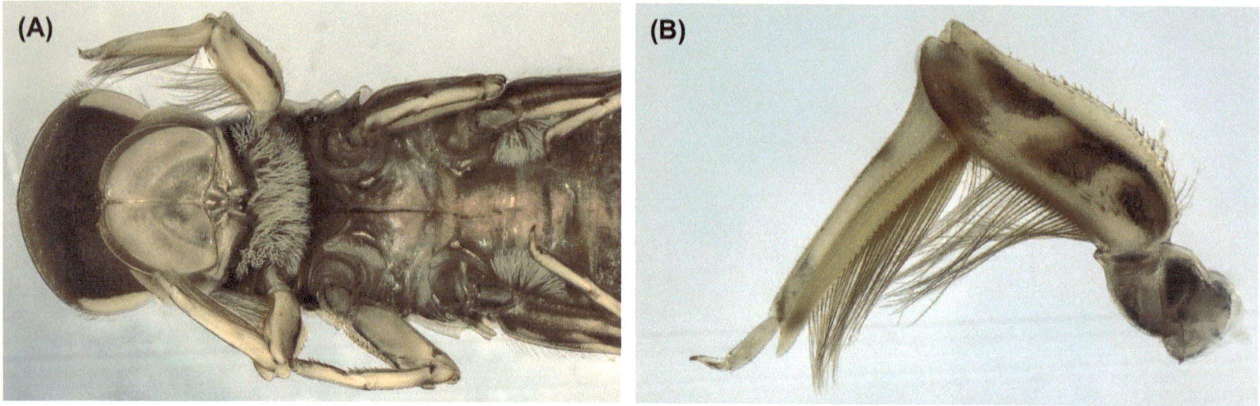

FIGURE 3.8 Nymphal structures of *Oligoneuria* (Oligoneuriidae): (A) head and thorax, ventral view; and (B) fore leg.

4(1') Mandibular tusks median margin distal third with one or two prominent tubercles; mandibles outer margin with few long setae (Figs. 3.30 B, C); abdominal gill I with a single lamella 5

4' Mandibular tusks median margin with prominent basal or subbasal tubercle (rarely absent), few to many apical crenulations; mandibles outer margin with numerous setae (Fig. 3.30 A); abdominal gill I bifurcated .... *Campsurus*

5(4') Mandibular tusks median margin with two tubercles (submedian and subapical) (Fig. 3.30 B); distal projection of foretibia-tarsus 2/5 the length of claw .... *Tortopus*

5' Mandibular tusks median margin with a single subapical tubercle (Fig. 3.30 C); distal projection of foretibia-tarsus length 2/3 claw .... *Tortopsis*

Oligoneuriidae: Genera

Half of this family's diversity comes from the Neotropics, where 30 species in five genera have been reported. *Homoeoneuria* are psammophilous, while the remaining Neotropical genera are clingers, living on wood debris, overhanging branches, dead leaves, or rocks. *Lachlania* and *Homoeoneuria* are the only genera with a distribution that extends from South to Central and North America.

1 Epimera and episterna not projected laterally (Fig. 3.31); abdominal segments with posterolateral projection restricted to segments VIII to IX (Fig. 3.31) 2

1' Epimera and episterna projected laterally (Fig. 3.32); abdominal segments I–IX with posterolateral projections (Fig. 3.32) 3

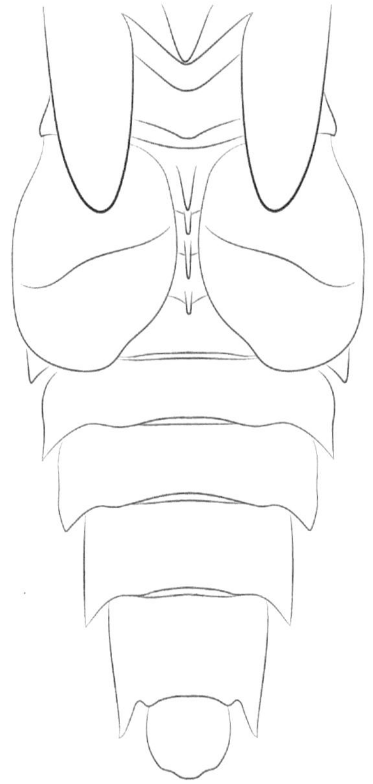

FIGURE 3.9 Fore wing pad and abdomen of *Siphlonella ventilans* Needham and Murphy (Oniscigastridae).

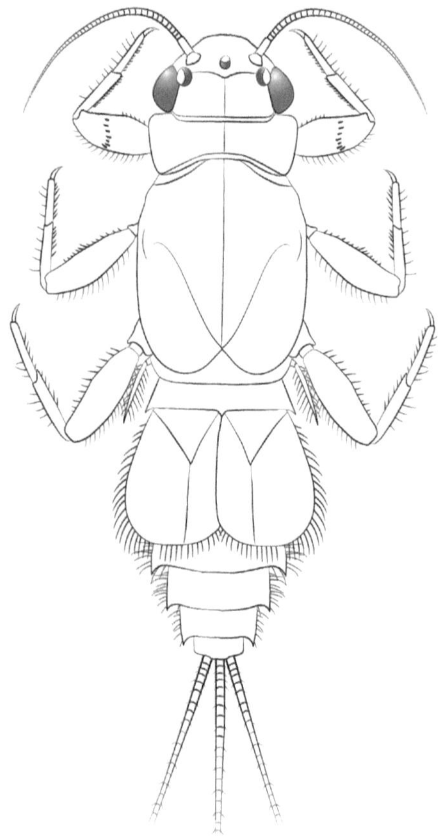

FIGURE 3.10 Habitus of *Caenis* nymph (Caenidae).

2(1) Head anteriorly with distinct pointed projection (Fig. 3.33); compound eyes clearly separated (Fig. 3.33); gills oval, broad (Fig. 3.31); mid and hind coxae short .... *Fittkauneuria*

2' Head without projection (Fig. 3.34 A); compound eyes almost contiguous (Fig. 3.34 A); gills lanceolate, narrow (Fig. 3.34 B); mid and hind coxae enlarged (Fig. 3.34 C) .... *Homoeoneuria*

3(1') Terminal filament absent (Fig. 3.32) .... *Lachlania*

3' Terminal filament present .... *Oligoneuria*

Caenidae: Genera

This family is cosmopolitan with approximately 240 species in 28 genera, 40 species in five genera are known from the Neotropics. The most common and abundant genus is *Caenis*, with nearly 30 species. *Allorethocus* and *Brasilocaenis* Puthz are endemic to South America, *Insulibrachys* to Cuba, while *Latineosus* is present in South and Central America. *Caenis* is usually associated with roots and leaf debris, and is common in stagnant water, but can also be found in depositional areas of lotic systems. Leaf debris in rivers is a common habitat for *Brasilocaenis*, while sandy river bottom, especially when covered with a thin silt layer, is habitat for *Allorethocus*.

1 Three ocellar tubercles (Figs. 3.36, 37); foreleg clearly shorter than mid and hind legs 2

1' Ocellar tubercles absent (Fig. 3.35); foreleg similar in length to mid and hind legs (Fig. 3.35) 4

2(1) Ocellar tubercles conical and without dense, long setae (Fig. 3.37) 3
[continental South America]

2' Ocellar tubercles broadly cylindrical, with apex somewhat truncate, and with apical portion covered with dense, long setae (Fig. 3.36) .... *Insulibrachis*
[Caribbean]

3(2) Maxillary palp article II length >2.0× article I (Fig. 3.38 A); posterolateral projection VI moderately curved medially but not nearly parallel with posterior margin of tergum VI (Fig. 3.38 B) .... *Alloretochus*

FIGURE 3.11 Habitus of *Melanemerella brasiliana* Ulmer (Melanemerellidae) nymph: (A) lateral view of living specimen; and (B) dorsal view.

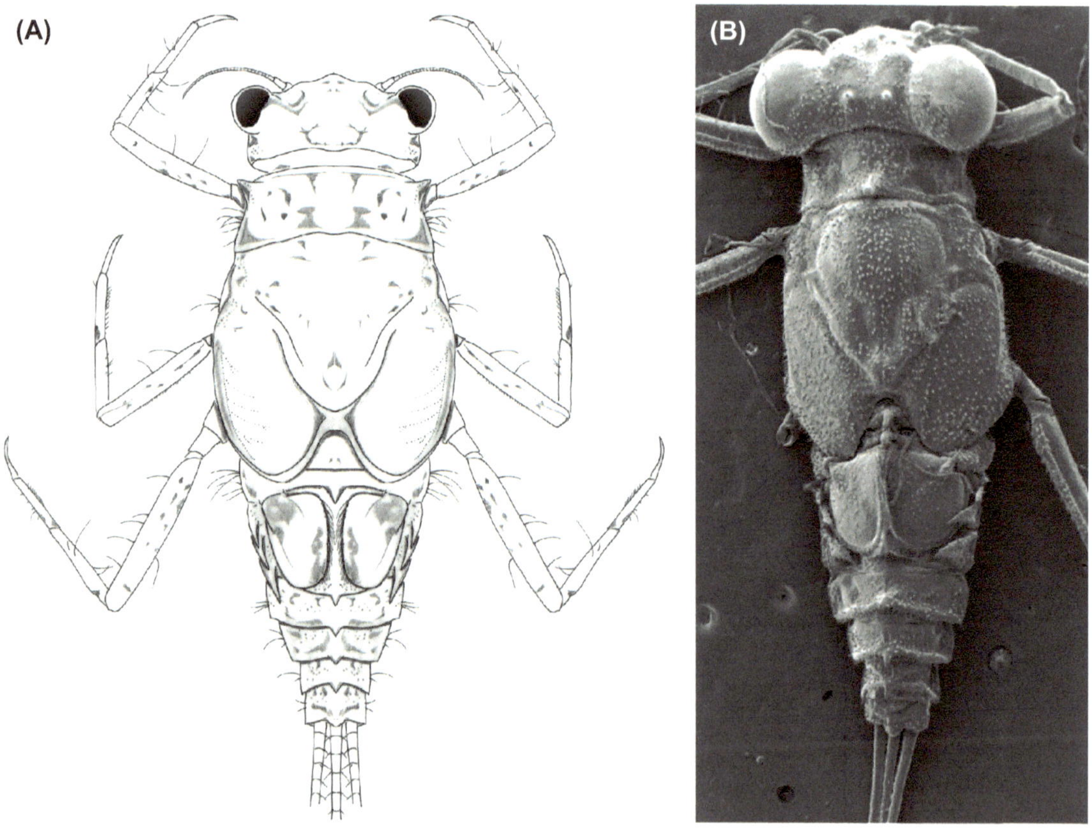

FIGURE 3.12 Habitus of *Coryphorus aquilus* Peters (Coryphoridae) nymph: (A) dorsal view of female, drawing; and (B) dorsal view of male, SEM picture.

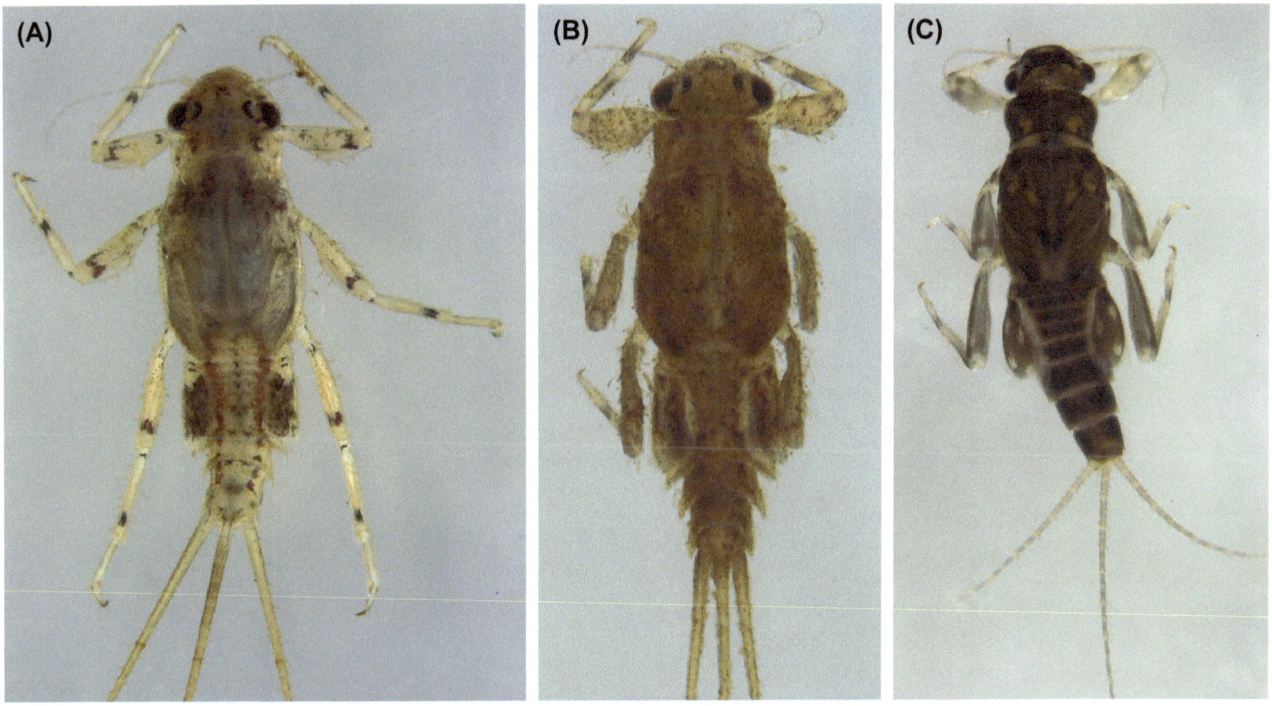

FIGURE 3.13 Habitus of Leptohyphidae nymphs: (A) *Tricorythodes*; (B) *Cabecar*; and (C) *Vacupernius*.

FIGURE 3.14 Habitus of a living nymph of *Chiloporter eatoni* (Lestage) (Ameletopsidae).

FIGURE 3.15 Head of nymph of *Hermanella mazama* (Nascimento, Mariano, and Salles) (Leptophlebiidae), dorsal view.

FIGURE 3.16 Habitus of a living nymph of *Hylister* (Leptophlebiidae).

FIGURE 3.19 Habitus of a living nymph of *Metamonius anceps* (Eaton) (Nesameletidae).

FIGURE 3.17 Habitus of a living nymph of *Fittkaulus* (Leptophlebiidae).

FIGURE 3.20 Habitus of a living nymph of *Cloeon smaeleni* Lestage (Baetidae).

FIGURE 3.18 Head of nymph of *Epeorus* (Heptageniidae), lateral view.

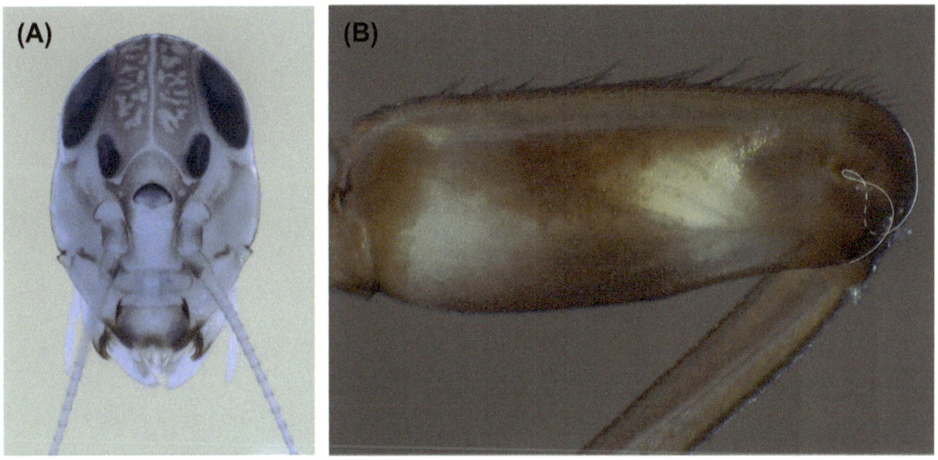

FIGURE 3.21 Nymphal structures of Baetidae: (A) head of *Cryptonympha*, frontal view; and (B) anterior surface of leg of *Camelobaetidius rufiventris* Boldrini and Salles.

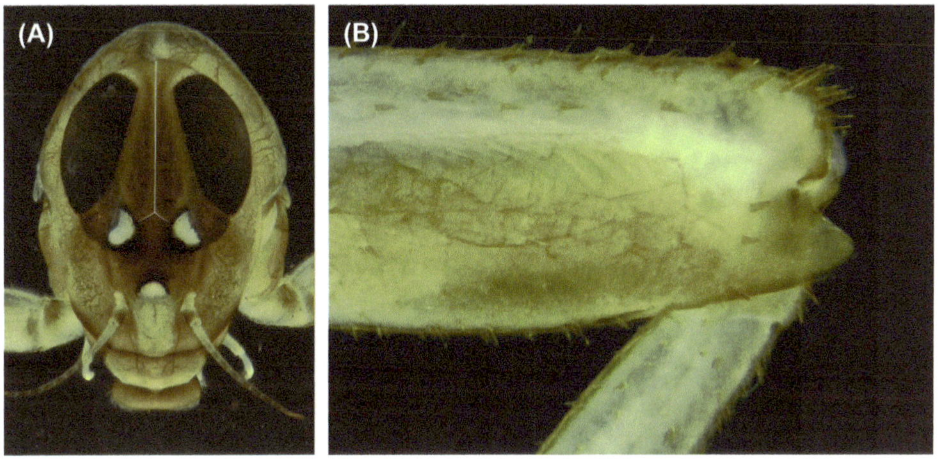

FIGURE 3.22 Nymphal structures of *Metamonius anceps* (Eaton) (Nesameletidae): (A) head, frontal view; and (B) anterior surface of leg.

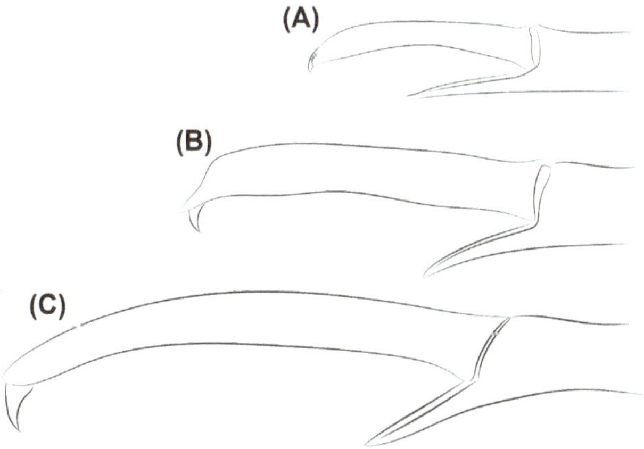

FIGURE 3.23 Partial view of fore tibiae and tarsi of Euthyplociidae nymphs: (A) *Euthyplocia*; (B) *Mesoplocia*; and (C) *Campylocia*.

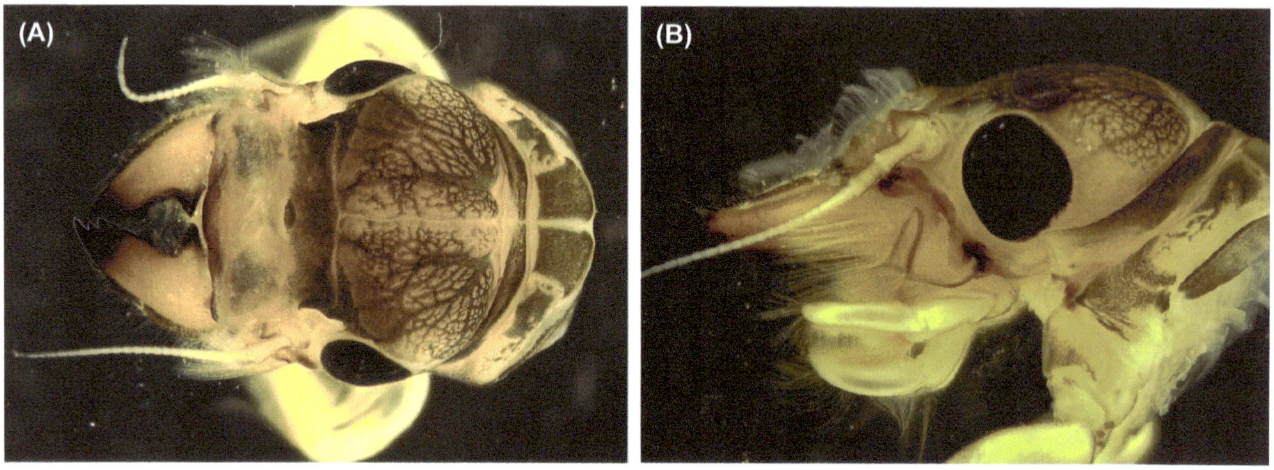

FIGURE 3.24 Nymphal head of *Asthenopus* (Polymitarcyidae): (A) dorsal view; and (B) lateral view.

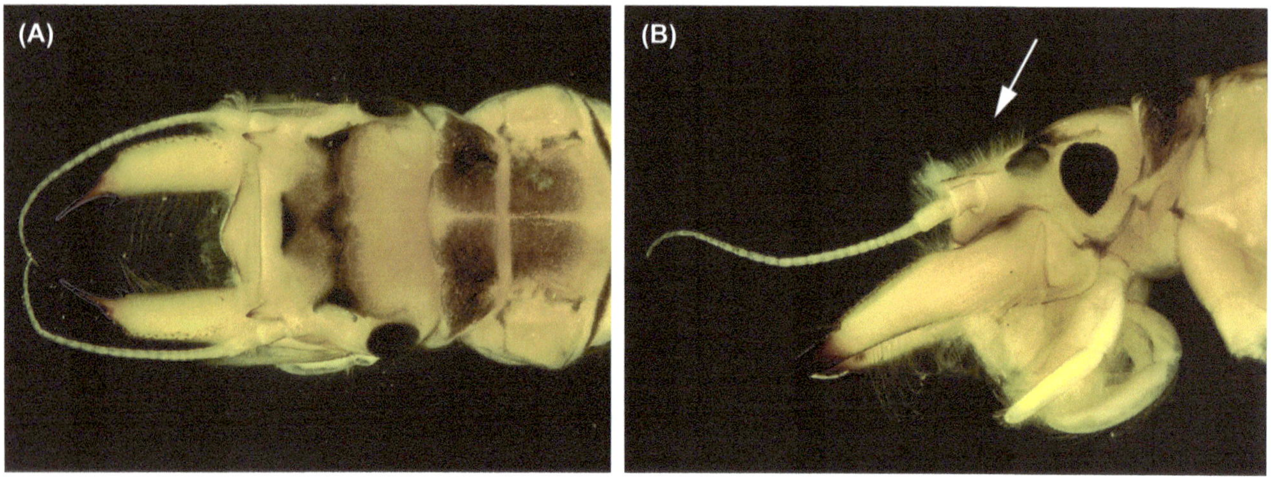

FIGURE 3.25 Nymphal head of *Tortopsis* (Polymitarcyidae): (A) dorsal view; and (B) lateral view. *White arrow* showing patch of dense setae.

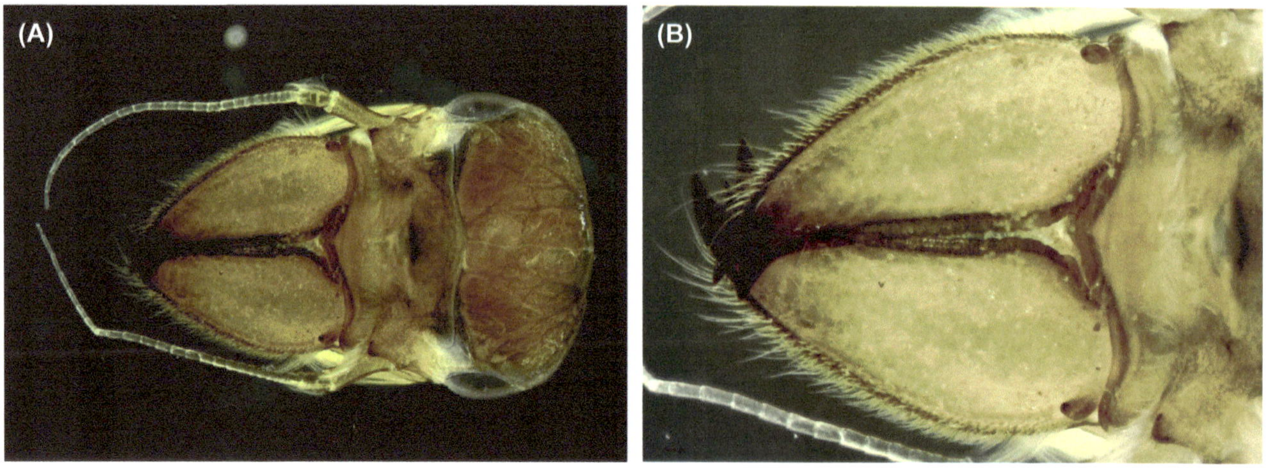

FIGURE 3.26 Nymphal head of *Hubbardipes* (Polymitarcyidae): (A) dorsal view; and (B) detail of mandibular tusks, dorsal view.

FIGURE 3.27 Nymphal head of *Asthenopus* (Polymitarcyidae), dorsal view. *White arrows* showing subbasal tubercle.

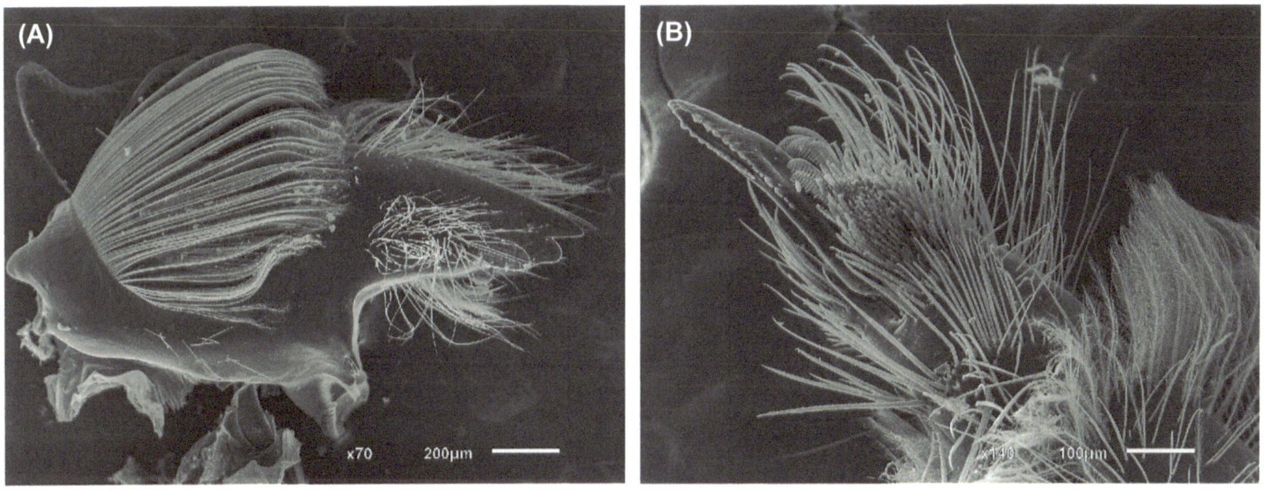

FIGURE 3.28 Nymphal structures of *Asthenopodes* (Polymitarcyidae), SEM pictures: (A) right mandible; and (B) fore leg.

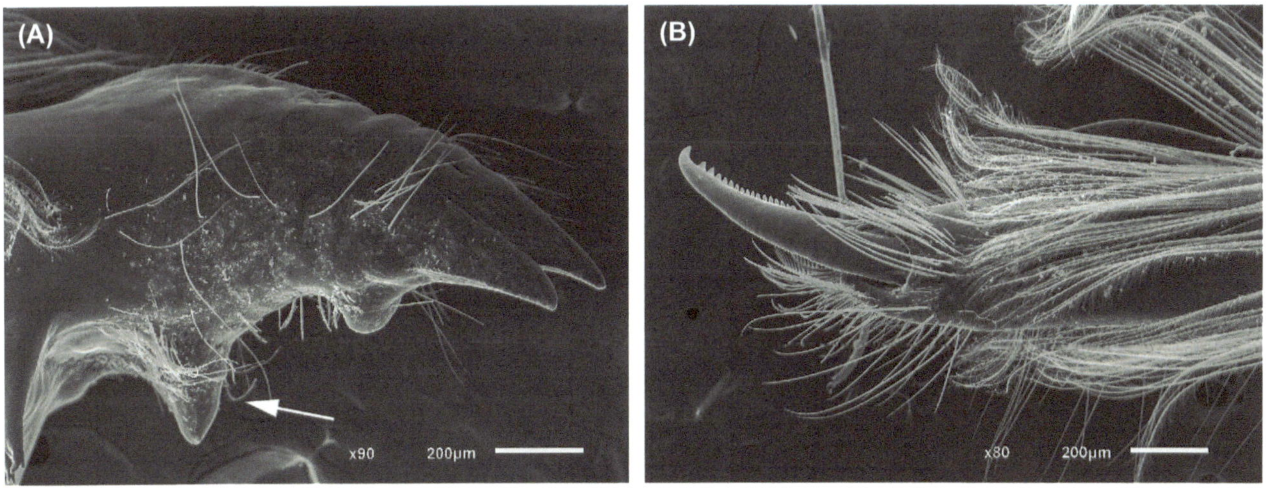

FIGURE 3.29 Nymphal structures of *Asthenopus* (Polymitarcyidae), SEM pictures: (A) right mandible; and (B) fore leg. *White arrow* showing subbasal tubercle.

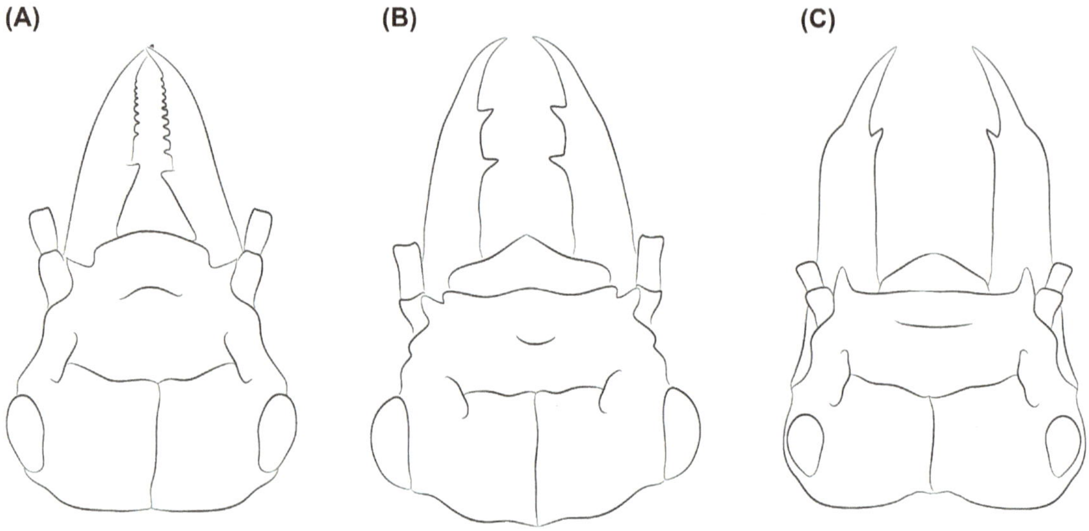

FIGURE 3.30 Schematic drawing of Polymitarcyidae heads: (A) *Campsurus*; (B) *Tortopus*; and (C) *Tortopsis*.

FIGURE 3.31 Habitus of *Fittkauneuria* (Oligoneuriidae) nymph. *Black arrows* showing posterolateral projections of abdominal segments VIII and IX.

FIGURE 3.32 Habitus of *Lachlania* (Oligoneuriidae) nymph. *White arrows* showing lateral projections of epimera and episterna.

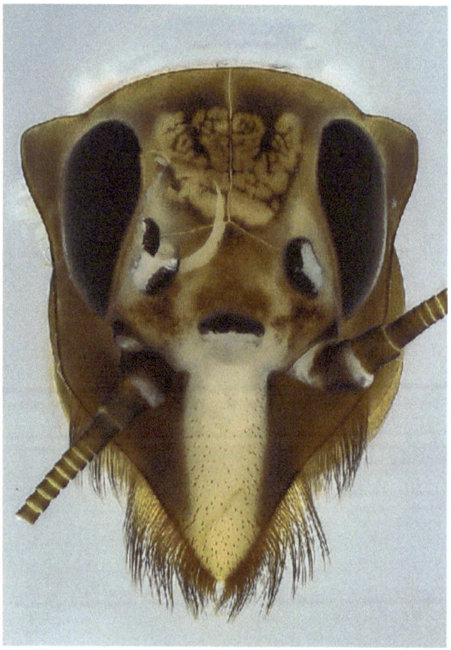

FIGURE 3.33 Head of *Fittkauneuria* (Oligoneuriidae) nymph, frontal view.

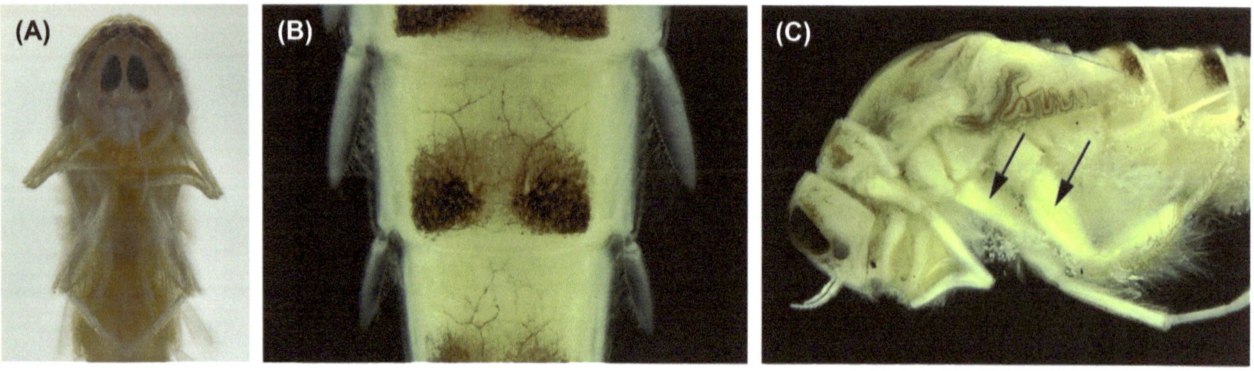

FIGURE 3.34 Nymphal structures of *Homoeoneuria* (Oligoneuriidae): (A) frontal view of head and partial view of thorax; (B) abdominal terga VI, VII and VIII; and (C) lateral view of head, thorax, and abdominal segments I and II. *Black arrows* showing elongate coxae.

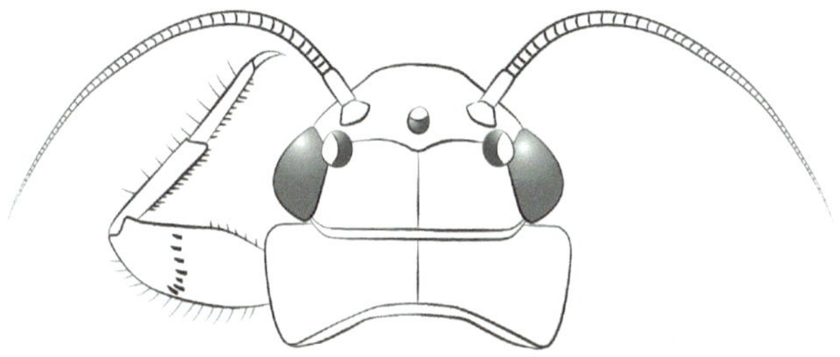

FIGURE 3.35 Head, prothorax and left fore leg of *Caenis* (Caenidae) nymph.

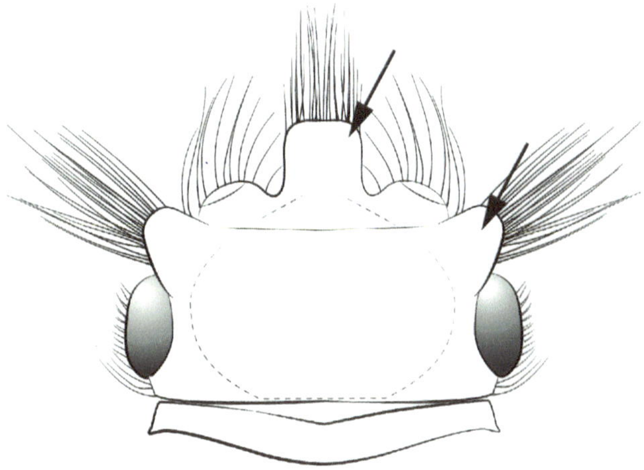

FIGURE 3.36 Head of *Insulibrachis* (Caenidae) nymph. *Arrows* showing ocelar tubercles. *After Sun & McCafferty (2008).*

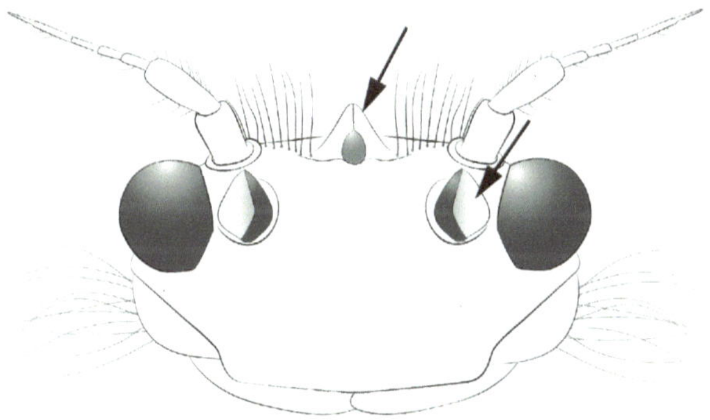

FIGURE 3.37 Head of *Latineosus sigillatus* (Molineri) (Caenidae) nymph. *Arrows* showing ocelar tubercles.

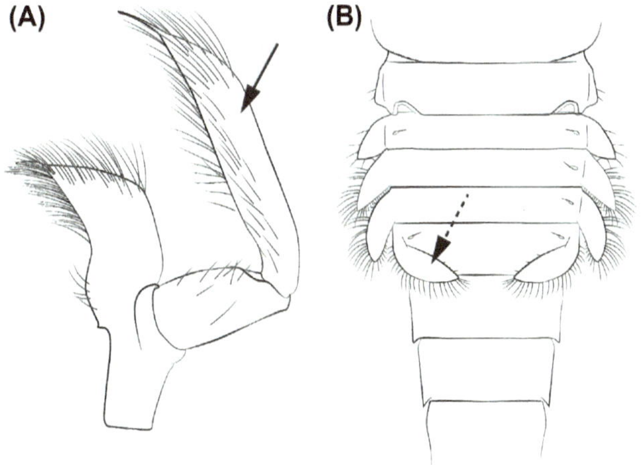

FIGURE 3.38 Nymphal structures of *Allorethocus* (Caenidae): (A) maxilla; and (B) abdominal segments I to VIII. *Solid arrow* showing second segment of maxillary palp, *dashed arrow* showing posterolateral projection of abdominal segment VI. *After Sun & McCafferty (2008).*

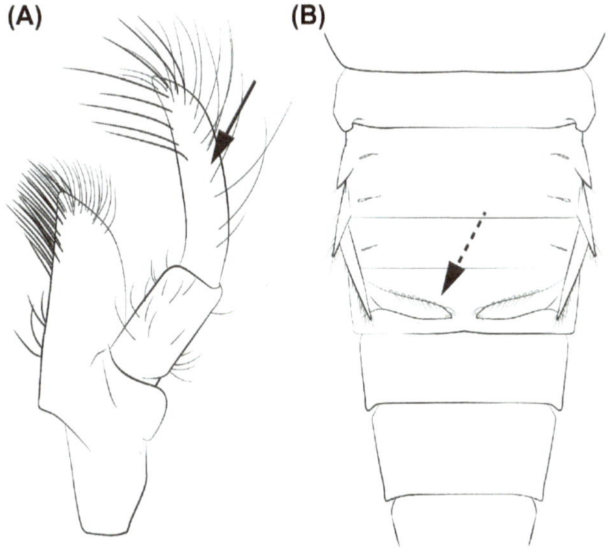

FIGURE 3.39 Nymphal structures of *Latineosus* (Caenidae): (A) maxilla; and (B) abdominal segments I to VIII. *Solid arrow* showing second segment of maxillary palp, *dashed arrow* showing posterolateral projection of abdominal segment VI. *After Sun & McCafferty (2008).*

3' Maxillary palp article II 1.5–1.8× length of article I (Fig. 3.39 A); segment VI posterolateral projection of distally nearly parallel with posterior margin of tergum VI (Fig. 3.39 B) .... *Latineosus*

4(1') Coxae of mid and hind leg with or without a semicircular projection (Fig. 3.40; operculate gill without short and robust spines (Fig. 3.42); segments IV–VII posterolateral projection short, not reaching half of following segment .... *Caenis*

4' Coxae of mid and hind leg with triangular projection (Fig. 3.41); operculate gill dorsal surface with short and robust spines (Fig. 3.43); segments IV–VII posterolateral projection long, reaching half of following segment .... *Brasilocaenis*

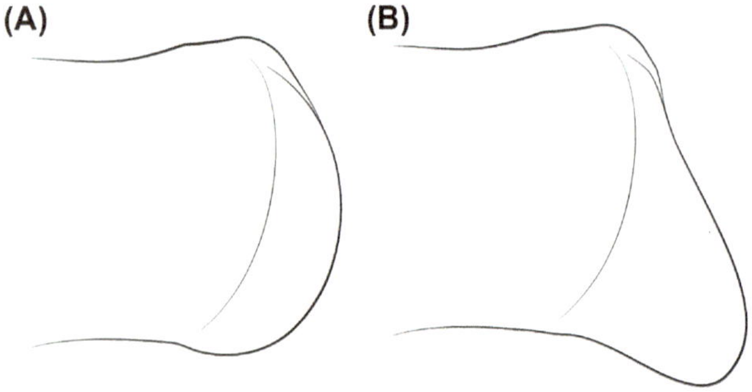

FIGURE 3.40 Mesocoxal processes of *Caenis* (Caenidae) nymphs: (A) *C. fittkaui* Malzacher; and (B) *C. pflugfelderi* Malzacher. *Adapted from Malzacher (1986).*

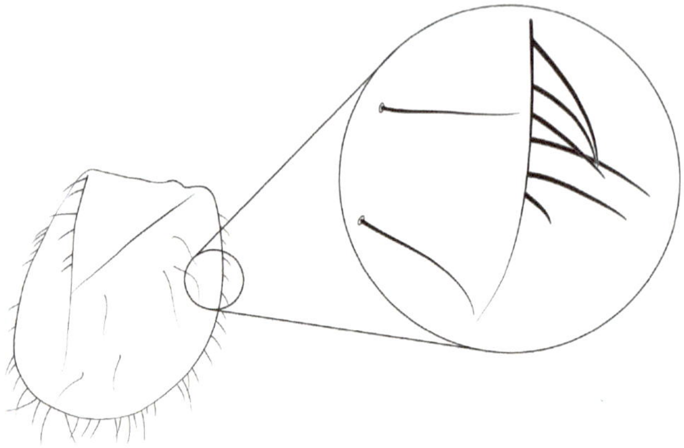

FIGURE 3.42 Operculate gill of *Caenis* (Caenidae) nymph.

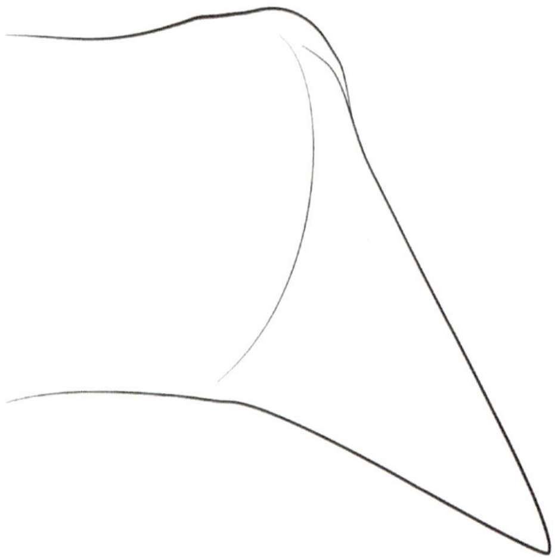

FIGURE 3.41 Mesocoxal processes of *Brasilocaenis irmleri* Puthz (Caenidae) nymph. *Adapted from Malzacher (1986).*

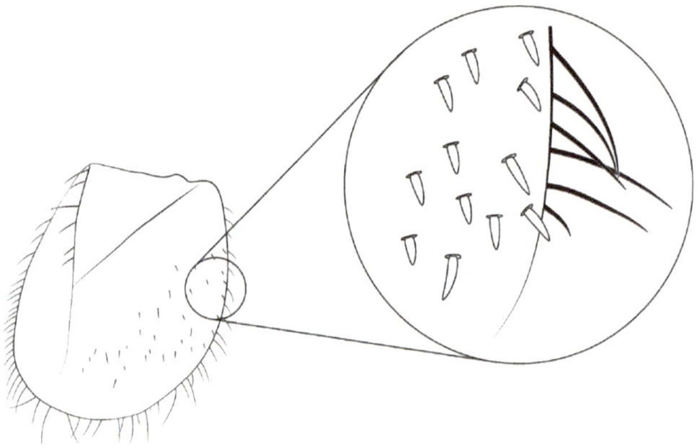

FIGURE 3.43 Operculate gill of *Brasilocaenis* (Caenidae) nymph.

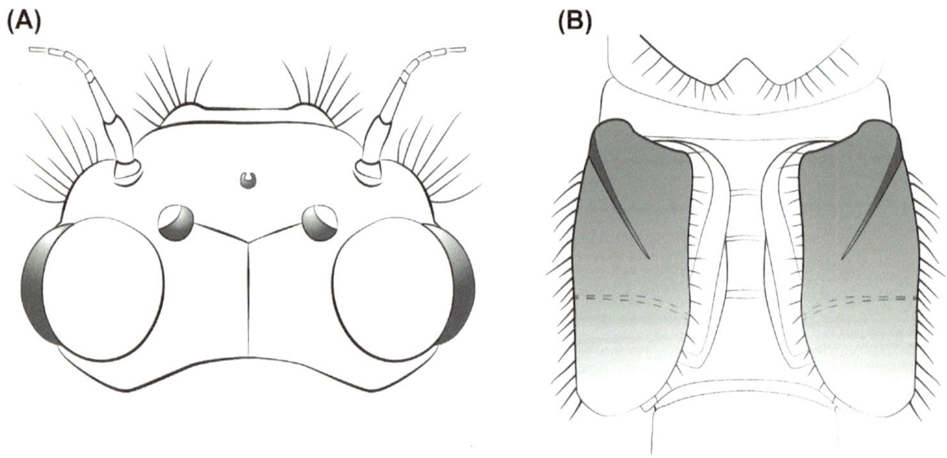

FIGURE 3.44 Nymphal structures of *Leptohyphodes inanis* (Pictet) (Leptohyphidae): (A) head of male; and (B) abdominal segments I to VII.

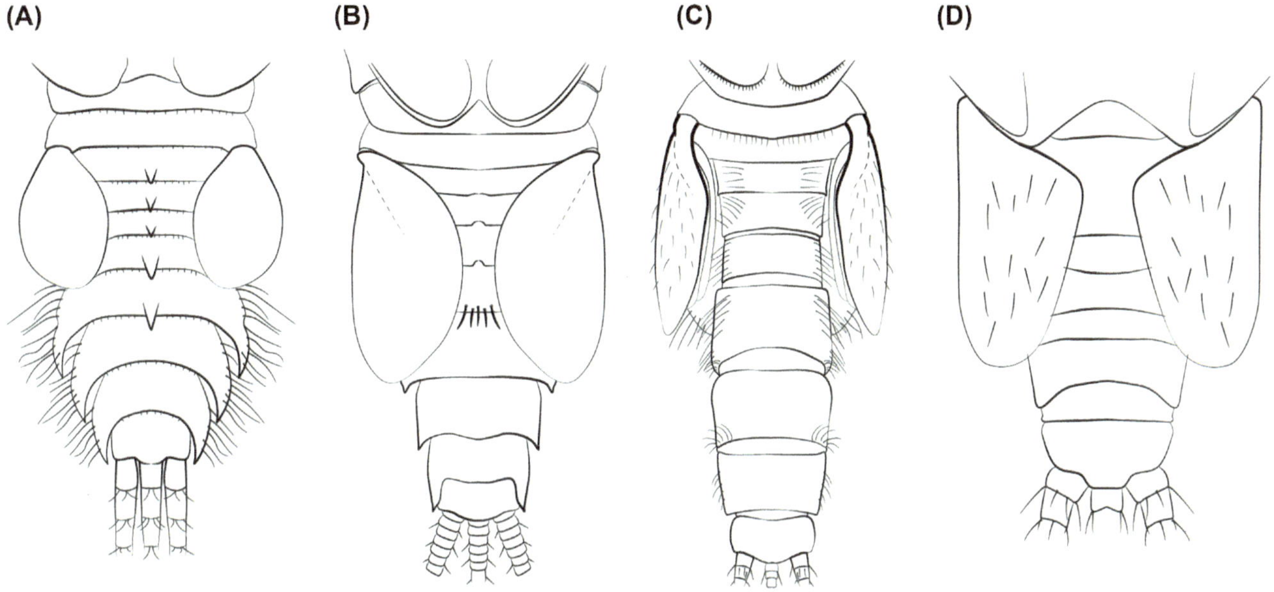

FIGURE 3.45 Abdomen of Leptohyphidae nymphs, dorsal view: (A) *Tricorythopsis*; (B) *Ableptemetes*; (C) *Haplohyphes*; and (D) *Tricorythodes*.

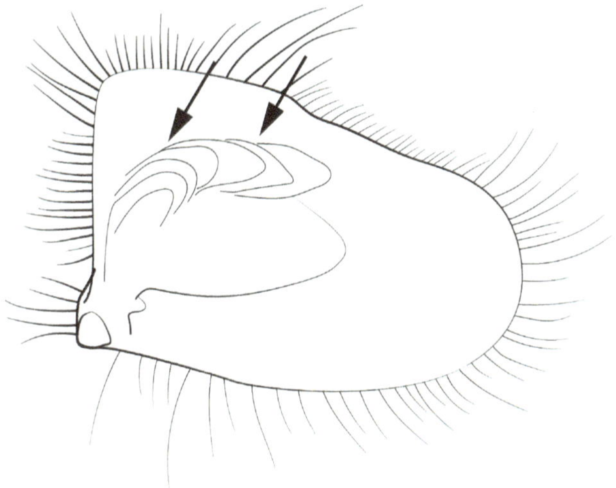

FIGURE 3.46 Operculate gill of of *Leptohyphodes inanis* (Pictet) (Leptohyphidae), ventral view. *Arrows* showing imbricated lobes.

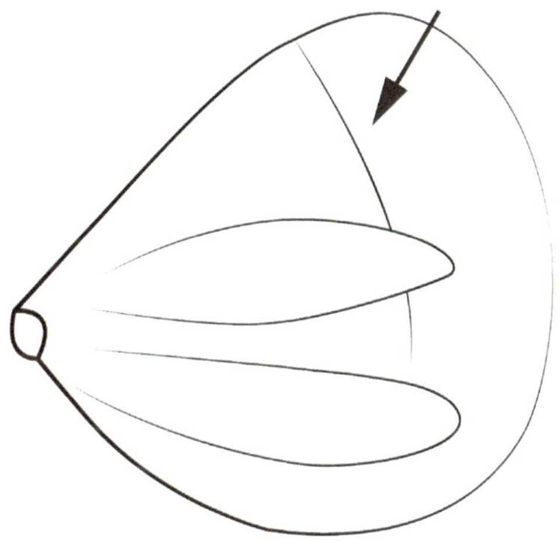

FIGURE 3.47 Operculate gill of of *Amanahyphes* (Leptohyphidae), ventral view. *Arrow showing* transverse submedian line.

Leptohyphidae: Genera

This New World family is one of the most diverse. Of the 150 species, nearly 130 species have been described from the Neotropics. Some of the 15 genera that comprises the family, 11 are endemic to the Neotropics. The distribution of some Neotropical genera, however, is much more restricted. *Leptohyphodes* (Atlantic Forest), *Amanahyphes* (Amazon Basin) and *Haplohyhes* (Andes). Nymphs are typical crawlers and can be found in a wide variety of habitats, from slow to fast currents, from organic to inorganic substrata, but always restricted to lotic systems.

1 Eyes of male enlarged, sometimes divided (Fig. 3.44 A); operculate gills subquadrate to semitriangular, meeting or almost meeting medially (Fig. 3.44 B); gills present on abdominal segments II–V 2

1' Eyes of male small, blackish, and not divided; operculate gills triangular, semitriangular, or oval, never meeting medially (Fig. 3.45); gills present on abdominal segments II–VI 3

2(1) Operculate gills semitriangular and without transverse line (Fig. 3.44 B); gill ventral lamellae with margins fringed with small imbricated lobes (Fig. 3.46); eyes of male mature nymph with whitish or yellowish upper portion .... *Leptohyphodes*

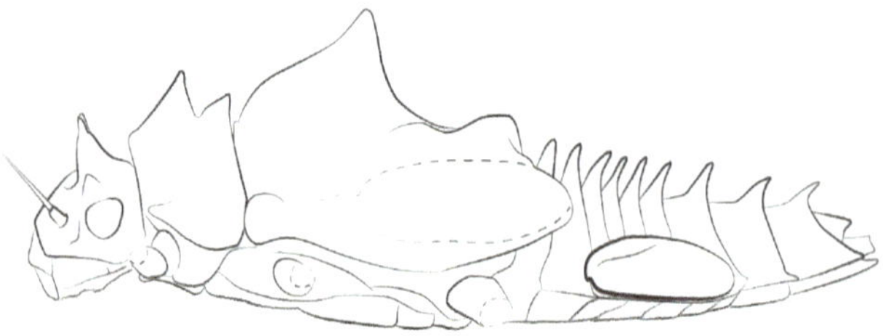

FIGURE 3.48 Nymph of *Loricyphes froehlichi* Molineri & Mariano (2015) (Leptohyphidae), lateral view. *After Molineri & Mariano (2015).*

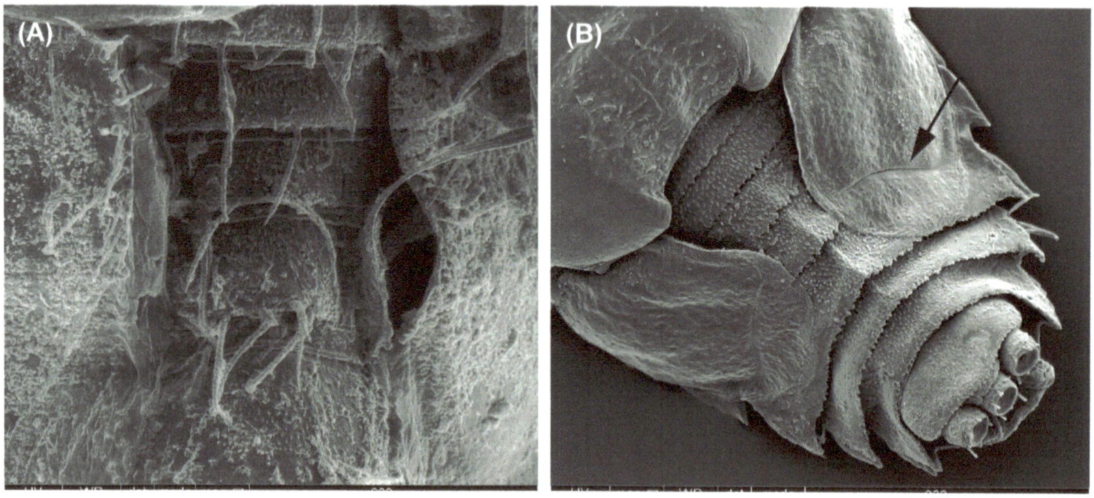

FIGURE 3.49 Abdominal terga and operculate gills of Leptohyphidae nymphs, SEM pictures: (A) *Tricorythodes*; (B) *Tricorythopsis*.

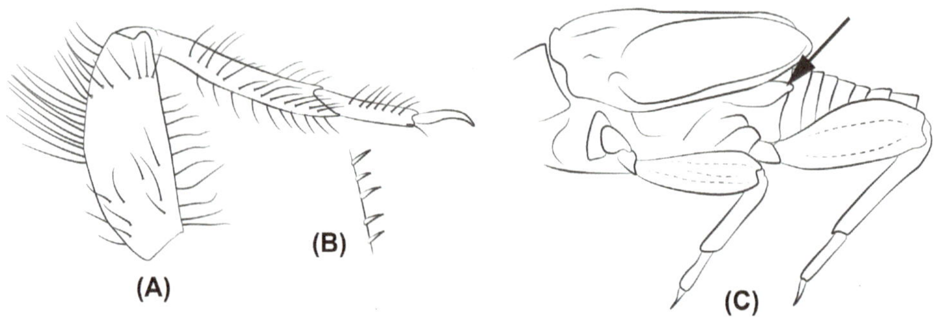

FIGURE 3.50 Nymphal structures of *Haplohyphes* (Leptohyphidae): (A) fore leg; (B) detail of medial margin of femur; and (C) partial view of thorax and abdomen, lateral view. *Arrow showing* hind wing pad.

2'	Operculate gills subquadrate to fusiform and with a transverse submedian line (Fig. 3.47); gills not fringed (Fig. 3.47); eyes of male mature nymph blackish	*Amanahyphes*
3(1)	Tubercles absent, rarely present but not in all three tagma at the same time (except one species of *Leptohyphes*, but tubercles are blunt and paired in the abdomen)	4
3'	Head, thorax and abdomen with acute tubercles (Fig. 3.48)	*Loricyphes*
4(3)	At least abdominal terga I–V with smooth posterior margin (Fig. 3.49 A)	5
4'	Abdominal terga I–X with tiny spines along posterior margin (Fig. 3.49 B)	8
5(4)	Abdomen base much wider than apex (Fig. 3.45 D); profemora anterior margin without spines (Fig. 3.51 A), if present, setaform (Fig. 3.51 B) or spatulate (Fig. 3.51 C); hind wing pads absent	6
5'	Abdomen base slightly wider than apex (Fig. 3.45 C); profemora anterior margin with long setae and stout spines on leading edge (Figs. 3.50 A, B); hind wing pads present (Fig. 3.50 C)	*Haplohyphes*

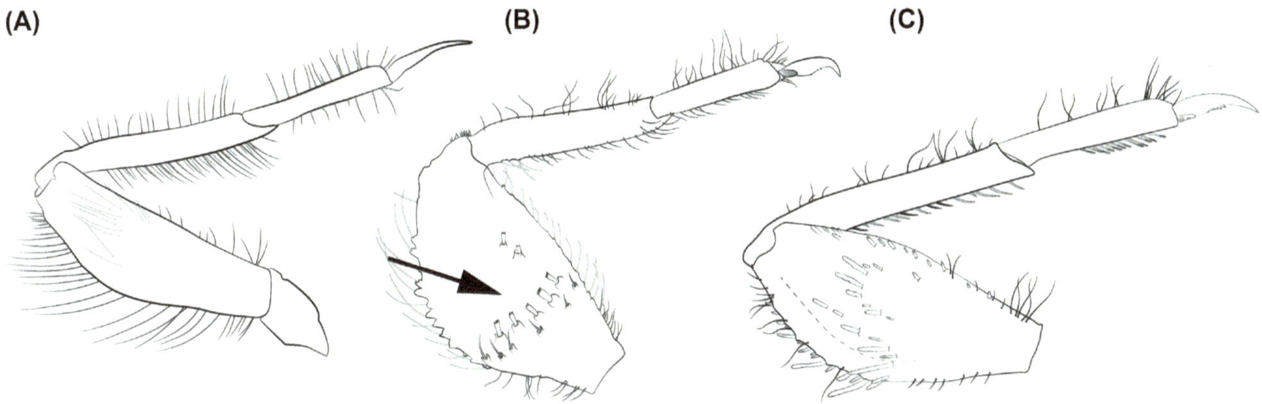

FIGURE 3.51 Fore legs of Leptohyphidae nymphs: (A) *Tricorythodes*; (B) *Cabecar*; and (C) *Macunahyphes*.

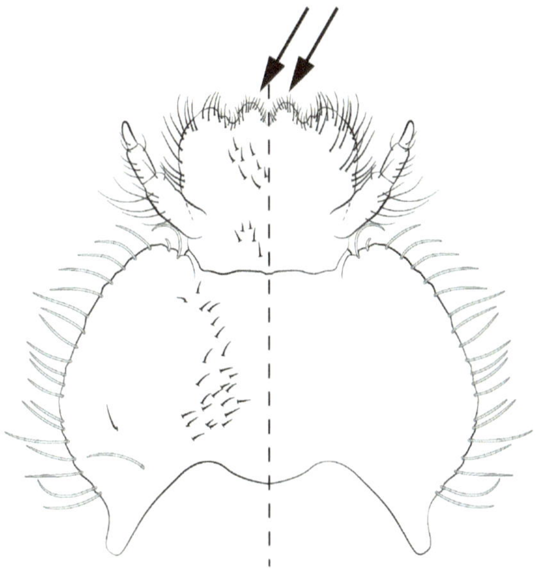

FIGURE 3.52 Labium of *Cabecar* (Leptohyphidae). *Arrows* showing glossae.

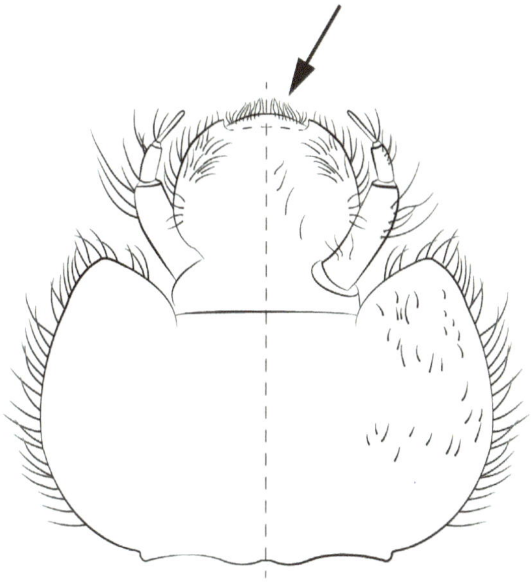

FIGURE 3.53 Labium of *Macunahyphes* (Leptohyphidae). *Arrows* showing glossae. *After Dias et al. (2005).*

6(5) Femora with relatively short spatulate setae (Figs. 3.51 B, C) 7

6' Femora with thin and long setae only (Fig. 3.51 A) .... *Tricorythodes*

7(6) Pronotum and mesonotum with anterolateral projections; middle and hind femora with a longitudinal row of 5–9 spatulate setae; glossae small but separate and distinguishable (Fig. 3.52) .... *Cabecar*

7' Pronotum and mesonotum without anterolateral projections; middle and hind femora with numerous spatulate setae on the dorsal face; glossae fused (Fig. 3.53) .... *Macunahyphes*

8(4) Operculate gills oval or semioval (Figs. 3.54 B–D) 9

8' Operculate gills subtriangular (Fig. 3.54 A); rare .... *Ableptemetes* [Mexico and Central America]

9(8) Operculate gills oval or suboval but never with weak transverse line (Figs. 3.54 C, D); operculate gill with three lamella (Figs. 3.54 C, D) 10

9' Operculate gills oval frequently with a median or subapical weak transverse line (Figs. 3.49 B, 54 B) (except *T. rondoniensis* and *T. spongicola*); operculate gill with five lamella (Fig. 3.54 B) .... *Tricorythopsis*

10(9) Operculate gills without basal spine (Fig. 3.54 D); caudal filaments not as below; dorsum of middle and hind femora dorsally without longitudinal ridge or transverse row of spines near base (Fig. 3.57) 11

10' Operculate gills with basal spine (Fig. 3.54 C); caudal filaments with darker articles basally (female) or medially (male) (Fig. 3.55); middle and hind femora dorsum generally with a longitudinal ridge and with a transverse of spine row near base (Fig. 3.56) .... *Leptohyphes*

11(10) Profemora dorsally distal 1/3 or more with transverse dorsal row of spatulate setae (Fig. 3.59) 12

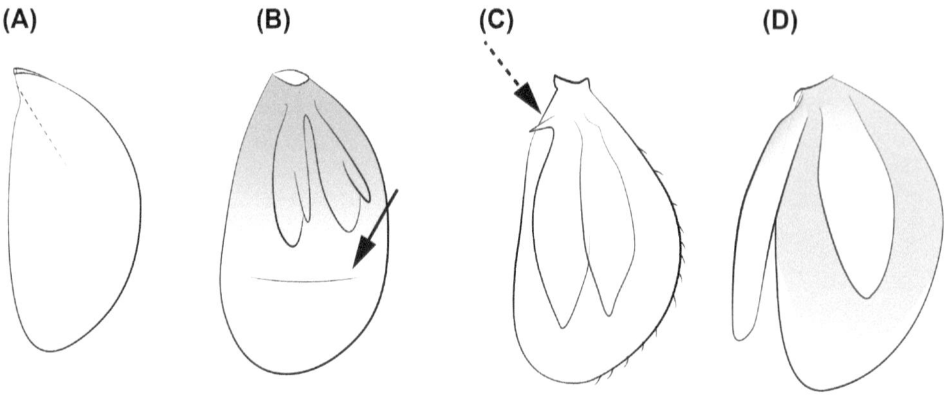

FIGURE 3.54 Operculate gills of Leptohyphidae nymphs, ventral view: (A) *Ableptemetes*; (B) *Tricorythopsis*; (C) *Leptohyphes*; and (D) *Traverhyphes*.

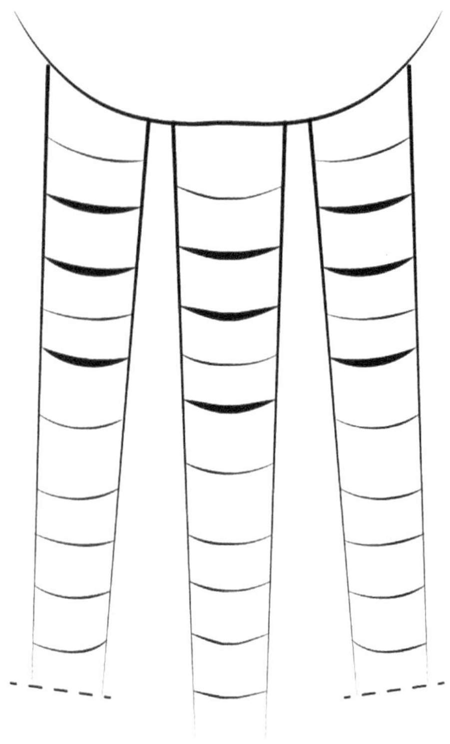

FIGURE 3.55 Caudal filaments of *Leptohyphes* (Leptohyphidae) female nymph.

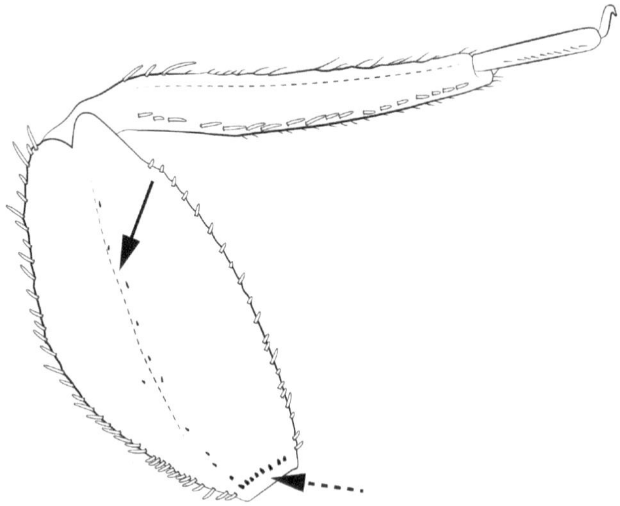

FIGURE 3.56 Hind leg of *Leptohyphes* (Leptohyphidae) nymph. *Solid arrow* showing longitudinal ridge. *Dashed arrow* showing transversal row of spines near base.

11'	Profemora dorsally with medial transverse row of spatulate setae (Fig. 3.58)	*Vacupernius*
12(11)	Body relatively stout (Fig. 3.60); maxillary palpi biarticulate (Figs. 3.61 A, B); operculate gills ovoid and relatively small, not reaching posterior half of abdominal segment VII (Fig. 3.60); gill formula: 3/3/3/3/1–2	13
12'	Body relatively elongated (Fig. 3.62); maxillary palpi bi- or triarticulate (Fig. 3.63); operculate gills ovoid to subquadrate and relatively large, at least reaching posterior half of abdominal segment VII (Fig. 3.62); gill formula: 3/3–4/3–4/3/1–2	14
13(12)	Abdominal gill VI with two lamellae (Fig. 3.64 A); maxillary palpi generally with apical seta (Fig. 3.61 C); fore tarsal claw with a subapical single row of four denticles	*Allenhyphes* (in part)
13'	Abdominal gill VI with one lamella (Fig. 3.64 B); maxillary palpi without apical seta (Fig. 3.61 B); fore tarsal claw with a subapical single row of 6–7 denticles	*Yaurina*
14(12)	Maxillary palpi triarticulate and relatively large (Fig. 3.63 B)	15
14'	Maxillary palpi biarticulate and relatively small (Figs. 3.63 C–E)	*Traverhyphes* (in part)
15(14)	Profemora dorsal spatulate setae not bifid and with smooth apex (Fig. 3.65 C); female mature nymph sterna	16

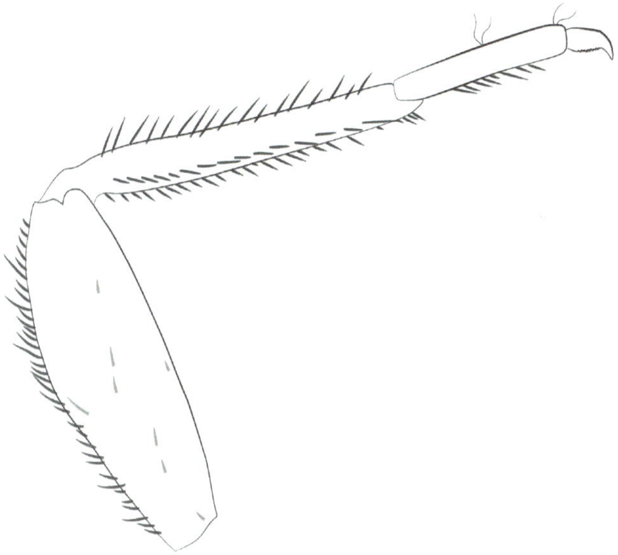

FIGURE 3.57 Hind leg of *Lumahyphes* (Leptohyphidae) nymph.

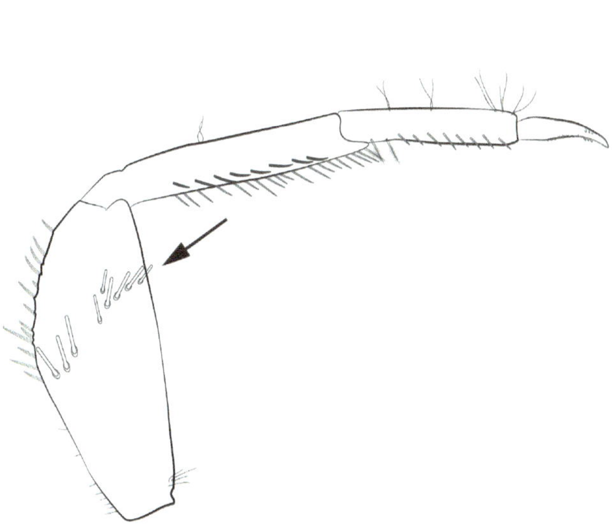

FIGURE 3.59 Fore leg of *Lumahyphes* (Leptohyphidae) nymph. *Solid arrow* showing transverse dorsal row of spatulate setae on femur.

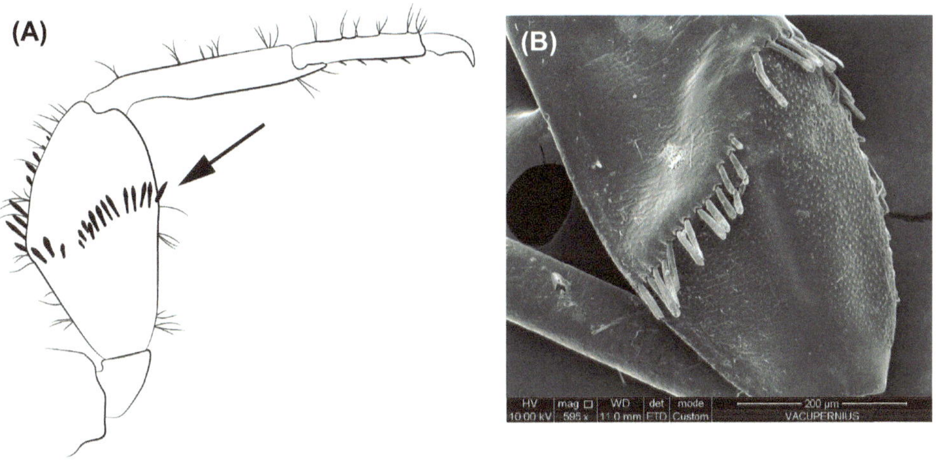

FIGURE 3.58 Fore leg of *Vacupernius* (Leptohyphidae) nymph. *Solid arrow* showing transverse dorsal row of spatulate setae on femur.

15' Profemora dorsal spatulate setae bifid or serrate (Figs. 3.65 A, B); female mature nymph generally with a median black macula between abdominal sterna VII and VIII .... *Lumahyphes*

16(15) Maxillary palpi with small apical seta (Fig. 3.63 B) .... *Traverhyphes* (in part)

16' Maxillary palpi without apical seta (Figs. 3.61 C) .... *Allenhyphes* (in part)

Ameletopsidae: Genera

This family has only seven species; three species in two genera are known from the Neotropics, exclusively from the Andean-Patagonian region: *Chaquihua penai* Demoulin, *Chaquihua bullocki* (Navás), and *Chiloporter eatoni* Lestage. Nymphs are carnivorous and crawlers, although able to perform fast swimming movements (Domínguez et al., 2006).

1 Gills composed of dorsal lamellar and ventral fringed portions (Fig. 3.14) .... *Chiloporter eatoni* Lestage

1' Gills composed of only dorsal lamella (Fig. 3.66) .... *Chaquihua*

FIGURE 3.60 Habitus of *Allenhyphes* (Leptohyphidae) nymph.

FIGURE 3.62 Habitus of *Traverhyphes* (Leptohyphidae) nymph.

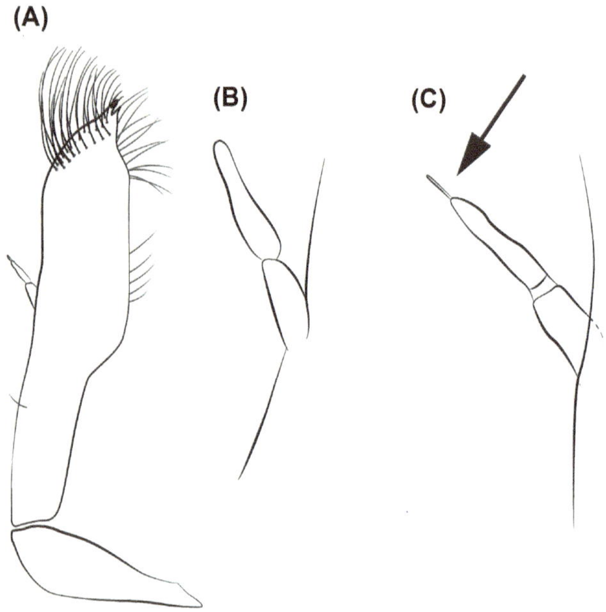

FIGURE 3.61 Maxillae of *Allenhyphes-Yaurina* (Leptohyphidae) nymph: (A) maxilla of *Allenhyphes*; (B) maxillary palp of two articles, without apical seta of *Yaurina*; and (C) maxillary palp of two articles, with apical seta of *Allenhyphes* (*solid arrow*).

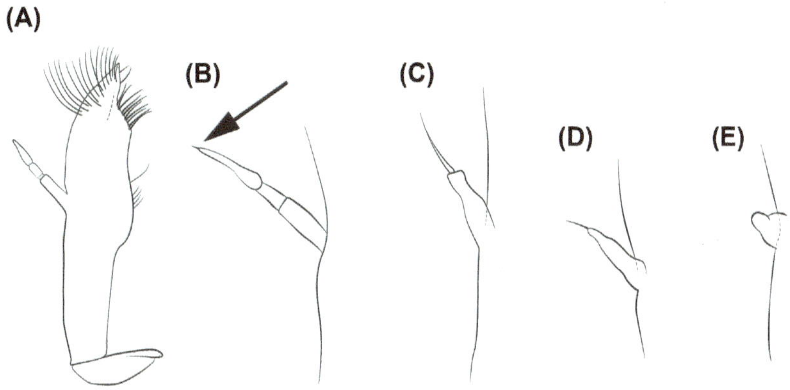

FIGURE 3.63 Maxillae of *Traverhyphes* (Leptohyphidae) nymph: (A) maxilla; (B) maxillary palp of three articles, with apical seta; (C) maxillary palp of two articles, with apical seta (*solid arrow*); (D) maxillary palp of two articles, with apical seta; and (E) maxillary palp of two articles, without apical seta.

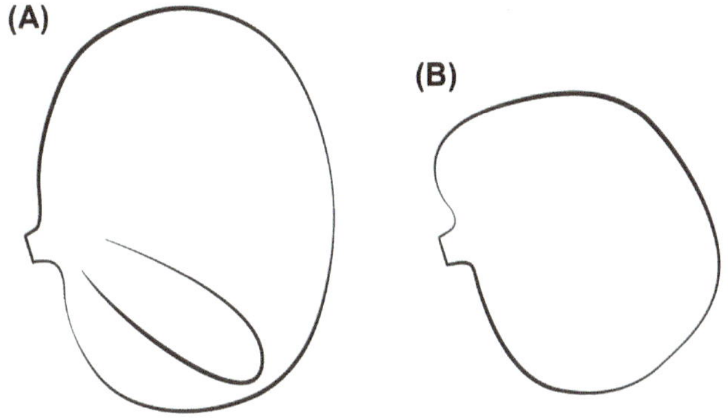

FIGURE 3.64 Abdominal gill VI of Leptohyphidae nymph: (A) *Allenhyphes*; and (B) *Yaurina*.

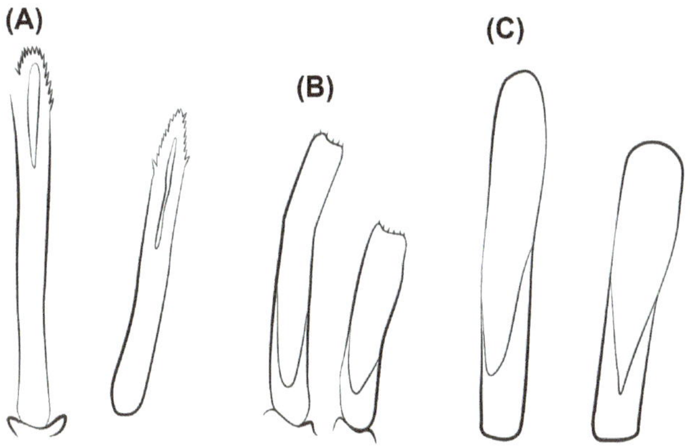

FIGURE 3.65 Detail of spatulate setae on dorsum of fore femora of Leptohyphidae nymph: (A) and (B) *Lumahyphes*; and (C) *Traverhyphes*.

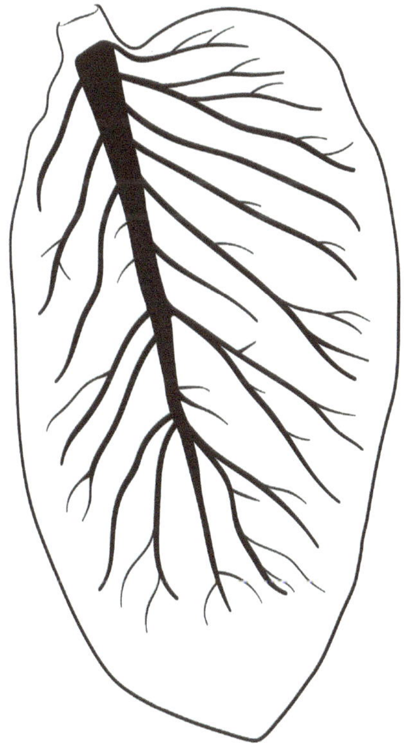

FIGURE 3.66 Abdominal gill of *Chaquihua* (Ameletopsidae) nymph.

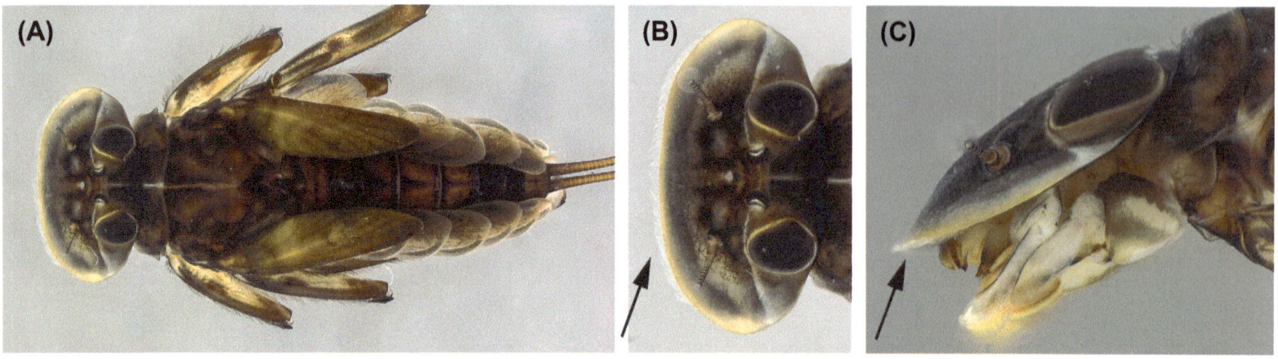

FIGURE 3.67 Nymph of *Epeorus* (Heptageniidae): (A) habitus; (B) detail of head in dorsal view; and (C) detail of head in ventral view. *Solid arrow* showing dense fringe of setae.

FIGURE 3.68 Nymph of *Rithrogena* (Heptageniidae), ventral view.

Heptageniidae: Genera

Modified from Webb & McCafferty (2008). This family has around 600 species: five species in four genera have been reported from Central America. *Stenacron* is also reported from southern Florida (Berner & Pescador, 1988).

1	Median caudal filament well developed (Fig. 3.68); head anterior margin without dense fringe of setae	2
1'	Median caudal filament vestigial (Fig. 3.67 A); head anterior margin with dense fringe of setae (Fig. 3.67 B) (reported from Colombia in error)	*Epeorus*
2(1)	First pair of gills do not meet or overlap ventrally; maxillary palp not enlarged (Fig. 3.69 A)	3
2'	First pair of gills meet or overlap ventrally to form friction disk (Fig. 3.68); maxillary palp enlarged (Fig. 3.69 B)	*Rhithrogena*

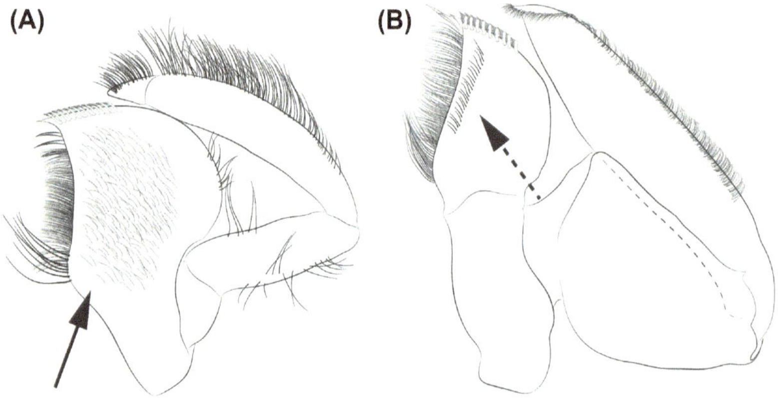

FIGURE 3.69 Maxillae of Heptageniidae nymphs: (A) *Ecdyonurus*; and (B) *Epeorus*.

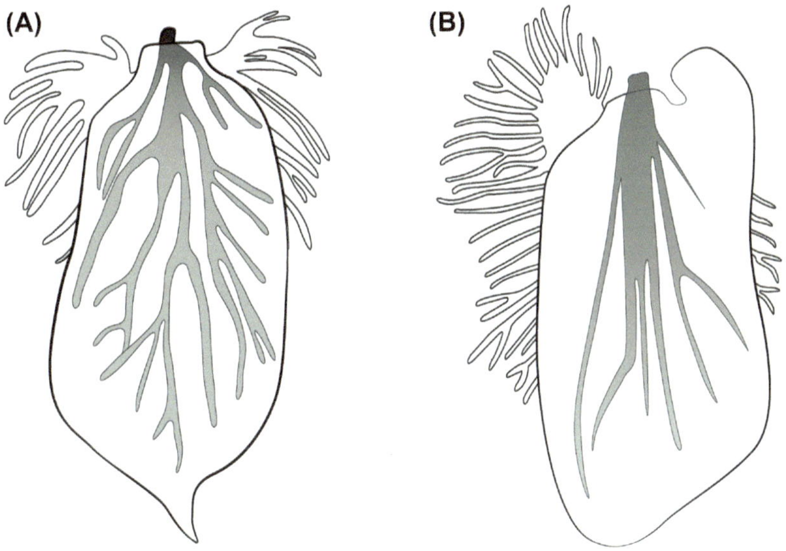

FIGURE 3.70 Gills of Heptageniidae nymphs: (A) *Stenacron*; and (B) *Mccaffertium*.

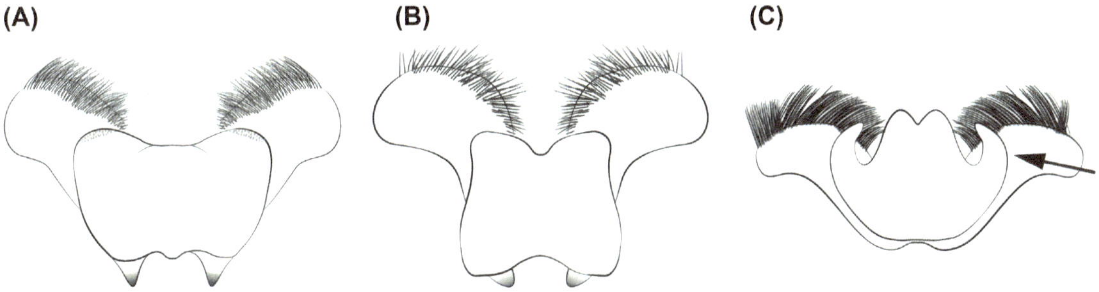

FIGURE 3.71 Hypopharynx of Leptophlebiidae nymphs: (A) *Fittkaulus*; (B) *Terpides*; and (C) *Ecuaphlebia*.

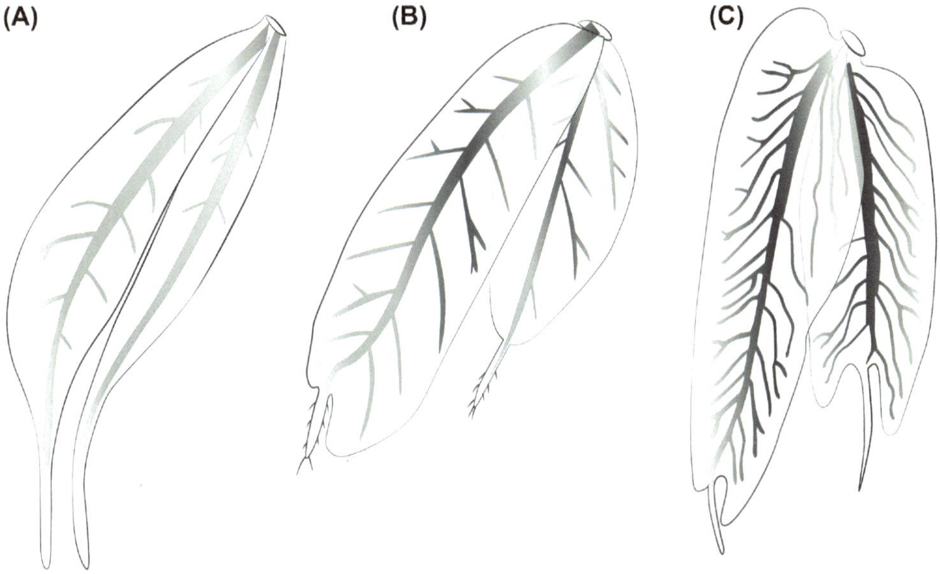

FIGURE 3.72 Gills of Leptophlebiidae nymphs: (A) *Terpides*; (B) *Fittkaulus*; and (C) *Tikuna*.

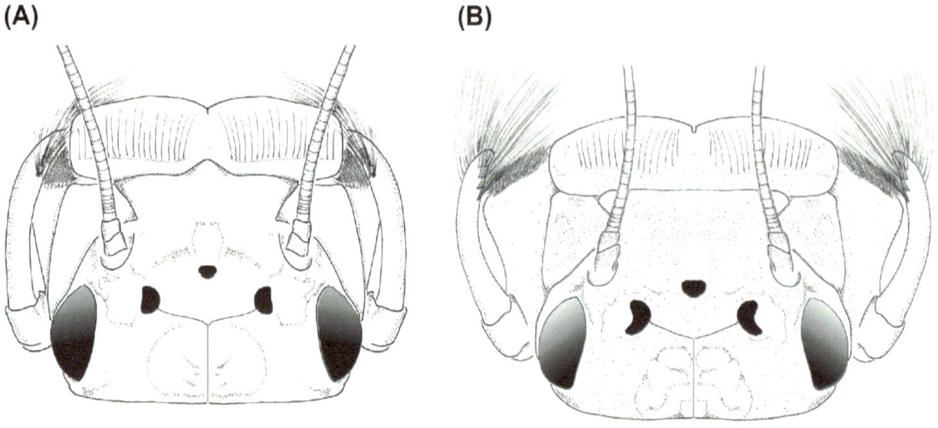

FIGURE 3.73 Head, in dorsal view, of Leptophlebiidae nymphs: (A) *Needhamella*; and (B) *Hylister.*

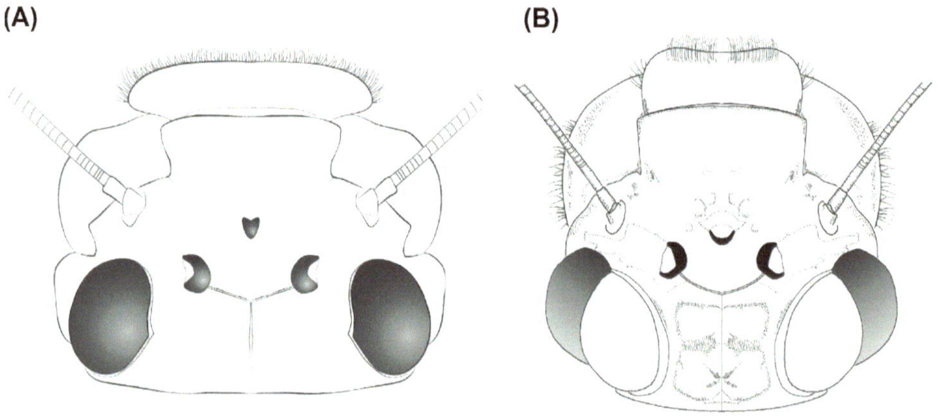

FIGURE 3.74 Head, in dorsal view, of Leptophlebiidae nymphs: (A) *Thraulodes*; and (B) *Massartella.*

3(2)	Maxillae ventrally with setal row (as in Fig. 3.69 B); gill one lamellae of variable, but never reniform	4
3'	Maxillae ventrally with scattered setae (Fig. 3.69 A); gills one lamellae usually reniform	*Ecdyonurus*
4(3)	Gill lamellae I to VI acutely pointed (Fig. 3.70 A); maxillae galea-lacinia crown with comb setae	*Stenacron*
4'	Gill lamellae I to VI truncate (Fig. 3.70 B); maxillae galea-lacinia crown with or without comb setae	*Mccaffertium*

Leptophlebiidae: Genera

Modified from Domínguez et al. (2006). This is one of the most diverse families of Ephemeroptera, with around 270 species and 50 genera recorded for the Neotropics (almost a third of the total number of genera and species). Of these genera, 75% are endemic to this region. Four of them are endemic to the Caribbean and 34 to South America. Most of the nymphs are prognathous, but a few are hypognathous. They have herbivorous mouthparts and are considered collectors-gatherers, collector-filterers, or shredders-detritivores. The nymphs can be found from the sea level, up to 4,500 m. Some genera that live at sea level in Patagonia, but are restricted to higher and higher altitudes toward the Andes in Northern Argentina, Bolivia, and Peru. Most of the genera present in the austral part of South America are more related to taxa from Australia and New Zealand, than to others present in the same continent. Nymphs are predominantly crawlers, with a somewhat flattened body, while a few genera with a streamlined body and abundant caudal filament setae are excellent swimmers (member of the Terpides lineage).

1	Hind wing pads present	2
1'	Hind wing pads absent	48
2(1)	Body hydrodynamic (Fig. 3.17); head hypognathous (Fig. 3.17); caudal filaments with dense row of setae (Fig. 3.17); hypopharynx lingua without lateral projections (Figs. 3.71 A, B)	3
2'	Body flattened (Fig. 3.16); head prognathous (Figs. 3.15, 16); caudal filaments with spines or setae, but not forming swimming setae; hypopharynx lingua with lateral projections (Fig. 3.71 C)	5

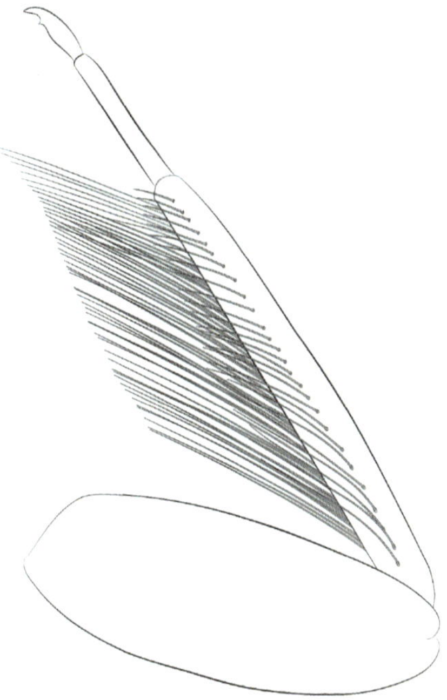

FIGURE 3.75 Fore leg, in dorsal view, of *Hagenulus caligatus* (Eaton) nymph. *After Peters (1971).*

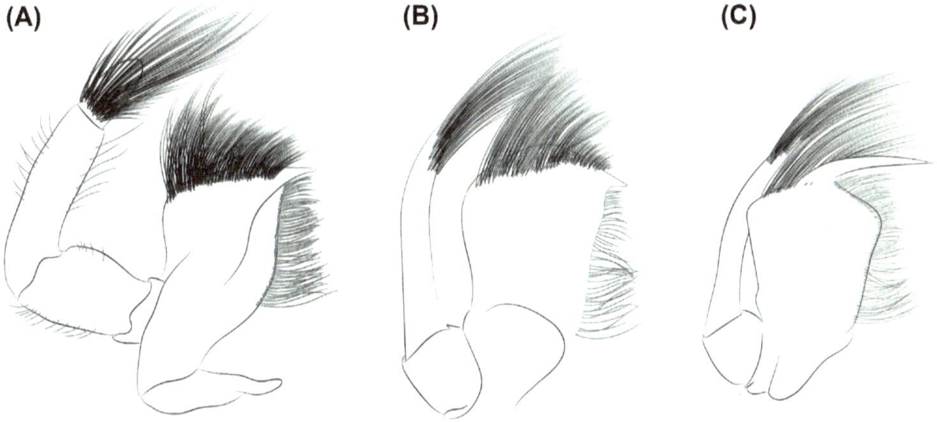

FIGURE 3.76 Maxillae of Leptophlebiidae nymphs: (A) *Hagenulus caligatus* (Eaton); (B) *Hylister*; (C) *Traverella*.

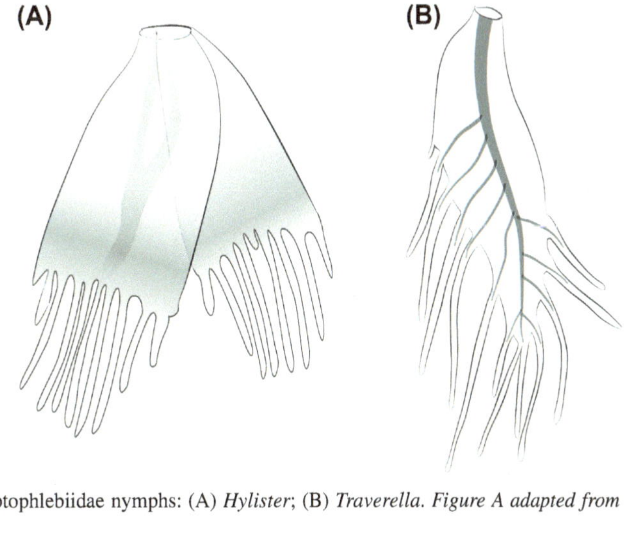

FIGURE 3.77 Gills of Leptophlebiidae nymphs: (A) *Hylister*; (B) *Traverella*. *Figure A adapted from Domínguez & Flowers (1989).*

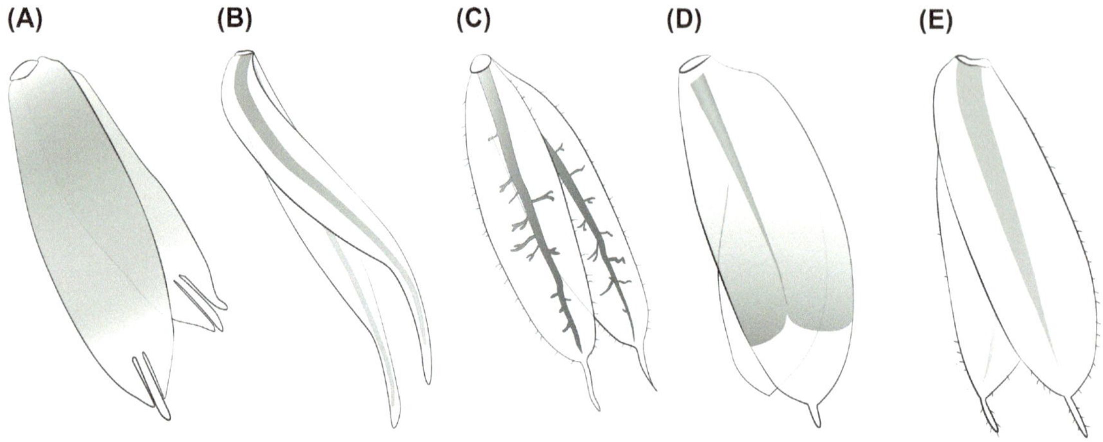

FIGURE 3.78 Gills of Leptophlebiidae nymphs: (A) *Hermanella*; (B) *Leentvaaria*; (C) *Hydrosmilodon*; (D) *Paramaka*; and (E) *Needhamella*. *Figures adapted from: (A) Domínguez & Flowers (1989); (B) Domínguez et al. (2001); (C) Flowers & Domínguez (1992); (D) Blanco-Belmonte et al. (2003); and (E) Flowers & Domínguez (1992).*

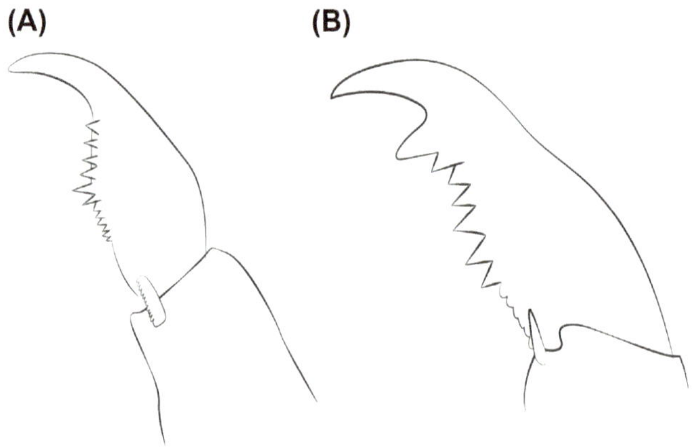

FIGURE 3.79 Tarsal claws of Leptophlebiidae nymphs: (A) *Traverella*; and (B) *Hylister*.

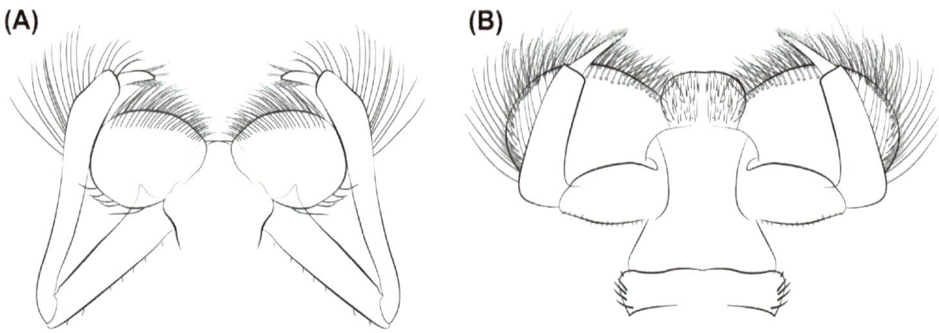

FIGURE 3.80 Labia of Leptophlebiidae nymphs: (A) *Leentvaaria*; and (B) *Paramaka*.

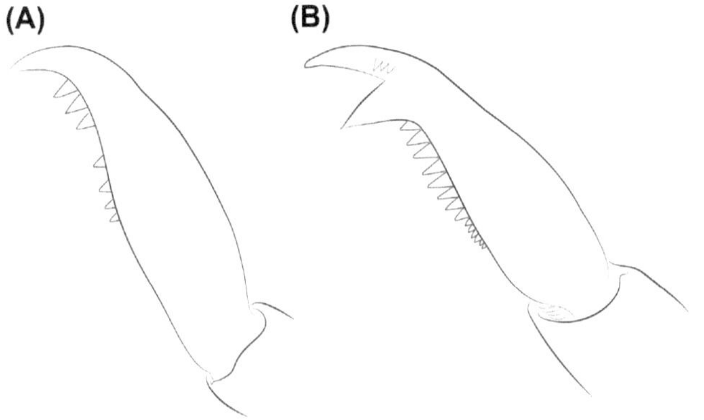

FIGURE 3.81 Tarsal claws of Leptophlebiidae nymphs: (A) *Hydromastodon*; and (B) *Hydrosmilodon*.

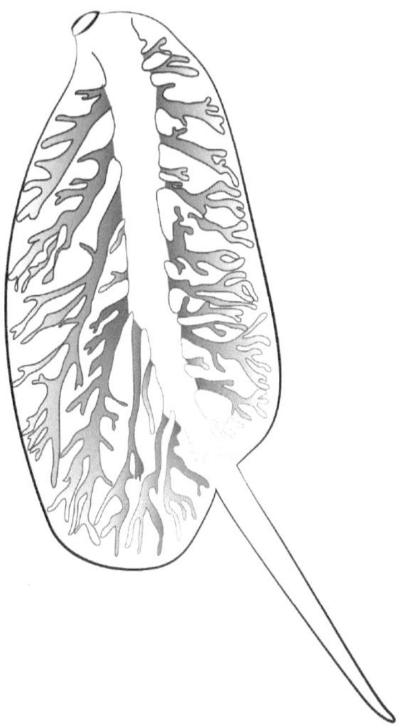

FIGURE 3.82 Gill of *Magallanella* (Leptophlebiidae) nymph. *Adapted from Pescador & Peters (1980).*

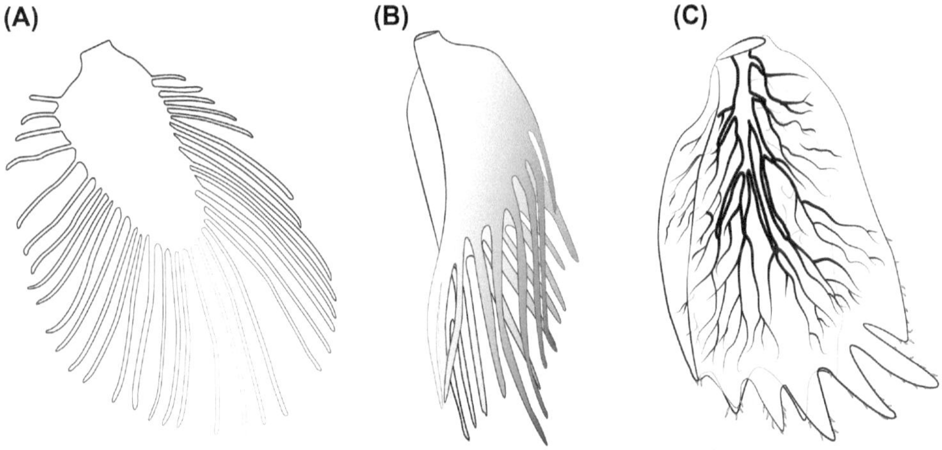

FIGURE 3.83 Gills of Leptophlebiidae nymphs: (A) *Ulmeritoides*; (B) *Atopophlebia*; and (C) *Dactylophlebia*.

3(2) Gills lamellar, gills II to VI dorsal portions ending in a filament and with 1 or 2 lateral lobes (Figs. 3.72 B, C)........ 4

3' Gills lamellar, tapering gradually toward apex and ending in single filament (Fig. 3.72 A)........ *Terpides*

4(3) Gills II–VI ventral portion without developed lobes (Fig. 3.72 B)........ *Fittkaulu'*

4' Gills II–VI ventral portion with apical internal and external lobes (Fig. 3.73 C)........ *Tikuna*

5(2') Labrum as wide as, or wider than, head (Figs. 3.73 A, B, 75)........ 6

5' Labrum narrower than head (Figs. 3.74 A, B)........ 14

6(5) Labrum anteromedian emargination absent; maxillary palp article I poculiform, attached to side of maxilla (Figs. 3.76 B, C)........ 7

6' Labrum anteromedian emargination present; maxillary palp article I not poculiform nor attached at base of maxilla (Fig. 3.76 A) .... *Hagenulus* (in part)

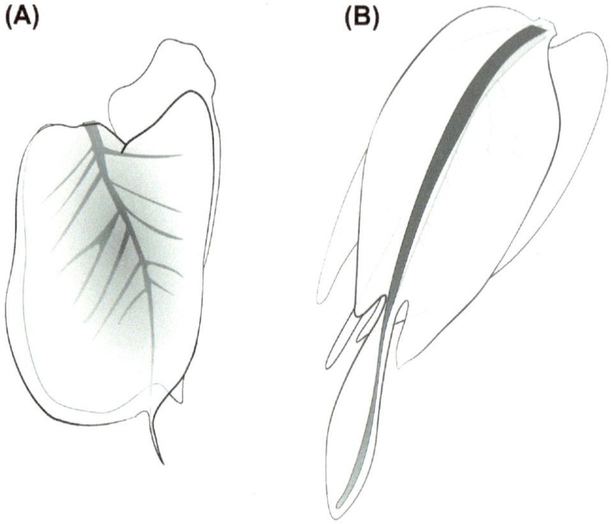

FIGURE 3.84 Gills of Leptophlebiidae nymphs: (A) *Massartella*; and (B) *Choroterps*.

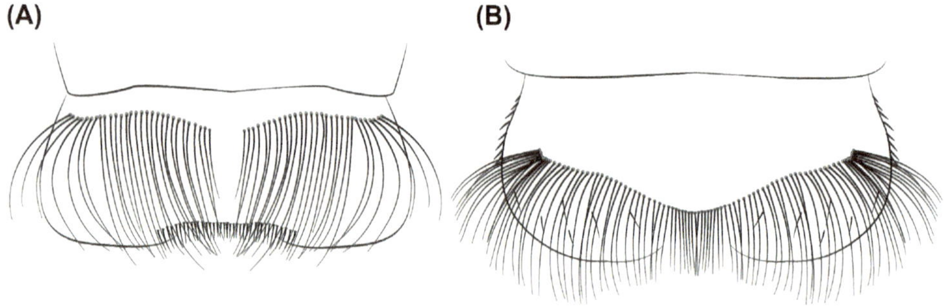

FIGURE 3.85 Labra of Leptophlebiidae nymphs: (A) *Ulmeritus*; and (B) *Ulmeritoides*.

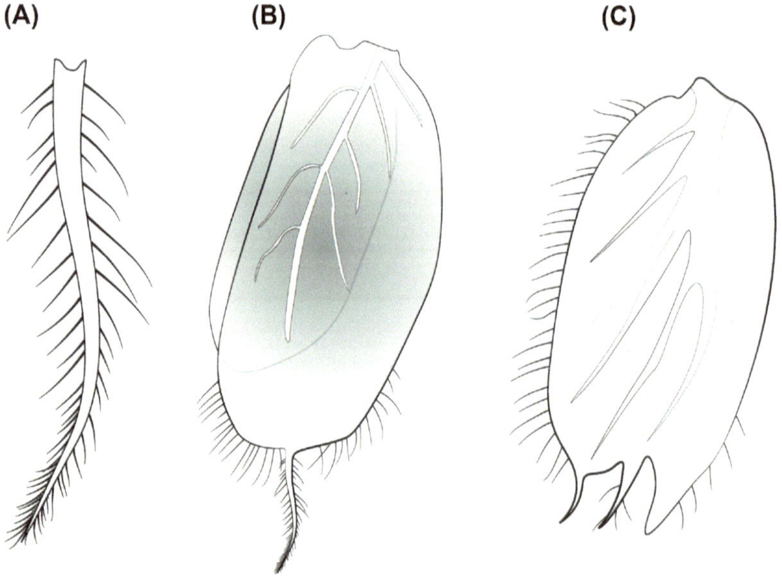

FIGURE 3.86 Gills of Leptophlebiidae nymphs: (A) Gill VII of *Hapsiphlebia*; and (B) Gill IV of *Hapsiphlebia*; (C) Gill IV of *Rhigotopus*. *Figures adapted from: (A-B) Pescador & Edmunds (1972); and (C) Miserendino (1996).*

7(6) Gills with six or more digitiform projections (Fig. 3.77) ... 8

7' Gills with three or fewer digitiform projections (can be absent) (Fig. 3.78) ... 9

8(7) Gills oval, fringed with filaments (Fig. 3.77 B); tarsal claws denticle equal size (Fig. 3.79 A) ... *Traverella*

8' Gills rather quadrangular, truncated apically, with 6–10 digitiform projections (Fig. 3.77 A); tarsal claws with much larger apical denticle (Fig. 3.79 B) ... *Hylister*

9(7') Labial palpi article II length <3× article III (Fig. 3.80 B) ... 10

9' Labial palpi article II length ~6× article III (Fig. 3.80 A) ... *Leentvaaria*

10(9) Some or all gills with single median digitiform projection (Figs. 3.78 B–E) ... 11

10' Gills ending in lateral lobes and longer median process (Fig. 3.78 A) ... *Hermanella*

11(10) Maxillary tusk 1/2 or more width of apex of maxilla (Fig. 3.76 C) ... 12

11' Maxillary tusk approximately 1/4 or less width of apex of maxilla (Fig. 3.76 B) ... 13

12(11) Tarsal claws with subequal denticles, in two different rows (Fig. 3.81 A) ... *Hydromastodon*

12' Tarsal claws with subapical denticle much larger than the others, with one row of denticles (Fig. 3.81 B) ... *Hydrosmilodon*

13(11') Gills of segment VII vestigial; apical process of gills variable (Fig. 3.78 D), small or absent on some gills ... *Paramaka*

13' Gills similar in form on abdominal segments I–VII, dorsal and ventral portions always with digitiform process (Fig. 3.78 E) ... *Needhamella*

14(5') Gills on abdominal segments II–VI composed by two narrow-to-broad lamellae (one ventral and one dorsal portion) ... 15

14' Each abdominal gill composed by single ovoid lamella, ending abruptly in a thick digitiform projection (Fig. 3.82) ... *Magallanella*

15(14) Gill margins with five or more filaments (Fig. 3.83) ... 16

15' Gill margins with zero to three filaments (Figs. 3.84, 86) ... 19

16(15) Entire gill margins fringed with filamentous projections (Fig. 3.83 A) ... 17

16' Filamentous projections only present on the distal portion of the gill (Figs. 3.83 B, C) ... 18

17(16) Row of long dorsal setae of labrum divided, without setae medially (Fig. 3.85 A) ... *Ulmeritus*

17' Row of long dorsal setae of labrum entire, forming a single sinusoidal row (Fig. 3.85 B) ... *Ulmeritoides*

18(16') Gills I–VI with filamentous projections on apical margin (Fig. 3.83 B), gill VII reduced to pair of filaments; margins of mandibles fringed with long setae, setae denser on apical ½ ... *Atopophlebia*

18' Gills I–VII similar, with filamentous projections on apical margin (Fig. 3.83 C); margins of mandibles with median tuft of setae, and some setae on basal ½ ... *Dactylophlebia*

19(15') Gills present on abdominal segments I–VII, gills variable but not as below (Figs. 3.84 B, 86) ... 20

19' Gills present on abdominal segments I–VI, broad, with (Fig. 3.84 A) or without an apical filament ... *Massartella*

20(19) Gills on abdominal segment I similarly shaped as other gills ... 21

20' Gills on abdominal segment I reduced to a single filament, gills on segments II–VII with three projections, dorsal portion with terminal filament long and expanded (Fig. 3.84 B) ... *Choroterpes*

21(20) Gills on abdominal segments VI–VII or VII vestigial, almost reduced to a single filament (Fig. 3.86 A); ventral portion smaller than dorsal portion (Figs. 3.86 B, C) ... 22

21' Gills on abdominal segments VI and VII with the same shape as gills on segments II–V (Fig. 3.87) ... 23

22(21) Gills on abdominal segments I–VI with three distal projections (Fig. 3.86 C), median projection sometimes a little longer; smaller ventral portion with similar projections ... *Rhigotopus*
[Andean-Patagonic Region]

22' Dorsal portion of gills on abdominal segments I–V wide, with narrow distal projection; small ventral portion without terminal projection (Fig. 3.86 B) ... *Hapsiphlebia*
[Andes]

23(21') Gills on abdominal segments I–VII variable, dorsal portion as large as, or larger than, ventral portion (Figs. 3.87 B-E) ... 24

23' Gills on abdominal segments I–VII oval, dorsal portion 3/4 as long and 1/2 as wide as ventral portion (Fig. 3.87 A) ... *Microphlebia*

24(23) Gills with long and narrow terminal projection, 4/5 to 1/2 as long as basal part; basal part wider, usually with lateral lobes (Figs. 3.87 B, C) ... 25

24' Distal projection of gills, if present, ≤1/3 as long as basal part (Figs. 3.87 D, E); basal part not as above ... 26

25(24) Posterolateral spines present on abdominal segments V or VI to IX; tarsal claw denticles gradually larger from the base to apex (as in Fig. 3.91 B) ... *Traverina*
[Caribbean]

25' Posterolateral spines present on abdominal segments II or III to IX; tarsal claw denticles of progressively larger apically, apical denticle often much larger (as in Fig. 3.81 B) ... *Miroculis*
[South America]

26(24') Posterolateral projections present on abdominal segments II or III–IX ... 27

26' Posterolateral projections present on abdominal segments V or VI or VII or VIII–IX ... 32

27(26) Labrum lateral margins divergent, forming lateral angles (Fig. 3.88 A) ... 28

27' Labrum lateral margins rounded (Fig. 3.88 B) ... 29

28(27) Labial and maxillary palpi strongly developed (Fig. 3.89 A); maxillary palp article III length 2× II, and labial palp article II length 2× article I ... *Segesta*

28' Labial and maxillary palpi not strongly developed (Fig. 3.89 B); maxillary palp article III shorter than II, and labial palp article I and II subequal ... *Thraulodes*

29(27') Mandible lateral margin glabrous, except for submedian tuft (Fig. 3.90 A) ... 30

29' Mandible lateral margin with setae on at least half of its length (Fig. 3.90 B) ... 31

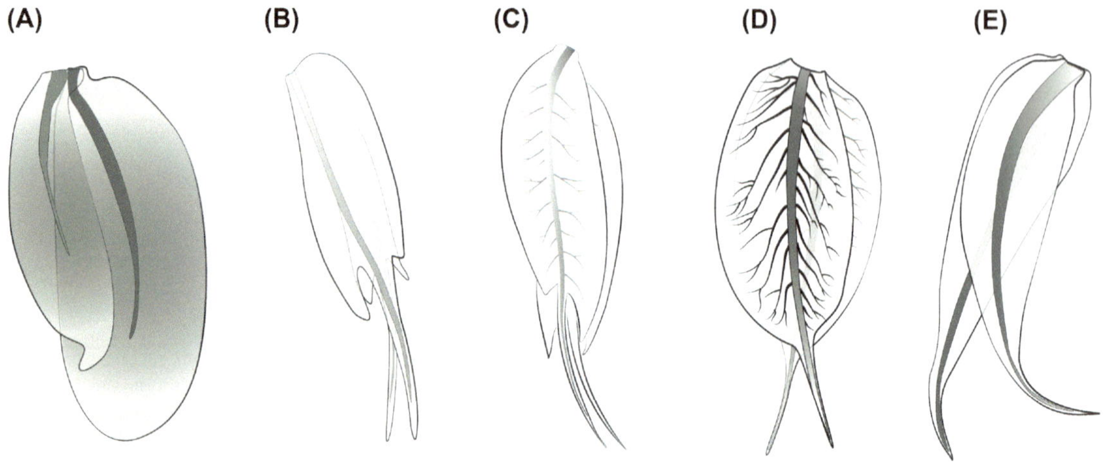

FIGURE 3.87 Gills of Leptophlebiidae nymphs: (A) *Microphlebia*; (B) *Miroculis*; (C) *Traverina*; (D) *Penaphlebia*; and (E) *Thraulodes*.

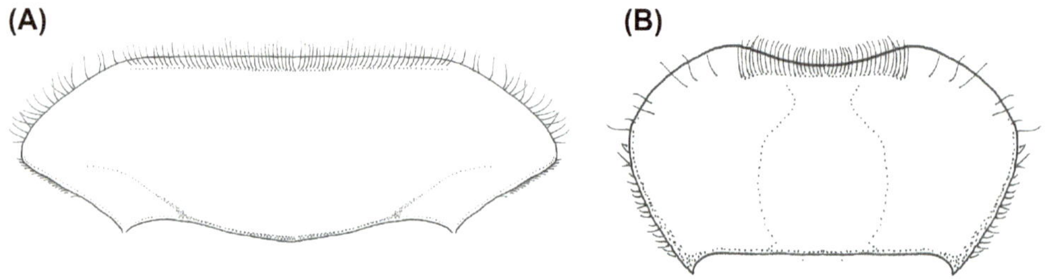

FIGURE 3.88 Labra of Leptophlebiidae nymphs: (A) *Thraulodes*; and (B) *Simothraulopsis*.

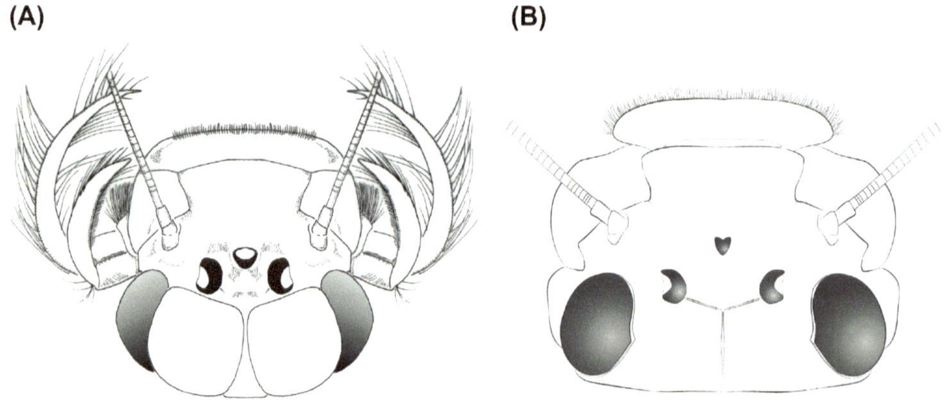

FIGURE 3.89 Head, in dorsal view, of Leptophlebiidae nymphs: (A) *Segesta*; and (B) *Thraulodes*.

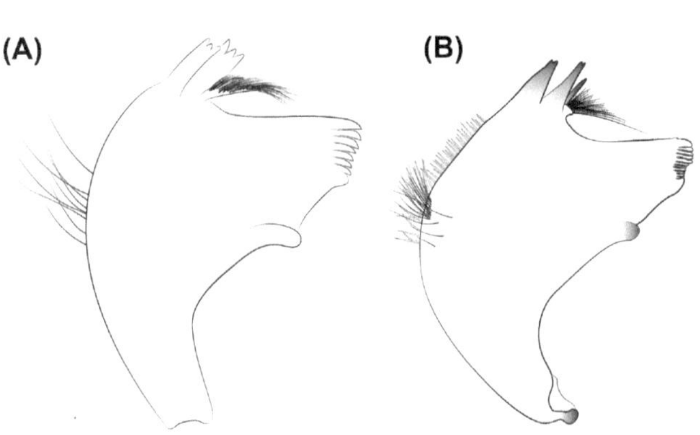

FIGURE 3.90 Left mandibles of Leptophlebiidae nymphs: (A) *Neohagenulus*; and (B) *Massartellopsis*. *Figures adapted from: (A) Peters (1971); and (B) Pescador & Peters (1987).*

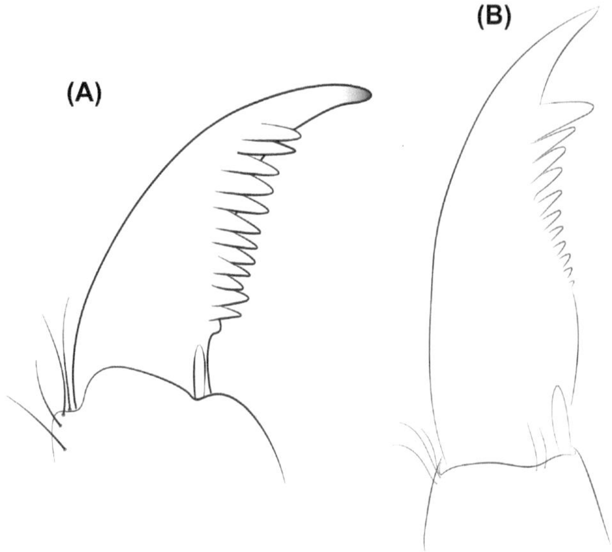

FIGURE 3.91 Tarsal claws of Leptophlebiidae nymphs: (A) *Secochela*; and (B) *Neohagenulus*. *Figures adapted from: (A) Pescador & Gonser (2001); and (B) Peters (1971).*

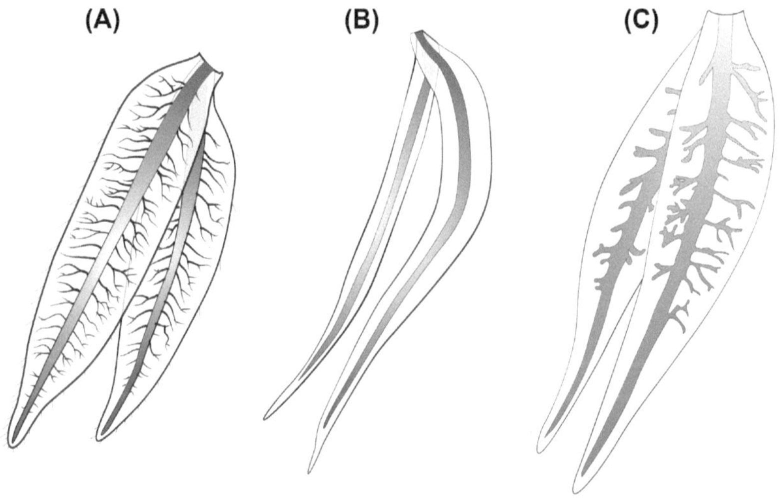

FIGURE 3.92 Gills of Leptophlebiidae nymphs: (A) *Massartellopsis*; (B) *Meridialaris*; and (C) *Meridialaris*. *Adapted from Pescador & Peters (1987).*

30(29)	Mandible lateral margin median tuft of setae very short; basal tarsal claw denticles unequal, remaining denticles gradually larger toward the apex (Fig. 3.91 A) .... *Secochela* [Southern Chile]
30'	Mandible lateral margin median tuft of setae long; tarsal claw denticles progressively larger apically (Fig. 3.91 B).......... *Neohagenulus* [Caribbean, Puerto Rico]
31(29')	Gills with main tracheal trunk and tracheal branches similarly pigmented (Fig. 3.92 A) .... *Massartellopsis*
31'	Gills with tracheal branches less pigmented than main tracheal trunk (Figs. 3.92 B, C) .... *Meridialaris*
32(26')	Abdominal gills variable, lanceolate to narrow (Fig. 3.94 A), apical process not as below; labium with glossa on same plane as, or dorsal to, paraglossa (Fig. 3.94 B) 33
32'	Abdominal gills wide and oval (width approximately 1/3 total length of gill) with apical process narrowed apically (Fig. 3.93 A); labium with glossa ventral to paraglossa (Fig. 3.93 B) .... *Penaphlebia*
33(32)	Posterolateral projections present on abdominal segments VIII–IX 34
33'	Posterolateral projections present on abdominal segments V or VI or VII–IX 38

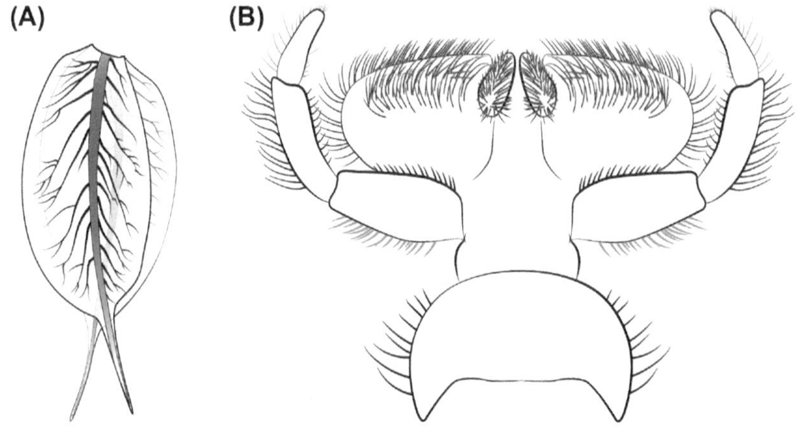

FIGURE 3.93 Nymphal structures of *Penaphlebia* (Leptophlebiidae): (A) gill; and (B) labium. *Adapted from Pescador & Peters (1991).*

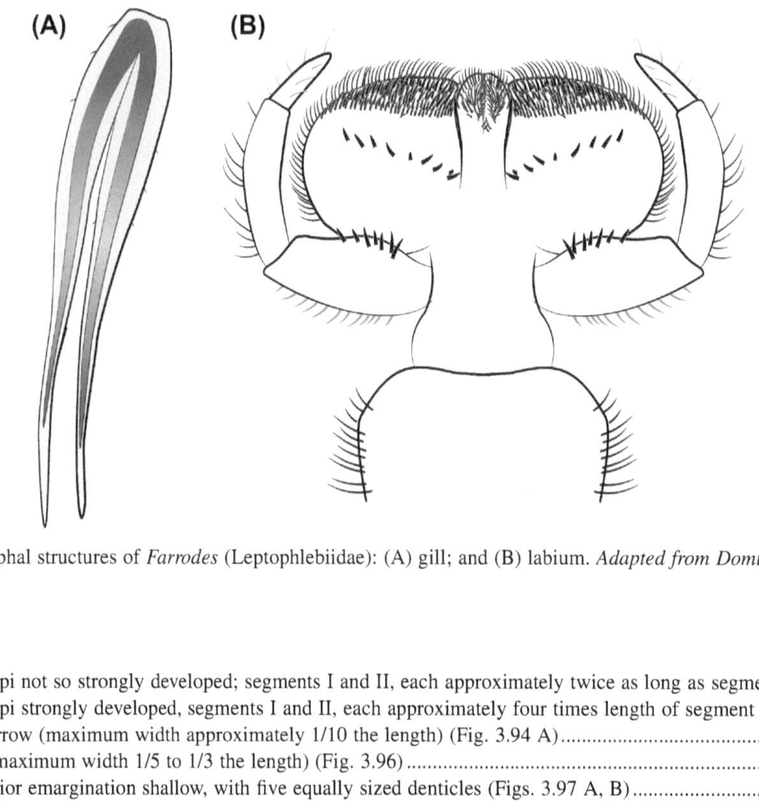

FIGURE 3.94 Nymphal structures of *Farrodes* (Leptophlebiidae): (A) gill; and (B) labium. *Adapted from Domínguez & Savage (1987).*

34(33)	Maxillary palpi not so strongly developed; segments I and II, each approximately twice as long as segment III	35
34'	Maxillary palpi strongly developed, segments I and II, each approximately four times length of segment III (Fig. 3.95)	*Lisetta*
35(34)	Gills very narrow (maximum width approximately 1/10 the length) (Fig. 3.94 A)	36
35'	Gills wider (maximum width 1/5 to 1/3 the length) (Fig. 3.96)	37
36(35)	Labrum anterior emargination shallow, with five equally sized denticles (Figs. 3.97 A, B)	*Farrodes*
36'	Labrum anterior emargination deep, margins rounded, with three weak crenulations (Fig. 3.97 C)	*Poranga*
37(35')	Widest section of gills approximately 1/5 of length (Fig. 3.96 A)	*Simothraulopsis* (in part)
37'	Widest section of gills approximately 1/3 of the length (Fig. 3.96 B)	*Homothraulus*
38(33)	Hind tibia not as below, cylindrical, subcylindrical or slightly flattened, dorsal surface never concave (Fig. 3.98 B)	39
38'	Hind tibia strongly flattened, dorsal surface concave (Fig. 3.98 A) [Caribbean, Cuba]	*Poecilophlebia*
39(38)	Tarsal claw denticles progressively larger toward apex (as in Fig. 3.91 B)	40

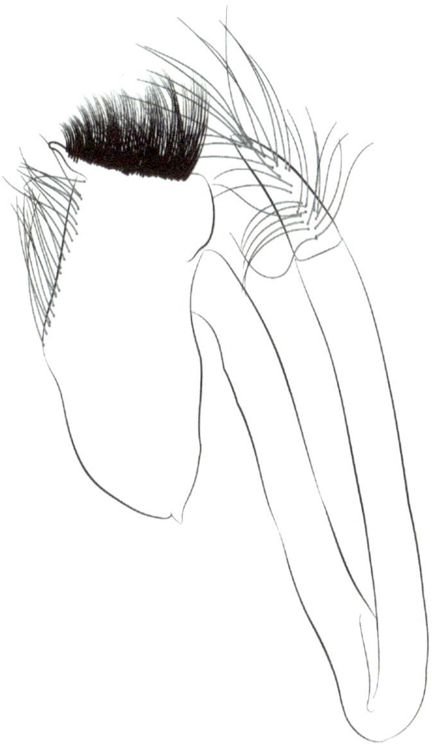

FIGURE 3.95 Maxilla of *Lisetta ernsti* Thomas and Dominique (Leptophlebiidae) nymph. *After Thomas et al. (2005).*

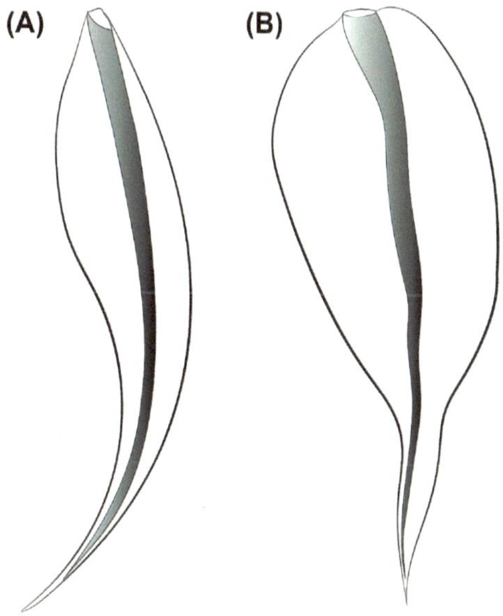

FIGURE 3.96 Gills of Leptophlebiidae nymphs: (A) *Simothraulopsis*; and (B) *Homothraulus*.

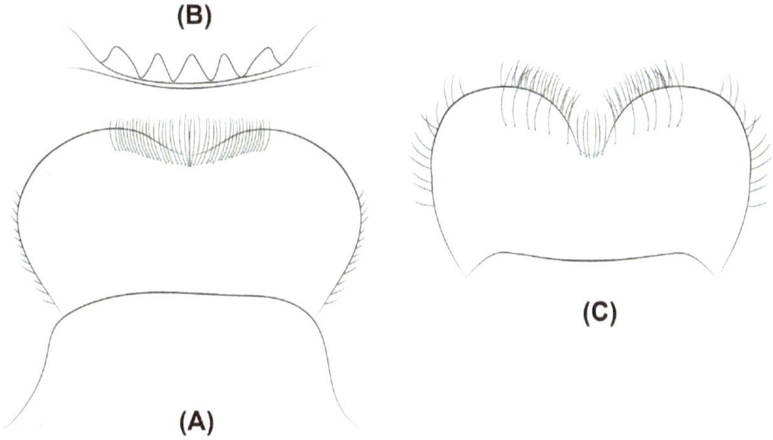

FIGURE 3.97 Labra of Leptophlebiidae nymphs: (A) *Farrodes*; (B) *Farrodes* (detail of anterior margin); and (C) *Poranga nessimiani* Gonçalves and Da-Silva.

39'	Tarsal claw denticles progressively larger toward claw apex, except apical denticle much larger (as in Fig. 3.81 B)	44
40(39)	Maxillary palp longer than maxilla (Fig. 3.99)	41
40'	Maxillary palp distinctly shorter than length of maxilla	42
41(40)	Maxillary tusk absent (Fig. 3.99 B); fore leg without rows of filtering setae.	*Demoulinellus*
41'	Maxillary tusk present (Fig. 3.99 A); fore leg with rows of filtering setae (as in Fig. 3.75)	*Hagenulus* (in part)
42(40')	Labrum anteromedial emargination broad, with denticles	43
42'	Labrum anteromedial emargination narrow, without denticles [Caribbean, Cuba]	*Turquinophlebia*

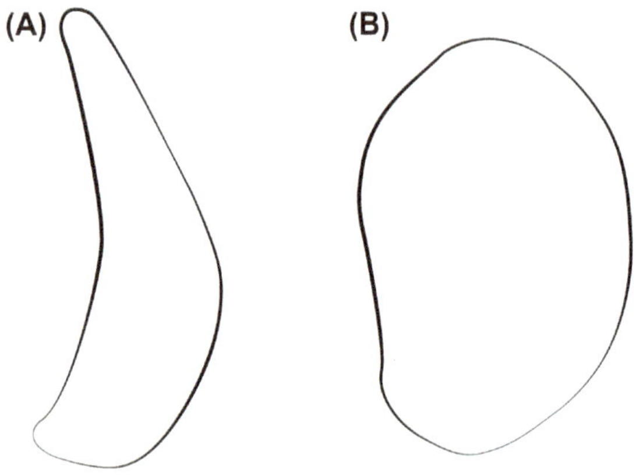

FIGURE 3.98 Cross section of hind tibiae of Leptophlebiidae nymphs: (A) *Poecilophlebia pacoi* Kluge; and (B) *Careospina baconaoi* Kluge. *Adapted from Kluge (1994).*

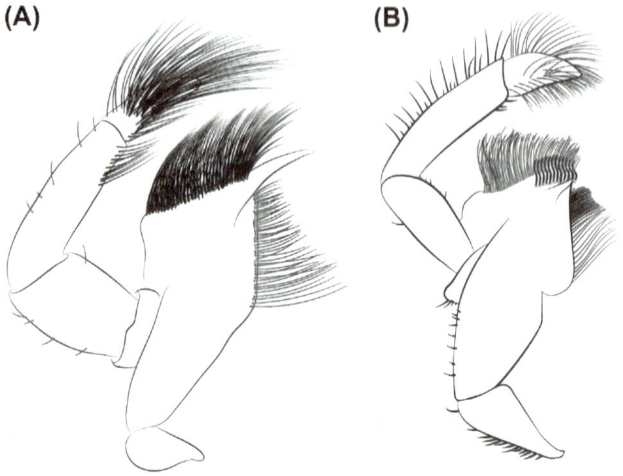

FIGURE 3.99 Maxillae of Leptophlebiidae nymphs: (A) *Hagenulus caligatus* (Eaton); and (B) *Demoulinellus. Figures adapted from: (A) Peters (1971); and (B) Pescador & Peters (1982).*

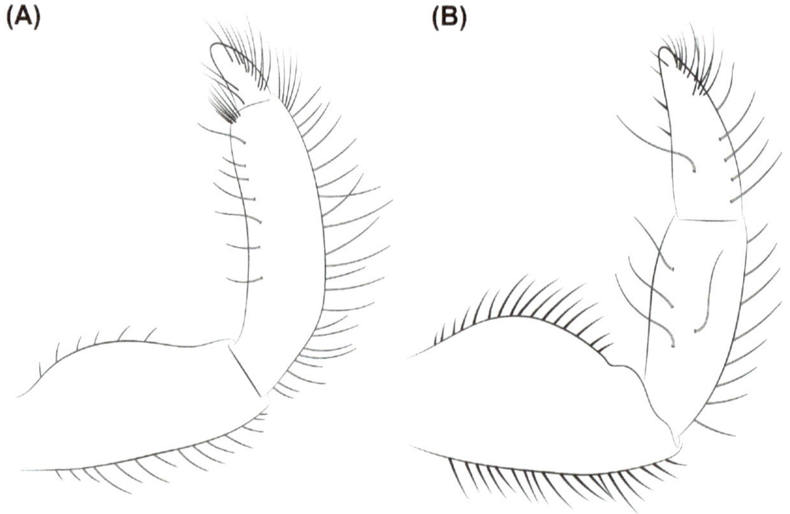

FIGURE 3.100 Labial palpi of Leptophlebiidae nymphs: (A) *Careospina*; and (B) *Nousia*.

43(42)	Labial palp article III slightly shorter or longer than article II (Fig. 3.100 B) [Andean-Patagonic Region]	*Nousia* (in part)
43'	Labial palp article III distinctly shorter than article II (approximately ¼) (Fig. 3.100 A) [Caribbean]	*Careospina*
44(39')	Maxilla galea-lacinia without tusk on apex	45
44'	Maxilla galea-lacinia with tusk on apex (Fig. 3.101)	*Gonserellus*
45(44)	Labrum anteromedian emargination never with two large denticles [continental South America]	46

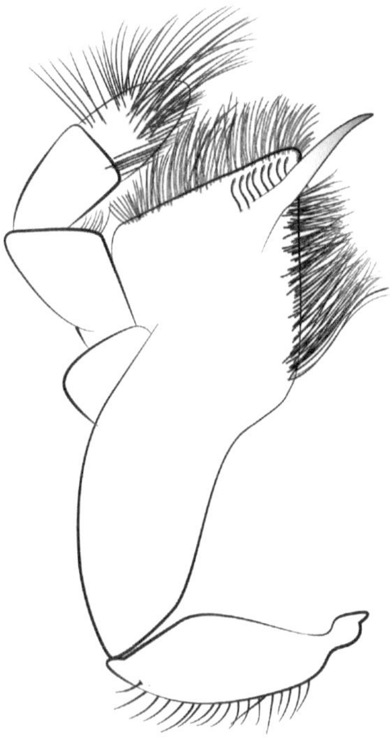

FIGURE 3.101 Maxilla of *Gonserellus* (Leptophlebiidae) nymph. *Adapted from Pescador & Peters (1985).*

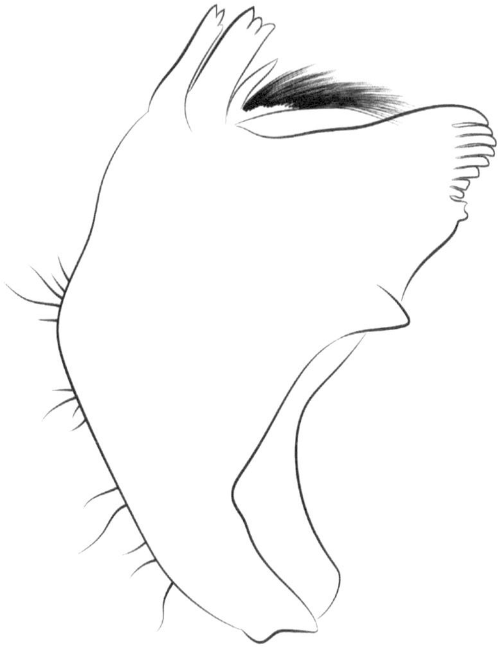

FIGURE 3.103 Left mandible of *Ecuaphlebia* (Leptophlebiidae) nymph. *Adapted from Peters (1969).*

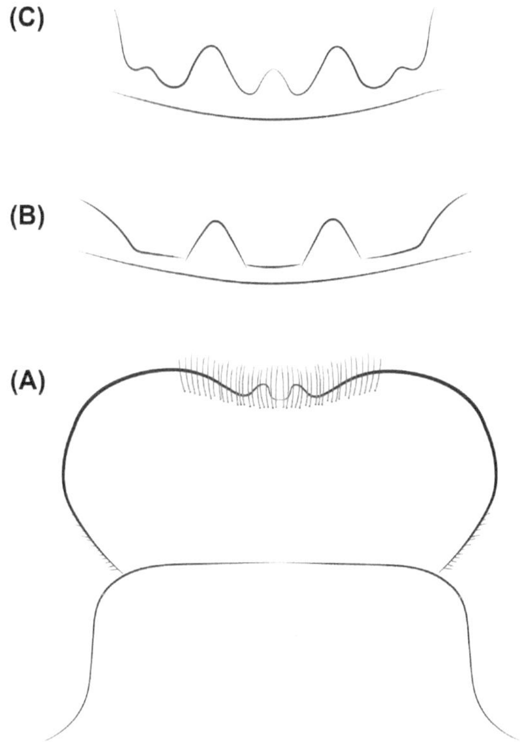

FIGURE 3.102 Labrum of *Borinquena* (Leptophlebiidae) nymph: (A) dorsal view; (B) and (C) detail of anterior margin.

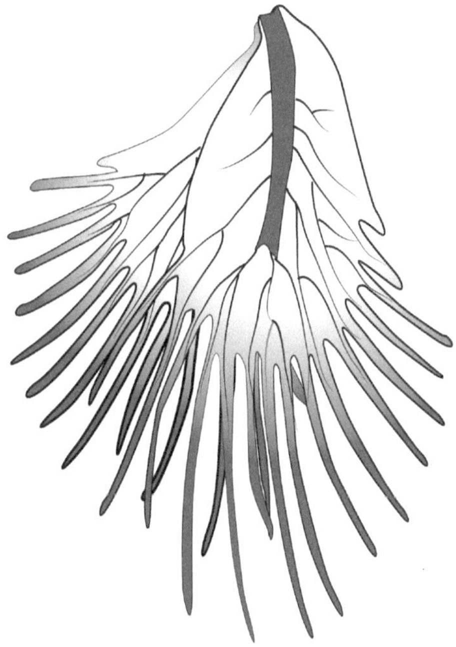

FIGURE 3.104 Gill of *Askola* (Leptophlebiidae) nymph.

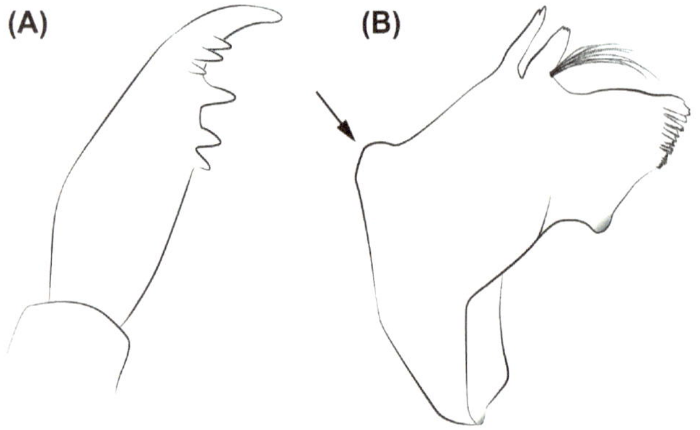

FIGURE 3.105 Nymphal structures of *Perissophlebiodes flinti* (Savage) (Leptophlebiidae): (A) tarsal claw; and (B) left mandible. *Adapted from Savage (1982).*

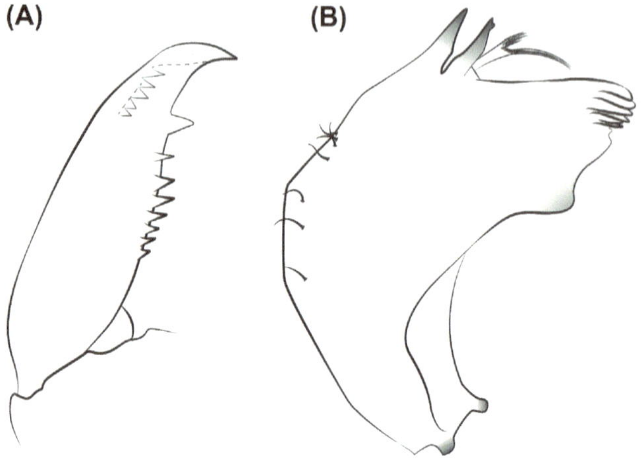

FIGURE 3.106 Nymphal structures of *Bessierus doloris* Thomas and Orth (Leptophlebiidae): (A) tarsal claw; and (B) left mandible. *After Thomas et al. (2001).*

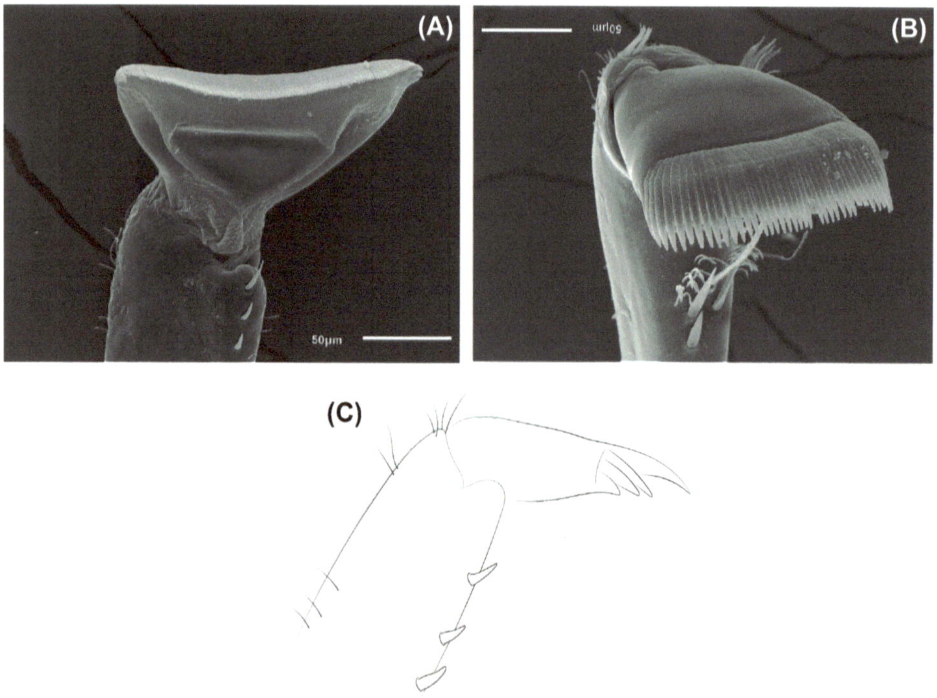

FIGURE 3.107 Tarsal claw of Baetidae nymphs: (A) *Tapajobaetis labiosus* Boldrini and Salles; (B) *Camelobaetidius*; and (C) *Corinella*. *After Boldrini et al. (2017).*

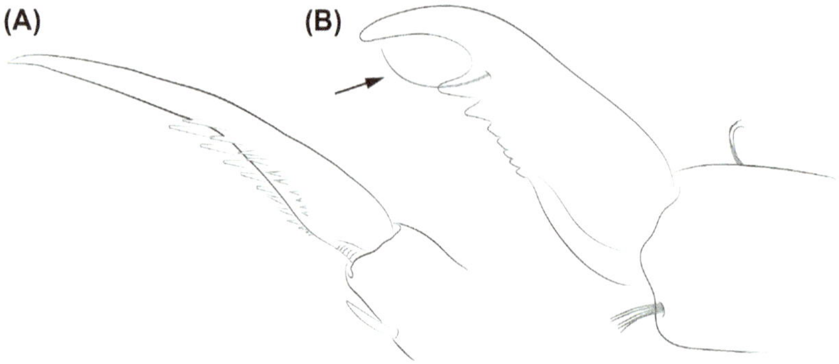

FIGURE 3.108 Tarsal claw of Baetidae nymphs: (A) *Waltzoyphius*; and (B) *Baetodes*.

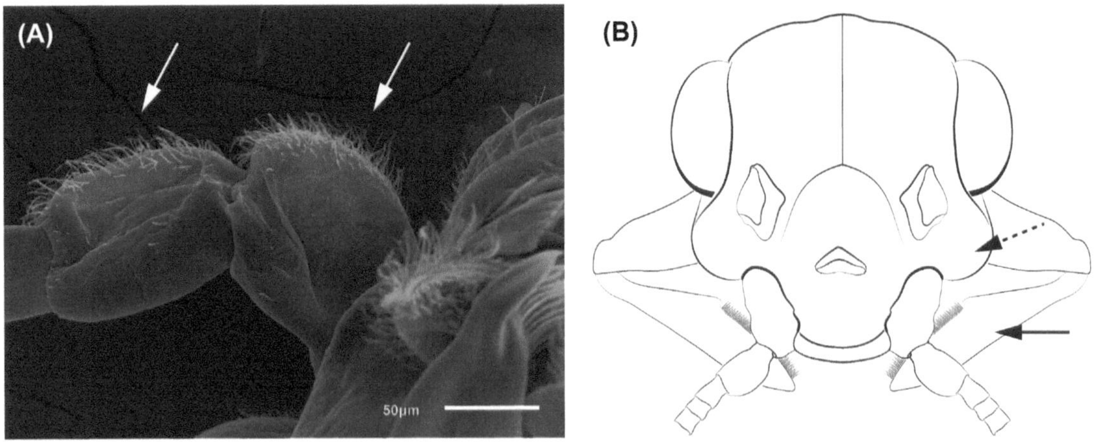

FIGURE 3.109 Head and appendages of *Tapajobaetis labiosus* Boldrini and Salles (Baetidae) nymph: (A) detail of scape and pedicel; and (B) frontal view of head. *White arrows* showing row of setae, *black solid arrow* showing labial palp, *black dashed arrow* showing gena. *After Boldrini et al. (2017).*

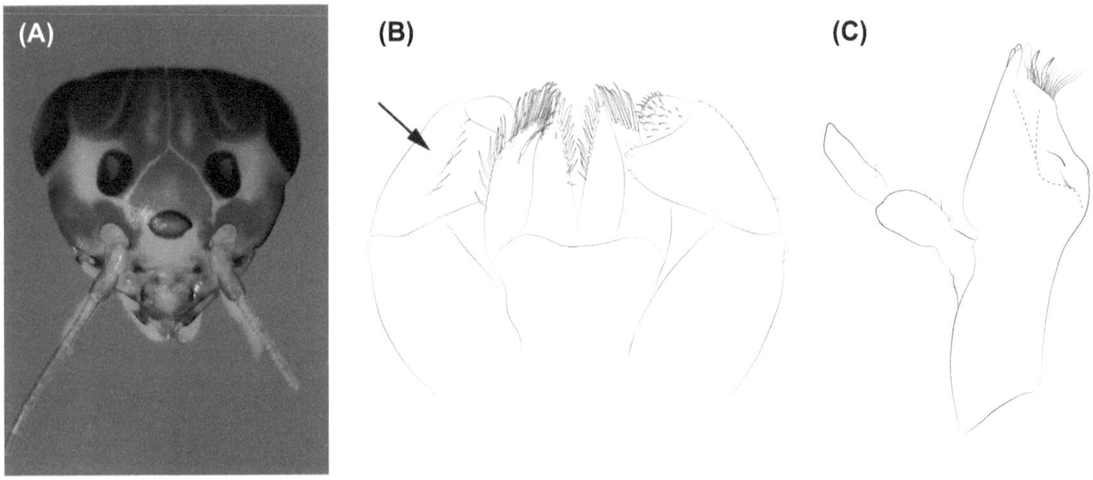

FIGURE 3.110 Head and appendages of *Camelobaetidius* (Baetidae) nymph: (A) frontal view of head; (B) labium; and (C) maxilla. *Black solid arrow* showing row of seta on dorsal surface labial palp segment II.

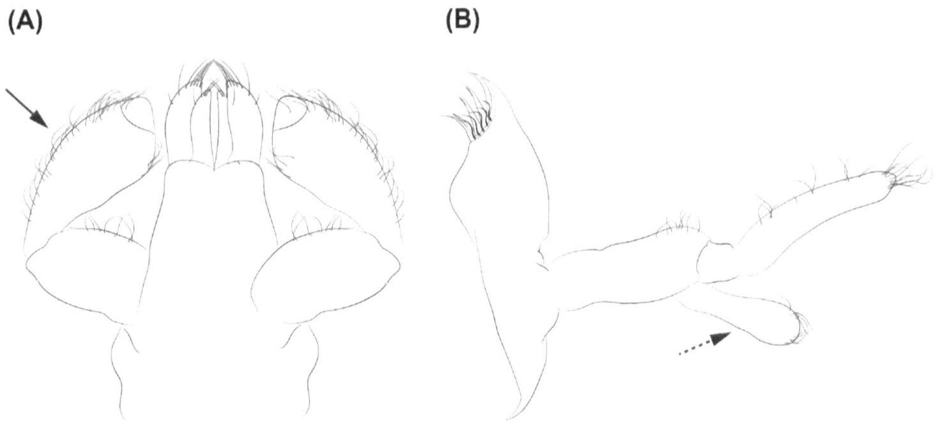

FIGURE 3.111 Selected mouth parts of *Corinella* (Baetidae): (A) labium; and (B) maxilla. *Black solid arrow* showing row of seta on outer margin of labial palp segment II. Black dashed arrow showing long papila.

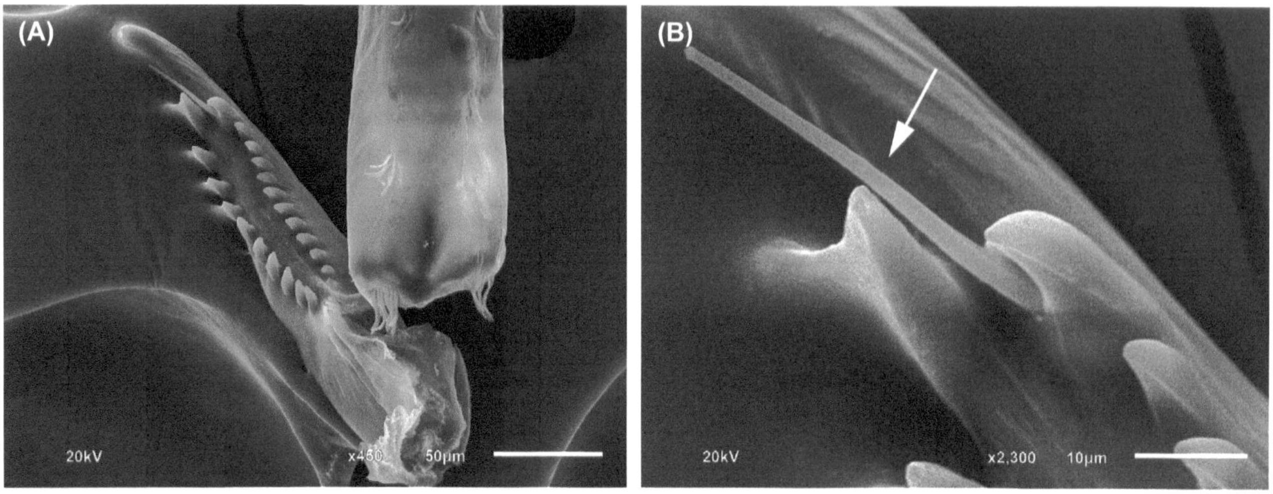

FIGURE 3.112 Tarsal claw of *Spiritiops* (Baetidae), SEM pictures: (A) general view; and (B) detail of apex. *White solid arrow* showing strong subapical seta.

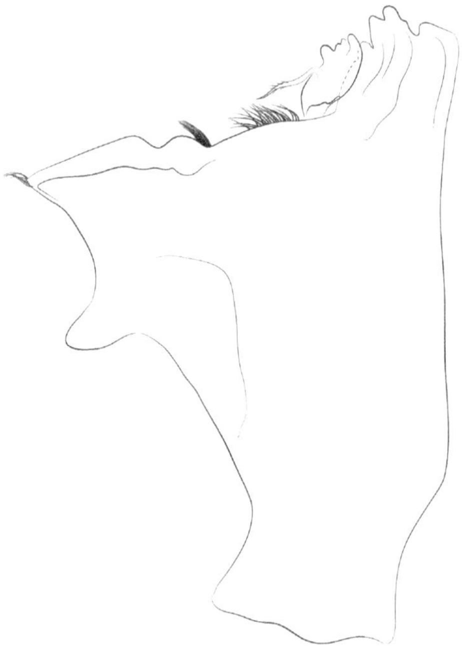

FIGURE 3.113 Right mandible of *Spiritiops* (Baetidae).

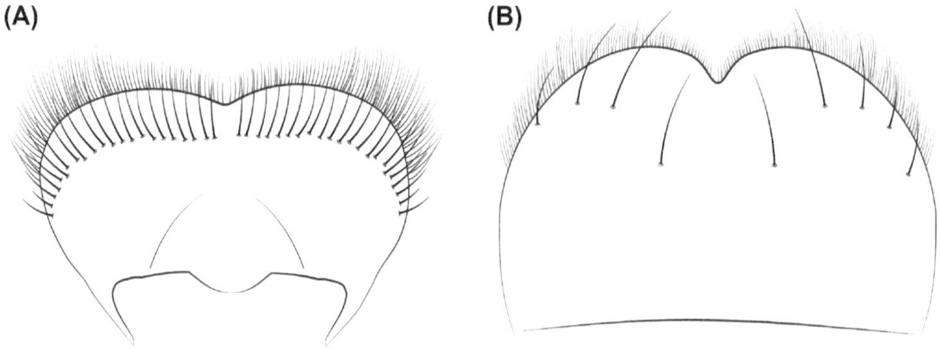

FIGURE 3.114 Labra of Baetidae nymphs: (A) *Moribaetis*; and (B) *Caribaetis*.

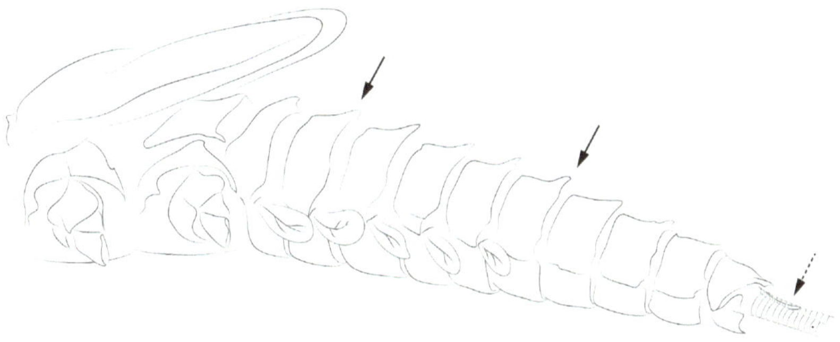

FIGURE 3.115 Lateral view of a nymph of *Baetodes* (Baetidae). *Black arrows* showing dorsal tubercle on abdomen, *black dashed arrow* showing reduced median terminal filament.

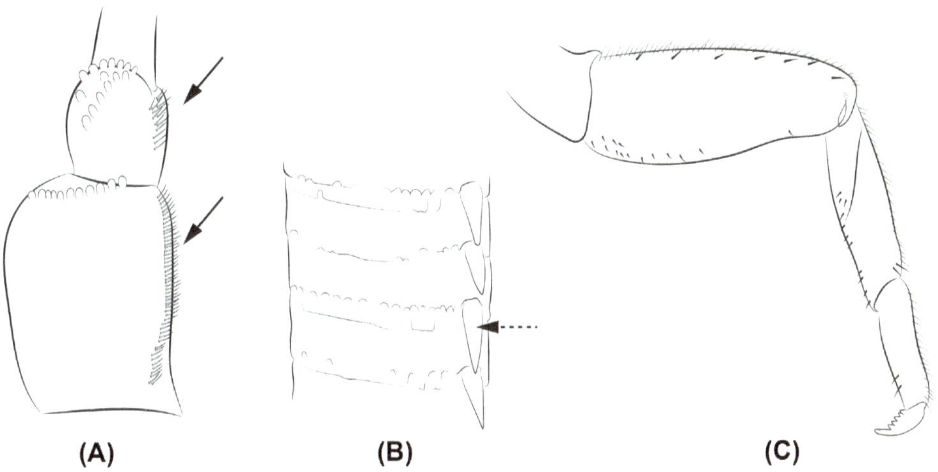

FIGURE 3.116 Nymphal structures of *Mayobaetis* (Baetidae): (A) scape and pedicel; (B) detail of cercus; and (C) fore leg. *Black solid arrows* showing row of setae, *black dashed arrow* showing strong spine.

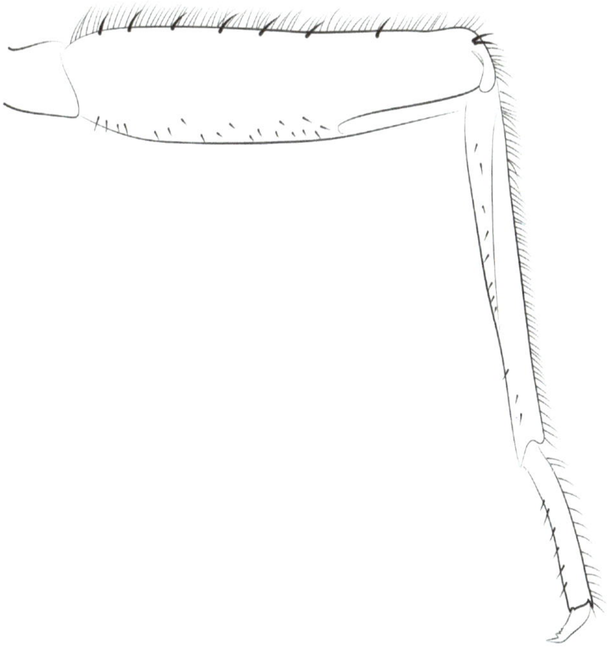

FIGURE 3.117 Fore leg of *Prebatodes* (Baetidae).

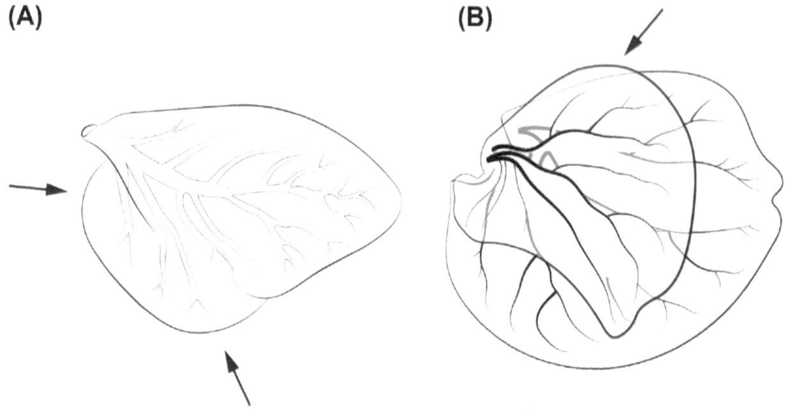

FIGURE 3.118 Gills of Baetidae nymphs: (A) *Callibaetoides caaigua* Cruz, Salles, and Hamada; and (B) *Cloeon smaeleni* Lestage.

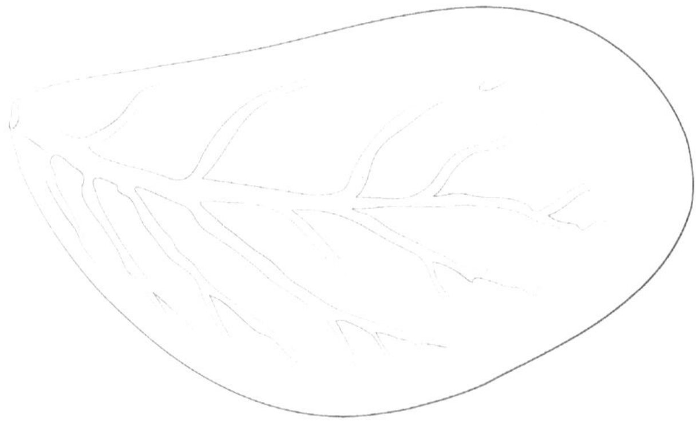

FIGURE 3.119 Gill of *Cloeodes* (Baetidae).

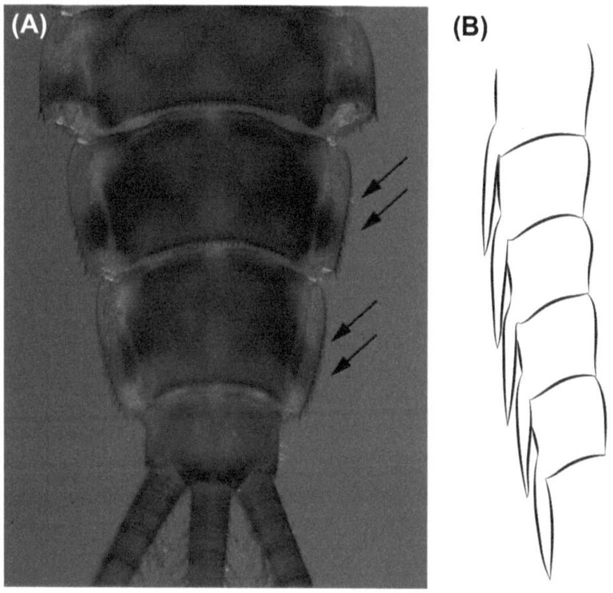

FIGURE 3.120 *Cloeon smaeleni* Lestage (Baetidae), nymphal structures: (A) dorsal view of abdominal segments VII to X; and (B) cercus. *Black solid arrows* showing spines on lateral margin.

45'	Labrum anteromedian emargination with two large denticles (one to three small denticles also present in some species) (Fig. 3.102) .... *Borinquena* [Caribbean]	
46(45)	Labial palp article III distinctly shorter than article II	47
46'	Labial palp article III slightly shorter or longer than article II (Fig. 3.100 B) .... *Nousia* (in part)	
47(46)	Mandible external margin glabrous, except for submedian tuft; gills wide, tapering abruptly on distal $^1/_3$ (Fig. 3.96 A) .... *Simothraulopsis* (in part) [Widespread]	
47'	Mandible external margin with scattered setae on basal half (Fig. 3.103); gills not so wide, gradually tapering toward apex ... *Ecuaphlebia* [Northern South America]	
48(1')	Gills narrow, tapering gradually toward apex (as in Fig. 3.94 A)	49
48'	Gills wide, apical 1/2 fringed with digitform processes (Fig. 3.104) .... *Askola*	
49(48)	Tarsal claws with two denticle rows (Figs. 3.105 A, 106 A); mandibles strongly curved (Figs. 3.105 B, 106 B)	50

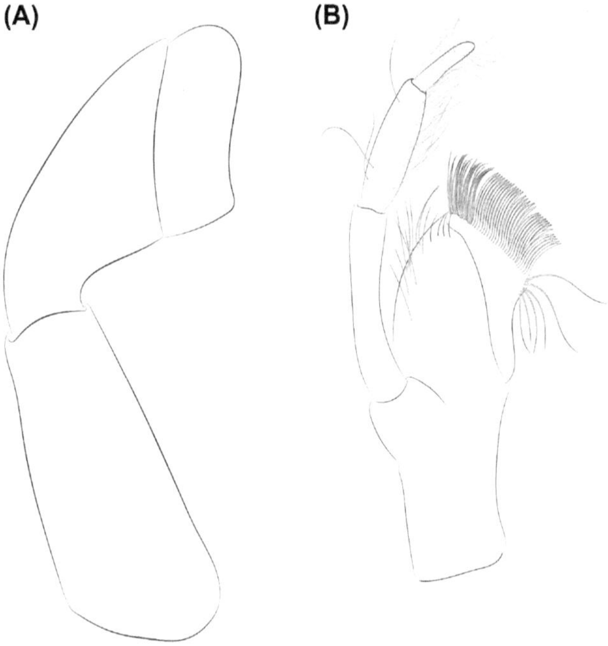

FIGURE 3.121 Mouth parts of *Callibaetoides caaigua* Cruz, Salles, and Hamada nymph (Baetidae): (A) labial palp; and (B) maxilla. *After Cruz et al. (2013).*

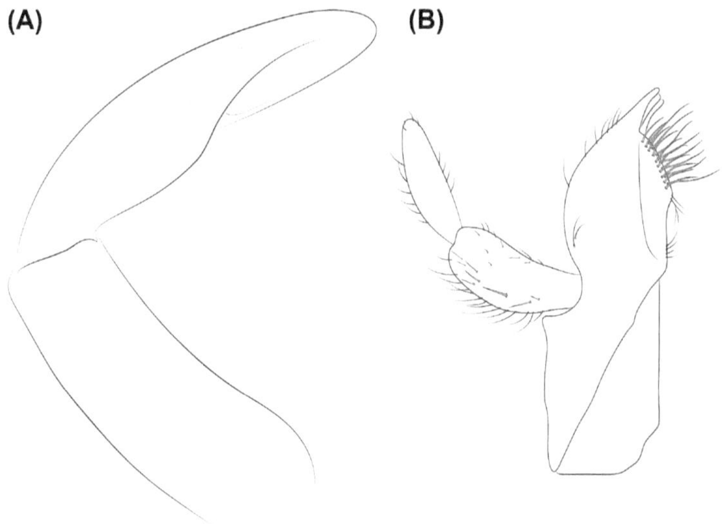

FIGURE 3.122 Mouth parts of *Callibaetis radiatus* Navás nymph (Baetidae): (A) labial palp; and (B) maxilla.

49' Tarsal claws with one denticle row, apical denticle much larger than others (Fig. 3.81B); mandibles smoothly curved as in Fig. 3.90. .... *Hagenulopsis*

50(49) Tarsal claws with three large medial denticles on median area (Fig. 3.105 A); mandible median area with round protuberance (Fig. 3.105 B) .... *Perissophlebiodes*

50' Tarsal claws with one submedial denticle (Fig. 3.106 A); mandible median area without protuberances (Fig. 3.106 B) .... *Bessierus*

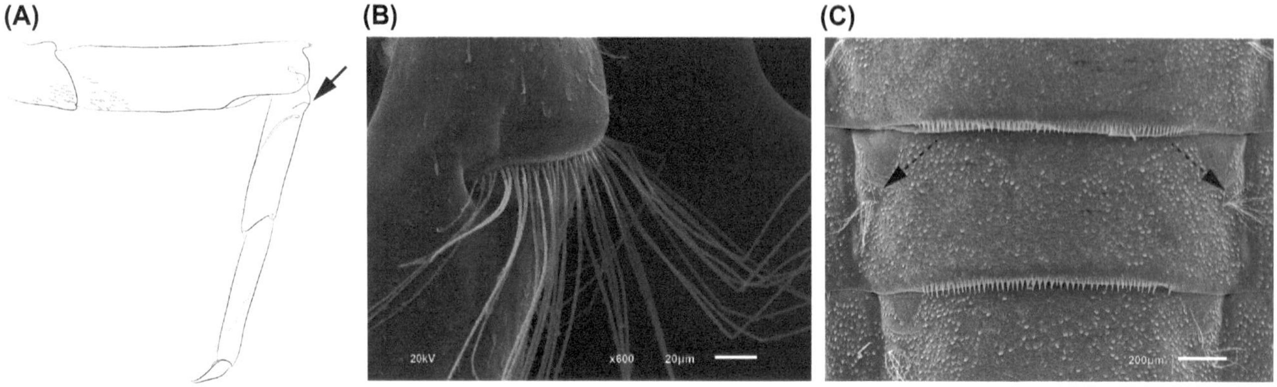

FIGURE 3.123 *Cloeodes* (Baetidae) nymphal structures: (A) fore leg; (B) base of tibia; and (C) abdominal sterna IV to VI. *Black solid arrow* showing subproximal arc of setae, *black dashed arrow* showing row of fine setae.

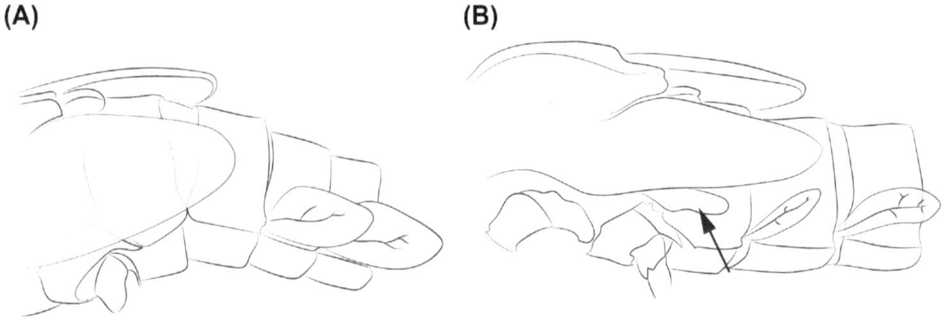

FIGURE 3.124 Partial view of thorax and abdomen of Baetidae nymphs: (A) *Americabaetis*; and (B) *Aturbina*. *Black solid arrow* showing hind wing pad.

FIGURE 3.125 Fore leg of *Apobaetis fiuzai* Salles and Lugo-Ortiz (Baetidae).

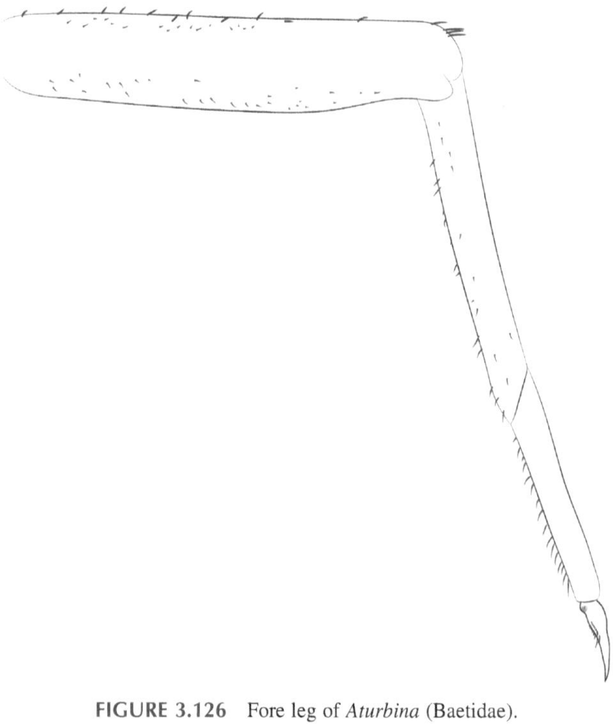

FIGURE 3.126 Fore leg of *Aturbina* (Baetidae).

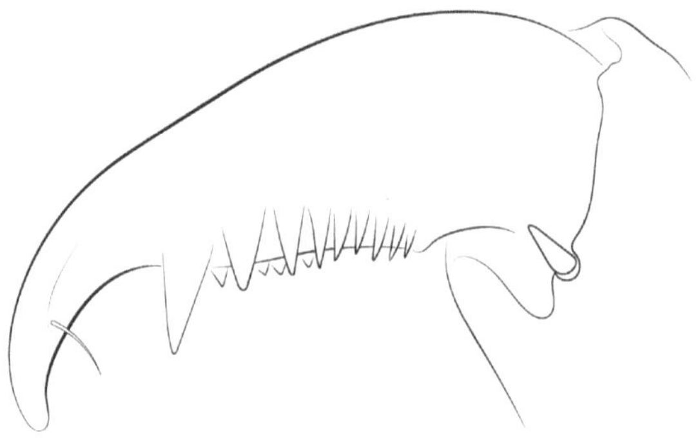

FIGURE 3.127 Tarsal claw of *Andesiops* (Baetidae).

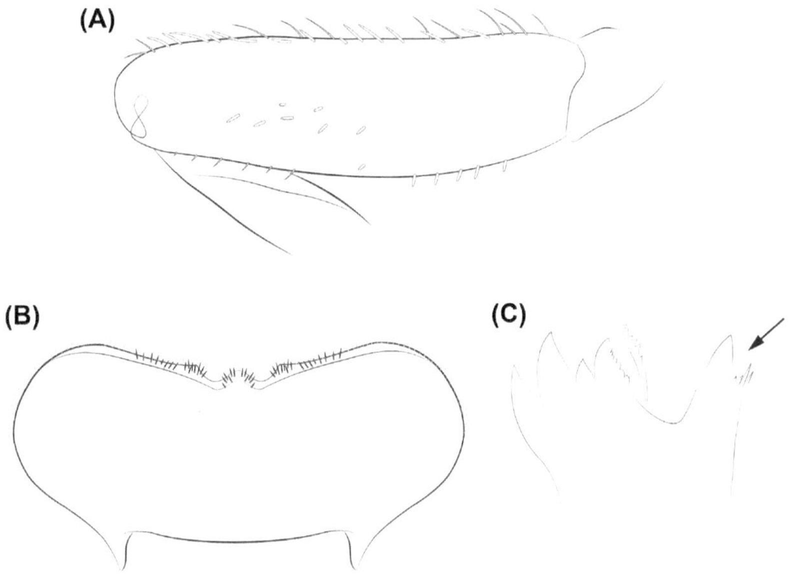

FIGURE 3.128 Nymphal structures of *Harpagobaetis gulosus* Mol (Baetidae): (A) partial view of fore leg; (B) labrum; and (C) distal half of left mandible.

Baetidae: Genera

This family has a thousand species distributed worldwide. The number of genera is also high (over 100), inferior only to Leptophlebiidae. The diversity of Baetidae in the Neotropics is higher than in other realms, with more than 260 species described in 32 genera. Twenty-one genera are endemic to South America, four to Central America and southern North America, while nine have a Panamerican distribution. Except for a few taxa, such as *Baetodes* and some species of *Camelobaetidius*, nymphs have a streamlined body and caudal filaments with well-developed swimming setae. Most genera inhabit lotic environments, but a few live in standing waters (e.g., *Callibaetis* and *Cloeon*).

1 Tarsal claws spatulate (denticles perpendicular to longitudinal axis of legs) (Fig. 3.107) 2

1' Tarsal claws pointed (denticles on the same axis of legs) (Fig. 3.108) 4

2(1) Scape and pedicel without row of setae; labial palpi and gena not as below (Fig. 3.110 A); tarsal claws with no more than 60 denticles (Figs. 3.107 B, C) 3

2' Scape and pedicel with row of setae (Fig. 3.109 A); labial palpi robust and elongated (Fig. 3.109 B); gena protruding (Fig. 3.109 B); tarsal claws with more than 100 denticles (Fig. 3.107 A) .... *Tapajobaetis*

3(2) Labial palpi scattered with short setae (Fig. 3.110 B); labial palpi article II with a dorsal setal row (Fig. 3.110 B); maxillary palp most often without papilla (Fig. 3.110 C) .... *Camelobaetidius*
[Widespread]

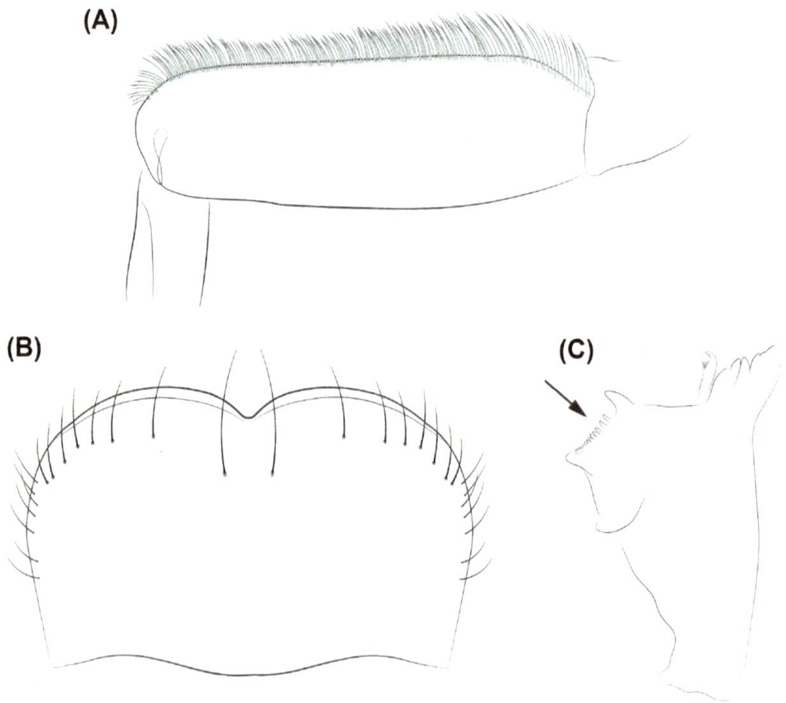

FIGURE 3.129 Nymphal structures of *Andesiops* (Baetidae): (A) partial view of fore leg; (B) labrum; and (C) distal half of left mandible.

FIGURE 3.130 Fore leg of *Varipes helenae* Salles and Batista (Baetidae).

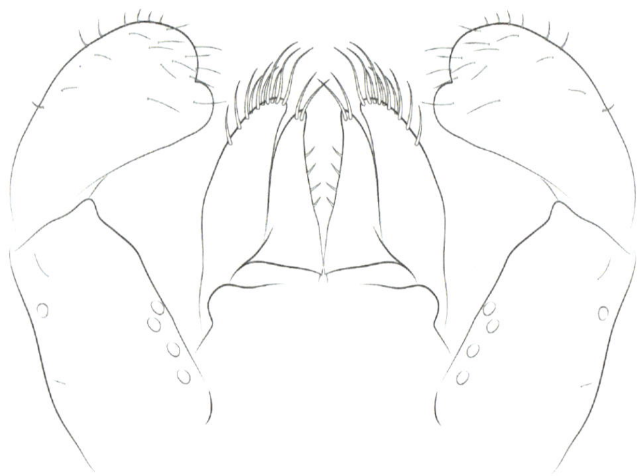

FIGURE 3.131 Labium of *Varipes helenae* Salles and Batista (Baetidae).

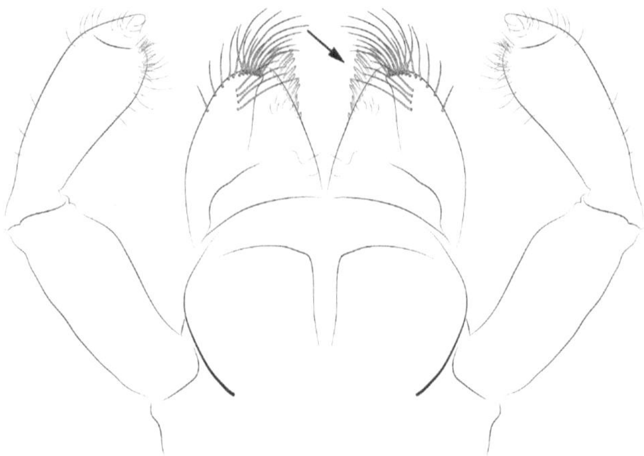

FIGURE 3.132 Labium of *Rivudiva minantenna* Lugo-Ortiz and McCafferty (Baetidae).

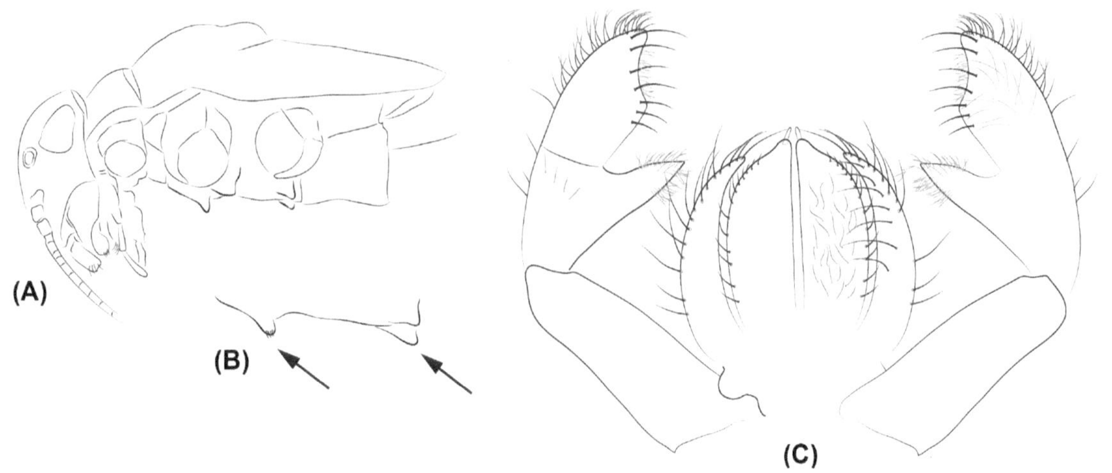

FIGURE 3.133 Nymphal structures of *Waltzoyphius fasciatus* McCafferty and Lugo-Ortiz (Baetidae): (A) head and thorax, lateral view; (B) detail of meso- and meta-sterna; and (C) labium. *Black solid arrow* showing ventral projections.

3' Labial palpi scattered with fine, long setae (Fig. 3.111 A); labial palpi article II without a dorsal setal row (Fig. 3.111 A); maxillary palp always with a long papilla (Fig. 3.111 B); rare ..... *Corinnella*

4(1') Tarsal claws with strong subapical seta (or setae) (Figs. 3.108 B, 112) 5

4' Tarsal claws without strong subapical seta (fine apical seta may be present) (Fig. 3.108 A) 11

5(4) Median terminal filament almost as long as cerci; antennal pedicel without row or tuft of setae 6

5' Median terminal filament reduced (less than ½ length of cerci); antennal pedicel with row or tuft of setae (as in Fig. 3.109 A) 8

6(5) Tarsal claws with one row of denticles (Fig. 3.108 B); right mandible prostheca without medial process (as in Figs. 3.142, 148 A) 7

6' Tarsal claws with two rows of denticles (Fig. 3.112); right mandible prostheca of with plumose medial process (Fig. 3.113).. *Spiritiops*

7(6) Labrum distally expanded, dorsal surface basomedially raised (Fig. 3.114 A) ..... *Moribaetis*

7' Labrum not expanded distally, dorsal surface flat (Fig. 3.114 B) ..... *Caribaetis*

8(5') Gills dorsal, present on segments I or II to VII; abdominal terga without tubercles 9

8' Gills ventral, present on segments I to V (Fig. 3.115); abdominal terga usually with tubercles (Fig. 3.115) ..... *Baetodes*

9(8) Medial terminal filament articulate; gill I present, reduced 10
[Andes]

9' Medial terminal filament uniarticulate; gill I absent ..... *Lugoiops*
[Central America]

10(9) Scape and pedicel dorsoventrally flattened (Fig. 3.116 A); scape length 3x pedicel (Fig. 3.116 A); cerci lateral margin with strong spines (Fig. 3.116 B); legs robust (Fig. 3.116 C) ..... *Mayobaetis*

10' Scape and pedicel cylindrical or subcylindrical, scape length subequal to pedicel; cerci without strong lateral spines; legs slender (Fig. 3.117) ..... *Prebaetodes*

11(4') Gills with recurved flaps (Fig. 3.118) 12

11' Gills simple, without recurved flaps (Fig. 3.119) 14

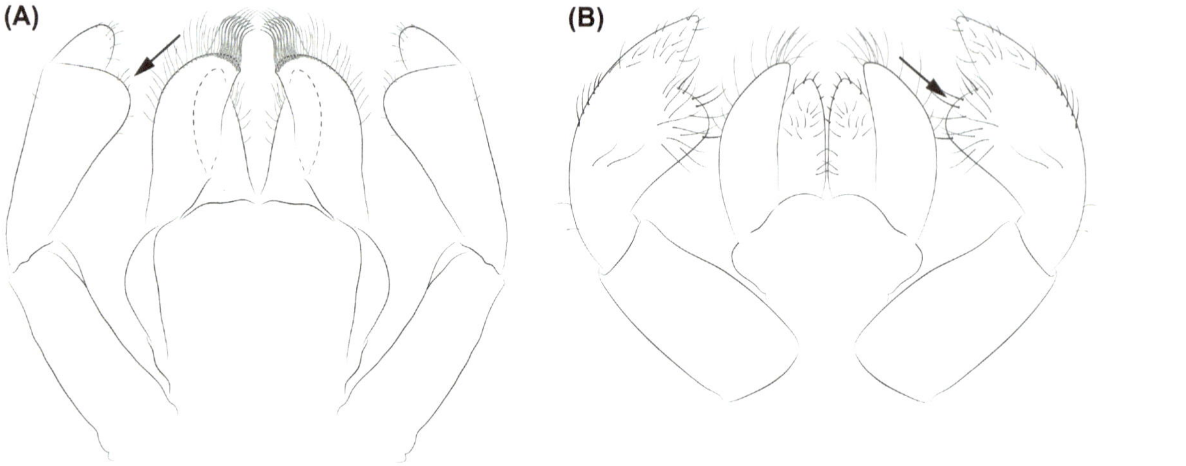

FIGURE 3.134 Labia of Baetidae nymphs: (A) *Tupiara ibirapitanga* Salles, Lugo-Ortiz, Franciscketti, and Da-Silva; and (B) *Paracloeodes atroari* Nieto and Salles.

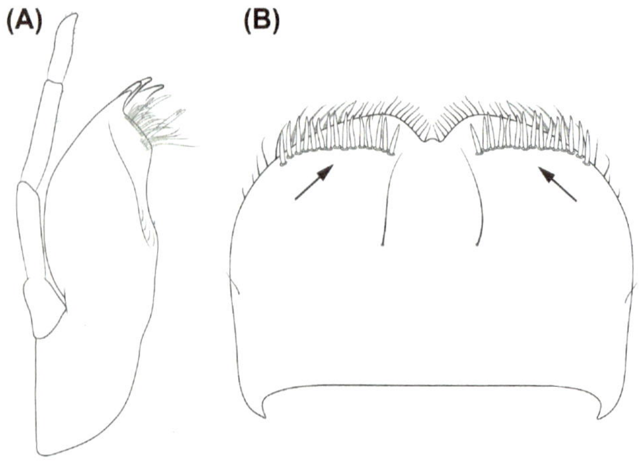

FIGURE 3.135 Mouth parts of *Aturbina* (Baetidae): (A) maxilla; and (B) labrum. *Black solid arrow* showing row of strong setae.

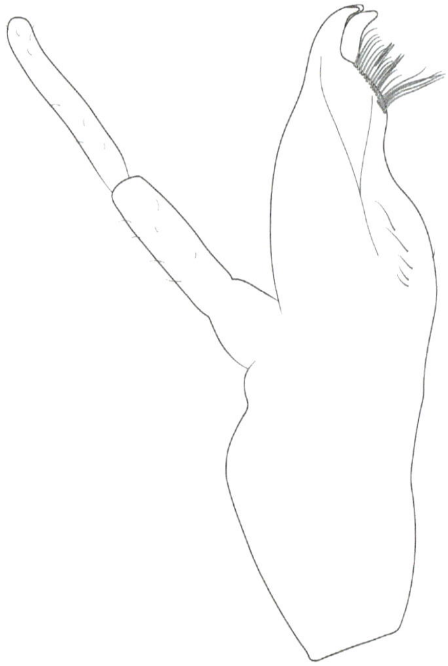

FIGURE 3.136 Maxilla of *Tupiara ibirapitanga* Salles, Lugo-Ortiz, Francischetti and Da-Silva (Baetidae).

12(11)	Abdominal segments lateral margin without spine row; lateral margin of cerci without strong spines on every article	13
12'	Abdominal segments VIII and IX lateral margin with spine row (Fig. 3.120 A); cerci distolateral margin with strong spines on every article (Fig. 3.120 B)	*Cloeon*
13(12')	Labial palp article III rectangular, distal margin almost straight (Fig. 3.121 A); maxillary palp tri-articulate (Fig. 3.121 B); body covered with scale and/or scale bases	*Callibaetoides*
13'	Labial palp article III concave (spoon shaped), distal margin rounded (Fig. 3.122 A); maxillary palp bi-articulate (Fig. 3.122 B); body not covered with scale and/or scale bases	*Callibaetis*
14(11')	Tibiae without subproximal arc of setae; abdominal sterna II to VI without lateral row of fine setae	15
14'	Tibiae with subproximal arc of setae (Figs. 3.123 A, B); abdominal sterna II to VI with lateral row of fine setae (Fig. 3.123 C)	*Cloeodes*
15(14)	Gill I present (Fig. 3.124 B); hind wing pads present or absent (Fig. 3.124 B)	16
15'	Gill I absent (Fig. 3.124 A); hind wing pads always absent (Fig. 3.124 A)	29

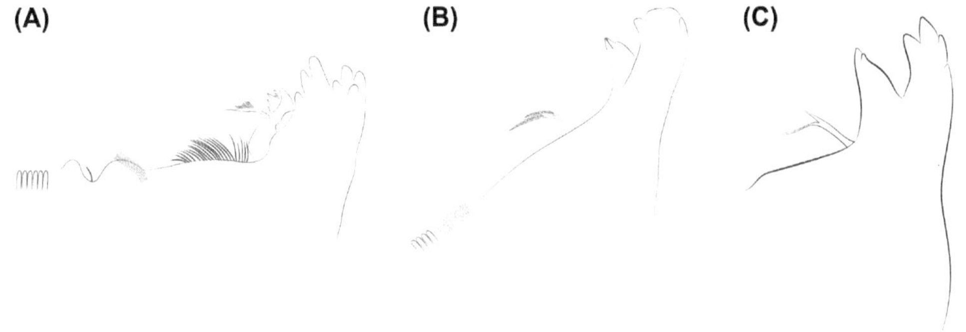

FIGURE 3.137 Distal half of right mandibles of Baetidae nymphs: (A) *Cryptonympha*; (B) *Tupiara ibirapitanga* Salles, Lugo-Ortiz, Francischetti, and Da-Silva; and (C) *Paracloeodes*.

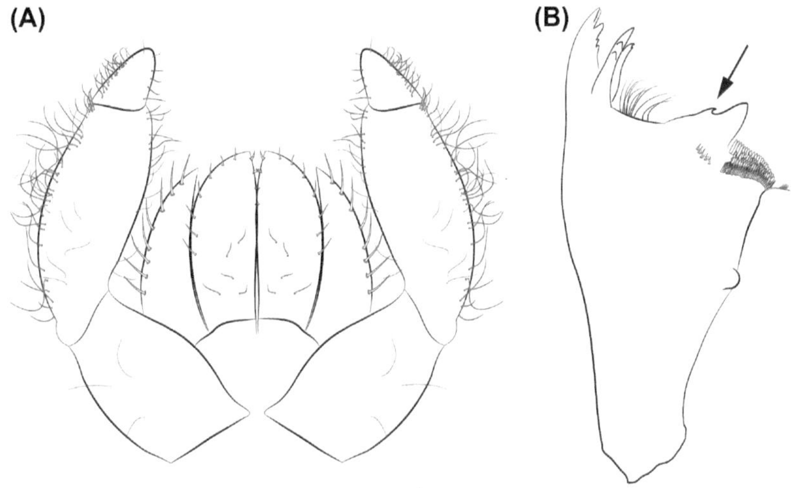

FIGURE 3.138 Mouth parts of *Adebrotus amazonicus* Lugo-Ortiz and McCafferty (Baetidae): (A) labium; and (B) left mandible. *Black solid arrow* showing row of sharp denticle.

16(15) Tarsal claws short, <0.5× the length of tarsi (Fig. 3.126) 17

16' Tarsal claws long, at least 0.75× the length of tarsi (Fig. 3.125)........ *Apobaetis*

17(16) Tarsal claws with two rows of denticles (although minute in some cases) (Figs. 3.108 A, 127)........ 18

17' Tarsal claws with one row of denticles (Fig. 3.108 B)........ 27

18(17) Tarsal claws with two subapical setae (Fig. 3.127)........ 19

18' Tarsal claws without subapical setae 20

19(18) Femora, dorsal margin with a row of robust spiniform setae (Fig. 3.128 A); labrum broad, at least 2× broader than long (Fig. 3.128 B); mandibles with mola adapted for preying (reduced and with pointed denticles) (Fig. 3.128 C)........ *Harpagobaetis*

19' Femora, dorsal margin with a row of long setae (Fig. 3.129 A); labrum 1.5× broader than long (Fig. 3.129 B); mandibles with mola not adapted for preying (with robust denticles) (Fig. 3.129 C)........ *Andesiops*

20(18') Legs anteriorly with conspicuous setae (strong and usually large) (Fig. 3.130)........ 21

20' Legs anteriorly without conspicuous setae (Figs. 3.125, 126)........ 22

21(20) Femora II and III proximal third with a transverse row of long spiniform setae (Fig. 3.130); labial palp article III rounded (Fig. 3.131); glossa apically with simple setae (Fig. 3.131)........ *Varipes*

21' Femora II and III without a transverse row of long spiniform setae; labial palp article III conical (Fig. 3.132); glossa apically with robust setae (Fig. 3.132)........ *Rivudiva*

22(20') Mesosternum, without triangular projection; labial palp article II with or without distomedial projection, but never shaped as below (Fig. 3.134)........ 23

22' Mesosternum, at least, with triangular projection (two projections on meta-sternum are also present in one species) (Figs. 3.133 A, B); labial palp article II with curved distomedial projection (Fig. 3.133 C)........ *Waltzoyphius*

23(22) Maxillary palpi biarticulate (Fig. 3.136); labrum without row of strong setae........ 24

23' Maxillary palp triarticulate (Fig. 3.135 A); labrum with row of strong setae close to distal margin (Fig. 3.135 B)........ *Aturbina*

24(23) Right prostheca slender, never denticulate (Figs. 3.137 B, C); incisors of both mandibles cleft to partially cleft (Figs. 3.137 B, C); right mandible without setae between incisors and mola (Figs. 3.137 B, C)........ 25

FIGURE 3.139 Distal half of left mandible of *Tupiara ibirapitanga* Salles, Lugo-Ortiz, Francischetti, and Da-Silva (Baetidae).

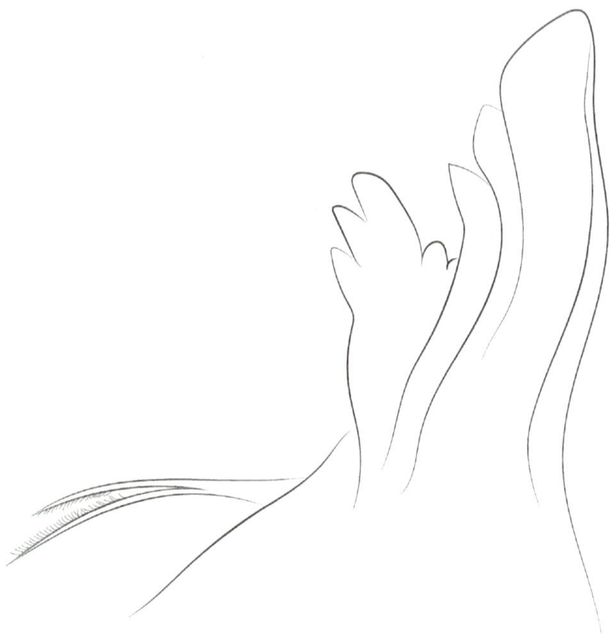

FIGURE 3.141 Distal half of right mandible of *Nanomis* (Baetidae).

FIGURE 3.140 Partial view of fore leg of *Baetis* (Baetidae). *Black solid arrow* showing villopore.

FIGURE 3.142 Distal half of right mandible of *Fallceon* (Baetidae).

24' Right prostheca robust, apically denticulate (Fig. 3.137 A); incisors of both mandibles fused (Fig. 3.137 A); right mandible with setae between incisors and mola (Fig. 3.137 A) .... *Cryptonympha*

25(24) Labial palp with distomedial process (Fig. 3.134); left mandible without a sharp denticle at the base of subtriangular process (Fig. 3.139) 26

25' Labial palp without distomedial process (Fig. 3.138 A); left mandible with a sharp denticle at the base of subtriangular process (Fig. 3.138 B); extremely rare .... *Adebrotus*

26(25) Mandibles with angle between incisors and mola obtuse (Figs. 3.137 B, 139); rare .... *Tupiara* [Atlantic Forest]

26' Mandibles with angle between incisors and mola around 90° (Fig. 3.137 C); common .... *Paracloeodes* [Widespread]

27(17') Villopore absent 28

27' Villopore (patch of small setae closely pressed together) present at base of femora (Fig. 3.140) .... *Baetis* [Central America. Nearctic]

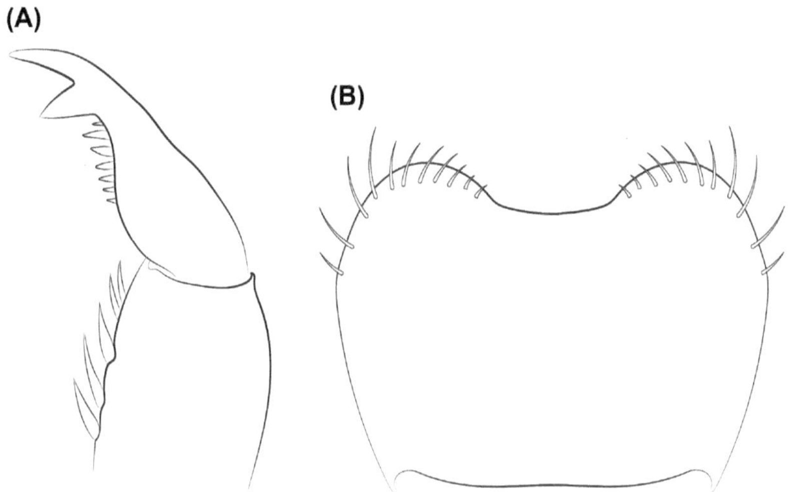

FIGURE 3.143 Nymphal structures of *Guajirolus rondoni* Salles (Baetidae): (A) fore tarsal claw; and (B) labrum.

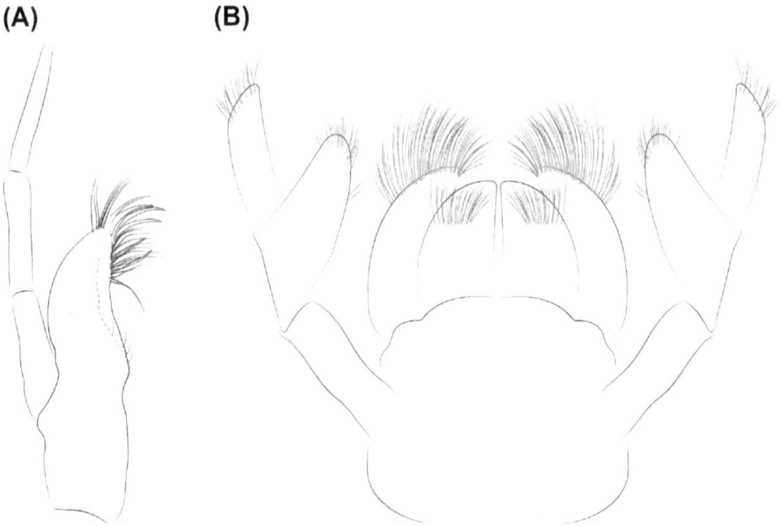

FIGURE 3.144 Mouth parts of *Guajirolus rondoni* Salles (Baetidae): (A) maxilla; and (B) labium.

28(27) Right prostheca slender, apically bifid (Fig. 3.141); right mandible without setae between incisors and mola (Fig. 3.141)......... *Nanomis* [Andes]

28' Right prostheca robust, apically denticulate (Fig. 3.142); right mandible with setae between incisors and mola (Fig. 3.142)..... *Fallceon* [Northern South America, Central America. Nearctic]

29(15') Tarsal claw with distal denticle much larger than others (Fig. 3.143 A); labrum distal margin with broad emargination on distal margin (Fig. 3.143 B); labial palp distal article elongate (at least 3× as long as broad) (Figs. 3.144 B, 145 B).. 30

29' Tarsal claw with distal denticle slightly larger or subequal to others; labrum distal margin with shallow or narrow emargination; labial pal distal article short (as long as broad).. 31

30(29) Maxillary palp article III slightly longer than articles I and II (Fig. 3.144 A); labial palp article II with broad medial projection (Fig. 3.144 B) .... *Guajirolus*

30' Maxillary palp article III more than 3× longer than articles I and II combined (Fig. 3.145 A); labial palp article II with long medial projection (Fig. 3.145 B).. *Chane*

31(29') Labrum semicircular (Fig. 3.147 A); paraglossa apex as broad as base (Figs. 3.147 B, 148 B, 149 B).. 32

31' Labrum trapezoidal (Fig. 3.146 A); glossa constricted in mid region (Fig. 3.146 B); paraglossa distally expanded (Fig. 3.146 B) .... *Tomedontus* [Amazon]

ORDER EPHEMEROPTERA

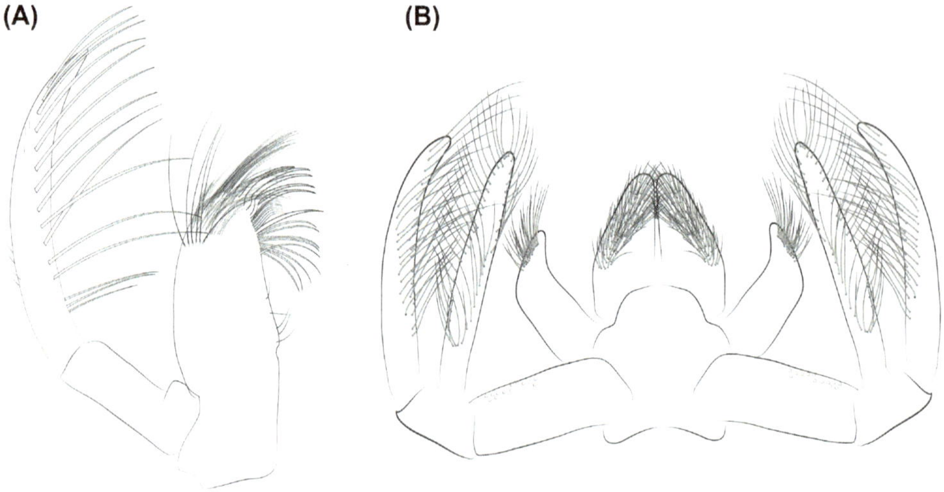

FIGURE 3.145 Mouth parts of *Chane baure* Nieto (Baetidae): (A) maxilla; and (B) labium.

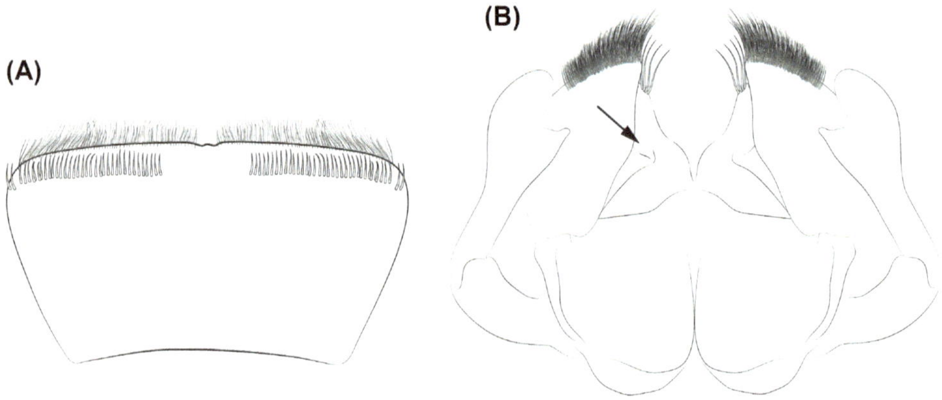

FIGURE 3.146 Mouth parts of *Tomedontus primus* Lugo-Ortiz and McCafferty (Baetidae): (A) labrum; and (B) labium.

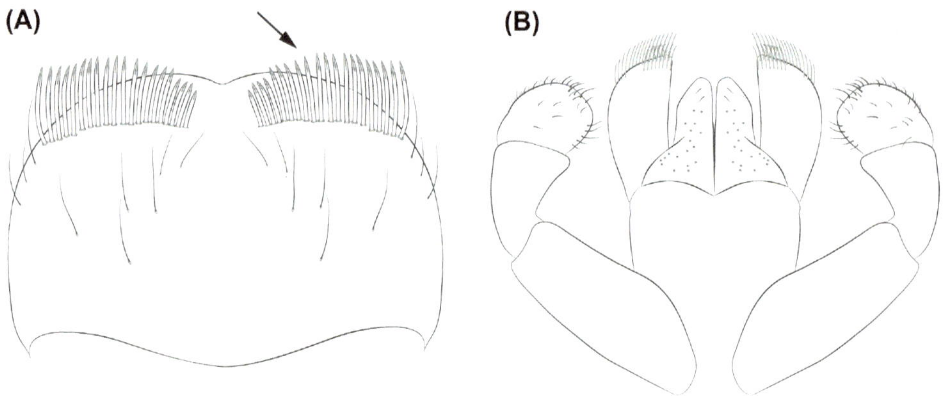

FIGURE 3.147 Mouth parts of *Zelusia principalis* Lugo-Ortiz and McCafferty (Baetidae): (A) labrum; and (B) labium.

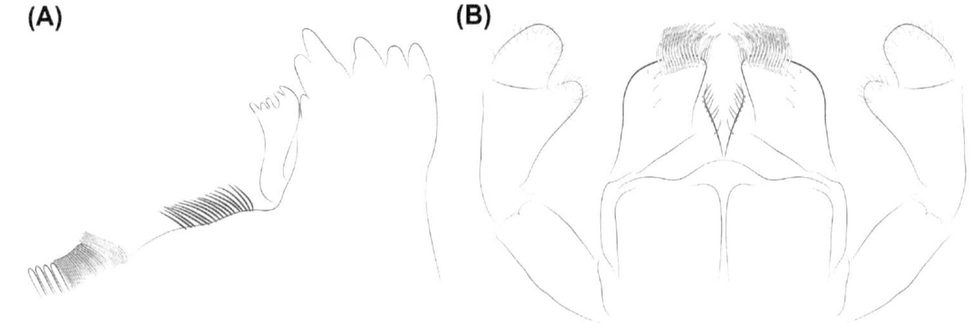

FIGURE 3.148 Mouth parts of *Americabaetis* (Baetidae): (A) distal half of right mandible; and (B) labium.

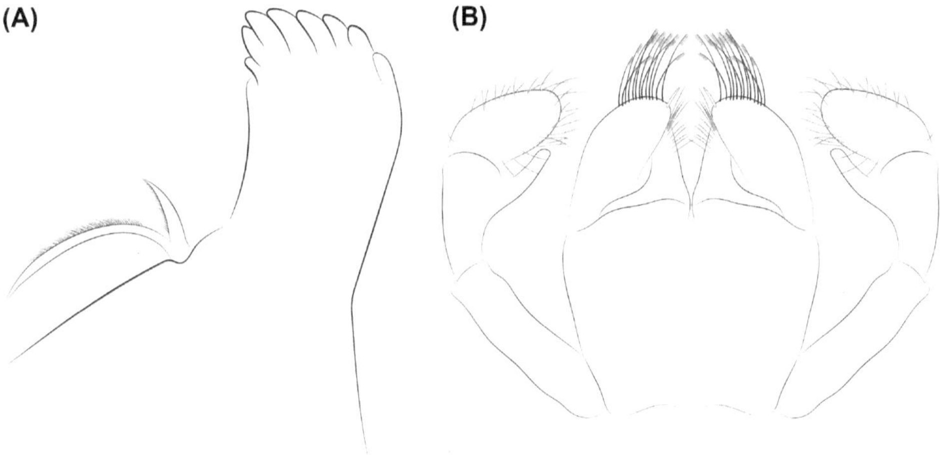

FIGURE 3.149 Mouth parts of *Parakari* (Baetidae): (A) distal half of right mandible; and (B) labium.

32(31) Labrum without distal row of strong setae (subdistal simple setae present); labial palp article II with distomedial projection (Figs. 3.148 B, 149 B) 33

32' Labrum with distal row of strong setae (Fig. 3.147 A); labial palp article II without distomedial projection (Fig. 3.147 B)........ *Zelusia*

33(32) Right prostheca robust, apically denticulate (Fig. 3.148 A); mandibles with setae between incisors and mola (Fig. 3.148 A); common .... *Americabetis* [Widespread]

32' Right prostheca slender, apically bifid (Fig. 3.149 A); mandibles without setae between incisors and mola (Fig. 3.149 A); rare .... *Parakari* [northern South America]

REFERENCES

Berner, L. & M.L. Pescador. 1988. The Mayflies of Florida. University Press of Florida.

Blanco-Belmonte, V., Ruaise, V. & J.G. Peters. 2003. The nymph of Paramaka Savage & Domínguez (Ephemeroptera: Leptophlebiidae: Atalophlebiinae). Pages 117–122 in: E. Gaino (ed.): Research update on Ephemeroptera and Plecoptera. University of Perugia, Perugia.

Cardoso, M.N., Y. Shimano, J.C. Nabout & J. Leandro. 2015. An estimate of the potential number of mayfly species (Ephemeroptera, Insecta) still to be described in Brazil. Revista Brasileira de Entomologia 59: 147–153.

Cruz, P.V., F.F. Salles & N. Hamada. 2013. A new genus and species of Baetidae (Insecta: Ephemerotera) from Brazil. Annales de Limnologie 49: 1–12.

Da-Silva, E.R. & F.F. Salles. 2012. Ephemeroptera Hyatt & Arms. Pages 231–244 *in*: J.A. Rafael, G.A.R. Melo, C.J.B. Carvalho, S.A. Casari & R. Constantino (eds.), Insetos do Brasil: Taxonomia e Diversidade. Holos, São Paulo.

Dias, L.G., F.F. Salles & C. Molineri. 2005. Macunahyphes: a New Genus for *Tricorythodes australis* (Ephemeroptera: Leptohyphidae). Annales de Limnologia 41: 195–201.

Domínguez, E. 1987. El género Thraulodes (Ephemeroptera: Leptophlebiidae) en la República Argentina. Acta Zoológica Lilloana 39: 47–65.

Domínguez, E. 1988. Ecuaphlebia: A new genus of Atalophlebiinae (Ephemeroptera: Leptophlebiidae) from Ecuador. Aquatic Insects 10: 227–235.

Domínguez, E. 1991. The status of the genus Ulmeritus (Ephemeroptera: Leptophlebiidae: Atalophlebiinae) and related taxa. Pages 157–168 in. J. Alba-Tercedor & A. Sanchez-Ortega (ed.), Overview and strategies of Ephemeroptera and Plecoptera. Sandhill Crane Press, Gainesville.

Domínguez, E. & H.M. Savage. 1987. Two new species of Farrodes from continental South America with comments on the distribution of faunal components in Argentina (Ephemeroptera: Leptophlebiidae). Studies on Neotropical Fauna and Environment 22: 43–52.

Domínguez, E. & R.W. Flowers. 1989. A revision of Hermanella and related genera (Ephemeroptera: Leptophlebiidae: Atalophlebiinae) from subtropical South America. Annals of the Entomological Society of America 82: 555–573.

Domínguez, E., Peters, W.L., Peters, J.G. & H.M. Savage. 1997. The imago of Simothraulopsis Demoulin with a redescription of the nymph (Ephemeroptera: Leptophlebiidae: Atalophlebiinae). Aquatic Insects 19: 141–150.

Domínguez, E., Ferreira, M.J. & C. Nieto. 2001. Redescription and phylogenetic relationships of Leentvaaria Demoulin (Ephemeroptera: Leptophlebiidae). Pages 313–320 in: E. Domínguez (ed.), Trends in research in Ephemeroptera and Plecoptera. Kluwer Academic/ Plenum, New York.

Domínguez, E., C. Molineri & C. Nieto. 2009. Capítulo 3: Ephemeroptera. Pages 55–93 *in*: E. Domínguez & H.R. Fernández (eds.), Macroinvertebrados bentónicos sudamericanos: sistemática y biología. Fundación Miguel Lillo, Tucumán.

Domínguez, E., M.D. Hubbard & W.L. Peters. 1992. Clave para Ninfas y Adultos de las Familias y Géneros de Ephemeroptera (Insecta) Sudamericanos. Biologia Acuatica No 16, La Plata, Argentina.

Domínguez, E., M.D. Hubbard & W.L. Peters. 1995. Capítulo "Ephemeroptera". Pages 1069–1089 *in*: E.C. Lopretto & G. Tell (eds.), Ecosistemas de Aguas Continentales, Metodologias para su estudio.

Domínguez, E., M.D. Hubbard, M.L. Pescador & C. Molineri. 2001. Capítulo "Ephemeroptera". Pages 17–53 *in*: H.R. Fernández & E. Domínguez (eds.), Guía para la determinación de los Artrópodos Sudamericanos. Serie Investigaciones.

Domínguez, E., C. Molineri, M.L. Pescador, M.D. Hubbard & C. Nieto. 2006. Ephemeroptera of South America. Pensoft Press, Sofia & Moscow.

Domínguez, E. & D.A. Dos Santos. 2014. Co-authorship networks (and other contextual factors) behind the growth of taxonomy of South American Ephemeroptera: A scientometric approach. Zootaxa 3754: 59–85.

Edmunds, G.F. Jr., S.L. Jensen & L. Berner. 1976. The Mayflies of North and Central America. University of Minnesota Press, Minneapolis.

Flowers, R.W. 1987. New species and life stages of Atopophlebia (Ephemeroptera: Leptophlebiidae: Atalophlebiinae). Aquatic Insects 9: 203–209.

Flowers, R.W. & E. Domínguez. 1992. New genus of Leptophlebiidae (Ephemeroptera) from Central and South America. Annals of the Entomological Society of America 85: 655–661.

Flowers, R.W. & C. De la Rosa. 2010. Ephemeroptera. Revista de Biologia Tropical 58: 63–93.

Gonçalves, I., D.M. Takiya, F.F. Salles, Peters, J.G. & J.L. Nessimian. 2017. Integrative taxonomic revision of *Campylocia* (mayflies: Ephemeroptera, Euthyplociidae). Systematics and Biodiversity 15: 564–581.

Hilsenhoff, W.L. 2001. Diversity and Classification of Insects and Collembola. Pages 661–731 *in*: J.H. Thorp & A.P. Covich (eds), Ecology and Classification of North American Freshwater Invertebrates. Academic press.

Lugo-Ortiz, C.R. & W.P. McCafferty. 1996. Phylogeny and classification of the Baetodes complex (Ephemeroptera: Baetidae), with description of a new species. Journal of the North American Benthological Society 15: 367–380.

Kluge, N. 1994. A revision of Leptophlebiidae of Cuba (Insecta, Ephemeroptera). Zoosystematica Rossica (1993) 2: 247–285.

Kluge, N. 2004. The Phylogenetic System of Ephemeroptera. Kluwer Academic Publishers, Dordrecht.

Malzacher, P. 1986. Caenidae aus dem Amazonasgebiet (Insecta, Ephemeroptera). Spixiana 9: 83–103.

McCafferty, P. & L.M. Jacobus. 2016. Mayfly central. Available at: https://www.entm.purdue.edu/mayfly/about.php.

McCafferty, W.P. 1998. Ephemeroptera and the Great American Interchange. Journal of the North American Benthological Society 17: 1–20.

McCafferty, W.P. & G.F. Edmunds. 1979. The higher classification of the Ephemeroptera and its evolutionary basis. Annals of the Entomological Society of America 72: 5–12.

McCafferty, W.P. & T.Q. Wang. 2000. Phylogenetic systematics of the major lineages of pannote mayflies (Ephemeroptera: Pannota). Transactions of the American Entomological Society 126: 9–101.

McCafferty, W.P. 1991. Toward a phylogenetic classification of the Ephemeroptera (Insecta): a commentary on systematics. Annals of the Entomological Society of America 84: 343–360.

Mercado, M., M. Maritza & E. Simon. 2004. Taxonomic revision of the genus *Metamonius* Eaton (Nesameletidae: Ephemeroptera), with notes on its biology and distribution. Studies on Neotropical fauna and environment 39: 149–157.

Miserendino, M.L. 1996. Primera descripción de la ninfa de Rhigotopus (Ephemeroptera: Leptophlebiidae) con datos sobre su biología. Revista de la Sociedad Entomológica Argentina 55: 21–24.

Molineri, C. & D. Emmerich. 2010. New species and new stage descriptions of *Campsurus major* species group (Polymitarcyidae: Campsurinae), with first report of silk-case construction in mayfly nymphs. Aquatic insects 32: 265–280.

Molineri, C. & E. Domínguez. 2003. Nymph and egg of *Melanemerella brasiliana* (Ephemeroptera: Ephemerelloidea: Melanemerellidae), with comments on its systematic position and the higher classification of Ephemerelloidea. Journal of the North American Benthological Society 22: 263–275.

Molineri, C. 2010. A cladistic revision of *Tortopus* Needham & Murphy with description of the new genus *Tortopus* (Ephemeroptera: Polymitarcyidae). Zootaxa 2481: 1–36.

Molineri, C., F.F. Salles & J.G. Peters. 2015. Phylogeny and biogeography of Asthenopodinae with a revision of *Asthenopus* reinstatement of *Asthenopodes* and the description of the new genera Hubbardipes and *Priasthenopus* (Ephemeroptera, Polymitarcidae). ZooKeys 478: 45–128.

Molineri, C. & R. Mariano. 2015. Description of *Loricyphes froehlichi*, a new genus and species of Leptohyphidae (Ephemeroptera) from São Paulo, Brazil. Annales de Limnologie 51: 323–328.

Ogden, T.H., J.L. Gattolliat, M. Sartori, A.H. Staniczek, T. Soldán & M.F. Whiting. 2009. Towards a new paradigm in mayfly phylogeny (Ephemeroptera): combined analysis of morphological and molecular data. Systematic entomology 34: 616–634.

Ogden, T.H. & M.F. Whiting. 2005. Phylogeny of Ephemeroptera (mayflies) based on molecular evidence. Molecular phylogenetics and evolution 37: 625–643.

Pescador, M.L. & T. Gonser. 2001. The nymph of Secochela illiesi (Ephemeroptera: Leptophlebiidae: Atalophlebiinae) from South America. Pages 347–352 in: E. Domínguez (ed.), Trends in research in Ephemeroptera and Plecoptera. Kluwer Academic/Plenum, New York.

Pescador, M.L. & W.L. Peters. 1980. Two new genera of cool-adapted Leptophlebiidae (Ephemeroptera) from southern South America. Annals of the Entomological Society of America 73: 332–338.

Pescador, M.L. & W.L. Peters. 1985. Biosystematics of the genus Nousia from southern South America (Ephemeroptera: Leptophlebiidae: Atalophlebiinae). Journal of the Kansas Entomological Society 58: 91–123.

Pescador, M.L. & W.L. Peters. 1987. Revision of the genera Meridialaris and Massartellopsis (Ephemeroptera: Leptophlebiidae: Atalophlebiinae) from South America. Transactions of the American Entomological Society 112: 147–189.

Pescador, M.L. & W.L. Peters. 1990. Biosystematics of the genus Massartella Lestage (Ephemeroptera: Leptophlebiidae: Atalophlebiinae) from South America. Aquatic Insects 12: 145–160.

Peters, W.L. 1969. Askola froehlichi a new genus and species from southern Brazil (Leptophlebiidae: Ephemeroptera). Florida Entomologist 52: 253–258.

Peters, W.L. & G.F. Edmunds Jr. 1972. A revision of the generic classification of certain Leptophlebiidae from southern South America (Ephemeroptera). Annals of the Entomological Society of America 65: 1398–1414.

Peters, W.L. & A.D. Harrison. 1974. Redescription of Terpides Demoulin from St. Vincent, West Indies (Ephemeroptera: Leptophlebiidae). Proceedings of the Entomological Society of Washington 76: 178–185.

Peters, J.G. & T. Soldan. 2005. Book review: Heckman C.H.W.: Encyclopedia of South American Aquatic Insects: Ephemeroptera. European Journal of Entomology 102: 791–791.

Peters, W.L. 1971. A revision of the Leptophlebiidae of the West Indies (Ephemeroptera). Smithsonian Contributions to Zoology 62: 1–48.

Savage, H.M. 1982. A curious new genus and species of Atalophlebiinae (Ephemeroptera: Leptophlebiidae) from the southern coastal mountains of Brazil. Studies on Neotropical Fauna and Environment 17: 209–217.

Savage, H.M. 1986. Systematics of the Terpides lineage from the Neotropics: Definition of the Terpides lineage, methods, and revision of Fittkaulus Savage & Peters. Spixiana 9: 255–270.

Savage, H.M. & W.L. Peters. 1983. Systematics of Miroculis and related genera from northern South America. Transactions of the American Entomological Society 108: 491–600.

Savage, H.M., R.W. Flowers. & W. Porras. 2005. Rediscovery of Choroterpes atramentum in Costa Rica, type species of Tikuna new genus (Ephemeroptera: Leptophlebiidae: Atalophlebiinae), and its role in the "Great American Interchange". Zootaxa 932: 1–14.

Sartori, M. & J.E. Brittain. 2015. Order Ephemeroptera. Pages 873–891 *in*: J.H. Thorp & D.C. Rogers (eds.), Thorp and Covich's Freshwater Invertebrates. Academic Press, Boston.

Sun, L. & W.P. McCafferty. 2008. Cladistics, classification and identification of the brachycercine mayflies (Insecta: Ephemeroptera: Caenidae). Zootaxa 1801: 1–239.

Thomas, A., K. Orth, V. Horeau & Y. Dominique. 2001. Les Ephémères de la Guyane Française. 3. *Bessierus doloris* n. gen. n. sp. [Ephemeroptera, Leptophlebiidae]. Ephemera (2000) 2: 49–57.

Thomas, A., Y. Dominique & K. Orth. 2005. Les Ephémères de la Guayne Française. 10. Description de *Lisetta ernsti* n. gen., n. sp., [Ephemeroptera, Leptophlebiidae]. Ephemera (2003) 5: 77–84.

Wang, T-Q. & W.P. McCafferty. 1996. New diagnostic characters for the mayfly family Baetidae (Ephemeroptera). Entomological News 107: 207–212.

Webb, J.M. & W.P. McCafferty. 2008. Heptageniidae of the world. Part II. Key to the genera. Canadian Journal of Arthropod Identification 7: 1–55.

Chapter 4

Order Plecoptera

Fernanda Avelino-Capistrano
Faculdades São José, and Laboratório de Entomologia, UFRJ, Rio de Janeiro, Rio de Janeiro, Brazil

Pablo Pessacq
Centro de Investigaciones Esquel de Montaña y Estepa Patagónicas (CIEMEP), Esquel, Chubut, Argentina

Leandro Silva Barbosa
Departamento de Entomologia, Museu Nacional, UFRJ, Rio de Janeiro, Brazil

Chapter Outline

INTRODUCTION

Plecoptera is a small neopteran order that probably emerged in the Upper Carboniferous (Grimaldi & Engel, 2005; Bethoux et al., 2011). The order is distributed across all continents except Antarctica and has about 4,000 species in 16 families (Fochetti & Tierno-de-Figueroa, 2008). In the Neotropical region, there are 542 described species (DeWalt et al., 2017) distributed in 6 families and 48 endemic genera (except for *Anacroneuria* which is also reported in the Nearctic Region; Stark et al., 2009; Vera, 2009; McCulloch et al., 2016). Adults are weak flyers and are found in riparian vegetation, typically at high elevations (up to 3,800–4,200 m) (Roback, 1980). However, they also occur at low elevations <100 m (e.g., *Anacroneuria* sp., Avelino-Capistrano, 2014) and even at sea level (*Klapopteryx kuscheli* Illies—Pessacq, pers. obs.). They can be easily found on substrates such as rocks, bryophytes, and litter in low-orders rivers with clean and flowing water. Most Plecoptera nymphs are very sensitive to environmental pollution (exceptions include *Anacroneuria*; Avelino-Capistrano, 2014) and are often used in freshwater monitoring programs (Callisto et al., 2001; Buss et al., 2002). However, much of those studies are based on family-level analyses (e.g., Junqueira & Campos, 1998; Miserendino & Pizzolón, 1999), in part because of the high diversity and lack of adequate keys to identify stoneflies at lower taxonomic levels. Some authors have associated nymphs with adults by rearing methods (e.g., Ribeiro, 2004; Ribeiro & Raphael, 2005; Pessacq & Omad, 2012), although in recent years, molecular tools have been successfully used for this purpose (e.g., Mynott et al., 2011; Tierno-de-Figueroa et al., 2011; Avelino-Capistrano et al., 2014, 2015). Keys to Neotropical nymphs have been developed by many authors (e.g., Benedetto, 1974; Romero, 2001), although the keys of Stark et al. (2009) are the most comprehensive. This chapter is based in part on this last work, with the addition of photographs, drawings, and the recently described genus *Pehuenioperla* (Vera, 2009).

Despite the improvement of breeding techniques, the number of nymphs described is still scarce compared to the

Thorp and Covich's Freshwater Invertebrates, Fourth Edition. https://doi.org/10.1016/B978-0-12-804223-6.00004-4

number of adult descriptions. In the Neotropical region, there are 542 described species, although only 100 nymphs are known (Table 4.1) (DeWalt et al., 2017). Only for some of the Patagonian taxa (Austroperlidae, Eustheniidae, Gripopterygidae, and Perlidae) the immature stage knowledge is more equated to the adult knowledge. This taxonomic gap in the group affects biodiversity conservation programs, either in the preparation of master plans for watersheds and conservation unit, as in biomonitoring programs.

The current classification of Plecoptera was established by Zwick (2000) based on a morphological phylogeny that proposed two suborders: Arctoperlaria and Antarctoperlaria. As the names suggest, the suborders are distributed between the two hemispheres, formed by the breakup of the supercontinent Pangaea (Zwick, 2000), with the Arctoperlaria, originated in Laurasia and currently distributed in the Northern Hemisphere and some regions of the Southern Hemisphere, and Antactoperlaria of Gondwanan origin, currently deployed in the Southern Hemisphere (except Africa and Madagascar) (Zwick, 2000). This hypothesis is supported by the recent work of McCulloch et al. (2016). The suborders divergence could have occurred between 121–110 Ma according to recent studies (Misof et al., 2014; McCulloch et al., 2016). The high endemism in southern hemisphere may be explained by the low dispersion abilities of stoneflies (McCulloch et al., 2016).

Suborder Arctoperlaria

Arctoperlaria is composed by two groups, each containing six families: Euholognatha (Capniidae, Leuctridae, Nemouridae, Notonemouridae, Scopuridae and Taeniopterygidae) and Systellognatha (Chloroperlidae, Perlidae, Perlodidae, Peltoperlidae, Pteronarcyidae and Styloperlidae) (Zwick, 2000). Only Perlidae and Notonemouridae are found in southern temperate regions, and the presence of these taxa in southern hemisphere could be a recent independent invasion from the north (Fochetti & Tierno-de-Figueiroa, 2008).

Suborder Antarctoperlaria

Antarctoperlaria consists of two superfamilies, both with two families: Eusthenioidea (Diamphipnoidae and Eustheniidae) and Gripopterygoidea (Austroperlidae and Gripopterygidae) (Zwick, 2000). The disjunct distribution of the Antactoperlaria in South America, Australia, and New Zealand has been suggested because of vicariance associated with Gondwanan fragmentation, and its absence from South Africa and India is explained by latter extinction (Fochetti & Tierno-de-Figueiroa, 2008).

LIMITATIONS

Studies of Neotropical stoneflies are mainly limited to adult descriptions. The first keys for South American stoneflies are those of Illies for Estheniidae and Diamphipnoidae (Illies, 1960), Gripopterygidae (Illies, 1963), and Perlidae (Illies, 1964), some of which are still useful for Patagonian stoneflies. The first key published for families and genera was Benedetto (1974); it was mainly based on classifications of Illies (1966) and Zwick (1973). The key of Romero (2001) also included families and nymphs, and updates the taxonomy of some genera. Heckman (2003) provided a key to South American stoneflies, with original plates, although he did not follow the taxonomy updates from Zwick (2000) and did not include some taxa previously described. Stark et al. (2009) provided an excellent update with a checklist of the Neotropical fauna. At family level, the most recent key for South American Gripopterygidae is McLellan & Zwick (2007). Regional keys to nymphs were also published by Froehlich (1984; southeastern Brazil), Bachmman (1995; Argentina), Hamada & Couceiro (2003; central Amazonia), Olifiers et al. (2004; Brazilian nymphs), Mugnai et al. (2010; Rio de Janeiro), Hamada & Silva (2014; Amazonas State) and Ribeiro & Gorayeb (2016). Froehlich (2010), in the catalog of Neotropical Plecoptera, presented the summary of the species, pointing out which nymphs are associated with the adults of each species.

Due to poor knowledge of immature stages and the high species diversity, some authors use specific regional keys to name nymphs without adult association (e.g., *Anacroneuria* VZ-10 in Stark, 1995). It is also relatively simple to identify stoneflies at the species level for certain families with low diversity, especially Estheniidae, Perlidae (in part), Austroperlidae, and with some difficulty for Gripopterygidae. Correlations of nymphs with their adult stage is still poor for Diamphipnoidae and Notonemouridae, despite the low diversity of these taxa. Recent works with Brazilian nymphs (Avelino-Capistrano et al., 2011a, b, Ribeiro & Gorayeb, 2016) correlated nymphs with adults using features such as color, shape of the mouthparts and legs chaetotaxy, and cerci. However, Avelino-Capistrano (2014) observed a great polymorphism in nymphs of *Kempnyia* and *Anacroneuria*.

TERMINOLOGY AND MORPHOLOGY

The basic morphology of nymphs is shown in Fig. 4.1 A–H. In this work terminology of many anatomic nymph's, parts are present like acronyms or abbreviations to facilitate the comprehension of the structures in the pictures. This section is a summary of the acronyms and abbreviations used in the Figs. 4.1–48.

TABLE 4.1 Taxa, Diversity and Number of Described Nymphs of Neotropical Plecoptera.

Taxon	Distribution	Species	Described Nymph
Austroperlidae			
Andesobius McLellan, 2001b	Argentina, Chile	1	1
Klapopteryx Navás, 1928	Argentina, Chile	2	2
Penturoperla Illies, 1960	Argentina, Chile	1	1
Diamphipnoidae			
Diamphipnoa Gerstaecker, 1873	Argentina, Chile	4	2
Diamphipnopsis Illies, 1960	Argentina, Chile	2	1
Eustheniidae			
Neuroperla Illies, 1960	Argentina, Chile	1	1
Neuroperlopsis Illies, 1960	Argentina, Chile	1	1
Gripopterygidae			
Alfonsoperla McLellan & Zwick, 2007	Argentina,Chile	1	1
Andiperla Aubert, 1956	Argentina, Chile	1	1
Andiperlodes Illies, 1960	Argentina, Chile	2	1
Antarctoperla Illies, 1963	Argentina, Chile	2	2
Araucanioperla Illies, 1963	Argentina, Chile	2	1
Aubertoperla Illies, 1963	Argentina, Chile	2	2
Ceratoperla Illies, 1963	Argentina, Chile	2	2
Chilenoperla Illies, 1963	Argentina, Chile	6	3
Claudioperla Illies, 1963	Chile and Argentina to Ecuador, Colombia	1	1
Falklandoperla McLellan, 2001	Malvinas Islands	1	1
Gripopteryx Pictet, 1841	Brazil, Argentina, Paraguai e Uruguai	17	11
Guaranyperla Froehlich, 2001	Brazil	3	1
Limnoperla Illies, 1963	Argentina, Chile	1	1
Megandiperla Illies, 1960	Chile	1	1
Neopentura Illies, 1965	Argentina, Chile	1	1
Notoperla Enderlein, 1909	Argentina, Chile	7	5
Notoperlopsis Illies, 1963	Argentina, Chile	1	1
Paragripopteryx Enderlein, 1909	Argentina, Brazil, Uruguay	14	7
Pehuenioperla Vera, 2009	Chile	1	1
Pelurgoperla Illies, 1963	Argentina, Chile	1	1
Plegoperla Illies, 1963	Chile	2	0
Potamoperla Illies, 1963	Argentina, Chile	1	1

(*Continued*)

TABLE 4.1 Taxa, Diversity and Number of Described Nymphs of Neotropical Plecoptera.—cont'd

Taxon	Distribution	Species	Described Nymph
Rhithroperla Illies, 1963	Argentina, Chile	2	1
Senzilloides Illies, 1963	Argentina, Chile	1	1
Gripopterygidae			
Teutoperla Illies, 1963	Argentina, Chile	4	2
Tupiperla Froehlich, 1969	Argentina, Brazil, Paraguay	18	1
Uncicauda McLellan & Zwick, 2007	Chile	2	1
Notonemouridae			
Austronemoura Aubert, 1960	Argentina, Chile	10	0
Neofulla Claassen, 1936	Argentina, Chile	3	1
Neonemura Navás, 1919	Argentina, Chile	4	0
Udamocercia Enderlein, 1909	Argentina, Chile	3	1
Perlidae			
Anacroneuria Klapálek, 1909	From Arizona (EUA) to northeast Argentina, Trinidad and Tobago	344	17
Enderleina Jewett, 1960	Brasil, Venezuela	7	2
Inconeuria Klapálek, 1916	Peru, Chile	3	1
Kempnyella Illies, 1964	Argentina, Chile	2	1
Kempnyia Klapálek, 1914	Brazil	38	10
Klapálekia Claassen, 1936	Colombia	1	0
Macrogynoplax Enderlein, 1909	Brazil, Guyana, Peru, Suriname, Venezuela, Colombia	15	4
Nigroperla Illies, 1964	Chile	1	1
Onychoplax Klapálek, 1916	Brazil (?)	1	0
Pictetoperla Illies, 1964	Argentina, Chile	2	2

Cephalic Structures

Antenna (ATN): Filiform, measuring up to 1/3 the size of the body. Some nymphs have many bristles covering the antennal flagellum (e.g., *Potamoperla*) and may be covered with mud (*Perlugoperla*).

Condilus (CD): Crticulation area between the mandible and the head.

Epicranial Suture (ES): Used in taxonomy of Brazilian Perlidae. In nymphs of *Kempnyia* and *Macrogynoplax*, the ES is curved between paired ocelli (ESC; Fig. 4.12); in *Anacroneuria* nymphs, this suture appears below the paired ocelli (Figs. 4.1 A, 8).

Frontoclypeus (FR): Area between coronal suture and labrum. In some species, the color patterns of FR can be used to distinguish nymph species (e.g., *Anacroneuria*; see Ribeiro & Gorayeb, 2016). The **M-shaped line (MSL)** is a color pattern in "M," "U," "V" or "W" form and is formed by clears areas in FR area, sometimes it is present only like a low relief form.

Mouth Parts: Plecoptera mouth parts were separate into two groups by Stewart & Stark (2002): Type I (with well-developed molar area on the mandible and the maxillary lacinia and galea adapted for brushing or scraping) and Type II (with more elongate lacinia usually

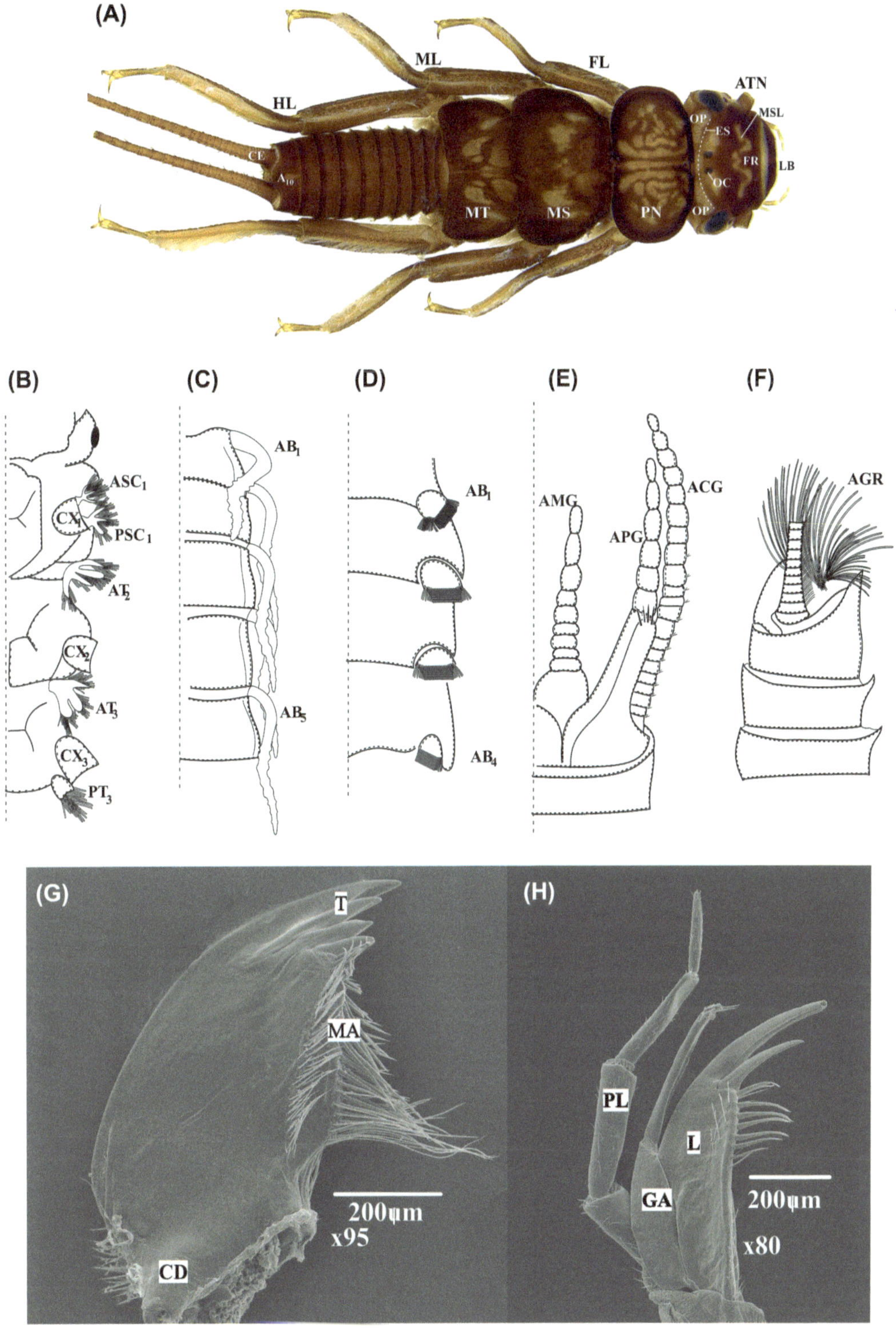

FIGURE 4.1 Plecoptera nymph. (A) *Anacroneuria* sp. dorsal habitus (LB: labrum; FR: frontoclypeus; MSL: M-shaped line; ATN: antenna; ES: epicranial suture; OC: ocellus; OP: occiput; PN: pronotum; FL: foreleg; MS: mesothorax; ML: midleg; MT: metatorax; HL: hind leg; A10: abdominal segment 10; CE: cercus). (B) Branched thoracic gills of *Anacroneuria* sp. (Ventral View: ASC1: anterior pronotal supracoxal; AT2: anterior mesothoracic; AT3: anterior metathoracic; CX: coxa; PSC1: posterior pronotal supracoxal; PT3: posterior metathoracic). (C) *Neuroperla* sp. abdominal gills (AB1 and AB5). (D) *Diamphipnoa annulata* (Brauer) abdominal gill remnants (AB1 to AB5). (E) *Klapopteryx sp*. anal gills. (ACG: anal cercal gill; AMG: anal medial gill; APG: anal paraproctal gill). (F) *Gripopteryx* sp. anal gill rosette (AGR). *Anacroneuria* sp. (G) Mandible (MA: molar area; CD: condyle; T: teeth). (H) Maxilla (GA: galea; L: lacinia; PL: maxillary palpi). *(F) Redrawn from Stark et al. (2009).*

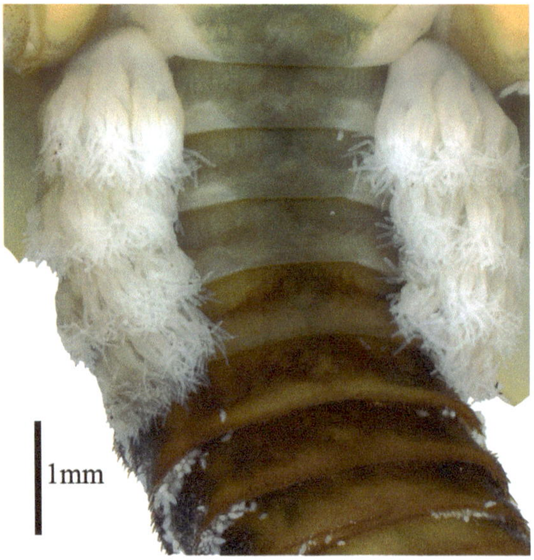

FIGURE 4.2 *Diamphipnoa helgae* Illies. Abdomen in ventral view. Ventrolateral gills on abdominal segments I–V. *Photo courtesy of Neusa Hamada.*

FIGURE 4.3 *Kempnyia* sp., thoracic segments in lateral view.

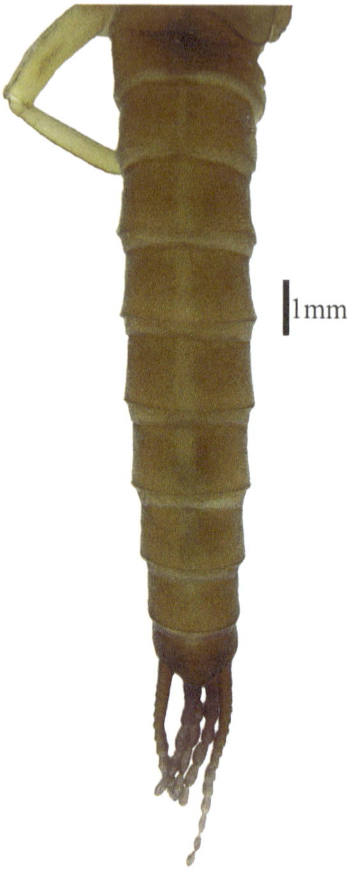

FIGURE 4.4 *Klapopteryx* sp. Abdomen in dorsal view. Abdomen and anal region.

FIGURE 4.5 *Diamphinoa helgae* Illies. Posterior leg in lateral view. Leg with fringe of bristles.

bearing two long, sharp teeth, and mandibles that lack a molar area). Here, these acronyms are used to designing the mouth parts: **galea (GA): lacinia (L)**; **labrum (LB)**; **molar area (MA)**; **palpus (PL)**; **teeth (T)**.

Ocellus (OC): When they are present, two (paired) or three ocelli can be found; the paired ocelli are localized above ES; the third ocellus is localized anteriorly and sometimes it is barely visible.

Occiput (OP): Area behind the ES; sometimes, bristles could cover this area (e.g., *Udamocercia*).

Thoracic Structures

Anterior Pronotal Supracoxal Gill ($ASC_{1\text{-}3}$): Gills that might arise anterior to coxae (Fig. 4.1 B). In *Kempnyia* and *Macrogynoplax* nymphs these gills (ASC_{1-3}) are present, but in *Anacroneuria*, only gill ASC_1 is present.

Anterior Metathoracic Gills (AT_{1-3}): Gills that arise on anterior ventral margins of each segment.

Coxa (CX): Similar to other insects.

Legs: Foreleg (FL): First leg; **Midleg (ML):** Second leg; **Hind Leg (HL):** Third leg. In general, the stoneflies legs are similar between families and genera, but a raptorial

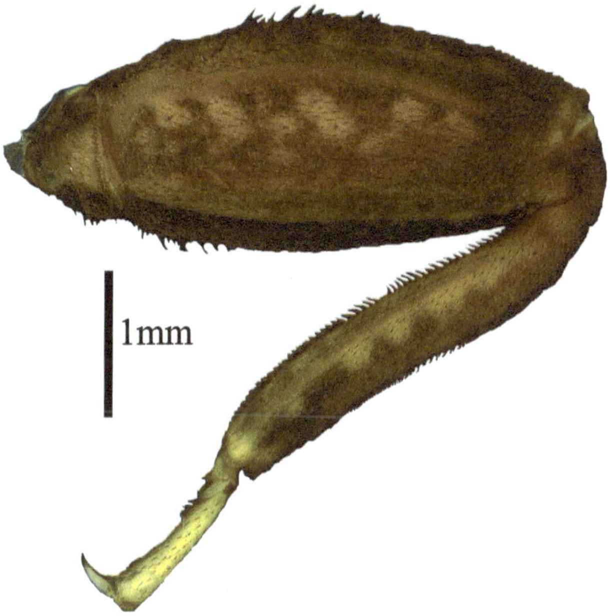

FIGURE 4.6 *Diamphipnopsis* sp. Posterior leg in lateral view. Leg without fringe of bristles.

FIGURE 4.7 *Neuroperla schedingi* (Navás), dorsal habitus.

FIGURE 4.8 *Anacroneuria* sp., head and pronotum.

FIGURE 4.9 *Macrogynoplax* sp., raptorial leg.

FIGURE 4.10 *Macrogynoplax* sp., dorsal habitus.

FIGURE 4.11 *Kempnyia* sp., leg in lateral view.

FIGURE 4.13 *Kempnyia* sp., dorsal habitus.

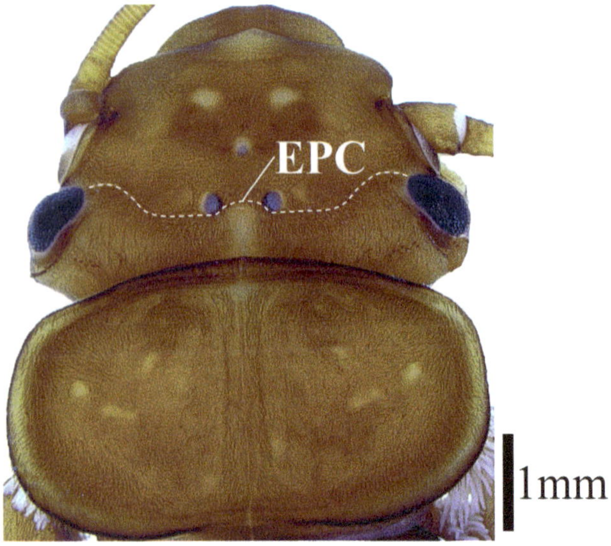

FIGURE 4.12 *Kempnyia* sp. Head and pronotum. EPC = epicranial suture curved. Bar = 1 mm.

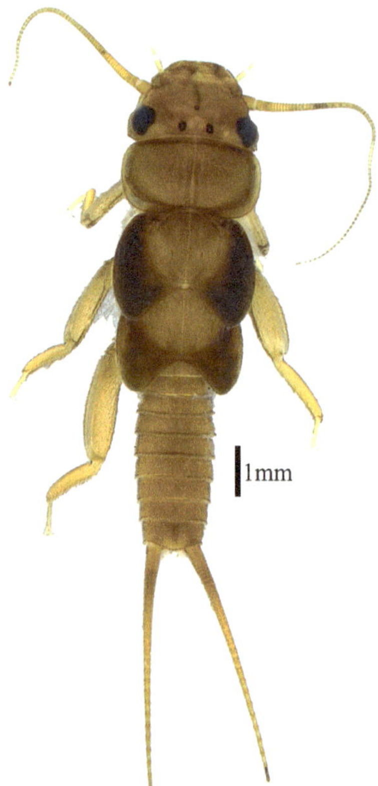

FIGURE 4.14 *Enderleina* sp., dorsal habitus. *Photo courtesy of Neusa Hamada.*

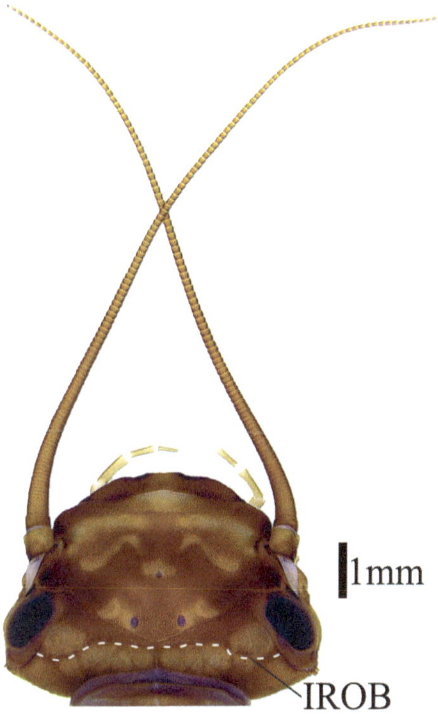

FIGURE 4.15 *Nigroperla costalis* Illies, head in dorsal view. IROB = irregular row of occipital bristles.

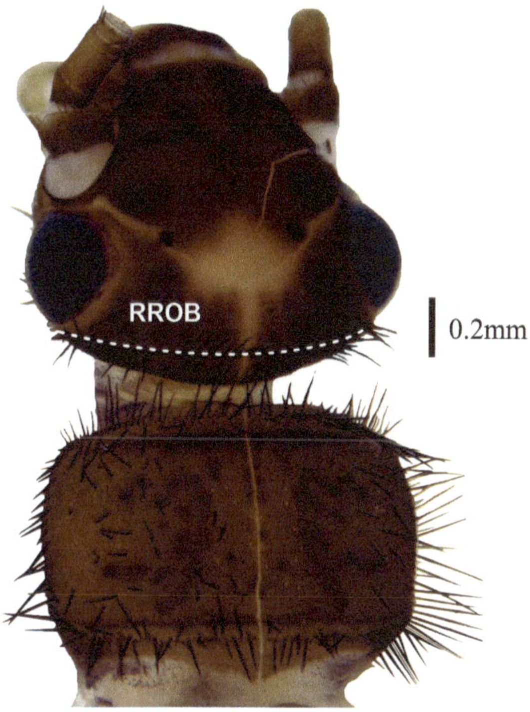

FIGURE 4.17 *Udamocercia* sp., head and pronotum in dorsal view. RROB = regular row of occipital bristles.

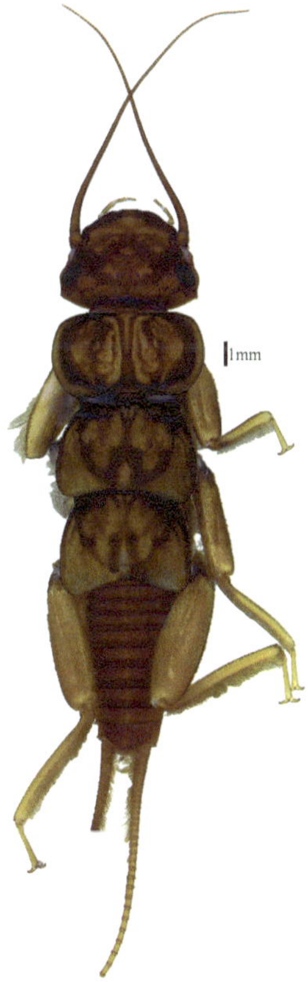

FIGURE 4.16 *Nigroperla costalis* Illies, dorsal habitus.

FIGURE 4.18 *Udamocercia* sp., dorsal habitus.

FIGURE 4.19 *Austronemoura* sp., dorsal habitus.

foreleg is present in *Macrogynoplax* nymphs. Bristles can cover all legs (e.g., *Udamocercia*), but in general, a fringe of bristles is found covering the dorsal margin of femur and tibia (e.g., *Macrogynoplax*). A ventrodistal spine in femora is present (sometimes small) in nymphs of *Tupiperla* and *Guaranyperla*.

Mesothorax (MS): The second thoracic segment. During nymph development, the pterothecae (or wingpads) grows, increasing in each ecdysis, forming the future wings.

Metatorax (MT): The third thoracic segment. During nymph development, the pterothecae (or wingpads) grows, increasing in each ecdysis, forming the future wings.

Pronotum (PN): The first thoracic segment. This may be covered by spines (e.g., *Gripopteryx*), bristles (e.g., *Udamocercia*), or with lateral extensions (e.g., *Guaranyperla*). The color pattern is sometimes used in species descriptions.

Posterior Metathoracic Gill (PT): Gills that arise on posterior ventral margin of the metathoracic segment.

Posterior Pronotal Supracoxal Gill (PSC_{2-3}): Gills that arise on posterior ventral margins of each segment.

Abdominal Structures

Abdominal Segment 10 (A_{10}): Is the last segment of the abdomen and may end in an apical projection (e.g., *Gripopteryx*), with a rounded shape (e.g., *Chilenoperla, Falklandoperla*), or in a triangular shape (e.g., *Klapopteryx*).

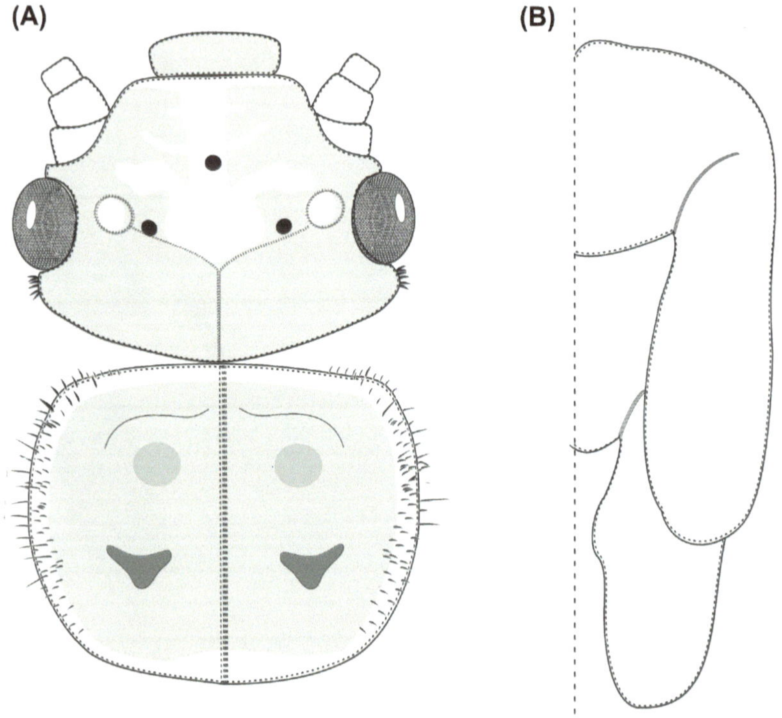

FIGURE 4.20 *Neonemura* sp. (A) Head and pronotum in dorsal view. (B) Nymphal wingpads. *(B) Redrawn from Stark et al. (2009; Figs. 32 E–F).*

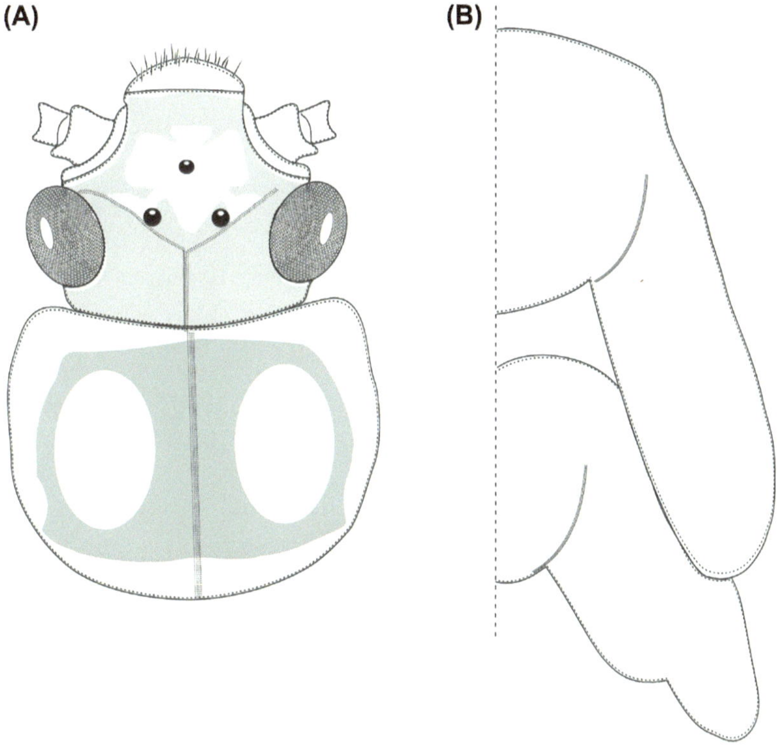

FIGURE 4.21 *Neofulla areolata* (Navás). (A) Head and pronotum in dorsal view. (B) Nymphal wingpads. *(B) Redrawn from Stark et al. (2009; Figs. 32 E–F).*

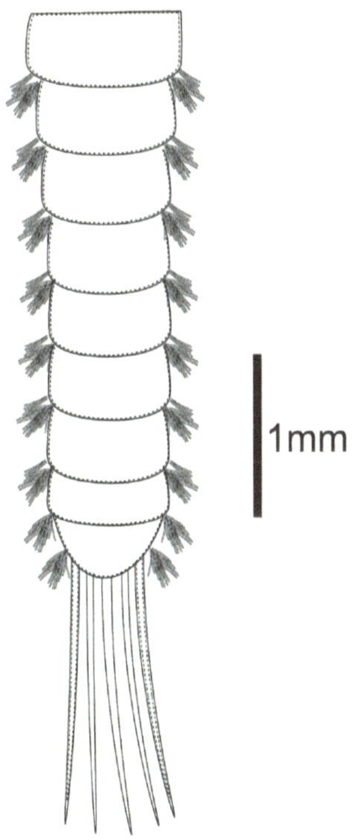

FIGURE 4.22 *Penturoperla barbata* Illies, abdomen, in dorsal view. *Redrawn from McLellan (2001; Fig. 29).*

FIGURE 4.23 *Klapopteryx* sp., head and pronotum in dorsal view.

FIGURE 4.24 *Andiperla willinki* Aubert, "the Patagonian Dragon." Live nymph on Perito Moreno Glacier (Santa Cruz, Argentina), in dorsal view. *Photo courtesy of Jacint Cerdá Moles.*

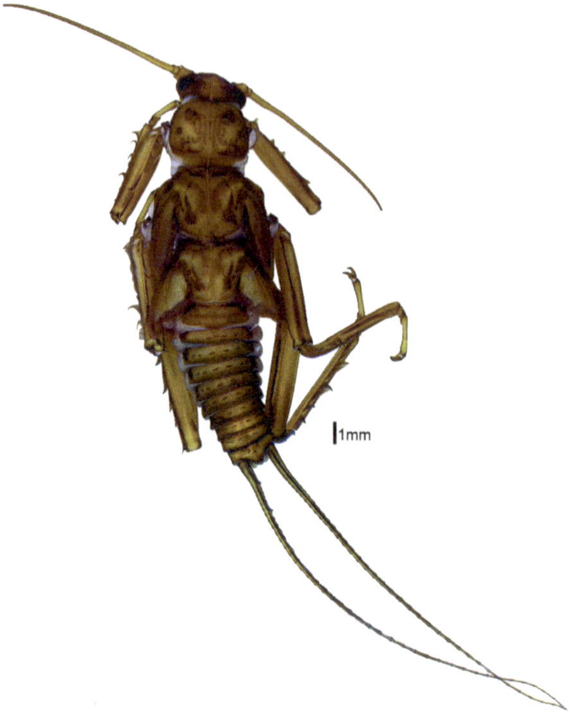

FIGURE 4.26 *Notoperla* sp., dorsal habitus. Bar = 1 mm.

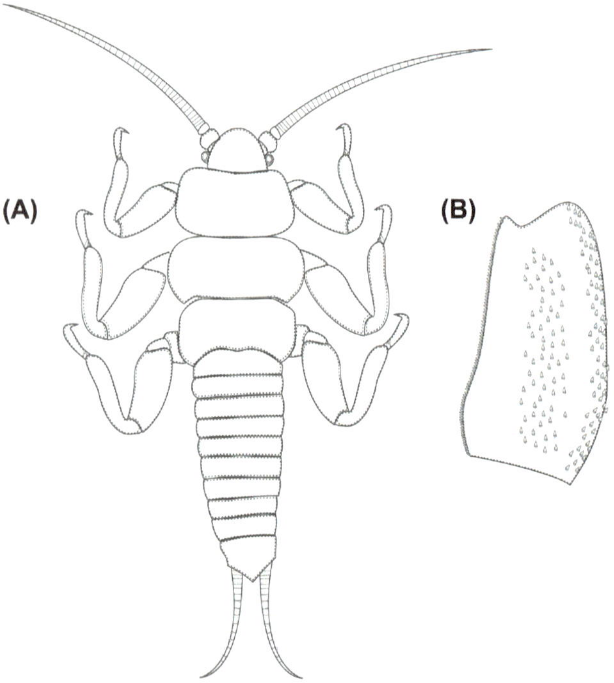

FIGURE 4.25 *Andiperlodes tehuelche* Pessacq. (A) Femora in dorsal view. (B) Dorsal habitus. *(B) Redrawn from Pessacq (2009; Figs. 4 E and 3).*

FIGURE 4.27 *Falklandoperla kelper* McLellan, dorsal habitus.

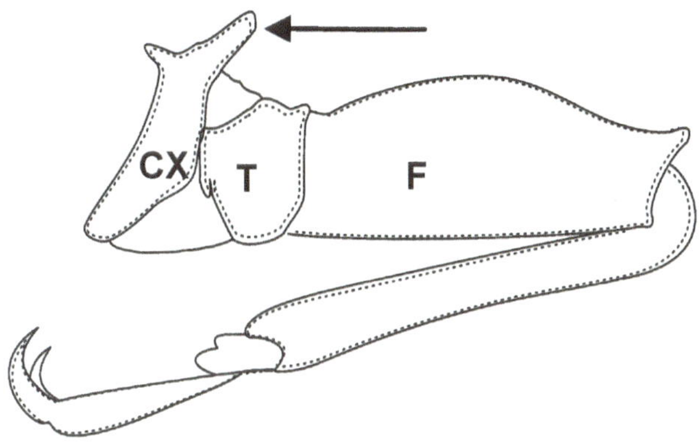

FIGURE 4.28 *Notoperla* sp., coxa projection on oblique anterior view (CX = coxa; T = trochanter; F = femur). *Redrawn from McLellan & Zwick (2007; Fig. 49).*

FIGURE 4.29 *Neopentura semifusca* Illies, dorsal habitus. APS = abdomen paired spines.

Abdominal Gills (AB_{1-5}): _resent in nymphs of Eustheniidae (on the five or six basal abdominal segments; Fig. 4.1 C) and Diamphipnoidae (on the four basal abdominal segments; Fig. 4.1 D).

Anal Gills: Anal Cercal Gill (ACG), Anal Medial Gill (AMG), and Anal Paraproctal Gill (APG): Present in Austroperlidae nymphs. In Neotropical nymphs, only three gills are present with one located on each paraproct and the third, medial gill is attached to underside of dorsoapical projection of segment X (Fig 4.1 E).

Anal Gills Rosette (AGR): This thick cluster of gills filaments in the anal region (Fig. 4.1F) is generally found in Gripopterygidae nymphs, except in *Notoperla*. Sometimes these gills can be retracted into rectal cavity.

Cercus (CE): Two long and whiplike appendages divided into numerous segments. However, some genera have reduced cerci (e.g., *Ceratoperla*, *Neopentura*, *Plegoperla*).

MATERIAL PREPARATION AND PRESERVATION

Plecoptera nymphs are usually preserved in 70 or 80 percent alcohol. For the use in molecular work, absolute or 96 percent alcohol is used, to preserve the DNA. Storage in freezers also reduces the DNA degradation process. For the nymph identification, in general, the genitalia or paraprocts are not used as they are not fully formed, and other characters of the external morphology (e.g., habitus, color, number of bristles, the mouthparts form) are used for nymph's classification (Dorvillé & Froehlich, 1999; Avelino-Capistrano et al., 2011a, b).

Rearing of nymphs is essential for species-level identification. In many studies, the association of nymphs and adults is conducted through the rearing of immatures. However, this is not easy because the nymphs are very sensitive to changes in temperature, current, and oxygen dissolved in the water. Some species do not resist the transfer from the natural environment to laboratory. Many authors have developed methodologies for rearing Plecoptera nymphs, and some have success rearing them in cages placed in the river where they were collected (e.g., Ribeiro, 2003 and Ribeiro & Raphael, 2005, Pessacq & Omad, 2012). But, in some cases, it is impossible or impractical to use this technique, especially where fast moving water (e.g., during storms) can drag down the cages. To avoid these problems, Kapoor (1972) and Novaes et al. (2012) developed an apparatus with a tank fed by a pump with continuous flow of water. Other artificial breeding techniques in laboratory have used polystyrene boxes (Ribeiro, 2004; Avelino-Capistrano, 2010), glass jars (Dorville, 1998), or aquariums (Avelino-Capistrano et al., 2011a, b, 2015) provided with aerators. Vera (e.g., 2006a, b, c; 2008) successfully reared several species in laboratory using refrigerated water from the rivers where the specimens were collected, an aerator, and replacement of water every one or two weeks. According to Poulton (2005), the successful rearing depends on several conditions, such as the functional food type of the species reared and the remaining time of growth until emergence, besides the water's physicochemical and behavior conditions. Due the taxonomic impediment in the nymph identification and its association to adults, molecular markers, particularly genes composing the mitochondrial DNA, have been used to solve these questions (e.g., Avelino-Capistrano et al., 2014, 2016; Hebert et al., 2003a, b).

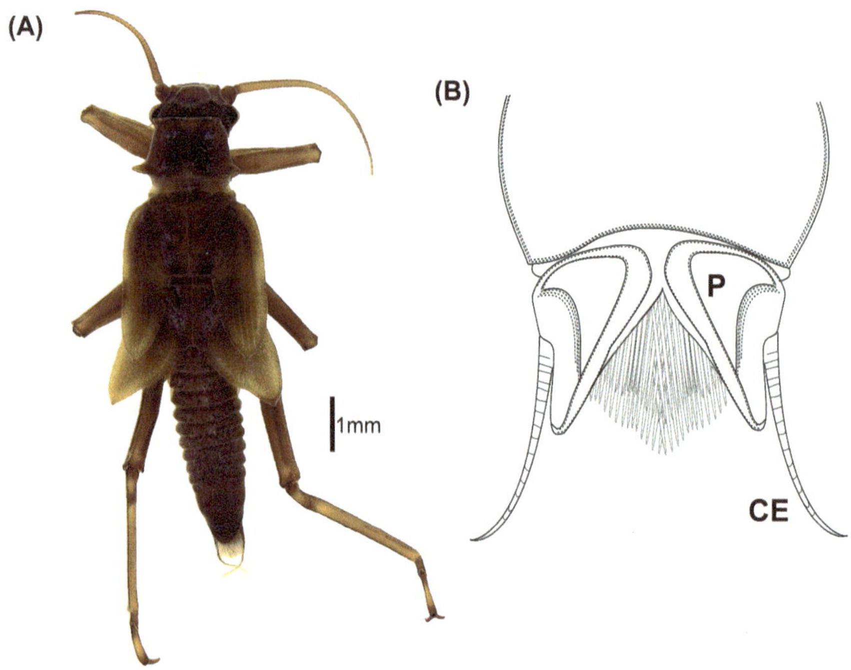

FIGURE 4.30 *Ceratoperla* sp. (A) dorsal habitus. PPS = pronotum paired spines. (B) Abdomen in ventral view. CE = cerci. P = paraprocts.

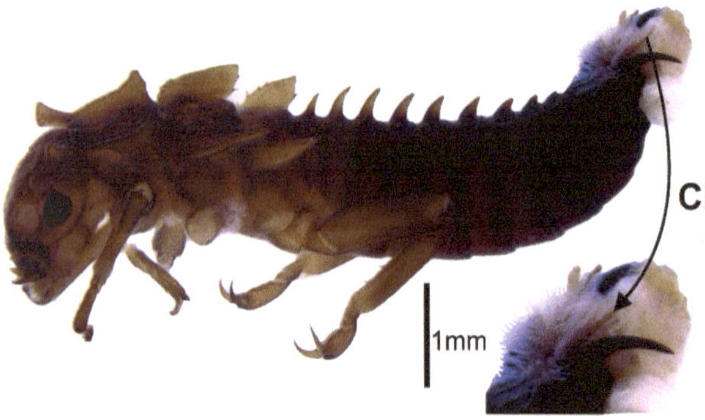

FIGURE 4.31 *Neopentura semifusca* Illies, habitus in lateral view.

FIGURE 4.32 *Pelurgoperla personata* Illies, dorso-lateral habitus.

KEYS TO PLECOPTERA

This key is adapted from Benedetto (1974), Romero (2001), and Stark et al. (2009).

Plecoptera: Families

1 Abdominal segments with ventrolateral gills (Figs. 4.1 C–D, 2) 2

1' Abdominal segments without ventrolateral gills, but these may be present on the thorax or anal region 3

2(1) Abdominal segments I–IV with ventrolateral gills (Figs. 4.1 D, 2); gills branched with thick trunks, not beaded, fringed along posteroventral or apical margins .... **Diamphipnoidae [p. 134]**

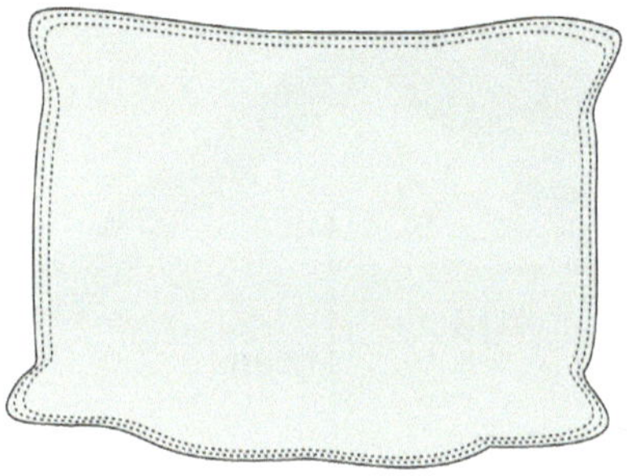

FIGURE 4.33 *Antarctoperla* sp. Illies, 1963. Pronotum (redrawn from Illies, 1963 - Fig. 32).

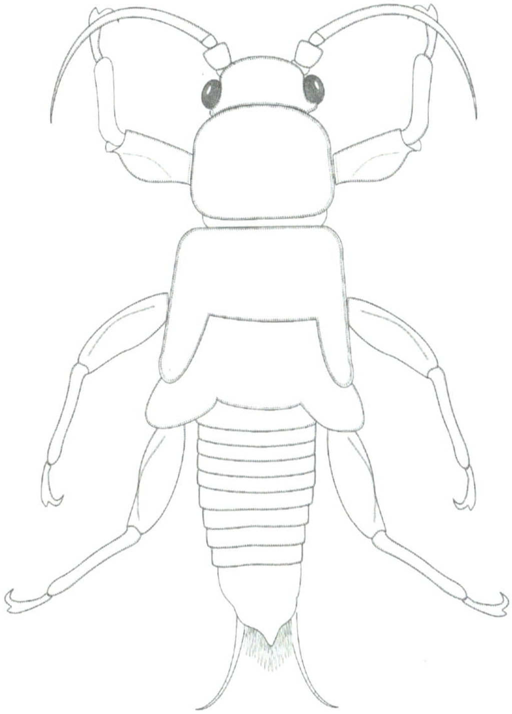

FIGURE 4.34 *Alfonsoperla* sp. McLellan & Zwick, 2007. Dorsal habitus (redrawn from McLellan & Zwick, 2007 – Fig. 11).

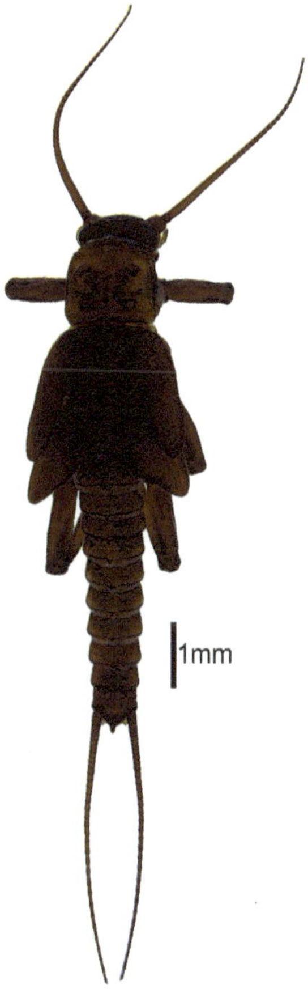

FIGURE 4.35 *Senzilloides panguipulli* (Navás). Habitus.

2' Abdominal segments I–V or I–VI with ventrolateral gills (Fig. 4.1 C); gills composed of beaded single filaments .... **Eustheniidae [p. 137]**

3(1) Thoracic segments without gills........ 4

3' Thoracic segments with branched gills around coxae (Figs. 4.1 B, 3)........ **Perlidae [p. 137]**

4(3) Abdominal segments VII–IX fused into complete rings without pleural membrane; furcal pits usually located far from spinasternum; anal gills usually present (Figs. 4.1 E–F) but sometimes contracted inside abdomen, absent in *Notoperla* (Gripopterygidae) 5

4' Abdominal segments VII–IX separated into terga and sterna by pleural membrane; furcal pits located near spinasternum, or obliterated; anal gills absent........ **Notonemouridae [p. 137]**

5(4) Cerci short with apical segments beaded; anal gills composed of three terminal beaded filaments (Figs. 4.1 E, 4) .... **Austroperlidae [p. 138]**

5' Cerci usually long, apical segments not beaded; anal gills in a tuft of cyllindrical filaments (Fig. 4.1 F), or absent (*Notoperla*) .... **Gripopterygidae [p. 138]**

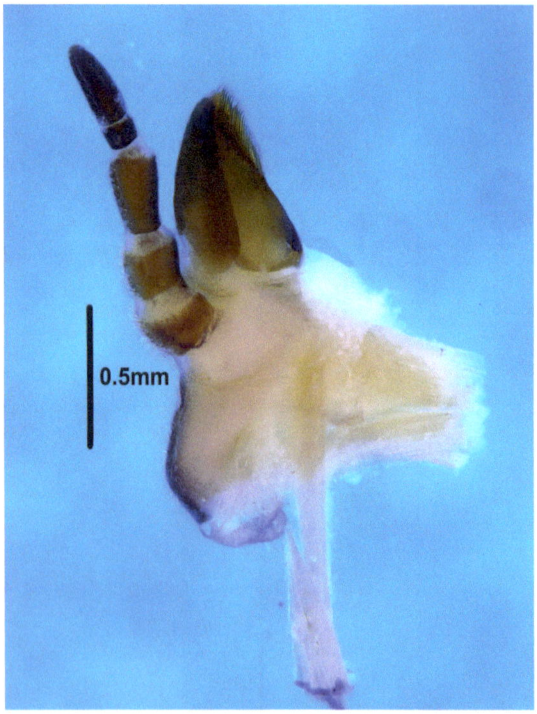

FIGURE 4.36 *Gripopteryx* sp. Pictet. Left maxilla.

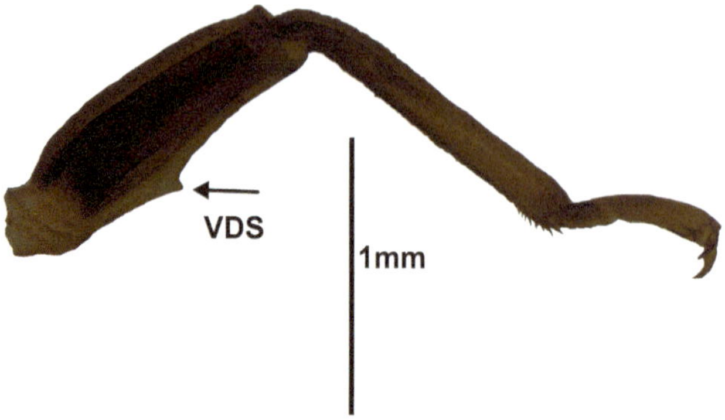

FIGURE 4.38 *Tupiperla* sp. Froehlich. Leg in lateral view. VDS = femora ventrodistal spine.

FIGURE 4.37 *Gripopteryx* sp. Pictet. Habitus in lateral view.

FIGURE 4.39 *Paragripopteryx* sp. Enderlein. Habitus.

Plecoptera: Diamphipnoidae: Genera

The biology of the group is poorly known, with aquatic nymphs reported shreddering woody debris (Wantzen & Wagner, 2006). Omad et al. (2015) reported *Diamphipnopsis samali* Illies feeding on wood and fungal hyphae, and with a clear preference for waterfalls and wood habitats. Distribution: Southern Argentina and Chile.

1	Legs and cerci medial margins with well developed setal fringes (Fig. 4.5); galea with dense apical hair brush	*Diamphipnoa*
1'	Legs and cerci medial margins without setal fringes (Fig. 4.6); galea without dense hair brush	*Diamphipnopsis*

FIGURE 4.40 *Guaranyperla* sp. Froehlich. Head and pronotum. PLE = pronotum lateral extensions.

FIGURE 4.42 *Claudioperla tigrina* (Froehlich). Habitus.

FIGURE 4.41 *Tupiperla* sp. Froehlich. Habitus.

FIGURE 4.43 *Rhithroperla* sp. Illies. Habitus.

FIGURE 4.44 *Chilenoperla* sp. Illies. Habitus.

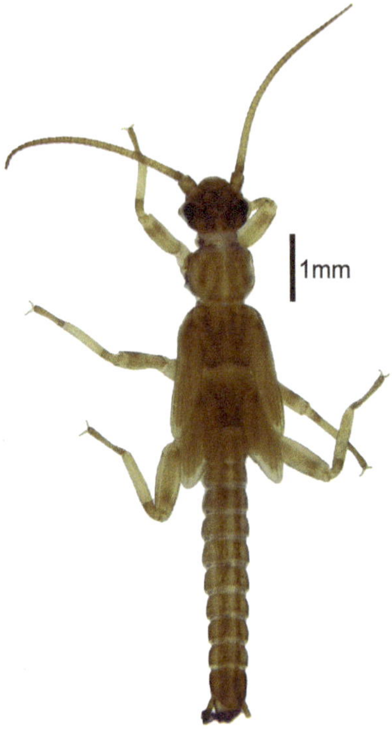

FIGURE 4.46 *Limnoperla jafueli* (Navás). Habitus.

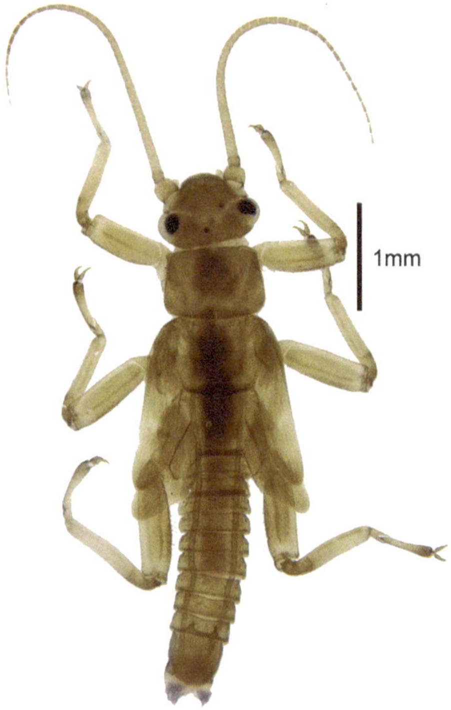

FIGURE 4.45 *Aubertoperla* sp. Illies. Habitus.

Plecoptera: Eustheniidae: Genera

Tillyard (1921) established this family to include a large group of Australian and South American stoneflies, with two subfamilies: Eustheniinae and Stenoperlinae. Only two genera are reported from the Neotropical region: *Neuroperla* and *Neuroperlopsis*. The nymphs are poorly known, but studies conducted in Australia show the species require two to three years to emerge, and they have an extended emergence (Hynes & Hynes, 1975). Distribution: Argentina, Chile. Australia, New Zealand, and Tasmania.

1 Abdominal segments I–V with gills (Fig. 4.1 C); gills distinctly beaded .... *Neuroperla* (Fig. 4.7)
1' Abdominal segments I–VI with gills (Fig. 4.1 C); gills not distinctly beaded........ *Neuroperlopsis*

Plecoptera: Perlidae: Genera

This is the most speciose family, with 51 genera and more than 900 species with a wide distribution, occurring through the entire Neotropical Region, except Tierra del Fuego. Perlid nymphs are the best studied in the Neotropics, and association of these insects with litter has been recorded by many authors (Batista et al., 2001; Henriques-Oliveira & Nessimian, 2010; Avelino-Capistrano et al., 2017). Many studies of association have been performed on adults and nymphs, and several authors have developed regional keys to immature stages at species level (Ribeiro & Rafael, 2007, 2009; Ribeiro & Gorayeb, 2016). Despite these efforts, nymphs of *Klapálekia* and *Onychoplax* are still unknown. Nevertheless, biology data is still sparse, and few studies have addressed feeding (Dorvillé & Froehlich, 2001; Docille et al., 2016) and emergence behaviors (Hynes, 1948; Turcotte & Harper, 1982; Froehlich, 1991; Avelino-Capistrano et al., 2015). Nymphs are active predators, but Dorvillé & Froehlich (2001) found a variety of plants detritus, algae, and fungi among gut contents. Distribution: Widely distributed in Neotropics (in all South America, especially in southernmost region). Neartic, Ethiopian, Paleartic, and Oriental regions (including some Indonesian-Malaysian islands).

1 Two ocelli (Fig. 4.8)........ 2
1' Three ocelli, anterior ocellus sometimes minute........ 6
2(1) Fore femora normal (Fig. 4.11); maxillary palpi only slightly longer than lacinia (Fig. 4.1 H)........ 3
2' Fore femora raptorial (Fig. 4.9); maxillary palpi much longer than lacinia........ *Macrogynoplax* (Fig. 4.10)
3(2) Gill PSC [2, 3] present........ 4
3' Gill PSC [2, 3] absent (Fig. 4.1 B)........ *Anacroneuria* (Fig. 4.8)
4(3) Gill ASC [2, 3] present........ 5
4' Gill ASC [2, 3] absent........ *Pictetoperla* (in part)
5(4) Gill ASC [1, 2, 3] with a single trunk; epicranial suture normal........ *Inconeuria*
5' Gill ASC [1, 2, 3] with a double trunk; epicranial suture curved forward between posterior ocelli (Fig. 4.12)........ *Kempnyia* (in part) (Fig. 4.13)
6(1) Anal and ASC [2, 3] gills present........ 7
6' Anal and ASC [2, 3] gills absent........ 9
7 (6) Ocelli normal; at least basal cercal segments with dense setal fringe; some gills, including ASC [1] and AT [2, 3] with double or triple trunks........ 8
7' Ocelli minute and widely spaced; cerci without dense setal fringe; all thoracic gills with single trunks........ *Enderleina* (Fig. 4.14)
8 (7) Epicranial suture curved forward between posterior ocelli (Fig. 4.13); gill ASC [2,3] with double trunks (Fig. 4.3)........ *Kempnyia* (in part)
8' Epicranial suture normal; gill ASC [2,3] with single trunks........ *Kempnyella*
9(6) Medial lacinial margin with a short row of ca. 6 setae extending from subapical tooth base toward lacinial base; occiput with irregular row of short peg-like sensilla (Fig. 4.15)........ *Nigroperla* (Fig. 4.16)
9' Medial lacinial margin with a small cluster of 2–4 setae at base of subapical tooth; occiput without irregular row of short peg-shaped sensilla........ *Pictetoperla* (in part)

Plecoptera: Notonemouridae: Genera

The family has 23 genera and 115 species, with gondwanian distribution. Only four genera are recorded from the Neotropical region. According to Stark et al. (2009), little information is available about Neotropical Notonemouridae, unlike New Zealand species. Species of *Austronemoura* were recorded from low altitudes (≥200 m a.s.l.) (Wais, 1984) to high altitudes (>1.200 m als.) and low order streams in Patagonian (Modenutti et al., 1998). Distribution: Southern Argentina and Chile.

1 Pronotal margins and abdominal terga posterior margins fringed with long, strong setae (Fig. 4.17)........ 2
1' Pronotal margins and abdominal terga posterior margins without setae or short and weakly developed........ 3
2(1) Occipital setal row present (Fig. 4.17); abdominal and pronotal setae stiff, dark (red or brown) and rather uniform in appearance........ *Udamocercia* (Fig. 4.18)
2' Occipital setal row absent; abdominal and pronotal setae of mixed length and thickness........ *Austronemoura* (Fig. 4.19)
3(1) Lateral pronotal disk with patches of short setae (Fig. 4.20 A); hind wingpads parallel to body axis (Fig. 4.20 B)........ *Neonemura*
3' Lateral pronotal disk bare (Fig. 4.21 A); hind wingpads strongly divergent to body axis (Fig. 4.21 B)........ *Neofulla*

Plecoptera: Austroperlidae: Genera

McLellan (2001) revised the Neotropical Austroperlidae, and three genera and four species are recorded in this region (Stark *et al.*, 2009). Austroperlid nymphs are ubiquitous in streams on both sides of the Southern Andes (Benedetto, 1974). Modenutti et al. (1998) recorded *Klapopteryx kusheli* Illies nymphs in Patagonian low-order streams. Mancilla et al. (2009) found a significant relationship between Austroperlid nymphs and rivers localized in areas covered by native vegetation. Distribution: Southern Argentina and Chile. Australia, New Zeland, and Tasmania.

1 Abdomen clothed with spatulate setae along tergal margins; gill segments beaded; fourth maxillary papal article equal in length to segment III 2
1' Abdomen clothed with long setae along posterior margins of terga (Fig. 4.22); gill segments not distinctly beaded; fourth maxillary papal article shorter than segment III .... *Penturoperla*
2(1) Frons anterior margin marked with pale V-shaped line with base forward of anterior ocellus; interocellar region pale, in contrast to anterior and lateral pigment areas on head; abdomen length $\approx$1.0–1.3× thorax .... *Andesobius*
2' Frons anterior margin uniformly pigmented, without pale V (Fig. 4.23); interocellar region not contrasting in pigmentation with anterior and lateral areas of frons; abdomen length $\approx$1.25–2.50x thorax .... *Klapopteryx*

Plecoptera: Gripopterygidae: Genera

Twenty-seven genera are reported from in the Neotropical region. Gripopterygidae nymphs are generally scrapers, feeding on periphyton and on aged fallen leaves (Stark *et al.*, 2009). In many areas of South America, nymphs are recorded in mountainous areas, especially small and cold rivers (Lecci & Froehlich, 2011; Avelino-Capistrano & Nessimian, 2013, 2014). In Patagonia, nymphs are found from higher to lower elevations (Modenutti et al., 1998; Wais, 1984). *Andiperlodes holdgatei* Illies and *Plegoperla* nymphs are still unknown. It must be noted that the placement of *Andiperlodes tehuelche* Pessacq in this genus is provisional (Pessacq, 2009). Distribution: South America, Australia, New Zeland, and Tasmania.

1 Occelli absent 2
1' Ocelli present, may be indistinct 4
2(1) Cerci longer than tergum X; body length $\leq$12 mm 3
2' Cerci very short (shorter than tergum X); body length long, $\geq$15 mm .... *Megandiperla*
3(2) Terrestrial nymphs, living on glaciers (Fig. 4.24) .... *Andiperla*
3' Aquatic nymphs, living on small streams on the Patagonian steppe; femora (Fig. 4.25 A) distinctly expanded on the ventral margin of basal third (Fig. 4.25 B) .... *Andiperlodes tehuelche*
4(1) Pterothecae absent 5
4' Pterothecae present 6
5(4) Supranal gill tuft absent (Fig. 4.26) .... *Notoperla* (in part)
5' Supranal gill tuft present (Fig. 4.27) .... *Falklandoperla*
6(4) Supranal gill tuft present, coxae simple 7
6' Supranal gill tuft absent; coxae with dorsal projection (Fig. 4.28) ... . *Notoperla* (in part)
7(6) Cerci short, with $\leq$20 cercomeres 8
7' Cerci with >20 cercomeres 14
8(7) Paired spines present dorsally on abdomen (Fig. 4.29) 9
8' Abdomen without paired dorsal spines 10
9(8) Pronotum with strong spines, anterior pair submarginal (Fig. 4.30 A); paraprocts not sharply pointed; cerci about twice as long as paraprocts (Fig. 4.30 B) .... *Ceratoperla*
9' Pronotum corners extended into spines; paraprocts sharply pointed, may be hooked; cerci shorter than paraprocts (Fig. 4.31) .... *Neopentura*
10(8) Nymphs brownish, not covered with mud that sticks to long steae 11
10' Nymphs pale, covered with mud that sticks to long hairs (Fig. 4.32) .... *Pelurgoperla/Chilenoperla* (in part)
11(10) Maxillary palpi distinctly 5-articulate; abdominal tergum X rounded posteriorly or with a small point 12
11' Maxillary palpi 4-articulate; tergum X ending in a pronounced point 13
12(11) Pronotum with a rounded corner projections (Fig. 4.33); paraprocts ending in a small point .... *Antarctoperla*
12' Pronotal corners extending into spines; paraprocts ending in long scletotized point .... *Araucanioperla*
13(11) Femora with a midlength dorsal bulge; pronotum wider than head; tibial spurs absent; anterior wingpads dorsal (Fig. 4.34) .... *Alfonsoperla*
13' Femora simple; pronotum slightly narrower than head; tibial spurs present; anterior wingpads projected laterally .... *Uncicauda*
14(6) Mediodorsal spines or tubercules present in terga I–X 15
14' Abdominal terga lacking spines or tubercules 16
15(14) Laciniae with dorsal fringe but no distal comb of strong bristles .... *Senzilloides* (Fig. 4.35) [Patagonia]
15' Laciniae with distal comb of strong bristles (Fig. 4.36) .... *Gripopteryx* (Fig. 4.37) [Eastern South America]
16(14) Eastern South America, from central Brazil highlands to Paraguay, northeastern Argentina and Uruguay 17
16' Patagonian region and along the Andes north to Colombia 19

17(16)	Femora with a ventrodistal spine, small in some nymphs (Fig. 4.38)	18
17'	Femora without spines	*Paragripopteryx* (Fig. 4.39)
18(17)	Thorax with large lateral extensions (Fig. 4.40)	*Guaranyperla*
18'	Thorax normal (Fig. 4.41)	*Tupiperla*
19(16)	Maxillary palps apparently tetramerous	20
19'	Maxillary palps clearly pentamerous.	21
20(19)	Nymph uniform brown, antennae less ½ as long as body (Fig. 4.42)	*Claudioperla*
20'	Nymph yellowish brown, legs mostly with dark bands; antennae less 2/3 as long as the body (Fig. 4.43)	*Rhithroperla*
21(19)	Flagellum of antennae hairy	22
21'	Flagellum of antennae not hairy; large, mature nymphs more than 12 mm long	*Notoperlopsis*
22 (21)	Body covered by hairs, size variable	23
22'	Body, cerci, and legs covered by many hairs; smaller, mature nymphs less than 12 mm long	*Potamoperla*
23(22)	Margins of pronotum not as above; base of antennae not hairy	24
23'	Margins of pronotum and posterior borders of abdominal terga with row of short bristles; base of antennae hairy (Fig. 4.44)	*Chilenoperla puerilis*
24(23)	Paraprocts expanded in ventral view concealing bases of cerci	25
24'	Paraprocts not laterally expanded, cercus base in ventral view exposed	26
25(24)	Paraprocts tongue-shaped; nymph stout, brown (Fig. 4.45)	*Aubertoperla*
25'	Paraprocts trapezoidal shape with the apex rounded and directed dorsally	*Pehuenioperla*
26(24)	Meso- and metanotum hind margins angularly excised (Fig. 4.46)	*Limnoperla*
26'	Meso- and metanotum hind margins straight	*Teutoperla*

ACKNOWLEDGMENTS

We appreciate the help of a number of colleagues in providing recent and older references on this insect order and in collecting nymphs. We also thank UFRJ for supporting during this work. Thanks to Jacint Cerdá Moles for the photo of *Andiperla willnki* in Glaciar Perito Moreno (Santa Cruz, Argentina) and to Ben Price from National History Museum (London) to lended material of *Falklandoperla*. We appreciate use of the Electron Microscopy Platform from Fundação Oswaldo Cruz where the scanning electron micrographs (SEMs) of mouthparts of *Anacroneuria* nymph were made.

REFERENCES

Avelino-Capistrano, F. 2010. Plecoptera (Insecta) da Estação Biológica de Santa Lúcia, Espírito Santo, Brasil: distribuição espacial, sazonal e preferência por substratos. Dissertação de Mestrado, Universidade Federal Rural do Rio de Janeiro, Seropédica, Brasil. 76 pp.

Avelino-Capistrano, F. 2014. Taxonomia integrativa de Plecoptera (Insecta) da bacia do Rio Macaé, Rio de Janeiro, Brasil. Tese de Doutorado, Universidade Federal Rural do Rio de Janeiro, Seropédica, Brasil. 233 pp.

Avelino-Capistrano, F. & J.L. Nessimian. 2013. A new species and new records of Gripopterygidae (Plecoptera) from the Serra dos Orgãos, Rio de Janeiro State, Brazil. Zootaxa 3683: 185–191.

Avelino-Capistrano, F. & J.L. Nessimian. 2014. Novos registros da fauna de Plecoptera (Insecta) para o Estado do Espírito Santo, Brasil. Boletim do Museu de Biologia Mello Leitão 33: 5–18.

Avelino-Capistrano, F., G.L. Almeida & L.S. Barbosa. 2017. Diversidade e distribuição de ninfas de Plecoptera (Insecta) por substratos em Riachos do Parque Estadual da Pedra Branca, Rio de Janeiro, Brasil. EntomoBrasilis 10: 76–81.

Avelino-Capistrano, F., L.S. Barbosa & D.M. Takiya. 2016. Description of a new *Kempnyia* Klapálek from Brazil (Plecoptera: Perlidae) with life stages associated using DNA barcodes. Zootaxa 4079: 372–380.

Avelino-Capistrano, F., L.S. Barbosa & G.L. Almeida. 2011a. The nymph of *Anacroneuria debilis* (Pictet, 1841) (Plecoptera: Perlidae). Zootaxa 2882: 59–63.

Avelino-Capistrano, F.S., L.S. Barbosa & G.L. Almeida. 2011b. Complementary descriptions of *Kempnyia gracilenta* (Enderlein, 1909) and *Kempnyia reticulata* (Klapálek, 1916) (Plecoptera: Perlidae). Illiesia 7: 142–147.

Avelino-Capistrano, F., L.S. Barbosa & G.L. Almeida. 2015. Notas na criação de *Anacroneuria debilis* (Pictet, 1841) (Plecoptera: Perlidae) no Rio de Janeiro, Brasil. Revista Ciência Atual 6: 02–07.

Avelino-Capistrano, F., J.L. Nessimian, J.R. Santos-Mallet & D.M Takiya. 2014. DNA-based identification and descriptions of immatures of *Kempnyia* Klapálek (Insecta: Plecoptera) from Macaé River Basin, Rio de Janeiro State, Brazil. Freshwater Science 331: 325–337. DOI: https://doi.org/10.1086/675226.

Bachmann, A.O. 1995. Insecta Plecoptera. Pages 1093–1111 *in*: E. Lopretto & G. Tell (eds.), Ecossistemas de Aguas Cotinentales, metodologias para su studio. Tomo III. Ediciones Sur, La Plata, Argentina.

Baptista, D.F., Buss, D.F., Dorvillé, L.F.M. & J.L. Nessimian. 2001. Diversity and habitat preference of aquatic insects along the longitudinal gradient of the Macaé River Basin, Rio de Janeiro, Brazil. Revista Brasileira de Biologia 61: 249–258.

Benedetto, L. 1974. Clave para la determinación de los plecópteros sudamericanos. Studies on Neotropical Fauna and Environment 9: 141–170.

Bethoux, O., Y. Cui, B. Kondratieff, B. Stark & D. Ren. 2011. At last, a Pennsylvanian stem-stonefly (Plecoptera) discovered. BMC Evolutionay Biology 11: 248.

Buss, D.F., D.F. Baptista, M.P. Silveira, J.L. Nessimian & L.F.M. Dorvillé. 2002. Influence of water chemistry and environmental degradation on macroinvertebrate assemblages in a river basin in south-east Brazil. Hydrobiologia 481: 125–136.

Callisto, M., M. Moretti & M. Goulart. 2001. Macroinvertebrados bentônicos como ferramenta para avaliar a saúde de riachos. Revista Brasileira de Recursos Hídricos 6: 71–82.

DeWalt, R.E., M.D. Maehr, U. Neu-Becker & G. Stueber. 2017. Plecoptera Species File Online. Version 5.0/5.0. [retrieval date]. <http://Plecoptera.SpeciesFile.org>.

Docille, T., D.C.O. Rosa, R. Figueiró & J. Nessimian. 2016. Urbanization alters the flow of energy through stream food webs. Insect Conservation and Diversity. DOI: 10.1111/icad.12176.

Dorvillé, L.F.M. 1998. Caracterização Taxonômica e Biologia de uma Espécie de *Kempnyia* Klapálek (Plecoptera, Perlidae) em um riacho da Floresta da Tijuca, Rio de Janeiro (RJ). Mestrado em Ciências Biológicas (Zoologia). Museu Nacional, Universidade Federal do Rio de Janeiro. 120 pp.

Dorvillé, L.F.M. & C.G. Froehlich. 1999. Additional characters to distinguish the nymphs of the perlid genera from southeastern Brazil (Insecta, Plecoptera). Aquatic Insects 21: 281–284.

Dorvillé, L.F.M. & C.G. Froehlich. 2001. Description of the nymph of *Kempnyia tijucana* Dorvillé and Froehlich (Plecoptera, Perlidae), with notes on its development and biology. Trends in Research in Ephemeroptera and Plecoptera 13: 385–392.

Enderlein, G. 1909. Klassifickation der Plecopteropteren, sowie Diagnosen neur Gattungen und Arten. Zoologischer Anzeiger 34: 385–419.

Fochetti, R. & J.M. Tierno-de-Figueroa. 2008. Global diversity of stoneflies (Plecoptera; Insecta) in freshwater. Hydrobiologia 595: 365–377.

Froehlich. C.G. 1984. Brazilian Plecoptera 4. Nymphs of Perlidae genera from south-eastern Brazil. Annales de Limnologie 20: 43–48.

Froehlich, C.G. 1991. Flight Periods of *Kempnyia* and *Macrogynoplax* (Plecoptera: Perlidae) in southeastern Brazil. Pages 353–358 *in*: J. Alba-Tecedor & A. Sanches-Ortega (eds.), Overview and Strategies of Ephemeroptera and Plecoptera, Sandhill Crane Press, Gainesville.

Froehlich, C.G. 2010. Catalogue of Neotropical Plecoptera. Illiesia 6: 118–205.

Grimaldi, D. & M.S. Engel. 2005. Evolution of the insects. Cambrige University Press.

Hamada, N. & S.R.M. Couceiro. 2003. An illustrated key to nymphs of Perlidae (Insecta, Plecoptera) genera in Central Amazonia, Brazil. Revista Brasileira de Entomologia 47: 477–480.

Hamada, N. & J.O. Da Silva. 2014. Ordem Plecoptera. Pages 258–288 *in*: N. Hamada, J.L. Nessimian & R.B. Querino (eds.), Insetos Aquáticos na Amazônia Brasileira: taxonomia, biologia e ecologia. Editora INPA, Manaus.

Hebert, P.D.N., A. Cywinska, S.L. Ball & J.R. deWaard. 2003a. Biological identifications through DNA barcodes. Proceedings of the Royal Society of London B: Biological Sciences 270: 313–321.

Hebert, P.D.N., S. Ratnasingham & J.R. deWaard. 2003b. Barcoding animal life: cytochrome c oxidase subunit 1 divergences among closely related species. Proceedings of the Royal Society of London B: Biological Sciences 270: DOI: 10.1098/rsbl.2003.0025.

Heckman, C.W. 2003. Encyclopedia of South America aquatic insects: Plecoptera. Illustrated keys to known families, genera and species in South America. Kluwerr Academic Publishers, Dordrecht, Boston, London.

Henriques-Oliveira, A.L. & J.L. Nessimian. 2001. Aquatic macroinvertebrate diversity and composition in streams along an altitudinal gradient in Southeastern Brazil. Biota Neotropica 10: http://www.biotaneotropica.org.br/v10n3/en/abstract?article+bn02010032010.

Hynes, H.B.N. 1948. The nymph of *Anacroneuria aroucana* Kimmins (Plecoptera, Perlidae). Proceedings of the Royal Entomology Society of London 23: 105–110.

Hynes, H.B.N. & M.E. Hynes. 1975. Life histories of many of stoneflies (Plecoptera) of south-eastern mainland Australia. Australian Journal of Marine and Freshwater Research 26: 113–153.

Illies, J. 1960. Penturoperlidae, eine neue Plecopterenfamilie. Zoologischer Anzeiger 164: 26–41.

Illies, J. 1963. Revision der südamerikanischen Gripopterygidae (Plecoptera). Mitteilungen der Schweizerischen Entomologischen Gesellschaft 36: 145–248.

Illies, J. 1964. Südamerikanische Perlidae (Plecoptera), besonders aus Chile und Argentinien. Beiträge zur Neotropischen Fauna 3: 207–233.

Illies, J. 1966. Katalog der rezenten Plecoptera. Lieferung 82. Das Terreich, Walter de Gruyter & Co., Berlin.

Junqueira, V.M. & S.C.M. Campos. 1998. Adaptation of the "BMWP" method for water quality evaluation to Rio das Velhas watershed (Minas Gerais, Brazil). Acta Limnologica Brasiliensia 10: 125–135.

Kapoor, N.N. 1972. Rearing and maintenance of Plecoptera nymphs. Hydrobiologia 40: 51–53.

Klapálek, F. 1909. Vorläufige Bericht über exotische Plecopteren. Wiener Entomologische Zeitung, 28: 215–232.

Lecci, L.S. & C.G. Froehlich. 2011. Taxonomic revision of *Gripopteryx* (Pictet, 1841) (Plecoptera: Gripopterygidae). Zootaxa 2792: 1–21.

Mancilla, G., C. Valdovinos, M. Azocar, P. Jorquera & R. Figueroa. 2009. Efecto Del reemplazo de La vegetación nativa de ribera sobre La comunidad de macroinvertebrados bentônicos em arroyos de climas templados, Chile central. Hidrobiológica 19: 193–203.

McCulloch, G.A., G.P. Wallis & J.M. Waters. 2016. A time-calibrated phylogeny of southern hemisphere stoneflies: testing for Gondwanan origins. Molecular Phylogenetics and Evolution 96: 150–160.

McLellan, I.D. 2001. A revision of South American Austroperlidae (Plecoptera). Aquatic Insects 23: 233–251.

McLellan, I.D. & P. Zwick. 2007. New species of and keys to South American Gripopterygidae (Plecoptera). Illiesia 3: 20–42.

Miserendino, M.L. & L.A. Pizzolón. 1999. Rapid assessment of river water quality using macroinvertebrates: a family level biotic index for the Patagonic Andean Zone. Acta Limnologica Brasiliensia 11: 137–148.

Misof, B., S. Liu, K. Meusemann, R.S. Peters, A. Donath, C. Mayer, P.B. Frandsen, J. Ware, T. Flouri, R.G. Beutel, O. Niehuis, M. Petersen, F. Izquierdo-Carrasco, T. Wappler, J. Rust, A.J., Aberer, U. Aspöck, H. Aspöck, D. Bartel, A. Blanke, S. Berger, A. Böhm, T.R. Buckley, B. Calcott, J. Chen, F. Friedrich, M. Fukui, M., Fujita, C. Greve, P. Grobe, S. Gu, Y. Huang, L.S. Jermiin, A.Y. Kawahara, L. Krogmann, M. Kubiak, R. Lanfear, H. Letsch, Y. Li, Z. Li, J. Li, H. Lu, R. Machida, Y. Mashimo, P. Kapli, D.D. McKenna, G. Meng, Y. Nakagaki, J.L. Navarrete-Heredia, M. Ott, Y. Ou, G. Pass, L. Podsiadlowski, H. Pohl, B.N. vonReumont, K. Schütte, K. Sekiya, S. Shimizu, A. Slipinski, A. Stamatakis, W. Song, X. Su, N.U. Szucsich, M. Tan, X. Tan, M. Tang, J. Tang, G. Timelthaler, S. Tomizuka, M. Trautwein, X. Tong, T. Uchifune, M.G. Walzl, B.M. Wiegmann, J. Wilbrandt, B. Wipfler, T.K.F. Wong, Q. Wu, G. Wu, Y. Xie, S. Yang, Q. Yang, D.K. Yeates, K. Yoshizawa, Q. Zhang, R. Zhang, W. Zhang, Y. Zhang, J. Zhao, C. Zhou, L. Zhou, T. Ziesmann, S. Zou, Y. Li, X. Xu, Y. Zhang, H. Yang, J. Wang, J. Wang, K.M. Kjer & X. Zhou. 2014. Phylogenomics resolves the timing and pattern of insect evolution. Science 346: 763. DOI: 10.1126/science.1257570.

Modenutti, B.E., E. Balseiro, M.C. Diéguez, C. Queimaliños & R. Albariño. 1998. Heterogenity of fresh-water Patagonian ecossystems. Ecología Austral 8: 155–165.

Mugnai, R., J.L. Nessimian & D. Baptista. 2010. Manual de Identificação de Macroinvertebrados Aquáticos do Estado do Rio De Janeiro. 1ª Edição. Editora Techinical Books.

Mynott, J., J. Webb & P. Suter. 2011. Adult and larval associations of the alpine stonefly genus *Riekoperla* McLellan (Plecoptera: Gripopterygidae) using mitochondrial DNA. Invertebrate Systematics 25: 11–21.

Novaes, M.C., P.C. Bispo & J. Higuti. 2012. A description of the nymph of *Anacroneuria ofaye* Froehlich (Plecoptera: Perlidae) and an apparatus for rearing Neotropical stonefly species. Zootaxa 3547: 71–77.

Olifiers, M.H., L.F.M. Dorvillé, J.L. Nessimian & N. Hamada. 2004. A. key to Brazilian genera of Plecoptera (Insecta) based on nymphs. Zootaxa 651: 1–15.

Omad, G., P. Pessacq & L.B. Epele. 2015. Spatial distribution, feeding and length mass relationship of *Diamphipnopsis samali* (Plecoptera, Diamphipnoidea) in a North Patagonia Andean stream, Argentina. Revista de la Sociedad Entomológica Argentina 74: 27–35.

Pessacq, P. 2009. A new Gripopterygidae (Plecoptera) species from southern Patagonia. Zootaxa 2264: 51–57.

Pessacq, P. & G. Omad. 2012. Description of the larva and redescription of the adult male of *Diamphipnoa helgae* Illies, 1960 (Plecoptera: Diamphipnoidae). Zootaxa 3341: 64–68.

Poulton, B. 2005. Methods for collection and rearing of stoneflies. Pages 34–36 *in*: R.E. DeWalt, B.C. Poulton, K. Alexander & D. Heimdal (org.), Plecoptera Technical Workshop. Available in: http://plsa.inhs.uiuc.edu/plecoptera/pdf/NABS_Plecoptera_Workshop_2005.pdf.

Ribeiro, J.M.F. 2004. Plecoptera (Insecta) adultos da Reserva Ducke. Dissertação de Mestrado. Programa de Pós-Graduação em Entomologia, Instituto de Pesquisas da Amazônia, 81 pp.

Ribeiro, J.M.F. & I.S. Gorayeb. 2016. Key to Plecoptera nymphs from the Brazilian Amazon (Insecta). Zootaxa 4208: 433–335.

Ribeiro, J.M.F. & J.A. Rafael. 2005. A key to adult Amazonian stonefly genera with new geographical records of *Enderleina* Jewett for Brazil and first description of the *E. froehlichi* Ribeiro-Ferreira female (Insecta: Plecoptera). Zootaxa 1096: 61–67.

Ribeiro, J.M.F. & J.A. Rafael. 2007. Description of one new species and a key to adults of *Macrogynoplax* Enderlein (Plecoptera, Perlidae) from Reserva Florestal Adolpho Ducke, Amazonas, Brazil. Zootaxa 1511: 17–28.

Ribeiro, J.M.F. & J.A. Rafael. 2009. Redescription of the species of *Anacroneuria* Klapálek known from the Reserva Florestal Adolpho Ducke, Amazonas, Brazil with a neotype designation, and a key to adult males, Amazonas, Brazil. (Plecoptera, Perlidae). Zootaxa 2004: 1–15.

Roback, S.S. 1980. III. Plecoptera nymphs. Pages 176–217 *in*: S.S. Roback, O.S. Flint, N. Nieser & P.J. Splangler (eds.), Results of the Catherwood Bolivian-Peruvian altiplano expedition Part I. Aquatic insects except Diptera: Proceedings of the Academy of Natural Science of Philadelphia.

Romero, V.F. 2001. Plecoptera. Pages 93–109 *in*: H.R. Fernandez & E. Dominguez (eds.), Guia para La determination de los Artrópodos Bentónicos Sudamericanos. Universidad Nacional de Tucumán, Tucumán.

Stark, B.P. 1989. The genus *Enderleina* (Plecoptera: Perlidae). Aquatic Insects 11: 153–160.

Stark, B.P. 1995. New species and records of *Anacroneuria* (Klapálek) from Venezuela (Insecta, Plecoptera, Perlidae). Spixiana 18: 211–249.

Stark, B.P., C.G. Froehlich & M.C. Zúñiga. 2009. South American Stoneflies (Plecoptera). Volume 5. Aquatic Biodiversity in Latin America. Pensoft, Sofia-Moscow.

Tierno-de-Figueroa, J.M., B. Gaetani, J.M. Luzón-Ortega, M.J. López-Rodrigues & R. Fochetti. 2011. On the identity of *Isoperla curtata* (Plecoptera: Perlodidae): behavioural and molecular approaches show the existence of two separate species. Zootaxa 3000: 49–58.

Tillyard, R.J. 1921. A new classification of the order Perlaria. Canadian Entomologist 53: 35–44.

Trucotte, P. & P.P. Harper. 1982. Drift patterns in high Andean stream. Hydrobiologia 89: 141–151.

Vera, A. 2006a. Una nueva especie de *Teutoperla* Illies, 1963 para Chile: *Teutoperla maulina* n. sp. (Plecoptera:, Gripopterygidae), con comentarios sobre su biología. Acta Entomológica Chilena 30: 29–38.

Vera, A. 2006b. Una nueva especie de Gripopterygidae de Chile, *Potamoperla testacea* n.sp. (Insecta: Plecoptera). Gayana 70: 168–175.

Vera, A. 2006c. Redescubrimiento de *Neopentura semifusca* Illies 1965 (Plecoptera: Gripopterygidae), descripción del imago macho, redescripción de la hembra y la ninfa. Revista de la Sociedad Entomológica Argentina 65: 69–77.

Vera, A. 2008. Una nueva especie de *Chilenoperla* (Plecoptera: Gripopterygidae) y las consecuencias taxonómicas del descubrimiento de su ninfa. Gayana 72: 144–156.

Vera, A. 2009. *Pehuenioperla llaima* a new genus and species of Gripopterygidae (Plecoptera) from South America. Revista de la Sociedad Entomológica Argentina 68: 317–327.

Wais, I.R. 1984. Two Patagonian basins – Negro (Argentina) and Valdivia (Chile) – as habitats for Plecoptera. Annals of Limnology 20: 115–122.

Wanzen, K.M. & R. Wagner. 2006. Detritus processing by invertebrate shredders: a neotropical-temperate comparision. Journal of the North American Benthological Society 25: 216–232.

Zwick, P. 1973. Insecta: Plecoptera. Phylogenetisches System und Katalog. Das terreich 94. Walter de Gruyter, Berlin.

Zwick, P. 2000. Phylogenetic System and Zoogeography of the Plecoptera. Annual Review Entomology 45: 709–746.

Chapter 5

Order Orthoptera

Ana Lúcia Nunes Gutjahr
Department of Natural Sciences, State University of Pará, Belém, Pará, Brazil

Carlos Elias de Souza Braga
Department of Natural Sciences, State University of Pará, Belém, Pará, Brazil

Chapter Outline

INTRODUCTION

The Orthoptera encompasses insects of different forms, habits and behavior and has its highest expression in the locusts that are cosmopolitan and are well represented in the Amazon region. Orthopterans have more than 28,000 valid species (Cigliano et al.. 2018) and therefore are considered the seventh largest group of Insecta and the second most diverse of hemimetabolous insects. This order has representatives of great ecological and economic importance and their species can be classified by their feeding habits: phytophagous (most), predators, and omnivores (Nunes-Gutjahr, 2009; Buhl et al., 2011).

The superfamily Acridoidea, which are the true locusts (Amédégnato & Devriese, 2008; Nunes-Gutjahr & Braga, 2012), includes terrestrial species using the most varied environments and plant debris as well as semiaquatic species (Amédégnato, 2003). Currently 11 families are known, with 43 subfamilies and 8,039 valid species, of which only one family (Acrididae), three subfamilies (Leptysminae, Pauliniinae, and Marelliinae), and 47 species are considered as semiaquatic. The life cycles of the aquatic species are associated with emergent aquatic plants or with herbaceous or graminacea of wetlands. Most species are host specific for feeding and oviposition on host plants.

Most semiaquatic locusts live and develop in association with tropical aquatic macrophytes in wetlands and rivers, including in the Amazon basin (Nunes et al., 1992) or are found within low-lying areas that are periodically flooded, including those within forests. While macrophytes are important for environmental balance, they can also become a big problem when subjected to environmental change. Therefore, it is important to the plant communities and their associated aquatic fauna to understand the environmental impacts of changing aquatic environments. In this context, the locusts living in aquatic weeds can be good allies in the biological control of these plants (Braga et al., 2013), as is the orthopteran *Cornops aquaticum* (Bruner) that develops in Pontederiaceae colonies and is not considered a kind of plague cultivar (Centre for Overseas Pest Research – COPR, 1982).

Species from semiaquatic habits have morphological and physiological adaptation to the aquatic environment.

Thorp and Covich's Freshwater Invertebrates, Fourth Edition. https://doi.org/10.1016/B978-0-12-804223-6.00005-6

FIGURE 5.1 A Lemenaceae (Arales) population in body of fresh water in a forest environment.

Such adjustments are essential to facilitate movement in the water, as these locusts can swim and dive. A good example of this adaptation is related to adult *Stenacris fissicauda fissicauda* (Bruner) (Acrididae, Leptysminae) which may remain submerged for a period of 4–7 min and their nymphs for up to 1 min (Nunes, 1989). This ability to dive is considered a kind of exit strategy to avoid predation, especially by birds in colonies of macrophytes, where these insects live (Nunes, 1989).

Among the different taxonomic groups of Orthoptera, the Acridomorpha of Acridoidea and Tetrigoidea superfamilies have semiaquatic species or have close relationship with freshwater, although Amédégnato & Devriese (2008) admitted that most species of Tetrigoidea occur in limno-terrestrial habitats. However, some species of Tetrigoidea depend on aquatic habitats and can swim and feed on aquatic weeds, as well as lichens, mosses, and debris (Bhalerao et al., 1987; Kocárek et al., 2008). In the Amazon region it is common to find Tetrigoidea species in wet areas inside the forests with intense populations of Lemnaceae (Fig. 5.1), on which they feed (field observation). Other representatives of Orthoptera, such as some species of Gryllidae and Tettigoniidae, living in association with colonies of aquatic weeds are also good swimmers.

Vegetation Used by Semiaquatic Locusts

The semiaquatic locusts are usually associated with a variety of plant species, including floating vegetation (macrophytes) and fields of graminacea or other plants adapted to the aquatic environment. They use this vegetation as a host plant when they have specific dietary or oviposition requirements. They can also use temporary or secondary host plants when necessary. The use of temporary host plants is common in regions with extreme seasonal water fluctuations, as occurs in the flooded areas of Central Amazon, where water levels may fluctuate up to 10 m over a period of five to six months (Junk, 1980).

The locust-macrophyte associations in the Neotropics may occur between a species of grasshopper and more than one plant and even more than one species of locusts exploiting a single plant species, as can be seen in the species *Paulinia acuminata* (Pauliniinae) and macrophytes *Pistia stratiotes* and *Salvinia auriculata*; *Marellia remipes* (Marelliinae) with *Salvinia auriculata* and *wide Nymphaea; C. aquaticum* (Leptysminae) and different species of Pontederiaceae of *Eichhornia* genres (*E. crassipes and E. azurea*) and Pontederia (*P. cordata and P. rotundifolia*); *S. f. fissicauda and Tucayaca gracilis* (both Leptysminae), living in sympatric in Poaceae *Paspalum repens* colonies (Nunes et al., 1992; Nunes & Adis, 1994; Bownes et al., 2010; Franceschini et al., 2011 Capello et al., 2012; Braga et al., 2011; Oliveira et al., 2015).

Adaptations to the Aquatic Environment

The morphological adaptations of semiaquatic species of locusts include fusiform body shape, well-developed muscles on the third pair of femurs and enlarged hind tibia (paddle-shaped, adapted for swimming), thorns, and modified tibial spurs bearing hydrophobic bristles (Roberts, 1978; Amédégnato & Devriese, 2008) (Fig. 5.2). The Acrididae family includes the best semiaquatic representatives where morphological characteristics are quite evident.

When submerged, locusts use a plastron for aeration similar to those employed by certain beetles and water

FIGURE 5.2 Aspect of the hind tibia of a semiaquatic locust (Acrididae).

bugs of Hydrophilidae and Naucoridae families, respectively. In *S. f. fissicauda* (Acrididae: Leptysminae), the plastron is characterized as an elongated air bubble which is located at the side of the body between the tegminas and the abdomen or between the hind femoral and abdomen (most common in nymphs) (Nunes, 1989). Diving not only provides access to some food types but also is a strategy in *C. aquaticum* and *P. acuminata* to escape predators (Lhano, 2002).

Most locust species oviposit in the soil, however semiaquatic species can oviposit on the host plant surface or inside tissues (Nunes, 1989; Nunes et al., 1992; Amédégnato, 2003; Amédégnato & Devriese, 2008). The species that perform endophytic oviposition (inside the plant tissue) (Fig. 5.3 A) have the female valves indented, adapted to cut the tissue of their host macrophytes (Amédégnato & Devriese, 2008), as noted in *S. f. fissicauda* and *C. aquaticum*. When ovipositing on plant surfaces, the eggs are placed at the bases of the grooves or slits leaf petiole plants (Fig. 5.3 B), as observed in *Tetrataenia surinama* (Linnaeus), or they may lay their eggs underwater by sticking groups of eggs on roots of host macrophytes, with this type of oviposition characteristic of *P. acuminata* species mainly Salvinaceae (Fig. 5.3 C).

Another adaptation of these locusts is related to morphological and chromatic mimicry of their host plants which can be observed for locusts associated with populations of floating grasses (Poaceae and Cyperaceae) of *Stenacris* genera and *Cylindrotettix*.

LIMITATIONS

The existing knowledge on semiaquatic Acridoidea is limited, and current studies mostly concern the host specificity, food consumption, general ecology, and life cycle, and this research is limited to only a few species. However, taxonomical research of semiaquatic groups is mostly restricted to descriptions of species and their revisions, but even this research is limited. This is understandable considering that the taxonomists who study Acridoidea are dedicated almost exclusively to specific groups and to terrestrial species in general.

We see the need for a publication that compiles all available information about these locusts. Such a publication on semiaquatic and terrestrial Acridoidea should describe standard protocols for locust inventories. The absence of such a publication prevents development of easily comparable studies in many different locations, contributing to a huge bottleneck in the knowledge of the dynamics of these insects.

Another major problem that hinders bioecological studies and biodiversity inventories is the lack of an adequate identification key for semiaquatic Acridoidea. Finally, we need to overcome the lack of training of specialists for this taxonomic group of insects.

TERMINOLOGY AND MORPHOLOGY

The semiaquatic locusts are morphologically very similar to terrestrial locusts, however, their enlarged hind tibia facilitates swimming. Morphological structures in the

FIGURE 5.3 Aspects of egg mass of semiaquatic locust eggs (Acrididae): (A) endophytic posture of eggs; (B) in petiole slit of plant leaves; (C) in the roots of aquatic plants.

head, thorax, and abdomen are important for taxonomic identification of the subfamilies, tribes, the genera, and species of semiaquatic locusts. Only adults have taxonomic value in identifying genera and species, however, the immature can be used for identification of higher taxonomic levels.

The locusts' head has great taxonomic importance, in relation to changes in its shape, color, structures, and adornments. The frons may be flat or convex; the fastigium is a prolongation of the vertex and forehead, located between the compound eyes, which may be absent and is formed and with variable size when it is present; the locusts have three ocelli; the compound eyes can be globular or elongated and close or far apart (interocular space) (Fig. 5.8); the foveolas (Fig. 5.4) or folds are depressions located in the upper part of the head; the antennae can be filiform, ensiformis, or subensiformis (Fig. 5.9). The mouthparts are termed opisthognathous when directed backward or hypognathous (orthognatous) when directed vertically.

The thorax is one of the most important segments taxonomically, as colors of the prothorax, mesothorax, metathorax, and appendages (wings and legs) are quite variable with blotches, bands, stripes, or features in suprageneric through specific taxa. The amount and disposition of colors on the side, middle, and back are important along with the presence or absence of wrinkles and calluses. The pronotum is the largest and most developed thoracic sclerite and is important for identification of subfamilies, tribes, genera, and species. It is divided into two portions, one anterior, the prozone, and another posterior, the metazone (Fig. 5.4). The proesternal spine or tubercle (ventral bulge prothorax) is important to distinguish the three subfamilies. In the Leptysminae, it can be high or elongated (width equal to or less than the height), espiniforme, or stellular. In the Marellinae and Pauliniinae it is low (greater width height) or compressed frontally and subrectangular. Legs have three tarsomeres; the first two are ambulatory (used to walk), while the third pair is the largest, most developed, and adapted for

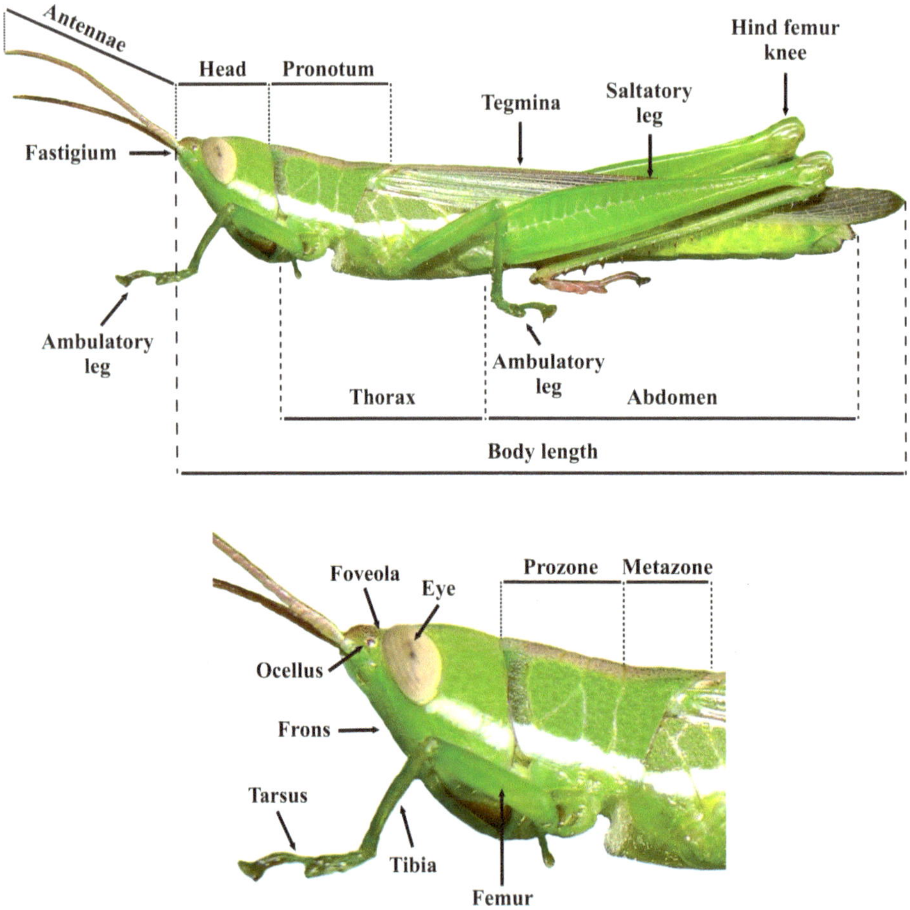

FIGURE 5.4 Representation of the external morphological structures of a semiaquatic grasshopper (Acrididae).

jumping (saltatory leg) (Fig. 5.4). In these semiaquatic species, these legs enable the locust to swim because the tibias are expanded and remiform with hydrophobic bristles (Fig. 5.7 B). The first pair of wings is coriaceous (tegmina) and the second is membranous. Wings are always present and be brachypterous (not exceeding the last abdominal segment) or macropterous (extending beyond the abdominal end). The wings may vary in color with different arrangements of the ribs, which are important taxonomic characters. In Leptysminae, the wings are absent (apterous) in the terrestrial tribe Chloropseustini.

The abdomen is sessile and has 10 to 11 metameres (segments). The eardrum (hearing organ) is on the first abdominal segment in the three semiaquatic subfamilies (Leptysminae, Marellinae, and Pauliniinae) but is absent in the terrestrial Chloropseustini tribe of Leptysminae. The last abdominal segment is of greatest taxonomic importance because of the presence of either male (aedeagus or phallus) or female (ovipositor) genitalia and accessory structures, especially the highly variable subanal and subgenital plates (Fig. 5.6). In females the ovipositor valves are variable; and in most semiaquatic species they are heavily indented and adapted to endophytic oviposition (inside the leaf petiole) or smooth and adapted to epiphytic oviposition (under leaves, petioles or fold petiole, or the stems of host plants). The cercus of females can be conical or cylindrical. In males, phallic complex, which is formed by three characteristic structures or sclerites (endophallus, ectophallus, and epiphallus) vary in size and shape. The subanal plate can be ornamented (wrinkles or calluses) with different pigmentation among species. The cercus of the males can be tapered, cylindrical, or curved as in Leptysminae.

MATERIAL PREPARATION AND PRESERVATION

Semiaquatic and terrestrial locusts are prepared and preserved in the same way. They must be preserved dry, or mounted using stainless steel entomological pins and dried in an oven for at least 48 h (Almeida & Câmara, 2008) at a temperature of 50–60°C. The pin should pierce the posterior portion of the pronotum, immediately to the right of the body midline (Triplehorn & Johnson, 2011). This preparation procedure should be followed very soon after collection to avoid the rapid decomposition that can occur in the tropics.

Locusts that cannot be immediately transported to the laboratory should remain in wet beds (temporary holding containers with paper tissues soaked with ethyl acetate on the bottom of the container) to avoid hardening or they may be placed in bags or paper envelopes for a natural sun drying. Upon arriving at the lab, this dry material should be placed in a rehydration moist chamber to soften the locust and allow proper assembly without breaking the specimen. Another form of temporary preservation of locusts is to place them in plastic bags in a freezer where they will retain their color and not dehydrate. After

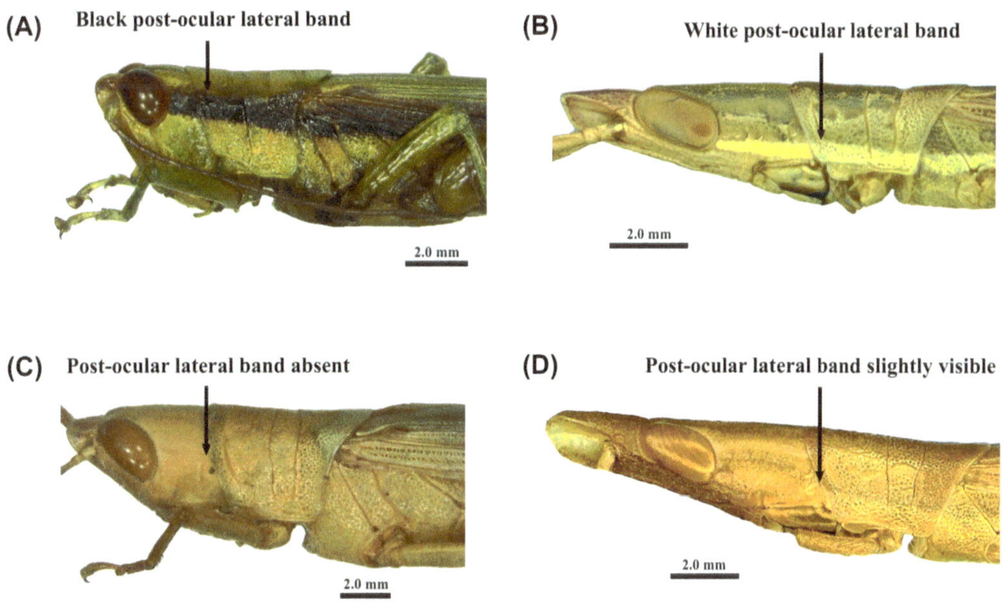

FIGURE 5.5 Chromatic characteristics of semiaquatic grasshoppers (Acrididae): (A) black postocular band; (B) white postocular band; (C) postocular band absent; and (D) poorly visible postocular band.

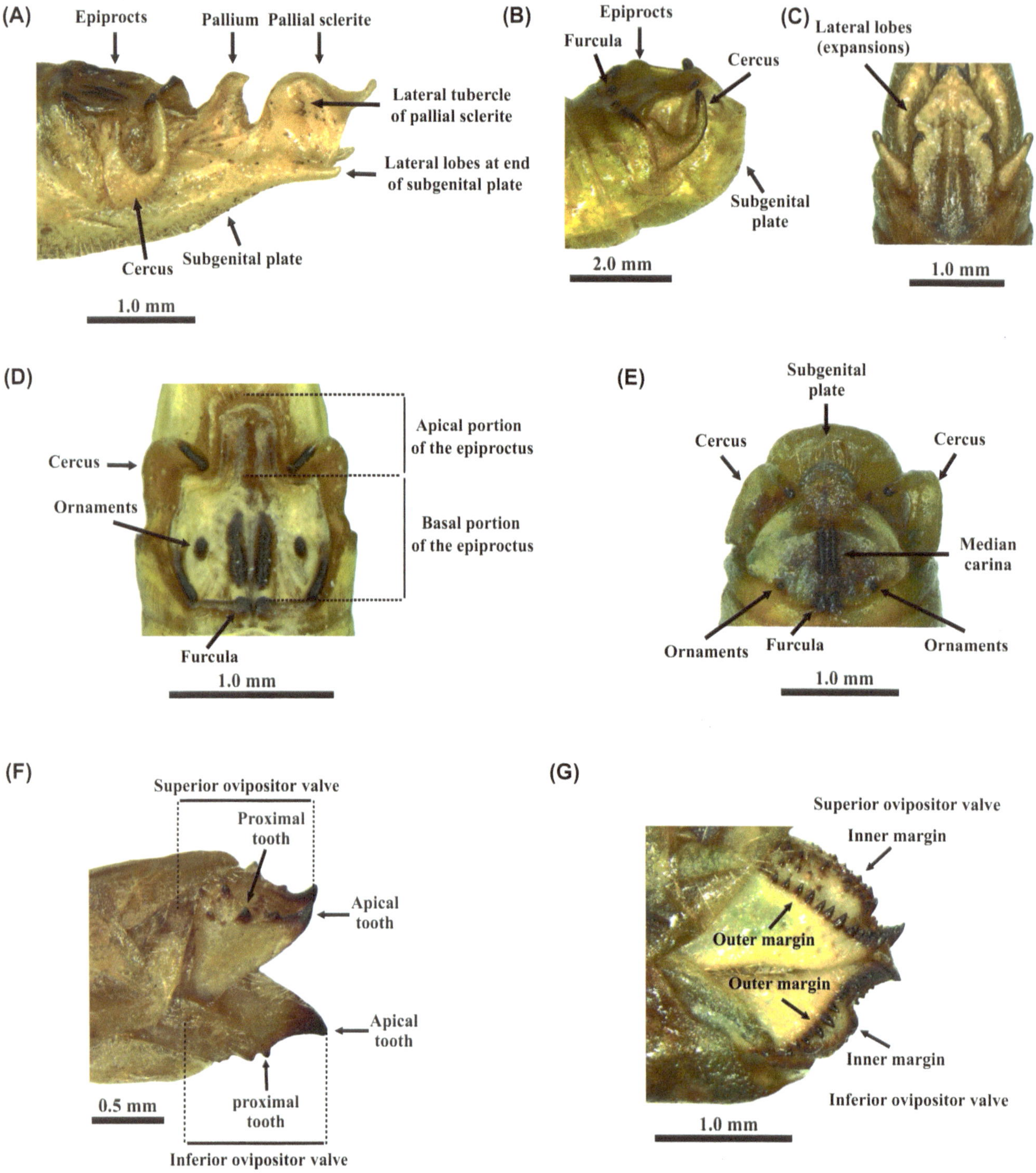

FIGURE 5.6 Morphological aspects of terminalia structures of semiaquatic grasshoppers (Acrididae): (A–B) male, in lateral view; (C–E) male, in dorsal view; and (F–G) female, in lateral view.

thawing and pinning the specimen, it will appear as if it has just been collected.

The locusts should not be preserved in a humid environment (80% alcohol) because they will lose their coloring, which is a very important taxonomic characteristic. Only immature specimens, which are not used in the specific identifications can be preserved in alcohol.

Label the pinned locust with the origin data of the specimens (country, state, county, and city), date,

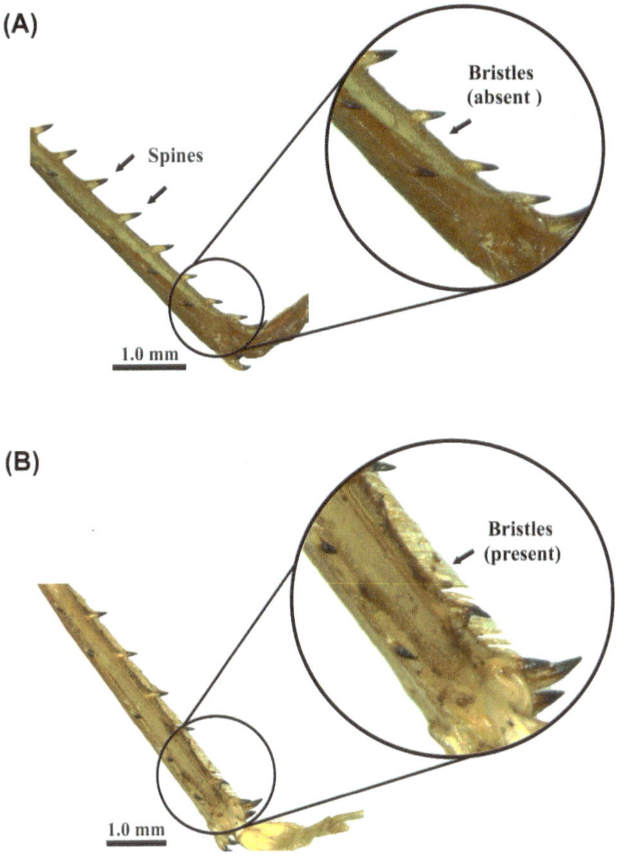

FIGURE 5.7 Aspect of the internal and external margins of the saltatory leg tibia of semiaquatic grasshoppers (Acrididae): (A) absence of a row of short bristles; and (B) presence of a row of short bristles.

collector, collection method, and other field information, such as geographic coordinates and the locust's habitat characteristics and collection environment, as well as other relevant information. These labels must be made of hard white paper (Canson type, 120 g weight) and preferably be printed on a laser printer, so that the letters will not fade. Handwritten labels are no longer considered acceptable.

After the locusts are prepared and preserved, the final packaging should be in an entomological drawer that needs to be kept in dry and dark conditions, as moisture enables the development of fungi and light contributes to fading chromatic specimens, respectively. We emphasize the need to keep into entomological drawers, a container with mothballs to chase insects away that feed on the collectible specimens as the ones belonging to Psocoptera orders (Liposcelidae), Hymenoptera (Formicidae), and Coleoptera (Dermestidae), and also to contribute decrease of humidity inside the entomological drawer.

KEYS TO SEMIAQUATIC GRASSHOPPERS

The identification keys for Neotropical semiaquatic grasshopper species is based on taxonomic characteristics common to all locusts and some exclusive to semiaquatic groups. The semiaquatic grasshoppers are those that live in association with floating aquatic vegetation (aquatic macrophyte colonies) and riparian vegetation, where they develop their life cycle and can swim and dive with great ease (Nunes, 2011), or those living in vegetation located in

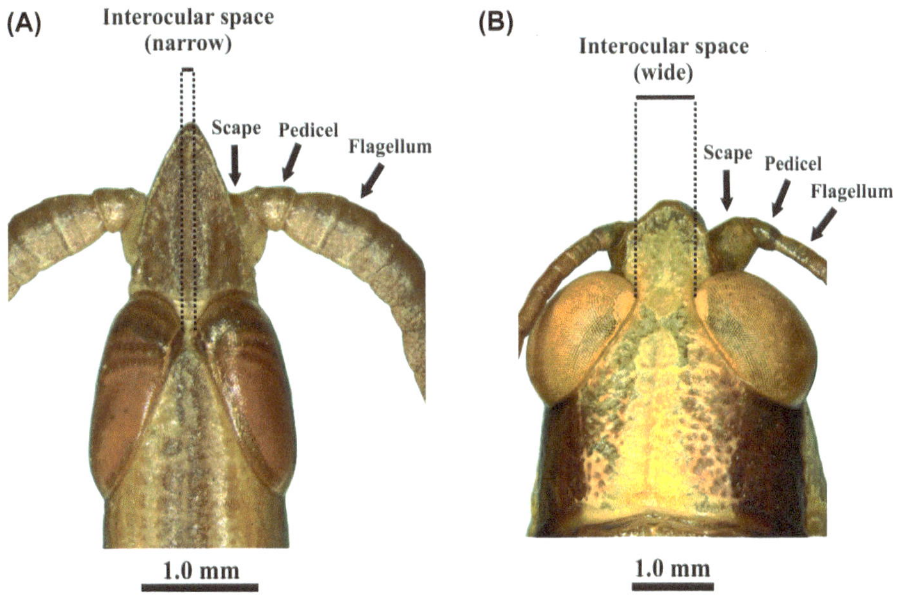

FIGURE 5.8 Space between compound eyes (interocular space) of semiaquatic grasshoppers (Acrididae): (A) narrow; and (B) wide.

FIGURE 5.9 Types of semiaquatic grasshopper (Acrididae) antennae: (A) filiform; (B) ensiform (articles of the antennal flagellum flattened until the 1/3 posterior area); and (C) subensiform (articles of the antennal flagellum flattened near the base).

shallow areas (puddles), which are very humid or are flooded periodically (Nunes-Gutjahr & Braga, 2014). Life in those habitats requires specific anatomical structures, which are emphasized in Figs. 5.4–10. Keys are applicable to subfamilies, tribes, genera, and species of semiaquatic grasshoppers classified in the Acridomorpha infraorder, Acridoidea superfamily, Acrididae family and Marelliinae, Pauliniinae, and Leptysminae subfamilies. Regarding these taxa, it should be clarified that the some of the tribes, genera, or species of the mentioned subfamilies, although recognized as semiaquatic, have terrestrial habits and, therefore, will not be treated in the keys. Species belonging to the Tetrigidae family will also not be addressed, since most species are classified as limno-terrestrial, although some exhibit swimming adaptations and are found on the banks of rivers and lakes. The genera belonging to some monospecific subfamilies and tribes (that present only one species) will be presented directly in the keys when entering the dichotomies referring to these taxa.

This key covers 3 subfamilies, 2 tribes, 17 genera, and 47 species.

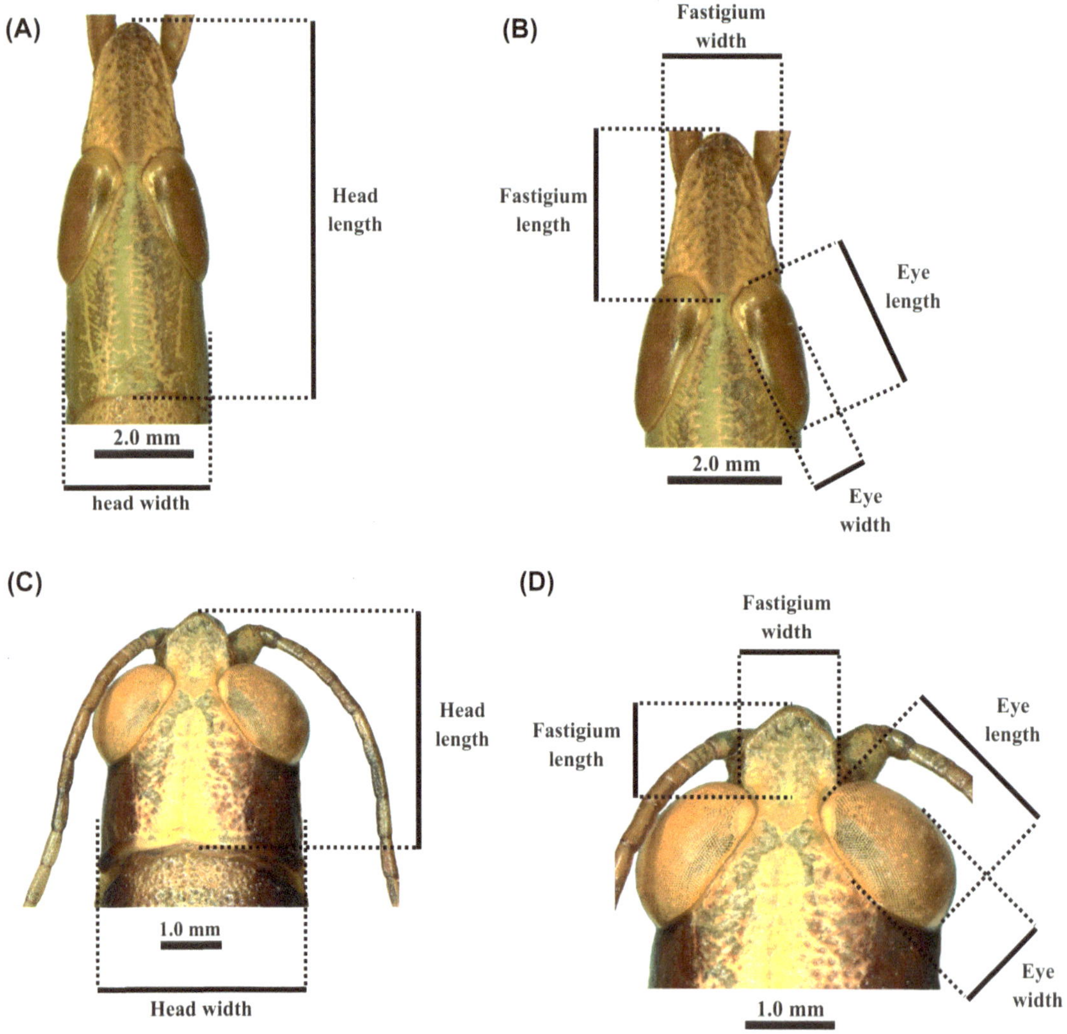

FIGURE 5.10 Head morphology of semiaquatic grasshoppers (Acrididae): (A) long head; (B) long fastigium; (C) short head; and (D) short fastigium.

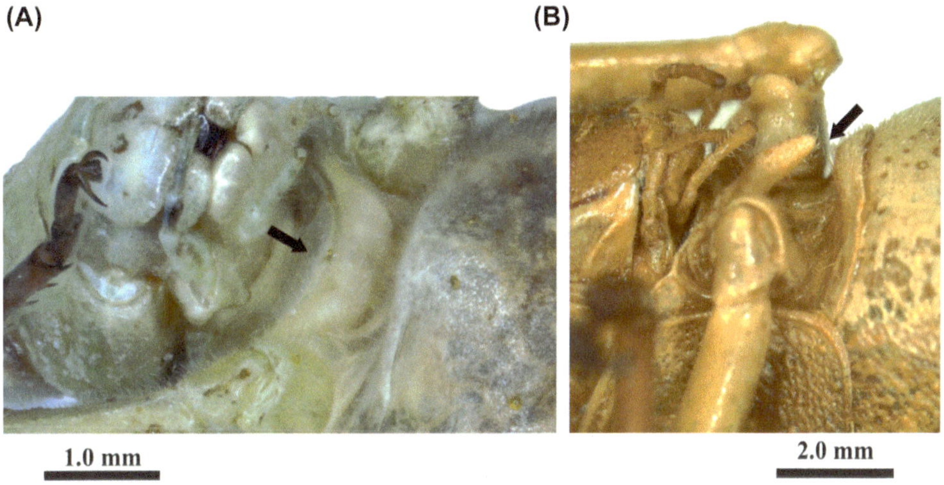

FIGURE 5.11 Morphology of the prosternal tubercle of semiaquatic grasshoppers (Acrididae): (A) low prosternal tubercule (ventral view); and (B) high prosternal tubercule (latero-ventral view).

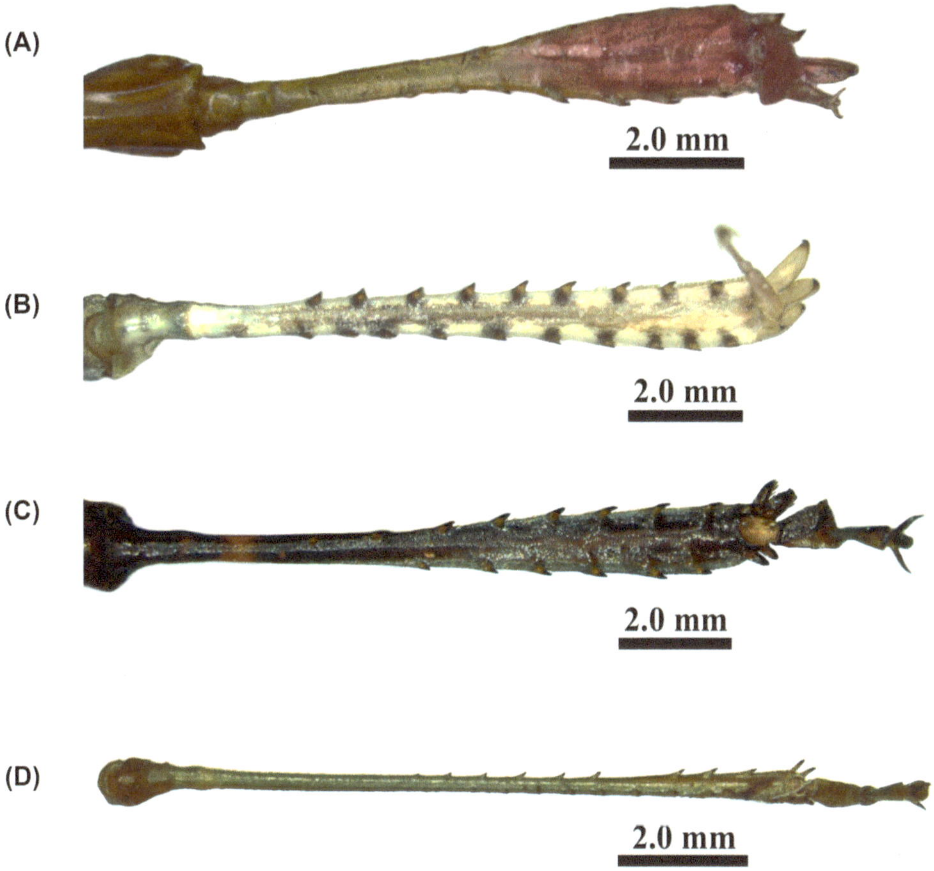

FIGURE 5.12 Morphology of the saltatory legs of semiaquatic grasshoppers (Acrididae): (A–C) strongly enlarged hind tibia; and (D) slightly enlarged hind tibia.

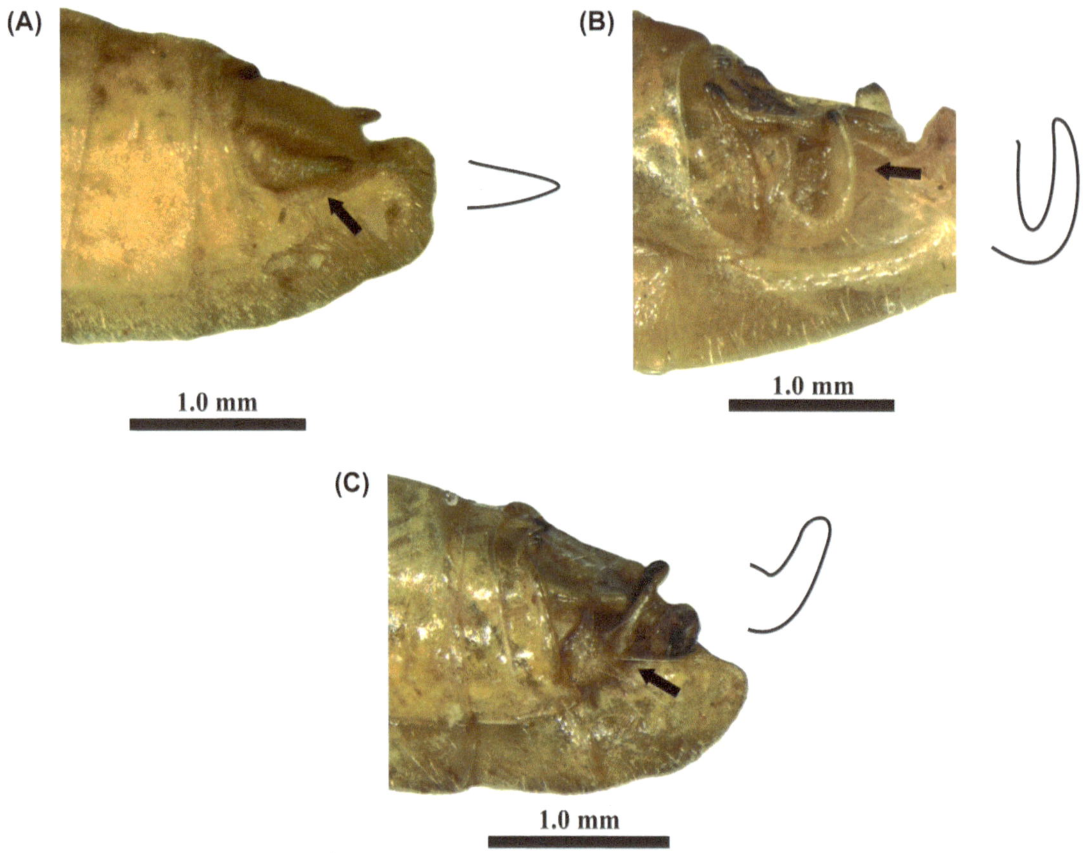

FIGURE 5.13 Cerci morphology of semiaquatic grasshopper (Acrididae) genitalia: (A) conical (nonrecurved) cerci; and (B, C) upwardly curved cerci.

Orthoptera: Acrididae: Subfamilies

1. Prosternal tubercle low, anteriorly compressed and subrectangular (width > breadth) (Fig. 5.11 A); hind tibia strongly dilated (Figs. 5.12 A, B); males with conical (nonrecurved) cerci (Fig. 5.13 A) 2

1' Prosternal tubercle high, pointed, or varied (width < height) (Fig. 5.11 B); hind tibia strongly (Fig. 5.12 C) or slightly enlarged (Fig. 5.12 D); males with dorsally curved cerci (Figs. 5.13 B, C) (except for *Xenismacris*) .... **Leptysminae [p. 152]**

2(1) Hind tibia enlarged from the posterior third (apical 1/3) (Fig. 5.12 A); male subgenital plate long (width < length) (Fig. 5.14 A), conical without lateral lobes (Fig. 5.14 B); male epiprocts hastate, with two lateral lobes (Fig. 5.14 C); female subgenital plate long (Fig. 5.14 D)........ Marelliinae, one species: *Marellia remipes* Uvarov

2' Hind tibia gradually enlarged from the base (apical 2/3) (Fig. 5.12 B); short subgenital plate in males (width subequal to length) (Fig. 5.14 E), trapezoidal with two lateral lobes (Fig. 5.14 F); epiprocts without lateral lobes in males (Fig. 5.14 G); short subgenital plate in females (Fig. 5.14 H) .... Pauliniinae, one species: *Paulinia acuminata* (De Geer)

Orthoptera: Leptysminae: Tribes

1 Black postocular lateral band present (except *Haroldgrantia*) (Fig. 5.15 A); males cerci curved (except *Xenismacris*) never hamulate (Fig. 5.13 C); male subgenital plate short (length subequal to width) (Fig. 5.15 B) .... **Tetrataeniini [p. 153]**

1' Black postocular lateral band absent (Fig. 5.15 C); males with dorsally curved cerci, hamulate (Fig. 5.13 B); male subgenital plate long (length > width) (Fig. 5.15 D)........ **Leptysmini [p. 160]**

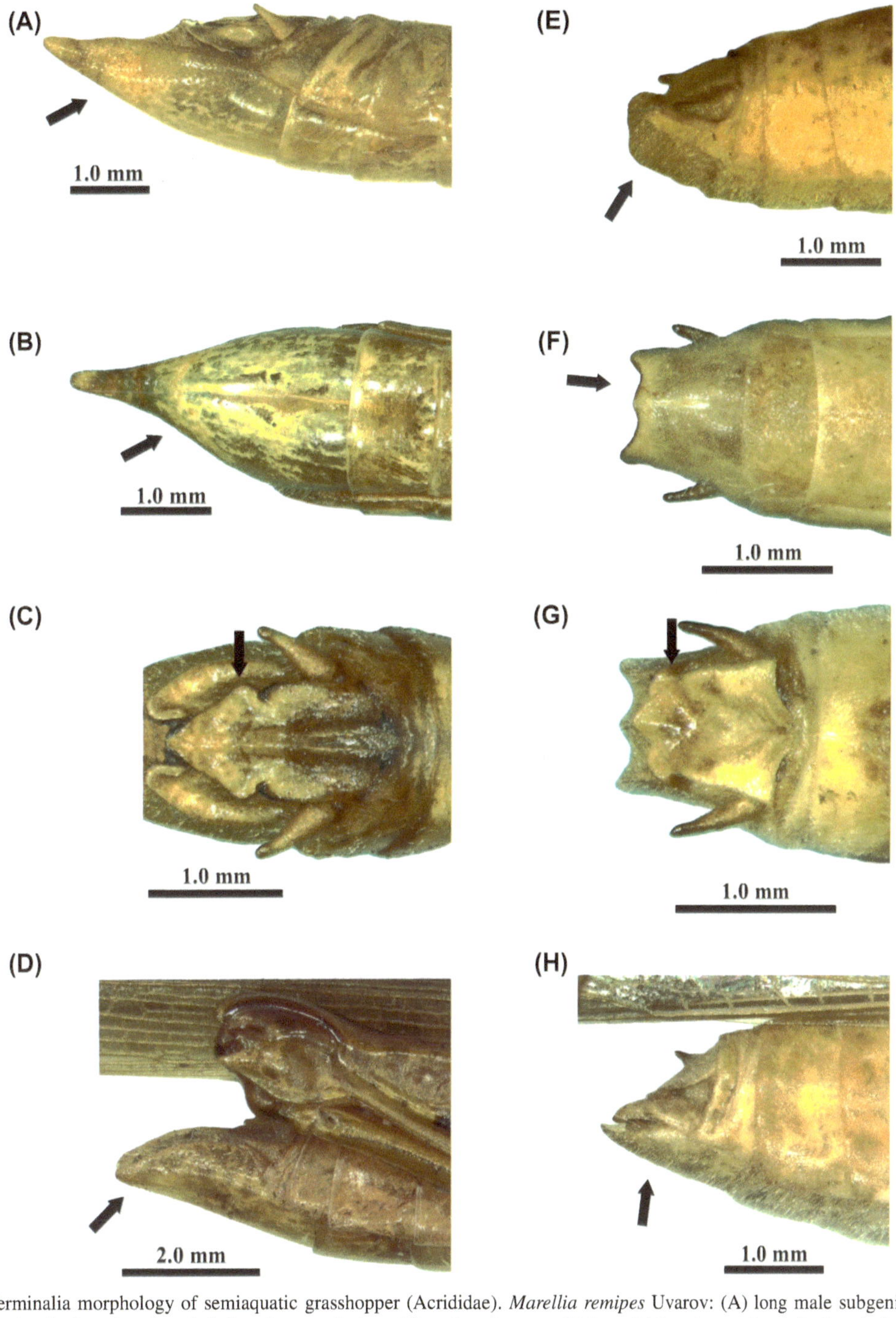

FIGURE 5.14 Terminalia morphology of semiaquatic grasshopper (Acrididae). *Marellia remipes* Uvarov: (A) long male subgenital plate, in lateral view; (B) triangular (conical) male subgenital plate, in ventral view; (C) male epiprocts with lateral lobes (expansions), in dorsal view; (D) long female subgenital plate. *Paulinia acuminate* (De Geer: (E) short male subgenital plate, in lateral view; (E) subsquare (trapezoidal) male subgenital plate, in ventral view; (G) male epiprocts without lateral lobes, in dorsal view; and (H) short female subgenital plate, in lateral view.

Orthoptera: Leptysminae: Tetrataeniini: Genera

1	Males with dorsally curved cerci (Fig. 5.16 A)	2
1’	Males with conical cerci (nonrecurved) (Fig. 5.16 B)	***Xenismacris* [p. 157]**
2(1)	Black postocular lateral band present (Fig. 5.17 A)	3

FIGURE 5.15 Morphological characteristics of the semiaquatic Leptysminae (Acrididae) tribes: Tetrataeniini: (A) presence of a black postocular band (lateral view); and (B) short subgenital plate (ventral view); Leptysmini: (C) absence of a black postocular band (lateral view); and (D) long subgenital plate (ventral view).

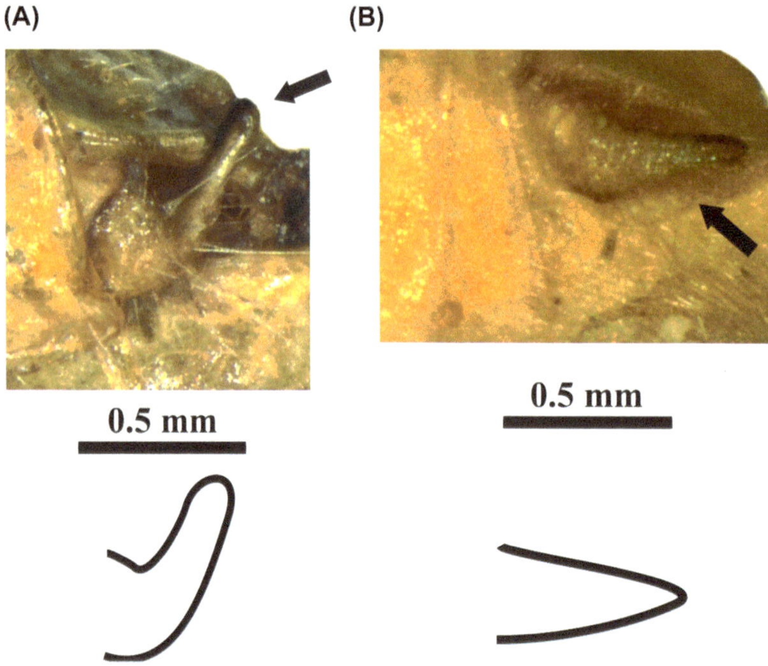

FIGURE 5.16 Morphological Tetrataeniini (Acrididae) cerci aspects: (A) upwards recurved cerci in lateral view; and (B) conical (nonrecurved) cerci in lateral view.

2' Black postocular lateral band absent (Fig. 5.17 B) .... *Haroldgrantia lignosa* Carbonell, Ronderos, and Mesa

3(1) Well-developed wings (macroptera) in both sexes, surpassing abdominal segment five (Fig. 5.18 A); superior and inferior ovipositor valves serrated (Fig. 5.18 B).. 4

3' Male brachipterous, wings not exceeding abdominal segment five (Fig. 5.18 C), winged females; superior ovipositor valves slightly serrated and inferior valves smooth (nonjagged) (Fig. 5.18 D) .... *Eumastusia koebelei* (Rehn)

4(3) Frons straight or convex in lateral view (Fig. 5.19 A) 5

4' Frons concave in lateral view (Fig. 5.19 B) .... ***Stenopola* [p. 157]**

5(4) Both male and female genitalia length > width, in lateral view (Figs. 5.20 A, B); male median femur with apical spiniform process (Fig. 5.21 A); male cerci slightly curved upwards (Fig. 5.21 B)... *Tetrataenia surinama* (Linnaeus)

5' Both male and female genitalia length < width, in lateral view (breadth subequal to length) (Figs. 5.20 C, D); male mesofemur without an apical spiniform process (Fig. 5.21 C); male cerci strongly curved dorsally (Fig. 5.21 D) .... ***Cornops* [p. 159]**

FIGURE 5.17 Morphological Tetrataeniini (Acrididae) aspects: (A) presence of a black postocular band (lateral view), *Haroldgrantia lignosa* Carbonell, Ronderos, and Mesa; and (B) absence of a black postocular band (lateral view).

FIGURE 5.18 Morphological Tetrataeniini (Acrididae) characteristics: (A) well-developed wings (macropterous grasshoppers), lateral view; (B) serrated upper and lower valves, lateral view; (C) under-developed wings (brachypterous grasshoppers), dorsolateral view; and (D) slightly serrated upper valves and flat inferior valves, lateral view.

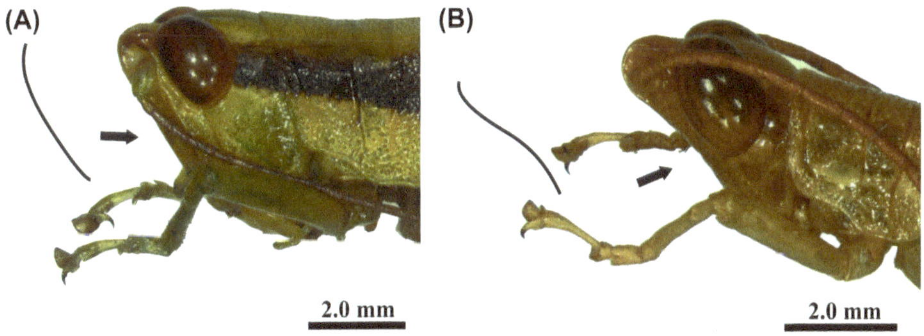

FIGURE 5.19 Head morphology of semiaquatic grasshoppers (Acrididae), in lateral view: (A) head with a straight or convex forehead; and (B) head with a concave forehead.

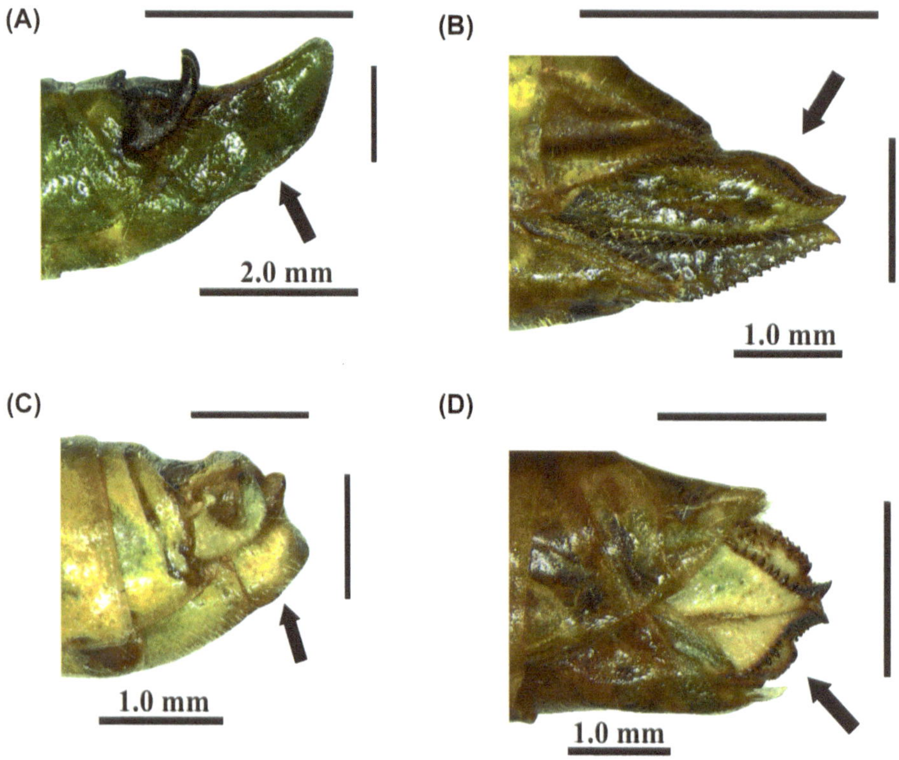

FIGURE 5.20 Morphology of Tetrataeniini (Acrididae) terminalia. *Tetrataenia* (A) male external genitalia; (B) female external genitalia. *Cornops* (C) male external genitalia; and (D) female external genitalia.

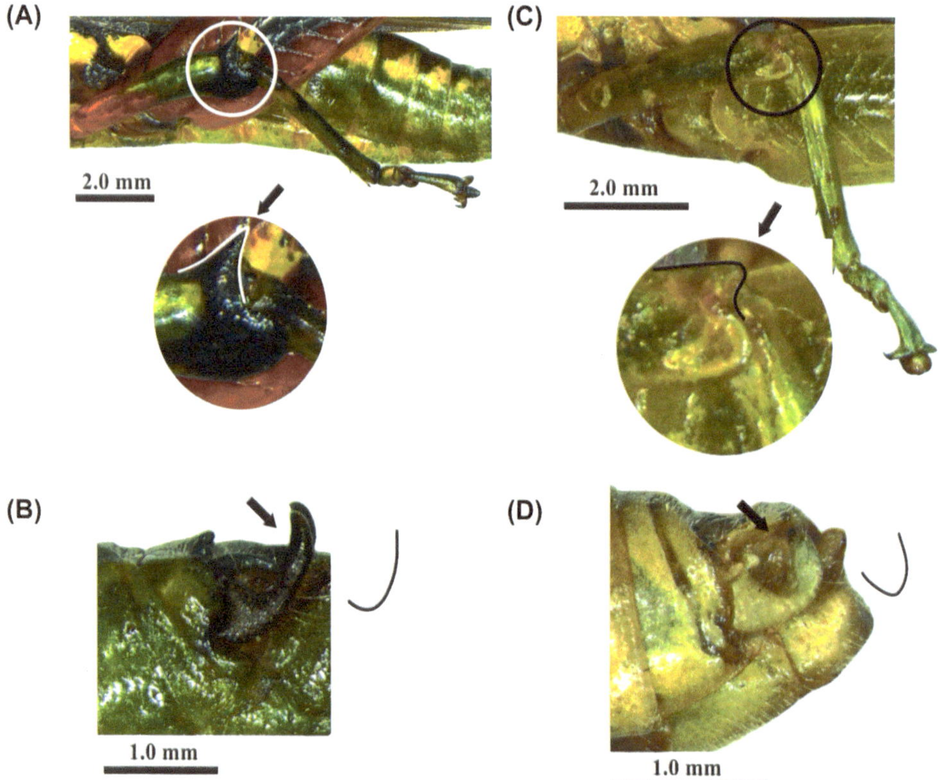

FIGURE 5.21 Morphological aspects of Tetrataenini (Acrididae) median legs and male cerci. *Tetrataenia*: (A) spinous process present at the apex of the median femur; (B) slightly upwards recurved cerci. *Cornops*: (C) Spinous process absent at the apex of the median femur; and (D) strongly upwards recurved cerci.

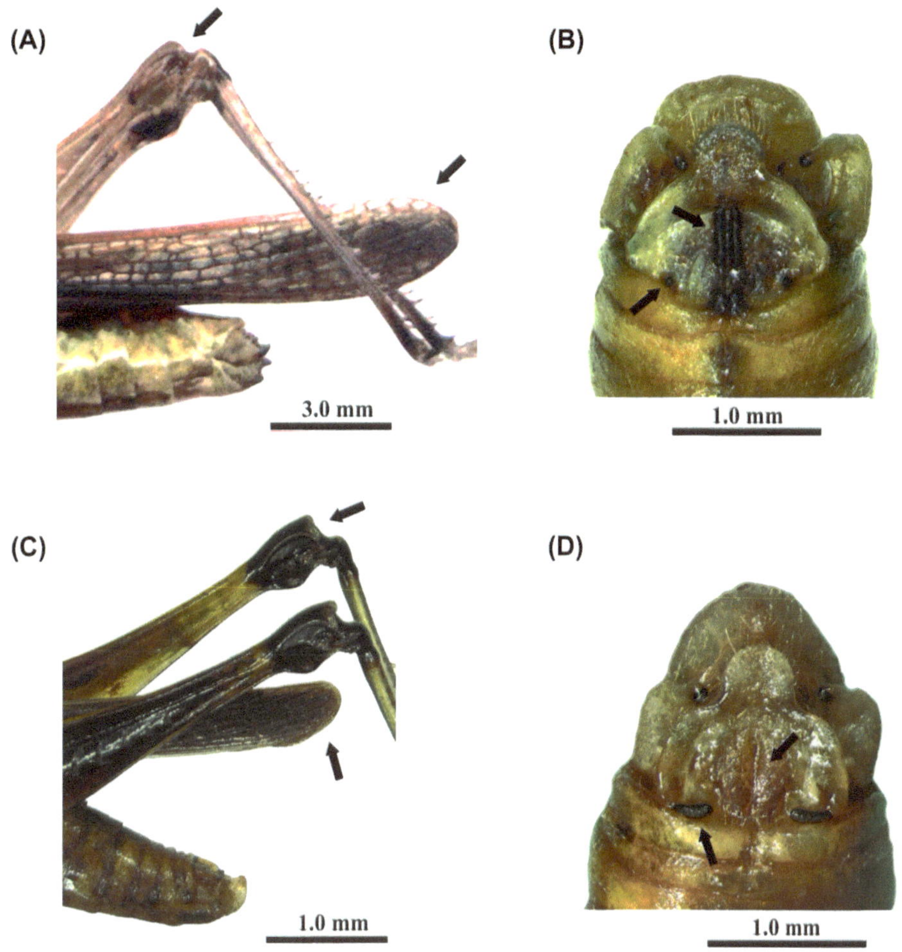

FIGURE 5.22 Morphological aspects of *Cornops*. *Cornops* spp.: (A) light brown hind knee (*top arrow*) and tegmina exceeding the knee (*bottom arrow*), in lateral view; (B) epiprocts with median carina (*top arrow*) and with a circular callosity in the basal portion (*bottom arrow*), in dorsal view. *Cornops brevipenne* Roberts and Carbonell; (C) black hind knee (*top arrow*) and tegmina do not exceed the knee (*bottom arrow*), in lateral view; and (D) epiprocts without median carina (*top arrow*) and with a cylindrical callosity in the basal portion (*bottom arrow*), in dorsal view.

Orthoptera: Leptysminae: Tetrataeniini: *Xenismacris*: Species

1 Fastigium short (smaller than the head width) (Figs. 5.10 C, D); interocular space wide (subequal or greater than antennal pedicel width) (Fig. 5.8 A); male epiprocts without lateral spininiform expansions (Figs. 5.6 C–E) ... 2

1' Fastigium long (≥ head width) (Figs. 5.10 A, B); interocular space narrow (< antennal pedicel width) (Fig. 5.8 A); male epiprocts with lateral spininiform expansions (Figs. 5.6 C–E) ... *Xenismacris cyanoptera* (Gerstaecker)

2(1) Antennae filiform (Fig. 5.9 A); male epiprocts displaying two small circular ornaments near the apical portion (Figs. 5.6 C–E) ... *Xenismacris filicornis* (Descamps & Amédégnato)

2' Antenna ensiform (Fig. 5.9 B); male epiprocts ornate, displaying two small parallel spots near the base and a large triangular ornamentation in the center-apical region (Figs. 5.6 C–E) ... *Xenismacris aetoma* (Roberts & Carbonell)

Orthoptera: Leptysminae: Tetrataeniini: *Stenopola*: Species

1 Antennae ensiform (Fig. 5.9 B); red blotches on the dorsal portion of the fastigium, on the head and between the hind tibia spines ... *Stenopola nigricans* (Roberts & Carbonell)

1' Filiform antennae (Fig. 5.9 A); head, fastigium, and hind tibia without red pigmentation ... *Stenopola puncticeps* (Stål)

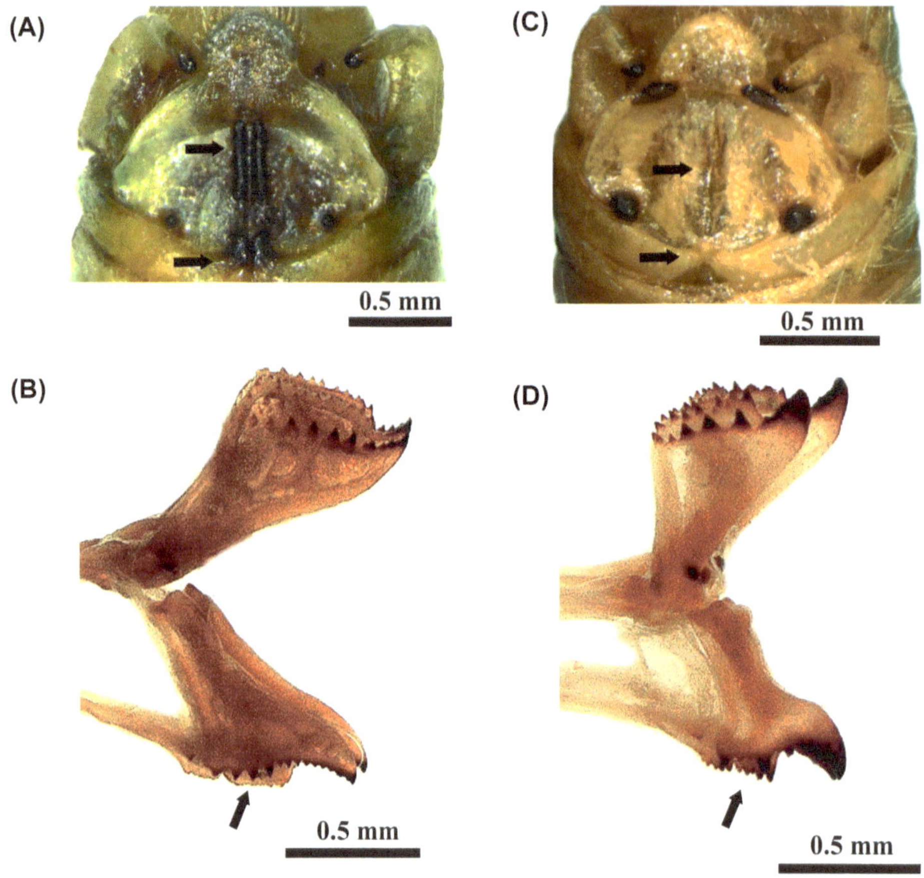

FIGURE 5.23 Morphological aspects of *Cornops* (Acrididae: Leptysminae). *Cornops aquaticum* (Bruner): (A) male epiprocts, in dorsal view, with black medium carina (*top arrow*) and terminalia, in dorsal view, with a black furcula (*bottom arrow*); (B) external margin of the lower valves of the ovipositor with uniform-sized teeth, in lateral view. *Cornops paraguayense* (Bruner): (C) male epiprocts with unpigmented medium carina (*top arrow*) and furcula (*bottom arrow*), in dorsal view; and (D) external margin of the lower oviductor valves with two or more relatively large-sized teeth, in lateral view.

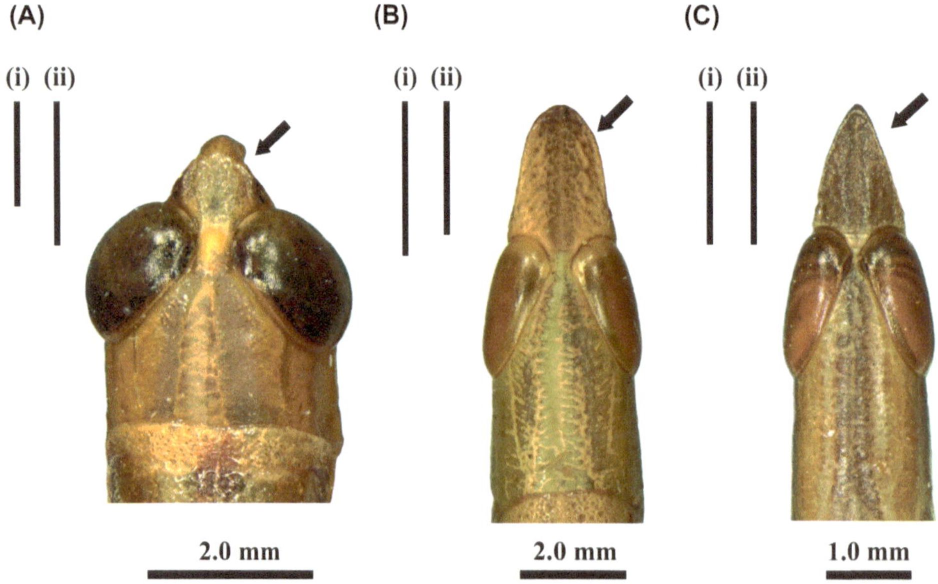

FIGURE 5.24 Leptysmini (Acrididae) morphological head aspects, in dorsal view: (A) short fastigium; (B) long fastigium with rounded apex; and (C) long fastigium with triangular apex. I = fastigium length, ii = compound eye length.

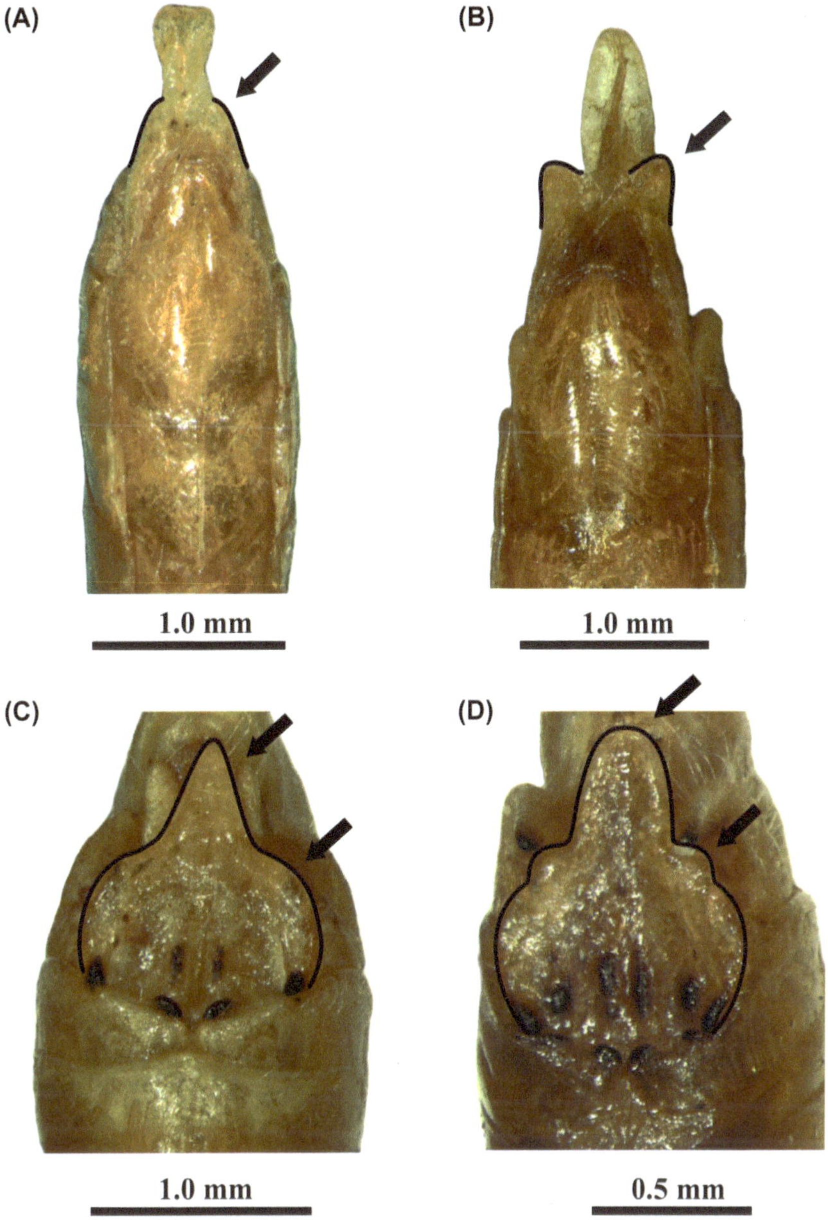

FIGURE 5.25 Morphological aspects of male Leptysmini (Acrididae) subgenital plates (ventral view) and epiprocts (dorsal view): (A) subgenital plate without lateral lobes; (B) subgenital plate with lateral lobes; (C) epiprocts with a laterally rounded basal portion (*bottom arrow*) and triangular apical portion (*top arrow*); and (D) epiprocts with a laterally lobed basal portion (*bottom arrow*) and a round apical portion (*upper arrow*).

Orthoptera: Leptysminae: Tetrataeniini: *Cornops*: Species

1 Wings and hind tibia green or green-blue; hind femur knee light brown or green (Fig. 5.22 A); male epiprocts pigmented, with a pair of median carina and with two circular black callosities in the basal portion (Fig. 5.22 B); tegmina long, exceeding hind femur (Fig. 5.22 A) 2

1' Wings and hind tibia dark blue; hind femur knee black apically (Fig. 5.21 C); male epiprocts nonpigmented, without median carina and with two cylindrical black callosities in the basal portion (Fig. 5.22 D); tegmina short, not exceeding hind femur (Fig. 5.22 C) .... *Cornops brevipenne* (Roberts & Carbonell)

2(1) Male epiprocts median carinae and furcula with black pigmentation (Fig. 5.23 A); inferior oviductor valves lateral margin with uniform-sized teeth (Fig. 5.23 B) .... *Cornops aquaticum* (Bruner)

2' Male epiprocts median carinae and furcula unpigmented (Fig. 5.23 C); inferior oviductor valves lateral margin with two or more relatively large teeth (Fig. 5.23 D) .... *Cornops paraguayense* (Bruner)

FIGURE 5.26 Morphological aspects of Leptysmini (Acrididae) saltatory legs (hind leg): (A) red or orange-red; and (B) green.

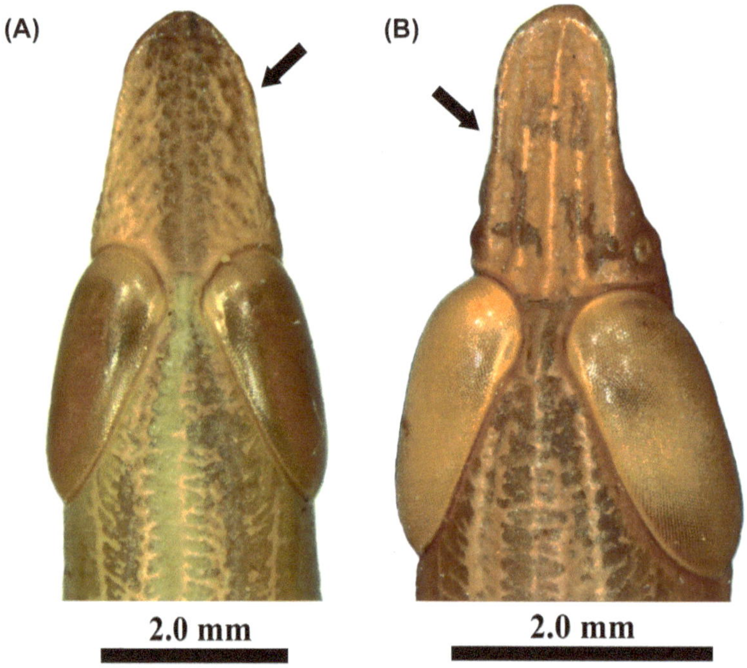

FIGURE 5.27 morphological Leptysmini (Acrididae) head aspects (dorsal view): (A) fastigium without longitudinal carina; and (B) fastigium with longitudinal carina.

Orthoptera: Leptysmini: Genera

1	Fastigium short (<composite eyes) (Fig. 5.24 A); antennae filiform (except for *Tucayaca*) (Fig. 5.9 A)	2
1'	Fastigium long (>composite eyes) (Figs. 5.24 B–C); antennae ensiform (Fig. 5.9 B) or subensiform (Fig. 5.9 C)	5
2 (1)	Male subgenital plates simple, without lateral lobes (Fig. 5.25 A)	3
2'	Male subgenital plates of various shapes, with lateral lobes (Fig. 5.25 B)	*Stenacris*
3 (2)	Antennae filiform (Fig. 5.9 A); epiprocts with basal portion laterally rounded and apical portion triangular (Fig. 5.25 C)	*Belosacris*
3'	Antennae subensiform (Fig. 5.9 C); epiprocts with basal portion laterally lobed and apical portion rounded (Fig. 5.25 D)	*Tucayaca*
4 (1)	Fastigium apex rounded (Fig. 5.24 B)	5
4'	Fastigium apex triangular (Fig. 5.24 C)	*Leptysma*
5 (4)	Fastigium dorsal portion lacking longitudinal carinae (Fig. 5.27 A)	6
5'	Fastigium dorsal portion with 2–3 longitudinal carinae (Fig. 5.27 B)	*Leptysmina*
6 (5)	Antennae subensiform (Fig. 5.9 C)	7
6'	Antennae ensiform (Fig. 5.9 B)	*Carbonellacris grossa* (Bruner)

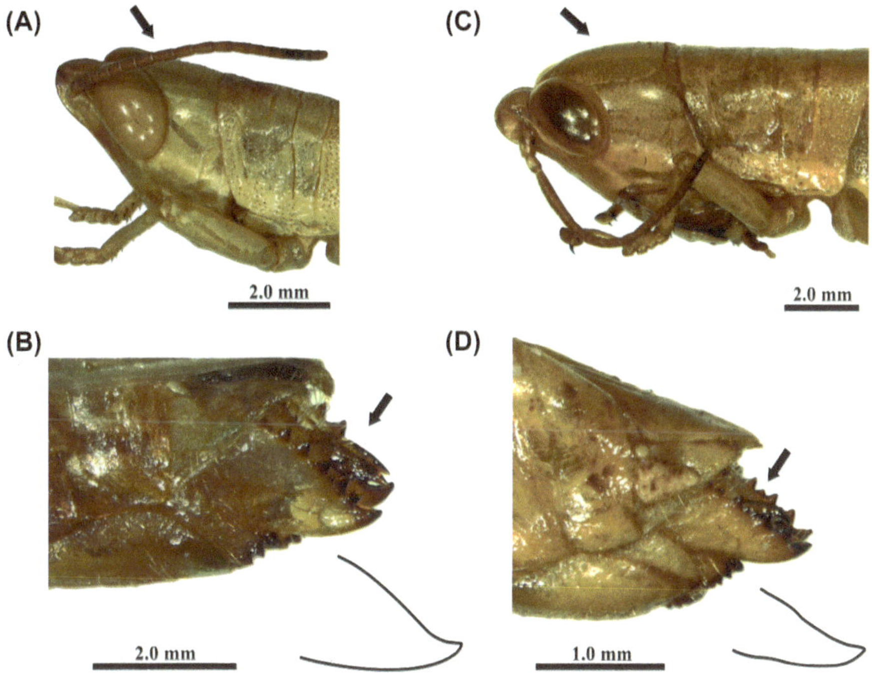

FIGURE 5.28 Morphological aspects of the head and ovipositor valves in *Stenacris* (Acrididae: Leptysminae), in lateral view: (A) normal head (nonglobular); (B) shortened superior ovipositor valves; (C) Globular shaped head; (D) elongated superior ovipositor valves.

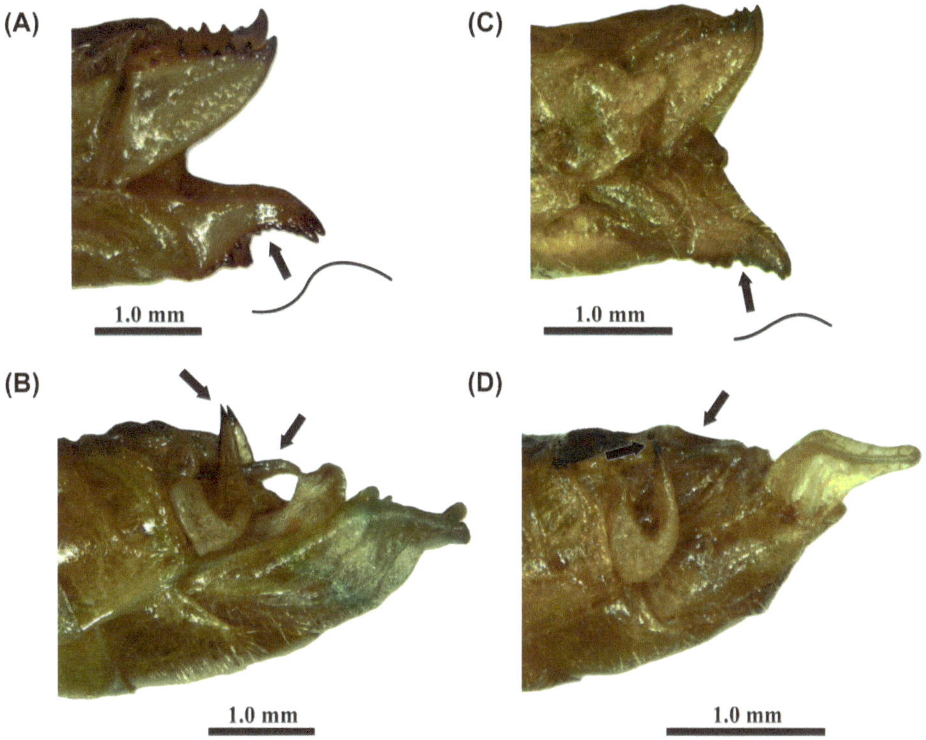

FIGURE 5.29 External genitalia morphology of *Stenacris* (Acrididae: Leptysminae), in lateral view: (A) heavily arched inferior ovipositor valve; (B) male cerci with pointed apex (*upper arrow*), extending beyond the lateral border of the epiprocts (*top arrow*); (C) almost straight or weakly arched inferior ovipositor valve; and (D) male cerci with rounded apex (*bottom arrow*), not extending beyond the lateral border of the epiprocts (*top arrow*).

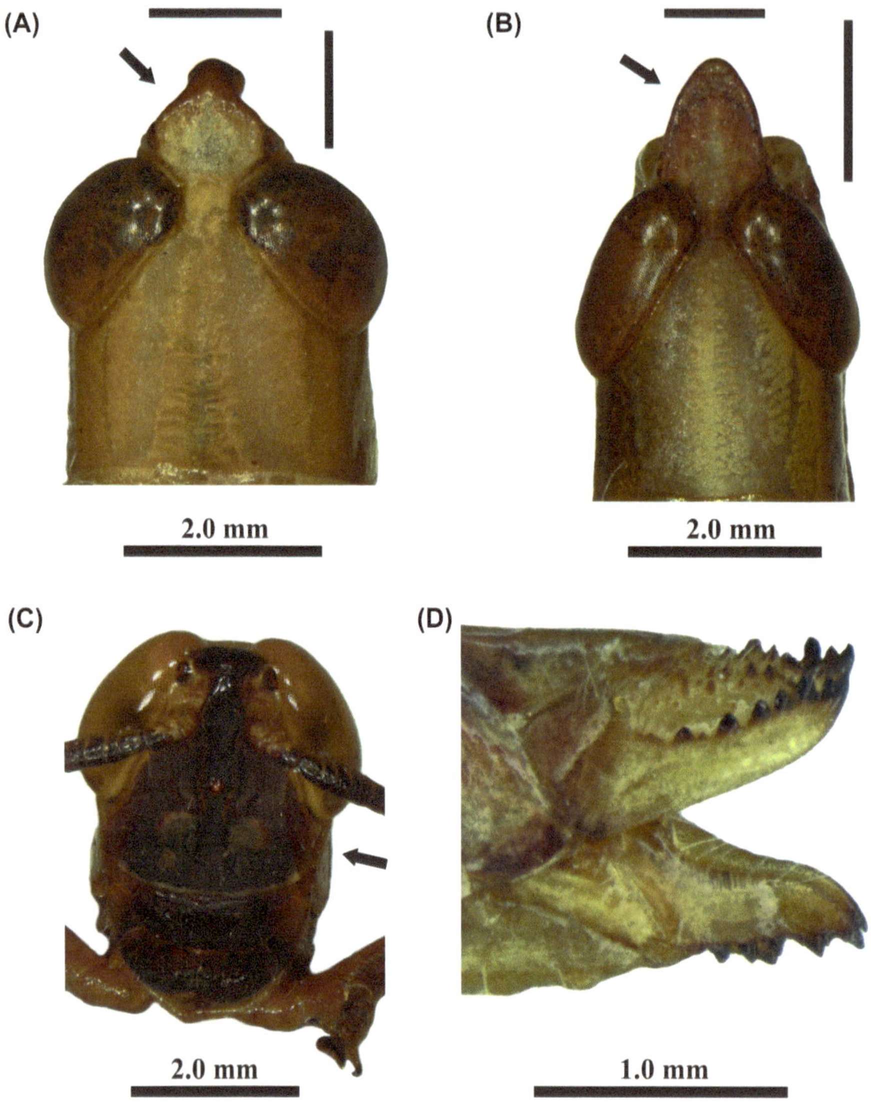

FIGURE 5.30 Morphological aspects of *Tucayaca* (Acrididae: Leptysminae): (A) widened fastigium, in dorsal view; (B) elongated fastigium, in dorsal view; and (C) dark brown face, frontal view; (D) ovipositor valves with relatively similar sized teeth.

7 (6)	Hind tibia lateral and medial margins without a dense row of short bristles (Fig. 5.7 A)	8
7'	Hind tibia lateral and medial margins each with a dense row of short bristles (Fig. 5.7 B)	*Columbacris caudata* Bruner
8 (7)	Superior ovipositor valves with a single large tooth on the dorsal outer portion (Figs. 5.6 F, G); male abdomen apex with a robust, strongly prominent palial sclerite and with a lateral tubercle (Fig. 5.6 A)	*Cylindrotettix*
8'	Superior ovipositor valves with three large teeth on the dorsal outer portion (Figs. 5.6 F, G); male abdomen apex with an obscure palial sclerite and without lateral tubercles (Fig. 5.6 B)	*Seabratettix angustus* Roberts

Orthoptera: Leptysmini: *Stenacris*: Species

1	Head normal (Fig. 5.28 A); total body length (fastigium anterior margin to tegmina apex) < 32 mm in males and 39 mm in females; shortened superior ovipositor valves (length subequal to height) (Fig. 5.28 B)	2
1'	Head globular (Fig. 5.28 C); total body length (fastigium anterior margin to tegmina apex) > 32 mm in males and 39 mm in females; elongated superior ovipositor valves (length > height) (Fig. 5.28 D)	*Stenacris megacephala* Bruner
2 (1)	Aedeagus strongly recurved	3
2'	Aedeagus straight	4
3 (2)	Male subgenital plate with lateral lobes long and digitiform (Fig. 5.6 A); male cerci with apex truncated	*Stenacris vitreipennis* (Marschall)

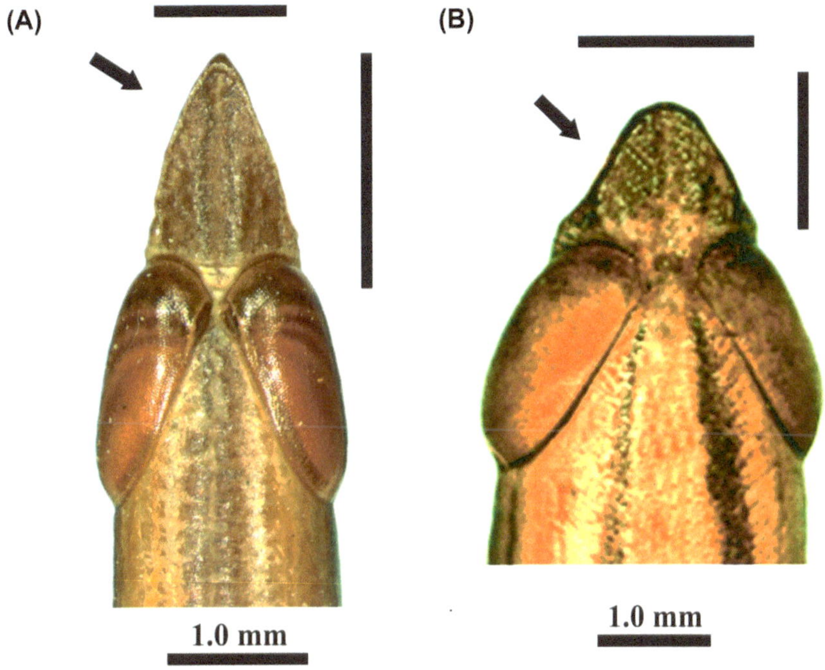

FIGURE 5.31 Head morphology, in dorsal view of *Leptysma* (Acrididae: Leptysminae): (A) elongated fastigium, in dorsal view, of *Leptysma marginicollis* (Serville); and (B) widened fastigium.

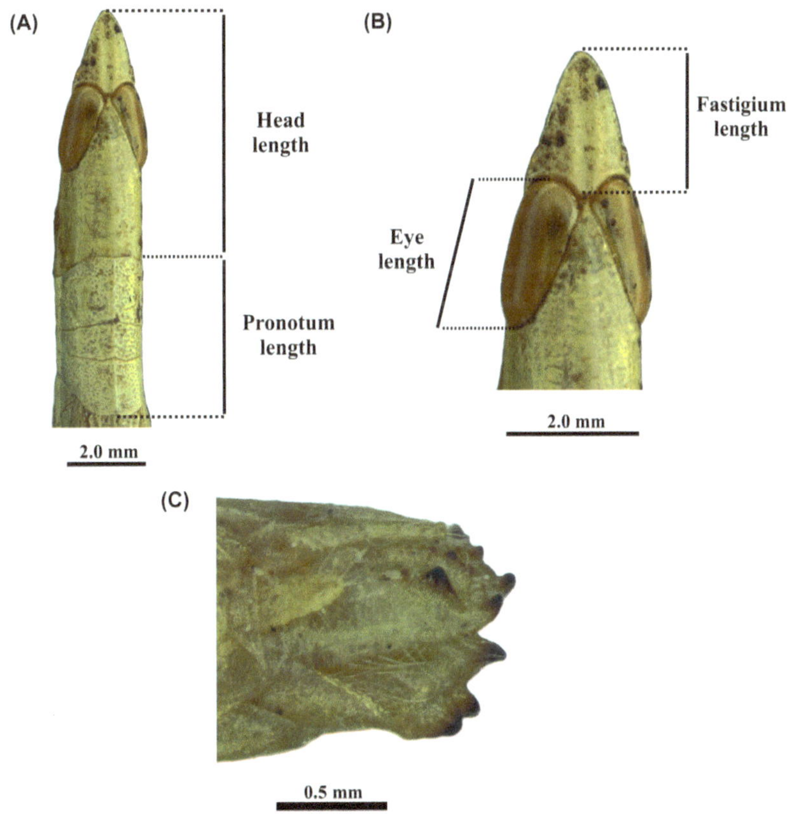

FIGURE 5.32 External morphology of the head, pronotum (dorsal view) and ovipositor (lateral view) of *Leptysma* (Acrididae: Leptysminae): (A) head and pronotum; (B) fastigium and compound eye; and (C) ovipositor valve with fused teeth.

FIGURE 5.33 Male terminalia of *Cylindrotettix* (Acrididae: Leptysminae), in lateral view. Featured structures: *a*, apex of the subgenital plate; *b*, base of the subgenital plate; *e*, epiprocts (supraanal plate); *ll*, lateral lobes of the subgenital plate; *lt*, lateral tubercle of the palial sclerite; *p*, pallium; *ps*, palial sclerite; *sp*, subgenital plate.

FIGURE 5.34 Male terminalia of *Cylindrotettix* (Acrididae: Leptysminae), in dorsal view. Featured structures: *e*, epiprocts (supraanal plate); *lt*, lateral tubercle of the palial sclerite; *p*, pallium; *ps*, palial sclerite; *sp*, subgenital plate.

3' Male subgenital plate with small, nondigitiform, lateral lobes (Fig. 5.6 A); male cerci with apex conical .... *Stenacris caribea* (Rehn and Hebard)

4 (2') Head to the thorax bearing a white lateral band always present in male and present or absent in female (Fig. 5.5) 5

4' Head to the thorax lacking white lateral band in both sexes (Fig. 5.5). .... *Stenacris minor* (Bruner)

5 (4) Inferior ovipositor valve lower margin strongly arched in lateral view (Fig. 5.29 A); male cerci long with apex pointed, projecting well above epiprocts lateral border (Fig. 5.29 B) .... *Stenacris fissicauda* (Bruner)

5' Inferior ovipositor valve lower margin nearly straight or weakly arched in lateral view (Fig. 5.29 C); male cerci shorter, apex rounded, not projecting above epiprocts lateral border (Fig. 5.29 D) .... *Stenacris xanthochlora* (Marschall)

Orthoptera: Leptysmini: *Belosacris*: Species

1 3 Body without a lateral white band (Fig. 5.5); hind tibia red or orange-red (Fig. 5.26 A) .... *Belosacris coccineipes* (Bruner)

1' Body with a lateral white band (Fig. 5.5); hind tibia green (Fig. 5.26 B) .... *Belosacris stali* (Bruner)

Orthoptera: Leptysmini: *Tucayaca*: Species

1 Fastigium widened, with width larger than length, in lateral view (Fig. 5.30 A) 2

1' Fastigium elongated, with length larger than width, in lateral view (Fig. 5.30 B) 3

2 (1) Hind tibia grayish-green; frons dark brown or red (Fig. 5.30 C) .... *Tucayaca biserrata* Roberts

2' Hind tibia bluish; frons green .... *Tucayaca coeruleipes* Roberts

3 (1) Superior and inferior ovipositor valves with tooth similar in size or proportionally the same 4

3' Superior and inferior ovipositor valves with proximal tooth larger than the others teeth (Fig. 5.30 D) .... *Tucayaca aquatica* Bruner

4 (3) Fastigium length and head width ratio between 0.40–0.46 mm (Figs. 5.10 A, C); body length 21.5–29.0 mm (Fig. 5.4) .... *Tucayaca gracilis* (Giglio-Tos)

4' Fastigium length and head width ratio between 0.29–0.39 mm (Figs. 5.10 A, C); body length 19.0–21.0 mm (Fig. 5.4) .... *Tucayaca parvula* Roberts

Orthoptera: Leptysmini: *Leptysma*: Species

1 Fastigium longer than wide (Fig. 5.31 A); head larger than the pronotum, in dorsal view (Fig. 5.32 A) 2

1' Fastigium wide, almost as wide as long (Fig. 5.31 B); head and pronotum equal in length, in dorsal view (Fig. 5.32 A) .... *Leptysma marginicollis* (Serville)

2 (1) Fastigium shorter or equal to eye length, in dorsal view (Fig. 5.32 B) 3

2' Fastigium longer than eyes, in dorsal view (Fig. 5.32 B) .... *Leptysma argentina* Bruner

3 (2) Antennae subensiform (Fig. 5.9 C) 4

3' Antennae ensiform (Fig. 5.9 B) .... *Leptysma intermedia* Bruner

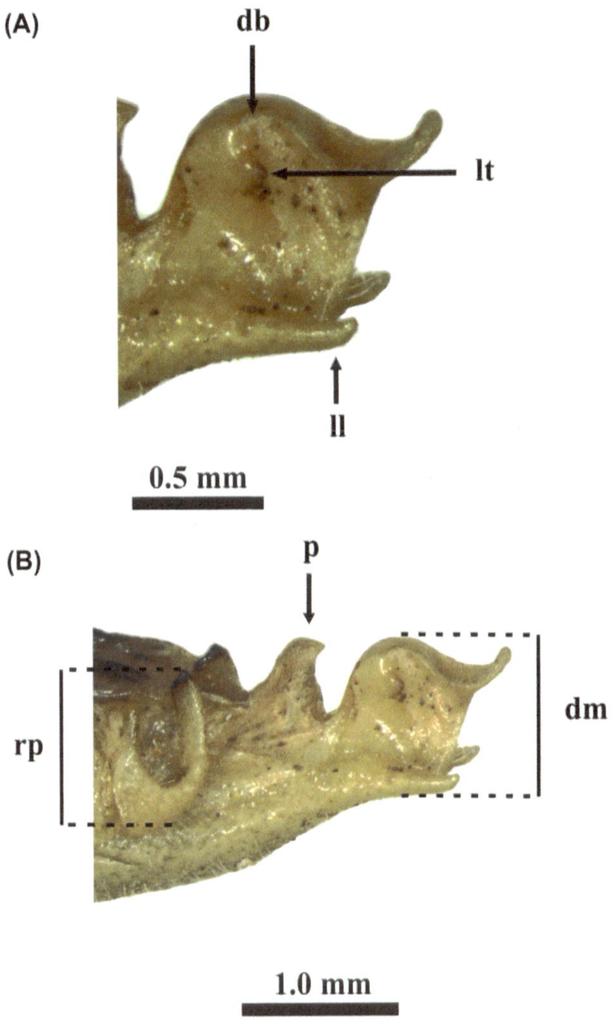

FIGURE 5.35 Male terminalia of *Cylindrotettix* (Acrididae: Leptysminae), in lateral view: (A) *db*, dorsal border of the palial sclerite; *ll*, lateral lobes of the subgenital plate, *lt*, lateral tubercle of the palial sclerite; and (B) *dm*, distance between the dorsal margin of the palial sclerite and the inferior margin of the subgenital plate; *p*, pallium; *rp*, length dimensions of the curved cerci portion.

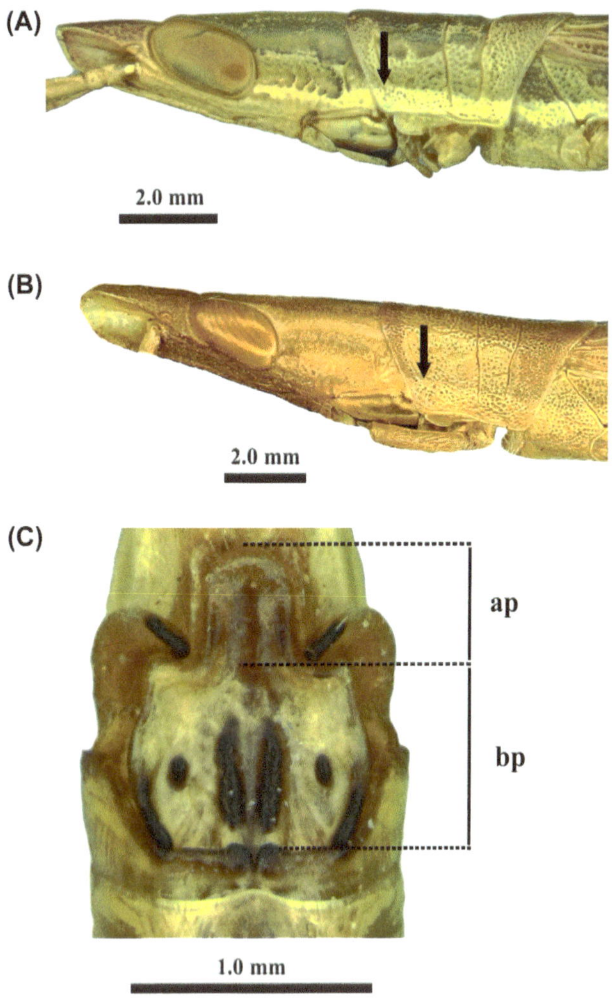

FIGURE 5.36 Male head and pronotum (lateral view) and, epiroproct (dorsal view) of *Cylindrotettix* (Acrididae: Leptysminae): (A) white postocular lateral band visible; (B) white postocular lateral band absent or obscured; and (C) epiprocts division, with limitation of the apical and basal portion: *ap*, apical portion of the epiprocts and *bp*, basal portion of the epiprocts.

4 (3) Ovipositor valve external margin with small and subequal teeth (Fig. 5.32 C) .... *Leptysma tainan* Rehn & Hebard

4' Ovipositor valve external margin with fused teeth, large and irregular in size (Fig. 5.32 C) .... *Leptysma filiformis* (Serville)

Orthoptera: Leptysmini: *Leptysmina*: Species

1 Fastigium long and thin (length > width) (Fig. 5.10) 2

1' Fastigium short and wide (width subequal to length) (Fig. 5.10) .... *Leptysmina pallida* Giglio-Tos

2(1) Body total length, from anterior margin of fastigium to the apex of the tegmina in males ≤4.0 cm and in females up to 4.5 cm (Fig. 5.4) .... *Leptysmina gracilis* Bruner

2' Body total length, from anterior margin of fastigium to the apex of the tegmina in males >4.0 cm and in females >4.5 cm (on average, 5.0 cm) (Fig. 5.4) .... *Leptysmina amazonica* Carbonell & Ronderos

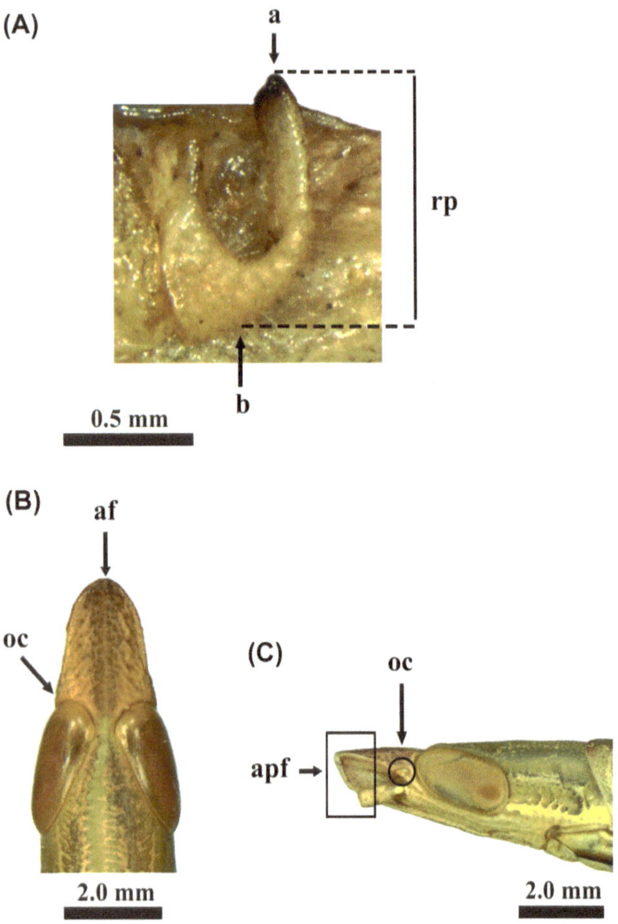

FIGURE 5.37 *Cylindrotettix* (Acrididae: Leptysminae) external morphology. Featured structures: (A) cerci, lateral view; (B) head, in dorsal view; and (C) head, in lateral view. *a*, apex of the curved portion of cerci; *af*, fastigium apex; *apf*, Zfastigium apical region; *b*, base of the curved portion of cerci, *oc*, lateral ocelli; *rp*, recurved portion of the cerci.

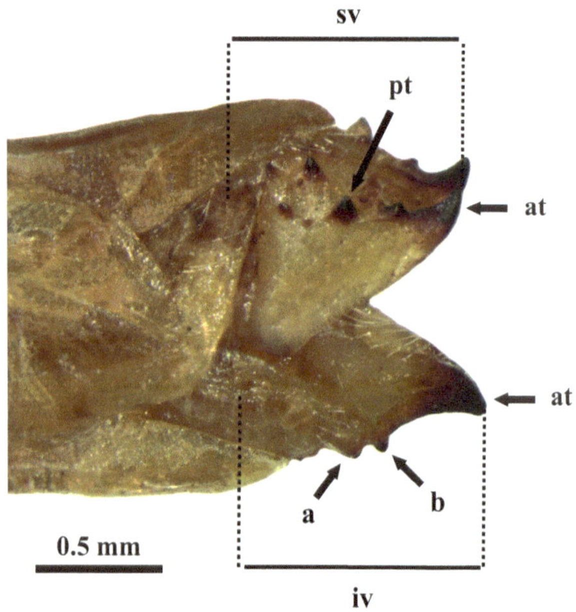

FIGURE 5.38 Female terminalia of *Cylindrotettix* (Acrididae: Leptysminae), in lateral view. Featured structures: upper and lower ovipositor valves; apical and proximal valve teeth; *a*, second proximal tooth; *at*, apical tooth; *b*, first proximal tooth; *iv*, inferior ovipositor valve; *pt*, proximal tooth; *sv*, superior ovipositor valve.

Orthoptera: Leptysmini: *Cylindrotettix*: Species

1 Pallial sclerite in lateral view ornamented by a lateral tubercle located at the same distance or anterior to subgenital plate apex (Figs. 5.33–34) 2

1' Pallial sclerite in lateral view ornamented by a lateral tubercle posterior to subgenital plate apex (Figs. 5.33–34) .... *Cylindrotettix insularis* Bruner

2 (1) Pallial sclerite dorsal border with a lateral tubercle located below the dorsal margin in lateral view (Fig. 5.35 A) 3

2' Pallial sclerite dorsal border with a lateral tubercle in lateral view (Fig. 5.35 A) .... *Cylindrotettix santarosae* Roberts

3 (2) White postocular lateral band clearly visible (Fig. 5.36 A) 4

3' White postocular lateral band absent or obscured (Fig. 5.36 B) .... *Cylindrotettix uniformis* (Bruner)

4 (3) Distance between pallial sclerite dorsal margin and the subgenital plate inferior margin smaller than the length of the cerci recurved portion (Fig. 5.35 B); pallial sclerite lateral tubercle small and strongly pointed (Figs. 5.33–34) 5

4' Distance between pallial sclerite dorsal margin and the subgenital plate inferior margin greater than the length of the cerci recurved portion (Fig. 5.35 B); pallial sclerite lateral tubercle large and robust, slightly pointed (Figs. 5.33–34) .... *Cylindrotettix obscurus* (Thunberg)

5 (4) Epiprocts apical portion shorter than basal portion in dorsal view (Fig. 5.36 C); pallium not near pallial sclerite (Figs. 5.33, 35 B); cerci recurved portion with apex as broad as or wider than base (Figs. 5.37 A) 6

5' Epiprocts apical portion the same size or larger than basal portion in dorsal view (Fig. 5.36 C); pallium near or extending dorsally over palial sclerite (Figs. 5.33, 35 B); cerci recurved portion with apex narrower than base (Fig. 5.37 A) 8

6(5) Fastigium in dorsal view tapering moderately from lateral ocelli to a recurved, enlarged apex (Fig. 5.37 B); fastigium apical part in lateral view not or slightly curved (Fig. 5.37 C) 7

6' Fastigium in dorsal view tapering evenly from lateral ocelli to a triangular apex (Fig. 5.37 B); fastigium apical part in lateral view generally curved ventrally (Fig. 5.37 C)........ *Cylindrotettix attenuatus* Roberts

7 (6) Cingulate lobes extending around the aedeagus well above the lateral lobe extremities and almost at the base of the apical processes; aedeagus apical processes extended outward forming a small, poculiform structure; subgenital plate lobes with apex broadly rounded (Figs. 5.33, 35 A)........ *Cylindrotettix riverae* Roberts

7' Cingulate lobes extending along the posterior side of the aedeagus approximately to the level or well below the lateral lobe ends; aedeagus apical processes simple, unstriated, or poculiform; subgenital plates slightly rounded or tapered (Figs. 5.33, 35 A)........ *Cylindrotettix dubius* Roberts

8 (5) Subgenital plate apex with pronounced lateral lobes (Figs. 5.33, 35 A); inferior ovipositor valve external margin with only one proximal tooth (Fig. 5.38); superior valve without small between the proximal and apical teeth (Fig. 5.38)........ 9

8' Subgenital plate apex without lateral lobes (Figs. 5.33, 35 A); inferior ovipositor valve external margin e with two proximal teeth (Fig. 5.38); superior valve with one or more small teeth between the proximal and apical teeth (Fig. 5.38)........ *Cylindrotettix chacoensis* Roberts

9 (8) Epiprocts apical portion longer than basal portion (Fig. 5.36 C); pallium extending over dorsal part of pallial sclerite (Figs. 5.33, 35 B)........ *Cylindrotettix elongatus* Roberts

9' Epiprocts narrow portion length subequal to basal portion (Fig. 5.36 C); pallium extending to the anteriodorsal part of pallial sclerite (Figs. 5.33, 35 B)........ *Cylindrotettix dorsalis* (Burmeister)

ACKNOWLEDGMENTS

We thank Dr. Neusa Hamada for the invitation to prepare the Orthoptera chapter of this book.

REFERENCES

Almeida, A.V. & C.A.G. Câmara. 2008. Distribution of grasshoppers (Othoptera: Acridoidea) in the Tapacurá ecological station (São Lourenço da Mata, PE/Brazil). Brazilian Journal Biology 68: 21–24.

Amédégnato, C. 2003. Chapter 22: Microhabitat distribution of forest grasshoppers in the Amazon. Pages 237–255 in: Y. Basset, V. Novotny, S.E. Miller & R.L. Kitching (eds.), Arthropods of tropical forests: spatio-temporal dynamics and resource use in the canopy. Cambridge University Press, Cambridge.

Amédégnato, C. & H. Devriese. 2008. Global diversity of true and pygmy grasshoppers (Acridomorpha, Orthoptera) in freshwater. Hydrobiologia 595: 535–543.

Bhalerao, A.M.N., M. Naidu & S.Y. Paranjape. 1987. Some observations on the nutrition reproduction correlation in grouse locusts (Orthoptera: Tetrigidae). Proceedings: Animal Sciences 96: 323–327.

Bownes, A., M.P. Hill & M.J. Byrne. 2010. Assessing density–damage relationships between water hyacinth and its grasshopper herbivore. Entomologia Experimentalis et Applicata 137: 246–254.

Braga, C.E., A.L. Nunes, J.W. Morais & J. Adis. 2011. Fenologia de Cornops aquaticum (Bruner 1906) (Orthoptera: Acrididae) associado a Eichhornia crassipes (Mart.) Solms (Pontederiaceae) em um Lago de Várzea na Amazônia Central, Brasil. Revista de la Sociedad Entomológica Argentina 70: 185–196.

Braga, C.E., A.L. Nunes-Gutjahr, J.W. Morais & J. Adis. 2013. Avaliação do potencial do gafanhoto Cornops aquaticum (Orthoptera) como agente de controle Biológico de Eichhornia crassipes (Pontederiaceae). Interciencia 38: 590–596.

Buhl, J., G.A. Sword, F.J. Clissold & S.J. Simpson. 2011. Group structure in locust migratory bands. Behavioral Ecology and Sociobiology 65: 265–273.

Capello, S., M. Marchese & M.L. Wysiecki. 2012. Feeding habits and trophic niche overlap of aquatic Orthoptera associated with macrophytes. Zoological Studies 51: 51–58.

Centre for Overseas Pest Research (COPR). 1982. The Locust and Grasshopper Agricultural Manual. Published by The Centre for Overseas Pest, Research, London, England. 690 pp.

Cigliano, M.M., H. Braun, D.C. Eades & D. Otte. 2018. Orthoptera Species File. Version 5.0/5.0. http://Orthoptera.SpeciesFile.org.

Franceschini, C., M.L. Wysiecki, A. Poi de Neff, M.E. Galassi & F.S. Martinez. 2011. The role of the host-specific grasshopper Cornops aquaticum (Orthoptera: Acrididae) as consumer of native Eichhornia crassipes (Pontederiaceae) floating meadows. Revista de Biología Tropical 59: 1407–1418.

Junk, W.J. 1980. Áreas inundáveis – Um desafio para limnologia. Acta Amazonica 10: 775–795.

Kocárek, P., S. Grucmanová, Z. Filipcová, L. Bradová, V. Plásek & J. Holusa. 2008. Bryophagy in the groundhopper Tetrix ceperoi (Orthoptera: Tetrigidae): analysis of alimentary tract contents. Scripra Facultatis Rerum Naturalium Environmental Changes Universitatis Ostraviensis 186: 348–352.

Lhano, M.G. 2002. Aspectos biológicos e ecológicos de Cornops aquaticum (Bruner, 1906) (Orthoptera: Acrididae) em Eichhornia azurea (Swartz) Kurth (Pontederiaceae) no Pantanal de Poconé, Mato Grosso. M.Sc. Thesis, Institute of Biosciences, Federal University of Mato Grosso. Cuiabá, Brazil. 123 pp.

Nunes, A.L. 1989. Estudo sobre o ciclo de vida e fenologia de Stenacris fissicauda fissicauda (Bruner, 1908) (Orthoptera-Acrididae) em um lago de várzea da Amazônia Central, Brasil. M.Sc. Thesis, National Institute of Amazon Researches, Manaus, Brazil. 122 pp.

Nunes, A.L. & J. Adis. 1994. Comportamento populacional de Tucayaca gracilis (Giglio-Tos, 1897) (Orthoptera-Acrididae) frente a oscilação do nível d'água na várzea da Amazônia Central. Boletim do Museu Paraense Emílio Goeldi 10: 211–224.

Nunes, A.L., J. Adis & J.A. Nunes de Mello. 1992. Estudo sobre o ciclo de vida e fenologia de Stenacris fissicauda fissicauda (Bruner, 1908) (Orthoptera: Acrididae) em um lago de várzea da Amazônia Central, Brasil. Boletim do Museu Paraense Emílio Goeldi 8: 349–374.

Nunes-Gutjahr, A.L. 2009. Chapter 28: Os gafanhotos Acridoidea. Pages 451–460 in: P.L.B. Lisboa (ed.), Caxiuanã: desafios para a conservação de uma Floresta Nacional na Amazônia. Museu Paraense Emilio Goeldi, Belém, PA.

Nunes-Gutjahr, A.L. & C.E. Braga. 2012. Gafanhotos da Flona Caxiuanã: Guia prático. Museu Paraense Emilio Goeldi, Belém, PA. 62 pp.

Oliveira, A.F., A.L. Nunes-Gutjahr & C.E. Braga. 2015. Aceitabilidade alimentar do gafanhoto Cornops aquaticum (Bruner, 1906) (Orthoptera: Acrididae) na Amazônia oriental, Brasil. Boletim do Museu Paraense Emílio Goeldi 10: 267–277.

Roberts, H.R. 1978. A revision of the tribe Leptysmini except the genus Cylindrotettix (Orthoptera: Acrididae: Leptysminae). Proceedings of the Academy of Natural Sciences of Philadelphia 129: 33–69.

Triplehorn, C.A. & N.F. Johnson. 2011. Introdução ao Estudo dos Insetos. Tradução da 7ª Edição de Borror and Delong's Introduction to the study of insects. Cengage Learning, São Paulo, SP, BR, 816 pp.

Chapter 6

Order Blattodea

Sonia Maria Lopes
Laboratório de Biodiversidade e Sistemática de Blattaria, Departamento de Entomologia, Museu Nacional, Universidade Federal do Rio de Janeiro, Rio de Janeiro, Rio de Janeiro, Brazil

Chapter Outline

INTRODUCTION

The Blattodea includes cockroaches and termites (Beccaloni & Eggleton, 2011), and it is composed of three superfamilies (Blaberoidea, Blattoidea, and Corydioidea), three epifamilies (Blattoidae, Cryptocercoidae, and Termitoidae), nine families (Blaberidae, Ectobiidae, Blattidae, Lamproblattidae, Tryonicidae, Cryptocercidae, Termitidae, Corydiidae, and Nocticolidae), and 24 subfamilies (Beccaloni, 2014). The suborder Blattaria (cockroaches), within Blattodea, has approximately 4500 species in 460 genera worldwide (Rentz, 2014), with the largest number of species reported from the Neotropics because of its relatively large area of tropical forest. Blattaria is one of the oldest insect in the fossil record, dating back to the Cretaceous (Bell et al., 2007). Fossil aquatic cockroaches are found from the Mesozoic, while extant species of aquatic cockroaches are recorded from south and southeastern Asia and Japan (Annandale 1900, 1906, Green, 1902; Shelford, 1907, 1909; Takahashi, 1921, Nesemann et al., 2010).

Blaberidae is divided into 12 subfamilies and 135 genera; however, the relatively few semiaquatic species are found only within the subfamily Epilamprinae (tribes Epilamprini [Figs. 6.1–2] and Poeciloderrhini) (Nessimian, 2014) and those members of the family Ectobiidae (Fig. 6.3) that inhabit bromeliads (Rocha e Silva-Albuquerque & Lopes, 1976). Blaberidae is the most advanced among the Blattodea, but they are morphologically very similar with only minor variations in their genitalia and proventriculus (Mckittrick, 1964). Females have an enlarged atrium in which eggs are incubated; most species are ovoviviparous and rarely viviparous; they lack an egg case, ootheca (Lopes & Oliveira, 2013).

Blattaria is important to ecosystem functioning, particularly in the fragmentation and decomposition of organic matter process and consequent release of nutrients (Bell et al., 2007). On a global scale, cockroaches are responsible for breaking down complex carbohydrates and extracting minerals from decaying matter, allowing these nutrients to return to the ecosystem (Lavelle, 2002). In addition to their role in recycling, cockroaches are also known for their ability to carry bacteria and fungi, thus functioning as vectors of some diseases (Fotedar et al., 1991).

There are two categories of semiaquatic cockroaches: (1) species associated with phytotelmata such as bromeliads, where nymphs and adults can be found; and (2) species associated with lakes, rivers, and small streams, as well as phytotelmata. In the last category, nymphs are often found in the semiaquatic environment, while adults occupy the nearby trees. These species can live on the water surface, submerging to escape predators or when searching for food. Approximately 34 species in 18 genera of semiaquatic cockroaches are included in Blaberidae (Epilamprinae), and Ectobiidae are known from bromeliads in Central and South America (Roth & Willis, 1960; Albuquerque & Lopes, 1976).

Thorp and Covich's Freshwater Invertebrates, Fourth Edition. https://doi.org/10.1016/B978-0-12-804223-6.00006-8

FIGURE 6.1 Blaberidae, Epilamprinae (Blattodea: Blattaria) male, dorsal habitus (41 mm).

FIGURE 6.2 Blaberidae, Epilamprinae (Blattodea: Blattaria) male, dorsal habitus (54 mm).

LIMITATIONS

The great diversity of cockroaches in tropical forests is relatively unexplored by entomologists (Bell et al., 2007). Although the ecological associations of modern cockroaches should be well known from direct observation, most species are still little more than names on museum specimens, and our knowledge of them is fragmentary. Aquatic forms might have been simply overlooked, because little research has been done on this group, and this makes it difficult to develop identification keys. Furthermore, the color of semiaquatic immature forms is so similar to the substrate color that it is difficult to collect and to associate them with the adult stage.

FIGURE 6.3 Ectobiidae (Blattodea: Blattaria), male, dorsal habitus (17 mm).

TERMINOLOGY AND MORPHOLOGY

Cockroaches are dorsoventrally flattened insects with long, filiform antennae, and mandibulated, ventrally projecting mouthparts. The prothorax has an enlarged, shield-like pronotum that often covers the head, and the meso- and metathorax are rectangular and subequal. The forewings are sclerotized tegmina protecting membranous hind wings folded fan-like beneath. Hind wings often may be reduced or absent and, if present, characteristically have many vein branches and a large anal lobe. The legs may be spiny, and the tarsi are five-articulate. Abdomen has 10 visible segments in the male, and segments may have tergal modifications. The subgenital plate usually bears two styles (Figs. 6.4, 5 B, 6 B, 8 B), which cover well-developed asymmetrical genitalia and the reduced 11th segment. The supranal plate has a pair of cerci, typically with many articulated (Figs. 6.5 A, 6 A, 8 A). The female ovipositor valves are small and concealed beneath tergum 10 (Gullan & Cranston, 2014).

MATERIAL PREPARATION AND PRESERVATION

Cockroaches are best collected manually, with the aid of gloves and forceps, and then preserved in glass containers with 70% ethanol. When specimens are collected from

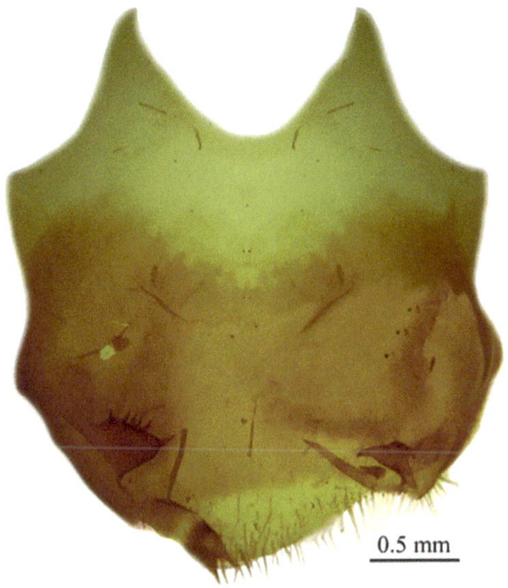

FIGURE 6.4 Ectobiidae (Blattodea: Blattaria): male subgenital plate.

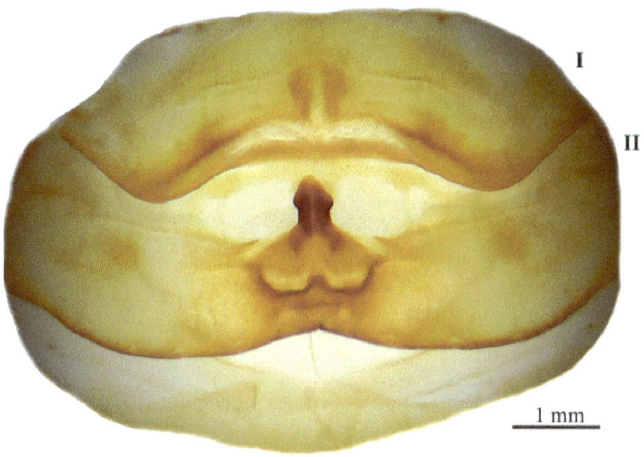

FIGURE 6.7 Blaberidae, Epilamprinae, Epilamprini (Blattodea: Blattaria) male: tergal modification on abdominal segments I and II.

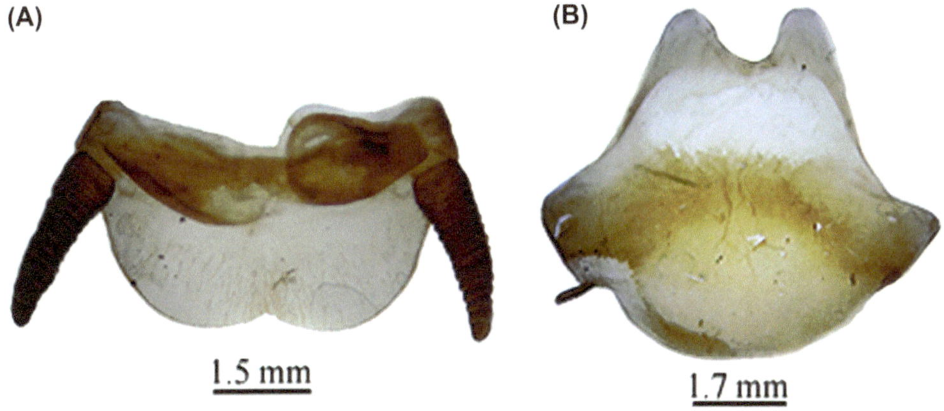

FIGURE 6.5 Blaberidae (Blattodea: Blattaria) male: (A) supranal and (B) subgenital plates.

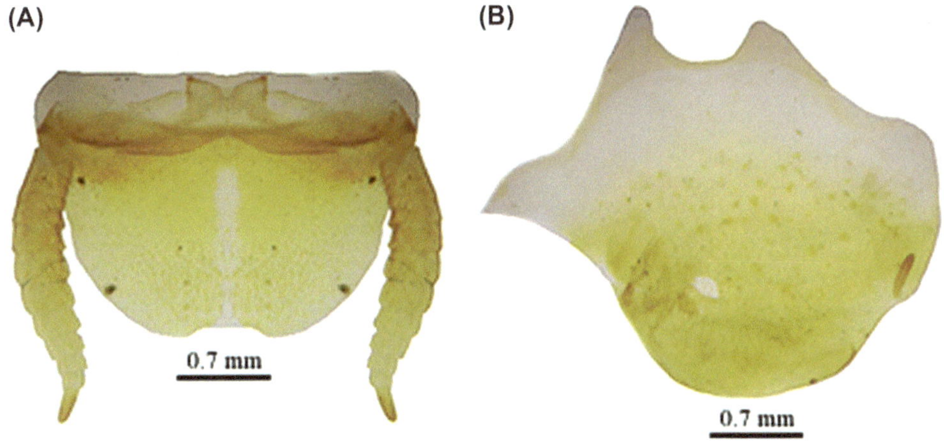

FIGURE 6.6 Blaberidae, Epilamprinae (Blattodea: Blattaria) male: (A) supranal and (B) subgenital plates.

FIGURE 6.8 Blaberidae, Epilamprinae, Poeciloderrhini (Blattodea: Blattaria) male: (A) supranal and (B) subgenital plates.

bromeliads, care needs to be taken to preserve this fragile microenvironment. After being collected, specimens should be mounted on entomological pins number 1 or 2. If they were preserved in ethanol, they need to be dried over paper towel before being pinned.

The genital plates are observed after dissection of the last portion of the abdomen of males and females. The terminal region of the abdomen is left submerged for about 24 h in 10% cold potassium hydroxide. The genital plates are then dissected, washed in distilled water for 5 min, and dehydrated in a series of increased ethanol concentrations. After observation, the genitalia should be stored in glycerin in a microvial attached to the specimen (Gurney et al., 1964). The nomenclature of the genital structures and the taxonomic classification follow Roth (2003).

KEYS TO BLATTARIA

This key applies only to those families with semiaquatic species.

Blattaria: Families

1 Occurring in many aquatic environments; body length 22–77 mm; femur I without spines, if spines present they are dissimilar between anterio- and posterioventral margins .... **Blaberidae (Epilamprinae) [p. 172]**

1' Occuring in bromeliads: body length 8–24 mm; femur I spinose, with similar spines on anterio- and posterioventral margins .... Ectobiidae

Blaberidae: Epilamprinae: Tribes

1 Male abdomen segments I and II without tergal modifications; subgenital plate asymmetric with styles subequal (Fig. 6.6 B) .... Epilamprini

1' Male abdomen segments I and II with distinct tergal modifications (Fig. 6.7); subgenital plate asymmetric with one style much smaller than the other (Fig. 6.8 A) .... Poeciloderrhini

ACKNOWLEDGMENTS

I thank Dr. Neusa Hamada for the invitation to prepare the Blattodea chapter of this book. I also appreciate the contributions of Dr. Janet Read for help with the English translation of this chapter and Prof. Edivar Heeren de Oliveira for suggested revisions.

REFERENCES

Annandale, N. 1900. Notes on Orthoptera in the Siamese Malay State. Entomological Records 12: 75.

Annandale, N. 1906. Notes on the freshwater fauna of India. No. III. An Indian aquatic cockroach and beetle larva. Journal & Proceedings of the Asiatic Society of Bengal, New Series 2: 105–107.

Beccaloni, G.W. 2014. Blattodea species file online. Version 5.0/5.0. World Wide Web electronic publication. http://Blattodea.SpeciesFile.org.

Beccaloni, G.W. & P. Eggleton. 2011. Order Blattodea Brunner von Wattenwyl, 1882. Pages 199–200 *in*: Z.-Q. Zhang (ed.), Animal biodiversity: An outline of higher-level classification and survey of taxonomic richness. Zootaxa 3148: 1–237.

Bell, W.J., L.M. Roth & C.A. Nalepa. 2007. Cockroaches. Ecology, Behavior and Natural History. The Johns Hopkins University Press, Baltimore.

Fotedar, R., U.B. Shriniwas & A. Verma. 1991. Cockroaches (*Blattella germanica*) as carriers of microorganisms of medical importance in hospitals. Epidemiology and Infection 107: 181–187.

Green, E.E. 1902. Aquatic Orthoptera in Ceylon. Entomological Monthly Magazine 38: 214–215.

Gullan, P.J. & P.S. Cranston. 2014. The Insects. An Outline of Entomology. Willey Blackwell.

Gurney, A.B., I.P. Kramer & G.C. Steyskal. 1964. Some Techniques for the Preparation, Study and Storage in Microvials of Insect Genitalia. Annals of the Entomological Society of America 57: 240–242.

Kier, G., H. Kreft, T.M. Lee, W. Jetz, P.L. Ibisch, C. Nowichi & W. Barthlott. 2009. A global assessment of endemism and species richness across island and mainland regions. Proceedings of the National Academy of Sciences 106: 9322–9327.

Lavelle, P. 2002. Functional domains in soils. Ecological Research 17: 441–450.

Lopes, S.M. & E.H. Oliveira. 2013. New group, new species and new records from Brazil of the atropos, brasilianus and giganteus groups (Blattaria, Blaberidae, Blaberinae). http://www.biotaneotropica.org.br/v13n3.

Mckittrick, F.A. 1964. Evolutionary studies of cockroaches. Memoir by Cornell University. Agricultural Experiment Station 389: 1–197.

Nesemann, H., R.D.T. Shah, D.N. Shah & S. Sharma 2010. First records of *Rhicnoda natatrix* and *Rhicnoda rugosa* (Blattodea: Blaberidae) from Nepal and India (Maharashtra) with notes on habitat quality. Journal of Threatened Taxa 2: 648–652.

Nessimian, J.L. 2014. Ordem Blattodea. Pages 288–295 *in*: N. Hamada, J.L. Nessimian & R.B. Querino (eds.), Insetos Aquáticos na Amazônia brasileira: taxonomia, biologia e ecologia. Manaus. Ed. INPA.

Rentz, D. 2014. A guide to the cockroaches of Australia: CSIRO Publishing, Collingwood.

Rocha e Silva-Albuquerque, I. & S.M.R. Lopes. 1976. Blattaria de bromélia (Dictyoptera). Revista Brasileira de Biologia 36: 837–901.

Roth, L.M. 1970. The male genitalia of Blattaria. II. *Poeciloderrhis* spp (Blaberidae: Epilamprnae). Psyche 77: 104–119.

Roth, L.M. 2003. Systematics and Phylogeny of Cockroaches (Dictyoptera, Blattaria). Oriental Insects, 37: 1–186. https://doi.org/10.1080/003053 16.2003.10417344.

Shelford, R. 1907. Aquatic Cockroaches. Zoologist 1907: 221–226.

Roth, L.M. & E.R. Willis. 1960. The biotic associations of cockroaches. Smithsonian Miscellaneous Collections 141: 1–470.

Shelford, R. 1909. Notes on some amphibious cockroaches. Records of the Indian Museum Calcutta 3: 125–127.

Takahashi, R. 1921. Orthoptera and water. Transactions of the Sapporo Natural History Society 8: 85–100.

Chapter 7

Order Hemiptera

Felipe Ferraz Figueiredo Moreira

Laboratório de Biodiversidade Entomológica, Fundação Oswaldo Cruz, Rio de Janeiro, Rio de Janeiro, Brazil

Higor Daniel Duarte Rodrigues

Museu de Zoologia, Universidade de São Paulo, São Paulo, São Paulo, Brazil

Robert W. Sites

Enns Entomology Museum, Division of Plant Sciences, University of Missouri, Columbia, MO, United States

Isabelle da Rocha Silva Cordeiro

Laboratório de Biodiversidade Entomológica, Fundação Oswaldo Cruz, Rio de Janeiro, Rio de Janeiro, Brazil

Oséias Martins Magalhães

Laboratório de Biodiversidade Entomológica, Fundação Oswaldo Cruz, Rio de Janeiro, Rio de Janeiro, Brazil

Chapter Outline

INTRODUCTION

Hemiptera is the largest and most diverse order among hemimetabolous insects, with about 89,000 species distributed worldwide. All Neotropical aquatic taxa are in the suborder Heteroptera. The heteropterans are further divided into seven infraorders: Enicocephalomorpha, Dipsocoromorpha, Gerromorpha, Nepomorpha, Leptopodomorpha, Cimicomorpha, and Pentatomomorpha (Štys & Kerzhner, 1975).

Most true bugs are terrestrial, but members of the infraorders Gerromorpha, Nepomorpha, and Leptopodomorpha are primarily associated with freshwater bodies, with some taxa tolerant of, or restricted to salt water, or even secondarily terrestrial (Schuh & Slater, 1995). Our chapter concerns representatives of only these three infraorders of Heteroptera and excludes all other groups of Hemiptera. These three together include more than 4,800 species, but are far less diverse than their terrestrial relatives. Leptopodomorpha is the least species rich, with about 380 species, whereas Gerromorpha has over 2,100, and Nepomorpha more than 2,300 (Polhemus & Polhemus, 2008).

Gerromorpha classification is still based mostly on Andersen (1982), who included semiaquatic bugs in four superfamilies and eight families: Mesovelioidea (Mesoveliidae),

Thorp and Covich's Freshwater Invertebrates, Fourth Edition. https://doi.org/10.1016/B978-0-12-804223-6.00007-X

Hebroidea (Hebridae), Hydrometroidea [Paraphrynoveliidae + (Macroveliidae + Hydrometridae)], and Gerroidea [Hermatobatidae + (Veliidae + Gerridae)]. There has been a progressive abandonment of the superfamilies, some of which are now considered paraphyletic based on recent phylogenetic analyses (Damgaard, 2008, 2012). For this reason, we consider only eight families is this chapter, not the superfamilies.

Nepomorpha classification has undergone more changes at the family level than has Gerromorpha. Among the more important classifications are the studies of Popov (1971), Mahner (1993), and Hebsgaard et al. (2004). The first two were not based on phylogenetic analyses, making the results of Hebsgaard et al. (2004) currently more relevant. Hebsgaard et al. (2004) combined morphological characters with molecular data and supported the monophyly of the 11 families of Nepomorpha. They considered Nepoidea (Belostomatidae + Nepidae) as basal and the sister group of the remaining families: Corixoidea (Corixidae), Aphelocheiroidea (Potamocoridae + Aphelocheiridae), Naucoroidea (Naucoridae), Ochteroidea (Ochteridae + Gelastocoridae), Notonectoidea (Notonectidae), and Pleoidea (Helotrephidae + Pleidae). There are some exceptions in the classification of Corixidae (*sensu lato*), because some researchers consider the two subfamilies Micronectinae and Diaprepocorinae as distinct families (e.g., Nieser, 2002; Chen et al., 2005). More recently, these two lineages frequently have been reported in the literature as families, and results from a preliminary analysis based on morphological and molecular characters support the familial status of Micronectidae (Tinerella, 2008).

The relationships within Leptopodomorpha have not changed since the publication of Schuh & Polhemus (1980), a study based on morphological characters in which they proposed Saldidae + Aepophilidae as the sister group of Omaniidae + Leptopodidae (Weirauch & Schuh 2011). The study of evolutionary relationships among genera and species within these three infraorders is still incipient, particularly among Neotropical taxa (Barbosa & Rodrigues, 2015; Melo, 2015; Moreira, 2015). The latest compilation available indicates that the Neotropical Region is represented by more than 45 genera and 500 species of Gerromorpha, more than 50 genera and 700 species of Nepomorpha, and fewer than 10 genera and 40 species of Leptopodomorpha (Polhemus & Polhemus, 2008). The vast majority of these are related to inland water bodies, with only a few species from terrestrial, coastal, or open ocean environments (Barbosa & Rodrigues, 2015; Melo, 2015; Moreira, 2015).

LIMITATIONS

Despite a rich research history (e.g., Champion, 1898–1901; Nieser, 1975; Polhemus, 1997; see also Woodruff, 1958, 1963; Ruhoff, 1968; Polhemus, 2015), there remain many understudied geographical areas and taxonomic groups of aquatic and semiaquatic bugs in the Neotropics, and dozens of undescribed species are known to exist (Polhemus & Polhemus, 2007). This is the result not only of difficult or dangerous access to some regions, but also from a general lack of local taxonomists and the overwhelming bureaucracy that limits or restricts collecting and studying the biodiversity in some countries. This perspective has been slowly changing since the early 2000s by the rise of prolific cooperative research groups interested in aquatic and semiaquatic bugs in Mexico, Costa Rica, Colombia, Brazil, and Argentina (Moreira et al., 2015; Sites & Reynoso-Velasco, 2015; Moreira, 2015).

Several taxonomic revisions and identification keys have been published in the last decade (e.g., Nieser & Chen, 2008; Ribeiro & Estévez, 2009; Polhemus & Polhemus, 2010; Mazzucconi, 2011; Barbosa et al., 2012; Morales-Castaño & Castro-Vargas, 2013; Rodrigues et al., 2014; Aristizábal et al., 2015; Reynoso-Velasco et al, 2016; Floriano et al., 2017). However, there has been a remarkable increase in the publication of erroneous information. Most of these consist of lists with dubious records included in ecological and other nontaxonomic publications, but there are also poorly prepared faunal studies and species descriptions, which is of much greater concern. We refrain from reviewing those publications here, but we concur with the frequent criticism of the widely used series *Encyclopedia of South American Aquatic Insects*, including Heckman's (2011) book on Heteroptera. This series has many problems regarding nomenclature, species compilation, distribution lists, misinterpretation of characters, lack of specimen examination, and construction of identification keys without a critical analysis. Such problems have been reported for various volumes of this series, including Ephemeroptera (see Flowers, 2004; Peters & Soldán, 2005), Odonata (see Carvalho & Pinto, 2007; Muzón, 2007), and Plecoptera (see Kondratieff, 2004; Soldán, 2005).

TERMINOLOGY AND MORPHOLOGY

- Acetabulum: Place of insertion of coxae in the thorax.
- Antennomere: Each article of an antenna.
- Apterous: Individual with fore- and hindwings completely absent.
- Arolium: Unpaired medial structure that arises between the bases of the pretarsal claws.
- Brachypterous: Individual with fore- and hindwings short, with clavus and corium of forewing fused.
- Bucculae: Posteroventral projections of the head capsule in the shape of vertical plates.
- Carina: Ridge or keel on the integument.
- Clavus: Portion of the forewing lying adjacent to the scutellum and containing the first and second anal veins.
- Connexivum: Lateral margin of the abdomen formed by laterotergites and laterosternites.
- Corium: Portion of the forewing separated from the clavus by the claval sulcus, which runs obliquely from

the basal articulation of the wing toward the posterodistal margin.

- Embolium: In the forewing, the submarginal part of the costal margin of the corium situated proximally to the costal fracture.
- Genae: Portion of the head of Corixidae below the compound eye.
- Grasping comb: Comb formed by a row of short, stout spines in the distal portion of the protibia of most male Veliidae.
- Hypo-ocular suture: Suture posteroventral to the compound eye in Corixidae.
- Interocular: Space between the compound eyes.
- Laterosternite: Lateral subdivisions of the abdominal sterna of pregenital segments; sclerite lateral to mediosternite.
- Laterotergite: Lateral subdivisions of the abdominal terga; sclerite lateral to mediotergite.
- Macropterous: Individual with fore- and hindwings elongate, clavus, and corium of forewing distinct, membrane well-developed.
- Membrane: Portion of the forewing distal to the clavus and corium, with or without veins or closed cells.
- Pala: Modified distal tarsal article of the foreleg of Corixidae.
- Palm: Portion of the corixid pala between upper and lower rows of palmar hairs.
- Respiratory siphon: Paired distal structures of abdominal tergum VIII, forming a tube in adults of Belostomatidae and Nepidae, used to obtain atmospheric air.
- Rostral groove: Formed by the well-developed bucculae that conceal the basal articles of the rostrum.
- Scutellum: Dorsal plate of the mesothorax, usually with triangular shape.
- Strigil: Structure of rows of comb-like teeth on abdominal tergum VI of male Corixidae, used for stridulation.
- Tarsal formula: Number of tarsomeres of each leg (fore-middle-hind).
- Tarsomere: Each article of a tarsus.

MATERIAL PRESERVATION AND PREPARATION

Preservation can be done in two ways, where the material is either pinned or kept in 70–80% ethyl alcohol, unless the specimens are to be used for molecular studies, in which case they should be kept in absolute ethyl alcohol. In general, specimens have longer archival longevity if pinned and kept in museum environmental conditions. However, for some families, especially Corixidae and Notonectidae, it is recommended to keep them in alcohol, because dry specimens become very fragile over the years, which can result in damage of structures and loss of important information. The precise placement of the pin through the specimen is very important and varies greatly depending on the taxon. In general, members of Heteroptera are pinned through the right side of the scutellum to avoid obliterating diagnostic structures on the midline. If the scutellum is very small, it is recommended to pin through the base of the forewing, also on the right side. Except for large Gerromorpha (e.g., some genera of Gerridae), where it is possible to pin the specimen, the most appropriate technique for this infraorder is to glue the specimen to a triangular paper point (small elongate triangular mounts—*sensu* Schuh & Slater, 1995). The small portion at the apex of the triangle must be slightly bent, and a sufficient amount of white glue applied to this fold in order to hold the specimen after drying. It is best to keep the appendages close to the body to minimize the likelihood of breakage.

In nearly all genera of aquatic Heteroptera, it is necessary to examine the genitalia to identify species. Material in alcohol is easier because the specimen is soft; however, in pinned specimens it is necessary to soften the insect's body before dissection. The pinned specimen can be relaxed by immersion in 70% ethyl alcohol for 6–10 h, with larger specimens requiring up to 24 h. A faster option is to heat water to a temperature of approximately 80°C and then immerse the specimen for a few minutes. To examine taxonomic features, it is generally not necessary to clear the genitalia. However, if necessary, a cold solution of 10% KOH can be used for approximately 10 h; small species will require less time.

KEYS TO FRESHWATER HEMIPTERA

Hemiptera: Infraorders

1 Antennae longer than head, inserted in front of eyes and plainly visible from above (Figs. 7.1 C, 7.2 G, 7.3 B); water surface or shore bugs 2

1' Antennae shorter than head, inserted beneath eyes, not visible from above (Figs. 7.1 A, H) except in Ochteridae (Figs. 7.1 B, 7.4 A); truly aquatic or shore bugs........ **Nepomorpha [p. 179]**

2(1) Metacoxae large, transverse, broadly joined to thoracic pleura (Fig. 7.1 D); forewing membrane usually with four or five similar closed cells (Fig. 7.1 F) .... Leptopodomorpha, one family: **Saldidae [p. 204]**

2' Metacoxae small, conical or cylindrical, freely rotatory in acetabula (Fig. 7.1 E); forewing membrane without cells or with dissimilar cells (Fig. 7.1 G) .... **Gerromorpha [p. 207]**

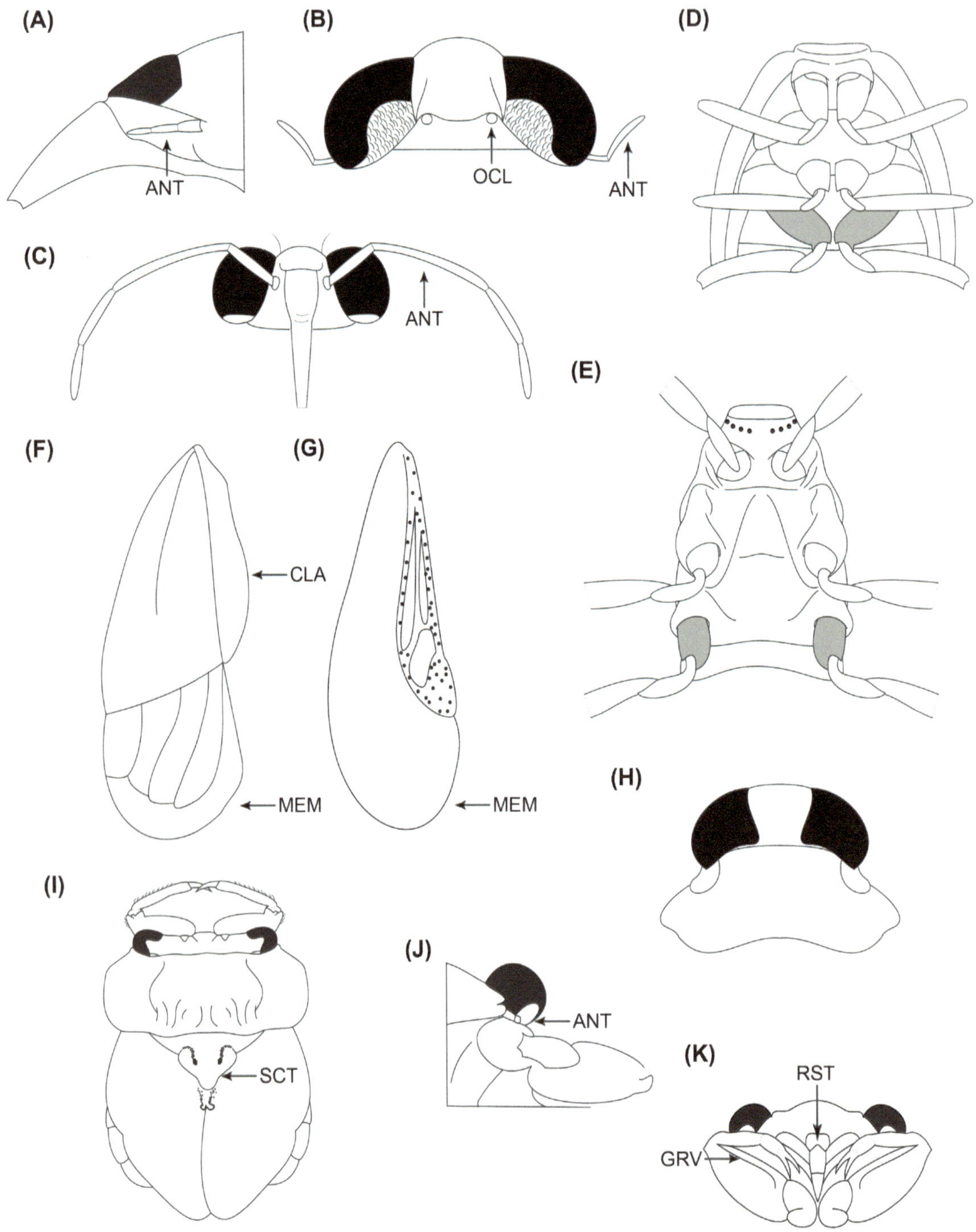

FIGURE 7.1 Naucoridae: (A) head, ventral view. Ochteridae: (B) head, dorsal view. Saldidae: (C) head, ventral view; (D) thorax, ventral view, metacoxae in gray. Macroveliidae: (E) thorax, ventral view, metacoxae in gray. Saldidae: (F) forewing. Mesoveliidae: (G) forewing. Notonectidae: (H) head and pronotum, dorsal view. *Nerthra*: (I) habitus, dorsal view. Gelastocoridae: (J) head and prothorax, ventral view; and: (K) head and prothorax, ventral view. Abbreviations: *ANT*, antenna; *CLA*, clavus; *GRV*, groove; *MEM*, membrane; *OCL*, ocellus; *RST*, rostrum; *SCT*, scutellum.

Hemiptera: Nepomorpha: Families

1 Ocelli present (Fig. 7.1 B); riparian habitat 2
1' Ocelli absent (Fig. 7.1 H); aquatic habitat 3
2(1) Antennae shorter than head, not visible dorsally (Figs. 7.1 I, J); rostrum short, reaching procoxae (Fig. 7.1 K); scutellum tumid (Fig. 7.1 I); profemora enlarged, with a groove for reception of tibiae (Fig. 7.1 K) .... **Gelastocoridae [p. 179]**
2' Antennae filiform, usually visible dorsally (Fig. 7.4 A); rostrum long, reaching at least to metacoxae (Fig. 7.4 B); scutellum flat (Fig. 7.4 A); all legs cursorial, profemora not enlarged (Fig. 7.5) .... **Ochteridae [p. 179]**
3(1) Rostrum cylindrical to conical, distinctly articulate, without transverse striations (Figs. 7.4 B, E); protarsi not spatulate, cylindrical, with one or more articles and distinct pretarsal claws (Figs. 7.4 F, K) 4
3' Rostrum triangular, apparently not articulate, with transverse striations (Figs. 7.4 C, 7.6 A); protarsi uniarticulate, modified, spatulate, with a fringe of long setae and without distinct pretarsal claws (Fig. 7.4 D) .... **Corixidae (*sensu lato*) [p. 183]**
4(3) Hemelytral membrane with venation (Figs. 7.4 G, I); respiratory siphon present at end of body (Figs. 7.19, 7.20 C, G) 5
4' Hemelytral membrane without venation (Figs. 7.4 H, 7.7 F, G); respiratory siphon absent from end of body 6
5(4) All tarsi uniarticulate (Fig. 7.4 J); hindlegs without fringe of setae (Fig. 7.7 A); respiratory siphon filiform, not retractile and usually long (Figs. 7.4 I, 7.7 C) .... **Nepidae [p. 187]**
5' Tarsi bi- or triarticulate (Fig. 7.4 K), rarely protarsi uniarticulate; hindlegs with fringe of setae (Fig. 7.7 B); siphon flattened, retractile and short (Fig. 7.7 D) .... **Belostomatidae [p. 189]**
6(4) Body with dorsal surface flat to slightly convex (Figs. 7.4 H, 7.7 G, 7.8, 7.9); forelegs retentorial (Figs. 7.4 H, 7.7 E, H) 7
6' Body with dorsal surface usually convex (Figs. 7.7 F, K, 7.10, 7.11, 7.12, 7.13 A, F); forelegs not or only weakly retentorial (Figs. 7.7 K, 7.13 B) 8
7(6) Antennae long, extending beyond lateral margins of head (Fig. 7.7 H); profemora not or slightly enlarged (Fig. 7.7 H); male genitalia always symmetrical (Fig. 7.7 I); body length $\leq$4 mm .... Potamocoridae, one genus: *Potamocoris*
7' Antennae short, not exceeding lateral margins of head in dorsal view (Fig. 7.1 A); profemora greatly enlarged (Fig. 7.7 E); male genitalia usually asymmetrical (Fig. 7.7 J); body length >4 mm .... **Naucoridae [p. 191]**
8(6) Body globose, with dorsal surface punctuate (Figs. 7.11, 7.12, 7.13 A, F), length <4 mm; legs subequal in length (Figs. 7.11, 7.12, 7.13 B); metapretarsal claws well developed (Figs. 7.11, 7.13 B) 9
8' Body elongate, without punctuations on dorsal surface (Figs. 7.7 F, K, 7.10), length $\geq$4 mm; hindlegs longer than fore- and middle legs (Figs. 7.7 K, 7.10); metapretarsal claws inconspicuous (Figs. 7.7 L, 7.10) .... **Notonectidae [p. 196]**
9(8) Head fused with pronotum, although suppressed suture sometimes apparent (Figs. 7.11, 7.13 A, C); antennae uniarticulate in brachypterous forms and biarticulate in macropterous forms (Fig. 7.13 D); rostrum quadriarticulate (Fig. 7.13 E) .... **Helotrephidae [p. 202]**
9' Head not fused with pronotum (Figs. 7.12, 7.13 F); antennae and rostrum triarticulate (Fig. 7.13 G) .... **Pleidae [p. 202]**

Hemiptera: Nepomorpha: Gelastocoridae: Genera

Gelastocoridae (Fig. 7.14), commonly known as toad bugs, occur worldwide with approximately 100 species (Cassis & Silveira, 2001). Phylogenetic studies of Nepomorpha have frequently associated Gelastocoridae with Ochteridae forming a monophyletic clade, the superfamily Ochteroidea (Rieger, 1976; Mahner, 1993; Hebsgaard et al., 2004; Hua et al. 2009; Li et al., 2014). Both families have riparian habitat preferences and strongly asymmetrical male genitalia (Polhemus & Polhemus, 2012).

1 Pronotum rectangular, generally with anterior and posterior margins subequal in width (Fig. 7.13 H); profemora subtriangular, very broad at base, about as wide as long (Fig. 7.13 J); protarsi fused with tibiae, bearing one claw (Fig. 7.13 J); color usually uniform .... *Nerthra*
1' Pronotum with anterior margin distinctly narrower than posterior margin (Fig. 7.13 I); profemora moderately incrassate, about half as wide as long (Fig. 7.13 K); protarsi not fused with tibiae, bearing two claws (Fig. 7.13 K); color heterogeneous, pruinose (Fig. 7.14) .... *Gelastocoris*

Hemiptera: Nepomorpha: Ochteridae: Genera

Ochteridae (Fig. 7.5), commonly known as velvety shore bugs, is a predominantly tropical family with only three genera and approximately 60 described species. The body length ranges from 4.0 to 9.5 mm and they generally have a pruinose appearance with dark coloration and light markings on the pronotum and forewings. Ochterids are noted for their saldid-like appearance, but the short antennae inserted ventrally distinguish them from Saldidae (Schuh & Slater, 1995). Adults fly quickly when disturbed and are not easily collected.

1 Eyes separated from vertex of head by channels adjacent to their mesal margins (Fig. 7.13 L); forewing with 4–6 cells in apical three-fourths of membrane (Fig. 7.15 A); body length 3.30–6.30 mm (compare Figs. 7.15 B, C) .... *Ochterus*
1' Vertex of head lacking channels adjacent to mesal margins of eyes (Fig. 7.13 M); forewing with 10–12 cells in apical three-fourths of membrane (Fig. 7.5); body length 6.20–8.20 mm (compare Figs. 7.15 B, C) .... *Ocyochterus*

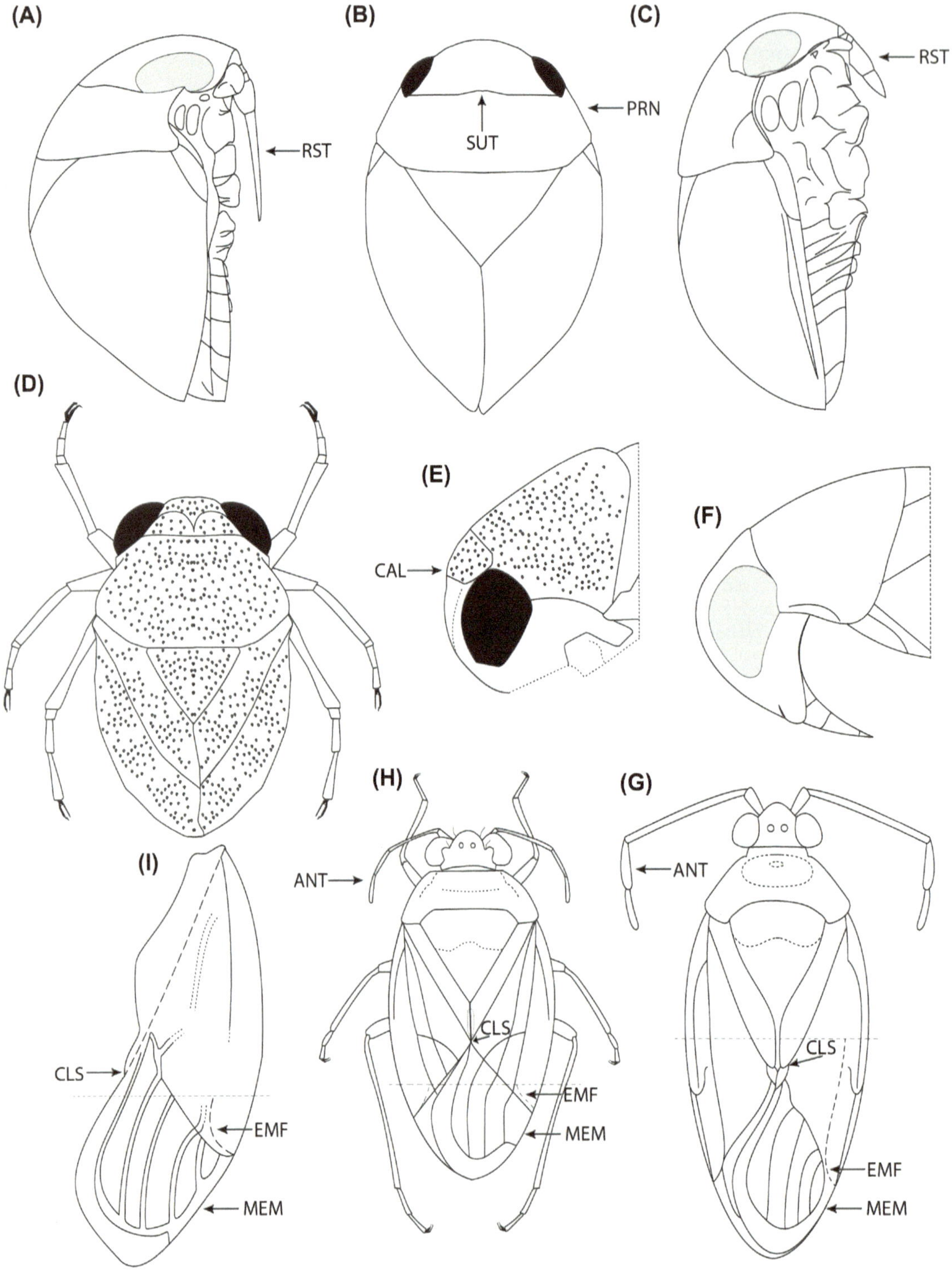

FIGURE 7.2 *Paratrephes*: (A) habitus, lateral view; (B) habitus, dorsal view. *Neotrephes*: (C) habitus, lateral view. *Heteroplea:* (D) habitus, dorsal view; (E) head and part of thorax, lateral view. *Neoplea:* (F) head and part of thorax, lateral view. *Pentacora*: (G) habitus, dorsal view. *Saldula*: (H) habitus, dorsal view. *Pseudosaldula*: (I) forewing. Abbreviations: *ANT*, antenna; *CAL*, callus; *CLS*, end of claval suture; *EMF*, embolar fracture; *MEM*, membrane; *PRN*, pronotum; *RST*, rostrum; *SUT*, cephalonotal suture.

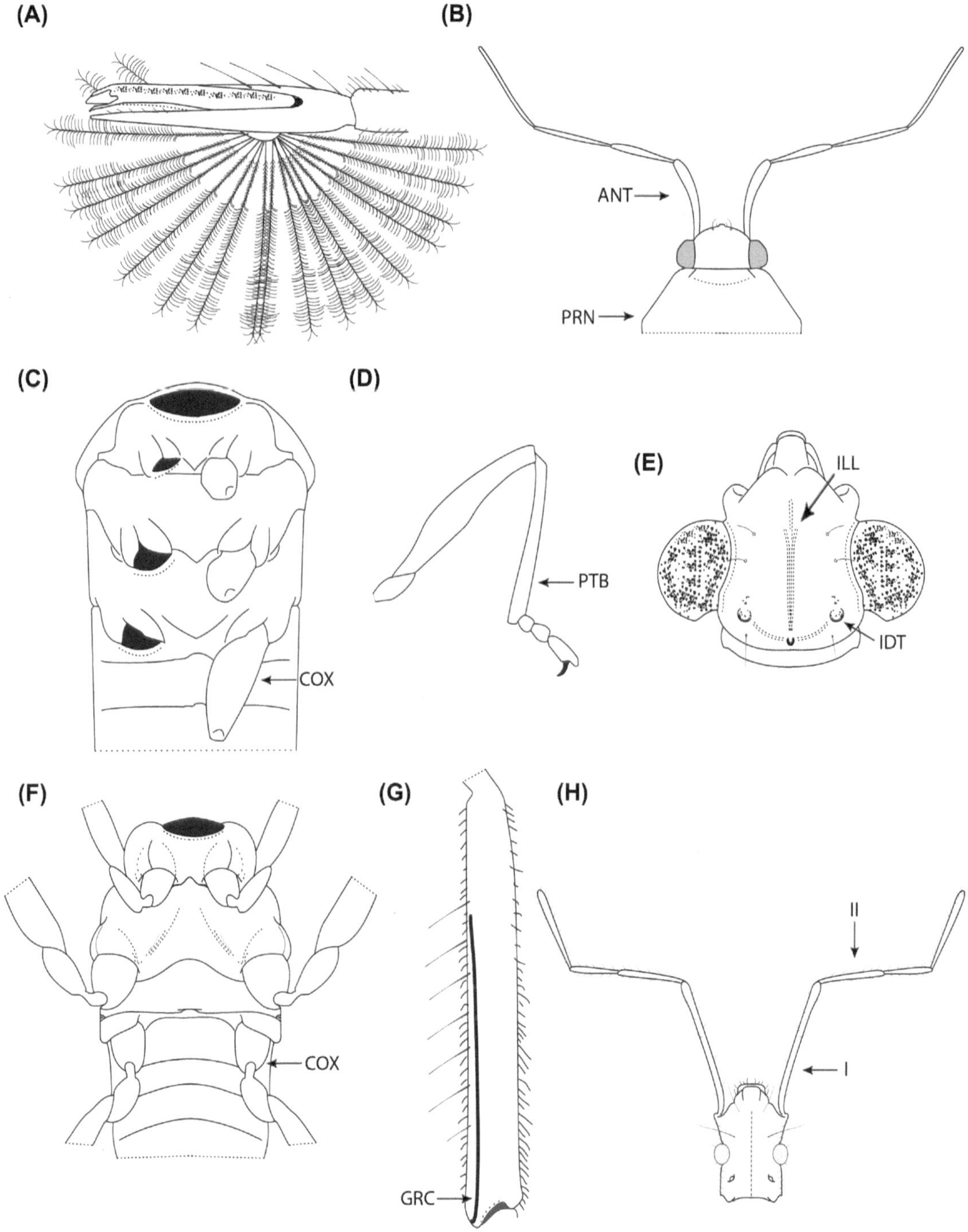

FIGURE 7.3 *Rhagovelia*: (A) apex of mesotarsus. *Mesoveloidea*: (B) head and part of pronotum, dorsal view. Mesoveliidae: (C) thorax and part of abdomen, ventral view. *Mesoveloidea*: (D) foreleg. *Microvelia*: (E) head, dorsal view. *Rhagovelia*: (F) thorax and part of abdomen, ventral view. *Angilia*: (G) protibia. *Veliometra*: (H) head, dorsal view. Abbreviations: *ANT*, antenna; *COX*, coxa; *GRC*, grasping comb; *I*, antennomere 1; *II*, antennomere 2; *IDT*, indentation; *ILL*, impressed longitudinal line; *PRN*, pronotum; *PTB*, protibia.

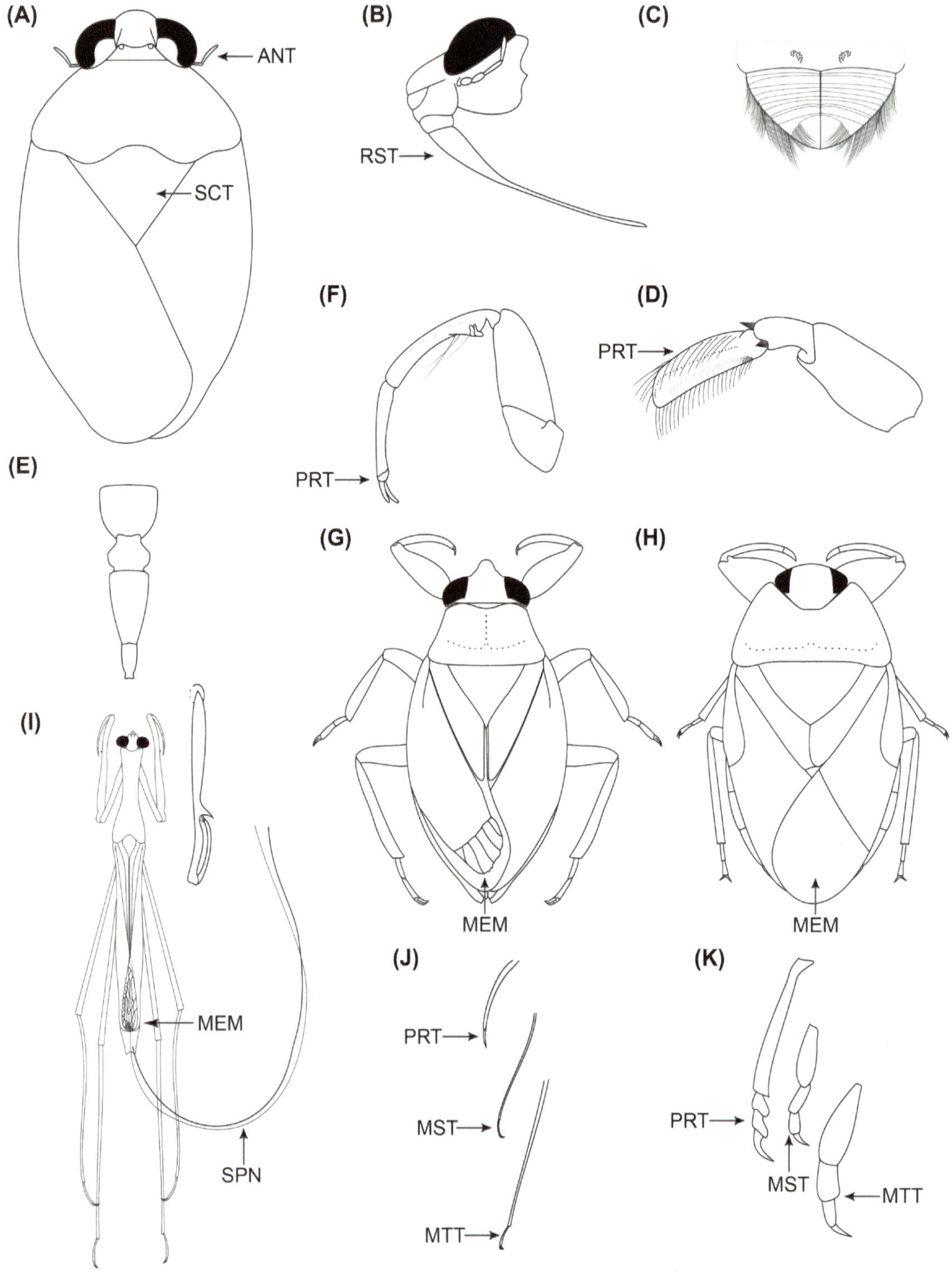

FIGURE 7.4 Ochteridae: (A) habitus dorsal view; (B) head, lateral view. Corixidae: (C) rostrum; (D) foreleg. Notonectidae: (E) rostrum; (F) foreleg. *Belostoma*: (G) habitus, dorsal view. Naucoridae: (H) habitus, dorsal view. *Ranatra*: (I) habitus, dorsal view; (J) apex of legs. Belostomatidae: (K) apex of legs. Abbreviations: *ANT*, antenna; *MEM*, membrane; *MST*, mesotarsus; *MTT*, metatarsus; *PRT*, protarsus; *RST*, rostrum; *SCT*, scutellum; *SPN*, siphon.

FIGURE 7.5 *Ocyochterus* (Ochteridae): habitus, dorsal view.

Hemiptera: Nepomorpha: Corixidae (*Sensu Lato*): Genera

This key is modified from Hungerford (1948). Corixidae (Fig. 7.16), commonly known as water boatmen, is the family of Nepomorpha with the greatest diversity and has approximately 560 species in 35 genera and six subfamilies: Corixinae, Cymatiainae, Diaprepocorinae, Heterocorixinae, Micronectinae, and Stenocorixinae (Hungerford, 1948). Two of these subfamilies (Micronectinae and Diaprepocorinae) are considered as families by some authors (e.g., Nieser, 2002; Chen et al., 2005). Here we treat all as subfamilies of Corixidae because robust phylogenetic studies testing the monophyly of the groups are lacking. Females are difficult to identify because of the lack of interspecific characters and should be matched with males based partly on dorsal color patterns (Hungerford, 1948).

1 Pronotum not covering scutellum (Fig. 7.15 D) 2
1' Pronotum almost entirely covering scutellum (Fig. 7.15 E) 4
2(1) Mesosternum without carina (Fig. 7.15 G) 3
2' Mesosternum with a distinct carina (Fig. 7.15 F) .... *Synaptogobia*
3(2) Antennae triarticulate (Fig. 7.15 H); metaxyphus short (Fig. 7.15 I) .... *Tenagobia*
3' Antennae uniarticulate (Fig. 7.15 J); metaxyphus long (similar to Fig. 7.15 F) .... *Monogobia*
4(1) Head in lateral view with posterior margin of eye almost straight; hypo-ocular suture arising about midway along posterior margin of eye (Fig. 7.6 B) 5
4' Head in lateral view with posterior margin of eye concave; hypo-ocular suture arising from posteroventral angle of eye (Fig. 7.6 A). .... *Heterocorixa*
5(4) Protarsi with apical claw well-developed (Figs. 7.6 C, D); pala of both sexes narrowly digitate (Figs. 7.6 C, D); males with row of stout, short pegs along upper edge of palm (Fig. 7.6 D); frons in both sexes usually with a tomentose patch (Fig. 7.6 E) 6
5' Protarsi with apical claw spiniform, usually resembling spines along lower margin of palm (Fig. 7.6 F); pala of various shapes, not narrowly digitate (Fig. 7.6 G); males with row of pegs usually well above fringe of setae along upper edge of palm (Fig. 7.6 G); frons with setae not in a tomentum, often absent (Fig. 7.6 H) 7
6(5) Male abdomen sinistral, strigil absent (Fig. 7.6 I); female face slightly concave, abdomen slightly asymmetrical .... *Neocorixa*
6' Male abdomen dextral, strigil present (Fig. 7.6 J); female face not concave, abdomen symmetrical .... *Graptocorixa*
7(5) Male protibiae not produced over pala (Fig. 7.6 G) 8
7' Male protibiae distinctly produced over pala (Fig. 7.6 K) 12
8(7) Male vertex not produced or rounded, frons flat; male pala without deep transverse groove on dorsal surface (Fig. 7.6 M); females with costal margin of forewings not thickened at basal third 9

8'	Male vertex conically produced anterad (acuminate), frons deeply concave (Fig. 7.6 L); male pala nearly divided by deep transverse groove on dorsal surface (Fig. 7.6 G); females with costal margin of forewings with an elongate-ellipsoid, polished thickening at basal third	*Ramphocorixa*
9(8)	Eye posteromesal angle not acutely produced (Fig. 7.17 C); pronotum and hemelytra rastrate; male profemora not enlarged (Fig. 7.17 D)	10
9'	Eye posteromesal angle acutely produced (Fig. 7.17 A); pronotum and hemelytra smooth; male profemora greatly enlarged (Fig. 7.17 B)	*Krizousacorixa*
10(9)	Protibiae without apical comb of spines (Fig. 7.17 G); males with dextral asymmetry and with strigil (Fig. 7.17 H) or with sinistral asymmetry and without strigil (Fig. 7.17 I)	11
10'	Protibiae with apical comb of spines (Fig. 7.17 E); males with dextral asymmetry and without strigil, although strigilar pedicel present (Fig. 7.17 F)	*Morphocorixa*
11(10)	Male pala with one row of pegs (Fig. 7.17 J); metaxyphus short, triangular (Figs. 7.15 G, 7.17 K)	*Sigara*
11'	Male pala with two rows of pegs (Fig. 7.17 L); metaxyphus long, apically rounded	*Neosigara*
12(7)	Eye posteromesal angle obtusely angulate (Fig. 7.18 B)	13
12'	Eye posteromesal angle sharply right-angled (Fig. 7.18 A), occasionally slightly produced	*Trichocorixella*
13(12)	Ventral surface of metafemora pubescent only on basal third of anterior margin (Fig. 7.18 C)	14
13'	Ventral surface of metafemora pubescent at least on one-half of anterior margin	15
14(13)	Apices of clavi not exceeding a line drawn through nodal furrows of hemelytra in females and slightly exceeding in males (Fig. 7.18 D); tibiae strongly produced apically over pala (Fig. 7.18 E); males with sinistral asymmetry, strigil on the left (Fig. 7.18 F)	*Trichocorixa*
14'	Clavi distinctly exceeding a line drawn through nodal furrows of hemelytra in both sexes (Fig. 7.18 G); tibiae slightly produced over pala (Fig. 7.18 H); males with dextral asymmetry, strigil on the right (Fig. 7.18 I)	*Ectemnostega*
15(13)	Males without strigil (Fig. 7.18 J)	*Centrocorisa*
15'	Males with strigil (Fig. 7.18 K)	*Corisella*

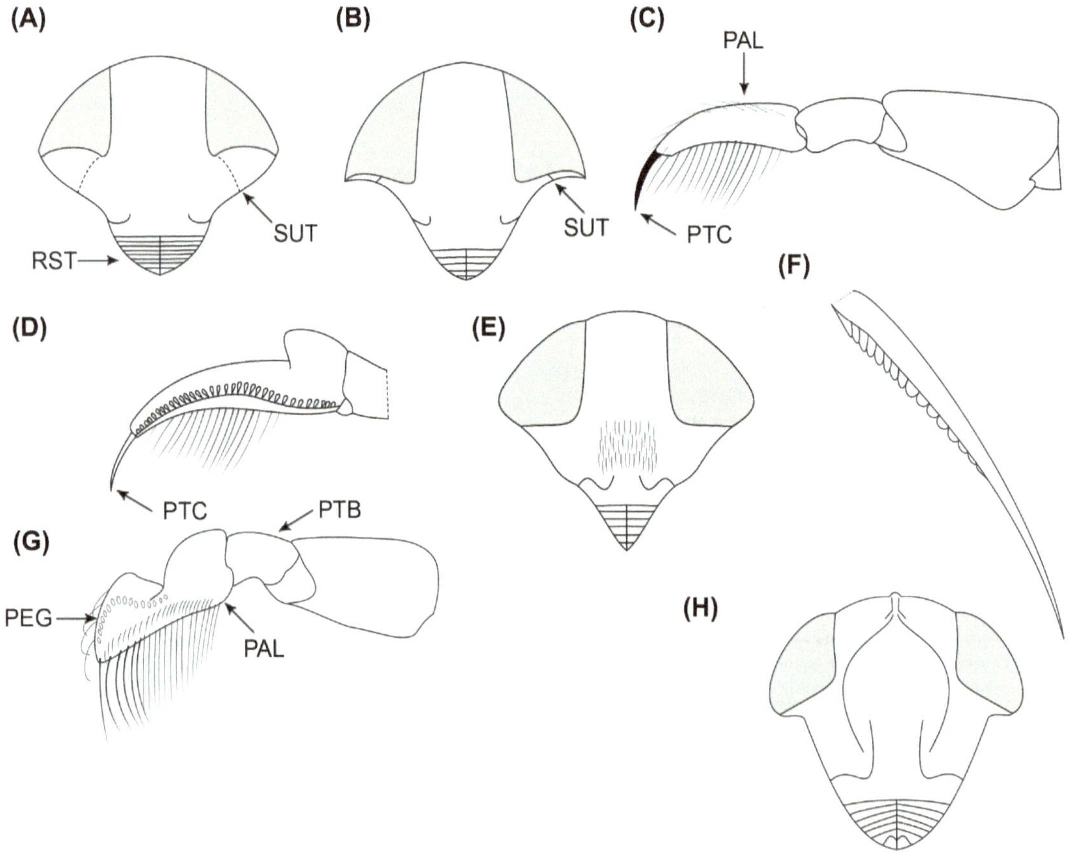

FIGURE 7.6 A. *Heterocorixa*: (A) head, frontal view. Corixinae: (B) head, frontal view. *Graptocorixa*: (C) male foreleg; (D) male pala; (E) head, frontal view. *Ramphocorixa*: (F) protarsal claw; (G) male foreleg; (H) head, frontal view. *Neocorixa*: (I) male abdomen, dorsal view. *Graptocorixa*: (J) male abdomen, dorsal view. *Trichocorixella*: (K) male foreleg. *Ramphocorixa*: (L) head, frontal view. *Morphocorixa*: (M) male pala. Abbreviations: *PAL*, pala; *PEG*, row of pegs; *PTB*, protibia; *PTC*, protarsal claw; *RST*, rostrum; *SUT*, hypo-ocular suture; *STG*, strigil; *VRT*, vertex.

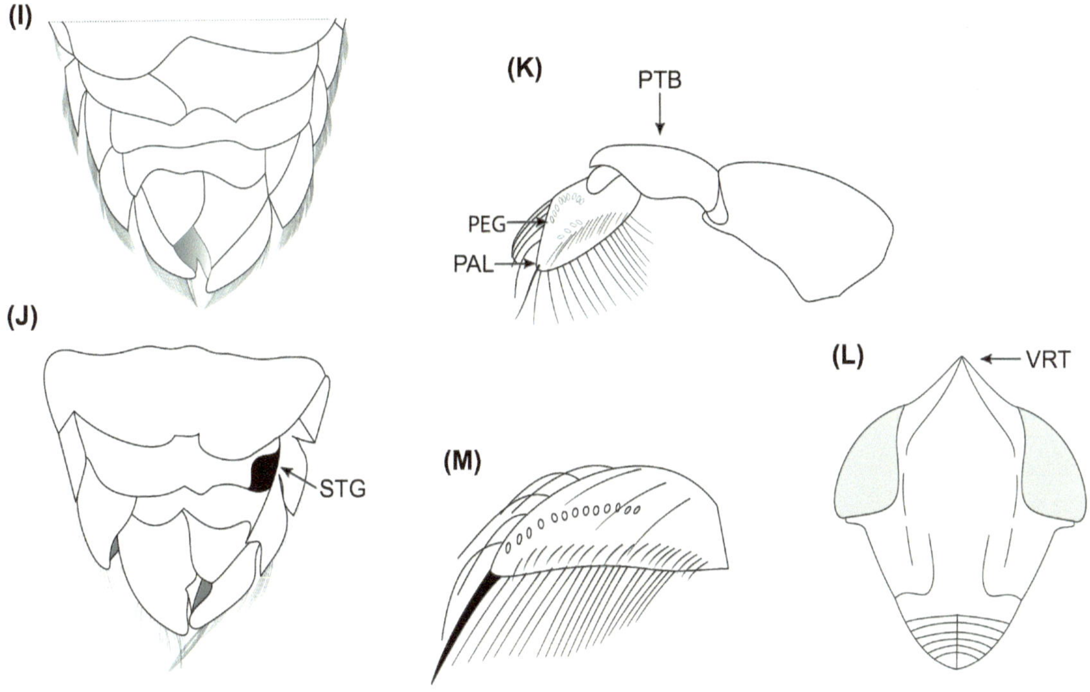

FIGURE 7.6 (Continued)

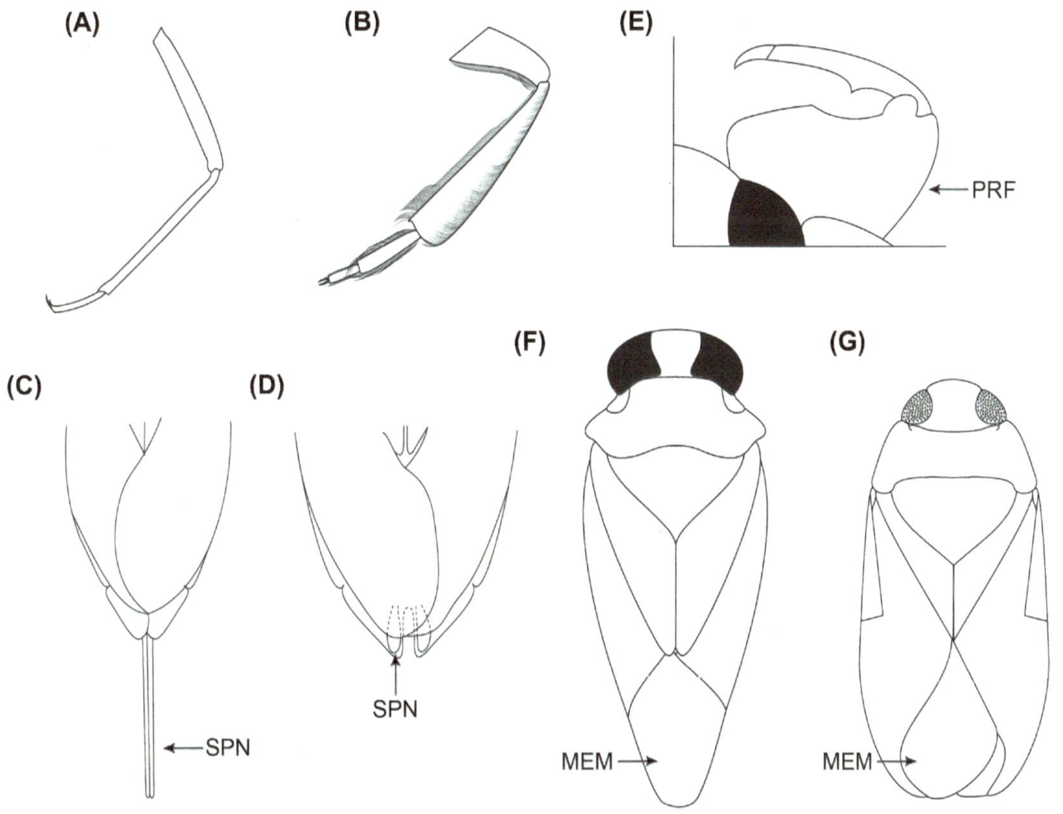

FIGURE 7.7 Nepidae: (A) hindleg. Belostomatidae: (B) hindleg. Nepidae: (C) apex of body, dorsal view. Belostomatidae: (D) apex of body, dorsal view. Naucoridae: (E) part of head and foreleg, dorsal view. Notonectidae: (F) habitus, dorsal view. *Potamocoris*: (G) habitus, dorsal view; (H) head and prothorax, ventral view; (I) apex of body, male, ventral view. Naucoridae: (J) apex of body, male, ventral view. Notonectidae: (K) habitus, dorsal view; and (L) apex of hindleg. Abbreviations: *ANT*, antenna; *MEM*, membrane; *MTC*, metapretarsal claw; *PRF*, profemur; *SPN*, siphon.

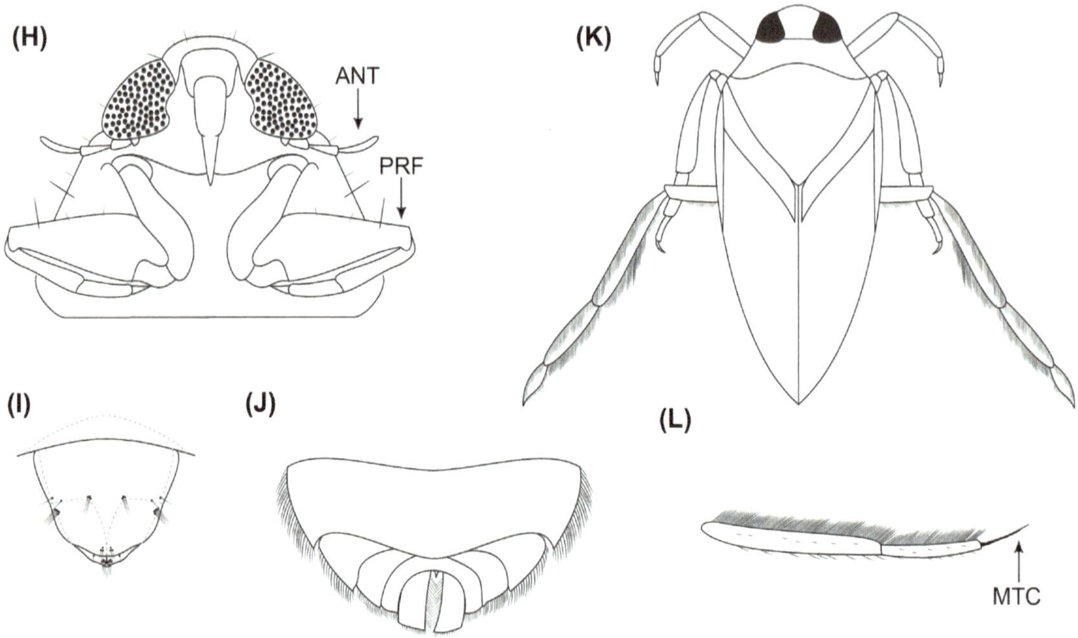

FIGURE 7.7 (Continued)

FIGURE 7.8 *Potamocoris* (Potamocoridae): habitus, dorsal view.

FIGURE 7.9 *Ambrysus* (Naucoridae): habitus, dorsal view.

FIGURE 7.11 *Neotrephes* (Helotrephidae): habitus, dorsal view.

FIGURE 7.10 *Enitharoides* (Notonectidae): habitus, dorsal view.

FIGURE 7.12 *Neoplea* (Pleidae): habitus, lateral view.

Hemiptera: Nepomorpha: Nepidae: Genera

Nepidae (Fig. 7.19) are commonly known as water scorpions, which is derived from the nonretractile respiratory siphons arising from the apex of their abdomen. The respiratory siphons can be as long as or longer than the body length. Water scorpions have raptorial forelegs and are generally narrow and elongated; although a few genera are oval in shape. This family is divided into two subfamilies, Nepinae and Ranatrinae, and contains approximately 225 species in 14 genera (Keffer, 2004). Nepids generally inhabit lentic environments or banks of streams, usually associated with aquatic vegetation.

1	Body flattened (Fig. 7.20 C); anterolateral corners of pronotum wider than head (including eyes) (Fig. 7.20 D)	2
1'	Body cylindrical (Figs. 7.4 I, 7.19, 7.20 A); anterolateral corners of pronotum not wider than head (including eyes) (Fig. 7.20 B).	*Ranatra*
2(1)	Body and pronotum elongate (Fig. 7.20 E); pronotum longer than wide, length/width ratio 5–6 (Fig. 7.20 F)	*Curicta*
2'	Body oval (Fig. 7.20 G); pronotum quadrate, length/width ratio ≤3 (Fig. 7.20 D)	*Telmatotrephes*

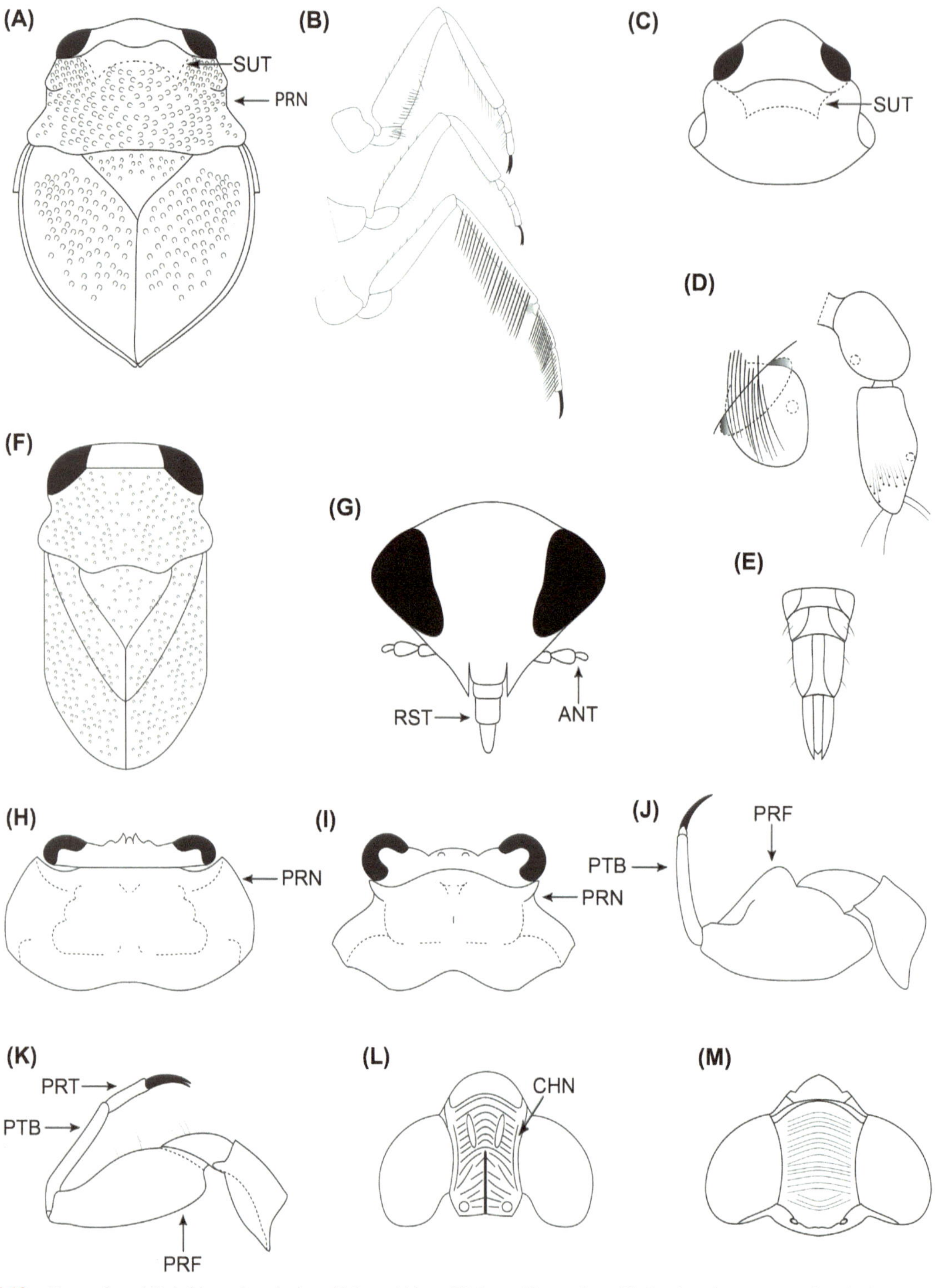

FIGURE 7.13 *Neotrephes*: (A) habitus, dorsal view. Helotrephidae: (B) legs. *Neotrephes*: (C) head and pronotum, dorsal view. Helotrephidae: (D) uniarticulate (left) and biarticulate (right) antennae; (E) rostrum. Pleidae: (F) dorsal view; (G) head, frontal view. *Nerthra*: (H) head and pronotum, dorsal view. *Gelastocoris*: (I) head and pronotum, dorsal view. *Nerthra*: (J) foreleg. *Gelastocoris*: (K) foreleg. *Ochterus*: (L) head, dorsal view. *Ocyochterus*: (M) head, dorsal view. Abbreviations: *ANT*, antenna; *CHN*, channel; *PRF*, profemur; *PRT*, protarsus; *PRN*, pronotum; *PTB*, protibia; *RST*, rostrum; *SUT*, cephalonotal suture.

FIGURE 7.14 *Gelastocoris* (Gelastocoridae): habitus, dorsal view.

Hemiptera: Nepomorpha: Belostomatidae: Genera

Belostomatidae (Fig. 7.21), also known as giant water bugs, toe biters, or electric-light bugs, are the largest Heteroptera and range from 10 to 100 mm in body length. They are typically associated with lentic environments, usually in areas with aquatic vegetation, although they also can be found in the vegetated margins of streams. Belostomatidae is divided into three subfamilies: Belostomatinae, Horvathiniinae, and Lethocerinae (Lauck & Menke, 1961; Štys & Jansson, 1988), and contains more than 140 species in 11 genera (Estévez & Ribeiro, 2011). Species of Belostomatinae have back-brooding behavior in which the females lays eggs on the forewings of the males. Belostomatids can be recognized by the body large, oval to elongate oval, the forelegs are retentorial with the femur enlarged and an opposable tibia, and abdominal tergum VIII is modified into a pair of strap-like retractable respiratory siphons.

1	Rostrum with article 1 short, wider than long, and clearly shorter than article 2 (Fig. 7.20 H); abdominal sterna 5 and 6 subdivided laterally by suturiform folds (Fig. 7.20 I); laterosternite spiracles on or adjacent to mesal margins (Fig. 7.20 I)	2
1'	Rostrum with article 1 much longer than its greatest width, subequal in length to article 2 (Fig. 7.20 J); abdominal sterna not subdivided by a suture (Fig. 7.20 K); laterosternites with submedial spiracles (Fig. 7.20 K)	4
2(1)	Body elongate (Fig. 7.22 A); three protarsomeres (often appearing as two laterally), bearing one long claw (Fig. 7.22 C); metatibiae and -tarsi thinly compressed, much more dilated than mesotibiae and -tarsi (Fig. 7.22 E)	3
2'	Body oval (Fig. 7.22 B); two protarsomeres (often appearing as one), bearing two very short, vestigial claws (Fig. 7.22 D); meso- and metatibiae and -tarsi similar, narrow, flattened, but not broadly dilated (Fig. 7.22 F)	*Horvathinia*
3(2)	Profemora with medial setal pads with two symmetrical sulci (Fig. 7.22 G); metafemora pads with one sulcus; laterosternites 2 and 3 lateral borders narrowed, almost straight (Fig. 7.22 I)	*Lethocerus*
3'	Profemora with medial setal pads without sulci (Fig. 7.22 H); metafemora pads with a deep cleft, but not divided; laterosternites 2 and 3 lateral borders not narrowed, arcuate, following abdomen lateral margin (Fig. 7.22 J)	*Benacus*
4(1)	Forewing membrane distinctly reduced, its greatest width ≤ that of clavus (Fig. 7.22 L); most membrane cells small or poorly defined, or short, rectangular, and of various lengths (Fig. 7.22 L)	5
4'	Forewing membrane well developed, its greatest width > that of clavus (Fig. 7.22 K); most membrane cells equal in length, as long, narrow rectangles (Fig. 7.22 K)	*Belostoma*
5(4)	Abdomen lateral margins not smooth, interrupted between segments (scaliform laterotergites) (Fig. 7.23 A)	*Weberiella*
5'	Abdomen lateral margins smooth, not scaliform (Fig. 7.23 B)	*Abedus*

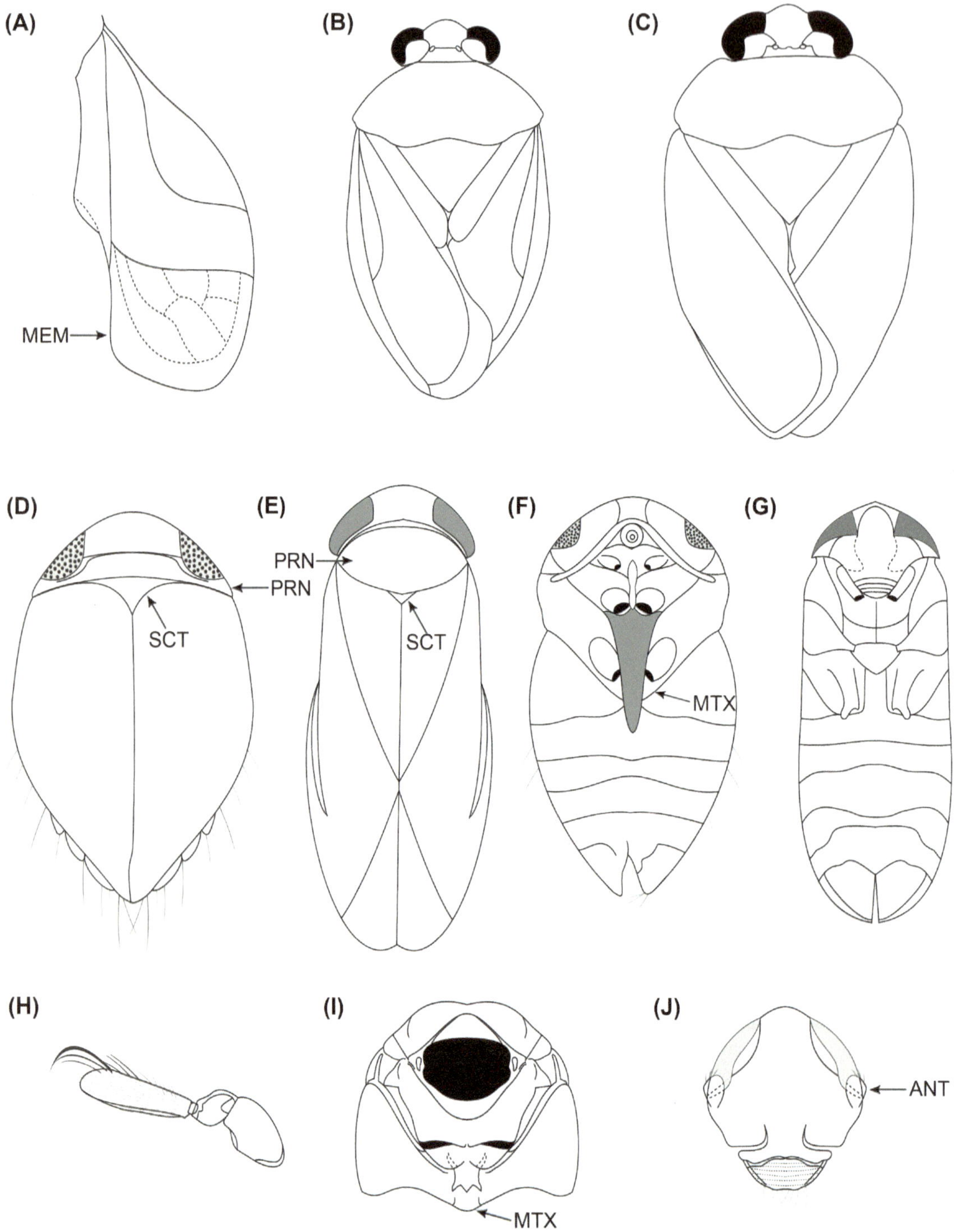

FIGURE 7.15 *Ochterus*: (A) forewing; (B) habitus, dorsal view. *Ocyochterus*: (C) habitus, dorsal view. *Synaptogobia*: (D) habitus, dorsal view. *Neosigara*: (E) habitus, dorsal view. *Synaptogobia*: (F) habitus, ventral view, mesosternal carina in gray. *Sigara*: (G) habitus, ventral view. *Tenagobia*: (H) antenna; (I) pterothorax, ventral view. *Monogobia*: (J) head, ventral view. Abbreviations: *ANT*, antenna; *MEM*, membrane; *MTX*, metaxyphum; *PRN*, pronotum; *SCT*, scutellum.

FIGURE 7.16 *Hesperocorixa* (Corixidae): habitus, dorsal view.

Hemiptera: Nepomorpha: Naucoridae: Genera

Naucoridae (Fig. 7.9), commonly known as creeping water bugs or saucer bugs, is divided into five subfamilies: Cheirochelinae, Cryphocricinae, Laccocorinae, Limnocorinae, and Naucorinae (Štys & Jansson 1988); all except Cheirochelinae occur in the Neotropics. The family has a worldwide distribution with 39 genera and more than 400 species (Sites, 2015; Reynoso-Velasco & Sites, 2016). They are small- to medium-sized insects, with body ovoid to slightly elongate and 4–20 mm long; the body is dorsoventrally flattened, the foreleg is retentorial with an enlarged femur; the middle and hindlegs have pale swimming hairs; and the membrane of the forewing does not have veins. This family can be found in a wide range of habitats, including riffles, vegetated pond margins, algal mats, brackish water, and the vertical rock faces and splash zones of waterfalls.

1	Propleura extending mesad, covering posterior part of prosternum (prosternellum) (Fig. 7.23 C) 2
1'	Prosternum exposed posteriorly, without mesal extensions of propleura (Fig. 7.23 D) 5
2(1)	Abdominal sterna margined by a wide glabrous band (Fig. 7.23 E, left) 3
2'	Abdominal sterna pubescent to or nearly to lateral margins (Fig. 7.23 E, right) .... *Ambrysus*
3(2)	Pronotum anterior margin straight or slightly concave behind interocular space (Fig. 7.23 G) 4
3'	Pronotum anterior margin deeply emarginate behind interocular space (Fig. 7.23 F) .... *Cataractocoris*
4(3)	Head oriented vertically (Fig. 7.23 H); male profemora excavate distally, protibiae with prominent tooth (Fig. 7.23 I); forewing with patches of black setae (Fig. 7.23 J) .... *Hygropetrocoris*
4'	Head oriented horizontally; male profemora not excavate distally, protibiae without tooth (Fig. 7.24 A); forewing without patches of black setae (Fig. 7.24 B) .... *Carvalhoiella*
5(1)	Head front folded posteroventrally such that labrum is set back from anterior margin (Fig. 7.24 C); propretarsi with two claws (Fig. 7.24 D); male pro- and mesotibiae ventrally with well-developed tomentose patch (weakly developed in females) (Fig. 7.24 J) 6
5'	Head front not folded posteroventrally such that labrum is at anterior margin; propretarsi with a single claw; pro- and mesotibiae without tomentose patch in both sexes 8
6(5)	Pronotum posterior angles not directed posteriorly (Fig. 7.24 G); hemelytra with fully developed overlapping membranes (Fig. 7.24 H) 7
6'	Pronotum posterior angles obtuse, but directed posteriorly (Fig. 7.24 E); hemelytra with membranes reduced and not overlapping (Fig. 7.24 F) .... *Decarloa*
7(6)	Metatibiae with circlet of stout, divergent, apical, and subapical spines, not arranged in a distinct row (Fig. 7.24 I); females with two tarsomeres on foreleg (Fig. 7.24 J) .... *Ctenipocoris*
7'	Metatibiae ventrally with row of five apical, stout, subparallel spines (Fig. 7.24 K); females with one tarsomere on foreleg .... *Interocoris*
8(5)	Meso- and metasterna without prominent median carinae; eyes with mesal margins usually parallel or converging anteriorly (Fig. 7.25 C) 9
8'	Meso- and metasterna with prominent median carinae (Fig. 7.25 A); eyes with mesal margins diverging anteriorly (Fig. 7.25 B) .... *Limnocoris*
9(8)	Pronotum anterior margin deeply emarginated behind interocular space (Figs. 7.25 C, D) 10
9'	Pronotum anterior margin straight or slightly concave behind interocular space (Fig. 7.25 E) 11
10(9)	Pronotum lateral margins crenulate (Figs. 7.25 C, F); female mediosternite 6 symmetrical; brachypterous forms with abdominal terga 3–5 distinctly separated (Fig. 7.25 G) .... *Cryphocricos*
10'	Pronotum lateral margins smooth (Fig. 7.25 H); female mediosternite 6 asymmetrical; brachypterous forms with abdominal terga 3–5 (male) or 3–6 (female) fused (Fig. 7.25 I) .... *Procryphocricos*
11(9)	Mesofemora greatly flattened dorsoventrally, posterodorsal margins extended into thickened, lamellar projections with a fine brush of setae; meso- and metatarsi longer than corresponding tibiae .... *Placomerus*
11'	Mesofemora not flattened or margined as described above; meso- and metatarsi shorter than corresponding tibiae .... *Pelocoris*

ORDER HEMIPTERA

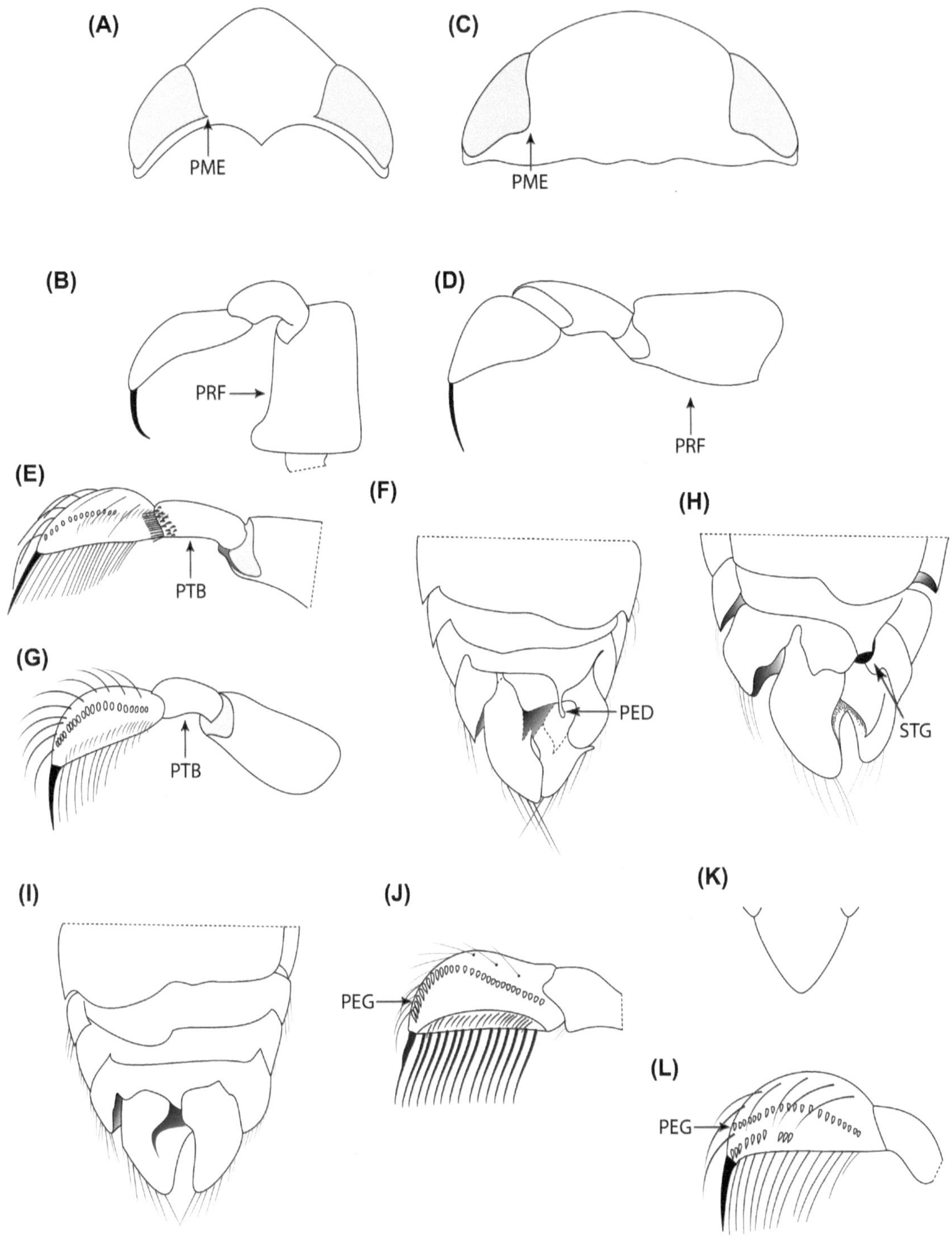

FIGURE 7.17 *Krizousacorixa*: (A) head, dorsal view; (B) male foreleg. *Ectemnostega*: (C) head, dorsal view. *Trichocorixa*: (D) male foreleg. *Morphocorixa*: (E) male foreleg. *Morphocorixa*: (F) male abdomen, dorsal view. *Neosigara*: (G) male foreleg; (H) male abdomen, dorsal view; (I) male abdomen, dorsal view. *Sigara*: (I) male pala; (K) metaxyphus. *Neosigara*: (L) male pala. Abbreviations: *PED*, strigilar pedicel; *PEG*, row of pegs; *PME*, posteromesal angle of eye; *PRF*, profemur; *PTB*, protibia; *STG*, strigil.

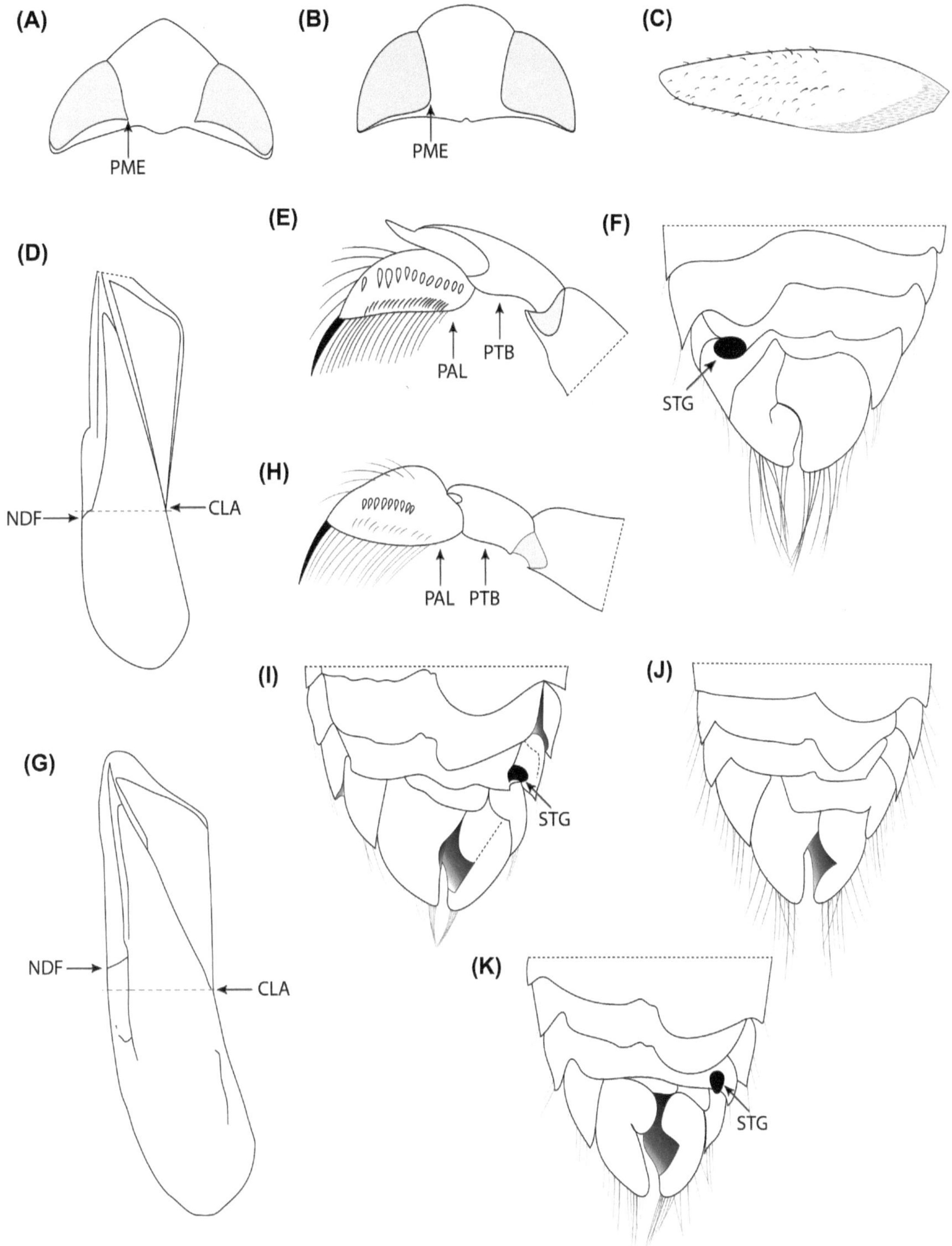

FIGURE 7.18 A. *Trichocorixella*: (A) head, dorsal view. *Trichocorixa*: (B) head, dorsal view. *Ectemnostega*: (C) metafemur, pubescent area in gray. *Trichocorixa*: (D) forewing, female. *Trichocorixa*: (E) male foreleg; (F) male abdomen, dorsal view. *Ectemnostega*: (G) female forewing; (H) male foreleg; (I) male abdomen, dorsal view. *Centrocorisa*: (J) male abdomen, dorsal view. *Corisella*: (K) male abdomen, dorsal view. Abbreviations: *CLA*, apex of clavus; *NDF*, nodal furrow; *PAL*, pala; *PME*, posteromesal margin of eye; *PTB*, protibia; *STG*, strigil.

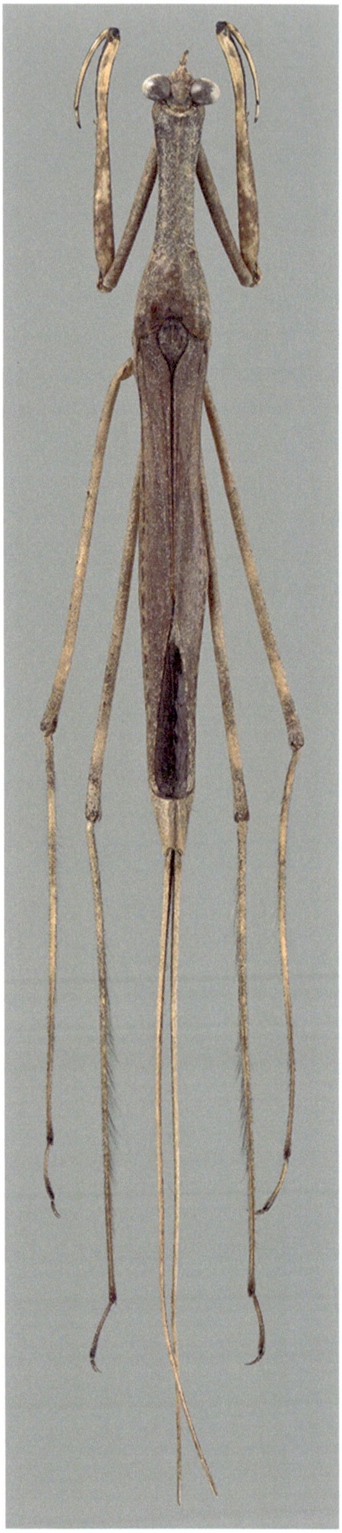

FIGURE 7.19 *Ranatra* (Nepidae): habitus, dorsal view.

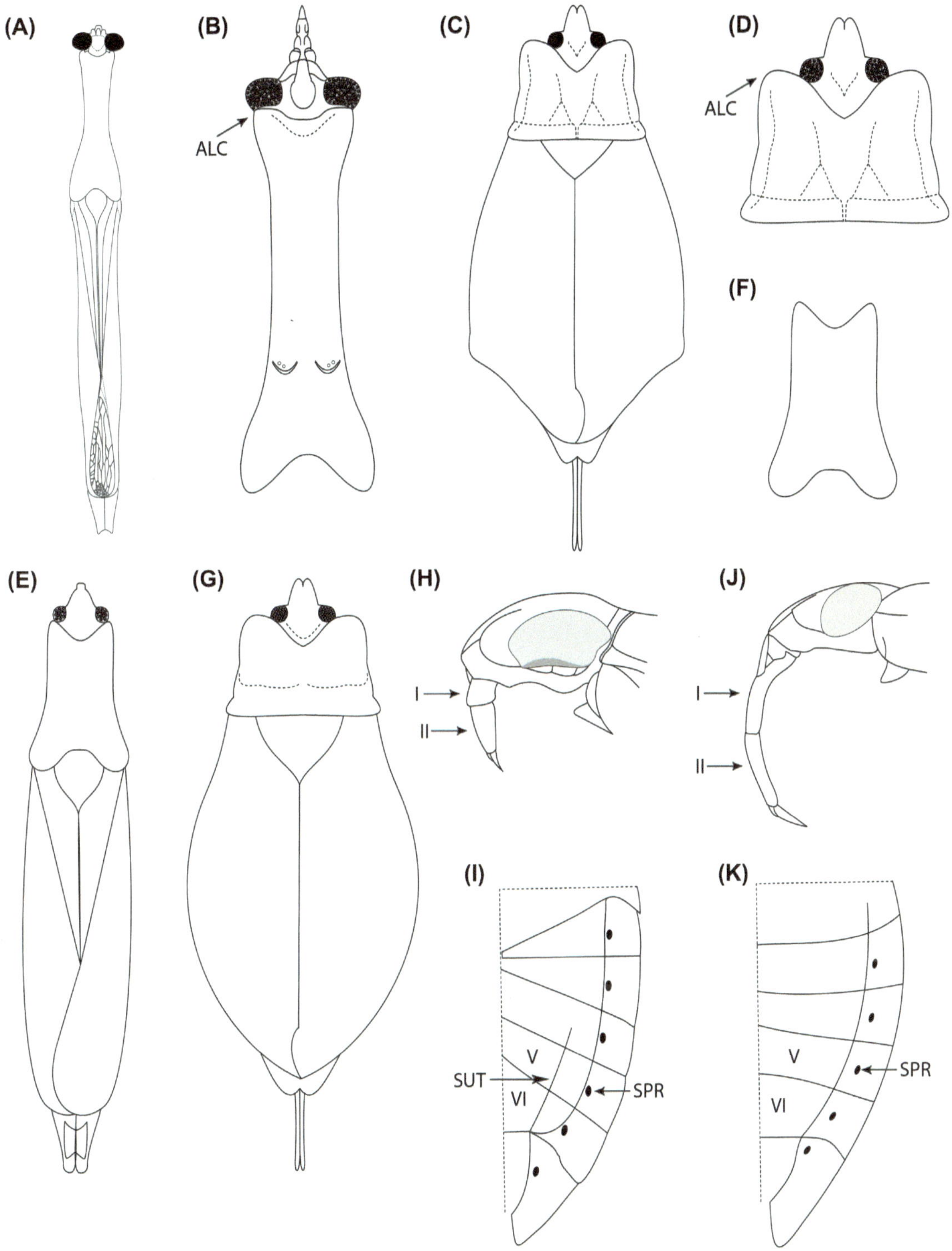

FIGURE 7.20 *Ranatra*: (A) habitus, dorsal view; (B) head and pronotum, dorsal view. *Telmatotrephes*: (C) habitus, dorsal view; (D) head and pronotum, dorsal view. *Curicta*: (E) habitus, dorsal view; (F) pronotum, dorsal view. *Telmatotrephes*: (G) habitus, dorsal view. *Horvathinia*: (H) head, lateral view; (I) abdomen, ventral view. *Belostoma*: (J) head, lateral view; (K) abdomen, ventral view. Abbreviations: *ALC*, anterolateral corner of pronotum; *I*, article 1 of rostrum; *II*, article 2 of rostrum; *V*, abdominal sternum 5; *VI*, abdominal sternum 6; *SPR*, abdominal spiracle; *SUT*, suturiform fold.

FIGURE 7.21 *Belostoma* (Belostomatidae): habitus, dorsal view.

Hemiptera: Nepomorpha: Notonectidae: Genera

Notonectidae (Fig. 7.10), commonly known as backswimmers, is a family of elongate and fusiform water bugs, ranging in length from 4 to 15 mm. Notonectids swim upside down, have reverse countershading, and use the remiform hindlegs to propel the body through the water. Members of this family are predaceous, feeding on small arthropods and even small fish. They occur in a variety of freshwater habitats, including pools, ponds, lakes, and banks of streams, and have a worldwide distribution. They are divided into two subfamilies, Notonectinae and Anisopinae, with approximately 400 species in 11 genera, four of which occur in the Neotropical region (Polhemus & Polhemus, 2008). Members of the subfamily Anisopinae have hemoglobin, which is concentrated in the abdomen (mainly in segments III–VII), and enable the insects to maintain neutral buoyancy in the water column (Schuh & Slater, 1995). All species have paired claws well-developed on the fore- and middle legs and reduced on the hindlegs. They lack ocelli and forewing veins, and the abdominal sternum has a median carina (Barbosa & Rodrigues, 2015).

1 Hemelytral commissure without setae-lined pit (Figs. 7.26 A, F, H); male protibiae without stridulatory comb (Fig. 7.26 B), abdominal tergum 7 caudosinistral margin without spine 2

1' Hemelytral commissure anteriorly with setae-lined pit (Fig. 7.25 J); male protibiae with stridulatory comb (Fig. 7.25 K), abdominal tergum 7 caudosinistral margin with a spine (Fig. 7.25 L) .... *Buenoa*

2(1) Eyes separated (Figs. 7.26 E, H); clavi distal ends without spine-like process (Fig. 7.26 F); mesofemora with preapical spur (Fig. 7.26 G) 3

2' Eyes contiguous posteromesally (Fig. 7.26 A); clavi distal ends with spine-like process (Fig. 7.26 C); mesofemora without preapical spur (Fig. 7.26 D) .... *Martarega*

3(2) Pronotum lateral margins foveate (Fig. 7.26 H); male genital capsule asymmetrical (Fig. 7.26 I) .... *Enitharoides*

3' Pronotum lateral margins not foveate (Fig. 7.26 J); male genital capsule symmetrical (Fig. 7.26 K) .... *Notonecta*

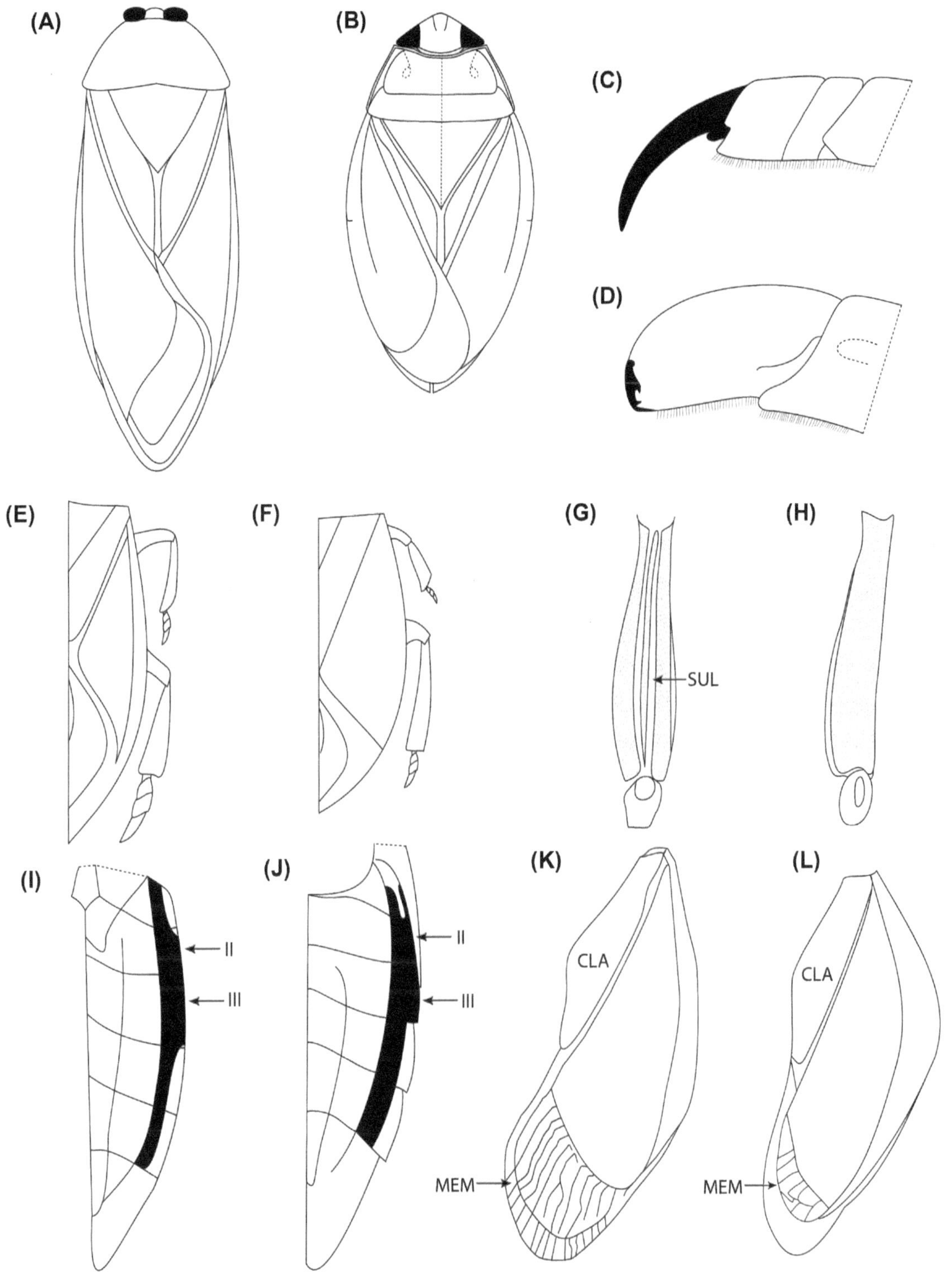

FIGURE 7.22 *Lethocerus*: (A) habitus, dorsal view. *Horvathinia*: (B) habitus, dorsal view. *Lethocerus*: (C) protarsus. *Horvathinia*: (D) protarsus. *Lethocerus*: (E) part of thorax, abdomen, midlegs, and hindlegs, dorsal view. *Horvathinia*: (F) part of thorax, abdomen, midlegs, and hindlegs, dorsal view. *Lethocerus*: (G) profemur, ventral view. *Benacus*: (H) profemur, ventral view. *Lethocerus*: (I) abdomen, ventral view, stripe of hydrophobic setae in black. *Benacus*: (J) abdomen, ventral view, stripe of hydrophobic setae in black. *Belostoma*: (K) forewing. *Weberiella*: (L) forewing. Abbreviations: *CLA*, clavus; *II*, abdominal laterosternite 2; *III*, abdominal laterosternite 3; *MEM*, membrane; *SUL*, sulcus].

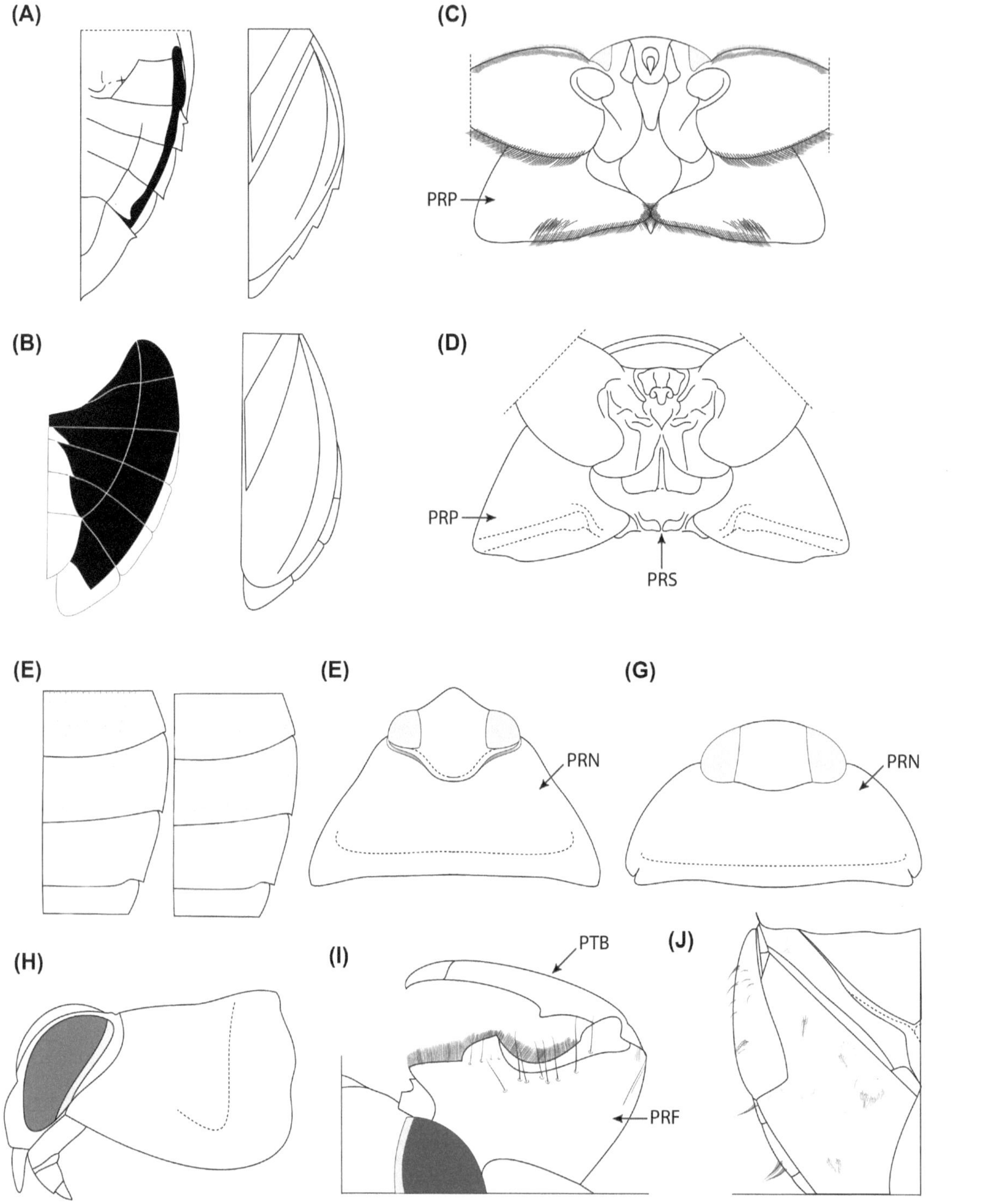

FIGURE 7.23 *Weberiella*: (A) ventral view (left) and dorsal view (right). *Abedus*: (B) ventral view (left) and dorsal view (right). *Ambrysus*: (C) head and prothorax, ventral view. *Ctenipocoris*: (D) head and prothorax, ventral view. *Carvalhoiella* (left) and *Ambrysus* (right): (E) abdomen, ventral view. *Cataractocoris*: (F) head and pronotum, dorsal view. *Carvalhoiella*: (G) head and pronotum, dorsal view. *Hygropetrocoris*: (H) head and prothorax, lateral view; (I) part of head and foreleg, dorsal view. *Hygropetrocoris*: (J) part of thorax and abdomen, dorsal view. Abbreviations: *PRF*, profemur; *PRN*, pronotum; *PRP*, propleuron; *PRS*, prosternum; *PTB*, protibia].

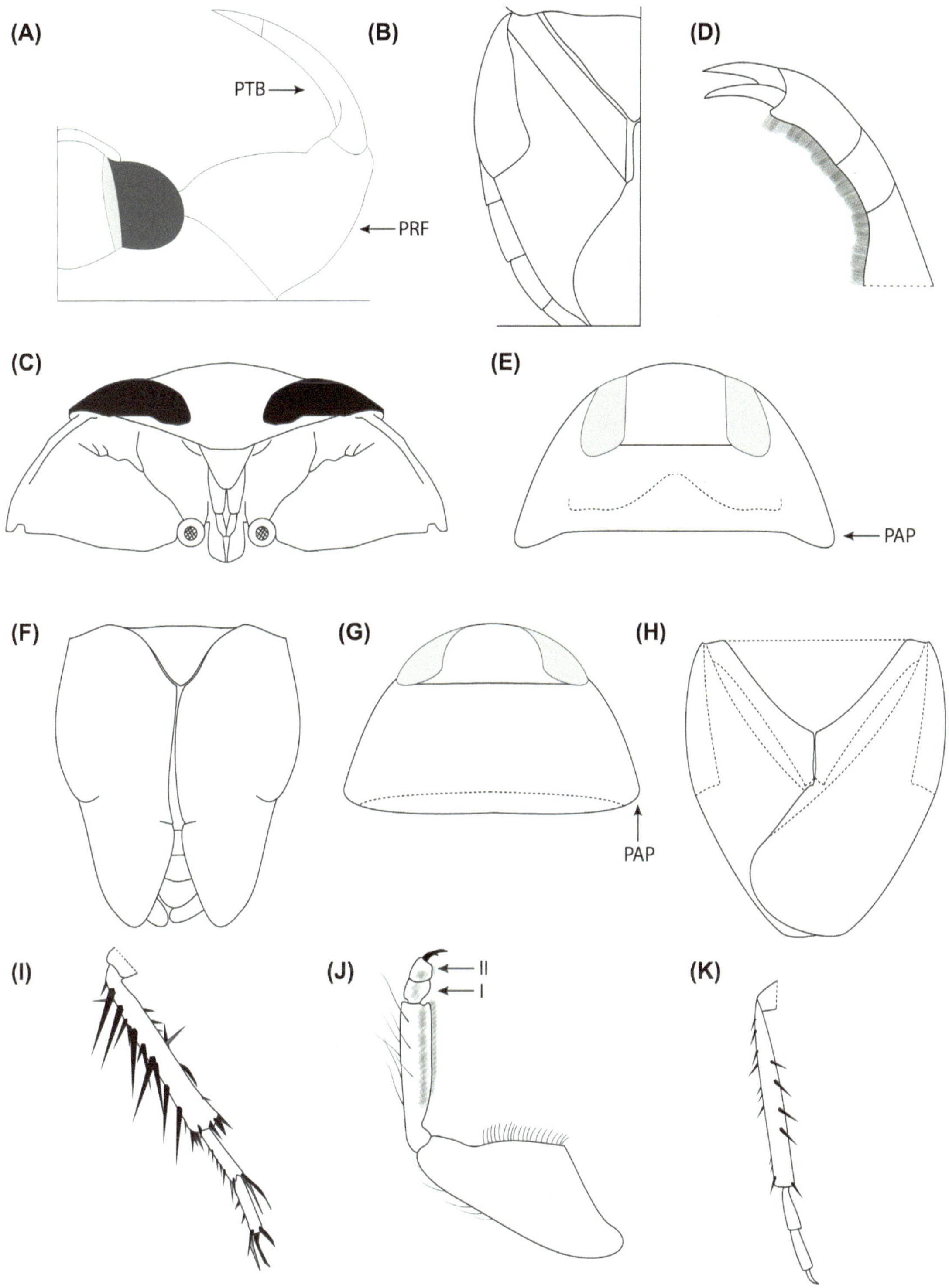

FIGURE 7.24 *Carvalhoiella*: (A) part of head, pronotum, and foreleg, dorsal view; (B) part of thorax and abdomen, dorsal view. *Ctenipocoris*: (C) head and prothorax, ventral view; (D) apex of foreleg. *Decarloa*: (E) head and pronotum, dorsal view; (F) part of thorax and abdomen, dorsal view. *Ctenipocoris*: (G) head and pronotum, dorsal view; (H) part of thorax and abdomen, dorsal view; (I) apex of hindleg; (J) foreleg. *Interocoris*: (K) apex of hindleg. Abbreviaitons: *I*, tarsal article 1; *II*, tarsal article 2; *PAP*, posterior angle of pronotum; *PRF*, profemur; *PTB*, protibia.

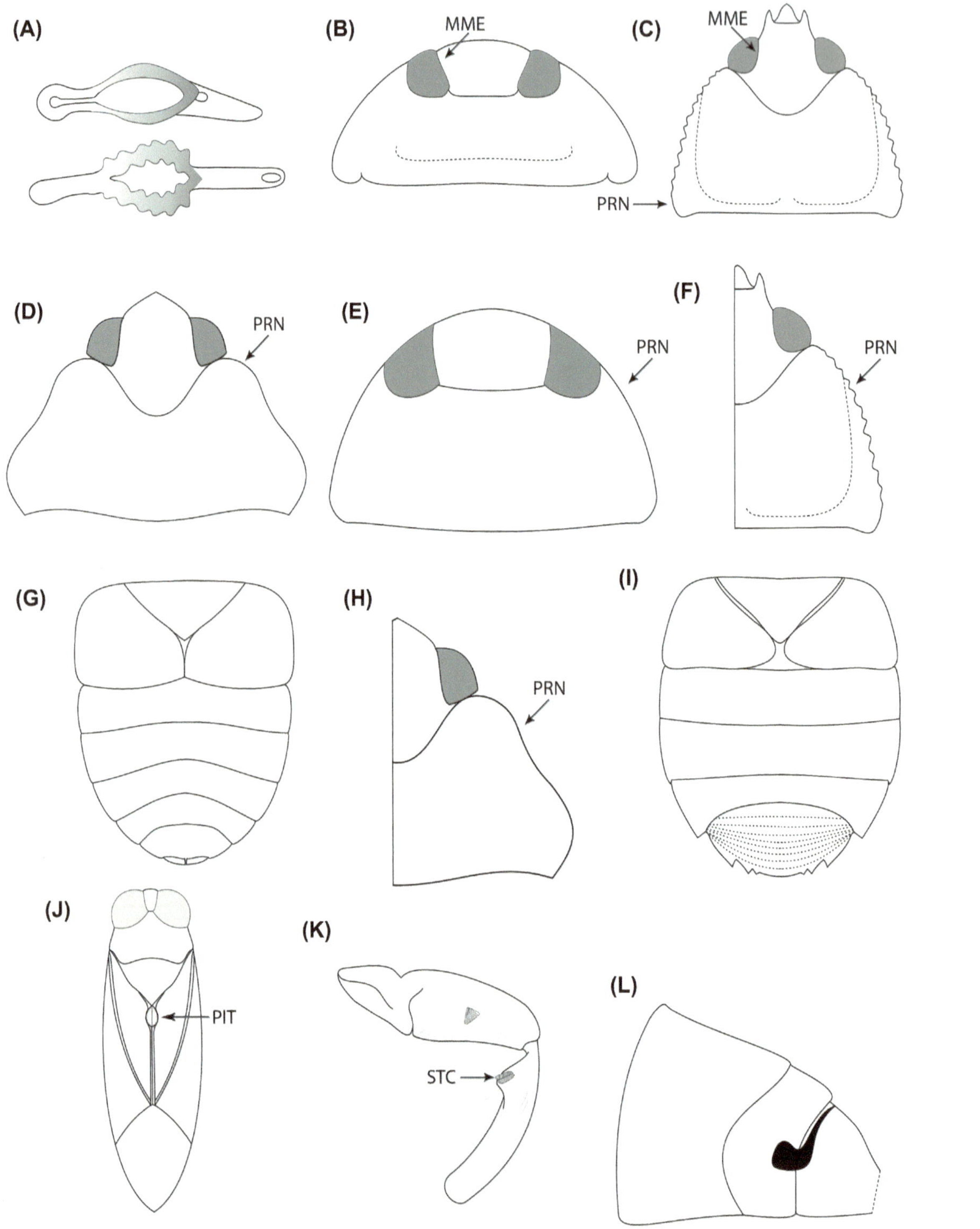

FIGURE 7.25 *Limnocoris*: (A) mesosternal carina (top), metasternal carina (bottom); (B) head and pronotum, dorsal view. *Cryphocricos*: (C) head and pronotum, dorsal view. *Procryphocricos*: (D) head and pronotum, dorsal view. *Placomerus*: (E) head and pronotum, dorsal view. *Cryphocricos*: (F) head and pronotum, dorsal view; (G) part of thorax and abdomen, dorsal view. *Procryphocricos*: (H) head and pronotum, dorsal view; (I) part of thorax and abdomen, dorsal view. *Buenoa*: (J) habitus, dorsal view; (K) male foreleg; (L) apex of abdomen, lateral view, spine in black. Abbreviations: *MME*, mesal margin of eye; *PIT*, hair-lined pit; *PRN*, pronotum; *STC*, stridulatory comb.

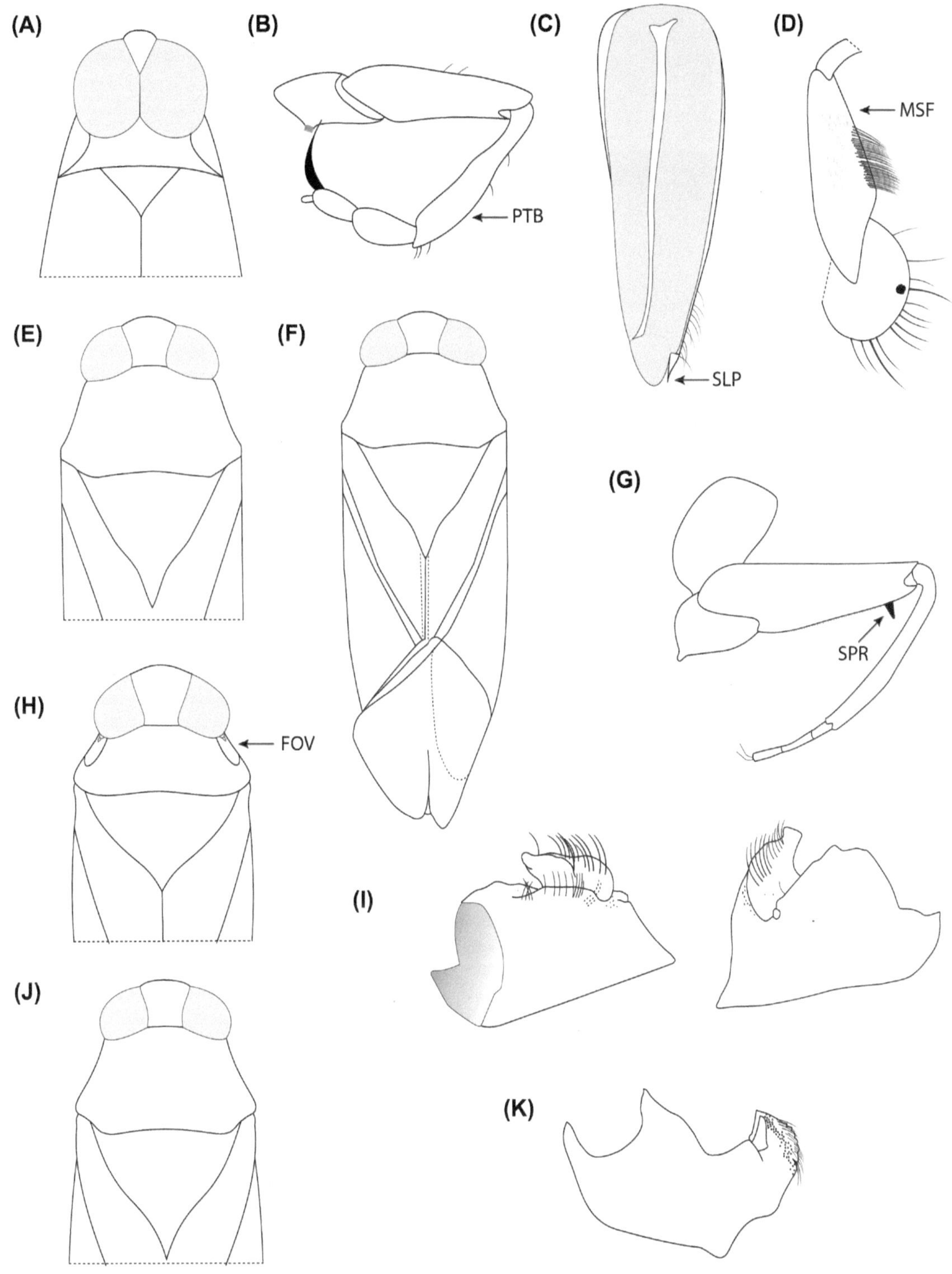

FIGURE 7.26 *Martarega*: (A) head and part of thorax, dorsal view; (B) male foreleg; (C) forewing; (D) midleg. *Notonecta*: (E), head and part of thorax, dorsal view; (F) habitus, dorsal view; (G) midleg. *Enitharoides*: (H) head and part of thorax, dorsal view; (I) male genital capsule, left side (left) and right side (right) views. *Notonecta*: (J) head and part of thorax, dorsal view; (K) male genital capsule. Abbreviations: *FOV*, fovea; *MSF*, mesofemur; *PTB*, protibia; *SLP*, spine-like process; *SPR*, spur.

FIGURE 7.27 Saldidae: habitus, dorsal view.

Hemiptera: Nepomorpha: Helotrephidae: Genera

Helotrephidae (Fig. 7.11) are small water bugs ranging in length from 1.0 to 4.0 mm. This family together with Pleidae forms the superfamily Pleoidea, and both probably evolved from pleid-like ancestors (Zettel, 1997). The body is generally globular with the dorsal surface covered by rounded punctuations, the head is fused with the prothorax, the antennae are generally uni- or biarticulate, the forewings are elytriform without venation, the sterna of the thorax and the anterior abdominal segments (mainly II–V) have a median carina, and the male genital capsule is rotated 90° (Nieser & Chen, 2002). Currently, Helotrephidae has 20 genera and approximately 171 species worldwide (Papácek & Zettel, 2005), and the greatest diversity of this family is in the Old World (Sites & Polhemus 2001). The family is represented in the Neotropics by only the subfamily Neotrophinae and two genera. Brachypterous specimens are more common than the macropterous form, and in some species it is the only form known. They can be found in a great variety of environments, including rock pools, waterfalls, water tanks, banks of streams, and temporary water bodies in arid zones (Papácek & Zettel, 2005).

1	Rostrum long, extending to middle of metacoxae, fourth article nearly four times as long as third (Fig. 7.2 A); pronotum lateral margins weakly sinuate, punctation weak (Fig. 7.2 B); cephalonotal suture nearly straight behind eyes (Fig. 7.32 B)	*Paratrephes*
1'	Rostrum short, not reaching mesocoxae, third and fourth articles subequal in length (Fig. 7.2 C); pronotum lateral margins strongly sinuate, punctation strong (Figs. 7.11, 7.13 A); cephalonotal suture in brachypterous form distinctly curved posteriorly behind eyes (Figs. 7.13A, C)	*Neotrephes*

Hemiptera: Nepomorpha: Pleidae: Genera

Pleidae (Fig. 7.12), also known as pygmy backswimmers, are small bugs ranging from 1.5 to 3.5 mm long. They have a naviform to semiglobular body, with the dorsal surface covered by rounded punctuations, and the forewings are coriaceous without membrane. The head is immobile relative to the thorax; however, they are not fused as in Helotrephidae. The ovipositor is well developed and with distinct teeth. The genera are uniform in morphology, and the differences between them are based on the number of tarsomeres (Cook, 2011). As in Helotrephidae, pleids also have the body shape related to the habit of swimming upside down. Four genera and almost 40 described species are known worldwide. Pleids are typical inhabitants of stagnant waters, usually with aquatic vegetation (Chen et al., 2005), and feed on small invertebrates such as mosquito larvae and ostracods (Schuh & Slater, 1995). In summer temperatures, some species of pleids may remain submerged for more than 10 h without replenishing their air bubble (Andersen & Weir, 2004).

1	Vertex without callus (Fig. 7.2 F); tarsal formula 3-2-3 or 3-2-2	2
1'	Vertex with distinct callus (Figs. 7.2 D, E); tarsal formula 3-3-3	*Heteroplea*
2(1)	Tarsal formula 3-2-3; abdominal sterna 2–5 with median carinae	*Neoplea*
2'	Tarsal formula 3-2-2; abdominal sterna 2–6 with median carinae	*Paraplea*

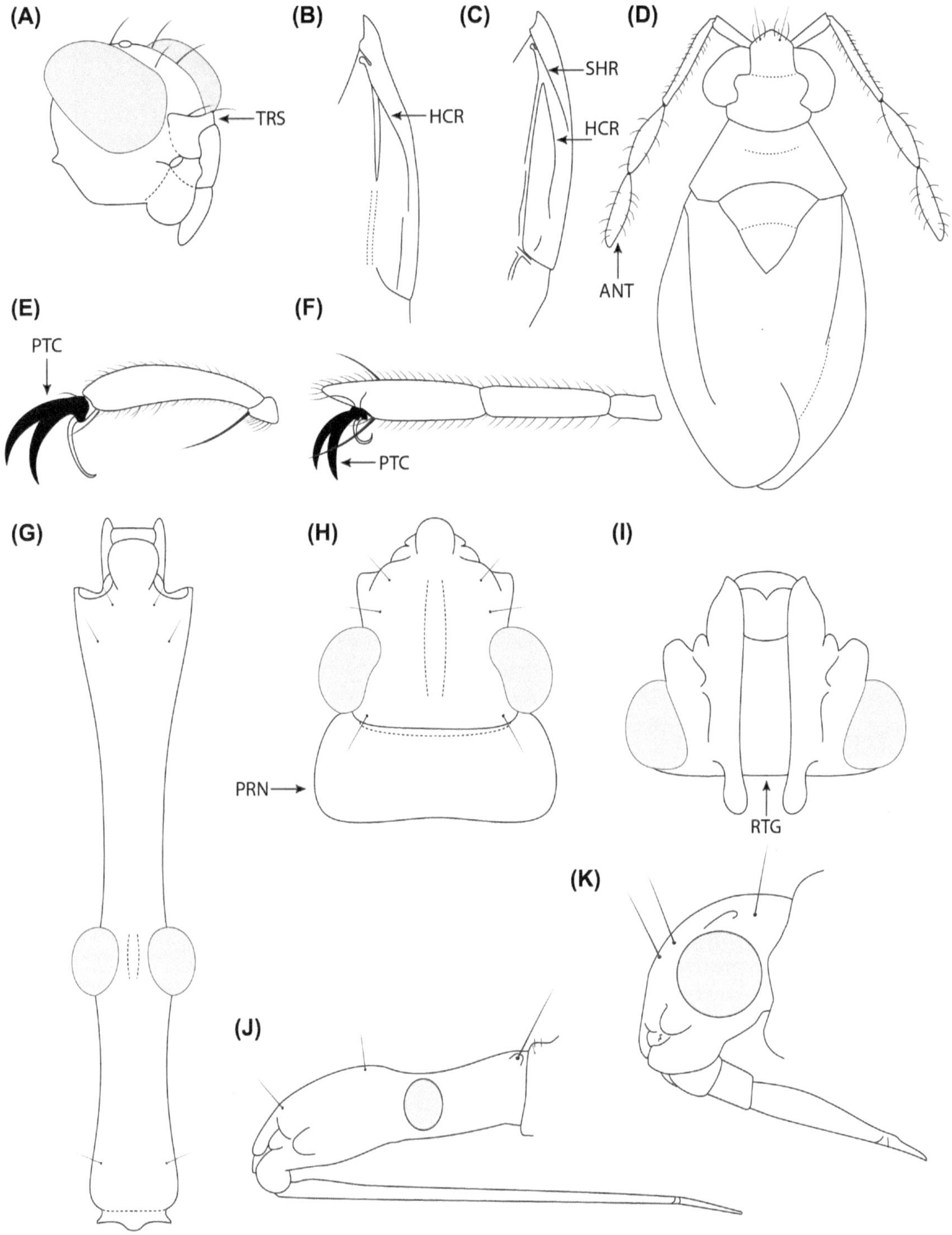

FIGURE 7.28 *Saldula*: (A) head, frontolateral view. *Rupisalda*: (B) forewing, hypocostal region. *Saldula*: (C) forewing, hypocostal region. *Ioscytus*: (D) forewing, hypocostal region. *Hyrcanus*: (E) protarsus. *Mesoveloidea*: (F) protarsus, dorsal view. *Hydrometra*: (G) head, dorsal view. *Mesovelia*: (H) head and pronotum, dorsal view. *Timasius*: (I) head, ventral view. *Limnobatodes*: (J) head and part of prothorax, lateral view. *Steinovelia*: (K) head and part of prothorax, lateral view. Abbreviations: *ANT*, antenna; *HCR*, hypocostal ridge; *PRN*, pronotum; *PTC*, pretarsal claw; *RTG*, rostral groove; *SHR*, secondary hypocostal ridge; *TRS*, transversal swelling.

FIGURE 7.29 *Hebrus* (Hebridae): habitus, dorsal view.

FIGURE 7.30 *Mesovelia* (Mesoveliidae): habitus, dorsal view.

Infraorder Leptopodomorpha

Members of Leptopodomorpha (Fig. 7.27) are commonly known as shore bugs because most species inhabit lake shores, beaches, and stream banks (Polhemus, 1985). This infraorder includes approximately 400 species divided into four extant families: Aepophilidae, Leptopodidae, Omaniidae, and Saldidae. Schuh & Polhemus (1980) considered Saldidae as the sister group of Aepophilidae, and Omaniidae as the sister group of Leptopodidae. Most leptopodomorphan species belong to Saldidae, which is distributed worldwide and has 29 genera and approximately 340 species (Polhemus & Polhemus, 2008). In the Neotropical region, 41 species have been recorded, one of which belongs to Leptopodidae and the other 40 are placed in Saldidae. The body is generally ovate, but some species are nearly globose or elongate and parallel-sided (Schuh & Slater, 1995). The head has at least three pairs of trichobothria and the eyes are usually large and sometimes reniform, occupying most of the side of the head. The antennae are longer than the head with four antennomeres. Wing polymorphism is common with some species having two forms of the forewing, although macroptery is the most common wing form in saldids. In macropterous specimens, the veins form 3–5 closed cells. The legs usually are slender, short, with spines on the tibiae and femora, and are adapted for hopping or jumping. This group is uncommon in scientific collections because they are easily overlooked by nonspecialists (Melo, 2015).

Hemiptera: Leptopodomorpha: Saldidae: Genera

1	Forewings with short embolar fracture, not reaching more than halfway to end of claval suture (Figs. 7.2 H, I); membrane usually with four cells (Fig. 7.2 H), except five in *Pseudosaldula* (Fig. 7.2 I) 2
1'	Forewings with long embolar fracture reaching at least to end of claval suture (Fig. 7.2 G); membrane with five cells (Fig. 7.2 G) .... *Pentacora*
2(1)	Head with a transverse swelling (Fig. 7.28 A) 3
2'	Head without a transverse swelling .... *Oiosalda*
3(2)	Forewing membrane with four cells (Fig. 7.2 H) 4
3'	Forewing membrane with five cells (Fig. 7.2 I) .... *Pseudosaldula*
4(3)	Hypocostal ridge complex, secondary hypocostal ridge present (Fig. 7.28 C) 5
4'	Hypocostal ridge simple, secondary hypocostal ridge absent (Fig. 7.28 B) .... *Rupisalda*
5(4)	Antennae relatively slender, antennomeres III–IV not thicker than distal end of II (Fig. 7.2 H); secondary hypocostal ridge without strigil; metafemora distal portion without plectrum 6
5'	Antennae relatively thick, antennomeres III–IV thicker than distal end of II (Fig. 7.28 D); secondary hypocostal ridge with strigil; metafemora distal portion with plectrum .... *Ioscytus*
6(5)	Body usually longer than 3.5 mm, if shorter then pronotum anterior margin wider than collar; corium veins more or less distinct .... *Saldula*
6'	Body usually shorter than 3.5 mm; pronotum anterior margin usually narrower than collar; corium veins indistinct .... *Micracanthia*

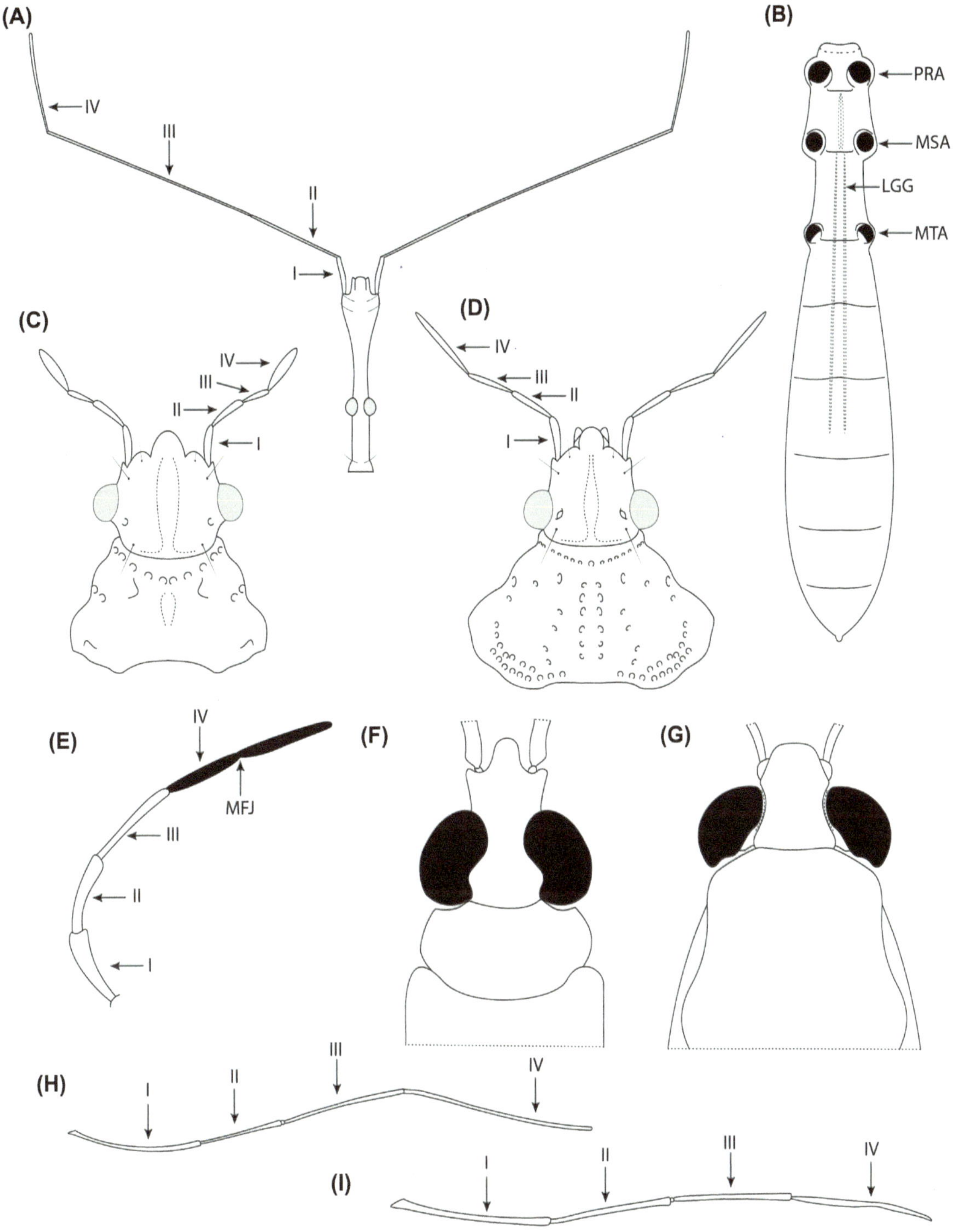

FIGURE 7.31 *Bacillometroides*: (A) head, dorsal view; (B) thorax and abdomen, ventral view. *Merragata*: (C) head and pronotum, dorsal view. *Lipogomphus*: (D) head and pronotum, dorsal view. *Hebrus*: (E) antenna, superficially divided antennomere four in black. *Cylindrostethus*: (F) head and part of thorax, dorsal view. *Brachymetra*: (G) head and part of thorax, dorsal view. *Tachygerris*: (H) antenna. *Eurygerris*: (I) antenna. Abbreviations: *I*, antennomere 1; *II*, antennomere 2; *III*, antennomere 3; *IV*, antennomere 4; *LGG*, longitudinal groove; *MFJ*, membranous false joint; *MSA*, mesoacetabulum; *MTA*, metacetabulum; *PRA*, proacetabulum].

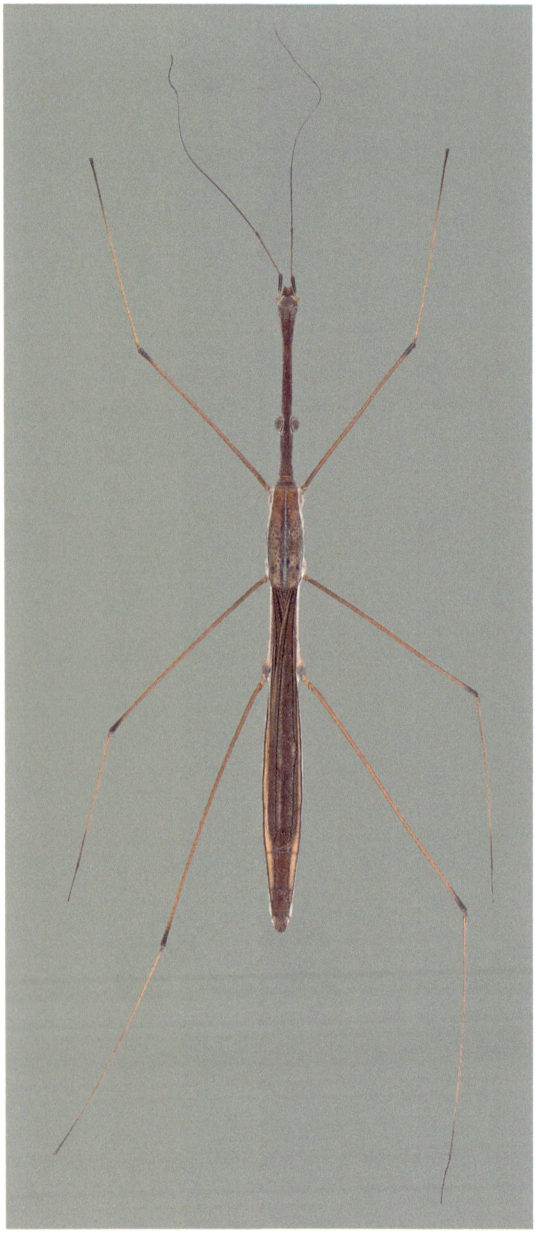

FIGURE 7.32 *Hydrometra* (Hydrometridae): habitus, dorsal view.

FIGURE 7.33 *Halobatopsis* (Gerridae): habitus, dorsal view.

Hemiptera: Gerromorpha: Families

1	Pretarsal claws apical (Fig. 7.28 E)	2
1'	Pretarsal claws preapical (Fig. 7.28 F)	4
2(1)	Body, legs, and antennae not as below (Figs. 7.29, 7.30); eyes located close to thorax anterior margin (Figs. 7.3 B, 7.28 H, 7.31D)	3
2'	Body long and almost always slender (Fig. 7.32); eyes located far from thorax anterior margin (Figs. 7.28 G, 7.32; antennae and legs long and very slender (Figs. 7.31 A, 7.32)	**Hydrometridae [p. 207]** (in part)
3(2)	Body compact and robust (Fig. 7.29); head ventral surface with a deep rostral groove (Fig. 7.28 I); tarsal formula 2-2-2	**Hebridae [p. 207]**
3'	Body delicate, elongated (Fig. 7.30); head ventral surface without deep rostral groove; tarsal formula 3-3-3	Mesoveliidae (in part), one genus: *Mesovelia*
4(1)	Head not elongated (Fig. 7.28 K); eyes located close to thorax anterior margin (Fig. 7.28 K)	5
4'	Head elongated (Fig. 7.28 J); eyes located far from thorax anterior margin (Fig. 7.28 J)	Hydrometridae (in part), one genus: *Limnobatodes*
5(4)	Tarsal formula 1-2-2 or 3-3-3; if 2-2-2, then mesotarsi deeply cleft and bearing a swimming fan (Fig. 7.3 A)	6
5'	Tarsal formula 2-2-2; mesotarsi never deeply cleft and bearing a swimming fan	**Gerridae [p. 211]**
6(5)	Head dorsum without impressed longitudinal line or pair of subbasal indentations (Fig. 7.3 B); paired coxae inserted close to each other (Fig. 7.3 C); male protibiae without grasping comb (Fig. 7.3 D)	Mesoveliidae (in part), one genus: *Mesoveloidea*
6'	Head dorsum usually with an impressed longitudinal line and a pair of subbasal indentations (Fig. 7.3 E); paired coxae widely separated from each other (Fig. 7.3 F); male protibiae usually with apical grasping comb (Fig. 7.3 G)	**Veliidae [p. 213]**

Hemiptera: Gerromorpha: Hydrometridae: Genera

Hydrometridae is represented by more than 125 freshwater species in eight genera (Polhemus & Polhemus, 2008, 2010). These bugs (Fig. 7.32) are almost always slender, 2.7–22.0 mm long, the eyes are located far from the anterior margin of the thorax, and the antennae and legs are very long and slender. The pretarsal claws are apical in the subfamilies Heterocleptinae and Hydrometrinae, but preapical in Limnobatodinae (Andersen, 1982; Moreira, 2015).

1	Pretarsal claws apical (Fig. 7.28 E)	2
1'	Pretarsal claws preapical (Fig. 7.28 F)	*Limnobatodes*
2(1)	Body length ≥6 mm; antennomere I subequal to or shorter than II, only slightly passing head apex (Fig. 7.31 A)	3
2'	Body length 3–5 mm; antennomere I much longer than II, distinctly passing head apex (Fig. 7.3 H)	*Veliometra*
3(2)	Meso- and metasterna with longitudinal grooves (Fig. 7.31 B)	4
3'	Meso- and metasterna without longitudinal grooves	*Hydrometra*
4(3)	Antennomere IV bulbous distally; coxae equally spaced	*Bacillometra*
4'	Antennomere IV filamentous, not enlarged distally (Fig. 7.31 A); mesocoxae farther from metacoxae than from procoxae (Fig. 7.31 B)	*Bacillometroides*

Hemiptera: Gerromorpha: Hebridae: Genera

Hebridae (Fig. 7.29) is composed of nine genera and more than 220 described species. Individuals are usually are collected at the margins of freshwater bodies, although one Asian genus has subaquatic representatives (Polhemus & Polhemus, 2008; Moreira 2015). These insects are small, range from 1.3 to almost 4.0 mm, and the scutellum of winged forms is reduced to a narrow transverse plate posterior to the pronotum. They can be recognized by the compact and robust body, the ventral surface of the head has a deep rostral groove, article IV of the rostrum is about half as long as III, the metanotum of winged forms has a median subtriangular elevation, and the tarsal formula is 2-2-2 with the claws always apical (Andersen, 1982; Moreira, 2015).

1	Antennae as long as maximum pronotum width, antennomeres slender, antennomere I shorter than IV (Fig. 7.31 D)	2
1'	Antennae shorter than maximum pronotum width, antennomeres thick, antennomeres I and IV subequal in length (Fig. 7.31 C)	*Merragata*
2(1)	Antennae appearing pentarticulate because of a membranous false joint on antennomere IV (Fig. 7.31 E)	*Hebrus*
2'	Antennae clearly quadriarticulate, false joint absent (Fig. 7.31 D)	*Lipogomphus*

Hemiptera: Gerromorpha: Mesoveliidae: Genera

Mesoveliids (Fig. 7.30) comprise 46 extant species in 12 genera that occupy freshwater environments, except for a few humid terrestrial or marine taxa (Damgaard et al., 2012). The body length ranges from 1.2 to 4.5 mm and adults can be winged or wingless. They are the only gerromorphans with a typical heteropteran scutellum in winged forms, which can be triangular or posteriorly rounded. The coxae are inserted close to the ventral midline, all tarsi are triarticulate with claws inserted either apically or preapically, and the ovipositor is well-developed with long and serrate gonapophyses for insertion of eggs into plant tissue (Andersen, 1982; Damgaard et al., 2012; Moreira, 2015).

1	Monomorphic macropterous; pretarsal claws preapical (Fig. 7.28 F)	*Mesoveloidea*
1'	Wing dimorphic (macropterous or apterous); pretarsal claws apical (Fig. 7.28 E)	*Mesovelia*

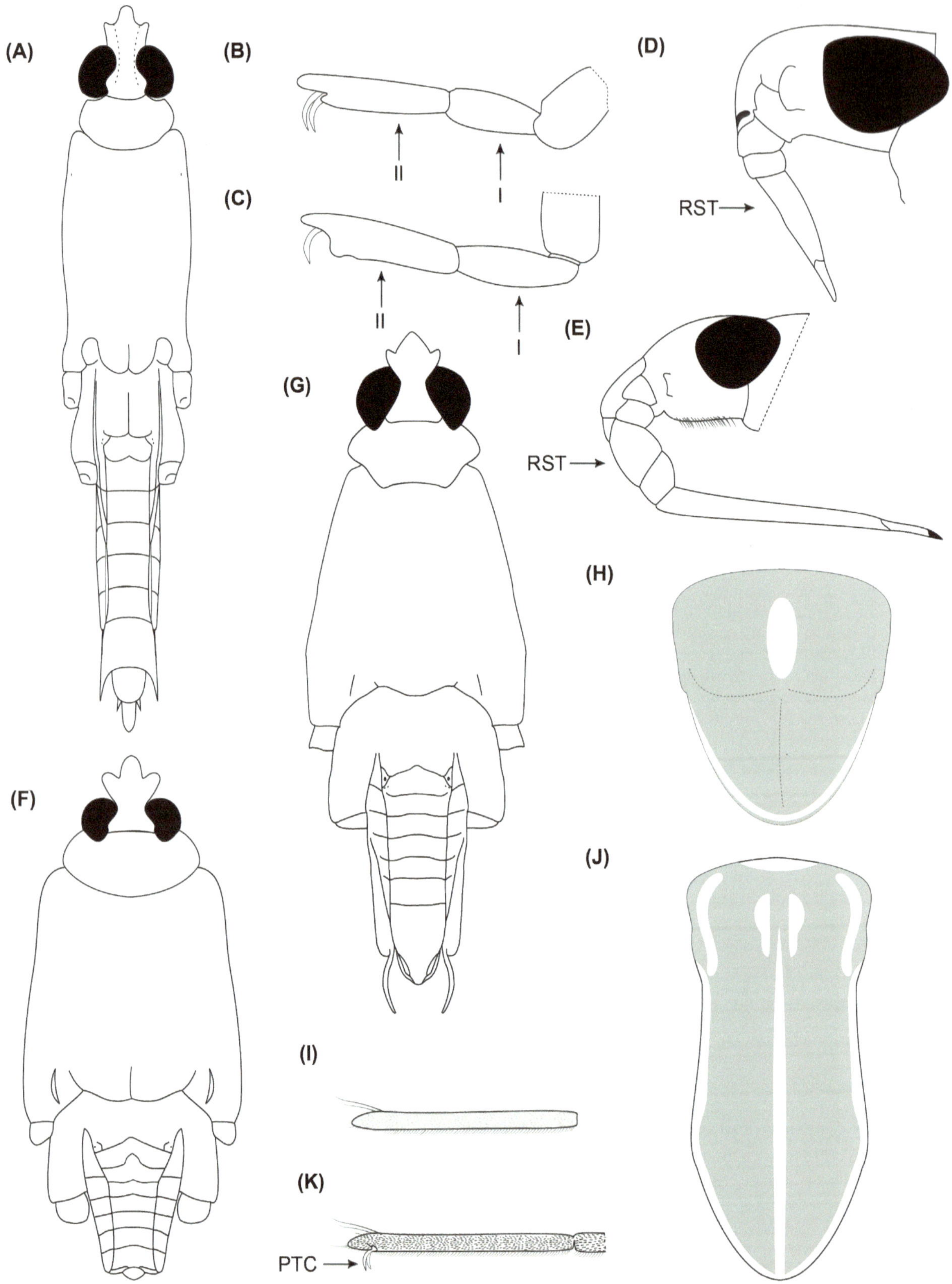

FIGURE 7.34 *Cylindrostethus*: (A) dorsal view. *Limnogonus*: (B) apex of foreleg. *Aquarius*: (C) apex of foreleg. *Potamobates*: (D) head, lateral view. *Limnogonus*: (E) head, lateral view. *Platygerris*: (F) dorsal view. *Potamobates*: (G) dorsal view. *Neogerris*: (H) pronotum, dorsal view; (I) apex of mesotarsus. *Limnogonus*: (J) pronotum, dorsal view; (K) apex of mesotarsus. Abbreviations: *I*, tarsal article 1; *II*, tarsal article 2; *PTC*, pretarsal claw; *RST*, rostrum.

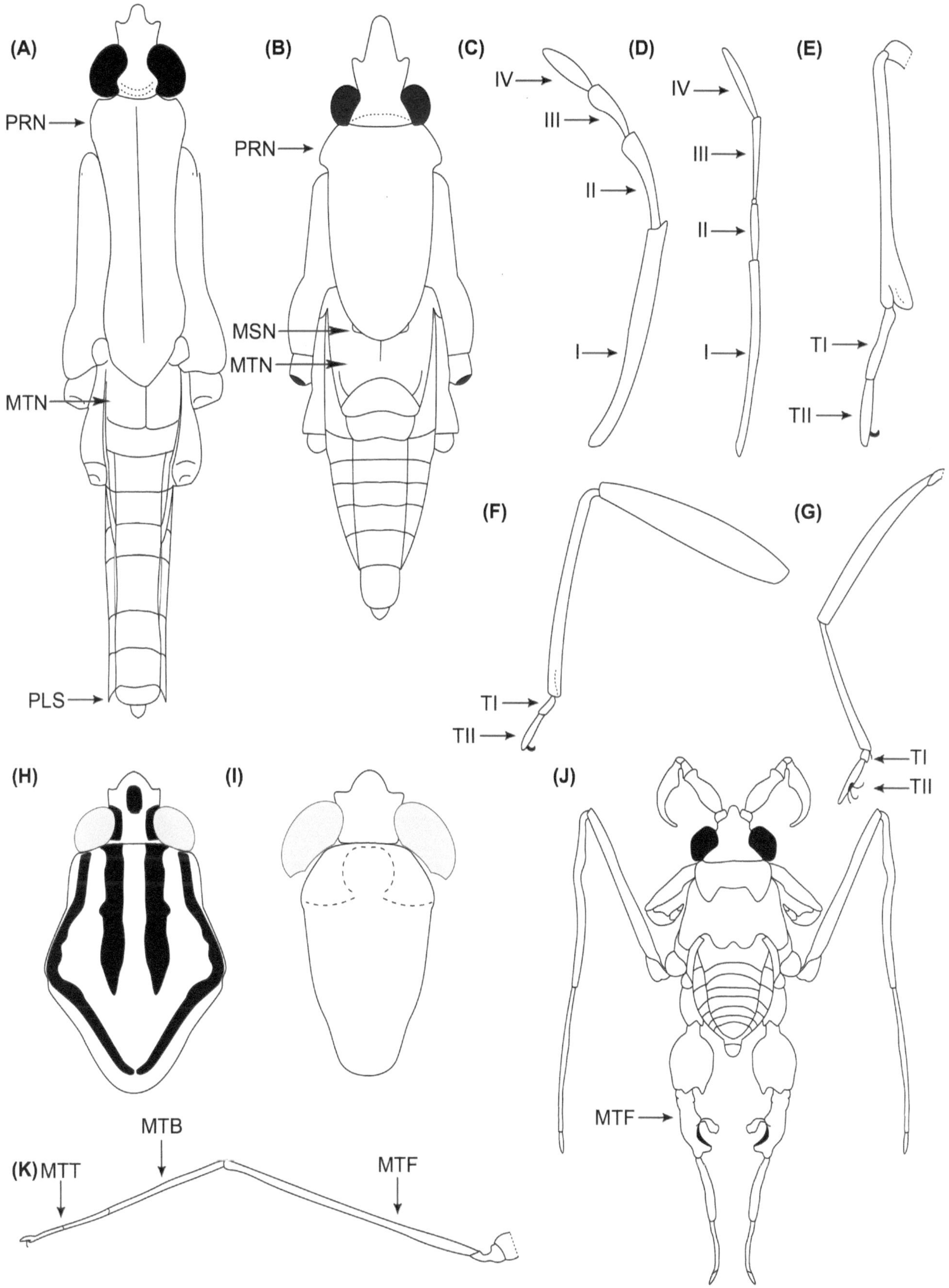

FIGURE 7.35 *Aquarius*: (A) dorsal view. *Eurygerris*: (B) dorsal view. *Metrobates*: (C) male antenna. *Charmatometra*: (D) antenna; (E) apex of foreleg. *Brachymetra*: (F) foreleg. *Trepobatoides*: (G) foreleg. *Eobates*: (H) head and pronotum, dorsal view. *Brachymetra*: (I) head and pronotum, dorsal view. *Rheumatobates*: (J) male, dorsal view. *Trepobates*: (K) hindleg. Abbreviations: *I*, antennomere 1; *II*, antennomere 2; *III*, antennomere 3; *IV*, antennomere 4; *MSN*, mesonotum; *MTB*, metatibia; *MTF*, metafemur; *MTN*, metanotum; *MTT*, metatarsus; *PLS*, posterolateral spine; *PRN*, pronotum; *TI*, tarsal article 1; *TII*, tarsal article 2.

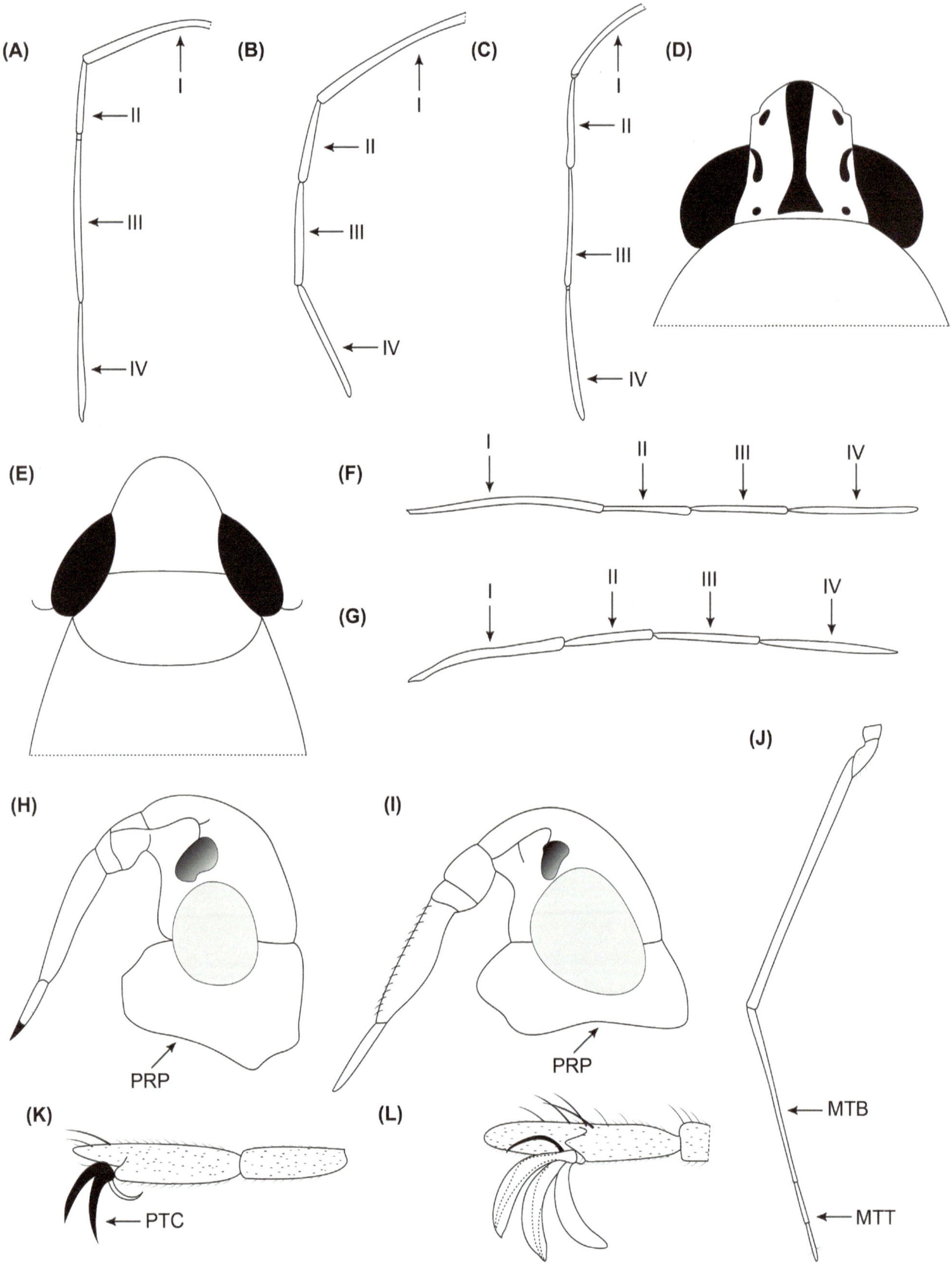

FIGURE 7.36 *Telmatometra*: (A) antenna. *Trepobates*: (B) antenna. *Telmatometroides*: (C) antenna; (D) head and part of pronotum; dorsal view. *Lathriobatoides*: (E) head and part of thorax, dorsal view. *Trepobatoides*: (F) antenna. *Ovatametra*: (G) antenna. *Trepobates*: (H) head and prothorax, lateral view. *Halobatopsis*: (I) head and prothorax, lateral view; (J) hindleg. *Microvelia*: (K) mesotarsus. *Xiphovelia*: (L) apex of mesotarsus. Abbreviations: *I*, antennomere 1; *II*, antennomere 2; *III*, antennomere 3; *IV*, antennomere 4; *MTB*, metatibia; *MTT*, metatarsus; *PRP*, propleura; *PTC*, pretarsal claw.

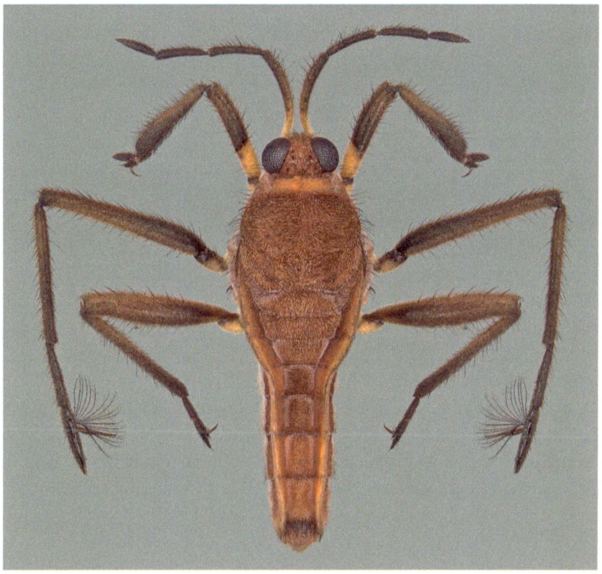

FIGURE 7.37 *Rhagovelia* (Veliidae): habitus, dorsal view.

Hemiptera: Gerromorpha: Gerridae: Genera

Gerridae (Fig. 7.33) is a large and morphologically diverse family with over 60 genera and 750 described species (Polhemus & Polhemus, 2008). Gerrids can be found in many kinds of habitats occupied by semiaquatic bugs, from humid terrestrial environments to the sea, and it is the only group of insects to colonize the open ocean (Cheng, 1976). The body size of gerrids ranges from some of the smallest in Gerromorpha (1.5 mm) to the by far largest (36.0 mm). They can be identified by the head not distinctly elongated, the eyes are located close to the anterior margin of the thorax, the mesocoxae are located much closer to the metacoxae than to the procoxae, the metafemur is long and generally surpasses the apex of abdomen, and the tarsi are biarticulate with preapical pretarsal claws (Andersen, 1982; Moreira, 2015).

1	Eyes reniform, emarginated on mesal margin (Fig. 7.31 F)	2
1'	Eyes entire, not emarginated (Fig. 7.31 G)	9
2(1)	Antennae much shorter than body; antennomere IV not longest (Fig. 7.31 I)	3
2'	Antennae almost as long as or longer than body; antennomere IV longest (Fig. 7.31 H)	*Tachygerris*
3(2)	Body short and robust, at most 5.5 times longer than its greatest width	4
3'	Body very long and cylindrical, at least 6.5 times longer than its greatest width (Fig. 7.34 A)	*Cylindrostethus*
4(3)	Forelegs with tarsomere I about half length of tarsomere II (Fig. 7.34 B)	5
4'	Forelegs with tarsomeres I and II subequal in length (Fig. 7.34 C)	8
5(4)	Antennomere IV curved; rostrum short, not surpassing prosternum posterior margin (Fig. 7.34 D)	6
5'	Antennomere IV straight; rostrum long, surpassing prosternum posterior margin (Fig. 7.34 E)	7
6(5)	Body strongly flattened; abdomen relatively short (Fig. 7.34 F)	*Platygerris*
6'	Body not strongly flattened; abdomen relatively long (Fig. 7.34 G)	*Potamobates*
7(5)	Pronotum anteriorly with a large, central, light macula (Fig. 7.34 H); mesotarsi without pretarsal claws (Fig. 7.34 I)	*Neogerris*
7'	Pronotum anteriorly with two relatively small, elongated, light maculae (Fig. 7.34 J); mesotarsi with pretarsal claws (Fig. 7.34 K)	*Limnogonus*
8(4)	Pronotum completely covering mesonotum (Fig. 7.35 A); laterotergites with small posterolateral spines (Fig. 7.35 A)	*Aquarius*
8'	Pronotum leaving at least small lateral portions of mesonotum exposed (Fig. 7.35 B); laterotergites without posterolateral spines (Fig. 7.35 B)	*Eurygerris*
9(1)	Antennomere II subequal in length or shorter than III; antennomeres II–III of males not distally widened (Fig. 7.35 D)	10
9'	Antennomere II longer than III; antennomeres II–III of males distally widened (Fig. 7.35 C)	*Metrobates*
10(9)	Forelegs with tarsomere I shorter than or equal to II (Fig. 7.35 F)	11
10'	Forelegs with tarsomere I longer than II (Fig. 7.35 E)	*Charmatometra*
11(10)	Forelegs with tarsomere II not longer than two times length of I (Fig. 7.35 F)	12
11'	Forelegs with tarsomere II longer than two times length of I (Fig. 7.35 G)	13

12(11)	Pronotum yellowish with several distinct black stripes (Fig. 7.35 H)	*Eobates*
12'	Pronotum orange to brown, usually only with lateral margins darkened, but sometimes with a median black stripe (Fig. 7.35 I)	*Brachymetra*
13(11)	Mesofemora shorter than mesotibiae and metafemora; metafemora never modified (Fig. 7.35 K)	14
13'	Mesofemora longer than mesotibiae and metafemora; metafemora modified in males of some species (Fig. 7.35 J)	*Rheumatobates*
14(13)	Antennomere III 10–40% longer than I (Fig. 7.36 A)	15
14'	Antennomere III 40–80% length of I (Fig. 7.36 B)	17
15(14)	Antennomere III shorter than two times length of II (Fig. 7.36 C)	16
15'	Antennomere III longer than two times length of II (Fig. 7.36 A)	*Telmatometra*
16(15)	Interocular space with a dark longitudinal stripe (Fig. 7.36 D)	*Telmatometroides*
16'	Interocular space without dark longitudinal stripe (Fig. 7.36 E)	*Lathriobatoides*
17(14)	Antennomere I at most as long as II–III together (Fig. 7.36 G)	18
17'	Antennomere I much longer than II–III together (Fig. 7.36 F)	*Trepobatoides*
18(17)	Mesotibiae almost as long as or slightly longer than length of body	19
18'	Mesotibiae distinctly shorter than length of body	*Ovatametra*
19(18)	Eyes in lateral view not extending beyond half of propleura (Fig. 7.36 H); metatibiae distinctly shorter than two times length of metatarsi (Fig. 7.35 K)	*Trepobates*
19'	Eyes in lateral view extending beyond half of propleura (Fig. 7.36 I); metatibiae longer than two times length of metatarsi (Fig. 7.36 J)	*Halobatopsis*

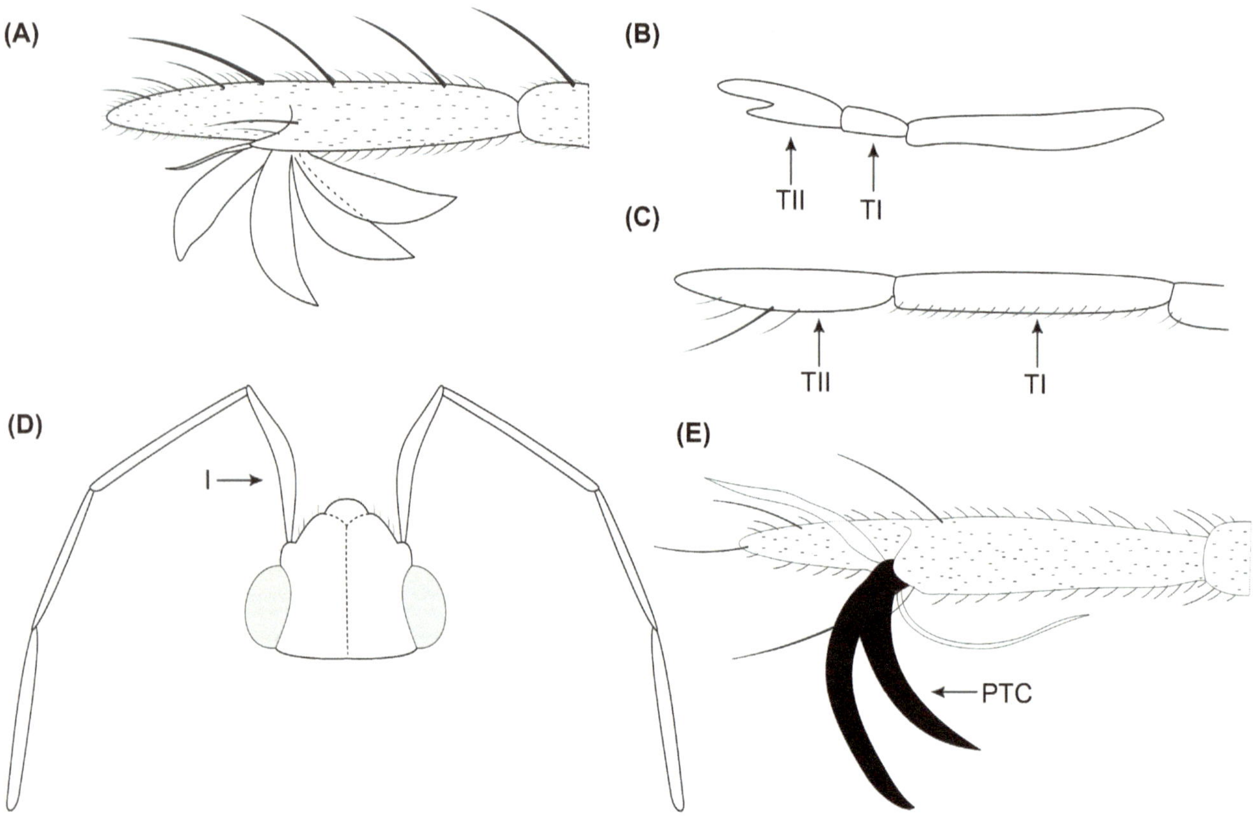

FIGURE 7.38 *Husseyella*: (A) apex of mesotarsus. *Euvelia*: (B) apex of midleg, pretarsal structures omitted. *Husseyella*: (C) apex of midleg, pretarsal structures omitted. *Aegilipsovelia*: (D) head, dorsal view. *Aegilipsovelia*: (E) apex of mesotarsus. *Microvelia*: (F) head, dorsal view. *Stridulivelia*: (G) lateral view. *Veloidea*: (I) apex of mesotarsus. Abbreviations: *I*, antennomere 1; *PTC*, pretarsal claw; *SGG*, shining glabrous groove; *TI*, tarsal article 1; *TII*, tarsal article 2.

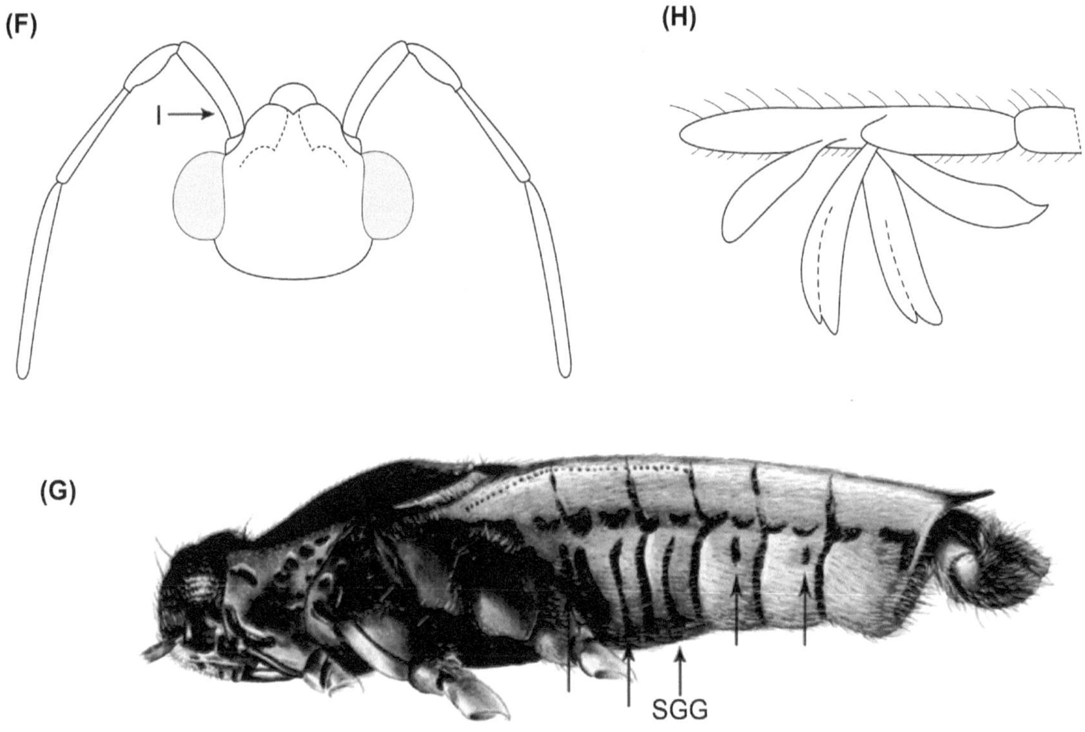

FIGURE 7.38 (Continued)

Hemiptera: Gerromorpha: Veliidae: Genera

Veliidae (Fig. 7.37) is the largest Gerromorpha family, currently including more than 60 genera and 960 described species, most of which are associated with freshwater environments, whereas the others are either secondarily terrestrial or occupy coastal marine habitats (Polhemus & Polhemus, 2008; Moreira, 2015). Veliids can be from 1.0 to 10.5 mm long, and members of different subfamilies are usually quite distinct from one another. All have the coxae of the same pair of legs widely separated, the pronotum of macropterous forms is extended posteriorly covering the rest of the nota, and the pretarsal claws are inserted preapically; none of these features are synapomorphic (Andersen, 1982; Moreira, 2015).

1 Tarsal formula 1-2-2 2
1' Tarsal formula 2-2-2 or 3-3-3 6
2(1) Mesotarsi with three or four foliform structures preapically (modified pretarsal claws and ventral arolium; Figs. 7.36 L, 7.38 A) 3
2' Mesotarsi with falcate pretarsal claws and setiform arolium (Fig. 7.36 K) 5
3(2) Mesotarsi with four foliform structures subapically (Fig. 7.38 A) 4
3' Mesotarsi with three foliform structures subapically (Figs. 7.36 L) .... *Xiphovelia*
4(3) Tarsomere II of middle legs longer than I, widened and deeply cleft subapically (Fig. 7.38 B) .... *Euvelia*
4' Tarsomere II of middle legs shorter than I, not widened and deeply cleft subapically (Fig. 7.38 C) .... *Husseyella*
5(2) Antennomere I at least twice as long as distance from base of antennae to apex of head (Fig. 7.38 D); pretarsal claws of middle and hindlegs very long, subequal in length to tarsomere I (Fig. 7.38 E) .... *Aegilipsovelia*
5' Antennomere I and pretarsal claws of middle and hindlegs shorter (Figs. 7.36 K, 38 F) .... *Microvelia*
6(1) Tarsal formula 3-3-3; mesotarsi and pretarsal structures not as below (Figs. 7.39 A, B) 7
6' Tarsal formula 2-2-2 or 3-3-3; mesotarsi deeply cleft, with pretarsal claws and arolia modified as a swimming fan (Fig. 7.3 A) .... *Rhagovelia*
7(6) Abdominal sterna without lateral grooves 8
7' Abdominal sterna with lateral shining glabrous grooves (Fig. 7.38 G) .... *Stridulivelia*
8(7) Meso- and metatarsi with falcate pretarsal claws and setiform arolia (Figs. 7.39 A, B) 9
8' Meso- and metatarsi each with four leaflike structures (modified pretarsal claws and arolia) arising from a short cleft at middle of last tarsomere (Fig. 7.38 H) .... *Veloidea*
9(8) Last tarsomere of each leg not expanded, subcylindrical; tarsomere II of middle legs at least 2.0 times length of I (Fig. 7.39 B) 10
9' Last tarsomere of each leg laterally expanded, suboval; tarsomere II of middle legs at most 1.5 times length of I (Fig. 7.39 A) .... *Oiovelia*

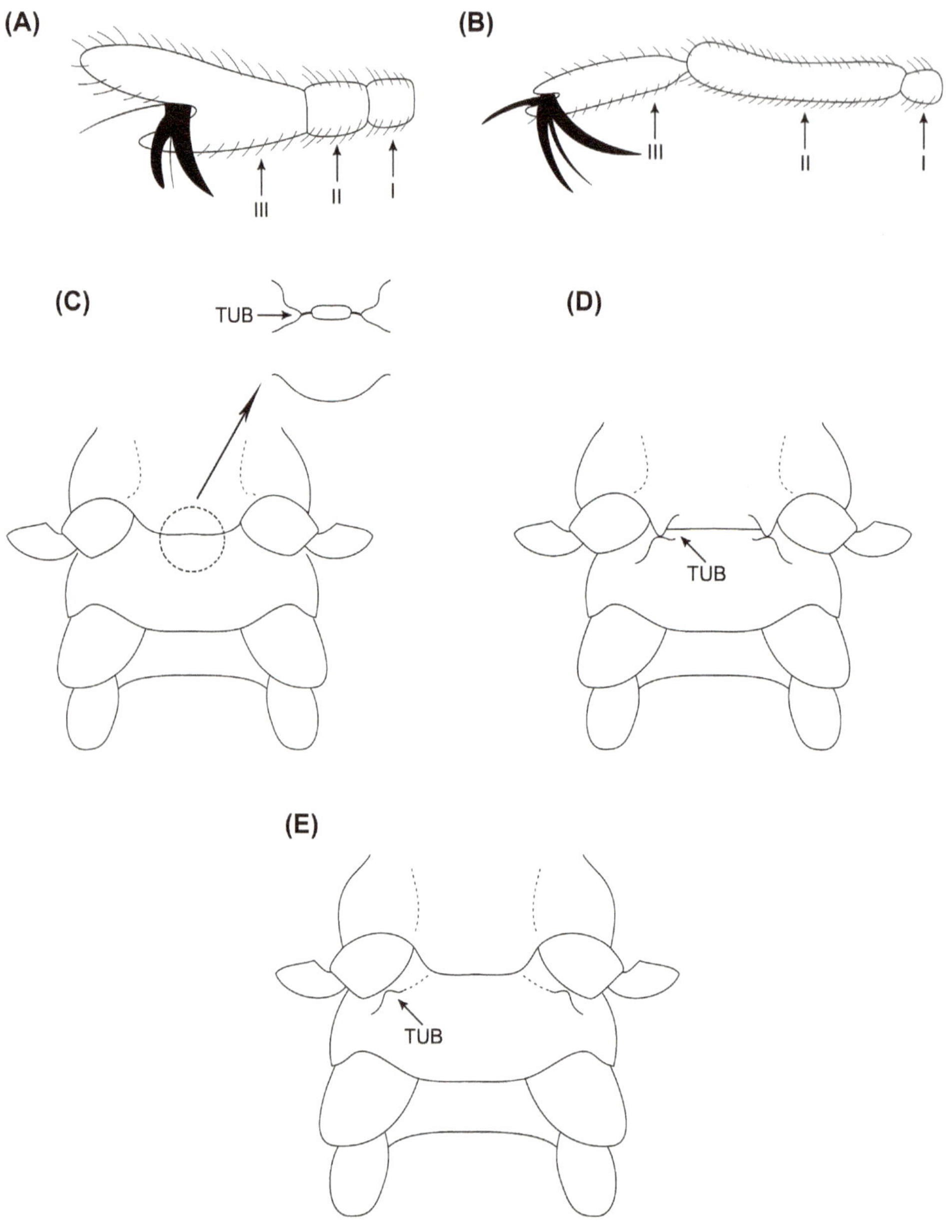

FIGURE 7.39 *Oiovelia*: (A) mesotarsus. *Paravelia*: (B) mesotarsus. *Paravelia*: (C) part of thorax and abdomen, ventral view, and detail of central tubercles. *Platyvelia*: (D) part of thorax and abdomen, ventral view. *Steinovelia*: (E) part of thorax and abdomen, ventral view. Abbreviations: *I*, tarsal article 1; *II*, tarsal article 2; *III*, tarsal article 3; *TUB*, tubercle.

10(9) Metasternum with a pair of anterolateral tubercles near mesoacetabula (Figs. 7.39 D, E) 11
10' Tubercles of meso- and metasternum, if present, located centrally (Fig. 7.39 C) (in some species the mesoacetabula are slightly expanded mesally, but do not form distinct tubercles) .... *Paravelia*
11(10) Mesoacetabula prolonged into tubercles, positioned in front of metasternal tubercles (Fig. 7.39 D) .... *Platyvelia*
11' Mesoacetabula not prolonged into tubercles, only metasternal tubercles present (Fig. 7.39 E) .... *Steinovelia*

ACKNOWLEDGMENTS

Financial support was provided by the Fundação de Amparo à Pesquisa do Estado do Rio de Janeiro as a grant to F.F.F.M. (process #210.508/2016), the Fundação de Amparo à Pesquisa do Estado de São Paulo as scholarship for H.D.D.R. (process #2013/16654−0), and the Coordenação de Aperfeiçoamento de Pessoal de Nível Superior as scholarships for I.R.S.C. and O.M.M.

REFERENCES

Andersen, N.M. 1982. The semiaquatic bugs (Hemiptera, Gerromorpha). Phylogeny, adaptations, biogeography and classification. Entomonograph 3: 1–455.

Andersen, N.M. & T.A. Weir. 2004. Mesoveliidae, Hebridae, and Hydrometridae of Australia (Hemiptera: Heteroptera: Gerromorpha), with a reanalysis of the phylogeny of the semiaquatic bugs. Invertebrate Systematics 18: 467–522.

Aristizábal, H., C.F.B. Floriano, F.F.F. Moreira & P.C. Bispo. 2015. New species and notes on Microveliinae from northern South America (Hemiptera: Heteroptera: Veliidae). Zootaxa 3980: 591–596.

Barbosa, J.F., J.R.I. Ribeiro & J.L. Nessimian. 2012. A synopsis of *Martarega* White, 1879 (Insecta: Heteroptera: Notonectidae) occurring in the Brazilian Amazonia, with descriptions of three new species. Journal of Natural History 46: 1025–1057.

Barbosa, J.F. & H.D.D. Rodrigues. 2015. The true water bugs (Nepomorpha). Pages 159–199 *in*: A.R. Panizzi & J. Grazia (eds.), True bugs (Heteroptera) of the neotropics. Springer Science+Business Media, Dordrecht, The Netherlands.

Carvalho, A.L. & A.P. Pinto. 2007. Book review: Heckman, C.W. 2006. Encyclopedia of South American aquatic insects: Odonata - Anisoptera. Dordrecht, Springer, viii+725pp. ISBN-10 1-4020-4801-7. Lundiana 8: 157–159.

Cassis, G. & R. Silveira. 2001. A revision and phylogenetic analysis of the *Nerthra alaticollis* species-group (Heteroptera: Gelastocoridae: Nerthrinae). Journal of the New York Entomological Society 109: 1–46.

Champion, G. C. 1898–1901. Hemiptera-Heteroptera. Biologia Centrali-Americana, Rhynchota 2: 1–416.

Chen, P.P., N. Nieser & H. Zettel. 2005. The aquatic and semi-aquatic bugs (Heteroptera: Nepomorpha & Gerromorpha) of Malesia. Fauna Malesiana Handbook 5: 1–546.

Cheng, L. 1976. Marine insects. North Holland Publishing Company, Amsterdam. 581 pp.

Cook, J.L. 2011. A new genus and species of Pleidae (Hemiptera) from Venezuela, with notes on the genera of Pleidae. Zootaxa 3067: 26–34.

Damgaard, J. 2008. Phylogeny of the semiaquatic bugs (Hemiptera-Heteroptera, Gerromorpha). Insect Systematics and Evolution 39: 431–460.

Damgaard, J. 2012. What do we know about the phylogeny of the semiaquatic bugs (Hemiptera: Heteroptera: Gerromorpha)? Entomologica Americana 118: 81–98.

Damgaard, J., F.F.F. Moreira, M. Hayashi, T.A. Weir & H. Zettel. 2012. Molecular phylogeny of the pond treaders (Insecta: Hemiptera: Heteroptera: Mesoveliidae), discussion of the fossil record and a checklist of species assigned to the family. Insect Systematics and Evolution 43: 175–212.

Estévez, A.L. & J.R.I. Ribeiro. 2011. *Weberiella* De Carlo, 1966 (Insecta: Heteroptera: Belostomatidae) revisited: redescription with a key to the genera of Belostomatidae and considerations on back brooding behavior. Zoologischer Anzeiger 250: 46–54.

Floriano, C.F.B. & F.F.F. Moreira. 2015. A new species of *Rhagovelia* Mayr, 1865 (Hemiptera: Heteroptera: Veliidae) from Brazil. Zootaxa 4018: 437–443.

Flowers, R.W. 2004. Book review: Encyclopedia of South American aquatic insects: Ephemeroptera. Illustrated keys to known families, genera and species in South America. C. W. Heckman. Kluwer Academic Publishers, Dordrecht/Boston/London. 2002. 419 pp (hardcover). ISBN 1-4020-0775-2. Journal of the North American Benthological Society 23: 152–153.

Hebsgaard, M.B., N.M. Andersen & J. Damgaard. 2004. Phylogeny of true water bugs (Nepomorpha: Hemiptera–Heteroptera) based on 16S and 28S rDNA and morphology. Systematic Entomology 29: 488–508.

Heckman, C.W. 2011. Encyclopedia of South American aquatic insects: Hemiptera - Heteroptera. Illustrated keys to known families, genera, and species in South America. Springer Science+Business Media, Dordrecht, The Netherlands.

Hua, J., M. Li, P. Dong, Y. Cui, Q. Xie & W. Bu. 2009. Phylogenetic analysis of the true water bugs (Insecta: Hemiptera: Heteroptera: Nepomorpha): evidence from mitochondrial genomes. BMC Evolutionary Biology 9: 134.

Hungerford, H.B. 1948. The Corixidae of the Western Hemisphere (Hemiptera). University of Kansas Science Bulletin 32: 5–827.

Keffer, S.L. 2004. Morphology and evolution of waterscorpion male genitalia (Heteroptera: Nepidae). Systematic Entomology 29: 142–172.

Kondratieff, B.C. 2004. Book reviews: The Encyclopedia of South American Aquatic Insects: Plecoptera Charles W. Heckman Kluwer Academic Publishers, Dordrecht 2003, 329 pp. ISBN: 1-4020-1520-8. American Entomologist 50: 236–237.

Lauck, D.R. & A.S. Menke. 1961. The higher classification of the Belostomatidae (Hemiptera). Annals of the Entomological Society of America 54: 644–657.

Li, T., J. Hua, A.M. Wright, Y. Cui, Q. Xie, W. Bu & D.M. Hillis. 2014. Long-branch attraction and the phylogeny of true water bugs (Hemiptera: Nepomorpha) as estimated from mitochondrial genomes. BMC Evolutionary Biology 14: 99.

Mahner, M. 1993. Systema cryptoceratum phylogeneticum (Insecta, Heteroptera). Zoologica, Originalabhandlungen aus dem Gesamtgebiet der Zoologie 48(143): i–ix + 1–302.

Mazzucconi, S.A. 2011. A new species of *Martarega* White, 1879 from South America (Hemiptera: Heteroptera: Notonectidae), with an identification key to all described species of the genus. Aquatic Insects 33: 113–126.

Melo, A. L. 2015. Shore bugs (Saldidae). Pages 203–213 *in*: A.R. Panizzi & J. Grazia (eds.), True bugs (Heteroptera) of the neotropics. Springer Science+Business Media, Dordrecht, The Netherlands.

Morales-Castaño, I.T. & M.I. Castro-Vargas. 2013. A new species and new records of *Tachygerris* Drake, 1957 (Hemiptera: Heteroptera: Gerridae) from Colombia. Zootaxa 3616: 277–283.

Moreira, F.F.F. 2015. The semiaquatic gerromorphans. Pages 113–156 *in*: A.R. Panizzi & J. Grazia (eds.), True bugs (Heteroptera) of the neotropics. Springer Science+Business Media, Dordrecht, The Netherlands.

Moreira, F.F.F., B. Pacheco-Chaves, M. Springer & I.R.S. Cordeiro. 2015. Two new species of *Rhagovelia* (Hemiptera: Heteroptera: Veliidae) from Costa Rica, with a key and new records from the country. Zootaxa 3980: 477–500.

Muzón, J. 2007. Comentário bibliográfico: Charles W. Heckman. 2006. Encyclopedia of South American Aquatic Insects: Odonata - Anisoptera. Illustrated Keys to Known Families, Genera, and Species in South America. Springer, Dordrecht, The Netherlands. viii+ 725 pp. Revista de la Sociedad Entomológica Argentina 66: 193–194.

Nieser, N. 1975. The water bugs (Heteroptera: Nepomorpha) of the Guyana Region. Studies on the Fauna of Suriname and Other Guyanas 16: 1–308.

Nieser, N. 2002. Guide to aquatic Heteroptera of Singapore and Peninsular Malaysia. IV. Corixoidea. The Raffles Bulletin of Zoology 50: 263–274.

Nieser, N. & P.P. Chen. 2002. Six new species of *Neotrephes* China, 1936 (Heteroptera: Helotrephidae) from Brazil, with a key to Neotropical Helotrephidae. Lundiana 3: 31–40.

Nieser, N. & P.P. Chen. 2008. A new species of *Tenagobia* Bergroth from Bolivia, with notes on the subgenus *Fasciagobia* Nieser, 1977 (Heteroptera, Nepomorpha: Micronectidae). Folia Entomologica Hungarica 69: 5–13.

Papácek, M. & H. Zettel. 2005 (2004). Helotrephidae of the World (Hemiptera: Heteroptera: Nepomorpha): checklist and bibliography. Acta Societatis Zoologicae Bohemicae 68:99–108.

Peters, J.G. & T. Soldán. 2005. Book review: Heckman Ch.W.: Encyclopedia of South American Aquatic Insects: Ephemeroptera. Kluwer Academic Publishers, Dordrecht/Boston/London, 2002, viii + 419 pp. ISBN 1-4020-0775-2. European Journal of Entomology 102: 791–792.

Polhemus, D.A. 1997. Systematics of the genus *Rhagovelia* Mayr (Heteroptera: Veliidae) in the Western Hemisphere (exclusive of the *angustipes* complex). Entomological Society of America, Lanham, MD.

Polhemus, D.A. 2015. Dr. John T. Polhemus, 1929–2013: summary of collecting expeditions, scientific publications, and taxonomic names proposed. Proceedings of the Entomological Society of Washington 117: 529–591.

Polhemus, J.T. 1985. Shore bugs (Heteroptera, Hemiptera; Saldidae): a world overview and taxonomy of Middle American forms. The Different Drummer, Englewood.

Polhemus, J.T. & D.A. Polhemus. 2007. Global trends in the description of aquatic and semiaquatic Heteroptera species, 1758–2004. Tijdschrift voor Entomologie 150: 271–288.

Polhemus, J.T. & D.A. Polhemus. 2008. Global diversity of true bugs (Heteroptera; Insecta) in freshwater. Hydrobiologia 595: 379–391.

Polhemus, J.T. & D.A. Polhemus. 2010. *Bacillometroides*, a new genus of Hydrometridae (Heteroptera) for three previously described species from South America. Entomologica Americana 116: 58–63.

Polhemus, J.T. & D.A. Polhemus. 2012. Guide to the aquatic Heteroptera of Singapore and peninsular Malaysia. IX. Infraorder Nepomorpha, families Ochteridae and Gelastocoridae. The Raffles Bulletin of Zoology 60: 343–359.

Popov, Y.A. 1971. [Historical development of the Hemiptera infraorder Nepomorpha (Heteroptera)] (in Russian). Trudy Paleontologicheskogo Instituta 129: 1–230.

Reynoso-Velasco, D. & R.W. Sites. 2016. Revision of the *Ambrysus guttatipennis* Stål species complex (Heteroptera: Naucoridae: Cryphocricinae) with the descriptions of six new species from Mexico. Zootaxa 4154: 201–250.

Reynoso-Velasco, D., R.W. Sites, & R. Novelo-Gutiérrez. 2016. The *Ambrysus* Stål (Heteroptera: Naucoridae: Cryphocricinae) of Mexico: Revision of the subgenus *Syncollus* La Rivers. Zootaxa 4126: 44–76.

Ribeiro, J.R.I. & A.L. Estévez. 2009. The small species of *Belostoma* Latreille (Heteroptera, Belostomatidae). III. A revision of *oxyurum* group, with a new species from Brazil and description of the male of *B. noualhieri* Montandon. Revista Brasileira de Entomologia 53: 207–215.

Rieger, C. 1976. Skelett und Muskulatur des Kopfes und Prothorax von *Ochterus marginatus* Latreille. Zoomorphologie 83: 109–191.

Rodrigues, H.D.D., F.F.F. Moreira, N. Nieser, P.P. Chen, A.L. Melo, K. Dias-Silva & N.F.S. Giehl. 2014. The genus *Paravelia* Breddin, 1898 (Hemiptera: Heteroptera: Veliidae) in Brazil, with descriptions of eight new species. Zootaxa 3784: 1–47.

Ruhoff, F.A. 1968. Bibliography and index to scientific contributions of Carl J. Drake for the years 1914–1967. United States National Museum Bulletin 267:1–81.

Schuh, R.T. & J.T. Polhemus. 1980. Analysis of taxonomic congruence among morphological, ecological, and biogeographical data sets for the Leptopodomorpha (Hemiptera). Systematic Zoology 29: 1–26.

Schuh, R.T. & J.A. Slater. 1995. True bugs of the world (Hemiptera: Heteroptera). Classification and natural history. Cornell University Press, Ithaca, NY. 336 pp.

Sites, R.W. 2015. New taxa of Cryphocricinae (Heteroptera: Naucoridae) from the Guiana Shield: *Hygropetrocoris* Sites, n. gen. and two new species of *Ambrysus* Stål. Zootaxa, 4033: 427–438.

Sites, R.W. & J.T Polhemus. 2001. Distribution of Helotrephidae (Heteroptera) in Thailand. Journal of the New York Entomological Society, 109: 372–391.

Sites, R.W. & D. Reynoso-Velasco. 2015. Review of the *Ambrysus stali* La Rivers species complex (Heteroptera: Nepomorpha: Naucoridae) with the description of a new species from Mesoamerica. Zootaxa 4018: 279–281.

Soldán, T. 2005. Book review: Heckman Ch.W.: Encyclopedia of South American Aquatic Insects: Plecoptera. Kluwer Academic Publishers, Dordrecht/Boston/London, 2003, viii + 329 pp. ISBN 1-4020-1520-8. Price USD 197.00. European Journal of Entomology 102: 800.

Štys, P. & A. Jansson. 1988. Check-list of recent family-group and genus-group names of Nepomorpha (Heteroptera) of the world. Acta Entomologica Fennica 50: 1–44.

Štys, P. & I. Kerzhner. 1975. The rank and nomenclature of higher taxa in recent Heteroptera. Acta Entomologica Bohemoslovaca 72: 65–79.

Tinerella, P.P. 2008. Taxonomic revision and systematics of New Guinea and Oceania pygmy water boatmen (Hemiptera: Heteroptera: Corixoidea: Micronectidae). Zootaxa 1797: 1–66.

Weirauch, C. & R.T. Schuh. 2011. Systematics and evolution of Heteroptera: 25 years of progress. Annual Review of Entomology 56: 487–510.

Woodruff, L.C. 1958. Herbert Barker Hungerford. University of Kansas Science Bulletin 38: i-xiv.

Woodruff, L.C. 1963. Herbert Barker Hungerford, 1885–1963. Journal of the Kansas Entomological Society 36: 197–199.

Zettel, H. 1997. One new genus and two new species of Helotrephidae (Insecta: Heteroptera) from India, with notes on the phylogeny of the family. Annalen des Naturhistorischen Museums in Wien 99: 83–95.

Chapter 8

Order Megaloptera

Adrian Ardila-Camacho

Laboratorio de Entomología, Centro de Investigaciones en Microbiología y Parasitología Tropical – CIMPAT, Universidad de los Andes, Bogotá, Colombia

Atilano Contreras-Ramos

Departamento de Zoología, Instituto de Biología, Universidad Nacional Autónoma de México, Ciudad Universitaria, Ciudad de México, México

Chapter Outline

INTRODUCTION

Megaloptera belongs to an ancient neuropteroid lineage within the Holometabola (Aspöck et al., 2001; Grimaldi & Engel, 2005; Misof et al., 2007; Aspöck et al., 2012). Estimates of the oldest megalopteran fossils of the order are from the Permian, with a few records from the Mesozoic, while most fossil megalopterans are of Cenozoic age (Jepson & Heads, 2016). The phylogenetic position of Megaloptera within Neuropterida has been uncertain, but recent morphological and molecular studies place Raphidioptera as sister to Megaloptera + Neuroptera (Haring & Aspöck, 2004; Aspöck & Aspöck, 2008; Wang et al., 2012; Zhao et al., 2014; Liu et al., 2015c). Megaloptera is divided into two families, Sialidae (alderflies) and Corydalidae, comprised of the subfamilies, Corydalinae (dobsonflies) and Chauliodinae (fishflies). At present, the order includes around 380 species (Wang et al., 2012; Zhao et al., 2014; Liu et al., 2015a, c). Sialidae is present in all biogeographic realms, except tropical Africa and portions of the Oriental region, and it comprises around 75 species in eight genera (Liu et al., 2015a); Chauliodinae contains 128 species in 18 genera sparsely distributed in Asia, Australia, North and South America, South Africa, and Madagascar (Liu et al., 2012; Cardoso-Costa et al., 2013). For its part, Corydalinae includes around 143 species in 9 genera, distributed in the Americas, South Africa, and the Oriental region (Cover & Resh, 2008; Contreras-Ramos, 2011).

Megaloptera are distinguished from other neuropteroids by the presence of a prognathous adult head, broad anal area of the hindwing, and exclusively aquatic larval stages (New & Theischinger, 1993; Liu et al., 2015c). Larval Megaloptera are dorsoventrally flattened, campodeiform (Fig. 8.1), with well-developed chewing mouthparts and movable labrum (Fig. 8.2 D), antennae short, and six stemmata present (Beutel & Friedrich, 2008). Adults of Sialidae are small sized (forewing $\leq$15 mm), without ocelli, prothorax wider than long, and legs with fourth tarsomere bilobed. Adults of Corydalidae are medium to large sized (forewing $\geq$15 mm), with dorsal ocelli present and fourth tarsomere unmodified. Dobsonflies have two postocular spines developed, male genitalia with gonostyli, 9 as claspers and 10 lobiform (New & Theischinger, 1993; Contreras-Ramos, 1998). Fishflies have postocular plane reduced, usually specialized antennae, three crossveins in the radial field, and gonocoxite and gonostyli, nine reduced (Liu & Yang, 2006).

The Neotropical Megaloptera currently consist of 2 genera and 9 species of Sialidae, 5 genera and 10 species and subspecies of Chauliodinae, and 3 genera and 56

ORDER MEGALOPTERA

Thorp and Covich's Freshwater Invertebrates, Fourth Edition. https://doi.org/10.1016/B978-0-12-804223-6.00008-1

FIGURE 8.1 Corydalidae larvae in their natural environment: (A) *Neohermes filicornis* (Banks); and (B) *Corydalus* sp. *(A) Photo courtesy of Matthew Cover; (B) photo by A. Ardila-Camacho.*

species of Corydalinae. Megaloptera larvae generally inhabit lotic ecosystems, from large rivers to small streams and creeks (New & Theischinger, 1993). Corydalidae larvae are generally found in concealed places like submerged vegetation, under logs and stones, in ponds, and/or "riffle" zones, while Sialidae prefer small streams with soft substrates (Rasmussen & Pescador, 2002). Because of their predatory habits, larval Megaloptera are important elements to aquatic ecosystems; they may be considered major predators in the absence of predatory fishes. Megaloptera larvae have functional spiracles, however they are generally associated with well oxygenated and cold-water bodies (Costa et al., 2006), so they may be considered indicators of good water quality. Their life cycle generally lasts from one to three years in the Nearctic, with at least 11 larval instars; a similar pattern should be expected in the Neotropics, depending on climatic conditions (e.g., latitude and elevation). Because of a holometabolous life cycle, with long-lived aquatic larvae and terrestrial adults, the Megaloptera play an important role in the cycling of energy between aquatic and terrestrial ecosystems (New & Theischinger, 1993; Cover et al., 2015). High numbers, potentially in high densities, of generalist predacious larvae, make Megaloptera an important insect group for the conservation of hydric sources.

Sialidae

Neotropical alderflies are seldom encountered in field work, which may reflect low population numbers and patchy distribution (Contreras-Ramos, 2008). Egg masses are uncovered and unilaminar, composed of several hundreds of eggs arranged in rows; they are laid on stems, plants, and stones near water (New & Theischinger, 1993; Rasmussen & Pescador, 2002; Bentes et al., 2014). Life histories of Neotropical species of Sialidae are virtually unknown, but data on Nearctic species indicate up to 10 larval instars during one or two years (Contreras-Ramos, 2009). Studies in the Neotropics indicate that larval Sialidae may be associated to small creeks with low flow velocity, soft and muddy bottoms, and availability of decaying plant material (Contreras-Ramos, 2008; Clavier et al., 2009; Azevêdo, 2009; Archangelsky et al., 2017).

Corydalidae

The biology of Corydalinae is the best known among the family Corydalidae and Megaloptera as a whole. Egg masses of *Corydalus* Latreille, 1802 are three-layered and have a covering of a white or yellowish and hard substance, giving it an appearance of bird feces. Such a substance is produced by a gland located in the end of the female abdomen and has a protective function. Egg masses are usually found during rainy season on stones, logs, or leaves that overhang the water, but oviposition far from the water has been recorded (Mayorga et al., 2017). After hatching, first instar larvae fall or crawl to the water. Larval Corydalinae are found in intermittent or perennial lotic ecosystems, they are generally monovoltine, although there are species that could be from bivoltine to semivoltine, depending on the environmental conditions (Contreras-Ramos, 1998, 2009). In the tropical regions, the larvae may pass through around 11 instars in about one year, while the cycle is generally longer in the Nearctic species. As in

FIGURE 8.2 Corydalidae larva morphology (Corydalinae, *Chloronia* sp.): (A) habitus, dorsal view; (B) habitus, ventral view; (C) thorax, dorsal view; (D) head, dorsal view; (E) head, ventral view; (F) macro- and microsetae of abdominal tergite 5; (G) microsetae of abdominal sternite 5; (H) mandible; (I) spiracle of abdominal segment 1; and (J) abdominal segments 9 and 10, dorsal view.

other Megaloptera, dobsonfly larvae are generalist predators on other invertebrates, they are large sized insects when reached the mature stage (in many *Corydalus* species >60 mm of body length). These larvae are more active at night, but during the day and even most of the night they ambush their prey below stones, logs, submerged vegetation, and other concealed places (Hayashi & Nakane, 1989). When mature larvae are ready to pupate (prepupae), they leave the water during the night searching for an appropriate place, often near the water habitat. Then, the prepupa builds a pupal chamber generally under a large stone or log in a relatively moist place. After a few days, a decticous and exarate pupa emerges, which despite its vulnerable condition, it is capable of a defensive behavior. In around 12 days imaginal hatching occurs, and the adult leaves the pupal chamber, extends its wings, and achieves cuticle hardening.

At present, except for *Neohermes filicornis* (Banks, 1903), species of Chauliodinae in the Neotropics are poorly studied; it is necessary to describe larvae of the Chilean and Brazilian species and to develop ecological and biological studies in these regions. However, in North America larval Chauliodinae may be found from intermittent to perennial streams, their life cycle is generally long, and periods ranging from one to five years for the complete larval cycle have been reported (Cover et al., 2015). Egg masses of Nearctic species of *Neohermes* Neohermes Banks 1908 and *Protochauliodes* van der weele, 1906 are single-layered and have a thin and soft covering, light brown in color in *Neohermes* (Baker & Neunzig, 1968; Evans & Neunzig, 1996; Contreras-Ramos, 2009).

LIMITATIONS

Besides several studies on morphology and ecology (Azevêdo, 2003, 2009), immature stages in the Neotropics have received little attention. Also, comprehensive genus level diagnoses have not been prepared. Available genus-level keys (Contreras-Ramos & Harris, 1998; Contreras-Ramos, 2009; Azevêdo & Hamada, 2014), are generally restricted to a specific region, and thus do not include all Neotropical genera. The established morphological characters for delimitation of some genera might be shared with other species that were not included in these studies. For example, submental projections, mandible shape, scars, and color pattern of head and pronotum, as well as macrosetae of abdominal terga of *Platyneuromus* van der Weele, 1909 are quite similar to some South American species of *Corydalus*. Taking in consideration this conservative morphology, Contreras-Ramos & Harris (1998) suggested that an exploration of setal characters and other microscopic traits for both specific and generic delimitation are necessary. As mentioned in other sections, many species are not associated with the adults and remain undescribed. At present, five *Corydalus* species, two of *Ilyobius* Enderlein, 1910 and only one in the genera *Platyneuromus*, and *Chloronia* Banks, 1908 have detailed descriptions (Penny & Flint, 1982; Contreras-Ramos & Harris, 1998; Azevêdo,

FIGURE 8.3 Sialidae, *Ilyobius* Larvae: (A) habitus, dorsal view; and (B) habitus, ventral view. *Photos courtesy of Neusa Hamada.*

FIGURE 8.4 Habitus of larval Corydalidae. *Corydalus* sp.: (A) dorsal view; and (B) ventral view. *(B) Photos courtesy of Neusa Hamada.*

2003, 2009; Azevêdo & Hamada, 2006, 2007; Archangelsky et al., 2017). Often preservation methods and collecting data are not adequate; consequently, morphological studies and distributional information are limited, challenging further studies on larvae.

FIGURE 8.5 Habitus of larval Corydalidae (preserved specimens), *Corydalus* spp.: (A) *Corydalus armatus* Hagen, 1861; and (B) *Corydalus* sp. *(B) Photos by A. Ardila-Camacho.*

TERMINOLOGY AND MORPHOLOGY

Morphological terminology generally follows that of New & Theischinger (1993), Contreras-Ramos & Harris (1998), Azevêdo & Hamada (2007, 2014), and Beutel & Friedrich (2008). Terminology of setal characters is based on Contreras-Ramos & Harris (1998).

Larvae of Sialidae are small insects (0.8–12 mm of body length), while Corydalidae larvae are generally large sized (20–80 mm of body length). Head is broad, strongly sclerotized, flattened or convex, robust, and prognathous, sometimes with muscle attachment marks (Figs. 8.8 C–E). Six well-developed stemmata are present, arranged either as two vertical rows of three ocelli in Sialidae, or as a pentagon with five peripheral and a single central stemma in Corydalidae. Antennae are short with four articles in Sialidae (Fig. 8.8 F) and Chauliodinae, and five in Corydalinae (Fig. 8.2 D), the second flagellomere has specialized setae in the apical portion. The epicraneal suture is conspicuous, Y-shaped; the clypeus is traverse, subrectangular; labrum is broad or narrow, often medially produced; gena is sometimes angulate. The mandibles are well developed and strong, with a tapering and sharply pointed apical tooth, and from two to four preapical teeth (Fig. 8.2 H); the maxilla is elongate, stipes cylindrical with a small galea and 4- or 5-articulate apical palpus; lacinia is reduced (Fig. 8.2 E), somewhat conspicuous in Sialidae (Fig. 8.6 A). Labium has small glossae and fleshy lateral paraglossae; labial

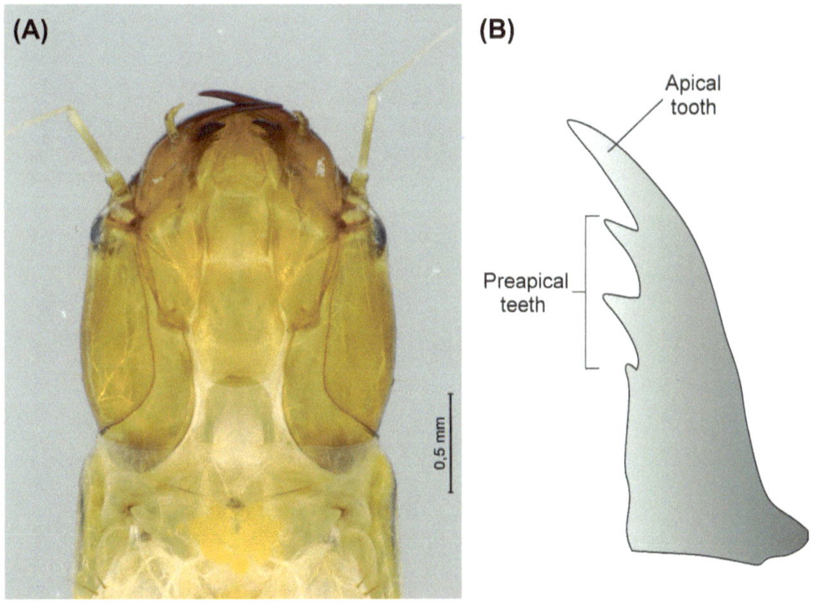

FIGURE 8.6 Mouth pieces of larval Sialidae: (A) *Ilyobius* sp., head in ventral view; and (B) *Protosialis americana* (Rambur, 1842), mandible. *(A) Photo courtesy of Neusa Hamada; (B) after Bowles & Sites (2015), modified.*

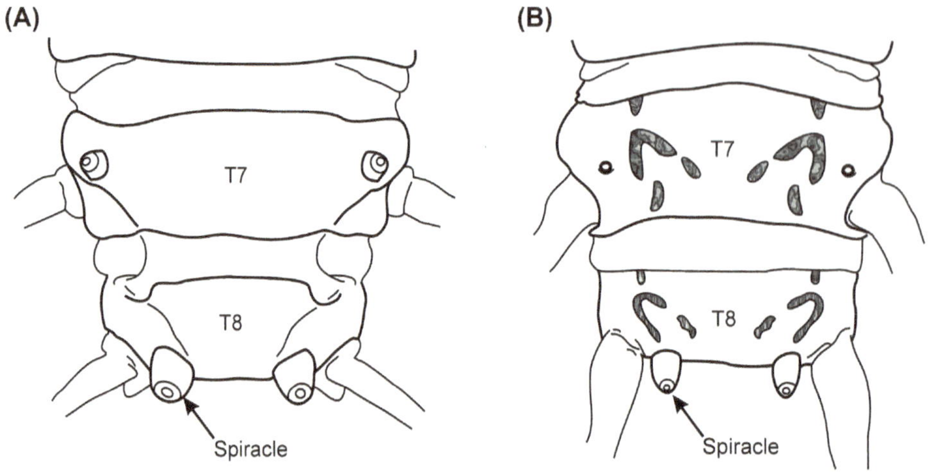

FIGURE 8.7 Diagnostic features of Neotropical larval Chauliodinae: (A) *Archichauliodes* sp., abdominal segments 7 and 8, dorsal view; and. (B) *Protochauliodes* sp., abdominal segments 7 and 8, dorsal view. *(B) After Flint (1973), modified.*

FIGURE 8.8 Cephalic and thoracic patterns of Neotropical Corydalinae and Sialidae: (A) *Chloronia* sp.; (B) *Chloronia* sp.; (C) *Corydalus* sp.; (D) *Corydalus armatus* Hagen, 1861; (E) *Corydalus* sp.; and (F) *Ilyobius* sp. *(F) Photo courtesy of Neusa Hamada.*

palpus is 3-articulate; the mentum is subquadrate to broadly rounded, sometimes divided into two plates (Fig. 8.2 E); the submentum is developed, in Corydalidae with submental projections present (Fig. 8.2 E); the gular plate is short or elongate. The prothorax is longer than meso- and metathorax together; pronotum is subquadrate (Figs. 8.2, 8), generally with muscle-attachment marks (Figs. 8.8 C–E), strongly sclerotized; pleural area is semimembranous; basisternum is strongly sclerotized. Meso- and metathorax are similar in size and shape, with nota strongly sclerotized (Fig. 8.2 C), and sternal area semimembranous. Legs are similar in size and shape, with foreleg somewhat shorter than mid- and hindleg; tarsi are not divided into tarsomeres, and bear two tarsal claws (Figs. 8.3–4). Abdomen is soft: 1.5 times as long as head and thorax together, dorsoventrally flattened (Figs. 8.2–5); segments 1–7 (Sialidae) or 1–8 (Corydalidae), bearing a pair of long lateral filaments. In Sialidae they are 5-articulate, except by the first one, which is 4-articulate (Fig. 8.3). In Corydalidae they are 2-articulate, although subdivision is not so well defined (Figs. 8.4, 5). Terga and filaments are densely covered with decumbent microsetae and numerous or sparse erect macrosetae (Figs. 8.2 F), often with different morphologies, which are distinguishing characters for the taxonomy of the group (Figs. 8.9, 8.10), these setae are also extended to the pleural area or even to thoracic nota (Fig. 8.2 C). Unmodified spiracles are situated laterally on each side of abdominal terga 1–8 in Corydalidae and Sialidae (Fig. 8.2 I), and 1–7 in Chauliodinae (Fig. 8.1 A). Corydalinae has ventral tufts of tracheal gills near the lateral filaments base (Fig. 8.2 B). Abdominal segments 8 and 9 (Sialidae) or only 9 (Corydalidae) are unornamented (Figs. 8.3–5). In Sialidae, a long and tapered caudal filament is located on the last abdominal segment, while Corydalidae has a pair of prolegs equipped each with a filament and paired hamulate claws (Fig. 8.2 J). Chauliodinae presents a pair of spiracles on the abdominal segment 8th; in some genera, they are modified as long respiratory tubes (Fig. 8.7).

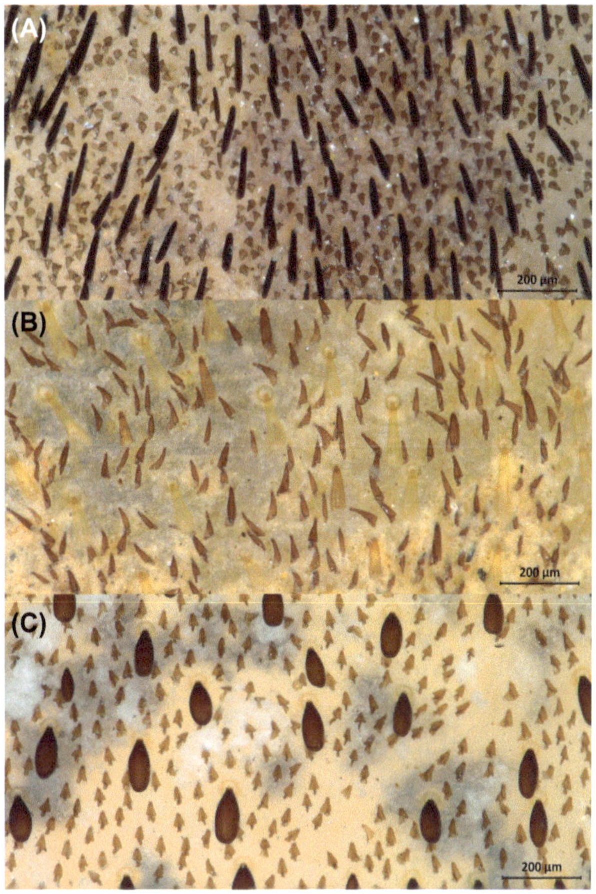

FIGURE 8.9 Macro- and microsetae on abdominal terga of Neotropical Corydalinae: (A) *Chloronia* sp.; (B) *Platyneuromus soror* (Hagen, 1861); and (C) *Corydalus* sp.

MATERIAL PREPARATION AND PRESERVATION

The larva can be fixed and preserved in different ways. In the field, the larvae can be orally injected using a tuberculin syringe filled with either acid alcohol (nine parts of 80% ethyl alcohol and one part of glacial acetic acid) or buffered 10% formalin or other formalin-based fixative (e.g., Kahle's fluid) (Contreras-Ramos & Harris, 1998; Contreras Ramos, 1999; Rasmussen & Pescador, 2002). In the laboratory, they may be anesthetized at −4°C, and then fixed with the same procedure. Overinjection must be avoided, or otherwise the normal morphology can be altered (Contreras Ramos, 1999). In this process, the larvae need to be carefully handled to keep the body stretched and straight, avoiding their contraction or coiling while the tissues are fixed by the killing agent. Larvae are maintained in the same fixative solution for at least 24 h, and then transferred to 70–80% ethyl alcohol for permanent preservation. Preservative fluid must be periodically replaced, especially after collection. Vials containing the specimens should be adequately labeled with collecting information and stored in cabinets to avoid the damaging effect of the light (Rasmussen & Pescador, 2002). This method may be followed for young and mature larvae. First, instar larvae may be killed in water just below boiling point, then they should be transferred to 70–80% ethyl alcohol. For slide preparation, first, instars should be treated over night with 10% potassium hydroxide (KOH) at room temperature (Contreras-Ramos & Harris, 1998), then rinsed with distilled water; they should be mounted in slides with a

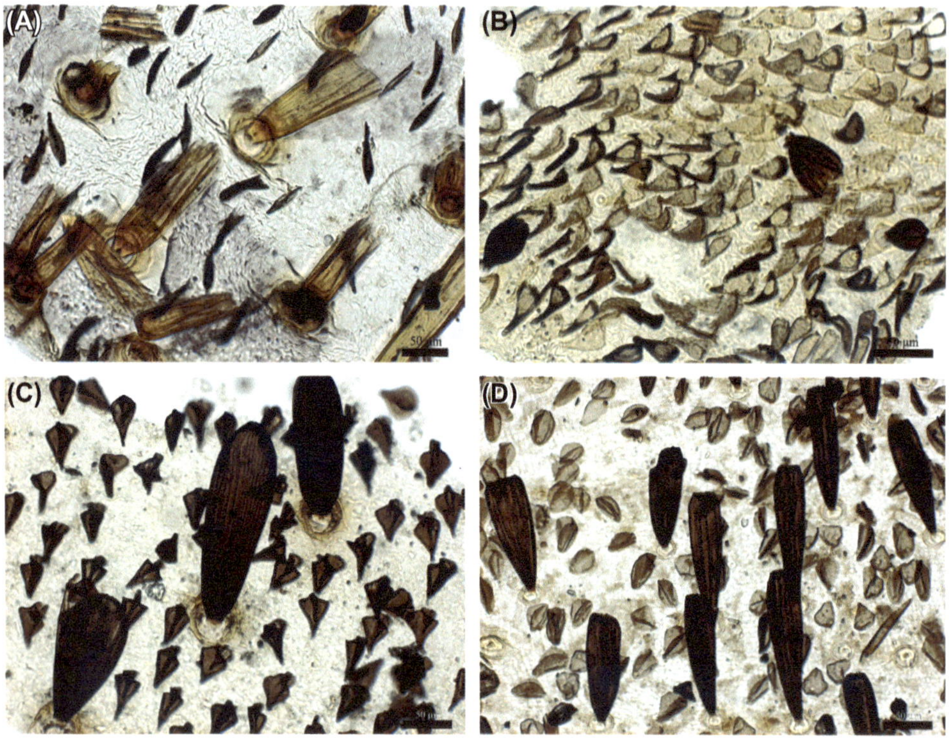

FIGURE 8.10 Macro- and microsetae on abdominal terga of Neotropical Corydalinae: (A) *Corydalus* sp.; (B) *Corydalus* sp.; (C) *Corydalus flavicornis* Stitz, 1914; and (D) *Chloronia* sp. Scale bars 50 μm.

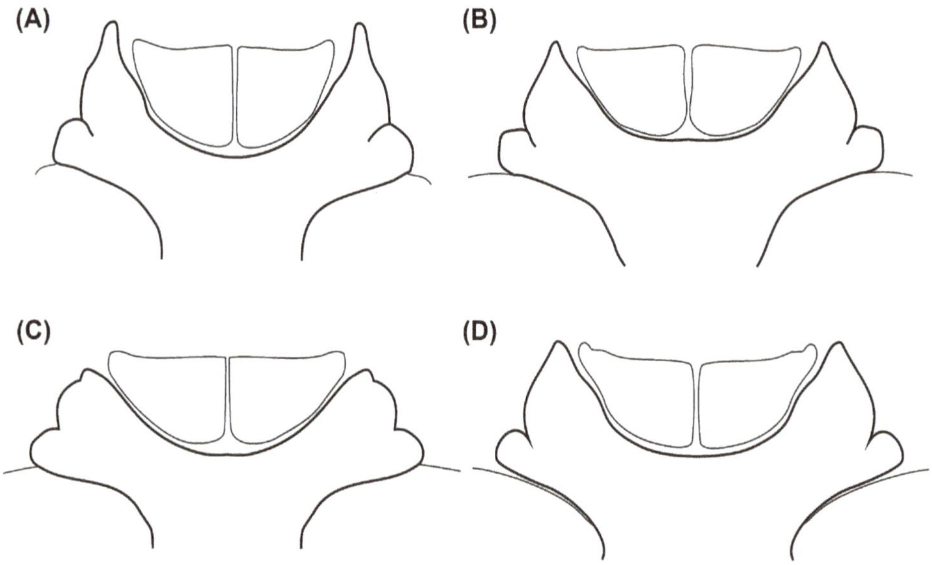

FIGURE 8.11 Distal part of submentum and mental plates of Neotropical Corydalinae: (A) *Corydalus ignotus* Contreras-Ramos; (B) *Platyneuromus soror* (Hagen, 1861); (C) *Corydalus armatus* Hagen, 1861; and (D) *Corydalus* sp.

water-soluble medium such as Hoyer's or Euparal, or they may be treated for Canada Balsam (e.g., Wirth & Marston, 1968). Setae on abdominal terga of mature larvae may be observed under compound microscope, either on temporary or permanent slides; a small portion of the cuticle should be removed with a thin scalpel or micro-dissecting scissors under a dissecting stereomicroscope. The portion of cuticle, if necessary, may be subjected to a 10% KOH treatment (Contreras-Ramos & Harris, 1998). First instar and cuticle portions may also be treated for Scanning Electron Microcopy (e.g., Contreras-Ramos & Harris, 1998).

KEYS TO LARVAL MEGALOPTERA

This key may be used to identify Neotropical genera, however *Nothochauliodes* Flint, 1983 from Chile and *Puri* Cardoso-Costa, Azevêdo & Ferreira-Jr, 2013 from Brazil were excluded, because their larval stages are unknown.

Megaloptera: Families

1 Abdomen with eight pairs of lateral filaments (Figs. 8.4, 5); segment 10 with a pair of anal prolegs (terminal filament absent; Figs. 8.2, 4–5) .... **Corydalidae [p. 225]**

1' Abdomen with seven pairs of lateral filaments (Fig. 8.3); segment 10 with long terminal filament (anal prolegs absent) .... **Sialidae [p. 226]**

Megaloptera: Corydalidae: Subfamilies

1 Abdomen with ventral gill tufts (Figs. 8.2 B, 4 B) .... **Corydalinae [p. 225]**

1' Abdomen without ventral gill tufts .... **Chauliodinae [p. 225]**

Megaloptera: Corydalidae: Corydalinae: Genera

Neotropical Megaloptera are mainly represented by this subfamily. The Amazon region, Andean mountain ranges and their valleys, and the Guiana shield are the areas of highest diversity (Contreras-Ramos, 2005; Cover & Resh, 2008).

1 Head and pronotum mottled, from light to dark brown with conspicuous light muscle scars (Figs. 8.8 C–E); macrosetae of abdominal terga with variable coloration and morphology 2

1' Head and pronotum from pale brown to dark reddish brown or dark brown, without muscle scars (Figs. 8.8 A, B), pronotum generally with four dark spots, sometimes obscure or fused, forming two lateral dark stripes (Figs. 8.8 A, B); macrosetae of abdominal terga dark, elongate and widened (Fig. 8.9 A), with deep longitudinal furrows (Fig. 8.10 D) .... *Chloronia*
[Mexico to Brazil, Bolivia and Peru, Antilles]

2(1) Abdominal terga macrosetae yellowish clear, either tubular with widened, striated and open apex, or slender, smooth, and with inconspicuous opening (Fig. 8.9 B); submental projections long reaching or passing anterior margin of mental plates (Fig. 8.11 B) .... *Platyneuromus*
[Mexico to Panama]

2' Abdominal terga macrosetae either dark, inflated and striated (Figs. 8.9 C, 10 B, C); or clear, elongate with dense or slight striations (Fig. 8.10 A) or completely smooth; Central American species with short submental projections not reaching or passing anterior margin of mental plates (Fig. 8.11 C); South American species either with short submental projections not reaching or passing anterior margins of mental plates or long, reaching or passing anterior limits of mental plates (Figs. 8.11 A, C, D); abdominal terga with yellowish clear macrosetae sometimes present (Fig. 8.10 A) .... *Corydalus*
[Neotropical. Nearctic]

Megaloptera: Corydalidae: Chauliodinae: Genera

Chauliodinae, or fishflies, is a geographically restricted group, with the highest species richness in the Oriental realm (Liu & Yang, 2006). They are also present in Australia and New Zealand, but the New World and South Africa are the regions with the highest generic diversity (Liu & Yang, 2006). *Puri* Cardoso-Costa, Azevêdo, and Ferreira-Jr was recently described from southeastern Brazil (Cardoso-Costa et al., 2013). Larval descriptions are unavailable, but keys, diagnosis and illustrations for *Archichauliodes* and *Protochauliodes* are provided by Flint (1973). Azevêdo & Hamada (2008, 2014) provided illustrations of an undetermined Chauliodine from São Paulo, Brazil, that may correspond to the larva of *Puri*.

1 Abdominal segment 8 spiracles at the end of a respiratory tube 2

1' Abdominal segment 8 spiracles sessile .... *Neohermes*
[Mexico. Nearctic]

2(1) Lateral margins of head rounded; muscle scars distinct and pale; abdominal spiracles large (Fig. 8.7 A) .... *Archichauliodes*
[Chile]

2' Lateral margins of head straight; muscle scars dark and concolorous; abdominal spiracles small (Fig. 8.7 B) .... *Protochauliodes*
[Chile]

Megaloptera: Sialidae: Genera

Sialidae is represented by two genera in the Neotropics, *Protosialis* van der Weele, 1909, with a single species from Cuba, and *Ilyobius* Enderlein, 1910 (Liu et al., 2015a, b), widely distributed across the Neotropics. Currently, *Ilyobius* comprises eight species distributed from Mexico to Argentina (Contreras-Ramos, 2005, 2008; Liu et al., 2015a, b; Archangelsky et al., 2017).

1 Mandibles with three preapical teeth (Fig. 8.6 B)........ *Protosialis*
[Cuba]

1' Mandibles with two preapical teeth (Fig. 8.6 A)........ *Ilyobius*
[Widespread]

ACKNOWLEDGMENTS

Sincere thanks to Neusa Hamada, Instituto Nacional de Pesquisas da Amazônia, Manaus, Brazil for the invitation to participate with this chapter and for providing high resolution images of larval Sialidae. Thanks to Matthew Cover, Department of Biological Sciences, California State University, California, USA for providing images of Chauliodinae larvae. The fisrt author gives his sincere acknowledgment to Departamento de Ciencias Biológicas, Universidad de los Andes, for the financial support, to Alexander García of the Universidad Distrital of Bogotá for his hospitality in the Kumangui Research Laboratory, and to Julián Arias Pineda, Angela Aristizábal, Diego Gómez, Daniela García and Juan Carlos Quijano of the Universidad de los Andes, for their support collecting dobsonfly larvae and for their help during field trips.

REFERENCES

Archangelsky, M., P. Pessacq & M. Berrondo. 2017. Description of the larva of *Ilyobius chilensis* (McLachlan) (Megaloptera: Sialidae) and notes on the adult morphology. Zootaxa 4318(1): 177–186.

Ardila-Camacho, A. 2014. A new species of Corydalus Latreille, 1802 (Megaloptera, Corydalidae) and first record of C. clavijoi Contreras-Ramos, 2002 and C. nubilus Erichson, 1848 from Colombia. Zootaxa 3811: 107–118.

Aspöck, U. & H. Aspöck. 2008. Phylogenetic relevance of the genital sclerites of Neuropterida (Insecta: Holometabola). Systematic Entomology 33: 97–127.

Aspöck, U., J. Plant & H.L. Nemeschkal. 2001. Cladistic analysis of Neuropterida and their systematic position whitin Neuropterida (Insecta: Holometabola: Neuropterida: Neuroptera). Systematic Entomology 26: 73–86.

Aspöck, U., E. Haring & H. Aspöck. 2012. The phylogeny of the Neuropterida: long lasting and current controversies and challenges (Insecta: Endopterygota). Arthropod Systematics & Phylogeny 70: 119–129.

Azevêdo, C.A.S. 2003. Taxonomia e bionomia de imaturos de Megaloptera (Insecta) na Amazônia Central, Brasil. Master's thesis. Programa de Pós-Graduação em Biologia Tropical e Recursos Naturais, Entomologia, Universidade Federal do Amazonas, Instituto Nacional de Pesquisas da Amazônia, Manaus, Amazonas, Brazil. 159 pp.

Azevêdo, C.A.S. 2009. Taxonomia, bionomia e estrutura da comunidade de larvas de Megaloptera (Insecta) em igarapés no estados do Amazonas e Roraima, Brasil. Ph. D. Thesis, Programa de Pós-Graduação em Biologia Tropical e Recursos Naturais, Entomologia, Universidade Federal do Amazonas, Instituto Nacional de Pesquisas da Amazônia, Manaus, Amazonas, Brazil. 115 pp.

Azevêdo, C.A.S. & N. Hamada. 2006. Description of last-instar larva of Corydalus nubilus Erichson, 1848 (Megaloptera: Corydalidae) and notes on its bionomics. Zootaxa 1177: 57–68.

Azevêdo, C.A.S. & N. Hamada. 2007. Description of the larvae of Corydalus batesii MacLachlan and C. ignotus Contreras-Ramos (Megaloptera: Corydalidae) with notes on life history and behavior. Zootaxa 1631: 33–45.

Azevêdo, C.A.S. & N. Hamada. 2008. Megaloptera. Pages 1–12 *in*: Guia on-line: C.G. Froehlich (org.), Identificação de larvas de Insetos Aquáticos do Estado de São Paulo. Available at: http://sites.ffclrp.usp.br/aguadoce/guiaonline.

Azevêdo, C.A.S. & N. Hamada. 2014. Ordem Megaloptera. Pages 335–342 *in*: N. Hamada, J.L. Nessiman & R.B. Querino (eds.), Insetos Aquáticos na Amazônia brasileira: taxonomia, biologia e ecologia. Editora INPA, Manaus, Brasil.

Baker, J.R. & H.H. Neunzig. 1968. The egg masses, eggs, and first-instar larvae of eastern North American Corydalidae. Annals of the Entomological Society of America 61: 1181–1187.

Bentes, S.P.C., N. Hamada & R.L. Ferreira-Keppler. 2014. Caracterização morfológica de ovos de insetos aquáticos e seus habitats na Amazônia central, Brasil. Pages 51–68 *in*: N. Hamada, J.L. Nessiman & R.B. Querino (eds.), Insetos Aquáticos na Amazônia brasilera: taxonomia, biologia e ecologia. Editora INPA, Manaus, Brasil.

Beutel, R.G. & F. Friedrich. 2008. Comparative study of larval head structures of Megaloptera (Hexapoda). European Journal of Entomology 105: 917–938.

Cardoso-Costa, G., C.A.S. Azevêdo & N. Ferreira, Jr. 2013. New genus and new species of Chauliodinae (Insecta: Megaloptera: Corydalidae) from Brazil. Zootaxa 3613: 391–399.

Clavier, S., A. Contreras-Ramos & L. Guillemet. 2009. First report of Protosialis Weele, 1909 in French Guiana (Megaloptera, Sialidae). Ephemera 11: 135–140.

Contreras-Ramos, A. 1998. Systematics of the dobsonfly genus Corydalus (Megaloptera, Corydalidae). Thomas Say Publications in Entomology: Monographs. Entomological Society of America, Lanham, Maryland. 360 pp.

Contreras-Ramos, A. 1999. Métodos para estudios en sistemática de Megaloptera (Insecta: Neuropterida) con base en morfología. Dugesiana 6: 1–15.

Contreras-Ramos, A. 2002. Six new species of dobsonflies from Venezuela (Megaloptera: Corydalidae: Corydalinae). Aquatic Insects 24: 55–75.

Contreras-Ramos, A. 2005. Recent accounts on the systematics and biogeography of Neotropical Megaloptera (Corydalidae, Sialidae). Pages 67–72 *in*: R.A. Pantaleoni, A. Letardi & C. Corazza, (eds.),

Proceedings of the Ninth International Symposium on Neuropterology (20–23 June 2005). Annali del Museo Civico di Storia Naturale di Ferrara, Ferrara, Italy.

Contreras-Ramos, A. 2008. Notes on some Neotropical alderflies (Sialidae: Megaloptera). Annals of the Entomological Society of America 101: 808–814.

Contreras-Ramos, A. 2009. Megaloptera. Pages 233–245 *in*: E. Domínguez & H.R. Fernández (eds.), Macroinvertebrados bentónicos sudamericanos –Sistemática y biología–Fundación Miguel Lillo, Tucumán, Argentina.

Contreras-Ramos, A. 2011. Phylogenetic review of dobsonflies of the subfamily Corydalinae and the genus Corydalus Latreille (Megaloptera: Corydalidae). Zootaxa 2862:1–38.

Contreras-Ramos, A. & S.C. Harris. 1998. The immature stages of Platyneuromus (Corydalidae), with a key to the genera of larval Megaloptera of Mexico. Journal of the North American Benthological Society 17: 489–517.

Costa, C., S. Ide & E.C. Simonka. 2006. Megaloptera. Pages 87–92 *in*: C. Costa & C.E. Simonka (eds.), Insectos Inmaduros: metamorfosis e identificación. 3rd Edition. M3M: Monografias Tercer Milenio 5: 87–92.

Cover, M.R. & V.H. Resh. 2008. Global diversity of dobsonflies, fishflies, and alderflies (Megaloptera; Insecta) and spongillaflies, nevrorthids, and osmylids (Neuroptera; Insecta) in freshwater. Hydrobiologia 595: 409–417.

Cover, M.R., J.H. Seo & V.H. Resh. 2015. Life history, burrowing behavior, and distribution of Neohermes filicornis (Megaloptera: Corydalidae), a long-lived aquatic insect in intermittent streams. Western North American Naturalist 75: 474–490.

Evans, E.D. & H.H. Neunzig. 1996. Megaloptera and aquatic Neuroptera. Pages 298–308 *in*: R.W. Merritt & K.W. Cummins (eds.), An introduction to the aquatic insects of North America. 3rd edition. Kendall/Hunt Pub. Co.: Dubuque. xiii + 862 pp.

Flint, O.S. Jr. 1973. The Megaloptera of Chile (Neuroptera). Revista Chilena de Entomología 7: 31–45.

Flint, O.S. Jr. 1992. A review of the genus Chloronia in Costa Rica, with the description of two new species (Neuropterida: Megaloptera: Corydalidae). Proceedings of the Biological Society of Washington 105: 801–809.

Glorioso, M.J. & O.S. Flint, Jr. 1984. A review of the genus Platyneuromus (Insecta: Neuroptera: Corydalidae). Proceedings of the Biological Society of Washington 97: 601–614.

Grimaldi, D. & M.S. Engel. 2005. Evolution of the Insects. Cambridge University Press, New York. 755 pp.

Haring, E. & U. Aspöck. 2004. Phylogeny of the Neuropterida: a first molecular approach. Systematic Entomology 29: 415–430.

Hayashi, F. & M. Nakane. 1989. Radio tracking and activity monitoring of the dobsonfly larva, Protohermes grandis (Megaloptera: Corydalidae). Oecologia 78: 468–472.

Jepson, J.E. & S.W Heads. 2016. Fossil Megaloptera (Insecta: Neuropterida) from the Lower Cretaceous Crato Formation of Brazil. Zootaxa 4098: 134–144.

Liu, X.-y. & D. Yang. 2006. Phylogeny of the subfamily of Chauliodinae (Megaloptera: Corydalidae), with description of a new genus from the Oriental Realm. Systematic Entomology 31: 652–670.

Liu, X.-y., Y.-j. Wang, C.-k. Shih, D. Ren & D. Yang. 2012. Early evolution and historical biogeography of fishflies (Megaloptera: Chauliodinae): implications from a phylogeny combining fossil and extant taxa. PLoS ONE 7: 1–12.

Liu, X.-y., F. Hayashi & D. Yang. 2015a. Phylogeny of the family Sialidae (Insecta: Megaloptera) inferred from morphological data, with implications for generic classification and historical biogeography. Cladistics 31: 18–49.

Liu, X.-y., F. Hayashi & D. Yang. 2015b. Taxonomic notes of the Neotropical alderfly genus Ilyobius Enderlein, 1910 (Megaloptera, Sialidae), with description of a new species. Deutsche Entomologische Zeitschrift, Berlin 62: 55–63.

Liu, X.-y., Y. LÜ, H. Aspöck, D. Yang & U. Aspöck. 2015c. Homology of the genital sclerites of Megaloptera (Insecta: Neuropterida) and their phylogenetic relevance. Systematic Entomology 41: 256–286.

Mayorga, A., R. Barba-Álvarez, & A. Contreras-Ramos. 2017. Notes on sex ratio and oviposition of Platyneuromus honduranus (Navás, 1928) in Chiapas, Mexico (Megaloptera, Coryalidae). American Midland Naturalist 178: 284–289.

Misof, B., O. Niehuis, I. Bischoff, A. Rickert, D. Erpenbeck & A. Staniczek. 2007. Towards an 18S phylogeny of hexapods: Accounting for group-specific character covariance in optimized mixed nucleotide/doublet models. Zoology 110: 409–429.

New, T.R. & G. Theischinger. 1993. Megaloptera, Alderflies and Dobsonflies. Handbuch der Zoologie, Vol. 4 (Arthropoda: Insecta), Part 33. Walter de Gruyter, Berlin. 97 pp.

Penny, N.D. & O.S. Jr. Flint. 1982. A revision of the genus Chloronia (Neuroptera: Corydalidae). Smithsonian Contributions to Zoology 348: 1–27.

Rasmussen, A.K. & M.L. Pescador. 2002. A guide to the Megaloptera and aquatic Neuroptera of Florida. Department of Environmental Protection, Division of Water Resource Management, Tallahassee, Florida. 45 pp.

Wang, Y.-y., X.-y. Liu, S.L. Winterton & D. Yang 2012. The first mitochondrial genome for the fishfly subfamily Chauliodinae and implications for the higher phylogeny of Megaloptera. PLoS One 7: 1–13.

Wirth, W.W. & N. Marston. 1968. A method for mounting small insects on microscope slides in Canada Balsam. Annals of the Entomological Society of America 61: 783–784.

Zhao, C.-j., X.-y. Liu & D. Yang. 2014. Wing base structural data support the sister relationship of Megaloptera and Neuroptera (Insecta: Neuropterida). PLoS One 9: 1–23.

Chapter 9

Order Neuroptera

Caleb C. Martins
Laboratório de Morfologia e Evolução de Diptera, Departamento de Biologia, FFCLRP, Universidade de São Paulo, Ribeirão Preto, São Paulo, Brazil

Adrian Ardila-Camacho
Laboratorio de Entomología, Centro de Investigaciones en Microbiología y Parasitología Tropical – CIMPAT, Universidad de los Andes, Bogotá, Colombia

Chapter Outline

INTRODUCTION

The order Neuroptera has approximately 4,500 described species grouped into 17 families distributed among the suborders Nevrorthiformia, Myrmeleontiformia, and Hemerobiiformia. Nevrorthiformia includes only Nevrorthidae and is considered a sister clade to the other two suborders. Myrmeleontiformia has five families generally distributed in two clades: (1) Nemopteridae + Psychopsidae; and (2) Nymphidae + (Myrmeleontidae + Ascalaphidae). The other 11 families belong to the suborder Hemerobiiformia, the most heterogeneous group whose monophyly is not yet sufficiently established (Haring & Aspöck, 2004; Aspöck & Aspöck, 2008; Winterton et al., 2010; Zimmermann et al., 2011).

Neuropterans have highly conserved morphologies among the holometabolous orders and include insects with diaphanous wings streaky with reticulate venation. Adults range from 2 to 150 mm in length, have compound eyes, antennae with multiple flagellomeres, mouth parts with developed jaws, and four membranous wings with numerous cross veins (mainly along the costal area). In most cases, the radial sector has parallel branches. Among the Neuroptera families, both fore- and hindwings have similar shape, size, and venation except Nemopteridae (their hind wings are long and thin) and some Hemerobiidae (hindwings smaller than forewings) (Triplehorn & Johnson, 2005; Freitasde & Penny, 2012).

Neuropteran larvae are generally predatory, although in some Ithonidae and Polystoechotidae, the hypothesis of a phytosuccivorous feeding mode has been well-supported (Gallard, 1932; Aspöck et al., 2001; Grebennikov, 2004). They can be campodeiform, fusiform or scarabaeiform and up to 30 mm in length when fully developed. The head is typically prognathic with up to seven stemmata at each side, and the antennae are most often filiform. The mandible and maxilla are furrowed and juxtaposed forming a tube through which the prey body fluid is ingested. The labial palpus may have two to five articles, and maxillary palpi are absent. The prothorax is large with two dorsal sclerites, and the meso- and metathorax bear small dorsal sclerites. Legs generally are well developed except in stages two and three of Mantispidae and Berothidae. The first tarsus is articulated. The abdomen is 10-segmented, without cerci, and the terminal segment has an adhesive organ used for substrate attachment. Internally, the mid- and hindgut are closed. Larvae have eight Malpighian tubules with six adhered to the hindgut and modified as silk organs in the last larval instar (New, 1989; Tauber, 1991; Costa et al., 2006; Freitasde & Penny, 2012).

Thorp and Covich's Freshwater Invertebrates, Fourth Edition. https://doi.org/10.1016/B978-0-12-804223-6.00009-3

The Sisyridae and Nevrorthidae have larvae that develop in the aquatic environment, and the Osmylidae has some species associated with aquatic habitats. The Sisyridae has a worldwide distribution, while Nevrorthidae is restricted to the Mediterranean, Japan, China, Taiwan, and Australia. Osmylidae has a nearly global distribution, with the noteworthy exception of North and Central America (Aspöck & Aspöck, 2008; Cover & Resh, 2008; Shepard & Contreras-Ramos, 2009; Yang et al., 2010; Ardila-Camacho & Noriega, 2014). We report here only data on Osmylidae and Sisyridae, because Nevrorthidae does not occur in the Neotropics.

Sisyridae

Sisyridae, commonly known as spongillaflies, is a small family with worldwide distribution and approximately 60 described species, 30 of which occur in the Neotropics. Two genera, *Sisyra* Burmeister, 1839 and *Climacia* McLachlan, 1869 (Cover & Resh, 2008), occur in the Neotropical region. Their eggs are generally deposited in groups of 2–5, but there are reports of groups of 22. The eggs are elongate with a short micropilar region, and are generally whitish in color and covered with thin silk web produced by the female; however, they can appear darker when the embryo is developed (Brown, 1952; Eliott, 1996; New, 1991; Hamada et al., 2014a, 2014b).

The larvae have three instars, and only the second and third (Figs. 9.1–2) have abdominal ventral gills. After hatching, the larvae drop to the water and swim by flexing the body until they find a freshwater sponge (Porifera: Spongillidae). Larvae feed throughout this stage using their long, thin, styliform mouthparts. Larvae also have been shown to feed on Bryozoa (Weissmair & Waringer, 1994). The last larval instar swims actively at night to the shoreline, where they emerge and seek cryptic sites away from the water to construct their pupal cocoons (Pupedis, 1980; New, 1991). The exarate pupae do not have mobile appendages; antennae are inserted dorsal to the eye and fixed along the anterior wing margin. The mandibles are sclerotized (Brown, 1952; Hamada et al., 2014b).

FIGURE 9.1 Habitus of *Sisyra* Burmeister, 1839 (Sisyridae) larva: (A) dorsal view; and (B) lateral view. Bar = 0.5 mm.

FIGURE 9.2 Habitus of *Climacia* McLachlan, 1869 (Sisyridae) larva: (A) dorsal view; and (B) lateral view. Bar = 0.5 mm.

Adults are up to 8 mm in length, similar to Hemerobiidae but differ in having only one branch of Rs from R in the forewings compared to two or more branches of Rs from R in Hemerobiidae. The head is hypognathous with filiform antennae, large compound eyes, vertex convex, and the ocelli are absent. Wings are oval with many primary veins forked at the wing margins, but the transverse costal veins are not forked. The wings also bear many macrotrichia, mainly along the alar veins. The wings of *Sisyra* are generally brown, while those of *Climacia* usually have a light yellow base color with often prominent maculations (Brown, 1952; New, 1991; Freitasde & Penny, 2012). The abdomen is 10-segmented and weakly sclerotized. The female ovipositor is tubular in shape and extends dorsally while at rest. The tenth sternite of the male is modified into a pair of claspers.

Osmylidae

Currently there are about 200 described species distributed among eight subfamilies (Eidoporisminae, Gumillinae, Kempyninae, Osmylinae, Porisminae, Protosmylinae, Spilosmylinae, and Stenosmylinae). In South America, 19 species are distributed among four subfamilies and five genera: Gumillinae (*Gumilla* Navás, 1912), Stenosmylinae (*Isostenosmylus* Krüger, 1913 and *Phymatosmylus* Adams, 1969), Kempyninae (*Kempynus* Navás, 1912), and Protosmylinae (*Paryphosmylus* Krüger, 1913) (Aspöck & Aspöck, 2008; Shepard & Contreras-Ramos, 2009; Yang et al., 2010; Ardila-Camacho & Noriega, 2014; Martins et al., 2016).

FIGURE 9.3 Habitus of Osmylidae (Neuroptera) larva: (A) dorsal view; and (B) lateral view. Bar = 1 mm.

Osmylid eggs are deposited in groups of up to 12 on vegetation, tree trunks, or stones (New, 1991; Beutel et al., 2010). Immature stages are long and fusiform with long tapering straight jaws (Fig. 9.3). The larvae of some genera of the subfamilies Osmylinae and Kempyninae are strongly associated with freshwater habitats and are considered to be amphibious. Although, osmylids are not effective swimmers, they often have been observed to crawl under water and have been collected in benthic samples in streams and springs (Cover & Bogan, 2014). The body surface has a hydrophobic cuticle that forms a plastron when immersed in the water (Wichard et al., 2002). Water-dependent species live along margins of creeks and rivers and only go to the water to feed on soft aquatic insect prey, especially chironomid larvae (Tjeder, 1957; Shepard & Contreras-Ramos, 2009).

Adults are medium to large insects with forewings of up to 30 mm. Their heads are hypognathous, usually bearing three ocelli, and have filiform antennae that are generally less than half of forewing length. The prothorax is longer than wide and possesses odor glands. The wings are broad with the apex ovate or falcate and contain intricate venation and trichosors along the margins. The abdomen has the ectoproct dorsally fused with *callus cerci* arranged in a rosette. The male ninth tergite is unmodified and the ninth gonocoxites are external. The complex of tenth gonocoxites resembles paired claspers or an unpaired sclerite. Females have two spermathecae, and the ninth gonocoxite is elongated and equipped with terminal gonostyli (Tjeder, 1957; Aspöck & Aspöck, 2008; Ardila-Camacho & Noriega, 2014; Martins et al., 2016).

LIMITATIONS

Larvae of most species and several genera of Osmylidae remain unknown. Similarly, although the larvae for *Climacia* and *Sisyra* are known from the Neotropics (Pupedis 1980; Hamada et al. 2014b), larvae for most species have not been associated in this geographic realm. The lack of knowledge about larvae for these two families precludes preparation of comprehensive species keys. Additionally, while the biology of the Sisyridae is well known, that of the Osmylidae is not.

The Australian and Mediterranean osmylid faunas are better known compared to the Neotropical fauna, largely due to the comprehensive works of New (1983, 1986), Cambulat (2013) and Monserrat (2014). Several studies have addressed the adult Osmylidae of the Neotropics (Navás, 1912, 1928; Krüger, 1913; Kimmins, 1940; Adams, 1969, 1971; Oswald, 1994; Ardila-Camacho & Noriega 2014; Martins et al., 2016), but the larvae of different genera or species were largely not addressed in these studies.

This lack of knowledge about the species of Sisyridae and Osmylidae in the Neotropics is directly related to the current low number of researchers, mainly taxonomists, who study these families.

TERMINOLOGY AND MORPHOLOGY

Sisyridae larvae have three instars, but only the second and third instars have abdominal gills. The third instar larva is fusiform, up to 4 mm long, and its coloration may be white, brown, or green depending on their food (Figs. 9.1–2). The last instar larva has a small, oval head with six lateral stemmata (Figs. 9.1–2), filiform, and multi-articulated antennae, and elongated mouthpieces with stylets shape held together and forming a tube; maxillary and labial palps are absent (Figs. 9.4 A, B). The thoracic segments are wider than long and each of them has a pair of developed legs, with a one-segmented tarsus and an apical thin claw. The abdomen is soft, with closed spiracles and tracheal gills present on the venter of abdominal segments 1–7 (Fig. 9.4 C). Silk produced in the Malpighian tubules is secreted through a spinneret, located on the last abdominal segment, and is used for the construction of the cocoon (Pupedis, 1980). Tubercles with bristles are found on the sides and dorsum of the thorax and abdomen. The locations of bristles and tubercle structures are important for the identification of sisyrid genera: *Climacia* has the tergite of the sixth abdominal segment with a median tubercle that is larger than the two lateral tubercles, while *Sisyra* has a lateral tubercle that is larger than the other two. Additionally, *Climacia* has a bristle in the median tubercle that is longer and thinner compared to the bristles present on the other two lateral tubercles, while *Sisyra* has a longer and thinner bristle located in one of the lateral tubercles.

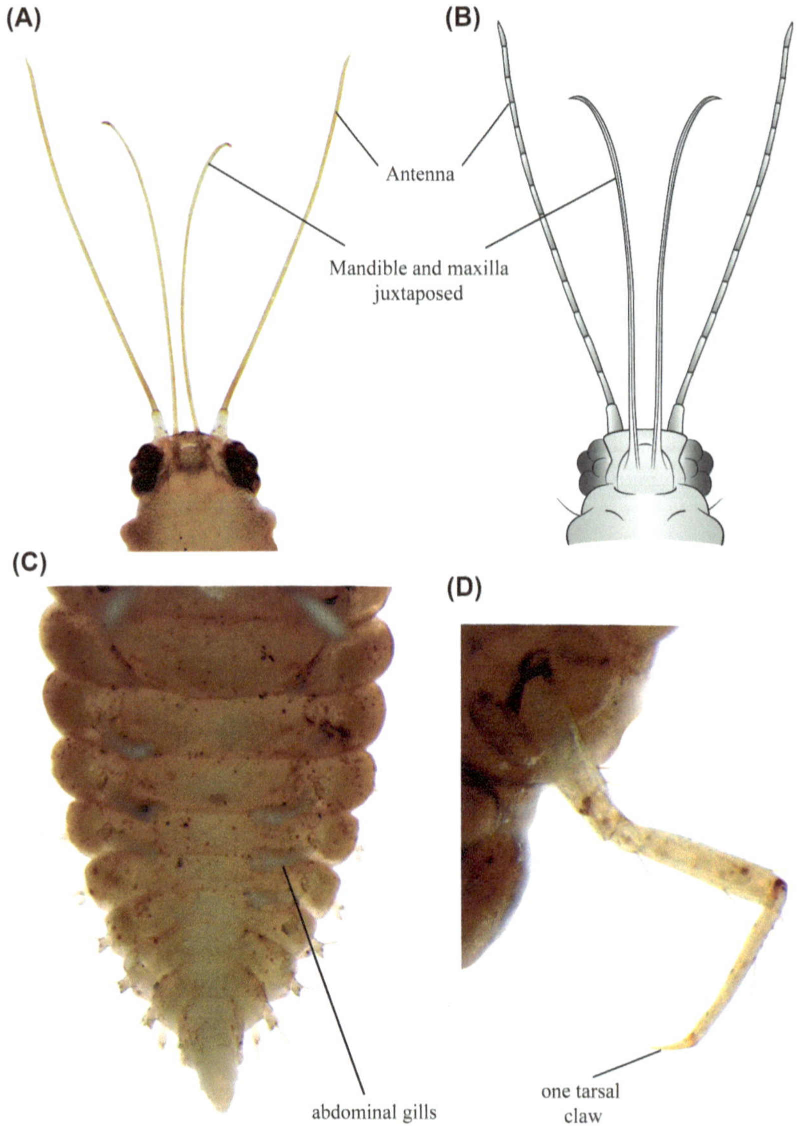

FIGURE 9.4 Sisyridae (Neuroptera) larva: (A, B.) head, ventral view; and (C). abdomen, ventral view. (D) Legs, lateral view.

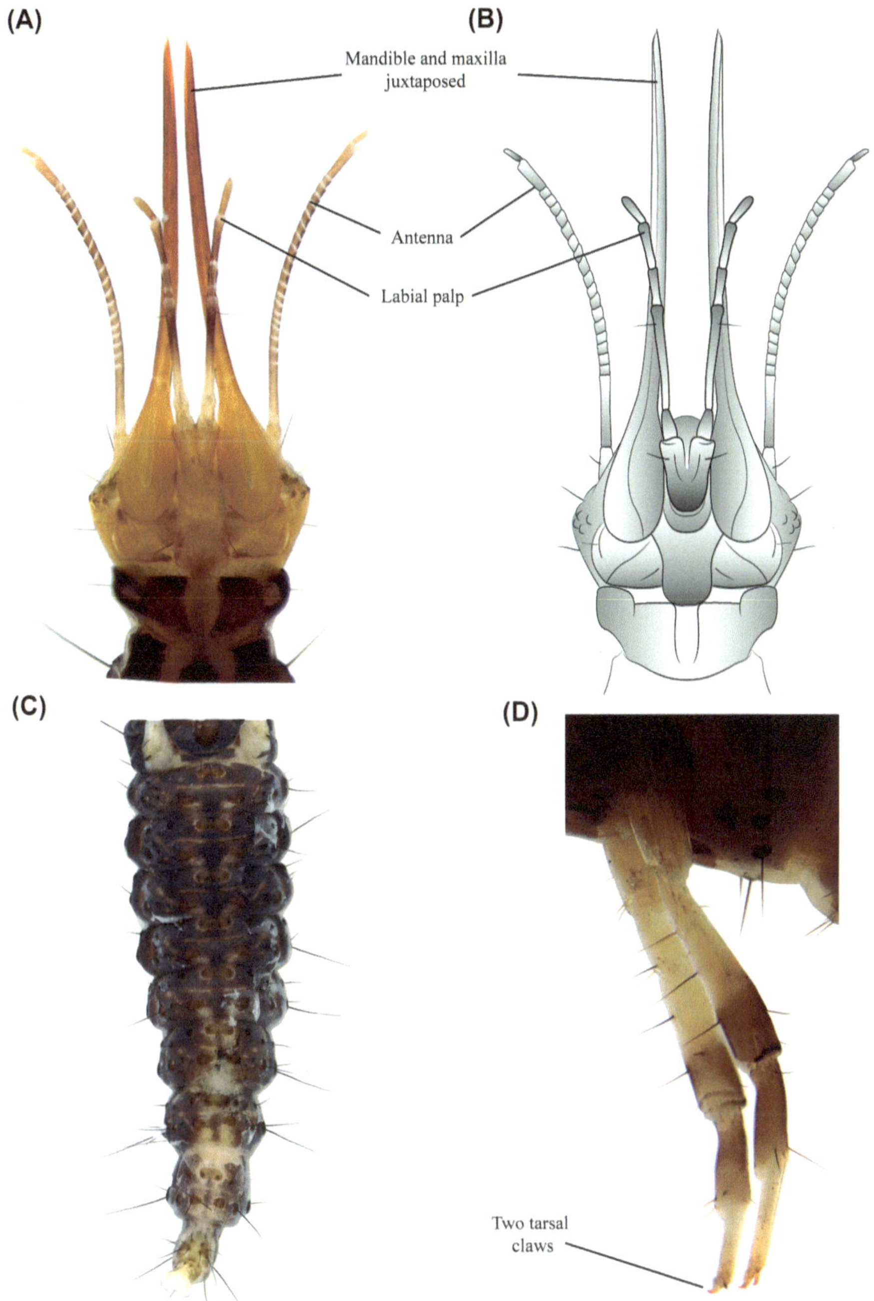

FIGURE 9.5 Osmylidae (Neuroptera) larva: (A, B) head, ventral view; (C) abdomen, ventral view; and (D) legs, lateral view.

Also, *Climacia* has a pair of bristles on the dorsal regions of eighth abdominal tergite, while the dorsal regions of eighth abdominal tergite of *Sisyra* has no bristles (Pupedis, 1980; Hamada et al., 2014b).

Osmylid larvae have three instars also; but unlike Sisyridae, they do not have abdominal gills. The third instar larvae are campodeiform, brown or light yellow in color, and up to 7 mm in length. The head is small and oval with six lateral stemmata, long filiform antennae (Fig. 9.3). The mouthparts are very elongate reaching twice the head length and are held together forming a tube. Maxillary palpi are absent, and the labial palpi are as long as the antenna (Figs. 9.4 A, B). The larval prothorax is as long as wide, and the meso- and metathorax are wider than long. The thoracic tergites are more heavily sclerotized than the abdominal tergites, and each segment has a pair of well-developed legs bearing a single article tarsus with two apical curved claws. The abdomen is soft, without gills but

FIGURE 9.6 *Climacia* McLachlan, 1869 (Sisyridae): larval abdomen, dorsal view.

FIGURE 9.7 *Sisyra* Burmeister, 1839 (Sisyridae): larval abdomen, dorsal view.

with functional spiracles, and the tenth abdominal segment has two reversible prolegs that aid locomotion. The entire surface of the sclerites is covered with long, thick, and straight setae (New, 1991; Cover & Resh, 2008). The larvae of many Neotropical species are unknown, and thus it is not possible to identify them to genus level.

MATERIAL PREPARATION AND PRESERVATION

Whenever possible, specimens should be examined alive and then studied after preservation to appreciate the consequence of fixation on body shape. Specimens should be handled with care because they have soft, fragile bodies, and bristles can be broken during handling. Specimens are best examined using a stereomicroscope, and they should be placed in a petri plate and covered with 70% ethanol. Larvae should be preserved in 70% ethanol containing glycerin. Alternatively, some specimens can be fixed using aldehydes (glutaraldehyde 2% or formaldehyde 10% in buffer solution) and then preserved in 70% ethanol for morphological studies. Specimens preserved in 96% ethanol are also useful in molecular approaches.

KEY TO AQUATIC LARVAE OF NEUROPTERA

Neuroptera: Families

1 Labial palpi absent (Figs. 9.4 A, B); abdominal gills present (Fig. 9.4 C); one tarsal claw present (Fig. 9.4 D) .... **Sisyridae**

1' Labial palpi present (Figs. 9.5 A, B); abdominal gills absent (Fig. 9.5 C); two tarsal claws present (Fig. 9.5 D) .... Osmylidae

Sisyridae: Genera

Pupedis (1980), Hamada et al. (2014b) and Heckman (2017) provided important keys to larvae of Neotropical Sisyridae genera.

1 Abdominal segment VI tergite with median tubercle larger than the two lateral tubercles; median tubercle bristle longer and thinner than lateral tubercles bristles; abdominal segment VIII tergite dorsally with a pair of bristles (Fig. 9.6) .... *Climacia*

1' Abdominal segment VI tergite with lateral tubercle larger than median tubercle; lateral tubercle bristle longer and thinner than lateral tubercle bristles; abdominal segment VIII tergite dorsally without a pair of bristles (Fig. 9.7) .... *Sisyra*

ACKNOWLEDGMENTS

We are grateful to Dr. David E. Bowles of Missouri State University, Missouri, USA, James H. Thorp, and Rogers D. Christopher for improving the English version and for his valuable comments and critical review of the chapter.

REFERENCES

Adams, P.A. 1969. A new genus and species of Osmylidae (Neuroptera) from Chile and Argentina, with a discussion of Planipennian genitalic homologies. Postilla 141: 1–11.

Adams, P.A. 1971. Variation and geographic distribution in some Argentine and Chilean Osmylidae, with a new species of Kempynus (Neuroptera). Bulletin of the Southern California Academy of Sciences 70: 45–49.

Ardila-Camacho, A. & J.A. Noriega. 2014. First record of Osmylidae (Neuroptera) from Colombia and description of two new species of Isostenosmylus Krüger, 1913. Zootaxa 3826: 315–328.

Aspöck, U. & H. Aspöck. 2008. Phylogenetic relevance of the genital sclerites of Neuropterida (Insecta: Holometabola). Systematic Entomology 33: 97–127.

Aspöck, U., J.D. Plant & H.L. Nemeschkal. 2001. Cladistic analysis of Neuroptera and their systematic position within the Neuropterida (Insecta: Holometabola: Neuropterida: Neuroptera). Systematic Entomology 26: 73–86.

Beutel, R.G., D. Zimmermann, M. Krauss, S. Randolf & B. Wipfler. 2010. Head morphology of Osmylus fulvicephalus (Osmylidae, Neuroptera) and its phylogenetic implications. Organisms Diversity and Evolution 10: 311–329.

Brown, H.P. 1952. The life history of Climacia areolaris (Hagen), a neuropterous 'parasite' of freshwater sponges. American Midland Naturalist 47: 130–160.

Cambulat, S. 2013. Redescription of Osmylus multiguttatus McLachlan, 1870 (Neuroptera: Osmylidae) with distributional remarks. Zootaxa 3741: 385–390.

Costa, C., S. Ide & C.E. Simonka. 2006. Neuroptera. Pages 93–98 *in*: C. Costa (ed.), Insectos Inmaduros: metamorfosis e identificación. M3M: Monografias Tercer Milenio, Sociedad Entomológica Aragonesa, Zaragoza, España.

Cover, M.R. & M.T. Bogan. 2014. Minor Insect Orders. Pages 1059–1072 *in*: J.H. Thorp & D.C. Rogers (eds.), Thorp and Covich's Freshwater Invertebrates, Fourth Edition Volume I: Ecology and General Biology. Elsevier, San Diego, United States.

Cover, M.R. & V.H. Resh. 2008. Global diversity of dobsonflies, fishflies, and alderflies (Megaloptera; Insecta) and spongillaflies, nevrorthids, and osmylids (Neuroptera; Insecta) in freshwater. Hydrobiologia 595: 409–417.

Elliott, J.M. 1996. British freshwater Megaloptera and Neuroptera: a key with ecological notes. Freshwater Biological Association Scientific Publication 54: 1–69.

de Freitas, S.. & N.D. Penny. 2012. Neuroptera. Pages 537–546 *in*: J.A. Rafael, G.A.R. Melo, C.J.B. Carvalho, S.A. Casari & R. Constantino (eds.), Insetos do Brasil: diversidade e taxonomia. Holos Editora, Ribeirão Preto, Brazil.

Gallard, L. 1932. Notes on the feeding habits of the brown moth-lacewing, Ithone fusca. Australian Naturalist 8: 168–170.

Grebennikov, V.V. 2004. Grub-like larvae of Neuroptera (Insecta): a morphological review of the families Ithonidae and Polystoechotidae and a description of Oliarces clara. European Journal of Entomology 101: 409–417.

Hamada, N., A.M.O. Pes & L.M. Fusari. 2014a. First records f Sisyridae (Neuroptera) in Rio de Janeiro State, Brazil, with bionomics notes on Sisyra panama. Florida Entomogical Society 97: 281–284.

Hamada, N., A.M.O. Pes & R. Boldrini. 2014b. Ordem Neuroptera, Família Sisyridae. Pages 343–348 *in*: N. Hamada, J.L. Nessimian & R.B. Querino (eds.), Insetos Aquáticos na Amazônia brasileira: taxonomia, biologia e ecologia. Editora do Instituto Nacional de Pesquisas da Amazônia, Manaus, Brazil.

Haring, E. & U. Aspöck. 2004. Phylogeny of the Neuropterida: a first molecular approach. Systematic Entomology 29: 415–430.

Heckman, C.W. 2017. Sisyridae. Pages 429–445. *in*: C.W. Heckman (ed.) Encyclopedia of South American Aquatic Insects: Neuroptera (including Megaloptera). Springer, Netherlands.

Kimmins, D.E. 1940. A revision of the osmylid subfamilies Stenosmylinae and Kalosmylinae (Neuroptera). Novitates Zoologicae 42: 165–201.

Krüger, L. 1913. Osmylidae. Beiträgezueiner Monographie der Neuropteren-Familie der Osmyliden. II. Charakteristik der Familie, Unterfamilien und Gattungen auf Grund des Geäders. Stettiner Entomologische Zeitung 74: 3–123.

Martins, C.C., Ardila-Camacho, A. & Aspöck, U. 2016. Neotropical osmylids (Neuroptera, Osmylidae): Three new species of Isostenosmylus Krüger, 1913, new distributional records, redescriptions, checklist and key for the Neotropical species. Zootaxa 4149: 1–66.

Monserrat, V.J. 2014. Los osmílidos de la Península Ibérica (Insecta: Neuropterida: Neuroptera: Osmylidae). Heteropterus Revista de Entomología 14: 55–72.

Navás, L. 1912. Insectos neurópteros nuevos o poco conocidos. Memorias de la Real Academia de Ciencias y Artes de Barcelona 10: 135–202.

New, T.R. 1983. A revision of the Australian Osmylidae: Kempyninae (Insecta: Neuroptera). Australian Journal of Zoology 31: 393–420.

New, T.R. 1986. A new Australian genus of Stenosmylinae (Neuroptera: Osmylidae). Systematic Entomology 11: 447–452.

New, T.R. 1989. Planipennia, Lacewings. Handbuch der Zoologie, Vol. 4 (Arthopoda: Insecta), Part 30. Walter de Gruyter, Berlin.

New, T.R. 1991. Neuroptera (lacewings). Pages 525–542 *in*: I.D. Naumann (ed.), The Insects of Australia. Vol. 1. 2nd Edition. Melbourne University Press, Melbourne, Australia.

Oswald, J.D. 1994. Two new South American species of the genus Kempynus Navás (Neuroptera: Osmylidae: Kempyninae). Proceedings of the Entomological Society of Washington 96: 367–372.

Pupedis, R.J. 1980. Generic differences among New World Spongilla-fly larvae and a description of the female of Climacia striata (Neuroptera: Sisyridae). Psyche 87: 305–314.

Shepard, W.D. & A. Contreras-Ramos. 2009. Neuroptera y Mecoptera. Pages 247–254 *in*: E. Domínguez & H.R. Fernández (eds.), Macroinvertebrados bentónicos sudamericanos. Sistemática y Biología. Fundación Miguel Lillo, Tucumán, Argentina.

Tauber, C.A. 1991. Order Neuroptera. Pages 126–143 *in*: F.W. Stehr (ed.), Immature insects. Vol. 2. Kendall/Hunt Publishing Co., Dubuque, Iowa.

Tjeder, B. 1957. Neuroptera-Planipennia. The Lace-wings of Southern Africa. 1. Introduction and families Coniopterygidae, Sisyridae, and Osmylidae. Pages 95–188 *in*: B. Hanström, P. Brinck & G. Rudebec (eds.). South African Animal Life. Vol. 4. Swedish Natural Science Research Council, Stockholm, Sweden.

Triplehorn, C.A. & N.F. Johnson. 2005. Chapter 27 Order Neuroptera: Alderflies, dobsonflies, fishflies, snakeflies, lacewings, antlions, and owlflies. Pages 469–480 *in*: C.A. Triplehorn & N. F. Johnson (eds.), Borror and DeLong's introduction to the study of insects. 7th ed. Thomson Brooks/Cole, Belmont, California.

Weissmair, W. & J. Waringer. 1994. Identification of the larvae and pupae of Sisyra fuscata (Fabricius, 1793) and Sisyra terminalis Curtis, 1854 (Insecta: Planipennia: Sisyridae), based on Austrian material. Aquatic Insects 16: 147–155.

Wichard, W., W. Arens & G. Eisenbeis. 2002. Biological Atlas of Aquatic Insects. Apollo Books, Stenstrup, Denmark.

Winterton, S.L., N.B. Hardy & B.M. Wiegmann. 2010. On wings of lace: phylogeny and Bayesian divergence time estimates of Neuropterida (Insecta) based on morphological and molecular data. Systematic Entomology 35: 349–378.

Yang, Q., V.N. Makarkin & D. Ren. 2010. Remarkable new genus of Gumillinae (Neuroptera: Osmylidae) from the Jurassic of China. Annals of the Entomological Society of America 103: 855–859.

Zimmermann, D., S. Randolf, B.D. Metscher & U. Aspöck. 2011. The function and phylogenetic implications of the tentorium in adult Neuroptera (Insecta). Arthropod Structure & Development 40: 571–582.

Chapter 10

Order Trichoptera

Ana Maria Pes
Coordenação de Biodiversidade, Instituto Nacional de Pesquisas da Amazônia, Manaus, Amazonas, Brazil

Ralph W. Holzenthal
Department of Entomology, University of Minnesota, St. Paul, MN, United States

Julieta V. Sganga
Departamento de Biodiversidad y Biología Experimental, Facultad de Ciencias Exactas y Naturales, Universidad de Buenos Aires, Buenos Aires, Buenos Aires, Argentina

Allan P.M. Santos
Departamento de Zoologia, Universidade Federal do Estado do Rio de Janeiro – UNIRIO, Rio de Janeiro, Rio de Janeiro, Brazil

Patrik Barcelos-Silva
Divisão do Curso de Pós-graduação em Entomologia, Instituto Nacional de Pesquisas da Amazônia, Manaus, Amazonas, Brazil

Lucas M. Camargos
Department of Entomology, University of Minnesota, St. Paul, MN, United States

Chapter Outline

INTRODUCTION

Caddisflies are among the most successful and diverse organisms in freshwater habitats, with about 15,000 described species around the world, divided into 49 extant families and 616 genera (Holzenthal et al., 2007, 2015). As in other insect orders, Trichoptera taxonomy is mainly based on adult features, but they are better known for the building behavior of their larvae. Trichoptera and the megadiverse Lepidoptera—moths and butterflies—are sister groups and constitute the superorder Amphiesmenoptera, characterized by many morphologic apomorphies (Kristensen, 1997), including the larvae having a modified salivary gland opening on the labium that produces silk. Additionally, molecular data consistently corroborate the monophyly of the superorder (Kjer et al., 2006; Misof et al., 2014). In Lepidoptera caterpillars, the silk is used mainly to spin cocoons; in caddisfly larvae the silk is used to build portable cases, commonly composed of mineral grains or leaves and twigs, or retreats with silken nets for food capture. Trichoptera diversity and success, in some part, are the results of the various ways in which silk is used among the different

Thorp and Covich's Freshwater Invertebrates, Fourth Edition. https://doi.org/10.1016/B978-0-12-804223-6.00010-X

groups, allowing larvae to exploit different kinds of available food resources and microhabitats (Mackay & Wiggins, 1979).

Immature stages of Trichoptera, from egg to pupa, are aquatic, and only the adults are terrestrial. Caddisfly larvae are especially abundant and diverse in running waters (lotic habitats), but they can be also found in standing water (lentic habitats), especially in temperate latitudes. At least one specie (*Phylloicus bromeliarum* Müller, 1880) is recorded living in water retained in bromeliad tanks. Trichoptera can occupy virtually every available substrate in rivers and streams, where they usually play a key role in energy flow (Resh & Rosenberg, 1984). In the Neotropics, some larvae are specialized to live in semiaquatic areas, such as rock surfaces above the water line, as seen in Xiphocentronidae and some Leptoceridae. Others are commonly found in madicolous habitats, with just a thin layer of water flowing over them, or in the spray zone of waterfalls, as in several Hydroptilidae genera.

After hatching and before pupating, caddisfly larvae complete five instars. The immature stages represent the longest part of the life cycle and are when they build their cases or retreats. In general, three groups can be defined according to larval construction behavior (Wiggins, 1996, 2004): fixed retreat-makers (Annulipalpia), portable case-makers, and closed cocoon-makers (Integripalpia). Retreat-making caddisflies usually build a fixed shelter of material from the substrate, also held together with silk, and from which they also spin a silken net to capture food particles from the water current (Wiggins, 2004). In some, the retreat itself is also the filter or capture structure. Philopotamids have a series of labial grooves associated with the silk gland opening that allow the larvae of this family to spin about 70 threads simultaneously, and they are used as a sac-like net with the smallest mesh opening known (0.5–5.5 microns), which enable them to feed on fine particulate organic matter, as well as associated bacteria and fungi (Wallace & Malas, 1976). Case-making caddisflies include those that typically build tubular, portable cases. Cases usually include material from the substrate, such as small fragments of leaves, sand grains, or small pebbles, twigs, or even empty mollusk shells. Silk acting as glue or mortar holds these components together. Although the portable case acts primarily as a camouflage or protective device against potential predators, it also has an important function in increasing respiratory efficiency (Wiggins, 1996). Thus, the case probably enabled caddisflies to exploit oxygen-poor habitats, such as some lentic waters (Wiggins, 2004; Malm et al., 2013). Closed cocoon-making caddisflies are a more heterogeneous group and include larvae that build a cocoon (closed to circulation of water) only for pupation (Wiggins, 2004). The cocoon-makers include larvae that build cases in all larval instars, only the last larval instar, and those that build no larval structures.

The Neotropical Region contains the second highest diversity of caddisflies species in the world, with a correspondent variety of types of retreats and portable cases. Only the Oriental Region has more species. Currently, about 3280 caddisfly species are known from the Neotropics (Holzenthal & Calor, 2017), but many new species continue to be described every year from the region. Forty-six new species were described for the Neotropical Region only in the first semester of 2017 (Barcelos-Silva et al., 2017; Cavalcante et al., 2017; Desidério et al., 2017; Dumas et al., 2017; Quinteiro & Holzenthal, 2017; Rocha et al., 2017; Souza & Santos, 2017 Valarino & Calor, 2017). In addition to providing descriptions of new caddisflies from the Neotropical Region, a more challenging task is to associate the immature stages and study their ecology and behavior. To date, 25 families and 155 genera have been recorded from the Neotropics; however, for 41 genera the immature stages remain completely unknown (Table 1), and knowledge of larvae and pupae at the species level is even less well known.

LIMITATIONS

Table 1 summarizes the current state of knowledge of the order in the Neotropical Region. The region covered by this chapter extends northward beyond the limits of the Neotropical Region, as defined by Wallace (1876). Here, we included all taxa from Mexico, including Nearctic northern Mexico, Central and South America, and the Antilles.

The literature for the Neotropical caddisfly fauna is extensive. Beginning with the earliest descriptions from the 1830s, a large number of works dealing with the description of new species, immature stages, and taxonomic revisions have been published. These works were compiled in the *Catalog of the Neotropical Caddisflies* (Flint et al., 1999b) that also provided a faunistic list of all the species known from the region until the year of its publication and a list of synonyms for all taxa and their distributions. A new catalog was recently published updating the literature through 2016 (Holzenthal & Calor, 2017). Flint et al. (1999a) provided a list of synonyms. Also, online catalogs have been created such as the "Trichoptera World Checklist" (Morse, 2016) and the "Taxonomic Catalog of the Brazilian Fauna" (Santos et al., 2016b) that compile updated data on the Neotropical caddisfly fauna. Another that was of fundamental importance to caddisfly literature was the "Trichopterorum Catalogus" (Fischer, 1960–1973b).

One of the greatest shortcomings of the Neotropical Trichoptera literature is the few number of works containing identification keys. Flint (1991, 1996) provided keys to adults of Neotropical families and genera but excluded taxa from the Chilean subregion. These taxa were treated later in the work of Angrisano & Sganga (2009) that included keys to adults, pupae, and larvae of South American families and genera. Other works that can be useful for the identification of the fauna, especially from Mexico, and to some extent Central America, are the chapters on caddisflies in the book *An Introduction to the Aquatic Insects of North America* and by the works of Wiggins on the North American larvae (Morse & Holzenthal, 1984, 2008; Wiggins, 1984, 1996, 2004, 2008). Other keys covering smaller areas were published by Posada-García & Roldán-Pérez (2003) for northwestern Colombia, by Springer (2006, 2010) for Costa Rica and Central America, Pes et al. (2005, 2014) for Amazonas state, Brazil, Calor (2007) for São Paulo state, Brazil, and Huamantinco & Ortíz (2010) for the Andes near Lima, Peru, and Central America (Springer, 2010). Although these works are very useful, an integrative work covering the whole territory under consideration is needed.

Knowledge on Neotropical caddisflies larvae is poor. As seen in Table 10.1, the immatures of some genera are completely unknown, and the diagnoses of others are based only on the description of one or two species, due to the difficulty of associating species of immatures with their respective adults.

TERMINOLOGY AND MORPHOLOGY

For information on morphologic terminology needed to use these identification keys, see Holzenthal et al. (2015) in the first volume of this work.

MATERIAL PREPARATION AND PRESERVATION

For methods for collecting, preserving, and assembling specimens for identification, see Holzenthal et al. (2015).

TABLE 10.1 List of Neotropical Families and Genera in Trichoptera Kirby, 1913, With Number of Recorded Species and Associated Larvae

Taxon	Number of Species	Number of Larvae
Suborder Annulipalpia Martynov, 1924		
Superfamily Philopotamoidea Stephens, 1829		
Family Philopotamidae Stephens, 1829	**377**	**14**
Alterosa Blahnik, 2005	39	NO
Chimarra Stephens, 1829	256	14
Chimarrhodella Lestage, 1925	12	NO
Sortosa Navás, 1918	20	0
Wormaldia McLachlan, 1865	50	0
Family Stenopsychidae Martynov, 1924	**3**	**NO**
Pseudostenopsyche Doehler, 1915	3	NO
Superfamily Psychomyioidea Walker, 1852		
Family Ecnomidae Ulmer, 1903	**59**	**1**
Austrotinodes Schmid, 1955	58	1
Chilocentropus Navás, 1934	1	NO
Family Polycentropodidae Ulmer, 1903	**283**	**7**
Cernotina Ross, 1938	68	1
Cyrnellus Banks, 1913	10	1
Nyctiophylax Brauer, 1865	4	0
Polycentropus Curtis, 1835	105	4

(Continued)

TABLE 10.1 List of Neotropical Families and Genera in Trichoptera Kirby, 1913, With Number of Recorded Species and Associated Larvae—cont'd

Taxon	Number of Species	Number of Larvae
Polyplectropus Ulmer, 1905	96	1
Family Pseudoneureclipsidae Ulmer, 1951	**4**	**1**
Antillopsyche Banks, 1941	4	1
Family Psychomyiidae Walker, 1852	**2**	**0**
Tinodes Curtis, 1834	2	0
Family Xiphocentronidae Ross, 1949	**60**	**4**
Cnodocentron Schmid, 1982	6	0
Machairocentron Schmid, 1982	6	1
Xiphocentron Brauer, 1870	48	3
Superfamily Hydropsychoidea Curtis, 1835		
Family Hydropsychidae Curtis, 1835	**478**	**40**
Blepharopus Kolenati, 1859	1	1
Calosopsyche Ross & Unzicker, 1977	13	2
Centromacronema Ulmer, 1905	17	NO
Cheumatopsyche Wallengren, 1891	6	0
Diplectrona Westwood, 1839	1	0
Hydropsyche Pictet, 1836	9	0
Leptonema Guérin-Méneville, 1843	125	3
Macronema Pictet, 1836	33	1
Macrostemum Kolenati, 1859	18	0
Palaehydropsyche Wichard, 1986	0	NO
Plectromacronema Ulmer, 1906	3	1
Plectropsyche Ross, 1947	3	NO
Pseudomacronema Ulmer, 1906	1	NO
Smicridea McLachlan, 1871	232	28
Streptopsyche Ross & Unzicker, 1977	5	NO
Synoestropsis Ulmer, 1905	10	1
***Suborder Integripalpia* Martynov, 1924**		
Superfamily Glossosomatoidea Wallengren, 1891		
Family Glossosomatidae Wallengren, 1891	**266**	**19**
Canoptila Mosely, 1939	2	NO
Cariboptila Flint, 1964	21	6
Culoptila Mosely, 1954	23	0
Glossosoma Curtis, 1834	1	0
Itauara Müller, 1888	22	1
Mastigoptila Flint, 1967	9	1
Merionoptila Schmid, 1959	1	NO

TABLE 10.1 List of Neotropical Families and Genera in Trichoptera Kirby, 1913, With Number of Recorded Species and Associated Larvae—cont'd

Taxon	Number of Species	Number of Larvae
Mortoniella Ulmer, 1906	97	4
Protoptila Banks, 1904	87	8
Scototrichia Mosely, 1934	1	NO
Tolhuaca Schmid, 1964	2	NO
Superfamily Hydroptiloidea Stephens, 1836		
Family Hydroptilidae Stephens, 1836	**931**	**60**
Acostatrichia Mosely, 1939	20	1
Alisotrichia Flint, 1964	58	1
Anchitrichia Flint, 1970	8	2
Angrisanoia Özdikmen, 2008	5	NO
Ascotrichia Flint, 1983	3	NO
Betrichia Mosely, 1939	10	NO
Bredinia Flint, 1968	16	2
Byrsopteryx Flint, 1981	16	4
Celaenotrichia Mosely	1	1
Cerasmatrichia Flint, Harris & Botosaneanu, 1994	8	1
Ceratotrichia Flint, 1992	5	NO
Costatrichia Mosely, 1937	14	NO
Dicaminus Müller, 1879	1	NO
Flintiella Angrisano, 1995	15	1
Hydroptila Dalman, 1819	72	4
Ithytrichia Eaton, 1873	2	1
Kumanskiella Harris & Flint, 1992	2	1
Leucotrichia Mosely, 1934	44	10
Mayatrichia Mosely, 1937	3	1
Mejicanotrichia Harris & Holzenthal, 1997	7	2
Metrichia Ross, 1938	109	8
Neotrichia Morton, 1905	138	2
Nothotrichia Flint, 1967	5	1
Ochrotrichia Mosely, 1934	160	0
Orinocotrichia Harris, Flint & Holzenthal, 2002	3	NO
Orthotrichia Eaton, 1873	3	0
Oxyethira Eaton, 1873	106	6
Peltopsyche Müller, 1879	6	2
Ragatrichia Oláh & Johanson, 2011	5	NO
Rhyacopsyche Müller, 1879	28	3

(Continued)

TABLE 10.1 List of Neotropical Families and Genera in Trichoptera Kirby, 1913, With Number of Recorded Species and Associated Larvae—cont'd

Taxon	Number of Species	Number of Larvae
Scelobotrichia Harris & Bueno, 1993	3	2
Taraxitrichia Flint & Harris, 1991	1	1
Tizatetrichia Harris, Flint & Holzenthal, 2002	1	NO
Tricholeiochiton Kloet & Hincks, 1944	1	0
Tupiniquintrichia Santos, Nessimian & Takiya, 2016	2	1
Zumatrichia Mosely, 1937	50	2
Superfamily Rhyacophiloidea Stephens, 1836		
Family Hydrobiosidae	**180**	**29**
Amphichorema Schmid, 1989	3	NO
Androchorema Flint, 1979	1	NO
Apatanodes Navás, 1934	2	1
Atopsyche Banks, 1905	129	10
Australobiosis Schmid, 1958	3	NO
Cailloma Ross & King, 1951	4	2
Clavichorema Schmid, 1955	7	1
Heterochorema Schmid, 1989	1	NO
Iguazu Ross & King, 1951	2	1
Isochorema Schmid, 1989	2	NO
Metachorema Schmid, 1957	2	1
Microchorema Schmid, 1955	4	1
Neoatopsyche Schmid, 1955	5	5
Neochorema Schmid, 1957	4	1
Neopsilochorema Schmid, 1955	1	1
Nolganema Navás, 1934	1	NO
Parachorema Schmid, 1957	1	NO
Pomphochorema Flint, 1969	1	NO
Pseudoradema Schmid, 1955	1	1
Rheochorema Schmid, 1955	4	2
Schajovskoya Flint, 1979	1	1
Stenochorema Schmid, 1955	1	1
Family Rhyacophilidae	**1**	**0**
Rhyacophila Pictet, 1834	1	0
Infraorder Brevitentoria		
Superfamily Leptoceroidea		
Family Atriplectididae	**2**	**2**
Neoatriplectides Holzenthal, 1997	2	2

TABLE 10.1 List of Neotropical Families and Genera in Trichoptera Kirby, 1913, With Number of Recorded Species and Associated Larvae—cont'd

Taxon	Number of Species	Number of Larvae
Family Calamoceratidae	**75**	**11**
Banyallarga Navás, 1916	17	1
Phylloicus Müller, 1880	58	10
Family Leptoceridae	**238**	**40**
Achoropsyche Holzenthal, 1984	1	NO
Amazonatolica Holzenthal & Pes, 2004	1	1
Amphoropsyche Holzenthal, 1985	15	2
Atanatolica Mosely, 1936	20	8
Brachysetodes Schmid, 1955	10	1
Grumichella Müller, 1879	13	3
Hudsonema Mosely, 1936	1	1
Mystacides Berthold, 1827	1	1
Nectopsyche Müller, 1879	49	12
Neoatrhripsodes Holzenthal, 1989	2	NO
Notalina Mosely, 1936	11	2
Oecetis McLachlan, 1877	69	4
Osflintia Calor & Holzenthal, 2008	1	0
Setodes Rambur, 1842	1	0
Triaenodes McLachlan, 1865	26	1
Triplectides Kolenati, 1859	17	4
Family Odontoceridae	**45**	**6**
Anastomoneura Huamantinco & Nessimian, 2004	1	1
Barypenthus Burmeister, 1839	1	1
Marilia Müller, 1880	43	4
Family Philorheithridae	**6**	**NO**
Mystacopsyche Schmid, 1955	2	NO
Psilopsyche Ulmer, 1907	4	NO
Superfamily Sericostomatoidea		
Family Anomalopsychidae	**28**	**7**
Anomalopsyche Flint, 1967	1	1
Contulma Flint, 1969	27	6
Family Helicophidae	**16**	**2**
Alloecentrellodes Flint, 1979	2	NO
Austrocentrus Schmid, 1964	3	1
Eosericostoma Schmid, 1955	2	1
Microthremma Schmid, 1955	8	NO

(*Continued*)

TABLE 10.1 List of Neotropical Families and Genera in Trichoptera Kirby, 1913, With Number of Recorded Species and Associated Larvae—cont'd

Taxon	Number of Species	Number of Larvae
Pseudosericostoma Schmid, 1957	1	NO
Family Helicopsychidae	**122**	**17**
Helicopsyche Siebold, 1856	122	17
Family Sericostomatidae	**8**	**3**
Grumicha Müller, 1879	1	1
Gumaga Tsuda, 1938	1	1
Notidobiella Schmid, 1955	6	1
Family Parasericostomatidae	**12**	**3**
Chiloecia Navás, 1930	1	NO
Myotrichia Schmid, 1955	1	1
Parasericostoma Schmid, 1957	10	2
Superfamily Tasimioidea		
Family Tasimiidae	**2**	**2**
Charadropsyche Flint, 1969	1	1
Trichovespula Schmid, 1955	1	1
Infraorder Plenitentoria		
Family Kokiriidae	**2**	**1**
Pangullia Navás	2	1
Superfamily Limnephiloidea		
Family Limnephilidae	**51**	**18**
Anomalocosmoecus Schmid, 1957	4	3
Antarctoecia Ulmer, 1907	2	2
Austrocosmoecus Schmid, 1955	1	1
Clistoronia Banks, 1916	1	0
Hesperophylax Banks, 1916	2	2
Limnephilus Leach, 1815	13	0
Metacosmoecus Schmid, 1955	1	1
Monocosmoecus Ulmer, 1906	6	2
Platycosmoecus Schmid, 1964	1	1
Verger Navás, 1918	20	6
Superfamily Phryganeoidea		
Family Lepidostomatidae	**28**	**5**
Lepidostoma Rambur, 1842	28	5
Total	**3280**	**292**
%	**100**	**9.0**

NO, larva unknown; *0*, the larvae is described in another region, but not to Neotropical species.

KEYS TO TRICHOPTERA

The keys below are to final instar larvae only. The following 14 families are limited in the keys because either they are monobasic or not all genera have been described.

Atriplectididae is a small and very peculiar family with four genera and only six species. The family shows a disjunct Gondwanan distribution, and in the Neotropics (northern Andes and southeast Brazil), it has one genus (*Neoatriplectides*) with two species (Holzenthal, 1997; Dumas & Nessimian, 2008). Larvae are usually found in shallow, lateral pools of small streams where they build somewhat flattened, portable cases of small rock fragments.

Ecnomidae has nine genera worldwide with about 400 species, *Austrotinodes* is the only genus in the Neotropics, and has 44 species (Holzenthal & Calor, 2017). Larvae are usually found in small streams, living under rocks or wood in running water, where they build elongate, very fragile shelters of silk covered by sand, but they also can be found in large rivers (Holzenthal & Calor, 2017). *Austrotinodes* larvae have a distinctive foretrochantin, longer than the coxa, strongly depressed, and remiform. In addition, all three thoracic nota are completely sclerotized.

Helicophidae is found in southern Chile and southwestern Argentina, with 16 species in five genera (Holzenthal & Calor, 2017). Larvae are usually found in clean small streams in forested areas, and they build tubular cases with sand grains or plant material. Of the five Neotropical genera, only *Austrocentrus* and *Eosericostoma* are described (Flint, 1992, 1997). *Austrocentrus* larvae have labrum with an anteromesal depression; mid and hind tarsal claws with short accessory setae ventrally, less than half the length of the claw and case cylindrical (Figs. 10.110 A–C). *Eosericostoma* larvae have labrum without an anterolateral depression, margin straight; mid and hind tarsal claws with long accessory setae, more than half the length of the claw and case with dorsal margin expanded laterally, covering the ventral margin (Figs. 10.111 A–G). *Alloecentrellodes*, *Microthremma*, and *Pseudosericostoma* larvae are unknown.

Helicopsychidae is currently composed of 270 described species (Holzenthal et al., 2011) distributed in two genera: the monotypic *Rakiura*, from New Zealand, and the cosmopolitan *Helicopsyche*, which is divided into six subgenera, two recorded from the Neotropics: the endemic *H.* (*Cochliopsyche*), with 16 species, and *H.* (*Feropsyche*), with 102 species (Holzenthal & Calor, 2017). Larvae are usually found on rocky bottoms of small- to medium-sized streams with moderate- to slow-flowing water. Helicopsychid larvae build a singular snail-shaped case with different sizes of mineral fragments that make them easily recognized from other Neotropical families.

Kokiriidae is a small family composed of nine species distributed in five genera; only one of them (*Pangullia*), with two species, is recorded from the Neotropical Region, with distribution limited to the Chilean subregion (Holzenthal & Calor, 2017). Rojas (2007) described the larva of one of the Chilean species: *Pangulia nea* Rojas, 2006.

Lepidostomatidae is widely distributed with about 400 described species, reaching its highest diversity in the Holoarctic, Oriental, and Afrotropical regions. It is poorly represented in the Neotropics, with only 28 species in *Lepidostoma*, which occurs in Mexico and Central America (Holzenthal & Calor, 2017). Larvae are usually found in small forest streams in leaf litter banks, generally above 1000 m above sea level, but they also occur in cool springs, lakes, and temporary streams. They build portable cylindrical cases that can be composed of sand grains or fragments of bark, twigs, and leaves (Flint et al., 1999b); some cases can be square in cross-section.

Parasericostomatidae was recently erected from Sericostomatidae (Johanson et al., 2016) to accommodate three Neotropical genera: *Parasericostoma* with 10 species, *Myotrichia*, and *Chiloecia*, both monotypic; all of them are recorded from southern Chile and northwestern Argentina. Larvae usually are collected from running, clean, unpolluted waters (Valverde & Miserendino, 1997; Albariño & Valverde, 1998). They build tubular, elongate, silken cases with inclusion of sand grains. There are no reliable characters to diagnose the genera, and the larva of *Chiloecia* is unknown.

Philopotamidae is found in running water in all zoogeographic regions; in the Neotropical Region, 377 species are reported (Holzenthal & Calor, 2017), in five genera. *Chimarra* is the most diverse, with about 250 species (66% belong to the Neotropical philopotamid fauna), and it is cosmopolitan. Larvae usually live under rocks or leaf litter in running water of moderate to fast flow and build a long, tubular retreat attached to the substrate, where they construct a net with very fine mesh composed by several layers of superimposed silk (Flint et al., 1999b). Identification of the immature stages to genus is complex because all known genera are very similar, with few diagnostic features. *Alterosa*, *Chimarrhodella*, and *Sortosa* larvae remain unknown. The genus *Chimarra* is characterized by presenting the prothoracic coxae with a long, slender, subapical, seta-bearing process (Figs. 10.89 D, 90 D). It is divided into four subgenera (*Chimarra, Chimarrita, Curgia*, and *Otarrha*), and *Chimarrita* larvae are very different from the other subgenera because it is very thin, the anterior margin of the frontoclypeus does not have the asymmetrical notch to the right of midline (Fig. 10.90 B), and the head has setae *no. 18* and *no. 22* strong and dark, positioned posteroventral and in a line (Fig. 10.90 C). *Wormaldia* is characterized by having prothoracic coxae without long, slender, subapical, seta-bearing process (Fig. 10.91 D) and foretrochantin forming a long and narrow process laterally that narrows toward the apex (Figs. 10.91 D, E). This genus occurs in Neotropical, Afrotropical, Nearctic, and Palaearctic regions. Larvae collected in the southeast region of Brazil possibly belong to *Alterosa* because they possess characteristics that distinguish them from the two known Neotropical genera in this region. These unassociated larvae are included in the generic key as "Genus A," and it is characterized by presenting the foretrochantin with a wide triangular base laterally and ventrally (Figs. 10.92 A, B).

Philorheithridae is a small family composed of 21 species in 17 genera with disjunct distribution in the Australasian and Neotropical regions, where six species distributed in two genera occur: *Mystacopsyche* and *Psilopsyche*, both endemic to the Chilean subregion (Holzenthal & Calor, 2017). There is no published information about immature stages of both genera in Neotropics, although they are known in the Australasian region. The Neotropical forms will likely be similar in morphology and behavior.

Pseudoneureclipsidae was recently established as a family distinct from Polycentropodidae. It is composed of two genera, the widespread *Pseudoneureclipsis*, with about 180 species, and the Neotropical *Antillopsyche*, with four extant species endemic to the Greater Antilles (Chamorro & Holzenthal, 2011). Larvae are usually found in small and clear water streams in backwater areas near the soil–water interface, where they build silken tubes attached to rocks (Holzenthal & Calor, 2017). Neotropical *Antillopsyche* larvae are typical psychomyioids, except for the presence of small, slightly sclerotized plates on all thoracic nota, and the delta-shaped foretrochantin separated from the episternum.

Psychomyiidae is a small, but widely distributed family, with about 460 known species; nevertheless, its occurrence in the Neotropical Region is notably lacking. In the New World, it is virtually exclusive to North America, with only two species belonging to *Tinodes* recorded at higher elevations in northwestern Mexico (Holzenthal & Calor, 2017). Their immature stages are usually found in flowing water, often on hydropetric surfaces, where they build elongate silken tubes with mineral fragments over rocky substrates (Holzenthal & Calor, 2017). Psychomyiid larvae share some similarities with others psychomyioids, but the hatchet-shaped foretrochantin, separated from the episternum, can distinguish them from the others.

Rhyacophilidae is a family of the Northern Hemisphere, extending south into India and southeastern Asia. *Rhyacophila rayneri* Ross is the only species known from the Nearctic part of Mexico; the family is included in the key because the distribution of the genus is in the limits of the regions Nearctic and Neotropical. The immature stages of this species are not known, but larvae of several species in this genus are described and are quite diverse in appearance. They are free-living predators, except for a few species that are phytophagous (Holzenthal & Calor, 2017).

Stenopsychidae has three species in *Pseudostenopsyche* reported for Chile; immatures from the Neotropical Region are unknown (Holzenthal & Calor, 2017). Reliable information about the immature stages and their biology are only available to Asian species, and therefore this family is not included in the key to family. The main characteristic used to identify larvae of this family is the presence of a pair of pointed projections on the forecoxae (Nozaki et al., 2016).

Xiphocentronidae is a small family with less than 200 described species placed in seven genera around the world. Three genera are recorded from the Neotropics, with a total of 58 species (Holzenthal & Calor, 2017). Immature stages are usually found in small streams on rocky substrate, often above the water surface and even in semiterrestrial habits, as described by Muñoz-Quesada & Holzenthal (1997). They build elongate, silken tubes covered with mineral fragments and debris or even a kind of portable case, as in *Xiphocentron moncho* Muñoz & Holzenthal. Xiphocentronid larvae are easily recognized by the fusion of the tibia and tarsus of the thoracic legs and the modified mesopleura, expanded anteriorly. Larvae of the three Neotropical genera are morphologically indistinguishable (Schmid, 1982; Pes et al., 2013).

Trichoptera: Families

1 Larva with portable cases of different forms, with fixed shelters, or free living; abdomen straight or slightly curved 2

1' Larva with portable case in the shape of a snail shell (Fig. 10.1 A); abdomen fixed in a spiral (Fig. 10.1 B) .... Helicopsychidae, one genus: *Helicopsyche* [Cosmopolitan]

2(1) Prothorax and mesothorax short, almost the same width; mesothorax unmodified; midlegs not modified as follows 3

2' Head, prothorax, and mesothorax anterior portion very narrow; mesothorax modified, extremely elongate, anterior portion with two pairs of dorsal plates and capable of retraction into body invagination; midlegs developed, robust, modified into a prehensile structure (Figs. 10.2 A–C) Atriplectididae, one genus: .... *Neoatriplectides* [Bolivia, Peru, Brazil]

3(2) Each thoracic notum completely covered by an entire sclerotized plate (Figs. 10.13 C, 14 C, 15 C, 18 B, 34 C) 4

3' Only pronotum completely covered by a sclerotized plate; meso- and metanotum may have sclerites of different sizes or may be completely membranous (Figs. 10.60 C, 86 A; and 10.108 C) 6

4(3) Abdominal segments ventral margin without filamentous gills (Figs. 10.3 A, 34 B, 38 B) 5

4' Abdominal segments ventral margin with filamentous gills (Figs. 10.15 A, 17 A, 19 A) .... **Hydropsychidae [p. 248]**

5(4) Foretrochantin longer than the foreleg coxa, flattened longitudinally and slightly curved (Fig. 10.3 B) .... Ecnomidae, one genus: *Austrotinodes* [Neotropics. Nearctic]

5' Foretrochantin shorter than the foreleg coxa, cylindrical and pointed apically .... **Hydroptilidae [p. 249]**

6(3) Prosternum with a horn (Fig. 10.56 D) 7

6' Prosternum without a horn (Fig. 10.107 C) 8

7(6) Antenna located midway between eye and cephalic capsule anterior margin (Figs. 10.46 D, 55 C); abdominal segment I with dorsomedial hump (Figs. 10.46 B, C, 49 B, 52 B, 55 E) .... **Limnephilidae [p. 251]**

7' Antenna located near eye anterior margin (Fig. 10.4 E); abdominal segment I without dorsomedial hump (Figs. 10.4 A–E) .... Lepidostomatidae, one genus: *Lepidostoma* [Central America. Cosmopolitan]

8(6) Antenna length $\leq 2\times$ width or inconspicuous; all legs of similar size; hindleg coxa turned ventrally 9

8' Antenna length $\sim 6\times$ width (Figs. 10.57 C, 63 C, 64 C); hindleg longer than others; hindleg coxa usually turned forward (Figs. 10.57 A, 62 E, 66 B) .... **Leptoceridae [p. 252]**

9(8) Labrum median region with no more than eight long pointed setae (Fig. 10.109 A) 10

9' Labrum with median transverse row of 8–32 long pointed setae (Figs. 10.70 D, 71 F) .... **Calamoceratidae [p. 253]**

10(9) Meso- and metanotum usually completely membranous (Figs. 10.86 A, 88 A, 93 A), if sclerites present, metanotum with lightly sclerotized, longitudinally elongate areas, length nearly $2\times$ width, with a noticeable sclerotized longitudinal bar (Fig. 10.5 A); anal prolegs elongate and detached from segment IX (Figs. 10.87 A, 88 A, 95 A, D); larva free living or building fixed shelters 11

10' Mesonotum usually with sclerites, or completely membranous (Figs. 10.97 A, B, 108 A, B, 113 F); metanotum with or without sclerites; if sclerites present, length subequal to width, without sclerotized longitudinal bar; anal prolegs broadly joined to segment IX (Figs. 10.107 A, 110 B); larva building portable cases 17

11(10) Meso- and metanotum completely membranous 12

11' Mesonotum with two lightly sclerotized plates; metanotum with a pair of dark longitudinal bars on lightly sclerotized areas (Fig. 10.5 A) .... Pseudoneureclipsidae, one genus: *Antillopsyche* [Caribbean (fossil from Mexico)]

12(11) Abdominal segment IX dorsal region with a sclerotized plate (Figs. 10.6 A, 78 A, 87 B) 13

12' Abdominal segment IX dorsal region without a sclerotized plate 14

13(12) Foreleg tibia and tarsus highly modified, can be appressed against femur ventral projection, forming a chelate structure with tarsal claw long and thin (Figs. 10.72 A, B, 78 C, 86 C) .... **Hydrobiosidae [p. 253]**

13' Foreleg with tibia, tarsus, and femur unmodified (Figs. 10.6 A–C) .... Rhyacophilidae, one genus *Rhyacophila* [Mexico. Nearctic]

14(12) Labrum sclerotized and rounded (Fig. 10.94 A) 15

14' Labrum membranous and T-shaped (Figs. 10.88 B, 89 B, 90 B), often retracted under frontoclypeus in preserved specimens .... Philopotamidae

15(14) Tibia and tarsus of all thoracic legs separated (legs 5-articulate); mesopleura unmodified (Figs. 10.92 B–D) 16

15' Tibia and tarsus of all thoracic legs fused (legs apparently 4-articulate) (Figs. 10.7 D–F); mesopleura extended anteriorly in form of a lobed process (Figs. 10.7 B, C) (Neotropical genera with larvae morphologically indistinguishable) .... Xiphocentronidae [Mexico to Argentina. Afrotropical, Nearctic, Oriental, Palearctic]

16(15) Foretrochantin separated from the episternum, extended as broadened, securiform lobe (Fig. 10.8 A) .... Psychomyiidae, one genus: *Tinodes* [Mexico. Widespread]

16' Foretrochantin fused with the episternum with acute apex, not expanded into a broadened lobe (Fig. 10.93 A) .... **Polycentropodidae [p. 254]**

17(10) Mesonotum dorsally with sclerotized plates covering at least half of segment; anal claw large, with or without accessory teeth; larva builds tubular portable cases (not domed), with anterior and posterior openings not directed ventrad 18

17' Mesonotum usually with three small sclerotized plates or completely membranous (Figs. 10.97 C, 100 D); anal claw with accessory teeth similar to the claw (Figs. 10.97 D, 100 C); larva builds portable cases in form of tortoise shell, with anterior and posterior openings directed ventrad (Figs. 10.97 A, 99 B, C, 100 A, 101 A) .... **Glossosomatidae [p. 255]**

18(17) Pronotum without a transverse carina, but anterolateral margin may have small pointed projections; anal claw without accessory teeth or with a distinct accessory hook, a single very small dorsal hook or a small accessory hook 20

18' Pronotum near posterior margin with a transverse carina, anterolaterally produced as a pointed or rounded lobe (Figs. 10.103 A, B, 104 A, B); anal claw with single basoventral tooth and with dorsal curved process bearing many secondary teeth or with a single dorsomesal comb of accessory teeth (Figs. 10.103 C, 104 C–E) .... **Anomalopsychidae [p. 256]**

20(19) Midleg tibia and tarsus fused (Figs. 10.9 C, 10 D); forelegs and midlegs raptorial (Figs. 10.9 B, C, 10 C, D) 21

20' Tibia and tarsus of midleg separated; fore- and midlegs unmodified 22

21(20) Foreleg tibia and tarsus fused (Fig. 10.9 B); mesonotum with plates lightly sclerotized (Fig. 10.9 A) .... Kokiriidae, one species: *Pangullia nea* Rojas [Chile]

21' Foreleg tibia and tarsus separated (Fig. 10.10 C); mesonotum with well-sclerotized plates (Figs. 10.10 B) .... Philorheithridae (unassociated larva) [Argentina. Australian]

22(20) Pronotum anterolateral margin usually rounded; if with a pointed process, metathorax covered by a large plate; abdominal segments VI to VIII posterodorsal region without setae 23

22' Pronotum anterolateral margin produced as a pointed process (Figs. 10.105 A, 106 A); metathorax with two small posterolateral sclerites; abdominal segments VI to VIII posterodorsal region with tufts of setae (Fig. 10.105 B) .... **Tasimiidae [p. 256]**

23(22) Foreleg flattened laterally (Figs. 10.11 G, 111 E, 112 D, 113 D, 114 G); foretrochantin conspicuous, elongate, about half coxa length (Figs. 10.111 E, 112 D, 113 D, 114 G); prosternum membranous 24

23' Foreleg subcylindrical, femur may be slightly flattened (Fig. 10.109 E); foretrochantin inconspicuous, not elongated, length <1/3 coxa; prosternum sclerotized (Figs. 10.107 E, 108 D) .... **Odontoceridae [p. 256]**

24(23) Mesonotum strongly sclerotized, completely covered with plates; *sa1* large, occupying at least half of metanotum (Figs. 10.11 C, 112 C, 113, 114 F) 25

24' Mesonotum slightly sclerotized, with plates covering at least anterior half; *sa1* small, occupying less than 1/6 of metanotum (Figs. 10.111 B, D) .... Helicophidae

25(24) Pronotum anterior half similar in texture to posterior half, not covered by setae (Figs. 10.11 C, D); foretrochantin subrectangular (Fig. 10.11 G) (Neotropical genera with larvae morphologically indistinguishable) .... Parasericostomatidae
[Argentina, Chile]

25' Pronotum anterior half with different texture of the posterior half, covered by many setae (Figs. 10.112 C,D, 113 E, F, 114 B, F); foretrochantin triangular or pointed (Figs. 112 C, 113E, 114F) .... **Sericostomatidae [p. 256]**

Trichoptera: Hydropsychidae: Genera

Hydropsychidae are the third most diverse family of the order with about 1500 described species. The family is cosmopolitan and is divided into five subfamilies, of which Diplectroninae, Hydropsychinae, Macronematinae, and Smicrideinae are recorded in the Neotropical Region, totaling 476 species in 15 genera (Holzenthal & Calor, 2017). They are primarily found in flowing waters, from small streams to large rivers, though some few species may be found in a lentic environment. Hydropsychid larvae are easily distinguished from the other families by having all thoracic nota completely sclerotized and possessing filamentous gills on the ventrolateral and ventral surfaces of all abdominal segments. Of the 15 genera, only *Pseudomacronema* and *Streptopsyche* remain without any information about immature stages. Immature stages of *Plectropsyche* and *Cheumatopsyche* are morphologically similar, and there are no characteristics to separate them (O. Flint & R. Holzenthal, Pers. Com. *in:* Flint et al. (1999b)).

1 Posterior ventral apotome very reduced or absent; if present, no more than half of the length of median ecdysial line 2

1' Posterior ventral apotome always present, at least half of the length of median ecdysial line (Fig. 10.12 B) .... *Diplectrona*
[Central America. Cosmopolitan]

2(1) Abdominal gills consisting of an elongate central stalk with numerous filaments arising more or less uniformly along the entire length (Figs. 10.15 A, 16 A, 17 A, 19 A) (Macronematinae) 3

2' Abdominal gills with or without central stalk with filaments arising independently from the abdomen; when central stalk present, filaments fewer than previous and not arising uniformly, frequently as an apical tuft (Figs. 10.21 A, 22 A) 9

3(2) Head and thorax as wide as long; forefemur usually unmodified, tibia and tarsus not prehensile; foretrochantin usually pointed or truncate 4

3' Head and thorax length $\geq 2\times$ width (Figs. 10.13 A–C); forefemur broad and expanded with ventral margin almost straight, forming a prehensile structure with tibia and tarsus (Fig. 10.13 B); foretrochantin modified in a broad triangular to quadrangular plate (Figs. 10.13 A, B) .... *Synoestropsis*
[Mexico to Argentina]

4(3) Abdominal segments II–VIII without paired ventrolateral pockets; forefemur unmodified 5

4' Abdominal segments II–VIII with paired ventrolateral pockets (Fig. 10.14 A) forefemur broad with ventral margin rounded (Figs. 10.14 A, D) .... *Plectromacronema*
[Mexico to Argentina]

5(4) Head with dorsal carina; labrum with paired basal sclerites (Figs. 10.15 C, 16 C) 6

5' Head without dorsal carina; labrum without paired basal sclerites (Fig. 10.18 C) 7

6(5) Sternum IX composed of short, paired plates, occupying less than half sternum length, with no more than five long setae (Fig. 10.15 B) .... *Blepharopus diaphanus* Kolenati
[Venezuela, Brazil, Argentina]

6' Sternum IX composed of large, paired plates, occupying at least half the sternum, length densely covered by long setae (Fig. 10.15 E). .... *Macrostemum*
[Neotropics. Afrotropical, Nearctic, Oriental]

7(5) Anal proleg very elongate with a median fold (Figs. 10.18 A, 19 A); abdominal gills consisting of two rows of lateral filaments arising from the central stalk (Figs. 10.18 A, 19 A) 8

7' Anal proleg short without median fold; abdominal gills consisting random of filaments randomly arising from the central stalk (Fig. 10.17 A) .... *Leptonema*
[Neotropics. Afrotropical]

8(7) Head ventrally with stridulatory lines present (Fig. 10.18 D); mesonotum with paired laterodorsal grooves (Fig. 10.18 D); foretrochantin bifid (Fig. 10.18 D) .... *Centromacronema*
[Mexico to Brazil]

8' Head ventrally without stridulatory lines (Fig. 10.19 C); mesonotum without laterodorsal grooves (Fig. 10.19 D); foretrochantin simple, truncate (Fig. 10.19 E) .... *Macronema*
[Mexico to Argentina]

9(2) Submentum notched anteriorly (Figs. 10.20 C, 21 C); foretrochantin forked (Figs. 10.20 E, 21 D) (Hydropsychinae, except *Calosopsyche*) 10

9' Submentum entire anteriorly (Fig. 10.23 C); foretrochantin simple (Fig. 10.23 D) 11

10(9) Prothorax venter posterior to prosternal plate with a pair of prominent sclerites in intersegmental fold (Fig. 10.20 D); frontoclypeus anterior margin usually straight to slightly convex (Fig. 10.20 B) .... *Hydropsyche*
[Mexico, Guatemala. Afrotropical, Nearctic, Oriental, Palearctic]

10' Prothorax venter posterior to prosternal plate, usually with minute pair of sclerites in intersegmental fold (Fig. 10.21 E); frontoclypeus anterior margin usually with a small median notch (Fig. 10.21 B) .... *Cheumatopsyche/Plectropsyche*
[Cosmopolitan]

11(9) Abdominal segments dorsally covered with decumbent, black, pointed and scattered, erect, long, black setae, without flattened and scaliform setae (Fig. 10.22 B); submentum lateral margin slightly convex, anterolateral surface with tuft of setae (Fig. 10.22 C); abdominal gills with a central stalk and few filaments (less than five) arising randomly from them (Fig. 10.22 D) .... *Calosopsyche*
[Caribbean, Central America]

11' Abdominal segments dorsally covered with flattened and scaliform setae; submentum lateral margin almost straight to slightly concave, anterolateral surface with few setae (Figs. 10. 23 B, 24 C); abdominal gills usually without central stalk, all filaments arising from base; when with central stalk, filaments (more than five) arising from the same length (Figs. 10.23 E, 24 A) .... ***Smicridea*** **[p. 249]**
[Mexico to Argentina]

Trichoptera: Hydropsychidae: *Smicridea*: Subgenera

1 Abdominal segment VIII ventral sclerite consisting of a single plate (Fig. 10.23 A) (except in *S.* (*S.*) *annulicornis* (Blanchard) and *S.* (*S.*) *frequens* (Návas), which has two, occurs in small streams to large rivers) .... *Smicridea*

1' Abdominal segment VIII ventral sclerite consisting of two separate small plates (Fig. 10.24 B) (usually present in large rivers) .... *Rhyachophylax*

Trichoptera: Hydroptilidae: Genera

Modified from Wiggins (1996) and Pes et al. (2005). Hydroptilidae are the most diverse family in Trichoptera, with more than 2,000 known species, 37 genera, including 940 species recorded from the Neotropical Region (Holzenthal & Calor, 2017). A characteristic shared by hydroptilids is heteromorphosis (or hypermetamorphosis), in which first to fourth larval instars are similar to small beetle larvae, acquiring the typical aspect of a caddisfly only in the fifth and final instar (Marshall, 1979; Wiggins, 2004). These early instars do not build a case, but the final, fifth instar does in almost all genera. Larvae of this family present a large diversity of case forms and behavior. This diversity includes species that build almost tubular cases, cases consisting of two lateral valves, and those that only build a shelter for pupation. However, a basic plan, in which the cases are composed of two lateral valves, can be seen even in tubular ones, in which a lateral fusion of two pieces is visible (Wiggins, 2004). The larvae of *Angrisanoia*, *Ascotrichia*, *Ceratotrichia*, *Costatrichia*, *Orinocotrichia*, and *Tizatetrichia* are unknown. Since Müller (1888), larvae of *Tupiniquintrichia* have not been collected, and very little is known about their morphology, except for the typical leucotrichiine cases. The larva and pupa of an undescribed species of *Betrichia* was described equivocally (A.M. Pes, unpublished data) as *Ceratotrichia* sp. in Amazonas state (Pes & Hamada, 2004).

1 Abdomen laterally compressed or almost cylindrical; larva with portable case; abdominal segments II–VII with or without tergites (Figs. 10.25 B, 27 A, 28 A, B, 29 A, B, 30 A, B, 31 A, B, 32 A, 33 A, B, 34 A, C, 35 A, E, 36 A, C, 37 A, D, 38 A, B) 2

1' Abdomen dorsoventrally depressed or with segments V and VI abruptly wider than segment IV (Figs. 10.42 B, 43 C, 45 B); larva free living or with a fixed retreat on rock, as a flat silken dome (Figs. 10.39 A, 41 B, 42 A, B, 44 A, 45 A); abdominal segments II–VII with tergites (Figs. 10.39 C, 40 A) 19

2(1) Pronotum divided longitudinally by middorsal ecdysial line, meso- and metanotum entire, without median ecdysial lines (Figs. 10.25 A, 26 A); abdominal segments with rectangular tergites covering most of dorsal area of the segment (Fig. 10.25 C) 3

2' All thoracic nota with middorsal ecdysial line (Figs. 10.31 C, 34 C); abdominal segments II–VII without tergites, or if present, tergites are small and circular, covering less than a quarter of the dorsal area of the segment (Fig. 10.34 C) 4

3(2) Pronotum with a small anteriolateral depression (Fig. 10.25 A) .... *Byrsopteryx*
[Mexico to Brazil]

3' Pronotum without lateral depression (Fig. 10.26 A) .... *Celaenotrichia*
[Argentina, Chile]

4(2) Anal prolegs elongate, projecting prominently from the body (Fig. 10.29 D); abdomen with a distinct lateral fringe of fine filaments (Fig. 10.29 B); portable case almost cylindrical or conical, with anterior ending wider than posterior (Figs. 10.27 A, 28 A, 29 A) 5

4' Anal prolegs short, not projecting prominently from the body (Figs. 10.30 B, 33 C, 34 B, 35 D, 36 B, 37 C, E, 38 B); abdomen usually without a lateral fringe; portable case of two lateral valves or, if cylindrical, with both anterior and posterior openings narrow (case fusiform) (Figs. 10.30 A, 31 A, B, 32 A, 33 A, 35 A, 36 A, 37 A, B, 38 A) 8

5(4) Abdominal segments well defined; portable case of sand grains, sponge spicules, algae, or sediments 6

5' Limits of each abdominal segment poorly defined; portable case of silk with or without mineral grains included, walls reinforced with transverse or longitudinal silken ridges (Fig. 10.27 A) .... *Mayatrichia*
[Ecuador. Nearctic]

6(5) Head slightly narrowing anteriorly, length ~2× width; antenna second article bifurcated, length 1/3 first article (Fig. 10.29 C); abdominal segment I with no sclerotized tergites; portable case of sand grains or organic materials (Fig. 10.29 A) 7

6' Head long and narrow, length ~3× width (Figs. 28 A, B); antenna second article length 1/2 first article, not bifurcated (Fig. 10.28 B); abdominal segment I with small tergites bearing stout setae (Fig. 10.28 C); portable case of sponge spicules (Fig. 10.28 A) *Taraxitrichia*
[Venezuela, unassociated larvae South and North of Brazil]

7(6) Thoracic nota margins with elongate setae (Figs. 10.29 A, B) ..... *Neotrichia*
[Neotropical]

7' Thoracic nota margins with short setae ..... *Kumanskiella*
[Greater Antilles]

8(4) Middle and hind legs distinctly longer than forelegs (Figs. 10.30 B, 31 C, 32 A) 9

8' Thoracic legs similar in length (Figs. 10.33 A, 34 B, 36 A) 12

9(8) Abdominal segments without prominent ventral or dorsal lobes; portable case of several forms (Figs. 10.31 A, 33 A, 34 A, 36 A).... 10

9' Abdominal segments with prominent ventral and dorsal lobes (Fig. 10.30 B); portable case of silk, compressed, anterior end with a small circular opening, posterior end open (Fig. 10.30 A) ..... *Ithytrichia*
[Mexico, Argentina. Nearctic]

10(9) Antenna article II without a basal seta (short or long in the 1st article) (Fig. 10.32 B); portable case of silk, usually with plant material incorporated, anterior opening not circular (Fig. 10.32 A) 11

10' Antenna article II with a long median seta (Fig. 10.31 D); portable case of transparent silk, narrowing anteriorly to a circular opening at the anterior end, resembling a bottleneck (Figs. 10.31 A, B) ..... *Oxyethira*
[Cosmopolitan]

11(10) Middle and hind legs length 3–4× longer than forelegs (this character from Old World species, the Neotropic larva unknown) ..... *Tricholeiochiton neotropicalis* Flint, 1991
[Brazil. Australia, Oriental, Palearctic]

11' Middle and hind legs length ≤2× forelegs (Fig. 10.32 A)] ..... *Flintiella*
[Mexico to Argentina]

12(8) Abdominal segment IX without filamentous gills arising from the lateral sclerites of anal prolegs (in some genera the medial filament may be present) 13

12' Abdominal segment IX with three long filamentous gills, one arising from dorsomedian area of the segment, other two near the lateral sclerites of anal prolegs (Fig. 10.33 B) ..... *Hydroptila*
[Cosmopolitan]

13(12) Portable case of two lateral valves, anterior and posterior openings not circular 14

13' Portable case fusiform, anterior and posterior ends with circular openings reinforced with silk, usually with a long stalk fixed to the substrate (Figs. 10.34 A, B) ..... *Rhyacopsyche*
[Mexico to Argentina]

14(13) Abdominal segments II–VII with dorsal chloride epithelia distinct or with small median tergites 15

14' Abdominal segments II–VII without dorsal chloride epithelia distinct or sclerotized median tergites 18

15(14) Meso- and metanotum without anterolateral lobules; abdominal segments II–VII with dorsal chloride epithelia rounded 16

15' Meso- and metanotum with anterolateral lobules; abdominal segments II–VII with dorsal chloride epithelia with or with small median tergites (Fig. 10.35 E) (probably associated with chloride epithelia) ..... *Ochrotrichia*
[Neotropics. Nearctic]

16(15) Thoracic legs with tarsal claw bearing a slender or stout basal seta, not foliaceous 17

16' Thoracic legs with tarsal claw bearing a long and foliaceous basal seta, only slightly smaller than the claw (Figs. 10.36 A, C) ..... *Bredinia*
[Mexico to Argentina]

17(16) Portable case of lateral silken valves, usually with filamentous algae, small pieces of briophytes incorporated, or covered by sand grains, in some species with two small dorsal chimneys (Figs. 10.37 A, B); middle and hind legs with tarsal claw lacking the basal seta (Figs. 10.37 G, H); abdominal segments II and IV usually with lateral humps ..... *Metrichia*
[Neotropics. Nearctic]

17' Portable case of lateral silken valves, with two small dorsal chimneys (Fig. 10.38 A); median and hind legs with tarsal claw bearing a short and blunt basal seta; abdominal segments II and IV without lateral humps (Fig. 10.38 B) ..... *Ragatrichia*
[South America]

18(14) Labrum asymmetrical; abdominal segments II and IV without lateral humps; portable case of silk with longitudinal ridges ..... *Orthotrichia*
[Cosmopolitan]

18' Labrum symmetrical; abdominal segments II and IV with small lateral humps; portable case of silk and filamentous algae (these characters from North America taxa, Neotropical larvae are unknown) ..... *Nothotrichia*
[Brazil, Chile, Costa Rica, Panama. Nearctic]

19(1) Abdomen strongly depressed dorsoventrally (Figs. 10.39 B, C, 40 A); abdominal segments V and VI equal or slightly wider than the anterior segments (Figs. 10.39 B, 40 A); abdominal segments II–VII with rectangular tergites covering most of the dorsal surface of the segments; larva free living, only building a fixed retreat immediately before pupation 20

19' Abdomen only slightly depressed dorsoventrally (Figs. 10.41 A, 43 B, 44 C); abdominal segments V and VI abruptly wider than segments I–IV (Figs. 10.41 A, 43 B); abdominal segments II–VII with small tergites, covering less than half of dorsal surface; dome-shaped fixed retreats (Figs. 10.41 B, 43 A, 44 A) 23

20(19) Thorax and abdomen with unmodified setae; femora without fimbriate setae 21
20' Thorax and abdomen dorsally with enlarged, mucronate setae (Figs. 10.39 B–E); femora with a fimbriate seta (Fig. 10.39 D) .... *Cerasmatrichia*
[Greater Antilles and northern South America]
21(20) Meso- and metanotum without median ecdysial line (Fig. 10.40 A) 22
21' Meso- and metanotum with median ecdysial line .... *Mejicanotrichia*
[Mexico, Guatemala]
22(21) Thoracic and abdominal tergites with lacunae or pores .... *Scelobotrichia*
[Mexico]
22' Thoracic and abdominal tergites without lacunae or pores (Fig. 10.40 A) .... *Alisotrichia*
[Neotropics. Nearctic]
23(19) Abdominal segments I–VII each with only one median tergite (Figs. 10.42 C, 43 B) 24
23' Abdominal segment I with tergite transverse and abdominal segments II–VII each with two small tergites (Fig. 10.41 A)...... *Betrichia*
[South America]
24(23) Foreleg femur with unmodified setae; pronotum anterolateral angle not produced; pupal case fixed to substrate ventrally (Figs. 10.43 A, 44 A) 25
24' Foreleg femur with an arborescent seta; pronotum anterolateral angle produced anteriorly (Fig. 10.42 C); pupal case with a long stalk fixed to substrate (Figs. 10.42 A, B) .... *Anchitrichia*
[Neotropics]
25(24) Each abdominal tergite with a pair of median circular lacunae (Fig. 10.45 C) 26
25' Abdominal tergites without lacunae (Fig. 10.43 B) (except larvae of the *Leucotricha brasiliana* Satter and Sykora with abdominal tergites with a pair of median circular lacunae) .... *Leucotrichia*
[Neotropics. Nearctic]
26(25) Head medially with a prominent dorsal tubercle (Fig. 10.44 C) .... *Peltopsyche*
[South America]
26' Head without prominent tubercles (Fig. 10.45 E); thoracic legs with tarsal claw bearing an enlarged basal seta, giving the appearance of a bifid tarsal claw (Fig. 10.45 D) .... *Zumatrichia/Acostatrichia*
[Neotropics. Nearctic]

Trichoptera: Limnephilidae: Genera

Modified from Flint (1982), Wiggins (1996), Morse & Holzenthal (2008), and Angrisano & Sganga (2009). Limnephilidae is a large family with about 1000 described species distributed in 93 genera around the world, and about 50 species in 10 genera in the Neotropics (Holzenthal & Calor, 2017). They are usually associated with water bodies in high elevations or in cool, southern temperate regions. Larvae may be found in small streams and creeks to large rivers, as well as lentic habitats such as lakes, ponds, and depositional areas in streams, where they are associated with rocky substrate, sand, or leaf litter. They can use a range of matter to build their tubular portable cases, such as plant or rock fragments, wood, sand, and sometimes even molluscan shells, small crustaceans, and other caddisfly cases (Holzenthal & Calor, 2017). All known larvae of Limnephilinae are from Nearctic region, and the characteristics used in the key are based on them. The Neotropical representatives are clearly related to the Nearctic fauna and probably share most of these characteristics.

1 Head with primary and secondary setae (Figs. 10.46 D, 48 D, E); mesonotum primary setal areas separate; mid- and hind femora ventral edge usually with two major setae (Figs. 10.48 G, H); abdominal segments with chloride epithelia present ventrally and laterally; abdominal segment IX dorsal sclerite sparsely setose (Figs. 10.46 B, C) 2
1' Head with primary setae only (Figs. 10.50 C, 52 D, 53 G, 54 C, 55 C, 56 D); mesonotum primary setal areas confluent; mid- and hind femora ventral edge with three or more major setae (Figs. 10.51 B, C, 52 H, 53 I, J, 54 I, J, 55 G, H, 56 F, G); abdominal segments with chloride epithelia present only ventrally; abdominal segment IX dorsal sclerite densely setose (Figs. 10.52 E, 53 D, 54 A) 4
2(1) Dorsal and ventral abdominal gills two or three branched (Fig. 10.48 C); segments II–VIII with lateral abdominal gills 3
2' Dorsal and ventral abdominal gills four branched or more (Figs. 10.46 B, C); segments II–V only with lateral abdominal gills (Figs. 10.46 B, C) .... *Hesperophylax*
[Mexico. Nearctic]
3(2) Abdominal segment I venter with more than 100 setae across all setal areas (Fig. 10.48 A); head and pronotum covered by small spines; anal proleg lateral sclerite with several short stout setae; mid femur with major setae positioned basal .... *Clistoronia graniculata* (Denning in Denning & Sykora, 1966)
[Mexico]
3' Abdominal segment I venter with less than 100 setae across all setal areas (Fig. 10.48 C); head and pronotum usually without small spines (Figs. 10.48 D, E); anal proleg lateral sclerite usually lacking short, stout setae (Figs. 10.48 B, C); mid femur with major setae positioned medially, equidistant from midpoint of femur (Fig. 10.48 G) .... *Limnephilus*
[Mexico to Panama. Holarctic]
4(1) First abdominal segment without sclerites; head and thoracic nota with simple setae; forefemur ventral margin basal half not expanded 5
4' First abdominal segment dorsally with four sclerites (Fig. 10.49 A); head and thoracic nota bearing large, spiniform setae; forefemur ventral margin with basal half expanded (Fig. 10.49 B) .... *Platycosmoecus beaumonti* (Schmid, 1958)
[Argentina, Chile]
5(4) Head without dorsal ridge or carina 6

5' Head with dorsal ridge or carina (Figs. 10.50 A, C) ..... *Anomalocosmoecus*
[Argentina, Bolivia, Colombia, Ecuador, Peru]

6(5) Tibiae smooth, with only an apical pair of mucronate setae, or with a fringe of short setae (Figs. 10.52 F–H, 53 H–J, 54 H–J, 55 F–H, 56 E–G) 7

6' Tibiae medial margins with two rows of irregular mucronate setae (Figs. 10.51 A–C) ..... *Antarctoecia nordenskioeldii* (Ulmer, 1905) in part
[Argentina, Bolivia]

7(6) Anal prolegs ventrally with setae on membrane (Figs. 10.52 B, C, E, 53 A, D) 8

7' Anal prolegs ventrally without setae (Figs. 10.54 D, 55 A) 9

8(7) Abdominal segments II–VII ventrally with ovoid rings outlining chloride epithelia, their width ~8× length (Fig. 10.52 C) ..... *Metacosmoecus nigrofasciatus* Schmid, 1955
[Chile]

8' Abdominal segments III–VII or IV–VII ventrally with ovoid rings outlining chloride epithelia, their width ~3 to 4× length (Figs. 10.53 A, C) ..... *Verger*
[Argentina, Chile]

9(7) Head, pro- and mesonotum without middorsal pale stripe (Figs. 10.55 C, E); head with distinctive muscle scars (Fig. 10.55 C); mid- and hind tibiae and tarsi without medial fringe of short setae (Figs. 10.55 G, H) 10

9' Head, pro- and mesonotum with a middorsal pale stripe (Figs. 10.54 A, C); head without distinctive muscle scars (Fig. 10.54 C); mid- and hind tibiae and tarsi medially often with a fringe of short setae (Figs. 10.54 I, J) ..... *Monocosmoecus*
[Argentina, Chile]

10(9) Forefemur width about half the length, medial margin with many pale setae (Fig. 10.55 F); mandibles toothed apically (Fig. 10.55 D) ..... *Austrocosmoecus hirsutus* Schmid, 1955
[Argentina, Chile]

10' Forefemur width about 1/3 of length, medial margin only with long, dark setae (Fig. 10.56 E); mandibles without teeth ..... *Antarctoecia brasiliensis* Huamantinco & Nessimian 2003, in part
[Brazil]

Trichoptera: Leptoceridae: Genera

Key modified from Wiggins (1996), Calor & Fröehlich (2008), Morse & Holzenthal (2008), and Angrisano & Sganga (2009). Leptoceridae is one of the largest families of the order (Malm & Johanson, 2011), with about 220 species in 16 Neotropical genera (Holzenthal & Calor, 2017), several of which are endemic. Larvae can be found in a variety of lotic and lentic environments and may use different organic and inorganic matter to build their tubular cases, but in some species, cases are made entirely of silk. Despite their large morphologic diversity, they can be recognized by their relatively elongate antennae and the position of hind coxa usually turned forward. Knowledge of the immature stage of the Neotropical representatives is relatively good, with only three genera *Achoropsyche* (widespread in South America), *Neoathripsodes* (mountains near Teresopolis, Brazil), and *Osflintia* (Peru) with the immature stages completely unknown.

1 Metanotum with three to six sclerotized plates (setal areas [*sa*] 1–3) (Figs. 10.57 C, 58 C, 59 D; 10 60 C, 61 C, 62 B); antennae long, prominent or short, inconspicuous 2

1' Metanotum entirely membranous, without sclerites (or with a pair of very small sclerites and/or setal areas represented by one or a few setae) (Figs. 10.63 C, 64 C, 65 C, 66 C, 67 C, 68 E, 69 D); antennae long, prominent, length about 6× width (Fig. 10.65 D) 7

2(1) Antennae long, prominent, length ~6× width (Fig. 10.65 D) 3

2' Antennae short, inconspicuous, length 2-3× width (Fig. 10.60 D) 5

3(2) Pronotum with anterior edge smooth (Fig. 10.58 D); hind tibia undivided; case constructed of plant fragments (Fig. 10.58 B) 4

3' Pronotum with anterior edge crenulate (Fig. 10.57 D); hind tibia divided (Fig. 10.57 B); case a hollowed out twig or discarded case of other caddisfly (Fig. 10.57 A); widespread ..... *Triplectides*
[Widespread. Australia]

4(3) Metanotal *sa*2 sclerites small, separate; *sa*1 sclerites large, *sa*3 sclerites intermediate in size (Fig. 10.58 C) ..... *Hudsonema flaminii* (Navás)
[Argentina, Chile]

4' Metanotum *sa*2 sclerites fused to form a single, small, medial sclerite; *sa*1 sclerites large, *sa*3 sclerites very small or represented by a few setae (Fig. 10.59 D) ..... *Notalina*
[Northern Andes, Brazil]

5(2) Hind tibia not expanded apicoventrally (Figs. 10.61 D, 62 C); case composed of mineral fragments, entirely of silk, or silk with mineral fragments embedded within (Figs. 10.61 A, 62 F–I) 6

5' Hind tibia broad, greatly expanded apicoventrally (Fig. 10.60 E); case tapered, light brown, composed entirely of transparent silk (Fig. 10.60 A) ..... *Amazonatolica hamadae,* Holzenthal & Pes
[Brazil]

6 Anal proleg accessory hook shorter than anal claw (Fig. 10.61 D); tarsal claws gradually curved (Fig. 10.61 F); cases tapered, long, constructed of light or dark silk, silk and small sand grains, or sand grains alone, sometimes with larger stones laterally, usually without apparent rings (Fig. 10.61 A) ..... *Atanatolica*
[Caribbean, Costa Rica to Brazil. Australia, Oriental]

6' Anal proleg accessory hook and anal claw equal (Fig. 10.62 D); tarsal claws abruptly curved (Fig. 10.62 E); cases tapered, constructed of light or dark silk, silk and small sand grains, or sand grains alone, usually with apparent rings (Figs. 10.62 F–I) .... *Grumichella*
[South America]

7(1) Maxillary palpi extending little, if any, beyond labrum (Fig. 10.68 C); mandibles short, wide, with teeth grouped close to apex around central concavity (Fig. 10.68 D); .... tarsal claws not as long and slender as previous (Fig. 10.68 B)

7' Maxillary palpi usually extended far beyond labrum anterior edge (Figs. 10.63 D, F); mandibles long and mucronate, with sharp apical tooth separated from remaining teeth (Fig. 10.63 E); tarsal claws long and slender (especially hind leg claws) (Fig. 10.63 E) .... *Oecetis*
[Cosmopolitan]

8(7) Anal prolegs without sclerotized spiny plates, although patches of spines or setae may be present (Fig. 10.68 C); widespread......... 9

8' Anal proleg with sclerotized, concave plate with marginal spines on each side of anal opening, and extended onto ventral lobe (Fig. 10.64 D) (the Neotropical species larvae are unknown; the characters are based on North American species) .... *Setodes*
[Hispaniola, Nearctic]

9(8) Foretibia and tarsus without styliform setae; *sa*3 with one to few setae (Fig. 10.68 E) 10

9' Foretibia and tarsus with subapical, ventral, flattened, styliform setae (Fig. 10.65 E); *sa*3 with patch of ca. 10–12 setae (Fig. 10.65 C). .... *Amphoropsyche*
[Antilles, northern South America]

10(9) Hind tibia subdivided or with a dividing constriction (Figs. 10.66 B, 67 B) 11

10' Hind tibia not subdivided, without constriction 12

11(10) Mandibles strongly asymmetrical (Fig. 10.66 E); hind legs usually with dense, long setal fringe (Fig. 10.66 D); slender case a spiral of plant pieces (Fig. 10.66 A) .... *Triaenodes*
[Cosmopolitan]

11' Mandibles only slightly asymmetrical (Fig. 10.67 D); hind legs with only few scattered long setae (Fig. 10.67 B); irregular case of plant and mineral materials, with twigs or conifer needles extending beyond ends (Fig. 10.67 A).................. *Mystacides alafimbriata* Hill-Griffin
[Mexico]

12(10) Abdominal segment IX dorsal plate and anal prolegs with long, thin setae only (Fig. 10.68 G); abdominal segment I lateral hump sclerite with round anterior portion and narrow, posterior extension, bearing a seta (Fig. 10.68 F); gills present or absent; cases constructed of various materials (Fig. 10.68 A) .... *Nectopsyche*
[Neotropics. Nearctic]

12' Abdominal segment IX dorsal plate and anal prolegs with thick, prominent setae (Fig. 10.69 E); abdominal segment I lateral hump angular (Fig. 10.69 C); gills absent; cases constructed of sand grains (Fig. 10.69 A)... *Brachysetodes*
[Argentina, Chile]

Trichoptera: Calamoceratidae: Genera

Key modified from Angrisano & Sganga (2009). Calamoceratidae is a small, widely distributed family with two genera reported from the Neotropics: *Banyallarga,* with 17 species, and *Phylloicus,* with 58 (Prather, 2002; 2003, Barcelos-Silva, et al., 2017). Larvae are frequently found in lateral pools and depositional areas of small streams and backwater areas, associated with leaf litter.

1 Pronotum anterior margin bearing about 15 long dark setae (Figs. 10.70 F, G); gular sclerite produced posteriorly, genae almost completely divided (Fig. 10.70 E); hindleg trochanter with midventral dense brush of short setae (Fig. 10.70 G); case cylindrical, constructed of sand and vegetal fragments (Fig. 10.70 A) .... *Banyallarga*
[Neotropics]

1' Pronotum anterior margin bearing about two to three long dark setae; gular sclerite reaching head anterior half (Fig. 10.71 C); hindleg trochanter without midventral dense brush of short setae (Fig. 10.71 D); case flattened, constructed of circular leaf fragments (Fig. 10.71 A) .... *Phylloicus*
[Neotropics. Nearctic]

Trichoptera: Hydrobiosidae: Genera

Key modified from Angrisano & Sganga (2009). Hydrobiosidae is widely distributed, with 22 recognized genera and 180 species (Holzenthal & Calor, 2017). Five of the 22 genera are not known in the immature stages. They do not build cases, but the final instar larva spins a strong, closed, osmotically semipermeable cocoon under a protective shelter of rock fragments for pupation. All known Hydrobiosidae larvae are predators. Larvae of the *Apatanodes* and *Neopsilochorema* are indistinguishable, just as adults are very similar (Angrisano, 1997). Larvae of the *Androchorema, Australobiosis, Heterochorema, Isochorema,* and *Nolganema* are unknown.

1 Foreleg femur wide, especially apically, without ventrodistal prolongation (Figs. 10.72 A, B, 73 B, 77 D) 2

1' Foreleg femur not particularly widened, with ventrodistal prolongation (Figs. 10.78 B, C, 79 C 82 B, C, 83 B, C, 84 B, 85 B, C, 86 B, 87 C) 7

2(1) Prosternum with sclerites; foreleg claw length subequal to tarsus, with pointed apex (Figs. 10.73 A, 74 A, 75 A, 76 A, 77 E) 3

2' Prosternum without sclerites, membranous (Fig. 10.72 D); foreleg claw thin, curved, length at least 3× tarsus, with rounded apex (Figs. 10.72 A–C) .... *Apatanodes/Neopsilochorema*
[Argentina, Chile]

3(2) Prosternum with three sclerites (one large mesal and two anterolateral small), or only one large mesal (Figs. 10.74 A, 75 A, 76 A, 77 E) 4

3' Prosternum with four sclerites (one large mesal, shorter than broad with irregular margins, one anterior long and narrow, and two small anteriolateral in oblique position) (Fig. 10.73 A) .... *Pseudoradema spinosissimum* Schmid [Chile]

4(3) Prosternum mesal sclerite large, nearly covering prosternum (Figs. 10.75 A, 76 A, 77 E); anal prolegs long, about 2.5 times longer than wide (Fig. 10.77 C) 5

4' Prosternum mesal sclerite wide, width >3× length, subrectangular, concave anteriorly, convex posteriorly (Fig. 10.74 A); anal prolegs short, length 2× width (Fig. 10.74 B) .... *Clavichorema* [Argentina, Chile]

5(4) Mesal sclerite posterior margin narrower than anterior margin, anteriolateral sclerites not lanceolate, or absent (Figs. 10.76 A, 77 E) 6

5' Mesal sclerite posterior margin wider than anterior margin, anteriolateral sclerites lanceolate (Fig. 10.75 A) .... *Neochorema sinuatum* Schmid, 1964 [Chile]

6(5) Prosternum mesal plate posteriolateral edges with paired, very small notches; anteriolateral sclerites absent (Fig. 10.76 A)........ *Iguazu* [Chile]

6' Prosternum mesal plate posteriolateral edges without paired notches; anteriolateral sclerites with rounded base and apex long and very slender (Fig. 10.77 E) .... *Rheochorema* [Argentina, Chile]

7(1) Foreleg tibia and tarsus partially or completely fused; forefemur apicoventral prolongation usually thin (Figs. 10.78 B, C, 79 C, 82 B, C, 83 B, C) 8

7' Foreleg tibia and tarsus not fused; forefemur apicoventral prolongation wide (Figs. 10.84 B, 85 B, C, 86 B, 87 C) 14

8(7) Prosternum membranous 9

8' Prosternum with sclerites (Figs. 10.79 B, 80 D, 82 A, 83 A) 10

9(8) Head posterior third with dark, narrow, longitudinal, narrow stripe (Fig. 10.78 A) .... *Stenochorema crassicostum* Schmid, 1955 [Chile]

9' Head posterior third without dark, longitudinal, narrow stripe .... *Schajovskoya neuquenensis* Flint, 1979 [Argentina, Chile]

10(8) Prosternum with only one small, triangular, mesal sclerite (Figs. 10.79 B, 80 D) 11

10' Prosternum with three sclerites, one mesal and two lateral (Figs. 10.82 A, 83 A) 13

11(10) Prosternal plate trapezoidal (Fig. 10.80 D) 12

11' Prosternal plate small, triangular (Fig. 10.79 B) .... *Cailloma* [Argentina, Bolivia, Chile, Ecuador, Peru]

12(11) Head, pronotum, and legs stramineous (Figs. 10.80 A–C) .... *Parachorema* [Argentina, Chile]

12' Head, pronotum, and legs dark brown (Figs. 10.81 A–C) .... *Amphichorema* [Argentina, Chile]

13(10) Mesal sclerite posterior margin convex, with central notch; lateral sclerite very small, ovoid, smaller than mesal sclerite (Fig. 10.82 A); tibia and tarsus fused internally and externally (Figs. 10.82 B, C) .... *Metachorema* [Argentina, Chile]

13' Mesal sclerite posterior margin straight; lateral sclerites rectangular, length subequal to mesal sclerite (Fig. 10.83 A); tibia and tarsus fused medially, separated laterally (Figs. 10.83 B, C) .... *Microchorema* [Chile]

14(7) Prosternum with more than one sclerite (Figs. 10.85 A, 86 A, 87 D, 88 C) 15

14' Prosternum with a single large central sclerite (Fig. 10.84 A) .... "Genus X" of Bravo & Angrisano (2005). [Chile]

15(14) Prosternum with mesal sclerite present (Figs. 10.86 A, 87 D, 88 C) 16

15' Prosternum with mesal sclerite absent (Fig. 10.85 A) .... *Pomphochorema chilensis* Flint [Chile]

16(15) Prosternum with mesal sclerite large, ovoid to hexagonal (Figs. 10.87 D, 88 C) 17

16' Prosternum with mesal sclerite short, arcuate (Fig. 10.86 A) .... "Genus Y" of Bravo & Angrisano (2005). [Argentina, Chile]

17(16) Prosternum mesal sclerite oval, anteriolateral sclerites three-pointed (Fig. 10.87 D) .... *Neoatopsyche* [Argentina, Chile]

17' Prosternum mesal sclerite approximately hexagonal; anterolateral sclerites subtriangular (Fig. 10.88 C) .... *Atopsyche* [Neotropics. Nearctic]

Trichoptera: Polycentropodidae: Genera

Polycentropodidae has five Neotropical genera and 283 species (Holzenthal & Calor, 2017); most are found in slow-flowing water, in lentic habits such as lakes, large rivers, pools, and backwater areas in small streams. Larvae build a kind of silken retreat, open at both ends, covered by debris, attached to rocky substrate or submerged wood. They are predators, feeding on other aquatic invertebrates (Holzenthal & Calor, 2017). We included here the larvae associated with *Nyctiophylax neotropicalis* Flint through immature rearing, but not yet formally described, as they are different morphologically from other widespread *Nyctiophylax*. Recent phylogenetic studies provided evidence that *N. neotropicalis* belongs to a different clade from the other *Nyctiophylax* (Chamorro & Holzenthal, 2011; Johanson et al., 2012); this conclusion is supported by larval morphology.

1 Anal claw medial margin with teeth or long spines (Figs. 10.93 F, G, 94 C, D)........ 2

1' Anal claw smooth, without teeth, or some species having a slightly serrated surface on the internal margin (Figs. 10.95 B, 96 E) 3

2(1) Anal claw with a distinct comb of teeth, length equals claw width at medial margin (Figs. 10.93 F, G); retreat long and narrow, of silk and sediments, underneath rocks, logs, and roots; larvae usually collected outside the retreat .... *Polyplectropus* [Cosmopolitan]

2' Anal claw medial margin basally with four spines (Figs. 10.94 C, D); retreat of silk, attached to rocks (Fig. 10.94 B) (only *N. neotropicalis* has associated larvae in Neotropics) .... *Nyctiophylax* [Nearly cosmopolitan]

3(1) Anal proleg dorsal plate between claw and lateral sclerite with two dark lines touching medially, making an "X" (Figs. 10.96 D, 97 C) 4

3' Anal proleg dorsal plate between claw and lateral sclerite with two dark lines that do not touch medially; claw medial margin smooth or slightly serrated (Figs. 10.95 B, C)........ *Cyrnellus* [Neotropics. Nearctic]

4(3) Head and pronotum muscle insertion spots light brown (Fig. 10.96 B); frontoclypeus ecdysial line with a slight median constriction; larva small, ~7 mm (Fig. 10.96 A)........ *Cernotina* [Neotropics. Nearctic]

4' Head and pronotum muscle insertion spots dark brown; frontoclypeus ecdysial line with a distinct median constriction (Figs. 10.97 A, B); larvae >15 mm (Fig. 10.97 A)........ *Polycentropus* [Cosmopolitan]

Trichoptera: Glossosomatidae: Genera

Glossosomatidae is cosmopolitan, with about 266 Neotropical species in 11 genera (Robertson & Holzenthal, 2013). Larvae usually live in cool running water, but in South America, *Protoptila* and *Mortoniella* also occur in warm flowing water, and they live on wood, leaves, and stones on the bottom. When feeding ceases during the last instar, the larva fastens the dorsal portion of the case, which resembles a dome, with silk to the substrate. Larvae are very small and with very pale thoracic sclerites; these are diagnostic at least for the Nearctic genera (Wiggins, 1996). There is very great variation among species within genera. Larvae of *Canoptila* (Brazil), *Merionoptila* (Argentina), *Scotiotrichia* (Argentina, Chile), and *Tolhuaca* (Chile) are unknown.

1 Mesonotum with three sclerites, metanotum with two sclerites; anal claw usually with one, two, three, or more accessory teeth pairs (Fig. 10.101 D) 2

1' Meso- and metanotum without sclerites (Fig. 10.98 C); anal claw with a pair of accessory teeth (Fig. 10.98 D)........ *Glossosoma* [Mexico. Holarctic, Oriental]

2(1) Tarsal claws with basal seta short, if long, never arising from basal process 3

2' Tarsal claws with basal seta long and slender, arising from basal process (Fig. 10.101 E) .... *Protoptila* [Neotropical. Nearctic]

3(2) Anal claw with one or two pairs of accessory teeth........ 4

3' Anal claw with three pairs of accessory teeth 6

4(3) Abdominal tergum IX posterior edge with different number of setae than follows........ 5

4' Abdominal tergum IX posterior edge with three pairs of long setae........ *Mastigoptila* [Argentina, Chile]

5(4) Tarsal claw hook length 2× base; basal process reduced, basal seta short, wide, and lightly pointed (Figs. 10.102 D–F) .... *Mortoniella* (in part) [Neotropical]

5' Tarsal claw hook length subequal to base; basal process reduced, basal seta short, wide, dull (Figs. 10.99 C, D) .... *Cariboptila* [Antilles, Mexico to Costa Rica. Nearctic]

6(3) Tarsal claw different from follows........ 7

6' Tarsal claw hook length 1.5× base; basal process large, basal seta short, slender, arising between process and hook (Fig. 10.102 E) .... *Mortoniella* (in part) [Ecuador, Peru]

7(6) Foreleg tibia with three plumose setae, mid- and hindleg tibiae with two plumose setae and one spine between them (Figs. 10.100 D, F) .... *Itauara* [Venezuela to Argentina]

7' Tibiae apicoventral margin with a stout spine........ *Culoptila* [Mexico to Costa Rica. Nearctic]

Trichoptera: Anomalopsychidae: Genera

Anomalopsychidae is endemic to the Neotropics, with species known from Costa Rica, the Andes from Colombia to Chile, and the mountains of southeastern Brazil. The family includes 28 species in two genera: *Anomalopsyche* and *Contulma* (Holzenthal & Flint, 1995; Holzenthal & Robertson, 2006; Holzenthal & Ríos-Touma, 2012). The larvae are possibly collector-scrapers; Holzenthal & Ríos-Touma (2012) observed their gut contents and found mostly diatoms. In the high Andean Paramo, larvae prefer small streams in forested areas and build a tubular case of sand grains.

1 Head and thorax with primary setae only (Fig. 10.103 B); mesonotum with dorsal plate only (*sa*1) (Fig. 10.103 B) .... *Anomalopsyche minuta* (Schmid)
[Chile]

1' Head and thorax with dense covering of short secondary setae and/or scales (Figs. 10.104 A, B); mesonotum with dorsal plate (*sa*1), and pair of smaller, lateral sclerites (*sa*3) or setal patches (Fig. 10.104 B) .... *Contulma*
[Costa Rica, to Brazil, Chile]

Trichoptera: Tasimiidae: Genera

Tasimiidae is a small family of which only the monotypic *Charadropsyche* and *Trichovespula* are recorded from the Neotropical Region, both limited to the Chilean subregion (Holzenthal & Calor, 2017). Larvae can be found in small, shallow streams where they cling to the faces of rocks (Holzenthal & Calor, 2017).

1 Pronotum dark reddish-brown, paler posteriorly, lateral margin developed into a flange produced into an anterolateral, pointed process (Fig. 10.105 A) .... *Charadropsyche penicillata* Flint, 1970
[Chile]

1' Pronotum uniformly pale, lateral margin with a small projection, but not a pointed process (Fig. 10.106 A) .... *Trichovespula macrocera* Schmid, 1955
[Chile]

Trichoptera: Odontoceridae: Genera

Odontoceridae has 46 described species in three genera in the Neotropics (Holzenthal & Calor, 2017). Larvae are usually found in slow to moderate running water of small- to medium-sized streams to large rivers, normally associated with bottoms of gravel or sand, where they are sometimes found buried in the substrate. They build portable tubular cases with sand grains and other mineral fragments.

1 Pronotum anterior angle without projection; larvae <20 mm (Figs. 10.108 B, C, 109 D) 2

1' Pronotum anterior angle with acute projection; larvae >20 mm (Figs. 10.107 A, D, E) .... *Barypenthus concolor* Burmeister, 1839
[Brazil]

2(1) Gular sclerite extending half head length, not separating the gena completely (Fig. 10.108 D); metanotum with only small sclerites (Fig. 10.108 C) (streams at elevations of 1860–2060 m) .... *Anastomoneura guahybae* Huamantico & Nessimian, 2004
[Brazil]

2' Gular sclerite reaching head posterior margin, completely separating gena (Fig. 10.109 C); metanotum usually with two rectangular dorsal plates (Fig. 10.109 D), ... *Marilia*
[Neotropics. Australian, Nearctic, Oriental]

Trichoptera: Sericostomatidae: Genera

Sericostomatidae contains eight Neotropical species in three genera (Holzenthal & Calor, 2017; Johanson et al., 2016). Larvae are usually found in small streams with moderate to fast water flow. They build tubular cases of small sand grains or completely of dark silk.

1 Mesonotum with *sa*3 not separate as discrete sclerite; case composed entirely of silk; if with sand grains, not very long and slender (Figs. 10.113 F, 114 E) 2

1' Mesonotum with *sa*3 separate as discrete sclerite (Fig. 10.112 D); case long and slender, of sand grains (Fig. 10.112 A) .... *Gumaga*
[Mexico. Nearctic]

2(1) Metanotum with dark sclerotized plate, anterior margin with about six setae (Figs. 10.113 E, F); foretrochantin triangular with pointed apex (Fig. 10.113 D); abdominal segment I with lateral hump small and prominent, delimited by two long and dark sclerites; case built entirely of dark silk (Fig. 10.113 A) .... *Grumicha grumicha* (Vallot, 1855)
[Argentina, Brazil]

2' Metanotum with a translucent sclerotized plate covered with many long and thin setae (Figs. 10.114 B, E, F); foretrochantin sub-quadrangular, with curved apex (Fig. 10.114 G); abdominal segment I with lateral hump small, not prominent, without sclerites; case built entirely of grains of sand (Fig. 10.114 A) .... *Notidobiella*
[Brazil, Chile, Ecuador]

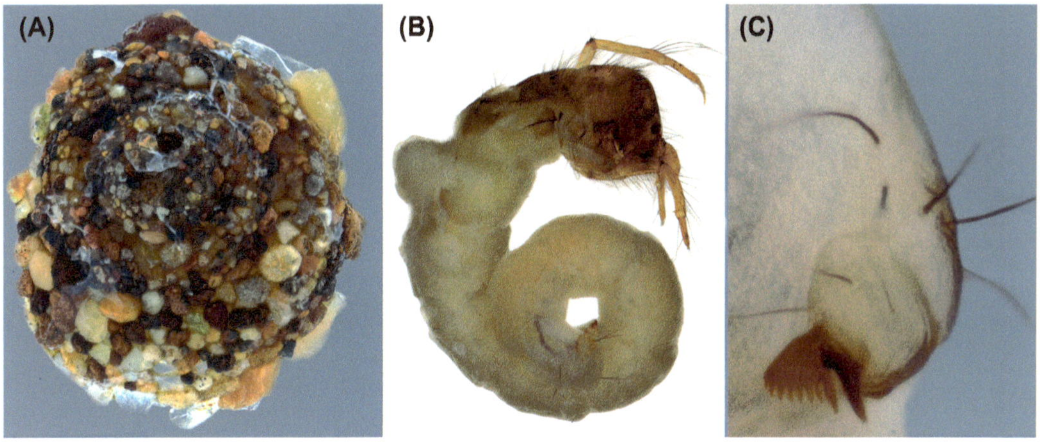

FIGURE 10.1 *Helicopsyche* sp. (Helicopsychidae): (A–B) case and larva, dorsal and lateral view (*From Pes et al. 2014*) and (C) anal claw.

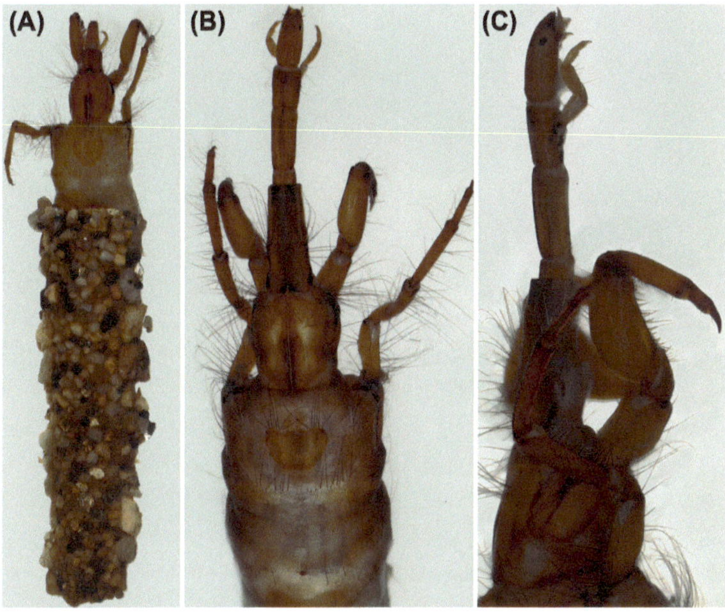

FIGURE 10.2 *Neoatriplectides desiderata* Dumas & Nessimian, 2008 (Atriplectididae): (A) case and larva; (B) thorax, dorsal view; and (C) thorax, dorsal view. (*Figs. B and C from Pes et al. 2014*).

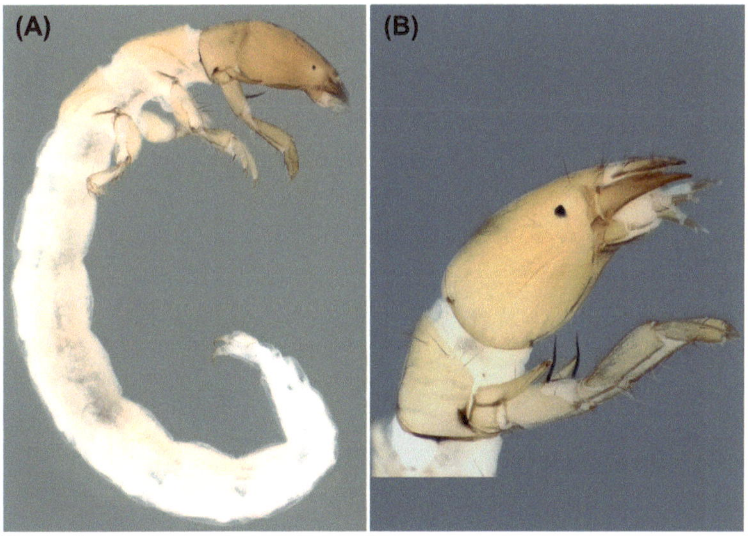

FIGURE 10.3 *Austrotinodes* sp. (Ecnomidae): (A) larva; (B) prothorax, lateral view.

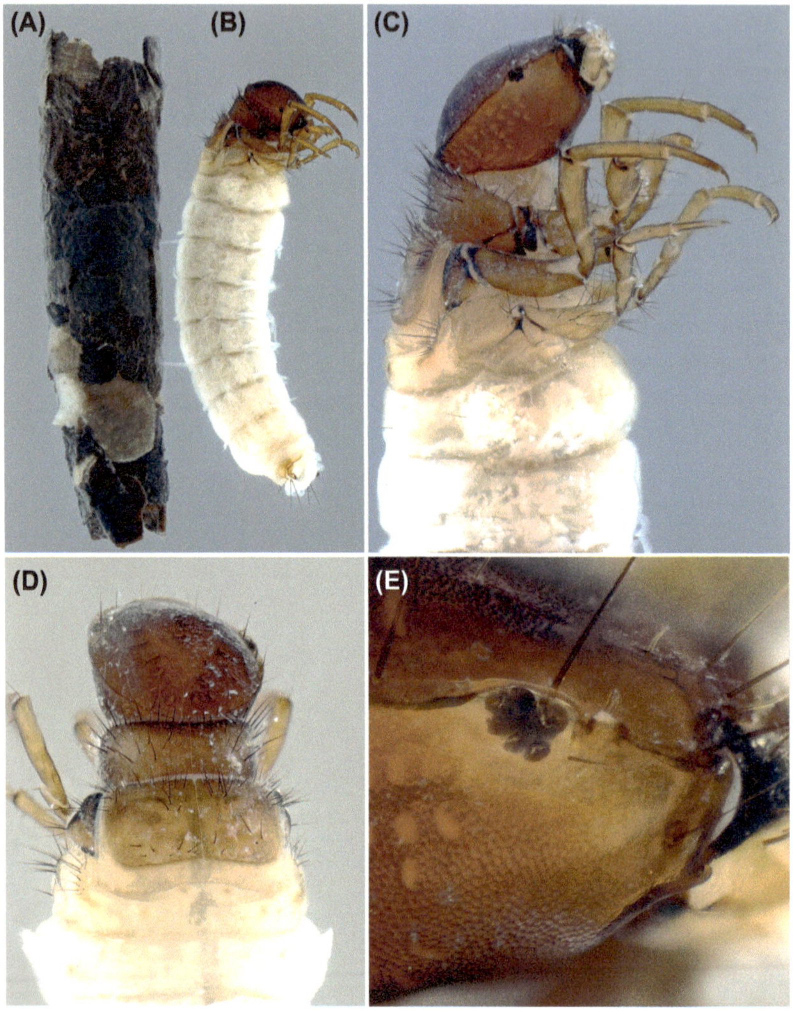

FIGURE 10.4 *Lepidostoma* sp. (Lepidostomatidae). (A–B) case and larva, lateral view; (C–D) head, thorax, lateral and dorsal view; and (E) head, lateral view.

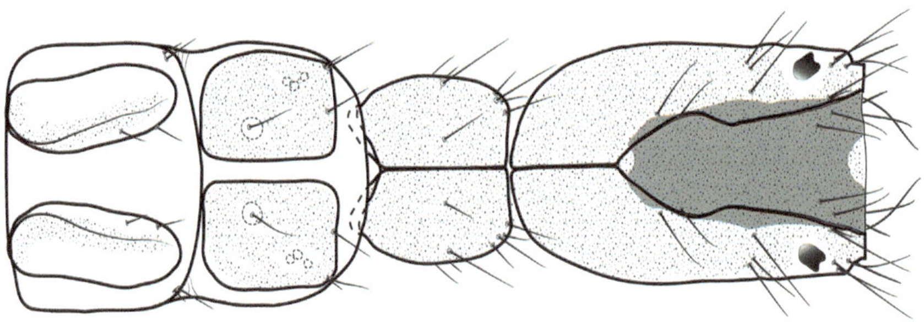

FIGURE 10.5 *Antillopsyche tubicola* Flint, 1964 (Pseudoneureclipsidae): thorax and head, dorsal view. *Modified from Flint (1964).*

FIGURE 10.6 *Rhyacophila* sp. (Rhyacophilidae): (A) larva, lateral view; (B) claw anal; and (C) foreleg.

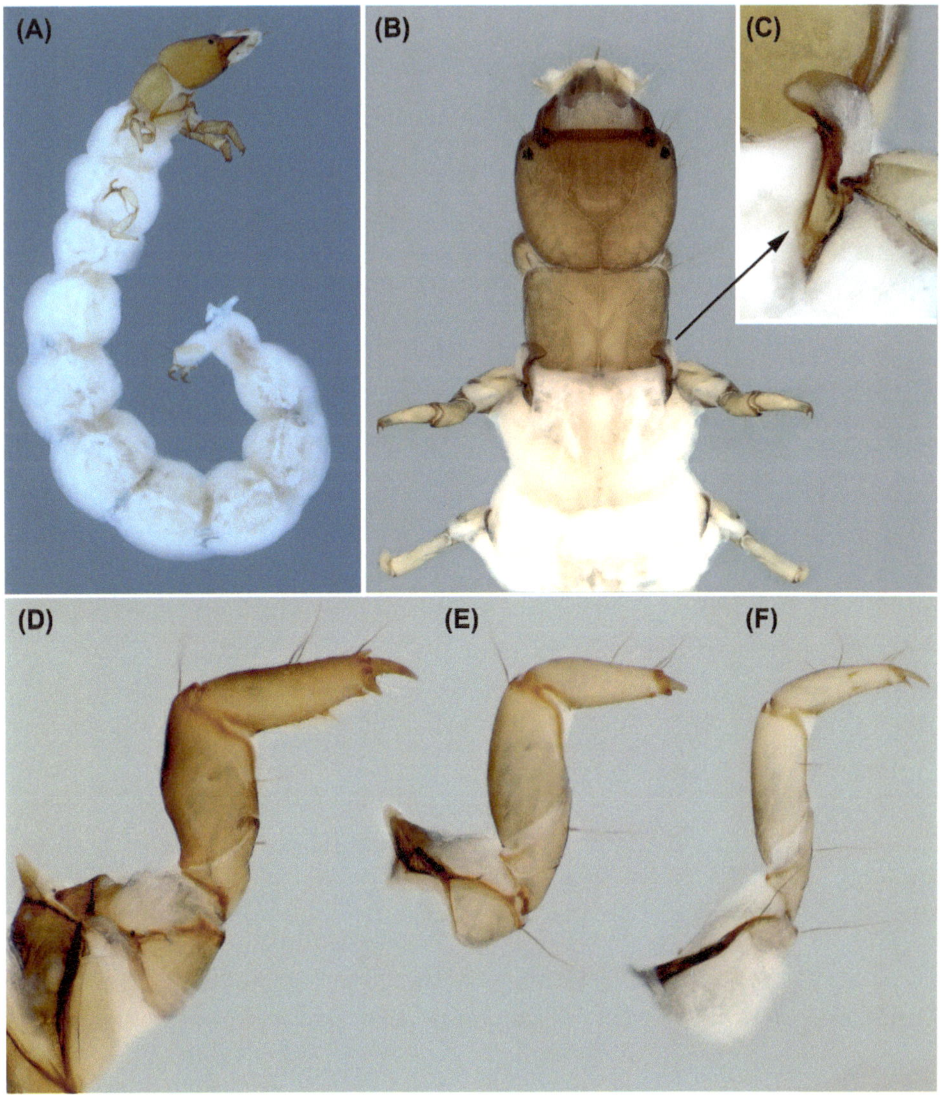

FIGURE 10.7 *Xiphocentron* sp. (Xiphocentronidae): (A) larva, lateral view; (B) head and prometanotum, dorsal view; (C) mesopleura, lateral view; and (D–F) legs.

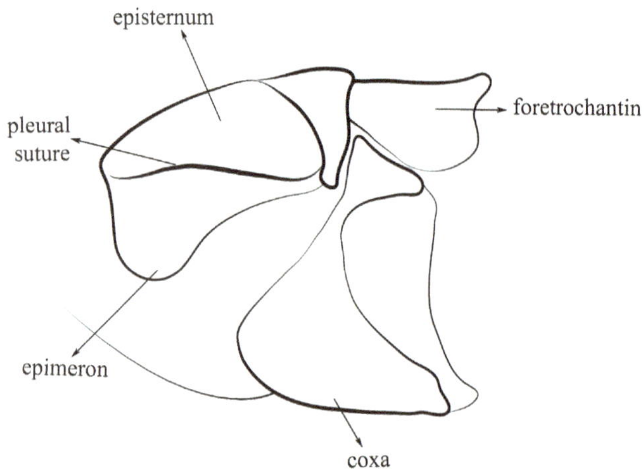

FIGURE 10.8 *Tinodes* sp. (Psycomiidae): forecoxa and foretrocantin.

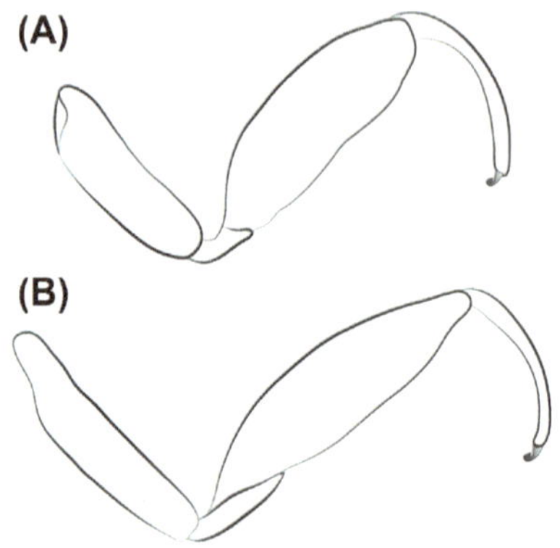

FIGURE 10.9 *Pangullia nea* Rojas, 2006 (Kokiriidae): (A) fore and (B) median legs.

FIGURE 10.10 Philorheithridae (of the Argentina): (A) case, lateral view; (B) thorax, lateral view; (C–E) legs.

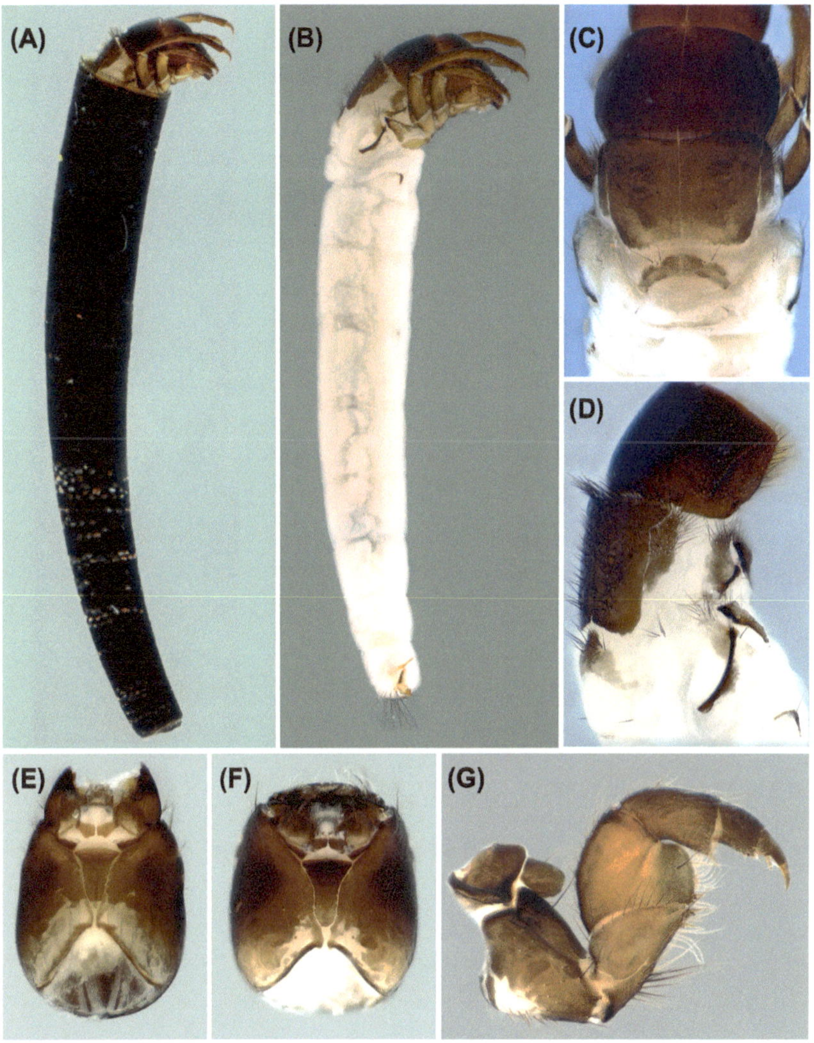

FIGURE 10.11 Parasericostomatidae. (A–B) Larva and case, lateral view (of the Chile); (C–D) thorax, dorsal and lateral view (of the Argentina); (E) head, ventral view (of the Argentina); (F) head, ventral view (of the Chile); and (G) foreleg.

FIGURE 10.12 *Diplectrona* sp. (Hydrosychidae): (A) larva, lateral view; and (B) head, ventral view.

FIGURE 10.13 *Synoestropsis* spp. (Hydropsychidae): (A) *Synoestropsis* sp. lateral view; (B) *S. furcata* Flint, 1974 head and prothorax, lateral view; and (C) *S. grisoli* Ulmer head and thorax, dorsal view.

FIGURE 10.14 *Plectromacronema computum* Ulmer, 1906 (Hydropsychidae): (A) Larva, lateral view and ventrolateral pockets magnified; (B) head, dorsal view; (C) thorax, dorsal view; and (D) thorax and head, lateral view.

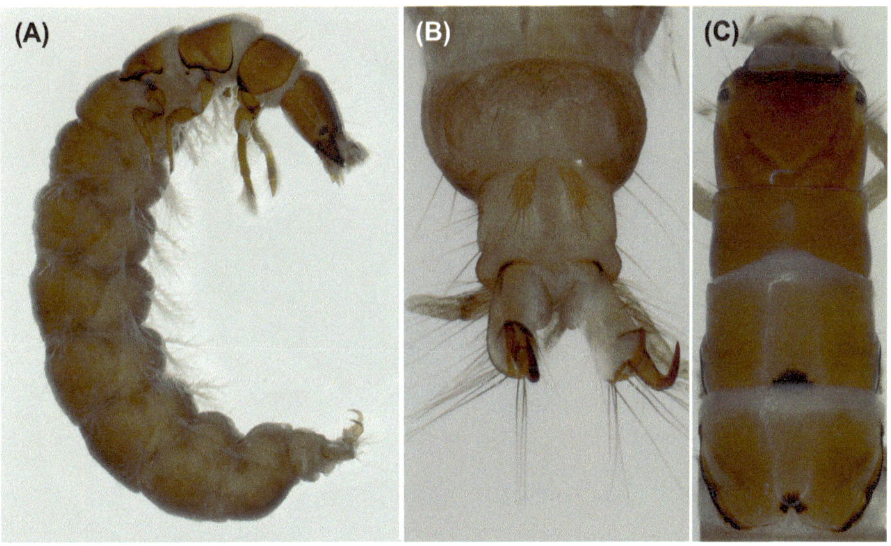

FIGURE 10.15 *Blepharopus diaphanus* Kolenati, 1959 (Hydropsychidae): (A) larva, lateral view; (B) abdominal segment VIII-IX, ventral view; and (C) head and thorax, dorsal view.

FIGURE 10.16 *Macrostemum* spp. (Hydropsychidae): (A) *M. arcuatum* (Erichson, 1848), larva, lateral view; (B) *M. hyalinum* (Pictet, 1836), head and thorax, dorsal view; (C–D) *M. hyalinum,* head, dorsal and ventral view; and (E) *M. erichisoni* (Banks, 1920), abdominal segment VIII-IX, ventral view.

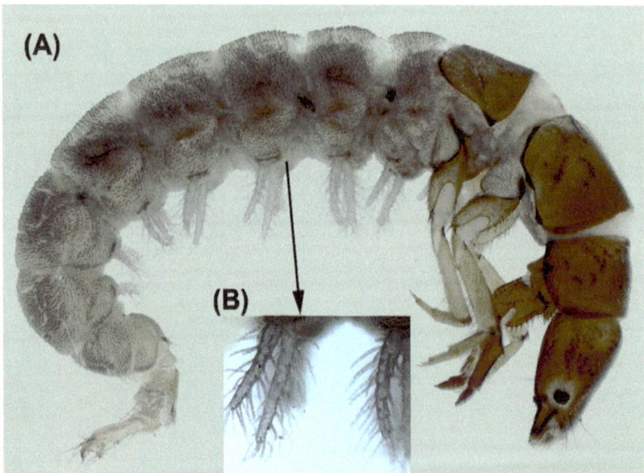

FIGURE 10.17 *Leptonema sparsum* (Ulmer, 1905) (Hydropsychidae): (A) larva (B) abdominal gills magnified, lateral view.

FIGURE 10.18 *Centromacronema obscurum* (Ulmer, 1905) (Hydropsychidae): (A) lateral view; (B) head and thorax, dorsal view; (C–D) head, dorsal and ventral view; and (E) foretrochantin and foreleg.

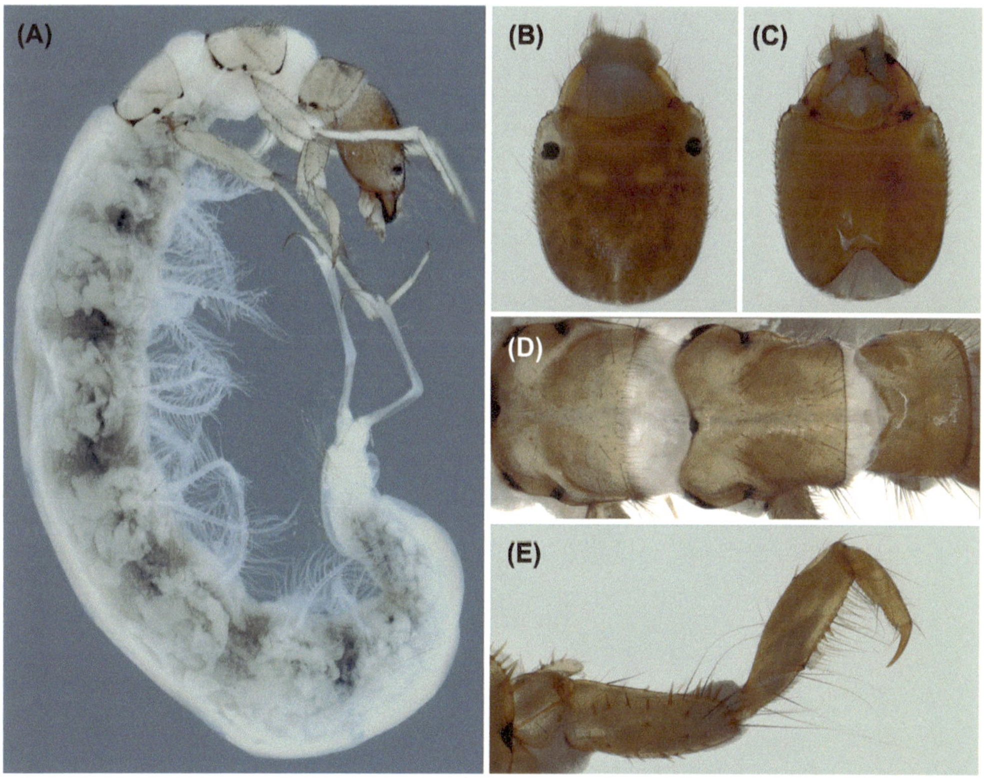

FIGURE 10.19 *Macronema bifidum* Flint, 1974 (Hydropsychidae): (A) Larva, lateral view; (B–C) head, dorsal and ventral view; (D) thorax, view dorsal; and (E) foretrochantin and foreleg.

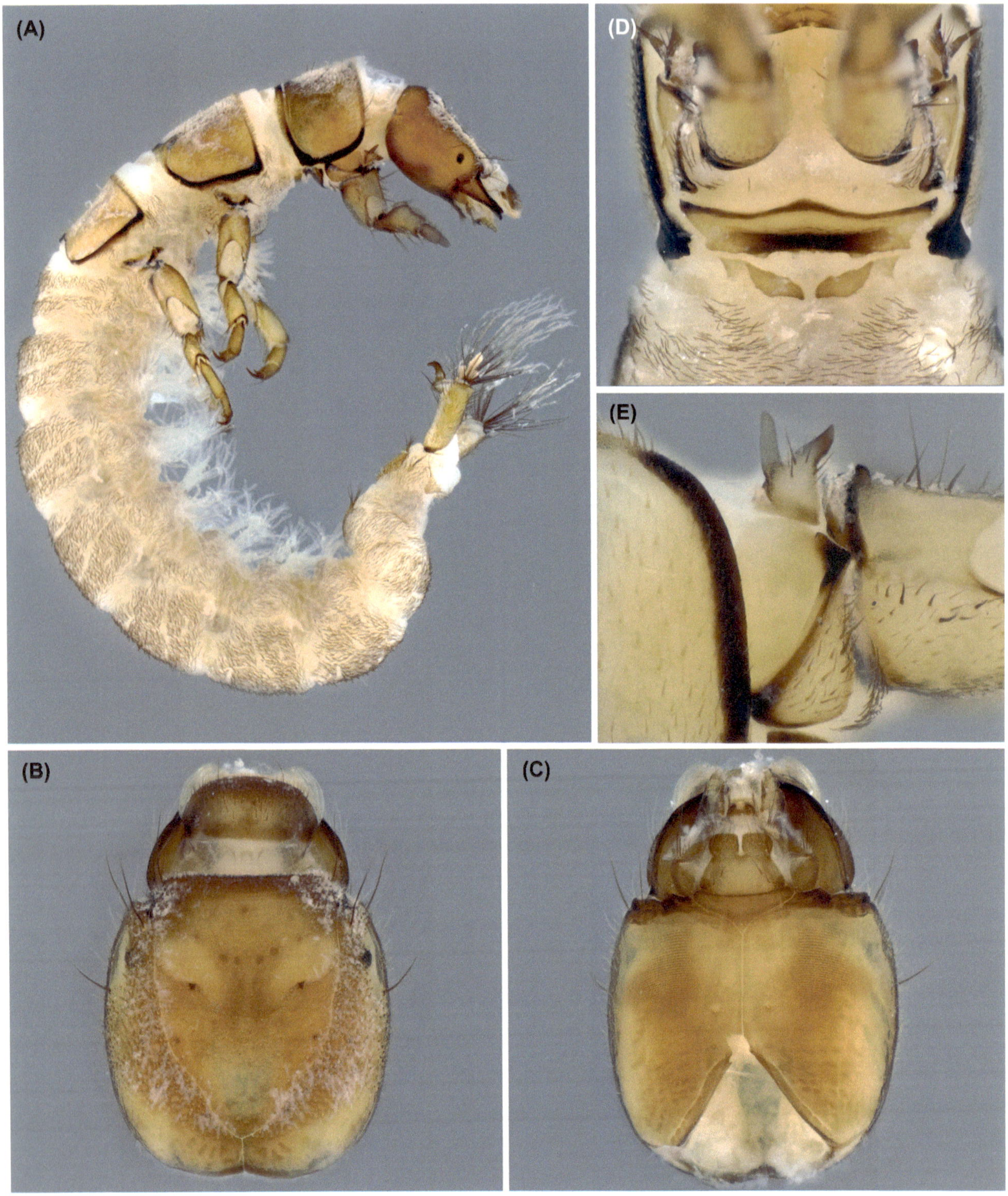

FIGURE 10.20 *Hydropsyche* sp. (Hydropsychidae): (A) lateral view; (B–C) head, dorsal and ventral view; (D) prothorax, ventral view; and (E) foretrochantin.

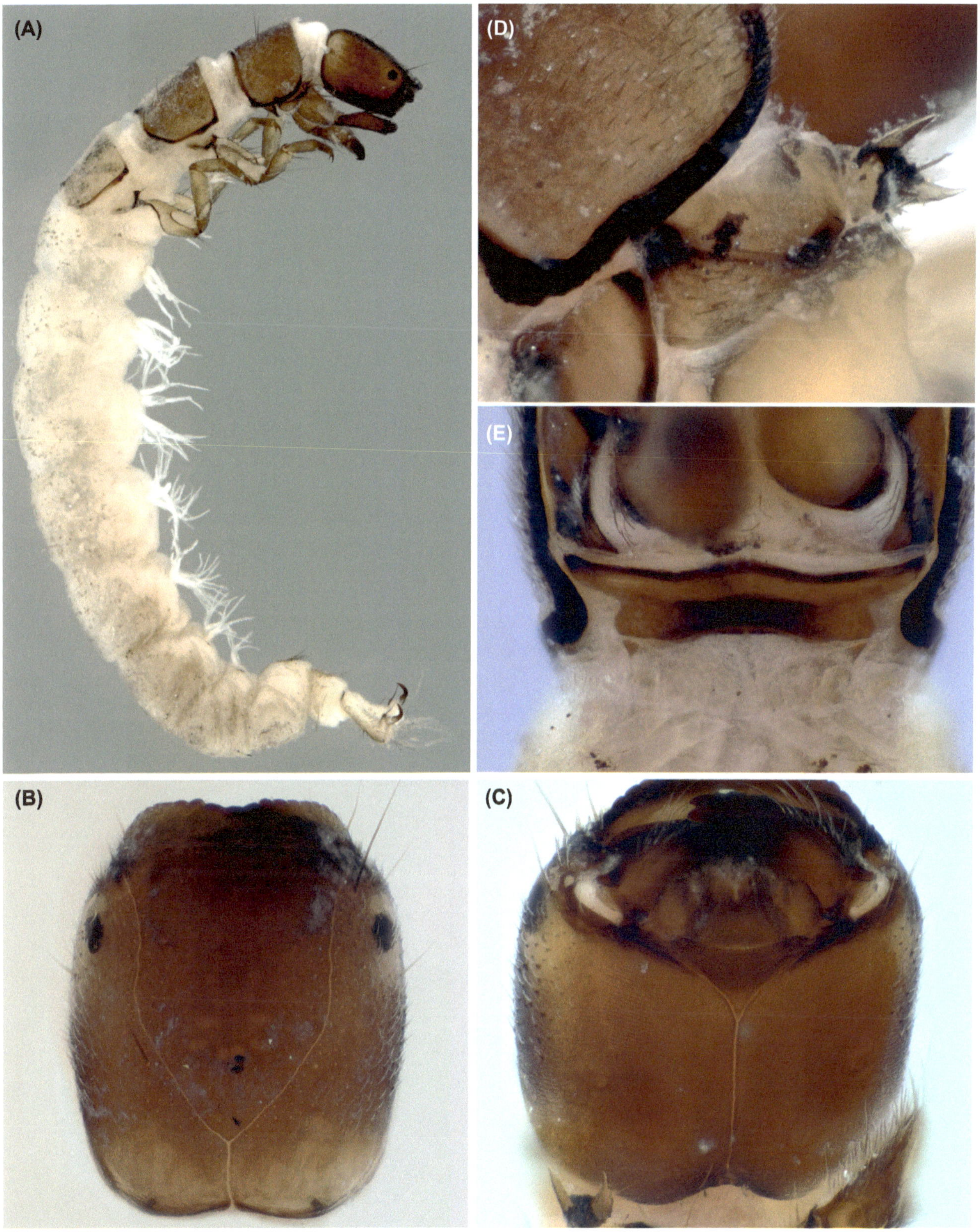

FIGURE 10.21 *Cheumatopsyche/Plectropsyche* (Hydropsychidae): (A) lateral view; (B-C) head, dorsal and ventral view; (D) foretrochantin; (E) prothorax, ventral view.

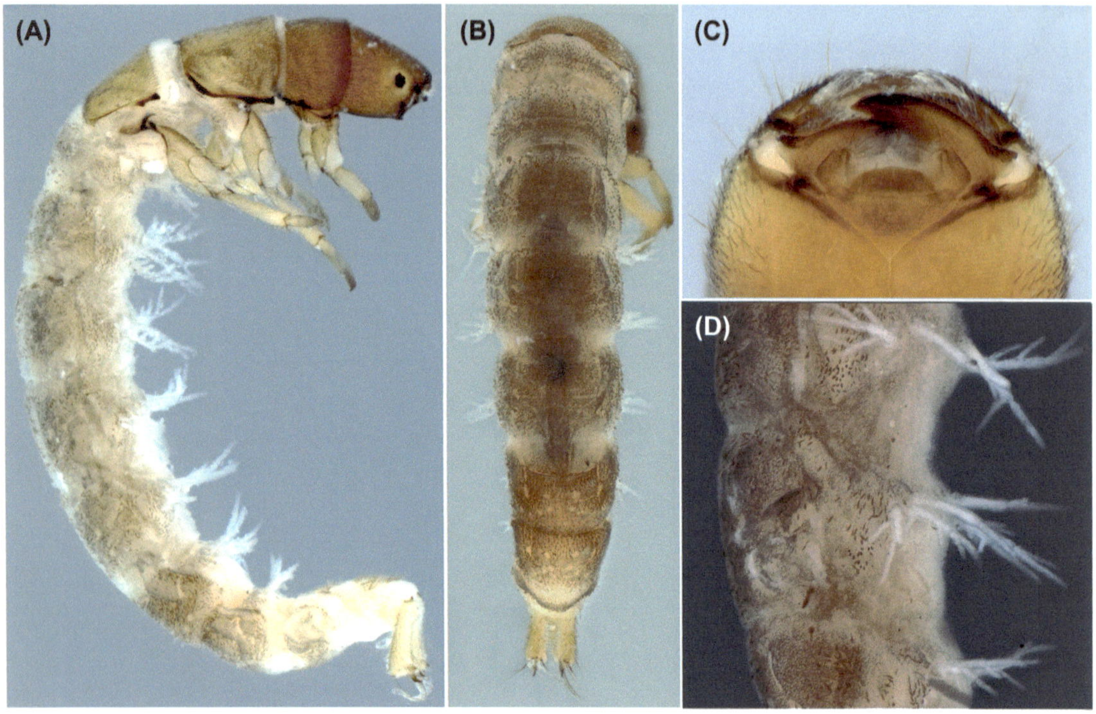

FIGURE 10.22 *Calosopsyche* sp. (Hydropsychidae): (A) lateral view; (B) abdomen, dorsal view; (C) head, mentum, and submentum, ventral view; and (D) abdominal segment IV-VI, lateral view.

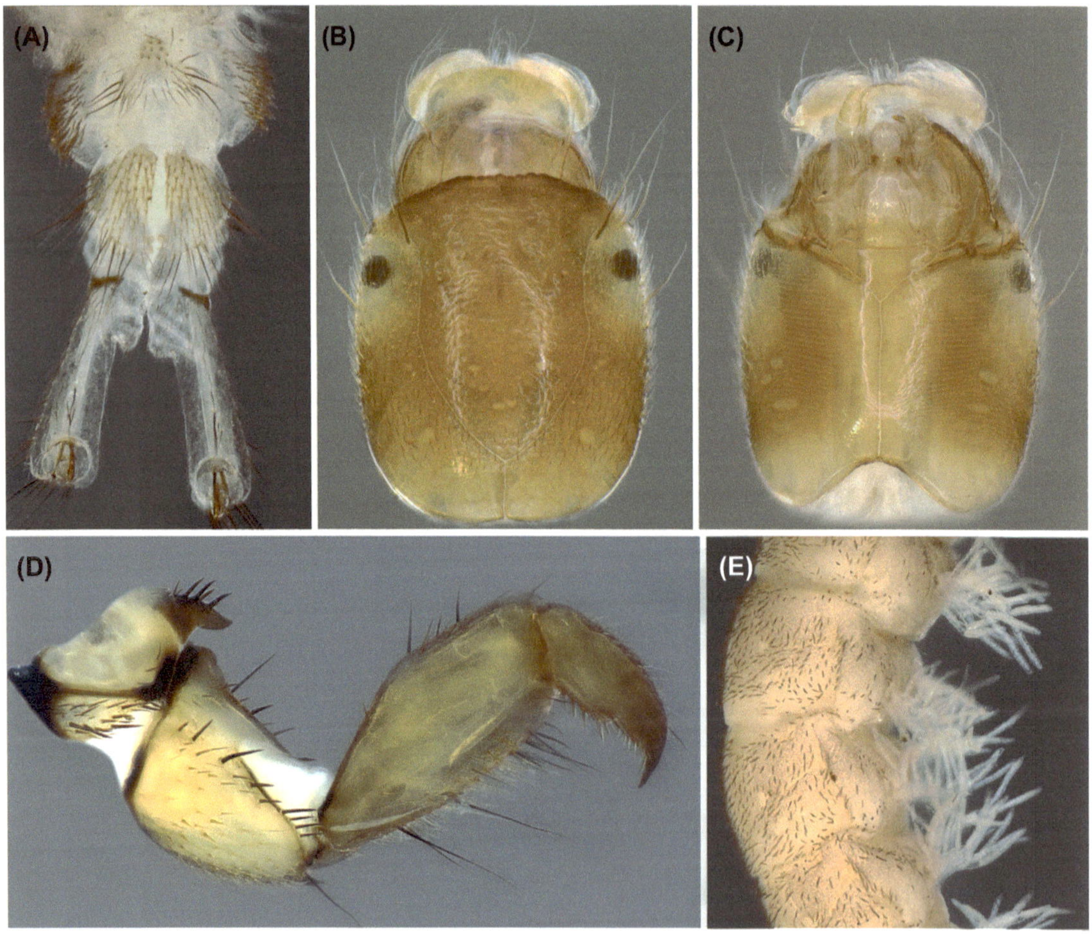

FIGURE 10.23 *Smicridea* (*S.*) *palifera* Flint, 1981 (Hydrosychidae): (A) Abdominal segment VIII-IX, ventral view; (B–C) head, dorsal and ventral view; (D) foretrochantin and foreleg; and (E) abdominal segment II-V, lateral view.

FIGURE 10.24 *Smicridea* spp. (Hydropsychidae): (A) *Smicridea* (*R.*) *caligata* Flint, 1974 larva, lateral view; (B) *S.* (*R.*) *appendiculata* Flint, 1972 abdominal segment VIII-IX, ventral view; and (C) head, dorsal view.

FIGURE 10.25 *Byrsopteryx espinosa* Harris & Holzenthal (Hydroptilidae): (A) larval habitus and case, lateral view; (B) larval habitus, dorsal view; (C) pro and mesothorax, dorsal view.

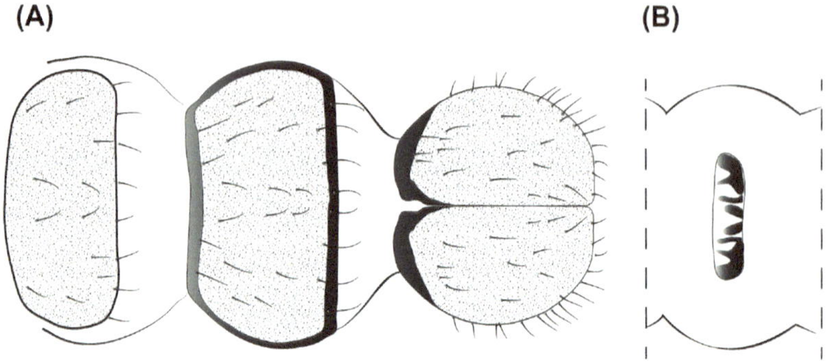

FIGURE 10.26 *Celaenotrichia edwardsi* Mosely, 1934 (Hydroptilidae). (A–B) thorax, and abdominal segment 1, dorsal view. *Modified from Harris & Flint (1993).*

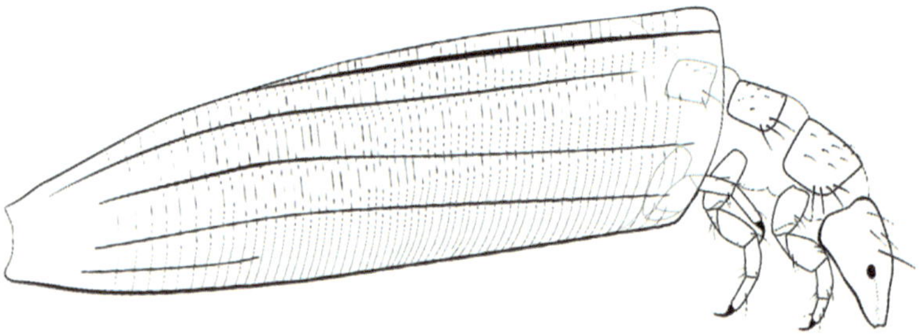

FIGURE 10.27 *Mayatrichia* sp. (Hydroptilidae): larva and case, lateral view.

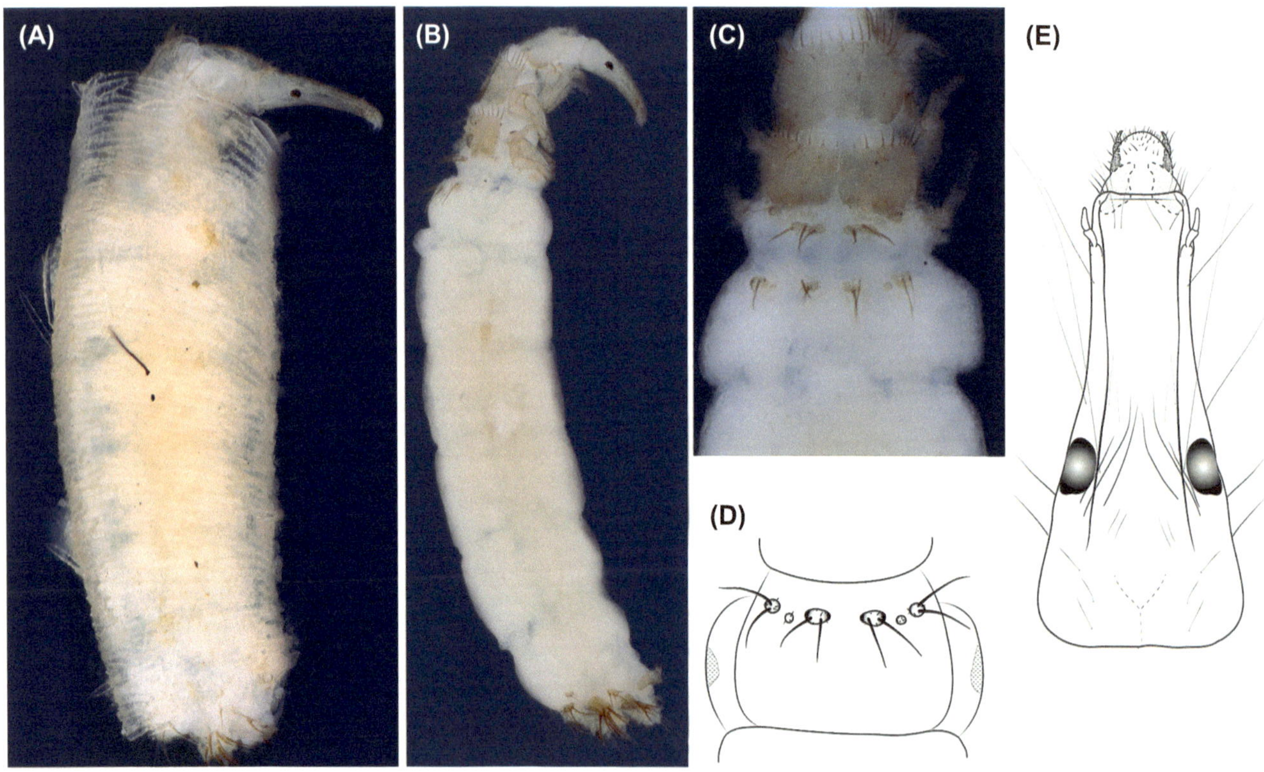

FIGURE 10.28 *Taraxitrichia* sp. (Hydroptilidae). (A–B) larva and case, lateral view; (C–D) abdominal segment I, dorsal view; and (E) head, dorsal view.

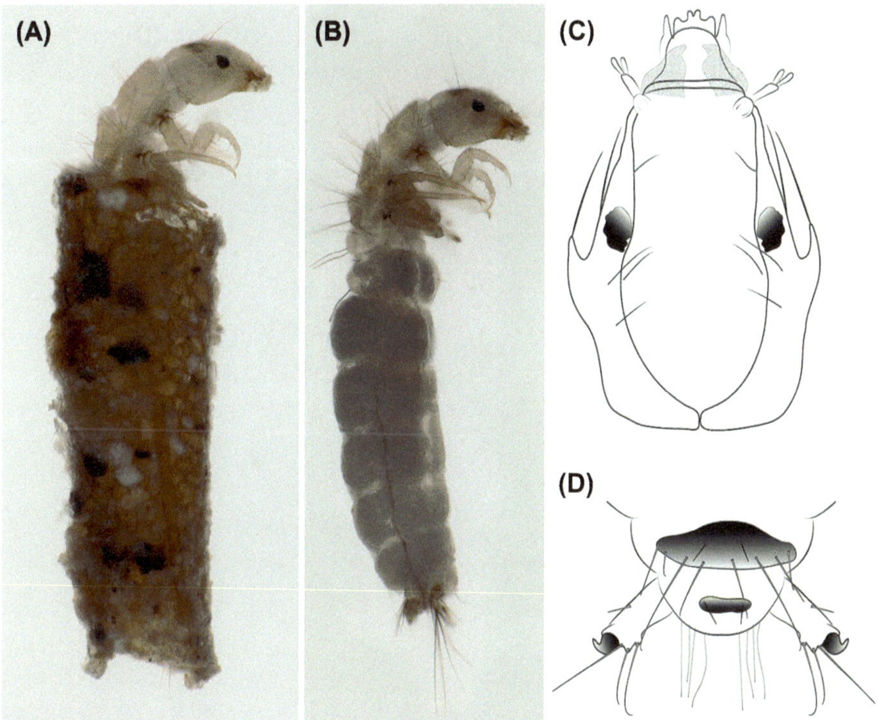

FIGURE 10.29 *Neotrichia* sp. (Hydroptilidae). (A–B) larva and case, lateral view (*From Pes et al. 2014*); (C) head, dorsal view; and (D) abdominal segments IX and X, dorsal view.

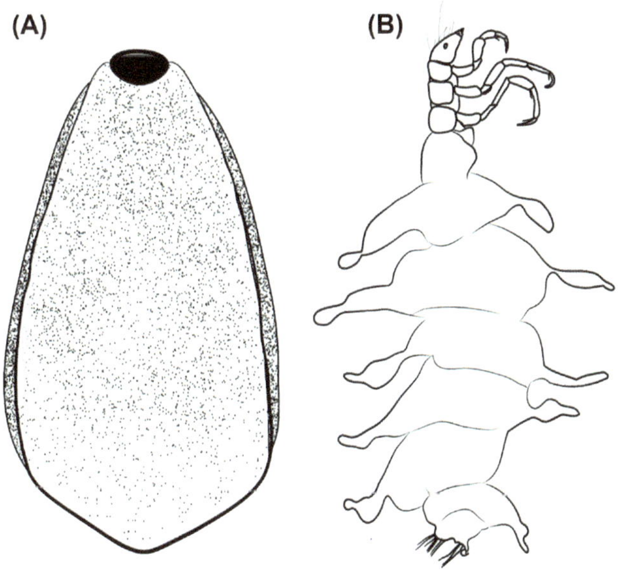

FIGURE 10.30 *Ithytrichia* sp. (Hydroptilidae): case; larva, lateral view.

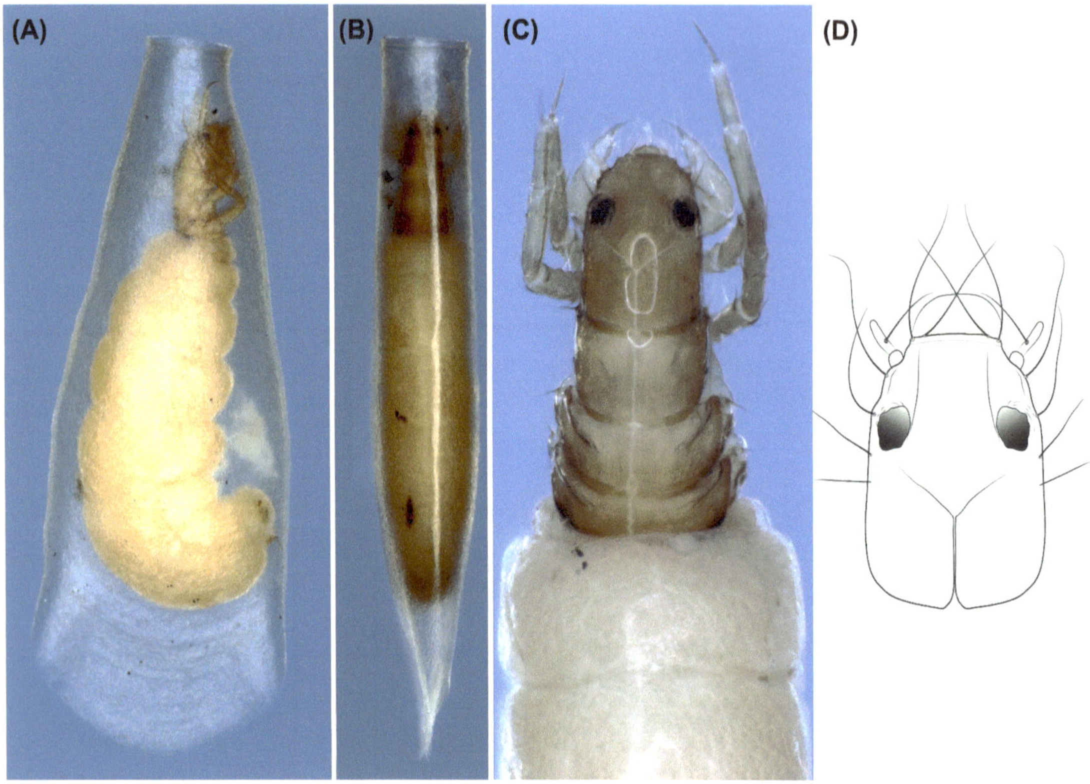

FIGURE 10.31 *Oxyethira* sp. (Hydroptilidae). (A–B) larva and case, lateral and dorsal view; (C) head and thorax, dorsal view; and (D) head, dorsal view.

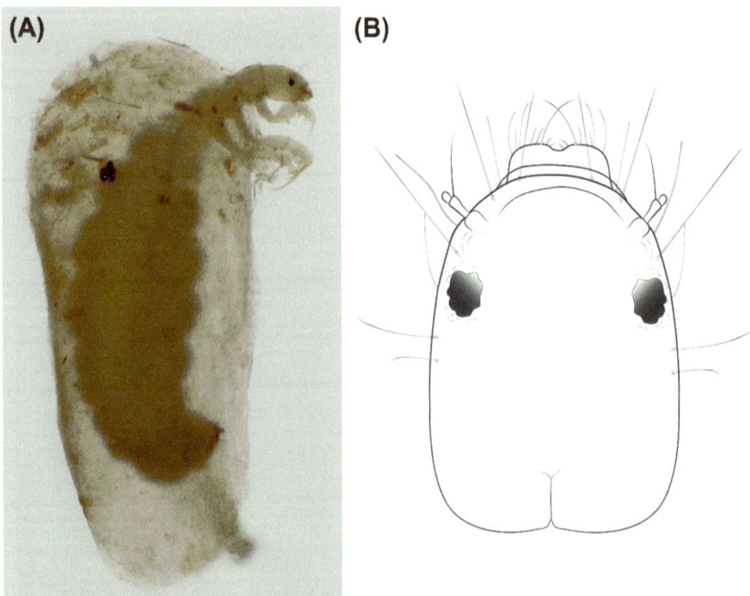

FIGURE 10.32 *Flintiella* sp. (Hydroptilidae): (A) larva and case, lateral view (*From Pes et al. 2014*); and (B) head, dorsal view.

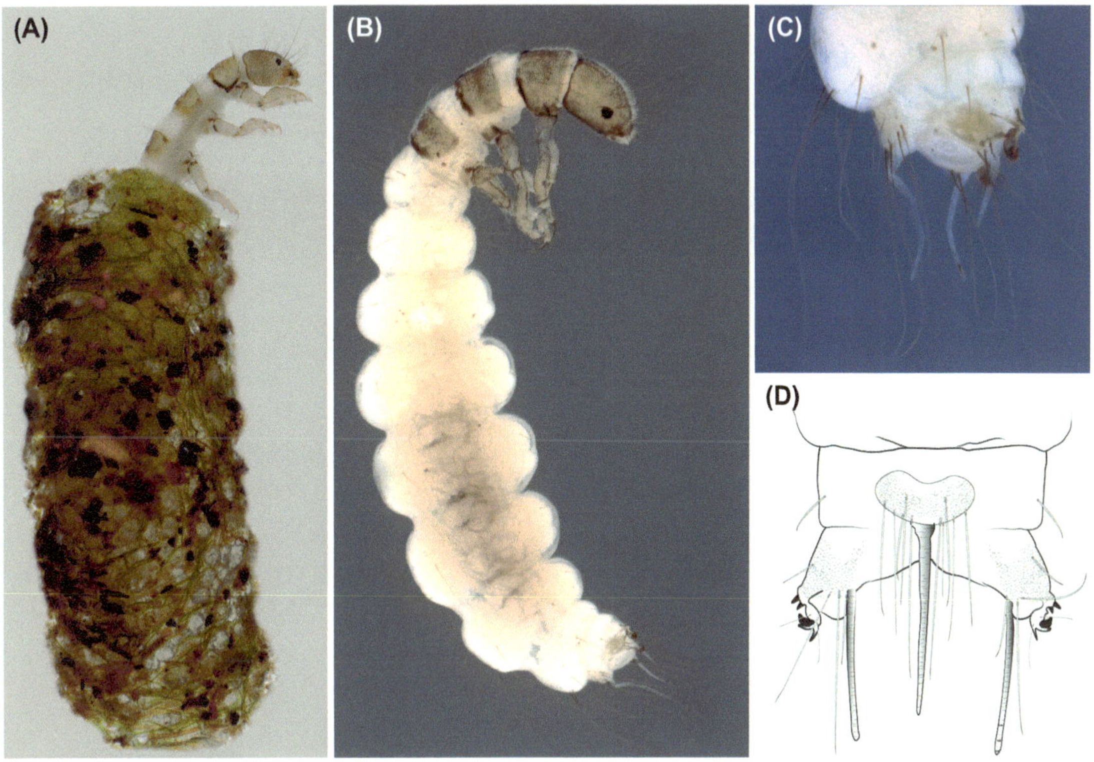

FIGURE 10.33 *Hydroptila* sp. (Hydroptilidae): (A–B) larva and case, lateral view; and (C–D) abdominal segments IX and X, lateral and dorsal view.

FIGURE 10.34 *Rhyacopsyche* sp. (Hydroptilidae): (A) case with larva inside, lateral view. (*From Pes et al. 2014*); (B) larval habitus, lateral view; (C) head, thorax and partial view of abdomen, in dorsal view.

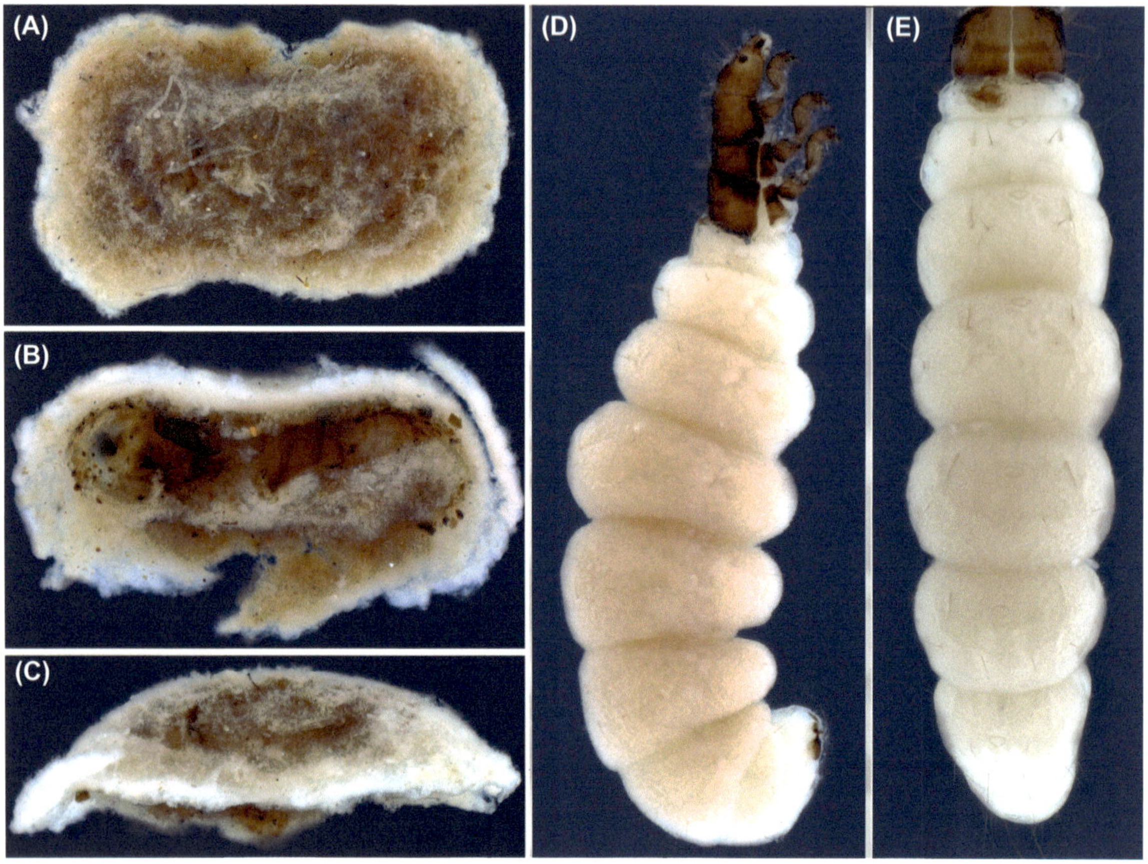

FIGURE 10.35 *Ochrotrichia* sp. (Hydroptilidae): (A–C) case, dorsal, ventral, and lateral view; (D) larva, lateral view; (E) abdominal segments I–VII, dorsal view.

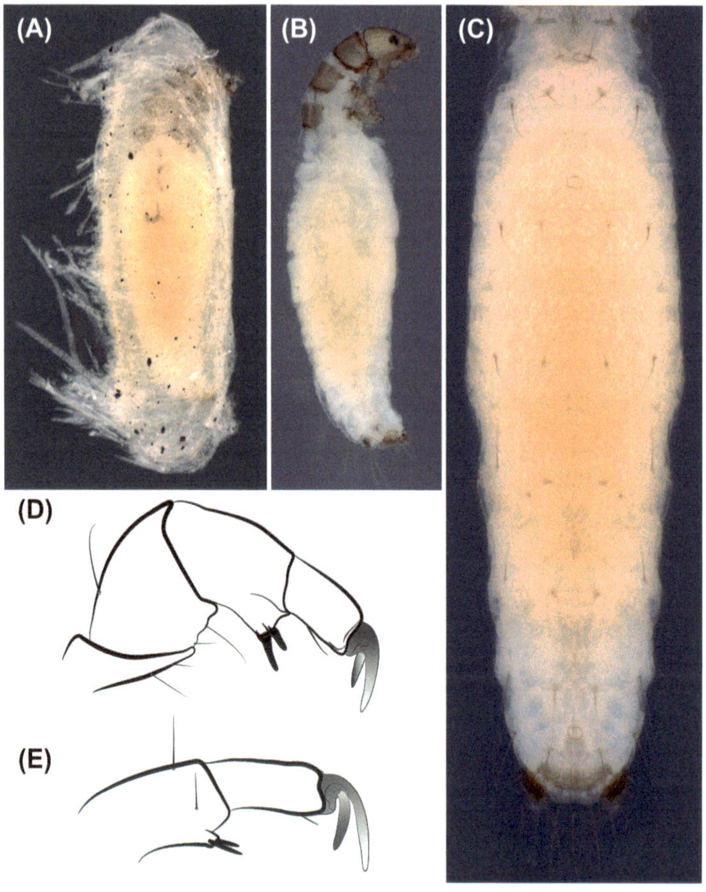

FIGURE 10.36 *Bredinia* sp. (Hydroptilidae): (A–B) larva and case, lateral view; (C) abdominal segments I–IX, dorsal view; and (D–E) fore- and hindlegs.

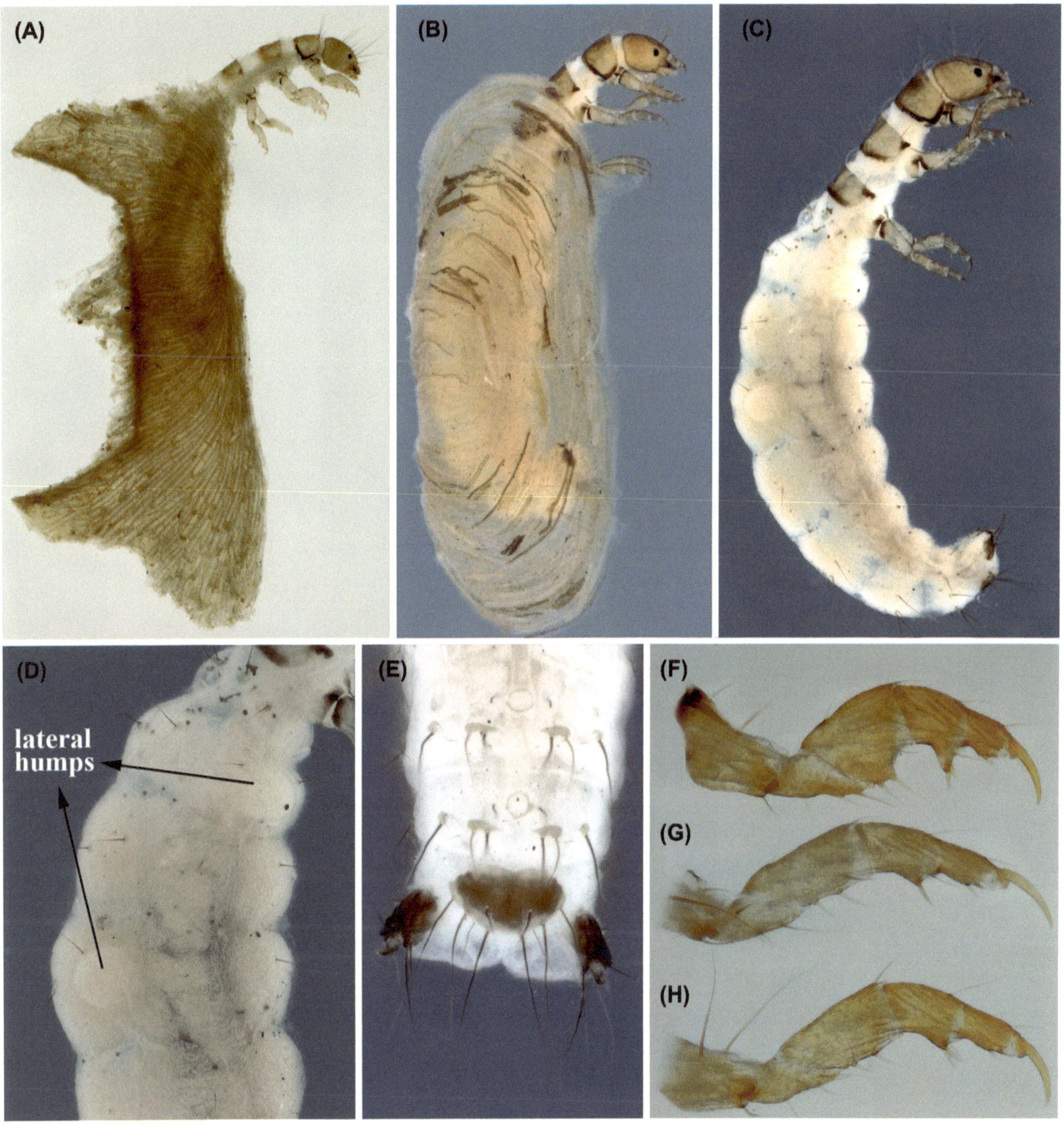

FIGURE 10.37 *Metrichia* spp. (Hydroptilidae): (A–C) larvae and cases, lateral view; (D) abdominal segments I–V, lateral view; (E) abdominal segments VIII–X, dorsal view; and (F–H) legs.

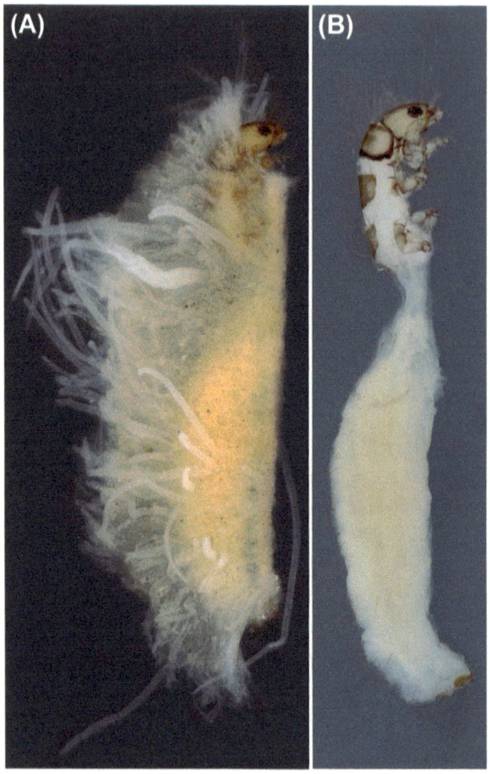

FIGURE 10.38 *Ragatrichia yatay* (Angrisano, 1989) (Hydroptilidae): (A–B) larva and case, lateral view.

FIGURE 10.39 *Cerasmatrichia* spp. (Hydroptilidae): (A) larva and case, dorsal view; (B) larva, lateral view; (C) larva, dorsal view; (D) thorax and abdominal segments I–V; dorsal view; and (E) head and prothorax, dorsal view. *(Figs. B, C, D from Pes et al. 2014).*

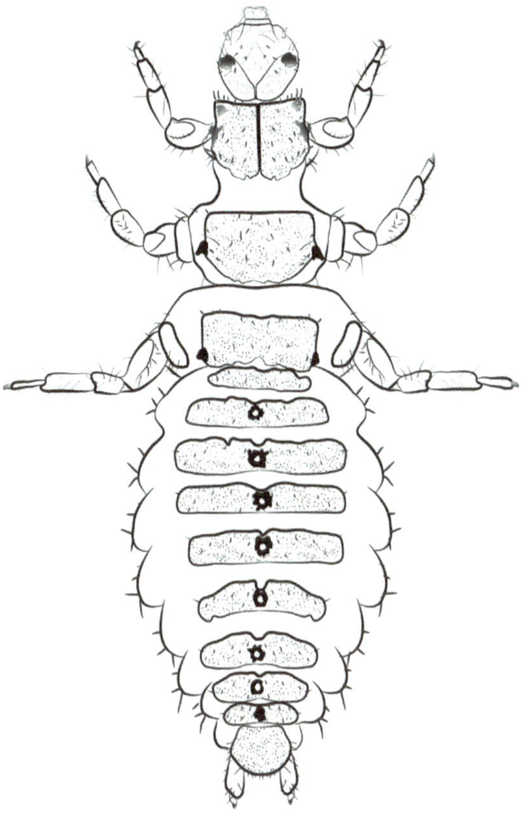

FIGURE 10.40 *Alisotrichia* sp. (Hydroptilidae): larva, dorsal view.

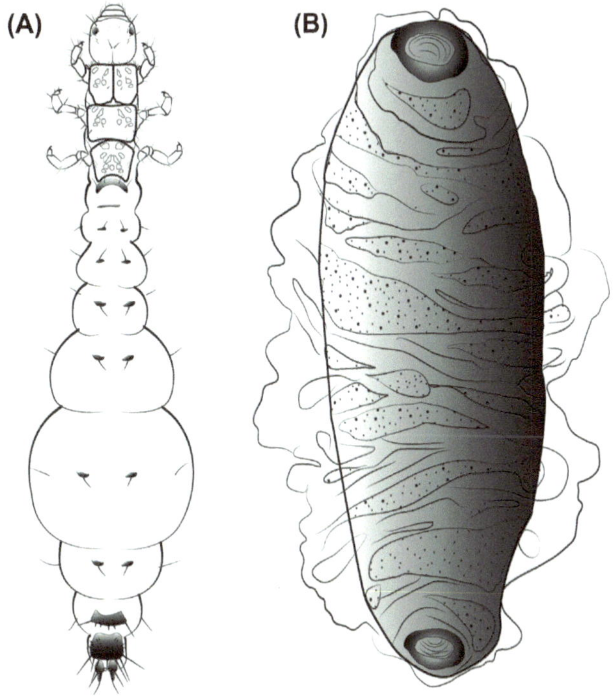

FIGURE 10.41 *Betrichia* sp. (Hydroptilidae): (A) larva, dorsal view; and (B) case, dorsal view.

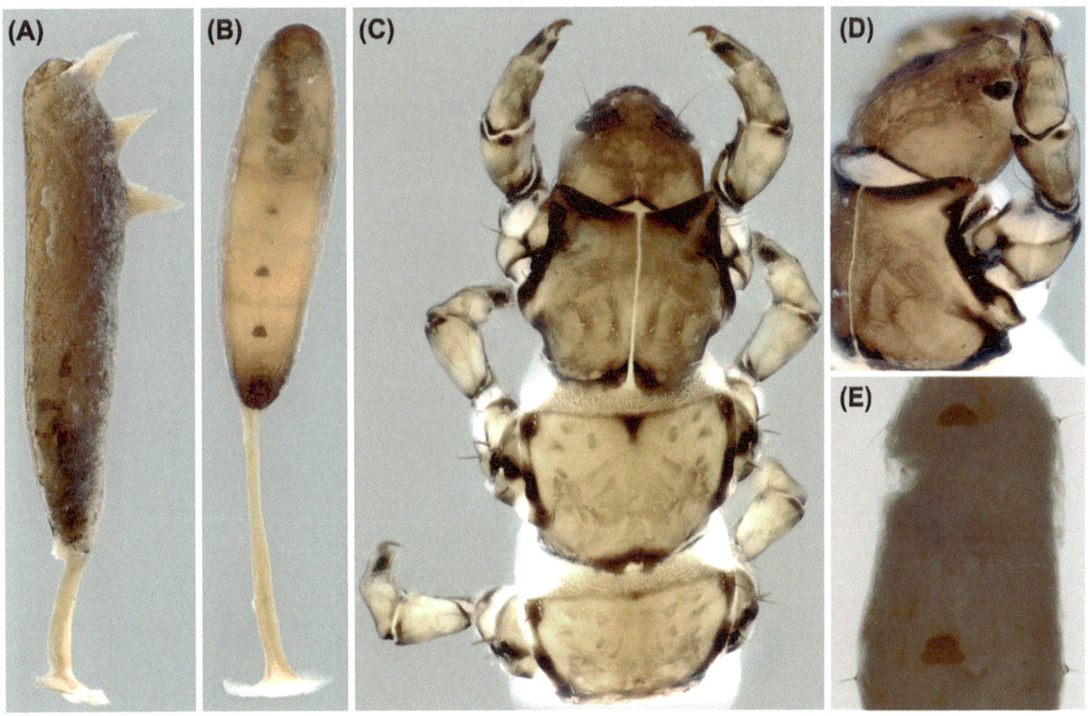

FIGURE 10.42 *Anchitrichia* sp. (Hydroptilidae). (A–B) cases of the pupa; (C) thorax and head, dorsal view; (D) pronotum, laterodorsal view; and (E) abdominal segments V–VI, dorsal view.

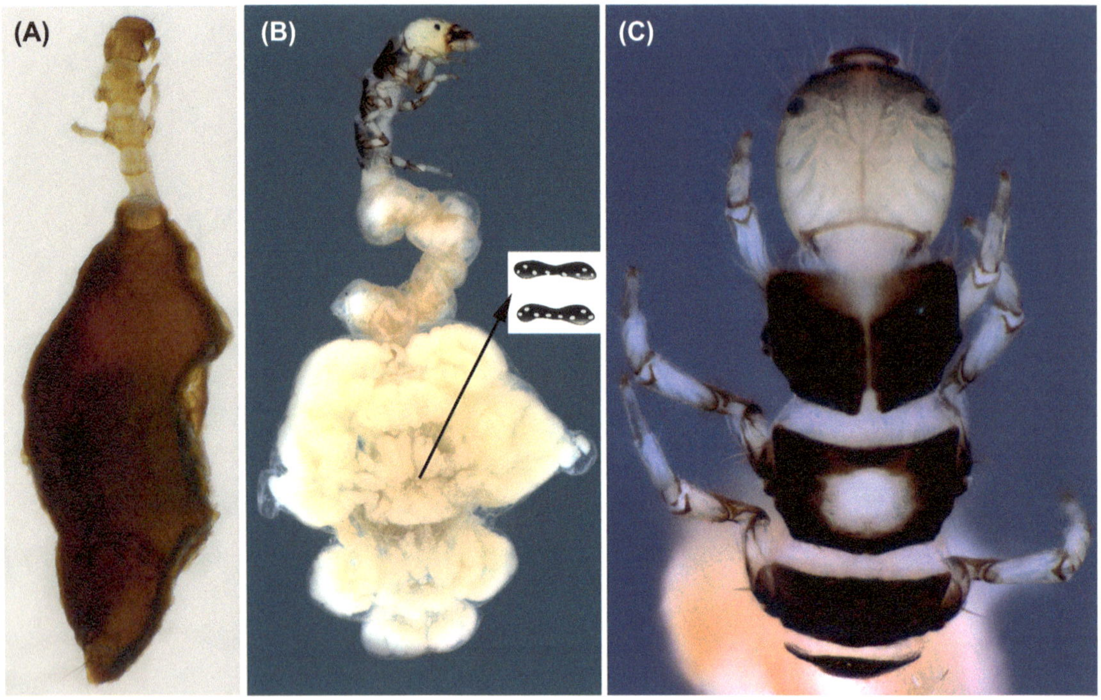

FIGURE 10.43 *Leucotrichia* spp. (Hydroptilidae): (A) larva and case, dorsal view; (B) larva, laterodorsal view; and (C) thorax and head, dorsal view.

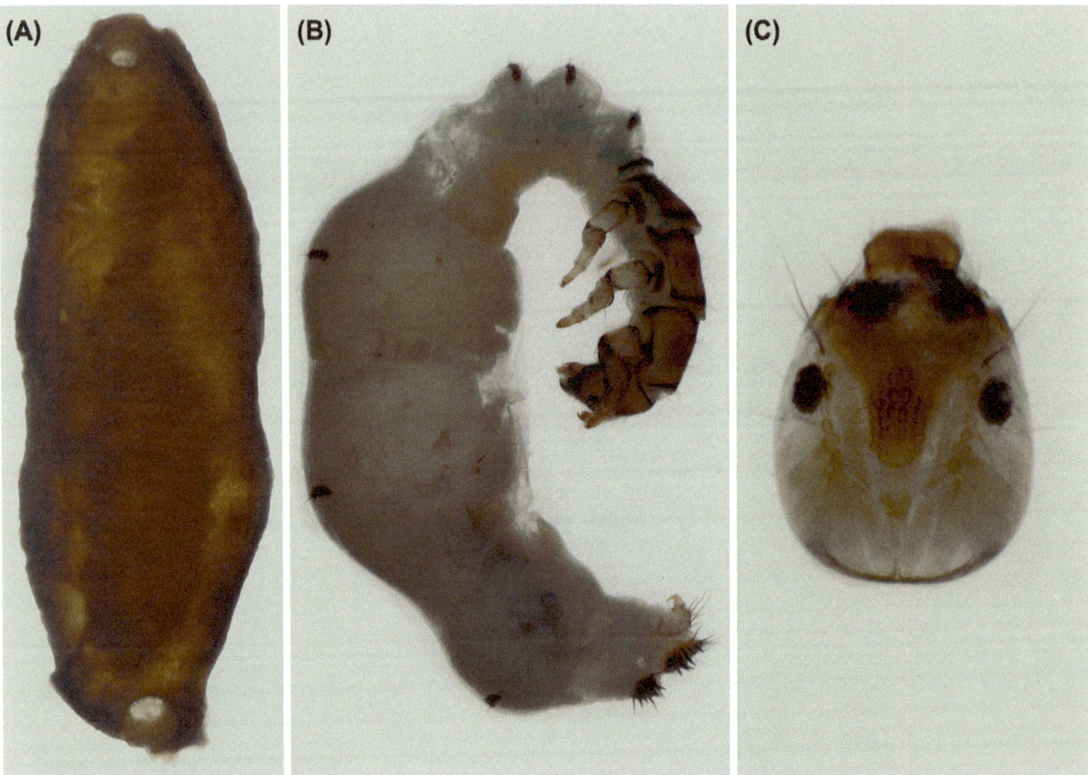

FIGURE 10.44 *Peltopsyche* sp. (Hydroptilidae): (A) case, dorsal view; (B) larva, lateral view; and (C) head, dorsal view.

FIGURE 10.45 *Acostatrichia/Zumatrichia* sp. (Hydroptilidae): (A) cases; (B) larva, lateral view; (C) abdominal segments V–VI, dorsal view; (D) foreleg; and (E) head, dorsal view.

FIGURE 10.46 *Hesperophylax* sp. (Limnephilidae): (A) larva and case, lateral view; (B–C) abdomen, lateral and dorsal view; (D) head, dorsal view; (E) midleg.

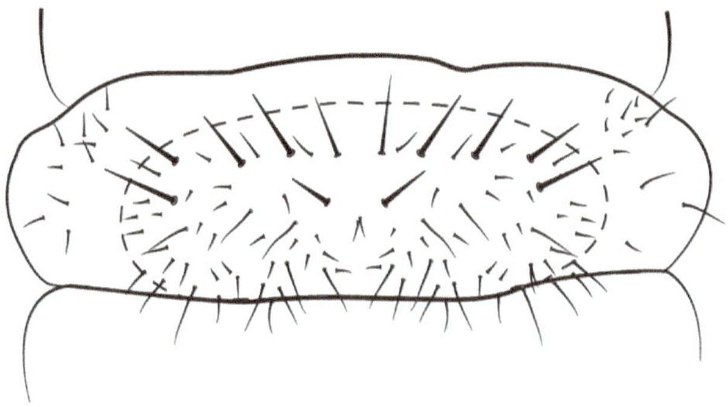

FIGURE 10.47 *Clistoronia* sp. (Limnephilidae): (A) abdominal segment I, ventral view. *Modified from Morse and Holzenthal (2008).*

FIGURE 10.48 *Limnephilus* sp. (Limnephilidae): (A) larva and case, lateral view; (B) VII–IX abdominal segments, lateral view; (C) abdomen, ventral view; (D) head, dorsal view; (E) thorax, dorsal view; and (F–H) legs.

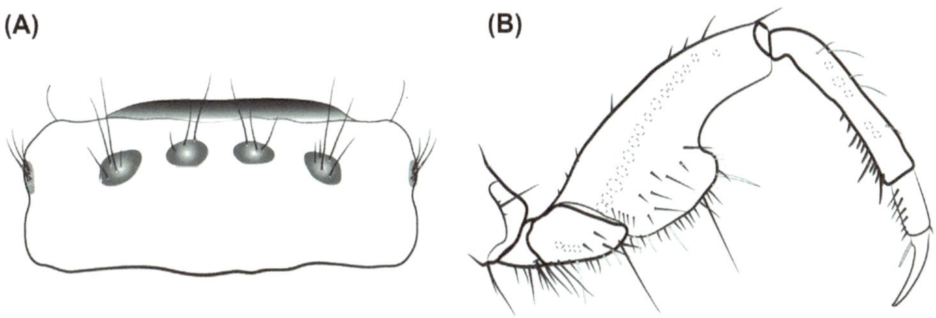

FIGURE 10.49 *Platycosmoecus beaumonti* (Schmid, 1958) (Limnephilidae): (A) first abdominal segment, ventral view; and (B) foreleg, lateral view. *Modified from Flint (1982).*

FIGURE 10.50 *Anomalocosmoecus* sp. (Limnephilidae): (A–B) larva and case, lateral view; and (C) head, sublateral view.

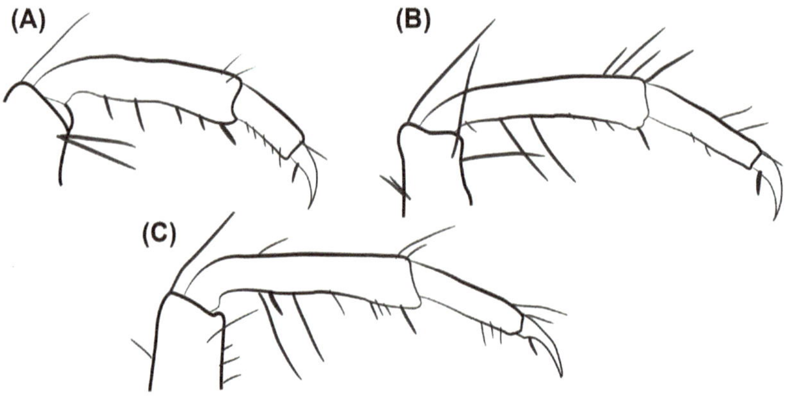

FIGURE 10.51 *Antarctoecia* (Limnephilidae): (A–B) tibia and tarsus of the legs. *Modified from Flint (1982).*

FIGURE 10.52 *Metacosmoecus nigrofasciatus* Schmid, 1955 (Limnephilidae): (A–B) case and larva, lateral view; (C) abdominal segments, ventral view; (D) head, dorsal view; (E) abdominal segments VIII–X, dorsal view; and (F–H) legs.

FIGURE 10.53 *Verger* spp. (Limnephilidae): (A) larva, ventral view; (B) thorax and first abdominal segments, lateral view; (C) detail chloride epithelia of the VI and VII segments, ventral view; (D) abdominal segments VII–X, lateral view; (E–F) larvae and cases, lateral view; and (G) head, dorsal view; (H–J) legs.

FIGURE 10.54 *Monocosmoecus* spp. (Limnephilidae): (A) larva, dorsal view; (B) detail chloride epithelia of the V–VII segments; (C) head, dorsal view; (D) abdomen, ventral view; (E–J) cases; and (H–I) legs.

FIGURE 10.55 *Austrocosmoecus hirsutus* Schmid, 1955 (Limnephilidae): (A) larva, lateral view; (B) case; (C) head, dorsal view; (D) mandibles, sublateral view; (E) thorax, dorsal view; and (F–H) legs.

FIGURE 10.56 *Antarctoecia brasiliensis* Hamantinco & Nessimian, 2003 (Limnephilidae): (A–B) larva and case, lateral view: (C) abdomen, ventral view; (D) head and prosternum, sublateral view; and (E–F) legs.

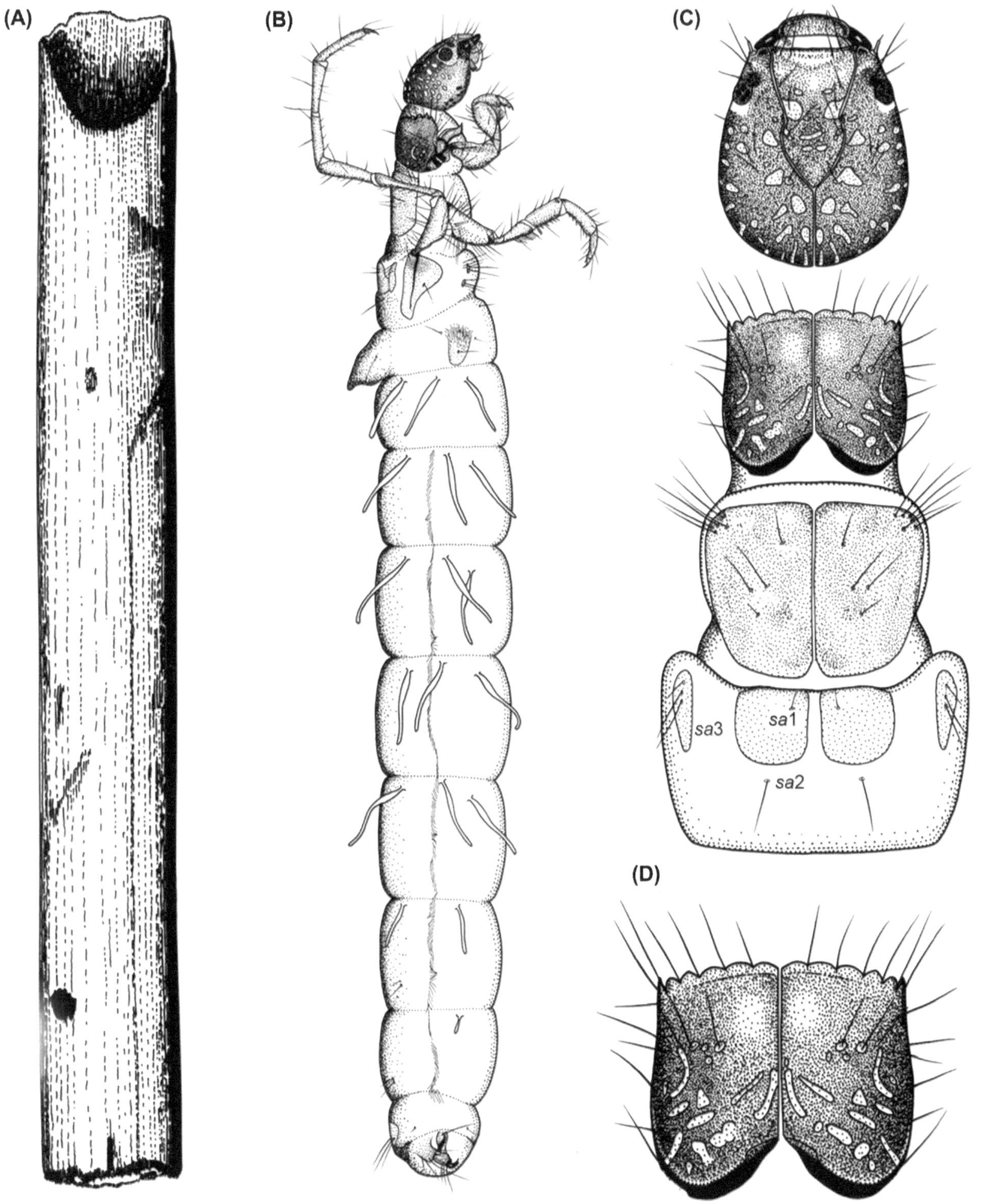

FIGURE 10.57 *Triplectides* sp. (Leptoceridae): (A–B) case and larva, lateral view; (C) head and thorax, dorsal view; and (D) pronotum, dorsal view.

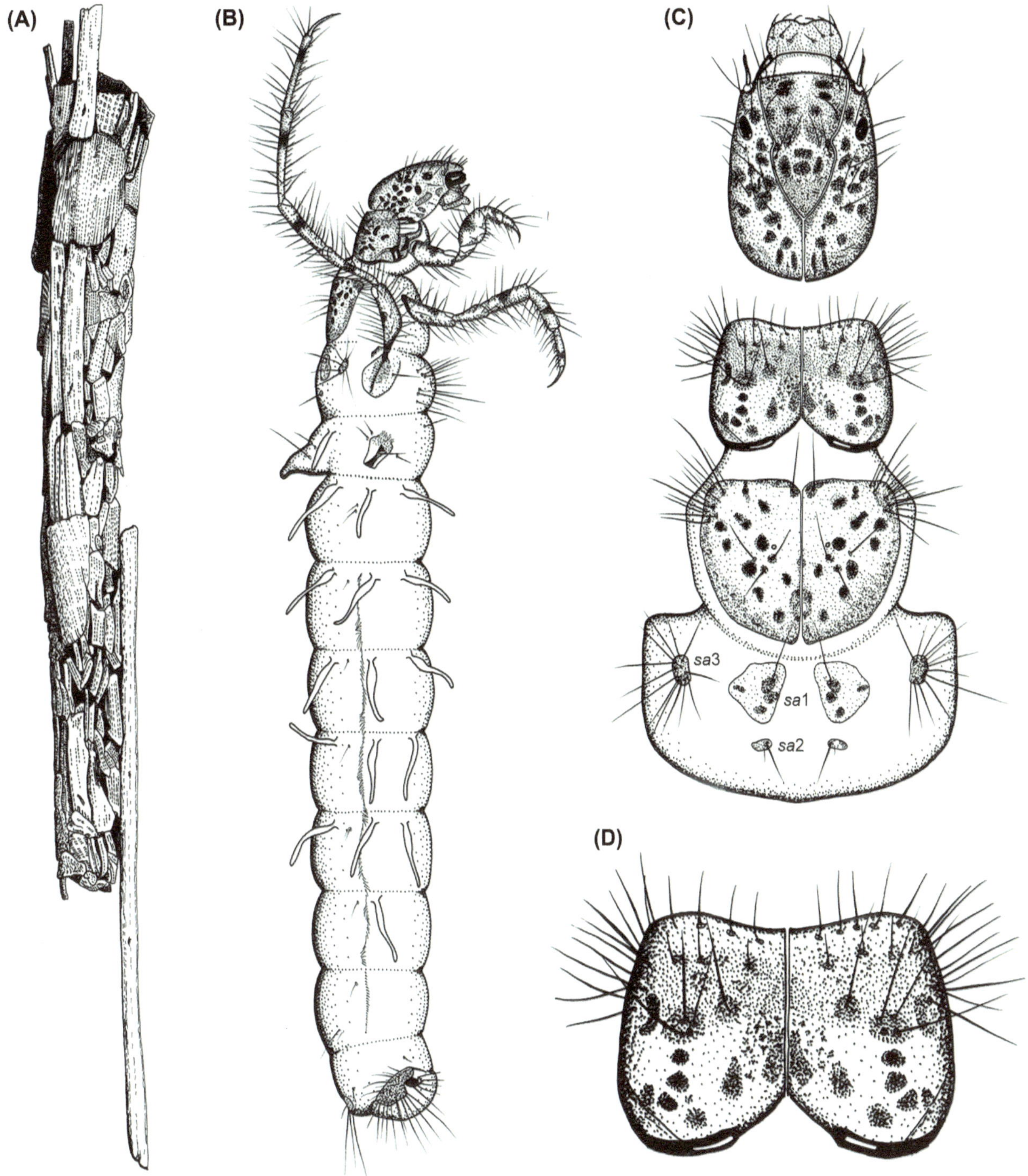

FIGURE 10.58 *Hudsonema flaminii* (Navás, 1926) (Leptoceridae): (A–B) case and larva, lateral view: and (C) head and thorax, dorsal view; and (D) pronotum, dorsal view. *From Holzenthal (1986c).*

FIGURE 10.59 *Notalina* sp. (Leptoceridae): (A–B) case and larva, lateral view; (C) head and thorax, dorsal view; and (D) thorax, dorsal view.

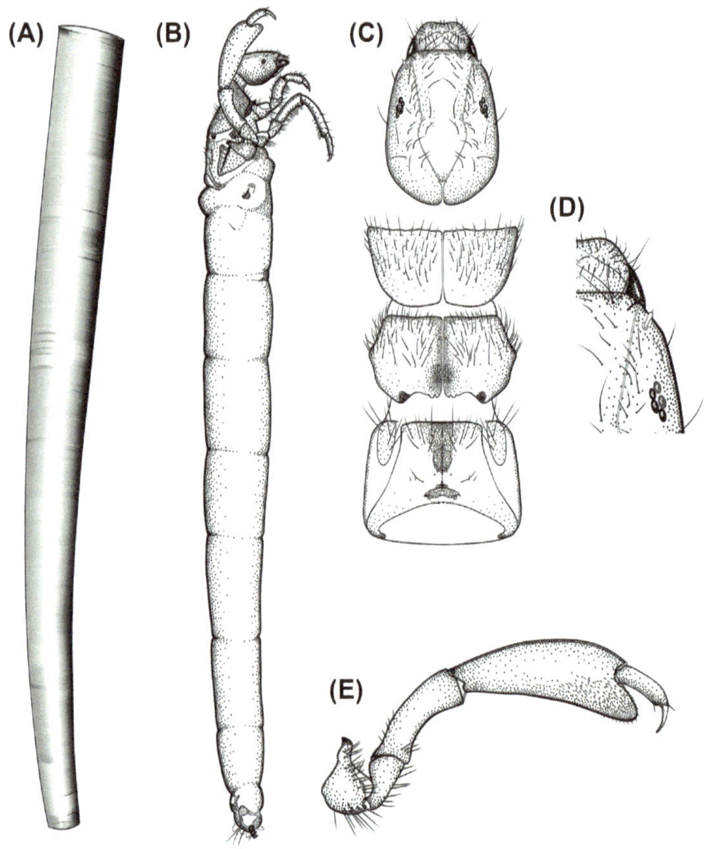

FIGURE 10.60 *Amazonatolica hamadae* Holzenthal & Pes, 2004 (Leptoceridae): (A–B) case and larva, lateral view: (C) head and thorax, dorsal view; (D) antennae; and (E) hind leg. *From Holzenthal & Pes (2004)*

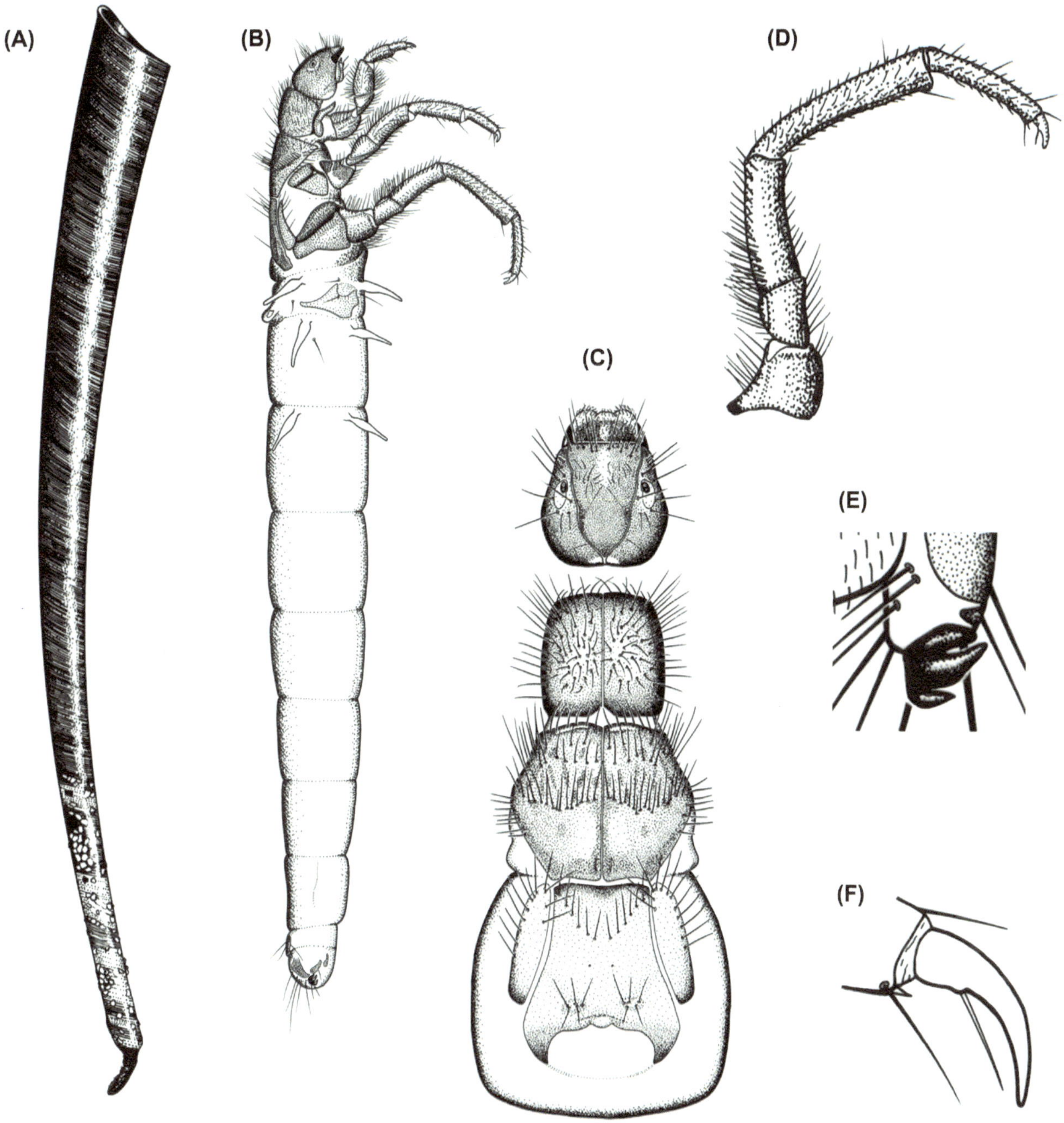

FIGURE 10.61 *Atanatolica flinti* Holzenthal, 1988 (Leptoceridae). (A–B) case and larva, lateral view: (C) head and thorax, dorsal view; (D) hind leg; (E) anal claw, lateral view; (F) tarsal claw. *From Holzenthal (1988b).*

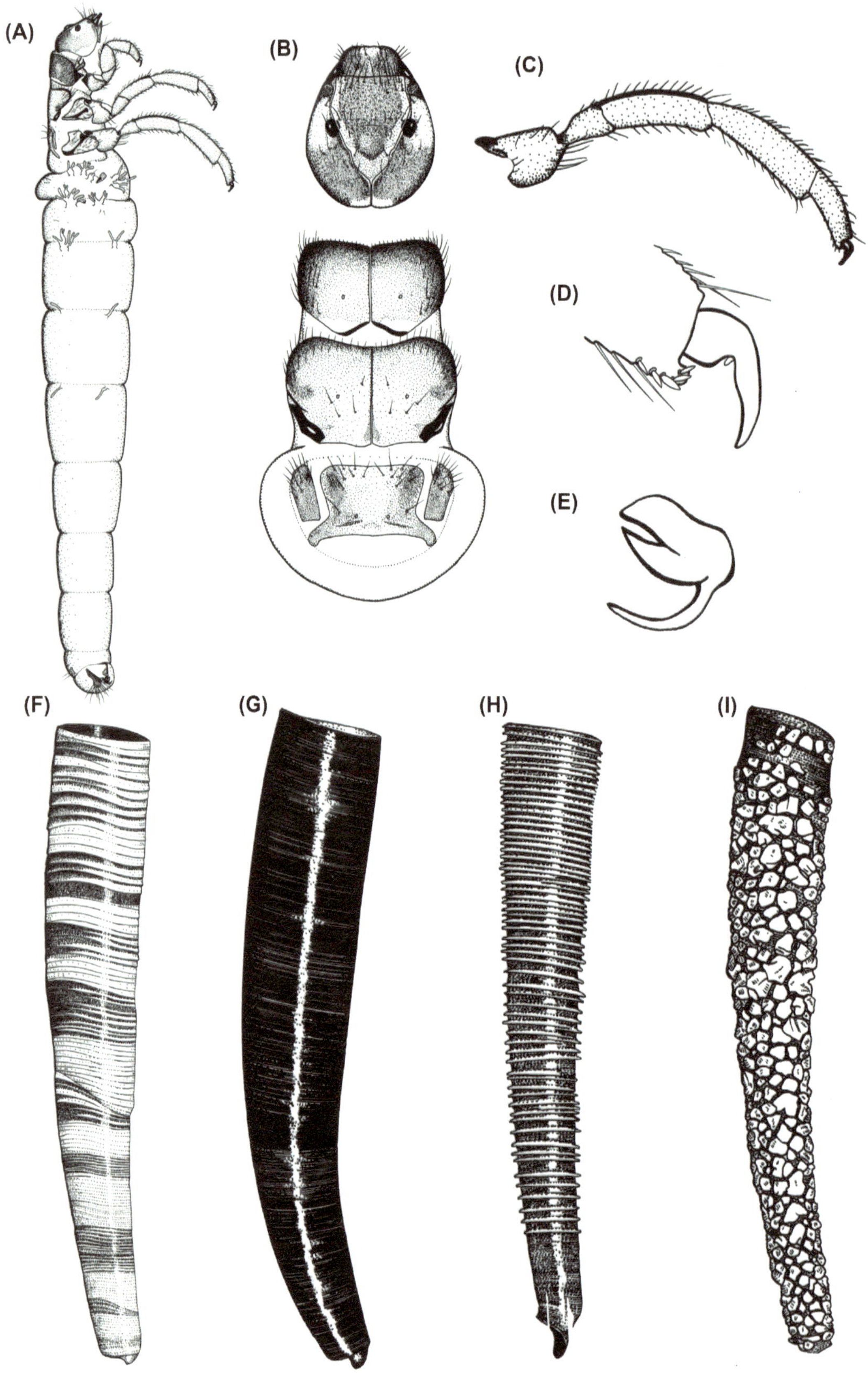

FIGURE 10.62 *Grumichella* spp. (Leptoceridae): (A) larva, lateral view; (B) head and thorax, dorsal view; (C) hind leg; (D) tarsal claw; (E) anal claw; (F–I) cases. *From Holzenthal (1988b).*

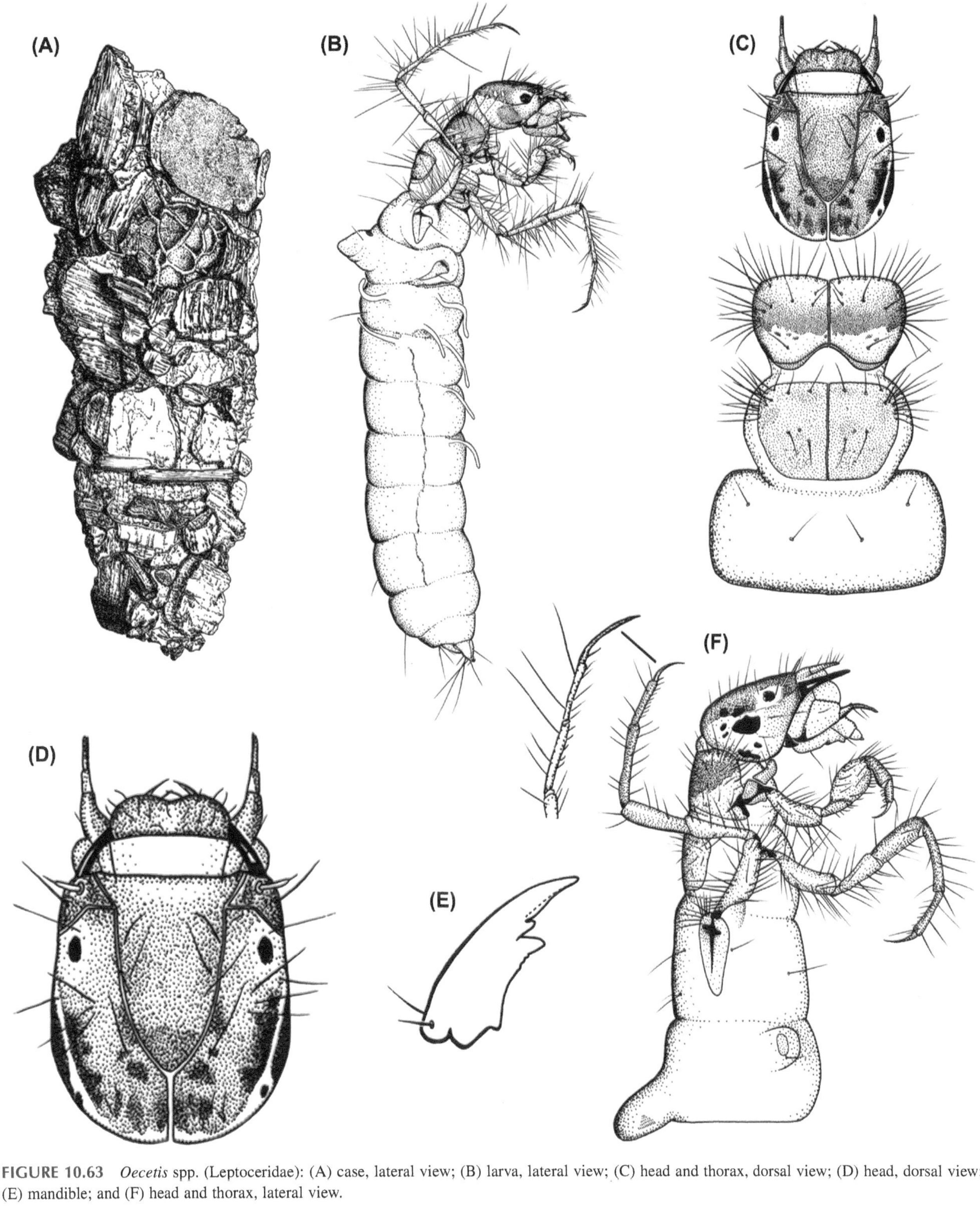

FIGURE 10.63 *Oecetis* spp. (Leptoceridae): (A) case, lateral view; (B) larva, lateral view; (C) head and thorax, dorsal view; (D) head, dorsal view; (E) mandible; and (F) head and thorax, lateral view.

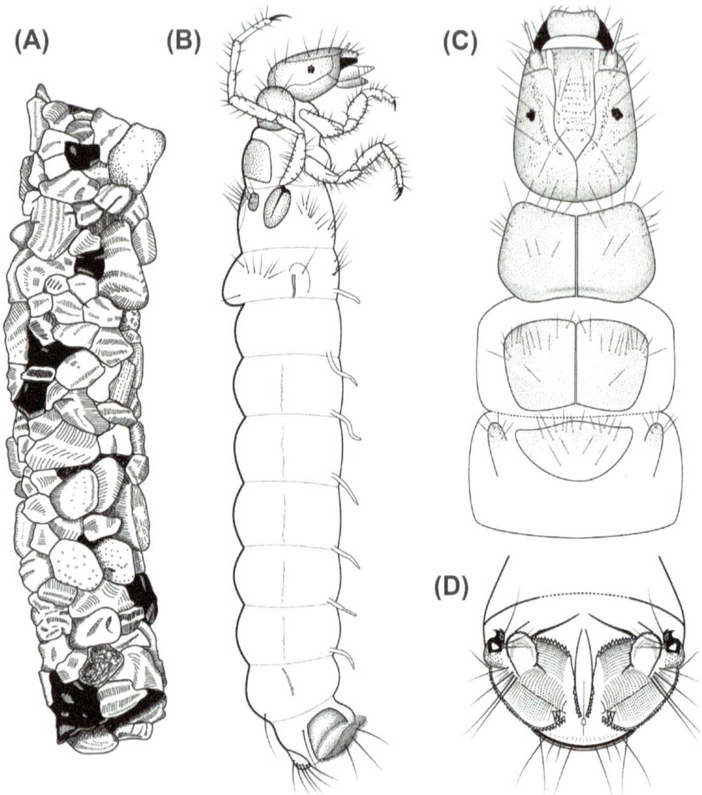

FIGURE 10.64 *Setodes* sp. (Leptoceridae): (A) case, lateral view; (B) larva, lateral view: (C) head and thorax, dorsal view; and (D) terminal abdominal segment, posterior view.

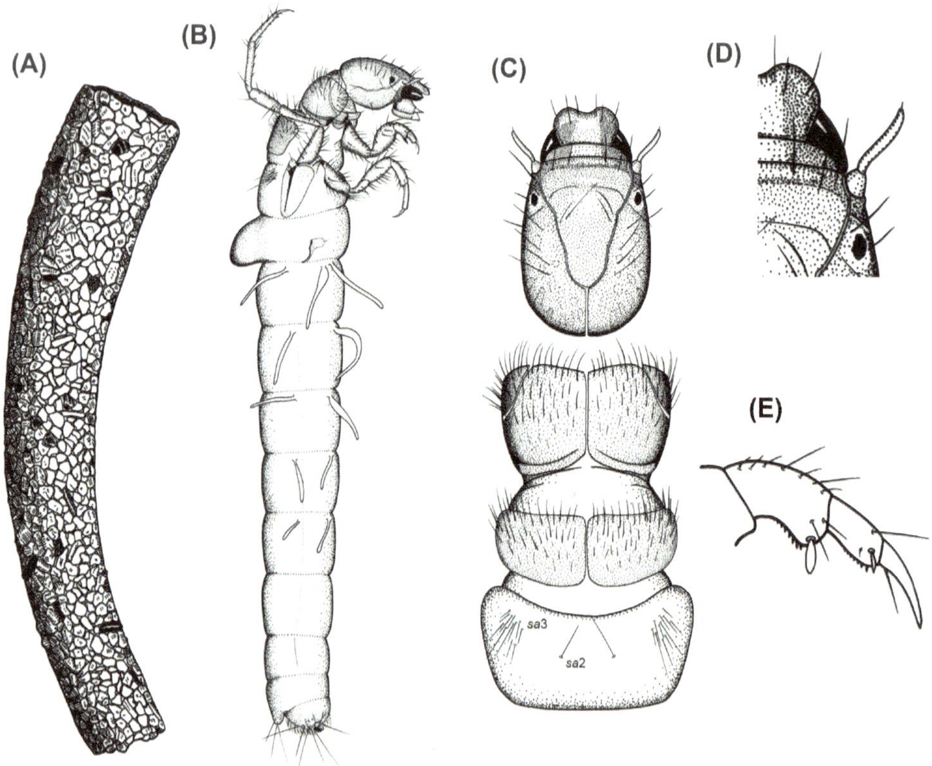

FIGURE 10.65 *Amphoropsyche* sp. (Leptoceridae): (A) case, lateral view; (B) larva, lateral view: (C) head and thorax, dorsal view; (D) antennae; and (E) tarsus. *From Holzenthal (1986b).*

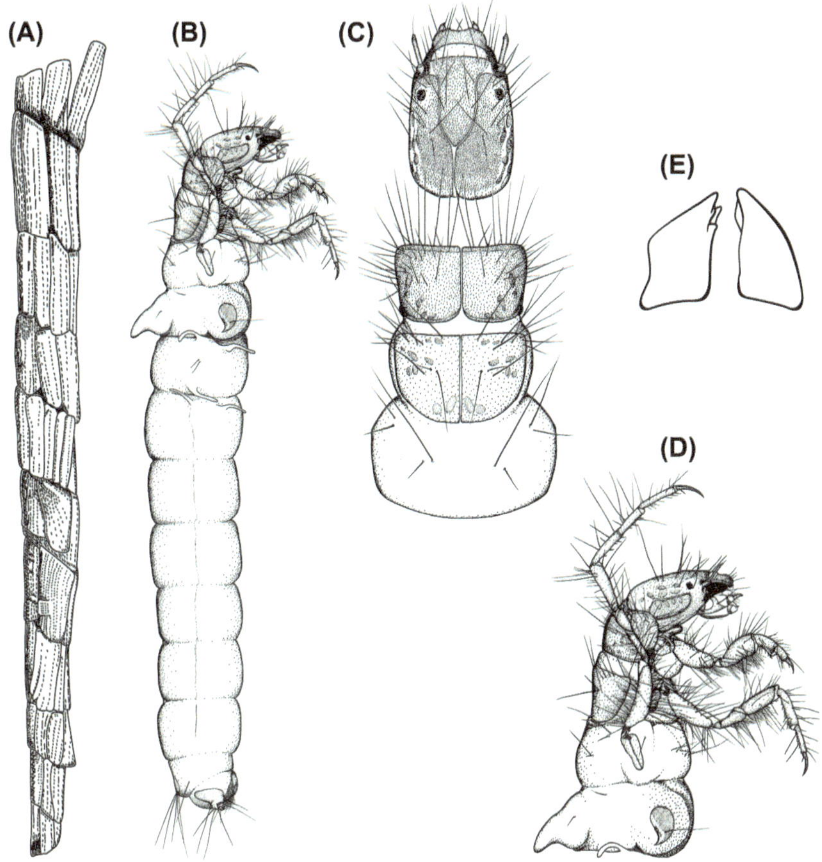

FIGURE 10.66 *Triaenodes* sp. (Leptoceridae): (A) case, lateral view; (B) larva, lateral view: (C–D) head and thorax, dorsal and lateral view; and (E) mandibles. *Figs. B, C–D from Holzenthal & Andersen (2004).*

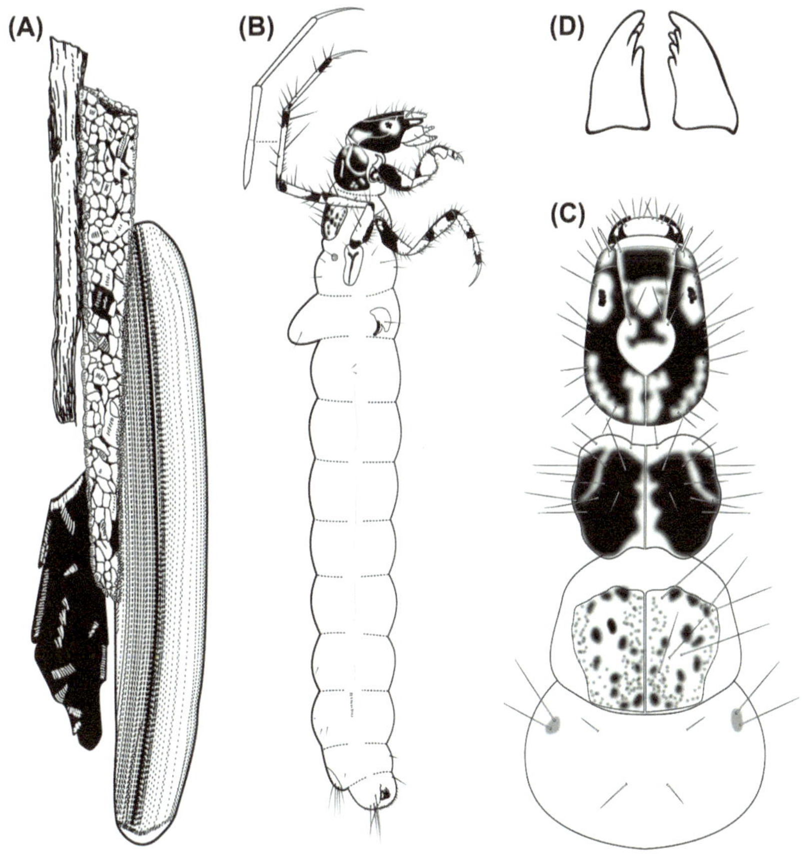

FIGURE 10.67 *Mystacides sepulchralis* (Walker, 1852) (Leptoceridae): (A) case, lateral view; (B) larva, lateral view: (C) head and thorax, dorsal view; and (D) mandibles.

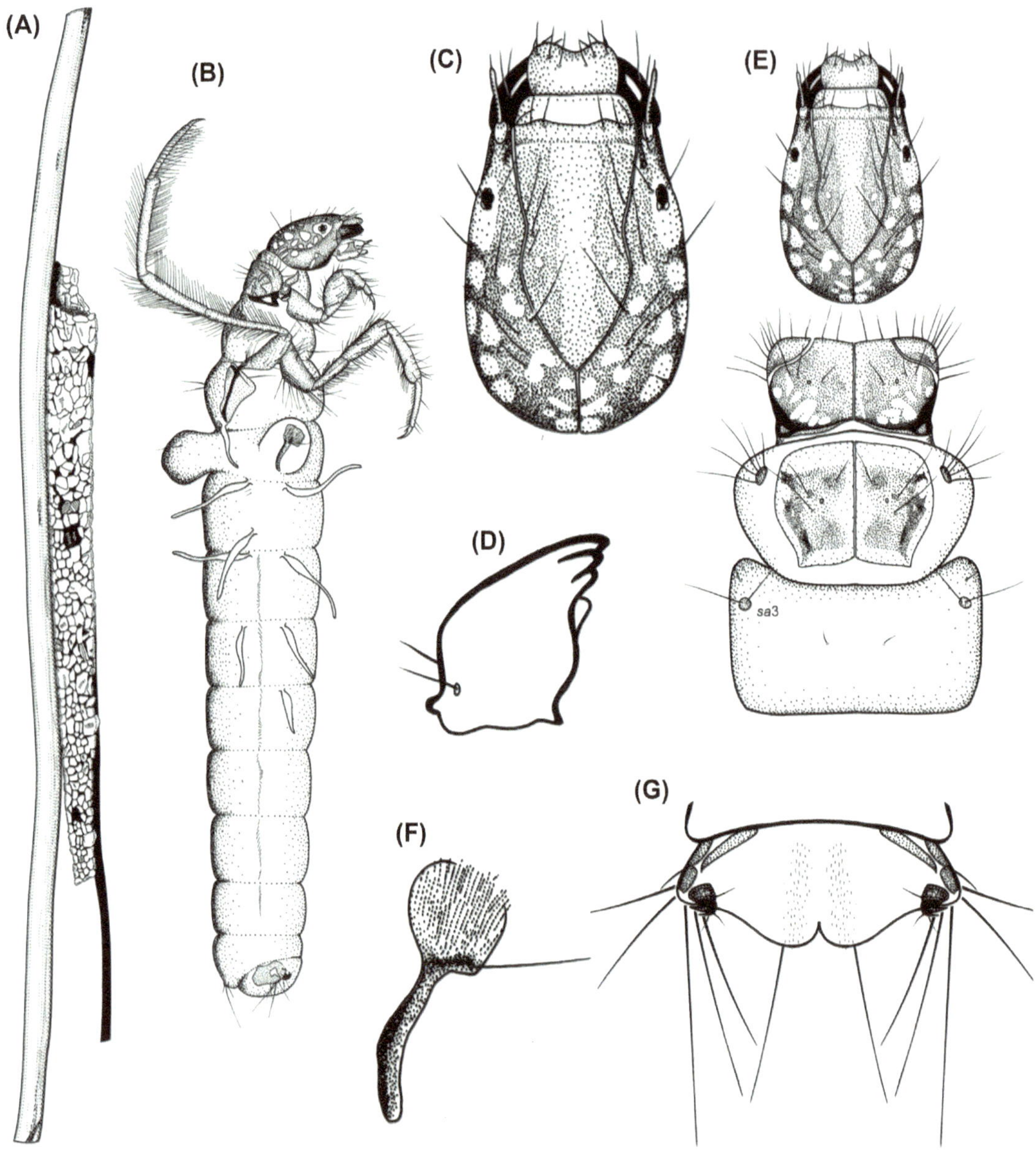

FIGURE 10.68 *Nectopsyche* spp. (Leptoceridae): (A) case, lateral view; (B) larva, lateral view; (C) head, dorsal view; (D) mandible; (E) head and thorax, dorsal view; (F) lateral hump sclerite of abdominal segment I; and (G) abdominal segment IX dorsal plate and anal prolegs, dorsal view. *Figs. B–G from Holzenthal (1995).*

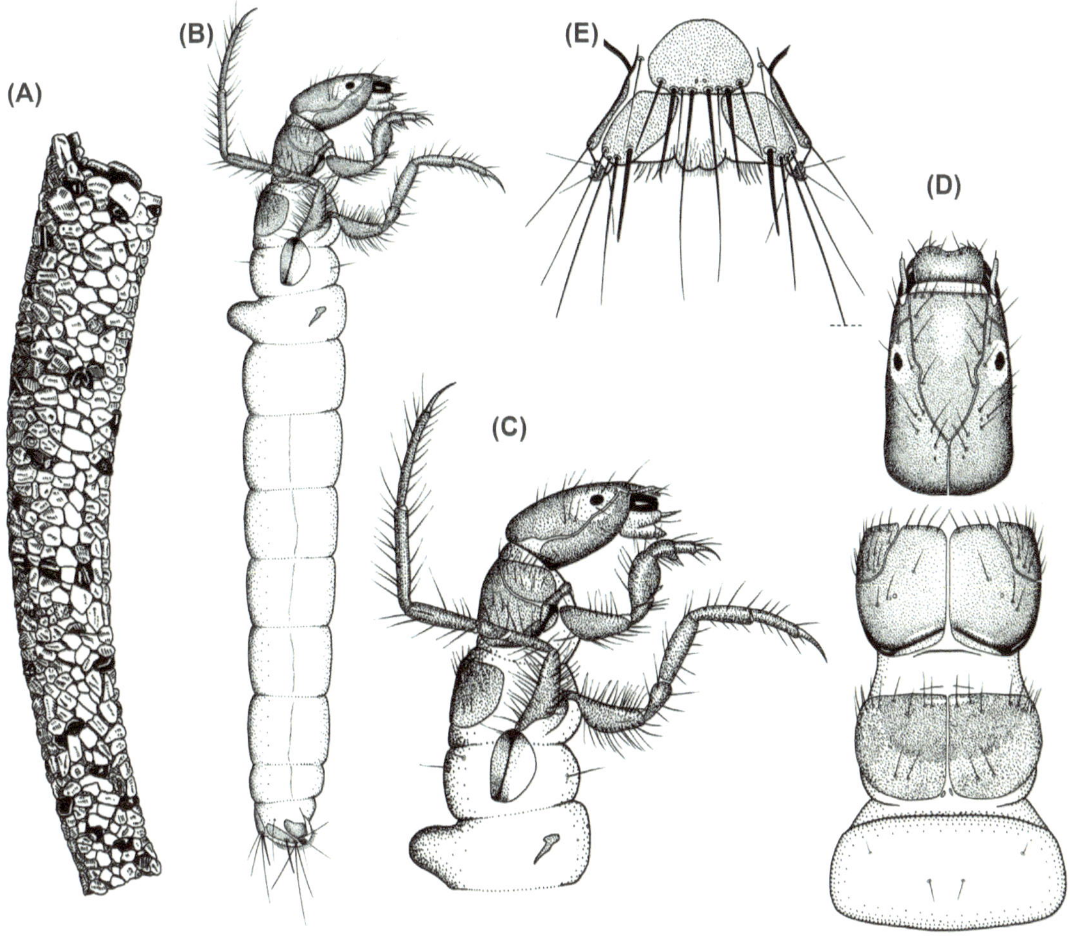

FIGURE 10.69 *Brachysetodes* (Leptoceridae): (A) case, lateral view; (B) larva, lateral view (C–D); head and thorax, lateral and dorsal view; and (E) abdominal segment IX dorsal plate and anal prolegs, dorsal view. *From Holzenthal (1986a).*

FIGURE 10.70 *Banyallarga* sp. (Calamoceratidae): (A) larva and case, dorsal view; (B—C) head, dorsal and ventral view; (D) labrum; (E) thorax, dorsal view; (F) thorax, lateral view; and (G) trochanter of the hindleg.

FIGURE 10.71 *Phylloicus* spp. (Calamoceratidae): (A) case, ventral view; (B) larva, dorsal view; (C) pronotum, dorsal view; (D) hindleg; (E, F) head, dorsal and ventral view; and (G) labrum.

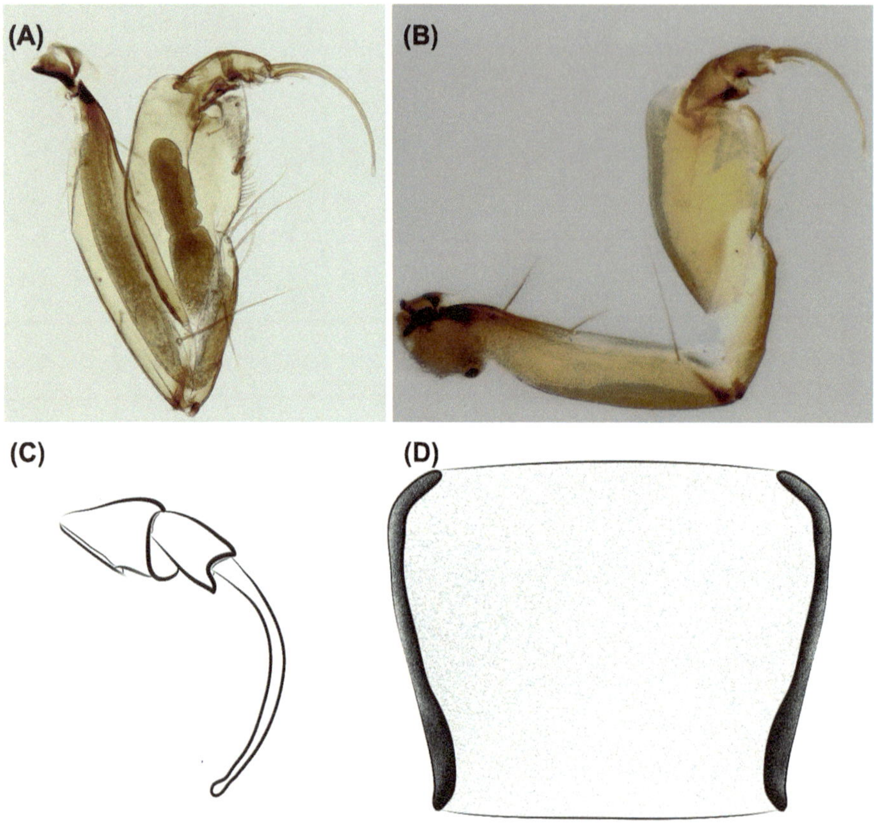

FIGURE 10.72 *Apatanodes/Neopsilochorema* sp. (Hydrobiosidae): (A–C) forelegs; and (D) prosternum. *Modified from Angrisano (1997).*

FIGURE 10.73 *Pseudoradema spinosissimum* Schmid, 1955 (Hydrobiosidae): (A) prosternum; and (B) foretibia and tarsus. *Modified from Bravo & Angrisano (2005).*

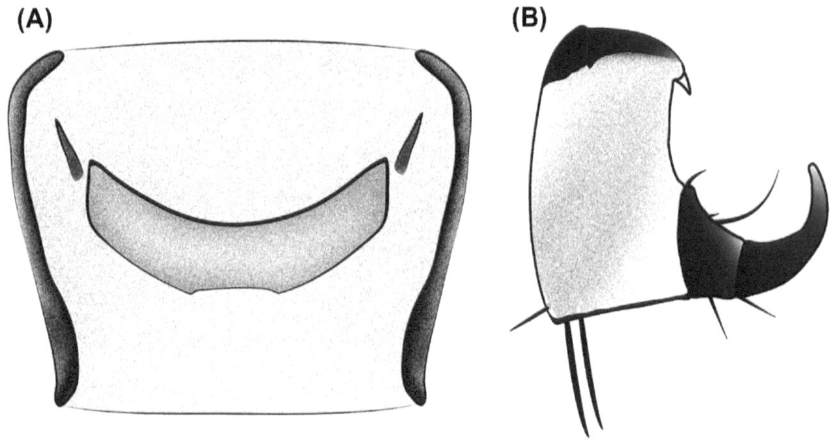

FIGURE 10.74 *Clavichorema* sp. (Hydrobiosidae): (A) prosternum; and (B) anal claw. *Modified from Bravo & Angrisano (2005).*

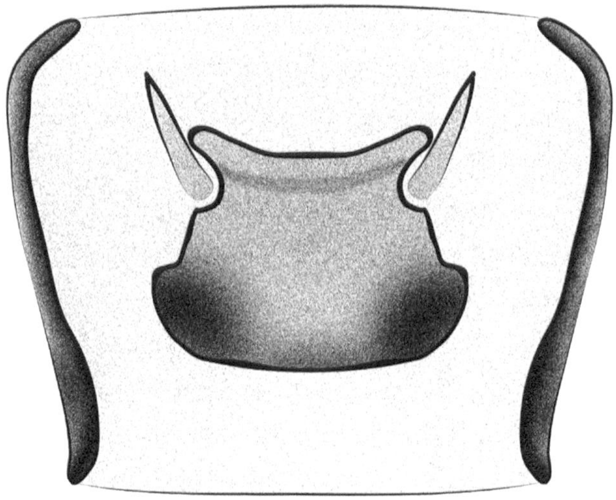

FIGURE 10.75 *Neochorema sinuatum* Schmid, 1964 (Hydrobiosidae): prosternum. *Modified from Bravo & Angrisano (2004a).*

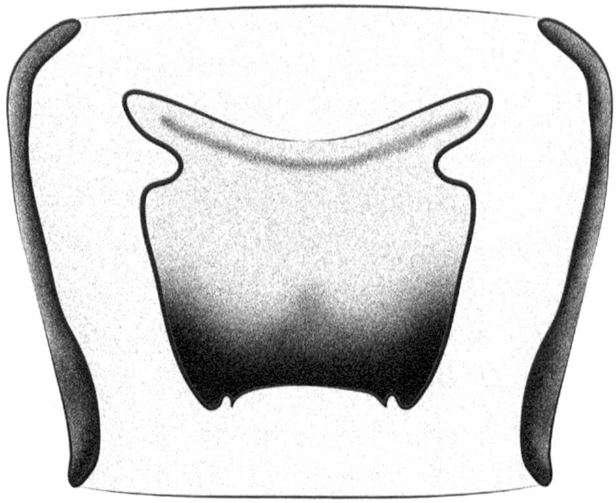

FIGURE 10.76 *Iguazu* sp. (Hydrobiosidae): prosternum. *Modified from Angrisano (2001).*

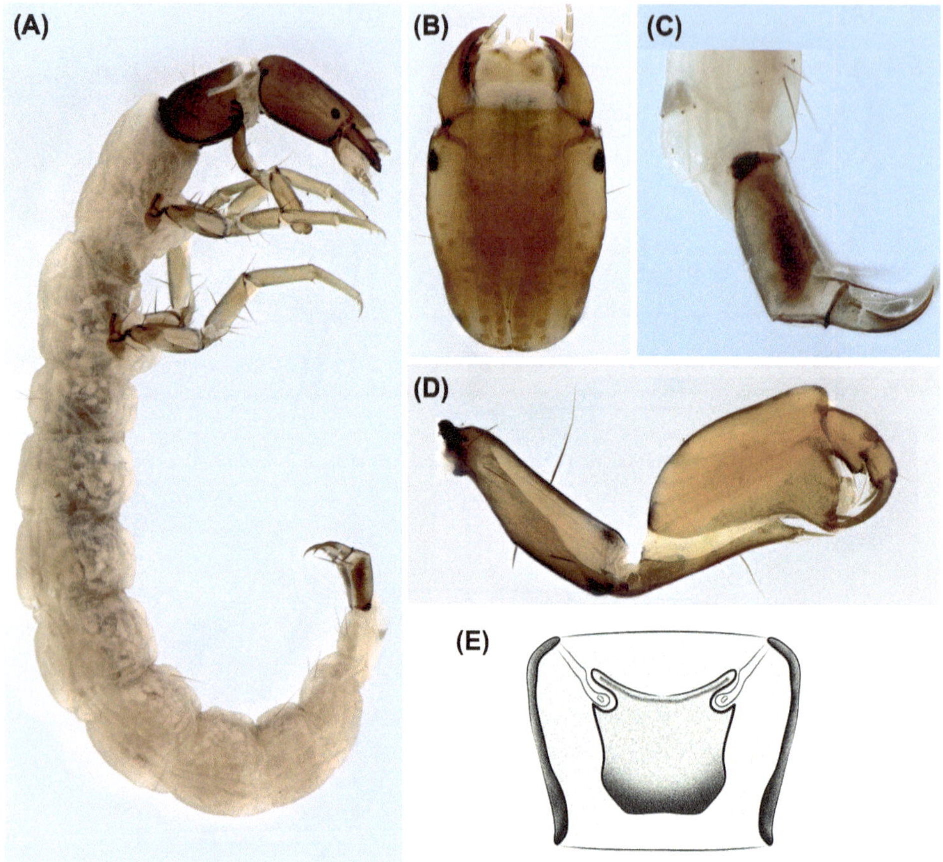

FIGURE 10.77 *Rheochorema* sp. (Hydrobiosidae): (A) larva, lateral view; (B) head, dorsal view; (C) anal claw; (D) foreleg; and (E) prosternum.

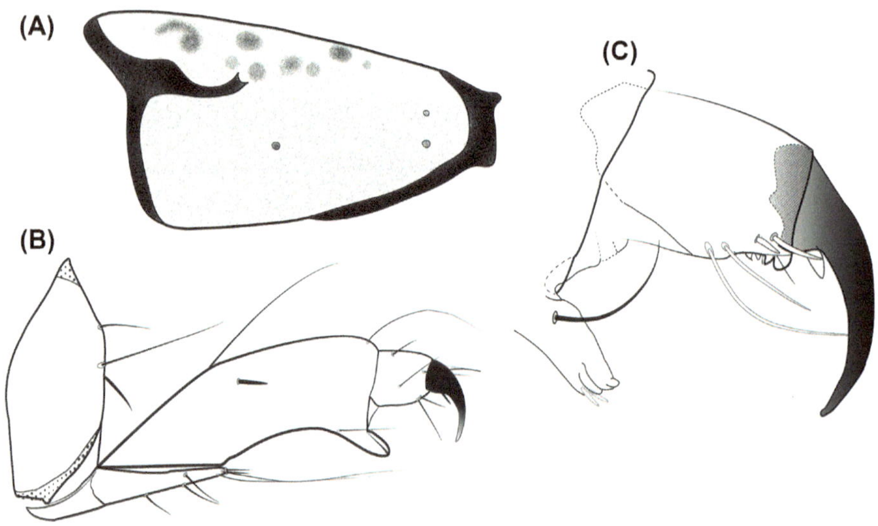

FIGURE 10.78 *Stenochorema* sp. (Hydrobiosidae): (A) head, lateral view; and (B–C) foreleg. *Modified from Angrisano (2000).*

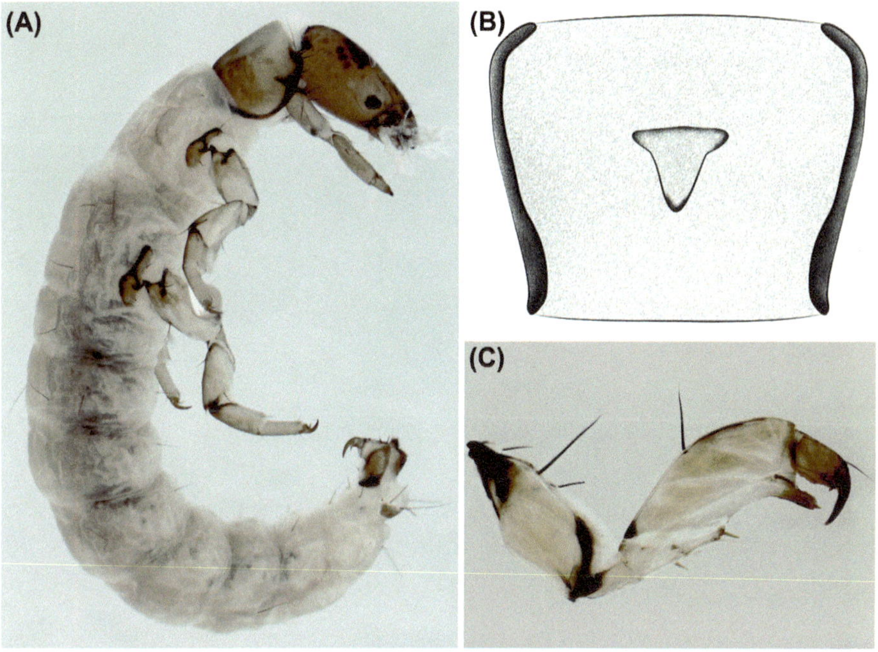

FIGURE 10.79 *Cailloma* sp. (Hydrobiosidae): (A) larva, lateral view; (B) prosternum; and (C) foreleg.

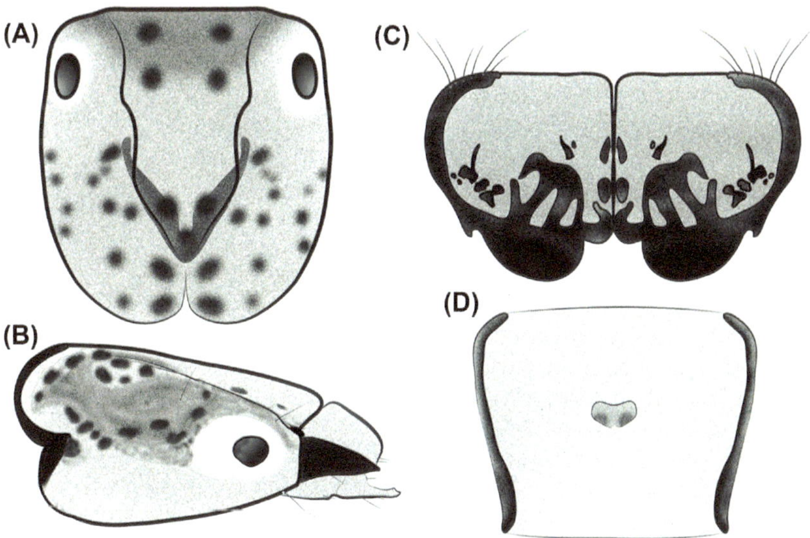

FIGURE 10.80 *Parachorema bifidum* Schmid, 1957 (Hydrobiosidae): (A–B) head, dorsal and lateral view; (C) pronotum, dorsal view; and (D) prosternum. *Modified from Bravo & Angrisano (2004b).*

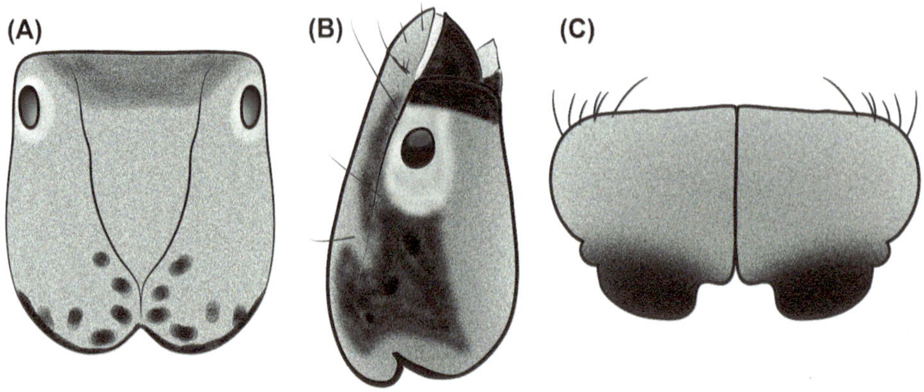

FIGURE 10.81 Larva of the *Amphichorema costiferum* (Flint, 1969) (Hydrobiosidae): (A–B) head, dorsal and lateral view; and (C) pronotum, dorsal view. *Modified from Bravo & Angrisano (2004b).*

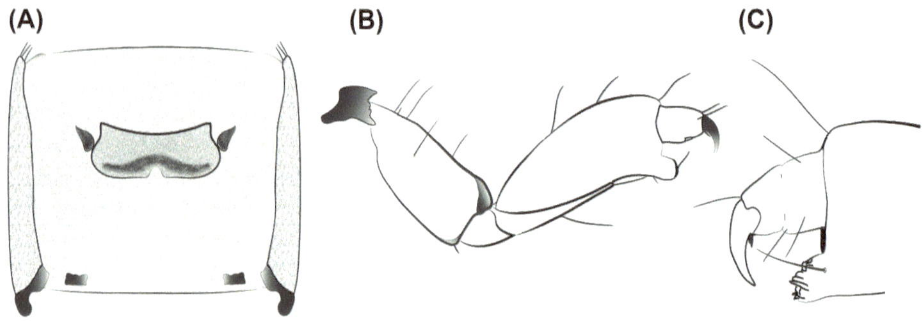

FIGURE 10.82 *Metachorema griseum* Schmid, 1957 (Hydrobiosidae): (A) prosternum; and (B–C) foreleg external view and foreleg apex internal view. *Modified from Bravo & Angrisano (2003).*

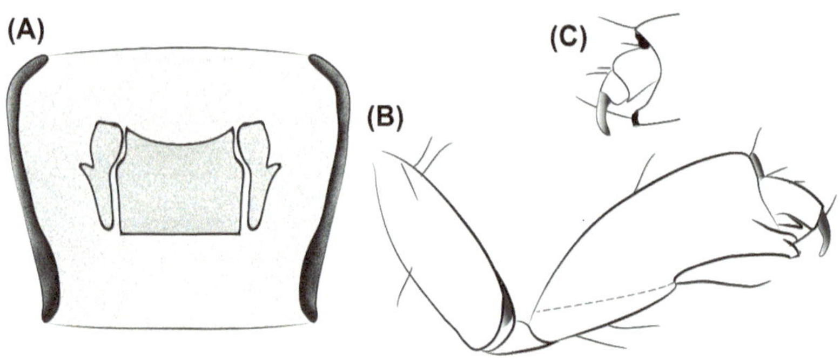

FIGURE 10.83 *Microchorema extensum* Schmid 1964 (Hydrobiosidae): (A) prosternum; and (B–C) foreleg, external view and foreleg apex, internal view. *Modified from Bravo & Angrisano (2004a).*

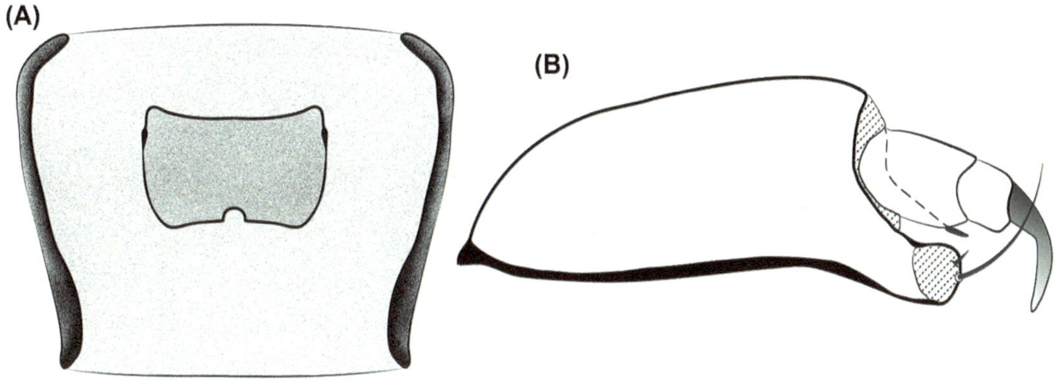

FIGURE 10.84 Genus X (Hydrobiosidae): (A) prosternum; and (B) foreleg. *Modified from Bravo & Angrisano (2004a).*

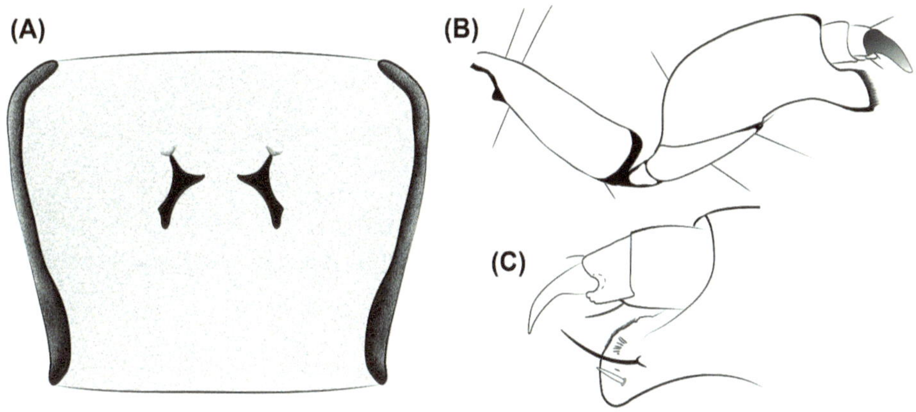

FIGURE 10.85 *Pomphochorema chilensis* Flint (Hydrobiosidae): (A) prosternum; and (B–C) foreleg, external view and foreleg apex, internal view. *Modified from Bravo & Angrisano (2001/2004).*

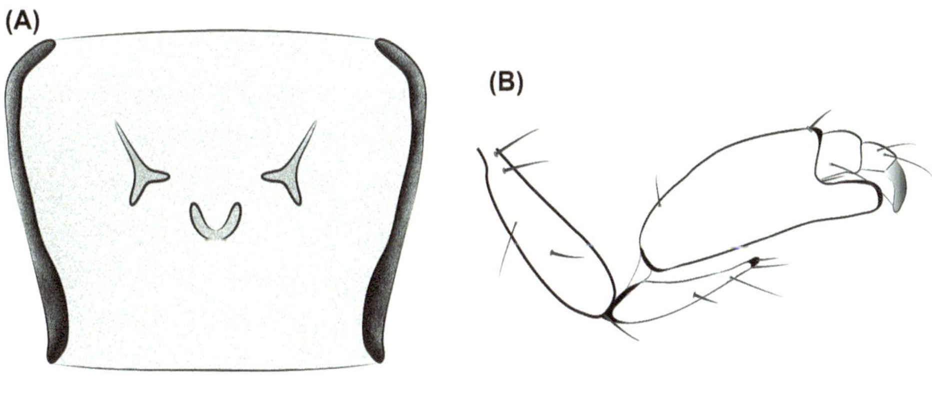

FIGURE 10.86 Genus Y (Hydrobiosidae): (A) prosternum; and (B) foreleg, external view. *Modified from Bravo & Angrisano (2004a).*

FIGURE 10.87 *Neoatopsyche* sp. (Hydrobiosidae): (A) larva, view lateral; (B) head, view dorsal; (C) foreleg, external view; and (D) prosternum.

FIGURE 10.88 *Atopsyche* sp. (Hydrobiosidae): (A) larva, lateral view; and (B) prosternum.

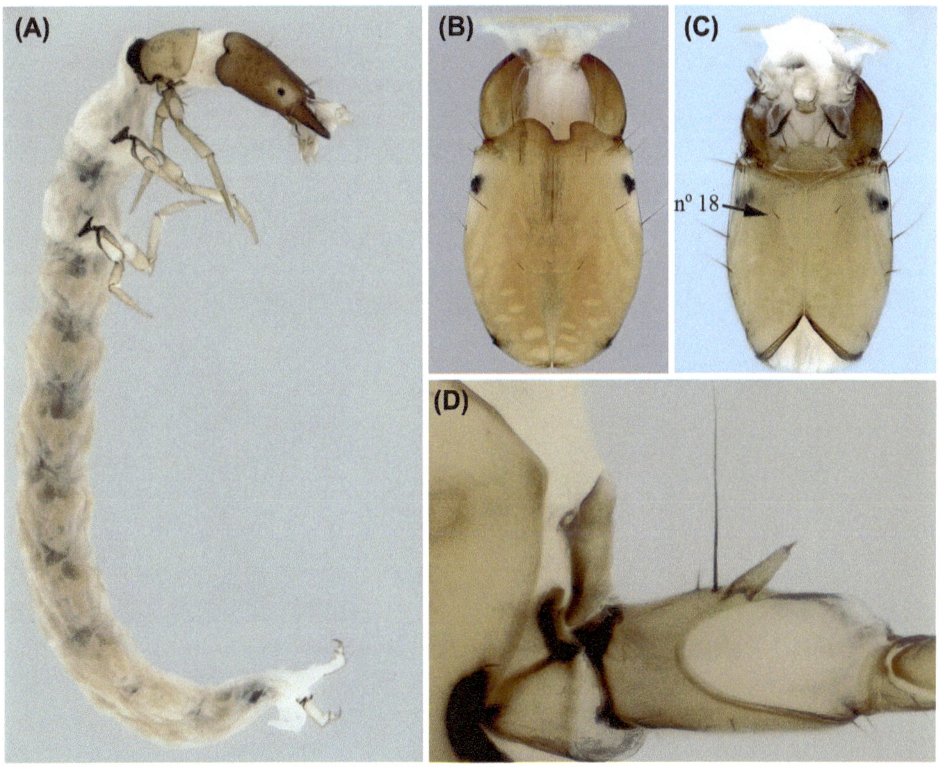

FIGURE 10.89 *Chimarra* spp. (Philopotamidae): (A) larva, lateral view; (B–C) head, dorsal and ventral view; and (D) foretrochantin and forecoxa.

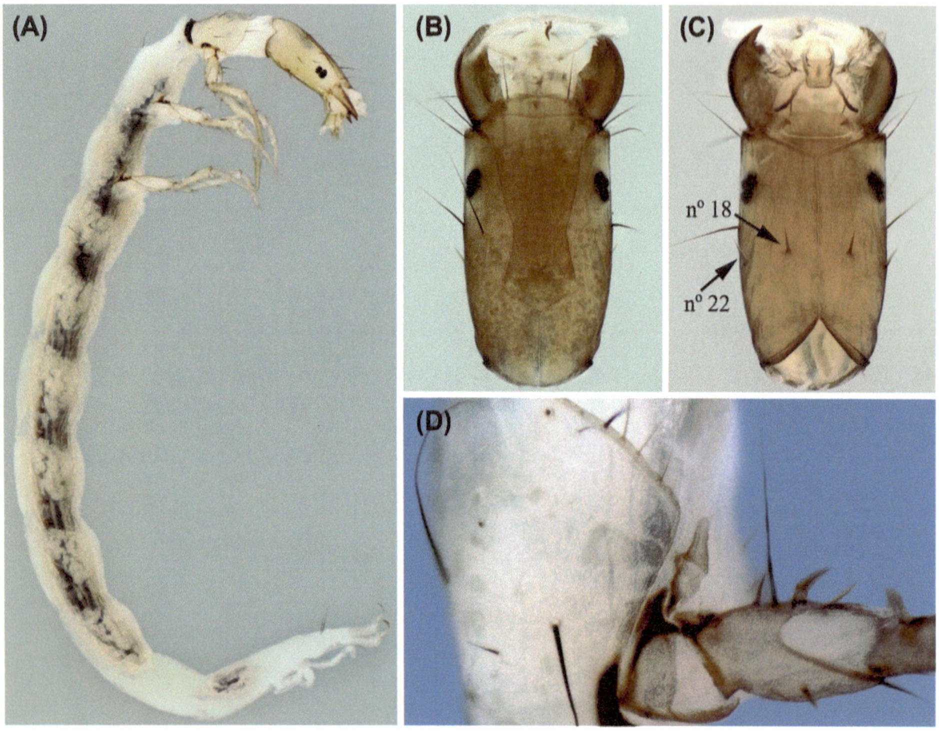

FIGURE 10.90 *Chimarra* (*Chimarrita*) spp. (Philopotamidae): (A) *C.* (*C.*) *simpliciforma* Flint, 1971 larva, lateral view; (B) *C.* (*C.*) *akantha* Blahnik, 1997 foretrochantin and forecoxa; and (C–D) head, dorsal and ventral view.

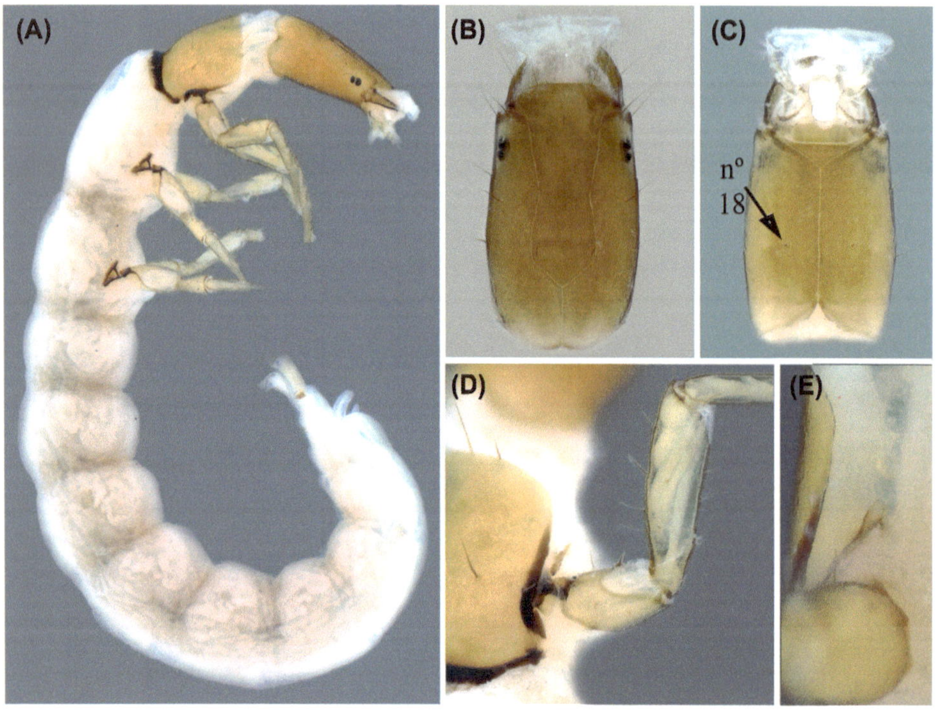

FIGURE 10.91 *Wormaldia* sp. (Philopotamidae): (A) larva, lateral view; (B–C) foretrochantin, forecoxa, lateral view and foretrochantin, ventral view; and (D–E) head, dorsal and ventral view.

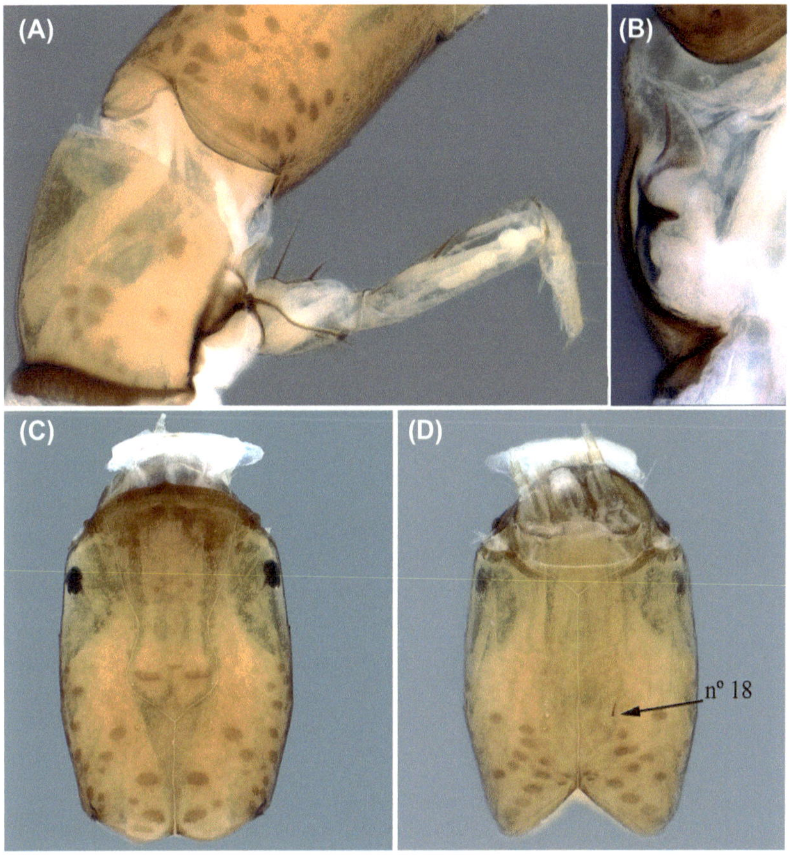

FIGURE 10.92 Genus A (Philopotamidae): (A–B) foretrochantin, forecoxa, lateral view and foretronchantin, ventral view; and (C–D) head, dorsal and ventral view.

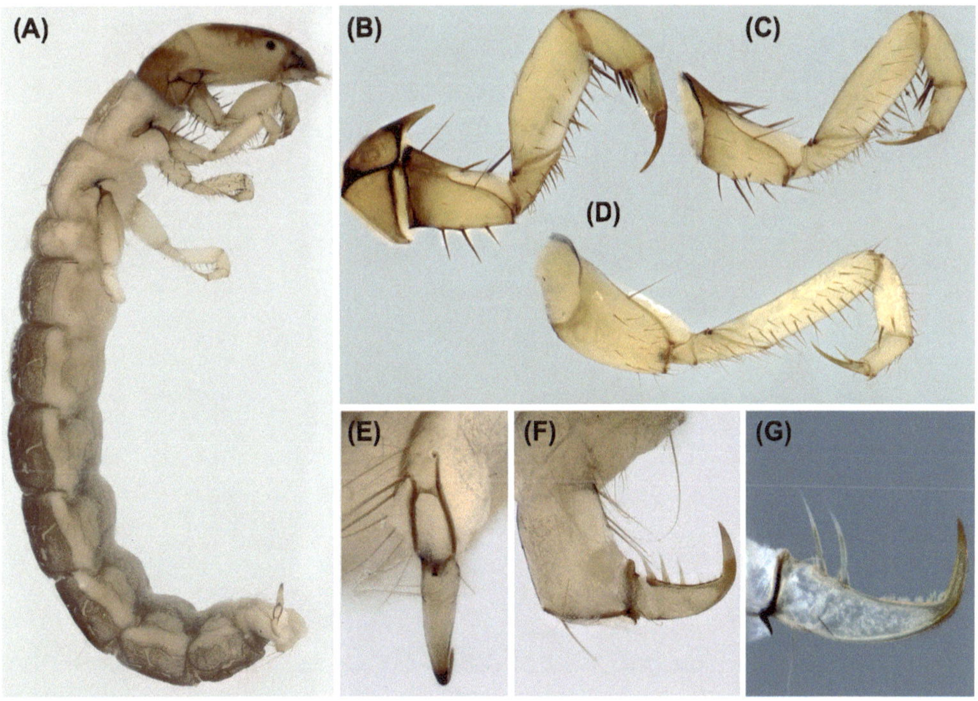

FIGURE 10.93 *Polyplectropus* sp. (Polycentropodidae): (A) larva, view lateral; (B–D) legs lateral view; and (E–F) plate of anal proleg, dorsal and proleg claw, lateral view.

FIGURE 10.94 *Nyctiophylax neotropicalis* Flint, 1971 (Polycentropodidae). (A—B) larva and case, dorsal view; and (C—D) plate of anal proleg, dorsal and sublateral view.

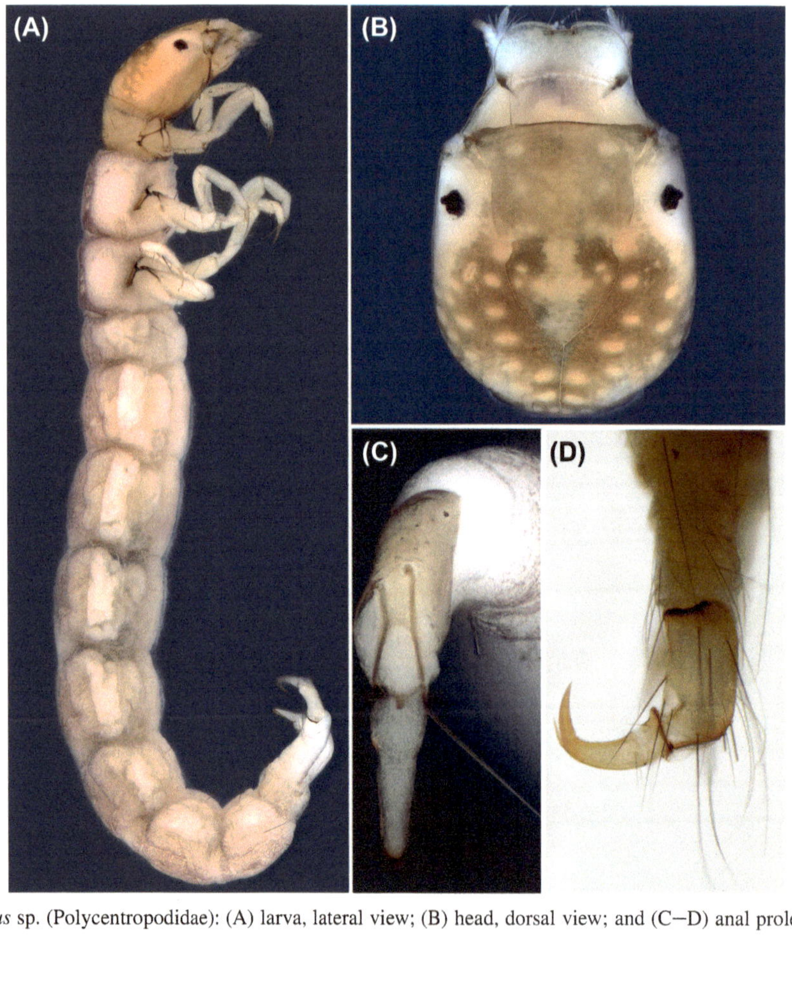

FIGURE 10.95 *Cyrnellus* sp. (Polycentropodidae): (A) larva, lateral view; (B) head, dorsal view; and (C—D) anal proleg, dorsal and lateral view.

FIGURE 10.96 *Cernotina* sp. (Polycentropodidae): (A) larva, lateral view; (B) pronotum and head, dorsal view; (C) pronotum and head, lateral view; and (D) anal proleg, dorsal view.

FIGURE 10.97 *Polycentropus* sp. (Polycentropodidae): (A) larva, lateral view; (B) pronotum and head, dorsal view; and (C) anal proleg, dorsal view.

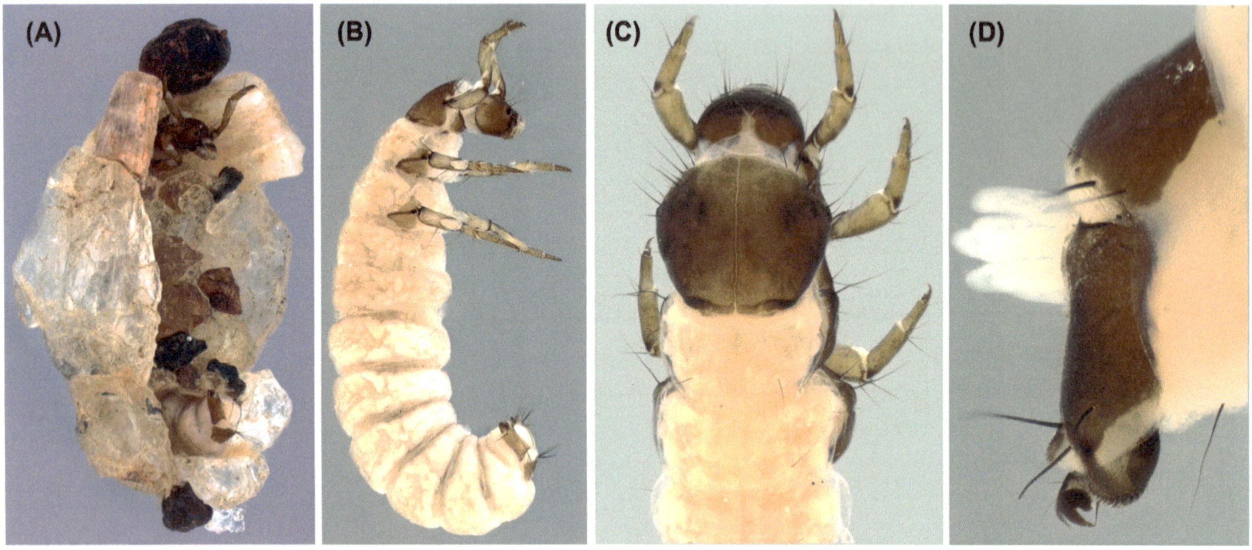

FIGURE 10.98 *Glossosoma* sp. (Glossosomatidae): (A) larva and case, ventral view; (B) larva, lateral view; (C) thorax, dorsal view; and (D) anal proleg, lateral view.

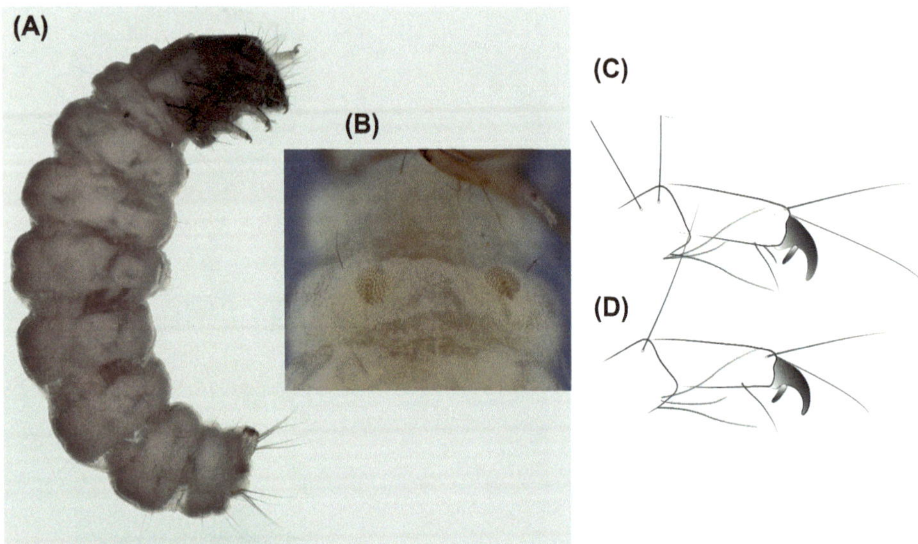

FIGURE 10.99 *Cariboptila pedophila* (Flint, 1964) (Glossosomatidae): (A) larva, lateral view; (B–C) pro- and midlegs, lateral view; and (D) abdominal sternum II.

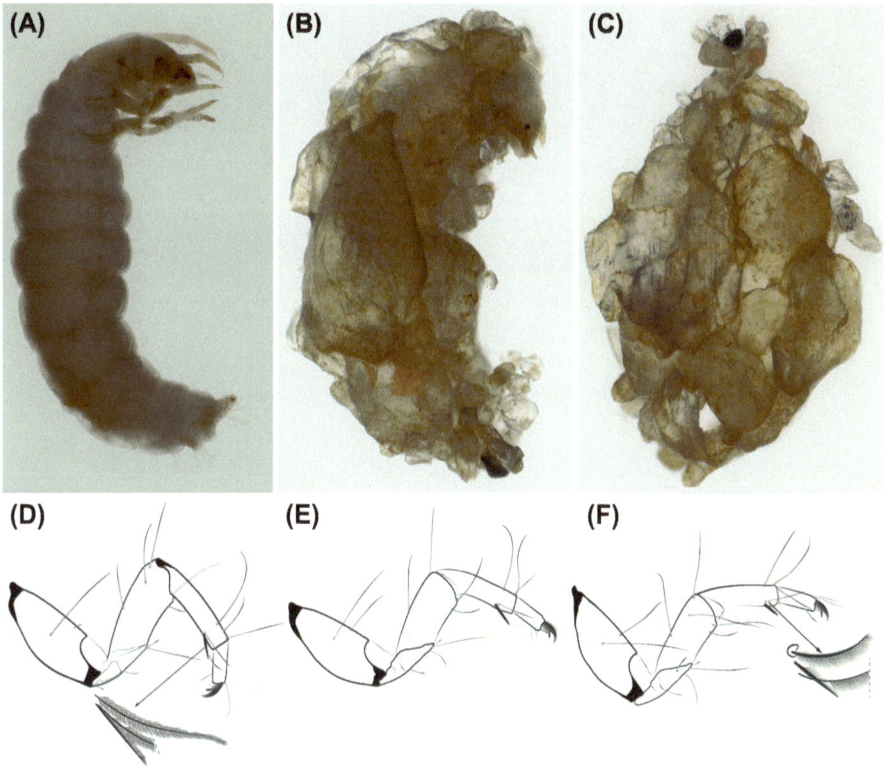

FIGURE 10.100 *Itauara aff. guyanensis* (Glossosomatidae): (A) larva, lateral view; (B–D) legs, lateral view (apicoventral tibial feathered setae, of the fore- and hindlegs); and (E–F) case, lateral and dorsal view.

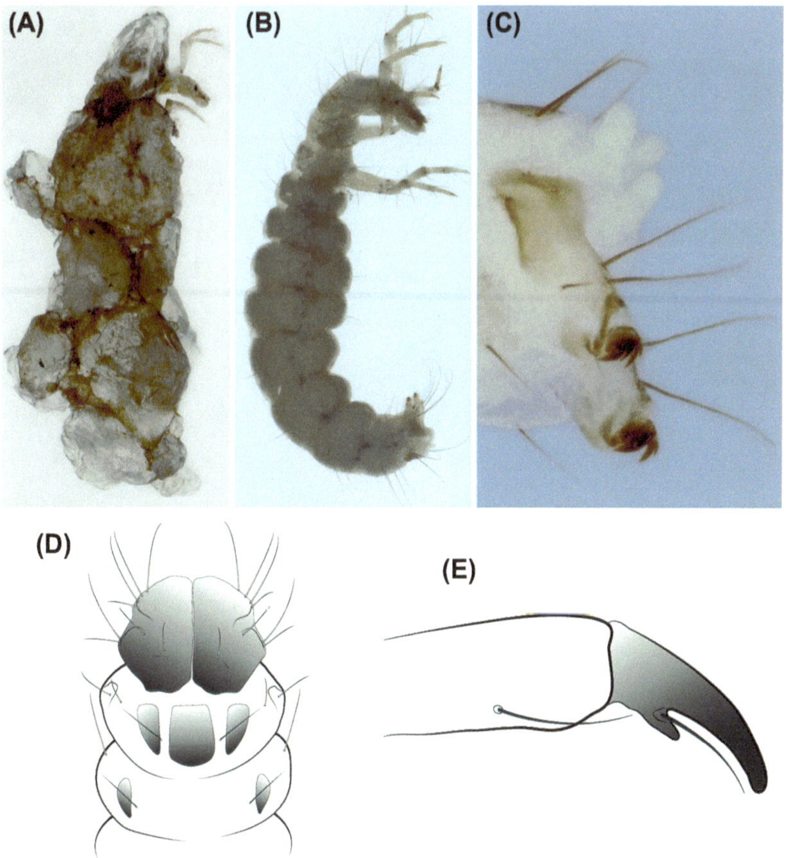

FIGURE 10.101 *Protoptila* sp. (Glossosomatidae). (A–B) case and larva, lateral view; (C) thorax, dorsal view; (D) anal proleg, lateroventral view; (E) tarsal chaw of the foreleg.

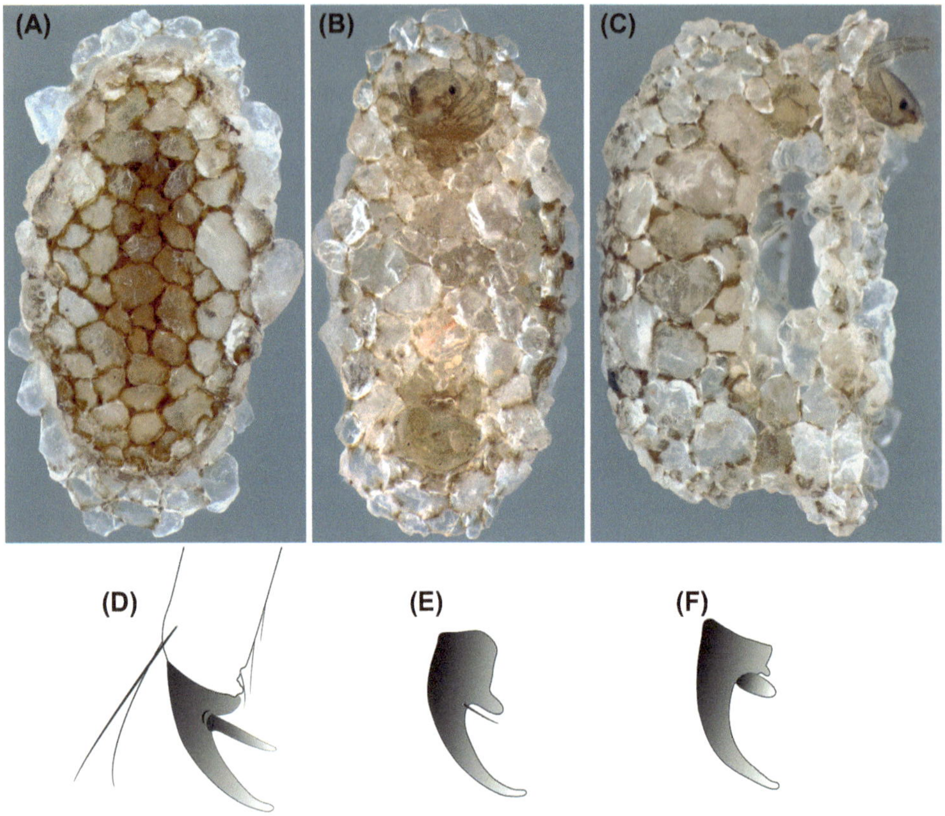

FIGURE 10.102 *Mortoniella* spp. (Glossosomatidae). *Mortoniella* sp.: (A–C) case, dorsal, ventral and lateral view; (D) tarsal claw, hindleg; *M. aries* (Flint): (E) tarsal claw; *M. apiculata* Flint; and (F) tarsal chaw.

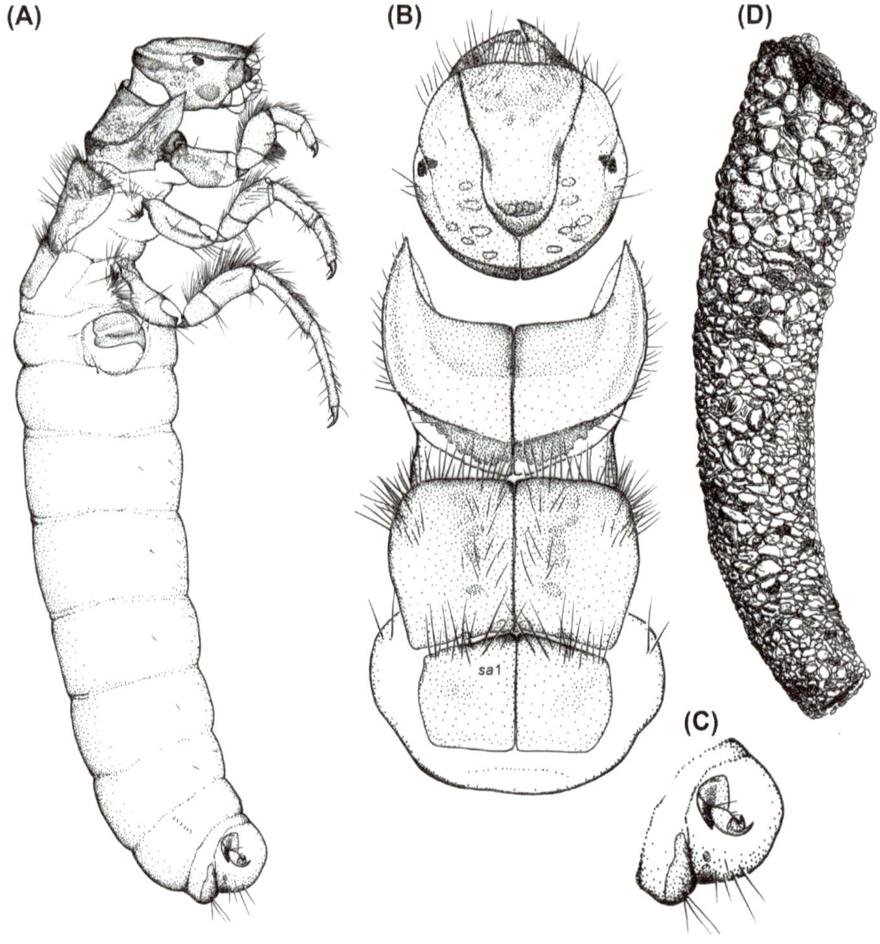

FIGURE 10.103 *Anomalopsyche* sp: (Anomalophsychidae): (A) larva, lateral view; (B) head and thorax, dorsal view; (C) anal proleg, lateral view; and (D) case, lateral view.

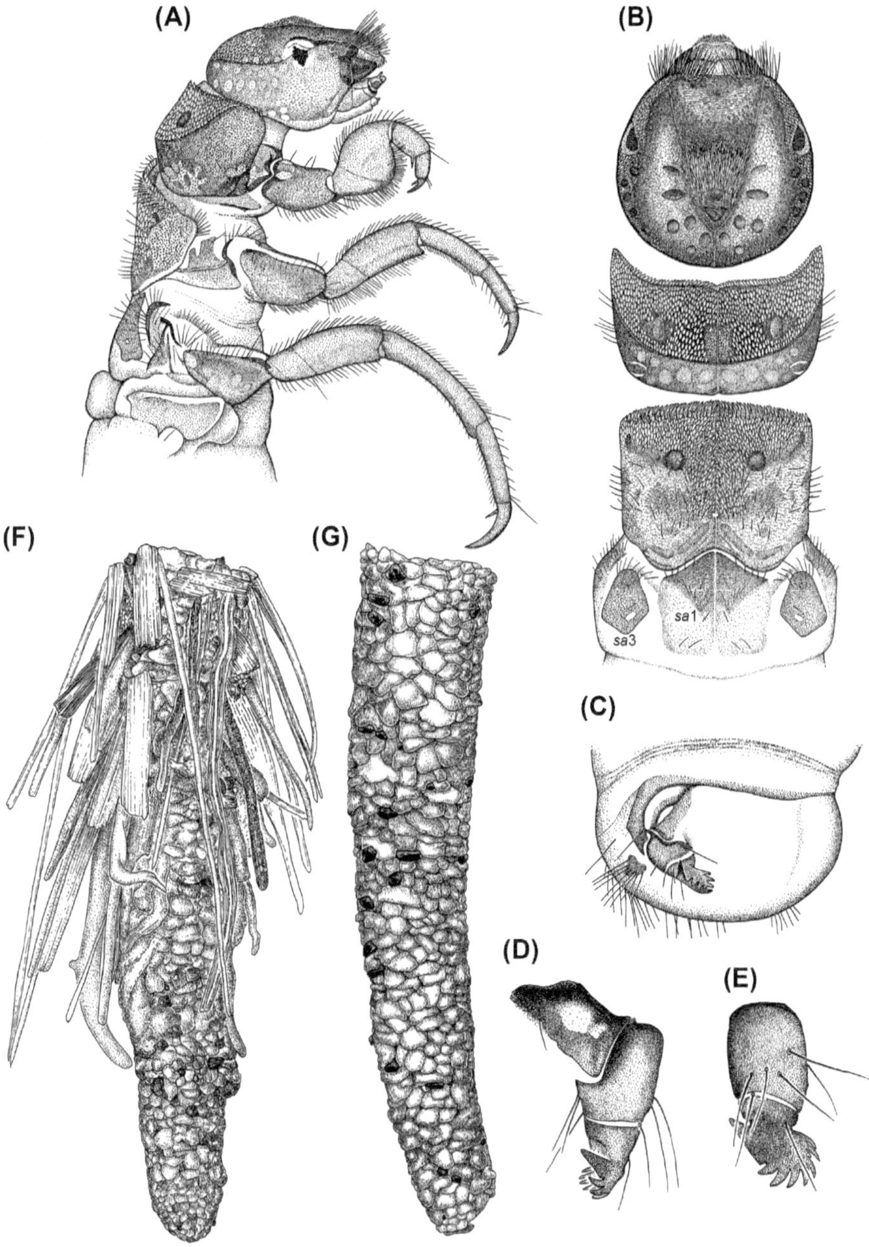

FIGURE 10.104 *Contulma* spp: (Anomalopsychidae). *C. paluguillensis* Holzenthal & Ríos-Touma: (A–C) larva, lateral view, head and thorax, dorsal view, and abdominal segment IX and anal proleg, oblique lateral; *C. nevada* Holzenthal & Flint, 1995; (D–E) anal proleg; *C. penai* Holzenthal & Flint; and (F–G) case, lateral view.

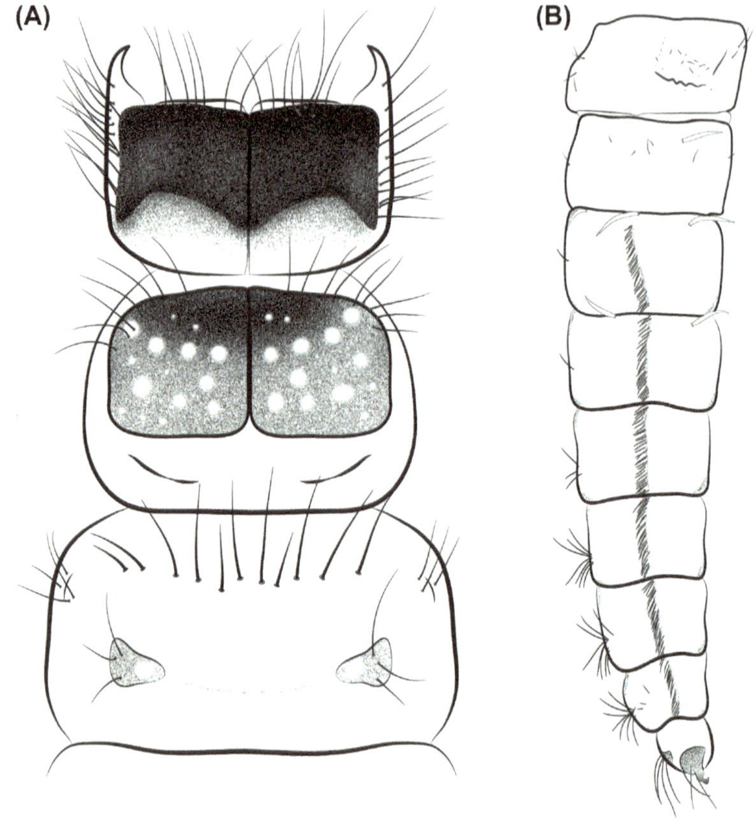

FIGURE 10.105 *Charadropsyche penicillata* Flint, 1972 (Tasimiidae): (A) thorax, dorsal view; and (B) abdomen, lateral view. *Modified from Flint (1999).*

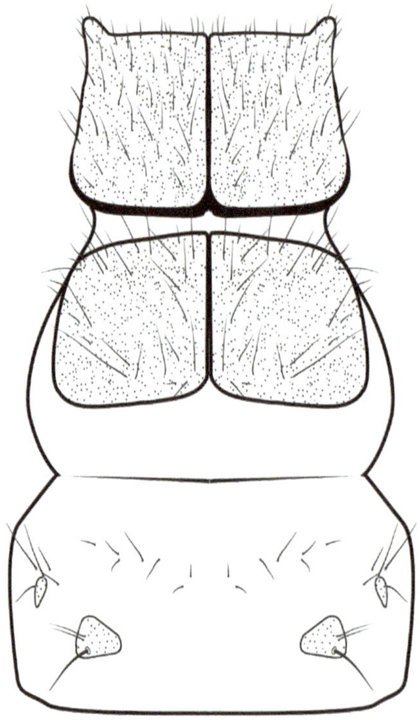

FIGURE 10.106 *Trichovespula macrocera* Schmid, 1955 (Tasimiidae): thorax, dorsal view. *Modified from Flint (1967).*

FIGURE 10.107 *Baryphentus concolor* Burmeister, 1839 (Odontoceridae): (A) larva and case, lateral view; (B–C) head, dorsal and ventral view; and (D–E) thorax, dorsal and ventral view.

FIGURE 10.108 *Anastomoneura guahybae* Hauamantico & Nessimian, 2004 (Odontoceridae): (A) case; (B) larva, lateral view; (C) head and thorax, I II abdominal segments, dorsal view; and (D) head, promesothorax, ventral view.

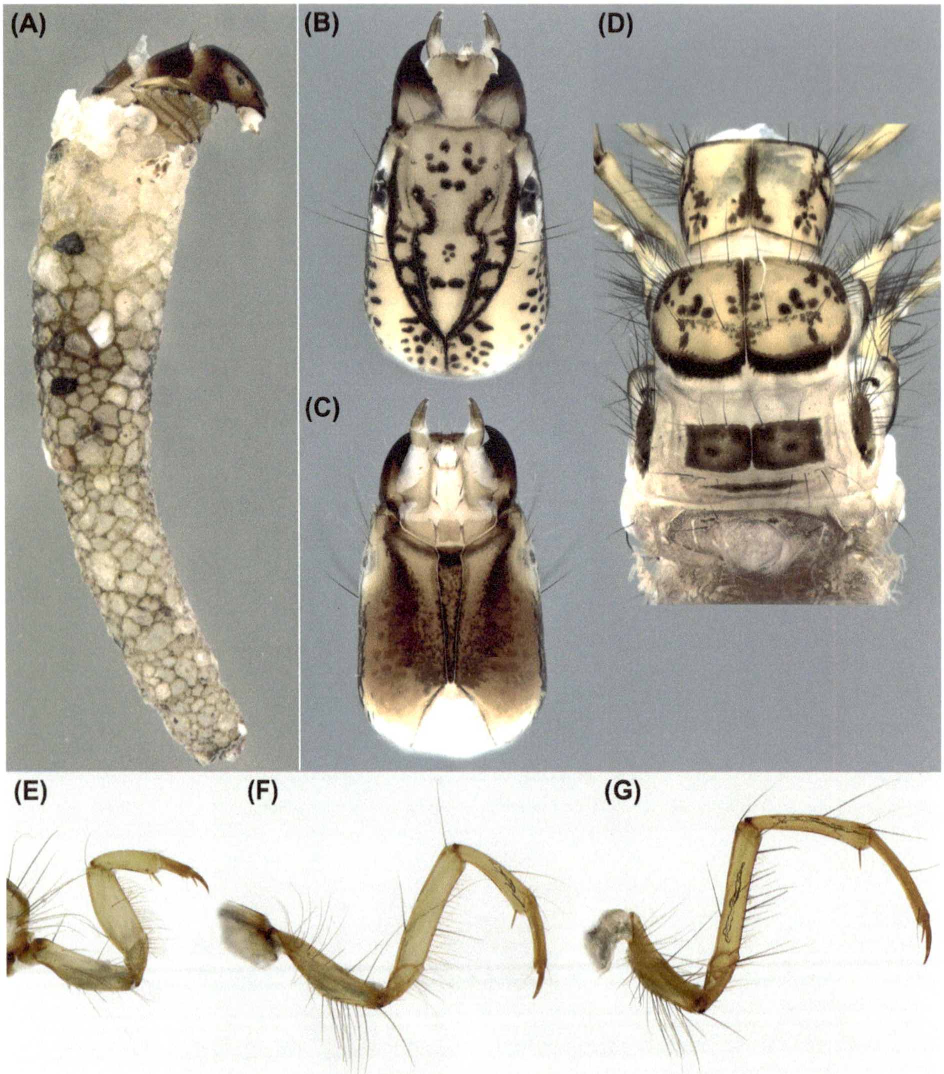

FIGURE 10.109 *Marilia* sp. (Odontoceridae): (A) larva and case, lateral view; (B–C) head, dorsal and ventral view; and (D) thorax, dorsal view; (E–G) legs.

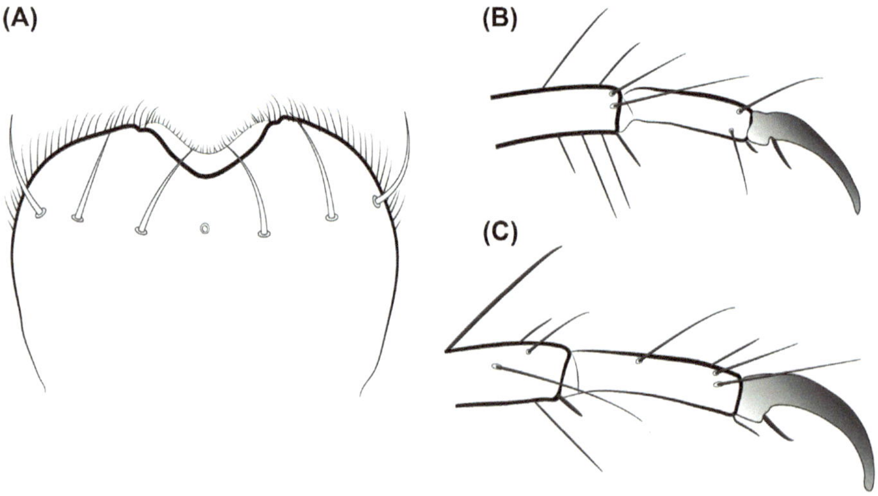

FIGURE 10.110 *Austrocentrus valgiformis* Flint 1997 (Helicophidae): (A) labrum; and (B–C) tarsal chaw of the mid and hindlegs. *Modified from Flint (1997).*

FIGURE 10.111 *Eosericostoma* sp. (Helicophidae): (A) larva and case, ventral view; (B) larva, dorsal view; (C) thorax, dorsal view; (D) labrum; and (E–G) mid and hindlegs.

FIGURE 10.112 *Gumaga* sp. (Sericostomatidae): (A–B) case and larval habitus lateral view; and (C–D) head and thorax, dorsal and lateral view.

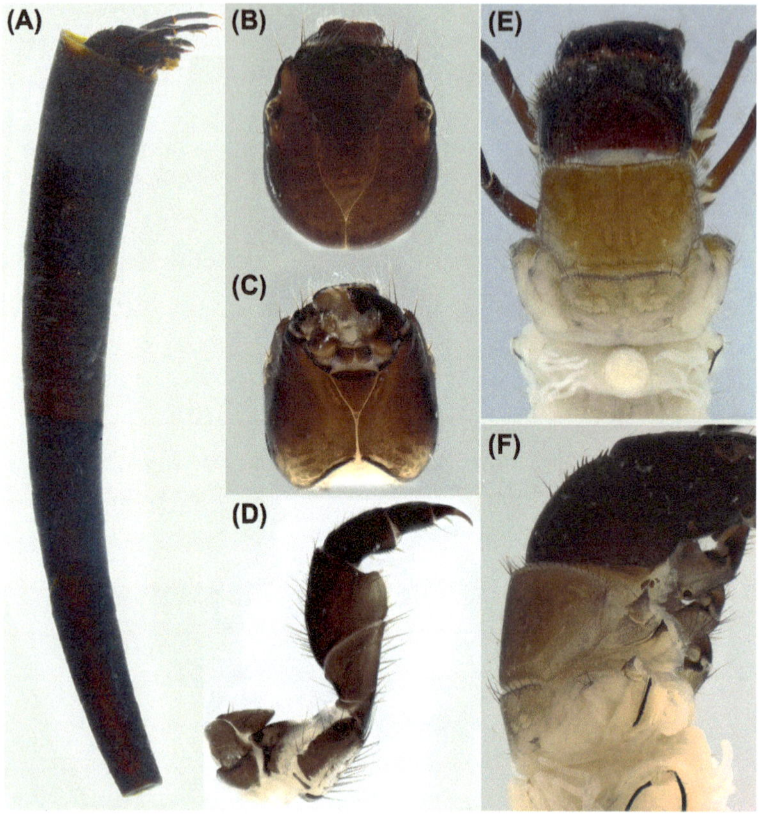

FIGURE 10.113 *Grumicha* sp. (Sericostomatidae): (A) larva and case, lateral view; (B–C) head and thorax, dorsal and lateral view; (D–E) head, dorsal and ventral view; and (F) foreleg and foretrochantin.

FIGURE 10.114 *Notidobiella* sp. (Sericostomatidae): (A–B) larva and case, lateral view; (C–D) head and thorax, dorsal and lateral view; (E–F) head, dorsal and ventral view; and (G) foreleg and foretrochantin.

ACKNOWLEDGMENTS

We would like to thank Alice Wells, Fresia Ester Rojas A., and Takao Nozaki for providing information and images of some families, Jorge L. Nessimian UFRJ, by the loan of some larvae to be photographed. Thanks to Neusa Hamada and Jeane M.C. do Nascimento for collecting larvae of Hydrobiosidae and Limnephilidae from the Andes region.

REFERENCES

Albariño, R.J. & A.C. Valverde. 1998. Hábito alimentario del estado larval de *Parasericostoma cristatum* (Trichoptera: Sericostomatidae). Revista de la Sociedad Entomológica Argentina 57: 131–135.

Angrisano, E.B. 1997. Contribution to the knowledge of the larvae of Hydrobiosidae. I. *Neopsilochorema tricarinatum* and *Austrochorema rectispinum*. Pages 15–17 *in*: Holzenthal, R.W. & O.S Flint, Jr. (eds.), Proceedings of the 8th International Symposium on Trichoptera. Ohio Biological Survey, Columbus, Ohio.

Angrisano, E.B. 2001a. Description of the preimaginal stages of *Stenochorema crassicostum* Schmid (Trichoptera, Hydrobiosidae). Aquatic Insects 24: 289–295.

Angrisano, E.B. 2001b. Las larvas de *Rheochorema* e *Iguazu* (Trichoptera, Hydrobiosidae). Revista de la Sociedad Entomológica Argentina 60: 195–202.

Angrisano, E.B. & J.V. Sganga. 2009. Capitulo 9: Trichoptera. Pages 255–307 *in*: Domínguez, E.E. & H.R. Fernández (eds.), Macroinvertebrados Bentónicos Sudamericanos. Sistemática y Biología. Fundación Miguel Lillo, Tucumán, Argentina.

Barcelos-Silva, P., G.R. Desidério & A.M. Pes. 2017. A new species of *Phylloicus* Müller 1880 (Trichoptera: Calamoceratidae) from Espírito Santo State, southeastern Brazil. Zootaxa 4238: 287–292. https://doi.org/10.11646/zootaxa.4238.2.11.

Bentes, S.P.C., A.M.O. Pes, N. Hamada & R.L. Ferreira-Keppler. 2008. Larvas de *Synoestropsis* sp. (Trichoptera: Hydropsychidae) são predadoras? Acta Amazonica 38: 579–581. https://doi. org/10.1590/S0044-59672008000300023.

Bravo, W. & E.B. Angrisano. 2003a. Los estadios preimaginales de *Pomphochorema chilense* Flint, 1969 (Trichoptera, Hydrobiosidae). Physis [2001] 59: 37–41.

Bravo, W. & E.B. Angrisano. 2003b. Contribution to the knowledge of the preimaginal stages of Neotropical Hydrobiosidae, *Metachorema griseum* Schmid (Trichoptera). Aquatic Insects 25: 203–209.

Bravo, W. & E.B. Angrisano. 2004a. Descripción de los estados preimaginales de cuatro géneros de Hydrobiosidae (Trichoptera) neotropicales. Revista de la Sociedad Entomológica Argentina 63: 97–105.

Bravo, W. & E.B. Angrisano. 2004b. The larvae and pupae of *Amphichorema costiferum* (Flint 1969) and *Parachorema bifidum* Schmid 1957 (Trichoptera: Hydrobiosidae). Journal of the Washington Academy of Sciences 90: 20–27.

Bravo, W. & E.B. Angrisano. 2005. Contribución al conocimiento de los Hydrobiosidae: las larvas de *Clavichorema* y *Pseudoradema* (Trichoptera: Hydrobiosidae). Folia Entomológica Mexicana 44: 19–26.

Calor, A.R. 2007. Trichoptera. *in*: Guia on-line de Identificação de larvas de Insetos Aquáticos do Estado de São Paulo. Disponível em: http://sites.ffclrp.usp.br/aguadoce/index_trico.

Calor, A.R. & R.W. Holzenthal. 2008. Phylogeny of Grumichellini Morse, 1981 (Trichoptera: Leptoceridae) with the description of a new genus from southeastern Peru. Aquatic Insects 30: 245–259.

Cavalcante, B.M. da S., L. L. Dumas & J. L. Nessimian. 2017. Two new species of *Austrotinodes* Schmid (Insecta, Trichoptera, Ecnomidae) from southeastern Brazil. Zootaxa 4263: 395–400. https://doi.org/10.11646/zootaxa.4263.2.13.

Chamorro, M.L. & R.W. Holzenthal. 2010. Phylogeny of Polycentropodidae Ulmer, 1903 (Trichoptera, Annulipalpia, Psychomyioidea) inferred from larval, pupal and adult characters. Invertebrate Systematics 25: 219–253.

Dean, J.C., R.M. St. Clair & D.I. Cartwright. 2004. Identification keys to Australian families and genera of caddis-fly larvae (Trichoptera). Cooperative Research Centre for Freshwater Ecology Identifcation Guide 50: 131 pp.

Desidério, G.R., P. Barcelos-Silva, A.M. Pes, W.R.M. de Souza, & C.A.S. de Azevêdo. 2017. Caddisflies (Insecta: Trichoptera) from Maranhão State, Northeast Region, Brazil: A new species, checklist, and new geographical records. Zootaxa 4221: 151–171. https://doi.org/10.11646/zootaxa.4221.2.1.

Dumas, L.L. & J.L. Nessimian. 2008. A new species of *Neoatriplectides* Holzenthal, 1997 (Insecta: Trichoptera: Atriplectididae), from Brazil, including description of the pupa of the genus. Zootaxa 1773: 63–68.

Fischer, F.C.J. 1960. Necrotaulidae, Prosepididontidae, Rhyacophilidae. Trichopterorum Catalogus 1. Nederlandsche Entomologische Vereeniging, Amsterdam, iv + 168 pp.

Fischer, F.C.J. 1961 Philopotamidae, Hydroptilidae, Stenopsychidae. Trichopterorum Catalogus 2. Nederlandsche Entomologische Vereeniging, Amsterdam, iv + 190 pp.

Fischer, F.C.J. 1962. Polycentropodidae, Psychomyidae. Trichopterorum Catalogus 3. Nederlandsche Entomologische Vereeniging, Amsterdam, vi + 236 pp.

Fischer, F.C.J. 1963. Hydropsychidae, Arctopsychidae. Trichopterorum Catalogus 4. Nederlandsche Entomologische Vereeniging, Amsterdam, vi + 225 pp.

Fischer, F.C.J. 1964. Phryganeidae, Limnocentropodidae, Molannidae. Trichopterorum Catalogus 5. Nederlandsche Entomologische Vereeniging, Amsterdam, vi + 214 pp.

Fischer, F.C.J. 1965. Calamoceratidae, Philorheithridae, Odontoceridae, Leptoceridae Pars 1. Trichopterorum Catalogus 6. Nederlandsche Entomologische Vereeniging, Amsterdam, vi + 242 pp.

Fischer, F.C.J. 1966. Leptoceridae Pars 2. Trichopterorum Catalogus 7. Nederlandsche Entomologische Vereeniging, Amsterdam, iii + 163 pp.

Fischer, F.C.J. 1967. Goeridae, Limnephilidae Pars 1. Trichopterorum Catalogus 8. Nederlandsche Entomologische Vereeniging, Amsterdam, vi + 263 pp.

Fischer, F.C.J. 1968. Limnephilidae Pars 2. Trichopterorum Catalogus 9. Nederlandsche Entomologische Vereeniging, Amsterdam, iii + 363 pp.

Fischer, F.C.J. 1969. Limnephilidae Pars 3. Trichopterorum Catalogus 10. Nederlandsche Entomologische Vereeniging, Amsterdam, iii + 332 pp.

Fischer, F.C.J. 1970. Philanisidae, Lepidostomatidae, Brachycentridae, Beraeidae, Sericostomatidae, Thremmatidae, Helicopsychidae. Trichopterorum Catalogus 11. Nederlandsche Entomologische Vereeniging, Amsterdam, vi + 316 pp.

Fischer, F.C.J. 1971. Supplement to Vol. I and II. Trichopterorum Catalogus 12. Nederlandsche Entomologische Vereeniging, Amsterdam, vii + 311 pp.

Fischer, F.C.J. 1972a Supplement to Vol. III and IV. Trichopterorum Catalogus 13. Nederlandsche Entomologische Vereeniging, Amsterdam, vii + 172 pp.

Fischer, F.C.J. 1972b. Supplement to Vol. V, VI, and VII. Trichopterorum Catalogus 14. Nederlandsche Entomologische Vereeniging, Amsterdam, vii + 154 pp.

Fischer, F.C.J. 1973a. Supplement to Vol. VIII, IX, X and XI. Trichopterorum Catalogus 15. Nederlandsche Entomologische Vereeniging, Amsterdam, vii + 166 pp.

Fischer, F.C.J. 1973b. Index Volumes I-XV. Trichopterorum Catalogus. Nederlandsche Entomologische Vereeniging, Amsterdam, vi + 214 pp.

Flint, O.S, Jr. 1981. Studies of Neotropical caddisflies, XXVII: Anomalopsychidae, a new family of Trichoptera. Pages 75–85 *in*: Moretti, G.P. (ed.), Proceedings of the 3rd International Symposium on Trichoptera. Dr. W. Junk Publishers, The Hague.

Flint, O.S, Jr. 1982a. Studies of Neotropical caddisflies, XXX: larvae of the genera of South American Limnephilidae (Trichoptera). Smithsonian Contributions to Zoology 355: 1–30.

Flint, O.S, Jr. 1982b. Studies of Neotropical caddisflies, XXXIV: the genus *Plectromacronema* (Trichoptera: Hydropsychidae). Proceedings of the Biological Society of Washington 96: 225–237.

Flint, O.S., Jr. 1991. Studies of Neotropical caddisflies, XLV: The taxonomy, phenology, and faunistics of the Trichoptera of Antioquia, Colombia. Smithsonian Contributions to Zoology 520: 1–113.

Flint, O.S., Jr. 1992. Studies of Neotropical caddisflies, XLIX: The taxonomy and relationship of the genus *Eosericostoma*, with descriptions of the immature stages (Trichoptera: Helicophidae). Proceedings of the Biological Society of Washington 105: 494–510.

Flint, O.S., Jr. 1996. The Trichoptera collected on the expeditions to Parque Manu, Madre de Dios, Peru. Pages 369–430 *in*: Wilson, D.E. & A. Sandoval (eds.), Manu: The biodiversity of southeastern Peru. Smithsonian Institution Press. Washington, DC. USA.

Flint, O.S., Jr. 1997. Studies of Neotropical caddisflies, LIV: the Patagonian genus *Austrocentrus*, with the description of its immature stages (Trichoptera: Helicophidae). Pages 99–108 *in*: Holzenthal, R.W. & O.S. Flint, Jr. (eds.), Proceedings of the 8th International Symposium on Trichoptera, 1995. Ohio Biological Survey, Columbus, USA.

Flint, O.S. Jr, R.W. Holzenthal & S.C. Harris. 1999a. Nomenclatural and systematic changes in the Neotropical caddisflies. Insecta Mundi 13: 73–84.

Flint, O.S. Jr, R.W. Holzenthal & S.C. Harris. 1999b. Catalog of the Neotropical Caddisflies (Insecta: Trichoptera). Ohio Biological Survey. Columbus. USA. 239 pp.

Harris, S.C & O.S. Jr. Flint. 1993. Studies of Neotropical caddisflies, XLVIII; the larva of *Celaenotrichia edwardsi* Mosely, with an assessment of the genus (Trichoptera: Hydroptilidae). *In:* Otto, C. (Ed) Proceedings of the 7th International Symposium on Trichoptera. Backhuys Publishers, Leiden, The Netherlands, 101–106.

Holzenthal, R.W. 1986a. Studies in Neotropical Leptoceridae (Trichoptera), IV: a revision of Brachysetodes Schmid. Transactions of the American Entomological Society 111: 407–440.

Holzenthal, R.W. 1986b. Studies in Neotropical Leptoceridae (Trichoptera), a new species of *Amphoropsyche*, with a redescription of the immature stages of A. insularis (Flint). Annals of the Entomological Society of America 79: 251–255. https://doi.org/10.1093/aesa/79.1.251

Holzenthal, R.W. 1986c. Studies in Neotropical Leptoceridae (Trichoptera), VI: immature stages of *Hudsonema flaminii* (Návas) and the evolution and historical biogeography of Hudsonemini (Triplectininae). Proceedings of the Entomological Society of Washington 88: 268–279.

Holzenthal, R.W. 1988a. Systematics of Neotropical Triplectides (Trichoptera: Leptoceridae). Annals of the Entomological Society of America 81: 187–208. https://doi.org/10.1093/aesa/81.2.187

Holzenthal, R.W. 1988b Studies in Neotropical Leptoceridae (Trichoptera), VIII: the genera *Atanatolica* Mosely and Grumichella Müller (Triplectidinae: Grumichellini). Transactions of the American Entomological Society 114: 71–128.

Holzenthal, R.W. 1995. The caddisfly genus *Nectopsyche*: new gemma group species from Costa Rica and the Neotropics (Trichoptera: Leptoceridae). Journal of the North American Benthological Society 14: 61–83. https://doi.org/10.2307/1467725

Holzenthal, R.W. 1997. The caddisfly (Trichoptera) family Atriplectididae in the Neotropics. Pages 157–165 *in*: Holzenthal, R.W. & O.S., Jr. Flint (eds.), Proceedings of the 8th International Symposium on Trichoptera, 1995. Ohio Biological Survey, Columbus, USA.

Holzenthal, R.W. & T. Andersen. 2004. The caddisfly genus *Triaenodes* in the Neotropics (Trichoptera: Leptoceridae). Zootaxa 511: 1–80. https://doi.org/10.11646/zootaxa.511.1.1.

Holzenthal, R.W. & R.J. Blahnik. 2010. Systematics of the Neotropical caddisfly genus Notidobiella Schmid (Trichoptera, Sericostomatidae), with the description of 3 new species. ZooKeys 71: 23–47.

Holzenthal, R.W. & O.S., Jr. Flint. 1995. Studies of Neotropical caddisflies, LI: systematics of the Neotropical caddisfly genus Contulma (Trichoptera: Anomalopsychidae). Smithsonian Contributions to Zoology 575: 1–59.

Holzenthal, R.W. & A.R. Calor. 2017. Catalog of the Neotropical Trichoptera (Caddisflies). ZooKeys 654: 1–566. https://doi.org/10.3897/zookeys.654.9516.

Holzenthal, R.W. & A.M.O. Pes. 2004. A new genus of long-horned caddisfly from the Amazon basin (Trichoptera: Leptoceridae: Grumichellini). Zootaxa 621: 1–16. https://doi.org/10.11646/zootaxa.621.1.1

Holzenthal, R.W. & D.R. Robertson. 2006. Four new species of Contulma from South America (Trichoptera: Anomalopsychidae). Zootaxa 1355: 49–59.

Holzenthal, R.W. & B. Ríos-Touma. 2012. *Contulma paluguillensis* (Trichoptera: Anomalopsychidae), a new caddisfly from the high Andes of Ecuador, and its natural history. Freshwater Science 31: 442–450.

Holzenthal, R.W. & O.S., Jr. Flint. 1995. Studies of Neotropical caddisflies, LI: systematics of the Neotropical caddisfly genus *Contulma* (Trichoptera: Anomalopsychidae). Smithsonian Contributions to Zoology 575: 1–59.

Holzenthal, R.W. & D.R. Robertson. 2006. Four new species of *Contulma* from South America (Trichoptera: Anomalopsychidae). Zootaxa 1355: 49–59.

Holzenthal, R.W. & R.J. Blahnik. 2010. Systematics of the Neotropical caddisfly genus *Notidobiella* Schmid (Trichoptera, Sericostomatidae), with the description of 3 new species. ZooKeys 71: 23–47.

Holzenthal, R.W., R.J. Blahnik, A.P. Prather & K.M. Kjer. 2007. Order Trichoptera Kirby, 1813 (Insecta), caddisflies. Zootaxa 1668: 639–698.

Holzenthal, R.W., J.C. Morse & K.M. Kjer. 2010. Order Trichoptera Kirby, 1813. In: Zhang, Z.-Q. (ed.), Animal Biodiversity: An Outline of Higher-level Classification and Survey of Taxonomic Richness. Zootaxa 3148: 209–210.

Holzenthal, R.W., R.E. Thomson & B. Ríos-Touma. 2015. Chapter 38: Order Trichoptera. Pages 965–1002 *in*: Thorp, J. and D.C. Rogers (eds.), Ecology and General Biology, 4th ed: Thorp and Covich's Freshwater Invertebrates. Academic Press.

Huamantinco, A.A. & J.L. Nessimian. 2004. A New Neotropical Genus and Species of Odontocerinae (Trichoptera: Odontoceridae) from Southeastern Brazil. Aquatic Insects 26: 281–288.

Huamantinco, A.A. & W. Ortiz. 2010. Clave de géneros de larvas de Trichoptera (Insecta) de la Vertiente Occidental de los Andes, Lima, Perú. Revista Peruana de Biologia 17: 75–80.

Johanson, K.A., T. Malm & M. Espeland. 2016. Molecular phylogeny of Sericostomatoidea (Trichoptera) with the establishment of three new families. Systematic Entomology 42: 1–27. https://doi.org/10.1111/syen.12209.

Johanson, K.A., T. Malm, M. Espeland & E. Weingartner. 2012. Phylogeny of the Polycentropodidae (Insecta: Trichoptera) based on protein-coding genes reveal non-monophyletic genera. Molecular Phylogenetics and Evolution 65: 126–135. https://doi.org/10.1016/j.ympev.2012.05.029.

Kjer, K.M., J.A. Thomas, X. Zhou, P.B. Frandsen, E. Scott & R.W. Holzenthal. 2016. Progress on the phylogeny of caddisflies (Trichoptera). Pages 216–224 *in*: Vshivkova, T. & J.C. Morse (eds.), ZooSymposia (Proceedings of the 14th International Symposium on Trichoptera). Aukland, NZ, Magnolia Press.

Kristensen, N.P. 1997. Early evolution of the Trichoptera + Lepidoptera lineage: phylogeny and the ecological scenario. Mémoires Museum National d'Histoire Naturelle 173: 253–271.

Mackay, R.J & G.B. Wiggins. 1979. Ecological diversity of Trichoptera. Annual Review of Entomology 24: 185–208.

Malicky, H. 2001. Notes on the taxonomy of *Rhadicoleptus, Ptilocolepus, Pseudoneureclipsis*. Braueria 28: 19–20.

Malicky, H. 2005. Ein kommentiertes Verzeichnis der Köcherfliegen (Trichoptera) Europas und des Mediterrangebiets. Linzer Biologische Beiträge 37Ç 533–596.

Malm, T. & K.A. Johanson. 2010. A new classification of the long-horned caddisflies (Trichoptera: Leptoceridae) based on molecular data. BMC Evolutionary Biology, 11, 17 pp. DOI:10.1186/1471-2148-11-10.

Malm, T., K.A. Johanson & N. Wahlberg. 2013. The evolutionary history of Trichoptera (Insecta): a case of successful adaptation to life in freshwater. Systematic Entomology 38: 459–473.

Marshall, J.E. 1979. A review of the genera of the Hydroptilidae (Trichoptera). Bulletin of the British Museum (Natural History). Entomology 39: 135–239.

Misof, B., S. Liu, K. Meusemann, R.S. Peters, A. Donath, C. Mayer, P.B. Frandsen, J. Ware, T. Flouri, R.G. Beutel, O. Niehuis, M. Petersen, F., Izquierdo-Carrasco, T. Wappler, J. Rust, A.J. Aberer, U. Aspöck, H. Aspöck, D. Bartel, A. Blanke, S. Berger, A. Böhm, T.R. Buckley, B. Calcott, J. Chen, F. Friedrich, M. Fukui, M. Fujita, C. Greve, P. Grobe, S. Gu, Y. Huang, L.S. Jermiin, A.Y. Kawahara, L. Krogmann, M. Kubiak, R. Lanfear, H. Letsch, Y. Li, Z. Li, J. Li, H. Lu, R. Machida, Y. Mashimo, P. Kapli, D.D. McKenna, G. Meng, Y. Nakagaki, J.L. Navarrete-Heredia, M. Ott, Y. Ou, G. Pass, L. Podsiadlowski, H. Pohl, B.M. von Reumont, K. Schütte, K. Sekiya, S. Shimizu, A. Slipinski, A. Stamatakis, W. Song, X. Su, N.U. Szucsich, M. Tan, X. Tan, M. Tang, J. Tang, G. Timelthaler, S. Tomizuka, M. Trautwein, X. Tong, T. Uchifune, M.G. Walzl, B.M. Wiegmann, J. Wilbrandt, B. Wipfler, T.K. Wong, Q. Wu, G. Wu, Y. Xie, S. Yang, Q. Yang, D.K. Yeates, K. Yoshizawa, Q. Zhang, R. Zhang, W. Zhang, Y. Zhang, J. Zhao, C. Zhou, L. Zhou, T. Ziesmann, S. Zou, Y. Li, X. Xu, Y. Zhang, H. Yang, J. Wang, J. Wang, K.M. Kjer, & X. Zhou. 2014. Phylogenomics resolves the timing and pattern of insect evolution. Science 346: 763–767.

Morse, J.C. & R.W. Holzenthal. 1984. Trichoptera genera. Pages 312–347 *in*: Merritt, R.W. & K.W. Cummins (eds.), An introduction to the Aquatic Insects of North America. Kendall/Hunt Publ. Comp.

Morse, J.C. & R.W. Holzenthal. 2008. Chapter 18, Caddisfly genera. Pages 481–552 *in*: Merritt, R.W., K.W. Cummins & M.A. Berg (eds.), An Introduction to the Aquatic Insects of North America, 4th ed. Kendall/Hunt, Dubuque, Iowa.

Morse, J.C. (ed.) 2016. Trichoptera World Checklist. http://entweb.clemson.edu/database/trichopt/index.htm.

Muñoz-Quesada, F. & R.W. Holzenthal. 1997. A new species of *Xiphocentron (Antillotrichia)* from Costa Rica with semiterrestrial immature stage (Trichoptera: Xiphocentronidae). Pages 355–363 *in*: Holzenthal, R.W. & O.S. Flint, Jr. (eds.), Proceedings of the 8th International Symposium on Trichoptera. Ohio Biological Survey, Columbus, Ohio. USA.

Nozaki, T., R. Saito, N. Nishimura, H. Li-Peng & T. Koji. 2016. Larvae and females of two *Stenopsyche* species in Taiwan with redescription of the male of *S. formosana* (Insecta: Trichoptera). Zootaxa 4121: 485–494.

Paprocki, H, R.W. Holzenthal. 2002. A review of Brazilian genus *Barypenthus* Burmeister (Trichoptera: Odontoceridae). Proceedings of the 10th International Symposium on Trichoptera. Nova Supplementa Entomologica 15: 223–230.

Pes, A.M.O., N. Hamada. 2004. *Ceratotrichia* Flint, 1992 (Trichoptera: Hydroptilidae) larval and pupal description and new genus records for Brazil. Entomotropica 19: 31–37.

Pes, A.M., A.P.M. dos Santos, P. Barcelos-Silva & L.M. Camargos. 2014. Capítulo 23: Ordem Trichoptera. Pages 391–433 *in*: Hamada, N., J.L. Nessimian & R.B. Querino (eds.), Insetos Aquáticos na Amazônia brasileira: taxonomia, biologia e ecologia. Editora do INPA, Manaus, AM. Brazil.

Pes, A.M.O., N. Hamada & J.L. Nessimian. 2005. Chaves de identificação de larvas para famílias e gêneros de Trichoptera (Insecta) da Amazônia Central, Brasil. Revista Brasileira de Entomologia 49: 181–204.

Posada-García, J.A. & G. Roldán-Perez. 2003. Clave ilustrada y diversidad de las larvas de Trichoptera en el nor-occidente de Colombia. Caldasia 25: 169–192.

Prather, A.L. 2003. Revision of the Neotropical caddisfly genus *Phylloicus* (Trichoptera: Calamoceratidae). Zootaxa 275: 1–214. https://doi.org/10.11646/zootaxa.275.1.1.

Prather, A.L. 2004. Revision of the Neotropical caddisfly genus *Banyallarga* (Trichoptera: Calamoceratidae). Zootaxa 435: 1–76. https://doi.org/10.11646/zootaxa.435.1.1.

Resh, V.H. & D.M. Rosenberg. 1984. The ecology of aquatic insects. Praeger Publishers, New York. 625 pp.

Quinteiro, F.B. & R.W. Holzenthal. 2017. Fourteen new species of *Oecetis* McLachlan, 1877 (Trichoptera: Leptoceridae) from the Neotropica region. PeerJ. 5: 3753:1–35. https://doi.org/10.7717/peerj.3753.

Ríos-Touma, B., A.C. Encalada & N. Prat. 2012. Oviposition of aquatic insects in a tropical high altitude stream. Environmental Entomology 41: 1322–1331. https://doi.org/10.1603/EN12037.

Robertson, D.R. & R.W. Holzenthal. 2013. Revision and phylogeny of the caddisfly subfamily Protoptilinae (Trichoptera: Glossosomatidae) inferred from adult morphology and mitochondrial DNA. Zootaxa 3723: 1–99. https://doi.org/10.11646/zootaxa.3723.1.1.

Rocha, I.C., L.L. Dumas & J.L. Nessimian. 2017. Description of two new species of *Xiphocentron* Brauer, 1870 (Trichoptera: Xiphocentronidae) from southeastern Brazil. Tropical Zoology. 30:1–8. 8 https://doi.org/10.1080/03946975.2017.1362867.

Rojas, F.E. 2007. Estados preimaginales de *Pangullia nea* Rojas, 2006 (Trichoptera: Kokiriidae) de Chile. Acta Entomologica Chilena 31: 23–30.

Santos, A.P.M., J.L. Nessimian & D.M. Takiya. 2016a. Revised classification and evolution of leucotrichiine microcaddisflies (Trichoptera: Hydroptilidae) based on morphological and molecular data. Systematic Entomology 41: 458–480. DOI: 10.1111/syen.12168.

Santos, A.P.M., A.R. Calor, L.L. Dumas, A.M.O. Pes, W.R.M. Souza & A.L. Henriques-Oliveira. 2016b. Trichoptera In Taxonomic Catalog of the Brazilian Fauna. (URL: http://fauna.jbrj.gov.br/fauna/listaBrasil.

Souza, W.R.M., A.P.M. Santos. 2017. Taxonomic study of the genus *Oxyethira* Eaton 1873 (Trichoptera: Hydroptilidae) from Northeast Brazil: Eleven new species and distributional records. Zootaxa 4236: 484–506. https://doi.org/10.11646/zootaxa.4236.3.4.

Springer, M. 2006. Clave taxonomica para larvas de las familias del orden Trichoptera (Insecta) de Costa Rica. Revista de Biologia Tropical. 54: 273–286.

Springer, M. 2010. Trichoptera. Revista de Biologia Tropical. 58: 151–198.

Valverde, A.C. & M.L. Miserendino. 1997 Los estados inmaduros de *Parasericostoma ovale* (Trichoptera: Sericostomatidae). Revista de la Sociedad Entomológica Argentina 56: 33–37.

Vilarino, A. & A.R. Calor. 2017. Trichoptera of Serra da Jibóia, Bahia, Brazil: new species of Helicopsyche (Helicopsychidae) and new records. Zootaxa 4311: 507–522. https://doi.org/10.11646/zootaxa.4310.4.4.

Wallace, J.B. & D. Malas. 1976. The fine structure of capture nets of larval Philopotamidae (Trichoptera), with special emphasis on *Dolophilodes distinctus*. Canadian Journal of Zoology 54: 1788–1802.

Wallace, A.R. 1876. The Geographical Distribution of Animals: With a Study of the Relations of Living and Extinct Faunas as Elucidating the Past Changes of the Earth's surface. Vol. 2, Macmillan, London. 607 pp.

Wiggins, G.B. 1984. Trichoptera. Pages 271–310 *in*: Merritt, R.W. & K.W. Cummins (eds.), An introduction to the Aquatic Insects of North America. Kendall/Hunt Publ. Comp. Dubuque, Iowa. USA.

Wiggins, G.B. 1996. Larvae of the North American caddisfly genera (Trichoptera), second ed. University of Toronto Press, Toronto, Canada. 457 pp.

Wiggins, G.B. 2004. Caddisflies: the underwater architects. University of Toronto Press, Toronto, Canada. 292 pp.

Wiggins, G.B & D.C. Currie. 2008. Chapter 17. Trichoptera families. Pages 439–480 *in*: Merritt, R.W., K.W. Cummins & M.B. Berg (eds.), An introduction to the aquatic insects of North America. Kendall/Hunt Publishing Co., Dubuque, Iowa.

Wiggins, G.B. & W. Wichard. 1989. Phylogeny of pupation in Trichoptera, with proposals on the origin and higher classification of the order. Journal of the North American Benthological Society 8: 260–276.

Williams, N.E. & G.B. Wiggins. 1981. A proposed setal nomenclature and homology for larval Trichoptera. Pages 421–429 *in*: Moretti, G.P. (ed.), Proceedings of the 3rd International Symposium on Trichoptera. The Hague: Dr W. Junk Publishers. The Netherlands.

Chapter 11

Order Lepidoptera

Márlon Breno Graça

Coordenação de Biodiversidade, Curso de Pós-Graduação em Entomologia, Instituto Nacional de Pesquisas da Amazônia, Manaus, Amazonas, Brazil

M. Alma Solis

MAS, Systematic Entomology Laboratory, USDA, National Museum of Natural History, Washington, DC, United States

Chapter Outline

INTRODUCTION

Regarded as the third most diverse group of insects (van Nieukerken et al., 2011), the Lepidoptera (Greek: *lepidos* = scale, *pteron* = wing) are commonly known as butterflies and moths. Currently, there are over 157,000 recognized species within 15,578 genera worldwide. Lepidoptera is mainly the paraphyletic grouping of the moths, leaving the butterflies with around 12% of the known diversity (van Nieukerken et al., 2011).

Lepidopterans are widespread and have colonized nearly all types of environments. The vast majority are terrestrial organisms, but some species inhabit during at least one life stage water bodies or plants that occur in water. As early-lineage larvae began to use green plants parts as food resources, the radiation of the angiosperms likely led to coevolution and the development of adaptive strategies to exploit host plants, including those of aquatic and semiaquatic habits (Mey & Speidel, 2008).

Studies on aquatic and semiaquatic moths in the Neotropics are extremely sparse, so they are constantly overlooked as members of freshwater communities. Still, the Neotropical region is regarded as home of higher diversity of freshwater Lepidoptera than any other zoogeographic region (Mey & Speidel, 2008). Currently, these water-related moths have been identified as belonging to eight families: Cosmopterigidae, Momphidae, Tortricidae, Pyralidae, Crambidae, Sphingidae, Erebidae, and Noctuidae.

Cosmopterigidae

Commonly known as cosmet moths, cosmopterigids are small moths (wingspan ranging from 6–12 mm) that occur worldwide and comprise 135 genera and 1,792 species (Sinev, 2002), and 166 are known from the Neotropical region (Koster, 2010). Larvae can be recognized by the crochets of the prolegs, which are disposed in a biordinal circle. Their feeding habits include leaf mining, leaf rolling, boring stems, and roots of plants, others feed on flowers and flower heads, mycelia, or can be predators of scale insects and spider's eggs (Common, 1990). They feed mostly on herbs in the Convolvulaceae, Cyperaceae, and Poaceae families, and numerous species are regarded as crop pests (Hodges, 1999). Leaf miners build blotch-type

Thorp and Covich's Freshwater Invertebrates, Fourth Edition. https://doi.org/10.1016/B978-0-12-804223-6.00011-1

mines and may switch or stay all juvenile stage in a single mine (Koster, 2010). Pupation takes place in silken chambers, in or outside the mines (Koster, 2010). Among the species reported from the Neotropics, very few display aquatic habits. The widely distributed *Cosmopterix attenuatella* (Walker) mine leaves of the aquatic sedge *Scirpus* (Koster, 2010), and the Galápagos endemic *C. galapagosensis* Landry is a leaf miner of *Eleocharis mutata* (L.) Roem and Schult (Cyperaceae) (Landry, 2001; Koster, 2010).

Momphidae

Momphidae comprises relatively slender-bodied and medium-sized moths with wingspan ranging from 7 to 21 mm. The family is relatively small, represented by six genera and 115 species in the world (van Nieukerken et al., 2011). The larva has T1 with two lateral setae, crochets on prolegs in uniordinal transverse circles (Hodges, 1999), and are generally leaf miners, but some species feed inside stems and seeds. The only known momphid on an aquatic plant reported from the Neotropical region is *Mompha orfilai* Bourquin, a leaf miner of *Ludwigia longifolia* (DC) H. Hara, the long-leaf willow primrose, in Argentina (Romero & Navarro, 2009).

Tortricidae

Tortrix moths are widely distributed, fairly robust microlepidopterans, whose wingspan varies from 7 mm and may reach 40 mm (Horak, 1999). Currently, 10,387 species are known from the world, with the New World tropics comprising most of the species richness (Gilligan et al., 2014). The larvae of tortrix moths have the crochets in circular disposition, the setae SD1 of the eighth abdominal segment is anterior to the spiracle (Horak. 1999, Gilligan & Epstein, 2009). and some bear an anal comb on the final segment. The common name of the family (leafrollers) refers to the habit of the larvae of many species of folding leaves to create shelters. Besides leaf consuming, tortricids display different feeding strategies, such as gall-making, flower-feeding, litter-feeding, and predation (Horak & Brown, 1991; Brown et al., 2008).

The associations between species of tortricids from the Neotropical region and aquatic plants are yet to be confirmed. However, in other regions, *Bactra* is reported to be associated with *Bolboschoenus*, *Cyperus*, and *Scirpus* (Cyperaceae) of aquatic environments, and *Clepsis*, with over 60 species reported from South and Central America, grows on *Scirpus* as well (Brown et al., 2008). At least two North American species of *Choristoneura* feed on *Typha* (Typhaceae), even though *Choristoneura argentifasciata* Heppner, from the Florida Everglades, has not been confirmed as a cattail feeder (Brown et al., 2008). *Typha* may also host *Archips* (Lange, 1978), but the host plant of *Archips biforata* Meyrick from Brazil is not definitively known. Despite the unconfirmed semiaquatic status of these taxa, we have included Tortricidae here due to their considerable species richness in the Neotropical region and therefore their potential of discovery on aquatic plants from this region.

Pyralidae

In this chapter, we treat Pyralidae in the strict sense (Solis & Mitter, 1992; Regier et al., 2012; Nuss et al., 2003–2015). Van Nieukerken et al. (2011) recognized 1,055 genera and 5,921 species of pyralids worldwide, and species richness from the Neotropical region was last estimated to be ~1.100 species (Munroe et al., 1995; Solis, 1997). Pyralid moth larvae nearly always possess a ring-shaped sclerotized plate surrounding the SD1 setae on the eighth abdominal segment and on the mesothorax (Phycitinae), metathorax (Chrysauginae) or A1 (Galleriinae) as well (Minet, 1985; Solis & Mitter, 1992; Solis, 1999). Their feeding breadth comprises leaf mining, leaf rolling, leaf webbing, stem boring, root feeding, seed feeding, bee-wax consuming, and even predation of scale insects (some Phycitinae) (Solis, 1997). The pupal stage occurs in silk sealed cocoons made by larvae.

From the five subfamilies of Pyralidae, the speciose Phycitinae is the only one reported to have a Neotropical aquatic representative, *Arcola malloi* Pastrana, the alligator weed stemborer. It has been introduced to other countries from Argentina for the biological control of alligator weed, *Alternanthera philoxeroides* (Mart.) Griseb. (Amaranthaceae) (Pastrana, 1961; Center et al., 1999). The larvae grow up to 19 mm, spin a silk cocoon, and conceal themselves in cavities within the plant stem.

Crambidae

This family is the most diverse of the pyraloids, with 9,655 species recognized in 1,020 genera (van Nieukerken et al., 2011) and 13 subfamilies (Regier et al., 2012). Eleven subfamilies in the Crambidae are Neotropical in distribution; but in this chapter, we will focus on the five subfamilies, Acentropinae, Crambinae, Pyraustinae, Schoenobiinae, and Spilomelinae, that have Neotropical aquatic/semiaquatic species.

Subfamily Acentropinae

In all of Lepidoptera, this subfamily has the most representatives with morphological adaptations for living underwater during at least one life stage (Lange, 1978; Solis, 2008). Its known diversity is over 700 species worldwide, and its greatest richness is in the Neotropical region (Mey & Speidel, 2008). Classically, the acentropines were divided into two tribes, Argyractini and Nymphulini; but since this classification has no phylogenetic support (Munroe & Solis, 1999), our chapter does not recognize this division.

Females often lay the eggs on rock surface, as in *Petrophila*, or in submerged or underwater vegetation, as in *Elophila*. Acentropinae larvae utilize many mechanisms for oxygen intake, including direct breathing through tracheal or blood gills in species of *Argyractis, Petrophila*, and *Parapoynx*. Species that lack gills in all life stages, such as those of *Elophila*, often resort to plastron-like devices using either hydrophobic cuticular setae, or bubbles in interstitial spaces of the host plant or cases made with the host plant (Solis, 2008). Larvae can be associated with submerged or emerged hydrophytes by leaf mining (first instar *Parapoynx*), consuming leaves externally (*Elophila* and later instar *Parapoynx*), or root predating (*Neargyractis*), as well as feed on algae and diatoms scraped from rocks (*Petrophila*) (Lange, 1978; Solis, 2008; Habeck & Cuda, 2009). Additionally, there are records of species predating blackflies in Brazil (Gorayeb & Pinger, 1978). Pupae of Acentropinae possess several shared unique characters, for example, the protruding spiracles on the second to fourth abdominal segments (Speidel, 1984; Yoshiyasu, 1985) that support the group's monophyly. Pupation occurs in or out of the aquatic environment, inside silken cocoons with reduced or protruding spiracles, and can consist of two cocoon layers covered with holes to permit water flowing, as in *Petrophila* (Solis, 2008).

There are 18 genera of Acentropinae known from the Neotropical region (Munroe, 1995; Mey & Speidel, 2008). The following genera have described aquatic stages: *Argyractis, Elophila, Neargyractis, Oxyelophila, Parapoynx, Petrophila* and *Usingeriessa* (Solis et al., 2018). The life cycles of *Argyractoides, Chrysendeton, Cryptocosma, Giorgia, Nymphulodes, Oligostigma, Oligostigmoides, Pseudolithosia*, and *Synclitodes* are unknown.

Subfamily Crambinae

Crambines comprise 1,987 species worldwide (Nuss et al., 2003–2015). In the Neotropical region, *Xubida* contains at least one freshwater species, *Xubida infusellus* (Walker), which can be found from Guatemala to Argentina, including Trinidad Island (De Loach et al., 1980). The larvae bore into many aquatic weeds in the family Pontederiaceae, such as *Eichhornia azurea* (Sw.) Kunth, *E. heterosperma* Alexander, *E. paniculata* (Spreng.) Solms-Laub, *Pontederia cordata* L. and *P. rotundefolia* L. (De Loach et al., 1980) that exhibit severe damage from the larval tunneling (Julien et al., 2001). After hatching from the egg, the first instar larva begins mining the leaf and then it girdles the petiole, leaving a characteristic yellow ring (Julien et al., 2001).

Subfamily Pyraustinae

The Pyraustinae *sensu stricto* (excluding the Spilomelinae) contains 1,413 valid species of grass moths (Munroe, 1995; Solis & Maes, 2002). *Ostrinia penitalis* (Grote) occurs from Canada to Brazilian Amazon (Mutuura & Munroe, 1970). In the United States (Florida) is known to grow on floating plants in the Polygonaceae (*Polygonum* spp.), Nelumbonaceae (*Nelumbo lutea* Willd.), and Nymphaeaceae (*Nuphar*) (Heppner & Habeck, 1976; Center et al., 1999). The female of this species deposits masses of eggs on the upper surface of the host leaves and newly hatched larvae promptly attach silk threads to the leaf as anchor points to avoid being dislocated by wind and wave action. Young larvae feed on the leaf lamina, but then move into the petiole, where they pupate (Center et al., 1999). *Neohelvibotys pelotasalis* (Capps) is found on alligator weed from Brazil and Uruguay (Romero & Navarro, 2009).

Subfamily Schoenobiinae

This subfamily is composed of 201 species worldwide (Nuss et al., 2003–2015). Larvae may be borers or build retreats made of leaves. The larval stage is usually a borer of marsh-living grasses in the Poaceae (Munroe & Solis, 1999). In the Neotropics, Schoenobiinae is represented by ten genera (Munroe, 1995; Nuss et al., 2003–2015), with *Rupela* occurring from Mexico to Argentina, including Caribbean islands. Further information on the rice sweetheart, *Rupela albina* Becker and Solis, the most common species of this genus, can be found in Passoa & Habeck (1987).

Subfamily Spilomelinae

Formerly included in the Pyraustinae, Spilomelinae was raised to the subfamily status based on morphological characters (Minet, 1981); but according to Solis & Maes (2002), its current grouping represents a polyphyletic clade and demands further systematic investigation. As recognized now, it stands as the largest subfamily of crambid moths with around 3,700 species worldwide (Solis & Maes 2002; Nuss et al., 2003–2015).

Neotropical spilomelines include the lesser Canna leafroller *Geshna cannalis* (Quaintance) that feeds on *Canna flaccida* Salisb. (Cannaceae) and *Thalia geniculate* L. (Marantaceae). This species builds a shelter by rolling leaves of the host to feed on the leaf dorsal epidermis and parenchyma, whereas the ventral epidermis is left untouched (Center et al., 1999). *Niphograpta albiguttalis* Warren, an important biological control agent of the invasive water hyacinth *Eichhornia crassipes* (Mart.) Solms. Female *N. albiguttalis* prefer to lay their eggs on damaged plants (De Loach & Cordo, 1978). The larvae burrow into intact leaf petioles, where they feed and grow up to 2 cm, as of the last instar. Juveniles of *N. albiguttalis* might be confused with those of fellow spilomeline *Samea multiplicalis* (Guenée). *Niphograpta albiguttalis* is larger (~18 mm when fully grown) and has conspicuous brown spots and orange/dark brown head, while *S. multiplicalis* is smaller (~13 mm when fully grown) and has pale brownish head and inconspicuous brown spots (Center et al., 1999). Distribution coverage of *S. multiplicalis* is from warmer regions of North America to Argentina, and besides *E. crassipes*, it grows on *Pistia stratiotes* L., *Lemna* (Araceae), on waterferns *Salvinia*, and on the mosquito fern *Azolla caroliniana* Willd. (Salviniaceae). *Nomophila indistinctalis* (Walker), whose larvae feed on *Alternanthera philoxeroides* (Amaranthaceae), occurs in Brazil (Romero & Navarro, 2009).

Sphingidae

The sphinx moths are among the most conspicuous lepidopterans known, and the family contains 1,463 valid species in 206 genera (van Nieukerken et al., 2011), most of which are entirely terrestrial. Nonetheless, one species, the banded sphinx moth *Eumorpha fasciatus* (Sulzer), is recorded to grow on the aquatic *Ludwigia octovalvis* (Jacq.) Raven and very likely on other members of the evening primrose family (Onagraceae) (Center et al., 1999). The larvae can be of two color-morphs: pink-colored caterpillars that are mainly found on the plant stem, and green-colored caterpillars that dwell under the leaves. For pupation, they leave the host plant and pupate in the soil (Center et al., 1999). Unlike other sphinx moths, the proboscis of *E. fasciatus* is fused to the body during the pupal stage (Hodges, 1971; Center et al., 1999).

Erebidae

Erebidae is currently the most speciose family in the Lepidoptera, with approximately 24,600 species recognized worldwide (van Nieukerken et al., 2011) and consists of medium to large moths. Classic families Arctiidae and Lymantriidae are now subordinated within Erebidae (Zahiri et al., 2012). Only one genus in the Arctiinae, *Paracles*, is reported to live on Neotropical freshwater habitats (Mey & Speidel, 2008, Romero & Navarro, 2009). *Paracles* is widely distributed throughout Central and South America, and four species have aquatic stages. These larvae possess hydrophobic setae on the epidermis, which enable them to breathe by means of a plastron, colonize light-streamed water bodies, where they feed on submerged plants, and swim with undulatory movements (Adis, 1983; Mey & Speidel, 2008; Meneses et al., 2013). This genus is associated with *E. crassipes*, but its diet breadth is quite wide (Perkins, 1974). *Paracles laboulbeni* (Bar) is reported from Surinam and southern Brazil (Adis, 1983). The caterpillars are black, with dorsal hairs reddish or bright rufous, and reported to grow on *Mayaca fluviatilis* Aubl. (Mayacaceae). More recently, *Paracles klagesi* (Rothschild) was recorded from an ecotone in Brazil, living among *Tonina fluviatilis* Aubl. (Eriocaulaceae) and other macrophytes in the Nymphaeaceae (Meneses et al., 2013). However, the caterpillar fed on *Elodea canadensis* Michx. (Hydrocharitaceae) and *Cabomba* sp. (Cabombaceae) as well (Meneses et al., 2013). *Paracles burmeisteri* (Berg), known for associating with pondweed *Potamogeton* (Potamogetonaceae) and *Paracles azollae* (Berg) have been collected in Brazilian savanna, but without notes on the aquatic habits (Ferro & Teston, 2009).

Noctuidae

Noctuids used to compose the largest family within Lepidoptera, but recent systematic studies utilizing molecular characters have elevated subgroups, such as the Erebidae, to family status and placed subgroups, such as the arctiines, into Erebidae (Zahiri et al., 2012; Goldstein, in press). Now owlet moths are represented by 11,772 species worldwide (van Nieukerken et al., 2011). Despite the family's species richness, few noctuids are regarded as aquatic, especially in the Neotropics. From *Bellura*, at least two aquatic species occur in the Neotropical region: *B. densa* (Walker) and *B. obliqua* (Walker) (Solis, 2008). *Bellura densa* lives on *Pontederia* and *Eichhornia* (Pontederiaceae), while *B. obliqua* feeds on cattail species *Typha angustifolia* L., *T. domingesis* Pers. and *T. latifolia* L. (Typhaceae) (Vogel & Oliver, 1969; Center et al., 1999). Females of the latter species lay several-layered egg masses of up to 40 eggs, covered with a thick yellowish substance made of exudates and body scales, on the surface of cattail leaves. First instar larvae are leafminers, but from the second instar on, they bore into the stalk and sometimes reach the rhizome, blocking flower formation. They pupate inside chambers lined with loose silk strands, but do not spin cocoons. Other *Bellura* species occurring in the Neotropics, whose immatures have not been investigated, are *B. matanzasensis* (Dyar) from Cuba and *B. pleostigma* (Dyar) from Tabasco, Mexico. *Nonagria* also contains aquatic species on other continents, but the two species recorded from Ecuador have not been confirmed to have aquatic or semiaquatic juveniles (Romero & Navarro, 2009).

LIMITATIONS

Moths are among the most diversified forms of insects known and, conversely, their world is yet to be fully explored. Studies on tropical moths have been somewhat scarce, either because of insufficient sampling or by neglecting of the collected material (Gaston & May, 1992). Kristensen et al. (2007) highlighted the geographical bias on immature knowledge and the discrepancy between New World and European approaches by lepidopterists. With respect to Neotropical micro-moths, which compose a great deal of semiaquatic/aquatic lepidopterans, most studies on basic biology and life history are conducted on species of agricultural importance or potential for biological control of aquatic weeds (see Center et al., 1999; Julien et al., 2001). Although there has been a growing interest in aquatic moths from the New World, there is a definite lacuna of reliable information. Many examples of disputed records are present, mainly involving either misidentification of species, such as an aquatic cosmopterigid from São Paulo, Brazil (Kikuchi & Uieda, 1998), or doubtful association with aquatic habitats, e.g. for *Stigmella guittonae* (Bourquin), *Coleophora haywardi* Pastrana, and *C. versurella* Zeller (Pastrana, 2004) (see the warning in Shapiro, 2008). In these particular cases, we consulted J. R. Stonis and J.-F. Landry—specialists in Nepticulidae and Coleophoridae, respectively—who contend that *S. guittonae*, *C. haywardi*, and *C. versurella* do not seem to have relationships with aquatic plants (pers. comm.). Consequently, we have excluded them to avoid duplicating uncertain records.

One of the big limitations of any biologist who endeavors to study fundamental and applied ecology of Neotropical aquatic moths is the dearth information on life history. Even for the principal lineage of aquatic Lepidoptera, the Acentropinae, larval and pupal descriptions are not available for most tropical species (Mey & Speidel, 2008; Romero & Navarro, 2009). In some groups, the concealed behavior of larvae (life miners, leaf rollers, borers) makes locating the immature stages challenging, and their occasional minute size does not seem very inviting. Another challenge is that adult moths must be collected and preserved in a dry mode to conserve wing patterns of systematic value, unlike most of the aquatic insects, which are preserved in liquid (Mey & Speidel, 2008). Proper processing of the material requires singular methods and instruments that are not often part of the armory of an aquatic entomologist. As a consequence, aquatic moths are often more deeply investigated by specialized lepidopterists, rather than limnologists, thus receiving relatively less attention in studies of freshwater assemblages (Mey & Speidel, 2008). These shortcomings have impeded aquatic biologists from advancing further on questions related to large-scale ecological patterns, historical biogeography, endemic groups, and phylogeography of aquatic moths.

Although there have been efforts to solve the puzzle of aquatic moths, the overall literature on aquatic moths is very sparse and relatively old, with most larval descriptions and associations dating from before the 1990s. Hence, they are rather hard for aquatic entomologists and ecologists to find (Mey & Speidel, 2008). Regarding the Neotropical region, the work of Romero & Navarro (2009) and Nessimian et al. (2014) exemplify the attempts to create a compilation of information on aquatic moths in South America and Brazil, respectively. Both of these book chapters present a dichotomous key that greatly influenced and served as background, alongside Solis (2008), for the key we present in the following pages. We adapted them to cover the entire Neotropical region, while adding new groups (Sphingidae), removing others (Nepticulidae, Coleophoridae), and updating the current systematic status of some (e.g., Erebidae, Momphidae). Also, due to the lack of more refined larval morphological descriptions as well as species distributional patterns, we were not able to cross the genus boundary.

TERMINOLOGY AND MORPHOLOGY

Lepidopterans are holometabolous insects, and the egg, larva, and pupa of some species may be found associated with water. For more detailed information on this subject, see Triplehorn & Johnson (2005), Dias (2006) and Nessimian et al. (2014).

The eggs greatly vary in shape: elliptical, spherical, hemispherical, flattened, slightly cuboid, and asymmetrical. The chorion may be bare or ridged, have punctuations, with transverse or longitudinal projections. The micropilar area is on the upper pole of elongated eggs and on the external margin of flattened ones.

The larvae (caterpillars, Fig. 11.1) possess a rounded, sclerotized head, whose posterior edge is sometimes retracted into the prothorax. The upturned Y-shaped epicranial suture divides the head into three main portions: two epicranial regions (lateral) and one frons (central). Each epicranium contains six lateral simple eyes (stemmata), five of them scattered in semicircle and one more ventral, near the antenna socket. The antennae are very short, with three segments. The frons is laterally bordered by the adfrontal area. The clypeus is located below the frons, and is inferiorly bordered by a membranous anteclypeus. The anteclypeus articulates with the bilobed labrum, the latter partially covering the strong pair of toothed mandible at rest. On the inferior portion of the head, each maxilla bears one small maxillary palp, and the spinneret is located on the anterior part of the labium, between the maxillae. The mouthparts may be oriented downwards (hypognathous head) or forwards (prognathous head, in mining larvae).

The thorax has three segments: prothorax, mesothorax, and metathorax. Each one of them bears a pair of segmented legs with an apical tarsal claw. The prothorax has a dorsal sclerotized plate called prothoracic shield and a pair of lateral spiracles. In some groups (some Crambidae), there are filamentous gills on the meso- and metathorax.

The abdominal tagma is divided in ten segments (A1–A10). The spiracles are present from A1 to A8, and the last pair may be larger than the others. A9 and A10 are

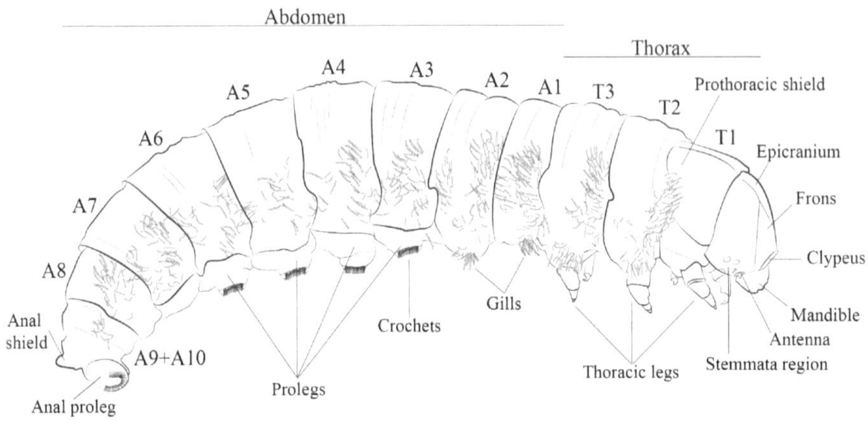

FIGURE 11.1 Body plan of an aquatic moth larva with filamentous gills.

fused, forming the anal segment, lack spiracles and possess a sclerotized dorsal plate named anal shield. The abdomen may or may not have developed abdominal appendages termed prolegs, which are ventral fleshy protuberances that aid in locomotion. In general, prolegs grow from A3–A6 and A9 + A10 and generally possess terminal crochets, whose size and disposition carries great systematic value. Most lepidopterans have crochets in a single series (uniserial), while they may all be of the same size (uniordinal), or alternate between two (biordinal) or three sizes (triordinal). The crochets may be scattered in a complete circle, an internal or external interrupted ellipse (mesal and lateral penellipse, respectively), or an ellipse interrupted in both margins (transverse bands). Also, they can consist of longitudinal series: if the middle crochets are longer than those of the edges, this is a heteroideous mesoseries; if they are of the same size (uniordinal) or alternate between long and small ones (biordinal), this is a homoideous mesoseries. Some groups possess differentiated structures in the abdomen, such as horn-like or lobe-shaped protuberances of sphinx moths and the anal comb in the Tortricidae (Common, 1990).

The pupa in Lepidoptera is often obtect, meaning that the appendages are tightly attached to the pupa's body, and adecticous, lacking functional mandibles. They may or may not be enclosed in a cocoon, usually made of silk and surrounding sediment material (leaves, bark fragments, etc.). The body is generally cylindrical and some imagoes' structures (mouthparts, antennae, wings, legs) can be visualized through the pupal cuticle. Pairs of spiracles are present in the mesothorax and in abdominal segments A10–A8; in A3–A4, they can be greatly enlarged in aquatic forms. In the posterior edge of A10, the cremaster is composed of a series of hooks or spikes that aid the attachment of the pupa to the substrate. In groups in which the pupae are not fixed inside the shelter or cocoon, the cremaster lacks such attachment devices.

MATERIAL PREPARATION AND PRESERVATION

Depending on the life stage in which the moths are sampled, different procedures of preparation and preservation should be employed to facilitate further identification. Even though the book is concentrated on aquatic stages, we feel it is indispensable that aquatic entomologists, limnologists and ecologists are aware of how to preserve terrestrial adults for the sake of obtaining correct associations and identifications of species.

Immature: After collecting, larvae and pupae should be conserved in liquid solutions. Ethanol (96%) is the most common solution utilized, but formalin (4%–10%) is also used (Oliveira & Pes, 2014). The specimens can be fixed together with their freshwater substrates if the material cannot be sorted in the field (Oliveira & Pes, 2014). Importantly, before placing larvae in alcohol, one should briefly boil them to avoid further complete darkening of the material (Winter, 2000). We highly recommend that biologists place part of their material in absolute ethanol (~99%) without boiling, for further molecular purposes. In such case, the solution must be changed at least three times and kept at −20°C out of light reach (Oliveira & Pes, 2014). Plastic vials can be used to keep the samples, but make sure that there is a complete sealing to avoid evaporation of the preserving fluid (Oliveira & Pes, 2014).

Imago: We advise aquatic entomologists to sacrifice adult micro-moths by putting them for around 1 min inside small killing vials containing a volatile chemical, such as ethyl acetate (70%), which is less harmful for humans than cyanide. Ethyl acetate acts very rapidly, and long exposure to it might damage color pattern. If there is a refrigerator or freezer, one can also do it by freezing the insect. These two methods avoid collectors applying mechanical pressure onto the moths and occasionally injury important structures. For stouter moths, such as *Paracles* and some Noctuidae, freezing can also be an alternative, but we usually sacrifice them by injecting a few drops of aqueous ammonia ($NH_{3(aq)}$) or alcohol into the thorax. Hence, syringes and hypodermic needles are important tools if the aquatic entomologists wish to collect adult moths. After being stunned, medium and large-sized adults should be temporarily kept in paper envelopes for transportation to the collector's lab (Triplehorn & Johnson, 2005), but minute species (<10 mm of wingspan) are better kept inside small plastic tubes. Many biologists permanently leave the material inside envelopes, especially if there is limited space in the collection, but in most cases adults are pinned—in the center of mesothorax—and spread (Triplehorn & Johnson, 2005). Medium and large-sized species can be pinned as the traditional manner, but minute ones are better double-mounted with a minuten pin and spread with a small blow to separate the wings (Triplehorn & Johnson, 2005). They must be preferably mounted within less than 24 h after being killed or preserved at low temperatures if mounting cannot be done promptly (Oliveira & Pes, 2014).

KEYS TO LARVAL LEPIDOPTERA

Families

1 Crochets variable; A8 without lobiform protuberance; body without annulets 2

1' Crochets A3–A6 in biordinal mesoseries (Fig. 11.2); A8 with dorsal lobiform protuberance sometimes containing a small eyespot; usually 6–8 annulets per segment; associated with plants in the evening primrose family (Onagraceae) .... Sphingidae

2(1) A8, spiracles dorsally or laterally located; A8, if spiracles are lateral then subdorsal setae (SD1) dorsal to spiracle; A9, dorsal setae (D2) on separate pinacula; A10 never with anal comb 3

2' A8, spiracles laterally located; A8, subdorsal setae (SD1) anterior to spiracle; A9, dorsal setae (D2) on a single pinaculum; A10 contains an anal comb in some cases (Fig. 11.3) .... Tortricidae

3(2) Metathorax (T3), distance between the coxae (x) wider than the coxa width (y) (Fig. 11.4) 4

3' Metathorax (T3), distance between the coxae smaller than the coxa width 5

4(3) Crochets A3–A6 in uniordinal circle; associated with water primrose (Onagraceae) .... Momphidae

4' Crochets A3–A6 in biordinal circle; associated with aquatic sedges (Cyperaceae) .... Cosmopterigidae

5(3) Body not covered with protuberant verrucae 6

5' Body covered with protuberant verrucae, bearing bunches of setae (Fig. 11.5) .... Erebidae

6(5) Crochets A3–A6 bi or triordinal; mature larvae <20 mm long 7

6' Crochets A3–A6 uniordinal; mature larvae usually 50–70 mm long .... **Noctuidae** **[p. 332]**

7(6) Mesothorax and A8, base of subdorsal setae (SD1) surrounded by a sclerotized ring; A9, lateral group (L) with three setae (Fig. 11.6) .... Pyralidae

7' Mesothorax and A8, base of subdorsal setae (SD1) not surrounded by a sclerotized ring; A9, lateral group (L) with one seta (Fig. 11.7) .... **Crambidae** **[p. 333]**

FIGURE 11.2 Sphinx moth (Sphingidae) larva proleg. Crochets in biordinal mesoseries. *Arrows* indicate the two rows of the crochet series.

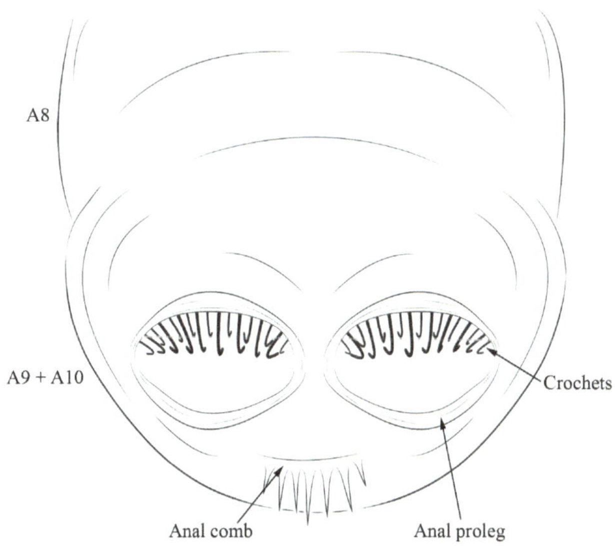

FIGURE 11.3 Tortricid moth (Tortricidae) larva, posterior end. Schematic ventral view showing the anal comb in A10. *Courtesy of Todd Gilligan (Animal and Plant Health Inspection Service, USA).*

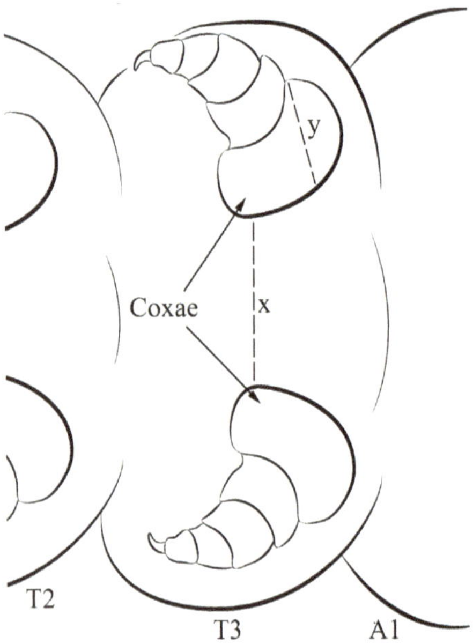

FIGURE 11.4 *Cosmopterix pulchrimella* Chambers (Cosmopterigidae) larva, metathorax, ventral view. Schematic ventral view showing the distance between the metathoracic coxae (x) compared with the coxa width (y). *Courtesy of Willem Ellis (Zoölogisch Museum Amsterdam, The Netherlands).*

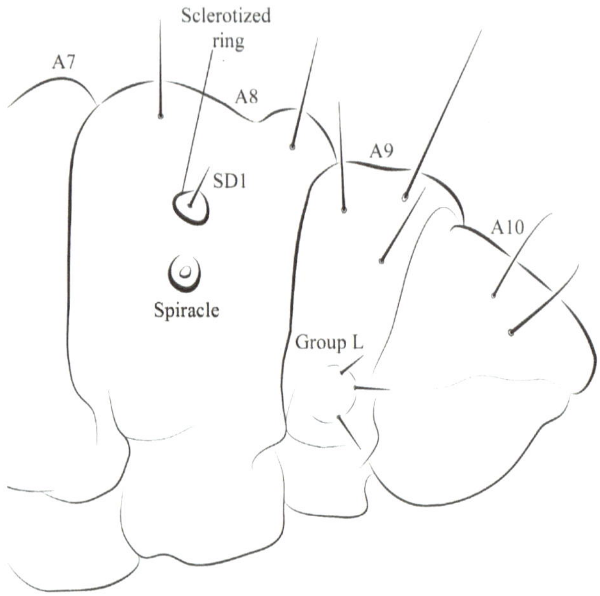

FIGURE 11.6 Phycitinae (Pyralidae) larva. Schematic lateral view of the terminal segments, with the sclerotized ring around seta SD1 of A8 and the lateral group of A9 evidenced.

FIGURE 11.5 *Paracles* (Erebidae) larva from Brazilian Amazon: (A) lateral; and (B) dorsal view.

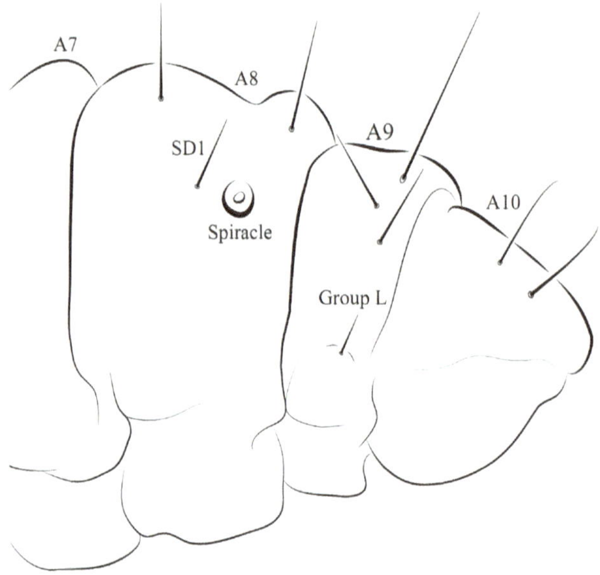

FIGURE 11.7 Crambinae (Crambidae) larva. Schematic lateral view of the terminal segments, showing that SD1 seta in A8 is not surrounded by a sclerotized ring and A9 lateral group contains a single seta.

Noctuidae: Genera

The key includes two genera whose larvae occur in freshwater environments.

1 Crochets A3–A6 in incomplete circle (Fig. 11.8 A); A8 spiracles located dorsally (Fig. 11.8 B), specialized in breathing above waterline .... *Bellura*

1' Crochets A3–A6 in mesoseries; A8 spiracles not specialized .... *Nonagria*

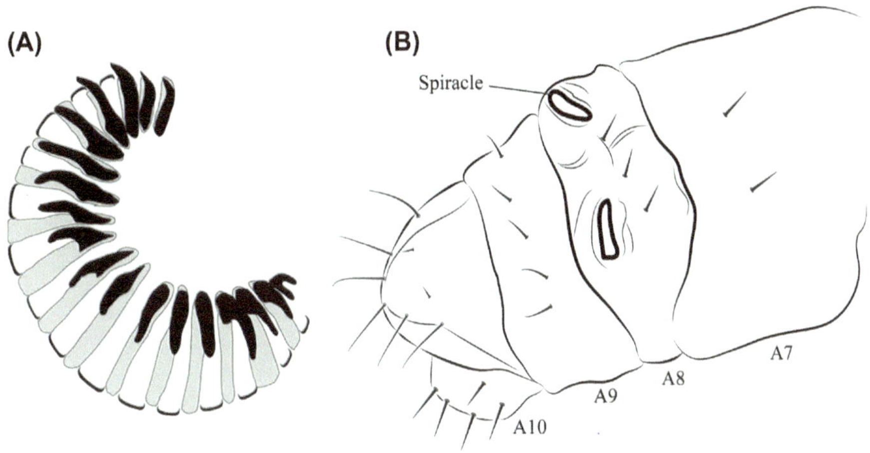

FIGURE 11.8 *Bellura* sp. (Noctuidae) larva: (A) crochets (black) in uniordinal incomplete circle; and (B) spiracles of A8 located dorsally for above-water breathing. *Modified from Solis (2008).*

FIGURE 11.9 *Parapoynx* (Crambidae: Acentropinae) larva. Habitus, lateral view with insert showing thorax and abdominal branched gills.

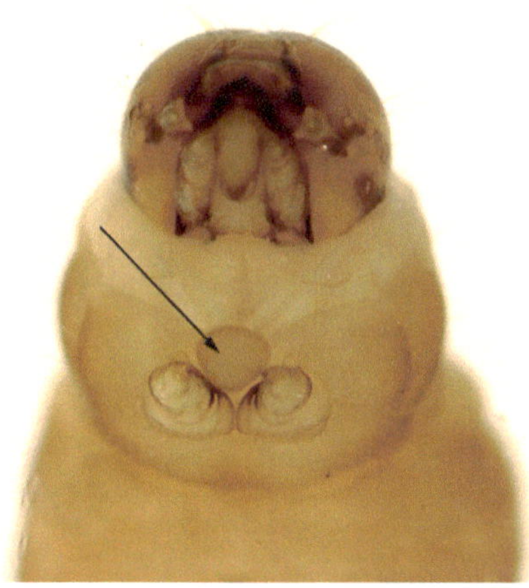

FIGURE 11.10 Schoenobiinae (Crambidae) larva. Ventral view of head and thorax, *arrow* indicates the prothoracic membranous sac.

Crambidae: Subfamilies

The key does not include the Spilomelinae, as there are no immature characters that support the monophyly of the group (Allyson, 1984); thus it is difficult to distinguish the larvae.

1 Body without filamentous gills 2
1' Body with filamentous gills (Figs. 11.1, 9) .... **Acentropinae [p. 334]**(in part)
2(1) Crochets A3–A6 in ellipse or transverse bands........ 3
2' Crochets A3–A6 in circle or penellipse........ 4
3(2) Prothorax, membranous sac anterior to coxae present (Fig. 11.10)........ Schoenobiinae
3' Prothorax, membranous sac anterior to coxae absent .... **Acentropinae [p. 334]**(in part)
4(2) Mesothorax, one transverse plate posterior to the dorsal pinacula (Fig. 11.11); Crochets A3–A6 in circle (Fig. 11.12) .... Crambinae
4' Mesothorax, two transverse plates posterior to the dorsal pinacula (Fig. 11.13); Crochets A3–A6 in mesal penellipse (Fig. 11.14)........ Pyraustinae

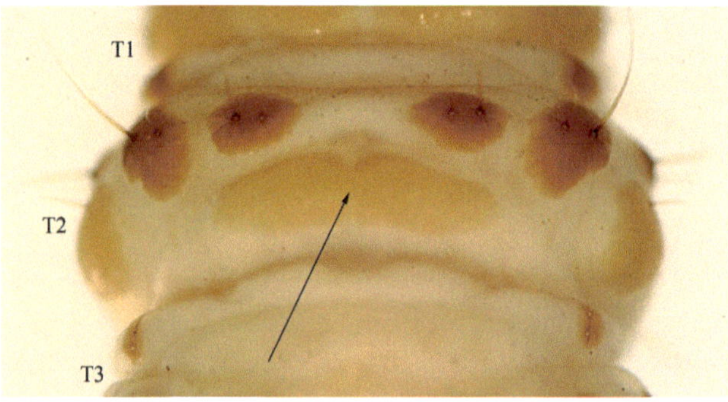

FIGURE 11.11 Crambinae (Crambidae) larva. Dorsal view of the thorax, *arrow* indicates the single transverse plate posterior to the dorsal setae.

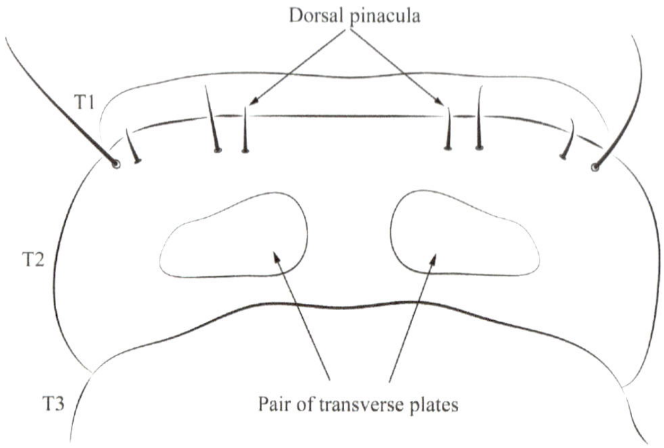

FIGURE 11.13 Pyraustinae (Crambidae) larva. Schematic dorsal view of the thorax, with two transverse plates posterior to the dorsal pinacula.

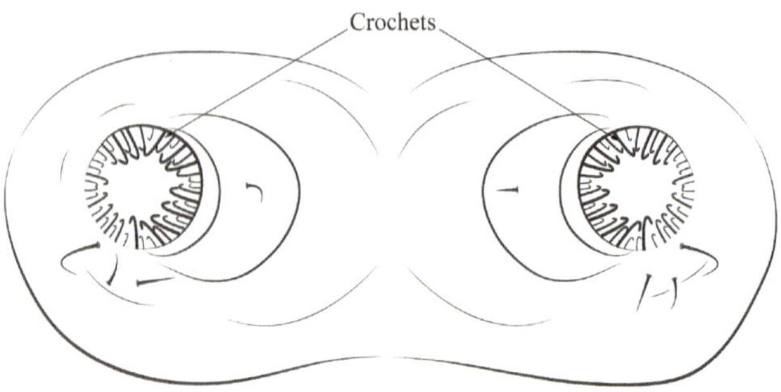

FIGURE 11.12 Crambinae (Crambidae) larva, prolegs. Ventral schematic figure, showing the abdominal crochets in complete circle.

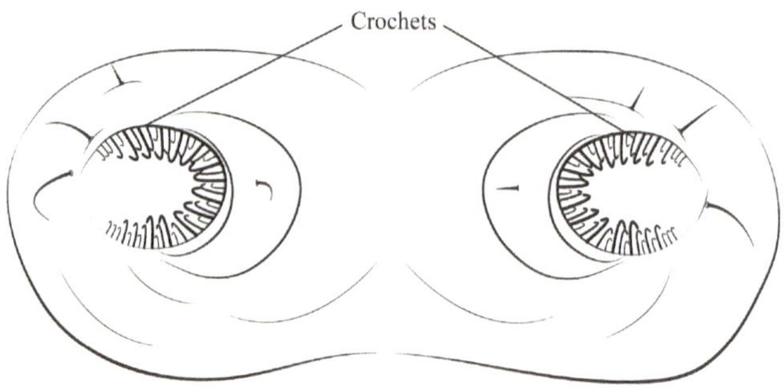

FIGURE 11.14 Pyraustinae (Crambidae) larva. Schematic ventral view of the abdominal crochets in triordinal mesal penellipse.

Crambidae: Acentropinae: Genera

1	Crochets A3–A6 in ellipse; body with filamentous gills (Fig. 11.1, 9)	2
1'	Crochets A3–A6 in biordinal transverse bands (Fig. 11.15), body without filamentous gills	*Elophila*
2(1)	All gills simple, not branched, or with only a few branched (Fig. 11.16); crochets variable	3
2'	Gills branched (Fig. 11.9); A3–A6 crochets in bi- or triordinal ellipse (Fig. 11.17)	*Parapoynx*
3(2)	Head hypognathous, not flattened; gills widely distributed	4
3'	Head prognathous, flattened dorso-ventrally; gills laterally and uniformly distributed (Fig. 11.18)	*Petrophila*

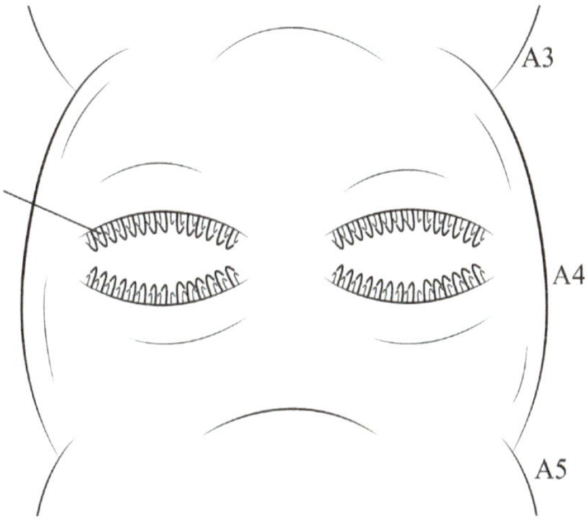

FIGURE 11.15 *Elophila* (Crambidae: Acentropinae) larva. Schematic representation of the crochets in biordinal transverse bands.

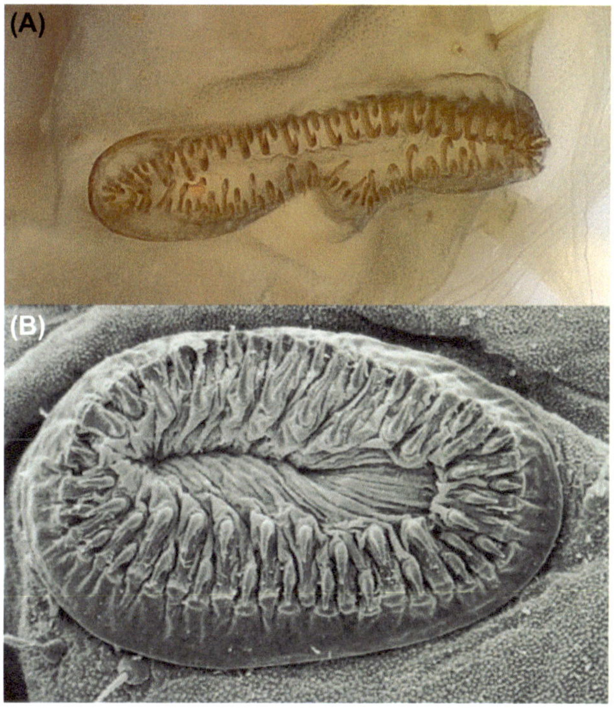

FIGURE 11.17 *Parapoynx* sp. (Crambidae: Acentropinae) larva. Proleg crochets in biordinal ellipse; (A) Stereomicroscopy; (B) Scanning Electron Microscopy. *(B) Modified from Solis (2008).*

FIGURE 11.16 Lepidoptera larva. Lateral view, with simple gills.

FIGURE 11.18 *Petrophila* sp. (Crambidae: Acentropinae) larva: habitus: (A) lateral; and (B) dorsal view.

4(3)	A6, ventral tubercles absent; one group of subventral gills on each segment	5
4'	A6, ventral tubercles present (Fig. 11.19); one or two groups of subventral gills on each segment	*Argyractis*
5(4)	A1–A7, two sets of 3–4 dorsal gills in a transverse row; two groups of subdorsal gills in clumps	*Oxyelophila*
5'	Without dorsal gills on any segment; two groups of subdorsal gills in a transverse row	*Neargyractis*

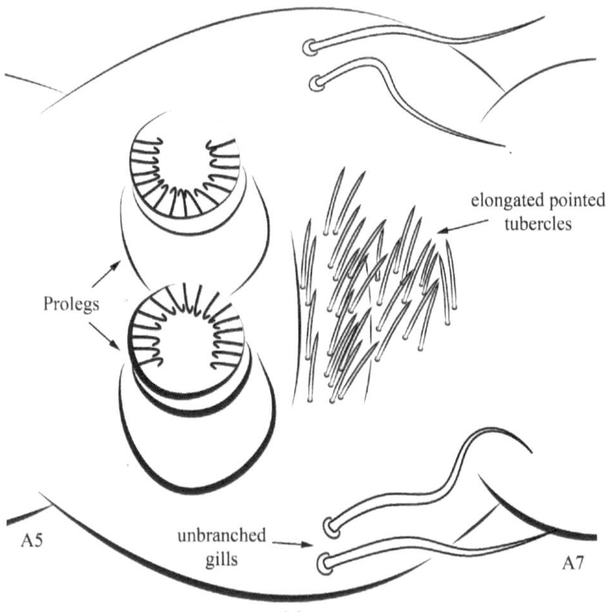

FIGURE 11.19 *Argyractis* sp. (Crambidae: Acentropinae) larva. Lateroventral view of the sixth abdominal segment, showing the elongated pointed tubercles posterior to the prolegs.

ACKNOWLEDGMENTS

We thank Jonas Rimantas Stonis (Baltic-American Biotaxonomy Institute, Lithuania), Jean-François Landry (Canadian National Collection of Insects, Canada), Joaquin Baixeras (University of Valencia, Spain), and Richard Brown (Mississippi Entomological Museum, USA) for fruitful discussions via email, in which plenty of doubts were solved. Paul Goldstein (United States Department of Agriculture, USA) and Eugenie Phillips (Museo Nacional de Costa Rica, Costa Rica) provided information and comments on the Noctuoidea and Acentropinae, respectively. Diego Pádua, Jeane Nascimento, Leonardo Rossi, and Nikolas Cípola (National Institute for Amazonian Research, Brazil) helped with illustrations and photographs. Todd Gilligan (Animal and Plant Health Inspection Service, USA) and Willem Ellis (Zoölogisch Museum Amsterdam, The Netherlands) kindly allowed the usage of modified Tortricidae and Cosmopterigidae images, respectively, in our key.

REFERENCES

Adis, J. 1983. Eco-entomological observations from the Amazon. IV. Occurrence and feeding habits of the aquatic caterpillar *Palustra laboulbeni* Bar, 1873 (Arctiidae: Lepidoptera) in the vicinity of Manaus. Acta Amazonica 13: 31–36.

Allyson, S. 1984. Description of last-instar larvae of 22 species of North American Spilomelini (Lepidoptera: Pyralidae: Pyraustinae) with a key to species. Canadian Entomologist 116: 1301–1334.

Brown, J.G., G. Robinson & J.A. Powell. 2008. Food plant database of the leafrollers of the world (Lepidoptera: Tortricidae) (Version 2.0). http://www.tortricid.net/foodplants.asp.

Center, T.D., F.A. Dray, Jr., G.P. Jubinsky & M.J. Grodowitz. 1999. Insects and other arthropods that feed on aquatic and wetland plants. U.S. Department of Agriculture, Agricultural Research Service. Technical Bulletin 1870. 200 pp.

Common, I.F.B. 1990. Moths of Australia. Melbourne University Press, Melbourne, Australia. 544 pp.

De Loach, C.J. & H.A. Cordo. 1978. Life history and ecology of the moth *Sameodes albiguttalis*, candidate for biological control of water-hyancinth. Environmental Entomology 7: 309–321.

De Loach, C.J., H.A. Cordo, R. Ferrer & J. Runnacles. 1980. *Acigona infusella*, a potential biological control agent for water hyacinth: observations in Argentina (with descriptions of two new species of *Apanteles* by L. De Santis). Annals of the Entomological Society of America 73: 138–146.

Dias, M.M. 2006. Lepidoptera. Pages 175–204 *in*: C. Costa, S. Ide & C.E. Simonka (eds.), Insetos imaturos: metamorfose e identificação. Holos, São Paulo, Brasil.

Ferro, V.G. & J.A. Teston. 2009. Composição de espécies de Arctiidae (Lepidoptera) no sul do Brasil: relação entre tipos de vegetação e entre a configuração espacial do hábitat. Revista Brasileira de Entomologia 53: 278–286.

Gaston, K.J. & R.M. May. 1992. Taxonomy of taxonomists. Nature 356: 281–282.

Gilligan, T.M. & M.E. Epstein. 2009. LBAM ID, Tools for diagnosing light brown apple moth and related western U. S. leafrollers (Tortricidae: Archipini). Colorado State University, California Department of Food and Agriculture, and Center for Plant Health Science and Technology, USDA, APHIS, PPQ.

Gilligan, T.M., J. Baixeras, J.W. Brown & K.R. Tuck. 2014. T@RTS: online world catalogue of the Tortricidae (ver. 3.0). Available from http://www.trotricid.net.

Gilligan, T.M. & Passoa, S.C. 2014. LepIntercept: An identification resource for intercepted Lepidoptera larvae. Available from http://idtools.org/id/leps/lepintercept/index.html.

Goldstein, P.Z. In press. Diversity and significance of Lepidoptera: A phylogenetic perspective *in*: R. Foottit and P. Adler (eds.) Insect Biodiversity: Science and Society. John Wiley & Sons.

Gorayeb, I.S. & R.R. Pinger. 1978. Detecção de predadores naturais das larvas de *Simulium fulvinotum* Cerq. & Mello, 1968 (Diptera, Nematocera). Acta Amazonica 8: 629–637.

Habeck, D.H. & J.P. Cuda. 2009. Waterlily leafcutter, *Elophila obliteralis* (Walker) (Insecta: Lepidoptera: Crambidae: Acentropinae). EENY 424, University of Florida, Gainesville, FL.

Hepnner, J.B. & D.H. Habeck. 1976. Insects associated with *Polygonum* (Polygonaceae) in North Central Florida. I. Introduction and Lepidoptera. The Florida Entomologist 59: 231–239.

Hodges, R.W. 1971. Sphingoidea. The Moths of America North of Mexico. Fasc. 21. E.W. Classey Limited and R.B.D. Publications Inc. London, UK. 158 pp.

Hodges, R.W. 1999. The Gelechioidea. Pages 131–158 *in*: N.P. Kristensen (ed.), Handbook of zoology, Lepidoptera, moths and butterflies, Vol. 1: Evolution, systematics, and biogeography. Walter de Gruyter, Berlin, NY.

Horak, M. 1999. The Tortricoidea. Pages 299–215 *in*: N.P. Kristensen (ed.), Handbook of zoology, Lepidoptera, moths and butterflies, Vol. 1: Evolution, systematics, and biogeography. Walter de Gruyter, Berlin, NY.

Horak, M. & R.L. Brown. 1991. Taxonomy and phylogeny. Pages 23–50. *in*: L.P.S. van der Geest and H.H. Evenhuis (eds.), Tortricoid pests, their biology, natural enemies and control. Elsevier Science Publishers, Amsterdam, Netherlands.

ORDER LEPIDOPTERA

Julien, M.H., M.W. Griffiths & J.N. Stanley. 2001. Biological control of water hyacinth. The moths *Niphograpta albiguttalis* and Xubida infusellus: biologies, host ranges, and rearing, releasing and monitoring techniques. ACIAR Monograph No. 79.

Kikuchi, R.M. & V.S. Uieda. 1998. Composição da comunidade de invertebrados de um ambiente lótico tropical e sua variação especial e temporal. Pages 157–173 *in*: J.L. Nessimian & A.L. Carvalho (eds.), Ecología de Insetos Acuáticos. Séries Oecologia Brasiliensis, vol. V. PPGE-UFRJ, Rio de Janeiro, Brazil.

Koster, J.C. 2010. The genera *Cosmopterix* Hübner and *Pebobs* Hodges in the New World with special attention to the Neotropical fauna (Lepidoptera: Cosmopterigidae). Zoologische Mededelingen Leiden 84: 251–575.

Kristensen, N.P., M. J. Scoble & O. Karsholt. 2007. Lepidoptera phylogeny and systematics: the state of inventorying moth and butterfly diversity. Zootaxa 1668: 699–747.

Landry, B. 2001. The Cosmopterigidae (Lepidoptera) of the Galápagos Islands, Ecuador. Revue Suisse de Zoologie 108: 513–539.

Lange, W.H. 1978. Aquatic and semiaquatic Lepidoptera. Pages 187–201 *in*: R.W. Merritt & K.W. Cummings (eds.), An introduction to the aquatic insects of North America, 1st edition. Kendall/Hunt Publishing Company, Dubuque, IA.

Meneses, A.R., M.V.O. Bevilaqua, N. Hamada & R.B. Querino. 2013. The aquatic habit and host plants of *Paracles klagesi* (Rothschild) (Lepidoptera, Erebidae, Arctiinae) in Brazil. Revista Brasileira de Entomologia 57: 350–352.

Mey, W. & W. Speidel. 2008. Global diversity of butterflies (Lepidoptera) in freshwater. Hydrobiologia 595: 521–528.

Minet, J. 1981. Les Pyraloidea et leurs principales divisions systemmatiques. Bulletin de la Société Entomologique de France 86: 262–280.

Minet, J. 1985. Étude morphologique et phylogénétique des organs tympaniques des Pyraloidea. 2-Pyralidae, Crambidae, première partie (Lepidoptera Glossata). Annales de la Société Entomologique de France 21, 69–86.

Munroe, E. 1995. Crambidae. Pages 34–79 *in*: J.B. Heppner (ed.), Atlas of Neotropical Lepidoptera checklist: part 2; Hyblaeoidea, Pyraloidea, Tortricoidea. Scientific Publishers, Gainesville, FL.

Munroe, E. & M.A. Solis. 1999. The Pyraloidea. Pages 233–256 *in*: N.P. Kristensen (ed.), Handbook of zoology, Lepidoptera, moths and butterflies, Vol. 1: Evolution, systematics, and biogeography. Walter de Gruyter, Berlin, NY.

Munroe, E., V.O. Becker, J.C. Shaffer, M. Shaffer & M.A. Solis. 1995. Pyralidae. Pages 80–105 *in*: J.B. Heppner (ed.), Atlas of Neotropical Lepidoptera checklist: part 2; Hyblaeoidea, Pyraloidea, Tortricoidea. Scientific Publishers, Gainesville, FL.

Mutuura, A. & E.G. Munroe, 1970. Taxonomy and distribution of the European corn borer and allied species: genus *Ostrinia* [Hübner] (Lepidoptera: Pyralidae). Memoirs of the Entomological Society of Canada 71: 1–112.

Nessimian, J.L., E.R. Da-Silva & L.B.N. Coelho. 2014. Ordem Lepidoptera. Pages 435–446 *in*: N. Hamada, J.L. Nessimian & R.B. Querino (eds.), Insetos aquáticos na Amazônia brasileira: taxonomia, biologia e ecologia. Editora do INPA, Manaus, Brazil.

van Nieukerken, E.J., L. Kaila, I.J. Kitching, N.P. Kristensen, D.C. Lees, J. Minet, C. Mitter, M. Mutanen, J.C. Regier, T.J. Simonsen, N. Wahlberg, S.-H. Yen, R. Zahiri, D. Adamski, J. Baixeras, D. Bartsch, B.Å. Bengtsson, J.W. Brown, S.R. Bucheli, D.R. Davis, J. De Prins, W. De Prins, M.E. Epstein, P. Gentili-Poole, C. Gielis, P. Hättenschwiler, A. Hausmann, J.D. Holloway, A. Kallies, O. Karsholt, A.Y. Kawahara, S. (J.C.) Koster, M.V. Kozlov, J.D. Lafontaine, G. Lamas, J.-F. Landry, S. Lee, M. Nuss, K.-T. Park, C. Penz, J. Rota, A. Schintlmeister, B.C. Schimdt, J.-C. Sohn, M.A. Solis, G.M. Tarmann, A.D. Warren, S. Weller, R.V. Yakovlev, V.V. Zolotuhin & A. Zwick. 2011. Order Lepidoptera Linnaeus, 1758. Pages 212–221 *in*: Z.-Q. Zhang (ed.) Animal biodiversity: An outline of higher-level classification and survey of taxonomic richness. Zootaxa 3148.

Nuss, M., B. Landry, R. Mally, F. Vegliante, A. Tränkner, F. Bauer, J. Hayden, A. Segerer, R. Schouten, H. Li, T. Trofimova, M.A. Solis, J. De Prins & W. Speidel. 2003–2015. Global Information System on Pyraloidea. Available from http://www.pyraloidea.org.

Oliveira, V.C. & A.M.O. Pes. 2014. Inventário da fauna de insetos aquaticos: coleta, preservação e criação. Pages 155–171 *in*: N. Hamada, J.L. Nessimian & R.B. Querino (eds.), Insetos aquáticos na Amazônia brasileira: taxonomia, biologia e ecologia. Editora INPA, Manaus, Brazil.

Passoa, S. 1988. Systematic position of *Acentria ephemerella* (Denis & Schiffermüller), Nymphulinae and Schoenobiinae based on morphology of immature stages (Pyralidae). Journal of the Lepidopterists' Society 42: 247–262.

Passoa, S. & D.H. Habeck. 1987. A description of the larva and pupa of *Rupela albinella*, a pest of rice in latin America (Lepidoptera: Pyralidae: Schoenobiinae). The Florida Entomologist 70: 368–378.

Pastrana, J. A. 1961. Una nueva Phycitidae (Lep.) parasite de la "lagunilla." Reporte Argentino Publicacion Technical 71, Instituto Nacional Technica Agorpecuar.

Pastrana, J.A. 2004. Los lepidópteros argentinos. Sus plantas hospedadoras y otros sustratos alimentícios. Editora da Sociedad Entomológica Argentina, Buenos Aires, Argentina. 350 pp.

Perkins, B.D. 1974. Arthropods that stress water hyacinth. Pest Articles and News Summaries 20: 304–314.

Regier, J.C., C. Mitter, M.A. Solis, J.E. Hayden, B. Landry, M. Nuss, T.J. Simonsen, S.-H. Yen, A. Zwick & M.P. Cummings. 2012. A molecular phylogeny for the pyraloid moths (Lepidoptera: Pyraloidea) and its implications for higher-level classification. Systematic Entomology 37: 635–656.

Romero, F. & F. Navarro. 2009. Lepidoptera. Pages 309–340 *in*: E. Dominguez & H.R. Fernandez (eds.), Macroinvertebrados bentónicos sudamericanos. Sistematica y biología. Fundación Miguel Lillo, Tucuman, Argentina.

Shapiro, A. 2008. Book review: Los lepidopteros argentinos: sus plantas hospedadoras y otros sustratos alimenticios. Journal of the Lepidoterists' Society 62: 114–115.

Sinev, S.Y. 2002. World catalogue of cosmopterigid moths (Lepidoptera: Cosmopterigidae). Katalog roskoshnykh uzkokrylykh molej (Lepidoptera: Cosmopterigidae). Trudy Zoologicheskogo Instituta 293: 1–183.

Solis, M.A. 1997. Snout moths: unraveling the taxonomic diversity of a speciose group in the Neotropics. Pages 231–242 *in*: M.L. Reaka-Kudla, D. Wilson & E.O. Wilson (eds.), Biodiversity II: understanding and protecting our biological resources. Joseph Henry Press, Washington, DC.

ORDER LEPIDOPTERA

Solis, M.A. 1999. Key to selected Pyraloidea (Lepidoptera) larvae intercepted at U.S. Ports of entry: Revision of Pyraloidea in "Keys to some frequently intercepted lepidopterous larvae" by D. M. Weisman, 1986. Proceedings of the Entomological Society of Washington 101: 645–686.

Solis, M.A. 2008. Aquatic and semiaquatic Lepidoptera. Pages 553–569 *in*: R.W. Merritt, K.W. Cummins & M.B. Berg (eds.), An introduction to the aquatic insects of North America, 4th edition. Kendall/Hunt Publishing Company, Dubuque, IA.

Solis, M.A. & C. Mitter. 1992. Review and preliminary phylogenetic analysis of the subfamilies of the Pyralidae (sensu stricto) (Lepidoptera: Pyraloidea). Systematic Entomology 17: 79–90.

Solis, M.A. & K.V.N. Maes. 2002. Preliminary phylogenetic analysis of the subfamilies of Crambidae (Pyraloidea: Lepidoptera). Belgian Journal of Entomology. 4: 53–95.

Solis, M.A., N.E. Harms, E. Phillips-Rodriguez, S.J. Scheffer, M.L. Lewis, D.H. Janzen, W. Hallwachs & M.A. Metz. 2018. Aquatic larvae of two acentropines, *Usingeriessa onyxalis* (Hampson) and *Oxyelophila callista* (Forbes) (Lepidoptera: Crambidae). Proceedings of the Entomological Society of Washington. 120: 180–195.

Solis, M.A. & P. Tuskes. 2018. Two new species of *Petrophila* Guilding (Crambidae) from north central Arizona. Proceedings of the Entomological Society of Washington. In press.

Speidel, W. 1984. Revision der Acentropinae des palaearktischen Faunengebietes (Lepidoptera: Crambidae). Neue Entomologische Nachrichten 12: 1–157.

Triplehorn, C.A. & N.F. Johnson. 2005. Borror and Delong's introduction to the study of insects, 7th edition. Brooks/Cole Publishing, Belmont, CA. 864 pp.

Vogel, E. & A.D. Oliver, Jr. 1969. Life history and some factors affecting the population of *Arzama densa* in Louisiana. Annals of the Entomological Society of America 62: 749–752.

Winter, W.D. 2000. Basic techniques for observing and studying moths & butterflies. Memoirs of the Lepidopterists' Society. 5: 1–444.

Yoshiyasu, Y. 1985. A systematic study of the Nymphulinae and the Musotiminae of Japan (Lepidoptera: Pyralidae). Scientific Report of Kyoto Prefectural University 37: 1–162.

Zahiri, R., J.D. Holloway, I.J. Kitching, J.D. Lafontaine, M. Mutanen & N. Wahlberg. 2012. Molecular phylogenetics of Erebidae (Lepidoptera: Noctuoidea). Systematic Entomology 37: 102–124.

Chapter 12

Order Hymenoptera

Daniell R.R. Fernandes
Coordenação de Biodiversidade, Instituto Nacional de Pesquisas da Amazônia, Manaus, Amazonas, Brazil

Ranyse B. Querino
Empresa Brasileira de Pesquisa Agropecuária, Embrapa Meio-Norte, Teresina, Piauí, Brazil

Neusa Hamada
Coordenação de Biodiversidade, Instituto Nacional de Pesquisas da Amazônia, Manaus, Amazonas, Brazil

Chapter Outline

INTRODUCTION

The Hymenoptera have different ecological roles, such as pollinators, seed dispersers, and regulators of other arthropod populations (acting as predators and parasitoids). Most species are terrestrial, such as ants, bees, and wasps. However, a small and little-known group is adapted to the aquatic environment and, in some part of its biological cycle, is associated with water. In this chapter, the only Hymenoptera considered as "aquatic" are those that spend at least one part of their life in water associated with an aquatic host.

The immatures do not have any specialized structures for aquatic life. However, there are two characteristics of the adults that are known to be adapted to aquatic life: densely arranged short setae on the body and wings, that grants the wasps a hydrophobic characteristic as well maintaining an air layer around the body, and highly curved claws that allow them to hold on to the substrate so that the females are not carried away by the water currents while looking for their hosts (Bennett, 2008b).

All known aquatic Hymenoptera are parasitoids. Their natural history has interesting biological and ecological aspects (Fursov, 1995; Hagen, 1996; Querino & Hamada, 2014). These very minute wasps that parasitize a variety of aquatic hosts. They search the aquatic environment to oviposit on the immature stages (egg, larva, or pupa) of their hosts (Williams & Feltmate, 1992).

The Hymenoptera classification adopted here follows Aguiar et al. (2013), which recognizes 23 extant superfamilies. Traditionally, these superfamilies are organized into two groups, Symphyta (8 superfamilies) and Apocrita (15 superfamilies). There are no records of aquatic Symphyta. The Apocrita holds most of the Hymenoptera with a parasitoid habit. The superfamilies Chalcidoidea, Cynipoidea, Platygastroidea, Diaprioidea, and Ichneumonoidea have aquatic representatives (Table 12.1). In the Neotropical region, egg parasitoids were found in Eulophidae, Mymaridae, Trichogrammatidae (Chalcidoidea), and Platygastridae (Platygastroidea). Ichneumonidae (Ichneumonoidea) parasitize Lepidoptera larvae and/or pupae, while Braconidae (Ichneumonoidea), Diapriidae (Diaprioidea), and Figitidae (Cynipoidea) parasitize pupae and/or puparia of aquatic insects.

The aquatic Hymenoptera are present in all biogeographical regions, except for Antarctica (Bennett, 2008b). The most complete study with information on ecological relationships and distribution of aquatic Hymenoptera was elaborated by Bennett (2008a) for the North American fauna. Fursov (1995) presented data on the aquatic Chalcidoidea of Europe. Still, Hymenoptera is one of the

Thorp and Covich's Freshwater Invertebrates, Fourth Edition. https://doi.org/10.1016/B978-0-12-804223-6.00012-3

TABLE 12.1 Families and Genera of Aquatic Hymenoptera and Their Hosts in the Neotropical Region

Superfamily	Family	Genus	Host Order (Family)	Stages of Host	Distribution Country (State)	References
Chalcidoidea	Chalcididae	*Chalcis* Fabricius, 1787	Diptera (Stratiomyidae)	Eggs and Larvae	Mexico (Sonora)	Burks (1977)
	Eulophidae	*Ametalon* Ashmead, 1904	Odonata	Eggs	Brazil (Rio de Janeiro)	Lima (1960)
		Aprostocetus Westwood, 1833	Coleoptera Hemiptera Odonada	Eggs	Brazil (Amazonas, Rondônia)	Ramalheira et al. (2005)
	Mymaridae	*Anagrus* Haliday, 1833	Hemiptera Lepidoptera Odonata	Eggs	Brazil (Amazonas)	Triapitsyn et al. (2008) Feitosa et al. (2016)
		Gonatocerus Nees, 1834	Coleoptera	Eggs	Brazil (Amazonas)	Feitosa et al. (2006)
	Trichogrammatidae	*Hydrophylita* Ghesquière, 1946	Odonata	Eggs	Argentina; Brazil (Amazonas and Rio de Janeiro)	Querino & Pinto (2007)
		Pseudoligosita Girault, 1913	Odonata	Eggs	Brazil (Amazonas)	Querino & Hamada (2009)
Cynipoidea	Figitidae	Eucoilinae	Diptera	Pupae	Brazil (Amazonas)	Querino & Hamada (2014)
		Hexacola Foerster, 1869	Diptera (Ephydridae) Diptera (Chloropidae)	Pupae	Argentina (Buenos Aires, Cordoba, Misiones, Neuquen, and Rio Negro)	Gaddi et al. (2010) Reche & Gallardo (2015)
		Hydrelliaeucoila Díaz & Gallardo, 2009	Diptera (Ephydridae)	Pupae	Argentina (Buenos Aires)	Díaz et al. (2009)
Diaprioidea	Diapriidae	*Trichopria* Ashmead, 1893	Diptera (Syrphidae)	Pupae	Brazil (Amazonas)	Querino & Hamada (2014)
		Psychopria Masner & Garcia, 2002	Diptera (Psychodidae)	Pupae	Costa Rica (San José)	Masner & Garcia (2002)
Ichneumonoidea	Braconidae	*Chorebus* Haliday, 1833	Diptera (Ephydridae)	Pupae	Brazil (Amazonas)	Ramalheira et al. (2005)
	Ichneumonidae	*Tanychela* Townes, 1971	Lepidoptera (Pyralidae)	Larvae and pupae	Brazil (Rio de Janeiro); Mexico (Michoacán)	Dasch (1979); Resh & Jamieson (1988)
Platygastroidea	Platygastridae	Scelioninae	Coleoptera Hemiptera Odonada	Eggs	Brazil (Amazonas)	Querino & Hamada (2014)
		Thoron Haliday, 1833	Hemiptera (Nepidae)	Eggs	Brazil, Bolivia, Equador and Venezuela	Johnson & Masner (2004)

least-known aquatic insect groups, especially in the Neotropical region, where the few records and known associations are concentrated in Brazil. There are few studies on the biology and taxonomy of aquatic Hymenoptera in Brazil (Querino & Pinto, 2007; Triapitsyn et al., 2008; Querino & Hamada, 2009, 2014). This lack of knowledge impairs the prediction of how many aquatic Hymenoptera species exist, especially considering that many aquatic species belong to genera that are mainly terrestrial (Bennett, 2008b).

In the Neotropical region, the biology of aquatic Hymenoptera was documented for the first time in Brazil by Lima (1960). He reported three species of micro-hymenoptera that were reared from dragonfly eggs of the genus *Lestes* Leach, 1815 (Odonata), on leaf fragments of the white-garland lily (*Hedychium coronarium*), in Petrópolis, Rio de Janeiro state: *Ametalon lutzi* Lima (1960) (Eulophidae: Entedontinae), *Hydrophylita lestesi* Lima (1960) (Trichogrammatidae: Oligositinae), and *Monelata* sp. (Diapriidae). However, the latter species is a dubious record since diapriids are not known to be found in eggs.

Most records are of egg parasitoids. Among the Chalcidoidea, probably the species with the best-known biology is *Caraphractus cinctus* Walker, 1846 (Mymaridae) (Matheson & Crosby, 1912). This species mating occurs over or on the water surface, or over aquatic plants, and oviposition is on exposed eggs of Dytiscidae (Coleoptera) or in eggs oviposited inside plant tissue. Other aquatic chalcidoids belong to Eulophidae (eggs of Coleoptera and Odonata), Pteromalidae (eggs of Diptera, Coleoptera and Neuroptera), and Trichogrammatidae (eggs of Diptera, Hemiptera, Megaloptera, and Odonata).

In the Neotropical region, nine families and 14 genera of Hymenoptera are currently known to be associated with aquatic insects (Table 16.1).

Eulophidae. This is one of the largest families of Chalcidoidea and is probably the most-commonly collected Chalcidoidea family in all geographical regions. Currently, four subfamilies are recognized: Entedoninae, Euderinae, Eulophinae, and Tetrastichinae. The known aquatic species belong to Entedoninae and Tetrastichinae. There are five known genera and three species with aquatic records: *Aprostocetus* sp., *Horismenus mexicanus* Burks, 1968, *Mestocharis* sp., and *Tetrastichus polynemae* Ashmead, 1900 (Bennett, 2008a). In the Neotropical region there are *Ametalon lutzi* Lima (1960), parasitizing eggs of *Lestes* sp. (Odonata: Lestidae) Lima (1960), and *Aprostocetus* sp. parasitizing eggs of Odonata and Coleoptera, both recorded from Brazil (Querino & Hamada, 2014).

Mymaridae. Mymaridae is one of the most distinctive families of Chalcidoidea and is widely distributed. All Mymaridae are endoparasitoids of insect eggs, except for two species that parasitize larvae of Eulophidae (Huber et al., 2006a). The classification into subfamilies has been abandoned, and the genera are currently grouped based on the number of tarsal segments (e.g., Munro et al., 2011). Mymaridae species that attack eggs of Coleoptera, Hemiptera, and Odonata belong to the genera *Anagrus* Haliday, 1833, *Anaphes* Haliday, 1833, *Caraphractus* Walker, 1846, and *Polynema* Haliday, 1833 (Bennett, 2008a). The mymarids *Anagrus amazonensis* Triapitsyn, Querino & Feitosa, 2008, and *Gonatocerus* sp. are species that parasitize eggs of Odonata and Coleoptera in Brazil. There are no other records for the Neotropical region.

Trichogrammatidae. Members of this family are insect-egg endoparasitoids. Ten orders of insects are reported as hosts in the New World, Hemiptera being the most common association, followed by Coleoptera and Lepidoptera (Querino et al., 2012). Currently, two subfamilies are recognized: Oligositinae and Trichogrammatinae. The genera with known association with aquatic insects are *Hydrophylita* Ghesquière, 1946, *Lathromeroidae* Girault, 1912, *Prestwichia* Lubbock, 1864 and *Trichogramma* Westwood, 1833. For the Neotropical region, the following aquatic species are known: *Hydrophylita bachmanni* De Santis, 1964 in eggs of a Zygoptera (Odonata) species (De Santis 1964), *Hydrophylita lestesi* Lima (1960) in eggs of *Lestes* sp. (Lestidae, Odonata) Lima (1960), *Hydrophylita neusae* Querino & Pinto, 2007 in eggs of Zygoptera (Odonata), and *Pseudoligosita longifrangiata* Viggiani, 1981 in eggs of *Argia insipida* Hagen in Selys, 1865 (Odonata, Zygoptera).

Platygastridae. This taxon is a large group of parasitoids with a wide diversity of habitats and habits. They are predominantly endoparasitoids of eggs of several orders of insects; the known aquatic Platygastridae are parasitoids of Diptera, Hemiptera, and Odonata eggs. Several genera have been recorded attacking aquatic insects such as *Pseudanteris* Fouts, 1927, *Telenomus* Haliday, 1833, *Thoron* Haliday, 1833, *Thoronella* Masner, 1972, *Thoronidea* Masner & Huggert, 1979 and *Tiphodytes* Bradley, 1902 (Bennett, 2008a). To date, there have been few studies on aquatic Platygastridae from the Neotropical region. Querino & Hamada (2014) found Platygastridae parasitizing eggs of Coleoptera, Hemiptera, and Odonata; however, the authors could not determine the species or genera of these parasitoids.

Braconidae. Many aquatic species of Braconidae are endoparasitoids of leaf-mining insects, and the adults search the aquatic environment to oviposit on larvae and pupae of Ephydridae (Diptera) (Bennett, 2008a). In the Neotropical region, there are only records of *Chorebus* sp. (Alysiinae) parasitizing pupae of Ephydridae (Diptera) in Amazonian streams in the municipality (county) of Manaus, Brazil (Ramalheira et al., 2005).

Ichneumonidae. The aquatic Ichneumonidae are endoparasitoids of stem-mining caterpillars (Lepidoptera), of Pyralidae (Lepidoptera) pupae attached to rocks, and of Trichoptera pupae (Hagen, 1996). The only known aquatic genus in the Neotropical region is *Tanychela* Townes, 1971 (Cremastinae). *Tanychela aurea* Townes, 1971 was described based on material from the Rio Grande Dam (Rio de Janeiro state, Brazil), and Townes (1971) indicated that the genus was a likely parasitoid of aquatic insects and that it had very long tarsal claws (a characteristic currently known as typical of aquatic parasitoids). The same characteristic was reported by Bennett (2008b). *Tanychela pilosa* Dasch, 1979 was associated with the aquatic species *Petrophila confusalis* (Walker, 1866) (Lepidoptera: Pyralidae) in the Nearctic region (Resh & Jamieson, 1988), however this species also occurs in the Neotropical portion of Mexico (Michoacán) (Dasch, 1979).

Diapriidae. This is a cosmopolitan family with a great number of species, which can be found in humid and shaded areas such as forests and streams, where the diversity of taxa in this family is greatest. Based on the studies of Sharkey (2007) and Sharkey et al. (2012), Diapriidae is constituted of three subfamilies: Ambositrinae, Belytinae, and Diapriinae. The aquatic species of Diapriidae belong to Diapriinae, *Psychopria* Masner & García, 2002 and *Trichopria* Ashmead, 1893 being the only proven aquatic genera, parasitizing pupae of Coleoptera and Diptera (Bennett, 2008a; Masner & García, 2002). In the tropical region, *T. longicauda* (Kieffer, 1910) is known for parasitizing puparia of Syrphidae (Diptera) in Brazil (Querino & Hamada, 2014) and *P. hoguei* Masner & García, 2002 for parasitizing aquatic pupae of *Maruina* sp. (Diptera: Psychodidae) in Costa Rica (Masner & García, 2002).

Figitidae. Member in this large cosmopolitan family of Cynipoidea are mainly parasitoids of Diptera puparia, although they can also parasitize pupae of Neuroptera (Chrysopidae and Hemerobiidae) and Hymenoptera, with most being larva–pupa endoparasitoids. Eight subfamilies are currently recognized, with only Eucoilinae associated with larvae and pupae of aquatic insects (mainly Diptera). The known aquatic genera are *Hexacola* Foerster, 1869, *Kleidotoma* Westwood, 1833 (Bennett, 2008a), and *Hydrelliaeucoila* Díaz & Gallardo, 2009 (Diaz et al., 2009). *Hexacola* parasitizes pupae of *Lemnaphila neotropica* Lizarralde de Grosso, 1977 (Diptera: Ephydridae) (Gaddi et al., 2010), and *Hydrelliaeucoila* parasitizes pupae of *Hydrellia* sp. (Ephydridae), both in Argentina (Diaz et al., 2009).

LIMITATIONS

Overall, further understanding of the order is limited by the poorly known Neotropical diversity, the difficulty of accessing type-material, and the lack of taxonomical studies, taxon revisions, identification keys, and taxonomists. These factors are increased for aquatic Hymenoptera, which are basically unknown in this region.

Another great limitation is lack of data on bioecology. In addition to the association between host and parasitoid being difficult to ascertain precisely, the correct identification of the insect host is compromised because it cannot complete its cycle due to mortality caused by the parasitoid development.

TERMINOLOGY AND MORPHOLOGY

The external morphology of adult Hymenoptera is presented in general, adapted from Huber et al. (2006b), emphasizing the structures that are used for the identification of the families with aquatic representatives. Although the terminology used for the order varies among groups, recent initiatives, such as the Hymenoptera Anatomy Ontology (HAO), aim to alleviate this problem and standardize the nomenclature (Yoder et al., 2010; Seltmann et al., 2012). Any doubts concerning the terminology used should therefore be checked against HAO (http://hymao.org).

Head. The head is hypognathus. The anterior surface of the head, from the oral cavity until the anterior ocellus, and between the compound eyes, is the face. The face is usually subdivided in three areas: clypeus, face, and frons. The clypeus is a ventral area immediately above the oral cavity. The lateral margin of the clypeus includes the anterior tentorial pit, often inconspicuous, on each side. The face (*stricto sensu*) is the median area above (and sometimes on the sides of) the clypeus. The frons is the area dorsal to the face. The dorsal surface of the head, between the dorsal margin of the compound eyes, and including the ocelli, is the vertex. There are usually three ocelli present in the vertex and, laterally, a pair of well-developed compound eyes. Anteriorly, and usually between the compound eyes, are the toruli (alveoli), from which a multisegmented antenna projects. The posterior surface of the head is divided into five areas; however, not all of these are always present or conspicuous: Occiput, postocciput, postgena, gena, and hypostoma. The antennal flagellum is usually divided into two or more flagellomeres. The scapes are united on the front of the head by the toruli. The mouthparts surround the oral foramen and are composed of four elements. The elements, from the posterior to the anterior end, are: labrum (usually hidden behind the clypeus), a pair of mandibles, a pair of maxillae, and the labium. Both the maxillae and the labium have segmented palpi. The mouthparts of parasitoids are usually of the chewing type.

Mesosoma. It is covered dorsally by the mesonotum, which is divided into an anterior mesoscutum and a posterior mesoscutellum, these being separated by a transversal suture (the scutoscutellar suture). In many species of parasitoid Hymenoptera, there are a pair of lines or sulci on the mesoscutum, which are the notauli; these divide the sclerite into a median lobe and a pair of lateral lobes. The shape and presence/absence of notauli are characters used to separate some families. The tegulae are the sclerites immediately basal to the wings. The lateral areas of the scutellum, basal to the wings and covered by them when at rest, are the "axillae." The posterior part of the scutellum is the frenum; sometimes this term is used only for the carina, and then the area posterior to the carina is denominated the "frenal area." The metanotum area immediately posterior to the frenum is denominated the "dorsellum." In lateral view, the Hymenoptera pronotum may or may not reach the tegula (wing base), being separated from the latter by a sclerite derived from the mesopleuron (the prepectus), or by a lateral lobular expansion from the mesonotum. The leg consists of six segments, from the base toward the apex these are: coxa, trochanter, femur, tibia, tarsus, and pretarsus. The femur may have, basally, a partial second division, the trochantellus, which might seem like a second segment of the trochanter. Each tibia has one to two apical spurs and, in many Hymenoptera, one of these spurs is modified into a structure to clean the antenna. The tarsus has three to five tarsomeres, and each tarsomere may have a ventral pad, the plantula. The pretarsus is composed by two claws and their associated structures.

The wing venation of many parasitoid Hymenoptera is reduced and has a simplified nomenclature. From the base of the wing, the veins are: submarginal, marginal, and postmarginal; a stigmal vein branches posteriorly from the encounter of the marginal and postmarginal veins; the stigmal vein may form an apical thickening denominated "stigma," which often ends in a protuberance directed toward the postmarginal vein, the uncus.

Metasoma. In the Apocrita, the first abdominal segment is immobile and broadly connected to the rest of the thorax. The first segment is the propodeum, and its combination with the thorax is denominated the mesosoma. In Apocrita, the remaining abdominal segments are collectively called the metasoma. Hymenoptera females have an ovipositor, which may be hidden or exposed, depending on the species. In Aculeata, the ovipositor is modified for stinging. The male genitalia in Hymenoptera are usually inconspicuous, and may vary from simple tubular structures with reduced components to relatively complex and elaborate structures.

MATERIAL PREPARATION AND PRESERVATION

Aquatic parasitoid collection involves searching for the aquatic insect host. Since the species with known biology are endoparasitoids of eggs, larvae/nymphs, and pupae, these are collected and maintained under laboratory conditions to obtain parasitoids or the immature stages of the aquatic host.

Parasitoid specimens are fixed in 80% ethanol, except those intended for molecular studies, which are placed in absolute ethanol. Specimen preparation for identification involves either double mounting on a triangular/rectangular piece of paper, mounting on a pin or mounting in permanent microscope slides.

For Eulophidae, Platygastridae, Figitidae, and Diapriidae specimens, we recommend that the material be dried in a critical point drier (CPD), since in the traditional manner it is common for structures like the head, mesosoma, and metasoma to collapse during drying, thereby hindering (and sometimes making impossible) the identification of the specimen. Specimens larger than 3 mm can be dried following the protocol of Aguiar (2012), which consists in using a mini-air pump to dry the specimens, allowing for a final condition similar to the CPD. Other methods can be used to dry ethanol specimens, such as drying manually (Noyes, 1982), or using chemicals such as acetone (Truman, 1968) or xylene (Sabrosky, 1966).

The Chalcidoidea, mainly the Eulophidae, Mymaridae, and Trichogrammatidae, need to be prepared on microscope slides for species identification. Slides should be prepared with Canada balsam, following the methodologies of Platner et al. (1999) and Noyes (1982).

KEY TO AQUATIC HYMENOPTERA

Hymenoptera: Families

1	Antenna with 16 or more articles	2
1'	Antenna with 15 or less articles	3
2(1)	Forewing with one recurrent vein (m-cu) (Fig. 12.1)	Braconidae
2'	Forewing with two recurrent veins (m-cu and 2 m-cu) (Fig. 12.2)	Ichneumonidae
3(1)	Torulus directed anteriorly, at the same level as the frons (Fig. 12.4)	4
3'	Torulus directed upwards and on a distinct projection from the frons (best observed in lateral view) (Fig. 12.3)	Diapriidae
4(3)	Forewing with reduced venation, without closed cells, with venation usually restricted to the anterior margin of the wing or completely absent (Fig. 12.6)	5
4'	Forewing with less than three closed cells and a distinct triangular marginal cell (Fig. 12.5)	Figitidae
5(4)	Metasoma usually not flattened dorsoventrally (Fig. 12.9); ovipositor originating from the ventral surface of the metasoma; pronotum not reaching tegula in lateral view (Fig. 12.10)	6

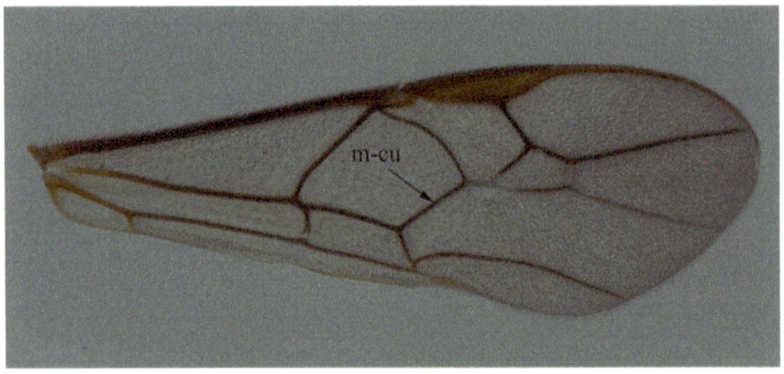

FIGURE 12.1 Braconidae. Forewing.

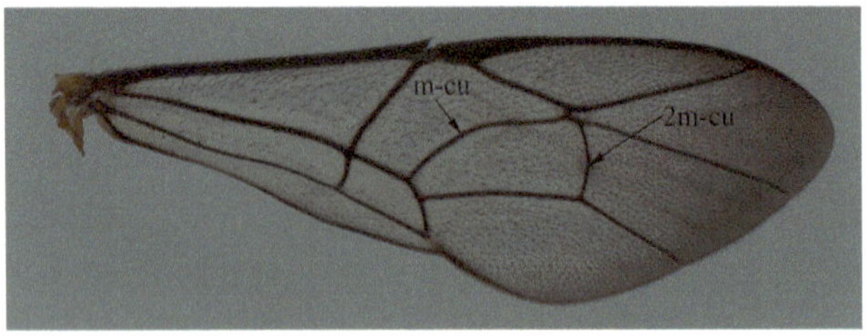

FIGURE 12.2 Ichneumonidae. Forewing.

FIGURE 12.3 Diapriidae. Lateral view of head.

FIGURE 12.4 Chalcididae. Lateral view of head.

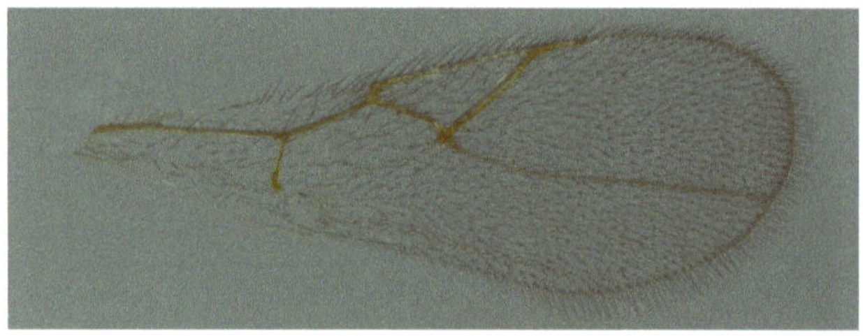

FIGURE 12.5 Figitidae. Forewing.

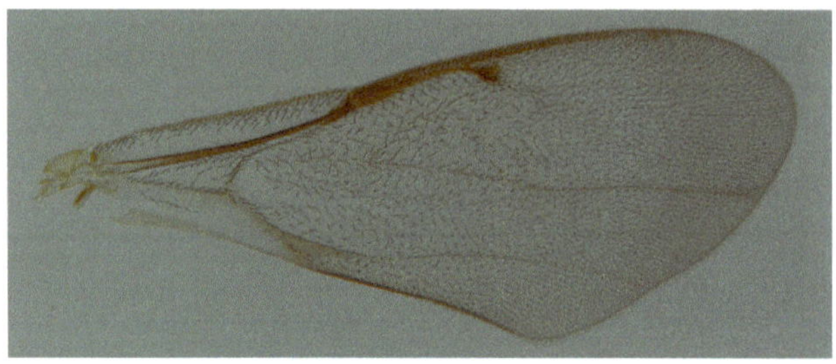

FIGURE 12.6 Chalcididae. Forewing.

5'	Metasoma usually flattened dorsoventrally (Fig. 12.7); ovipositor originating (externally) from the metasoma apex; pronotum reaching tegula in lateral view (Fig. 12.8) .... Platygastridae
6(5)	Hindfemur normal (Fig. 12.12) 7
6'	Hindfemur greatly enlarged (Fig. 12.11) .... Chalcididae
7(6)	Tarsi of four to five tarsomeres 8
7'	Tarsi of three tarsomeres .... Trichogrammatidae
8(7)	Head with dark lines forming an 'H' over the vertex and face (Fig. 12.13); fore- and hind wings usually with long marginal pile and hind wing petiolate; tarsus with four or five tarsomeres .... Mymaridae
8'	Head not as above, without dark lines forming an "H" (Fig. 12.14); hind wing not petiolate, with membranous area extending to the base; tarsus with four tarsomeres .... Eulophidae

FIGURE 12.7 Platygastridae. Lateral view of metasoma.

FIGURE 12.8 Platygastridae. Lateral view of mesosoma.

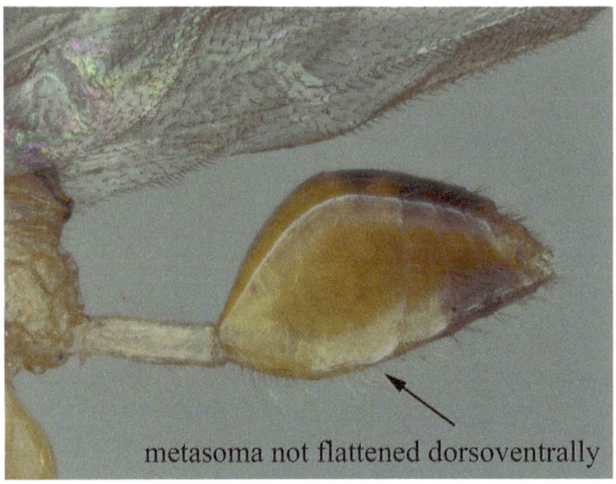

FIGURE 12.9 Chalcididae. Lateral view of metasoma.

FIGURE 12.10 Chalcididae. Lateral view of mesosoma.

FIGURE 12.11 Chalcididae. Hindfemur.

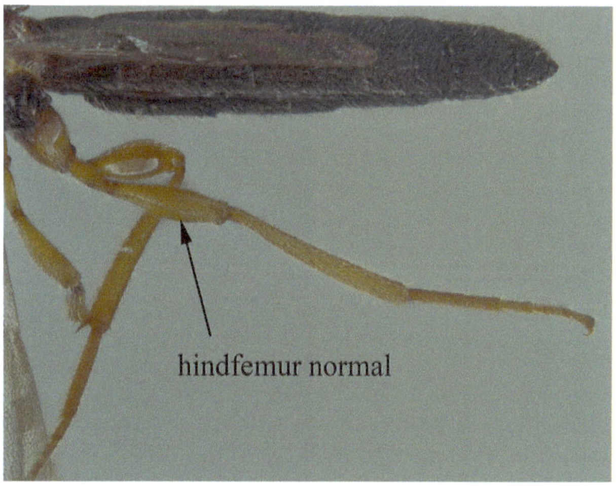

FIGURE 12.12 Platygastridae. Hindfemur.

FIGURE 12.13 Mymaridae. Frontal view of head.

FIGURE 12.14 Eulophidae. Frontal view of head.

REFERENCES

Aguiar, A.P. 2012. A technique to dry mount Hymenoptera (Hexapoda) from alcohol in a few seconds, and its application to other insect orders. Zootaxa 3412: 53–61.

Aguiar, A.P., A.R. Deans, M.S. Engel, M. Forshage, J.T. Huber, J.T. Jennings, N.F. Johnson, A.S. Lelej, J.T. Longino, V. Lohrmann, I. Mikó, M. Ohl, C. Rasmussen, A. Taeger & D.S.K. Yu. 2013. Order Hymenoptera. *in*: Z.-Q. Zhang (ed.), Zootaxa, 3703: 1–82. doi: 10.11646/zootaxa.3703.1.12a.

Bennett, A.M.R. 2008a. Aquatic Hymenoptera. Pages 673–682 *in*: R.W. Merritt, K.W. Cummins & M.B. Berg (eds.) An introduction to the Aquatic Insects of North America. Kendall-Hunt Publishing Company, Dubuque.

Bennett, A.M.R. 2008b. Global diversity of hymenopterans (Hymenoptera; Insecta) in freshwater. Hydrobiologia 595: 529–534. doi: 10.1007/s10750-007-9036-y.

Burks, B.D. 1977. The Mexican species of *Chalcis* Fabricius (Hymenoptera: Chalcididae). Proceedings of the Entomological Society of Washington 79: 383–399.

De Santis, L. 1964. Un nuevo tricogrammido argentino de biologia acuatica (Insecta Hymenoptera). Notas de Comision de Investigacion Cientifica 2: 3–7.

Dasch, C.E. 1979. Ichneumon-flies of America North of Mexico: 8. Subfamily Cremastinae. Memoirs of the American Entomological Institute 29: 1–702.

Diaz, N.B., F.E. Gallardo, A.L. Gaddi & G.C. Walsh. 2009. Description of a new genus and species of Eucoilinae (Hymenoptera: Cynipoidea: Figitidae) Parasitoid of Ephydridae (Diptera). Annals of The Entomological Society of America 102: 603–607. doi: 10.1603/008.102.0404.

Feitosa, M.C.B., R.B. Querino & N. Hamada. 2006. Mymaridae (Insecta: Hymenoptera) parasitóide de ovos de insetos aquáticos associados a macrófitas aquáticas de várzea e igarapés da Amazônia Central, Brasil in: XV Jornada de Iniciação Científica, 2006, Manaus. XV Jornada de Iniciação Científica. INPA, Manaus.

Feitosa, M.C.B., R.B. Querino & N. Hamada. 2016. Association of *Anagrus amazonenses* Triapitsyn, Querino & Feitosa (Hymenoptera, Mymaridae) with aquatic insects in upland streams and floodplain lakes in central Amazonia, Brazil. Revista Brasileira de Entomologia 60: 267–269. doi: 10.1016/j.rbe.2016.04.002.

Fursov, V.A. 1995. Review of European Chalcidoidea (Hymenoptera) parasitizing the eggs of aquatic insects. Bulletin of the Irish Biogeographical Society 18: 2–12.

Gaddi, A.L., N.B. Díaz & F.E Gallardo 2010. A new species of *Hexacola* Foerster (Hymenoptera: Figitidae), parasitoid of Lemnaphila neotropica Lizarralde de Grosso (Diptera: Ephydridae). Zootaxa 2622: 42–48.

Hagen, K.S. 1996. Aquatic Hymenoptera. Pages 474–483 *in*: R.W. Merritt & K.W. Cummins (eds.), An introduction to the aquatic insects of North America. Kendall/Hunt Publishing Company, Dubuque, Iowa.

Huber, J.T., Z. Mendel, A. Protasov & J. LaSalle. 2006a. Two new Australian species of *Stethynium* (Hymenoptera: Mymaridae), larval parasitoids of *Ophelimus maskelli* (Ashmead) (Hymenoptera: Eulophidae) on Eucalyptus. Journal of Natural History 40: 1909–1921.

Huber, J.T., M.J. Sharkey & F. Fernández. 2006b. Estructura y glosario. *in*: F. Fernández & M.J. Sharkey. Introducción a los Hymenoptera de la Región Neotropical. Sociedad Colombiana de Entomología and Universidad Nacional de Colombia, Bogotá.

Johnson, N.F. & L. Masner. 2004. The genus *Thoron* Haliday (Hymenoptera: Scelionidae), egg-parasitoids of waterscorpions (Hemiptera: Nepidae), with key to world species. American Museum Novitates 3452: 1–16. doi: 10.1206/0003-0082(2004)452<0001:TGTHHS>2.0.CO;2.

Lima, A. da C. 1960. Sôbre alguns microhimenópteros aquáticos (Chalcidoidea, Eulophidae e Trichogrammatinae). Arquivos do Instituto Biológico 27: 197–200.

Masner, L. & J.L. García. 2002. The genera of Diapriinae (Hymenoptera: Diapriidae) in the New World. Bulletin of the American Museum of Natural History 268: 1–138.

Matheson, R. & C.R. Crosby. 1912. Aquatic Hymenoptera in America. Annals of the Entomological Society of America 5: 65–67.

Munro, J.B., J.M. Heraty, R.A. Burks, D. Hawks, J. Mottern, A. Cruaud, J. Rasplus & P. Jansta. 2011. A Molecular Phylogeny of the Chalcidoidea (Hymenoptera). PLoS ONE 6(11): e27023. doi: 10.1371/journal.pone.0027023.

Noyes, J. 1982. Collecting and preserving chalcid wasps (Hymenoptera: Chalcidoidea). Journal of Natural History 16: 315–334. doi: 10.1080/00222938200770261.

Platner, J.D., R.K. Velten, M. Planoutene & J.D. Pinto. 1999. Slide-mounting techniques for *Trichogramma* (Trichogrammatidae) and other minute parasitic Hymenoptera. Entomological News 110: 56–64.

Querino, R.B. & N. Hamada. 2009. An Aquatic Microhymenopterous Egg-Parasitoid of *Argia insipida* Hagen in Selys (Odonata: Coenagrionidae) and Biological Observations in the Central Amazon, Brazil. Neotropical Entomology 38: 346–351. doi: 10.1590/S1519-566X2009000300008.

Querino, R.B. & N. Hamada. 2012. Hymenoptera. Pages 98–103 *in*: Hamada, N. & R.L. Ferreira-Keppler (eds.). Guia ilustrado de insetos aquáticos e semiaquáticos da Reserva Florestal Ducke, Manaus, Amazonas, Brasil. Editora da Universidade Federal do Amazonas, Manaus.

Querino, R.B. & N. Hamada. 2014. Ordem Hymenoptera. Pages 377–390 *in*: N. Hamada, J.L. Nessimian & R.B. Querino (eds.), Insetos aquáticos na Amazônia brasileira: taxonomia, biologia e ecologia. 1ed. Editora do INPA, Manaus.

Querino, R.B. & J.D. Pinto. 2007. A new *Hydrophylita* (Hymenoptera: Trichogrammatidae) from the Neotropics, with a key to species. Zootaxa 1437: 47–54.

Ramalheira, C.S., N. Hamada & R.B. Querino. 2005. Himenópteros parasitóides de ovos de insetos aquáticos associados a macrófitas de várzea e igarapés na Amazônia Central. Resumo Zoo-21. XIV Jornada de Iniciação Científica. INPA, Manaus.

Reche, A.V. & F.E. Gallardo. 2015. Revision of Argentinian species of *Hexacola* Foerster (Hymenoptera: Figitidae: Eucoilinae). Journal of Insect Biodiversity and Systematics 1(1): 77–86.

Resh, V.H. & W. Jamieson. 1988. Parasitism of the aquatic moth *Petrophila confusalis* (Lepidoptera: Pyralidae) by the aquatic wasp *Tanychela pilosa* (Hymenoptera: Ichneumonidae). Entomological News 99: 185–188.

Sabrosky, C.W. 1966. Mounting insects from alcohol. Bulletin of the Entomological Society of America 12(3): 349. doi: 10.1093/besa/12.3.349.

Seltmann, K., M. Yoder, I. Miko, M. Forshage, M. Bertone, D. Agosti, A. Austin, J. Balhoff, M. Borowiec, S. Brady, G. Broad, D. Brothers, R. Burks, M. Buffington, H. Campbell, K. Dew, A. Ernst, J. Fernandez-Triana, M. Gates, G. Gibson, J. Jennings, N. Johnson, D. Karlsson, R. Kawada, L. Krogmann, R. Kula, M. Ohl, C. Rasmussen, F. Ronquist, S. Schulmeister, M. Sharkey, E. Talamas, E. Tucker, L. Vilhelmsen, P. Ward, R. Wharton & A. Deans. 2012. A hymenopterists' guide to the Hymenoptera Anatomy Ontology: utility, clarification, and future directions. Journal of Hymenoptera Research 27: 67–88. doi: 10.3897/jhr.27.2961.

Sharkey, M.J. 2007. Phylogeny and classification of Hymenoptera. Zootaxa 1668: 521–548.

Sharkey, M.J., J.M. Carpenter, L. Vilhelmsen, J. Heraty, J. Liljeblad, A.P.G Dowling, S. Schulmeister, D. Murray, A.R. Deans, F. Ronquist, L. Krogmann & W.C. Wheeler. 2012. Phylogenetic relationships among superfamilies of Hymenoptera. Cladistics 28: 80–112. doi: 10.1111/j.1096-0031.2011.00366.x.

Townes, H.K. 1971. The genera of Ichneumonidae, part 4. Memoirs of the American Entomological Institute 18: 1–372.

Triapitsyn, S.V., R.B. Querino & M.C.B. Feitosa. 2008. A New Species of *Anagrus* (Hymenoptera: Mymaridae) from Amazonas, Brazil. Neotropical Entomology 37(6): 681–684. doi: 10.1590/S1519-566X2008000600009.

Truman, J.W. 1968. Acetone treatment for preservation of adult and larval mosquitoes. Annals of the Entomological Society of America 61(3): 779–780. doi: 10.1093/aesa/61.3.779.

Williams, D.D. & B.W. Feltmate. 1992. Aquatic Insects. CAB International, Wallingford.

Yoder, M.J., I. Mikó, K. Seltmann, M.A. Bertone & A.R. Deans. 2010. A gross anatomy ontology for Hymenoptera. PLoS One 5(12): e15991. doi: 10.1371/journal.pone.0015991.

Chapter 13

Order Mecoptera

Renato Jose Pires Machado
Instituto de Biociências, Universidade Federal do Mato Grosso, Cuiabá, Mato Grosso, Brazil

Chapter Outline

INTRODUCTION

Mecoptera is one of the smallest insect orders with approximately 600 extant species worldwide. These insects are commonly known as scorpionflies or hangingflies, based on the characteristics of the most abundant families, Panorpidae and Bittacidae, respectively. They are also famous for their complex mating behavior (Thornhill, 1976). In the Neotropical region, the order is represented by nearly 100 species divided into 12 genera and five families (Machado et al., 2013; Contreras-Ramos et al., 2014). All identification keys available for the Neotropical Mecoptera fauna are based exclusively on adults, and comprise all genera but only a small part of the species (Byers, 1989; Machado et al., 2009, 2013; Ascuntar-Osnas et al., 2016).

Panorpidae and Bittacidae are the most diverse families, comprising more than 95% of the known Neotropical species (Machado et al., 2013; Contreras-Ramos et al., 2014). Panorpidae is more abundant and diverse in the northern hemisphere, but species are also found in southern Mexico (Bicha, 2006; Byers, 2011; Contreras-Ramos et al., 2014). Bittacidae is dominant in the neotropics, with more than 70 widespread species (Machado et al., 2009; Contreras-Ramos et al., 2014). Larvae of both families are eruciform and always are associated with moist soils located very close to, but not in, water bodies (Setty, 1940).

Eomeropidae and Meropidae are the two least diverse families in the Neotropics. Adults live in soil (Palmer, 2010) and are known from one species each; Eomeropidae from Chile, and Meropidae from southeast Brazil (Machado et al., 2013). Meropid larvae remain unknown but they are probably terrestrial, living close to where adults are typically found (Pilgrim, 1972).

The last Neotropical family is the relict Nannochoristidae, another small family known by only one genus and eight species: four from southeast Australia, one from New Zealand, and three species from the temperate areas of southern Argentina and Chile: *Nannochorista andina* Byers, 1989, *N. edwardsi* Kimmins, 1929, and *N. neotropica* Navás, 1928.

Despite its low diversity, Nannochoristidae has received much more attention because of its unique characteristics. It is the only Mecoptera family with known aquatic larvae, and the habit and morphology of its species are key subjects in a continuous discussion about its systematic position (Fraulob et al., 2012). Some phylogenetic studies maintain that the family should be placed in a different order called Nannomecoptera (Hinton, 1981; Willmann, 1987; Kluge, 2003), while others have suggested that it should be placed as a sister group to Diptera (Beutel & Baum, 2008). However, more recent molecular data (Whiting, 2002; Wiegmann et al., 2009; Misof et al., 2014) and a few detailed larval morphological studies (Beutel et al., 2009; Fraulob et al., 2012) have placed Nannochoristidae as a basal lineage within Mecoptera.

The larva of Nannochoristidae was first discovered by Pilgrim (1972) for the single New Zealand species, *N. philpotti* (Tillyard, 1917), and still is the only larva formally described. Larvae of the Australian *N. dipteroides* Tylliard, 1917 was later included in the phylogenetic work of Fraulob et al. (2012). Kluge (2003) reported two larvae from Chile, but was unable to associate these specimens with any of the three known Neotropical species. Details of the habits and morphological features of the larvae are known in detail mainly by the New Zealand species but

Thorp and Covich's Freshwater Invertebrates, Fourth Edition. https://doi.org/10.1016/B978-0-12-804223-6.00013-5

their life styles are likely to be uniform among the whole family (Kluge, 2003; Beutel et al., 2009).

Larvae occur in fine organic deposits in shallow woodland streams and are generally found out of the current in backwater regions along stream margins (Pilgrim, 1972; Winterbourn, 1982; Winterbourn & Gregson, 1989). They are active predators on chironomid larvae and other small dipterans (Winterbourn, 1982). Larvae seem very active with sinuous abdominal movements and are supported by the legs (Pilgrim, 1972). The entire larval stage is aquatic and four instars with distinct sizes are present. By the end of the fourth instar, larvae become more sclerotized and start to show distinct colors and behaviors (Pilgrim, 1972). This stage, known as prepupa, does not feed and leaves the water for pupation, which occurs in damp vegetation or humid soil very close to the water (Pilgrim, 1972). The adults also occur along stream banks, and they probably have a liquid diet of nectar and fruit juice (Beutel & Baum, 2008; Palmer, 2010).

LIMITATIONS

As mentioned in the introduction, the larvae of the three Neotropical species remain undescribed and only the adults are known, and identified with the key provided by Byers (1989). However, it is likely that the Neotropical species inhabit environments that are very similar to the one described for *N. philpotti*, the New Zealand species. It is only with further collecting effort in areas of similar characteristics, near the places where the adults have been captured, that we will be able to better understand the life history of this relict group of aquatic insects.

TERMINOLOGY AND MORPHOLOGY

Larval morphological details are known only for *N. philpotti* and *N. dipteroides*, from New Zealand and Australia, respectively. Nevertheless, the two Chilean larvae mentioned by Kluge (2003) seem to be very similar to these larvae, and it is believed that their general morphology should be relatively constant throughout the family (Beutel et al., 2009). In general, the larvae are campodeiform, with a cylindrical and elongated body (Fig. 13.1 A); the head capsule (Figs. 13.1 B, C) is characterized by the presence of short antennae, large eyespots, and relatively large mandibles bearing a sharp apical tooth; the three segments of the thorax are very similar, bearing well-developed legs (Fig. 13.1 C); the abdomen is divided into ten cylindrical segments with a pair of terminal hooks, which are likely used to anchor the larvae in the substratum (Pilgrim, 1972; Melzer et al., 1994; Kluge, 2003; Beutel et al., 2009; Fraulob et al., 2012). Head and pronotum are

FIGURE 13.1 *Nannochorista* sp. (Mecoptera: Nannochoristidae) larva: (A) Habitus, lateral view; (B) head and thorax, dorsal view; (C) head and thorax, lateral view. Scale bars = 500 μm. *Photos courtesy of Elena Lukashevich and Dmitry Shcherbakov, Paleontological Institute, Moscow.*

dark and more sclerotized, while the abdomen is white and less sclerotized; the fourth instar length varies from 11 to 21 mm (Pilgrim, 1972). More morphological details of the larvae were provided by Beutel et al. (2009) and Fraulob et al. (2012) in a comprehensive description of the external and internal morphology of the head and abdomen, which also includes excellent electron microscopy images and robust phylogenetic discussions.

MATERIAL PREPARATION AND PRESERVATION

The larvae can be collected from the silt from the bottom of pools and quiet stream areas with the help of sieves with different mesh sizes, as described by Pilgrim (1972). Winterbourn (1982) also succeeded in collecting Nannochoristidae larvae in New Zealand with the substrate colonization method, using wooden blocks with longitudinal grooves that were left submerged in the streams. The larvae can be preserved in vials with 80% ethyl alcohol. For more detailed studies, the larvae can be clarified in 10% KOH and slide mounted.

ACKNOWLEDGMENTS

I would like to thank Elena Lukashevich and Dmitry Shcherbakov (Paleontological Institute, Moscow) for providing the images that illustrate this chapter. Thank you to Cecilia Simões for reviewing the first English version. The author also thanks the Brazilian National Council for Scientific and Technological Development (CNPq 400237/2017-2), for the support.

REFERENCES

Ascuntar-Osnas, O., D. Torres-Dominguez, P. Chacón de Ulloa & R.J.P. Machado. 2016. First record of the order Mecoptera (Insecta) for the Gorgona Island (Colombia), with the descriptions of a new species. Zootaxa 4097(2): 271–279.

Beutel, R.G. & E. Baum. 2008. A longstanding entomological problem finally solved? Head morphology of Nannochorista (Mecoptera, Insecta) and possible phylogenetic implications. Journal of Zoological Systematics and Evolutionary Research 46: 346–367.

Beutel, R.G., N.P. Kristensen & H. Pohl. 2009. Resolving insect phylogeny: the significance of cephalic structures of the Nannomecoptera in understanding endopterygote relationships. Arthropod Structure and Development 38: 427–460.

Bicha W. 2006. New scorpionflies (Mecoptera: Panorpidae) from Jalisco, Michoacán, and Oaxaca, Mexico. Proceedings of the Entomological Society of Washington 108: 24–34.

Byers, G.W. 1989. The Nannochoristidae of South America (Mecoptera). The University of Kansas Science Bulletin 54(2): 25–34.

Byers, G. W. 2011. Additions to the Mecoptera of Mexico. Journal of the Kansas Entomological Society 84: 1–11.

Contreras-Ramos, A., M.A. Sarmiento-Cordero & N.D. Penny. 2014. Biodiversidad de Mecoptera en México. Revista Mexicana de Biodiversidad 85: 339–344.

Fraulob, M., B. Wipfler, F. Hünefeld, H. Pohl & R.G. Beutel. 2012. The larval abdomen of the enigmatic Nannochoristidae (Mecoptera, Insecta). Arthropod Structure and Development 41: 187–198.

Hinton, H.E. 1981. Biology of Insect Eggs. Pergamon Press, Oxford.

Kluge, N. 2003. Larval leg structure of Nannochorista Tillyard, 1917 and characteristics of Mecoptera. Russian Entomological Journal 12: 1–6.

Machado, R.J.P., F.S.P. Godoi & J.A. Rafael. 2009. Neotropical Mecoptera (Insecta): new generic synonymies, new combinations, key to families and genera, and checklist of species. Zootaxa 2148: 27–38.

Machado, R.J.P., R. Kawada & J.A. Rafael. 2013. New continental record and new species of Austromerope (Mecoptera, Meropeidae) from Brazil. Zookeys 269: 51–65.

Melzer, R.R., H.F. Paulus & N.P. Kristensen. 1994. The larval eye of nannochoristid scorpionflies (Insecta, Mecoptera). Acta Zoologica 75: 201–208.

Misof, B., et al. 2014. Phylogenomics resolves the timing and pattern of insect evolution. Science 346: 763–767.

Palmer, C.M. 2010. Diversity of feeding strategies in adult Mecoptera. Terrestrial Arthropod Reviews 3: 111–128.

Pilgrim, R.L. 1972. The aquatic larva and the pupa of Choristella philpotti Tillyard, 1917 (Mecoptera: Nannochoristidae). Pacific Insects 14: 151–168.

Setty, L.R. 1940. Biology and morphology of some North American Bittacidae (Order Mecoptera). The American Midland Naturalist 23: 257–353.

Thornhill, R. 1976. Sexual selection and nuptial feeding behaviour in Bittacus apicalis (Insecta: Mecoptera). The American Naturalist 110: 529–548.

Whiting, M.F. 2002. Mecoptera is paraphyletic: multiple genes and phylogeny of Mecoptera and Siphonaptera. Zoologica Scripta 31: 93–104.

Wiegmann, B.M., M.D., Trautwein, J.W. Kim, B.K. Cassel, M.A. Bertone, S.L. Winterton & D.K. Yeates. 2009. Single-copy nuclear genes resolve the phylogeny of the holometabolous insects. BMC Biology 7: 34.

Willmann, R. 1987. The phylogenetic system of the Mecoptera. Systematic Entomology 12: 519–524.

Winterbourn, M.J. 1982. The invertebrate fauna of a forest stream and its association with fine particulate matter. New Zealand Journal of Marine and Freshwater Research 16: 271–281.

Winterbourn, M.J. & K.L.D. Gregson. 1989. Guide to the aquatic insects of New Zealand, Revised Edition. Bulletin of the Entomological Society of New Zealand 9: 1–95.

Section II.2

Order Odonata

Order Odonata

Pablo Pessacq
Centro de Investigaciones Esquel de Montaña y Estepa Patagónicas (CIEMEP), Esquel, Chubut, Argentina

Javier Muzón
Laboratorio de Biodiversidad y Genética Ambiental (BioGeA), Universidad Nacional de Avellaneda (UNDAV), Buenos Aires, Buenos Aires, Argentina

Ulisses Gaspar Neiss
Instituto de Criminalística, Departamento de Polícia Técnica-Científica, Manaus, Amazonas, Brazil

Chapter Outline

INTRODUCTION

Odonata, commonly known as damselflies and dragonflies, is one of the most ancient orders of Pterygota. It is comprised of cosmopolitan, hemimetabolous insects that are ubiquitous in most aquatic environments (including several brackish and terrestrial habitats). Due to their relatively big size and beauty, these insects have received adequate taxonomical attention, and currently there are approximately 6,000 identified species (Dijkstra et al., 2013). Another 1,000 to 1,500 species are expected to exist worldwide (Kalkman et al., 2008).

Although considered as a monophyletic group (e.g., Davis et al., 2011), the position of extant Odonata relative to Ephemeroptera and Neoptera and the validity of Paleoptera as a natural taxon are still debated (Hovmöller et al., 2002; Rehn, 2008; Odgen & Whiting, 2003). The higher relationships within the order, including the composition and relationships among superfamilies, is controversial and not fully resolved (see Bybee et al., 2008; Rehn, 2008; Trueman, 2007). For practical reasons, we follow the latest classification scheme (Dijkstra et al., 2013) which considers that: (1) Zygoptera is comprised of four superfamilies (Lestoidea, Platystictoidea, Calopterygoidea, and Coenagrionoidea); (2) Anisozygoptera includes only the superfamily Epiophlebioidea; and (3) Anisoptera is composed of five superfamilies (Aeshnoidea, Petaluroidea, Gomphoidea, Cordulegastroidea, and Libelluloidea). Except for Epiophlebioidea, all these superfamilies are represented in the Neotropical region. A detailed discussion of the phylogenetic of the order and the composition and systematics of the superfamily can be found in Suhling et al. (2015). Additional information is treated under each superfamily chapter.

The Neotropical region, with 1,768 species presently described (Neiss & Hamada, 2014), has the richest fauna of Odonata worldwide. Many new species and previously unknown immature stages are being described, but still, the larvae of only about 75% genera and 40% of the species are known (Garrison et al., 2006, 2010; von Ellenrieder, 2009). In the period of 2004–2009, 93 species and 11 genera, and the larval stage of 90 species were described (von Ellenrieder, 2009). Zygoptera is represented in this region by 10 families, 108 genera, and 974 species, while Anisoptera includes 9 families, 102 genera, and 794 species (Fleck, 2011; Fleck & El Adouzi, 2013; Neiss & Hamada, 2014).

Zygoptera has a very high level of endemicity. Indeed, 89 of the 103 genera of the group are endemic to the Neotropical region. Additionally, two families

Thorp and Covich's Freshwater Invertebrates, Fourth Edition. https://doi.org/10.1016/B978-0-12-804223-6.00014-7

(Dicteriadidae and Polythoridae) and many coenagrionid genera are also endemic.

Anisoptera also shows a high endemicity, with about 64% of the 59 currently described genera being endemic to the Neotropics, with the remarkable endemicity of Neopetaliidae in the subantarctic forests of southern South America, represented by the monotypic genus *Neopetalia*.

The traditional regionalization based on Wallace (1876) extends the Neotropical region from Mexico to central Argentina. Up to date classifications do not include Chile and part of Mexico and Argentina in this region (e.g., Morrone, 2014); but for practical reasons in the use of the keys, which are mainly oriented to Central and South American scientists and readers, we include here all the fauna of Argentina, Chile and Mexico, and will also refer to this area as the Neotropical region in a wide sense.

LIMITATIONS

In the recent years the publication of several keys for adults (e.g., Förster, 2001; Trapero & Naranjo, 2004; Lencioni, 2005, 2006) have made it easier to identify odonate taxa, especially generic keys for Anisoptera and Zygoptera of the New World (Garrison et al., 2006, 2010), but determination at the species level is still difficult and the Odonata literature is widely dispersed.

For the immature stages, there are several keys at the regional or country level: Amazonas (Neiss & Hamada, 2014), Brazil (Costa et al., 2004), Costa Rica (Ramírez, 2010), Mexico (Novelo-Gutiérrez, 1997) and Patagonia (Muzón et al., 2014). One particular problem is that larval descriptions are mainly based on the ultimate stadium, and the identification of early stages is almost impossible without rearing specimens, a difficult task for many seasonal and fast water species (Neiss & Hamada, 2014).

Given that only about 75% of the genera have been described and many descriptions are too brief or out of date, the keys presented in this book must be used with caution. The degree of knowledge for each family is described in their respective odonate chapters.

TERMINOLOGY AND MORPHOLOGY

Odonatologists have used the term larva(e) for the immature post-egg stages since the end of the 19th century (cf. Tillyard, 1917). Recently, some authors have advocated use of the term naiad for Odonata (Bybee et al., 2015); however, others disagreed (Muzón & Lozano, 2016; Rédei & Štys, 2016; Sahlén et al., 2016). In this chapter, the most widely use term larva is adopted, following the terminology proposed by Corbet (2002). Larvae of Odonata are mostly aquatic and active predators. They inhabit many different kinds of environments, and their morphology varies according to where they live. Details of Odonata's biology and morphology are discussed in Volume I (Thorp & Rogers, 2015) of this series.

Here we present a glossary which is fundamental for the correct understanding of the terminology used in the keys. Some structures are especially important for identifying larvae, including the head and its parts (e.g., antennae, labium, labial palps), abdominal spines or tubercles, caudal appendages, and anal pyramid, among others.

Glossary

Abdominal projections: Mainly in Anisoptera, larvae bear lateral spines and/or dorsal hooks, used as defense devices against predators like fishes. Their presence, size, and shape are useful for diagnostic purposes.

Anal pyramid: Epiproct and paraprocts are scaliform in Anisoptera (Fig. 14.2), resulting together with the cerci in a five-parts pyramid. At the base of this complex is the anus. Combination of contraction of rectal muscles and the anal pyramid are responsible for the "jet propulsion" of anisopterous larvae.

Caudal appendages: Also referred as caudal lamella, because of their nonappendicular origin. The epiproct and paraprocts are greatly enlarged in Zygoptera (Figs. 14.3, 4, 10, 11, 13) resulting in the dorsal (= epiproct) and lateral (= paraprocts) caudal appendages/lamellae. Dorsal and lateral appendages are quite similar in the same larva (differing mainly in chaetotaxy of dorsal and ventral margins), but they exhibit a wide range of differences among taxa. Caudal appendages give useful diagnostic characters, including shape, presence/absence and location of nodus, color pattern, tracheation pattern and chaetotaxy (i.e., series of spines/setae on dorsal and ventral margins). Corbet (1999), based on Tillyard (1917), gives a classification distinguishing different types, from which we use the following: (1) **saccoid or triquetral** (referred to as saccoid in this book): inflated, sac-shaped, sometimes with three carinae (e.g., Platystictidae, Fig. 14.11, most Megapodagrionidae, Fig. 14.4); (2) **triquetral**: composed of three lamellae, that in posterior view are T-shaped (e.g., Calopterygidae, Fig. 14.10, some Megapodagrionidae); and (3) **lamellate**: flattened, usually ovoid or lanceolate (dorsal one usually held in the vertical plane) (e.g., Coenagrionidae, Fig. 14.15, Lestidae, Fig. 14.13).

Caudal appendages nodus: Structure present at one margin of greatly enlarged caudal appendages in the Zygoptera, usually as a small cleft, that continues as a transversal line that divides the caudal appendage in a basal and a distal part; the distal one is less sclerotized (e.g., Fig. 14.6.27, 37). Usually the spines at the lamella's margin are reduced or absent distal to the nodus. In some species the nodus is slightly evident, and it is only indicated by a change in the stoutness of the marginal setae.

FIGURE 14.1 Gomphoidea, *Ebegomphus* sp., larval habitus, dorsal view.

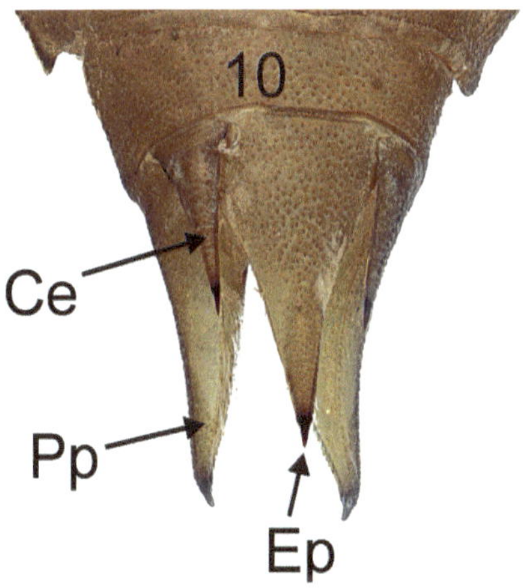

FIGURE 14.2 Gomphoidea, *Ophiogomphus susbehcha* Vogt & Smith, 1993. Larval abdominal segments 9–10. 10, abdominal segment 10; *Ce*, cercus; *Ep*, epiproct; *Pp*, Paraproct. *Pictures courtesy of Rodolfo Novelo-Gutierrez.*

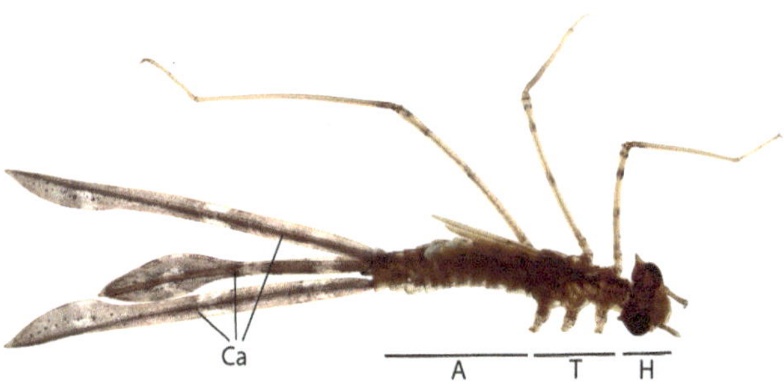

FIGURE 14.3 Calopterygoidea, *Megapodagrion megalopus* (Selys, 1862), larval habitus, dorsal view. *A*, abdomen; *Ca*, caudal appendages; *H*, head; *T*, thorax. *Original pictures by Neiss & Hamada (2014).*

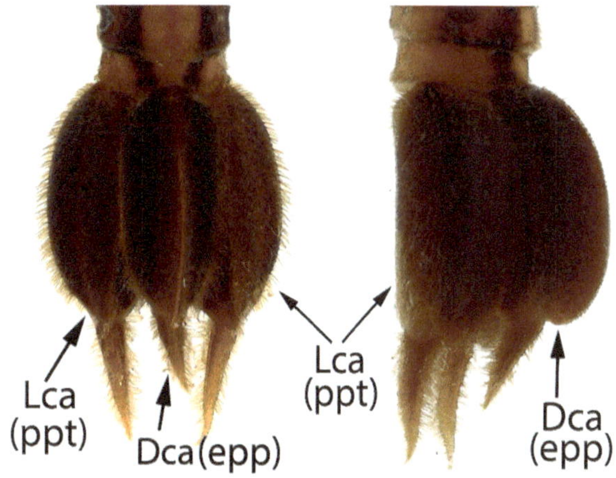

FIGURE 14.4 Calopterygoidea, *Paraphlebia zoe* Selys *in* Hagen, 1861. Larval caudal appendages, dorsal (left) and lateral (right) view. *Dca*, dorsal caudal appendages; *epp*, epiproct; *Lca*, lateral caudal appendages; *ppt*, paraprocts.

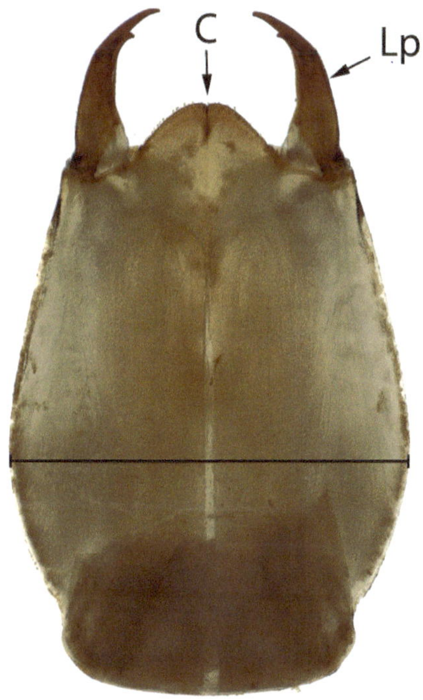

FIGURE 14.5 Platystictoidea, *Palaemnema brasiliensis* Machado, 2009. Larval prementum, dorsal view. *C*, prementum cleft; *Lp*, labial palp. *Black line* indicates maximum width.

FIGURE 14.6 Platystictoidea, *Palaemnema brasiliensis*. Larval labial palp dorsal view. *Eh*, end hook; *Mh*, movable hook. *Original pictures by Neiss & Hamada (2014).*

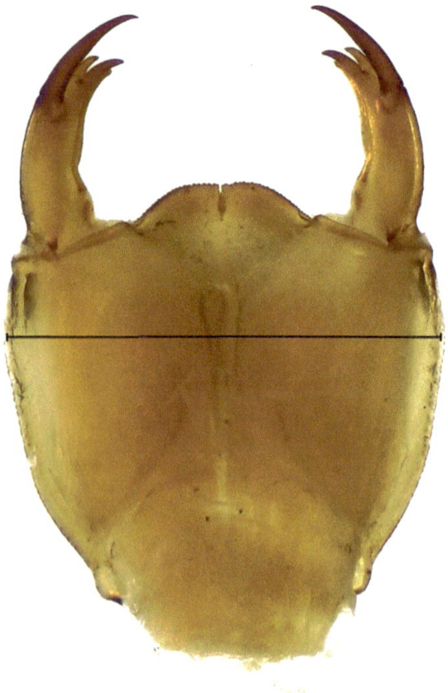

FIGURE 14.7 Coenagrionoidea, *Paraphlebia zoe*. Larva; prementum, dorsal view. *Black line* indicates maximum width.

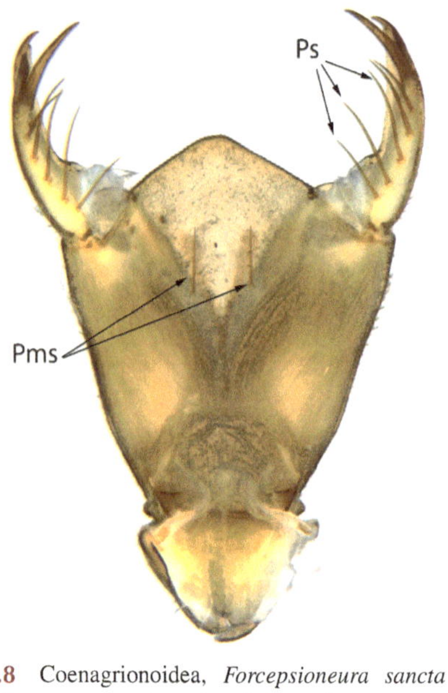

FIGURE 14.8 Coenagrionoidea, *Forcepsioneura sancta* (Hagen *in* Selys, 1860). Larva; prementum, dorsal view. *Pms*, premental setae; *Ps*, palpal setae.

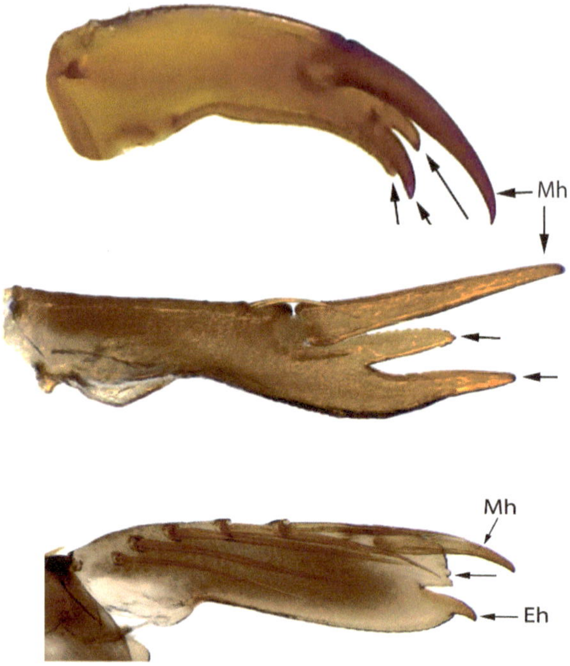

FIGURE 14.9 Coenagrionoidea, *Paraphlebia zoe* (above); *Argia* sp. (middle); *Psaironeura* sp. (below). Larval labial palps. *Arrows* indicate hooks. *Eh*, end hook; *Mh*, movable hook. *Original pictures by Neiss & Hamada (2014).*

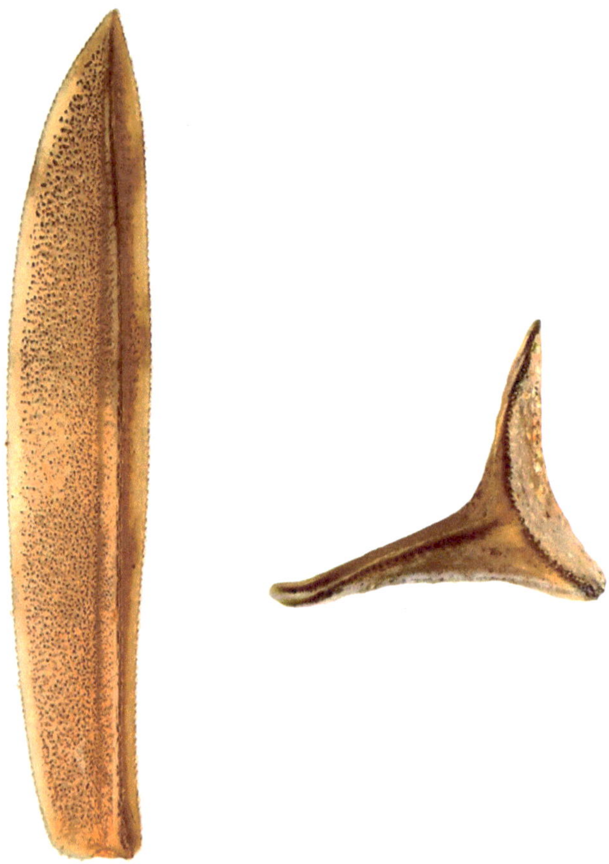

FIGURE 14.10 Calopterygoidea, *Hetaerina americana* (Fabricius, 1798). Larva lateral caudal appendage, dorsal (left) and posterior (right) view.

FIGURE 14.11 Platystictoidea, *Palaemnema brasiliensis*. Larval lateral caudal appendage, lateral (above) and dorsal (below) views.

FIGURE 14.12 Calopterygoidea, *Rimanella arcana* (Needham, 1933). Larval caudal appendages, dorsal view.

FIGURE 14.13 Lestoidea, *Perilestes attenuatus* Selys, 1886. Larval caudal appendage, lateral (above) and dorsal (below) view. *Original pictures by Neiss & Hamada (2014).*

FIGURE 14.14 Calopterygoidea, *Hetaerina* sp. larval habitus, dorsal view. *Original pictures by Neiss & Hamada (2014).*

FIGURE 14.15 Coenagrionoidea, *Psaironeura tenuissima* (Selys, 1886) larval habitus, dorsal view. *Original pictures by Neiss & Hamada (2014).*

Cerci: They are the 11th abdominal segment appendages. They are unsegmented, small, and little informative, except in several Coenagrionidae genera (e.g., *Enallagma*).

End hook: Hook present at the inner-distal margin of labial palp, usually well developed (e.g., most Coenagrionidae and Megapodagrionidae, Fig. 14.9), sometimes reduced (e.g., Austropetalidae), curved (e.g., most Coenagrionidae, Fig. 14.9), or straight (e.g., *Gynacantha*, *Staurophlebia*).

Labial palp: Modified raptorial palp (Figs. 14.5–9) which are very mobile and usually with an end hook and a movable hook (see in this glossary). These are highly variable in shape and relative size with the distal margin sometimes greatly lobulated (e.g., *Lestes*), with end hook and small teeth (e.g., *Psaironeura*, Fig. 14.9), or bare (e.g., *Triacanthagyna*). The inner margin may be crenulated or with small teeth (e.g., *Roppaneura*) or

ORDER ODONATA

bare (e.g., *Bromelagrion*), with (e.g., *Forcepsioneura*, *Psaironeura*, Figs. 14.8–9) or without palpal setae (e.g., *Palaemnema*, *Paraphlebia*, Fig. 14.6, 9) close to its outer margin.

Labium: The prehensile labium of Odonata larvae is unique due to its extendable capacity. It is integrated by a post and a prementum (Figs. 14.5, 7, 8, 27, 29) and two labial palps. The labial palp ends in a movable hook. In a resting position, the labium can reach posteriorly a different pair of coxae, or even abdominal S1 in some Lestidae. There are different labium types:

FIGURE 14.16 Lestoidea, *Perilestes solutus* Williamson & Williamson, 1924. Larval prementum, dorsal view. *C*, prementum cleft. *Original pictures by Neiss & Hamada (2014).*

FIGURE 14.17 Lestoidea, *Lestes* sp. larval prementum, dorsal view. *Original pictures by Neiss & Hamada (2014).*

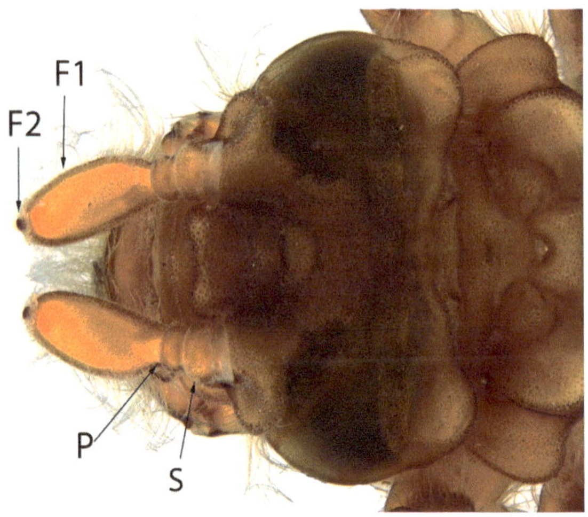

FIGURE 14.18 Gomphoidea, *Praeviosgomphus proprius* Belle, 1995. Larval head, dorsal view. *S*, scape; *F1–2*, flagellomeres 1–2; *P*, pedicel.

FIGURE 14.19 Cordulegastroidea, Cordullidae sp. Larval head dorsal view. *E*, scape *F1–6*, flagellomeres 1–6; *P*, pedicel.

FIGURE 14.20 *Phyllopetalia apollo* Selys, 1878. Detail of larval head, dorsal view. *F1–3*, flagellomeres 1–3.

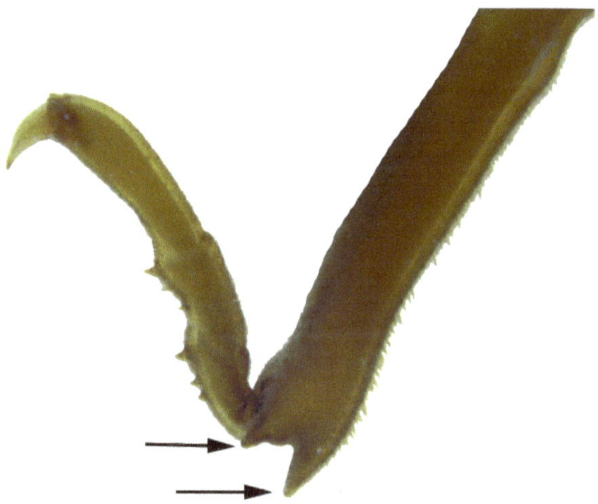

FIGURE 14.21 Petaluroidea, *Phenes raptor* Rambur, 1842. Larval leg. Arrows indicate tibial spines.

FIGURE 14.22 Aeshnoidea, *Rhionaeshna* sp. Larval tibia and tarsus.

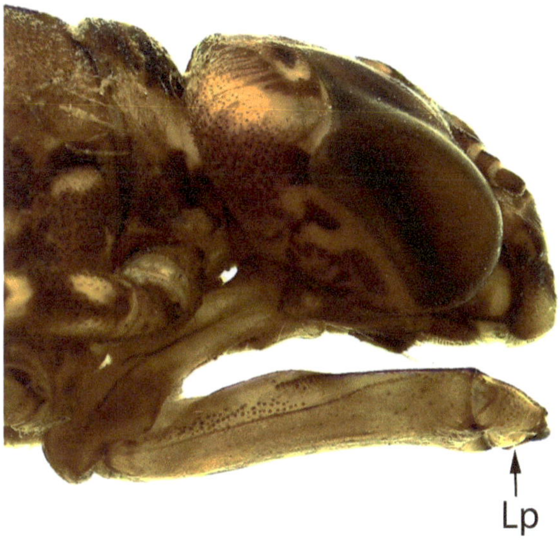

FIGURE 14.23 Aeshnoidea, *Rhionaeshna* sp. Larval head, lateral view. *Lp*, labial palp.

FIGURE 14.24 Libelluloidea, *Rialla villosa* (Rambur, 1842). Larval head, lateral view. *Lp*, labial palp.

FIGURE 14.25 Cordulegastroidea, *Cordulegaster diadema* Selys, 1868. Larval labial palp, inner view. *Mh*, movable hook.

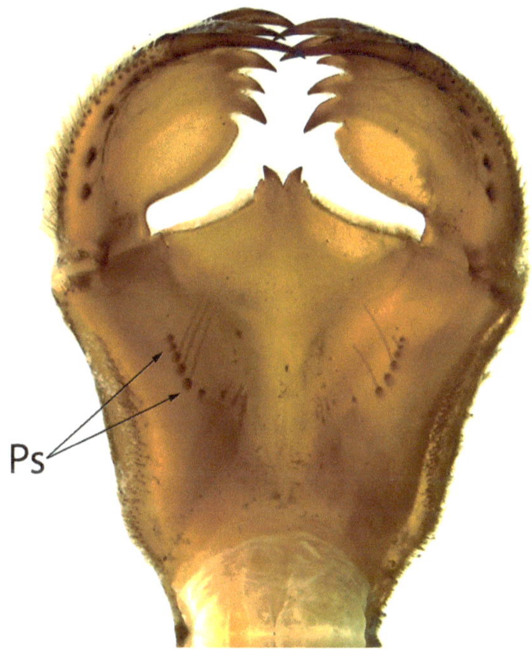

FIGURE 14.26 Cordulegastroidea, *Cordulegaster diadema*. Larval prementum, dorsal view. *Ps*, premental setae

broad and flat (e.g., most Zygoptera as in Figs. 14.7–8; and in Aeshnidae Fig. 14.23), extremely long, petiolated and scoop-shaped (Lestidae, Fig. 14.17), and short and scoop-shaped (e.g., Libellulidae and Cordulegastridae, Figs. 14.24, 27, 29) (Corbet, 1999). Besides these general distinctions, there is a great variation depending on feeding preferences or behavior. According to these preferences there are variations on the number and size of prementum and palpal setae, and different types of structures at the prementum anterior

FIGURE 14.27 Cordulegastroidea, *Neopetalia punctata* (Hagen in Selys, 1854). Larval prementum, dorsal view. Arrow indicates anterior margin with a cleft and dentiform median process

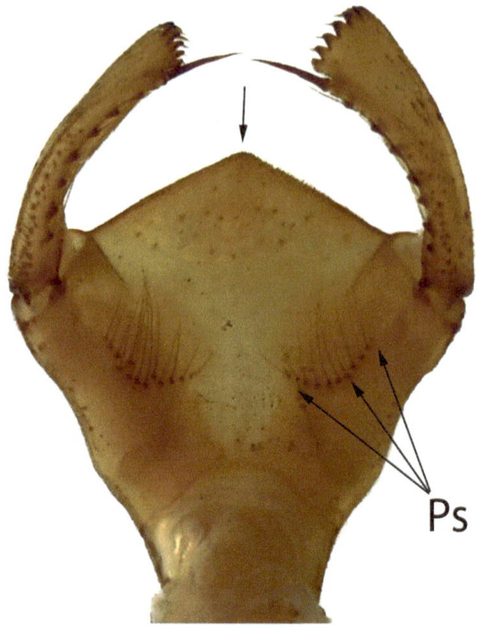

FIGURE 14.29 Libelluloidea, *Rialla villosa*. Larval prementum, dorsal view. *Ps*, premental setae. Arrow indicates absence of cleft and tooth-like median process

FIGURE 14.28 Libelluloidea, *Rialla villosa*. Labial palp.

margin (ligula) and palps' inner and distal margins (e.g., setae, spines, hooks, and crenulations).

Ligula: Anterior margin of prementum, with variable development and shape across the order. The ligula can be more or less triangular (Figs. 14.8, 29), convex, more or less rounded (Figs. 14.5, 7), with tooth-like median process (Fig. 14.26), and with (Figs. 14.5, 7) or without (Figs. 14.8, 29) a medial cleft.

Movable hook: Long, slightly curved hook (Figs. 14.6, 9, 25) developed at the outer distal margin of labial palp and articulated with it. The hook can be variable in length from about half to more than one and a half times the length of labial palp, and be bare or with raptorial setae (e.g., Lestidae).

Palpal setae: Setae developed close to the outer margin of labial palp. They vary in number and size (e.g., *Forcepsioneura*, *Psaironeura*, *Rialla*, Figs. 14.8, 9, 28) or can be absent (e.g., *Palaemnema*, *Paraphlebia*, Figs. 14.6, 9).

Prementum: Distal part of Odonata labium (e.g., Figs. 14.5, 7, 27) which bears the labial palps. It can be very long and slender (e.g., Lestidae, Fig. 14.17), more or less triangular (e.g., *Forcepsioneura*, Fig. 14.8), quadrangular (e.g., *Paraphlebia*, Fig. 14.7, Aeshnidae), etc. It bears (when present) premental setae on its dorsal surface. Its anterior margin is also known as ligula, may bear a prementum median cleft.

Prementum median cleft: Cleft present in the middle of labium distal margin. It can be absent (e.g., some Libellulidae, Coenagrionidae, Figs. 14.8, 29) or well developed (e.g., *Palaemnema*, *Paraphlebia*, *Perilestes*, Figs. 14.5, 7, 16).

TABLE 14.1 Genera of Coenagrionidae Previously Included Within Protoneuridae and Pseudostigmatidae

Genus	Protoneuridae	Pseudostigmatidae
Amazoneura	X	
Anomisma		X
Drepanoneura	X	
Epipleoneura	X	
Epipotoneura	X	
Forcepsioneura	X	
Idioneura	X	
Junix	X	
Lamproneura	X	
Mecistogaster		X
Megaloprepus		X
Microstigma		X
Neoneura	X	
Peristicta	X	
Phasmoneura	X	
Proneura	X	
Protoneura	X	
Psaironeura	X	
Pseudostigma		X
Roppaneura	X	

Premental setae: Setae developed at the dorsal surface of prementum. These vary in number and size (e.g., *Forcepsioneura*, *Cordulegaster*, *Rialla* Figs. 14.8, 26, 29), or can be absent (e.g., *Perilestes*, *Neopetalia*, Figs. 14.16, 27) (Table 14.1).

S1–9: Abbreviation for abdominal segments 1–9.

MATERIAL PREPARATION AND PRESERVATION

In Volume I (Thorp & Rogers, 2015) of this series, specimen preparation and preservation are discussed and to some extent, also in the odonate chapters of the present volume.

KEYS TO ODONATA

Superfamilies with low diversity and poorly represented in the Neotropical region (1–2 genera) do not require a separate chapter and a key of their own, therefore, they are included in the keys presented in this chapter.

Platystictoidea includes only one family, Platystictidae, distributed in the Oriental and Neotropical regions, where it has one genus (*Palaemnema*) and 43 species (Garrison et al., 2006). It is distributed in central and northern South America, but a few species enter North America (Garrison et al., 2006). The larva of six species are known, with all living in fast current streams with rocky bottom (Novelo-Gutiérrez and González-Soriano, 1986; Novelo-Gutiérrez, 2003; Neiss & Hamada, 2016).

Cordulegastroidea is a relatively small taxon, containing three families: Chlorogomphidae, Cordulegastridae, and Neopetaliidae, but only the last two are recorded in the Neotropical region. Cordulegastridae is a Holarctic family represented by 45 species in three genera. One genus (*Cordulegaster*) and 10 species are recorded in the New World, but only one enters Central America and is present from Mexico to Costa Rica. Larvae are characterized by living in forested streams where they bury themselves in soft sandy substrate (Garrison et al., 2006). Neopetaliidae is a monotypic family, endemic to the subantarctic forest of southern Argentina and Chile, with a wider distributional range in Chile than in Argentina (Muzón et al., 2014). The larva of *Neopetalia punctata* has been described (Carle & Louton, 1994), and it inhabits small streams within *Nothofagus* forests, burrowing shallowly in their muddy bottom (Carle & Louton, 1994).

Petaluroidea includes only the family Petaluridae which consist of very large dragonflies. It includes five genera and 11 species and has a transpacific distribution pattern, being present in Australia, New Zealand, North America and southern South America (Garrison et al., 2006). *Phenes raptor* is the only representative of the family in South America; it inhabits the subantarctic forest of southern Argentina and Chile (Muzón et al., 2014). Larvae of this species was described by Schmidt (1941) and Needham & Bullock (1943), they have terrestrial habits, being found in the moist forest ground and seepages (Garrison & Muzón, 1995). The biology of this species is well documented by Baird (2013).

Odonata: Suborders

1 Abdomen slender (Fig. 14.3, 15), ending in three large caudal appendages of variable shape (Figs. 14.3, 4, 10, 11, 15) .... **Zygoptera [p. 364]**

1' Abdomen robust (Fig. 14.1), ending in an anal pyramid (epiproct plus paraprocts) (Fig. 14.2) .... **Anisoptera [p. 364]**

Odonata: Zygoptera: Superfamilies

1 Prementum not distinctly widened on its basal half (Figs. 14.7, 8), labial palp with end hook and one or more additional teeth (Fig. 14.9), caudal appendages variable 2

1' Prementum distinctly widened on its basal half (Fig. 14.5), labial palp with only one robust and truncated tooth (end hook) (Fig. 14.6), caudal appendages saccoid (Fig. 14.11) .... Platystictoidea, one genus: *Palaemnema*

2(1) Caudal appendages lamellate, held nearly vertically (Figs. 14.13, 15), prementum cleft present (Lestoidea, Fig. 14.16) or entire (Coenagrionoidea, Fig. 14.8), labium may be petiolate and elongate, extending beyond metacoxae (Fig. 14.17) or may be normally developed 3

2' Caudal appendages triquetral (Fig. 14.10), saccoid (Figs. 14.4, 11) or greatly modified (Fig. 14.12), longer (Fig. 14.3) or shorter than body (Fig. 14.14). Prementum cleft, labium not elongate .... **Calopterygoidea** (Chapter 14.4, p. 499)

3(2) Prementum cleft, labium normal (Fig. 14.16) or petiolate and elongate, extending beyond metacoxae (Fig. 14.17) .... **Lestoidea** (Chapter 14.5, p. 469)

3' Prementum entire, labium normal (Fig. 14.8) .... **Coenagrionoidea** (Chapter 14.6, p. 475)

Odonata: Anisoptera: Superfamilies

1 Antennae with scape, pedicel, and 3–6 flagellomeres, the first < half antenna length (Figs. 14.19, 20) 2

1' Antennae with scape, pedicel, and two flagellomeres, the second reduced or vestigial, the first > half antenna length (Fig. 14.18) .... **Gomphoidea** (Chapter 14.2, p. 377)

2(1) Tibiae usually without robust apical spines (Fig. 14.22), if present (*Hypopetalia pestilens*), labial palp with several tooth on its inner margin (Fig. 14.14); body usually not robust and big; rarely terrestrial 3

2' Tibiae with robust apical spines for digging (Fig. 14.21); labial palp without tooth on its inner margin; body very robust length to >45 mm; terrestrial and aquatic .... Petaluroidea, Petaluridae, one species: *Phenes raptor* Rambur, 1842 [Patagonia]

3(2) Prementum spatulate, labial palp at rest covering anterior part of head (Fig. 14.24) 4

3' Prementum flat, labial palp not covering anterior part of head at rest (Fig. 14.23) .... **Aeshnoidea** (Chapter 14.1, p. 367)

4(3) Labial palpi with long and irregular teeth (Fig. 14.25), distal margin of prementum with a cleft and dentiform median process (Figs. 14.26, 27) .... **Cordulegastroidea [p. 364]**

4' Labial palps variable, usually without long and irregular teeth (Fig. 14.28); prementum distal margin entire and not dentiform (Fig. 14.29) .... **Libelluloidea** (Chapter 14.3, p. 399)

Odonata: Anisoptera: Cordulegastroidea: Families

1 Prementum and palpi without setae (Fig. 14.27) .... Neopetaliidae, one species: *Neopetalia punctata* (Hagen in Selys, 1854) [Patagonia]

1' Prementum and palpi with setae (Figs. 14.25, 26) .... Cordulegastridae, one genus; *Cordulegaster* [Costa Rica. Holarctic]

ACKNOWLEDGMENTS

We express our most sincere gratitude to Drs. Rodolfo Novelo-Gutiérrez, Enrique González-Soriano, and Natalia von Ellenrieder for the loan of larvae material and to Dr. Miguel Archangelsky for allowing us to use his microscope and digital camera. The Fig. 14.3, 6, 9, and 13–17 were originally published by Neiss & Hamada (2014) and were generously given by Editora do Instituto Nacional de Pesquisas da Amazônia, Manaus.

REFERENCES

Baird, I.R.C. 2013. Larval habitat and behaviour of Phenes raptor (Odonata: Petaluridae): a review of current knowledge, with new observations. International Journal of Odonatology 16:79–91.

Bybee, S.M., T.H. Ogden, M.A. Branham & M.F. Whiting. 2008. Molecules, morphology and fossils: a comprehensive approach to odonate phylogeny and the evolution of the odonate wing. Cladistics 24: 477–514.

Bybee, S.M., Q. Hansen, S. Büsse, H.M. Cahill Wightman & M. Branham. 2015. For consistency's sake: the precise use of larva, nymph and naiad within Insecta. Systematic Entomology 40: 667–670.

Carle, F.L. & J.A. Louton. 1994. The larva of Neopetalia punctata and establishment of Austropetaliidae Fam. Nov. (Odonata). Proceedings of the Entomological Society of Washington 96: 147–155.

Corbet, P.S. 1999. Dragonflies: Behavior and Ecology of Odonata. Harley Books, Colchester.

Corbet, P.S. 2002. Stadia and growth ratios of Odonata: a review. International Journal of Odonatology 5: 45–73.

Costa, J.M., L.O.I. Souza, A.M. Lourenço, B.B. Oldrini. 2004. Chave para as famílias e gêneros das larvas de Odonata citadas para o Brasil: Comentários e Registros Bibliográficos. Publicações Avulsas do Museu Nacional 99: 1–44.

Davis, R.B., D.B. Nicholson, E.L.R. Saunders & P.J. Mayhew. 2011. Fossil gaps inferred from phylogenies alter the apparent nature of diversification in dragonflies and their relatives. Evolutionary Biology 11: 252–261.

Dijkstra, K.-D. B., G. Bechly, S.N. Bybee, R.A. Dow, H.J. Dumont, N. von Ellenrieder, G. Fleck, R.W. Garrison, M. Hamalainen, V.J. Kalkman, H. Karube, M.L. May, A.G. Orr, D.R. Paulson, A.C. Rehn, G. Theischinger, J.W.H. Trueman, J. van Tol & J. Ware. 2013. The classification and diversity of dragonflies and damselflies (Odonata). Pages 36–45 in: Z.-Q. Zhang (ed.), Animal Biodiversity: An Outline of Higher-level Classification and Survey of Taxonomic Richness (Addenda 2013). Zootaxa 3703.

Fleck, G. 2011. Phylogenetic affinities of Petaluridae and basal Anisoptera families (Insecta: Odonata). Stuttgarter Beiträgezur Naturkunde A, NeueSerie 4: 83–104.

Fleck, G. & M. El Adouzi. 2013. The larva of the genus Palaeosynthemis Forster, 1903 (Odonata: Anisoptera: Synthemistidae) and a generic key to the larvae of non-New Caledonian Synthemistidae. Zootaxa 3619: 589–594.

Förster, S. 2001. The Dragonflies of Central America Exclusive of Mexico and the West Indies. 2nd edition. Gunnar Rehfeldt, Braunschweig, Germany.

Garrison, R.W. & J. Muzón. 1995. Collecting down at the other "down under". Argia 7: 23–26.

Garrison, R.W., N. von Ellenrieder, L.J.A. Louton. 2006. Dragonfly Genera of the New World – An illustrated and annotated key to the Anisoptera. The Johns Hopkins University Press, Baltimore.

Garrison, R.W., N. von Ellenrieder, L.J.A. Louton. 2010. Damselfly Genera of the New World: An Illustrated and Annotated Key to the Zygoptera. The Johns Hopkins University Press, Baltimore, Maryland.

Hovmöller, R., T. Pape & M. Källersjö. 2002. The Paleoptera problem: basal pterygote phylogeny inferred from 18S and 28S rDNA sequences. Cladistics 18: 313–323.

Kalkman, V.J., V. Clausnitzer, K-D.B. Dijkstra, A.G. Orr, D.R. Paulson & J. van Tol. 2008. Global diversity of dragonflies (Odonata) in freshwater. Hydrobiologia 595: 351–363.

Lencioni, F.A.A. 2005. Damselflies of Brazil - An illustrated identification guide - 1-Non-Coenagrionidae families. All Print Editora, São Paulo.

Lencioni, F.A.A. 2006. Damselflies of Brazil - An Illustrated identification guide - 2-Coenagrionidae. All Print Editora, São Paulo.

Morrone, J.J. 2014. Biogeographical regionalization of the Neotropical region. Zootaxa 3782: 001–110.

Muzón, J. & F. Lozano. 2016. Is it a question of consistency? A reply to Bybee et al. and Sahlén et al. Systematic Entomology 41: 695–696.

Muzón, J., P. Pessacq & F. Lozano. 2014. The Odonata (Insecta) of Patagonia: A synopsis of their current status with illustrated keys for their identification. Zootaxa 3784: 346–388.

Needham, J.G. & D.S. Bullock. 1943. The Odonata of Chile. Zoological Series of Field Museum of Natural History 24: 357–373.

Neiss, U.G. & N. Hamada. 2014. Ordem Odonata ([capítulo] 14). Pages 217–284 in: Hamada, N., J.L. Nessimian & R.B. Querino (eds.). Insetos aquáticos na Amazônia brasileira: taxonomia, biologia e ecologia. Editora do INPA, Manaus.

Neiss, U.G. & N. Hamada. 2016. Larva of Palaemnema brasiliensis Machado (Odonata: Platystictidae), from Amazonas, Brazil. Zootaxa 4078: 070–077.

Novelo-Gutiérrez, R. 1997. Clave para la separación de familias y géneros de las náyades de Odonata de México. parte 1. Zygoptera. Dugesiana 4: 1–10.

Novelo-Gutiérrez, R. 2003. The larva of Palaemnema domina Calvert, 1903 (Odonata: Platystictidae). Transactions of the American Entomological Society 129: 71–75.

Novelo-Gutiérrez, R. & E. González-Soriano. 1986. Description of the larva of Perissolestes magdalenae (Williamson & Williamson, 1924) (Zygoptera: Perilestidae). Odonatologica 15: 129–133.

Odgen, T.H. & M.F. Whiting 2003. The problem with "the Paleoptera Problem": sense and sensitivity. Cladistics 19: 432–442.

Ramírez, A. 2010. Odonata. Macroinvertebrados de agua dulce de Costa Rica I. Vol. I. Revista de Biología Tropical 58: 97–136.

Rédei, D. & P. Štys. 2016. Larva, nymph and naiad – for accuracy's sake. Systematic Entomology 41: 505–510.

Rehn, A.C. 2003. Phylogenetic analysis of higher-level relationships of Odonata. Systematic Entomology 28: 181–239.

Sahlén, G., F. Suhling, A. Martens, S.N. Gorb & O.M. Fincke. 2016. For consistency's sake? A reply to Bybee et al. Systematic Entomology 41: 307–308.

Schmidt, E. 1941. Petaluridae, Gomphidae und Petaliidae der Schönemannschen Sammlungaus Chile (Ordnung Odonata). Archiv für Naturgeschichte 10: 231–258.

Suhling, F., G. Sahlén, S. Gorb, V.J. Kalkman, K-D.B. Dijkstra & J. von Tol. 2015. Order Odonata, Chapter 35. Pages 893–932 in: J.H. Thorp & D.C. Rogers (eds.). Ecology and General Biology Thorp and Covich's Freswater Invertebrates, Vol. 1, 4th Edition, Academic Press-Elsevier.

Thorp, J.H. & D.C. Rogers (eds.). 2015. Ecology and General Biology. Vol. I in Thorp and Covich's Freshwater Invertebrates, 4th Edition, Academic Press-Elsevier.

Tillyard, R.J. 1917. The Biology of Dragonflies. Cambridge University Press, Cambridge.

Trapero-Quintana A.D. & J.C. Naranjo-López. 2014. Clave de identificación para los adultos de las especies del orden Odonata presentes en Cuba. Boletín de la Sociedad Entomológica Aragonesa 1: 171–180.

Trueman, J.W.H. 2007. A brief history of the classification and nomenclature of Odonata. Zootaxa 1668: 381–394.

von Ellenrieder, N. 2009. Databasing dragonflies: State of knowledge in the Neotropical region. Agrion 13: 58–72.

Wallace, A.R. 1876. The Geographical Distribution of Animals; With a Study of the Relations of Living and Extinct Faunas as Elucidating Past Changes of the Earth's Surface. Macmillan & Company, London.

Superfamily Aeshnoidea

Alcimar do Lago Carvalho
Departamento de Entomologia, Museu Nacional, UFRJ, Rio de Janeiro, Rio de Janeiro, Brazil

Pablo Pessacq
Centro de Investigaciones Esquel de Montaña y Estepa Patagónicas (CIEMEP), Esquel, Chubut, Argentina

Chapter Outline

INTRODUCTION

The superfamily Aeshnoidea, as currently considered (e.g., Dijkstra et al., 2013), is composed of two families with very unequal species richness and distribution. These are the highly diverse and worldwide Aeshnidae (469 species in 54 genera) and the Austropetaliidae, with its low diversity (10 species in 4 genera) and distribution restricted to the Australian region and southern South America. Previously, the families Gomphidae and Petaluridae were included within this superfamily, as shown in many traditional catalogs and general books on the order (e.g., Davies & Tobin, 1985; Silsby, 2001). In the most recent and comprehensive phylogenetic analyses of Anisoptera, the clade composed of Aeshnidae and Austropetaliidae is always justified as monophyletic (e.g., Bybee et al., 2008; Carle et al., 2015). In the analysis of Fleck et al. (2008), the Petaluridae was considered as its sister group. This hypothesis was confirmed afterwards based on larval characters (Fleck, 2011), with that monophyletic group consisting of three families named Siphonoprocta, which is a taxon not currently positioned in any classificatory hierarchical level.

The diversity of Aeshnidae in the New World is currently listed as 154 species in 20 genera (16 endemic). Fifteen genera are present in the Neotropics, with 13 predominantly having a Neotropical distribution. In total, 133 species can be considered truly Neotropical, corresponding to 28% of the diversity of the family. Descriptions of the ultimate instar larvae are available currently for 67 species, including representatives of all 15 Neotropical genera. Such literature is variable in quality and accuracy, in part because there are examples of questionable attributions for species and genera not reared in the laboratory from larvae to adult, hence, a revision of that literature is needed.

Austropetaliidae is a small family of large-sized and rare dragonflies distributed in South America in southern Chile and Argentina (Patagonia). The genera *Hypopetalia* and *Phyllopetalia* and seven species are endemic to the subantarctic forests of South America. Until now, no phylogenetic analyses on this small group have been conducted, but its monophyly and relation with Aeshnidae are clear, based on morphological (Garrison et al., 2006) and molecular data (Misof et al., 2001). South American austropetalids inhabit small streams and rivulets in *Nothofagus* (beech) forests. Use of semiterrestrial habitat have been reported by Carle (1996) and Baird (2013), while Pessacq & Brand (2009) reported larvae of *Hypopetalia* and *Phyllopetalia* in crevices of small falls, clinging under logs, or between pebbles and debris.

Aeshnidae is an important and cosmopolitan family of large-sized dragonflies, it is comprised of approximately 469 species in 54 genera, based on the World Odonata List (http://www.pugetsound.edu/academics/academic-resources/slater-museum/biodiversity-resources/dragonflies/world-odonata-list2/). The published numbers of Dijkstra et al.

Thorp and Covich's Freshwater Invertebrates, Fourth Edition. https://doi.org/10.1016/B978-0-12-804223-6.00015-9

(2013), 51 genera and 456 species, are long out of date since they were computed only to 2010 at publication. Since then three new genera and 29 species were described, and a number of synonyms were proposed.

Traditionally, aeshnid genera are arranged in five distinct groups at the tribe level and included within two subfamilies: Aeshnini, Anactini, and Gynacanthini within Aeshninae; and Brachytronini and Gomphaeschnini within Brachytroninae (e.g., Silsby, 2001). Although the family as a whole has been justified as monophyletic, analyses considering both morphology and molecular data (e.g., von Ellenrieder, 2002; Fleck et al., 2008; Carle et al., 2015) have shown that most of these taxa do not correspond to natural groups and need some revision. The morphological analysis of von Ellenrieder (2002) for the family is unique in its inclusion of larval information, abridging 10 of the 58 characters selected for the data matrix (ca. 17%). Few insights on the phylogeny and biogeography of some groups and genera were proposed based on the larvae (e.g., De Marmels, 2000; Fleck, 2011).

Ecologically, aeshnid larvae are primarily climbers, dwelling on vascular hydrophytes of mainly lentic habitats, such as lakes, swamps, marshes, etc. (Carvalho & Nessimian, 1998). Representatives found in lotic habitats tend to avoid riffles. The duration of the larval cycle of the species from Neotropics is practically unknown, with different taxa characterized by annual or even semiannual voltinism (see Carvalho, 1992; Santos, 1970; Santos, 1981). However, most species probably have much shorter cycles (Carvalho, 1989; Carvalho & Ferreira-Jr, 1989; Calil & Carvalho, 1999).

LIMITATIONS

There are few papers on the identification of the genera of Aeshnidae larvae from Neotropics. They are mainly regional, incomplete, and partially based on unreared larvae, and thus identified through supposition. Some of the comprehensive keys available, including most of the occurring genera, are those included in Carvalho (1989), Costa et al. (2004), Heckman (2006) and Neiss & Hamada (2014). Regarding Austropetaliidae, there is a key that includes the known larva from Patagonia that separates *Hypopetalia* from *Phyllopetalia* (Muzón et al., 2014), but an accurate generic characterization of the larva of *Phyllopetalia* is still subject to the description of the remaining species in the genus.

TERMINOLOGY AND MORPHOLOGY

The most important terminology for identifying these dragonflies are described in Chapter 14.0 in this book and Chapter 35 (Suhling et al., 2015) in Volume 1 of this series (Thorp & Rogers, 2015).

MATERIAL PREPARATION AND PRESERVATION

In Volume I (Thorp & Rogers, 2015) of this series specimen preparation and preservation are presented.

KEYS TO AESHNOIDEA

Distributional data for each genus in Neotropical region is indicated within brackets (based on Garrison et al., 2006).

Aeshnoidea: Families

1 Body surface granulated (covered with short scaliform setae composed of two parts) (Fig. 14.1.1); S1–10 with lateral lobes or very wide spines (Fig. 14.1.1); antennal flagellum with three articles (Fig. 14.1.2); labial palp with end hook poorly developed (Fig. 14.1.4) or indistinct (Fig. 14.1.3) .... **Austropetaliidae [p. 368]**
[Argentina, Chile]

1' Body surface not granulated (covered by cylindrical, simple setae), S6–9 (rarely S5–9, S5–10 or S7–9) with small and acute lateral spines (Figs. 14.1.7, 16, 18); antennal flagellum with four or five articles; labial palp with a small (Fig. 14.1.14) to a large end hook (Figs. 14.1.12, 14)........ **Aeshnidae [p. 369]**
[Widespread]

Aeshnoidea: Austropetaliidae: Genera

Only the larvae of *H. pestilens* McLachlan, 1870 and *P. apollo* Selys, 1878 are known with certainty (Schmidt, 1941; Pessacq & Brand, 2009), and the larva of *P. stictica* Hagen in Selys, 1858 was described by supposition (Schmidt, 1941). The larva of this family possesses distinctive lateral lobes of the abdominal segments, having a serrated aspect. The body surface is covered with scale-like setae and the prementum anterior margin possesses a deep and very narrow or closed cleft, structures that varies among genera and species.

1 Prementum with tuberculate processes adjacent to median cleft (Fig. 14.1.3); postclypeal knobs present; tibia apex with a dense concentration of setae (Fig. 14.1.5) .... *Phyllopetalia*
[Argentina, Chile]

1' Prementum tuberculate processes positioned some distance of median cleft (Fig. 14.1.4); postclypeal knobs absent; tibia apex without a dense concentration of setae .... *Hypopetalia*
[Argentina, Chile]

Aeshnoidea: Aeshnidae: Genera

Aeshnids are huge dragonflies with very characteristic larvae, relatively uniform in external morphology. Currently, descriptions of the ultimate instar larvae are available for 65 species, including representatives of all 16 genera occurring in Neotropics, covering ca. of 49% of the diversity. Most species can be diagnosed by the: angle between distal and internal borders of labial palp lamina forming an end hook; movable hook with a row of dorsal setae; ventral junction of metathoracic pleurae forming a discriminal line; epiproct apex excavated or truncated medially.

This key was designed mainly for ultimate instar larvae, in which wing pad length is always longer than the head width (Assis et al., 2000; Carvalho, 2007). When available, two characters for each step were selected, preferably based on structures easily observable in both sexes and related with distinct

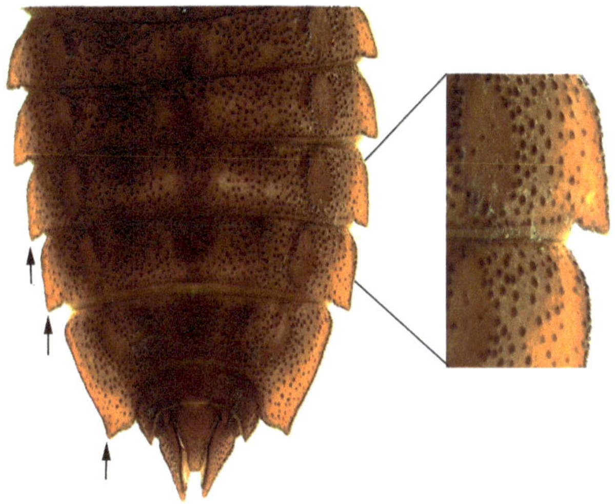

FIGURE 14.1.1 *Phyllopetalia apollo* Selys, 1878, larval abdomen S5–10, dorsal view. *Arrows* indicate lateral lobes. Detail shows granulated surface.

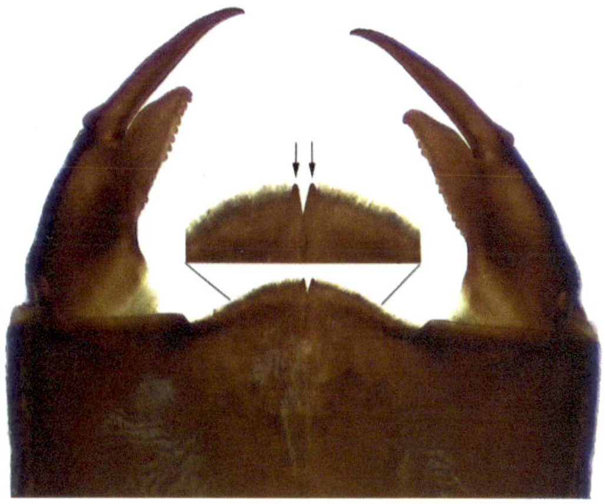

FIGURE 14.1.3 *Phyllopetalia Apollo* Selys, 1878, larval prementum, anterior margin and palps, dorsal view, and detail of prementum anterior margin. *Arrows* indicate tuberculate processes.

FIGURE 14.1.2 *Phyllopetalia Apollo* Selys, 1878, larval left antenna, dorsal view. F1–3: flagellomeres 1–3.

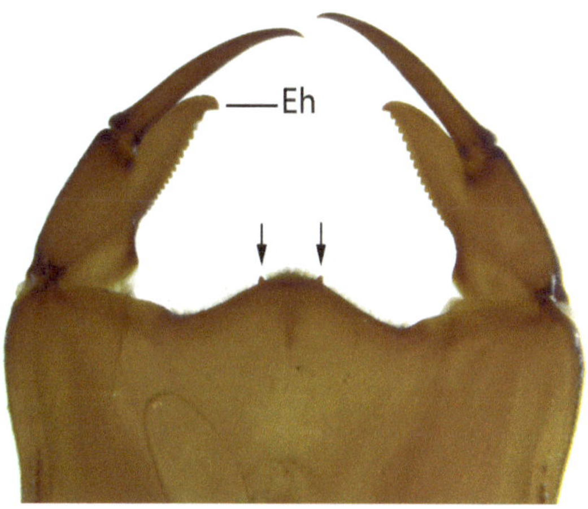

FIGURE 14.1.4 *Hypopetalia pestilens* McLachlan, 1870, larval prementum anterior margin and palps, dorsal view. *Eh*, end hook. *Arrows* indicate tuberculate processes.

body regions. When "ex." is indicated before a given figure citation, means that the image does not represent individuals included in the respective step of the key, but it undoubtedly exemplifies the condition of the character.

1 Femora basal half with a dorsal longitudinal line of small dark spots (Fig. 14.1.6); S10 with lateral spines (Fig. 14.1.7)...... 2

1' Femora basal half without a dorsal longitudinal line of small dark spots; S10 without lateral spines (ex. Figs. 14.1.16, 18)...... 3

2(1) Head and thorax dorsal surface without tubercles (Fig. 14.1.6); S9 and S10 without dorsal hooks (ex. Figs. 14.1.16, 18)...... *Neuraeschna*
[S Central America to N South America]

2' Head and thorax dorsal surface with tubercles (Fig. 14.1.8); S9 and S10 with small dorsal hooks...... *Staurophlebia*
[S Central America to N South America]

3(1) Head posteriolateral angles rounded (Fig. 14.1.9); prementum anterior margin on each side of median cleft, with a small spiniform or tuberculate process, smaller or barely longer than the fringe of piliform setae, sometimes poorly defined (ex. Fig. 14.1.10)...... 4

3' Head posteriolateral angles angulate (Fig. 14.1.11); prementum anterior margin on each side of median cleft, with an acute spiniform process, noticeably longer than the fringe of piliform setae (Fig. 14.1.12)...... *Coryphaeschna*
[Central America and Caribbean to N South America. Nearctic]

4(3) Palpal lobe dorsal surface with short or long setae organized in a row (Figs. 14.1.13–14)...... 5

4' Palpal lobe dorsal surface without setae or with small scattered ones, not organized in a row (ex. Fig. 14.1.12)...... 7

5(4) Prementum anterior margin with an obtuse tuberculate process on each side of the median cleft, distinctly shorter than the fringe of piliform setae, sometimes poorly defined (ex. Fig. 14.1.10); palpal setae long (at least the distal one), generally longer than width of the blade (Fig. 14.1.13)...... 6

5' Prementum anterior margin with an acute spiniform process on each side of the median cleft, slightly longer than the fringe of piliform setae; row of palpal setae composed by very short unities...... *Gynacantha* (in part)
[Central America and Caribbean to N South America. Nearctic]

6(5) Palpal setae slender, very unequal in length, diminishing gradually from distal to proximal (Fig. 14.1.13); S5 without lateral spines (ex. Fig. 14.1.16)...... *Gynacantha* (in part)
[Central America and Caribbean to N South America. Nearctic]

6' Palpal setae robust, nearly uniform in length, at least four distals (Fig. 14.1.14); S5 with short lateral spines (ex. Fig. 14.1.7)...... *Triacanthagyna*
[Central America and Caribbean to N South America. Nearctic]

7(4) Labium when folded reaching posteriorly at most the level of the second coxae (Fig. 14.1.15); S6 with short lateral spines (Fig. 14.1.16)...... 8

7' Labium when folded reaching posteriorly to the level of the third coxae or S1 (Fig. 14.1.17); S6 without lateral spines (Fig. 14.1.18)...... *Anax*
[Cosmopolitan]

8(7) Epiproct acuminate, apex pointed (Fig. 14.1.19)...... 9

8' Epiproct dorsal surface concave, with the apex truncate, trifid or bifid (Figs. 14.1.20–22, respectively)...... 11

9(8) Prothoracic supracoxal processes high, digitiform or pointed (Fig. 14.1.23); S5 with short lateral spines (ex. Fig. 14.1.7)...... 10

9' Prothoracic supracoxal processes low and blunt (Fig. 14.1.24); S5 without lateral spines (Fig. 14.1.16)...... *Rhionaeshna* (in part)
[Neotropics, except NE South America. Nearctic]

10(9) Prementum approximately as long as wide; movable hook similar in length to adjacent external margin of palp (ex., Fig. 14.1.14)...... *Allopetalia*
[Colombia, Ecuador, Peru, Venezuela, Chile]

10' Prementum distinctly longer than wider (Fig. 14.1.15); movable hook distinctly longer to adjacent external margin of palp (Fig. 14.1.25)...... *Racenaeschna*
[Venezuela]

11(8) Prementum anterior margin straight or slightly concave (Fig. 14.1.27); S5 with short lateral spines (ex., Fig. 14.1.7)...... 12

11' Prementum anterior margin convex (Fig. 14.1.10); S5 without lateral spines (ex., Fig. 14.1.16)...... 13

12(11) Movable hook similar in length to adjacent external margin of palp or shorter (ex. Fig. 14.1.14); cerci nearly as long as the half of epiproct (ex., Fig. 14.1.18)...... *Oplonaeschna*
[Mexico, Guatemala. Nearctic]

12' Movable hook distinctly longer than adjacent external margin of palp (Fig. 14.1.26); cerci nearly as long as the epiproct (Fig. 14.1.27)...... *Limnetron*
[Brazil, Peru, Argentina, Paraguay]

13(11) Cerci nearly as long as the epiproct or shorter (Fig. 14.1.28); epiproct with the apex variable, generally truncate with two or three short spines at apex (Figs. 14.1.20–21, respectively)...... 14

13' Cerci slightly longer than the epiproct (Fig. 14.1.29); epiproct apically bifid (Fig. 14.1.22)...... *Castoraeschna*
[Widespread in South America]

14(13) Cerci nearly as long as the epiproct (Fig. 14.1.20); paraprocts patterned with a subapical dark spot (Fig. 14.1.30)...... 15

14' Cerci distinct shorter than the epiproct (Fig. 14.1.28); paraprocts not patterned with a subapical dark spot (Fig. 14.1.31)...... *Aeshna*/*Andaeschna*/*Rhionaeschna* (in part)
[Mexico to Panama. Holarctic]/[northern South America]/[Widespread, except NE South America. Nearctic]

15(14) Prementum median cleft closed distally (Fig. 14.1.32); femora subapical dark ring when distinct, regular...... *Remartinia*
[Mexico to Argentina, except NE Brazil. Nearctic]

15' Prementum median cleft opened distally (Fig. 14.1.10); femora subapical dark ring encloses darker spots...... *Coryphaeschna*
[Central America to Argentina. Nearctic]

FIGURE 14.1.5 *Phyllopetalia apollo* Selys, 1878, larval distal tibia and tarsus, lateral view. *Arrows* indicate dense concentration of setae.

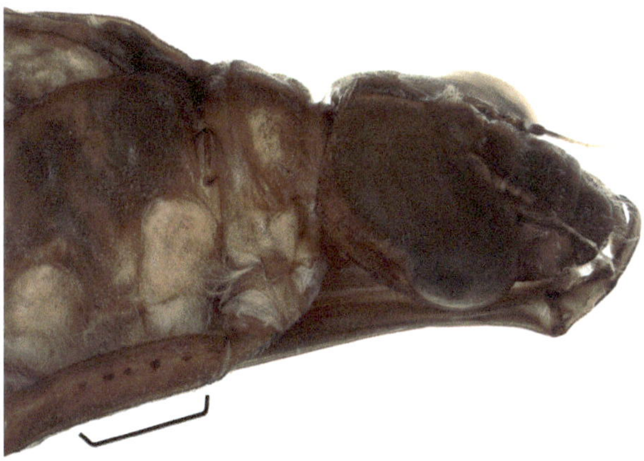

FIGURE 14.1.6 *Neuraeschna* sp., larval head and thorax, latero-dorsal view. *Arrows* indicate longitudinal line of small dark spots. *Original pictures by Neiss & Hamada (2014).*

FIGURE 14.1.7 *Staurophlebia reticulata* (Burmeister, 1839), larval abdomen S5–10 and anal appendages, dorsal view. *Arrows* indicate spines in segments 5 and 10.

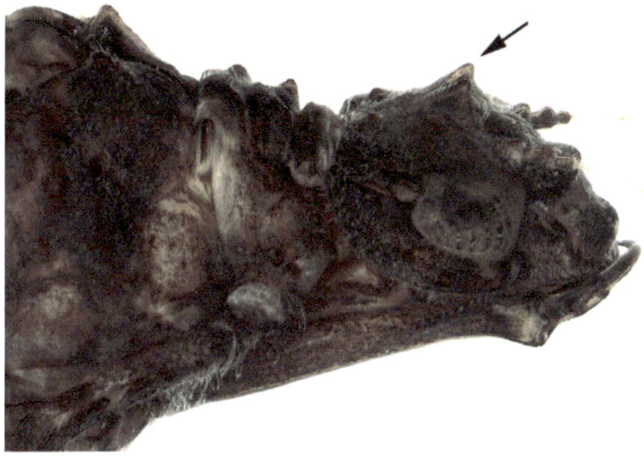

FIGURE 14.1.8 *Staurophlebia* sp., larval head and thorax, latero-dorsal view. *Arrow* indicate head tubercles. *Original pictures by Neiss & Hamada (2014).*

FIGURE 14.1.9 *Gynacantha gracilis* (Burmeister, 1839), larval head, dorsal view. *Original pictures by Neiss & Hamada (2014).*

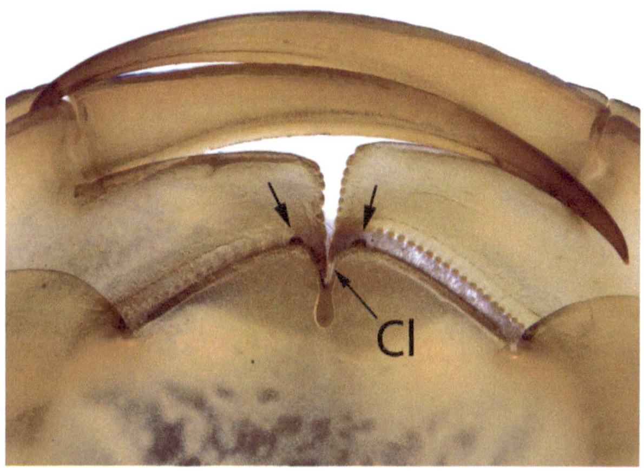

FIGURE 14.1.10 *Coryphaeschna adnexa* (Hagen, 1861), larval prementum anterior margin and palps, dorsal view. *Arrows* indicate tuberculate processes. *Cl*, prementum median cleft.

FIGURE 14.1.11 *Coryphaeschna* sp., larval head, dorsal view. *Original pictures by Neiss & Hamada (2014).*

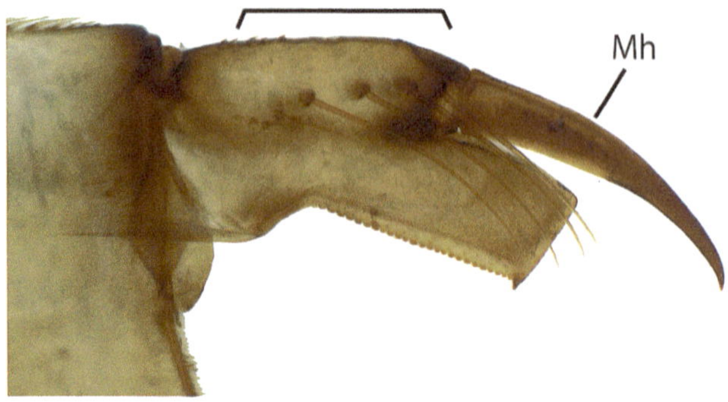

FIGURE 14.1.14 *Triacanthagyna* sp., larval left palp, dorsal view. *Arrows* indicate row of palpal setae. *Black line* indicates length of external margin of palp. *Mh*, movable hook. *Original pictures by Neiss & Hamada (2014).*

FIGURE 14.1.12 *Coryphaeschna* sp., larval prementum anterior margin and left palp, dorsal view. *Original pictures by Neiss & Hamada (2014).*

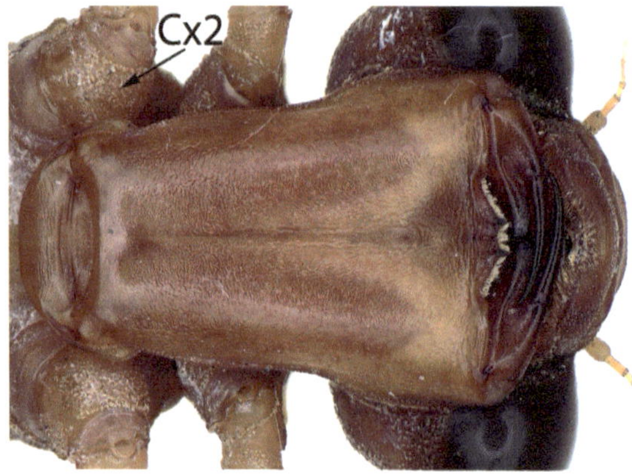

FIGURE 14.1.15 *Racenaeschna* sp., larval folded labium, ventral view. *Cx2*, coxa 2.

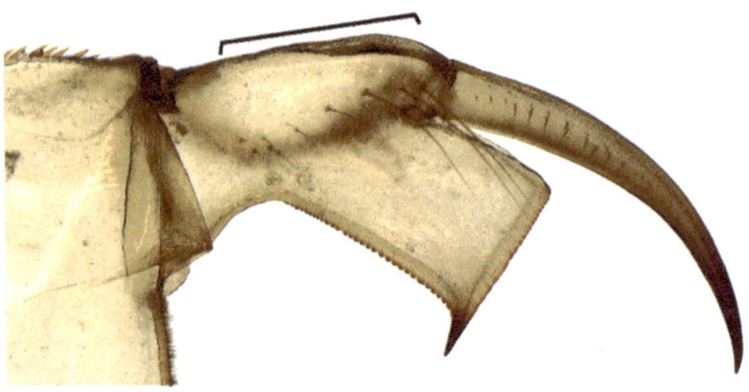

FIGURE 14.1.13 *Gynacantha* sp., larval left palp, dorsal view. *Arrows* indicate row of palpal setae. *Original pictures by Neiss & Hamada (2014).*

FIGURE 14.1.16 *Rhionaeschna itatiaia* (Carvalho & Salgado, 2004), larval abdomen S5–10 and anal appendages, dorsal view. *Arrow* indicate spine in abdominal segment 6. 5: abdominal segment 5, 10: abdominal segment 10.

FIGURE 14.1.17 *Anax amazili* (Burmeister, 1839), larval folded labium, ventral view. *Cx2*: coxa 2, *Cx3*: coxa 3.

FIGURE 14.1.18 *Anax amazili* (Burmeister, 1839), larval abdomen S5–10, dorsal view. *Ce*, cercus; *Ep*, epiproct.

FIGURE 14.1.19 *Rhionaeschna itatiaia* (Carvalho & Salgado, 2004), larval anal appendages, dorsal view. *Ep*, epiproct.

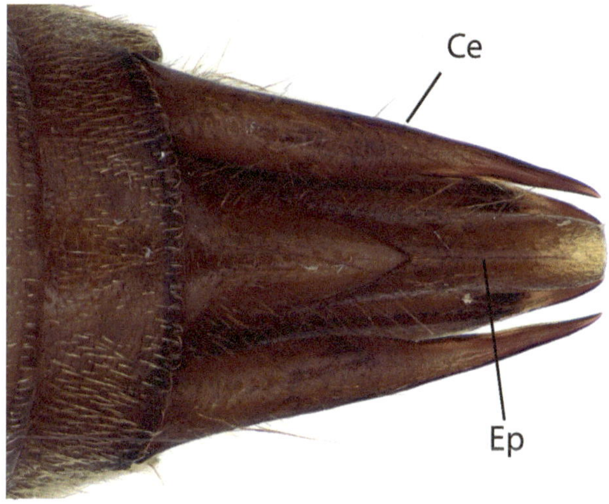

FIGURE 14.1.20 *Remartinia luteipennis luteipennis* (Burmeister, 1839), larval anal appendages, dorsal view. *Ce*, cercus; *Ep*, epiproct.

FIGURE 14.1.21 *Rhionaeschna brasiliensis* (von Ellenrieder & Costa, 2002), larval anal appendages, dorsal view.

FIGURE 14.1.22 *Castoraeschna castor* (Brauer, 1865), larval anal appendages, dorsal view.

FIGURE 14.1.23 *Limnetron debile* (Karsch, 1891), larval folded labium, ventral view. *Arrow* indicate prothoracic supracoxal processes.

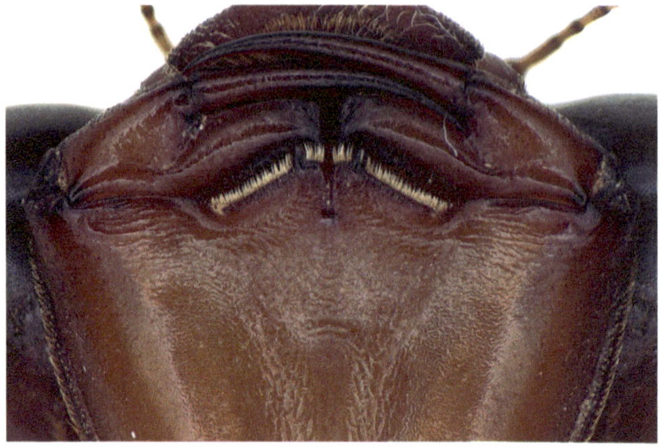

FIGURE 14.1.26 *Limnetron debile* (Karsch, 1891), larval prementum anterior margin and palps, ventral view.

FIGURE 14.1.24 *Rhionaeschna itatiaia* (Carvalho & Salgado, 2004), larval folded labium, ventral view. *Arrow* indicate prothoracic supracoxal processes.

FIGURE 14.1.27 *Limnetron debile* (Karsch, 1891), larval anal appendages, lateral view. *Ce*, cercus; *Ep*, epiproct.

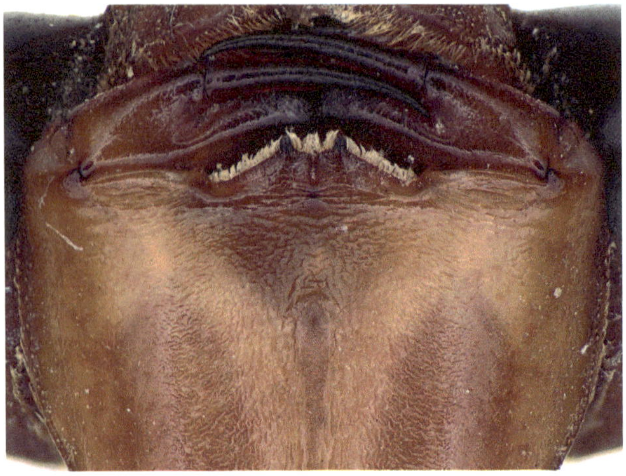

FIGURE 14.1.25 *Racenaeschna* sp., larval prementum anterior margin and palps, ventral view.

FIGURE 14.1.28 *Rhionaeschna brasiliensis* (von Ellenrieder & Costa, 2002), larval anal appendages, lateral view. *Ce*, cercus; *Ep*, epiproct.

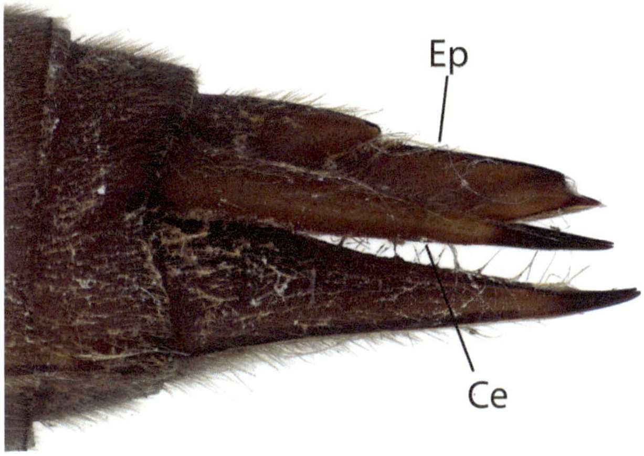

FIGURE 14.1.29 *Castoraeschna castor* (Brauer, 1865), larval anal appendages, lateral view. *Ce*, Cercus; *Ep*, Epiproct.

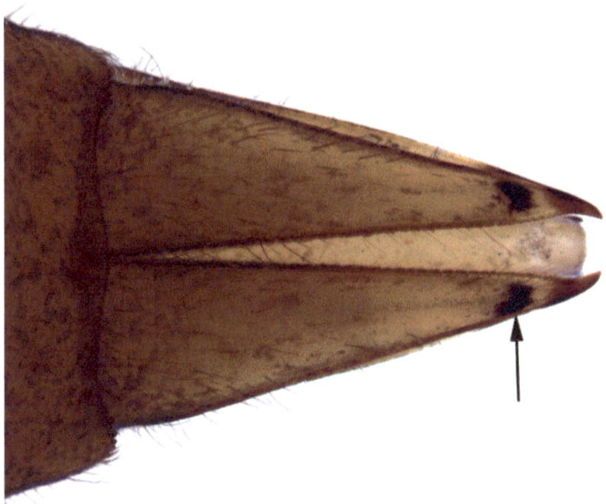

FIGURE 14.1.30 *Coryphaeschna adnexa* (Hagen, 1861), larval paraprocts, ventral view. *Arrow* indicate subapical dark spot.

FIGURE 14.1.31 *Rhionaeschna brasiliensis* (von Ellenrieder & Costa, 2002), larval paraprocts, ventral view.

FIGURE 14.1.32 *Remartinia luteipennis luteipennis* (Burmeister, 1839), larval prementum anterior margin and palps, ventral view. *Arrow* indicate prementum median cleft.

ACKNOWLEDGMENTS

Figs. 14.1.7, 10, and 15–32 were obtained with the aid of Marcus Vinícius Oliveira de Almeida, using the Stereomicroscope Leica M205 coupled to digital camera Leica DFC450C and the automontage software Leica Application Suite version 4.8.0 (Projeto Informatização da Coleção Entomológica do Museu Nacional/UFRJ, SIBBR/CNPq proc. 405588/2015-1). Figs. 14.1.6, 8, 9, and 11–14 were originally published by Neiss & Hamada (2014) and were generously provided by Editora do Instituto Nacional de Pesquisas da Amazônia, Manaus.

REFERENCES

Assis, J.C.F., A.L. Carvalho & L.F.M. Dorvillé. 2000. Aspects of larval development of Limnetron debile (Karsch), in a mountain stream of Rio de Janeiro State, Brazil (Anisoptera: Aeshnidae). Odonatologica 29: 151–155.

Baird, I.R.C. 2013. Larval habitat and behaviour of Phenes raptor (Odonata: Petaluridae): a review of current knowledge, with new observations. International Journal of Odonatology 16: 79–91.

Bybee, S.M., T.H. Ogden, M.A. Branham & M.F. Whiting. 2008. Molecules, morphology and fossils: a comprehensive approach to odonate phylogeny and the evolution of the odonate wing. Cladistics 24: 477–514.

Calil, E.R. & A.L. Carvalho. 1999. Descrições da larva de último estádio e do adulto de Triacanthagyna septima (Selys, 1857) (Odonata, Aeshnidae), com notas sobre a biologia da espécie. Revista Brasileira de Entomologia 43: 73–83.

Carle, F.L. 1996. Revision of Austropetaliidae (Anisoptera: Aeshnoidea). Odonatologica 25: 231–259.

Carle, F.L., K.M. Kjer & M.L. May. 2015. A molecular phylogeny and classification of Anisoptera (Odonata). Arthropod Systematics & Phylogeny 73: 281–301.

Carvalho, A.L. 1989. Description of the larva of Neuraeschna costalis (Burmeister), with notes on its biology, and a key to the genera of Brazilian Aeshnidae larvae (Anisoptera). Odonatologica 18: 325–332.

Carvalho, A.L. 1992. Aspectos da biologia de Coryphaeschna perrensi (McLachlan, 1887) (Odonata, Aeshnidae), com ênfase no período larval. Revista Brasileira de Entomologia 36: 791–802.

Carvalho, A.L. 2007. Recomendações para a coleta, criação e colecionamento de larvas de Odonata. Arquivos do Museu Nacional 65: 3–15.

Carvalho, A.L. & N. Ferreira, Jr. 1989. Descrição da larva de Gynacantha mexicana Selys, 1869, e notas sobre sua biologia (Odonata, Aeshnidae). Revista Brasileira de Entomologia 33: 413–419.

Carvalho, A.L. & J.L. Nessimian. 1998. Odonata do Estado do Rio de Janeiro, Brasil: hábitats e hábitos das larvas. Pages 3–28 *in*: J.L. Nessimian & A.L. Carvalho. (eds.), Ecologia de Insetos Aquáticos. Series Oecologia Brasiliensis, vol. V. PPGE-UFRJ, Rio de Janeiro.

Costa, J.M., L.O. Irineu de Souza & B.B. Oldrini. 2004. Chave para identificação das famílias e gêneros das larvas conhecidas de Odonata do Brasil: comentários e registros bibliográficos (Insecta, Odonata). Publicacões avulsas do Museu Nacional 99: 1–44.

Davies, D.A.L. & P. Tobin. 1985. The dragonflies of the world: a systematic list of extant species of Odonata. Vol. 2. Anisoptera. Societas Internationalis Odonatologica Rapid Communications (Suppl.) 5: X+1–151.

De Marmels, J. 2000. The larva of Allopetalia pustulosa Selys, 1873 (Anisoptera: Aeshnidae), with notes on aeshnoid evolution and biogeography. Odonatologica 29: 113–128.

Dijkstra, K.-D.B., G. Bechly, S.M. Bybee, R.A. Dow, H.J. Dumont, G. Fleck, R.W. Garrison, M. Hämäläinen, V.J. Kalkman, H. Karube, M.L. May, A.G. Orr, D.R. Paulson, A.C. Rehn, G. Theischinger, J.W.H. Trueman, J. van Tol, N. von Ellenrieder & J. Ware. 2013. The classification and diversity of dragonflies and damselflies (Odonata). Zootaxa 3703: 36–45.

von Ellenrieder, N. 2002. A phylogenetic analysis of the extant Aeshnidae (Odonata: Anisoptera). Systematic Entomology 27: 437–467.

Fleck, G., B. Ullrich, M. Brenk, C. Wallnisch, M. Orland, S. Bleidissel & B. Misof. 2008. A phylogeny of anisopterous dragonflies (Insecta, Odonata) using mtRNA genes and mixed nucleotide/doublet models. Journal of Zoological Systematics and Evolutionary Research 46: 310–322.

Fleck, G. 2011. Phylogenetic affinities of Petaluridae and basal Anisoptera families (Insecta: Odonata). Stuttgarter Beiträge zur Naturkunde A, Neue Serie 4: 83–104.

Garrison, R.W., N. von Ellenrieder & J.A. Louton. 2006. Dragonfly genera of the New World: an illustrated and annotated key to the Anisoptera. The Johns Hopkins University Press, Baltimore.

Heckman, C.W. 2006. Encyclopedia of South American Aquatic Insects: Odonata-Anisoptera. Illustrated keys to known families, genera, and species in South America. Springer, Dordrecht.

Misof B., A.M. Rickert, T.R. Buckley, G. Fleck & K.P. Sauer. 2001. Phylogenetic signal and its decay in mitochondrial SSU and LSU rRNA gene fragments of Anisoptera. Molecular Biology and Evolution 18(1): 27–37.

Muzón, J., P. Pessacq & F. Lozano. 2014. The Odonata (Insecta) of Patagonia: A synopsis of their current status with illustrated keys for their identification. Zootaxa 3784: 346–388.

Neiss, U.G. & N. Hamada. 2014. Ordem Odonata ([capítulo] 14). Pages 217–284 *in*: N. Hamada, J.L. Nessimian & R.B. Querino (eds.), Insetos aquáticos na Amazônia brasileira: taxonomia, biologia e ecologia. Editora do INPA, Manaus.

Pessacq, P. & C. Brand. 2009. Description of the larva of Phyllopetalia apollo Selys and redescription of that of Hypopetalia pestilens McLachlan (Anisoptera: Austropetaliidae). Odonatologica 38: 235–246.

Santos, N.D. 1970. Contribuição ao conhecimento da fauna do Estado da Guanabara, 72. Descrição da ninfa de Castoraeschna castor (Brauer, 1865) Calvert, 1952 (Odonata: Aeshnidae). Atas da Sociedade de Biologia do Rio de Janeiro 13: 47–48.

Santos, N.D. 1981. Odonata. Pages 64–85 *in*: S.H Hurlbert, G. Rodríguez & N.D. Santos (eds.), Aquatic Biota of Tropical South America. Part 1: Arthropoda. San Diego State University, San Diego.

Schmidt, E. 1941. Petaluridae, Gomphidae und Petaliidae der Schönemannschen Sammlung aus Chile (Ordnung Odonata). Archiv für Naturgeschichte 10: 231–258.

Silsby, J. 2001. Dragonflies of the World. CSIRO Publishing, Collingwood.

Suhling, F., G. Sahlén, S. Gorb, V.J. Kalkman, K-D.B. Dijkstra & J. von Tol. 2015. Order Odonata, Chapter 35. Pages 893–932 *in*: J.H. Thorp & D.C. Rogers (eds.), Ecology and General Biology, Thorp and Covich's Freshwater Invertebrates, Vol. 1, 4th Edition, Academic Press-Elsevier.

Thorp, J.H. & D.C. Rogers (eds.). 2015. Ecology and General Biology Thorp and Covich's Freswater Invertebrates, Vol. 1, 4th Edition, Academic Press-Elsevier.

Chapter 14.2

Superfamily Gomphoidea

Rodolfo Novelo-Gutiérrez
Instituto de Ecología, A. C., Xalapa, Veracruz, México

Alonso Ramírez
Department of Environmental Sciences, University of Puerto Rico, San Juan, Puerto Rico

Enrique González-Soriano
Instituto de Biología, Departamento de Zoología, Universidad Nacional Autónoma de México, Ciudad de México, México

SUPERFAMILY GOMPHOIDEA

Chapter Outline

INTRODUCTION

Gomphoidea is currently considered to be composed of the single family Gomphidae. This is supported by recent phylogenetic analyses (Bybee et al., 2008; Dumont et al., 2010) and is adopted by the latest classificatory scheme (Dijkstra et al., 2013). The family comprises a group of large-sized, worldwide distributed anisopterans. With approximately 1000 species, it is the second largest family of the suborder Anisoptera.

Although this group does not have a common name in Spanish, the term "libélulas topo" (Pérez-Gutiérrez et al., 2013) seems to be appropriate, as it refers to a biological feature of the larva of many species, which routinely burrow in the sand or mud of rivers and streams. Gomphidae represents between 34–35% (277–286 spp.) of all known Neotropical anisopterans (816 spp.) (Suhling et al., 2015, González-Soriano unpublished data).

The Gomphidae is a relatively well-known group of anisopterans in terms of alpha taxonomy. Former classificatory schemes relied only on wing venation (e.g., Williamson, 1920; Belle, 1979; Carle, 1986). Early attempts to resolve its phylogenetic classification (e.g., Carle, 1986) were unsuccessful (Garrison et al., 2006). Belle (1996) was the first to classify South American gomphids, compiling all the scattered literature and providing references and illustrations of wings for all known South American species. Studies have found support for the monophyly of the family (e.g., Bybee et al., 2008; Dumont et al., 2010). Bybee et al. (2008) and Dumont et al. (2010) favored the recognition of the superfamily Gomphoidea. However, with no extensive molecular phylogenetic reconstruction available yet, the validity of any classification remains untested (Dijkstra et al., 2013).

Most species inhabit running waters. Their habitats include, among others, banks of fine sand, mud and black detritus on small shallow forested streams (e.g., *Epigomphus*), muddy or sandy banks with decayed leaves on streams and rivers (*Phyllogomphoides*) or muddy edges in sluggish streams and lentic waters (*Aphylla*). Gomphid larvae can be classified ecologically (see also Table 14.2.1; and based in part on Corbet, 1999) as: (1) sluggish hiders - slow movement individuals that conceal themselves usually under a cover of leaf and/or leaf detritus combined with fine sediment such as silt and mud; (2) shallow burrowers - active individuals that conceal themselves in coarse or fine gravel, sand, and silt, adopting a position such that only

TABLE 14.2.1 List of Genera of Gomphidae From the Neotropical Region, With Notes on the Larval Microhabitats

Genus	Habitat	Reference	Notes
Agriogomphus	Among detritus, leaf particles in muddy edges with slow current in small streams	Belle (1966), Novelo-Gutiérrez (1989)	Sluggish hiders
Aphylla	Muddy edges of streams and ponds	Needham (1940)	Deep burrowers
Archaeogomphus	Probably in black detritus in muddy edges with slow current in small streams	Belle (1992)	Sluggish hiders
Brasiliogomphus	Larva unknown		
Cacoides	In leafy detritus over sandy substrate, streams	Belle (1970)	Sluggish hiders?
Cyanogomphus	Larva unknown		
Desmogomphus	Among coarse sand and pebbles in areas of flow	Neiss & Hamada (2014)	Shallow burrowers?
Diaphlebia	Larva unknown		
Dromogomphus	Small, slow flowing, coastal streams with low turbity, fine sediments, and emergent vegetation	Abbot (2005)	Shallow burrowers
Ebegomphus	Black detritus in muddy edges with slow current in small streams	Neiss & Hamada (2014)	Sluggish hiders?
Epigomphus	Areas of fine sand, mud, and black detritus at the edges of small shaded streams	Ramírez (1996), Novelo-Gutiérrez et al. (2015, 2016)	Sluggish hiders
Erpetogomphus	Sand banks, coarse gravel, in runs or slow water movement in small to large sandy and rocky streams or rivers	Novelo-Gutiérrez (2002a,b, 2005) Ramírez (1996)	Shallow burrowers
Gomphoides	Larva unknown	Belle (1992)	
Gomphurus	Areas of silt with slow flow	Needham et al. (2014)	Shallow burrowers
Idiogomphoides	Larva unknown	Belle (1992)	
Melanocacus	Among leaf detritus in sandy bottom streams	Garrison et al. (2006)	Sluggish hiders?
Neogomphus	Shallow areas of slow flow with pebbles in mountain streams	Belle (1992)	Shallow burrowers?
Octogomphus	Among leaf detritus in edges of small rocky or muddy forested streams	Needham et al. (2014) Paulson (2009)	Sluggish hiders
Ophiogomphus	Sand banks, coarse gravel, in runs or slow water movement in small to large sandy and rocky streams or rivers	Paulson (2009)	Shallow burrowers
Perigomphus	Coarse sand and gravel. In runs or slow water in small forested streams	Ramírez pers. obs.	Shallow burrowers
Peruviogomphus	Lakes and oxbow lakes in lowlands	Garrison et al. (2006)	
Phanogomphus	Areas of silt with slow flow	Needham et al. (2014)	Shallow burrowers
Phyllocycla	Sand and fine gravel banks at the edge of small shallow streams	Needham (1940)	Deep burrowers
Phyllogomphoides	Black detritus and litter at slow flowing or still water at the edge of lotic waters	Novelo-Gutiérrez (1993)	Sluggish hiders

TABLE 14.2.1 List of Genera of Gomphidae From the Neotropical Region, With Notes on the Larval Microhabitats—cont'd

Genus	Habitat	Reference	Notes
Praeviogomphus	Sandy substrate in shallow pools in first order streams	Carvalho (2000)	Shallow burrowers
Progomphus	Sandy areas of permanent streams	Belle (1991)	Rapid shallow burrowers
Stylurus	Weedy rivers, streams, and lakes with moderate to little current	Abbot (2005)	Shallow burrowers
Tibiagomphus	Areas of fine sediments in stream edges, forested streams	Rodrigues Capítulo (1984)	Shallow burrowers
Zonophora	Leaf accumulations in depositional areas, sometimes with roots	Neiss & Hamada (2014)	Sluggish hiders?

We have adopted the definitions provided by Corbet (1999) with some modifications in order to include most of the genera known as larvae:
Sluggish hiders: Slow movement individuals that conceal themselves usually under a cover of leaf and/or leaf detritus combined with fine sediment such as silt and mud.
Shallow burrowers: Active individuals that conceal themselves in coarse or fine gravel, sand, and silt, adopting a position such that only "the antennae and the caudal appendages project above the sediment in which the rest of the body lies buried" (Corbet, 1999).
Deep burrowers: Larvae which abdominal segment 10 has been extremely modified as an elongated tube, which enable them to maintain respiratory contact with sediment-free water above it, while they keep deeply buried.

"the antennae and the caudal appendages project above the sediment in which the rest of the body lies buried" (Corbet, 1999); and (3) deep burrowers - larvae, which abdominal segment 10 has been extremely modified as an elongated tube that enables them to maintain respiratory contact with sediment-free water above it while they keep deeply buried. Larval development is poorly known. In general, larvae develop slowly, taking from one to several years to complete development, as in *Asiagomphus pryeri* (Selys) (Aoki, 1999) and *Phyllogomphoides* sp (Novelo-Gutiérrez, 1993). A remarkable exception is *Paragomphus genei* (Selys) which completes development in as few as 60 days in Namibia (Suhling & Martens, 2007).

In the Western Hemisphere, the diversity is higher in the Neotropical region (29 genera, 287 species) than in the Nearctic region (14 genera, 101 species) (Paulson & González-Soriano, 1994; González-Soriano & Novelo-Gutiérrez, 2007; Meurgey, 2013; Paulson, 2014, 2016a, b; Garrison & von Ellenrieder, 2016). Eighteen genera (see Table 14.2.1) (with 84 species) are completely restricted to the Neotropics. The most speciose genera are *Progomphus* (69 species), *Phyllogomphoides* (46), *Phyllocycla* (31), *Epigomphus* (31), *Aphylla* (24), and *Erpetogomphus* (23). Four of the most species-rich genera *Aphylla, Phyllocycla, Phyllogomphoides*, and *Progomphus* (143 species) are mostly distributed in the Neotropics, but some species enter into the Nearctic region. On the other hand, most *Erpetogomphus* are distributed in Mexico, although some have penetrated into the United States and Central America (Garrison, 1994). *Dromogomphus, Gomphurus, Octogomphus, Ophiogomphus, Phanogomphus*, and *Stylurus* are more speciose in the Nearctic (72 species), than in the Neotropical region (8).

LIMITATIONS

The larval stage of many species still remain to be described in part because Neotropical Gomphidae are never abundant and are hard to find, even as adults, especially inhabitants of the rainforest. Even the habitats of many groups remain to be discovered and properly characterized. More research is needed in the family as a whole and the larval stages in particular, as the Neotropical region is among the most diverse biogeographical areas of the world. Adult identification for species is reasonably covered in recent keys (i.e., Förster, 2001; Garrison et al., 2006). In contrast, generic identifications for immature stages are far from adequate. There are a few regional keys for Mexico (Novelo-Gutiérrez, 1997, 2005), Central America (Ramírez, 1996), the Amazon (Neiss & Hamada, 2014), Brazil (Costa et al. 2004), and Patagonia (Muzón et al., 2014). However, available keys are focused on last instar larvae, and the identification of early stages is impossible without rearing specimens, a difficult task for many seasonal and fast water species (Neiss & Hamada, 2014). We also need revisions of some of the old larval descriptions, which are often too brief and incomplete and thus often of little use for actual comparisons.

TERMINOLOGY AND MORPHOLOGY

The most important terminology for identifying these dragonflies are described in Chapter 14. Larvae share the following features: the body abundantly setose; antennae, four articulated with the third antennal article the largest and fourth article minute or vestigial; head depressed and wider than long; prementum flat, ligula without median cleft, apical margin with piliform setae; molar crest of mandible movable; stiff legs, being scarcely opposable, short, and setose; claws simple, with pulvilliform empodium; male larva gonapophyses lacking; and female larva gonapophyses rudimentary or vestigial.

MATERIAL PREPARATION AND PRESERVATION

Material preparation and preservation is similar to other groups included in Odonata (see Chapter 14), but transportation of living specimens may require small containers with wet detritus or with some sediment such as sand (as in the particular case for *Progomphus* spp.) that should be placed on coolers with several larvae each.

KEYS TO GOMPHIDAE

The larvae of 116 species (41%) have been described, with 24 of them (8%) described by supposition. At genus level, the larvae of 26 out of the 29 genera (90%) are known. Larvae can be easily identified by the fourth antennal article being minute or vestigial. *Aphylla* and *Phyllocycla*, have the tenth abdominal segment tubular and greatly elongated; this characteristic allows the larva to inhale silt-free water while buried.

Larvae of two genera are not described: *Brasiliogomphus* (species known only from the holotype, from São Paulo, Brazil), and *Cyanogomphus* (southernmost Brazil and adjacent Argentina). The larva of *Diaphlebia* is not formally described; but based on material in the collection of the authors, it is included in the key.

Gomphoidea: Gomphidae: Genera

Distributional data presented in this section is modified from Garrison et al. (2006).

1	Abdominal segment 10 enlarged as a tube (Fig. 14.2.1), at least twice as long as S8 + 9; lateral spines lacking or very reduced 2
1'	Abdominal segment 10 not enlarged as a tube, usually of the same length than preceding segments, occasionally longer than S9; lateral spines variable 3
2(1)	Ligula hemispherical, strongly produced, labial palp medial margin lacking teeth (Fig. 14.2.2) .... *Phyllocycla* [Neotropics. Nearctic]
2'	Ligula convex, poorly developed, labial palp medial margin with strong, robust teeth (Fig. 14.2.3) .... *Aphylla* [Neotropics. Nearctic]
3(1)	Wing sheaths moderately (Fig. 14.2.4) to strongly divergent (Fig. 14.2.5) 4
3'	Wing sheaths parallel along midline (Fig. 14.2.6), posterior wing sheaths sometimes convergent at apices (Fig. 14.2.7) 10
4(3)	Abdomen with dorsal protuberances on at least some segments; prementum anterior margin (ligula) and sterna variable 5
4'	Abdomen lacking dorsal protuberances; prementum anterior margin (ligula) poorly developed with truncated teeth only medially (Fig. 14.2.8); sterna 2 and 9 divided into three sternites, sterna 3–8 divided into five sternites .... *Epigomphus* [Widespread]
5(4)	Wing sheaths moderately divergent (Fig. 14.2.4); labial palp moderately (Fig. 14.2.9) to strongly incurved (Fig. 14.2.10), apex acute 6
5'	Wing sheaths strongly divergent (Fig. 14.2.11); labial palp medial margin slightly incurved, apex rounded (Fig. 14.2.12) to obtuse (Fig. 14.2.13) 7
6(5)	Lateral spines on S5–9, widely divergent and slightly upturned, simple on all or simple on S5 and with a subapical, pale, S6–9 with medial lobe (Fig. 14.2.14); ligula convex at sides and slightly concave at middle (Fig. 14.2.10) .... *Idiogomphoides* [Guyana, Surinam to Brazil]
6'	Lateral spines on S7–9, not as above; ligula widely convex or slightly concave medially .... *Phyllogomphoides* [Widespread. Nearctic]

FIGURE 14.2.1 *Phyllocycla* sp.: larval dorsal habitus.

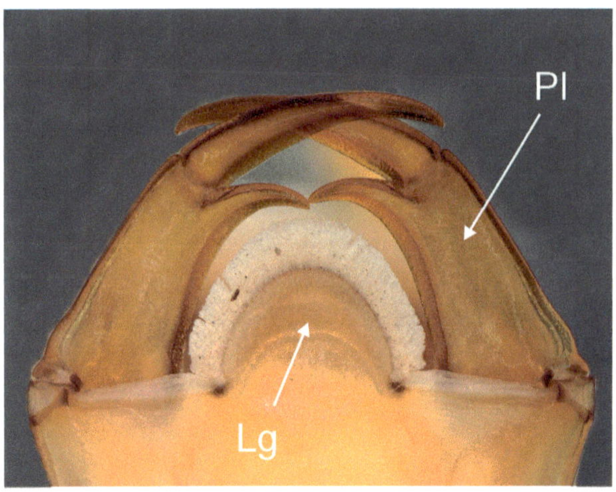

FIGURE 14.2.2 *Phyllocycla* sp.: details of larval prementum. *Pl*, palp; *Lg*, ligula.

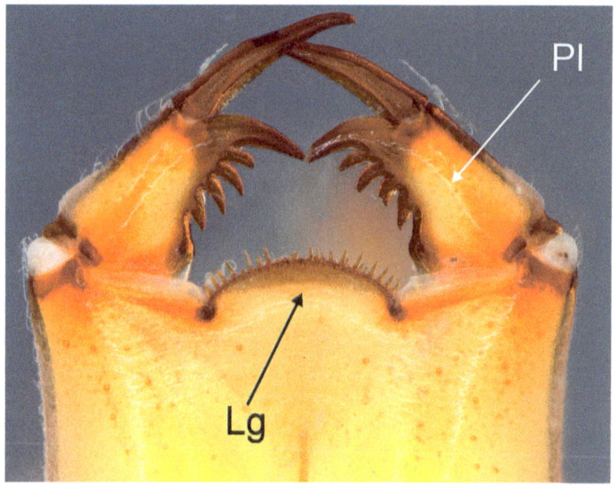

FIGURE 14.2.3 *Aphylla protracta* (Hagen in Selys): details of larval prementum. *Pl*, palp; *Lg*, ligula.

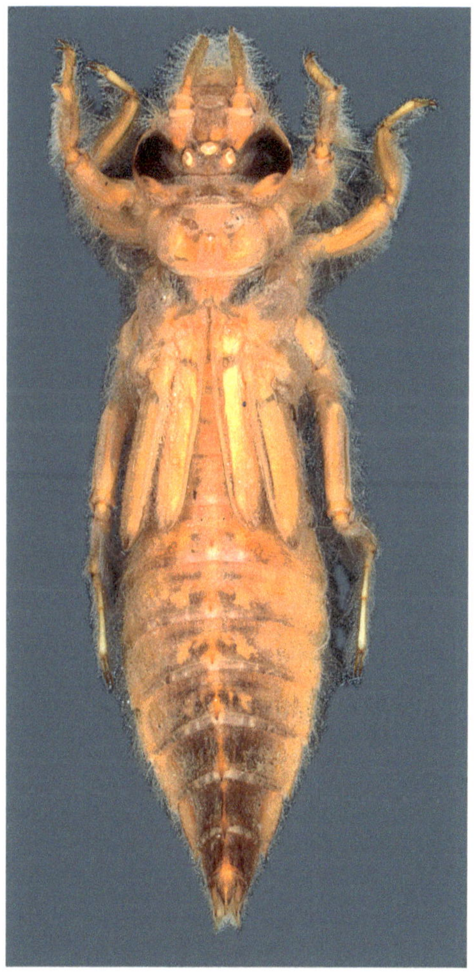

FIGURE 14.2.4 *Phyllogomphoides pacificus* (Selys): larval dorsal habitus showing wing sheaths moderately divergent.

FIGURE 14.2.5 *Epigomphus subsimilis* Calvert: larval dorsal habitus showing wing sheaths strongly divergent.

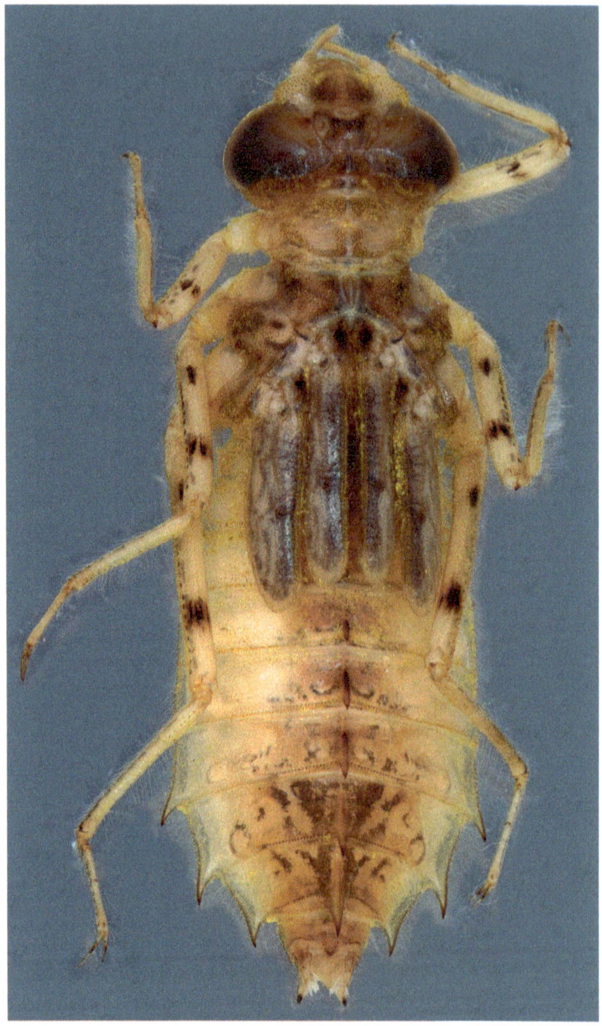

FIGURE 14.2.6 *Archaeogomphus* sp.: larval dorsal habitus showing wing sheaths parallel.

7(5)	Mesocoxae and procoxae more or less at the same distance at their bases; ligula variable; antennomere IV length <0.20× antennomere III; pro- and mesotibiae with burrowing hooks; pro- and meso tarsal claws of ordinary divergent type; cerci length ≥0.66× epi- and paraprocts	8
7'	Mesocoxae closer at bases than procoxae (Fig. 14.2.15); ligula very prominent (Fig. 14.2.12); antennomere IV length 0.25–0.50× antennomere III, this last usually covered by spiniform setae (Fig. 14.2.16); pro- and mesotibiae without burrowing hooks; pro- and meso tarsal claws cheliform (Fig. 14.2.17); cerci notably shorter than epi- and paraprocts [Widespread. Nearctic]	*Progomphus* (in part)
8(7)	Cerci shorter than epi- and paraprocts	9
8'	Caudal appendages subequal (Fig. 14.2.18) [Mexico to Colombia, Venezuela. Nearctic]	*Erpetogomphus*
9(8)	Paraproct length usually 3× of S10 middorsal, cerci length ~0.75× epi- and paraprocts (Fig. 14.2.19); ligula convex, edged with row of piliform setae, and many blunt, close-set small teeth (Fig. 14.2.20) [Mexico. Holarctic]	*Ophiogomphus*
9'	Paraprocts length ≤2× of S10 middorsal, cerci length usually 0.66× epi- and paraprocts (Fig. 14.2.21); ligula slightly convex or strongly developed (*P. tibialis*), edged with row of piliform setae, and a pair of median small teeth (Fig. 14.2.22) or lacking (*P. geijskesi*) [Widespread. Nearctic]	*Progomphus* (in part)
10(3)	Tarsal formula 2-2-2 (all tarsi about the same length)	11
10'	Tarsal formula 2-2-3 (distal tarsus longer than basal and median tarsi)	13
11(10)	Abdomen semicircular, S10 sunken into S9 (Figs. 14.2.7, 25); ligula strongly convex, labial palp medial margin concave (Fig. 14.2.23); mature larvae >20 mm	12

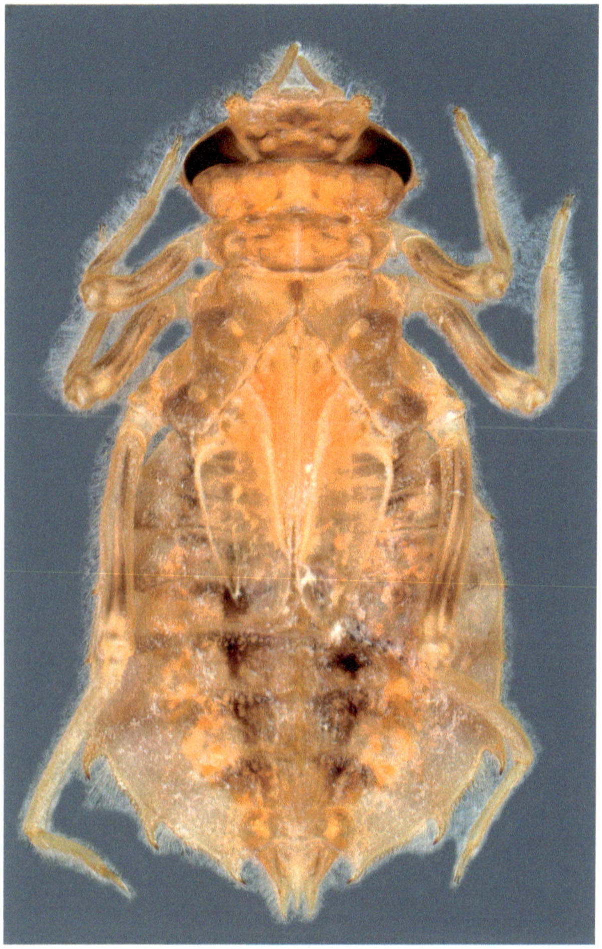

FIGURE 14.2.7 *Cacoides* sp.: larval dorsal habitus showing wing sheaths parallel, with posterior wing sheaths convergent at apex.

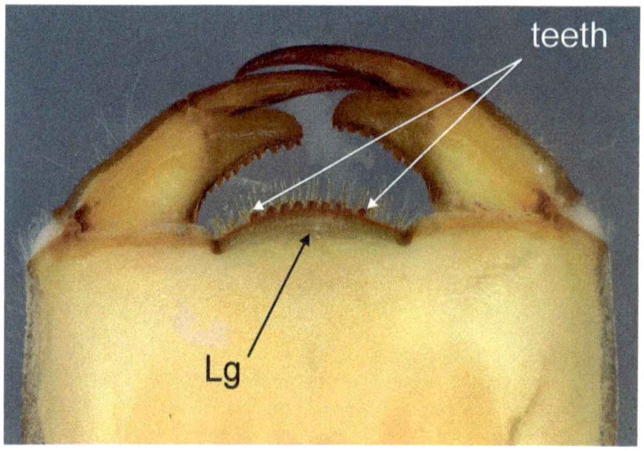

FIGURE 14.2.8 *Epigomphus crepidus* Kennedy: details of larval prementum showing teeth only on the middle part of ligula (Lg).

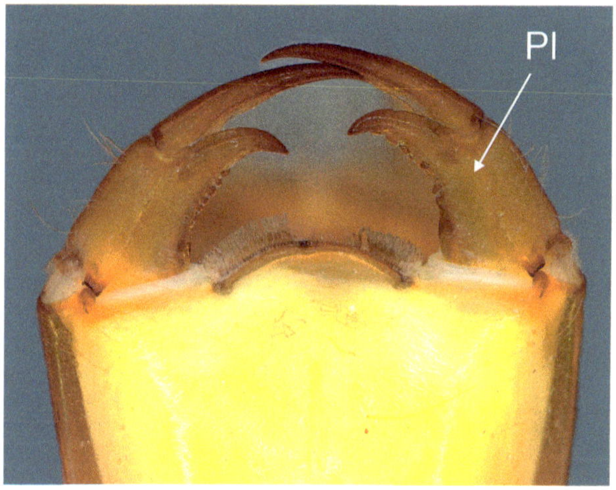

FIGURE 14.2.9 *Phyllogomphoides pacificus* (Selys): details of larval prementum showing ligula and palpi moderately incurved (Pl).

11' Abdomen enlarged, mostly parallel sided, S10 not sunken into S9 (Fig. 14.2.6); ligula and labial palp medial margin straight (Fig. 14.2.24); mature larvae no more than 14 mm .... *Archaeogomphus*
[Mexico to Argentina]

12(11) S7 lateral spines strongly developed and dorsally carinate, straight and subperpendicular to abdomen (Fig. 14.2.25); hind femora medial surface coarsely granulose (Fig. 14.2.26); labial palp apical lobe rounded; occiput with 4, low rounded tubercles .... *Melanocacus*
[Venezuela, Surinam, Brazil]

12' S7 lateral spines incurved, not carinate (Fig. 14.2.7); hind femora medial surface not granulose; labial palp apical lobe tapering; occipital lobes slightly protruding dorsally; anterolateral corner of frons with a very large, rounded tubercle beset with strong, sharp, upcurved spines (Fig. 14.2.27) .... *Cacoides*
[NW South America]

13(10) Abdominal dorsal protuberances lacking or, if present, only as small, low, triangular spines on S7–9, S8–9, or S9; ligula poorly developed 14

13' Abdominal dorsal protuberances well developed on at least one of the abdominal segments 3–9; ligula poorly to moderately developed 22

14(13) Pro- and mesotibial burrowing hooks well developed; ligula with median teeth or lacking; labial palp neither strongly incurved nor truncate, medial margin more or less straight or slightly concave; sterna variable 15

14' Pro- and mesotibial burrowing hooks absent; ligula lacking teeth, only with stiff piliform setae, labial palp strongly incurved ending in a stout, sharp hook (Fig. 14.2.28) or with the apical margin truncate ending in a robust, conical tooth (Fig. 14.2.29); S9 with a low, triangular, dorsal protuberance (Fig. 14.2.30) or lacking; sterna 2–9 divided into three sternites .... *Stylurus*
[Mexico. Nearctic]

15(14) Abdomen with dorsal protuberances on S7–9 or S8–9; sterna 2–6 and 9 divided into three sternites, sterna 7–8 divided into five sternites 16

15' Abdomen lacking dorsal protuberances; sterna variable 17

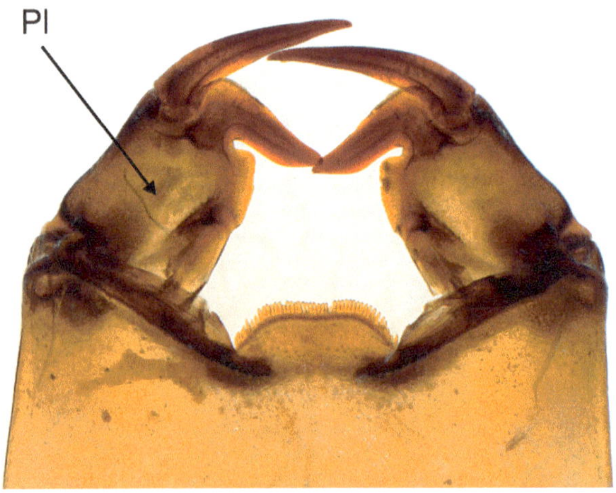

FIGURE 14.2.10 *Phyllogomphoides major* Belle: details of larval prementum showing ligula and palpi strongly incurved (Pl).

FIGURE 14.2.12 *Progomphus clendoni* Calvert: details of larval prementum showing ligula very prominent and palpi with inner margin slightly incurved and apex rounded.

FIGURE 14.2.11 *Progomphus clendoni* Calvert: larval dorsal habitus.

FIGURE 14.2.13 *Progomphus lambertoi* Novelo-Gutiérrez: details of larval prementum showing ligula very prominent and palpi with inner margin slightly incurved and apex obtusely pointed.

FIGURE 14.2.14 *Idiogomphoides* sp.: details of larval abdominal lateral spines on S7–8. Arrows indicate the subapical inner lobe.

FIGURE 14.2.16 *Progomphus clendoni* Calvert: dorsal view of larval head and pronotum showing the proportions of 3rd and 4th antennomeres, as well as the cover of bristle-like setae on 3rd antennomere. 3rd: third antennomere, 4th: fourth antennomere.

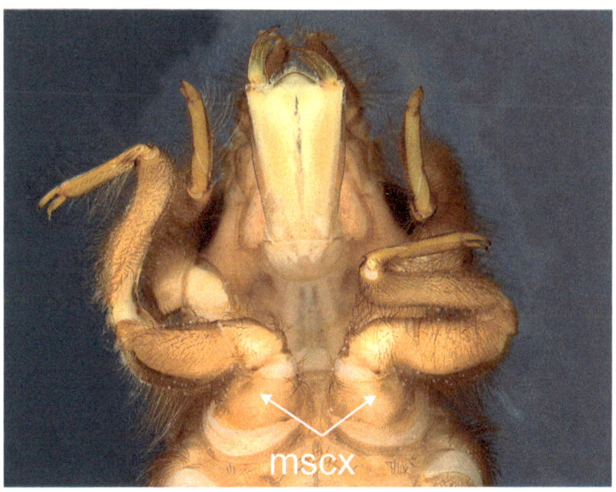

FIGURE 14.2.15 *Progomphus clendoni* Calvert (ventral), showing larval mesocoxae (mscx) close to each other.

FIGURE 14.2.17 *Progomphus lambertoi* Novelo-Gutiérrez: details of larval midleg, showing cheliform claws (Cl).

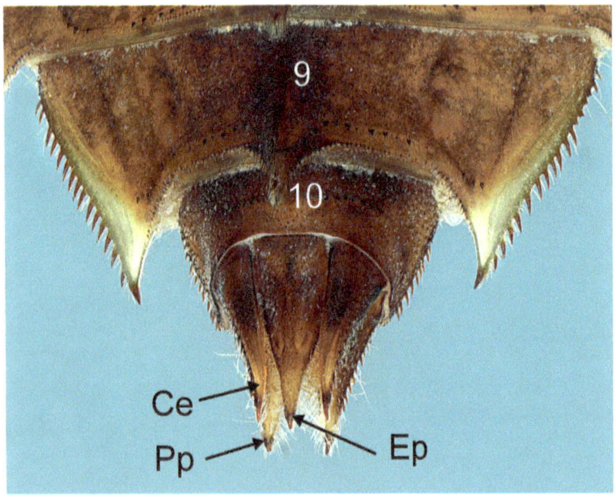

FIGURE 14.2.18 *Erpetogomphus sabaleticus* Williamson: details of larval tergites 9–10 and caudal appendages, dorsal view. *Ce*, cercus; *Ep*, epiproct; *Pp*, paraproct.

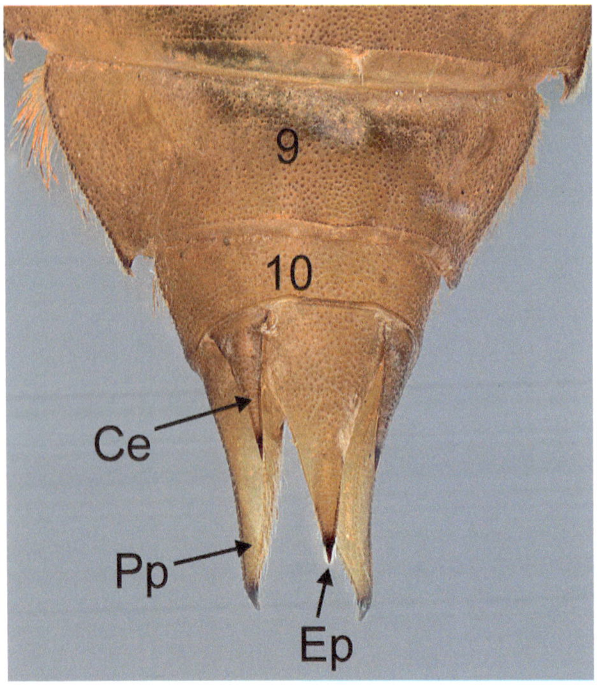

FIGURE 14.2.19 *Ophiogomphus susbehcha* Vogt & Smith: details of larval abdominal tergites 9–10 and caudal appendages, dorsal view. *Ce*, cercus; *Ep*, epiproct; *Pp*, paraproct.

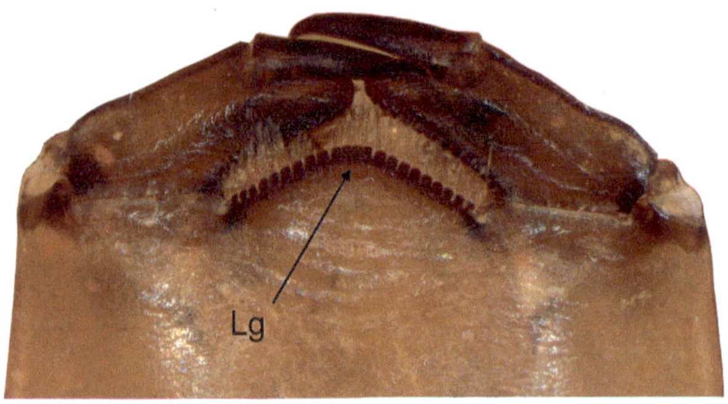

FIGURE 14.2.20 *Ophiogomphus susbehcha* Vogt & Smith: details of larval prementum showing the blunt, close-set small teeth on ligula (Lg) and palpi, ventral view.

FIGURE 14.2.21 *Progomphus pygmaeus* Selys: details of larval abdominal tergites 8–10 and caudal appendages, dorsal view. *Ce*, cercus; *Ep*, epiproct; *Pp*, paraproct.

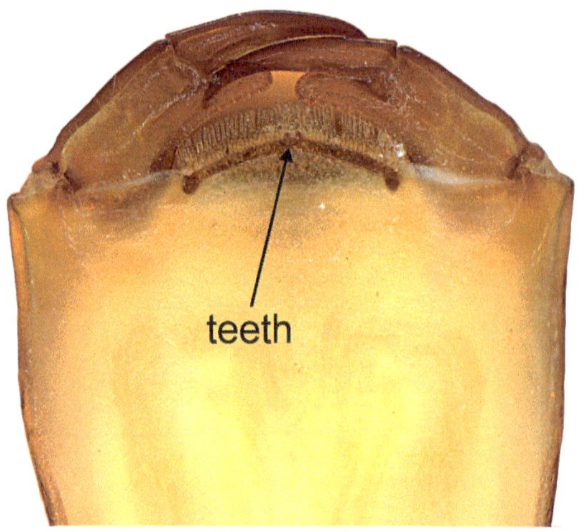

FIGURE 14.2.22 *Progomphus pygmaeus* Selys: details of larval prementum showing the two small teeth on ligula, ventral view.

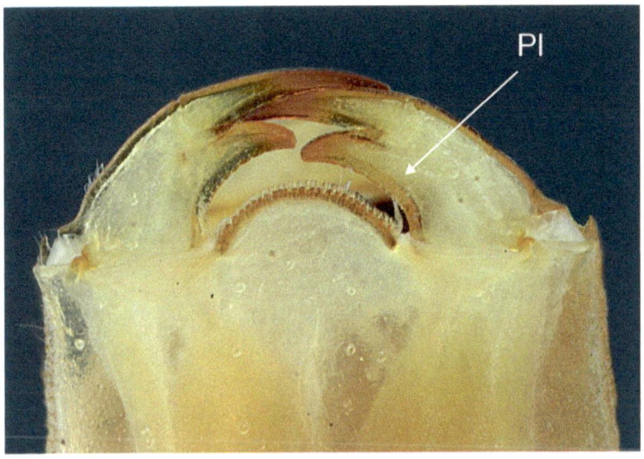

FIGURE 14.2.23 *Cacoides* sp.: details of larval prementum showing ligula strongly convex and palp (Pl) with inner margin concave.

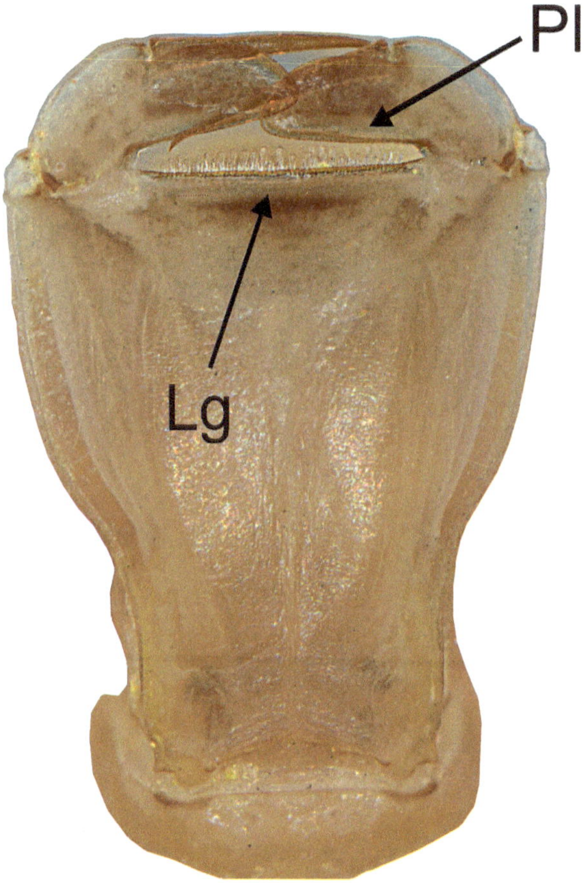

FIGURE 14.2.24 *Archaeogomphus* sp.: ventral view of larval prementum showing ligula (Lg) and internal margin of labial palp (Pl) straight.

FIGURE 14.2.25 *Melanocacus* sp.: larval dorsal habitus, abdominal tergites 6–10. See the lateral spines on S7 strongly developed and dorsally carinate.

FIGURE 14.2.26 *Melanocacus* sp.: details of larval hind femur (Hf) showing granuli (gr) on internal surface.

16(15) Ligula straight, apical margin with one tooth, palp medial margin with a row of sharp teeth (Fig. 14.2.31); abdominal dorsal protuberances on S7–9; S9 longer than preceding segments, lateral spines almost reaching posterior margin of S10; caudal appendages slightly longer than S10 (Fig. 14.2.32) .... *Dromogomphus*
[Mexico. Nearctic]

16' Ligula very slightly convex, lacking teeth, palp medial margin with a row of truncated teeth (Fig. 14.2.33); abdominal dorsal protuberances on S8–9; S9 slightly longer than preceding segments, lateral spines almost reaching the tips of paraprocts; caudal appendages shorter than S10 (Fig. 14.2.34) .... *Gomphurus*
[Mexico. Nearctic]

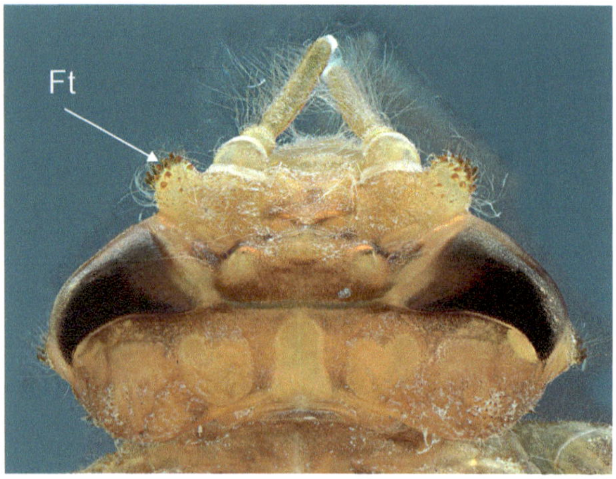

FIGURE 14.2.27 *Cacoides* sp.: details of head showing the anterolateral corner of frons with a rounded tubercle (Ft) beset with strong, sharp, upcurved spines.

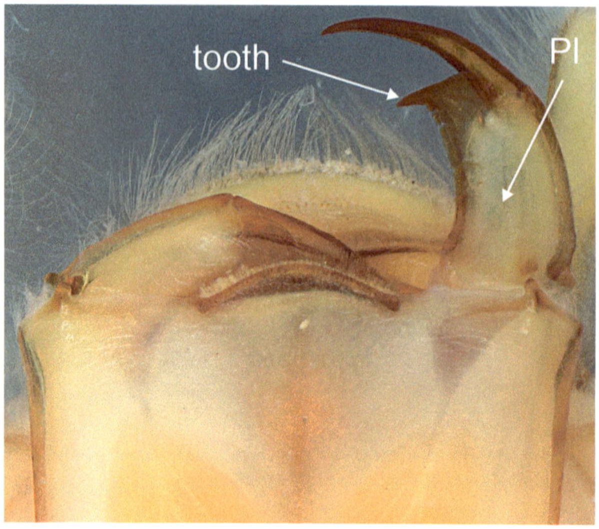

FIGURE 14.2.29 *Stylurus intricatus* (Hagen in Selys), details of larval prementum showing the ligula lacking teeth, and the labial palp (Pl) with the apical margin truncate ending in a robust, conical tooth, ventral view.

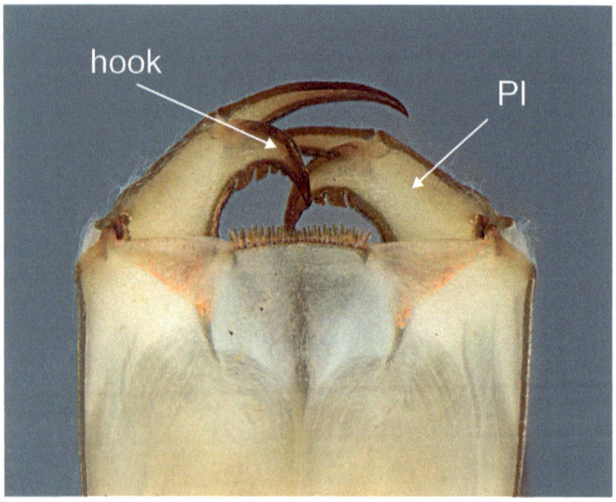

FIGURE 14.2.28 *Stylurus plagiatus* (Selys): details of larval prementum showing a poorly developed ligula without teeth, and a strongly incurved labial palp (Pl) ending in a stout, sharp hook, ventral view.

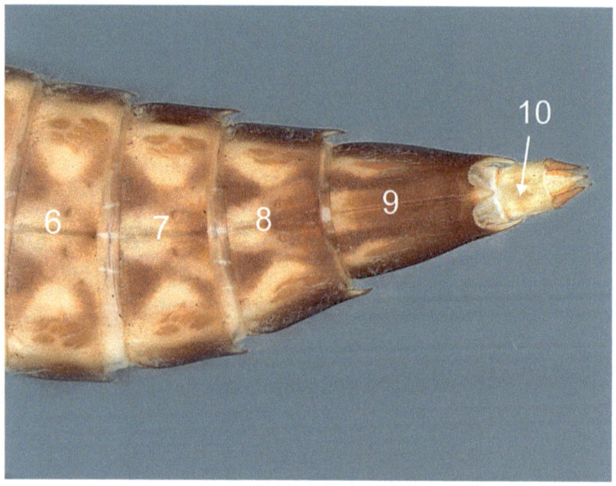

FIGURE 14.2.30 *Stylurus plagiatus* (Selys): details of larval abdominal tergites 6–10 showing the low, triangular, dorsal protuberance on S9, and caudal appendages, dorsal view.

17(15) Ligula with 2–5 teeth; dorsal length of S9 equal or shorter than preceding segments; caudal appendages variable 18

17' Ligula lacking teeth; S9 longer than preceding segments; caudal appendages shorter than S10 (Fig. 14.2.35).................. *Phanogomphus* [Mexico]

18(17) Epi- and paraprocts shorter than S8+9 together, bluntly or sharply pointed; paraprocts basointernal margin straight; 3rd antennomere length $\leq 2\times$ than its width or wider than long; S6– or S7–9 lateral spines variable; mature larva < 30 mm 19

18' Epi- and paraprocts as long as or longer than S8+9 together, sharply pointed, lateral spines only on S8–9, large, sharply pointed and upcurved (Fig. 14.2.36); paraprocts basointernal margin produced ventrally into a wide, convex lobe (Fig. 14.2.37); 3rd antennomere length $\geq 3\times$ times its widest part; mature larva 32–35 mm.. *Zonophora* [N South America. Paraguay and NE Argentina]

19(18) Third antennomere oval, longer than wide; frons straight or concave on anterior margin; ligula anterior margin with 3–5 small teeth; lateral spines on S6–, S7–, or S8–9 variable 20

19' Third antennomere transverse, subrectangular, medial margins straight where they meet, frons anterior margin convex (Fig. 14.2.38); ligula with a prominent denticle to each side of midline (Fig. 14.2.39); S6 ending in a mere angulation on lateral margins, those of S7–9 in a short blunt, spine (Fig. 14.2.40) .... *Perigomphus* [Costa Rica, Panamá, Colombia and Ecuador]

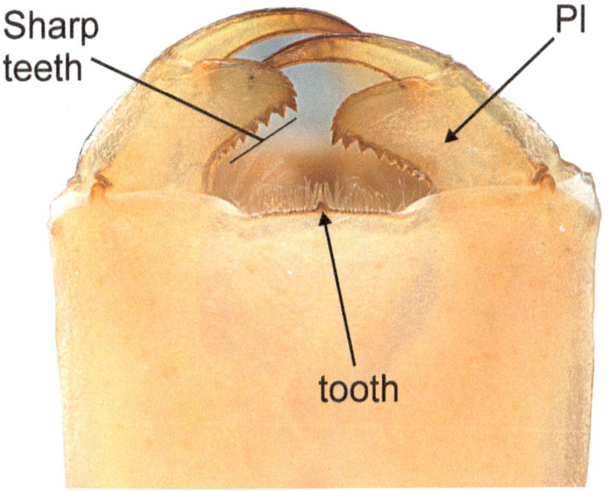

FIGURE 14.2.31 *Dromogomphus spoliatus* Hagen in Selys: details of larval prementum showing the straight ligula with one tooth at middle of its apical margin, and the internal margin of palp (Pl) with a row of sharp teeth.

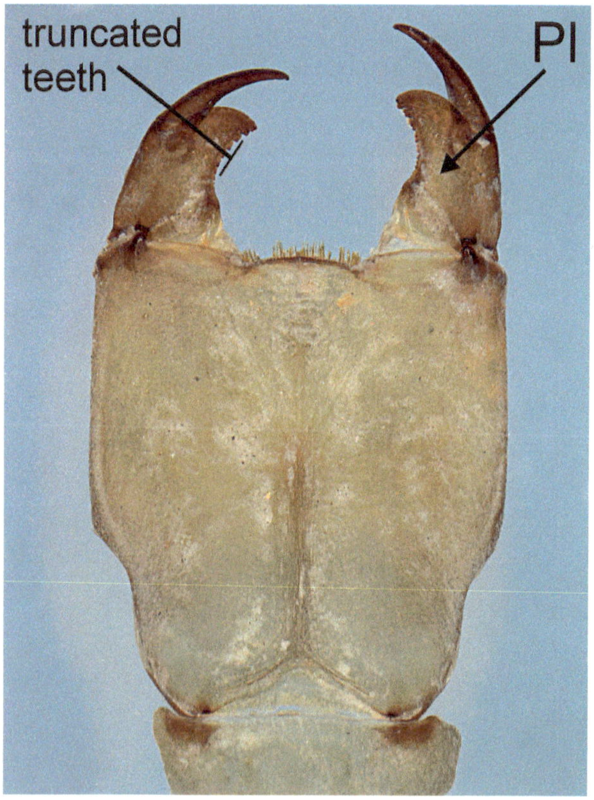

FIGURE 14.2.33 *Gomphurus gonzalezi* Dunkle: larval prementum showing the truncated teeth on internal margin of labial palp (Pl), ventral view.

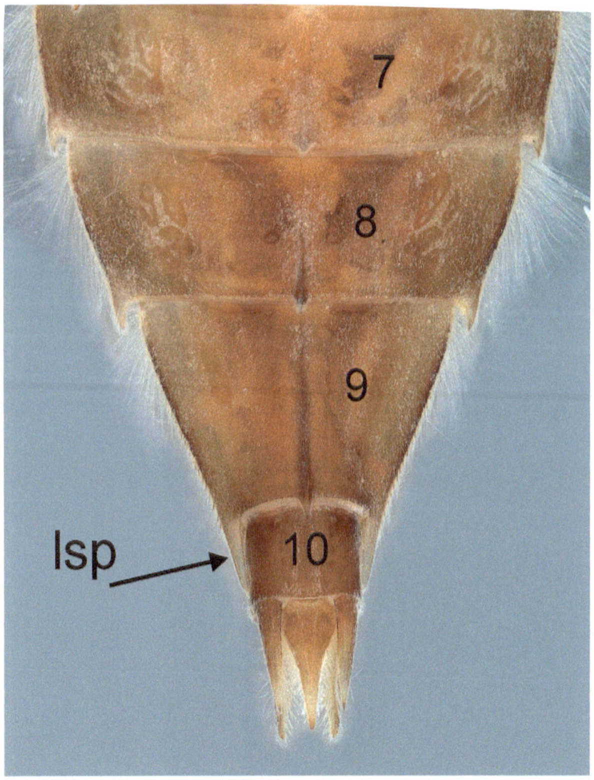

FIGURE 14.2.32 *Dromogomphus spoliatus* Hagen in Selys: details of larval abdomen showing the dorsal protuberances on S7–9, and the lateral spines (lsp) on S9 that almost reach the posterior margin of S10, dorsal view.

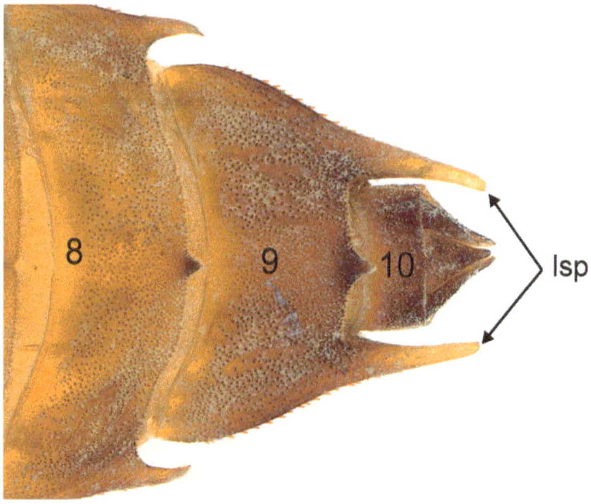

FIGURE 14.2.34 *Gomphurus gonzalezi* Dunkle: details of larval abdominal tergites 8–10 showing the dorsal protuberances on S8–9, and the lateral spines (lsp) on S9 that almost reach the apices of paraprocts, dorsal view.

20(19) Labial palp tapering apically, ligula with three small teeth (Fig. 14.2.41); prementum lateral margins smooth; lateral spines on S6–, S7–, or S8–9, bluntly pointed 21

20' Labial palp widely rounded apically, ligula with 4–5 small teeth, prementum lateral margins granulose (Fig. 14.2.42); lateral spines on S7–9 acutely pointed .... *Octogomphus* [Mexico. Nearctic]

21(20) S7–9 lateral spines upturned (Fig. 14.2.43); frons anterior margin concave (Fig. 14.2.44) .... *Praeviogomphus* [Brazil]

21' S6–, S7–, or S8–9 lateral spines straight (Fig. 14.2.45); frons anterior margin more or less sinuate (Fig. 14.2.46) .... *Neogomphus* [Chile, Argentina]

22(13') Pro- and mesotibial burrowing hooks absent 23

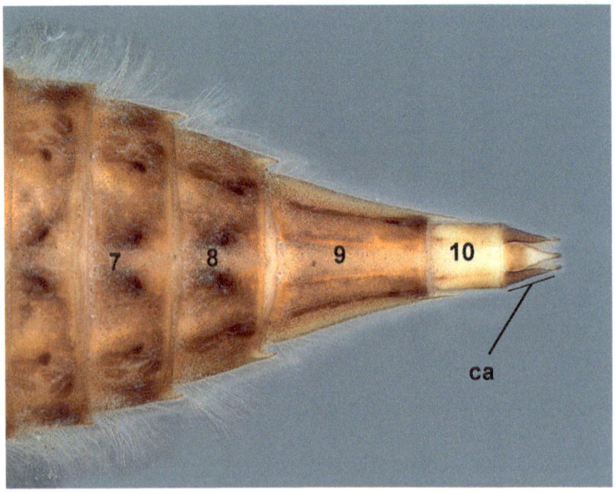

FIGURE 14.2.35 *Phanogomphus militaris* (Hagen in Selys): larval abdominal tergites 6–10 and caudal appendages (ca).

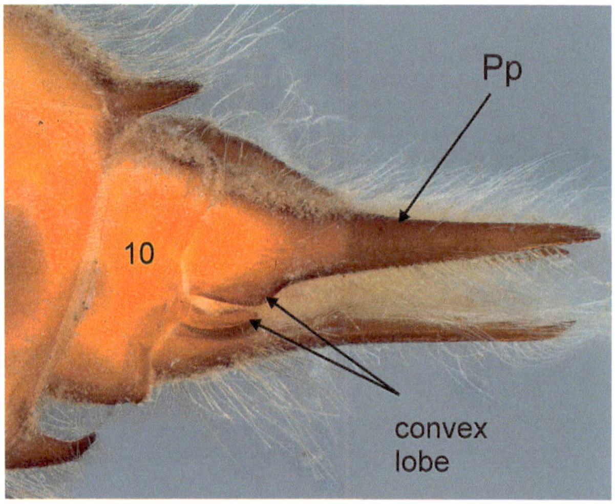

FIGURE 14.2.37 *Zonophora* sp.: larva lateroventral view of abdominal S10 and paraprocts (Pp) showing the ventral convex lobe of Pp.

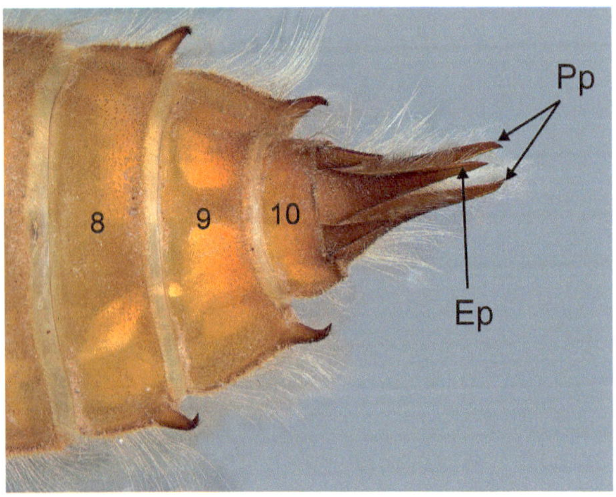

FIGURE 14.2.36 *Zonophora* sp.: larval abdominal tergites 8–10 showing the upcurved spines on S8–9, and the long epiproct (Ep) and paraprocts (Pp).

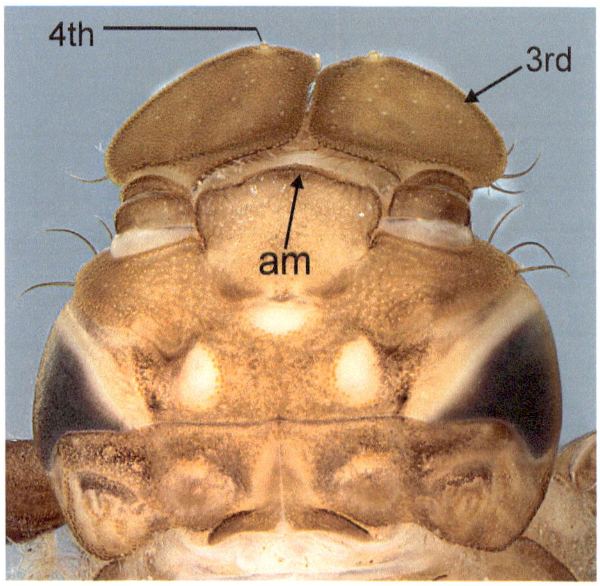

FIGURE 14.2.38 *Perigomphus pallidistylus* Belle: details of larval head showing the enlarged and vestigial 3rd and 4th antennomeres, respectively, and the convex anterior margin (am) of frons.

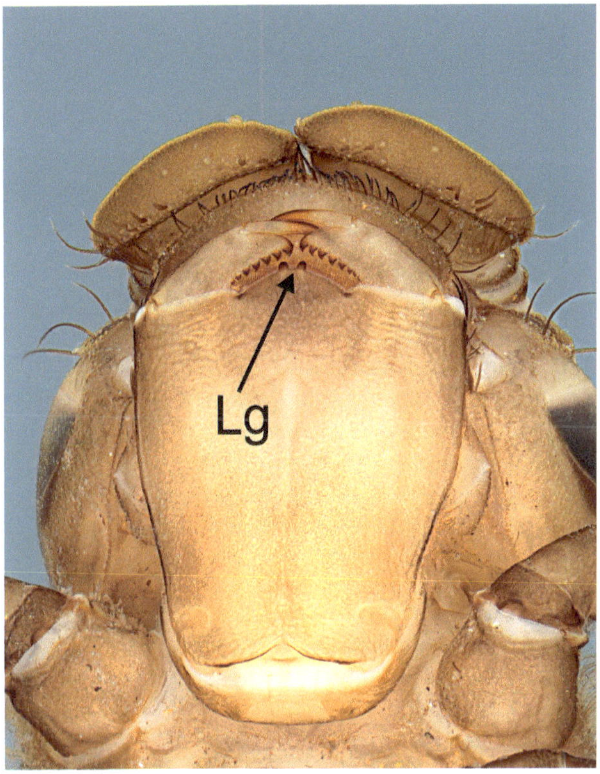

FIGURE 14.2.39 *Perigomphus pallidistylus* Belle: ventral view of larval prementum showing the two median teeth on ligula (Lg).

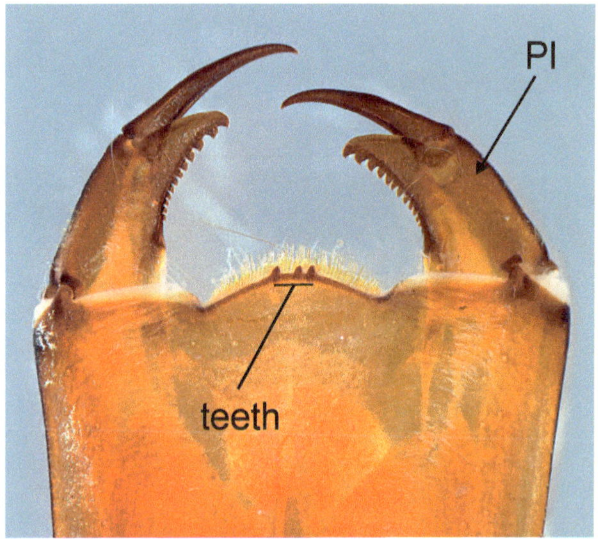

FIGURE 14.2.41 *Neogomphus edenticulatus* Carle & Cook: details of larval prementum showing ligula with three teeth at middle of its apical margin, and palpi (Pl) tapering apically with a row of sharp teeth, ventral view.

FIGURE 14.2.40 *Perigomphus pallidistylus* Belle, larval abdominal tergites 5–10 and caudal appendages. *Arrow* indicates the angulation at end of lateral margin of S6.

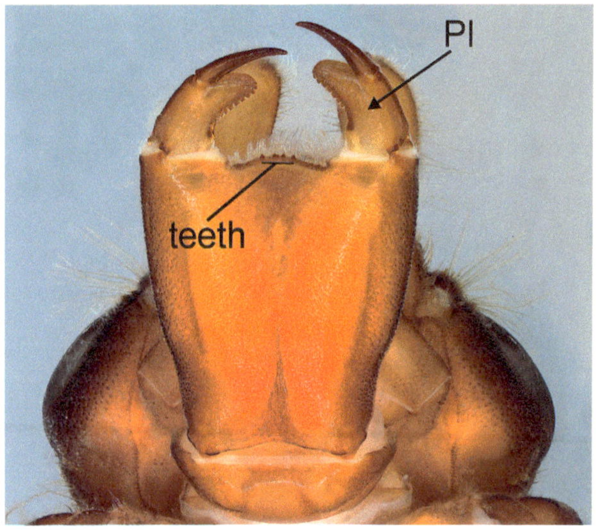

FIGURE 14.2.42 *Octogomphus specularis* (Hagen in Selys): details of larval prementum showing ligula with 4–5 teeth at middle of its apical margin, and palpi (Pl) widely rounded apically with a row of truncated teeth, ventral view.

22' Pro- and mesotibial burrowing hooks well developed 26

23(22) Body strongly depressed, abdomen semicircular (Fig. 14.2.47); 4th antennomere vestigial; prementum quadrate, labial palp medial margin smooth; S3–9 with lateral spines, dorsal protuberances variable, usually well developed on S5–9, that on S9 the largest and, in some cases, excessively enlarged 24

23' Body more or less cylindrical; antennomere IV short, digitiform; prementum subrectangular, labial palp medial margin toothed (Fig. 14.2.48); lateral spines on S7–9, dorsal protuberances on S3–9, gradually increasing in size rearwardly .... *Peruviogomphus* [Ecuador, Peru, Brazil]

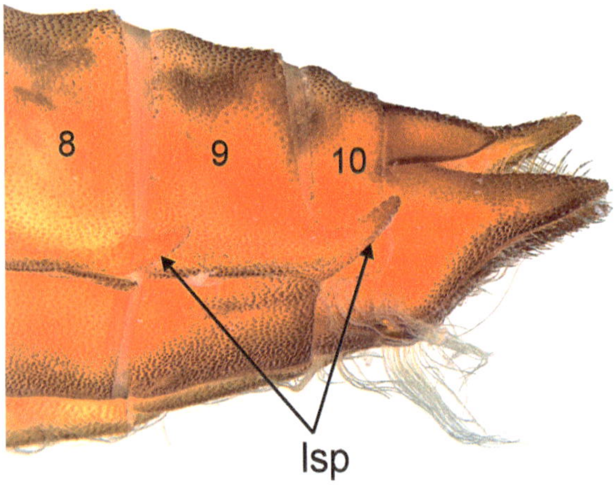

FIGURE 14.2.43 *Praeviogomphus proprius* Belle: details of larval abdominal S8–10 and caudal appendages, lateral view, showing the upturned lateral spines (lsp) on S8–9. *Photo courtesy of Pablo Pessacq.*

FIGURE 14.2.45 *Neogomphus edenticulatus* Carle & Cook: larval abdominal tergites 6–10 and caudal appendages.

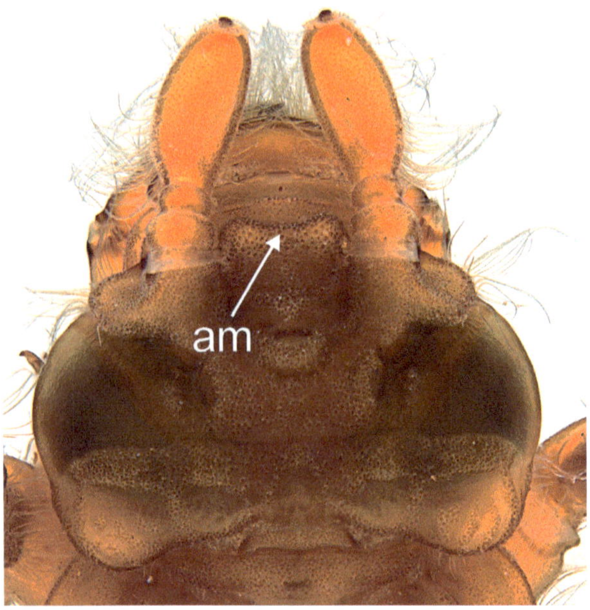

FIGURE 14.2.44 *Praeviogomphus proprius* Belle: details of larval head showing the anterior margin (am) of frons concave. *Photo courtesy of Pablo Pessacq.*

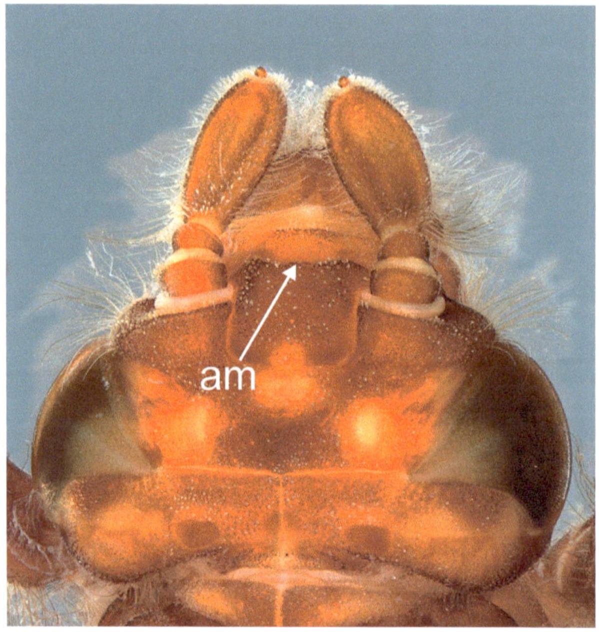

FIGURE 14.2.46 *Neogomphus edenticulatus* Carle & Cook: details of larval head showing the anterior margin (am) of frons undulate.

24(23) Sternum 9 divided into three sternites; dorsal protuberances variable; tibiae with or without brushes of long setae, if present, then only long on lateral surfaces 25

24' Sternum 9 divided into four sternites; dorsal protuberances on S6–9 subequal, almost vertical, that on S9 the highest (Fig. 14.2.49); femora and tibiae with brushes of long, abundant, stiff, black setae (Fig. 14.2.47) .... *Ebegomphus* [N South America]

25(24) S3–5 lateral spines strongly produced laterally (Fig. 14.2.50); lateral margins of S9–10 and caudal appendages forming a more or less continuous smooth outline; S5–9 dorsal protuberances large and well developed, on S9 excessively elongated, straight, and directed back obliquely, its posterior edge straight and not lying over S10 (Fig. 14.2.51); paraprocts almost twice longer than cerci...... *Tibiagomphus* [Brazil, Paraguay, Argentina, Uruguay]

FIGURE 14.2.47 *Ebegomphus* sp., larval dorsal habitus.

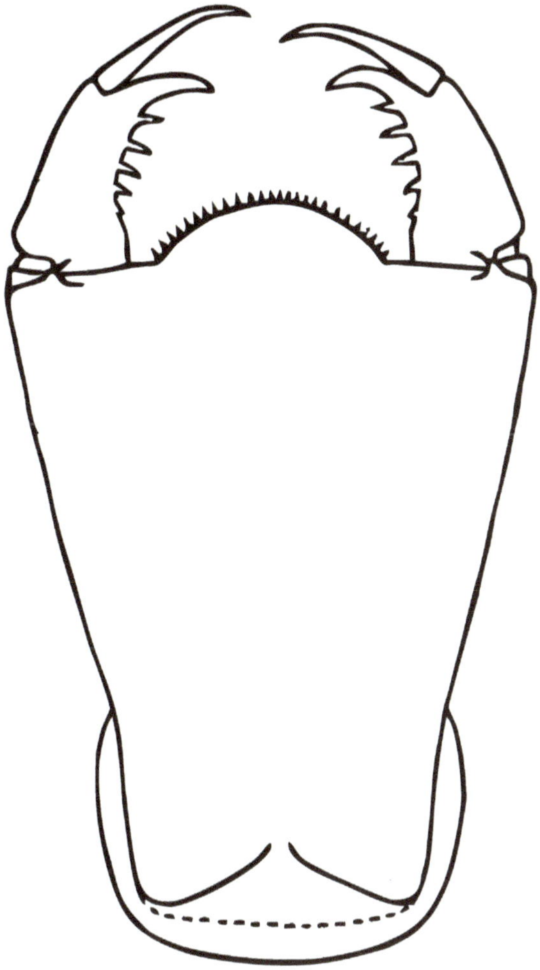

FIGURE 14.2.48 *Peruviogomphus* sp.: larval prementum showing the sharp teeth on internal margin of palpi, ventral view (after Belle, 1992).

25' S3–5 lateral spines not strongly produced laterally (Fig. 14.2.52); lateral margins of S9-10 and caudal appendages off set, not forming a continuous smooth outline; dorsal protuberances variable, if well developed on S6–9, then that of S9 the longest, acutely pointed, directed backward, its posterior edge concave distally (Fig. 14.2.53); if absent on S7–8, then that on S9 excessively elongated, tilted backward and lying over most of S10 (Fig. 14.2.54); paraprocts much less than twice longer than cerci ..... *Agriogomphus*
[Mexico to Peru and Brazil]

26(22) S10 shorter than S9; lateral spines of S9 large, surpassing posterior margin of S10; dorsal protuberances round or sharply pointed on S3– or S4–9; end hook shorter or of the same size than preceding teeth (Fig. 14.2.56) 27

26' S10 notably longer than S9; lateral spines of S9 short, not reaching the basal half of S10; dorsal protuberances on S3–9 sharply pointed and directed backward; end hook of labial palp long, larger than preceding teeth (Fig. 14.2.55) ..... *Gomphoides*
[Brazil, Paraguay, Argentina]

27(26) Third antennomere length 2–3x its widest part, narrow basally and widening distally; lateral spines on S4– or S6–9; dorsal protuberances sharply pointed and directed backward (Fig. 14.2.57); cerci parallel and sharply pointed ..... *Desmogomphus*
[Nicaragua to Colombia, Venezuela, Guyana and Surinam]

27' Third antennomere 5x longer than its widest part; lateral spines on S8–9, long, sharply pointed and upcurved (Fig. 14.2.58); dorsal protuberances almost vertical and rounded apically; tergites 4–8 with sublateral, rounded tubercles beset with stout, thick, sharp, curved spines (Fig. 14.2.59); cerci and paraprocts strongly divergent, cerci rounded apically, paraprocts sharply pointed, epiproct upturned ..... *Diaphlebia*?
[Venezuela, Brazil]

FIGURE 14.2.49 *Ebegomphus* sp.: larval abdominal tergites 6–10 showing dorsal protuberances subequal on S6–9, dorsolateral view.

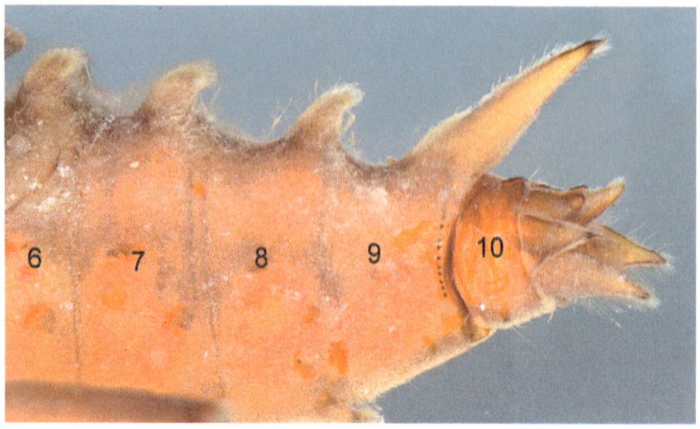

FIGURE 14.2.51 *Tibiagomphus uncatus* (Fraser): larval abdominal S6–10 showing dorsal protuberances subequal on S6–8, larger on S9, lateral view.

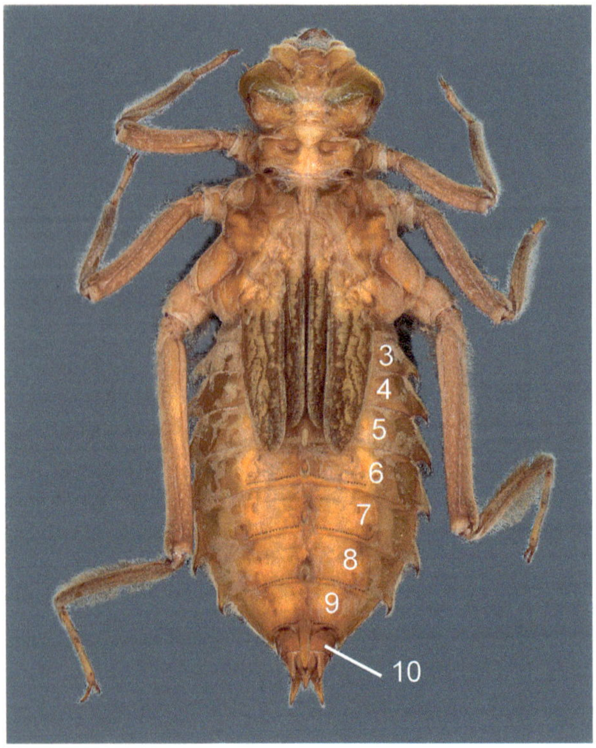

FIGURE 14.2.50 *Tibiagomphus uncatus* (Fraser): larval dorsal habitus. See lateral spines on S3–5 strongly produced laterally.

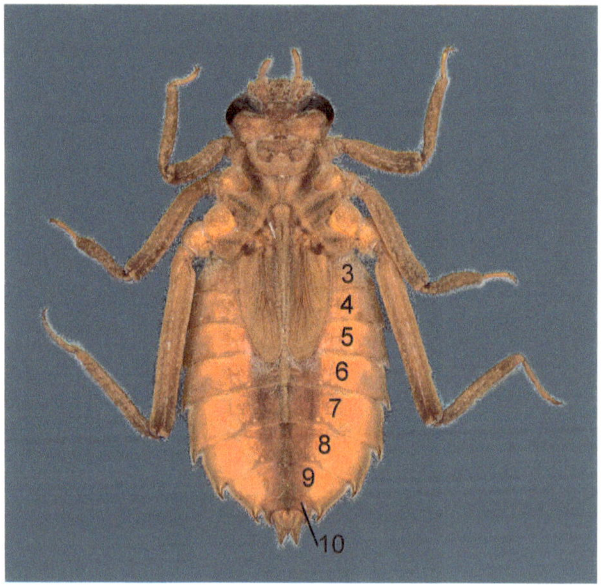

FIGURE 14.2.52 *Agriogomphus tumens* (Calvert): larval dorsal habitus, abdominal S3–10.

FIGURE 14.2.53 *Agriogomphus tumens* (Calvert): larval abdominal tergites 6–10 showing dorsal protuberances S6–9, lateral view.

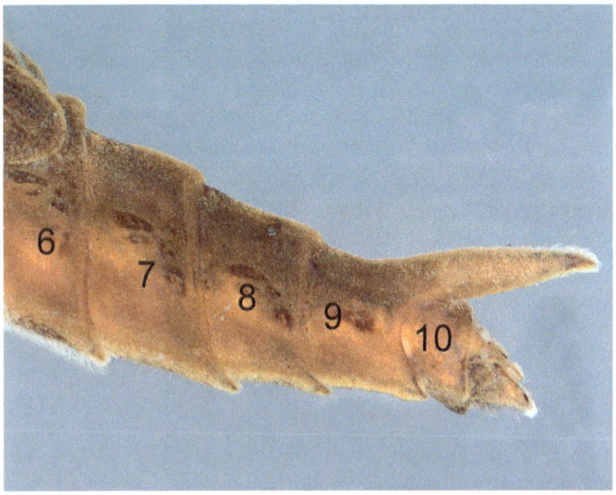

FIGURE 14.2.54 *Agriogomphus jessei* (Williamson): larval abdominal tergites 6–10 showing dorsal protuberances lacking on S7–8, and the large protuberance on S9 lying on S10, lateral view.

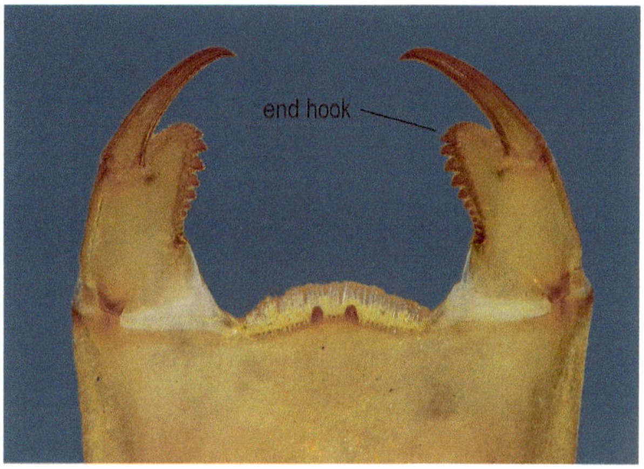

FIGURE 14.2.56 *Desmogomphus paucinervis* (Selys): details of larval prementum showing ligula with two median teeth, and palpi with end hook shorter than preceding teeth, ventral view.

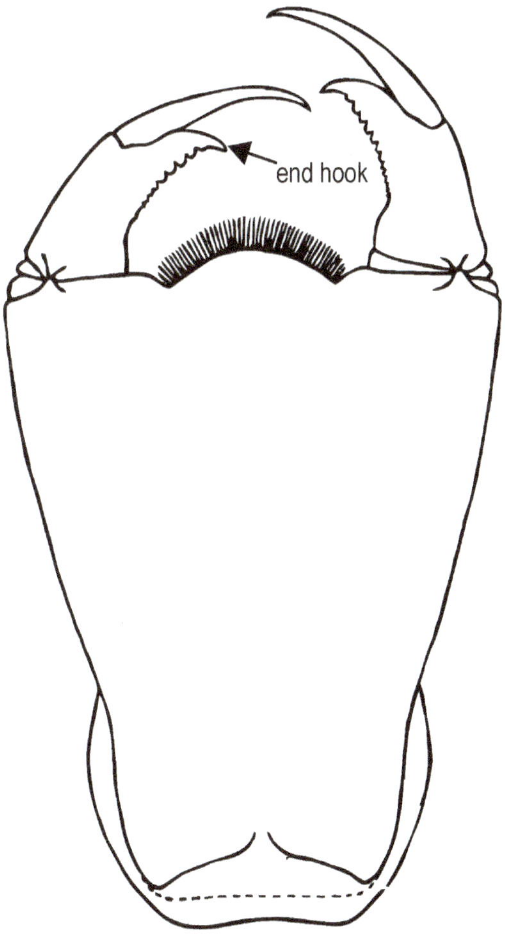

FIGURE 14.2.55 *Gomphoides infumata* (Rambur): ventral view of larval prementum. *After Belle, 1992).*

FIGURE 14.2.57 *Desmogomphus paucinervis* (Selys): larval dorsal habitus.

FIGURE 14.2.58 *Diaphlebia*? sp., larva dorsal habitus.

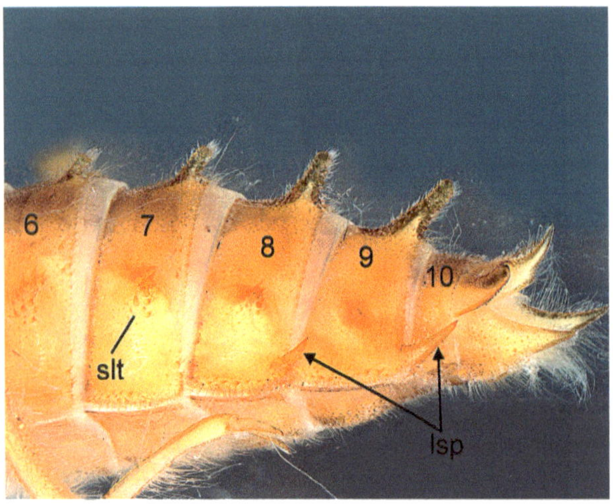

FIGURE 14.2.59 *Diaphlebia*? sp.: larval abdominal tergites 6–10 showing dorsal protuberances on S6–9, the spiny sublateral tubercles (slt), and lateral spines (lsp) on S8–9 directed upwards.

ACKNOWLEDGMENTS

The authors are very grateful to the following people for the loan/donation of larvae of Gomphidae essential for the achievement of this chapter: Luisa Fernanda Alvarez (Colombia), Robert W. Sites (USA), Ulisses G. Neiss (Brazil), Naturalis Biodiversity Center (Leiden), Pablo Pessacq (Argentina), Ken J. Tennessen (USA), Vanessa Amaya (Colombia), William D. Shepard (USA), Daniel Reynoso (México), and Yvonne von Nierop (The Netherlands). Special thanks are due to José Antonio Gómez (Mexico) who clean and edited the images accompanying the key.

REFERENCES

Abbot, J.C. 2005. Dragonflies and damselflies of Texas and the South – Central United States: Texas, Louisiana, Arkansas, Oklahoma, and New Mexico. Princeton University Press, Princeton.

Aoki, T. 1999. Larval development, emergence and seasonal regulation in *Asiagomphus pryeri* (Selys) (Odonata: Gomphidae). Hydrobiologia 394: 179–192.

Belle, J. 1966. Surinam dragon-flies of the *Agriogomphus* complex of genera. Studies on the Fauna of Suriname and other Guyanas 8: 29–60.

Belle, J. 1970. Studies on South America Gomphidae (Odonata) with special reference to the species from Surinam. Studies on the Fauna of Suriname and other Guyanas 11: 1–158.

Belle, J. 1979. An attempt to the subfamily classification of the Gomphidae based on some new interpretations of wing venation (Anisoptera). Odonatologica 8: 43–46.

Belle, J. 1991. The ultimate instar larvae of the Central American species of *Progomphus* Selys, with a description of *P. belyshevi* spec. nov. from Mexico (Anisoptera: Gomphidae). Odonatologica 20: 9–27.

Belle, J. 1992. Studies on ultimate instar larvae of neotropical Gomphidae, with the description of *Tibiagomphus* gen. nov. (Anisoptera). Odonatologica 21: 1–24.

Belle, J. 1996. Higher classification of the South American Gomphidae (Odonata). Zoologische Mededelingen, Leiden 70: 297–324.

Bybee, S.M., T.H Ogden, M.A. Branham, M.F. Whiting, 2008. Molecules, morphology and fossils: a comprehensive approach to odonate phylogeny and the evolution of the odonate wing. Cladistics 23: 1–38.

Carle, F.L. 1986. The classification, phylogeny and biogeography of the Gomphidae (Anisoptera). I. Classification. Odonatologica 15: 275–326.

Carvalho, A.L. 2000. Descriptions of the last instar larva and some structures in the pharate male adult of *Praeviogomphus proprius* Belle, 1995, with notes on the occurrence and taxonomic status of the species (Anisoptera: Gomphidae, Octogomphinae). Odonatologica 29: 239–246.

Corbet, P.S. 1999. Dragonflies-Behavior and Ecology of Odonata. Comstock Publishing Associates, Cornell University Press, Ithaca, New York.

Costa, J.M., L.O.I. Souza, A.M. Lourenco, & B.B. Oldrini. 2004. Chave para as famílias e gêneros das larvas de odonata citadas para o Brasil: Comentários e registros bibliográficos. Publicações Avulsas do Museu Nacional 99:1–44.

Dijkstra, K.-D. B., G. Bechly, S. M. Bybee, R.A. Dow, H.J. Dumont, G. Fleck, R.W. Garrison, M. Hamalainen, V.J. Kalkman, H. Karube, M.L. May, A.G. Orr, D.R. Paulson, A.C. Rehn, G. Theischinger, J.W.H. Trueman, J. van Tol, N. von Ellenrieder & J. Ware. 2013. The classification and diversity of dragonflies and damselflies (Odonata). Zootaxa 3703: 036–045.

Dumont H.J., H. Vierstraete & J.R. Vanfleteren. 2010. A molecular phylogeny of the Odonata (Insecta). Systematic Entomology 35: 618.

Förster, S. 2001. The dragonflies of Central America exclusive of Mexico and the West Indies. A guide to their identification, Second edition. Gunnar Reehfeldt, Braunschweig. Odonatological Monographs.

Garrison, R.W. 1994. A revision of the New World genus *Erpetogomphus* Hagen in Selys (Odonata:Gomphidae). Tijdschrift voor Entomologie 137: 173–269.

Garrison, R.W. & N. von Ellenrieder. 2016. A synonymic list of the New World Odonata (Revised version of 18 February 2016) Available at http://www.odonatacentral.org/docs/NWOL.pdf.

Garrison, R., N. von Ellenrieder & J.A. Louton. 2006. The dragonfly genera (Odonata:Anisoptera) of the New World: An illustrated and annotated key. The John Hopkins University Press.

González-Soriano, E. & R. Novelo-Gutiérrez. 2007. Odonata of México: revisited. Pages 105–136 *in*: B. K. Tyagi (ed.) Odonata: Biology of Dragonflies. Scientific Publishing (India).

Kalkman, V.J., V. Clausnitzer, K.D.B. Dijkstra, A.G. Orr, D.R. Paulson & J. van Tol. 2008. Global Diversity of dragonflies (Odonata) in freshwater. Hydrobiologia 595: 351–363.

Meurgey, F. 2013. A catalogue of the West Indian dragonflies (Insecta:Odonata). Annales de la Societe entomologique de France 49: 298–334.

Muzón, J., P. Pessacq & F. Lozano. 2014. The Odonata (Insecta) of Patagonia: A synopsis of their current status with illustrated keys for their identification. Zootaxa 3784: 346–388.

Needham, J.G. 1940. Studies on neotropical gomphinae dragonflies. Transactions of the American Entomological Society 65: 363–394.

Needham, J.G. 1944. Further studies on Neotropical gomphine dragonflies (Odonata). Transactions of the American Entomological Society 69: 171–224.

Needham, J.G., M.J. Westfall, Jr. & M.L. May. 2014. Dragonflies of North America. The Odonata (Anisoptera) Fauna of Canada, the Continental United States, Northern Mexico and the Greater Antilles. Scientific Publishers, Gainesville.

Neiss, U.G. & N. Hamada. 2014. Ordem Odonata, Chapter 14, pp. 217–284. *in*: N. Hamada, J.L. Nessimian & R.B. Querino (eds.), Insetos Aquáticos na Amazônia brasileira: taxonomia, biologia e ecologia. Instituto Nacional de Pesquisas da Amazônia, Manaus.

Novelo-Gutiérrez, R. 1989. The larva of *Agriogomphus tumens* Calvert (Anisoptera: Gomphidae). Odonatologica 18: 203–207.

Novelo-Gutiérrez, R. 1993. Four new larvae of *Phyllogomphoides* Belle from Mexico (Anisoptera: Gomphidae). Odonatologica 22: 17–26.

Novelo-Gutiérrez, R. 1997. Clave para la identificación de familias y géneros de las náyades de Odonata de México. Parte II. Anisoptera. Dugesiana 4: 31–41.

Novelo-Gutiérrez, R. 2002a. Larvae of the *ophibolus*-species group of *Erpetogomphus* Hagen in Selys from Mexico and Central America (Anisoptera: Gomphidae). Odonatologica 31: 35–46.

Novelo-Gutiérrez, R. 2002b. Two new Mexican larvae of *Erpetogomphus* Hagen in Selys (Odonata: Gomphidae). Journal of the New York Entomological Society 110: 370–375.

Novelo-Gutiérrez, R. 2005. Five new *Erpetogomphus* Hagen in Selys larvae from Mexico, with a key to the known species (Anisoptera: Gomphidae). Odonatologica 34: 243–257.

Novelo-Gutiérrez, R., J.A. Gómez-A. & S.G. Smith. 2015. Description of the larva of *Epigomphus crepidus* Kennedy, 1936 (Odonata: Gomphidae). Zootaxa 4027: 587–592.

Novelo-Gutiérrez, R., A. Ramírez & D. Delgado. 2016. The larvae of *Epigomphus jannyae* Belle, 1993 and *E. tumefactus* Calvert, 1903 (Insecta: Odonata: Gomphidae). PeerJ 4: e2338; DOI 10.7717/peerj.2338.

Paulson, D. 2009. Dragonflies and Damselflies of the West. Princeton University Press, Princeton.

Paulson, D. 2014. North American Odonata, the Odonata of North America. Available at http://www.pugetsound.edu/academics/academic-resources/slater-museum/biodiversity-resources/dragonflies/north-american-odonata/.

Paulson, D. 2016a. South American Odonata. (revised version of 14 June 2016) Available at http://www.pugetsound.edu/academics/academic-resources/slater-museum/biodiversity-resources/dragonflies/south-american-odonata/.

Paulson, D. 2016b. Middle American Odonata. (revised 2 June 2016). http://www.pugetsound.edu/academics/academic-resources/slater-museum/biodiversity-resources/dragonflies/middle-american-odonata/.

Paulson, D. & E. González-Soriano. Mexican Odonata, Odonata of Mexico by state. (compiled, 1994; last revised and updated 3 may 2016). http://www.pugetsound.edu/academics/academic-resources/slater-museum/biodiversity-resources/dragonflies/mexican-odonata/.

Pérez-Gutiérrez L., J. Montes-Fontalvo, M.I. Moreno & L.C. Gutiérrez. 2013. Libélulas de Colombia. Una guía de campo para su identificación. Editorial Universidad del Atlántico.

Ramírez, A. 1996. Six new dragonfly larvae of the family Gomphidae in Costa Rica, with a key to the Central American genera (Anisoptera). Odonatologica 25: 143–156.

Rodrígues-Capitulo, A. 1984. Una nueva especie del género *Cyanogomphus* Selys, incluyendo la decripción del último estadio preimaginal (Odonata: Gomphinae). Revista de la Sociedad Entomológica Argentina 43: 329–336.

Suhling, F. & A. Martens. 2007. Dragonflies and Damselflies of Namibia. Gamsberg Macmillan Publishers, Windhoek.

Suhling, F., G. Sahlén, S. Gorb, V.J. Kalkman, K-D.B. Dijkstra & J. van Tol. 2015. Order Odonata. Pages 893–932 *in*: Thorp, J & D.C. Rogers (eds.), Ecology and General Biology: Thorpand Covichs Freshwater Invertebrates, Academic Press.

Ware, J.L., E. Pilgrim, M.L. May, T.W. Donnelly & K. Tennessen. 2017. Phylogenetic relationships of North American Gomphidae and their close relatives. Systematic Entomology 42: 347–358.

Westfall, M.J. 1989. The larvae of *Desmogomphus paucinervis* (Selys, 1873) and *Perigomphus pallidistylus* (Belle, 1972) (Anisoptera:Gomphidae). Odonatologica 18: 99–106.

Williamson, E.B. 1920. A new gomphine genus from British Guiana with a note on the classification of the subfamily (order Odonata). Occasional Papers of the Museum of Zoology, University of Michigan 80: 1–11.

Odonata: Superfamily Libelluloidea

Ulisses Gaspar Neiss
Instituto de Criminalística, Departamento de Polícia Técnica-Científica, Manaus, Amazonas, Brazil

Gunther Fleck
Ver à soie Musée-Magnanerie, Lagorce, Ardèche, France

Pablo Pessacq
Centro de Investigaciones Esquel de Montaña y Estepa Patagónicas (CIEMEP), Esquel, Chubut, Argentina

Kenneth J. Tennessen
Florida State Collection of Arthropods, Gainesville, Florida, United States

Chapter Outline

INTRODUCTION

Libelluloidea, also called Eurypalpida, is the most diverse and speciose group of Anisoptera, as it includes almost 1,500 species in 195 genera. It has a cosmopolitan distribution and is present in all kinds of environments. The monophyly of the superfamily is strongly supported, and it represents the actual sister group of the Cordulegastroidea within the Cavilabiata (larvae with spoon-shaped mask). However, internal arrangements are still unresolved, and many different schemes have been proposed through the years (Ware et al., 2007; Bybee et al., 2008; Fleck et al., 2008b; Fleck, 2011; Dijkstra et al., 2013; Davis et al., 2011).

In our classificatory scheme, we follow Dijkstra et al. (2013); but in agreement with Fleck (2011) and Fleck & El Adouzi (2013), we recognize Synthemistidae as valid. Thus, four families are herein acknowledged: Synthemistidae, Macromiidae, Corduliidae, and Libellulidae. Corduliidae *sensu lato* (incorporating Synthemistidae, Macromiidae, Corduliidae, and several genera related to the Corduliidae), is a family concept still widely used, but has proved to be paraphyletic (Ware et al., 2007; Fleck et al., 2008b; Carle et al., 2015).

Systematics of Libelluloidea Families

Synthemistidae

It consists of two subfamilies: Synthemistinae, composed of 10 non-Neotropical genera, and Gomphomacromiinae comprised of three genera (Fleck, 2011; Fleck & El Adouzi, 2013), of which only *Gomphomacromia* is represented in the Neotropical region. This last genus consists of four species distributed along the Andes of South America from Ecuador to Southern Argentina and Chile (von Ellenrieder & Garrison, 2005). Only the larva of *G. paradoxa* Brauer, 1864 is known; but by supposition, it inhabits mountain streams, seepages, and bogs, and its larvae seem to be semiterrestrial (von Ellenrieder &

Thorp and Covich's Freshwater Invertebrates, Fourth Edition. https://doi.org/10.1016/B978-0-12-804223-6.00017-2

Garrison, 2005; Fleck pers. obs.). This genus can be recognized by the crenulations of the distal margin of labial palps, which are large but few in number (<6), and by the complete lack of raptorial setae and the absence of abdominal lateral spines.

Macromiidae

This is a small family of medium to large dragonflies. It is distributed nearly worldwide (Garrison et al., 2006). It is currently composed of 136 species (125 after Dijkstra et al., 2013) in four genera (Garrison et al., 2006). The monophyly of the family is supported by morphological (May, 1997) and molecular analyses (Ware et al., 2007). Only one genus, *Macromia*, and two species are represented in the Neotropical region, reaching northern Mexico. Larvae of both species are known. In the New World, larvae mainly inhabit running water with sandy bottoms and can be recognized by the unmistakable marked spider-like appearance due to extremely long legs and ovoid abdomen (shared with some Corduliidae) and by the conical projection in frons.

Corduliidae

This family is mainly distributed in the Holarctic region. In the Neotropics, it is represented by 50 species in eight endemic genera (with the exception of *Epitheca*, with a wider distribution). In a restricted sense, our proposed classification follows Dijkstra et al. (2013). Although the family seems to be monophyletic, its inner arrangement and exact composition is not fully resolved (Bechly, 1996, 2003; Fleck, 2012). *Lauromacromia* and *Neocordulia*, long considered as Corduliidae *sensu lato*, are now placed as genera *incertae sedis* since they cannot be placed satisfactorily in any family (Dijkstra et al., 2013). We have included them in the family key for practical reasons. *Neocordulia* is apparently closely related to the European genus *Oxygastra* (Fleck, 2002), which is placed by several authors in its own family Oxygastridae.

In the Neotropics, Corduliidae larvae preferentially inhabit cold or meso- or oligotrophic streams and lakes, mainly in mountain regions.

Libellulidae

Libellulidae, with 143 genera and more than 1,000 species, is the most diverse and ubiquitous family of Anisoptera. Its members, distributed worldwide, are quite variable in size and coloration, and inhabit a wide variety of environments. More than 390 species in 48 genera are known for the Neotropical region. The monophyly of the family seems to be strongly supported, but the internal arrangement and relationships among genera are still a matter of debate, and most traditional subfamilies are not supported by phylogenetic analyses (Fleck, 2004; Ware et al., 2007; Bybee et al., 2008; Fleck et al., 2008a, b; Dumont et al., 2010). In agreement with Artiss et al. (2001), we recognize *Plathemis* as a valid genus. Larvae virtually inhabit all freshwater aquatic ecosystems, including lotic and lentic systems as well as both permanent and temporary environments. A few species inhabit phytotelmata (e.g., some *Erythrodiplax* species and *Libellula herculea* Karsch, 1889; De Marmels, 1992; Haber et al., 2015), while *Erythrodiplax berenice* (Drury, 1773) is found in salt marshes (Dunson, 1980) and *Macrodiplax* can also be found in brackish water.

LIMITATIONS

Current knowledge of Libelluloidea immatures is still poor, many descriptions are old, based on association with only one exuvia, too brief, or lacking adequate descriptions and figures. Additionally, some genera need revision, since the taxonomy is essentially based on adults and not on larvae, which sometimes are difficult to separate (e.g., *Brechmorhoga*, *Macrothemis*, *Erythrodiplax*, *Crocothemis*, *Pachydiplax*, *Micrathyria*, *Nephepeltia*, *Anatya*, etc). However, there could be a considerable morphological variation among larvae of apparently monophyletic genera (*Aeschnosoma*, *Anatya*, *Libellula*, etc.). All this creates difficulties in larval identification, and thus on all disciplines that rely on it, such as ecology, biological assessments, behavior studies, etc.

There are several family level keys for different regions (e.g. De Marmels, 1990 for Venezuela; Carvalho & Calil, 2000; Costa et al., 2004 for Brazil; Heckman, 2006; von Ellenrieder & Garrison, 2009 for South America; Neiss & Hamada, 2014, for Brazilian Amazonia) and some genera keys, ranging from small regions or states (e.g., Carvalho et al., 2002; Muzón et al., 2014), countries (Novelo-Gutiérrez, 1997; Costa et al., 2004), and for South America as a whole (Heckman, 2006).

Given the poor knowledge of Libelluloidea immatures, the present key should be used with caution; and if it is possible, generic identity should be confirmed with original descriptions. Information on distribution of the Neotropical genera of Anisoptera can be found in Garrison et al. (2006).

KEYS TO LIBELLULOIDEA

Libelluloidea: Families

1 Labial palp distal margins with variable raptorial setae (Figs. 14.3.4, 5); pronotum not developed as a large shoulder pad (Figs. 14.3.6) (clade Trichodopalpida) 2

1' Labial palp distal margins without raptorial setae (Fig. 14.3.1); pronotum strongly developed laterally, with two large shoulder pads (Figs. 14.3.2, 3) ..... Synthemistidae, one genus: *Gomphomacromia* [Ecuador, Chile, Argentina]

2(1) Frons without a strong subpyramidal projection (Fig. 14.3.6) 3

2' Frons with a strong subpyramidal projection (like a horn) (Figs. 14.3.7, 8) ..... Macromiidae, one genus: *Macromia* [Mexico. Afrotropical, Australian, Nearctic, Oriental, Palaearctic]

3(2) Prementum ventrally without mediobasal longitudinal groove (Fig. 14.3.9) (except some *Perithemis* and *Tholymis*, where groove is weak, see general habitus of these larvae in Figs. 14.3.113, 88, respectively); labial palp distal margin dentation variable, can be weakly developed or obsolete (Fig. 14.3.10), if not obsolete, rarely with more than four raptorial setae on each tooth (Fig. 14.3.5); cerci usually shorter than paraproct, length often <3/5 paraprocts (Fig. 14.3.11) ..... **Libellulidae** [p. 401]

3' Prementum ventrally with an obvious well-marked mediobasal longitudinal groove (Fig. 14.3.12); labial palp distal margin crenation well-marked, separated by relatively deep notches, and usually with more than four raptorial setae on each tooth (except *Aeschnosoma* and *Navicordulia*) (Fig. 14.3.4); cerci usually long, length at least 3/5 paraprocts (Fig. 14.3.13) ..... **"Corduliidae"** [p. 401]

Libelluloidea: Corduliidae and Related *Incertae Sedis* Genera (*Neocordulia* and *Lauromacromia*): Genera

Larva of all genera, and 24 of the 50 Neotropical species of the family have been described. Although several genera possess a typical spider-like appearance, there are few morphological features that distinguish them from Libellulidae. These characters are the short longitudinal suture on the posterioventral surface of the prementum (Theischinger & Fleck, 2003) and in most genera, spatulate spines or setae in the frons. This key, modified from Fleck & Neiss (2012) and Carvalho et al. (2008), includes the genera *Lauromacromia* and *Neocordulia*, taxa *incertae sedis* previously considered as Corduliidae *sensu lato* (see Fleck, 2002; Dijkstra et al., 2013).

1 Some abdominal segments with middorsal hooks or protuberances (Fig. 14.3.14) 2

1' Abdominal middorsal hooks or protuberances absent (Fig. 14.3.15) 6

2(1) Some abdominal segments proximal to S8 with lateral spines (those of S5–S7 generally very distinct) (Fig. 14.3.16) 3

2' Abdominal segments proximal to S8 without lateral spines (Fig. 14.3.17) 4

3(2)

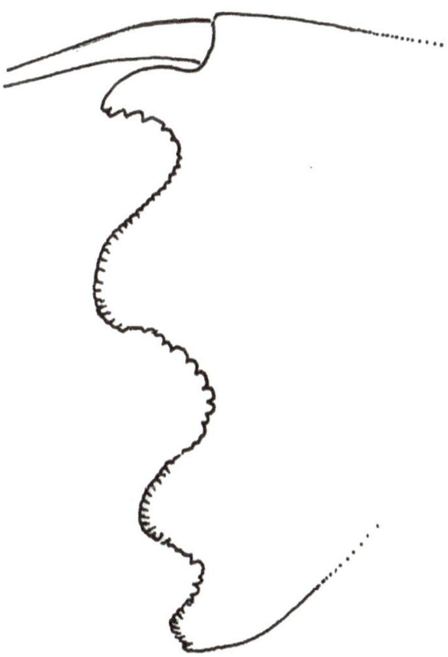

FIGURE14.3.1 *Gomphomacromia* sp., distal margins of labial palp, frontal view.

FIGURE 14.3.2 *Gomphomacromia* sp., pronotum, dorsal view. Arrows indicate the two large shoulder pad.

ODONATA: SUPERFAMILY LIBELLULOIDEA

FIGURE 14.3.3 *Gomphomacromia* sp., general habitus of larva, dorsal view.

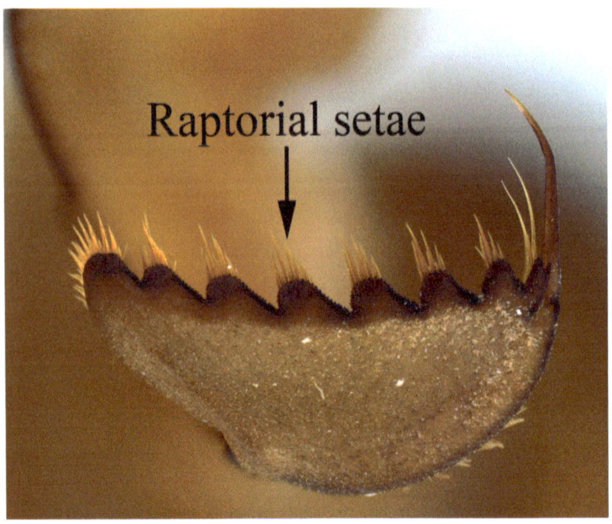

FIGURE 14.3.4 *Neocordulia* sp., distal margin of labial palp, frontal view. Arrow indicates raptorial setae.

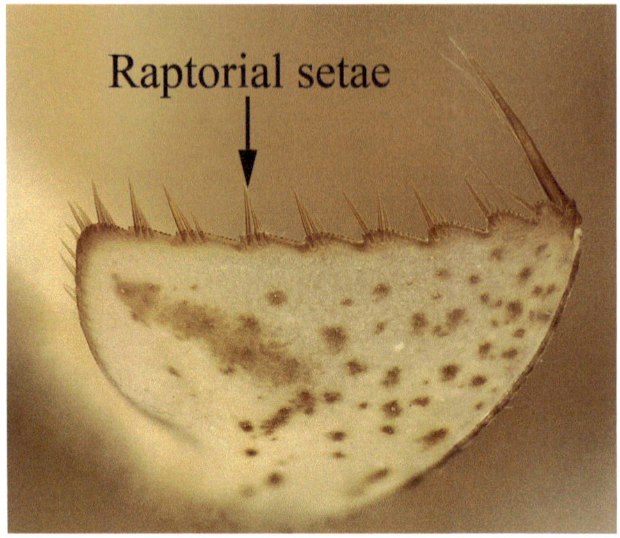

FIGURE 14.3.5 *Dythemis* sp., distal margin of labial palp, frontal view. Arrow indicate raptorial setae.

FIGURE 14.3.6 *Neocordulia* sp., pronotum and head, dorsal view. Arrows indicate pronotum not strongly developed laterally.

Occiput without conical protuberances (Fig. 14.3.18); frontal plate prominent and furnished with distinct spiniform and scaliform setae (Figs. 14.3.19, 20)........ *Lauromacromia*
[Venezuela, French Guiana, Brazil]

3' Occiput with a prominent pair of conical protuberances (Fig. 14.3.21), frontal plate more reduced and covered with minute spiniform setae (Figs. 14.3.22, 23)........ *Rialla*
[Chile and Argentina]

4(2) S9 lateral abdominal spines strongly developed and largely extending over anal pyramid apex, with length $\geq$ S9 middorsal length, and not convergent (Fig. 14.3.26) 5

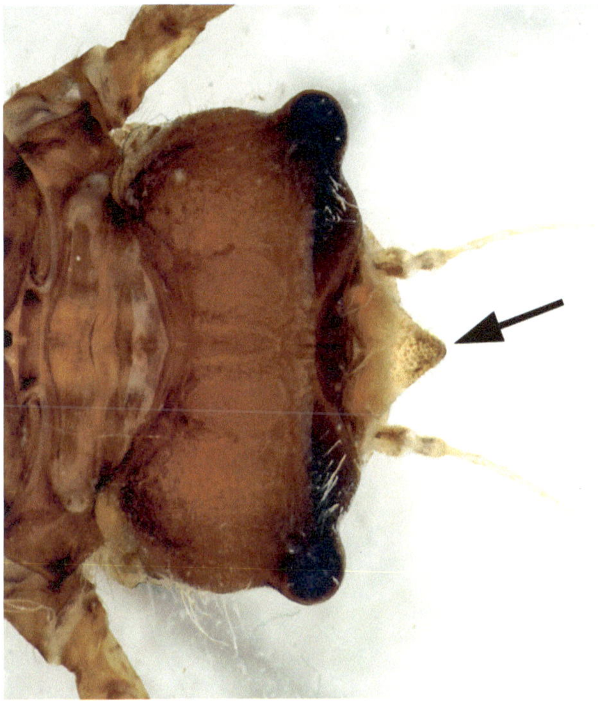

FIGURE 14.3.7 *Macromia anullata* Hagen, 1861, head, dorsal view. Arrow indicates subpyramidal projection on frons of head (like a horn).

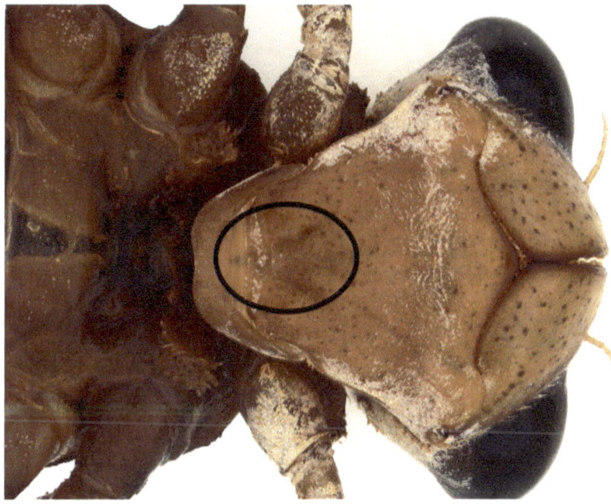

FIGURE 14.3.9 *Elasmothemis* sp., larval folded labium, ventral view. Circle indicate no mediobasal longitudinal groove.

FIGURE 14.3.8 *Macromia anullata*, general habitus of larva, dorsal view.

FIGURE 14.3.10 *Elasmothemis* sp., labial palp, inner view. Dentation of distal margin of labial palp obsolete.

FIGURE 14.3.11 *Pachydiplax* sp., apex of abdomen, dorsal view. Dark lines show cerci in comparison with paraprocts.

FIGURE 14.3.13 *Paracordulia* sp., apex of abdomen, dorsal view. Dark lines show cerci in comparison with paraprocts.

FIGURE 14.3.12 *Neocordulia* sp., larval folded labium, ventral view. Arrow indicates a well-marked mediobasal longitudinal groove.

FIGURE 14.3.14 *Paracordulia* sp., abdomen, dorso-lateral view. Arrows indicate middorsal hooks. *Original pictures by Neiss & Hamada (2014).*

4' S9 lateral abdominal spines not greatly developed, never extending over apex of the anal pyramid, with length < S9 middorsal length, and convergent (Figs. 14.3.24, 25) .... *Navicordulia*
[Venezuela, French Guiana, Brazil]

5(4) S8 and S9 lateral abdominal spines subequal and distinctly upturned in lateral view (Fig. 14.3.14); S8 and S9 lateral abdominal spines strongly divergent, those of S8 especially divergent (Figs. 14.3.26, 27) .... *Paracordulia*
[Amazon region]

FIGURE 14.3.15 *Neocordulia* sp., abdomen, lateral view. Arrows indicate absent of middorsal hooks or protuberances. *Original pictures by Neiss & Hamada (2014).*

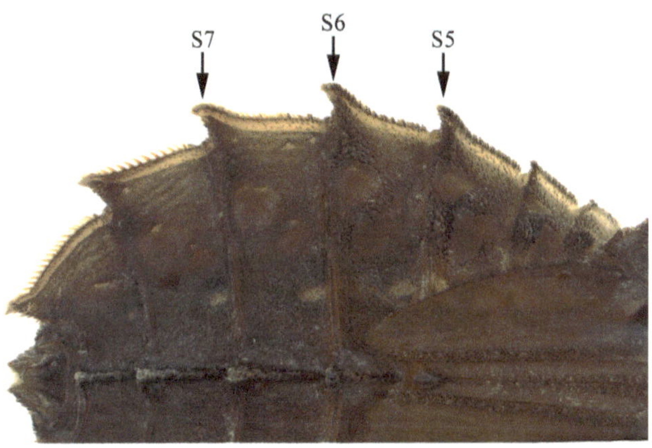

FIGURE 14.3.16 *Lauromacromia* sp., abdomen, dorsal view. Arrows indicate spine in abdominal segment 5 (S5), abdominal segment 6 (S6) and abdominal segment 7 (S7). *Original pictures by Neiss & Hamada (2014).*

FIGURE 14.3.17 *Aeschnosoma pseudoforcipula* Fleck, 2012, abdomen, dorsal view.

FIGURE 14.3.18 *Lauromacromia* sp., head, dorsal view. Circle indicates absence of conical protuberances on dorsal surface of occiput.

FIGURE 14.3.19 *Lauromacromia* sp., head, dorsal view, detail frontal plate (prominent and furnished with spinelike and scalelike setae).

FIGURE 14.3.20 *Lauromacromia* sp., general habitus of larva, dorsal view.

FIGURE 14.3.22 *Rialla* sp., head dorsal, detail frontal plate (more reduced and covered with minute spinifom setae).

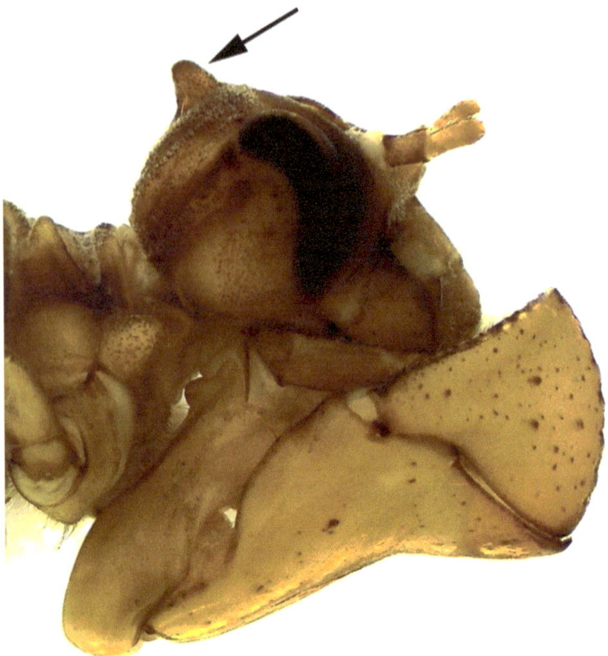

FIGURE 14.3.21 *Rialla villosa* Rambur, 1842, pronotum and head, lateral view. Arrow indicates a prominent conical protuberance on dorsal surface of occiput.

FIGURE 14.3.23 *Rialla* sp., general habitus of larva, dorsal view.

5' S8 lateral abdominal spines length <0.5× S9 spines, not distinctly upturned in lateral view (Fig. 14.3.28); S8 and S9 lateral spines not or only slightly divergent (Fig. 14.3.29)........ *Epitheca*
[Mexico. Nearctic, Palearctic]

6(1) S9 lateral abdominal spines poorly developed and never extending over anal pyramid apex (Fig. 14.3.34); occiput posterior and lateral margins not separated dorsally by a low tubercle made by a whorl of strong spiniform setae (Fig. 14.3.35, dark circles)........ 7

6' S9 lateral abdominal spines very strongly developed and largely extending over anal pyramid apex (Figs. 14.3.17, 30); occiput posterior and lateral margins separated dorsally by a low tubercle made by a whorl of strong spiniform setae (Figs. 14.3.31–33).... *Aeschnosoma*
[North of South America]

FIGURE 14.3.24 *Navicordulia nitens* De Marmels, 1991, apex of abdomen, dorsal view. Arrow indicates S9 lateral abdominal spine.

FIGURE 14.3.26 *Paracordulia* sp., abdomen, dorsal view. Arrow indicates S9 lateral abdominal spine. *Original pictures by Neiss & Hamada (2014).*

FIGURE 14.3.25 *Navicordulia nitens*, general habitus of exuvia, dorsal view.

FIGURE 14.3.27 *Paracordulia* sp., general habitus of exuvia, dorsal view. *Original pictures by Fleck & Neiss (2012).*

7(6) Antennae shorter than head middorsal length (Fig. 14.3.35, arrow); frontal plate prominent and with scaliform setae (Figs. 14.5.36, 37) .... *Neocordulia*
[Mexico to Paraguay and Brazil]

7' Antennae longer than head middorsal length; frontal plate not prominent and without scaliform setae (Fig. 14.3.38)..... *Cordulisantosia*
[Brazil; Rio de Janeiro, São Paulo]

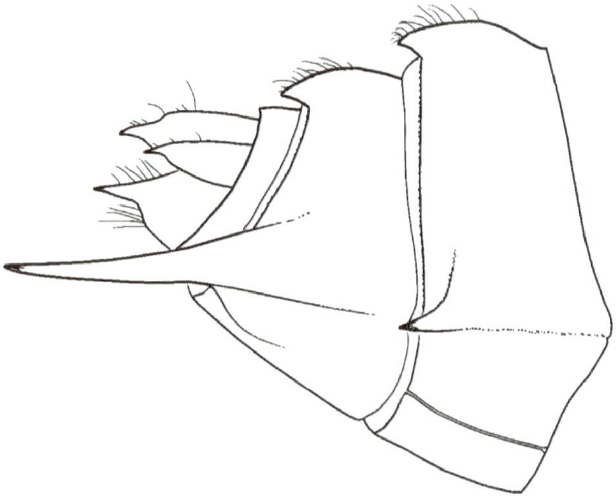

FIGURE 14.3.28 *Epitheca* sp., apex of abdomen, right lateral view (lateral spines on S8–9 not distinctly upturned in lateral view).

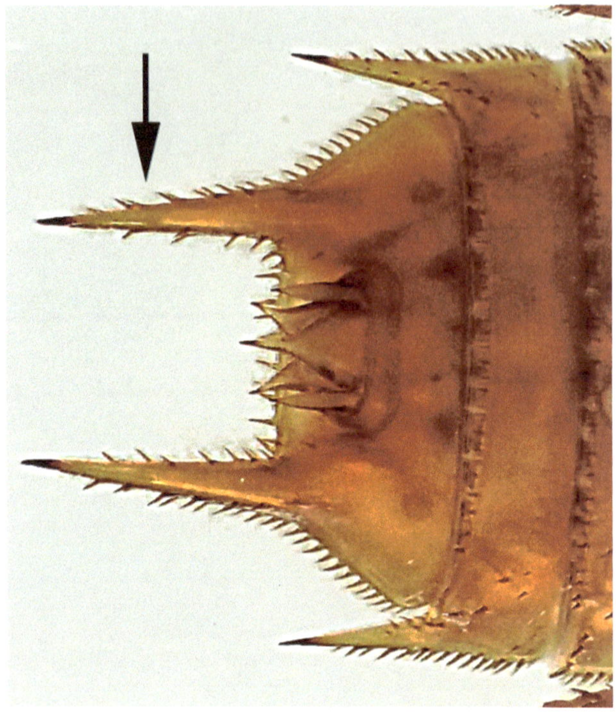

FIGURE 14.3.30 *Aeschnosoma* sp., apex of abdomen, dorsal view. Arrow indicates S9 lateral abdominal spine.

FIGURE 14.3.29 *Epitheca petechialis* Muttkowski, 1911, general habitus of larva, dorsal view.

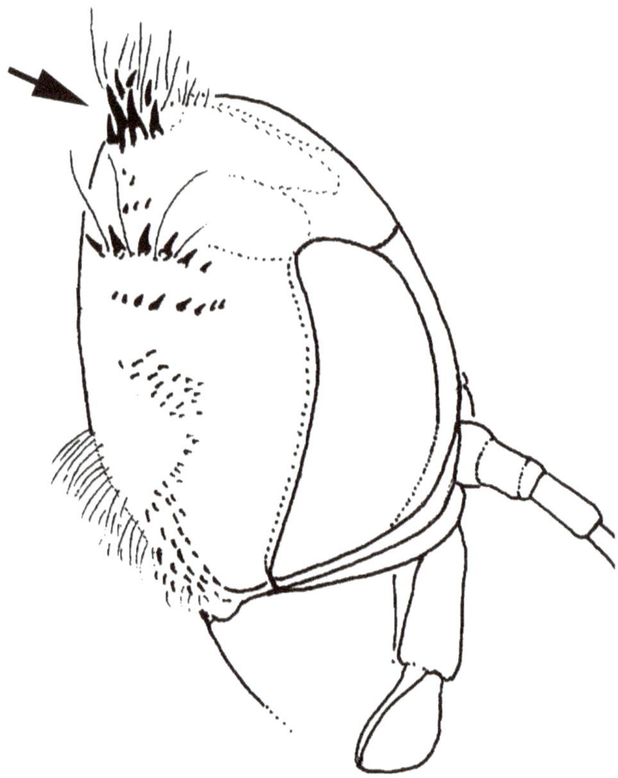

FIGURE 14.3.31 *Aeschnosoma* sp., head, lateral view. Arrow indicates a low tubercle made by a whorl of strong spinelike setae.

FIGURE 14.3.32 *Aeschnosoma* sp., general habitus of larva, dorsal view.

FIGURE 14.3.33 *Aeschnosoma pseudoforcipula* Fleck, De Marmels & Hamada, 2012, general habitus of larva, dorsal view.

FIGURE 14.3.34 *Neocordulia* sp., apex of abdomen, dorsal view. Arrow indicates S9 lateral abdominal spine. *Original picture by Neiss & Hamada (2014).*

FIGURE 14.3.35 *Neocordulia* sp., head, dorsal view. Circles shows posterior margin and lateral margins of occiput (without a low tubercle made by a whorl of strong spiniform setae). Arrow indicates the antenna (shorter than middorsal length of head).

FIGURE 14.3.36 *Neocordulia* sp., head, dorsal view, show frontal plate prominent and with scaliform setae.

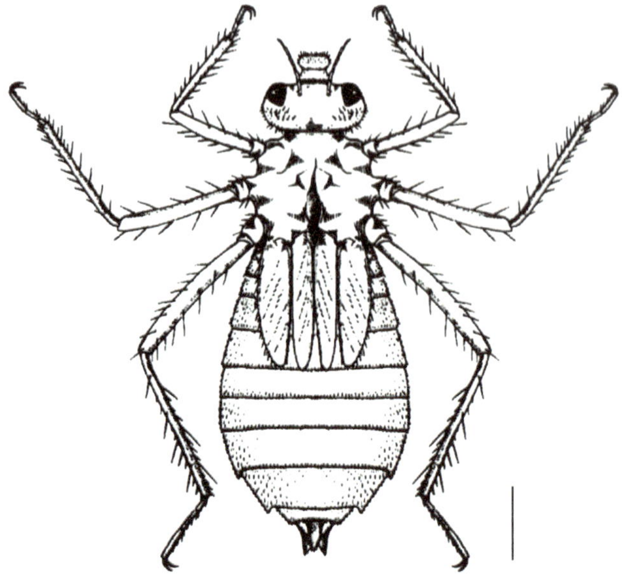

FIGURE 14.3.38 *Cordulisantosia machadoi* (Costa & Santos, 2000), general habitus of exuvia, dorsal view. *Modified from Costa & Santos (2000).*

FIGURE 14.3.37 *Neocordulia* sp., general habitus of larva, dorsal view.

FIGURE 14.3.39 *Macrothemis* sp., abdomen, dorsolateral view.

Libelluloidea: Libellulidae: Genera

The larvae are still undescribed for eight genera (*Antidythemis*, *Carajathemis*, *Edonis*, *Garrisonia*, *Misagria*, *Nothodiplax*, *Uracis*, and *Ypirangathemis*) and about 58% of the species. There are two typical morphologies: the "climbers/sprawlers" (also called "sympetrine" form) and the "burrowers" (also called "libelluline" form), which are characterized by smaller eyes, more squarish head, shorter legs, more elongated body, and hairier bodies than the climbers/spawlers form. Some climbers/spawlers genera with very long legs resemble the spider-like appearance of corduliids (e.g., *Argyrothemis* and

Oligoclada). The diversity of the family, the lack of knowledge for many species, and the nonmonophyly and similarities among several genera make the identification of several genera a challenge, and this key must be used carefully. Given the complexity of the group, habitus are provided for several genera to facilitate identification. The key is made for the ultimate instar. The presence of "+" between characters implies that all conditions must be filled; "*" indicates that exuviae must be carefully prepared for the specific characters. During emergence the thorax opens to allow the exit of the adult and could therefore be somewhat deformed. Exuviae always present divergent wing sheaths, whereas most living larvae have parallel wing sheaths.

1 Middorsal hooks or protuberances present on some abdominal segments, the middorsal hooks can be hidden by wing sheaths (e.g., *Pantala*), sediments, or detritus; long dorsal setae agglomerated as a brush must also not be confused with spines which are cuticular extensions (e.g., *E. berenice*) (Figs. 14.3.39–41) 2

1' No middorsal hooks or distinct protuberances of any abdominal segments (Fig. 14.3.42) 27

2(1) Lateral spines present on segments 6–9 (Fig. 14.3.43); abdominal segments at least partially fused 3

2' Lateral spines absent on segments 6–7 (present on segments 8–9 or lacking) (Fig. 14.3.44); abdominal segments free 4

3(2) S6 lateral spine very large, much longer than remaining ones, oriented perpendicular to main body axis; lateral spines decreasing in size from S6 to S9 (Figs. 14.3.45, 46) .... *Orionothemis*
[Brazil; Bahia]

FIGURE 14.3.40 *Brechmorhroga* sp., abdomen, lateral view.

3'

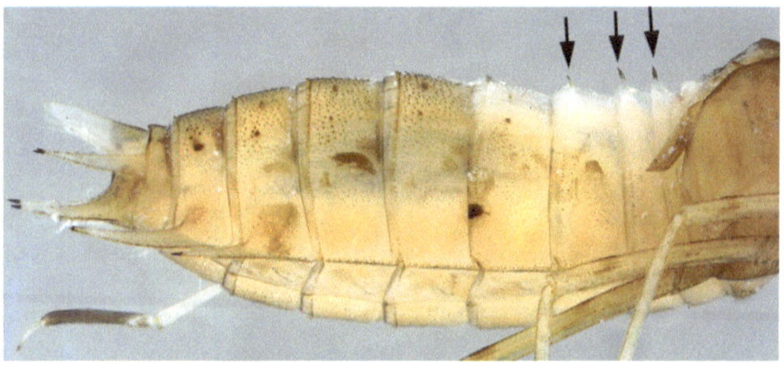

FIGURE 14.3.41 *Pantala* sp., abdomen, lateral view. Arrows indicate middorsal hooks on S2-4.

FIGURE 14.3.42 *Micrathyria* sp., abdomen, dorsolateral view.

FIGURE 14.3.43 *Elasmothemis* sp., abdomen, dorsal view of S5-10 (Arrows indicate lateral spines present on S6-9).

FIGURE 14.3.44 *Macrothemis* sp., abdomen, dorsal view. Arrows indicate lateral spines present on S8-9.

FIGURE 14.3.45 *Orionothemis felixorioni* Fleck, Hamada & Carvalho, 2009, abdomen, dorsal view (lateral spine on S6 very large).

FIGURE 14.3.46 *Orionothemis felixorioni*, general habitus of larva, dorsal view.

S6 lateral spine not very large, shorter than remaining ones, not oriented perpendicular to main body axis (Fig. 14.3.43); lateral spines not decreasing in size from S6 to S9 (size globally increase from S6 to S9 but S7 and S8 spines could be subequal (Figs. 14.3.43, 47)........ .... *Elasmothemis*
[Mexico to Argentina]

4(2) S9 without middorsal hook (Fig. 14.3.48, dark circle) 5

4' S9 with middorsal hook (Fig. 14.3.49) 13

5(4) S8 without middorsal hook (Fig. 14.3.48, dark circle) 6

5' S8 with middorsal hook (Fig. 14.3.50, white arrow) 9

6(5) Head not squarish as below, eyes globular or protruding laterally (Fig. 14.3.54); prementum distal margin not scalloped (Fig. 14.3.55).. 7

6' Head squarish, with long subparallel or parallel postoccular margins, eyes rather small in dorsal view placed anterolaterally, and slightly but distinctly protruding dorsally (Fig. 14.3.51); prementum distal margin distinctly scalloped with each scalloped separated by a small spine (Figs. 14.3.52, 53) .... *Plathemis*
[Mexico. Holarctic]

7(6) Elements of the anal pyramid shorter and not aciculate (Fig. 14.3.48, arrow); some middorsal hooks not hidden by wing sheaths and middorsal at least present on S3–7; combination of other characters not as below 8

7' All elements of the anal pyramid very long and aciculate (Fig. 14.3.56); middorsal hooks on S2–4 slender, erect, middorsal hidden by wing sheaths (Fig. 14.3.41); labial palp distal margin with deep crenations and rounded teeth + prementum distal margin without distinct microscopic spiniform setae (Fig. 14.3.55) + movable hook short and robust (Fig. 14.3.57) .... *Pantala*
[Neotropical. Pantropical]

8(7) S9 with lateral spines distinctly longer than middorsal length (Figs. 14.3.58, 60); eyes well protruding laterally (Fig. 14.3.59)........ .... *Celithemis*
[Cuba, Bahamas, Mexico. Holarctic]

8' S9 with lateral spines distinctly shorter than middorsal length (Fig. 14.3.48); eyes rounded, not greatly protruding laterally (Figs. 14.3.61, 62) .... *Paltothemis*
[Mexico to Panama. Nearctic]

9(5) Eyes not as below, larger, not distinctly produced dorsally (Fig. 14.3.65), S9 lateral spine length $\geq 0.5\times$ S9 middorsal length (Fig. 14.3.66) 10

9' Head squarish, with postoccular margins subparallel or parallel, eyes rather small in dorsal view, positioned anteriolaterally, slightly but distinctly protruding dorsally (Fig. 14.3.63); S9 lateral spine vestigial or very short, length $<1/3\times$ S9 middorsal length (Figs. 14.3.50, black arrow, 64) .... *Libellula* (in part)
[Mexico to Argentina. Holarctic]

FIGURE 14.3.47 *Elasmothemis* sp., general habitus of larva, dorsal view.

FIGURE 14.3.50 *Libellula herculea* Karsch, 1889, apex of abdomen, dorsal view. White arrow indicates middorsal hook present on S8. Dark arrow indicates lateral spine on S9.

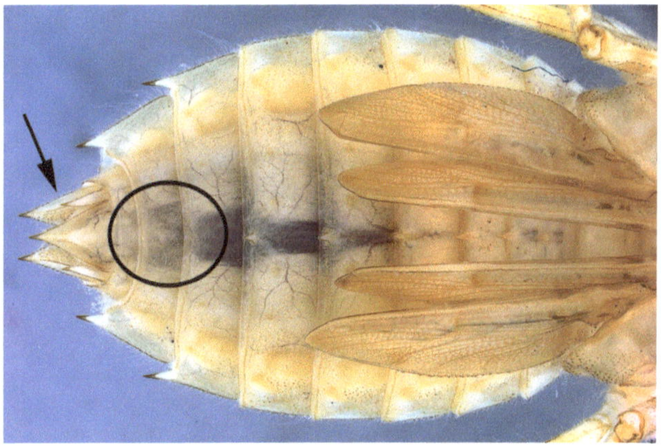

FIGURE 14.3.48 *Paltothemis* sp., abdomen, dorsal view. Circle indicates middorsal hook absent on S8 and S9. Arrow indicates anal pyramid.

FIGURE 14.3.49 *Tholymis citrina* Hagen, 1867, abdomen, dorsal view. Arrow indicates middorsal hook present on S9.

FIGURE 14.3.51 *Plathemis* sp., head, dorsal view.

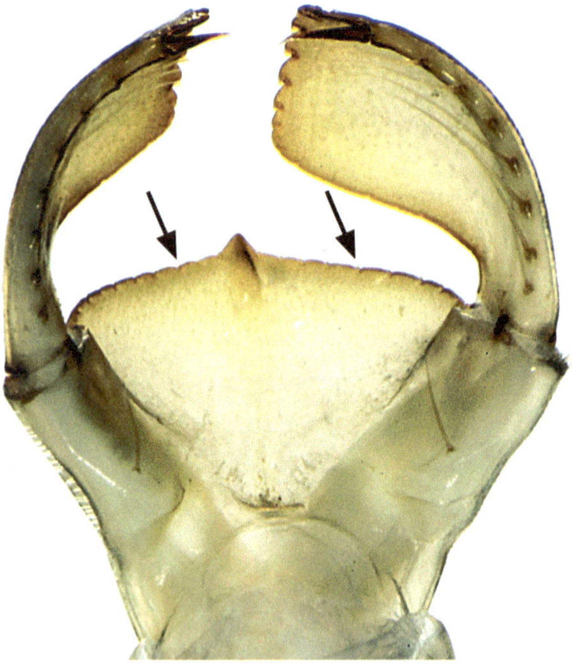

FIGURE 14.3.52 *Plathemis* sp., labium dorsal view. Arrows indicate distal margin of prementum distinctly scalloped.

FIGURE 14.3.54 *Pantala* sp., head, dorsal view.

FIGURE 14.3.53 *Plathemis* sp., general habitus of larva, dorsal view.

FIGURE 14.3.55 *Pantala* sp., labium, dorsal view. Dark arrow indicates distal margin of prementum not scalloped. White arrow indicates distal margin of labial palp with deep crenation.

FIGURE 14.3.56 *Pantala* sp., abdomen, dorsal view.

FIGURE 14.3.58 *Celithemis eponina* (Drury, 1773), apex of abdomen, dorsal view. Arrow indicates S9 abdominal lateral spine.

FIGURE 14.3.57 *Pantala* sp., general habitus of larva, dorsal view.

FIGURE 14.3.59 *Celithemis eponina*, head, dorsal view.

FIGURE 14.3.60 *Celithemis eponina*, general habitus of larva, dorsal view.

FIGURE 14.3.62 *Paltothemis* sp., general habitus of larva, dorsal view.

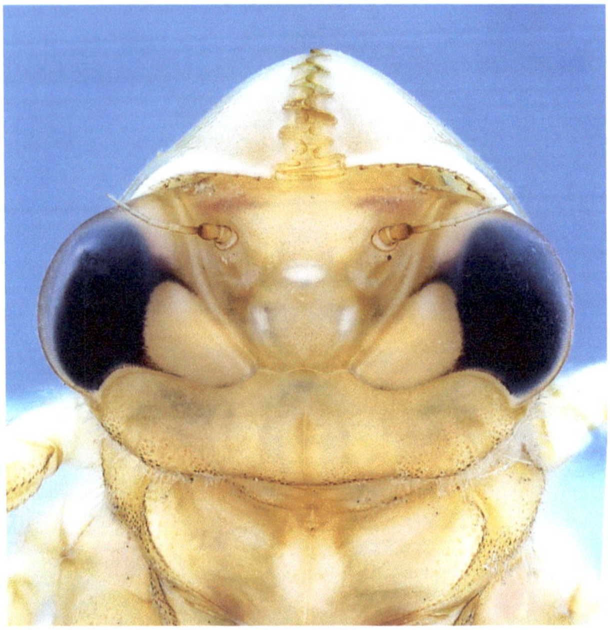

FIGURE 14.3.61 *Paltothemis* sp., head, dorsal view.

FIGURE 14.3.63 *Libellula herculea*, head, dorsal view.

FIGURE 14.3.64 *Libellula herculea*, general habitus of larva, dorsal view.

FIGURE 14.3.66 *Miathyria* sp., abdomen, dorsal view.

FIGURE 14.3.67 *Zenithoptera* sp., abdomen, dorsal view.

FIGURE 14.3.65 *Zenithoptera* sp., head, dorsal view.

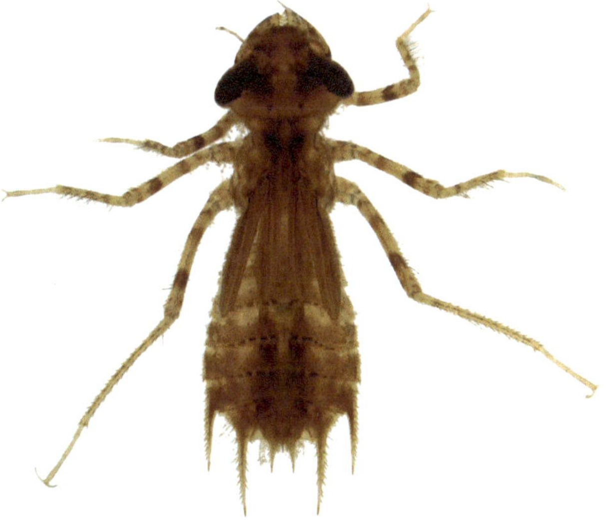

FIGURE 14.3.68 *Zenithoptera* sp., general habitus of larva, dorsal view.

FIGURE 14.3.69 *Macrodiplax balteata* (Hagen, 1861), apex abdomen, dorsal view.

FIGURE 14.3.70 *Macrodiplax balteata*, general habitus of larva, dorsal view.

10(9) S8–9 lateral spines not as developed as below, those on S8 shorter than middorsal length of S8–9 combined; S9 lateral spines not reaching or just reaching anal pyramid apex (Fig. 14.3.66) 11

10' S8–9 lateral spines very strongly developed, those on S8 longer than middorsal length of S8–9 combined; S9 lateral spine distinctly extending past anal pyramid apex (Figs. 14.3.67, 68) .... *Zenithoptera*
[Costa Rica to Paraguay and Argentina]

11(10) Epiproct in dorsal view distinctly shorter than paraprocts (Fig. 14.3.66); palpal setae generally 6–11, and premental setae generally 10–17 12

11' Epiproct in dorsal view as long as paraprocts (Fig. 14.3.69); palpal setae generally 11–12, and premental setae generally 16–21 (Fig. 14.3.70) .... *Macrodiplax*
[Bahamas, Antilles, Mexico, to Trinidad, Venezuela. Pantropical]

12(11) Palpal setae generally 8–11 (Fig. 14.3.71); epiproct about half the length of paraprocts or distinctly shorter ($\leq$0.55); abdomen well extended laterally, with S3 width/S6 width $<$0.75 (Figs. 14.3.72, 73) .... *Tauriphila*
[Mexico to Uruguay and Argentina, Caribbean, northern South America. Nearctic]

12' Palpal setae generally 6–7 (Fig. 14.3.74); epiproct length $\geq$0.67$\times$ paraprocts length (Fig. 14.3.66); abdomen not well extended laterally, with ratio S3 width/S6 width $>$0.85 (Fig. 14.3.75) .... *Miathyria*
[Caribbean, Mexico to Uruguay and Argentina. Nearctic]

13(4) Middorsal hooks without long lateral setae (Fig. 14.3.39); middorsal hooks on S2–9 or 3–9, exceptionally on S3–10 (*Tholymis*) with first two usually very small; palpal setae variable generally more than four, rarely four; wing sheaths generally not divergent* 14

13' S5–9 middorsal hooks with long lateral setae (Fig. 14.3.76); middorsal hooks present on S4–9; only four palpal setae; wing sheaths divergent (Fig. 14.3.77) .... *Argyrothemis*
[Amazon region]

14(13) Labial palp distal margin without digitiform teeth (Fig. 14.3.81); eyes not strongly protruding laterally with conical apex (Fig. 14.3.81) 15

14' Labial palp distal margin with at least some very deep, digitiform teeth (Fig. 14.3.78); eyes strongly protruding laterally and with conical apex (Figs. 14.3.79, 80) .... *Diastatops*
[Venezuela to Argentina and Uruguay]

15(14) Epiproct strongly elongated, in dorsal view length subequal to 2$\times$ or more basal width (Fig. 14.3.82) 16

15' Epiproct not so strongly elongated, in dorsal view length $<$2$\times$ basal width (Fig. 14.3.49) 17

16(15) Prementum distal margin broadly rounded and nearly glaborous (just a few minute, obscure, setae) (Fig. 14.3.83); palpal setae $\leq$ seven (Fig. 14.3.83); metafemora reaching S8 (Fig. 14.3.84) .... *Idiataphe*
[Neotropical, Caribbean. Nearctic]

FIGURE 14.3.71 *Tauriphila* sp., head and labium, dorsal view.

FIGURE 14.3.72 *Tauriphila* sp., abdomen, dorsal view. Dark lines indicate ratio width S3/width S6.

FIGURE 14.3.73 *Tauriphila* sp., general habitus of larva, dorsal view.

FIGURE 14.3.74 *Miathyria* sp., labium, dorsal view. Arrows indicate insertion of palpal setae.

16' Prementum distal margin normally shaped, triangular with small but distinct setae along border (Fig. 14.3.85); palpal setae ≥8 (Fig. 14.3.85); legs not so long, metafemora not reaching S8 (Fig. 14.3.86)........ *Brachymesia* [Neotropical, Caribbean. Nearctic]

17(15) S10 mid-dorsal hook absent (Figs. 14.3.39, 40)........ 18

17' S10 mid-dorsal hook small but distinct (in dorsal view as a small prominent flat protuberance) (Figs. 14.3.49, 87, 88)........ *Tholymis* [Neotropical. Pantropical]

FIGURE 14.3.75 *Miathyria* sp., general habitus of larva, dorsal view.

FIGURE 14.3.77 *Argyrothemis argentea*, general habitus of larva, dorsal view.

FIGURE 14.3.76 *Argyrothemis argentea* Ris, 1909, abdomen, lateral view. Arrows indicate long dorsal setae at the sides of S5- 9 middorsal hooks.

FIGURE 14.3.78 *Diastatops* sp., labium, dorsal view.

FIGURE 14.3.79 *Diastatops* sp., head, dorsal view.

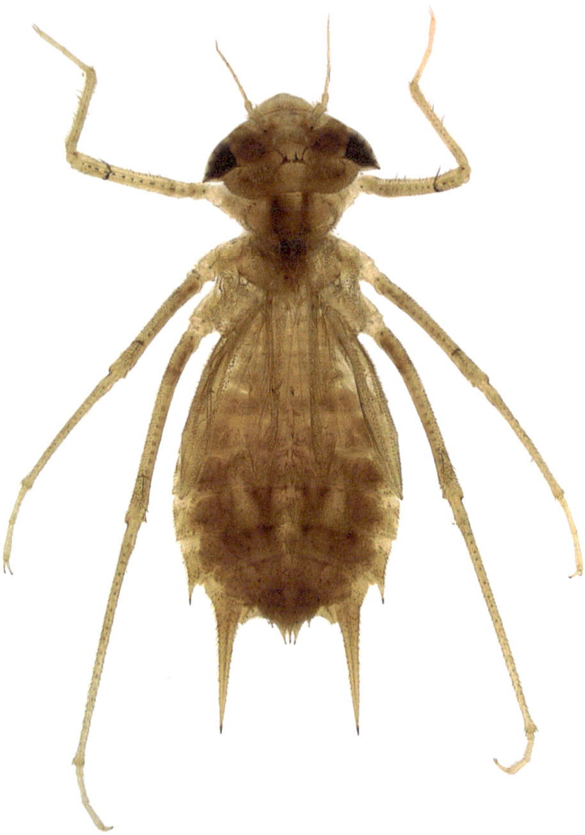

FIGURE 14.3.80 *Diastatops* sp., general habitus of larva, dorsal view.

FIGURE 14.3.81 *Idiataphe* sp., head and labium, dorsal view.

FIGURE 14.3.82 *Brachymesia* sp., abdomen, dorsal view. Arrow indicates the epiproct.

FIGURE 14.3.83 *Idiataphe* sp., labium, dorsal view.

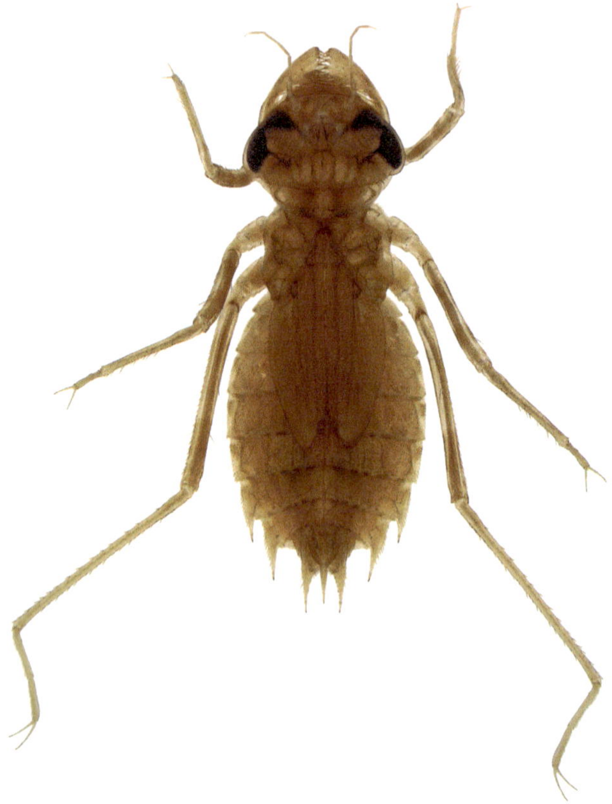

FIGURE 14.3.84 *Idiataphe* sp., general habitus of larva, dorsal view.

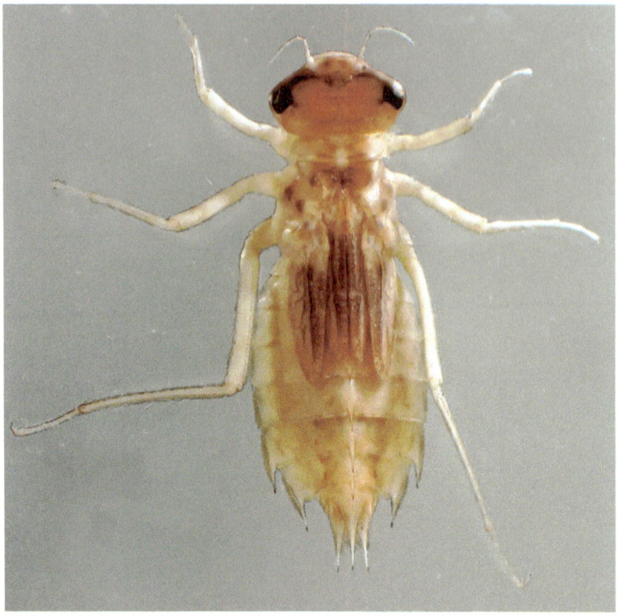

FIGURE 14.3.86 *Brachymesia* sp., general habitus of larva, dorsal view.

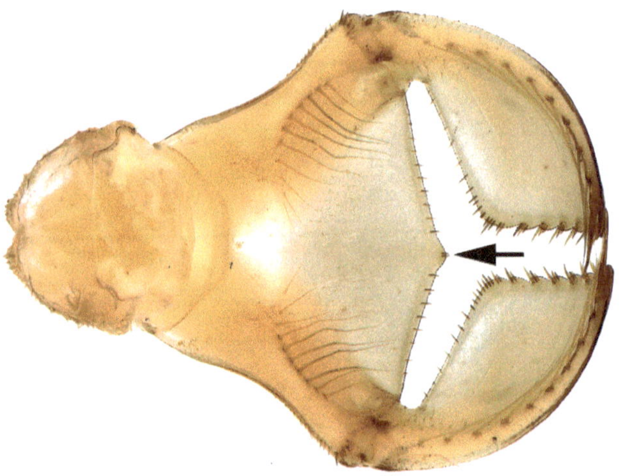

FIGURE 14.3.85 *Brachymesia* sp., labium, dorsal view. Arrow indicates distal margin of prementum (triangular).

FIGURE 14.3.87 *Tholymis citrina*, abdomen, lateral view. Arrow indicates middorsal hook present on S10.

18(17) S8–9 lateral spines length distinctly shorter than middorsal length of S8 and S9 together (Figs. 14.3.44, 48) 19

18' S8–9 lateral spines very long, those of S8 slightly divergent in dorsal view and as long as middorsal length of S8 and S9 together; those of S9 distinctly longer than middorsal length of S8 and S9 together (Figs. 14.3.89, 90) .... *Fylgia* [Amazon region]

19(18) Antennae not as below (Fig. 14.3.94); no tuft of specialized setae on occiput (Fig. 14.3.94) 20

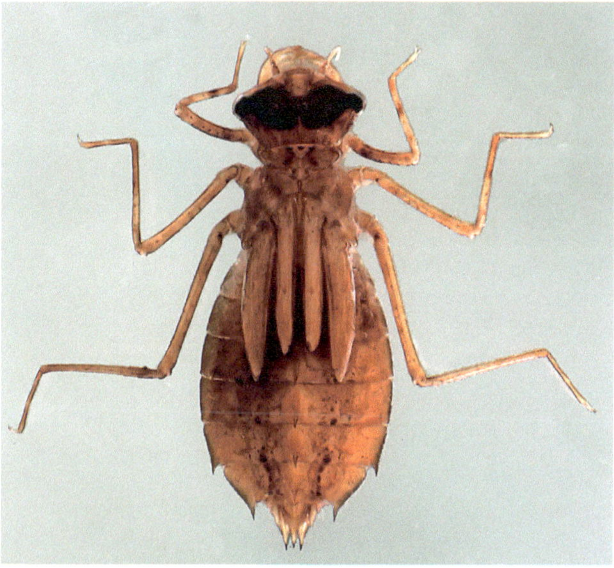

FIGURE 14.3.88 *Tholymis citrina*, general habitus of larva, dorsal view.

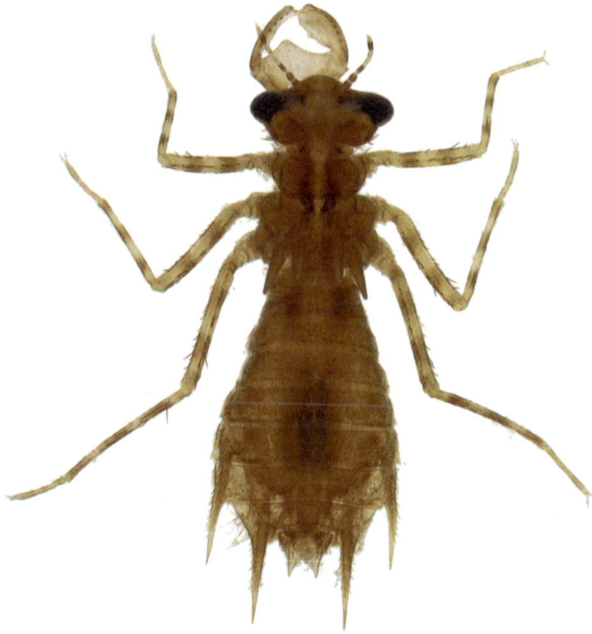

FIGURE 14.3.90 *Fylgia* sp., general habitus of larva, dorsal view.

FIGURE 14.3.89 *Fylgia* sp., abdomen, dorsal view.

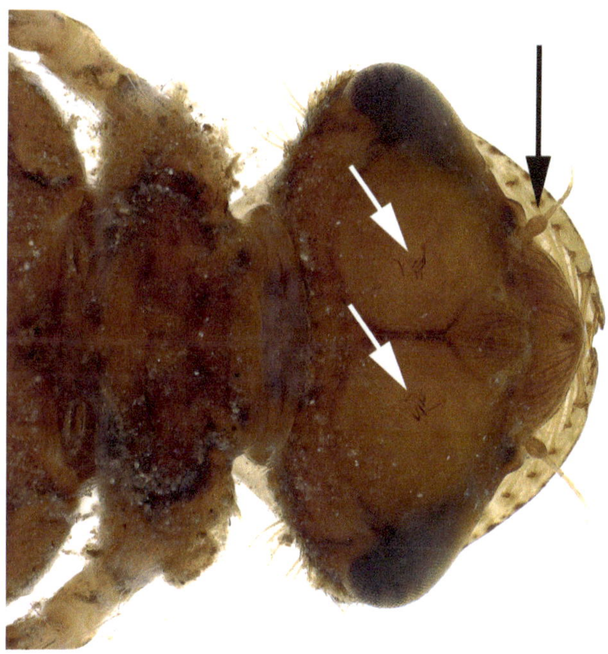

FIGURE 14.3.91 *Elga leptostyla* Ris, 1909, head, dorsal view. Dark arrow indicates antenna. White arrows indicate a pair of tuft of specialized dark setae on top of head.

19' Antennae length $\leq 0.2\times$ head width (Fig. 14.3.91) + antennomere three (first flagellum article) shorter than antennomere two (Fig. 14.3.92); a pair specialized, dark setal tufts on occiput (caution: these specialized setae are easily broken) (Figs. 14.3.91, 93) .... *Elga*
[Panama to Peru and Brazil]

20(19) Character combination not as below (Figs. 14.3.98, 99)........ 21

20' Eyes well produced dorsally (Fig. 14.3.95) + labial palps with only one raptorial seta per tooth (Figs. 14.3.96, 7)........ *Planiplax*
[Neotropical]

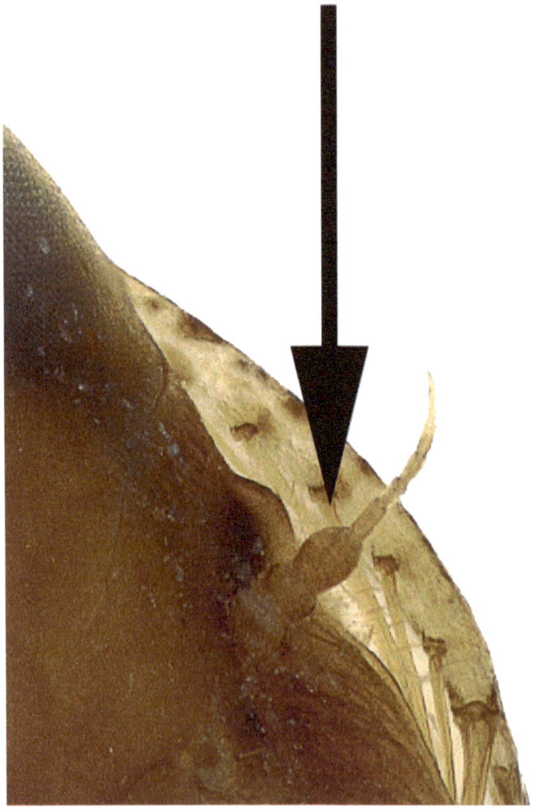

FIGURE 14.3.92 *Elga leptostyla*, detail of antenna, dorsal view. Arrow indicates antennal article 3.

FIGURE 14.3.94 *Perithemis* sp., head and antennae, dorsal view.

FIGURE 14.3.93 *Elga leptostyla*, general habitus of larva, dorsal view.

FIGURE 14.3.95 *Planiplax* sp., head and labium, frontal view.

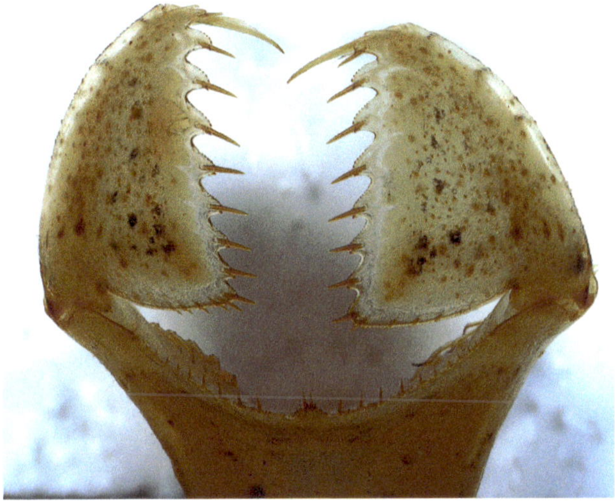

FIGURE 14.3.96 *Planiplax* sp., labial palps, frontal view (show one seta raptore per tooth).

FIGURE 14.3.98 *Perithemis* sp., head and labium, frontal view.

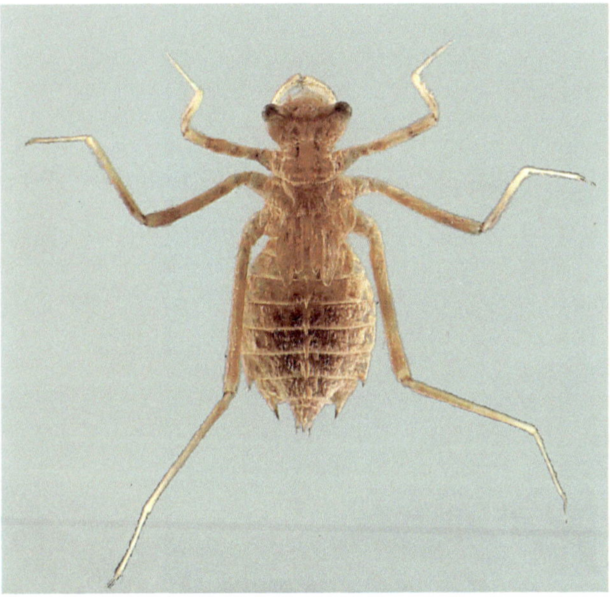

FIGURE 14.3.97 *Planiplax* sp., general habitus of larva, dorsal view.

FIGURE 14.3.99 *Dythemis* sp., labial palps, frontal view.

21(20) Metafemora not reaching or surpassing S9 (Fig. 14.3.118); prementum distal margin prominent, broadly convex, triangular, or shallow convex, but median portion of apex always with setae (Fig. 14.3.103) (caution: these setae can be broken, but their insertion should be still faintly visible); number of raptorial setae per distal palpal tooth variable 22

21' Metafemora reaching or surpassing S9 (Fig. 14.3.100); prementum distal margin convex or straight, with median portion slightly protruding, forming a kind of small subtriangular formation without setae (Fig. 14.3.101), labial palps generally with the five more dorsal dentations with one or two raptorial seta(e) (Figs. 14.3.100, 102) .... *Oligoclada*
[Guatemala to Argentina]

22(21) Head not enormous compared with body, width never >0.33× total body length (Fig. 14.3.118); eyes not strongly protruding laterally (Fig. 14.3.107); premental distal margin setae and raptorial setae generally spiniform; mental and palpal setae variable; labial palp distal dentation variable; propleuron chaetotaxy variable but generally not as below 23

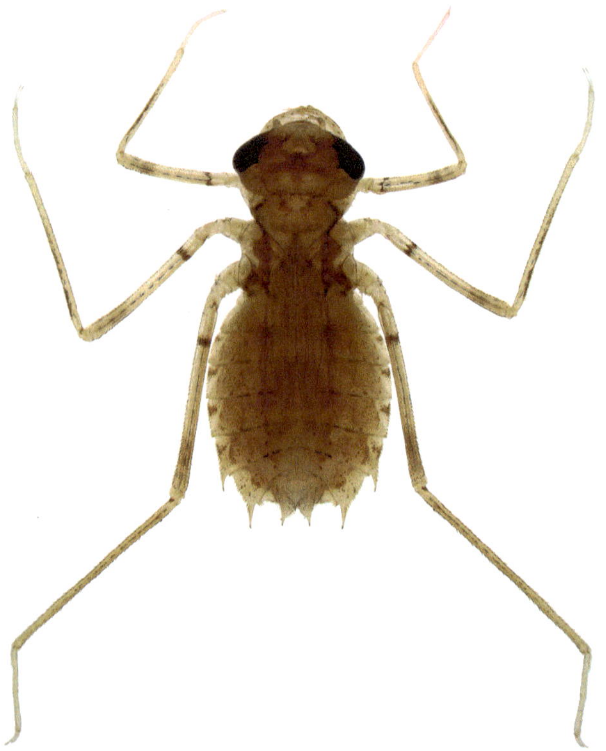

FIGURE 14.3.100 *Oligoclada* sp., general habitus of larva, dorsal view.

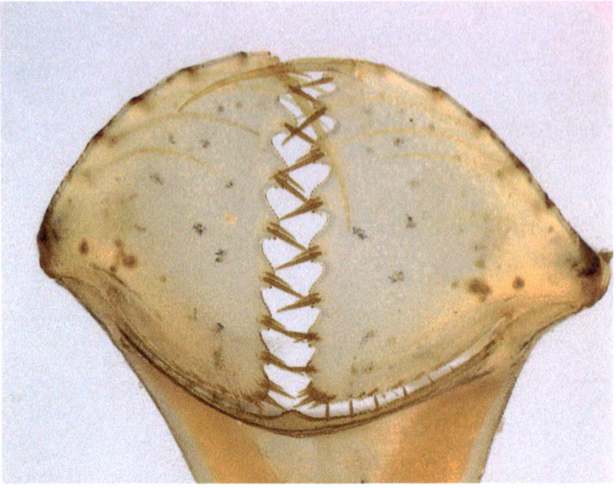

FIGURE 14.3.102 *Oligoclada* sp., labial palps, frontal view.

FIGURE 14.3.101 *Oligoclada* sp., labium, dorsal view. Arrow indicates apex of prementum deprived of setae.

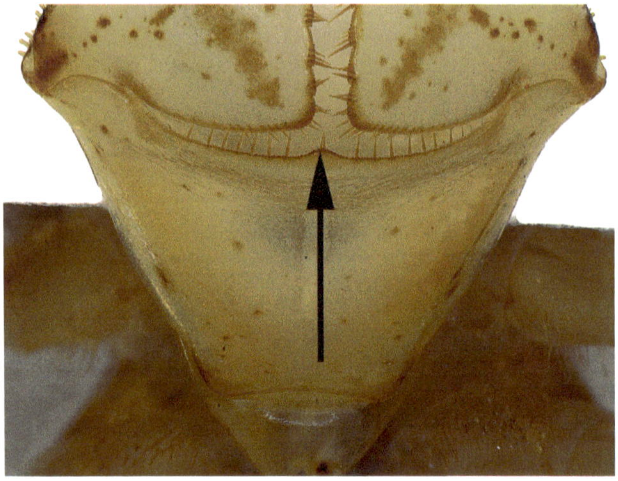

FIGURE 14.3.103 *Dythemis* sp., labium, fronto-ventral view. Arrow indicates apex of prementum with setae.

22' Head enormous compared to body, width distinctly >0.33× total body length (ratio head width/body length about 0.4) (Fig. 14.3.104); eyes enormous, well protruding laterally and globular (Fig. 14.3.105); palpal distal dentation raptorial setae and prementum distal margin setae with apex truncate or spatulate (Fig. 14.3.106); mental setae 6–8, palpal setae generally 5; labial palps distal margin dentation weak; propleuron without long filiform setae and with episternal spiniform setae small and robust (Fig. 14.3.104).......................... *Gynothemis* [Venezuela, Suriname, French Guiana, southeast Brazil]

23(22) Labial palps distal margin with marked dentation (Fig. 14.3.110), if distal palpal dentation present but not well-marked, then prementum distal margin setae and raptorial setae distinctly spatulate (*Brechmorhoga praedatrix* Calvert, 1909, Fig. 14.3.111)............................ 24

FIGURE 14.3.104 *Gynothemis* sp., general habitus of larva, dorsal view.

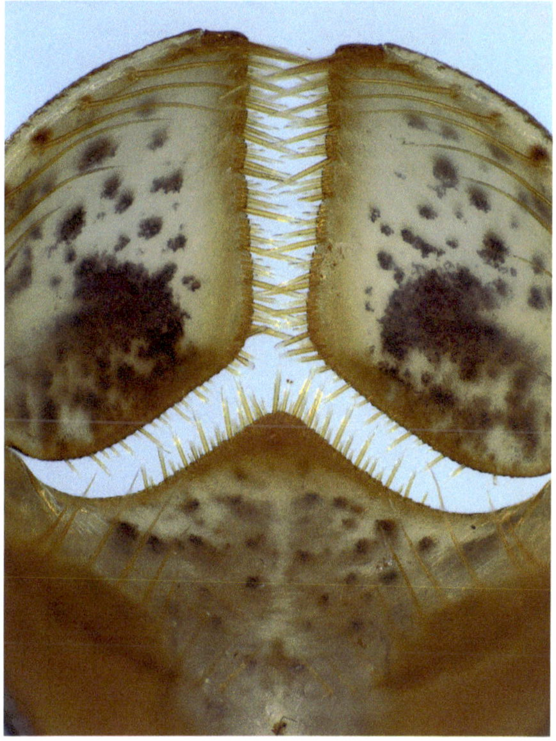

FIGURE 14.3.106 *Gynothemis* sp., distal prementum with labial palps, postero-dorsal view.

FIGURE 14.3.105 *Gynothemis* sp., head, dorsal view.

FIGURE 14.3.107 *Dythemis* sp., head, dorsal view.

FIGURE 14.3.108 *Dythemis* sp., dentation of labial palps, frontal view.

FIGURE 14.3.109 *Dythemis* sp., general habitus of larva, dorsal view.

FIGURE 14.3.110 *Brechmorhoga* sp., labial palps, frontal view.

23' Labial palp distal margin almost straight or with shallow crenulations (Fig. 14.3.108), palpal raptorial setae, and prementum distal margin setae spiniform (Figs. 14.3.108, 109)........ *Dythemis*
[Mexico to Argentina. Nearctic]

24(23) Ratio prementum width/head width >0.6; prementum/submentum articulation reaching mesocoxae (Fig. 14.3.114); antennae shorter as below, generally antennae length <1.5× distance between antennae bases (Fig. 14.3.115); eyes rather well developed, globular with an anterolateral position (anterior margin of the head not distinctly rounded in dorsal view) (Fig. 14.3.115) 25

24' Ratio prementum width/head width <0.6 (Figs. 14.3.98, 112); prementum/submentum articulation not reaching or hardly reaching level of mesocoxae (Fig. 14.3.112); antennae rather long, generally length >1.5× distance between antennae bases; eyes particularly small and with a rather marked lateral position (conferring to the head a distinct rounded anterior margin in dorsal view) (Figs. 14.3.94, 113) .... *Perithemis*
[Mexico to Uruguay and Argentina. Nearctic]

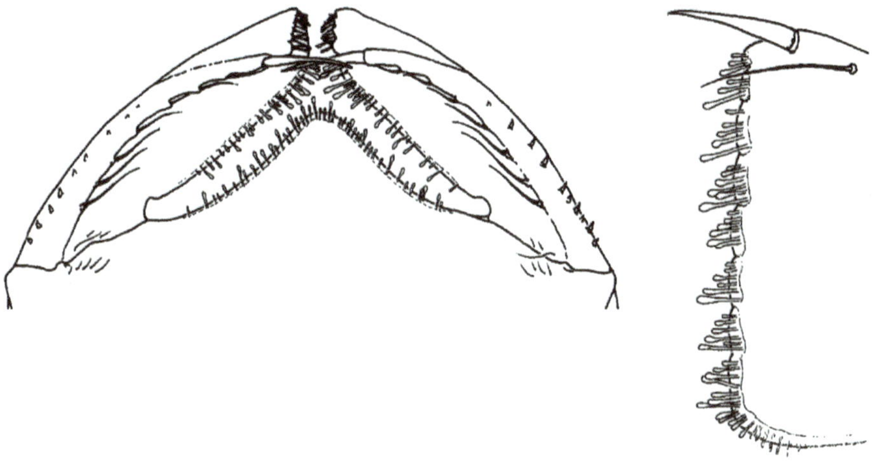

FIGURE 14.3.111 *Brechmorhoga praedatrix* Calvert, 1909, distal part of prementum in dorsal view and distal part of right labial palp in posterior view.

FIGURE 14.3.112 *Perithemis* sp., thorax, head and labium, ventral view.

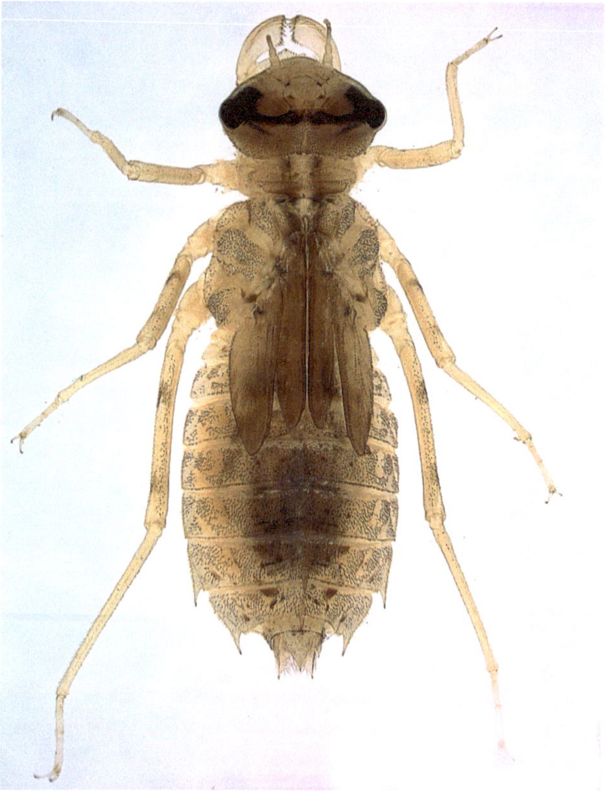

FIGURE 14.3.113 *Perithemis* sp., general habitus of larva, dorsal view.

25(24) Abdominal integument mostly granulose (covered with shaped setae) (Fig. 14.3.119); cercus length $\leq 0.5\times$ epiproct (Fig. 14.3.120); generally distal middorsal hooks not strongly developed dorsally, almost horizontal and dorsally flattened in lateral view (Fig. 14.3.40) 26

25' Abdominal integument mostly finely setose (covered with small spiniform setae or short soft setae) (Fig. 14.3.116); cercus length $\geq 0.5\times$ epiproct (Fig. 14.3.117); generally distal middorsal hooks, at least those of S6–8 well elevated and serrate (with well-rounded dorsal margin in lateral view) (Figs. 14.3.39, 118) .... *Macrothemis*
[Mexico to Uruguay and Argentina. Nearctic]

26(25) Wing sheaths divergent* (Fig. 14.3.121); middorsal S3 hook well developed, occupying entire dorsal length in lateral view (Figs. 14.3.121, 122) .... *Brechmorhoga*
[Caribbean, Mexico to Argentina. Nearctic]

FIGURE 14.3.114 *Macrothemis* sp., thorax, head and labium, ventral view.

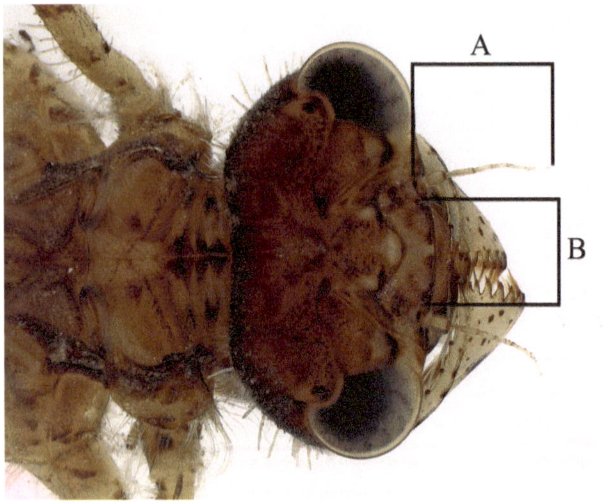

FIGURE 14.3.115 *Macrothemis* sp., head and antennae, dorsal view (Ratio A/B < 1.5).

FIGURE 14.3.116 *Macrothemis* sp., dorsal part of abdominal segments 6–7, right lateral view.

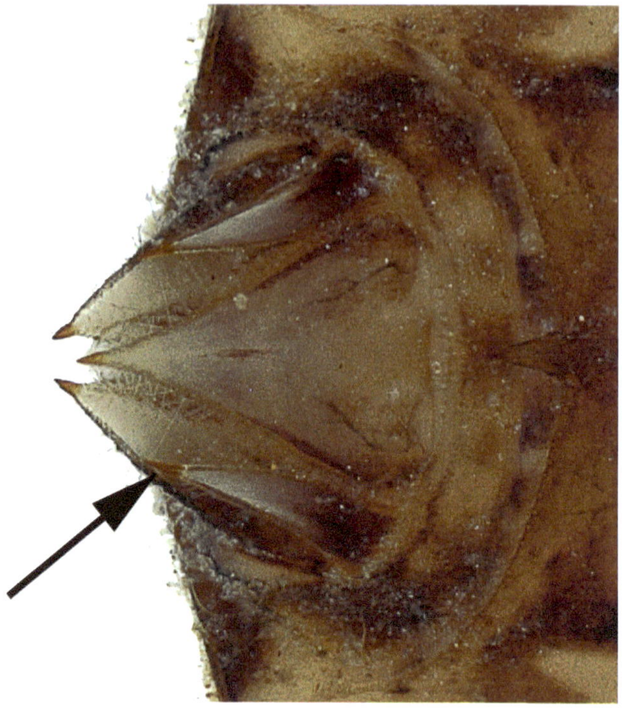

FIGURE 14.3.117 *Macrothemis* sp., anal pyramid, dorsal view. Arrow indicates the cercus.

FIGURE 14.3.118 *Macrothemis* sp., general habitus of larva, dorsal view.

FIGURE 14.3.119 *Brechmorhoga* sp., dorsal part of abdominal segments 6–7, right lateral view.

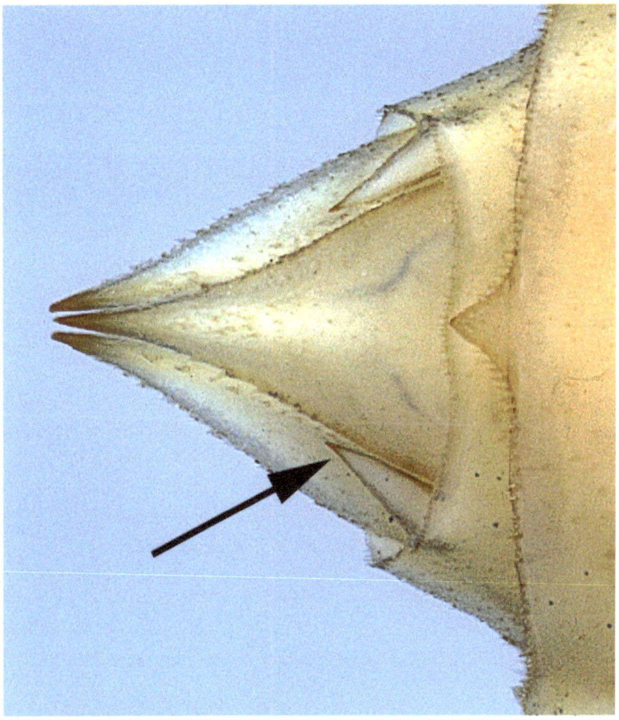

FIGURE 14.3.120 *Brechmorhoga* sp., anal pyramid, dorsal view. Arrow indicates the cercus.

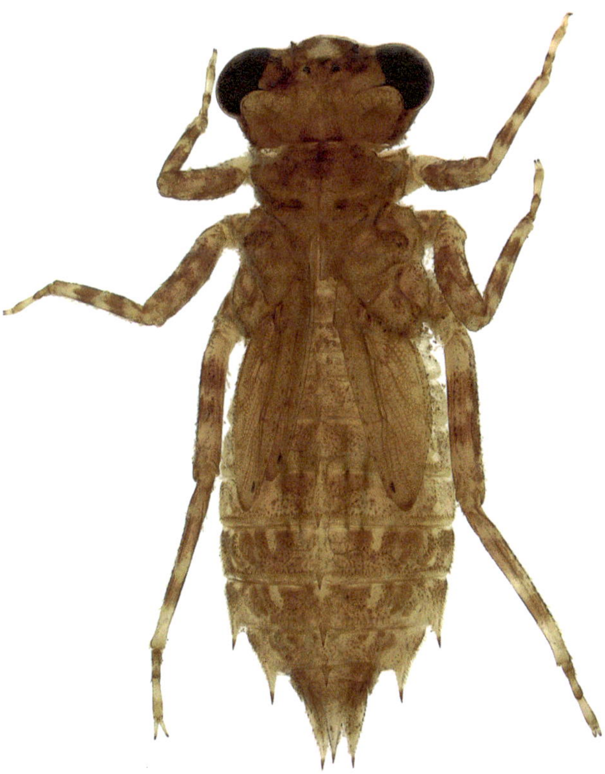

FIGURE 14.3.121 *Brechmorhoga* sp., general habitus of larva, dorsal view.

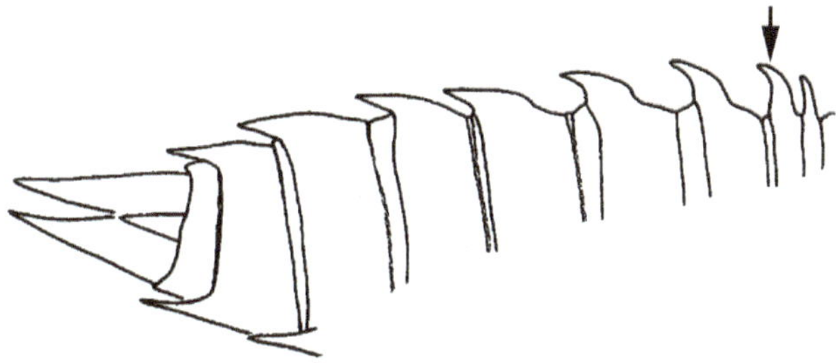

FIGURE 14.3.122 *Brechmorhoga praedatrix*, abdomen, right lateral view. Arrow indicates middorsal hook of the third abdominal segment.

FIGURE 14.3.123 *Scapanea frontalis* Burmeister, 1839, general habitus of larva, dorsal view.

FIGURE 14.3.124 *Scapanea frontalis*, abdomen, right lateral view. Arrow indicates middorsal hook of the third abdominal segment.

FIGURE 14.3.125 *Libellula saturata* Uhler, 1857, head, dorsal view.

FIGURE 14.3.127 *Libellula saturata*, head, frontal view.

FIGURE 14.3.126 *Orthemis* sp., head and labium, frontal view.

FIGURE 14.3.128 *Micrathyria* sp., head and labium, frontal view.

26' Wing sheaths parallel* (Fig. 14.3.123); S3 middorsal hook smaller, occupying the distal dorsal part in lateral view (Figs. 14.3.123, 124) .... *Scapanea*
[Cuba, Jamaica, Haiti, Dominica, Dominican Republic, Puerto Rico]

27(1) Head squarish, with long and subparallel or parallel postoccular margins (Fig. 14.3.125); eyes rather small in dorsal view in an anteriolateral position and protruding variably dorsally (Figs. 14.3.126, 127); larvae generally elongated and hairy (Figs. 14.3.134, 136)........ 28

27' Head and eyes not as above (Figs. 14.3.128–131); abdomen variable, often less elongated and less hairy (Fig. 14.3.146)........ 33

28(27) Eyes not strongly dorsally developed, only slightly protuberant dorsally (Fig. 14.3.127); prementum distal margin not distinctly scalloped (Fig. 14.3.135) 29

28' Eyes very strongly dorsally developed (Figs. 14.3.126, 132); generally distal margin of prementum distinctly scalloped and with a small spiniform seta present at each indentation (Figs. 14.3.133, 134)........ *Orthemis*
[Mexico to Chile and Argentina. Nearctic]

FIGURE 14.3.129 *Micrathyria* sp., head, right lateral view.

FIGURE 14.3.131 *Tramea* sp., head, dorsal view.

FIGURE 14.3.130 *Micrathyria* sp., head, dorsal view.

FIGURE 14.3.132 *Orthemis* sp., head, right lateral view.

29(28) Larvae <22 mm; *Dasythemis* and *Cannaphila*: these genera are closely related and the larvae extremely difficult to separate, and some specimens cannot be separated unambiguously; distribution may contribute to determination: *Dasythemis* occurs in Central South America from southern Venezuela and Colombia to Brazil and northernmost Argentina (absent in Chile); *Cannaphila* occurs in Mexico and Central America (including Antilles), Colombia, westernmost Venezuela, western Brazil (westernmost Amazonas, Acre, Rondonia), Ecuador, eastern Peru, central Bolivia, and northern Argentina 30

29' Larvae >23 mm (Fig. 14.3.136) with and without middorsal hooks, from Mexico south to North Argentina, only one species with middorsal hooks widely distributed in South America .... *Libellula* (in part)
[Mexico to Argentina. Holarctic]

30(29) Abdomen with lateral spines (Fig. 14.3.138) 31

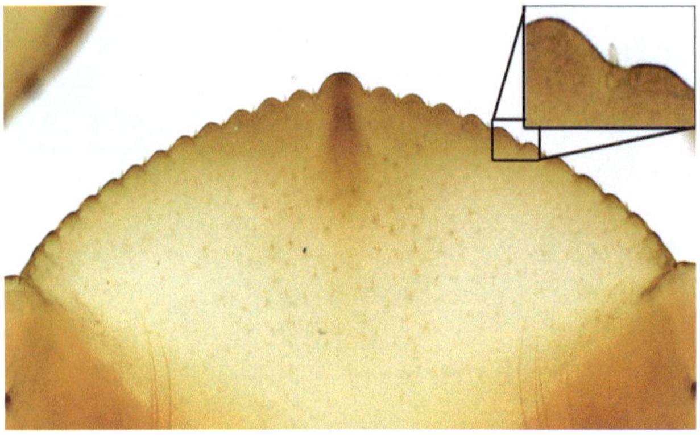

FIGURE 14.3.133 *Orthemis* sp., apex of prementum, dorsal view, and detail of scalloped margin showing minute seta

FIGURE 14.3.134 *Orthemis* sp., general habitus of larva, dorsal view.

FIGURE 14.3.135 *Libellula saturata*, apex of prementum, dorsal view.

FIGURE 14.3.136 *Libellula saturata*, general habitus of larva, dorsal view.

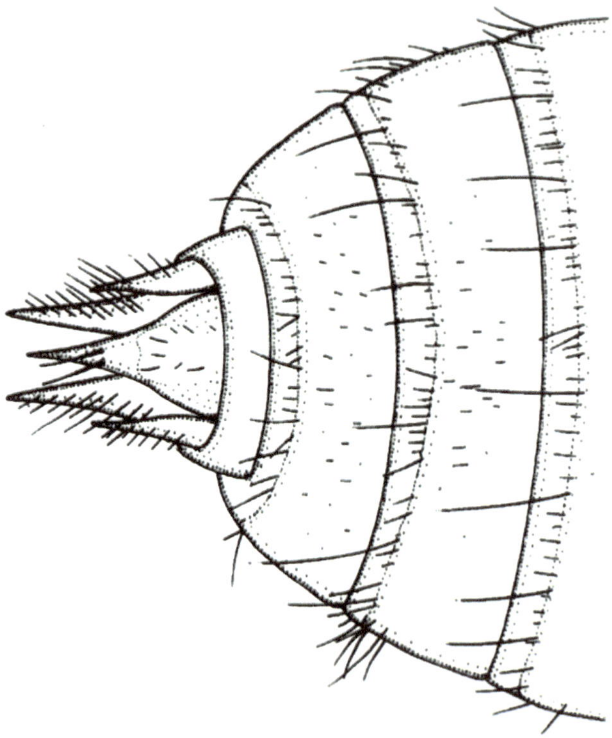

FIGURE 14.3.137 *Dasythemis mincki* Karsch, 1890, apex of abdomen, dorsal view. *Modified from Carvalho et al. (2002).*

30'	Abdomen without lateral spines (Fig. 14.3.137) .... *Dasythemis* (in part): *Dasythemis mincki* (Karsch, 1890)
31(30)	Generally four or five palpal setae and generally three premental long setae (Fig. 14.3.140); labial palp distal margin distinctly crenulate 32
31'	Generally six palpal setae and two premental long setae (Fig. 14.3.139); labial palp distal margin with obscure crenulations .... *Cannaphila* (in part): *C. insularis* Kirby, 1889
32(31)	Labial palp distal margin with about 12 crenulations .... *Dasythemis* (in part): *D. venosa* (Burmeister, 1839)
32'	Labial palp distal margin with about nine crenulations (Fig. 14.3.141) .... *Cannaphila* (in part): *C. vibex* (Hagen, 1861)
33(27)	Combination of characters not as below 34
33'	S8 lateral spines absent + S9 lateral spines very long, $\geq 2.5\times$ S9 middorsal length (Fig. 14.3.142) + hind legs very long, longer than body; metafemur reaching or extending over distal S9 (Fig. 14.3.143) .... *Rhodopygia*
34(33)	S8 lateral spines very long, $\geq$S8 middorsal length + S9 lateral spines very long, $>1.5\times$ S9 middorsal length (Fig. 14.3.144) 35
34'	Combination of characters not as above (including S8 without lateral spines) 36
35(34)	S8 lateral spines parallel or slightly convergent in dorsal view (Fig. 14.3.144); prementum distal margin broadly rounded (Fig. 14.3.145); cerci length $\geq 0.5\times$ paraprocts (Figs. 14.3.144, 146) .... *Tramea* [Mexico to Argentina. Nearctic, Pantropical]
35'	S8 lateral spines slightly divergent in dorsal view (Fig. 14.3.147); prementum distal margin almost triangular with straight margins (Fig. 14.3.148); cerci length $<0.33\times$ paraprocts (Figs. 14.3.147, 149) .... *Micrathyria* (in part) [*M. pseudeximia* Westfall, 1992, *M. ungulata* Föster, 1907] [Mexico to Argentina]
36(34)	Paraprocts straight or only moderately decurved at tip (Fig. 14.3.154, arrow); lateral spines variable; combination of others character not as below 37
36'	Paraprocts strongly decurved at tip (Fig. 14.3.150, arrow); S8 lateral spines absent, on S9 moderately developed, weak or absent (Fig. 14.3.151); larvae robust, short, not depressed dorsoventrally + robust, long hind legs + eyes globular and anteriorly developed + spiracles of pronotum on a strong elevation and largely exposed (for direct air breathing) (Figs. 14.3.152, arrow, 153) .... *Erythemis* [Mexico to Argentina]
37(36)	Larva not dorsoventrally depressed; pronotal shield distinctly elevated, making an angle $>55°$ with horizontal in lateral view (Figs. 14.3.155–157); S8 and S9 lateral spines absent or present but never strongly developed (those of S8 length generally $<0.5\times$ of middorsal S8 length, those of S9 length generally $\leq 0.75\times$ of S9 middorsal length (Fig. 14.3.158); head moderately transverse, eyes never strongly produced laterally (Fig. 14.3.159); epiproct length $\geq 0.8\times$ paracprocts in dorsal view; generally abdomen S1–5 dorsally not distinctly paler than S8–10 38

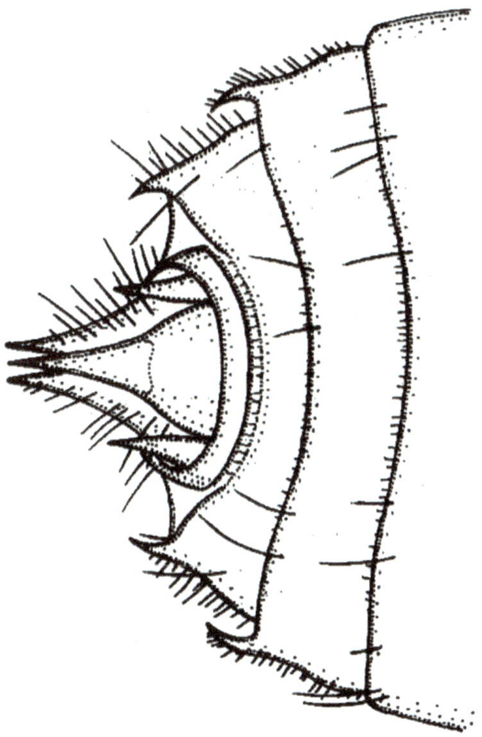

FIGURE 14.3.138 *Dasythemis venosa* (Burmeister, 1839), apex of abdomen, dorsal view. *Modified from Carvalho et al. (2002).*

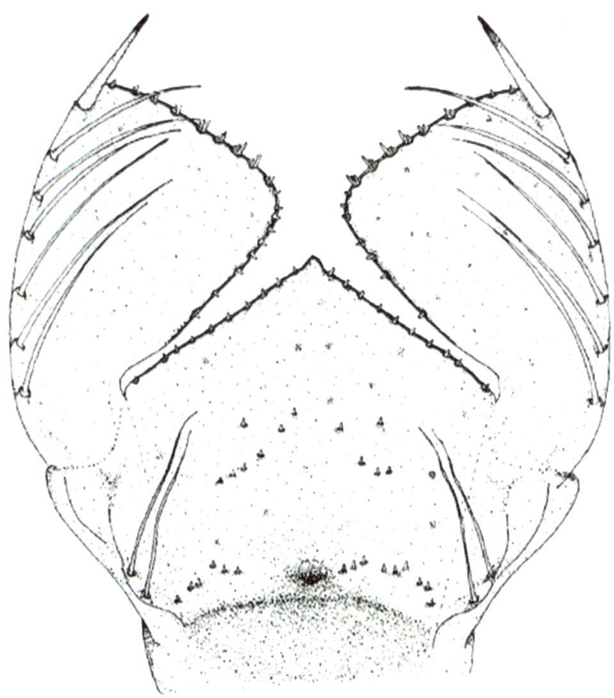

FIGURE 14.3.139 *Cannaphila insularis* Kirby, 1889, prementum with labial palps, dorsal view. *Modified from Trapero-Quintana & Araújo (2009).*

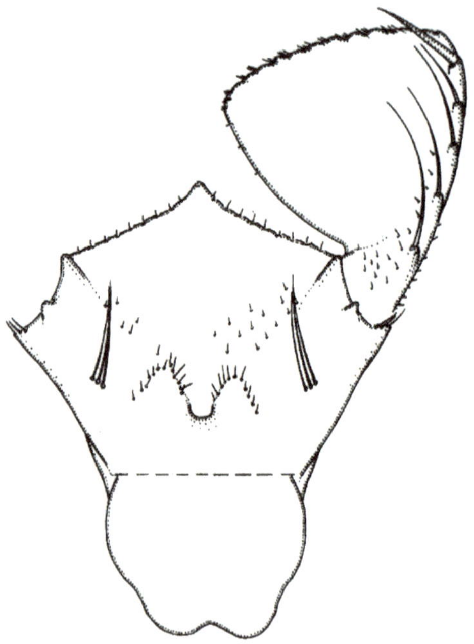

FIGURE 14.3.140 *Dasythemis venosa*, prementum with labial right palp, dorsal view. *Modified from Carvalho et al. (2002).*

FIGURE 14.3.141 *Cannaphila vibex* (Hagen, 1861), general habitus of larva, dorsal view.

37' Larva somewhat dorsoventrally depressed; pronotal shield not distinctly elevated (making an angle <50° with horizontal in lateral view) (Figs. 14.3.160, 161); pronotal shield may be subhorizontal in distal half (Fig. 14.3.162); S8 and S9 lateral spines present and of variable development but never weakly developed and generally rather well to strongly developed (with few exceptions S8 lateral spine length >0.33× and S9 lateral spine length >0.66× respective segment) (Fig. 14.3.163); head generally distinctly transverse with eyes often strongly produced laterally (Fig. 14.3.164); epiproct almost without exception (*Anatya* [0.85]) and possibly *Micrathyria stawiarskii* Santos, 1953 [0.85 drawing vs 0.67 text in the original description]) <0.8× paraprocts in dorsal view (Fig. 14.3.163); abdomen sometimes distinctly paler on dorsal S1–5 than S8–10; genera *Pachydiplax*, *Nephepeltia*, *Anatya*, and *Micrathyria*: these genera are closely related and the larvae are extremely difficult to separate; the characters given below relate to the anal pyramid, are quantitative, and based on a few specimens from the very few known larvae, therefore, some specimens could be misidentified; the best way to identify these larvae is by rearing 41

38(37) S7–9 ventral surface distal margin with a row of small, robust spiniform setae directed caudad (character needs to be examined on dry specimens) (Fig. 14.3.169); labial palp distal margin teeth each generally bearing one raptorial seta (Fig. 14.3.170) (except a few *Erythrodiplax*) (teeth may be shallow or obsolete but generally still separated by at least a weak notch); abdomen end generally somewhat abruptly rounded, epiproct length subequal to paraprocts (Fig. 14.3.171), or slightly shorter in dorsal view (ratio epiproct/paraproct most generally between 0.85 and 1) (Fig. 14.3.158); ultimate instar length variable 39

38' S7–9 ventral surface distal margin without a row of small and robust spiniform setae directed caudad (character needs to be controlled on dry specimens) (Fig. 14.3.165); labial palp distal margin each generally bearing two or three raptorial setae (Fig. 14.3.166); abdomen end generally smoothly tapering with epiproct slightly shorter than paraprocts in dorsal view (ratio epiproct/paraproct between 0.8 and 0.9) (Figs. 14.3.167, 168); ultimate instar length ≥16.5× mm .... *Sympetrum*
[Mexico to Venezula, Brazil, Chile, Argentina. Nearctic, Holarctic]

39(38) Ultimate instar larva length ≥15 mm + labial palp distal margin with only one raptorial seta per dentation; anal appendages apically not or only very slightly curved downward in lateral view; abdominal tergites surface (not distal margin) glabrous (small spines may be present, but glabrous aspect remains due to sparsely distributed spines and shining cuticle) (Fig. 14.3.173); from Mexico to Costa Rica and Cuba 40

39' Ultimate instar larva length ≤15 mm, rarely 15–18 mm, but if length ≥15 mm then labial palp distal margin teeth with two or three raptorial setae per dentation; anal appendages apically moderately but distinctly curved downward in lateral view (Fig. 14.3.154); abdominal tergites surface (not distal margin) often pilose (Fig. 14.3.172) .... *Erythrodiplax*
[Mexico to Chile, Argentina. Nearctic]

40(39) S8 and S9 lateral spine length generally >1/3× middorsal length of respective segment (Fig. 14.3.171) .... *Crocothemis*
[Cuba]

FIGURE 14.3.142 *Rhodopygia* sp., abdomen, dorsal view.

FIGURE 14.3.144 *Tramea* sp., apex of abdomen, dorsal view.

FIGURE 14.3.143 *Rhodopygia* sp., general habitus of larva, dorsal view.

FIGURE 14.3.145 *Tramea* sp., labium, dorsal view.

40' S8 and S9 lateral spine length generally $\leq 1/3$ middorsal length of respective segment (Fig. 14.3.174)................................... *Pseudoleon*
[Mexico to Costa Rica]

41(37) Generally S9 lateral spine length $\leq 0.75\times$ paraprocts and ending proximal to, or at level of epiproct apex (Figs. 14.3.163, 178, 180 C); eyes nonfunctional part posterior margin generally not underlined by a distinct black stripe or not as strongly marked as below (Fig. 14.3.164); last instar larva length variable (10–20 mm).. 42

41' S9 lateral spine $\geq 0.75\times$ paraprocts and > epiproct (about 15%–20% > epiproct length) (Fig. 14.3.175) + eyes nonfunctional part posterior margin underlined by a distinct black stripe (Figs. 14.3.176, arrows, 177) + ultimate instar larva length 15–20 mm.. *Pachydiplax*
[Mexico, Belize, Bahamas, Cuba]

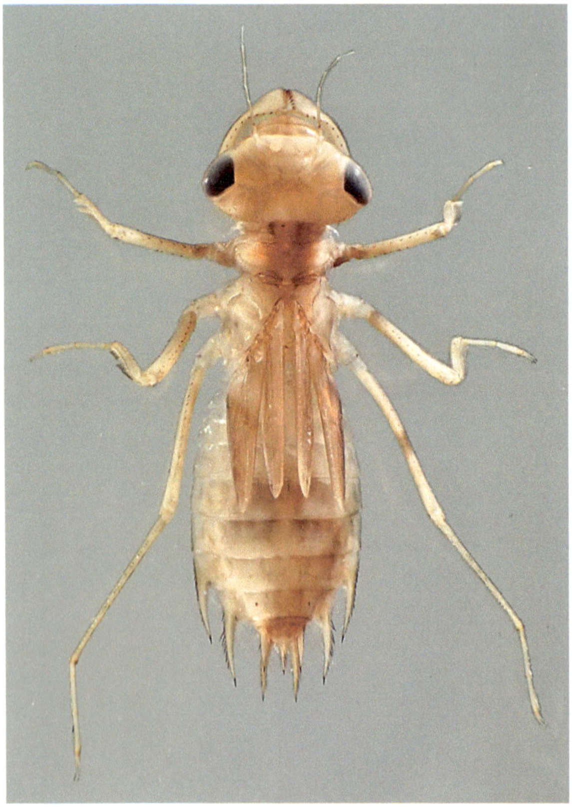

FIGURE 14.3.146 *Tramea* sp., general habitus of larva, dorsal view.

FIGURE 14.3.148 *Micrathyria pseudeximia*, labium, dorsal view.

FIGURE 14.3.147 *Micrathyria pseudeximia* Westfall, 1992, apex of abdomen, dorsal view.

FIGURE 14.3.149 *Micrathyria pseudeximia*, general habitus of larva, dorsal view.

FIGURE 14.3.150 *Erythemis* sp., abdomen, lateral view. Arrow indicates paraprocts.

FIGURE 14.3.151 *Erythemis* sp., abdomen, dorsal view.

FIGURE 14.3.153 *Erythemis* sp., general habitus of larva, dorsal view.

FIGURE 14.3.152 *Erythemis* sp., thorax and head, lateral view. Arrow indicates pronotal shield.

FIGURE 14.3.154 *Erythrodiplax* sp., abdomen, lateral view. Arrow indicates paraprocts.

FIGURE 14.3.155 *Sympetrum corruptum* (Hagen, 1861), thorax and head, lateral view. The figure shows the angle of pronotal shield made with the longitudinal body axis (here >50°).

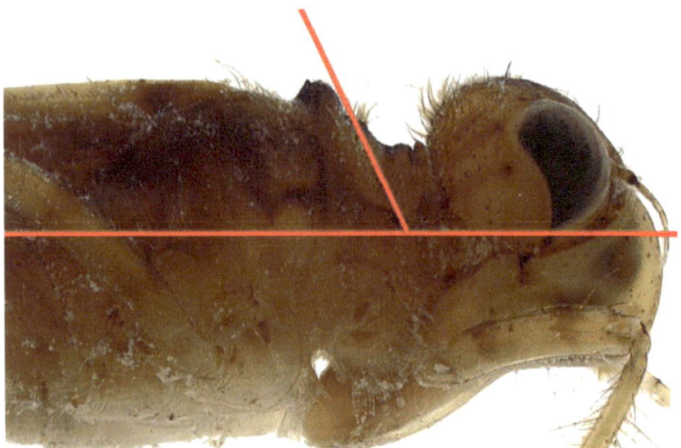

FIGURE 14.3.156 *Erythrodiplax* sp., thorax and head, lateral view. The angle of pronotal shield made with the longitudinal body axis = 65°.

FIGURE 14.3.157 *Erythrodiplax* sp., thorax and head, lateral view. The angle of pronotal shield made with the longitudinal body axis = 60°.

FIGURE 14.3.158 *Erythrodiplax* sp., apex of abdomen, dorsal view.

FIGURE 14.3.159 *Erythrodiplax* sp., head, dorsal view.

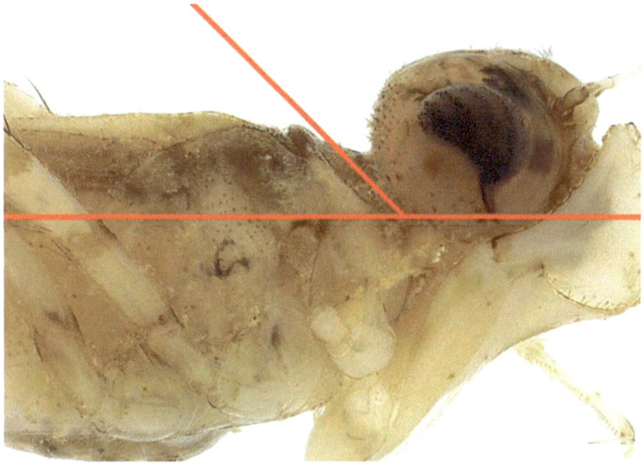

FIGURE 14.3.160 *Micrathyria* sp., thorax and head, lateral view. The angle of pronotal shield made with the longitudinal body axis = 45°.

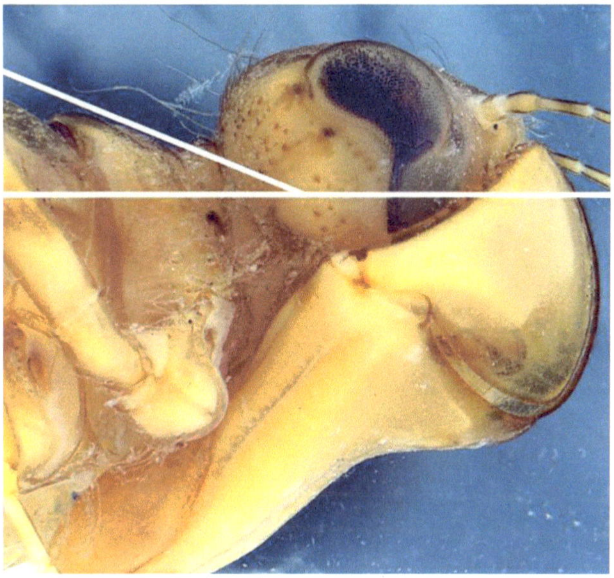

FIGURE 14.3.161 *Pachydiplax longipennis* (Burmeister, 1839), thorax and head, lateral view. The angle of pronotal shield made with the longitudinal body axis = 25°.

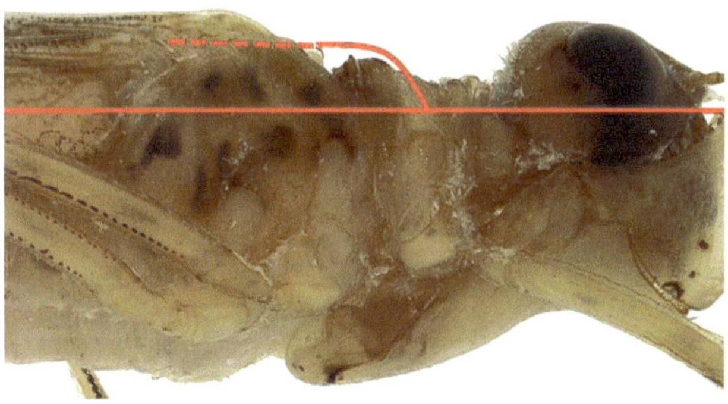

FIGURE 14.3.162 *Micrathyria* sp., thorax and head, lateral view. The angle of pronotal shield made with the longitudinal body axis <50° and is strongly flattened on its distal half (angle close to 0°).

FIGURE 14.3.163 *Micrathyria* sp., abdomen, dorsal view.

FIGURE 14.3.164 *Micrathyria* sp., head, dorsal view.

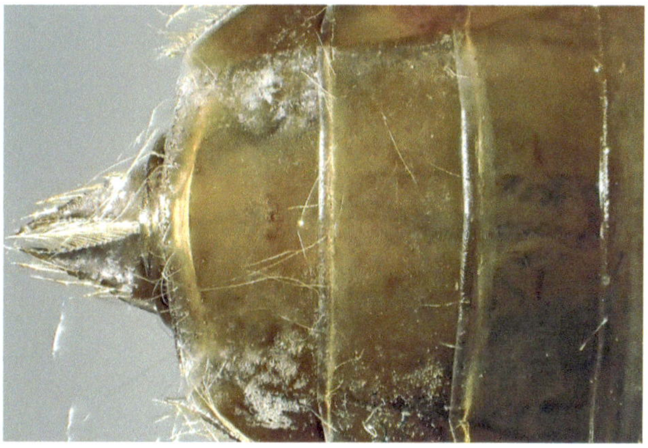

FIGURE 14.3.165 *Sympetrum corruptum*, apex of abdomen, ventral view.

FIGURE 14.3.166 *Sympetrum corruptum*, left labial palp, frontal view.

FIGURE 14.3.168 *Sympetrum corruptum*, general habitus of larva, dorsal view.

FIGURE 14.3.167 *Sympetrum corruptum*, abdomen, dorsal view.

FIGURE 14.3.169 *Crocothemis servilia* (Drury, 1773), apex of abdomen, ventral view. Arrows indicate a row of small spiniform setae directed caudad, present on ventral surface of distal margin of S7-9.

FIGURE 14.3.170 *Crocothemis servilia*, labial palps, frontal view.

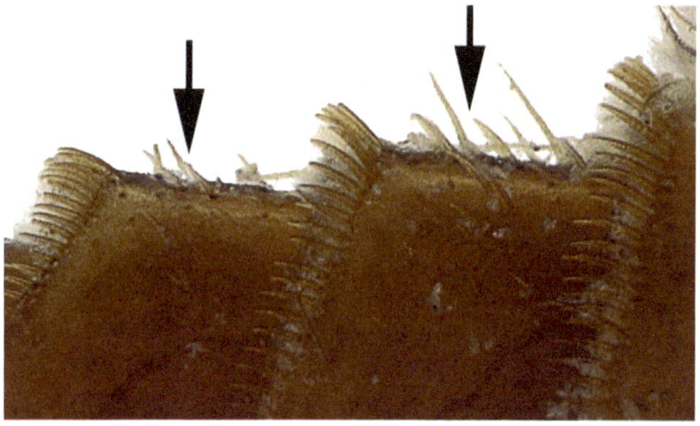

FIGURE 14.3.172 *Erythrodiplax* sp., dorsal part of two abdominal segments, right lateral view. Arrows indicate abdominal tergites surface often pilose.

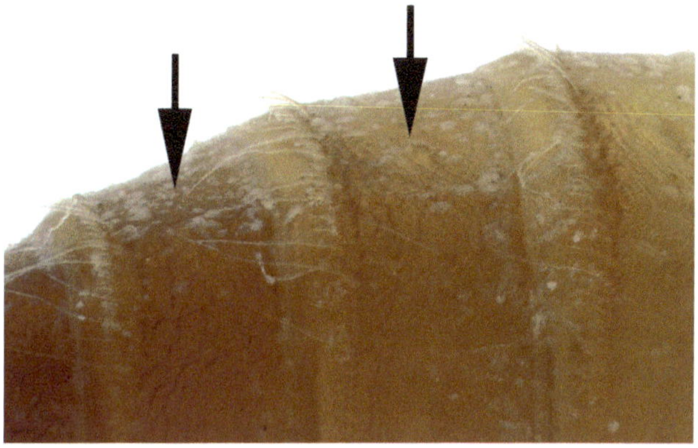

FIGURE 14.3.173 *Pseudoleon superbus* (Hagen, 1861), dorsal part of three abdominal segments, right lateral view. Arrows indicate abdominal tergites surface with glabrous aspect.

FIGURE 14.3.171 *Crocothemis servilia*, abdomen, dorsal view.

FIGURE 14.3.174 *Pseudoleon superbus*, abdomen, dorsal view.

FIGURE 14.3.175 *Pachydiplax longipennis*, abdomen, dorsal view.

FIGURE 14.3.176 *Pachydiplax longipennis*, head, dorsal view. Arrows indicate the long black stripes underlining posterior margin of the eyes.

FIGURE 14.3.177 *Pachydiplax longipennis*, general habitus of larva, dorsal view.

FIGURE 14.3.178 *Nephepeltia berlai* Santos, 1950, anal pyramid, dorsal view.

42(41)	Epiproct apex length >0.5× paraprocts in dorsal view (Fig. 14.3.180 C)	43
42'	Epiproct apex length 0.5× paraprocts in dorsal view (Figs. 14.3.178, 179) [Panama to Peru and Brazil]	*Nephepeltia*
43(42)	Epiproct apex length ≥0.8× paraprocts in dorsal view (Figs. 14.3.180 A, C) [Mexico to Brazil, Paraguay, Bolivia]	*Anatya*
43'	Epiproct apex length >0.5× and <0.8× paraprocts in dorsal view (Figs. 14.3.181, 182) [Mexico to central Argentina, Uruguay]	*Micrathyria*

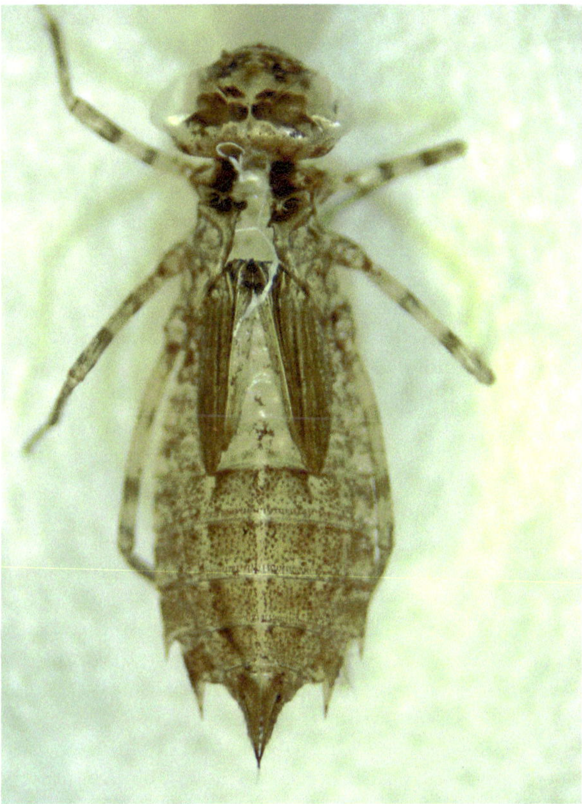

FIGURE 14.3.179 *Nephepeltia berlai*, general habitus of exuvia, dorsal view.

FIGURE 14.3.181 *Micrathyria* sp., anal pyramid, dorsal view.

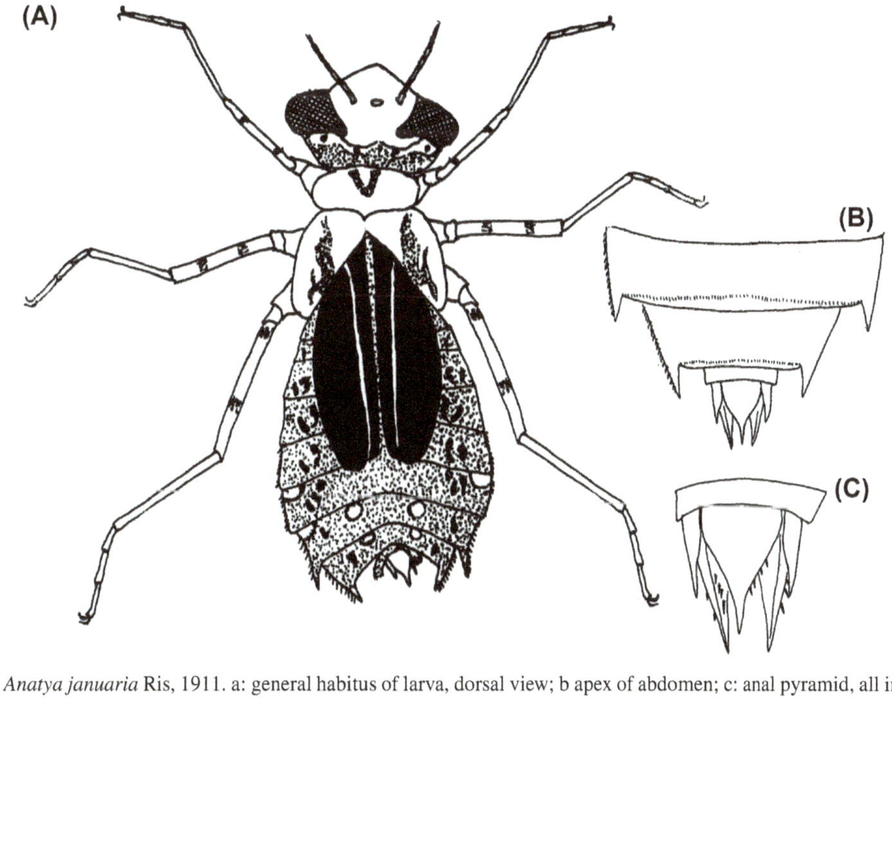

FIGURE 14.3.180 *Anatya januaria* Ris, 1911. a: general habitus of larva, dorsal view; b apex of abdomen; c: anal pyramid, all in dorsal view, based on Santos (1973).

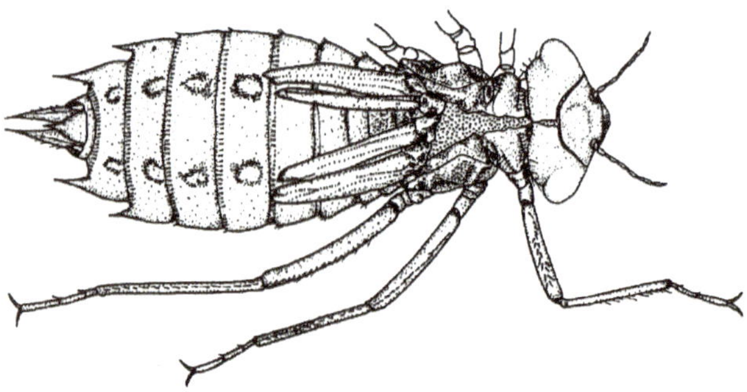

FIGURE 14.3.182 *Micrathyria longifasciata* Calvert, 1909, general habitus of exuvia, dorsal view. *Modified from Souza & Costa (2002).*

ACKNOWLEDGMENTS

We express our most sincere gratitude to Dr. Natalia von Ellenrieder and Rodolfo Novelo-Gutiérrez for the loan of material, and to Leo Rossy Ribeiro da Trindade for their help in making photographs using a Leica M165 with automontage software. Dr. Jürg DeMarmels provided the photos 14.3.24–25; Dr. Rodolfo Novelo-Gutiérrez provided the photos 14.3.70, 14.3.141, and 14.3.161 and several others valuable information about some libelulid larvae. Dr. Marina Schmidt Dalzochio provided the photos 14.3.147–149 and 14.3.178–179. The Figs. 14.3.14–16, 26 and 34 were originally published by Neiss & Hamada (2014) and were generously given by Editora INPA. Financial support was provided by PRONEX-CNPq-FAPEAM, MCT/INPA/PPI, PRO-Equipamentos/CAPES/INPA, and CNPq (process n. 307849/2014-7, to N. Hamada).

REFERENCES

Artiss T., T.R. Schultz, D.A. Polhemus & C. Simon. 2001. Molecular phylogenetic analysis of the odonate genera *Libellula, Ladona, and Plathemis* (Odonata: Libellulidae) based on mitochondrial cytochrome oxidase I and 16S rRNA sequence data. Molecular Phylogenetics and Evolution 18: 348–361.

Bechly, G. 1996. Morphologische Untersuchungen am Flügelgeäder der rezenten Libellen und deren Stammgruppenvertreter (Insecta; Pterygota; Odonata), unter besonderer Berücksichtigungder Phylogenetischen Systematik und des Grundplanes derOdonata. Petalura, Specialvolume 2: 1–402.

Bechly, G. 2003. Description of a new species of *Nannogomphus* (Insecta: Odonata: Nannogomphidae) from the Upper Jurassic Solnhofen Limestone in Germany. Stuttgarter Beiträge zur Naturkunde, Serie B 339: 1–6.

Bybee, S.M., T.H. Ogden, M.A. Branham & M.F. Whiting. 2008. Molecules, morphology and fossils: a comprehensive approach to odonate phylogeny and the evolution of the odonate wing. Cladistics 23: 1–38.

Carle, F.L., K.M. Kjer & M.L. May. 2015. A molecular phylogeny and classification of Anisoptera (Odonata). Arthropod Systematics & Phylogeny 73: 281–301.

Carvalho, A.L. & E.R. Calil. 2000. Chaves de identificação para as famílias de Odonata (Insecta) ocorrentes no Brasil, adultos e larvas. Papéis Avulsos de Zoologia 41: 223–241.

Carvalho, A.L., P.C. Werneck-de-Carvalho & E.R. Calil. 2002. Description of the larvae of two species of *Dasythemis* Karsch, with a key to the genera of Libellulidae occurring in the states of Rio de Janeiro and São Paulo, Brazil (Anisoptera). Odonatologica 31: 23–33.

Carvalho, A.L., L.G.V. Salgado & G. Fleck. 2008. Description of the larva of *Lauromacromia picinguaba* Carvalho, Salgado & Werneck-de-Carvalho 2004, with key to the genera of Corduliidae larvae occurring in South America. Zootaxa 1848: 57–65.

Costa, J.M., L.O. Irineu-de-Souza & B.B. Oldrini. 2004. Chave para identificação das famílias e gêneros das larvas conhecidas de Odonata do Brasil: comentários e registros bibliográficos (Insecta, Odonata). Publicações Avulsas do Museu Nacional 99: 1–44.

Davis, R.B., D.B. Nicholson, E.L.R. Saunders & P.J. Mayhew. 2011. Fossil gaps inferred from phylogenies alter the apparent nature of diversification in dragonflies and their relatives. Evolutionary Biology 11: 252–261.

De Marmels, J. 1990. Key to the ultimate larvae of the Venezuelan Odonate families. Opuscula Zoologica Fluminensia 50: 1–6.

De Marmels, J. 1992. Dragonflies (Odonata) from the Sierras of Tapirapecó and Unturán, in the extreme South of Venezuela. Acta Biologica Venezuelana 14: 57–78.

Dijkstra, K.-D.B. & V.J. Kalkman. 2012. Phylogeny, taxonomy and classification of European dragonflies and damselflies (Odonata): a review. Organisms Diversity & Evolution 12: 209–227.

Dijkstra, K.-D. B., G. Bechly, S.N. Bybee, R.A. Dow, H.J. Dumont, N. von Ellenrieder, G. Fleck, R.W. Garrison, M. Hämäläinen, V.J. Kalkman, H. Karube, M.L. May, A.G. Orr, D.R. Paulson, A.C. Rehn, G. Theischinger, J.W.H. Trueman, J. van Tol & J. Ware. 2013. The classification and diversity of dragonflies and damselflies (Odonata). Pages 36–45 *in*: Z.-Q. Zhang (ed.), Animal Biodiversity: An Outline of Higher-level Classification and Survey of Taxonomic Richness (Addenda 2013). Zootaxa 3703.

Dunson, W.A. 1980. Adaptations of nymphs of a marine dragonfly, *Erythrodiplax berenice*, to wide variations in salinity. Physiological Zoology 53: 445–452.

Dumont, H.J., A. Vierstraete & J.R. Vanfleteren. 2010. A molecular phylogeny of the Odonata (Insecta). Systematic Entomology 35: 6–18.

Fleck, G. 2002. Une larve d'Odonate remarquable de la Guyane française, probablement *Lauromacromia dubitalis* (Fraser, 1939) (Odonata, Anisoptera, 'Corduliidae'). Bulletin de la Société Entomologique de France 107: 223–230.

Fleck, G. 2004. The larva of the genus *Cyanothemis* Ris, 1915 (Odonata: Anisoptera: Libellulidae). Phylogenetic consequences. Annales de la Société Entomologique de France 40: 51–58.

Fleck, G. 2011. Phylogenetic affinities of Petaluridae and basal Anisoptera families (Insecta: Odonata). Stuttgarter Beiträgezur Naturkunde A, NeueSerie 4: 83–104.

Fleck, G. 2012. Preliminary notes on the genus *Aeschnosoma* Selys 1870 (Odonata: Anisoptera: Corduliidae s. str.). Annales de la SociétéEntomologique de France 48: 225–228.

Fleck, G.; Neiss, U.G. 2012. The larva of the genus *Paracordulia* Martin, 1907 (Odonata: Corduliidae *s.s.*) and a generic key to the larvae of Corduliidae *s.l.* occuring in South America. Zootaxa 3412: 62–68.

Fleck, G. & M. El Adouzi. 2013. The larva of the genus *Palaeosynthemis* Förster, 1903 (Odonata: Anisoptera: Synthemistidae) and a generic key to the larvae of non-New Caledonian Synthemistidae. Zootaxa 3619: 589–594.

Fleck, G., M. Brenk & B. Misof. 2008a. larval and molecular characters help to solve phylogenetic puzzles in the highly diverse dragonfly family Libellulidae (Insecta: Odonata: Anisoptera): The Tetrathemistinae are a polyphyletic group. Organisms, Diversity & Evolution 8: 1–16.

Fleck, G., B. Ullrich, M. Brenk, C. Wallnisch, M. Orland, S. Bleidissel & B. Misof. 2008b. A phylogeny of anisopterous dragonflies (Insecta, Odonata) using mtRNA genes and mixed nucleotide/doublet models. Journal of Zoological Systematics and Evolutionary Research 46: 310–322.

Garrison, R.W., N. von Ellenrieder & J.A. Louton. 2006. Dragonfly Genera of the New World. An Illustrated and Annotated Key to the Anisoptera. The Johns Hopkins University Press, Baltimore.

Haber, W.A., D.L. Wagner & C. De La Rosa. 2015. A new species of *Erythrodiplax* breeding in bromeliads in Costa Rica (Odonata: Libellulidae). Zootaxa 3947: 386–396.

Heckman, C. W. 2006. Encyclopedia of South American aquatic insects: Odonata – Anisoptera. Springer, Dordrecht.

May, M.L. 1997. Reconsideration of the status of the genera *Phyllomacromia* and *Macromia* (Anisoptera: Corduliidae). Odonatologica 26: 405–414.

Muzón, J., P. Pessacq & F. Lozano. 2014. The Odonata (Insecta) of Patagonia: A synopsis of their current status with illustrated keys for their identification. Zootaxa 3784: 346–388.

Neiss, U.G. & N. Hamada. 2014. Ordem Odonata. Pages 217–284 *in*: N. Hamada, J.L. Nessimian & R.B. Querino (eds.), Insetos aquáticos na Amazônia brasileira: taxonomia, biologia e ecologia. Editora INPA, Manaus.

Novelo-Gutiérrez, R. 1997. Clave para la determinación de familias y géneros de las náiades de Odonata de Mexico. Parte II. Anisoptera. Dugesiana 4: 31–40.

Theischinger G. & G. Fleck. 2003. A new character useful for taxonomy and phylogeny of Anisoptera (Odonata). Bulletin de la Société Entomologique de France 108: 409–412.

von Ellenrieder, N. & R.W. Garrison. 2005. Synopsis of the South American genus *Gomphomacromia* (Odonata: Gomphomacromiinae). International Journal of Odonatology 8: 81–96.

von Ellenrieder, N. & R.W. Garrison. 2009. Odonata. Pages 95–143 *in*: E. Domínguez & H.R. Fernández (eds.), Macroinvertebrados Bentónicos Sudamericanos. Fundación Miguel Lillo, Tucumán.

Ware, J.L., M.L. May & K. Kjer. 2007. Phylogeny of the higher Libelluloidea (Anisoptera: Odonata): an exploration of the most speciose superfamily of dragonflies. Molecular Phylogenetics and Evolution 45: 289–310.

Chapter 14.4

Superfamily Calopterygoidea

Danielle Anjos-Santos
Centro de Investigaciones Esquel de Montaña y Estepa Patagónicas (CIEMEP), Esquel, Chubut, Argentina

Ulisses Gaspar Neiss
Instituto de Criminalística, Departamento de Polícia Técnica-Científica, Manaus, Amazonas, Brazil

Pablo Pessacq
Centro de Investigaciones Esquel de Montaña y Estepa Patagónicas (CIEMEP), Esquel, Chubut, Argentina

SUPERFAMILY CALOPTERYGOIDEA

Chapter Outline

INTRODUCTION

The term Calopterygoidea has been traditionally used to refer to a very diverse assemblage of densely veined damselflies. Depending on the authors (e.g., Bechly, 1996; Rehn, 2003; Dumont et al., 2005), it refers to different concepts, including different families and genera. For example, Dumont et al. (2005) treated the Calopterygoid damselflies *sensu stricto* (Calopterygidae and Hetaerinidae) as the superfamily Calopterygoidea. The latest classificatory scheme (Dijkstra et al., 2013) considers the superfamily as composed by 10 families, five of which are represented in the Neotropical region (Amphipterygidae, Calopterygidae, Dicteriadidae, Megapodagrionidae and Polythoridae). This superfamily has worldwide distribution, being composed by 111 genera and 782 species (Dijkstra et al., 2013). The Neotropical region has 31 genera and 292 species (Garrison & von Ellenrieder, 2016), but this number is very likely to increase. Except for Dicteriadidae, there are many undescribed species in the remaining Calopterygoidea families (Garrison et al., 2010).

Several authors (Bybee et al., 2008; Carle et al., 2008; Dijkstra et al., 2014; Dumont et al., 2005; Rehn, 2003) included Calopterygoidea in morphological and molecular phylogenetic analyses, but its position within Zygoptera is still unclear. Dijkstra et al. (2013) concluded that "Calopterygoidea," by almost any definition, may ultimately prove to be paraphyletic, requiring the creation of further superfamilies. For the time being, however, we follow the classification proposed by Dijkstra et al. (2013).

The following families occur in the Neotropical region: Amphipterygidae, Calopterygidae, Dicteriadidae, Megapodagrionidae, and Polythoridae.

Amphipterygidae is a small family of medium-sized damselflies, distributed in Africa, Asia, and the New World (Garrison et al., 2010). Although it is currently composed of 15 species in four genera (Jocque & Argueta, 2014), the position and monophyly of this family is doubtful. Based on larval morphology, Novelo-Gutiérrez (1995) restricted the composition of this family to two subfamilies, as did Rehn (2003) in his cladistic analysis. Molecular phylogenies, however, are inconclusive, and this group may be polyphyletic (Bybee et al.,2008; Dumont et al., 2010). Two genera (*Amphipteryx* and *Rimanella*) with five species total, are endemic to the Neotropical region (Garrison et al., 2010, Jocque & Argueta, 2014), and one species of each genus has known larvae. Larvae of both

Thorp and Covich's Freshwater Invertebrates, Fourth Edition. https://doi.org/10.1016/B978-0-12-804223-6.00018-4

genera occur in similar habitats, and are found in small shallow creeks with gravel bottom and rapid flow, with some microhabitat preferences (for additional information see Novelo-Gutiérrez, 1995; De Marmels, 1989; Geijskes, 1940; Hamada & Oliveira, 2003).

Calopterygidae is composed of medium to very large damselflies. The 21 genera and 185 species currently recognized (Dijkstra et al., 2013) are distributed in all world regions, except Australia (Garrison et al., 2010). In the Neotropical region, 69 species in five genera are known, four of which are endemic (Garrison et al., 2010; Garrison & von Ellenrieder, 2016). Calopterygidae is considered monophyletic based on both molecular and morphological analyses (Bybee et al., 2008; Dumont et al., 2010; Rehn, 2003). Calopterygid larvae inhabit streams with rapid flow, where they are associated with macrophytes, roots, and detritus (Carvalho & Nessimian, 1998; Guillermo-Ferreira & Bispo, 2012; Neiss & Hamada, 2014).

Dicteriadidae is a small family of medium to large-sized damselflies. Two monotypic genera are endemic: *Dicterias atrosanguinea* Selys, 1853, which is restricted to the Amazon River basin, and *Heliocharis amazona* Selys, 1853, which is widely distributed in South America (Garrison et al., 2010). Both family monophyly and placement into the calopterygoid complex are corroborated by phylogenetic analysis based on adult and larval morphology (Rehn, 2003; Fleck et al., 2008, 2012; Dumont et al., 2010). *Dicterias* larvae occur in typical small blackwater sandy bottom streams of the Amazon forest, living on tree roots or aquatic vegetation in areas with moderate to rapid flow rates (Fleck et al., 2012; Neiss & Hamada, 2014). *Heliocharis* larvae are found in medium to large streams where they hang on submerged roots at the stream margin, in pools or in other areas with moderate to low flow (Geisjkes, 1986; Neiss & Hamada, 2014).

Megapodagrionidae is a heterogeneous family of medium- to large-sized damselflies, composed of 296 species in 42 genera (Dijkstra et al., 2013) distributed in tropical Africa, Australia, and the New World (Garrison et al., 2010). Currently, 147 species in 14 genera (all endemic) are known from the Neotropical region (Garrison & von Ellenrieder, 2016). According to several phylogenetic analyses, this family is clearly not monophyletic, containing at least six independent lineages (Rehn, 2003; Bybee et al., 2008; van Tol et al., 2009; Dumont et al., 2010, Kalkman et al., 2010). According to Dijkstra et al. (2014), recent molecular analyses (Bybee et al., 2008; Kalkman & Theischinger, 2013) suggest that several additional families may require description. Megapodagrionid larvae inhabit a wide variety of habitats, including streams of clear water with stony or sandy bottom, being found amongst aquatic macrophytes, roots, between detritus, and under stones. They can also be found in high altitudes streams, including several peculiar microhabitats such as the spray zone of waterfalls in forested streams (e.g., Calvert, 1915; De Marmels, 2001; Novelo-Gutiérrez, 2008; Tennessen, 2010; Neiss et al., 2011).

Polythoridae is an endemic family composed of medium to large-sized damselflies. This family has 60 species and seven genera (Garrison & von Ellenrieder, 2016), predominantly occurring in the Andean mountains and the Amazon Biome (Neiss & Hamada, 2014). The systematics are well resolved, and it is considered monophyletic in several analyses (Rehn, 2003; Bybee et al., 2008; Dijkstra et al., 2014). Most larvae are found in small to medium size forest streams with continuous flow, rocky substrates, and abundant detritus. Some genera also occur in mountain forest streams at elevations above 1000 m (De Marmels, 1982, 1995, 2007; Ramírez et al., 2011; Neiss & Hamada, 2014).

LIMITATIONS

Several ordinal level keys exist that include immature stages, but they are limited to certain regions or countries (e.g., Neiss & Hamada, 2014 for Amazonas; Costa et al., 2004 for Brazil; Novelo- Gutiérrez, 1997 for Mexico).

For the low diversity families (Amphipterygidae, Dicteriadidae), knowledge of the immature stages is limited. At the generic level, our knowledge of larvae in the remaining families of Calopterygoidea is relatively good, but as mentioned under each family, only a small portion of its species have been described (see Garrison et al., 2010). For the speciose and widely distributed genera *Hetaerina* and *Mnesarete* in Calopterygidae, separation of larvae to genera and species is often impossible (Garrison, 2006; Garrison et al., 2010; Neiss & Hamada, 2014). Larvae of *Heteragrion* and *Oxystigma* (Megapodagrionidae) and *Euthore* and *Polythore* (Polythoridae) are also very difficult to separate morphologically. Thus, these species are not separated in the key.

TERMINOLOGY AND MORPHOLOGY

The most important terminology for identifying these damselflies are described in Chapter 14.0 in this book and Chapter 35 (Suhling et al., 2015) in Volume 1 of this series (Thorp & Rogers, 2015).

MATERIAL PREPARATION AND PRESERVATION

In Volume I (Thorp & Rogers, 2015) of this series specimen preparation and preservation are presented.

KEYS TO CALOPTERYGOIDEA

Distributional data are from Garrison et al. (2010).

Calopterygoidea: Families

1 Antennal scape never longer than pedicel and flagellum together (Fig. 14.4.5), prementum without a very deep median cleft with long lateral lobes (Fig. 14.4.6) ... 2

1' Antennal scape longer than pedicel and flagellum together (at least 1.5 times longer) (Fig. 14.4.3), prementum with a very deep median cleft and long lateral lobes (Fig. 14.4.4)... **Calopterygidae [p. 451]**

2(1) Caudal appendages about as long as body length or longer (Fig. 14.4.1), dorsum and/or occipital region of head with tubercles or projections (Figs. 14.4.7–9) ... 3

2' Caudal appendages shorter than body (Fig. 14.4.2), if longer (*Rimanella*), strongly modified, long and slender and with abdominal terminal filamentous gill tuft; dorsum and occipital region of head without tubercles or projections (Figs. 14.10, 11), except for *Teinopodagrion*, with tubercles on the occiput (Fig. 14.4.12) ... 4

3(2) Posteriolateral margin of head with small tubercles; dorsum of head with two large projections (Fig. 14.4.7); caudal appendages with marginal thick setae, apex very sharp or ending in a spine (Fig. 14.4.13)... **Dicteriadidae [p. 452]**

3' Posteriolateral margin of head with well-developed projections (Figs. 14.4.8, 9); dorsum of head with two small tubercles or bare (Fig. 14.4.8); caudal appendages with marginal thinner setae, apex of caudal appendages not very sharp neither ending in a spine (Fig. 14.4.14) ... **Megapodagrionidae [p. 452]** (in part)

4(2) Abdominal terminal filamentous gill tuft absent (Figs. 14.4.17, 14.4.19, 20); caudal appendages not greatly modified, neither pedunculate nor covered with long flat setae ... 5

4' Abdominal terminal filamentous gill tuft present (Figs. 14.4.15, 16); caudal appendages greatly modified, lateral ones long and slender, with spiniform basal projections (*Rimanella*, Fig. 14.4.15) or pedunculate and covered with long flat setae (*Amphipteryx*, Fig. 14.4.16) ... **Amphipterygidae [p. 451]**

5(4) Abdomen with ventrolateral gills (Fig. 14.4.18); caudal appendages without spiniform seta, usually with elongated projections (Figs. 14.4. 17, 19)... **Polythoridae [p. 467]**

5' Abdomen with ventrolateral gills absent; caudal appendages with (Figs. 14.4.20, 21) or without (Fig. 14.4.22) rows of short spiniform setae ... **Megapodagrionidae [p. 452]** (in part)

Calopterygoidea: Amphipterygidae: Genera

Larvae known for both monotypic genera (Geijskes, 1940; Novelo-Gutierrez, 1995). The most striking larval characteristic is the modification of caudal appendages into two rigid processes that protects the filamentous gills.

1 Lateral caudal appendages $\geq$ body, divided into two parts: proximal (short, medially curved) and distal (long covered with thick short scaliform setae); median caudal appendage divided into four projections, that protect the gills (Fig. 14.4.15); head occipital region rounded with blunt projection (Fig. 14.4.23); antennae 8-articulate (Fig. 14.4.24) ... *Rimanella*
[Guiana shield in Venezuela, Surinam, and Brazil]

1' Lateral and median caudal appendages shorter than body, basally narrowed, resembling a pedicel, distally abruptly tumid and gradually tapering, ending in a long, blunt tip (Fig. 14.4.16); head occipital region quite rounded (Fig. 14.4.25), antennae 6-articulate (Fig. 14.4.26) ... *Amphipteryx*
[Guatemala, Honduras, Mexico]

Calopterygoidea: Calopterygidae: Genera

The larvae of four of the five genera and the larva of one third of the species are known (Garrison et al., 2010). The larvae are very peculiar: they have an elongated and slightly flattened body, the first antennomere very long, the prementum with deep median cleft and anteriorly projecting lateral lobes, the legs disposed laterally and bear triangular caudal appendages (De Marmels, 2007; Neiss & Hamada, 2014).

Unknown larva: *Bryoplathanon.*

1 First antennomere shorter than length of head and prothorax (Fig. 14.4.28) ... 2

1' First antennomere as long as length of head and prothorax (Fig. 14.4.27) ... *Iridictyon*
[Tepuis of Brazil, Guiana and Venezuela]

2(1) Abdominal segments 4–9 with posteriodorsal hooks, segment 10 posterior margin without mediodorsal spine (Fig. 14.4.29) ... *Ormenophlebia*
[Brazil, Colombia, Bolivia, Ecuador, Peru]

2' Abdominal segments with posteriodorsal hooks absent, if present, never only on segments 4–9 (Fig. 14.4.30), segment 10 posterior margin mediodorsal spine present or absent ... *Mnesarete/Hetaerina*
[South America]/[Neotropical. Nearctic]

Calopterygoidea: Dicteriadidae: Genera

Both monotypic genera have known larvae (Garrison et al., 2010; Fleck et al., 2012). The larvae possess long legs and very long caudal appendages (as long as body) armed with spiniform setae. The first antennomere is also long, the prementum possess a median cleft, and the head possess two prominent dorsal knobs (Geisjkes, 1986; Santos & Costa, 1988; Fleck et al., 2012).

1 Median caudal appendage apex ending abruptly in a sharp spur, directed ventrally in lateral view (Fig. 14.4.13, center); larva length without caudal appendages: ≤15 mm .... *Dicterias*
[Brazil along Amazonas river]
1' Median caudal appendage apex tapering gradually and directed ventrally (Fig. 14.4.13, above); larva length without caudal appendages >15 mm .... *Heliocharis*
[South America]

Calopterygoidea: Megapodagrionidae: Genera

The larvae of 13 of the 14 genera, and slightly more than 30% of the species are known. Larvae show great morphological variety, according to the group to which they belong and the habitat in which they live; some genera are characterized by head vertex and abdominal segments with tubercles, long legs, and caudal appendages as long as or longer than body size, while others are characterized by head and abdominal segments without tubercles, short legs, and short and inflated caudal appendages, armed with strong spines or variable pilosity along the edge (Neiss & Hamada, 2014).

Unknown larvae: *Archaeopodagrion.*

1 Occipital region of head with projections (Figs. 14.4.8, 9, 12) 2
1' Occipital region of head without projections (Figs. 14.4.5, 10, 11, 40, 43) 4
2(1) Abdomen without dorsal hooks, caudal appendages longer or equal than body (Fig. 14.4.1) 3
2' Abdomen with dorsal hooks (Fig. 14.4.31), caudal appendages shorter than body .... *Teinopodagrion*
[Argentina, Bolivia, Chile, Ecuador, Peru, Venezuela]
3(2) Caudal appendages very narrow (Fig. 14.4.14, below), head with two tubercles between the eyes, occipital projections dorsally directed (Fig. 14.4.9) .... *Allopodagrion*
[Argentina, Brazil]
3' Caudal appendages widened on its distal half (Fig. 14.4.14, above), head without two tubercles between the eyes (Fig. 14.4.8), occipital projections posteriorly directed (Fig. 14.4.8) .... *Megapodagrion*
[Brazil, Guiana, Peru, Venezuela]
4(1) Median caudal appendage larger than lateral ones (e.g., Fig. 14.4.32) 5
4' Median caudal appendage smaller or equal than lateral ones (Figs. 14.4.20, 21, 33) 6
5(4) Caudal appendages with well-developed dorsal ridges (Fig. 14.4.34), median caudal appendage almost circular (Fig. 14.4.34, rigth), labial palp with one seta close to movable hook .... *Dimeragrion*
[Northwest Brazil, Guiana, Venezuela]
5' Caudal appendages without well-developed dorsal ridges, median caudal appendage oval (Fig. 14.4.32), labial palp without one seta close to movable hook .... *Mesagrion*
[Colombia]
6(4) Caudal appendages with spines (a few ones in some *Heteragrion/Oxystigma* species) (Figs. 14.4.20, 21, 33) 7
6' Caudal appendages without spines (Fig. 14.4.22) 9
7(6) Labial palp with three teeth (Fig. 14.4.6) 8
7' Labial palp with two teeth (Fig. 14.4.35) .... *Thaumatoneura*
[Costa Rica, Panama]
8(7) Prothorax with prominent supracoxal processes (Fig. 14.4.36), median caudal appendage with well-developed dorsal and ventral ridges, terminal filament present only in lateral caudal appendages (Fig. 14.4.37) .... *Sciotropis*
[Coastal cordillera of Venezuela]
8' Prothorax without supracoxal processes (Fig. 14.4.38), all caudal appendages with well-developed dorsal and lateral ridges (Figs. 14.4.20, 21, 33), terminal filament, present in all caudal appendages (Figs. 14.4.21, 33), sometimes reduced (Fig. 14.4.20) *Heteragrion/Oxystigma*
[Widespread] [Brasil, Guianas, Venezuela]
9(6) Antennae length < head length (Fig. 14.4.39) 10
9' Antennae length ≥ head length (Fig. 14.4.40) 11
10(9) Prementum with a slightly curved row of very small transverse ridges on each side (Fig. 14.4.52), ligula as in Fig. 14.4.41 .... *Heteropodagrion*
[Colombia, Ecuador, Panama]
10' Prementum without transverse ridges on each side, ligula as in Fig. 14.4.42 .... *Paraphlebia*
[Central America]
11(9) Second antennomere longer than other segments (Fig. 14.4.43), terminal filament of median caudal appendage articulate (Fig. 14.4.44) .... *Hypolestes*
[Cuba, Dominican Republic, Jamaica]
11' Third antennomere longer than other segments (Fig. 14.4.45), terminal filament of median caudal appendage not articulate (Fig. 14.4.22) .... *Philogenia*
[Honduras to Bolivia]

FIGURE 14.4.1 *Megapodagrion megalopus* (Selys, 1862) (Megapodagrionidae), larval habitus, dorsolateral view. *Original pictures by Neiss & Hamada (2014).*

FIGURE 14.4.2 *Amphipteryx agrioides* Selys, 1853 (Amphipterygidae), larval habitus, dorsolateral view.

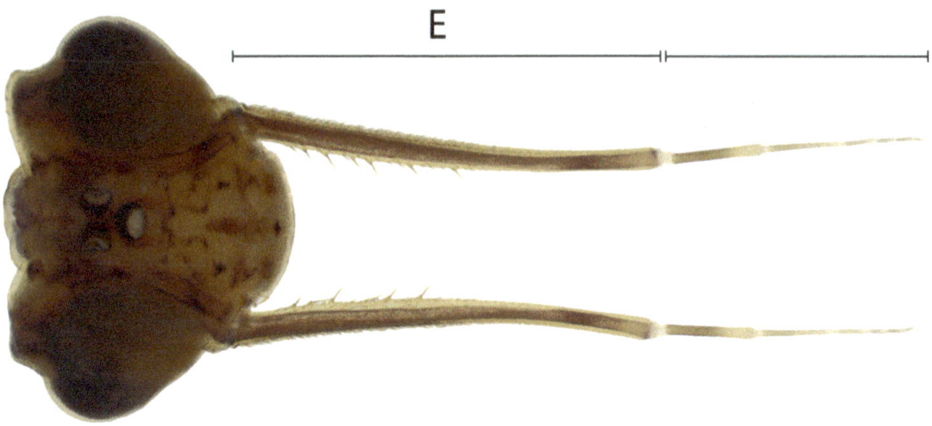

FIGURE 14.4.3 *Hetaerina* sp. (Calopterygidae), larval head, dorsal view. *E*, Escape (first antennal article). *Original pictures by Neiss & Hamada (2014).*

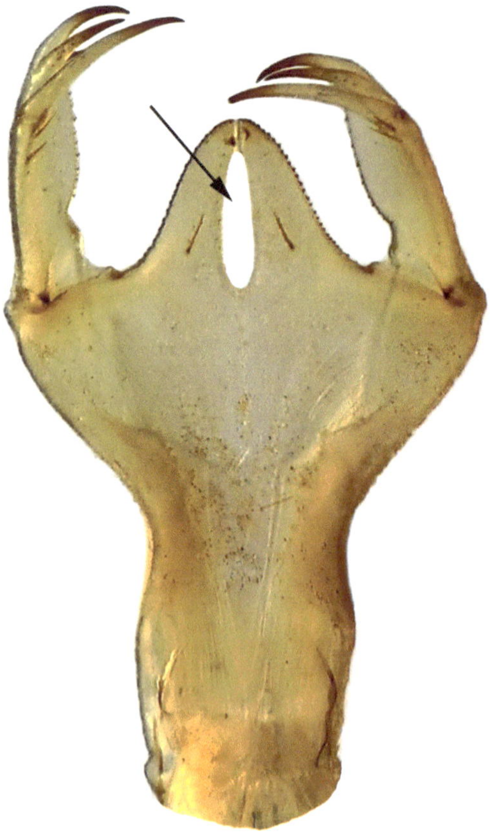

FIGURE 14.4.4 *Hetaerina americana* (Fabricius, 1798) (Calopterygidae), larval prementum, dorsal view. *Arrow* indicates premental cleft.

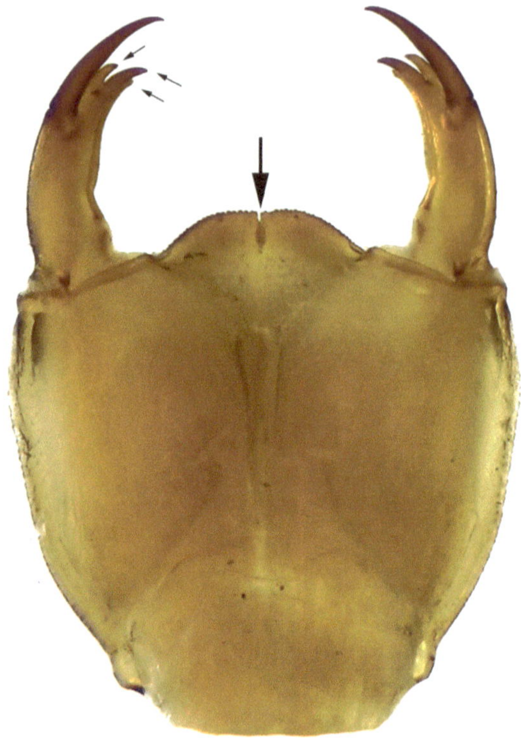

FIGURE 14.4.6 *Paraphlebia zoe* Selys *in* Hagen, 1861 (Megapodagrionidae), larval prementum, dorsal view. *Arrows* indicate premental cleft and palpal teeth.

FIGURE 14.4.5 *Heteragrion consors* Hagen *in* Selys, 1862 (Megapodagrionidae), larval head, dorsal view. *E*, Escape.

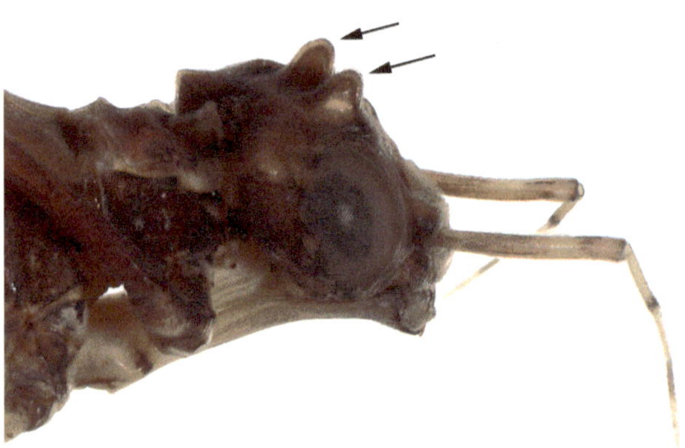

FIGURE 14.4.7 *Dicterias atrosanguinea* (Dicteriadidae), larval head, lateral view. *Arrows* indicate dorsum of head with large projections. *Original pictures by Neiss & Hamada (2014).*

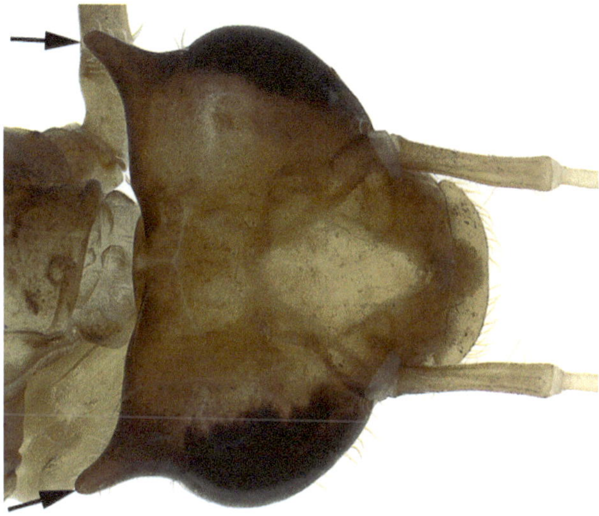

FIGURE 14.4.8 *Megapodagrion megalopus* (Megapodagrionidae), larval head, dorsal view. *Arrows* indicate posteriolateral margin of head with well-developed projections. *Original pictures by Neiss & Hamada (2014).*

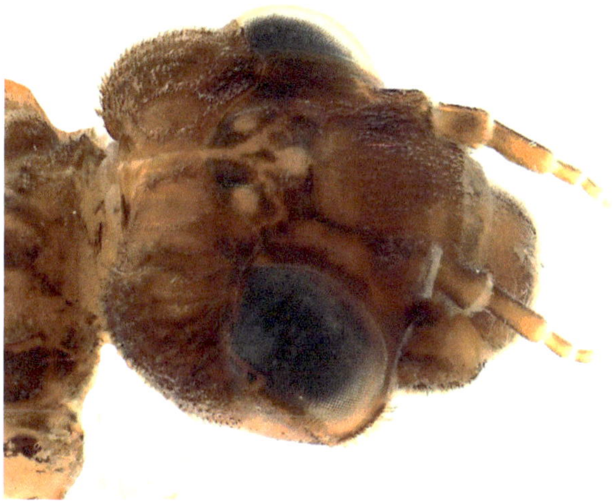

FIGURE 14.4.10 *Paraphlebia zoe* (Megapodagrionidae), larval head, dorsal view.

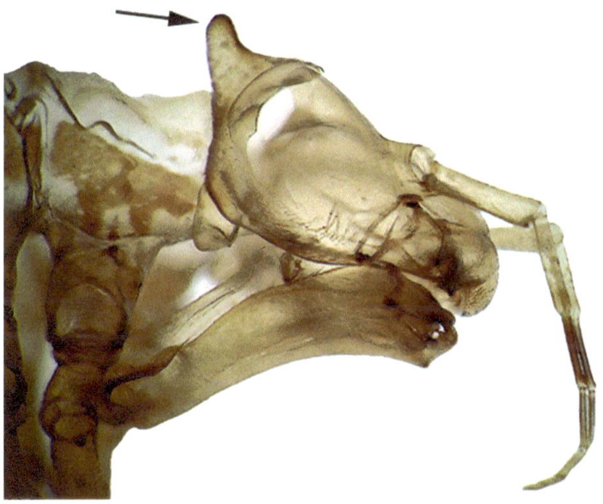

FIGURE 14.4.9 *Allopodagrion brachyurum* De Marmels, 2001 (Megapodagrionidae), larval head, lateral view. *Arrow* indicates posteriolateral margin of head with well-developed projections. *Original pictures by Neiss & Hamada (2014).*

FIGURE 14.4.11 *Heteragrion* sp. (Megapodagrionidae), larval head, dorsal view. *Original pictures by Neiss & Hamada (2014).*

FIGURE 14.4.12 *Teinopodagrion meridionale* De Marmels, 2001 (Megapodagrionidae), larval head, dorsolateral view. *Arrow* indicates tubercles on the occiput.

FIGURE 14.4.13 *Heliocharis amazona* (above), *Dicterias atrosanguinea* (center and below) (Dicteriadidae). Larva lateral caudal appendages, lateral view (above and center), caudal appendages, dorsal view (below). *Arrows* indicate marginal thick setae, and apex very sharp or ending in a spine. *Original pictures by Neiss & Hamada (2014).*

FIGURE 14.4.14 *Megapodagrion megalopus* (above), *Allopodagrion brachyurum* (below) (Megapodagrionidae). Larval caudal appendages, lateral view (above and center), caudal appendages, dorsal view (below). *Original pictures by Neiss & Hamada (2014).*

FIGURE 14.4.15 *Rimanella arcana* (Needham, 1933) (Amphipterygidae). Larval caudal appendages, dorsal view. *Arrow* indicates tuft of filamentous terminal gills. *Original pictures by Neiss & Hamada (2014).*

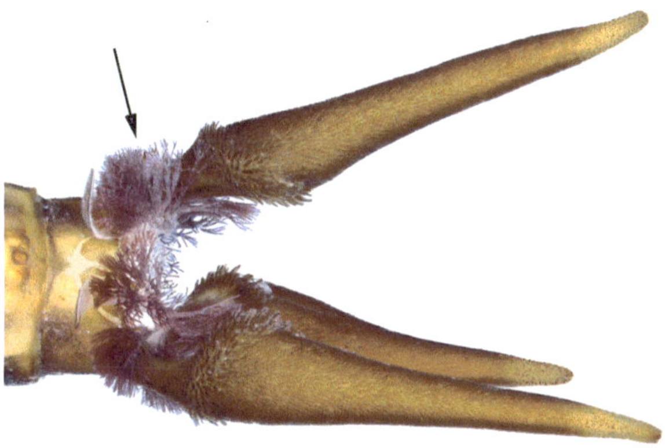

FIGURE 14.4.16 *Amphipteryx agrioides* (Amphipterygidae), larval caudal appendages, ventral view. *Arrow* indicates tuft of filamentous terminal gills.

FIGURE 14.4.17 *Cora marina* Selys, 1868 (Polythoridae), larval caudal appendages, ventral view. *Arrows* indicate elongated projections.

FIGURE 14.4.18 *Chalcopteryx scintillans* Mc Lachlan, 1870 (Polythoridae), larval abdomen, ventral view. *Arrows* indicate lateroventral abdominal gills. *Original pictures by Neiss & Hamada (2014).*

FIGURE 14.4.19 *Chalcopteryx scintillans* (Polythoridae). *Arrows* indicate elongated projections. *Original pictures by Neiss & Hamada (2014).*

FIGURE 14.4.20 *Heteragrion* sp. (Megapodagrionidae), larval caudal appendages, dorsal view. *Arrows* indicate spines and ridges. *Original pictures by Neiss & Hamada (2014).*

FIGURE 14.4.21 *Heteragrion* sp. (Megapodagrionidae), larval caudal appendages dorsal view. *Arrows* indicate spines.

FIGURE 14.4.22 *Philogenia mangosisa* Bick & Bick, 1988 (Megapodagrionidae), larval caudal appendages, lateral view. *Original pictures by Bybee & Tennessen (2008).*

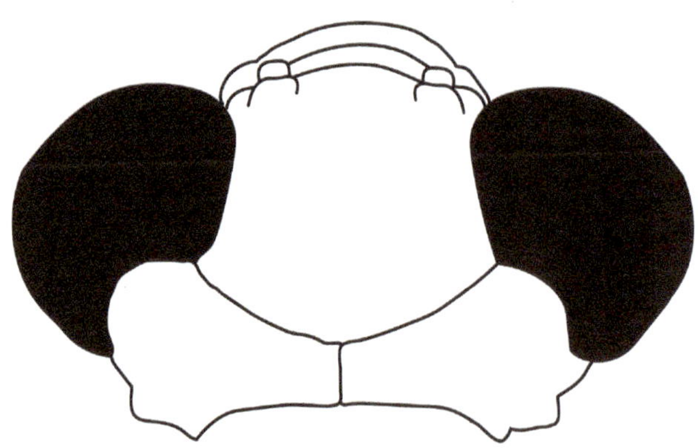

FIGURE 14.4.23 *Rimanella arcana* (Amphipterygidae), larval head, dorsal view. *Illustration adapted from Novelo-Gutiérrez (1995).*

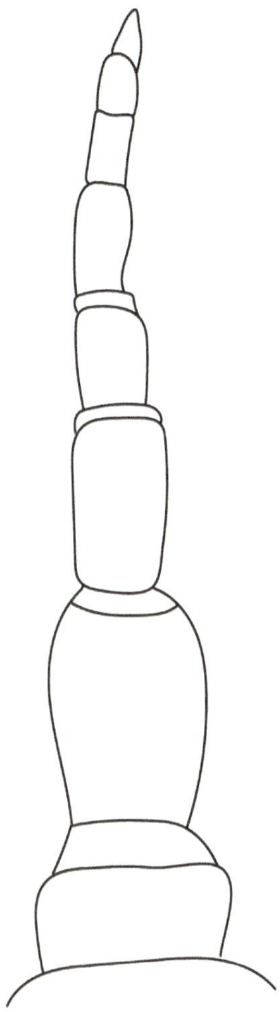

FIGURE 14.4.24 *Rimanella arcana* (Amphipterygidae), larval antenna, dorsal view. *Illustration adapted from Geijskes (1940).*

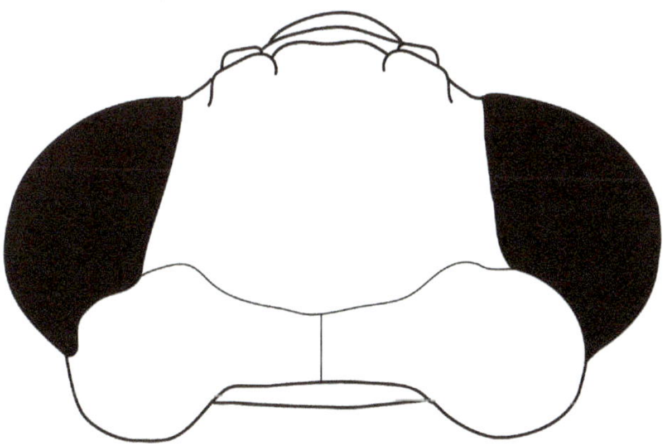

FIGURE 14.4.25 *Amphipteryx agrioides* (Amphipterygidae), larval head, dorsal view. *Illustration adapted from Novelo-Gutiérrez (1995).*

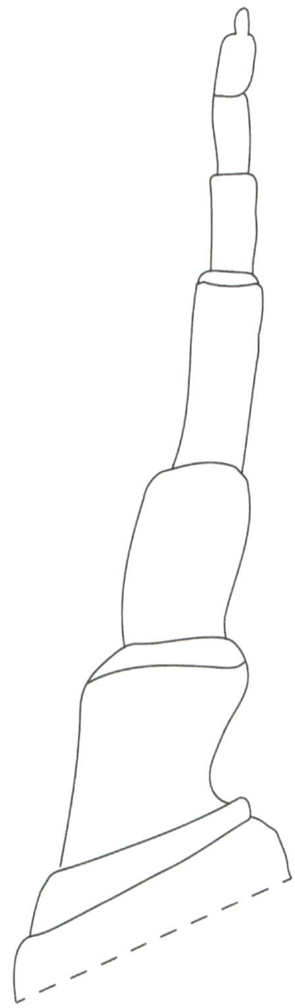

FIGURE 14.4.26 *Amphipteryx* sp. (Amphipterygidae), larval antenna, dorsal view. *Illustration adapted from Novelo-Gutiérrez (1995).*

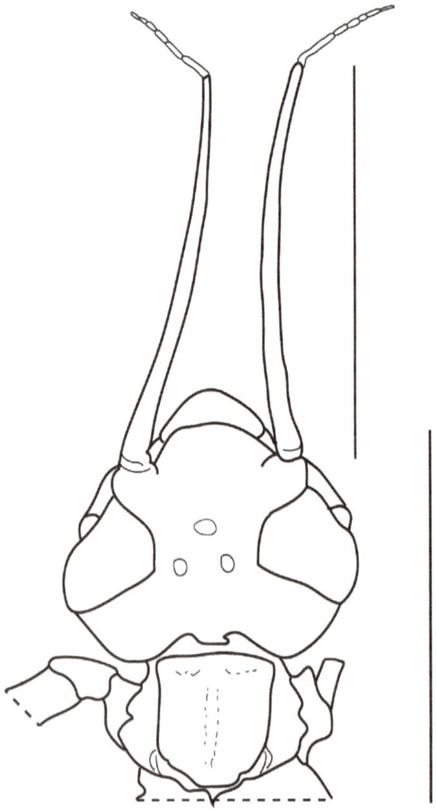

FIGURE 14.4.27 *Iridictyon trebbaui* Rácenis, 1968 (Calopterygidae), larval head and prothorax, dorsal view. *Illustration adapted from De Marmels (1992).*

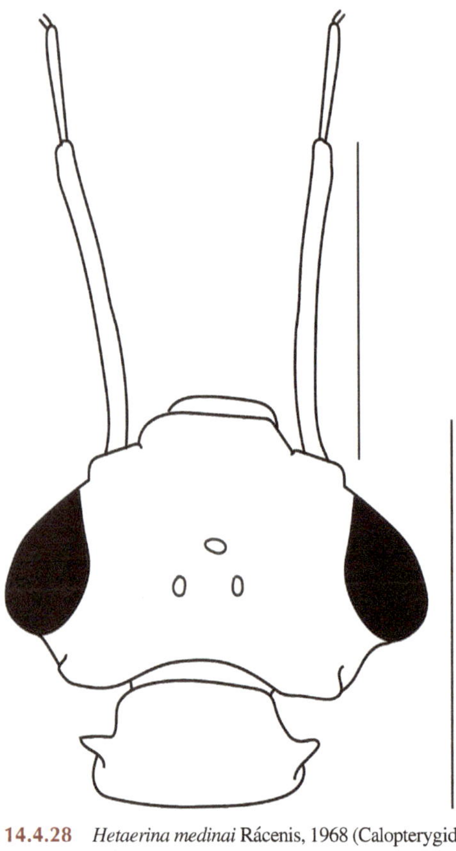

FIGURE 14.4.28 *Hetaerina medinai* Rácenis, 1968 (Calopterygidae), larval head and prothorax, dorsal view. *Illustration adapted from De Marmels (2007).*

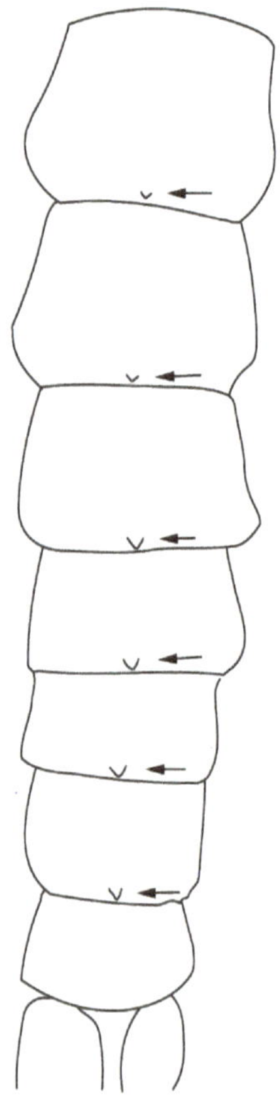

FIGURE 14.4.29 *Ormenophlebia imperatrix* (Mc Lachlan, 1878) (Calopterygidae) larva. *Arrows* indicate posterodorsal hooks. Abdomen dorsal view. *Illustration adapted from Garrison (2006).*

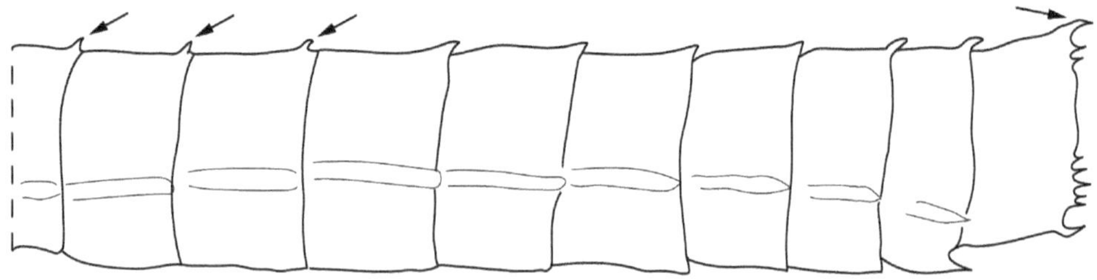

FIGURE 14.4.30 *Hetaerina mendezi* Jurzitza, 1982 (Calopterygidae). Male larva abdomen segments 1–10, lateral view. *Arrows* indicate firsts posterolateral hooks and mediodorsal spine. *Illustration adapted from von Ellenrieder (2007).*

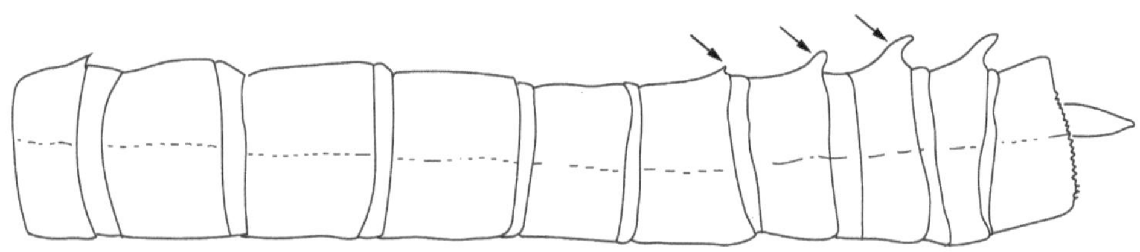

FIGURE 14.4.31 *Teinopodagrion meridionale* (Megapodagrionidae), larva abdomen, lateral view. *Arrows* indicate firsts posterolateral hooks. *Illustration adapted from von Ellenrieder (2006).*

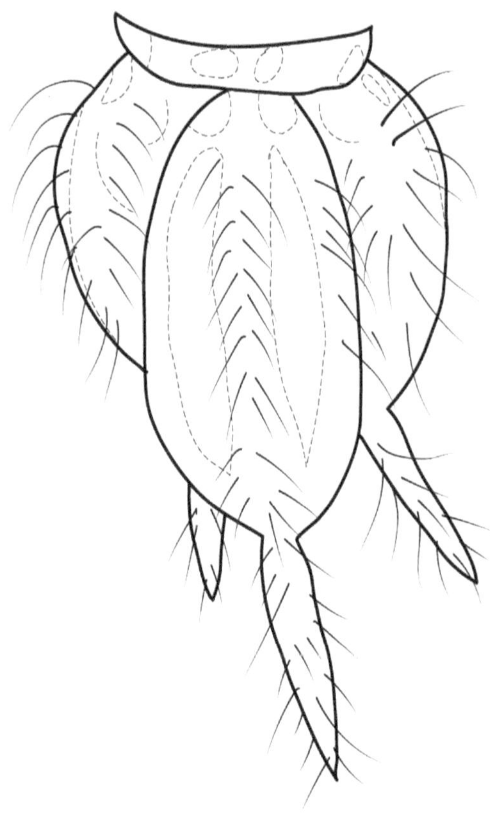

FIGURE 14.4.32 *Mesagrion leucorrhium* Selys, 1885 (Megapodagrionidae), larval caudal appendages dorsal view. *Illustration adapted from Pérez-Gutiérrez & Montes-Fontalvo (2011).*

FIGURE 14.4.33 *Heteragrion consors* (Megapodagrionidae), female larva caudal appendages, dorsal view.

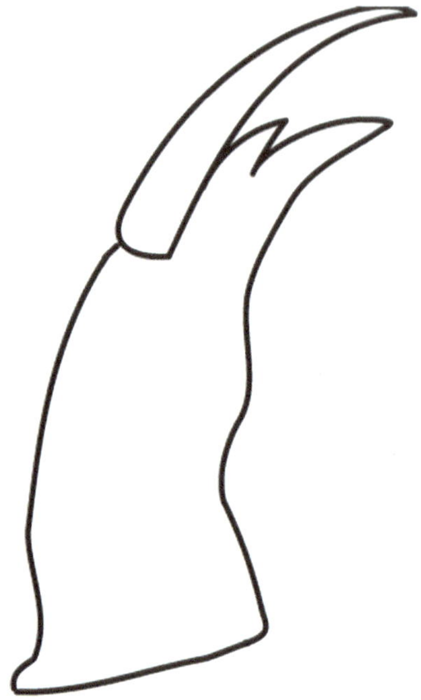

FIGURE 14.4.35 *Thaumatoneura* sp. (Megapodagrionidae), larval labial palp, dorsal surface view. *Illustration adapted from Calvert (1915).*

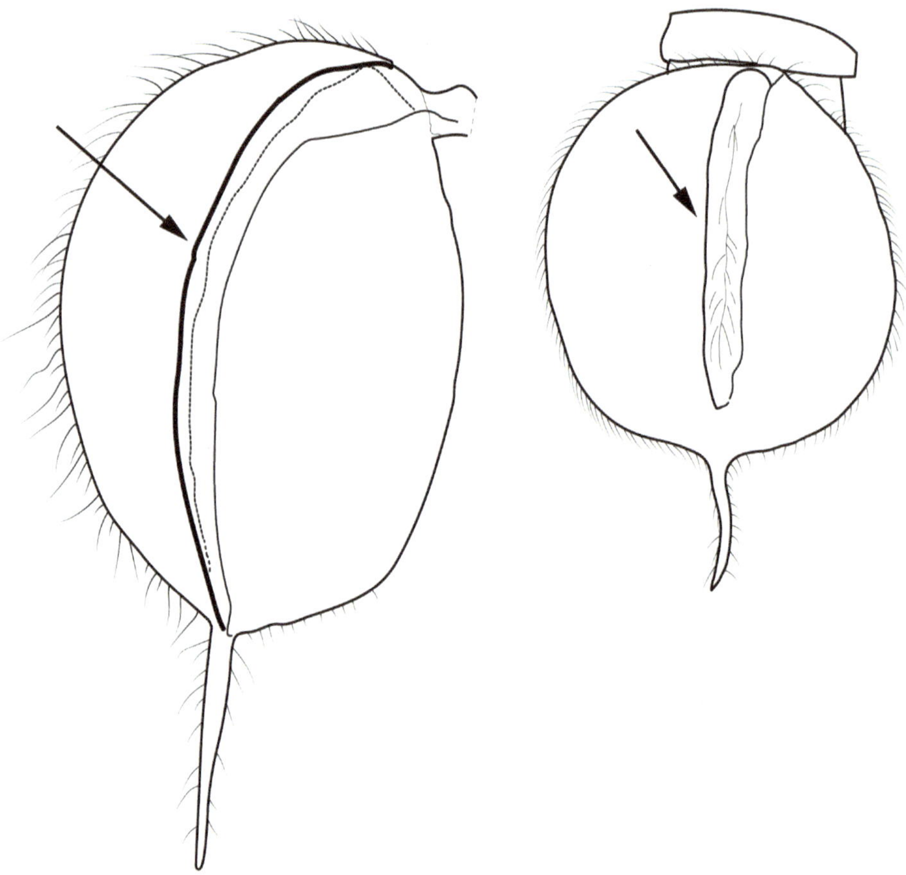

FIGURE 14.4.34 *Dimeragrion percubitale* Calvert, 1913 (Megapodagrionidae), larval left lateral caudal appendage, dorsal view (left), median caudal appendage, dorsal view (right), not to scale. *Arrows* indicate ridge. *Illustration adapted from De Marmels (1999).*

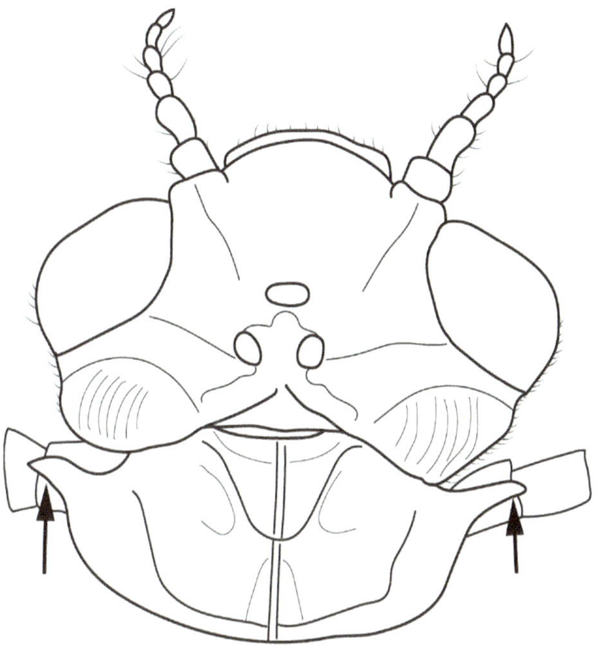

FIGURE 14.4.36 *Sciotropis cyclanthorum* Rácenis, 1959 (Megapodagrionidae), larva prothorax, dorsal view. *Arrows* indicate supracoxal processes. *Illustration adapted from De Marmels (2004).*

FIGURE 14.4.38 *Heteragrion consors* (Megapodagrionidae), larval head and prothorax, dorsal view.

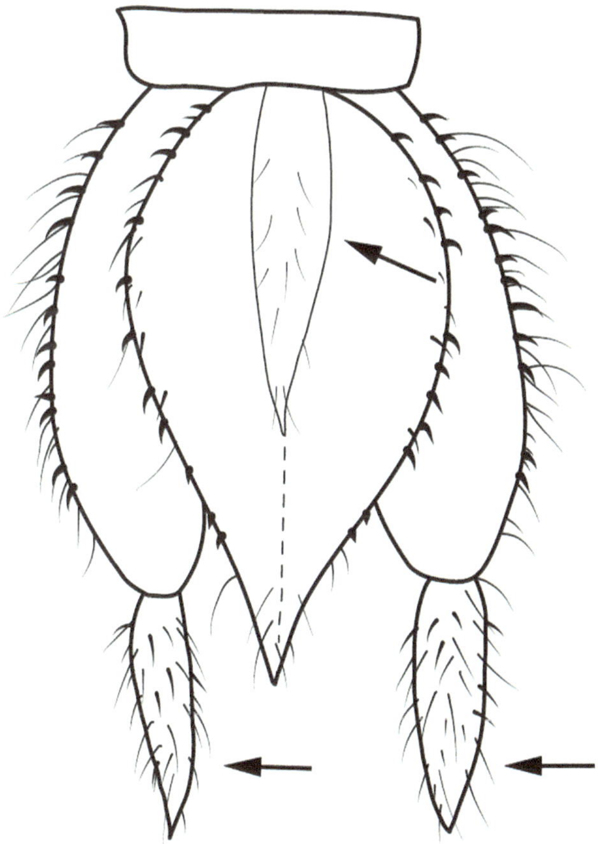

FIGURE 14.4.37 *Sciotropis cyclanthorum* (Megapodagrionidae), larval caudal appendages. *Arrows* indicate ridges and terminal filament. *Illustration adapted from De Marmels (2004).*

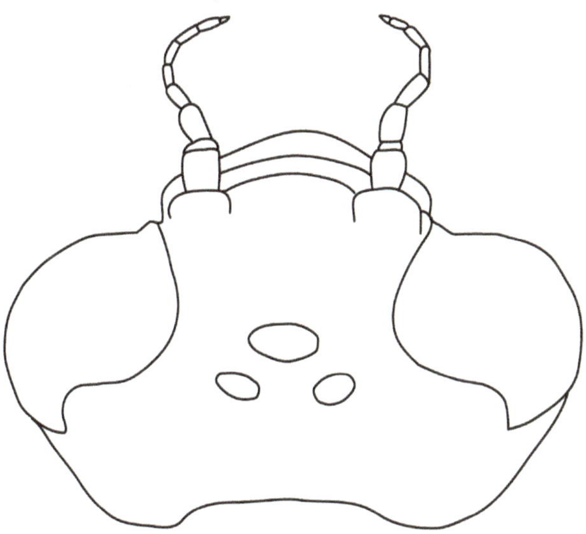

FIGURE 14.4.39 *Heteropodagrion sanguinipes* Selys, 1885 (Megapodagrionidae), larval head, dorsal view. *Illustration adapted from Tennessen (2010).*

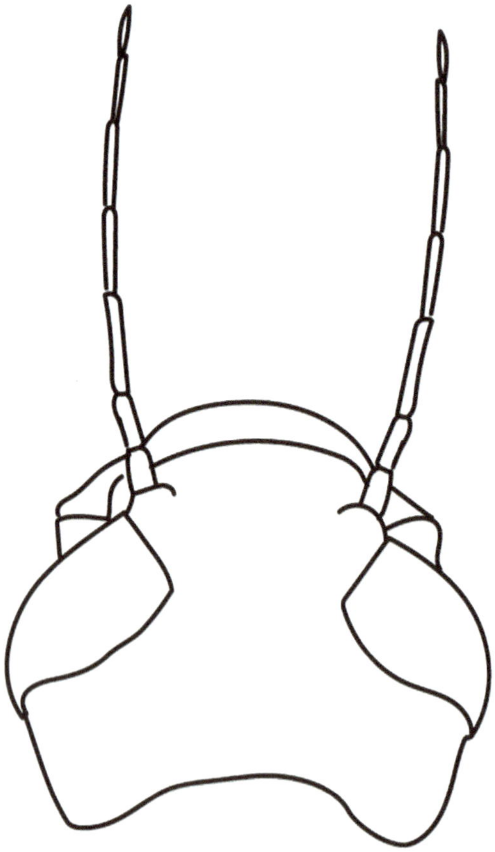

FIGURE 14.4.40 *Philogenia carrillica* Calvert, 1907 (Megapodagrionidae), larval head, dorsal view. *Illustration adapted from Ramírez & Novelo-Gutiérrez (1994).*

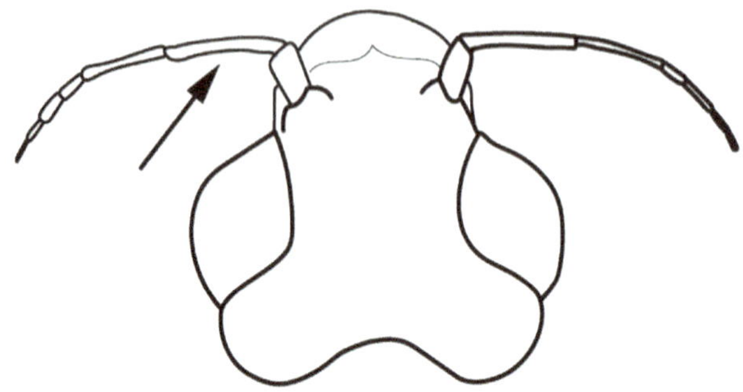

FIGURE 14.4.43 *Hypolestes clara* (Calvert, 1891) (Megapodagrionidae), larval head, dorsal view. *Arrow* indicates second antennomere. *Illustration adapted from Westfall (1987).*

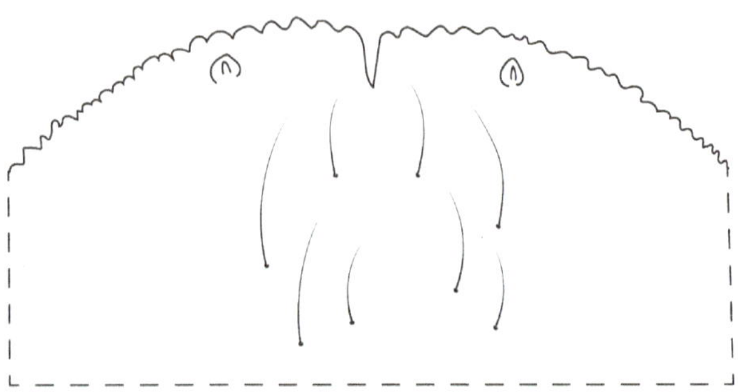

FIGURE 14.4.41 *Heteropodagrion sanguinipes* (Megapodagrionidae), larva prementum, dorsal view. *Illustration adapted from Tennessen (2010).*

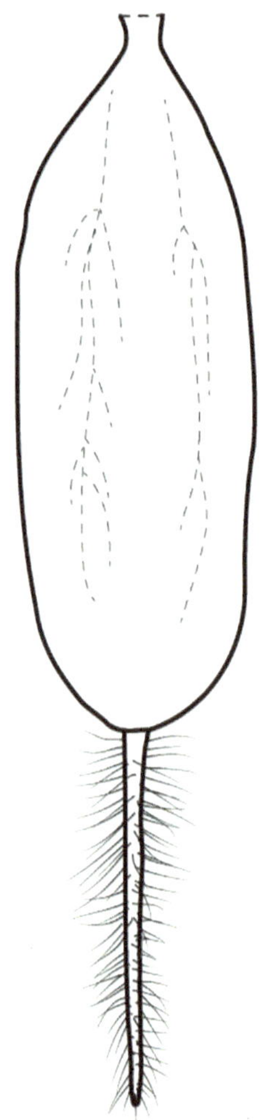

FIGURE 14.4.44 *Hypolestes trinitatis* (Gundlach, 1888) (Megapodagrionidae), larval median caudal appendage. *Illustration adapted from Alaya-Soto (1985).*

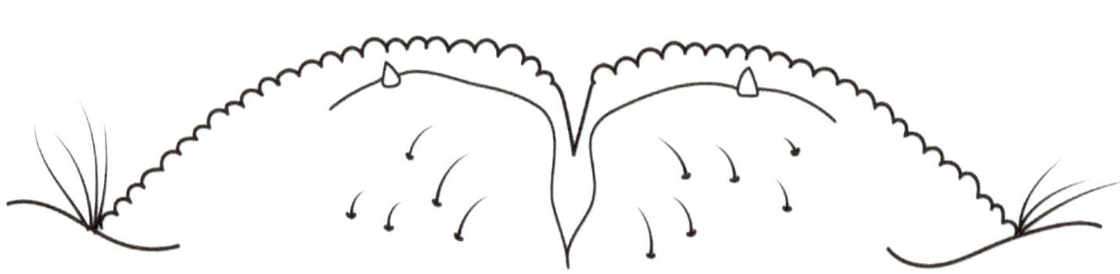

FIGURE 14.4.42 *Paraphlebia zoe* (Megapodagrionidae), larva prementum, dorsal view. *Illustration adapted from Novelo-Gutiérrez, 2008.*

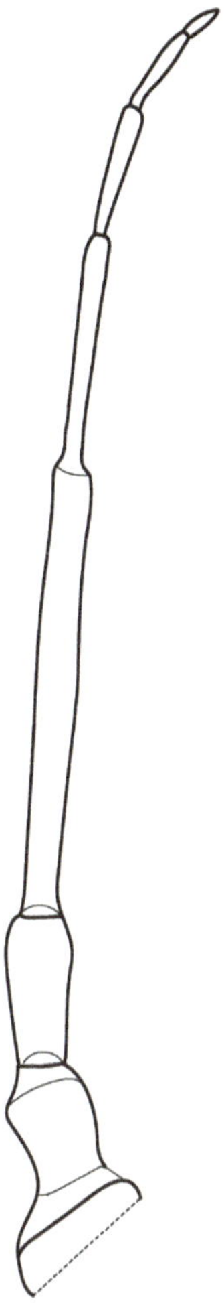

FIGURE 14.4.45 *Philogenia peacocki* Brooks, 1989 (Megapodagrionidae), larval antenna. *Illustration adapted from Ramírez & Novelo-Gutiérrez (1994).*

FIGURE 14.4.46 *Chalcothore montgomeryi* (Polythoridae), larva caudal appendages of female exuvia, lateral view. *Illustration adapted from De Marmels (1988).*

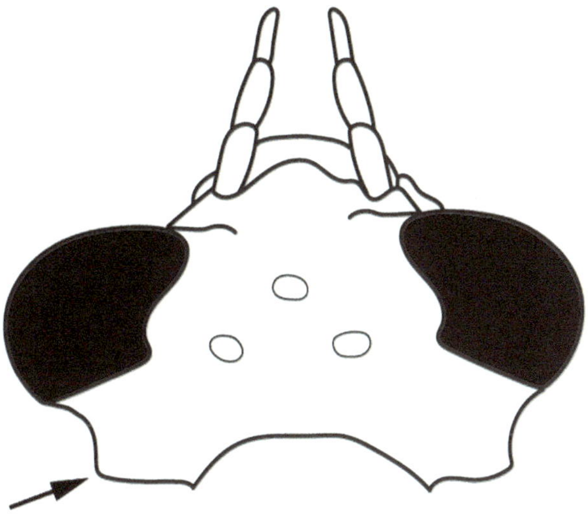

FIGURE 14.4.47 *Chalcothore montgomeryi* (Rácenis, 1968) (Polythoridae), larval head of female exuvia, dorsal view. *Arrow* indicates well pronounced angle. *Illustration adapted from De Marmels (1988).*

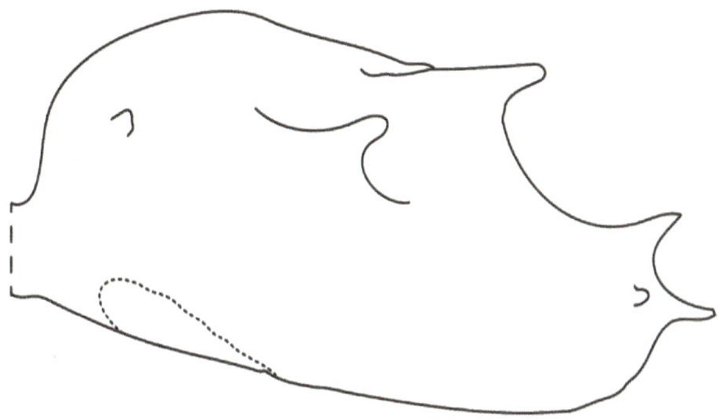

FIGURE 14.4.48 *Euthore fastigiata meridana* Selys, 1879 (Polythoridae). Larva left caudal appendages, lateral view. *Illustration adapted from De Marmels (1995).*

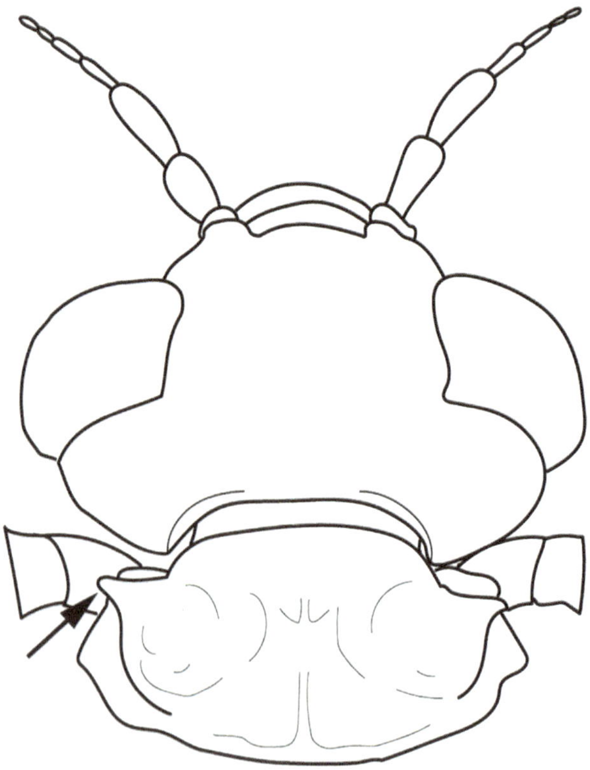

FIGURE 14.4.49 *Polythore spaeteri* Burmeister & Börzsöny, 2003 (Polythoridae), larval head and prothorax of the presumptive ultimate instar, dorsal view. *Arrow* indicates supracoxal process. *Illustration adapted from Etscher et al. (2006).*

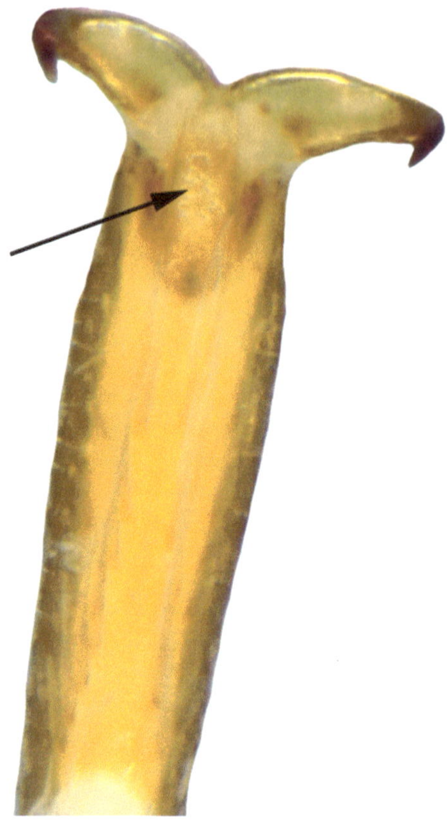

FIGURE 14.4.51 *Cora marina* Selys, 1868 (Polythoridae), larval right metatarsus, ventral view. *Arrow* indicates empodium-like structure. *Photo courtesy of Rodolfo Novelo-Gutierrez.*

FIGURE 14.4.50 *Chalcopteryx scintillans* (Polythoridae), larval head, dorsal view. *Illustration adapted from De Marmels (2007).*

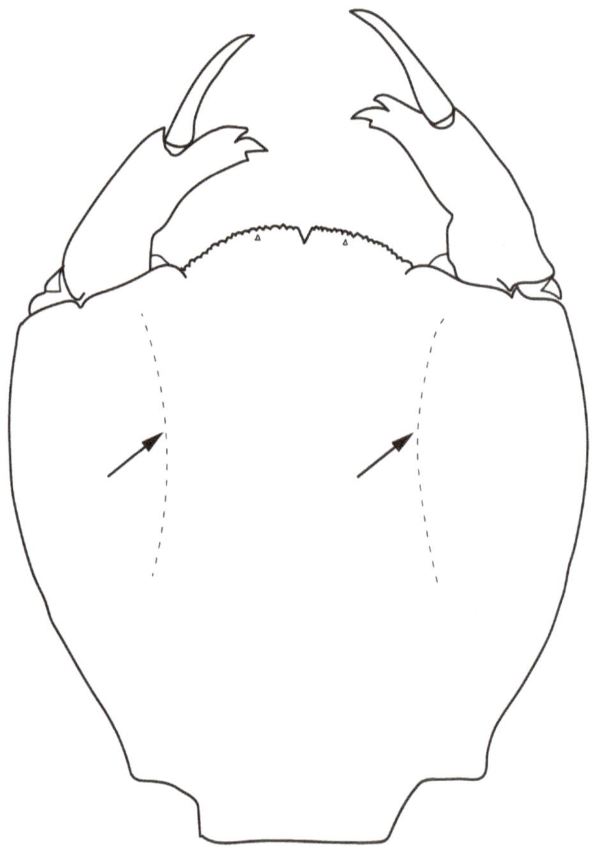

FIGURE 14.4.52 *Heteropodagrion sanguinipes* (Megapodagrionidae), larval prementum, dorsal view. *Arrows* indicate row of transverse ridges. *Illustration adapted from Tennessen (2010).*

Calopterygoidea: Polythoridae: Genera

The larva of five of the seven genera are known, but the knowledge of larva at species level is only 17% (Garrison et al., 2010). The larvae have ventrolateral filamentous abdominal gills between the segments (De Marmels, 1995, 2007; Neiss & Hamada, 2014). Furthermore, the caudal appendages are very globular with conical or digitiform projections, covered by thick squamiform bristles (Neiss & Hamada, 2014).

Unknown larvae: *Miocora* and *Stenocora*.

1 Caudal appendages with short or medium processes (less than half the length of the lamellae) (Fig. 14.4.48), occipital region of head rounded (Fig. 14.4.49), or angular (Fig. 14.4.50) 2

1' Caudal appendages with very long digitiform processes (approx. half the length of the caudal appendages, digitiform processes not included in caudal appendages length) (Fig. 14.4.46), occipital region of head with well pronounced angles (Fig. 14.4.47) .... *Chalcothore* [Venezuela]

2(1) Prothorax with supracoxal processes not prominent, occipital region of head rounded or angular (Fig. 14.4.50) 3

2' Prothorax with supracoxal processes prominent (Fig. 14.4.49), occipital region of head rounded (Fig. 14.4.49) .... *Euthore/Polythore* [Colombia, Ecuador, Guiana, Peru, Venezuela]/[Brazil, Bolivia, Guiana, Peru, Venezuela]

3(2) Median and lateral caudal appendages with no more than 3 protuberances (Fig. 14.4.19), simple claws without empodium structure, occipital region of head angular (Fig. 14.4.50) .... *Chalcopteryx* [Argentina, Brazil, Bolivia, Colombia, Paraguay, Peru, Venezuela]

3' Median and lateral caudal appendages with 4–6 protuberances (Fig. 14.4.17), simple claws with structure similar to empodium (Fig. 14.4.51), occipital region of head rounded (except *Cora cyane* Selys, 1853) .... *Cora* [Central America, Bolivia, Venezuela]

ACKNOWLEDGMENTS

We express our gratitude to Jürg De Marmels (MIZA) for critical reading of the keys, to Rodolfo Novelo-Gutiérrez (México) for the loan of larvae material and Miguel Archangelsky for allowing us to use his microscope and digital camera. The Figs. 14.1, 3, 7–9, 11, 13, 14, 18–20, originally presented by Neiss & Hamada (2014), were generously given by Editora do Instituto Nacional de Pesquisas da Amazônia, Manaus.

REFERENCES

Alaya-Soto, R. 1985. Descripción de la náyade de *Hypolestes trinitatis* Gundlanch (Odonata: Hypolestidae). Ciencias Biológicas 14: 111–113.

Bechly, G. 1996. Morphologische untersuchungen am Flugelge ader der rezenten libellen und deren stammgruppenvertreter (Insecta; Pterygota; Odonata) unter besonderer ber ucksichtigung der phylogenetischen systematik und des rundplanes der Odonata. Petalura, Special Volume 2: 1–402.

Bybee, S.M., T.H. Ogden, M.A. Branham & M.F. Whiting. 2008. Molecules, morphology and fossils: a comprehensive approach to odonate phylogeny and the evolution of the odonate wing. Cladistics 24: 477–514.

Bybee, S.M. & K.J. Tennessen. 2008. Description of the female and nymph of *Philogenia mangosisa* from southern Ecuador (Odonata: Megapodagrionidae). Zootaxa 1787: 63–68.

Calvert, P.P. 1915. Studies on Costa Rican Odonata. V. The waterfall-dwellers: *Thaumatoneura* imagos and possible male dimorphism. Entomological News 25: 337–348.

Carle, F.L., K.M. Kjer & M.L. May. 2008. Evolution of Odonata, with special reference to Coenagrionoidea (Zygoptera). Arthropod Systematics & Phylogeny 66: 37–44.

Carvalho, A.L. & J.L. Nessimian. 1998. Odonata do estado do RJ, Brasil: Hábitats e hábitos das larvas. Pages 157–173 *in:* J.L. Nessimian & A.L. Carvalho (ed.), Ecologia de insetos aquáticos. Oecologia Brasiliensis. Vol. 5. Rio de Janeiro, Programa de Pós-Graduação em Ecologia-Instituto de Biologia, Universidade Federal do Rio de Janeiro.

Costa, J.M., L.O.I. Souza, A.M. Lourenço & B.B. Oldrini. 2004. Chave para as famílias e gêneros das larvas de Odonata citadas para o Brasil: Comentários e Registros Bibliográficos. Publicações Avulsas do Museu Nacional 99: 1–44.

De Marmels, J. 1982. The genus *Euthore* Selys in Venezuela, with special notes on *Euthore fasciata fasciata* (Hagen, 1853) (Zygoptera: Polyrhoridae). Adv. Odonatology 1: 39–41.

De Marmels, J. 1988. Generic characters of *Chalcothore* De Marmels, 1985, with notes on the male of C. *montgomeryi* (Racenis, 1968) and a description of the larva (Zygoptera: Polythoridae). Odonatologica 17: 379–384.

De Marmels, J. 1989. Odonata or dragonflies from Cerro de la Neblina and the adjacent lowland between the rio Bária, the Casiquiare and rio Negro (Venezuela). I. Adults. Boletín of the Academia de las Ciencias Físicas, Matemáticas y Naturales 25: 11–78.

De Marmels, J. 1992. Caballitos del Diablo (Odonata) de las Sierras de Tapirapecó y Unturán, en el extremo sur de Venezuela. Acta Biológica Venezuélica 14: 57–78.

De Marmels, J. 1995. La larva de *Euthore fastigiata meridana* Selys, 1879 (Odonata:Polythoridae). Boletin de Entomología Venezolana 10: 131–134.

De Marmels, J. 1999. A new species of *Dimeragrion* Calvert 1913 from Pantepui, Venezuela (Odonata: Megapodagrionidae). Boletin de Entomología Venezolana 14: 27–36.

De Marmels, J. 2001. Revision of *Megapodagrion* Selys, 1886 (Insecta, Odonata: Megapodagrionidae). Dissertation zur Erlangung der naturwissen-schaftlichen Doktorwürde (Dr. sc. Nat.) vorgelegt der Matematisch-naturwissenschaftlichen Fakultät der Universität Zürich.

De Marmels, J. 2004. *Heteragrion makiritare* sp. nov., with descriptions of hitherto unknown females and larvae of other species from Venezuela (Odonata: Megapodagrionidae, Lestidae). International Journal of Odonatology 7: 439–458.

De Marmels, J. 2007. Thirteen new Zygoptera larvae from Venezuela (Calopterygidae, Polythoridae, Pseudostigmatidae, Platystictidae, Protoneuridae, Coenagrionidae). Odonatologica 36: 27–51.

Dijkstra, K-D.B., G. Bechly, S.M. Bybee, R.A. Dow, H.J. Dumont, G. Fleck., R.W. Garrison, M. Hämäläinen, V.J. Kalkman, H. Karube, M.L. May, A.G. Orr, D. Paulson, A.C. Rehn, G. Theischinger, J.W.H. Trueman, J. Van Tol, N. von Ellenrieder & J. Ware. 2013. The classification and diversity of dragonflies and damselflies (Odonata). Zootaxa 3703: 36–45.

Dijkstra, K-D.B., V.J. Kalkman, R.A. Dow, F.R. Stokvis & J. van Tol. 2014. Redefining the damselfly families: the first comprehensive molecular phylogeny of Zygoptera (Odonata). Systematic Entomology 39: 68–96.

Dumont, H.J., J.R. Vanfleteren, J.F. De Jonckheere & P.H.H. Weekers. 2005. Phylogenetic relationship, divergence time estimation, and global biogeographic patterns of calopterygoid damselflies (Odonata, Zygoptera) inferred from ribosomal DNA sequences. Systematic Biology 54: 347–362.

Dumont, H.J., H. Vierstraete & J.R. Vanfleteren. 2010. A molecular phylogeny of the Odonata (Insecta). Systematic Entomology 35: 6–18.

Etscher, V., M.A. Miller & E.-G. Burmeister. 2006. The larva of *Polythore spaeteri* Burmeister & Börzsöny, with comparison to other Polythorid larvae and molecular species assignment (Zygoptera: Polythoridae). Odonatologica 35: 127–142.

Fleck, G., M. Brenk & B. Misof. 2008. Larval and molecular characters help to solve phylogenetic puzzles in the highly diverse dragonfly family Libellulidae (Insecta: Odonata: Anisoptera): the Tetrathemistinae are a polyphyletic group. Organisms, Diversity and Evolution 8: 1–16.

Fleck, G., U.G. Neiss & N. Hamada. 2012. The larva of *Dicterias* Selys, 1853 (Odonata: Heliocharitidae [=Dicteriadidae]), and taxonomic and phylogenetic notes on Heliocharitidae. Zootaxa 3164: 32–40.

Garrison, R.W. 2006. A Synopsis of the genera *Mnesarete* Cowley, *Bryoplathanon* gen. nov., and *Ormenoplebia* [sic *Ormenoplebia*] gen. nov. (Odonata: Calopterygidae). Natural History Museum of Los Angeles County, Contributions in Science 506: 1–84.

Garrison, R.W., N. von Ellenrieder & J.A. Louton. 2010. Damselfly Genera of the New World: An Illustrated and Annotated Key to the Zygoptera. The Johns Hopkins University Press, Baltimore, Maryland.

Garrison, R. & N. von Ellenrieder. 2016. A Synonymic list of the New World Odonata (revised version of 18 February 2016) Available http://www.odonatacentral.org/docs/NWOL.pdf.

Geijskes, D.C. 1940. Notes on Odonata of Surinam I. *Rimanella arcana* Needham and its nymph (Odon. Zyg.). Revista de Entomologia 11: 173–179.

Geisjkes, D.C. 1986. The larva of *Dicterias cothurnata* (Föerster, 1906) (Zygoptera: Dicteriastidae). Odonatologica 15: 77–80.

Guillermo-Ferreira, R. & P.C. Bispo. 2012. Description of the larva of *Mnesarete pudica* (Hagen *in* Selys, 1853) (Odonata: Calopterygidae) and notes on known genera of South American Calopterygidae larvae. Zootaxa 3482: 77–81.

Hamada, N. & S.J. Oliveira. 2003. Food items of larvae of *Rimanella arcana* (Needham, 1933) (Odonata:Amphipterygidae) in Central Amazonia, Brazil. Entomotropica 18: 153–155.

Jocque, M. & I. Argueta. 2014. A new species in the genus *Amphipteryx* Selys, 1853 (Odonata, Amphipterygidae) from Pico Bonito National Park, Honduras. ZooKeys 408: 71–80.

Kalkman, V.J. & G. Theischinger. 2013. Generic revision of Argiolestidae (Odonata), with four new genera. International Journal of Odonatology 16: 1–52.

Kalkman, V.J., C.Y. Choong, A.G. Orr & K. Schütte. 2010. Remarks on the taxonomy of Megapodagrionidae with emphasis on the larval gills (Odonata). International Journal of Odonatology 13: 119–135.

Neiss, U.G., G.L. Fiorentin & J. De Marmels. 2011. The larva of *Allopodagrion brachyurum* De Marmels, 2001 (Odonata: Zygoptera: Megapodagrionidae) from Southern Brazil. Zootaxa 2836: 44–50.

Neiss, U.G. & N. Hamada. 2014. Ordem Odonata. Pages 216–284 *in:* N. Hamada, J.L. Nessimian & R.B Querino (eds.), Insetos aquáticos da Amazônia brasileira: taxonomia, biologia e ecologia. Editora do INPA, Manaus, Brazil.

Novelo-Gutiérrez, R. 1995. The larva of *Amphipteryx* and a reclassification of Amphipterygidae *sensu lato*, based upon the larvae (Zygoptera). Odonatologica 24: 73–87.

Novelo-Gutiérrez, R. 1997. Clave para la separación de familias y géneros de las náyades de Odonata de México. parte i. Zygoptera. Dugesiana 4: 1–10.

Novelo-Gutiérrez, R. 2008. Description of the larva of *Paraphlebia zoe* Selys in Hagen, 1861 (Odonata: Megapodagrionidae). Zootaxa 1876: 29–34.

Pérez-Gutiérrez, L.A. & J.M. Montes-Fontalvo. 2011. Rediscovery of *Mesagrion leucorrhinum* (Zygoptera: Megapodagrionidae): a "formal" description of female and ultimate stadium of larva with notes on habits. International Journal of Odonatology 14: 91–100.

Ramírez, A., M. Altamirando-Saavedra, P. Gutérrez-Fonseca & M. Springer. 2011. The neotropical damselfly genus *Cora*: New larval descriptions and a comparative analysis of larvae of known species (Odonata: Polythoridae). International Journal of Odonatology 14: 249–256.

Ramírez-Ulate, A. & R. Novelo-Gutiérrez. 1994. Megapodagrionidae (Odonata: Zygoptera) de Mexico y Centroamerica I. Las nayades de *Philogenia carrilica, P. peacocki* y *P. terraba*. Acta Zoológica Mexicana (nueva serie) 63: 61–73.

Rehn, A.C. 2003. Phylogenetic analysis of higher-level relationships of Odonata. Systematic Entomology 28: 181–239.

Santos, N.D. & J.M. Costa. 1988. The larva of *Heliocharis amazona* Selys, 1853 (Zygoptera: Heliocharitidae). Odonatologica 17: 135–139.

Suhling, F., G. Sahlén, S. Gorb, V.J. Kalkman, K-D.B. Dijkstra & J. von Tol. 2015. Order Odonata, Chapter 35. Pages 893–932 *in:* J.H. Thorp & D.C. Rogers (eds.). Ecology and General Biology Thorp and Covich's Freswater Invertebrates, Vol. 1, 4th Edition, Academic Press-Elsevier.

Tennessen, K.J. 2010. The madicolous nymph of *Heteropodagrion sanguinipes* Selys (Odonata: Megapodagrionidae). Zootaxa 2531: 29–38.

Thorp, J.H. & D.C. Rogers (eds.). 2015. Ecology and General Biology Thorp and Covich's Freswater Invertebrates, Vol. 1, 4th Edition, Academic Press-Elsevier.

van Tol, J., B.T. Reijnen & H.A. Thomassen. 2009. Phylogeny and biogeography of the Platystictidae (Odonata). *in:* van Tol, J., Phylogeny and biogeography of the Platystictidae (Odonata). PhD Thesis, Leiden University, Leiden.

von Ellenrieder, N. 2006. The larvae of *Teinopodagrion decipiens* De Marmels, 2001 and *T. meridionale* De Marmels, 2001 (Zygoptera: Megapodagrionidae). Odonatologica 35: 281–287.

von Ellenrieder, N. 2007. The larva of *Hetaerina mendenzi Jurzitza*, wirh comments on *H. rosea* Selys (Zygoptera: Calopterygidae). Odonatologica 26: 405–414.

Westfall, M.J. Jr. 1987. Order Odonata. Pages 95–117 in: Stehr, F.W. (Ed.), Immature Insects. Kendall/Hunt Publishing Company, Dubuque, Iowa.

Superfamily Lestoidea

Javier Muzón
Laboratorio de Biodiversidad y Genética Ambiental (BioGeA), Universidad Nacional de Avellaneda (UNDAV), Buenos Aires, Buenos Aires, Argentina

Chapter Outline

INTRODUCTION

The Lestoidea, known also as Lestinoidea (Bybee et al., 2008; Rehn, 2003), is a small taxon composed of medium to large damselflies encompassing approximately 21 genera and 210 species distributed worldwide (Dijkstra et al., 2013). It includes four families: the monotypic and Australian endemic Hemiphlebiidae, the speciose and cosmopolitan Lestidae and Synlestidae, and the Neotropical endemic Perilestidae (Dijkstra et al., 2014).

In the Neotropical region three families, five genera, and 62 species are reported, but this number is very likely to increase (Garrison et al., 2010). The Lestoidea has been proposed as the sister group of all the remaining Zygoptera (Bybee et al., 2008; Carle et al., 2008; Dumont et al., 2010; Davis et al., 2011).

In this group, larval caudal appendages are lamellate and held in a vertical plane. Most species occupy fast to moderate streams, except *Lestes*, which prefer temporary lentic environments.

Perilestidae is a small family composed of medium to large damselflies inhabiting low-order forest streams. The latest classification scheme (Dijkstra et al., 2013) considers the family as composed of two genera (*Perilestes* and *Perissolestes*), both endemic to South America. *Perilestes*, with seven species, is mainly distributed in the Amazonian basin (Williamson & Williamson, 1924; Kennedy, 1941a, b), and *Perissolestes*, composed of 11 species, exhibits a patchier distribution, being recorded in the tropical rainforest from southern Mexico, Guatemala, Costa Rica, Panama, Ecuador, Peru, Brazil, and French Guyana (Garrison et al., 2010). The known larvae are stream-dwellers (moderate lotic habitats, *sensu* Corbet, 1999), *Perilestes attenuatus* Selys, 1886, and *P. gracillimus* Kenedy, 1941 inhabits small forest streams with sandy beds, and can be found among leaf litter (Santos, 1969; Neiss & Hamada, 2010), whereas *Perissolestes magdalenae* (Williamson & Williamson, 1924) prefers backwaters and pools of forest streams (Novelo-Gutiérrez & González Soriano, 1986). The monophyly is still unresolved (Dijkstra et al., 2014). Based on morphological data, some authors consider it as composed of three genera, the Neotropical *Perilestes* and *Perissolestes* and the African *Nubiolestes* (Bechly, 1996, 2003; Rehn, 2003, Garrison et al., 2010). On the other hand, proposals based mainly on molecular data are more restrictive, excluding *Nubiolestes* and considering Perilestidae as a Neotropical endemic (Bybee et al., 2008, Dijkstra et al., 2014).

Synlestidae, also known as Chlorolestidae (Fraser, 1951; Davis & Tobin, 1984), is composed of medium to large and rather robust damselflies with strongly petiolate wings. The monophyly is still unresolved, and there is no consensus about the genera members (Dijkstra et al., 2014). According to several taxonomic classification proposals, Synlestidae is comprised of 32 to almost 40 species in eight

Thorp and Covich's Freshwater Invertebrates, Fourth Edition. https://doi.org/10.1016/B978-0-12-804223-6.00019-6

or nine genera, which have been recorded in all the regions of the world except for the Nearctic and the Pacific Ocean islands (Garrison et al., 2010; Kalkman et al., 2013; Dijkstra et al., 2014). The Neotropical region contains only the monotypic genus *Phylolestes*, which has been proposed as a Gondwanan relict by Westfall & May (2006). *Phylolestes ethelae* Christiansen was discovered in Haiti in 1934 by M. Bates and described in 1947 by K.A. Christiansen. It is endemic to forested mountain streams above 2000 m on the island of Hispaniola (Dominican Republic and Haiti). Its larva was described by Westfall (1976). The larva of *Philolestes* is characterized by the anterior margin of prementum with a median cleft, labial palp with terminal acute teeth, distinct white color pattern in legs, and S1 and caudal appendages foliate with rounded apex (Westfall, 1976; Westfall & May, 2006). It breeds in pools associated to clear and cold streams.

Lestidae is composed of large to medium damselflies, characterized by narrow and petiolated wings and male forcipated cerci. With more than 150 species in 9 genera, it is the most speciose family within Lestoidea, and it has been recorded in all the natural regions of the world except the Antarctica (Kalkman et al., 2013). There are two genera recorded in the Neotropics, the cosmopolitan *Lestes*, and the American endemic and rather robust *Archilestes*, but there are no strong differences between these two genera (Garrison, 1982). *Lestes* is represented in the Neotropical region by approximately 35 species, some of them reaching north to the Nearctic region. *Lestes undulatus* Say, 1840, has been recorded in northern Patagonia, being the southernmost lestid in South America (Muzón et al., 2014). *Archilestes* is composed of eight species, two of them, *A. grandis* (Rambur, 1842) and *A. californicus* McLachlan, 1895, are also distributed in the United States. The southernmost record is for *A. exoletus* (Hagen in Selys, 1862) in Misiones province, Argentina (Jurzitza, 1982). There are several contributions dealing with the Neotropics (Kennedy, 1942; Gloyd, 1980; Donnelly, 1981; Garrison, 1982; May, 1993; Belle, 1997; Muzón, 1993, 1994, 1997; Lencioni, 2005). There is consensus about the monophyly of Lestidae (Rehn, 2003; Bybee et al., 2008; Dumont et al., 2010; Dijkstra et al., 2014). Within *Archilestes* two subgenera have been proposed, *Superlestes* and *Cyptolestes* (Williamson, 1921; Donnelly, 1981), but there is a lack of good characters sustaining this proposal (Gloyd, 1980; Garrison, 1982). *Lestes* larvae are inhabitants of lentic waters, mainly temporary pools with abundant emergent vegetation. Larval cycles could be rapid, with three or more generations per year; delayed embryonic development (diapauses eggs) is common. *Archilestes* larvae are stream-dwellers (moderate lotic habitats). *Archilestes grandis* Rambur, 1842, exhibits delayed embryonic development.

LIMITATIONS

There are several regional keys that includes Lestoidea immature stages at generic level (e.g., Neiss & Hamada, 2014 for Amazonas, Costa et al., 2004 for Brazil), and there is a key for the known species of *Archilestes* (Dalzochio & Rodrigues, 2011).

Our knowledge of the immature stages of Perilestidae is restricted to three species: *Perilestes attenuatus* Selys,1886, *Perilestes fragilis* Hagen in Selys, 1862, and *Perissolestes magdalenae* (Williamson & Williamson) (Santos, 1969; Novelo-Gutiérrez & González Soriano, 1986; Neiss & Hamada, 2010), representing slightly more than 15% of the specific diversity. For Lestidae the situation is better, about 75% of the larva have been described for *Archilestes* (Kennedy, 1915; Needham, 1904; Novelo Gutierrez, 1994; Ramírez, 1994; De Marmels, 2004; Dalzochio & Rodrigues, 2011) and 33% for *Lestes* (e.g., Novelo-Gutiérrez & González Soriano, 1991, 2003; Muzón, 1993, 1997; Santos, 1972; De Marmels, 2004; Westfall & May, 2006; Muzón, & Weigel Muñoz, 2007).

TERMINOLOGY AND MORPHOLOGY

The most important terminology for identifying these damselflies are described in Chapter 14.0 in this book and Chapter 35 (Suhling et al., 2015) in Volume 1 of this series (Thorp & Rogers, 2015).

MATERIAL PREPARATION AND PRESERVATION

Specimen preparation and preservation are presented in Volume I (Thorp & Rogers, 2015) of this series.

KEYS TO THE SUPERFAMILY LESTOIDEA

Lestoidea: Families

1 Prementum not extending beyond metacoxa, not very narrow basally, flat (Fig. 14.5.2); labial palp and movable hook without setae (Fig. 14.5.5) 2

1' Prementum extending beyond metacoxa, very narrow basally, spatulate (Fig. 14.5.1); labial palp and movable hook with setae (Fig. 14.5.3) .... **Lestidae [p. 472]**

FIGURE 14.5.1 *Lestes* sp., larval prementum, dorsal view. *Original pictures by Neiss & Hamada (2014).*

FIGURE 14.5.2 *Perilestes solutus* Williamson & Williamson 1924, larval prementum, dorsal view. *Original pictures by Neiss & Hamada (2014).*

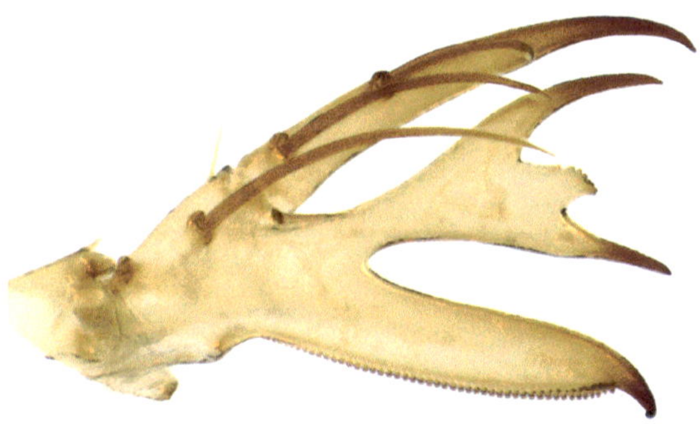

FIGURE 14.5.3 *Lestes* sp., larval labial palp, dorsal view. *Original pictures by Neiss & Hamada (2014).*

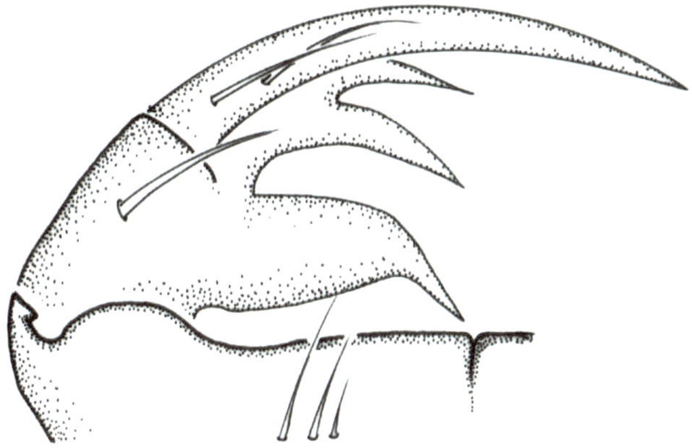

FIGURE 14.5.4 *Archilestes* sp., larval labial palp, dorsal view. *Original pictures by Neiss & Hamada (2014).*

FIGURE 14.5.5 *Perilestes solutus* larval labial palp, dorsal view. *Original pictures by Neiss & Hamada (2014).*

FIGURE 14.5.6 *Perilestes attenuatus* Selys, 1886 larval habitus, dorsal view. *Original pictures by Neiss & Hamada (2014).*

2(1) Abdomen S3–S10 with lateral spines (Fig. 14.5.6), S4–S10 with a mid-dorsal row of spines; labial palp with three apical hooks, the inner one smaller (Fig. 14.5.5) .... **Perilestidae [p. 472]**

2' Abdomen without lateral and dorsal spines; labial palp with two apical hooks; with a characteristic white pattern on first legs .... Synlestidae, one species: *Phylolestes ethelae* Christiansen [Hispaniola]

Lestoidea: Lestidae: Genera

1 Labial palp with two apical hooks and a serrated protuberance between them (Fig. 14.5.3)........ *Lestes*
[Cosmopolitan]

1' Labial palp with three apical hooks (Fig. 14.5.4)........ *Archilestes*
[Nearctic, Central America, Colombia, Venezuela, Brazil, Argentina]

Lestoidea: Perilestidae: Genera

1 S3, S4 lateral spines smaller than distal ones (Fig. 14.5.6)........ *Perilestes*
[North-central South America]

1' S3, S4 lateral spines similar as distal ones........ *Perissolestes*
[Central America, Colombia, French Guayana, Peru, Brazil, Bolivia]

ACKNOWLEDGMENTS

The figures, originally presented by Neiss & Hamada (2014), were generously given by the Editora do Instituto Nacional de Pesquisas da Amazônia, Manaus.

REFERENCES

Bechly, G. 1996. Morphologische Untersuchungen am Flügelgeäder der rezenten Libellen und deren Stammgruppenvertreter (Insecta; Pterygota; Odonata) unter besonderer Berücksichtigung der phylogentischen Systematik und des Grundplanes der Odonata. Petalura (Special Volume) 2: 1–402.

Bechly, G. 2003. Phylogenetic systematics of Odonata. *in*: M. Schorr & Lindeboom, M. (eds.), Dragonfly Research, Volume 1. Zerf-Tübingen (CD-ROM).

Belle, J. 1997. The genus *Lestes* (Odonata: Lestidae) Leach, 1815, in Surinam. Zoologische Mededelingen Leiden 71: 89–103.

Bybee, S.M., T.H. Ogden, M.A. Branham & M.F. Whiting. 2008. Molecules, morphology and fossils: a comprehensive approach to odonate phylogeny and the evolution of the odonate wing. Cladistics 24: 477–514.

Carle, F.L., K.M. Kjer & M.L. May. 2008. Evolution of Odonata, with special reference to Coenagrionoidea (Zygoptera). Arthropod Systematics & Phylogeny 66: 37–44.

Christiansen, K.A. 1947. A new genus and species of damselfly from southern Haiti (Odonata). Psyche 54: 256–262.

Corbet, P. 1999. Dragonflies: behavior and ecology of Odonata. Cornell University Press, Ithaca.

Costa, J.M., L.O.I. Souza, A.M. Lourenço & B.B. Oldrini. 2004. Chave para as famílias e gêneros das larvas de Odonata citadas para o Brasil: Comentários e Registros Bibliográficos. Publicações Avulsas do Museu Nacional 99: 1–44.

Dalzochio, M.S. & M.E. Rodrigues. 2011. Description of the larva of *Archilestes exoletus* (Hagen *in* Selys) (Odonata: Lestidae). Zootaxa 2756: 65–68.

Davis, R.B., D.B. Nicholson, E.L.R. Saunders & P.J. Mayhew. 2011. Fossil gaps inferred from phylogenies alter the apparent nature of diversification in dragonflies and their relatives. BMC Evolutionary Biology 11: 252–261. https://doi.org/10.1186/1471-2148-11-252.

Davis, D.A.L. & P. Tobin. 1984. The dragonflies of the world: A systematic list of the extant species of Odonata. Vol. 1. Zygoptera, Anisozygoptera. Societas Internationalis Rapid Communications (Supplement) 3: 1–127.

De Marmels, J. 2004. *Heteragrion makiritare* sp. nov., with descriptions of hitherto unknown females and larvae of other species from Venezuela (Odonata: Megapodagrionidae, Lestidae). International Journal of Odonatology 7: 439–458.

Dijkstra, K-D.B., B. Bechly, S.M. Bybee, R.A. Dow, H.J. Dumont, G. Fleck, R.W. Garrison, M. Hämäläinen, V.J. Kalkman, H.Karube, M.L. May, A.G. Orr, D.R. Paulson, A.C. Rehn, G. Theischinger, J.W.H. Trueman, J. van Tol, N. von Ellenrieder & J. Ware. 2013. The classification and diversity of dragonflies and damselflies (Odonata). Zootaxa 3703: 36–45.

Donnelly, T.W. 1981. A new species of *Archilestes* from Mexico and Central America, with further notes on the status of *Cyptolestes* Williamson (Odonata: Lestidae). Florida Entomologist 64: 412–417.

Donnelly, T.W. 1996. The status of *Lestes apollinaris* Navás and *L. henshawi* Calvert. Bulletin of American Odonatology 4: 69–74.

Dumont, H.J., A. Vierstraete & J.R. Vanfleteren. 2010. A molecular phylogeny of the Odonata (Insecta). Systematic Entomology, 35: 6–18. https://doi.org/10.1111/j.1365-3113.2009.00489.x.

Fraser, F.C. 1951. Outline of a new classification for the legion *Lestes* Selys (Order Odonata). Entomological News 62: 61–69.

Garrison, R.W. 1982. Archilestes neblina, a new damselfly from Costa Rica, with comments on the variability of *A. latialatus* Donnelly (Odonata: Lestidae). Occasional Papers of the University of Michigan 702: 1–12.

Garrison, R.W., N. von Ellenrieder & J.A. Louton. 2010. Damselfly genera (Odonata: Zygoptera) of the New World. An illustrated and annotated key to the Zygoptera. The Johns Hopkins University Press, Baltimore.

Gloyd, L.K. 1980. The taxonomic status of the genera *Superlestes* and *Cyptolestes* Williamson, 1921. (Odonata: Lestidae). Occasional Papers of the Museum of Zoology, University of Michigan 694: 1–3.

Jurzitza, G. 1982. Lista provisional de los odonatos del Parque Nacional Iguazú, Provincia de Misiones, República Argentina. Notulae Odonatologicae 1: 117–118.

Kalkman, V.J., V. Clausnitzer, K-D.B. Dijkstra, A.G. Orr, D.R. Paulson & J. van Tol. 2008. Global diversity of dragonflies (Odonata) in freshwater. Hydrobiologia 595: 351–363. https://doi.org/10.1007/s10750-007-9029-x.

Kennedy, C.H. 1915. Notes on the life history and ecology of the dragonflies (Odonata) of Washington and Oregon. Proceedings of the United States National Museum 49: 259–345.

Kennedy, C.H. 1941a. Perilestinae in Ecuador and Peru: revisional notes and descriptions (Lestidae: Odonata). Annals of the Entomological Society of America 34: 658–688.

Kennedy, C.H. 1941b. Perissolestes paprzyckii, a new perilestinae dragonfly from Peru (Lestidae: Odonata). Annals of the Entomological Society of America 34: 852–854.

Kennedy, C.H. 1942. *Lestes henshawi*, Ecuador, *L. urubamba*, Peru, and notes on other South American *Lestes* (Lestidae: Odonata). Revista de Entomologia 13: 274–290.

Lencioni, F.A.A. 2005. Damselflies of Brazil - An illustrated identification guide: 1- Non-Coenagrionidae families. All Print Editora, São Paulo.

May, M.L. 1993. *Lestes secula*, a new species of damselfly (Odonata: Zygoptera: Lestidae) from Panama. Journal of the New York Entomological Society 101: 410–416.

Muzón, J. 1993. *Lestes spatula* Fraser: description of the final larval instar and redescription of male and female adults (Zygoptera: Lestidae). Odonatologica 22: 443–454.

Muzón, J. 1994. *Lestes jurzitzai* spec. nov., a new damselfly from Rondonia, Brazil (Zygoptera: Lestidae). Odonatologica 23: 297–302.

Muzón, J. 1997. Redescripción de *Lestes auritus* y *L. paulistus* y descripción del último estadio larval de *L. undulatus* (Odonata: Lestidae). Revista de la Sociedad Entomológica Argentina 56: 159–166.

Muzón, J., P. Pessacq & F. Lozano. 2014. The Odonata (Insecta) of Patagonia: A synopsis of their current status with illustrated keys for their identification. Zootaxa 3784: 346–388.

Muzón, J. & S. Weigel Muñoz. 2007. Description of the final instar of *Lestes dichrostigma* Calvert (Zygoptera: Lestidae). Studies on Neotropical fauna and Environment 42: 235–239.

Needham, J.G. 1904. New dragonfly nymphs in the United States National Museum. Proceedings of the United States National Museum 27: 685–720.

Neiss, U.G. & N. Hamada. 2010. The larva of *Perilestes attenuatus* Selys, 1886 (Odonata: Perilestidae) from Amazonas, Brazil. Zootaxa 2614: 53–58.

Neiss, U.G. & N. Hamada. 2014. Ordem Odonata, Chapter 14, pages 217–284 *in:* Hamada, N., J.L. Nessimian & R.B. Querino (eds.), *Insetos Aquáticos na Amazônia brasileira: taxonomia, biologia e ecologia.* Editora do Instituto Nacional de Pesquisas da Amazônia, Manaus.

Novelo-Gutiérrez, R. 1994. La nayade de *Archilestes latialatus* Donnelly, 1981 (Zygoptera: Lestidae). Bulletin of American Odonatology 2: 1–7.

Novelo-Gutiérrez, R. & E. González Soriano. 1986. Description of the larva of *Perissolestes magdalenae* (Williamson and Williamson, 1924) (Zygoptera: Perilestidae). Odonatologica 15: 129–133.

Novelo-Gutiérrez, R. & E. González Soriano. 1991. Odonata de la Reserva de la Biósfera la Michila, Durango, México. Parte II. Náyades. Folia Entomológica Mexicana 81: 107–164.

Novelo-Gutiérrez, R. & E. González Soriano. 2003. The larva of *Lestes alfonsoi* González & Novelo (Zygoptera: Lestidae). Odonatologica 32: 289–294.

Ramírez, A. 1994. Descripción e historia natural de las náyades de odonatos de Costa Rica II. *Archilestes neblina* (Garrison, 1982) (Odonata, Lestidae) con una clave para las especies del género en Costa Rica. Folia Entomológica Mexicana 90: 9–16.

Rehn, A.C. 2003. Phylogenetic analysis of higher-level relationships of Odonata. Systematic Entomology 28: 181–239. https://doi.org/10.1046/j.1365-3113.2003.00210.x.

Santos, N.D. 1969. Contribução ao conhecimento da fauna do estado da Guanabara. 70. Descrição da ninfa de *Perilestes fragilis* Hagen *in* Selys, 1862 e notas sobre o imago (Odonata: Perilestidae). Atas da Sociedade de Biologia do Rio de Janeiro 12: 303–304.

Santos, N.D. 1972. Contribuição ao conhecimento da fauna do Estado da Guanabara. 79 Descrição da ninfa de *Lestes pictus* Selys, 1862 (Odonata: Lestidae). Atas da Sociedade de Biologia do Rio de Janeiro 15: 77–78.

Suhling, F., G. Sahlén, S. Gorb, V.J. Kalkman, K-D.B. Dijkstra & J. von Tol. 2015. Order Odonata, Chapter 35. Pages 893–932 *in:* Thorp, J.H. & D.C. Rogers (eds.). Ecology and General Biology Thorp and Covich's Freswater Invertebrates, Vol. 1, 4th Edition, Academic Press-Elsevier.

Thorp, J.H. & D.C. Rogers (eds.). Ecology and General Biology Thorp and Covich's Freswater Invertebrates, Vol. 1, 4th Edition, Academic Press-Elsevier.

Westfall, M.J. 1976. Taxonomic relationships of *Diceratobasis macrogaster* (Selys) and *Phylolestes ethelae* Christiansen of the West Indies as revealed by their larvae (Zygoptera: Coenagrionidae, Synlestidae). Odonatologica 5: 65–76.

Westfall, M.J., Jr. & M.L. May. 2006. Damselflies of North America. Revised edition. Scientific Publishers, Inc., Gainesville, Florida.

Williamson, E.B. 1921. Two new Neotropical genera of Lestinae (Odonata). Occasional Papers of the Museum of Zoology, University of Michigan 96: 1–9.

Williamson, E.B. & J.H. Williamson. 1924. The genus *Perilestes* (Odonata). Miscellaneous Publications, Museum of Zoology, University of Michigan 14: 1–39.

Chapter 14.6

Superfamily Coenagrionoidea

Federico Lozano
Laboratorio de Biodiversidad y Genética Ambiental (BioGeA), Universidad Nacional de Avellaneda (UNDAV), Buenos Aires, Buenos Aires, Argentina

Javier Muzón
Laboratorio de Biodiversidad y Genética Ambiental (BioGeA), Universidad Nacional de Avellaneda (UNDAV), Buenos Aires, Buenos Aires, Argentina

Danielle Anjos-Santos
Centro de Investigaciones Esquel de Montaña y Estepa Patagónicas (CIEMEP), Esquel, Chubut, Argentina

Pablo Pessacq
Centro de Investigaciones Esquel de Montaña y Estepa Patagónicas (CIEMEP), Esquel, Chubut, Argentina

Chapter Outline

INTRODUCTION

Coenagrionoidea, with approximately 170 genera and more than 1,700 species, is the largest superfamily within Zygoptera (Dijkstra et al., 2014). Several hypotheses have been proposed to explain its internal arrangement (O'Grady & May, 2003; Rehn, 2003; Carle et al., 2008), and the most recent molecular studies and classificatory scheme (Dijkstra et al., 2013, 2014) suggested that it is composed of three families: Isostictidae, Platycnemididae, and Coenagrionidae. Only the Coenagrionidae are represented within the Neotropical region. The Isostictidae, with 12 genera and 46 species is restricted to Australia, New Guinea, New Caledonia, and neighboring islands; and the Platycnemididae, with 40 genera and more than 400 species, is confined to the Old World.

Coenagrionidae is composed of small to large-sized, and usually colorful damselflies commonly known in Spanish as "*caballitos del diablo*" or "*alguaciles*." This is the most diverse family of Zygoptera; with 114 genera and 1,267 species, it represents 73% of all Coenagrionoidea (Dijkstra et al., 2014). Knowledge of the larvae is still very poor; to date, only 200 larvae have been described, which represents about 30% of the total known diversity. This number includes representatives of just 34 genera, approximately 50% of the generic diversity.

The internal topology of Coenagrionidae has long been discussed (see O'Grady & May, 2003). Recent studies suggest that two major groups are recovered, and according to Dijkstra et al. (2013) each group could be treated as a subfamily or even family. The Coenagrionidae *sensu* Dijkstra et al. (2014) implies the nonmonophyly of long-recognized families within Zygoptera. This has changed the macrosystematics of the order, especially within the Neotropical region where members of the Protoneuridae and Pseudostigmatidae are now included within Coenagrionidae (see Table 14.1). In this chapter, we follow the proposal of Dijkstra et al. (2014).

Members of this family are common inhabitants of all kinds of freshwater environments, such as shallow vegetated ponds and shaded streams (von Ellenrieder & Garrison, 2007; Muzón et al., 2010). Corbet (1999) included among the habitats of Coenagrionidae fast streams with boulders, slow moving streams with detritus, rivers, and permanent lotic environments. Additionally, larvae of all members of the former Pseudostigmatidae and a few other Coenagrionidae (e.g., *Bromeliagrion*, *Leptagrion*, *Roppaneura*)

Thorp and Covich's Freshwater Invertebrates, Fourth Edition. https://doi.org/10.1016/B978-0-12-804223-6.00020-2

inhabit phytotelmata (tree holes, bamboo internodes, and bromeliads, see Corbet, 1983; Fincke, 1984, 2005; Neiss et al., 2008) in Neotropical forests, and many other Neotropical species are found in temporary ponds. Most Coenagrionoidea larvae live among plants, e.g., submerged weeds, reeds, or grass-like vegetation, and some others may attach themselves to the roots of trees or live among coarse detritus (Suhling & Martens, 2007).

The Coenagrionidae in the Neotropics is represented by more than 650 species within 70 genera, of which 60 are endemic or mainly distributed in the Neotropical region (Garrison et al., 2010; Dijkstra et al., 2013). This number is continuously increasing; for example, in the last decade 121 new species and 11 new genera have been described. There are eight genera with more than 20 species each in the region: *Acanthagrion, Argia, Epipleoneura, Neoneura, Metaleptobasis, Oxyagrion, Protoneura,* and *Telebasis.* There are 17 monotypic genera (with only one species) in the region: *Anomisma, Apanisagrion, Dolonagrion, Enacantha, Fluminagrion, Hesperagrion, Hylaeonympha, Junix, Lamproneura, Leucobasis, Megaloprepus, Proneura, Roppaneura, Schistolobos, Telagrion, Tigriagrion,* and *Tukanobasis.* The remainder 45 genera have intermediary number of species.

In the Neotropics, five countries have more than 100 species recorded: Brazil (268), Venezuela (132), Peru (129), Mexico (111), and Colombia (106) (Paulson, 2014, 2016a, 2016b). This coincides with global diversity patterns. Diversity decreases southwards, with *Cyanallagma interruptum* (Selys) being the southernmost species of Coenagrionidae; it was recorded from Santa Cruz province, Argentina, in Patagonia (Muzón et al., 2014).

LIMITATIONS

Knowledge of the Coenagrionidae is still very incomplete and many larval descriptions need to be updated. There are several keys for the adults that include Coenagrionoidea (e.g., Lencioni, 2005, 2006; Garrison et al., 2010), but there are a few and only local keys for the larvae (von Ellenrieder & Garrison, 2007; Muzón et al., 2014; Neiss & Hamada, 2014) which makes larval identification very difficult.

Despite being the most diverse group within Zygoptera, larval morphology is strikingly uniform, which makes identification difficult, especially at the species level. Larval identification in most cases requires knowledge of either the local fauna from a thorough sampling of adults, or the collection and rearing of larvae to the adult.

Odonata larvae are described based on the morphology of the final instar (F0). It has been documented that morphological changes in structures used for identification occur during development (Corbet, 1995). This makes identification of younger instars almost impossible, especially within Coenagrionidae where the general morphology is pretty much uniform within the family.

TERMINOLOGY AND MORPHOLOGY

The most important terminology for identifying these damselflies are described in Chapter 14.0 in this book and Chapter 35 (Suhling et al., 2015) in Volume 1 of this series (Thorp & Rogers, 2015). Some of the main morphological features of the family include: head without tubercles; antennae with first article short; prementum either triangular or subrectangular with a variable number of premental setae (often the n + n notation is used to indicate number of setae on each side of the midline) and with ligula entire, with a variable number of small denticles; labial palps stout with a variable number of palpal setae, a strong and recurved movable hook, and a complex series of distal teeth; the mandibles have generally 4 or 5 incisors with the presence of an accessory tooth y and 2 or 3 molars (the notation of mandibular formula used to describe it follows Watson, 1956). The caudal appendages are extremely variable in morphology, and are used for both locomotion and to aid respiration, especially when there is low concentration of oxygen (Sesterhenn et al., 2013).

MATERIAL PREPARATION AND PRESERVATION

In Volume I (Thorp & Rogers, 2015) of this series specimen preparation and preservation are presented.

KEY TO COENAGRIONIDAE GENERA

Our knowledge on Neotropical coenagrionid larvae is currently restricted to 34 genera (48.57%) and approximately 210 species (31.81%). Larvae of this family share a trapezoidal labium which generally reaches coxae 2, a slender body with three lamellated caudal appendages (epiproct + paraprocts), held more or less in the vertical plane. Caudal appendages shape, presence/absence of nodus and spines, and number of premental and palpal setae, are the most commonly used characters to distinguish among genera and species, since they are extremely variable, and are widely used in the keys.

Genera previously assigned to Protoneuridae and Pseudostigmatidae are now included within Coenagrionidae (see Dijkstra et al., 2014). Unfortunately, this could be confusing for nonspecialists when tracking references. Therefore, the major changes, which include the inclusion of genera previously assigned to Protoneuridae and Pseudostigmatidae within Coenagrionidae, are summarized in Table 13.6.1.

Distributional data for each genus in Neotropical region is indicated between brackets (based on Garrison et al., 2010).

1	Premental setae absent (Fig. 14.6.1)	2
1'	Premental setae present (Fig. 14.6.2)	7
2(1)	Caudal appendages stalked and foliate, widening abruptly (Fig. 14.6.3); larvae inhabit phytotelmata	3
2'	Caudal appendages generally not as above, usually laminar and ending in a filamentous tip (Fig. 14.6.4); if stalked and foliate then larvae not in phytotelmata, usually in lotic environments [Neotropics. Nearctic]	*Argia*

FIGURE 14.6.1 *Argia serva* Hagen *in* Selys, 1865, labium, dorsal view.

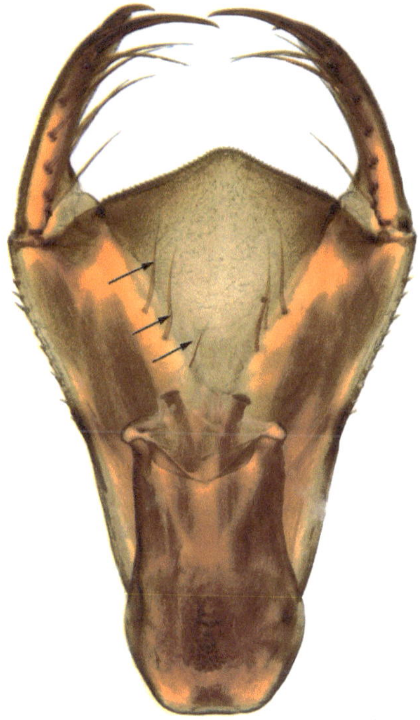

FIGURE 14.6.2 *Idioneura ancilla* Selys, 1860, labium, dorsal view. *Arrows* indicate premental setae.

FIGURE 14.6.3 *Mecistogaster* sp., caudal appendages, lateral view. *Original pictures by Neiss & Hamada (2014).*

3(2) Labial palp with five or more setae (Fig. 14.6.5) 4

3’ Labial palp with two setae (Fig. 14.6.6) .... *Platystigma*
[Neotropics]

4(3) Antennomere three length <2× second (Fig. 14.6.7) 5

4’ Antennomere three length >2× second (Fig. 14.6.8) .... *Microstigma*
[Neotropics. Nearctic]

5(4) Caudal appendages apically with a pale area (Figs. 14.6.9, 10) 6

5’ Caudal appendages apically without pale area (Fig. 14.6.3) .... *Mecistogaster*
[Neotropics]

6(5) Caudal appendages strongly petiolate (Fig. 14.6.9) .... *Megaloprepus*
[South central and Northeast South America]

6’ Caudal appendages weakly petiolate (Fig. 14.6.10) .... *Pseudostigma*
[Central America, Colombia]

FIGURE 14.6.4 *Argia* sp., caudal appendages, lateral view.

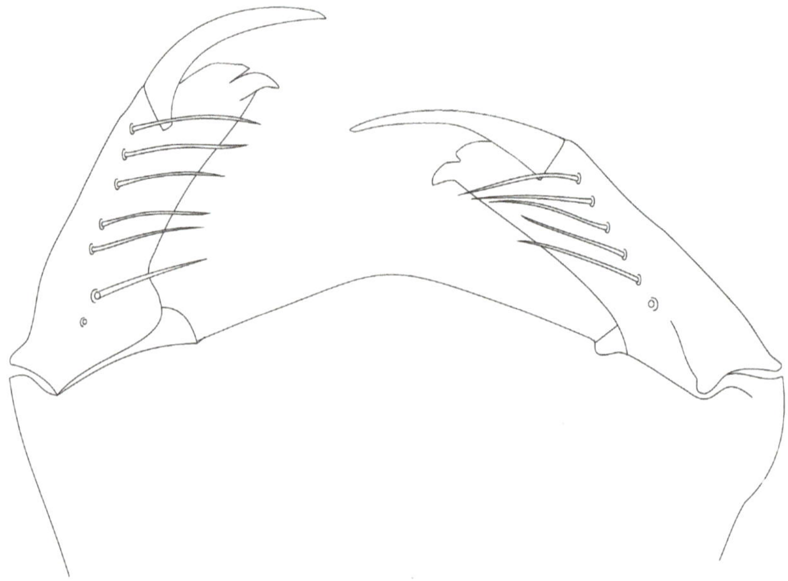

FIGURE 14.6.5 *Mecistogaster amalia* (Burmeister, 1839), labium, dorsal view. *Illustration adapted from Muzón et al. (2010).*

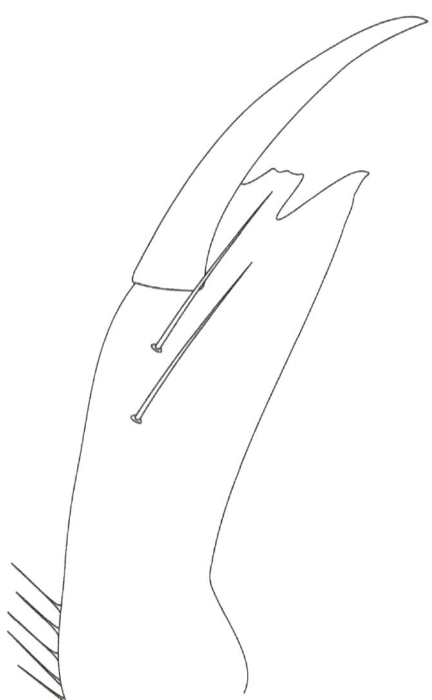

FIGURE 14.6.6 *Platystigma astictum* Machado & Lacerda, 2017, labial palp, dorsal view. Illustration adapted from Lencioni (2006).

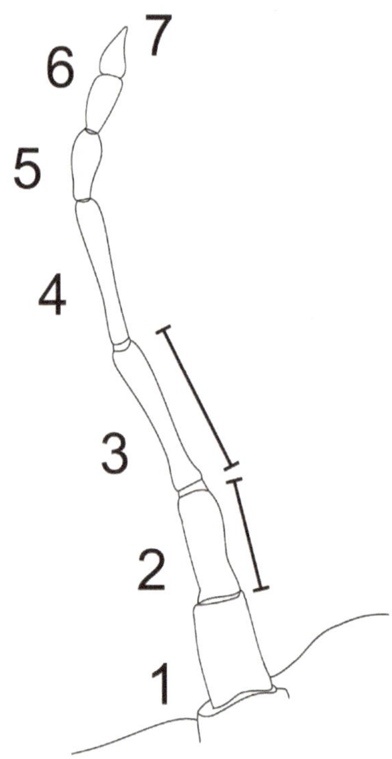

FIGURE 14.6.7 *Mecistogaster amalia*, antenna, dorsal view. Numbers indicates antennomeres. *Illustration adapted from Muzón et al. (2010).*

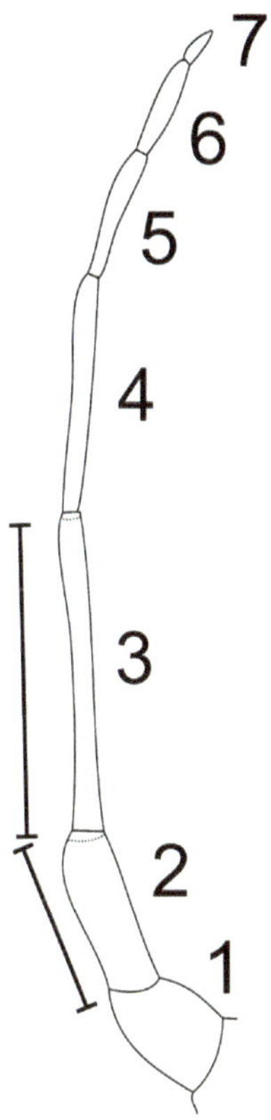

FIGURE 14.6.8 *Microstigma rotundatum* Selys, 1860, antenna, dorsal view. Numbers indicates antennomeres. *Illustration adapted from De Marmels (2007).*

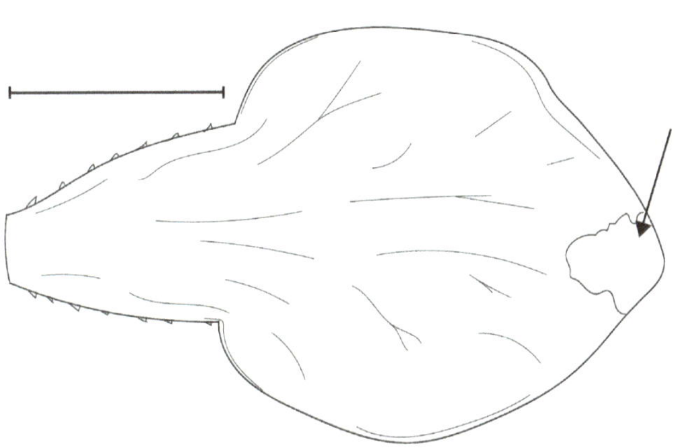

FIGURE 14.6.9 *Megaloprepus caerulatus* (Drury, 1782), caudal appendages, lateral view. Bar indicates petiole length; *arrow* indicate tip of caudal appendage with a pale area. *Illustration adapted from Hedström & Sahlén (2003).*

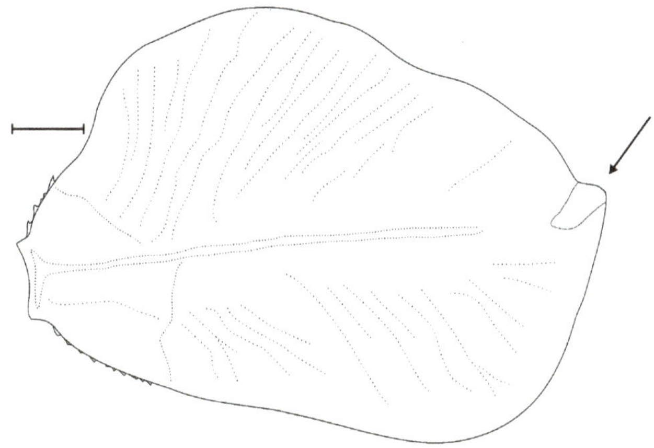

FIGURE 14.6.10 *Pseudostigma aberrans* Selys, 1860, caudal appendages, lateral view. Bar indicates petiole length; *arrows* indicate tips of caudal appendages with a pale area. *Illustration adapted from Novelo-Gutiérrez (1993).*

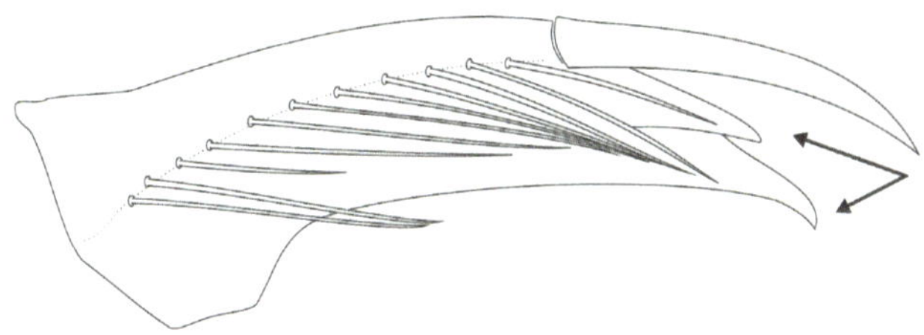

FIGURE 14.6.11 *Bromeliagrion rehni* Garrison *in* De Marmels & Garrison, 2005, labial palp, dorsal view. *Arrows* indicate palpal teeth. *Original pictures by Neiss & Hamada (2014).*

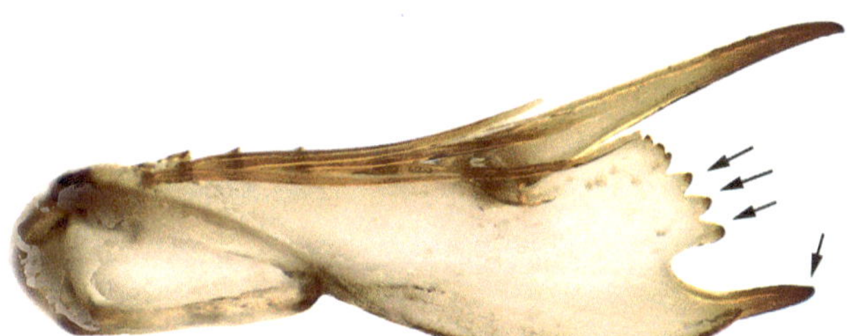

FIGURE 14.6.12 *Oxyagrion* sp., labial palp, dorsal view. *Arrows* indicate palpal teeth. *Original pictures by Neiss & Hamada (2014).*

7(1) Labial palp distal margin with two teeth (Fig. 14.6.11) 8
7' Labial palp distal margin smooth or with numerous small teeth and usually one larger tooth (Fig. 14.6.12) 13
8(7) Caudal appendages uniform in color, if color pattern is visible then without numerous dark spots that are larger near the margins (Fig. 14.6.13); premental setae variable 9
8' Caudal appendages with numerous dark spots, larger near the margins (Fig. 14.6.14); 3 + 3 premental setae .... *Diceratobasis* [Dominican Republic, Jamaica]
9(8) Labial palp with 10 or more setae (Fig. 14.6.11) 10
9' Labial palp with eight or nine setae (Fig. 14.6.15) .... *Leptagrion* (in part) [Brazil, Ecuador, Guianas, Peru, Surinam, Venezuela]

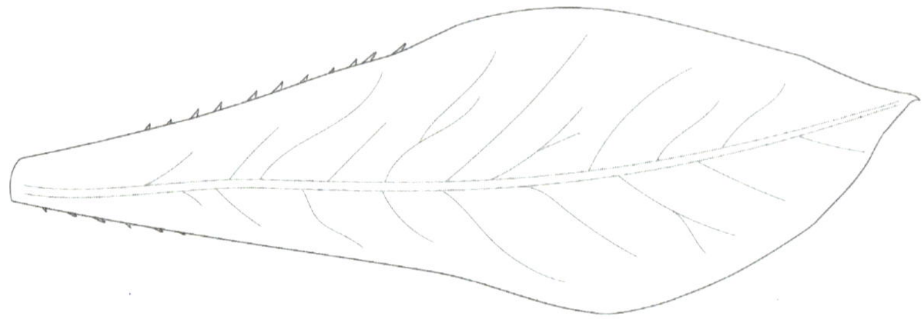

FIGURE 14.6.13 *Leptagrion capixabae* Santos, 1965, caudal appendages, lateral view. *Illustration adapted from Costa et al. (2009).*

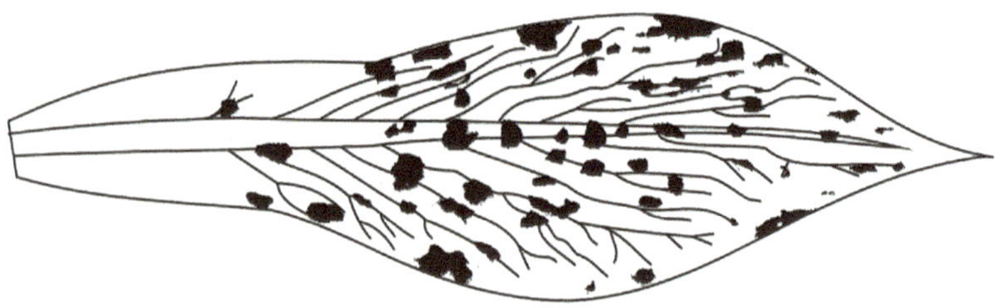

FIGURE 14.6.14 *Diceratobasis macrogaster* (Selys *in* Sagra, 1857), caudal appendages, lateral view. *Illustration adapted from Westfall (1976).*

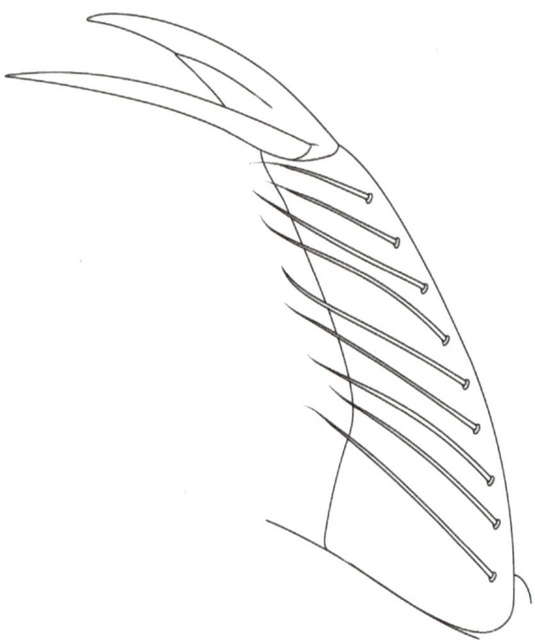

FIGURE 14.6.15 *Leptagrion dardanoi* Santos, 1968, labial palp, dorsal view. *Illustration adapted from Costa et al. (2009).*

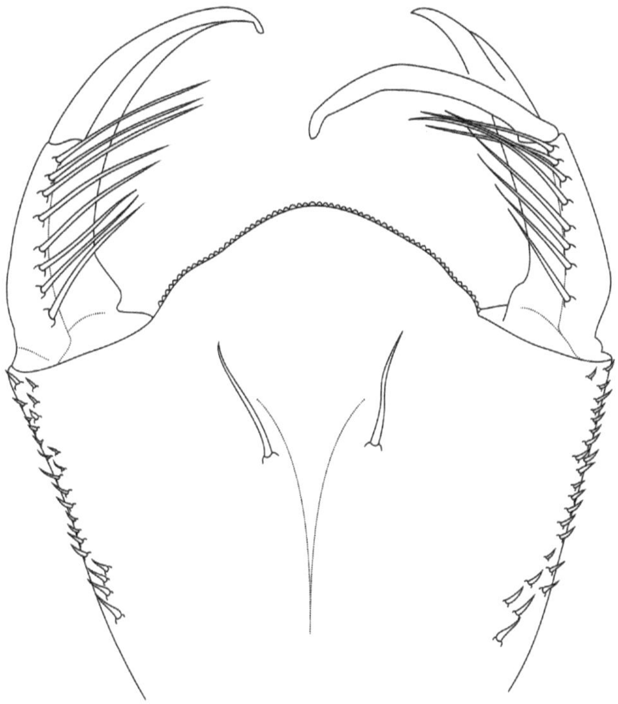

FIGURE 14.6.16 *Leptagrion andromache* Hagen *in* Selys, 1876, labium, dorsal view. *Illustration adapted from Muzon et al. (2009).*

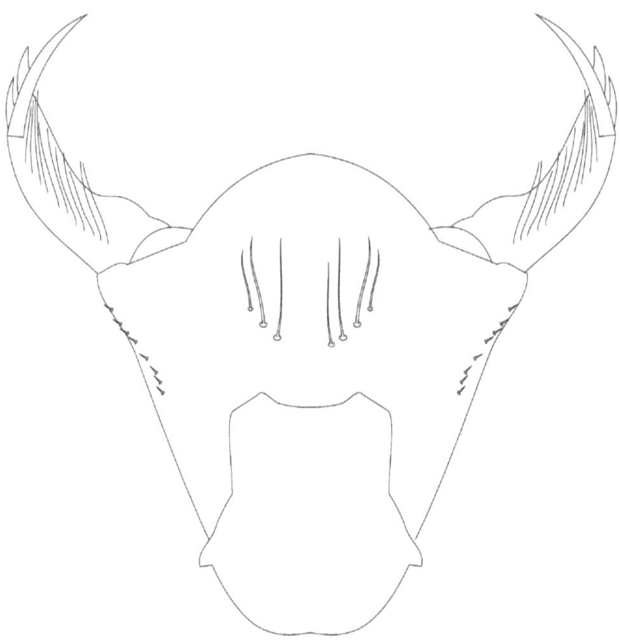

FIGURE 14.6.17 *Bromeliagrion rehni*, labium, dorsal view. *Original pictures by Neiss & Hamada (2014).*

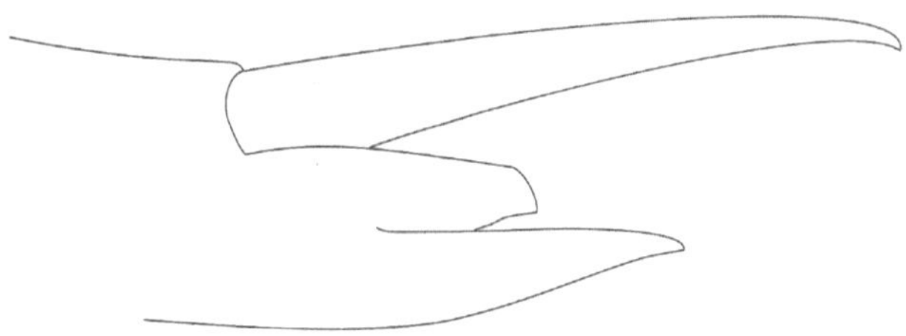

FIGURE 14.6.18 *Bromeliagrion fernandezianum* (Rácenis, 1958), labial palp, dorsal view. *Illustration adapted from De Marmels (1985).*

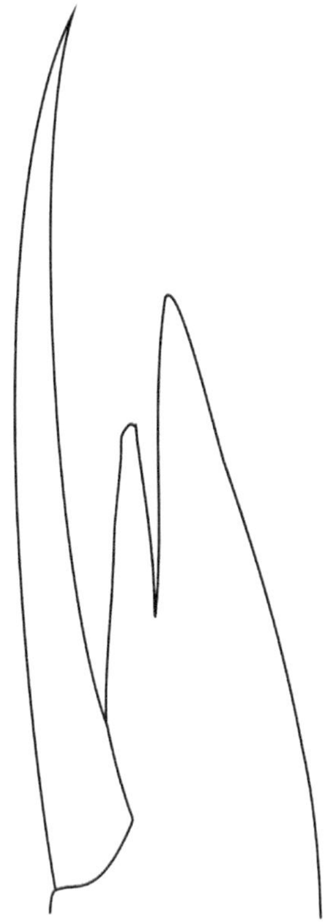

FIGURE 14.6.19 *Leptagrion andromache*, labial palp, dorsal view. *Illustration adapted from Muzón et al. (2009).*

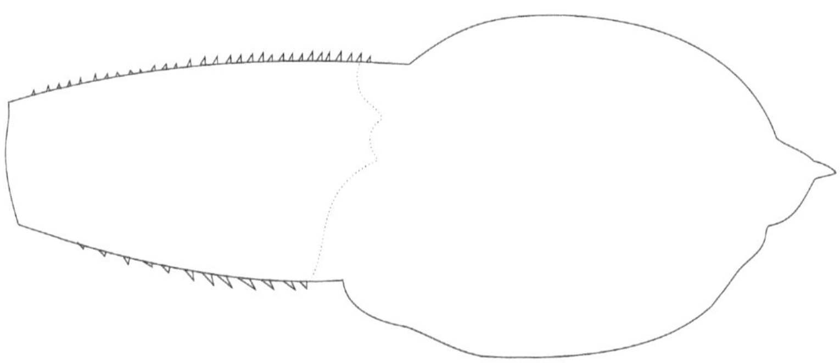

FIGURE 14.6.20 *Bromeliagrion fernandezianum*, caudal appendages, lateral view. *Illustration adapted from De Marmels (1985).*

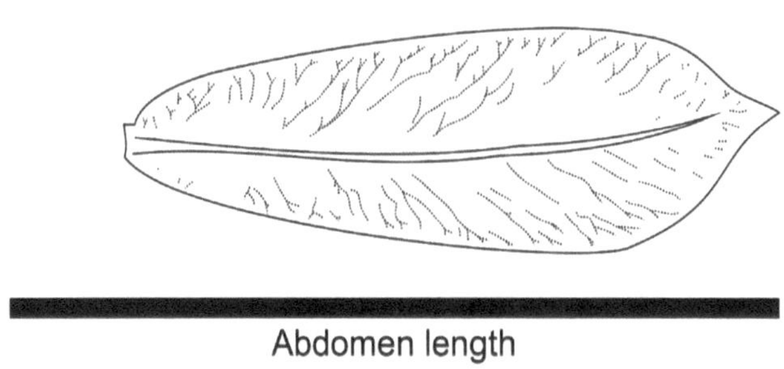

FIGURE 14.6.21 *Oxyagron ablutum* (Calvert, 1909), caudal appendages, lateral view in relation to abdomen length. *Illustration adapted from von Ellenrieder & Lozano (2008).*

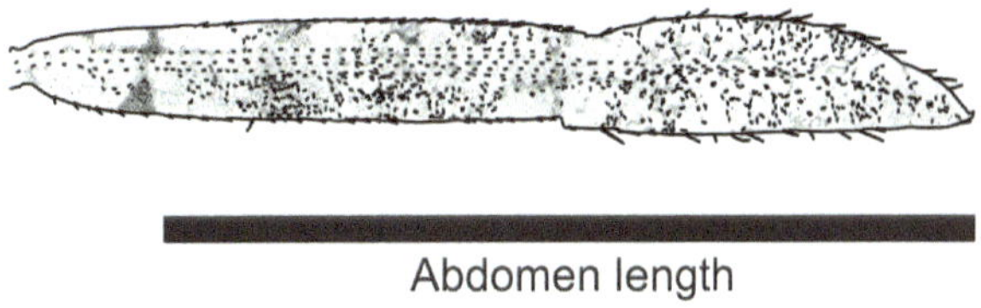

FIGURE 14.6.22 *Acanthagrion aepiolum* Tennessen, 2004, caudal appendages, lateral view in relation to abdomen length. *Illustration adapted from von Ellenrieder & Lozano (2008).*

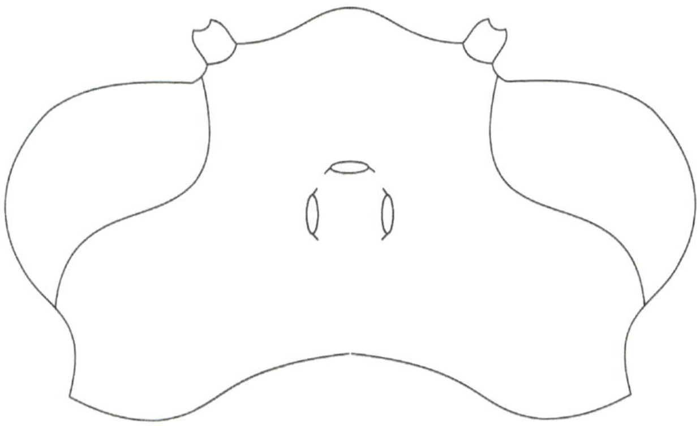

FIGURE 14.6.23 *Amphiagrion abbreviatum* (Selys, 1876), head, dorsal view. *Illustration adapted from von Cook & Antonelli (1969).*

FIGURE 14.6.25 *Roppaneura beckeri* Santos, 1966, labial palp, dorsal view, dorsal view.

FIGURE 14.6.24 *Idioneura* sp., head, dorsal view.

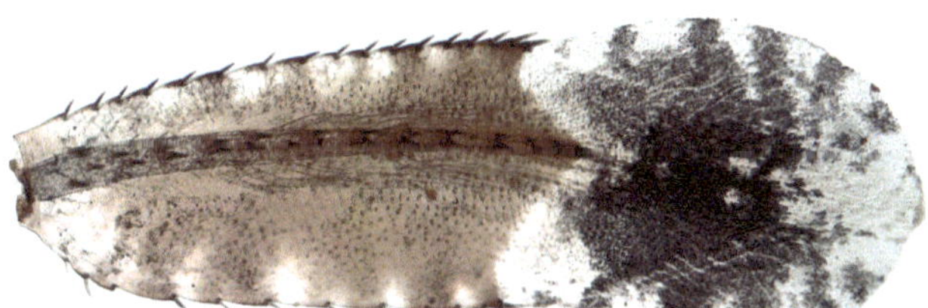

FIGURE 14.6.26 *Telebasis obsoleta* (Selys, 1876), caudal appendages, lateral view.

FIGURE 14.6.27 *Psaironeura* sp.,caudal appendages, lateral view.

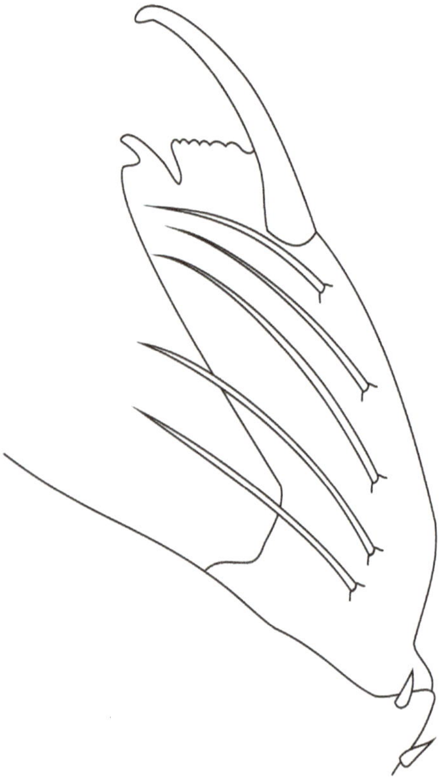

FIGURE 14.6.28 *Peristicta forceps* Hagen *in* Selys, 1860, labial palp, dorsal view. *Illustration adapted from Pessacq (2007).*

FIGURE 14.6.29 *Peristicta forceps*, caudal appendages, lateral view.

10(9)	Prementum with 1 + 1 primary setae (Fig. 14.6.16)	11
10'	Prementum with 3 + 3 primary setae (Fig. 14.6.17)	*Bromeliagrion* (in part)
11(10)	Teeth proximal to movable hook truncate (Fig. 14.6.18)	12
11'	Teeth proximal to movable hook pointed (Fig. 14.6.19) [Brazil]	*Leptagrion* (in part)
12(11)	Caudal appendages without transverse suture [Brazil]	*Leptagrion* (in part)
12'	Caudal appendages with transverse suture (Fig. 14.6.20)	*Bromeliagrion* (in part)
13(7)	Caudal appendages shorter than abdomen (Fig. 14.6.21)	14
13'	Caudal appendages as long as or longer than abdomen (Fig. 14.6.22) [Neotropics]	*Acanthagrion* (except for *A. fluviatile*)
14(13)	Postocular lobes rounded (Fig. 14.6.24)	15
14'	Postocular lobes angulate (Fig. 14.6.23) [Mexico. Nearctic]	*Amphiagrion*
15(14)	Labial palp distal margin with end hook but without defined smaller external teeth (Figs. 14.6.25, 30, 31)	16
15'	Labial palp distal margin with end hook and defined smaller external teeth (Fig. 14.6.12)	26
16(15)	Caudal appendages length $\leq 3\times$ width (Figs. 14.6.26, 29)	17
16'	Caudal appendages length $\leq 4\times$ width (Fig. 14.6.27)	19

FIGURE 14.6.30 *Psaironeura* sp., labial palp, dorsal view. *Original pictures by Neiss & Hamada (2014).*

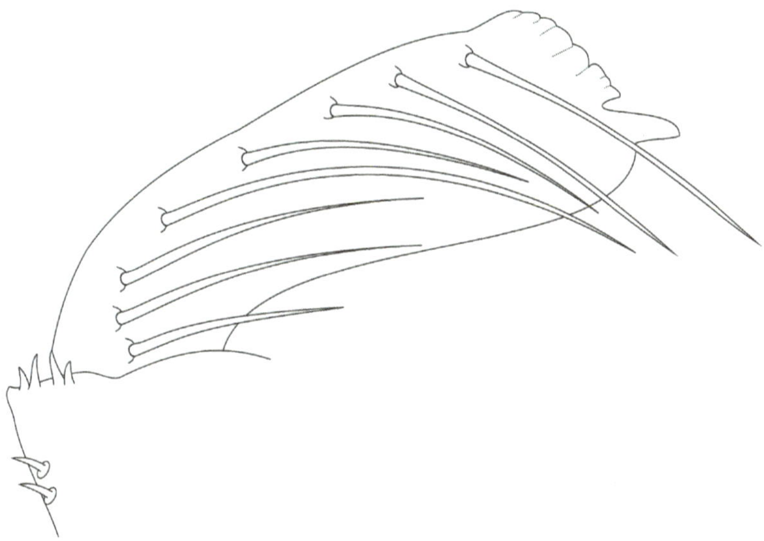

FIGURE 14.6.31 *Nehalennia minuta* (Selys *in* Sagra 1857), labial palp, dorsal view. *Original pictures by Neiss & Hamada (2014).*

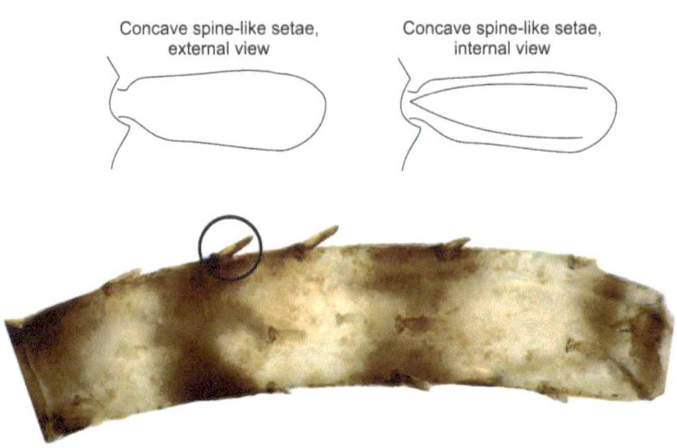

FIGURE 14.6.32 *Psaironeura tenuissima* (Selys 1886), femur, dorsal view. *Original pictures by Neiss & Hamada (2014).*

FIGURE 14.6.33 *Forcepsioneura sancta* (Hagen *in* Selys 1860), labial palp, dorsal view.

FIGURE 14.6.34 *Idioneura ancilla* Selys, 1860, caudal appendages, lateral view.

FIGURE 14.6.35 *Neoneura* sp., labium, dorsal view.

FIGURE 14.6.36 *Epipleoneura manauensis* Santos, 1964, caudal appendages, lateral view.

17(16) Labial palp with five to seven setae (Fig. 14.6.28) 18

17' Labial palp with eight or nine setae (Fig. 14.6.25), larva lives in axils of Umbelliferae .... *Roppaneura* [Brazil]

18(17) Caudal appendages with defined nodus (Fig. 14.6.26) .... *Telebasis* [Neotropics. Nearctic]

18' Caudal appendages with obscured nodus (Fig. 14.6.29) .... *Peristicta* [Argentina, Brazil, Paraguay, Uruguay]

19(16) Labial palp with three to six setae (Fig. 14.6.30) 20

19' Labial palp with seven setae (Fig. 14.6.31) .... *Nehalennia* [Neotropics. Nearctic]

FIGURE 14.6.37 *Neoneura* sp., caudal appendages, lateral view.

FIGURE 14.6.38 *Forcepsioneura sancta*, caudal appendages, lateral view.

FIGURE 14.6.39 *Epipleoneura manauensis*, labial palp, dorsal view. *Original pictures by Neiss & Hamada (2014).*

20(19) Occiput, head mid-dorsal region, femora, and abdomen with filiform or spiniform setae; labial palp usually with three to five setae (Fig. 14.6.33), five to six in *Idioneura* (Fig. 14.6.2) 21

20' Occiput, head mid-dorsal region, femora, and distal margin of S6–8 with concave spiniform setae (Fig. 14.6.32); labial palp with six setae (Fig. 14.6.30) .... *Psaironeura*
[Central America, Northern South America]

21(20) Prementum dorsally with one (Fig. 14.6.35) or two pairs of setae; labial palp with three to five (Fig. 14.6.33) setae; caudal appendages not as below 22

21' Prementum dorsal surface with three pairs of setae, although third pair may be reduced (Fig. 14.6.2); labial palp with five to six setae (Fig. 14.6.2); caudal appendages distinctly narrower basally, widened distally (Fig. 14.6.34) .... *Idioneura*
[Brazil]

22(21) Gills not divided (Fig. 14.6.36) 23

22' Gills divided as a rigid sclerotized basal half and a membranous, paler distal half (Fig. 14.6.37) 24

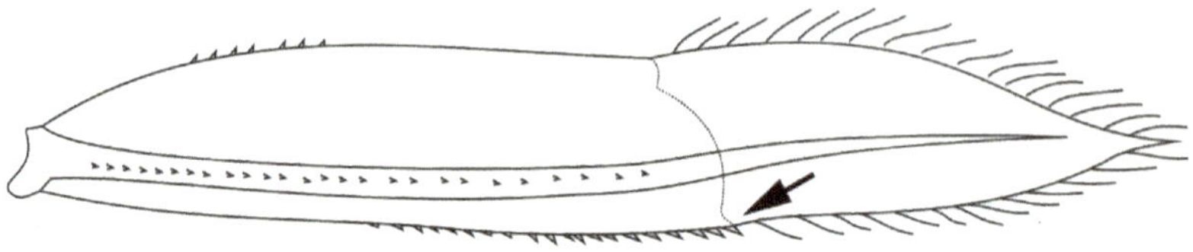

FIGURE 14.6.40 *Protoneura caligata* (Hagen *in* Selys, 1886), caudal appendages, lateral view. *Arrow* indicates nodus position. *Illustration adapted from Westfall (1964).*

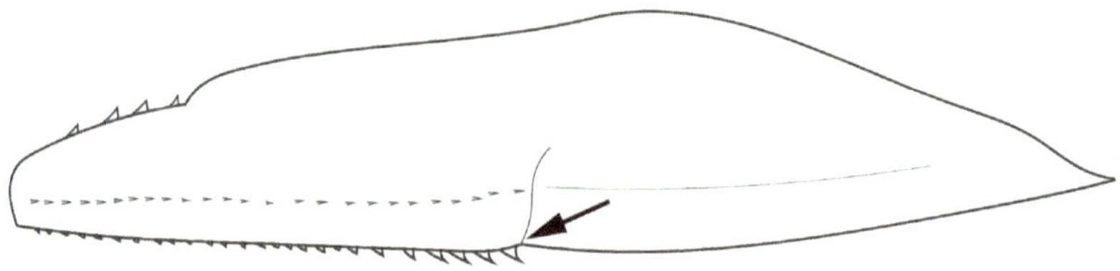

FIGURE 14.6.41 *Neoneura ethela* Williamson, 1917, caudal appendages, lateral view. *Arrow* indicates nodus position. *Illustration adapted from De Souza et al. (2012).*

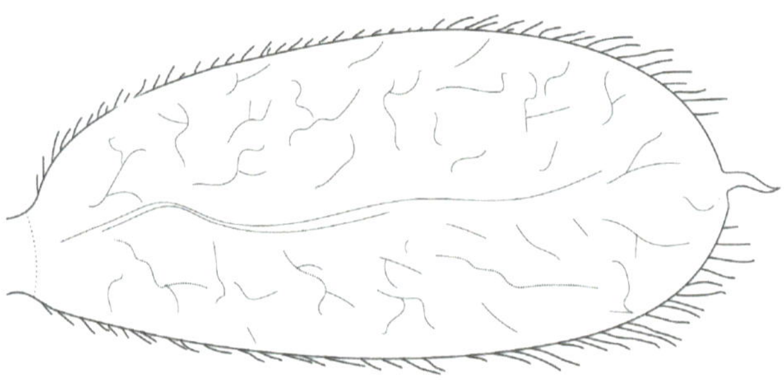

FIGURE 14.6.42 *Mesamphiagrion tamaense* (De Marmels, 1988), caudal appendages, lateral view. *Illustration adapted from von Ellenrieder & Garrison (2008).*

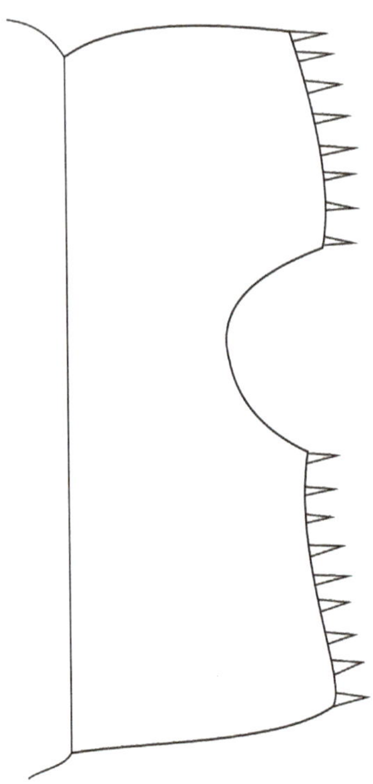

FIGURE 14.6.43 *Mesamphiagrion laterale* (Selys, 1876), S10 dorsal, view. *Illustration adapted from De Marmels (2007).*

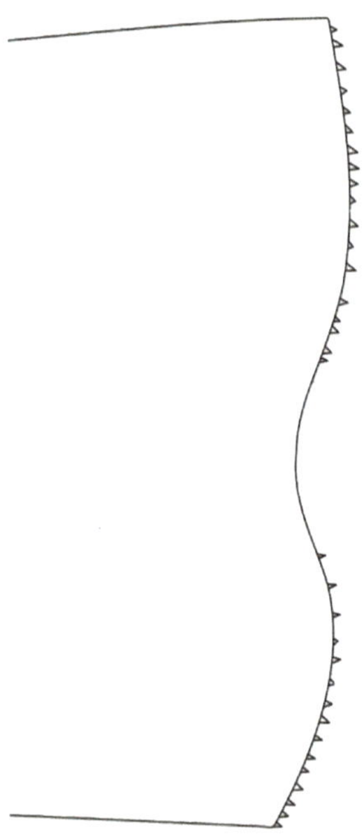

FIGURE 14.6.44 *Oreiallagma quadricolor* (Ris, 1918), S10 dorsal, view. *Illustration adapted from von Ellenrieder & Garrison (2008).*

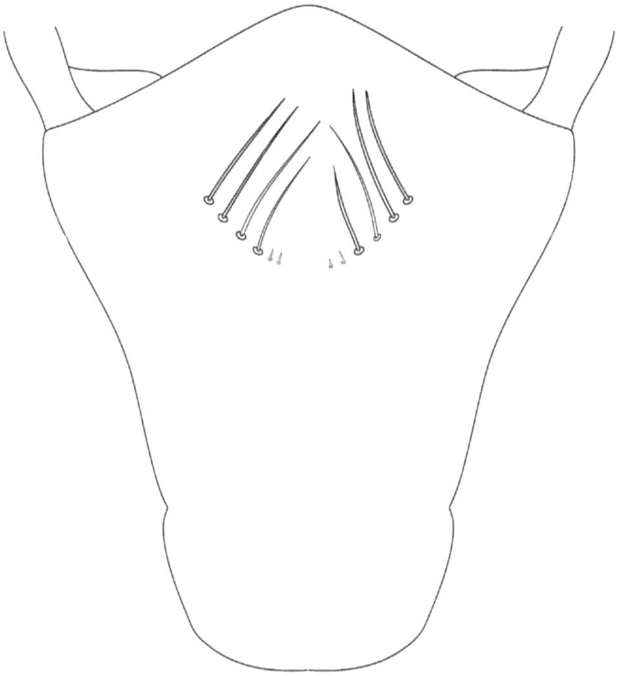

FIGURE 14.6.45 *Mesamphiagrion laterale*, labium, dorsal view. *Illustration adapted from De Marmels (2007).*

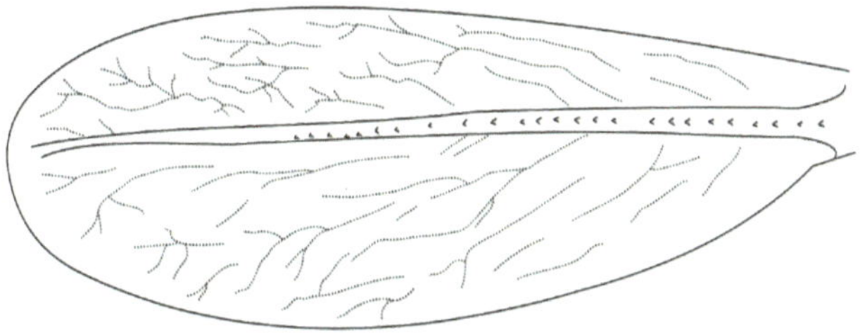

FIGURE 14.6.48 *Hesperagrion heterodoxum* (Selys, 1868), caudal appendages, view. *illustration adapted from Westfall & May (2006).*

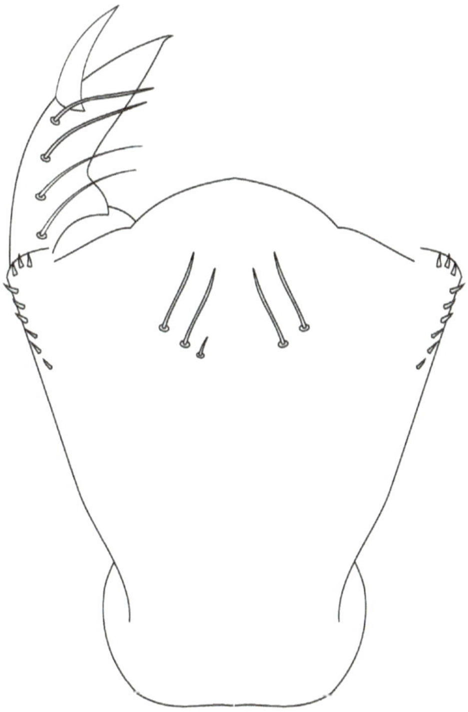

FIGURE 14.6.46 *Protallagma titicacae* (Calvert, 1909), labium, dorsal view. *Illustration adapted from Bulla (1972).*

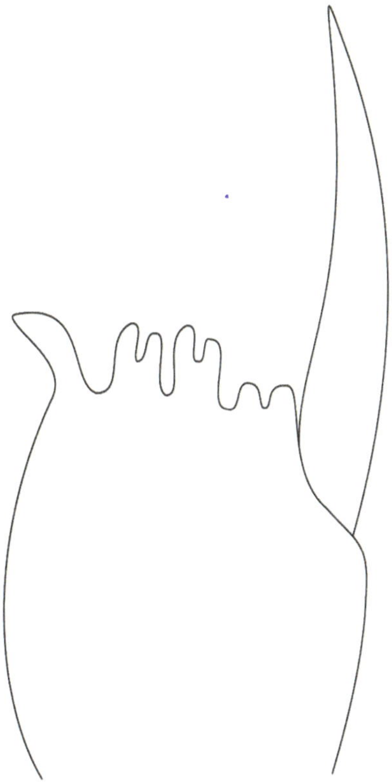

FIGURE 14.6.49 *Leptobasis* sp., labial palp, dorsal view. *Illustration adapted from Trapero Quintana & Naranjo López (2009).*

FIGURE 14.6.47 *Homeoura chelifera* (Selys, 1876), caudal appendages, lateral view.

23(22) Labial palp with five (Fig. 14.6.33) or six setae; lateral caudal appendage length > 7× width (Fig. 14.6.38) ... *Forcepsioneura*
[Brazil]

23' Labial palp with three or four setae (Fig. 14.6.39); lateral caudal appendage length < 6× width (Fig. 14.6.36) ... *Epipleoneura*
[Neotropics]

24(22) Labial palp with three or four (rarely five) setae; tibiae with one or no dark, transverse band; total length (excluding gills) more than 12 mm ... 25

24' Labial palp with five setae (except *P. romanae*, with six); tibiae with two or three dark, transverse bands; total length (excluding gills) less than 10 mm (except *P. calligata*, below) ... *Protoneura*
[Central America, Northern South America]

25(24) Lateral caudal appendages nodus located at about distal 3/5 (Fig. 14.6.40); lateral caudal appendages rigid portion dorsal margin with about 20 spiniform setae; median caudal appendage ventral margin with about 30 spiniform setae ... *Protoneura calligata*
[Cuba]

25' Lateral caudal appendages nodus located at about 1/2, 1/3 or 2/3 of appendages (Fig. 14.6.41); lateral caudal appendages rigid portion dorsal margin with less than 15 spiniform setae (Fig. 14.6.41); median caudal appendage ventral margin with less than 20 spiniform setae ... *Neoneura*
[Neotropics]

26(15) Caudal appendages ovoid, with at least one margin convex (Fig. 14.6.42) ... 27

26' Caudal appendages lanceolate, with margins subparallel at least along proximal third (Figs. 14.6.21, 47, 48) ... 29

27(26) S10 posterodorsal margin with an arcuate cleft; occipital lobes prominent (Fig. 14.6.43) ... 28

27' S10 posterodorsal margin entire; occipital lobes not prominent (Fig. 14.6.44) ... *Oreiallagma*
[Venezuela to Bolivia]

28(27) Palpal setae five or six; generally with 3 + 3 or 4 + 4 long premental setae (Fig. 14.6.45) ... *Mesamphiagrion*
[Colombia, Ecuador, Venezuela]

28' Palpal setae three or four; generally with 2 + 2 long premental setae (Fig. 14.6.46) ... *Protallagma*
[Argentina, Peru]

29(26) Caudal appendage apex usually acute (Figs. 14.6.21, 47), if rounded then setae on basal half stouter and more closely spaced than those more distal ... 30

29' Caudal appendage apex rounded (Fig. 14.6.48), marginal setae subequal in size and spacing along at least 9/10 of the gills *Hesperagrion*
[México. Nearctic]

30(29) Labial palp distal margin without bifid teeth (Fig. 14.6.12) ... 31

30' Labial palp distal margin with bifid teeth (Fig. 14.6.49) ... *Leptobasis*
[Central America, Northern South America, Greater Antilles. Nearctic]

31(30) Lateral caudal appendage apical sixth with a terminal angle of 70º or more (Fig. 14.6.50) ... 32

31' Lateral caudal appendage apical sixth with a terminal angle of 60º or less (Figs. 14.6.21, 47) ... 33

32(31) Labium with four premental setae (if five, one or two reduced) and five palpal setae (Fig. 14.6.51) ... *Anisagrion*
[Central America, Ecuador, Venezuela]

32' Labium with five to eight premental setae and six to eight palpal setae (Fig. 14.6.52) ... *Apanisagrion*
[México to Honduras. Nearctic]

33(31) Prementum with two or more setae on each side of midline (Fig. 14.6.53) ... 34

33' Prementum with one long seta on each side of midline (Fig. 14.6.54), may be accompanied by several short setae ... *Enallagma* (in part)
[Neotropics. Nearctic]

34(33) S2 through S8 with lateral carinae slightly concave in dorsal view and with prominent apices bearing two or more stout, curved setae ... 35

34' Abdominal segments with lateral carinae straight or slightly convex in dorsal view, without prominent apices and with setae, if present, no larger than the adjacent ones on the same segment ... 36

35(34) S9 with lateral carinae less prominent than those of preceding segments, and without stout setae; usually four palpal setae ... *Enallagma* (in part)
[Neotropics. Nearctic]

35' S9 with lateral carinae nearly as prominent as those of S8, and carinae of S8 and 9 both bearing stout setae; usually five palpal setae ... *Neoerythromma*

36(34) Total length of larvae excluding caudal appendages >9 mm ... 37

36' Total length of larvae excluding caudal appendages <6 mm ... *Argentagrion*
[Argentina, Bolivia, Brazil, Paraguay, Uruguay]

37(36) Eyes lacking pattern of lateral bands, but sometimes with dark spots ... 38

37' Eyes usually with a pattern of alternating light and dark lateral bands ... *Ischnura*
[Neotropics. Nearctic]

38(37) Species distributed in Central America ... 39

38' Species distributed in South America ... 40

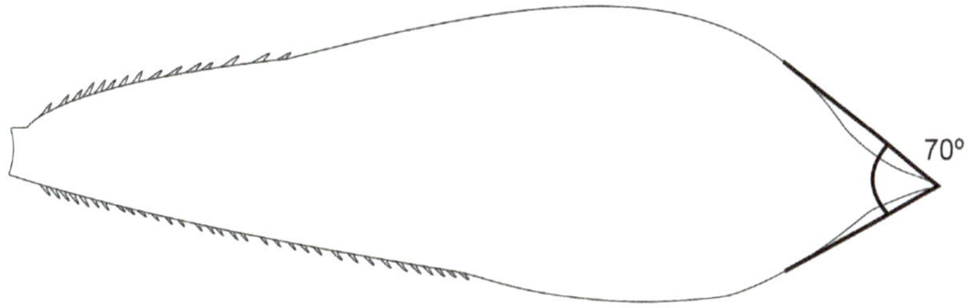

FIGURE 14.6.50 *Apanisagrion lais* (Brauer in Selys, 1876), caudal appendages, lateral view. *Illustration adapted from Novelo Gutiérrez (2010).*

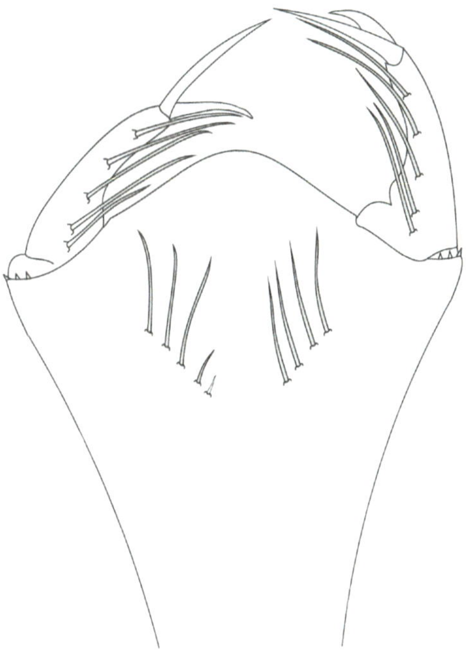

FIGURE 14.6.51 *Anisagrion inornatum* Selys, 1876, labium, dorsal view. *Illustration adapted from Tennessen (2012).*

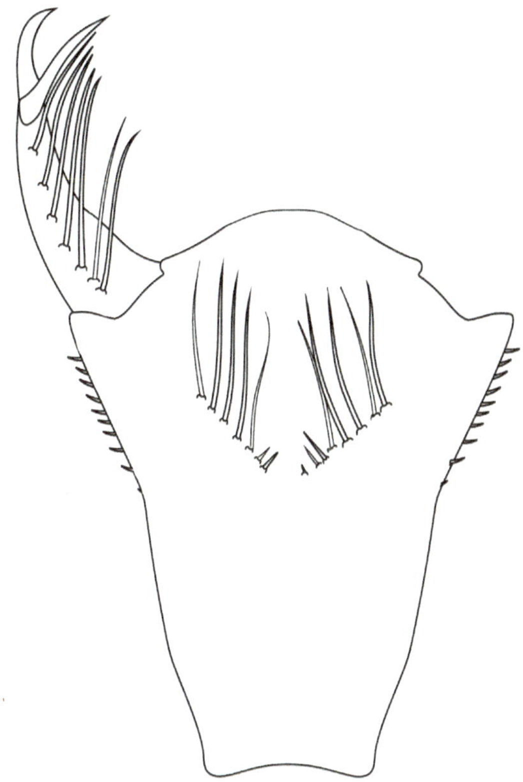

FIGURE 14.6.52 *Apanisagrion lais*, larval labium, dorsal view. *Illustration adapted from Novelo Gutiérrez (2010).*

39(38) Lateral carinae and general surface of abdomen (except dorsum of S9–10) without stout setae; antennal article III length 3x article I .... *Enacantha*
[Belize, Cuba, Guatemala, Mexico]

39' Lateral carinae and general surface of abdomen with stout setae; antennal article III length < 3× article I .... *Enallagma* (in part)
[Argentina, Bolivia, Brazil, Ecuador, Peru, Venezuela. Nearctic]

40(38) Labial palp medial margin crenulate (Fig. 14.6.55) (there are no generic characters which can be used to separate these genera; knowledge of local fauna is needed in order to be able to separate them) .... *Oxyagrion*/*Andinagrion*/*Homeoura*
[South America]/[Argentina and Uruguay]/[South America]

40' Labial palp medial margin not crenulate .... *Cyanallagma*
[Argentina, Brazil, Chile, Uruguay]

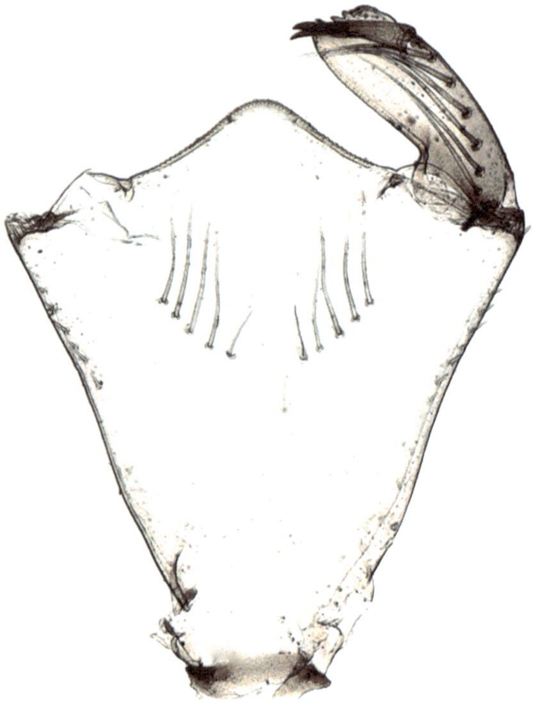

FIGURE 14.6.53 *Ischnura fluviatilis* Selys, 1876, labium, dorsal view.

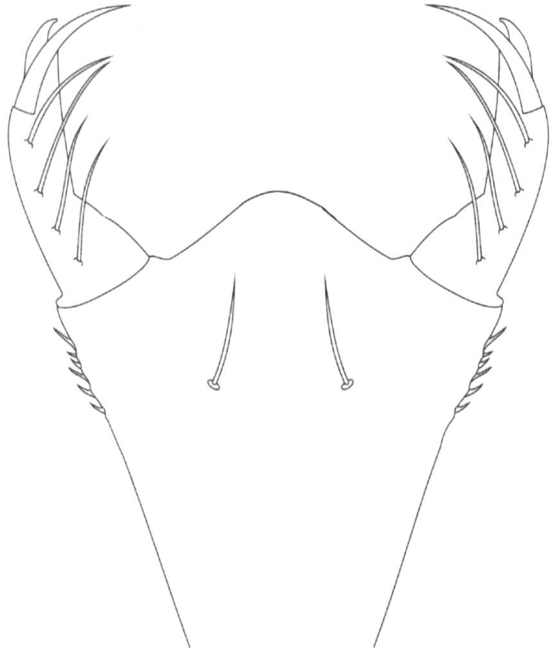

FIGURE 14.6.54 *Enallagma novahispaniae* Calvert, 1907, labium, dorsal view. *Illustration adapted from Novelo Gutierrez (2005).*

FIGURE 14.6.55 *Oxyagrion rubidum* (Rambur, 1842), labial palp, dorsal view.

ACKNOWLEDGMENTS

Figs. 14.6.3, 4, 14.6.12, 14.6.25–27, 14.6.30, 14.6.32, 14.6.36, 14.6.39 were originally presented by Neiss & Hamada (2014) and were generously given by Editora do Instituto Nacional de Pesquisas da Amazônia, Manaus. We express our most sincere gratitude to Dr. Natalia von Ellenrieder for the loan of material.

REFERENCES

Bulla, L.A. 1972. La ninfa de *Protallagma titicacae* (Calvert). Neotrópica 18: 127–133.

Carle, F.L., K.M. Kjer & M.L. May. 2008. Evolution of Odonata, with special reference to Coenagrionoidea (Zygoptera). Arthropod Systematics & Phylogeny 66: 37–44.

Cook, Jr. P.P. & A.L. Antonelli. 1969. The nymph of Amphiagrion abbreviatum. Annals of the Entomological Society of America. 62(2): 264–266.

Corbet, P.S. 1983. A Biology of Dragonflies. Classey, London.

Corbet, P.S. 1995. The larval stages of *Coenagrion mercuriale* (Charp.) (Odonata: Coenagrionidae). Proceedings of the Royal Entomological Society of London (A) 30: 115–126.

Corbet, P.S. 1999. Dragonflies: behavior and ecology of Odonata. Harley Books, Colchester.

Costa, J.M., B.B. Oldrini & D. Anjos-Santos. 2009. Eight new *Leptagrion* larvae from Brazil (Zygoptera: Coenagrionidae). Odonatologica 38: 93–111.

De Marmels, J. 1985. La nayade de *Leptagrion fernandezianum* Racenis, especie bromelícola (Odonata; Coenagrionidae) y consideraciones sobre la posible relación filogenética del género *Leptagrion* Selys. Boletín de Entomología Venezolana 4: 1–8.

De Marmels, J. 2007. Thirteen new Zygoptera larvae from Venezuela (Calopterygidae, Polythoridae, Pseudostigmatidae, Platystictidae, Protoneuridae, Coenagrionidae). Odonatologica 36: 27–51.

De Souza, L.O.I., M. Pepinelli & U.G. Neiss. 2012. The larva of *Neoneura ethela* Williamson, 1917 (Odonata: Protoneuridae). Zootaxa 3318: 63–67.

Dijkstra, K-D.B., G. Bechly, S.M. Bybee, R.A. Dow, H.J. Dumont, G. Fleck, R.W. Garrison, M. Hämäläinen, V.J. Kalkman, H. Karube, M.L. May, A.G. Orr, D.R. Paulson, A.C. Rehn, G. Theischinger, J.W.H. Trueman, J. van Tol, N. von Ellenrieder & J. Ware. 2013. The classification and diversity of dragonflies and damselflies (Odonata). *in*: Z.-Q. Zhang (eds.), Animal Biodiversity: An Outline of Higher-level Classification and Survey of Taxonomic Richness (Addenda 2013). Zootaxa 3703: 36–45.

Dijkstra, K-D.B., V.J. Kalkman, R.A. Dow, F.R. Stokvis & J. van Tol. 2014. Redefining the damselfly families: the first comprehensive molecular phylogeny of Zygoptera (Odonata). Systematic Entomology 39: 68–96.

Fincke, O.M. 1984. Giant damselflies in a tropical forest: reprodutive biology of Megaloprepus coerulatus with notes on Mecistogaster (Zygoptera: Pseudostigmatidae). Advances in Odonatology 2: 13–27.

Fincke, O.M., 2005. Use of forest and tree species, and dispersal by giant damselflies (Pseudostigmatidae): future prospects in fragmented forests. In: Cordero Rivera, A. (ed.) "Forests and dragonflies", Fourth WDA International Symposium of Odonatology, Pontevedra (Spain), July 2005, pp. 103–125.

Garrison, R.W., N. von Ellenrieder, L.J.A. Louton. 2010. Damselfly Genera of the New World: An Illustrated and Annotated Key to the Zygoptera. The Johns Hopkins University Press, Baltimore, Maryland.

Hedström, I. & G. Sahlén. 2003. An extended description of the larva of *Megaloprepus caerulatus* from Costa Rica (Odonata: Pseudostigmatidae), International Journal of Odonatology 6: 23–31.

Lencioni, F.A.A. 2005. Damselflies of Brazil - An illustrated identification guide - 1-Non-Coenagrionidae families. All Print Editora, São Paulo.

Lencioni, F.A.A. 2006. Damselflies of Brazil - An Illustrated identification guide - 2 –Coenagrionidae. All Print Editora, São Paulo.

Muzon J., S. Weigel Muñoz & R.E. Campos. 2009. Description of the bromeliad-dwelling final instar larva of *Leptagrion andromache* Hagen in Selys (Zygoptera: Coenagrionidae). Zootaxa. 2089: 65–68.

Muzón, J., S. Weigel Muñoz & R.E. Campos. 2010. The larva of *Mecistogaster amalia* (Odonata: Pseudostigmatidae). International Journal of Odonatology 13: 137–144.

Muzón, J., P. Pessacq & F. Lozano. 2014. The Odonata (Insecta) of Patagonia: A synopsis of their current status with illustrated keys for their identification. Zootaxa 3784: 346–388.

Neiss, U.G., F.A.A. Lencioni, N. Hamada & R.L. Ferreira-Kleppler. 2008. Larval redescription of Microstigma maculatum Hagen in Selys, 1860 (Odonata: Pseudostigmatidae) from Manaus, Amazonas, Brazil. Zootaza 1696: 57–62.

Neiss, U.G. & N. Hamada. 2014. Ordem Odonata ([capítulo] 14). Pages 217–284 *in*: N. Hamada, J.L. Nessimian & R.B. Querino (eds.), Insetos aquáticos na Amazônia brasileira: taxonomia, biologia e ecologia. Editora do INPA, Manaus.

Novelo Gutiérrez, R. 1993. La náyade de *Pseudostigma aberrans* Selys, 1860 (Odonata: Zygoptera: Pseudostigmatidae). Folia Entomológica Mexicana 87: 55–60.

Novelo Gutiérrez, R. 2005. La larva de *Enallagma novaehispaniae* Calvert, 1902 (Odonata: Zygoptera: Coenagrionidae). Folia Entomológica Mexicana 44: 219–224.

Novelo Gutiérrez, R. 2010. The larva of *Apanisagrion lais* (Brauer in Selys) (Zygoptera: Coenagrionidae). Odonatologica. 39: 259–264.

O'Grady, E. W. & M. L. May. 2003. A phylogenetic reassessment of the subfamilies of Coenagrionidae (Odonata: Zygoptera). Journal of Natural History 37: 2807–2834.

Paulson, D. 2014. North American Odonata, the Odonata of North America. Available at http://www.pugetsound.edu/academics/academic-resources/slater-museum/biodiversity-resources/dragonflies/north-american-odonata/.

Paulson, D. 2016a. South American Odonata. (revised version of 14 June 2016) Available at http://www.pugetsound.edu/academics/academic-resources/slater-museum/biodiversity-resources/dragonflies/south-american-odonata/.

Paulson, D. 2016b. Middle American Odonata. (revised 2 june 2016). http://www.pugetsound.edu/academics/academic-resources/slater-museum/biodiversity-resources/dragonflies/middle-american-odonata/.

Pessacq, P. 2007. *Peristicta aeneoviridis* Calvert, 1909 and P. forceps Hagen in Selys, 1860: redescriptions and a new synonymy (Zygoptera: Protoneuridae). Odonatologica. 36: 207–218.

Rehn, A.C. 2003. Phylogenetic analysis of higher-level relationships of Odonata. Systematic Entomology 28: 181–239.

Sesterhenn, T.M., E.E. Reardon & L.J. Chapman. 2013. Hypoxia and lost gills: respiratory ecology of a temperate larval damselfly. Journal of Insect Physiology 59: 19–25.

Suhling, F. & A. Martens. 2007. Dragonflies and Damselflies of Namibia. Gamsberg Macmillan, Namibia.

Suhling, F., G. Sahlén, S. Gorb, V.J. Kalkman, K-D.B. Dijkstra & J. von Tol. 2015. Order Odonata, Chapter 35. Pages 893–932 *in*: J.H. Thorp & D.C. Rogers (eds.). Ecology and General Biology Thorp and Covich's Freswater Invertebrates, Vol. 1, 4th Edition, Academic Press-Elsevier.

Tennessen, K. 2012. The nymph of *Anisagrion* Selys 1876, based on the discovery of *A. inornatum* (Selys, 1876) in Ecuador (Odonata: Coenagrionidae). Organisms Diversity & Evolution. 12: 297–300.

Thorp, J.H. & D.C. Rogers (eds.). 2015. Ecology and General Biology Thorp and Covich's Freswater Invertebrates, Vol. 1, 4th Edition, Academic Press-Elsevier.

Trapero Quintana, A. & C. Naranjo López. 2009. Clave para la identificación de especies de Odonata en estado larval de Cuba. Boletín Sociedad Entomológica Aragonesa. 44: 459–467.

von Ellenrieder N. & R.W. Garrison. 2007. Dragonflies and damselflies (Insecta: Odonata) of the Argentine Yungas: Species composition and identification. Scientific Reports 7. Publisher Società Zoologica La Torbiera, Italy.

von Ellenrieder, N. & R.W. Garrison. 2008. *Oreiallagma* gen. nov. with a redefinition of *Cyanallagma* Kennedy 1920 and *Mesamphiagrion* Kennedy 1920, and the description of *M. dunklei* sp. nov. and *M. ecuatoriale* sp. nov. from Ecuador (Odonata: Coenagrionidae). Zootaxa 1805: 1–51.

von Ellenrieder, N. & F. Lozano. 2008. Blues for the red *Oxyagrion:* a redefinition of the genera *Acanthagrion* and *Oxyagrion* (Odonata: Coenagrionidae). International Journal of Odonatology 11: 95–113.

Watson, M.C. 1956. The utilization of mandibular armature in taxonomic studies of anisopterous nymphs. Transactions of the American Entomological Society 81: 155–202.

Westfall, Jr, M.J. 1964. Notes on the Odonata of Cuba. Quarterly Journal of the Florida Academy of Sciences 27: 67–85.

Westfall Jr., M.J. 1976. Taxonomic relationships of *Diceratobasis macrogaster* (Selys) and Phylolestes ethelae Christiansen of the West Indies as revealed by their larvae (Zygoptera: Coenagrionidae, Synlestidae). Odonatologica 5: 65–76.

Westfall, Jr., M.J. & M.L. May. 2006. Damselflies of North America. Revised edition. Scientific Publishers, Inc., Gainesville.

Section II.3

Order Coleoptera

Chapter 15

Order Coleoptera: Introduction

Cesar J. Benetti

Coordenação de Biodiversidade, Instituto Nacional de Pesquisas da Amazônia, Manaus, Amazonas, Brazil

Mariano C. Michat

Laboratory of Entomology, IBBEA, CONICET-UBA., DBBE-FCEN, University of Buenos Aires, Buenos Aires, Argentina

Miguel Archangelsky

Laboratorio de Investigaciones en Ecología y Sistemática Animal (LIESA), Centro de Investigación Esquel de Montaña y Estepa Patagónica (CIEMEP, CONICET-UNPSJB), Esquel, Chubut, Argentina

Chapter Outline

ORDER COLEOPTERA: INTRODUCTION

INTRODUCTION

The beetles (order Coleoptera) comprise the largest group of insects in terms of species richness, with about 400,000 described species (Jäch & Balke, 2008) distributed in 211 families (Bouchard et al., 2011). Most species are terrestrial and only a minor percentage can be regarded as aquatic. In spite of this, Coleoptera represents one of the largest orders of aquatic invertebrates. Jäch & Balke (2008) summarized the diversity of the world's water beetles, estimating that about 18,000 species in 37 families have aquatic representatives.

In certain cases, it is difficult to define beetles as aquatic, riparian, or terrestrial. According to Jäch & Balke (2008) the difficulties in the ecological classification are related to different factors such as the amount of time they spend in contact with water, the degree of submergence, the water dependence, and the motivation for getting into contact with water. To facilitate their study, Jäch (1998) classified the beetles associated with aquatic habitats in six ecological groups: "True Water Beetles," "False Water Beetles," "Phytophilous Water Beetles," "Parasitic Water Beetles," "Facultative Water Beetles," and "Shore Beetles."

Most water beetles belong to families considered as "Predominantly aquatic" by Jäch & Balke (2008), i.e., those in which at least 50% of the species are aquatic. All the typical water beetle families are classified here, including "True Water Beetles" (Cneoglossidae, Dryopidae, Dytiscidae, Elmidae, Epimetopidae, Gyrinidae, Haliplidae, Helophoridae, Hydraenidae, Hydrochidae, Hydrophilidae, Hydroscaphidae, Lepiceridae, Lutrochidae, Meruidae, Noteridae, Spercheidae, Sphaeriusidae, Torridincolidae) and "False Water Beetles" (Psephenidae, Scirtidae). Other families have at least some representatives living in or associated with freshwater habitats, although most of the species are terrestrial. These include Chrysomelidae, Curculionidae, Lampyridae, Ptilodactylidae, Scarabaeidae, and Staphylinidae. The last group of families associated with water bodies is named "Shore beetles," and includes species that live at the margins of aquatic habitats (riparian species). These species depend on the aquatic environment to live, but in contrast to true water beetles they do not enter to water. Several species of the previously cited families can be classified here, in addition to most of the species of Georissidae, Heteroceridae, and Limnichidae.

Aquatic beetles live in almost every kind of freshwater environment, such as rivers, streams, lakes, ponds, pools,

Thorp and Covich's Freshwater Invertebrates, Fourth Edition. https://doi.org/10.1016/B978-0-12-804223-6.00021-4

ditches, wetlands, phytotelmata, and ground waters. Many species are benthic, although there are also neuston dwellers, such as adult Gyrinidae, species living exclusively on aquatic plants, and others that "walk" on the underside of the surface film (e.g., some Hydraenidae and Hydrophilidae). Several species are able to swim and dive (e.g., Dytiscidae, Hydrophilidae) and many others stay submerged for most of their life. Although larvae and adults of most water beetle families live in water, pupae are generally terrestrial, and pupation occurs near the water's edge.

As aquatic beetles have evolved from terrestrial ancestors, they have developed various strategies to obtain oxygen for their metabolic requirements. This has been accomplished by using their air-filled tracheal system as the structural plan for their aquatic respiratory systems. Different respiratory options evolved by water beetles were summarized in Eriksen et al. (1996). According to these authors, strategies include atmospheric breathers (e.g., Dytiscidae, Hydrophilidae), plant breathers (e.g., Chrysomelidae, Curculionidae), temporary air store (e.g., Dytiscidae, Noteridae, Haliplidae, Gyrinidae, Hydraenidae, Hydrophilidae), permanent air store (plastrons) (e.g., Curculionidae, Dryopidae, Elmidae, Hydraenidae, Hydrophilidae), spiracular gills (e.g., Hydroscaphidae, Psephenidae, Sphaeriusidae, Torridincolidae), and tracheal gills (e.g., Gyrinidae larvae). Respiration by diffusion of oxygen through the cuticle is also characteristic of several small-sized groups.

Aquatic beetles exhibit a variety of feeding habits (Beutel & Leschen, 2016). Members of Myxophaga (both larvae and adults) feed mainly on algae. Most Adephaga (larvae and adults) are carnivorous, either predaceous or scavengers (Dytiscidae, Noteridae, Gyrinidae). An exception within this group is Haliplidae in which adults are omnivorous and larvae are herbivorous. Members of Polyphaga, on their part, show various feeding strategies. Adults of Hydrophilidae are mostly scavengers, whereas larvae are generally predaceous. Members of several families such as Dryopidae, Elmidae, Hydraenidae, Hydrochidae, Limnichidae, Ptilodactylidae, and Spercheidae feed mainly on algae, decaying plant material, organic matter, or detritus, both as adults and/or larvae. Adults and larvae of Heteroceridae also appear to be detritus and algal feeders, although adults were also observed feeding on zooplankton and larvae consuming soft mud during the construction of their galleries. Finally, larvae of Psephenidae also feed on algae, whereas adults do not feed.

Water beetles are diverse both ecologically and in terms of life-history, and are functionally important in most inland waters, being involved in a range of ecosystem processes such as nutrient cycling and processing (Perissinotto et al., 2016). They have proven to be very useful as biological indicators and suitable for developing management criteria and selecting priority areas for aquatic conservation (Torres et al., 2012; Foster & Bilton, 2014).

Systematics

The order Coleoptera is comprised of four suborders, three of which have aquatic representatives: Myxophaga (77 described species, ca. 90% aquatic), Adephaga (ca. 30,000 described species, ca. 18% aquatic), and Polyphaga (ca. 370,000 described species, ca. 1.25% aquatic) (Jäch & Balke, 2008). Of the estimated global diversity of water beetles, 21% occurs in the Neotropical region.

In this chapter we follow the family classification utilized in Bouchard et al. (2011), with the exception of Hydrophiloidea, for which we follow Hansen (1999) (Epimetopidae, Georissidae, Helophoridae, Hydrochidae, and Spercheidae are considered separate families rather than subfamilies of Hydrophilidae). We also consider aquatic Coleoptera in a wide sense, i.e., including predominantly aquatic, not predominantly aquatic, and shore beetles. Therefore, the Neotropical region includes 30 families which belong to the suborders Adephaga (Dytiscidae, Gyrinidae, Haliplidae, Meruidae, Noteridae), Myxophaga (Hydroscaphidae, Lepiceridae, Sphaeriusidae, Torridincolidae) and Polyphaga (Chrysomelidae, Cneoglossidae, Curculionidae, Dryopidae, Elmidae, Epimetopidae, Georissidae, Helophoridae, Heteroceridae, Hydraenidae, Hydrochidae, Hydrophilidae, Lampyridae, Limnichidae, Lutrochidae, Psephenidae, Ptilodactylidae, Scarabaeidae, Scirtidae, Spercheidae, Staphylinidae).

The predominantly aquatic nonmonogeneric families are presented separately in the following chapters: Myxophaga (Hydroscaphidae and Torridincolidae); Adephaga (Dytiscidae); Adephaga (Gyrinidae, Haliplidae and Noteridae); Polyphaga (Hydrophilidae); Polyphaga (Dryopidae, Elmidae and Psephenidae); Polyphaga (Hydraenidae) and Polyphaga (Scirtidae).

Other families included in the keys, but not treated in detail, are summarized here:

Chrysomelidae: This family is predominantly terrestrial but includes some aquatic representatives. A few species live associated with leaves and roots of aquatic plants and riverbank vegetation. The genera *Agasicles* and *Lysathia* are known to live on aquatic vegetation.

Cneoglossidae: It is a small monogeneric family distributed from Mexico to Brazil. The only genus, *Cneoglossa*, includes eight species (Costa et al., 2005).

Curculionidae: This is the most speciose beetle family, with a majority of terrestrial species, although several genera include species associated with aquatic vegetation and some of them are true aquatic beetles, spending most of their life under water. Some species are important rice pests (e.g., *Oryzophagus oryzae* (Costa Lima, 1936)).

Epimetopidae: Known as hooded shore beetles, this family currently contains 66 described species in three genera: *Eumetopus* Balfour-Browne, 1949 (eight species from the Oriental Region); *Eupotemus* (two species from the Afrotropical Region) and *Epimetopus* (56 species from the New World) with only *E. arizonicus* Perkins occurring exclusively in the Nearctic Region (Ji & Jäch, 1998; Jäch, 2002; Skale & Jäch, 2003; Short & Fikáček, 2011; Perkins, 2012).

Georissidae: Known as minute mud-loving beetles, it is a monogeneric family that includes the genus *Georissus*. It comprises about 80 extant species and has a worldwide distribution (Hansen, 1999; Short & Fikáček, 2011). From the Neotropical region only two species are described: *G. humeralis* Pic (known from Argentina and Brazil) and *G. amrishi* Makhan (only known from Suriname), though the genus has been collected elsewhere.

Helophoridae: It contains the single genus *Helophorus* Fabricius with about 190 described recent species confined almost exclusively to the Palaearctic and Nearctic Regions (Hansen, 1991a, 1999; Angus, 1992; Short & Fikáček, 2011). The distribution of a few predominantly Nearctic species (*H. californicus* Smetana, *H. linearis* LeConte, *H. regularis* Sharp and *H. robertsi* Smetana) barely extend into northern Mexico and the Neotropical Region (Smetana, 1985; Hansen, 1999).

Heteroceridae: Both adults and larvae are shore beetles and live at the muddy margins of lotic and lentic freshwater environments, where they construct narrow tunnels in the wet sand or mud.

Hydrochidae: Known as elongated water scavenger beetles, hydrochids are cosmopolitan and currently comprise about 180 recent species described, all belonging to the genus *Hydrochus* (Hansen, 1999; Jäch & Balke, 2008). There are about 30 species described from the Neotropical region.

Lampyridae: This is mainly a terrestrial group, with only few groups (e.g., *Aspisoma*) living in association with aquatic environments such as floating plants. Larvae of a few species are truly aquatic, but adults are generally terrestrial.

Lepiceridae: Lepicerids are in a monogeneric family with three species. *Lepicerus inaequalis* Motschulsky and *L. pichilingue* Flowers, Shepard, and Mera are recorded on the Neotropical region, *L. bufo* (Hinton) is recorded only from Mexico in Sonora State. Interestingly, the family is also known from Burmese amber, suggesting it once had a much broader range (Kirejtshuk, 2009).

Limnichidae: Most species are riparian, living in wet margins of freshwater habitats, often very close to the water; this family also includes some genera associated with intertidal marine habitats (Hernando & Ribera, 2016). Some species might be regarded as aquatic.

Lutrochidae: This is a monotypic family with only a single described genus, *Lutrochus*, and 21 described species. Current distribution records for the family span from southern Canada (Roughley & Larson, 1991) to southern Brazil and Bolivia, with most of the described diversity in the Neotropics (Spangler et al., 2001).

Meruidae: It is a recently described family of aquatic Coleoptera with one described species, *Meru phyllisae* Spangler & Steiner only known from Venezuela.

Ptilodactylidae: Most species live on the shores of water bodies, but larvae of some species are known to be aquatic (e.g., *Anchytarsus*), while adults are terrestrial.

Scarabaeidae: This is a predominantly terrestrial group, with only few species that may spend up to several hours under water. Species of the genus *Chalepides* Casey, 1915, are associated with aquatic macrophytes, living in the plant axils.

Spercheidae: This monogeneric family is represented by the genus *Spercheus*. It comprises 18 described species and is distributed in all regions except the Nearctic (Archangelsky et al., 2016). From the Neotropical region only two species are known: *S. fimbriicollis* Bruch (from Argentina, Bolivia and Brazil) and *S. halophilus* Archangelsky (from Argentina).

Sphaeriusidae: This is another monogeneric family with about 20 species. *Sphaerius tropicus* Matthews is the only described species recorded on the Neotropical region (Reichardt, 1973; Beutel & Raffaini, 2003). Unidentified specimens were collected on Ecuador, Peru, and Venezuela (Beutel & Raffaini, 2003).

Staphylinidae: This family consists mostly of terrestrial insects with many riparian species and a few species that can be considered aquatic (e.g., *Thinobius*, associated with filamentous algae); some of them are found on the water surface (e.g., *Stenus*).

LIMITATIONS

The knowledge of the aquatic Coleoptera fauna in the Neotropical region is fragmentary. Certain groups are relatively well studied, others are known at local levels, and many others need to be reviewed. In addition, many areas, including the Amazon Basin and Central America, have not been properly surveyed. Adults are relatively well known in many families, especially in Adephaga, Hydrophiloidea, and Elmidae, but in other groups the knowledge is still poor (e.g., Limnichidae, Scirtidae, and Ptilodactylidae). With a few exceptions, knowledge of larval stages of Neotropical water beetles is poor, and this results in a lack of keys for identifying larvae of several families, although there is a recent surge of activity in describing water beetle larvae, especially within phylogenetic frameworks.

So far, there have been no attempts to develop comprehensive keys to aquatic Coleoptera of the Neotropical region. The most significant contribution is the chapter on Coleoptera published by Archangelsky et al. (2009), which includes keys to families and genera of South American aquatic beetles. Additionally, there are a variety of keys to water beetles of other biogeographic regions that may be useful, including those by Ciegler (2003), Bouchard (2004), White & Roughley (2008), and Epler (2010), which cover some of the North American fauna to family and/or genus level.

There are also some regional keys to identify families and/or genera. Among these are: Benetti et al. (2003) (Brazil, Adephaga), Manzo (2005) (South America, Elmidae), Manzo & Archangelsky (2008) (South America, Elmidae larvae), Benetti et al. (2006) and Segura et al. (2011a) (South Brazil, all families), Passos et al. (2007) and Segura et al. (2011b) (Southeast Brazil, Elmidae), Michat et al. (2008) (Argentina, Dytiscidae larvae), Arce-Pérez & Morón (2011) (Mexico, Hydrophiloidea), Libonatti et al. (2011) (Argentina, Dytiscidae), Barbosa et al. (2013) (Central Brazil, Elmidae), Michat & Archangelsky (2014) (Argentina, Gyrinidae), and Ferreira et al. (2014) (Brazilian Amazon, all families).

In addition, for different aquatic Coleoptera families, there are comprehensive global or continental studies which provide keys to genera: Brown (1970) (Dryopidae), Perkins (1980) (Hydraenidae), Brown (1981) (Elmidae), Hansen (1991a) (Hydrophilidae), Hansen (1991b) (Hydraenidae), Pederzani (1995) (Dytiscidae, Noteridae), Archangelsky (1997) (Hydrophiloidea larvae), Miller (2009) (Noteridae), Short et al. (2015) (Hydroscaphidae), and Miller & Bergsten (2016) (Dytiscidae).

References to recent revisions, regional species checklists, and descriptions of new species may be found in the treatments of individual families (see next chapters).

TERMINOLOGY AND MORPHOLOGY

Detailed accounts on the morphology of most water beetle families can be found in the Handbook of Zoology (Beutel & Leschen, 2016).

The smallest water beetles (e.g., Meruidae) are less than 1 mm in length, whereas the larger ones (some Dytiscidae and Hydrophilidae) are about 50 mm in length. Adults are recognized by the first pair of wings rigid (elytra) and the second pair of wings membranous and folded beneath the elytra at rest. In some species the hind wings can be reduced. The pronotum is large and the mesothorax and metathorax are covered by the elytra. The mouthparts are of the biting and chewing type, with the mandibles moving in a lateral plane. The maxillae are bilobed, usually with maxillary palps of four articles. The labium is well developed, usually with labial palps composed of three articles. The antennae consist of 11 or fewer articles and are variously shaped or modified. The legs may be adapted for swimming (especially the hind legs) and can be variously modified. The coxae are either free or fused to the sternum. The trochanter, femur, tibia, tarsus (with five or fewer segments) and paired claws are well developed.

Larvae of aquatic Coleoptera can be recognized by the presence of three pairs of segmented thoracic legs, and the absence of wing pads. They are generally elongate, rounded, or flattened in cross-section. The head capsule is well sclerotized, but the rest of the body is usually poorly sclerotized and soft (more strongly sclerotized in Byrrhoidea and some Adephaga). The mouthparts of most groups are of a biting and chewing type, but may be variously modified. The antennae are short (one to four segments) and the maxillary palps have 3-4 articles. The stemmata are usually present, in general up to six on each side of the head. A pair of well-developed legs is usually present on each of the three thoracic segments, which may occasionally be absent (for example in some Curculionidae). The abdomen is normally ten-segmented, but in many groups, there are nine or eight segments. The ninth segment usually carries a pair of urogomphi, segmented or not. Larvae of most species have spiracles on the mesothoracic and first eight abdominal segments, but these can be reduced in many groups. Abdominal tracheal gills are also variously developed in different groups.

Morphological terms used for the identification of aquatic Coleoptera vary slightly among families, although most of the terms are common. Figs. 15.1 A–D illustrate the main terms used for the identification of larvae and adults to family level. Some of the most commonly used terms related to aquatic Coleoptera are listed in this section (modified from CSIRO, 2001 and Epler, 2010). Additional terms are provided in the specific chapters.

Aedeagus (pl. Aedeagi): The male copulatory organ in Coleoptera, including median lobe and parameres.
Anal: In the direction of the anus, near the anus, or on the last abdominal segment.
Antennomere: A division of the antenna; "article."
Anterior: At or toward the front.

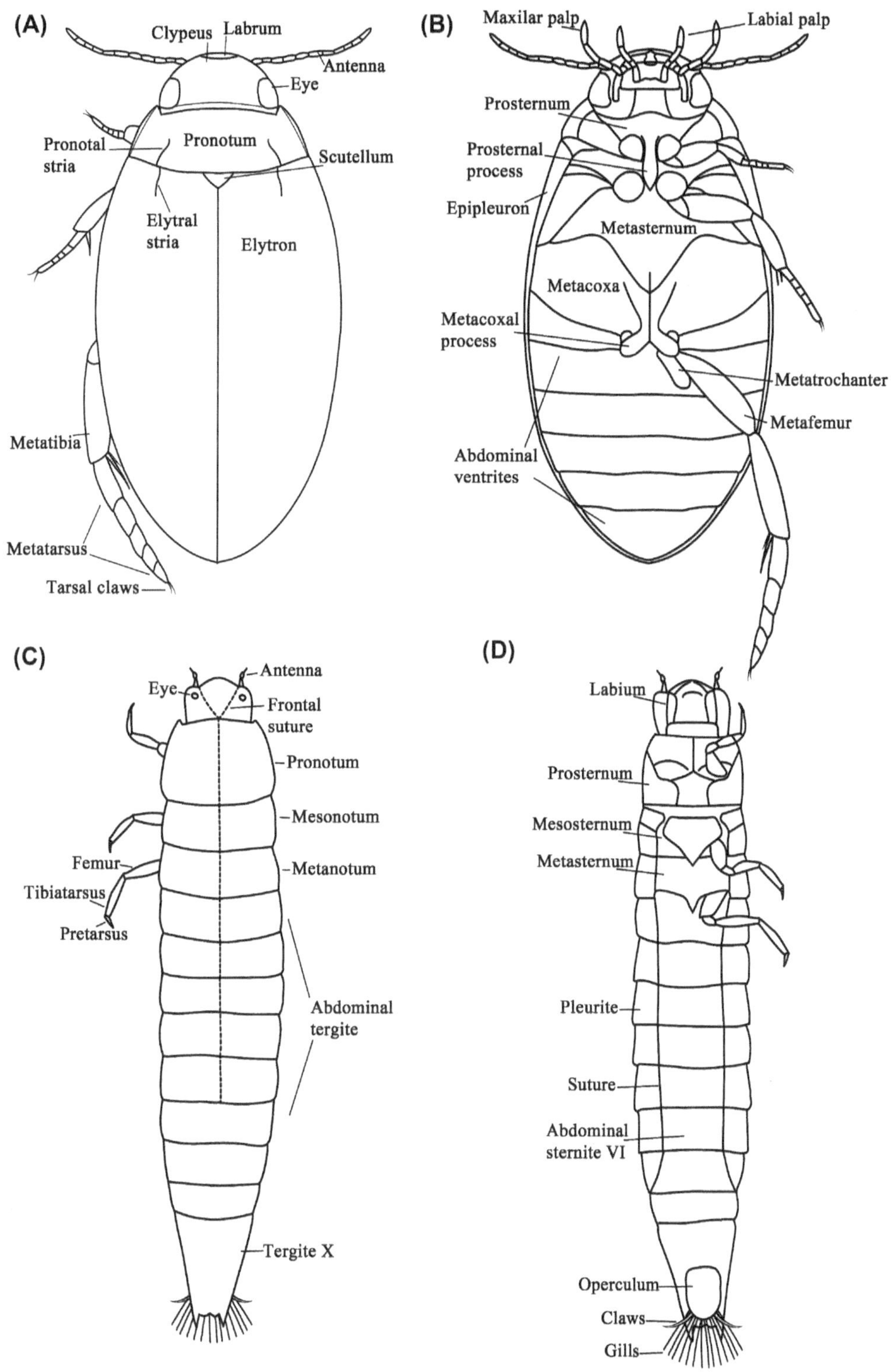

FIGURE 15.1 General characters of adults and larvae. Dytiscidae adult: (A) habitus, dorsal view, (B) habitus ventral view. Elmidae larva: (C) habitus, dorsal view, (D) habitus, ventral view. *(A) Modified from Larson et al. (2000). (B) Modified from Archangelsky et al. (2009).*

Apical: At or toward the apex.
Basal: At or toward the base, or closer to point of attachment.
Carina (pl. carinae): An elevated ridge or keel.
Carinate: Possessing a carina or carinae.
Cercus (pl. cerci): One of a pair of appendages originating from the eleventh abdominal segment, but usually visible as if on segment nine or ten (see also "urogomphus").
Clavate: Gradually thickening or widening distally.
Claw: A hooked structure on the distal end of the leg.
Club: Apical expanded articles of antenna.
Clypeus: The part of the head to which the labrum is attached anteriorly; it may be fused with the frons in a frontoclypeus or separated by a suture.
Coxa (pl. coxae): The proximal (basal) leg segment.
Crenulate: Having the margin very finely notched with rounded projections.
Cupule: Cup-shaped segment of antenna (antennomere 7 in hydrophilids).
Decumbent: Bent or curved downward.
Disc: The central portion of the pronotum or the combined elytra.
Distal: At or near the furthest end from the attachment of an appendage, opposite to proximal.
Dorsal: On the upper surface.
Elytron (pl. elytra): The modified, hardened fore pair of wings of beetles.
Emarginate: Notched along margin.
Epicranial line: A Y-shaped line of weakness on the vertex of the head.
Epimeron (pl. epimera): The posterior division of the thoracic pleuron, separated from the episternum by the pleural suture.
Epipleuron (pl. epipleura): The outer side of the elytron turned down on the side of the thorax and abdomen.
Explanate: Flattened.
Fascia (pl. fasciae): Transverse band.
Fasciate: Transversely banded.
Femur (pl. femora): The third article of the leg, following the coxa and trochanter.
Fimbriate: Having the edge bordered by slender processes; fringed.
Fore: Anterior, toward the head.
Frons: The single medio-anterior sclerite of the insect head, usually lying between the epicranium and the clypeus.
Frontoclypeus: The combined frons and clypeus on head.
Galea: Outer lobe of the maxilla.
Gena (pl. genae): On each side of the head, the part lying beneath the compound eye.
Glossa (pl. glossae): One of a pair of lobes on the inner apex of the prementum.
Gula: A ventromedian sclerotized plate on the head.
Hind: At or toward the posterior.
Instar: The growth stage between two successive molts.
Interval: Area between striae or rows of punctures on elytron.
Irrorate: Speckled with small spots referred to as "irrorations."
Keel: Longitudinal ridge.
Labial palp: A segmented appendage of the labium.
Labium: The "lower lip," forming the floor of the mouth, often with a pair of palps and two pairs of median lobes, derived from the sixth head segment.
Labrum: The "upper lip" forming the roof of the preoral cavity and mouth; derived from the first head segment.
Lacinia: Inner blade-like lobe of the maxilla.
Ligula: The glossae plus paraglossae of the prementum of the labium, whether fused or separate.
Mandible: The first pair of jaws; derived from the fourth head segment.
Maxilla (pl. maxillae): The second pair of jaws, derived from the fifth head segment.
Maxillary palp: A segmented sensory appendage borne on the stipes of the maxilla.
Median lobe: Central appendage of male genitalia; penis.
Mentum: The ventral fused plate derived from the labium.
Mesothorax: The second (and middle) segment of the thorax.
Metathorax: The third (and last) segment of the thorax.
Molar area: The grinding surface of the mandible.
Nasale: Anteromedial elongation of the larval frontoclypeus.
Notopleural suture: The suture separating the pronotum from the proepisternum.
Palpomere: Division of a palp/palpus; "article."
Paraglossa (pl. paraglossae): One of a pair of lobes distolateral on the prementum of the labium, lying outside the glossae, but mesal to the labial palps.
Pedicel: The second antennal article of adults.
Plica (pl. plicae): A fold, groove, or wrinkle.
Posterior: At or toward the rear.
Prementum: The free distal end of the labium, usually bearing the labial palps, glossae, and paraglossae.
Pronotum: The upper (dorsal) plate of the prothorax.

Prostheca: A movable blade on the inner surface of the mandible.
Prothorax: The first segment of the thorax.
Proximal: The closest, opposed to distal.
Pubescent: Clothed with fine setae.
Pygidium: Dorsal portion of the last visible abdominal segment.
Rugose: Wrinkled.
Scape: Basal article of antenna of adults.
Sclerite: A plate on the body wall surrounded by membrane or sutures.
Scutellum: The posterior third of the alinotum (either meso– or metanotum), lying behind the scutum.
Seta (pl. setae): A cuticular extension, a trichoid sensillum, also called a hair or macrotrichium.
Spiracle: An external opening of the tracheal system.
Spur: An articulated spine.
Stemma (pl. stemmata): The simple eyes, as found in larvae, sometimes aggregated into a more complex visual organ.
Sternite: A subdivision of a sternum.
Sternum (pl. sterna): The ventral face of thoracic (pro, meso and metasternum) and abdominal segments.
Stria (pl. striae): Impressed line.
Sulcus (pl. sulci): Groove (wider than a stria).
Suture: An external groove that may show the fusion of two plates.
Tarsomere: A subdivision of the tarsus.
Tarsus (pl. tarsi): The leg article distal to the tibia, comprising one to five tarsomeres.
Tegmen: The base of the male genitalia.
Tergite: Subdivision of tergum.
Tergum (pl. terga): The dorsal face of thoracic and abdominal segments.
Tibia (pl. tibiae): The fourth leg article, following the femur.
Tomentum: Area of short, dense, wooly setae.
Transverse: At right angles to the longitudinal axis.
Trochanter: The second leg article, following the coxa.
Trochantin: A small sclerite anterior to the coxa.
Tubercle, tubercule: A small knob-like or rounded protuberance.
Urogomphus (pl. urogomphi): Paired processes at posterior end of last abdominal segment, more common in immature stages, sometimes referred to as "cerci."
Venter: The under surface of the body.
Ventral: Toward or at the under surface.
Vertex: Dorsal portion of the head between the eyes, posterior to the frons.
Vitta (pl. vittae): Longitudinal stripe.

MATERIAL PREPARATION AND PRESERVATION

Methods used for the collection and preservation of water beetles are, in general, similar to those used for other freshwater invertebrates. For information regarding this topic, the reader is referred to Chapter 3 in Thorp & Rogers (2015). Since the adults of several groups are terrestrial, techniques used for the capture of beetles from terrestrial environments should also be considered; this is especially important to correctly associate aquatic immature stages and terrestrial adults.

Beetles, both larvae and adults, are usually preserved in 70–80% ethanol. However, if the purpose is to use the material for molecular analyses, it is recommended to preserve them in 96–100% ethanol. Adults can also be preserved dry, pinned/point mounted. Adult beetles are better pinned through the right elytron just anterior to the middle of the insect and just to the right of the suture. Beetles too small to be pinned can be glued to cardboard points.

For general observation of specimens, a stereomicroscope will suffice, but for examination of genitalia and larvae larger magnifications are required and the use of a compound microscope is needed. In some cases, it may be necessary to mount genitalia, larval pieces, or other body parts on microscope slides for high-powered observation. Slides may be temporary or permanent mounts. Temporary mounts can be made with water, alcohol, or glycerin. Permanent mounts can be made with different media, such as DMHF (dimethyl hydantoin formaldehyde), Hoyer's, Canada balsam, Euparal, or PVLG (polyvinyl-lacto-glycerol).

It is often necessary to remove or extrude the male genitalia for identification purposes. In the case of large beetles, the genitalia can often be extruded by squeezing the abdomen and the elytra. Removed genitalia can be mounted on microscope slides, placed in a microvial with glycerin or mounted with a pinned specimen, glued to a cardboard. To remove genitalia from glued, dried specimens they should be submerged in hot water for several minutes before attempting the extraction.

Larvae can be examined in temporary or permanent slides. The study of larvae may require a chaetotaxic analysis (the arrangement, nomenclature, and classification of setae). For this purpose, it is necessary to examine the specimens at high magnifications (for example 100× to 1000×).

Male genitalia and larvae may have to be cleared for proper observation. For that purpose, 10% potassium hydroxide (KOH) or lactic acid are recommended.

KEY TO AQUATIC COLEOPTERA FAMILIES

Family level keys are presented for both adults and larvae. Keys to eggs and pupae are not included.

Coleoptera: Adults

Modified from Archangelsky et al. (2009).

1 Compound eyes entire, rarely completely or partially divided (Epimetopidae); meso- and metathoracic legs not as below; antennae variable but not as below 2

1' Compound eyes completely divided in a dorsal and a ventral part (Fig. 15.2 I); meso- and metathoracic legs short and broad (Fig. 15.4 N); antennae short and broad, with scape asymmetrical and cup shaped, pedicel offset and subtriangular, and flagellum forming a short, medially-expanded unit (Fig. 15.3 A); example of habitus in Fig. 15.7 A .... **Gyrinidae (Chapter 15.2, p. 527)**

2(1) Metacoxae not as below; metatrochantera visible (Figs. 15.2 B, D) 3

2' Metacoxae expanded posteriorly in large rounded plates which cover half of the abdomen, the metatrochantera, and the basal half of the metafemora (Fig. 15.2 A); example of habitus in Fig. 15.7 B .... **Haliplidae (Chapter 15.2, p. 527)**

3(2) First abdominal ventrite completely divided by the metacoxae (Fig. 15.2 B); metacoxae fused to metasterna (Fig. 15.2 B) 4

3' First abdominal ventrite not completely divided by the metacoxae (Fig. 15.2 C); metacoxae not fused to metasterna (Fig. 15.2 C) 6

4(3) Metacoxae medial portions not produced ventrally in a platform (Fig. 15.2 B); scutellum visible (Fig. 15.7 E) or not 5

4' Metacoxae medial portions together produced ventrally in a platform with lateral margins nearly straight and triangular to rounded apical lobes that project externally to trochantera (Fig. 15.2 D); scutellum not visible (Fig. 15.7 C); example of habitus in Fig. 15.7 C .... **Noteridae (Chapter 15.2, p. 527)**

5(4) Total body length <1 mm; pronotum and elytra with strongly impressed punctation (Fig. 15.4 I); tarsal claws pectinate (Fig. 15.4 J); example of habitus in Fig. 15.7 D .... Meruidae

5' In almost all cases total body length greater than 1 mm; pronotum and/or elytra either smooth, striate, wrinkled, or variously punctate but with punctation not strongly impressed (Fig. 15.7 E); tarsal claws not pectinate (Fig. 15.4 L); example of habitus in Fig. 15.7 E .... **Dytiscidae (Chapter 15.3, p. 539)**

6(3) Head not prolonged in a snout; antennae not geniculate 7

6' Head prolonged in an elongate, drooping snout at least as long as broad (Fig. 15.2 J); antennae geniculate (Fig. 15.2 J); example of habitus in Fig. 15.7 F .... Curculionidae

7(6) Elytra leaving exposed at least three abdominal tergites (Figs. 15.7 G, H) 8

7' Elytra covering the abdomen completely or leaving exposed at most one abdominal tergite 9

8(7) Antennae composed of eight or less antennomeres; elytra exposing three to five abdominal tergites (Fig. 15.7 G); example of habitus in Fig. 15.7 G .... **Hydroscaphidae (Chapter 15.1, p. 519)**

8' Antennae composed of 11 (rarely 10) antennomeres; elytra exposing at least six abdominal tergites (Fig. 15.7 H); example of habitus in Fig. 15.7 H .... Staphylinidae

9(7) Size and shape variable; metacoxal plates absent (Fig. 15.2 C); four or more abdominal ventrites; meso- and metaventrite not as below 10

9' Total body length 0.5–1.2 mm, body convex; metacoxal plates large, covering the first abdominal ventrite (Fig. 15.2 E); three abdominal ventrites; meso- and metaventrite fused, metaventrite very large; example of habitus in Fig. 15.7 I .... Sphaeriusidae

10(9) Ultimate antennomere as long as the preceding three or four antennomeres combined (Figs. 15.3 B, C); total body length 1.0–2.7 mm 11

10' Ultimate antennomere at most as long as the preceding two antennomeres combined (e.g., Figs. 15.3 F, L); size variable 12

11(10) Antennae of nine antennomeres (Fig. 15.3 B); body flattened; example of habitus in Fig. 15.8 A .... **Torridincolidae (Chapter 15.1, p. 519)**

11' Antennae of four antennomeres (Fig. 15.3 C); body globular; example of habitus in Fig. 15.8 B .... Lepiceridae

12(10) Antennae short, inserted at compound eyes posterior margin, bearing a pubescent club, frequently asymmetrical and with basal antennomeres glabrous (Figs. 15.3 E–G) 13

12' Antennae not as above 19

13(12) Antennal club composed of three antennomeres (Figs. 15.3 F, G); five abdominal ventrites; total body length 1–50 mm 14

13' Antennal club composed of five antennomeres (Fig. 15.3 E); six or seven abdominal ventrites; total body length less than 2.5 mm; example of habitus in Fig. 15.8 C .... **Hydraenidae (Chapter 15.5, p. 557)**

14(13) Pronotum projected forward, totally or partly covering head (Figs. 15.8 D, E); first tarsomere very small (tarsi pseudotetramerous)..... 15

14' Pronotum only slightly projected forward, not or slightly covering head; tarsi composed of five tarsomeres (four tarsomeres in Spercheidae) 16

15(14) Pronotum lateral margins strongly expanded and bearing longitudinal carinae on disc (Fig. 15.4 A); elytra with longitudinal carinae and two longitudinal rows of punctures between two contiguous carinae (Fig. 15.8 D); compound eyes completely or partially divided; total body length up to 5 mm; example of habitus in Fig. 15.8 D .... Epimetopidae

15' Pronotum lateral margins not strongly expanded, pronotal disc smooth (Fig. 15.4 B); elytra lacking carinae, with rows of coarse punctures (Fig. 15.8 E); compound eyes entire; total body length <2 mm; example of habitus in Fig. 15.8 E .... Georissidae

16(14) Pronotum lacking grooves 17

16' Pronotum bearing five well-developed longitudinal grooves (Fig. 15.4 C); example of habitus in Fig. 15.8 F .... Helophoridae

17(16) Antennae before club with two or three glabrous antennomeres (Fig. 15.3 F); meso- and metatibiae lacking natatory setae; elytra with strongly impressed sculpture, generally covered by debris (Figs. 15.8 H, I); total body length <5 mm 18

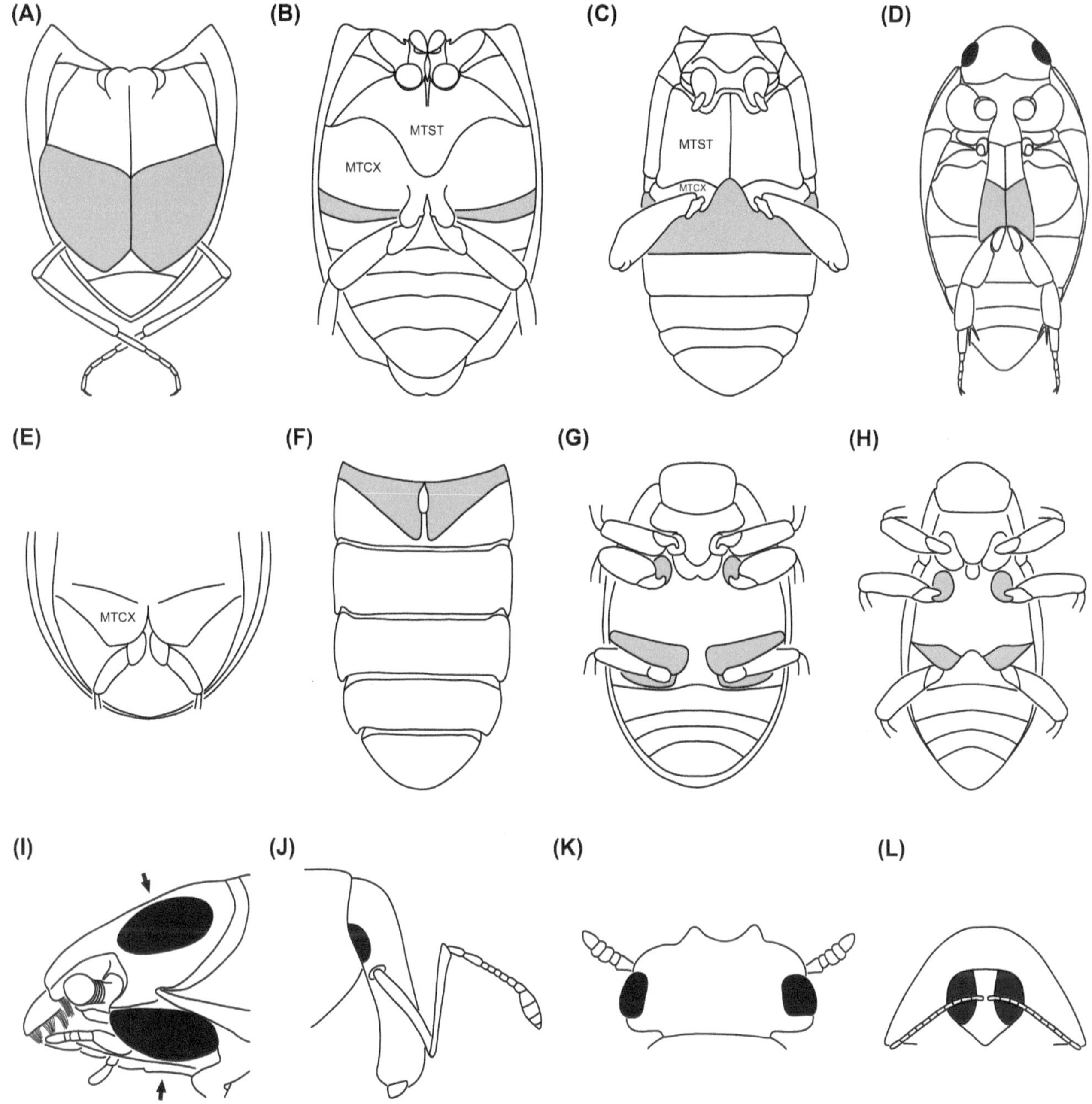

FIGURE 15.2 (Adults). Thorax and abdomen, ventral view: (A) Haliplidae, (B) Adephaga, (C) Polyphaga, (D) Noteridae, (E) Sphaeriusidae, (F) Cneoglossidae, (G) Lutrochidae, and (H) Elmidae. Head: (I) Gyrinidae, lateral view, *arrows* indicate compound eye, (J) Curculionidae, lateral view, (K) Spercheidae, dorsal view, and (L) Lampyridae, ventral view. Abbreviations: *MTCX* = metacoxa; *MTST* = metasternum.

17' Antennae before club with five glabrous antennomeres (Fig. 15.3 G), if only three glabrous antennomeres before club, meso- and metatibiae bearing long natatory setae (Fig 15.4 O); elytra generally smooth or with moderate sculpture; size variable; example of habitus in Fig. 15.8 G .... **Hydrophilidae (Chapter 15.4, p. 561)**

18(17) Body narrow, with iridescence in some regions (Fig. 15.8 H); pronotum lateral margins not dentate or fimbriate; labrum entire; tarsi of five tarsomeres; example of habitus in Fig. 15.8 H .... Hydrochidae

18' Body broad, lacking iridescence (Fig. 15.8 I); pronotum lateral margins dentate (Fig. 15.4 E) or fimbriate (Fig. 15.4 F); labrum medially indented (Fig 15.2 K); tarsi of four tarsomeres; example of habitus in Fig. 15.8 I .... Spercheidae

19(12) Tarsi pentamerous or tetramerous, but third tarsomere not as below (if pseudotetramerous, as in some Ptilodactylidae, the antennae are generally pectinate or serrate); antennae variable 20

19' Tarsi pseudotetramerous, i.e., composed of five tarsomeres but with third tarsomere broad and bilobed, concealing fourth tarsomere (Fig. 15.4 K); antennae long, filiform .... Chrysomelidae

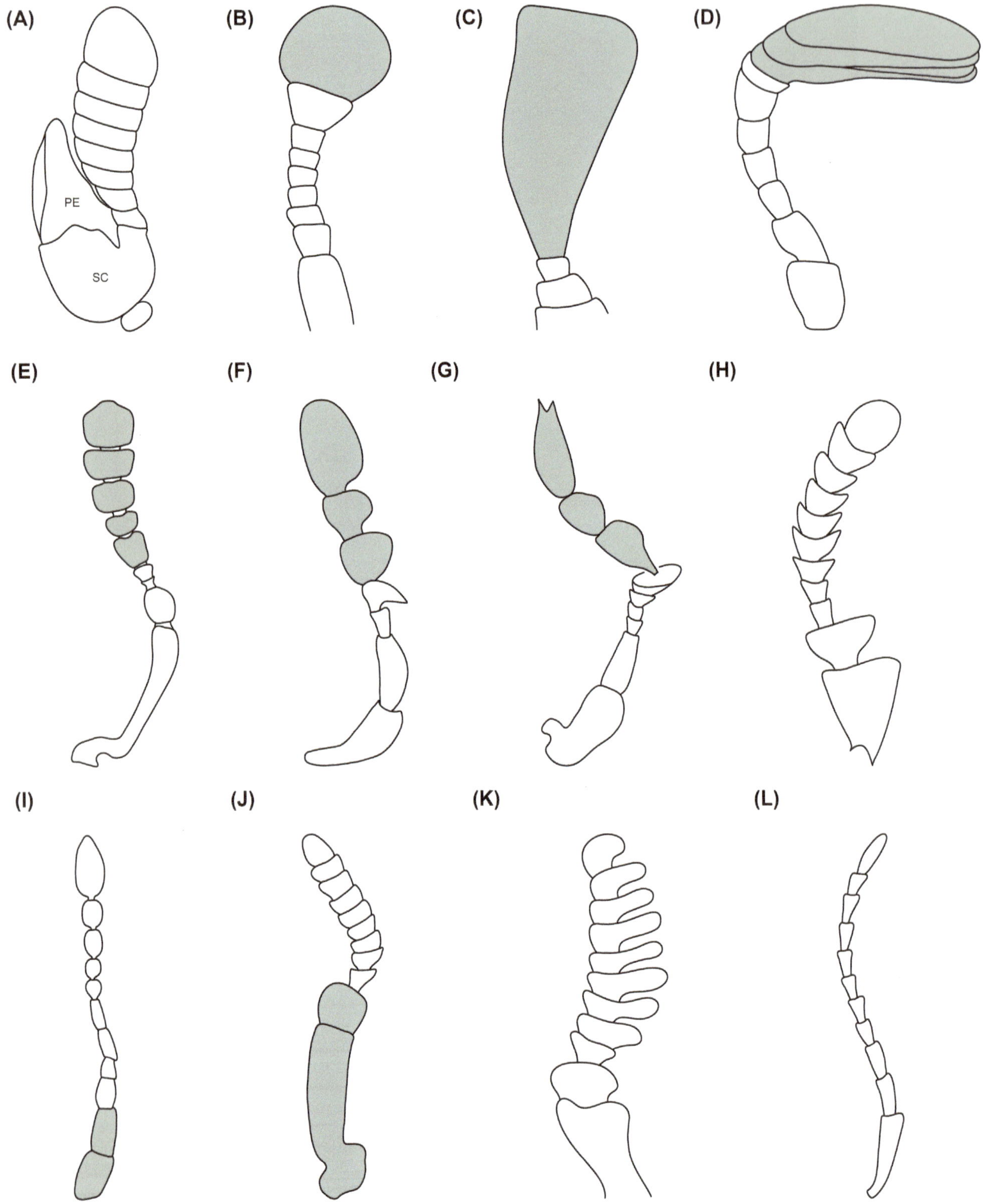

FIGURE 15.3 (Adults). Antenna: (A) Gyrinidae, (B) Torridincolidae, (C) Lepiceridae, (D) Scarabaeidae, (E) Hydraenidae, (F) Hydrochidae, (G) Hydrophilidae, (H) Heteroceridae, (I) Limnichidae, (J) Lutrochidae, (K) Dryopidae, and (L) Elmidae. Abbreviations: *PE* = pedicel; *SC* = scape.

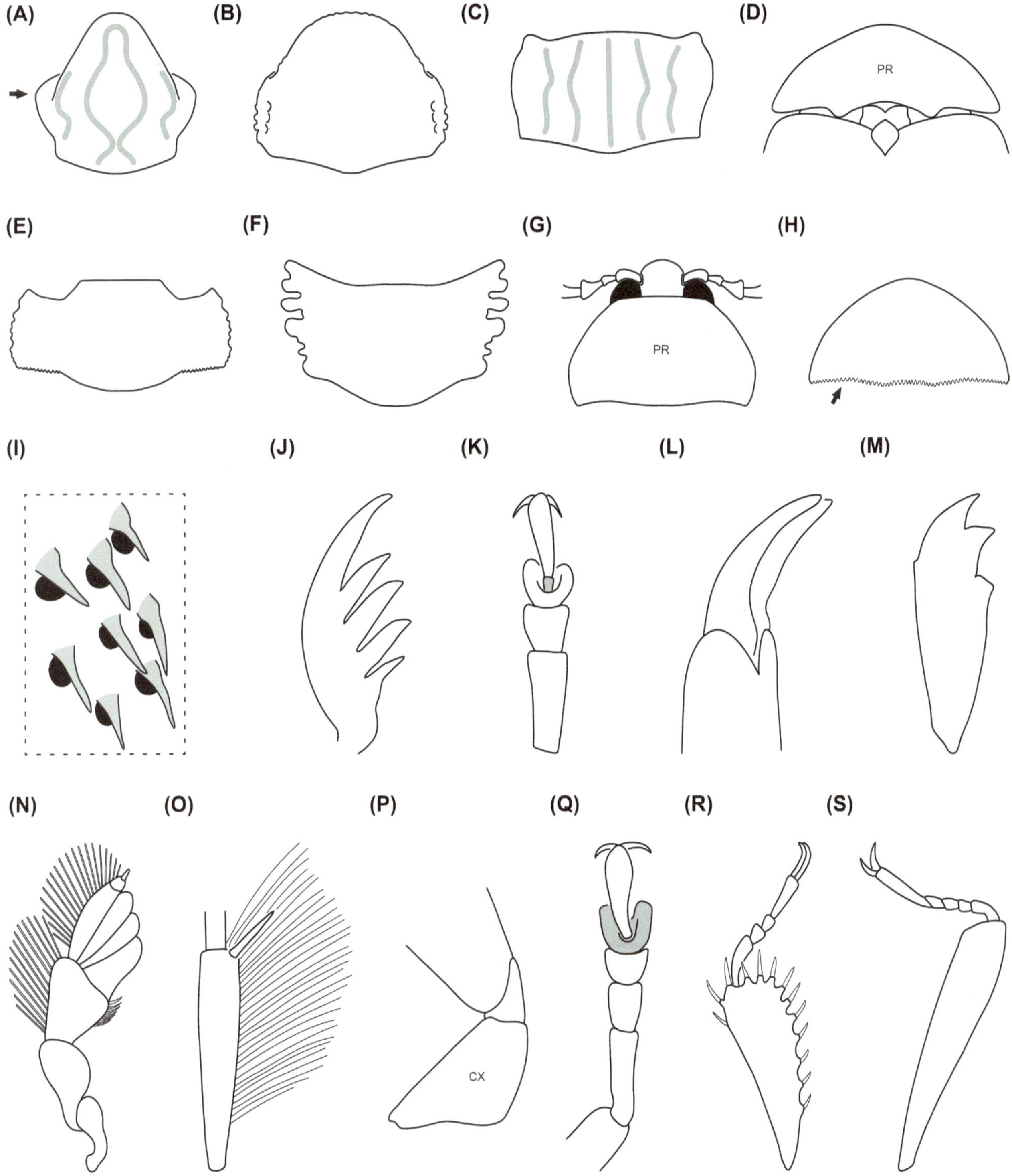

FIGURE 15.4 (Adults). Pronotum, dorsal view: (A) Epimetopidae, *arrow* indicates pronotal margin, (B) Georissidae, (C) Helophoridae, (D) Lampyridae, (E–F) Spercheidae, (G) Cneoglossidae, (H) Ptilodactylidae. Meruidae, (I) detail of elytrum; and (J) mesotarsal claw, lateral view. Chrysomelidae: (K) tarsomeres, dorsal view. Dytiscidae, (K) tarsal claws, posterior view. Scarabaeidae, (M) protibia, dorsal view. Gyrinidae, (N) metathoracic leg, ventral view. *Berosus* (Hydrophilidae), (O) mesothoracic leg, ventral view. Scirtidae, (P) prothoracic leg, anterior view; and (Q) tarsomeres, dorsal view. Heteroceridae: (R) prothoracic leg, posterior view. Elmidae: (S) mesothoracic leg, anterior view. Abbreviations: *CX* = coxa; *PR* = pronotum.

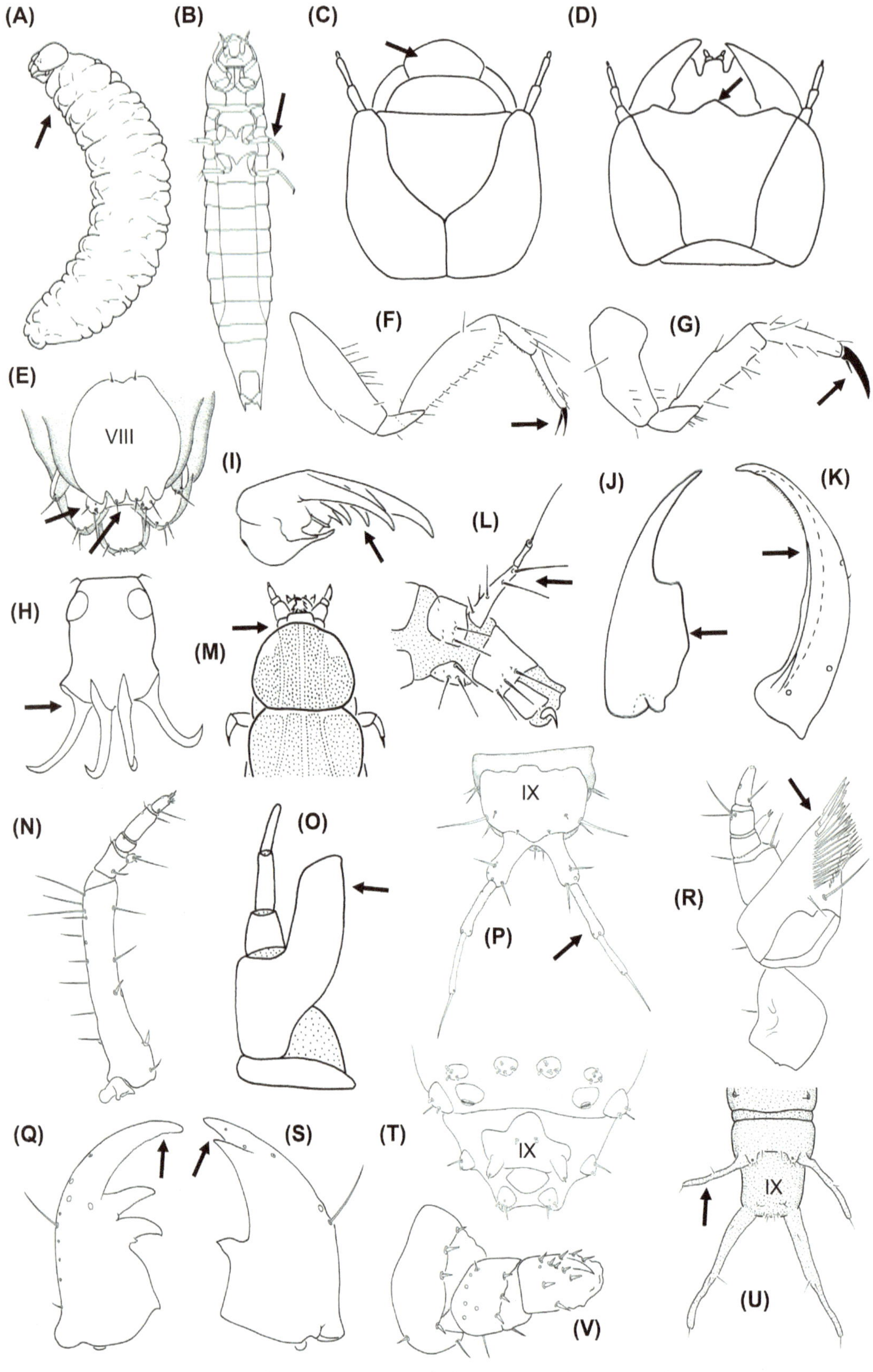

FIGURE 15.5 (Larvae). (A) Curculionidae, habitus, lateral view, *arrow* indicates absence of legs. (B) Elmidae, habitus, ventral view, *arrow* indicates thoracic legs. (C–D) Head, dorsal view, schematic drawings, *arrow* indicates labrum. (E) Hydrophilidae, abdominal segments VIII-IX, dorsal view; *arrow* indicates spiracular atrium. Leg, dorsal view: (F) Dytiscidae, (G) Hydrophilidae, *arrow* indicates pretarsal claws. (H) Gyrinidae, abdominal segment X, ventral view; *arrow* indicates hooks. (I) Meruidae, claws, lateral view. Mandible, dorsal view: (J) Noteridae, *arrow* indicates molar area, (K) Dytiscidae, *arrow* indicates sucking channel. (L) Hydraenidae, last abdominal segments, lateral view. (M) Lampyridae, head, dorsal view. Stipes, dorsal view: (N) Hydrophilidae, (O) Schematic drawing, *arrow* indicates mala. (P) Helophoridae, abdomen posterior region, dorsal view, *arrow* indicates urogomphi. (Q) Hydrophilidae, mandible, dorsal view, *arrow* indicates distal end. Spercheidae, (R) stipes, dorsal view; *arrow* indicates lacinia; (S) mandible, dorsal view, *arrow* indicates distal end. Last abdominal segments, ventral view: (T) Georissidae, (U) Epimetopidae, *arrow* indicates projection. (V) Georissidae, leg, dorsal view.

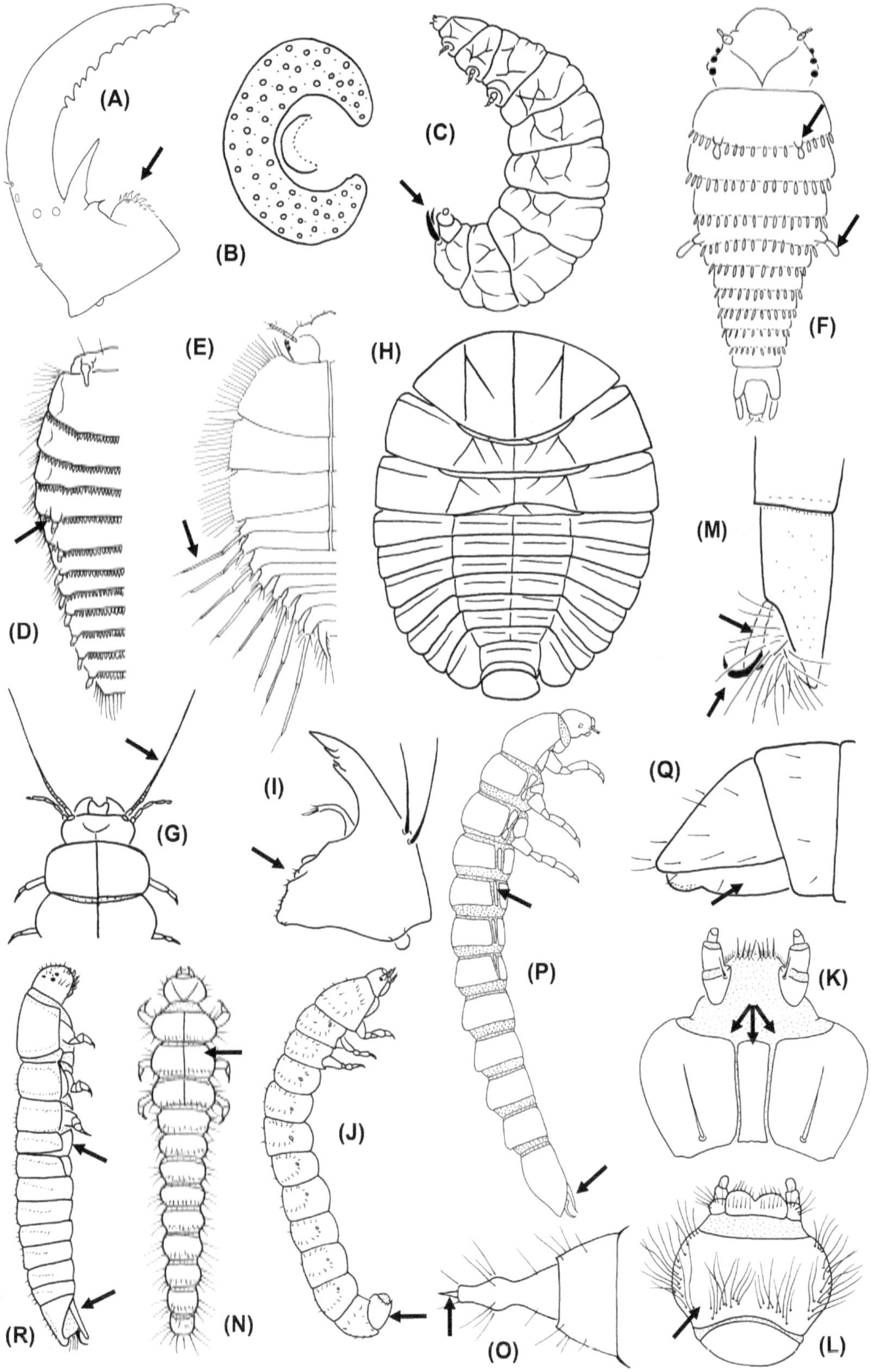

FIGURE 15.6 (Larvae). (A) Hydrochidae, mandible, dorsal view, *arrow* indicates pseudomolar area. (B) Scarabaeidae, spiracle, lateral view. (C) Chrysomelidae, habitus, lateral view, *arrow* indicates respiratory horns. Habitus, dorsal view: (D) Sphaeriusidae, *arrow* indicate abdominal processes; (E) Torridincolidae, *arrow* indicate abdominal processes; (F) Hydroscaphidae, *arrow* indicates digitiform process. (G) Scirtidae, head, pronotum, and mesonotum, dorsal view, *arrow* indicates antenna. (H) Psephenidae, habitus, dorsal view. (I) Hydraenidae, mandible, dorsal view, *arrow* indicates molar area. (J) Ptilodactylidae, habitus, lateral view, *arrow* indicates abdominal segment X. Labium, ventral view, *arrow* indicates mentum: (K) Ptilodactylidae, (L) Cneoglossidae. (M) Elmidae, last abdominal segments, lateral view, *arrow* indicates operculum. (N) Heteroceridae, habitus, dorsal view, *arrow* indicates mesothorax. (O) Lepiceridae, last abdominal segments, dorsal view, *arrow* indicates elongate process. (P) Elmidae, habitus, lateral view, *arrows* indicate pleural sclerites and operculum. (Q) Dryopidae, last abdominal segments, lateral view, *arrow* indicates operculum. (R) Lutrochidae, habitus, lateral view, *arrow* indicates operculum.

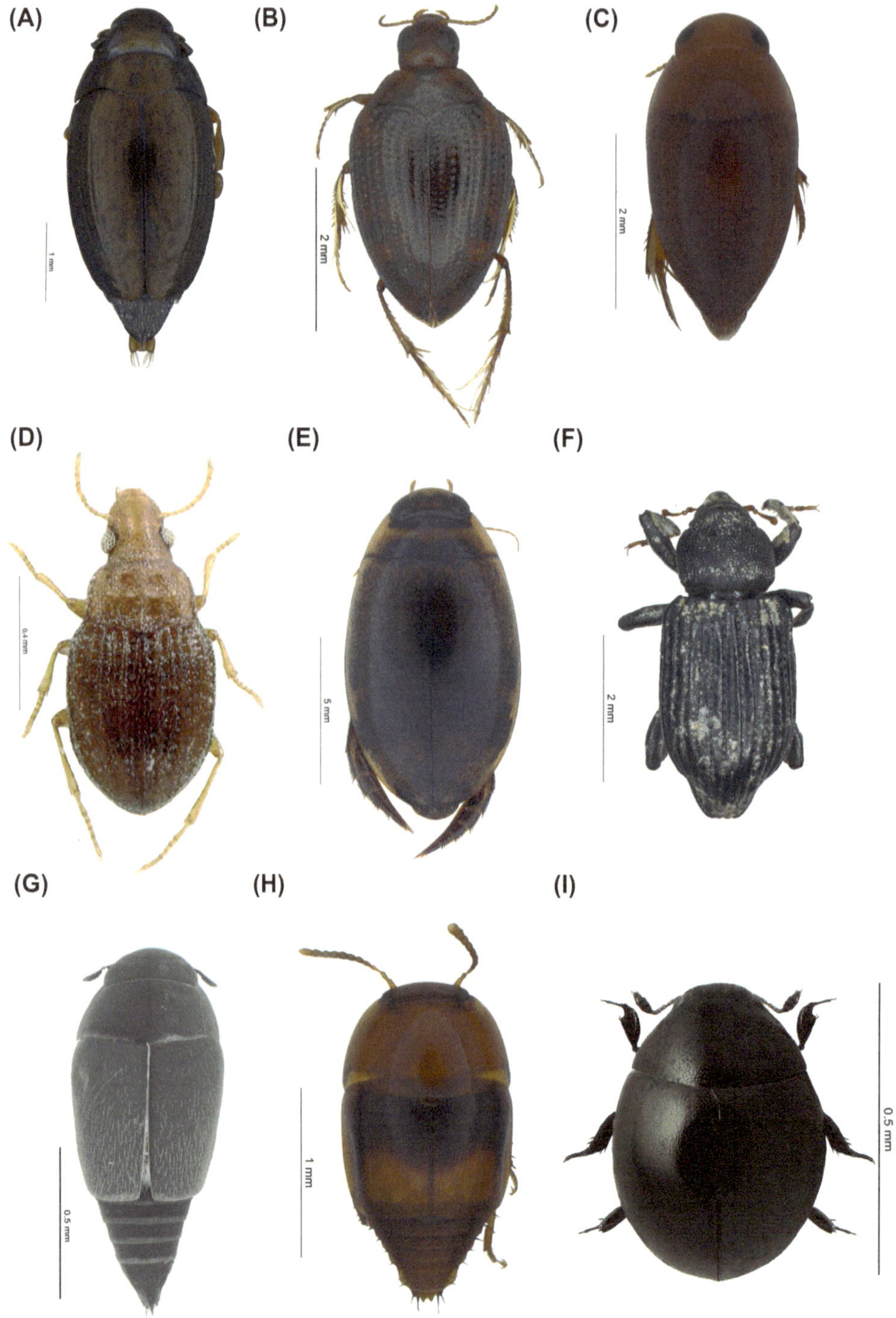

FIGURE 15.7 Adults, habitus, dorsal view. (A) Gyrinidae, (B) Haliplidae, (C) Noteridae, (D) Meruidae, (E) Dytiscidae, (F) Curculionidae, (G) Hydroscaphidae, (H) Staphylinidae, (I) Sphaeriusidae. *(D) After Beutel et al. (2006), with permission. (G) After Hall & Short (2010), with permission. (I) Lech Borowiec copyright.*

20(19) Antennae not as below; protibiae external margin lacking strong denticles (dentate and spinose in Heteroceridae, Fig. 15.4 R); pygidium not exposed; total body length generally <15 mm 21

20' Antennae short, laminate (i.e., distal articles transverse, projected mediad) (Fig. 15.3 D); protibiae external margin with few strong denticles (Fig. 15.4 M); pygidium exposed; total body length >15 mm .... Scarabaeidae

21(20) Pronotum projected forward, totally or partly covering head (Fig. 15.9 A); compound eyes large and prominent (Fig. 15.2 L); metacoxae lacking projections or grooves 22

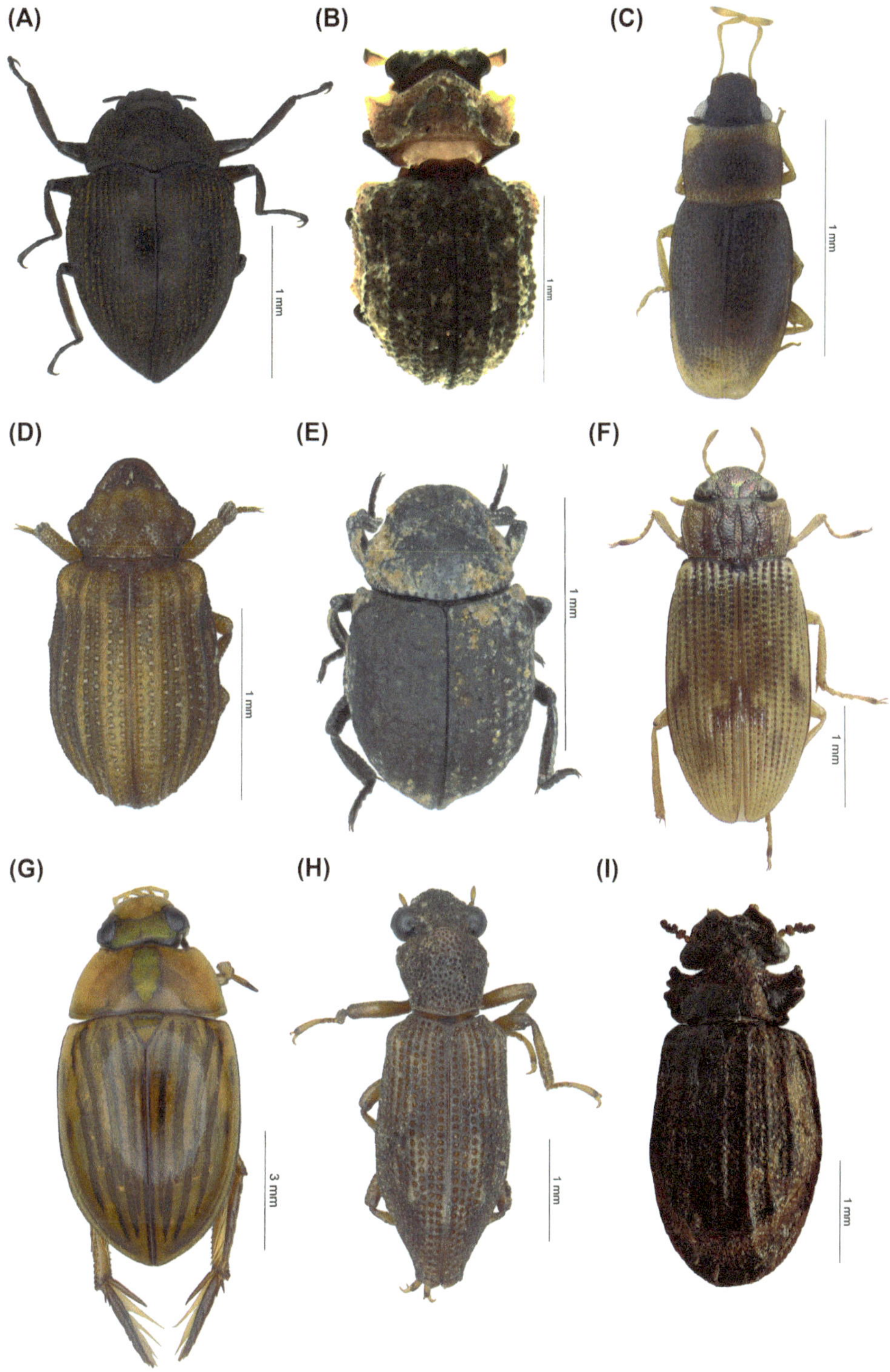

FIGURE 15.8 Adults, habitus, dorsal view: (A) Torridincolidae, (B) Lepiceridae, William Shepard copyright, (C) Hydraenidae, (D) Epimetopidae, (E) Georissidae, (F) Helophoridae, Robert Angus copyright (G) Hydrophilidae, (H) Hydrochidae, (I) Spercheidae.

21' Pronotum not projected forward, not covering head (except some Ptilodactylidae); compound eyes normally developed; metacoxae with projection that covers trochanter base (except some Scirtidae), sometimes further developed in a large groove that holds femur 23

22(21) Pronotum projected forward, completely covering head (Fig. 15.4 D); seven abdominal ventrites, may bear light-emitting organs .... Lampyridae

22' Pronotum projected forward, partly covering head (Fig. 15.4 G); five abdominal ventrites, first grooved for reception of metacoxae (Fig. 15.2 F); light-emitting organs absent; example of habitus in Fig. 15.9 A .... Cneoglossidae

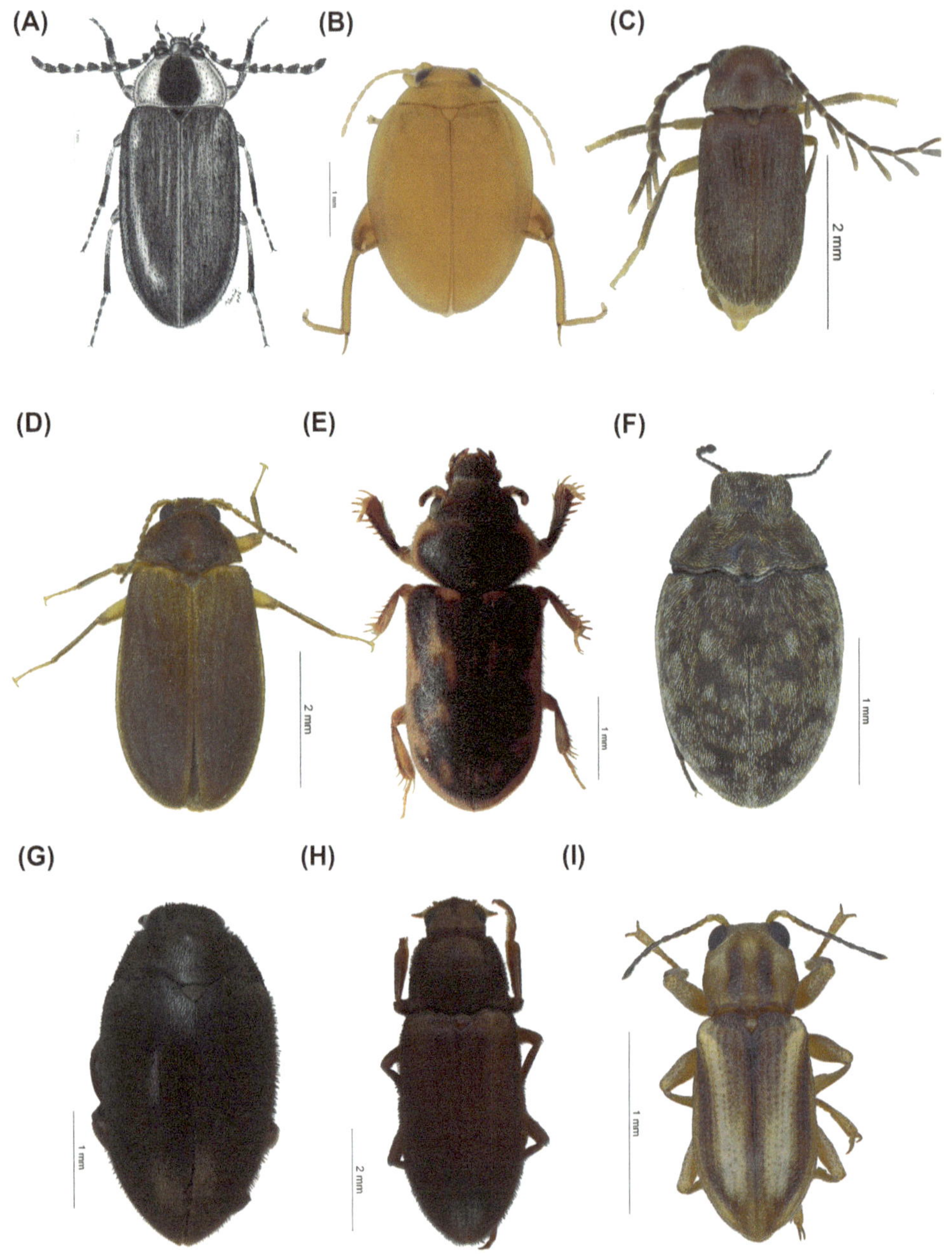

FIGURE 15.9 Adults, habitus, dorsal view: (A) Cneoglossidae, after Costa et al. (1999), (B) Scirtidae, (C) Ptilodactylidae, (D) Psephenidae, (E) Heteroceridae, (F) Limnichidae, (G) Lutrochidae, (H) Dryopidae, (I) Elmidae.

23(21) Procoxae shape variable, if conical, pronotum posterior margin crenulate (Fig. 15.4 H); tarsi pentamerous or tetramerous but never with fourth tarsomere bilobed 24

23' Procoxae more or less conical and prominent (Fig. 15.4 P); pronotum posterior margin not crenulate; tarsi pentamerous with fourth tarsomere bilobed (Fig. 15.4 Q); example of habitus in Fig. 15.9 B .... **Scirtidae (Chapter 15.7, p. 599)**

24(23) Procoxae more or less prominent and/or pronotum posterior margin crenulate (Fig. 15.4 H) 25

24' Procoxae transverse or rounded, not prominent; pronotum posterior margin not crenulate 26

25(24) Clypeus not protruding; labrum visible in anterior view; mandibles visible when closed; example of habitus in Fig. 15.9 C .... Ptilodactylidae

25' Clypeus protruding between antennae bases; labrum generally not visible in anterior view; mandibles not visible when closed; example of habitus in Fig. 15.9 D .... **Psephenidae (Chapter 15.6, p. 583)**

26(24) Pro- and mesotibiae narrow, not spinose (Fig. 15.4 S); antennae filiform or clubbed; tarsi generally pentamerous, last tarsomere longest (Fig. 15.4 S); claws generally large 27

FIGURE 15.10 Larvae, habitus: (A–G, (I), K) dorsal view; (H, J, L) lateral view: (A) Gyrinidae, (B) Noteridae, (C) Dytiscidae, (D) Hydrophilidae, (E) Torridincolidae, (F) Scirtidae, (G) Psephenidae, (H) Ptilodactylidae, (I) Cneoglossidae, after Costa et al. (1999), (J) Lutrochidae, (K) Elmidae, and (L) Dryopidae.

26' Pro- and mesotibiae broad, with external margin spinose (Fig. 15.4 R); antennae short and broad (Fig. 15.3 H); tarsi tetramerous, first and last tarsomeres longest (Fig. 15.4 R); claws slender; example of habitus in Fig. 15.9 E .... Heteroceridae

27(26) Metafemora retractile into a metacoxal groove (Fig. 15.2 G); mesocoxae strongly separated from each other, metacoxae closer to each other or contiguous (Fig. 15.2 G) 28

27' Metafemora not retractile into a metacoxal groove (Fig. 15.2 H); mesocoxae close to each other, if strongly separated from each other, metacoxae also separated (Fig. 15.2 H) 29

28(27) Antennomeres 1 and 2 combined representing less than 1/3 of total antennal length (Fig. 15.3 I); example of habitus in Fig. 15.9 F .... Limnichidae

28' Antennomeres 1 and 2 combined representing at least 1/3 of total antennal length (Fig. 15.3 J); example of habitus in Fig. 15.9 G .... Lutrochidae

29(27) Antennae short, last six to nine antennomeres forming a pectinate club (Fig. 15.3 K); example of habitus in Fig. 15.9 H .... **Dryopidae (Chapter 15.6, p. 583)**

29' Antennae long, slender, filiform, or clubbed, not pectinate (Fig. 15.3 L); example of habitus in Fig. 15.9 I .... **Elmidae (Chapter 15.6, p. 583)**

Coleoptera: Larvae

Modified from Archangelsky et al. (2009).

1 Thoracic legs absent or extremely reduced, difficult to see (at most with 1–2 min segments) (Fig. 15.5 A) 2

1' Legs with 3–6 segments (including pretarsus), easily seen (Fig. 15.5 B) 3

2(1) Body usually C-shaped (Fig. 15.5 A); legs absent; soft-bodied larvae usually lacking thoracic sclerites; head hypognathous; antennae short, composed of one antennomere; labrum free, not fused to clypeus (Fig. 15.5 C); abdomen lacking apical spiracular atrium .... Curculionidae

2' Body more or less elongated, never C-shaped; legs absent or reduced; larvae soft-bodied but with thoracic sclerites evident; head prognathous or hyperprognathous; three antennomeres; labrum fused to clypeus (Fig. 15.5 D); abdominal segments 8–9 forming a spiracular atrium (Fig. 15.5 E); subfamily Sphaeridiinae, associated with phytotelmata or riparian habitats .... **Hydrophilidae** (in part) **(Chapter 15.4, p. 561)**

3(1) Legs of six articles (including pretarsus) (Fig. 15.5 F); pretarsus usually with two claws (one claw in Haliplidae) 4

3' Legs of 3 or 5 articles; pretarsus with one claw (Fig. 15.5 G) 8

4(3) Abdomen apical segment without hooks; with or without tracheal gills 5

4' Abdomen apical segment (segment 10) bearing four strong hooks (Fig. 15.5 H); abdominal segments 1–9 with tracheal gills (Fig. 15.10 A); example of habitus in Fig. 15.10 A .... **Gyrinidae (Chapter 15.2, p. 527)**

5(4) Abdomen with 8–9 segments; pretarsus with two claws 6

5' Abdomen with 10 segments; pretarsus with one claw .... **Haliplidae (Chapter 15.2, p. 527)**

6(5) Abdomen with eight segments; claws simple, never pectinate 7

6' Abdomen with nine segments; claws pectinate (Fig. 15.5 I) .... Meruidae

7(6) Mandibles with well-developed molar area, without sucking channel (Fig. 15.5 J); example of habitus in Fig. 15.10 B .... **Noteridae (Chapter 15.2, p. 527)**

7' Mandibles falciform, lacking molar area (Fig. 15.5 K), with sucking channel (if absent, with strong denticles on medial margin); example of habitus in Fig. 15.10 C .... **Dytiscidae (Chapter 15.3, p. 539)**

8(3) Labrum fused to clypeus, forming a nasale or undivided clypeolabrum (Fig. 15.5 D) 9

8' Labrum free, not fused to clypeus (Fig. 15.5 C) 16

9(8) Body round to subcylindrical in cross-section, not strongly sclerotized, thoracic and abdominal sclerites less developed; head, at rest, not covered by pronotum; mandibles not channeled; urogomphi usually present (Fig. 15.5 L); abdominal segment 10, when present, not bearing retractile filaments 10

9' Body dorsoventrally flattened, strongly sclerotized, with well-developed thoracic and abdominal tergites; head, at rest, partially to totally covered by pronotum (Fig. 15.5 M); mandibles channeled; urogomphi absent; abdominal segment 10 bearing retractile filaments .... Lampyridae

10(9) Stipes lacking mala (Fig. 15.5 N) (present in Spercheidae); thoracic and abdominal spiracles biforous (in most cases the last abdominal pair is annular and concealed within a terminal spiracular atrium (Fig. 15.5 E)); urogomphi usually small, unarticulate (large and of three articles in Helophoridae); abdominal tergites usually strongly reduced 11

10' Stipes bearing a well-developed mala (Fig. 15.5 O); thoracic and abdominal spiracles annular; urogomphi well developed, of 1–3 articles (Fig. 15.5 L); abdominal tergites well developed .... Staphylinidae (in part)

11(10) Urogomphi short, unarticulate, difficult to see (Fig. 15.5 E) 12

11' Urogomphi long, of three articles (Fig. 15.5 P) .... Helophoridae

12(11) Stipes without lacinia (lacinia present but strongly reduced in Hydrochidae); mandible distal end single (Fig. 15.5 Q); each side of head usually with six stemmata (in some genera partially fused); gula absent (except in Hydrochidae) 13

12' Stipes with a well-developed lacinia (Fig. 15.5 R); mandible distal end bifid (Fig. 15.5 S); 5 stemmata on each side of head; gula present .... Spercheidae

13(12) Abdomen with nine complete segments (Figs. 15.5 T, U); segment 10 terminal, small but distinct; without spiracular atrium 14

13' Abdomen with eight complete segments; segments 9 and 10 modified and reduced, forming a spiracular atrium (Fig. 15.5 E) (except *Berosus*, without spiracular atrium but with long abdominal tracheal gills) 15

14(13) Legs reduced, of three articles (Fig. 15.5 V); terminal abdominal segments lacking fleshy projections (Fig. 15.5 T) ..... Georissidae

14' Legs normal, of five articles; abdominal segments 8 and 9 each with a pair of fleshy posterolateral projections (Fig. 15.5 U) ..... Epimetopidae

15(13) Mandibles with a spinose pseudomolar area at base (Fig. 15.6 A); gula well developed; antennae inserted close to anterolateral corner of head; lacinia small but distinct ..... Hydrochidae

15' Mandibles lacking pseudomolar area (Fig. 15.5 Q); gula reduced; antennae inserted further from anterolateral corner of head; lacinia absent; example of habitus in Fig. 15.10 D ..... **Hydrophilidae (in part) (Chapter 15.4, p. 561)**

16(8) Thorax and abdomen short and obese, lacking distinct sclerites 17

16' Thorax and abdomen variable in shape (cylindrical, dorsoventrally flattened or fusiform), but never obese, bearing distinct sclerites 18

17(16) Spiracles large, cribiform (Fig. 15.6 B); mandibles with a well-developed molar area; body length usually $\geq$10 mm ..... Scarabaeidae

17' Spiracles vestigial in abdominal segments 1–7, segment 8 with a pair of respiratory horns (Fig. 15.6 C); mandibles lacking molar area; larvae $<$10 mm ..... Chrysomelidae

18(16) At least some abdominal segments with elongate or digitiform processes (Figs. 15.6 D, E); antennae of two articles; larvae $\leq$2 mm 19

18' Abdominal segments lacking processes; antennae with three or more articles; size variable 21

19(18) Abdominal segments 1–8 with short or long processes present (Figs. 15.6 D, E) 20

19' Abdominal segments 1 and 8 with short digitiform processes present (Fig. 15.6 F) ..... **Hydroscaphidae (Chapter 15.1, p. 519)**

20(19) Abdominal processes long, articulated (Fig. 15.6 D); example of habitus in Fig. 15.10 E ..... **Torridincolidae (Chapter 15.1, p. 519)**

20' Abdominal processes short, not articulated (Fig. 15.6 D) ..... Sphaeriusidae

21(18) Antennae shorter than head, three antennomeres 22

21' Antennae longer than head, multiarticulate (Fig. 15.6 G); example of habitus in Fig. 15.10 F ..... **Scirtidae (Chapter 15.7, p. 599)**

22(21) Body shape variable, but never strongly flattened and wide; head and legs visible in dorsal view 23

22' Body strongly flattened and wide (Fig. 15.6 H); thoracic and abdominal segments broadly expanded, forming a shield; head and legs not visible in dorsal view; example of habitus in Fig. 15.10 G ..... **Psephenidae (Chapter 15.6, p. 583)**

23(22) Abdominal segment 9 bearing a pair of urogomphi (Fig. 15.5 L) (uni or bi-articulated) 24

23' Abdominal segment 9 lacking urogomphi 25

24(23) Mandibles falcate, without molar lobe ..... Staphylinidae (in part)

24' Mandibles short, triangular in shape, with well-developed molar lobe (Fig. 15.6 I) ..... **Hydraenidae (Chapter 15.5, p. 557)**

25(23) Abdomen 10 segmented, segment 10 usually not visible in dorsal view (Fig. 15.6 J) (some Ptilodactylidae with segment 10 partially fused to segment 9) 26

25' Abdomen clearly nine segmented 27

26(25) Mentum longitudinally subdivided in three parts (Fig. 15.6 K); example of habitus in Fig. 15.10 H ..... Ptilodactylidae

26' Mentum not subdivided (Fig. 15.6 L); example of habitus in Fig. 15.10 I ..... Cneoglossidae

27(25) Last abdominal segment without ventral operculum 28

27' Last abdominal segment bearing a ventral operculum which covers the anal region (Figs. 15.6 M, P–R) 29

28(27) Thorax much wider than abdomen; tergum IX short, rounded (Fig. 15.6 N) ..... Heteroceridae

28' Body parallel-sided, thorax not much wider than abdomen; tergum IX ending in an elongate process (Fig. 15.6 O) ..... Lepiceridae

29(27) Opercular chamber bearing two strong hooks and retractile gills (Figs. 15.6 M, P, R) 30

29' Opercular chamber without hooks or retractile gills (Fig. 15.6 Q) 31

30(29) Last abdominal segment rounded posteriorly; abdomen with pleural sclerites on first four segments or less (Fig. 15.6 R); example of habitus in Fig. 15.10 J ..... Lutrochidae

30' Last abdominal segment emarginated posteriorly (almost straight margin in some *Austrolimnius*); abdomen with pleural sclerites on at least first five segments (some *Luchoelmis* larvae, Fig. 15.6 P, with pleural sclerites on segments 1–4); example of habitus in Fig. 15.10 K ..... **Elmidae (Chapter 15.6, p. 583)**

31(29) Operculum reduced, may appear as a tenth abdominal segment; abdominal segments I–VII membranous ventrally ..... Limnichidae

31' Operculum well developed (Fig. 15.6 Q); abdominal segments 1–5 or 1–6 with small sclerotized sternites, segments 6–8 or 7–8 forming a ring; example of habitus in Fig. 15.10 L ..... **Dryopidae (Chapter 15.6, p. 583)**

ACKNOWLEDGMENTS

We are grateful to Andrew Short, Cleide Costa, Lech Borowiec, Robert Angus, Warren Steiner, and William Shepard for the images of some adult habitus. The work of Cesar J. Benetti was supported by CNPq and FAPEAM fellowships (processes 151461/2014-7 and 302031/2015-4). INPA/MCTI and CNPq provided financial support and laboratory infrastructure (PROTAX/CNPq - 440616/2015-8). The work of Mariano C. Michat was supported by CONICET under Grant PIP—2011—1087, by ANPCyT under Grant PICT—2014—0853, and by University of Buenos Aires under Grant UBACyT—20020150100170BA.

REFERENCES

Angus, R.B. 1992. Süsswasserfauna von Mitteleuropa, Bd. 20/10—12: Insecta, Coleoptera, Hydrophilidae, Helophorinae. Spectrum Akademischer Verlag, Heidelberg.

Arce- Pérez, R. & M.A. Morón. 2011. Sinopsis de los Hydrophiloidea de México (Coleoptera: Hydrophilidae, Helophoridae, Epimetopidae, Georissidae e Hydrochidae), con una clave para la identificación de los géneros. Revista Mexicana de Biodiversidad 82: 491—514.

Archangelsky, M. 1997. Studies on the biology, ecology, and systematics of the preimaginal stages of New World Hydrophiloidea (Coleoptera: Staphyliniformia). Bulletin of the Ohio Biological Survey (New Series) 12: 1—207.

Archangelsky, M., R.G. Beutel & A. Komarek. 2016. Chapter: Hydrophiloidea. Pages 231—272 *in*: R.G. Beutel & N.P. Kristensen (eds.), Handbook of Zoology, Arthropoda: Insecta Coleoptera, Beetles Volume 1: Morphology and Systematics, 2nd edition. Walter de Gruyter, Berlin, Germany, 684 pp.

Archangelsky, M., V. Manzo, M.C. Michat & P.L.M. Torres. 2009. Chapter 14: Coleoptera. Pages 411—468 *in*: E. Domínguez & H.R. Fernández (eds.), Macroinvertebrados bentónicos sudamericanos. Sistemática y biología. Fundación Miguel Lillo, Tucumán, Argentina.

Barbosa, F.F., A.S. Fernandes & L.G. Oliveira. 2013. Taxonomic key for the genera of Elmidae (Coleoptera, Byrrhoidea) occurring in Goiás State, Brazil, including new records and distributional notes. Revista Brasileira de Entomologia 57: 149—156.

Benetti, C.J., G.L. Fiorentin, J.A. Régil Cueto & U.G. Neiss. 2006. Chaves de identificação para famílias de coleópteros aquáticos ocorrentes no Rio Grande do Sul, Brasil. Neotropical Biology and Conservation 1: 24—28.

Benetti, C.J., J.A. Régil Cueto & G.L. Fiorentin. 2003. Gêneros de Hydradephaga (Coleoptera: Dytiscidae, Gyrinidae, Haliplidae, Noteridae) citados para o Brasil, com chaves para identificação. Biota Neotropica 3: 1—20.

Beutel, R.G. & G.B. Raffaini. 2003. First record of Spaeriusidae for Argentina (Coleoptera: Myxophaga). Koleopterologische Rundchau 73: 1—6.

Beutel, R.G. & R.A.B. Leschen. 2016. Handbook of Zoology, Volume 4/38: Coleoptera, Beetles, Volume 1: Morphology and Systematics (Archostemata, Adephaga, Myxophaga, Polyphaga partim). De Gruyter, Berlin, Germany. 684 pp.

Bouchard, P., Y. Bousquet, A.E. Davies, M.A. Alonso-Zarazaga, J.F. Lawrence, C.H.C. Lyal, A.F. Newton & C.A.M. Reid, M. Schmitt, S.A. Ślipiński, A.B.T. Smith. 2011. Family-group names in Coleoptera (Insecta). Zookeys 88: 1—972.

Bouchard, R.W. 2004. Chapter 12: Coleoptera (Aquatic beetles). Pages 141—158 *in*: R.W. Bouchard (ed.), Guide to aquatic invertebrates of the Upper Midwest. Water Resources Center, University of Minnesota, St. Paul, MN.

Brown, H.P. 1970. A key to the dryopid genera of the New World. Entomological News 81: 171—175.

Brown, H.P. 1981. Key to the world genera of Larainae (Coleoptera, Dryopoidea, Elmidae) with descriptions of the new genera from Hispaniola, Colombia, Australia and New Guinea. The Pan-Pacific Entomologist 57: 76—104.

Ciegler, J.C. 2003. Water Beetles of South Carolina (Coleoptera: Gyrinidae, Haliplidae, Noteridae, Dytiscidae, Hydrophilidae, Hydraenidae, Scirtidae, Elmidae, Dryopidae, Limnichidae, Heteroceridae, Psephenidae, Ptilodactylidae, and Chelonariidae). Biota of South Carolina, Vol. 3. Clemson University, Clemson, SC. 207 pp.

Costa, C., S.A. Vanin & S. Ide. 2005. Chapter: Cneoglossidae Guérin-Méneville. Pages 533—536 *in*: N.P. Kristensen & R.G. Beutel (eds.), Handbook of Zoology, Volume IV: Coleoptera, beetles Volume 1: Morphology and Systematics (Archostemata, Adephaga, Myxophaga, Polyphaga partim). De Gruyter, Berlin, Germany.

CSIRO 2001. Australian Aquatic Invertebrates. Available http://keys.lucidcentral.org/keys/lwrrdc/public/Aquatics/.

Epler, J.H. 2010. The water beetles of Florida. An identification manual for the families Chrysomelidae, Curculionidae, Dryopidae, Dytiscidae, Elmidae, Gyrinidae, Haliplidae, Helophoridae, Hydraenidae, Hydrochidae, Hydrophilidae, Noteridae, Psephenidae, Ptilodactylidae, and Scirtidae. Bureau of Laboratories, Florida Department of Environmental Protection, Tallahassee, FL. 414 pp.

Eriksen, C.H., G.A. Lamberti & V.H. Resh. 1996. Chapter: Aquatic insect respiration. Pages 29—40 *in*: R.W. Merritt & K.W. Cummins (eds.), An introduction to the aquatic insects of North America, 3rd ed. Kendall/Hunt Publ., Dubuque, Iowa.

Ferreira-Jr., N., B.H.L. Sampaio, A.S. Fernandes, B. Clarkson, R.B. Braga, M.I.S. Passos & A.D. Santos. 2014. Chapter 21: Ordem Coleoptera (coleos = invólucro, estojo; peteron = asa). Pages 349—375 *in*: N. Hamada, J.L. Nessimian & R.B. Querino (eds.), Insetos aquáticos na Amazônia brasileira: taxonomia, biologia e ecologia. Editora do Inpa, Manaus, Brazil.

Foster, G.N. & D.T. Bilton. 2014. Chapter: The conservation of predaceous diving beetles: knowns, unknowns and anecdotes. Pages 437—462 *in*: D.A. Yee (ed.), Ecology, Systematics, and Natural History of Predaceous Diving Beetles (Coleoptera: Dytiscidae). Springer, New York.

Hansen, M. 1991a. The hydrophiloid beetles. Phylogeny, classification and a revision of the genera. Biologiske Skrifter 40: 1—368.

Hansen, M. 1991b. A review of the genera of the beetle family Hydraenidae (Coleoptera). Steenstrupia 17: 1—52.

Hansen, M. 1999. World Catalogue of Insects 2: Hydrophiloidea (Coleoptera). Apollo Books, Stenstrup, Denmark. 416 pp.

Hernando, C. & I. Ribera. 2016. Chapter 19.5: Limnichidae. Pages 605—611 *in*: R.G. Beutel & R.A.B. Leschen (eds.), Handbook of Zoology, Volume 4/38: Coleoptera, Beetles, Volume 1: Morphology and Systematics (Archostemata, Adephaga, Myxophaga, Polyphaga partim). De Gruyter, Berlin, Germany.

Jäch, M.A. 1998. Annotated check list of aquatic and riparian/littoral beetle families of the world (Coleoptera). Water Beetles of China 2: 25—42.

Jäch, 2002 Jäch M.A. 2002. First record of Epimetopidae in Laos (Coleoptera: Epimetopidae). Koleopterologische Rundschau 72: 161—164.

Jäch, M.A. & M. Balke. 2008. Global diversity of water beetles (Coleoptera) in freshwater. Hydrobiologia 595: 419–442.

Ji, L. & M.A. Jäch. 1998. Chapter: Epimetopidae: Synopsis of the genus Eumetopus Balfour-Browne (Coleoptera). Pages 195–205 *in*: M.A. Jäch & J. JI (eds.), Water Beetles of China. Volume II. Zoologisch-Botanische Gesellschaft in Österreich & Wiener Coleopterenverein, Wien.

Kirejtshuk, A.G. 2009. A new genus and species of Sphaeriusidae (Coleoptera, Myxophaga) from Lower Cretaceous Burmese amber. Denisia 26: 99–102.

Larson, D.J., Y. Alarie & R.E. Roughley. 2000. Predaceous diving beetles (Coleoptera: Dytiscidae) of the Nearctic region, with emphasis on the fauna of Canada and Alaska. NRC Research Press, Ottawa, Canada. 982 pp.

Libonatti, M.L., M.C. Michat & P.L.M. Torres. 2011. Key to the subfamilies, tribes and genera of adult Dytiscidae of Argentina (Coleoptera: Adephaga). Revista de la Sociedad Entomológica Argentina 70: 317–336.

Manzo, V. & M. Archangelsky. 2008. A key to the known larvae of South American Elmidae (Coleoptera: Byrrhoidea), with a description of the mature larva of Macrelmis saltensis Manzo. Annales de Limnologie – International Journal of Limnology 44: 63–74.

Manzo, V. 2005. Key to the South America genera of Elmidae (Insecta: Coleoptera) with distributional data. Studies on Neotropical Fauna and Environment 40: 201–208.

Michat, M.C. & M. Archangelsky. 2014. Chapter: Gyrinidae. Pages 455–465 *in*: S. Roig-Juñent, L.E. Claps & J.J. Morrone (eds.), Biodiversidad de Artrópodos Argentinos, vol. 3. Facultad de Ciencias Naturales, Universidad Nacional de Tucumán, San Miguel de Tucumán, Argentina.

Michat, M.C., M. Archangelsky & A.O. Bachmann. 2008. Generic keys for the identification of larval Dytiscidae from Argentina (Coleoptera: Adephaga). Revista de la Sociedad Entomológica Argentina 67: 17–36.

Miller, K.B. & J. Bergsten. 2016. Diving Beetles of the World. Systematics and Biology of the Dytiscidae. Johns Hopkins University Press, Baltimore, MD. 336 pp.

Miller, K.B. 2009. On the systematics of Noteridae (Coleoptera: Adephaga: Hydradephaga): Phylogeny, description of a new tribe, genus and species, and survey of female genital morphology. Systematics and Biodiversity 7: 191–214.

Passos, M.I.S., J.L. Nessimian & N. Ferreira Junior. 2007. Chaves para identificação dos gêneros de Elmidae (Coleoptera) ocorrentes no Estado do Rio de Janeiro, Brasil. Revista Brasileira de Entomologia 51: 42–53.

Pederzani, F. 1995. Keys to the identification of the genera and subgenera of adults Dytiscidae (sensu lato) of the world (Coleoptera Dytiscidae). Atti della Accademia Roveretana degli Agiati 244, Ser. VII, 5–83.

Perissinotto, R., M.S. Bird & D.T. Bilton. 2016. Predaceous water beetles (Coleoptera, Hydradephaga) of the Lake St Lucia system, South Africa: biodiversity, community ecology and conservation implications. ZooKeys 595: 85–135.

Perkins, P.D. 1980. Aquatic beetles of the family Hydraenidae in the Western Hemisphere: classification, biogeography and inferred phylogeny (Insecta: Coleoptera). Quaestiones Entomologicae 16: 3–554.

Perkins, P.D. 2012. A revision of Epimetopus Lacordaire, the New World hooded shore beetles (Coleoptera: Epimetopidae). Zootaxa 3531: 1–95.

Reichardt, H. 1973. A critical study of the suborder Myxophaga, with a taxonomic revision of the Brazilian Torridincolidae and Hydroscaphidae (Coleoptera). Arquivos de Zoologia 24: 73–162.

Rogers, D.C. & J.H. Thorp. 2015. Chapter 3: Collecting, preserving and culturing invertebrates. Pages 57–62 *in*: J.H. Thorp & D.C. Rogers (eds.), Freshwater Invertebrates, 4th edition, Volume I, Ecology and General Biology. Academic Press, Academic Press, San Diego, CA.

Roughley, R.E. & D.J. Larson. 1991. Aquatic Coleoptera of Springs in Canada. Memoirs of the Entomological Society of Canada 123: 125–140.

Segura, M.O., F. Valente-Neto & A.A. Fonseca-Gessner 2011a. Chave de famílias de Coleoptera aquáticos (Insecta) do Estado de São Paulo, Brasil. Biota Neotropica 11: 393–412.

Segura, M.O., F. Valente-Neto & A.A. Fonseca-Gessner 2011b. Elmidae (Coleoptera, Byrrhoidea) larvae in the state of São Paulo, Brazil: Identification key, new records and distribution. ZooKeys 151: 53–74.

Short, A.E.Z. & M. Fikacek. 2011. World catalogue of the Hydrophiloidea (Coleoptera): additions and corrections II (2006–2010). Acta Entomologica Musei Nationalis Pragae 51: 83–122.

Short, A.E.Z., L.J. Joly, M. García, A.L. Wild, D.D. Bloom & D.R. Maddison. 2015. Molecular phylogeny of the Hydroscaphidae (Coleoptera: Myxophaga) with description of a remarkable new lineage from the Guiana Shield. Systematic Entomology 40: 214–229.

Skale, A. & M.A. Jäch. 2003. Chapter: A new species of Eumetopus Balfour-Browne, 1949 from Nepal (Insecta: Coleoptera: Epimetopidae). Pages 195–196 *in*: M. Hartmann & H. Baumbach (eds.), Biodiversität und Naturausstattung im Himalaya. Verein der Freunde & Förderer des Naturkundemuseums Erfurt, Erfurt.

Smetana, A. 1985. Revision of the subfamily Helophorinae of the Nearctic Region (Coleoptera: Hydrophilidae). Memoirs of the Entomological Society of Canada 131: 1–154.

Spangler, P.J., C.L. Staines, P.M. Spangler & S.L. Staines. 2001. A checklist of the Limnichidae and the Lutrochidae (Coleoptera) of the world. Insecta Mundi 15: 151–165.

Torres, P.L.M., M.C. Michat, M.L. Libonatti, L.A. Fernández, A. Oliva & A.O. Bachmann. 2012. Aquatic Coleoptera from Mburucuyá National Park (Corrientes Province, Argentina). Revista de la Sociedad Entomológica Argentina 71: 57–71.

White, D.S. & R.E. Roughley. 2008. Chapter 20: Aquatic Coleoptera. Pages 571–671 *in*: R.W. Merritt, K.W. Cummins & M.B. Berg (eds.), An Introduction to the Aquatic Insects of North America. Fourth Edition. Kendall/Hunt Publishing Co., Dubuque, IA. 1158 pp.

Chapter 15.1

Families Hydroscaphidae and Torridincolidae

Brunno H.L. Sampaio

Laboratório de Entomologia, Departamento de Zoologia, Instituto de Biologia, Universidade Federal do Rio de Janeiro, Rio de Janeiro, Rio de Janeiro, Brazil

Andrew E.Z. Short

Department of Ecology & Evolutionary Biology, Division of Entomology, Biology Institute, University of Kansas, Lawrence, KS, United States

Chapter Outline

INTRODUCTION

Hydroscaphidae and Torridincolidae belong to the suborder Myxophaga which is a small, rarely encountered suborder that is distributed worldwide except in Antarctica. The lineage consists of four families, all of which occur in the Neotropics: Lepiceridae, Hydroscaphidae, Sphaeriusidae, and Torridincolidae. Only two (Sphaeriusidae) from more than 80 species of Myxophaga are apparently not associated with aquatic environments (Beutel & Arce-Pérez, 2005). In this chapter, we will focus on the families Hydroscaphidae and Torridincolidae.

Hydroscaphidae (Skiff Beetles) have a broad but very patchy distribution in the Afrotropical, Nearctic, Neotropical, Oriental, and Palearctic regions. All four described genera are known from the Neotropics, and all but the widespread genus *Hydroscapha* are endemic to the region. Hydroscaphids are strongly associated with lotic habitats and may be found either in the gravel areas of streams or in nearby seepages.

Torridincolidae (Torrent Beetles) is the most diverse family and is, distributed in Africa, Asia, and South America. The greatest diversity is recorded in South America, mainly in Brazil, but there are also undescribed taxa in Venezuela (Short et al., 2015), Suriname and Guyana (Short, 2013, Short et al., 2016), and Paraguay (Shepard et al., 2013). Species are restricted to the hygropetric habitats such as seepages and along waterfalls, or in shallow streams and rivers flowing over rock. Larvae, pupae, and adults co-occur in the same habitat.

Systematics

With about 100 described species, Myxophaga is a very species-poor lineage of Coleoptera. It contains two extant superfamilies, the Lepiceroidea and Sphaeriusoidea. Lepiceroidea includes only Lepiceridae, while Sphaeriusoidea includes the three remaining families.

Hydroscaphidae has about 30 described species worldwide (Short et al., 2015) distributed among four genera: *Confossa*, *Scaphydra*, *Yara*, and *Hydroscapha*. The first three are endemic to the Neotropical region, while the latter is widespread with a single described species on Neotropics (*Hydroscapha perijaensis* Hall & Short) and undescribed species from Ecuador and Venezuela (Short et al., 2015).

Torridincolidae has almost 40 species in seven genera found mainly in the southern hemisphere. To date, three genera (all endemic) and 29 species have been recorded from the Neotropical region: *Claudiella*, *Iapir*, and *Ytu*. All currently described species are known from Brazil (Benetti & Hamada, 2016), but many undescribed taxa are recorded from Paraguay (Shepard et al., 2013) and Venezuela (Short et al., 2015).

Thorp and Covich's Freshwater Invertebrates, Fourth Edition. https://doi.org/10.1016/B978-0-12-804223-6.00022-6

In this chapter the family taxonomy follows Beutel (2005).

LIMITATIONS

The Myxophaga is very poorly known, including in the Neotropics. Their small size, cryptic habitats, and failure to be attracted to lights or other common mass-collecting techniques mean they are rarely encountered. The currently described diversity is likely significantly underestimated, including at the generic level.

The Neotropical fauna is still poorly known. The majority of the species descriptions are brief and lacking some details. The distribution of each species is in general restricted to type locality, though this is possibly due to their infrequent collection or study. Identification keys to genus as well as the species of *Yara* and *Confossa* are available (Reichardt, 1973; Short et al., 2010; Short et al., 2015). Molecular data is available for all genera of hydroscaphids, and relationships among the genera were recently studied (Short et al., 2015).

Torridincolidae is the most diverse family in the Neotropics, with three genera endemic to the region: *Iapir* and *Ytu* which have available keys to identification, and *Claudiella*, a thus far monotypic genus. Some species have recently been extensively described, but the remaining taxa are only briefly described. There are some phylogenetic hypotheses on the relationships among genera, based on larval morphology (Beutel et al., 1999) and larval and adult morphology (Beutel, 1999) on phylogenies of Myxophagan taxa. Some molecular data are available for a few taxa. Many new species have been collected in South America in the last decades and remain to be described (Shepard et al., 2013; Short et al., 2015; B.H.L. Sampaio, unpublished data).

TERMINOLOGY AND MORPHOLOGY

Hydroscaphidae, like other Myxophaga, are very small (0.9–2.0 mm), but are easily identified by the fusiform body shape, with short and truncate elytra on adults, exposing the terminal abdominal tergites. Compound eyes protrude only in *Confossa* species, antennae are short and clavate, with five antennomeres in *Scaphydra* and eight in other genera, and all tarsi contain three articles. Larvae are broad anteriorly, strongly narrowing posteriorly, with antennae of two articles, long lateral contact setae on tergites, and balloon-like spiracular gills (Beutel et al., 1999).

Torridincolidae adults are slightly longer than other the three families (1.1–2.7 mm), dark colored, usually black or brown, some of which bear iridescent scales. Surface smooth in some genera (*Delevea* and *Satonius*) or finely rugose, punctuate, and even carinate in other taxa. Short antennae with 11 (*Delevea* and *Satonius*) or nine antennomeres, tarsi pentamerous with a pair of claws, abdomen with four (*Claudiella* and *Iapir*) of five ventrites. Some species present sexual dimorphism on semi-lunar impression on ventrites (*Iapir*) or metallic sheen (*Ytu*). Larvae are usually fusiform and flattened, except from *Delevea*, which is oblong and subcylindrical (Beutel et al., 1999). Three stemmata on each side are found in *Claudiella*, *Iapir*, *Torridincola*, and *Ytu*, while the other genera present four stemmata on each side. All genera present antennae two-articulated. Tergites without contact setae only in *Iapir*, *Claudiella*, and *Torridincola*. Spiracular gills are segmented in all genera, inserted dorsally in *Incoltorrida* (B.H.L. Sampaio unpublished) and *Torridincola* and laterally in the other genera (Beutel et al., 1999).

MATERIAL PREPARATION AND PRESERVATION

Methods for collection and preservation of myxophagan beetles are similar to those used on other aquatic Coleoptera. Because of their minute size and specific habitat, these beetles are most efficiently collected manually in their habitat. In some cases, it is beneficial to scrub and collect the algae or debris from a seepage and then to examine it using a microscope in the lab. Special attention should be given to hygropetric environments and moist substrate near freshwater systems. Specimens may be either mounted and pinned or preserved in ethanol.

KEYS TO SUBORDER MYXOPHAGA

Keys to families are provided in Chapter 15.0.

Hydroscaphidae: Genera: Adults

1 Eyes not protruding beyond head outline (Fig. 15.1.2); elytra without dorsal sculpturing (Figs. 15.1.3 A, B); abdominal segments V–VII posterior margin without notches, or crenulate, and never bearing setal tufts (Fig. 15.1.2)........ 2

1' Eyes protruding laterally beyond head outline (Figs. 15.1.1 A, B); elytra dorsal sculpturing in the form of longitudinal carinae to coarse granulations (Figs 15.1.1 A, C); abdominal segments V–VII posterior margin lined with notches, which bear setal tufts (Figs. 15.1.1 A, D) .... *Confossa*

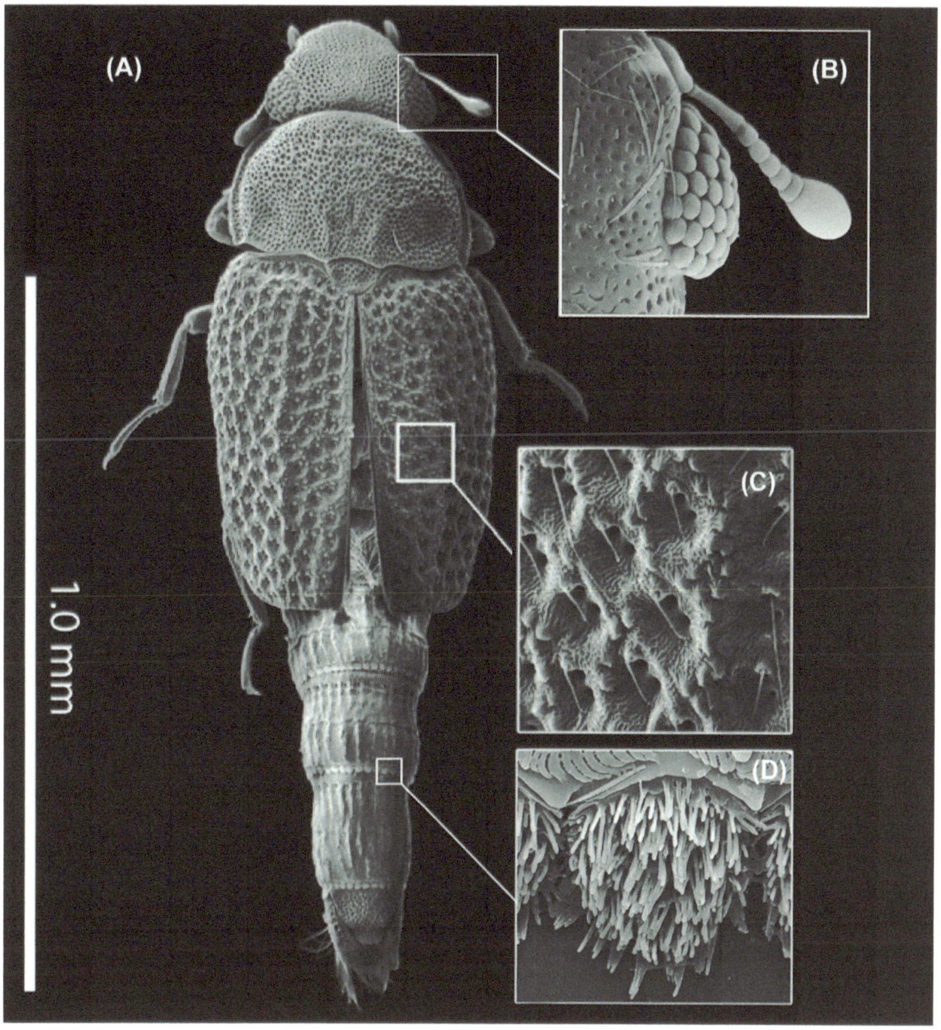

FIGURE 15.1.1 *Confossa sculptura* Short, Joly, Garcia & Maddison (Hydroscaphidae): (A) SEMs of dorsal habitus; (B) details of eye; (C) elytron sculpture; and (D) tuft-like setae on abdominal notches.

2(1) Antennae with eight articles; abdominal segments V–VII posterior margins smooth 3

2' Antennae with five articles (Fig. 15.1.2 C); abdominal segments V–VII posterior margins finely crenulate (Fig. 15.1.2 C).... *Scaphydra*

3(2) Metacoxae not covering hind femora (Fig. 15.1.2 B); metatrochanter posterior margin with a row of spiniform setae (Fig. 15.1.2 B); tergite IV without tomentum; dorsum densely setose (Fig. 15.1.3); mentum and submentum not fused (Fig. 15.1.2 B) .... *Yara*

3' Metacoxae expanded into large plates, covering hind femora (Fig. 15.1.2 A); metatrochanter posterior margin without a row of spiniform setae; tergite IV with tomentum; dorsum at most sparsely setose; mentum and submentum fused........ *Hydroscapha*

Torridincolidae: Genera: Adults

1 Pronotum posterior margin with two small medial incisions (Figs. 15.1.4 A, B); abdomen with four ventrites, last one with a semilunar impression (Fig. 15.1.5 A) 2

1' Pronotum posterior margin with a single small medial incision (Figs. 15.1.4 C, D); abdomen with five ventrites, last ventrite without impressions (Fig. 15.1.5 B) .... *Ytu*

2(1) Metatrochanter posterior margin with conspicuous dentiform projection (Fig. 15.1.6 A)........ *Claudiella*

2' Metatrochanter posterior margin without any projection or with a spiniform projection (Fig. 15.1.6 B) (except *Iapir quadridentatus* Pinto, Fonseca & Hamada) which bears a conspicuous dentiform projection)........ *Iapir*

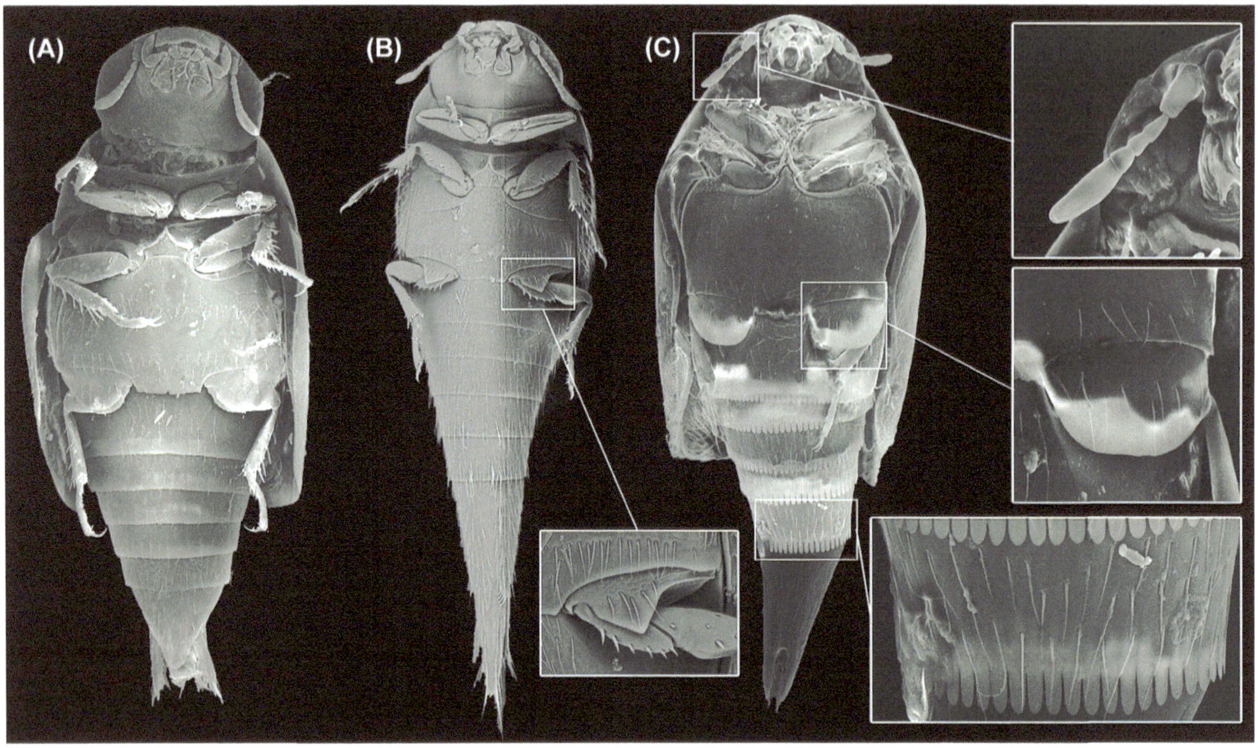

FIGURE 15.1.2 Hydroscaphidae: *Hydroscapha perijaensis* Hall & Short (Hydroscaphidae): (A) SEMs of ventral habitus; (B) *Yara maculata* Short, Joly & Garcia, detail of spine-bearing trochanter; and (C) *Scaphydra* sp., details of five-articulated antennae, femoral coxal plate, and denticulate posterior abdominal segment margins.

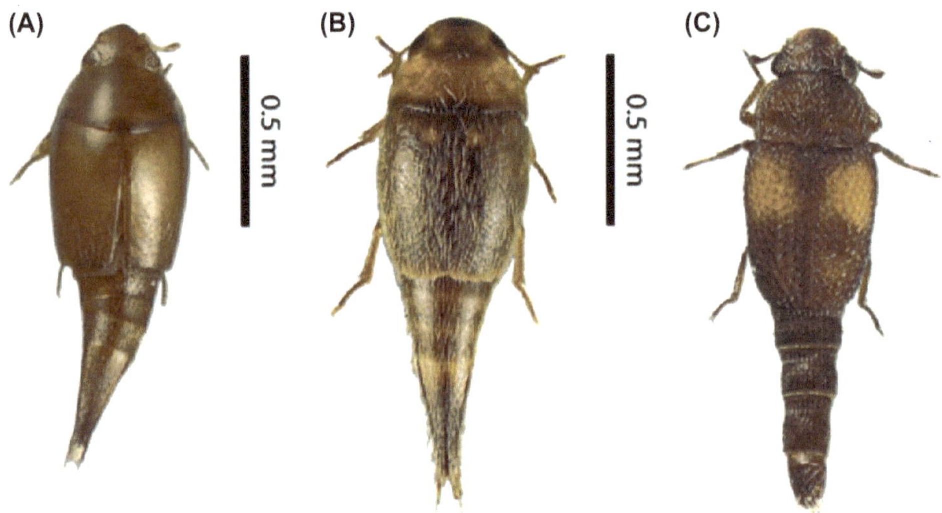

FIGURE 15.1.3 Hydroscaphidae (Coleoptera) habitus, in dorsal view: (A) *Scaphydra* sp.; (B) *Yara maculata* Short, Joly & Garcia; and (C) *Confossa sculptura* Short, Joly, Garcia & Maddison.

Torridincolidae: Genera: Larvae

Claudiella and *Iapir* immatures lack diagnostic characteristics (Vanin & Costa, 2001).

1	Thoracic segment lateral margins glabrous; tracheobranchia inserted sublaterally on anterior segments of abdomen (Figs. 15.1.7 A, B)	*Claudiella/Iapir*
1'	Thoracic segment lateral margins bearing contact setae; tracheobranchia inserted laterally on small projections on abdominal segments (Fig. 15.1.7 C)	*Ytu*

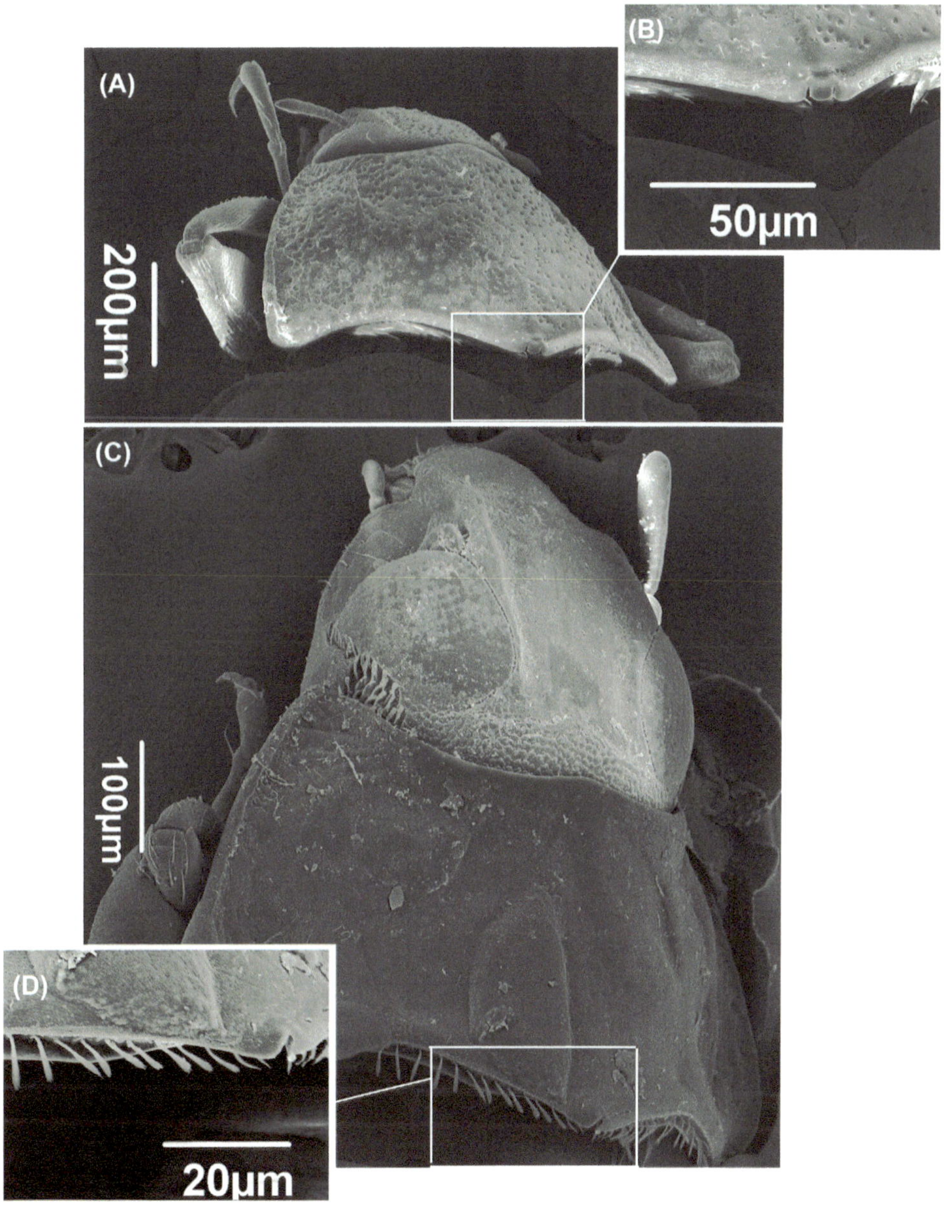

FIGURE 15.1.4 Torridincolidae (Coleoptera) adult, medial incision on posterior margin of pronotum: (A, B) *Iapir* sp.; and (C, D) *Ytu* sp.

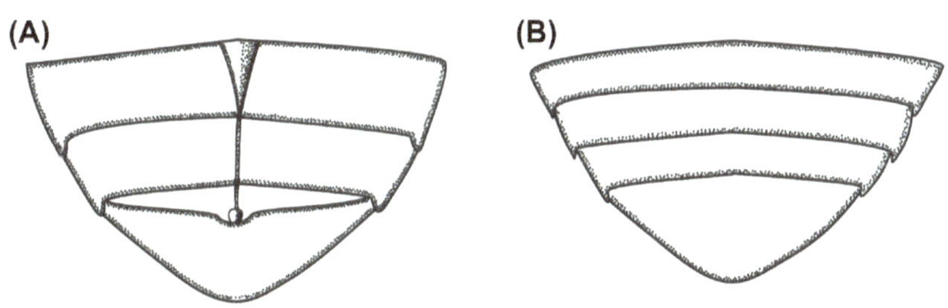

FIGURE 15.1.5 Torridincolidae (Coleoptera) adult, last abdominal ventrites: (A) *Iapir* sp., with semilunar impression on last ventrite; and (B) *Ytu* sp. without semilunar impression on last ventrite.

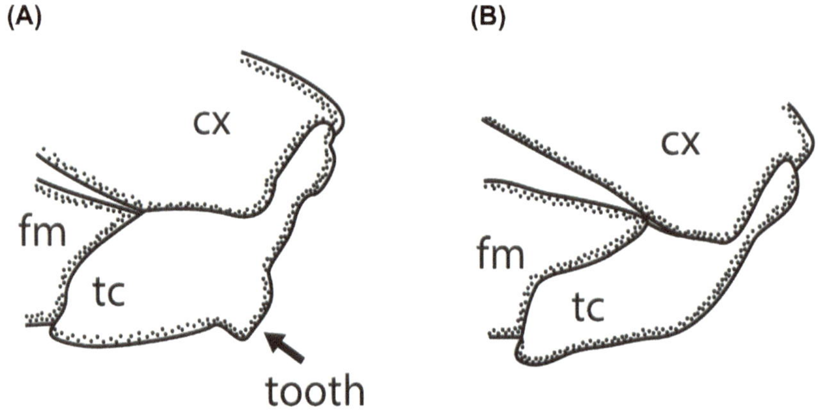

FIGURE 15.1.6 Torridincolidae (Coleoptera) adult, metatrochanter: (A) *Ytu* sp., with dentiform projection; and (B) *Iapir* sp., without projection. Arrow indicates tooth.

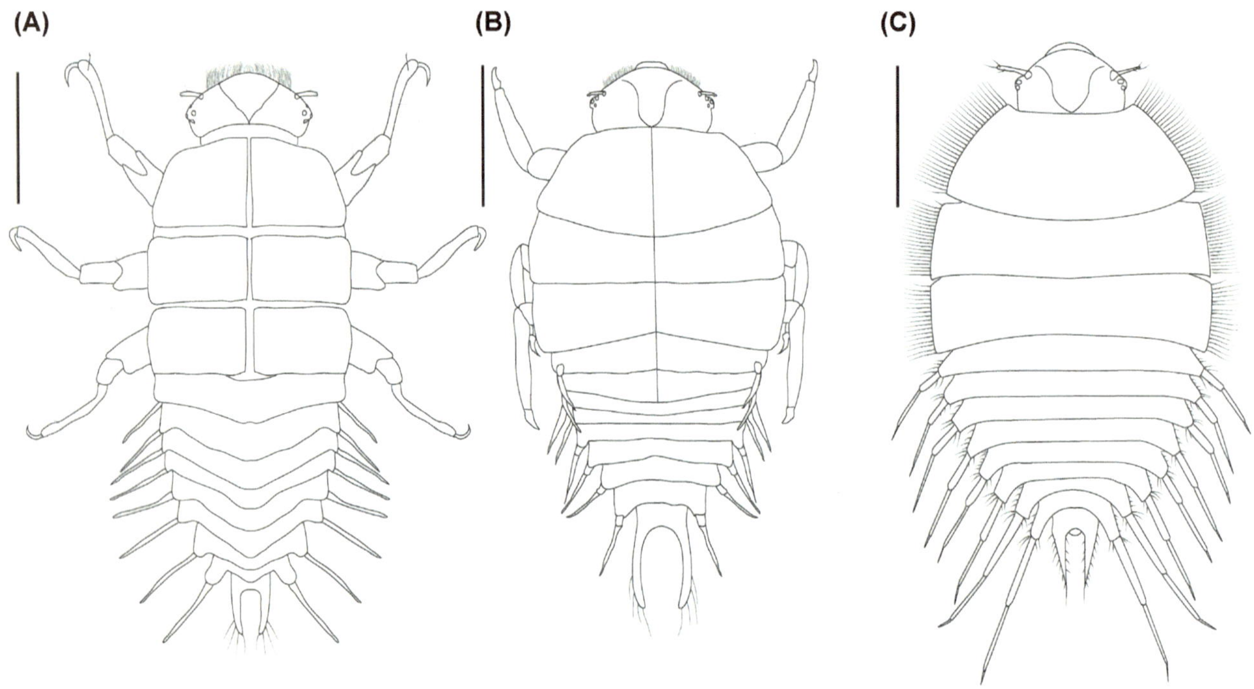

FIGURE 15.1.7 Torridincolidae (Coleoptera) larva, habitus in dorsal view: (A) *Claudiella* sp.; (B) *Iapir* sp.; and (C) *Ytu* sp. *(A) Modified from Vanin & Costa (2001). (B, C) Modified from Reichardt (1973).*

REFERENCES

Beutel, R.G. 1999. Phylogenetic analysis of Myxophaga (Coleoptera) with redescription of Lepicerus horni (Lepiceridae). Zoologischer Anzeiger 237: 291–308.

Benetti, C.J. & N. Hamada. 2016. Three new species of Claudiella Reichardt & Vanin, 1976 (Coleoptera, Torridincolidae) from Brazil. Zootaxa 4205: 151–161.

Beutel, R.G. 2005. 6. Myxophaga Crowson, 1955. Pages 43–52 *in*: N.P. Kristensen & R.G. Beutel (eds.), Handbook of Zoology, A Natural History of the Phyla of the Animal Kingdom, Volume 4 Arthropoda: Insecta, part 38. Walter de Gruyter, New York.

Beutel, R.G., D.R. Maddison & A. Haas. 1999. Phylogenetic analysis of Myxophaga (Coleoptera) using larval characters. Systematic Entomology 24: 171–192.

Beutel R.G. & R. Arce-pérez. 2005. 6.3 Spaeriusidae Erichson, 1845 (Jäch 1999) (=Microsporidae). Pages 48–49 *in*: N.P. Kristensen & R.G. Beutel (eds.), Handbook of Zoology, A Natural History of the Phyla of the Animal Kingdom, Volume 4 Arthropoda: Insecta, part 38. Walter de Gruyter, New York.

Reichardt, H. 1973. A critical study of the suborder Myxophaga, with a taxonomic revision of the Brazilian Torridincolidae and Hydroscaphidae (Coleoptera). Arquivos de Zoologia 24: 73–162.

Reichardt, H. 1976. Revision of the Lepiceridae (Coleoptera, Myxophaga). Papéis Avulsos de Zoologia, 30: 35–42.

Shepard, W.D., C.B. Barr & C.A. Aguilar-Julio. 2013. The occurrence of Torridincolidae (Coleoptera: Myxophaga) in Paraguay and a World checklist of species. Boletín del Museo Nacional de Historia Natural del Paraguay 17: 76–82.

Short, A.E.Z. 2013. Chapter 4. Aquatic Beetles of the Grensgebergte and Kasikasima Regions, Suriname (Insecta: Coleoptera). A Rapid Biological Assessment of the Upper Palumeu River Watershed (Grensgebergte and Kasikasima) of Southeastern Suriname. RAP Bulletin of Biological Assessment 67: 79–89.

Short, A.E.Z., L.J. Joly & M. García. 2010. Discovery of the skiff beetle genus Yara Reichardt and Hinton (Coleoptera: Myxophaga: Hydroscaphidae) in Venezuela, with description of a new species. The Coleopterists Bulletin 64: 151–156.

Short, A.E.Z., L.J. Joly, M. García, A.L. Wild, D.D. Bloom & D.R. Maddison. 2015. Molecular phylogeny of the Hydroscaphidae (Coleoptera: Myxophaga) with description of a remarkable new lineage from the Guiana Shield. Systematic Entomology 40: 214–229.

Chapter 15.2

Families Gyrinidae, Haliplidae, and Noteridae

Cesar J. Benetti
Coordenação de Biodiversidade, Instituto Nacional de Pesquisas da Amazônia, Manaus, Amazonas, Brazil

Mario Toledo
Museo di Storia Naturale, Parma, Italy

Daniara Colpani
Coordenação de Biodiversidade, Curso de Pós-Graduação em Entomologia, Instituto Nacional de Pesquisas da Amazônia, Manaus, Amazonas, Brazil

Bruno A.C. Guimarães
Laboratório de Entomologia, Departamento de Zoologia, Instituto de Biologia, Universidade Federal do Rio de Janeiro, Rio de Janeiro, Brazil

Chapter Outline

INTRODUCTION

The suborder Adephaga includes eight water beetle families, five of which known for the Neotropical Region: Dytiscidae, Meruidae, Gyrinidae, Haliplidae, and Noteridae. The last three families are treated in this chapter. In this chapter, we follow the family taxonomy utilized in Vondel & Spangler (2008), Miller (2009), Miller & Bergsten (2012), and Baca et al. (2017).

Gyrinidae

This is a worldwide distributed family, commonly known as "whirligig beetles," due to their habit of spinning in circles over the water surface. They are also known to have gregarious habits and form colonies over the water surface. Both adults and larvae are strictly aquatic and usually inhabit clear, clean waters, and they are more frequently found in the margins of lakes, ponds, and dams, although some species are much more common in running water (e.g., *Gyretes*), especially in standing areas of rivers and streams. Pupation occurs on the water shores, over sand or mud. Adults are predators of insects found on the superficial layer of the water. Larvae are predators that inhabit on the substrate and feed on a great variety of animals.

This family has about 750 described species (Jach & Balke, 2008), belonging to 13 genera, three tribes and thee subfamilies: Gyrininae, Heterogyrinae, and Spanglerogyrinae (Miller & Bergsten, 2012; Gustafson & Miller, 2013, Gustafson & Miller, 2017). In the Neotropical Region, they are only represented by the subfamily Gyrininae, with about 200 species in 5 genera.

Thorp and Covich's Freshwater Invertebrates, Fourth Edition. https://doi.org/10.1016/B978-0-12-804223-6.00023-8

Haliplidae

Members of this family can be found in every continent, although they are more diverse in northern temperate regions. Both adults and larvae live in a wide variety of freshwater habitats, but they are most frequently found in permanent ponds and lakes. They are commonly known as "crawling water beetles." Adults comb the surface, as they are not good swimmers. They are aquatic in all stages, except pupa, and the pupation occurs on the humid soil of water margins. Adults and larvae are phytophagous, shredders, and perforators.

About 200 described species in five genera (Jach & Balke, 2008) occur within this family. The Neotropics contains 56 species in two genera, *Haliplus* Latreille and *Peltodytes* Régimbart (Vondel & Spangler, 2008, Vondel, 2016b).

Noteridae

Beetles in this family are commonly called "burrowing water beetles" and are a relatively small, worldwide family occurring mainly in the tropics the highest diversity in this family occurs in the Neotropics. Members of Noteridae live mainly in lentic waters, rich in debris and vegetation, although several species are known to prefer lotic waters. Both adults and larvae are aquatic and, although very scarce information are currently available for most of the genera, pupation apparently occur in water within cocoons built among roots of water plants. Adult Noteridae have typically strong hind legs for digging in the mud of the bottom, whereas larvae (which are very poorly known) have stout bodies and burrowing habits, although in some genera it is known that larvae occur closer to the water surface. Both stages are probably largely carnivorous.

More than 260 species in this family occur in 17 genera, two subfamilies, and three tribes (Baca et al., 2014; Baca & Toledo, 2015; Baca et al., 2017). In the Neotropics two subfamilies, two tribes, nine genera, and 121 species are known (Nilsson, 2011).

LIMITATIONS

The fauna of aquatic Adephaga of the Neotropical Region is, in general, poorly known, although the level of treatment is uneven among the different families and genera.

Gyrinidae is one of the least known water beetle families in the Neotropics. Among the five genera recorded in this region, only two, *Dineutus* Macleay and *Enhydrus* Laporte, have been revised (Brinck, 1978; Gustafson & Miller, 2015). The other three, and most specious genera, are still poorly known. Several species of South American Gyrinidae are currently known only from old descriptions, sometimes incomplete, and in most cases without diagnostic figures, making their identification difficult. In general, adults are much better known than larvae, since the immature stages have been described for only a few species.

Haliplidae fauna of the Neotropical Region is more or less well-known, especially after the revision published by Vondel & Spangler (2008), which included all previous recorded species of this family from the Neotropicse and keys to genera and species. Unfortunately, very little is known about the immature stages of Neotropical Haliplidae.

Noteridae is a rather well-known family in the Neotropics, especially after publications of documents over the last few decades that revised, redefined, and newly described some genera and species. Nevertheless, the most specious genera, such as *Hydrocanthus* Say and *Suphisellus* Crotch, are still waiting for a review, and together they represent considerably over half the known taxa of Noteridae in this region. Immature stages of this family are practically unknown for the Neotropics.

Some papers covering the North American fauna (e.g., White & Roughley, 2008; Epler, 2010), provide useful keys, but they include only some of the Neotropical genera. Pederzani (1995) published an important global study on Dytiscidae (including Noteridae as subfamily) with a key to all world genera hitherto known. Unfortunately, figures were not included, making often proper identification of some genera difficult. In last years, a few regional works have been published, providing keys to genera: Benetti et al. (2003) to all known genera of Adephaga families in Brazil, and Ferreira-Jr. et al. (2014) to genera of all aquatic beetle families in Brazilian Amazonas. A key to genera (adults) is also present in the taxonomic revisions and phylogenetic studies of Miller (2009) for Noteridae, and Miller & Bergsten (2012) and Gustafson & Miller (2015) for Gyrinidae.

TERMINOLOGY AND MORPHOLOGY

Gyrinidae beetles are recognized by the split eyes and short, remiform middle, and hind legs. Adults have a shiny elliptical body, clearly hydrodynamic, black with shades of green, blue, golden, or copper. The shape of the eyes is one of the most distinctive character of this beetles, as they are divided in two parts (dorsal and ventral). Antennae are short, claviform, with 8–11 antennomeres fused in different degrees depending on the species. The length of the body can range from 3 to 15 mm, although most species range between 5 and 6 mm. The larvae undergo 2–3 instars. The most distinctive character of Gyrinidae larvae are the lateral gills of abdomen and the four anal claws. Larvae have narrow, falcate mandibles with internal perforation. The labrum is completely fused to the cephalic capsule. Legs have five segments and two claws. Additional information about morphology can be obtained in Beutel & Roughley (2016).

Adults of Haliplidae can be recognized by the enlarged metacoxal plates covering most of the hind legs and abdominal sternites. Antennae are long, filiform with 11 antennomeres. The legs are elongated and slender. They have an oval and glabrous body with colors that vary from completely testaceous to red-brown, usually with dark maculation on the elytra. Body length ranges from 2–5 mm. The larvae are very slender and undergo three instars. The legs have five segments and each tarsus bears only one claw. The mandibles are wide at the base and narrow at the apex, with internal perforation. The labrum is completely fused to the cephalic capsule. The tenth abdominal segment is much longer than the preceding one. More detailed accounts on the morphology of Haliplidae can be found in Vondel, 2016a.

Adults Noteridae are typically convex dorsally and flattened ventrally, with regularly oval body outline. The metacoxal process, which is slightly to visibly wide, combines with the medial part of metaventrite to form a sort of ventral platform ("noterid platform"), often in continuity with a large and wide prosternal process. Adults range from a bit more of 1 mm–8 mm but are more commonly 2.5–4 mm. These beetles are good and fast swimmers and have well developed natatory hind legs. Prothoracic legs are short and robust, with series of differently modified (depends on genera) stout spurs. In most genera of Noterinae, the protibiae bear a large and single, more or less curved, apical spur. This and a shift toward a more lateral position of the articulation of protarsus is probably an adaptation for borrowing in mud. Larval morphology of Noteridae is unknown, except for very few genera. In the Palaearctic genus *Noterus* Clairville, larvae are cylindrical, with short legs and conical last abdominal segment, bearing short urogompi. In the Nearctic and Neotropical genus *Suphis* Aubé, the body is globular, and the third antennal segment is over 10 times longer than the fourth segment. Larvae of Noteridae have short mandibles without suctorial channel. They have burrowing habits but in other genera larvae seem to live on submerged vegetation. For further information on this family see Holmen (1987); Miller (2009); Toledo (2010); Baca et al. (2017).

Most of the terms here used are explained in the general chapter on aquatic Coleoptera. Some additional terms are given in the next section.

MATERIAL PREPARATION AND PRESERVATION

Methods used for the collection and preservation of adephagous water beetle are similar to those used for other aquatic Coleoptera, which are explained in the general chapter. For most of the species of Noteridae, Haliplidae, and some Gyrinidae it may be important to concentrate efforts on collecting in lentic environments, such as lakes, ponds, dams, and pools. On the other hand, for collecting some genera such as *Gyretes* (Gyrinidae) or *Liocanthydrus* (Noteridae), it is important to collect in running waters. Collecting immature stages is also important, since the larvae of many species have not been described. The proper association between larvae and adults can be carried out by molecular techniques or by rearing the larvae.

KEYS TO GYRINIDAE, HALIPLIDAE, AND NOTERIDAE

Gyrinidae: Genera: Adults

Modified from Archangelsky et al. (2009) and Epler (2010).

1	Scutellum concealed with elytra closed (Fig. 15.2.1 A) 2
1'	Scutellum visible with elytra closed (Fig. 15.2.1 B) 3
2(1)	Pronotum and elytra lateral margins pubescent (Fig. 15.2.5 A); elytra shining, without striae; last abdominal ventrite conical; last two abdominal ventrites with median longitudinal row of long setae; example of habitus in Fig. 15.2.5 A .... *Gyretes* [Neotropical. Nearctic]
2'	Pronotum and elytra lateral margins without pubescence (Fig. 15.2.5 B); elytra with or without striae; last abdominal ventrite broadly rounded; last two abdominal ventrites glabrous, without median longitudinal row of long setae; example of habitus in Fig. 15.2.5 B .... *Dineutus* [Central America, Caribbean. Nearctic]
3(1)	Size >10 mm; elytron with or without longitudinal grooves or series of punctures; if elytral grooves present, with at most nine on each elytron (Fig. 15.2.5E) 4
3'	Size <9 mm; elytron with 11 longitudinal striae or series of punctures (reduced or obscured in some species) (Fig. 15.2.5 C) .... *Gyrinus* [Widespread]

4(3) Antenna flagellum with six antennomeres (Fig. 15.2.1 C); example of habitus in Fig. 15.2.5 D .... *Enhydrus* [Neotropical]

4' Antenna flagellum with nine antennomeres (Fig. 15.2.1 D); example of habitus in Fig. 15.2.5 E .... *Macrogyrus* (*Andogyrus*) [South America]

Gyrinidae: Genera: Larvae

Modified from Archangelsky et al. (2009), Epler (2010), and Michat et al. (2016).

1 Head with distinct neck (Fig. 15.2.1 E) 2

1' Head without distinct neck (Fig. 15.2.1 F) 4

2(1) Tracheal gills plumose; abdominal segment X without ventral spinulae (Fig. 15.2.1 H); maxilla first and second palpomeres and labial first palpomere without poriform additional structures 3

2' Tracheal gills not plumose; abdominal segment X with ventral spinulae (Fig. 15.2.1 G); maxilla first and second palpomeres and labial first palpomere bearing minute poriform additional structures; (based on instar I); example of habitus in Fig. 15.2.5 F .... *Enhydrus* [Neotropical]

3(2) Lacinia posterior margin dentate (Fig. 15.2.1 I) .... *Macrogyrus* (*Andogyrus*) [South America]

3' Lacinia posterior margin not dentate (Fig. 15.2.1 J) .... *Dineutus* [Central America, Caribbean. Nearctic]

4(1) Nasale with 2–4 teeth in a transverse row (Fig. 15.2.1 K) .... *Gyrinus* [Widespread]

4' Nasale truncate, without teeth (Fig. 15.2.1 L) .... *Gyretes* [Neotropical. Nearctic]

Haliplidae: Genera: Adults

Modified from Vondel & Spangler (2008) & Epler (2010).

1 Hind coxal plates leaving only last abdominal sternite exposed (Fig. 15.2.1 M); maxillary and labial palpi last article longer than penultimate segment .... *Peltodytes* [Cuba. Mexico. Widespread]

1' Hind coxal plates leaving last three abdominal sternites exposed (Fig. 15.2.1 N); maxillary and labial palpi last article shorter than penultimate segment; example of habitus in Fig. 15.2.5 G .... *Haliplus* [Cosmopolitan]

Haliplidae: Genera: Larvae

Modified from Epler (2010).

1 Each body segment with 1–3 dorsal pairs of long filaments (Fig. 15.2.1 O); last abdominal segment short; forelegs moderately to strongly chelate .... *Peltodytes* [Cuba. Mexico. Widespread]

1' Each body segment without dorsal filaments, short dorsal spines may be present (Fig. 15.2.1 P); last abdominal segment long; forelegs weakly chelate or nonchelate .... *Haliplus* [Cosmopolitan]

Noteridae: Genera: Adults

In part modified from Miller (2009).

1 Metacoxa and metaventrite not fused, suture entire (Figs. 15.2.2 B–H); protibia apically expanded or tapering; size >1.8 mm; Noterinae 2

1' Metacoxa and metaventrite fused laterally, suture obscured (Fig. 15.2.2 A); protibia apically expanded (Fig. 15.2.3 A); size <1.7 mm; example of habitus in Fig. 15.2.5 H; Notomicrinae .... *Notomicrus* [Neotropical. Nearctic. Oriental. Australian]

2(1) Protibia apically expanded or broadly rounded (Figs. 15.2.3 B, C); apical margin with several prominent spines; fringe of short and compact setae along lateral and apical margin absent 3

2' Protibia apically narrowly rounded, often tapering (Figs. 15.2.3 D–H); apical margin with only one (two in *Prionohydrus*) more or less large and robust, apically curved spur; fringe of short and compact setae along lateral margin present (less developed in *Prionohydrus*, missing in *Suphis*) 4

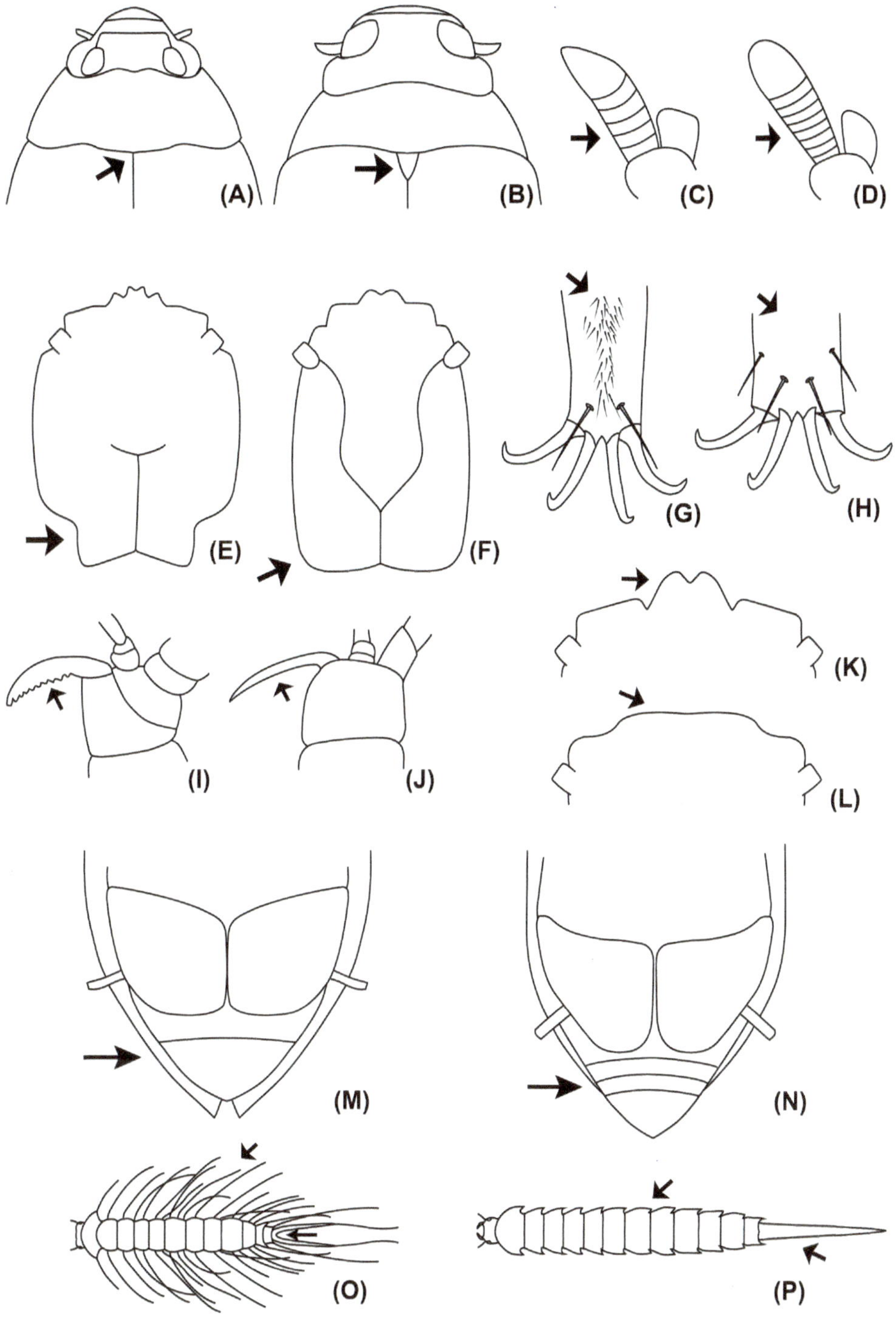

FIGURE 15.2.1 Gyrinidae adults: head and pronotum, dorsal view, (A) *Dineutus*, (B) *Gyrinus*, arrow indicates scutellum (A) concealed and (B) visible; antenna, dorsal view, (C) *Enhydrus*, (D) *Andogyrus*, arrow indicates flagellum. Gyrinidae larvae: head, dorsal view, (E) *Enhydrus*, (F) *Gyrinus*, arrow indicates neck; last abdominal segment, ventral view, (G) *Enhydrus*, (H) *Andogyrus*, *arrow* indicates spinulae; maxilla, ventral view, (I) *Andogyrus*, (J) *Dineutus*, arrow indicates posterior margin of lacinia; nasale, dorsal view, (K) *Gyrinus*, (L) *Gyretes*, arrow indicates teeth. Haliplidae adults, hind coxal plates and abdominal segments, ventral view: (M) *Peltodytes*, (N) *Haliplus*, arrow indicates abdominal sternites exposed. Haliplidae larvae, habitus, dorsal view: (O) *Peltodytes*, arrows indicate long filaments and last abdominal segment; (P) *Haliplus*, arrows indicate dorsal spines and last abdominal segment. *(G) Modified from Michat et al. (2016). (H) Modified from Archangelsky & Michat (2007). (N) Modified from Spangler (1991).*

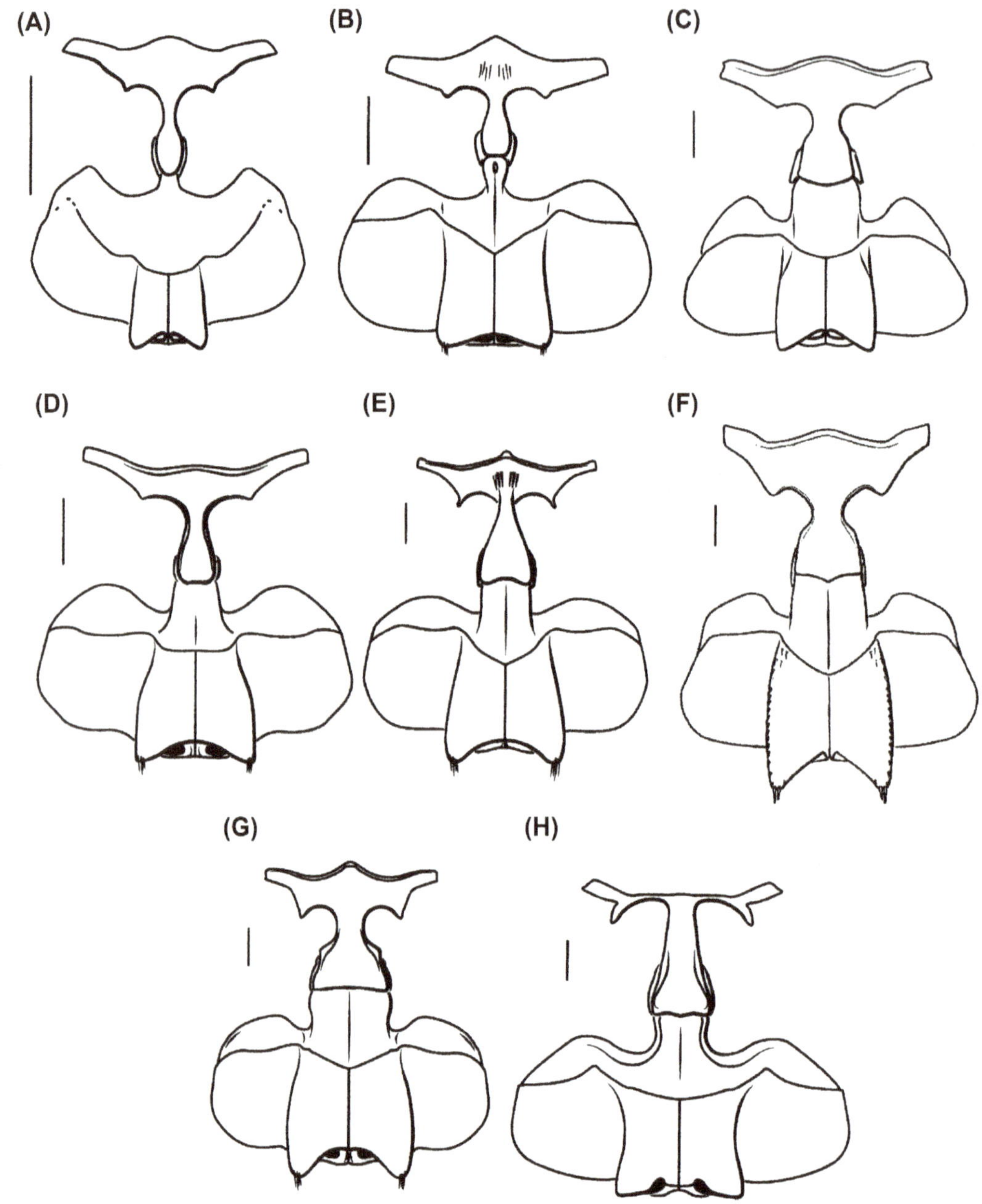

FIGURE 15.2.2 Prosternum, mesoventrite and metacoxae in Noteridae genera (adults): (A) *Notomicrus*; (B) *Suphisellus* in part (formally *Pronoterus*); (C) *Tonerus*; (D) *Mesonoterus*; (E) *Suphisellus* in part; (F) *Liocanthydrus*; (G) *Hydrocanthus*; (H) *Suphis*. Scale bars = 0.25 mm. *(H) From Miller (2009), modified.*

3(2) Prosternal process liguiform, with maximum width subapical, apex poorly truncate (Fig. 15.2.2 B); protibia with pronounced dorsoapical angle (Fig. 15.2.3 C)........ *Suphisellus* (in part)
[Argentina, Brazil, Guadalupe. Nearctic]

3' Prosternal process spatulate, broad with maximum width apically, apex visibly truncate (Fig. 15.2.2 C); protibia without visible angle, regularly rounded apically (Fig. 15.2.3 B) .... *Tonerus*
[Venezuela]

4(2) Protarsi inserted subapically on protibia (Figs. 15.2.3 D, E); protibia apically narrowly rounded, not tapering to apical spur, which is slender, weakly curved apically; prosternal process liguiform, apically somewhat rounded, or subtruncate (Fig. 15.2.2 D)........ 5

4' Protarsi inserted laterally on protibia (Figs. 15.2.3 F–H); protibia tapering to apical spur, which is robust and strongly curved; prosternal process spatulate, broad, apically truncate or subtruncate (Figs. 15.2.2 E–H)........ 6

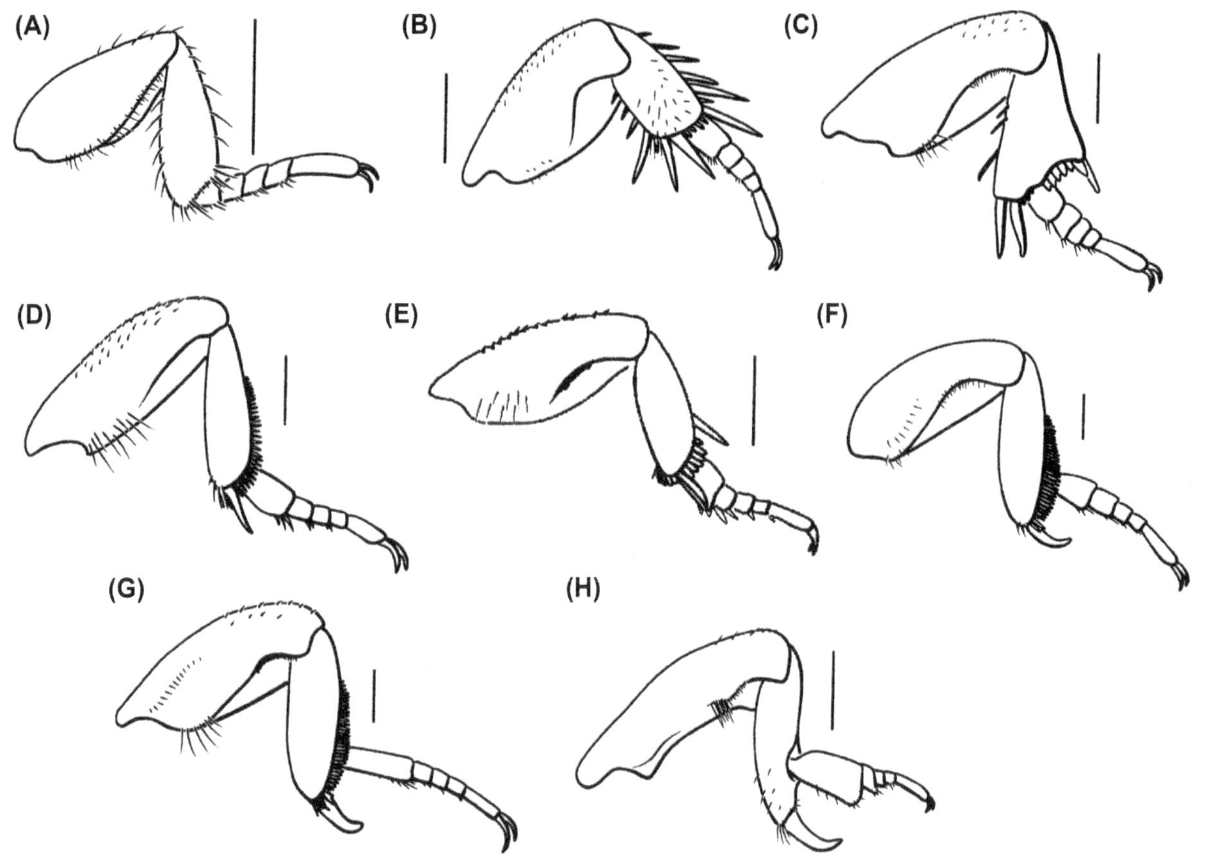

FIGURE 15.2.3 Left foreleg of Noteridae genera (adults): (A) *Notomicrus*; (B) *Tonerus;* (C) *Suphisellus* in part (formally *Pronoterus*); (D) *Mesonoterus*; (E) *Prionohydrus*; (F) *Suphisellus* in part; (G) *Hydrocanthus*; (H) *Suphis*. Scale bars = 0.10 mm. *(H) From Miller (2009), modified.*

5(4) Posterior meso- and metatibial spurs strongly serrate (Fig. 15.2.4 C); protibia with second, slightly smaller spur, poorly curved (Fig. 15.2.3 E); aedeagus median lobe tapering distally, apically rounded (Fig. 15.2.4 F); left lateral lobe without subapical constriction (Fig. 15.2.4 G) .... *Prionohydrus*
[Brazil, Venezuela]

5' Posterior meso- and metatibial spurs smooth; protibia without a second spur (Fig. 15.2.3 D); aedeagus median lobe apically wide, truncate (Fig. 15.2.4 H); left lateral lobe with a distinct subapical constriction (Fig. 15.2.4 I) .... *Mesonoterus*
[Neotropical. Nearctic]

6(4) Prosternal process width equal to subequal length (Figs. 15.2.2 F, G) 7

6' Prosternal process obviously longer than wide (Figs. 15.2.2 E, H) 8

7(6) Metatibial spurs finely but distinctly serrate (Fig. 15.2.4 A); meso- and metatibiae broad and flat, length 2× width; elytra uniformly dark or with dark wrinkles on paler background, always without pale spots; example of habitus in Fig. 15.2.5 I .... *Hydrocanthus*
[Neotropical. Nearctic]

7' Metatibial spurs smooth (Fig. 15.2.4 B); meso- and metatibiae length ≥3× width, not strongly flattened; elytra often (but not always) with clear pale spots on dark background .... *Liocanthydrus*
[South America]

8(6) Protibial lateral fringe present (Figs. 15.2.3 F, G); noterid platform extending onto metaventrite (Fig. 15.2.2 E); body elongate or less, but never strongly convex (Fig. 15.2.5 K) 9

8' Protibial lacking lateral fringe of short and compact setae (Fig. 15.2.3 H); noterid platform not extending onto metaventrite (Fig. 15.2.2 H); body very short, dorsally strongly convex; example of habitus in Fig. 15.2.5 J .... *Suphis*
[Neotropical. Nearctic]

9(8) Pronotum lateral bead narrow with fine crease at posterolateral angle extending anteriorly to about middle of pronotum (Fig. 15.2.4 D); example of habitus in Fig. 15.2.5 K .... *Suphisellus* (in part)
[Neotropical. Nearctic]

9' Pronotum with lateral bead broad, without fine crease at posterolateral angle (Fig. 15.2.4 E) .... *Canthysellus*
[Brazil, Colombia, Ecuador, Venezuela, Peru]

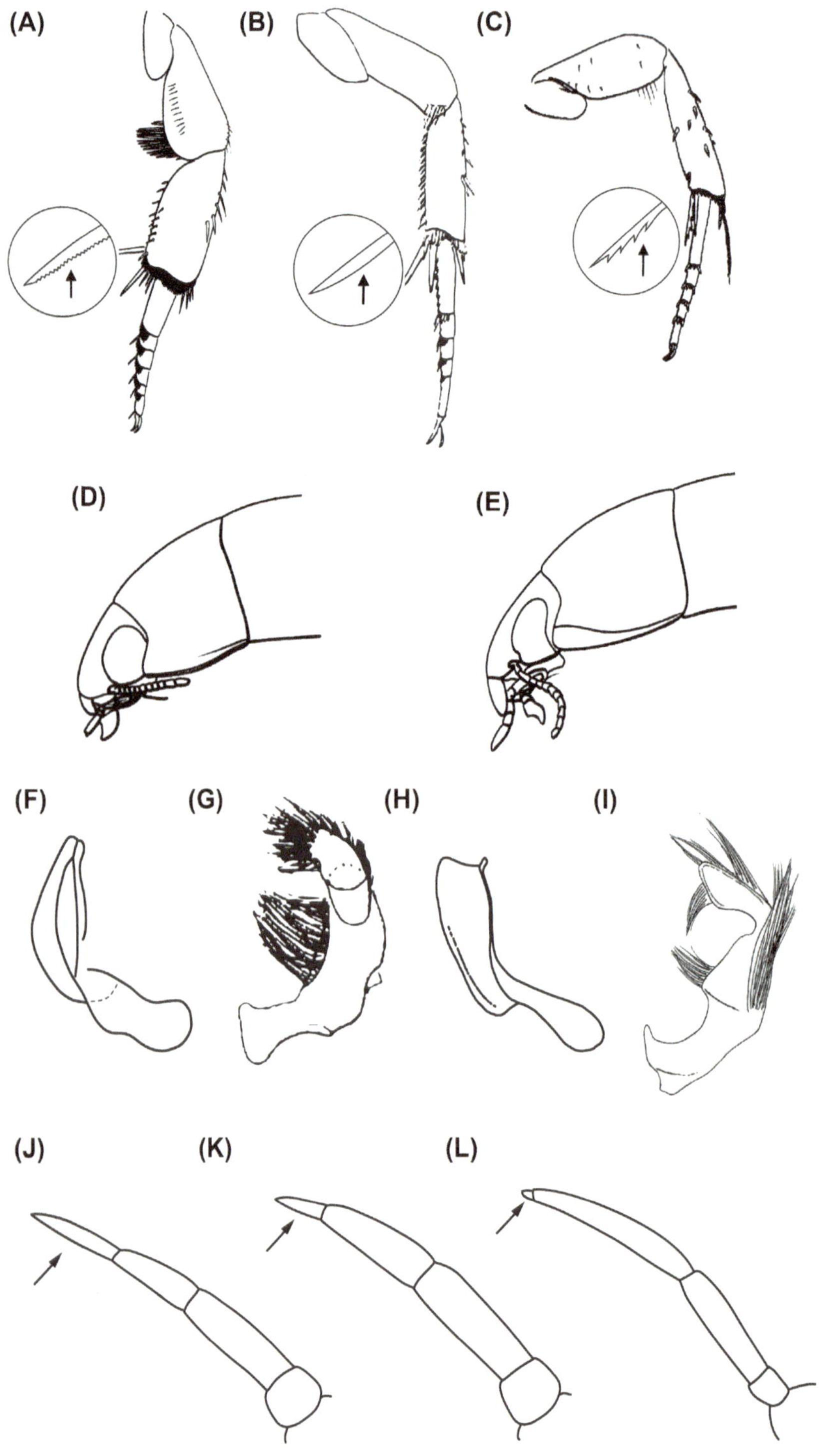

FIGURE 15.2.4 Noteridae genera (adults). Hindleg, in detail the outer tibial spur magnified: (A) *Hydrocanthus*; (B) *Liocanthydrus*; (C) *Prionohydrus*. Head and pronotum, left lateral view: (D) *Suphisellus*, (E) *Canthysellus*. Scale bar = 1.0 mm. Aedeagus: (F–H) median lobe; (G–I) left lateral lobe; (F, G) *Prionohydrus ubercornis* Gomez & Miller; (H, I) *Mesonoterus laevicollis* Sharp. Scale bar = 0.10 mm. Noteridae genera (larvae): (J) *Suphisellus*; (K) *Hydrocanthus*; (L) *Suphis*, arrow indicates antennal segment 3 (J–K) and 4 (L). *(A) From Miller (2009), modified. (B) From Baca et al. (2014). (C) From Gomez & Miller (2013). (E) From Miller (2009), modified. (F, G) From Gomez & Miller (2013).*

FIGURE 15.2.5 Aquatic Coleoptera, dorsal habitus. Gyrinidae adult: (A) *Gyretes*, (B) *Dineutus*, (C) *Gyrinus*, (D) *Enhydrus*, (E) *Andogyrus*. Gyrinidae larva: (F) *Enhydrus*. Haliplidae adult: (G) *Haliplus*. Noteridae adults: (H) *Notomicrus*, (I) *Hydrocanthus*, (J) *Suphis*, (K) *Suphisellus*. Noteridae larva: (L) *Hydrocanthus*.

Noteridae: Genera: Larvae

Modified from Spangler & Folkerts (1973) and Epler (2010). The larvae of several Neotropical genera are still unknown, so the present key should be considered very preliminary and only applicable to separate three of the known nine genera.

1 Third antennal article no longer than the fourth antennal article (Fig. 15.2.4 J); mandible with a stout preapical tooth .... *Suphisellus*
[Neotropical. Nearctic]

1' Third antennal article more than twice as long as the fourth antennal article (Fig. 15.2.4 K); mandible not strongly toothed................. 2

2(1) Body form globose; third antennal article about 12 times longer than the fourth antennal article (Fig. 15.2.4 L); mandible serrated........ .... *Suphis*
[Neotropical. Nearctic]

2' Body form cylindrical; third antennal article about three times longer than the fourth antennal article (Fig. 15.2.4 K); mandible smooth without inner teeth; example of habitus in Fig. 15.2.5 L........ *Hydrocanthus*
[Neotropical. Nearctic]

ACKNOWLEDGMENTS

We are grateful to Mariano Michat for his valuable comments on the Gyrinidae larvae key. The work of Cesar J. Benetti and Daniara Colpani was supported by CNPq and FAPEAM fellowships (processes 151461/2014-7, 307849/2014-7 and 302031/2015-4). INPA/MCTI and CNPq provided financial support and laboratory infrastructure (PROTAX/CNPq - 440616/2015-8). The work of Bruno A. C. Guimarães was supported by CNPq and Museu Nacional.

REFERENCES

Archangelsky, M., V. Manzo, M.C. Michat & P.L.M. Torres. 2009. Chapter 14: Coleoptera. Pages 411–468 *in*: E. Domínguez & H.R. Fernández (eds.), Macroinvertebrados bentónicos sudamericanos. Sistemática y biología. Fundación Miguel Lillo, Tucumán, Argentina.

Archangelsky, M. & M. Michat 2007. Morphology and chaetotaxy of the larval stages of Andogyrus seriatopunctatus Régimbart (Coleoptera: Adephaga: Gyrinidae). Zootaxa 1645: 19–33.

Baca, S.M., G. Gustafson, M. Toledo & K.B. Miller. 2014. Revision of the Neotropical burrowing water beetle genus Liocanthydrus Guignot (Coleoptera: Noteridae: Noterinae: Noterini) with the description of two new species. Zootaxa 3793: 231–246.

Baca, S.M. & M. Toledo. 2015. Canthysellus Baca and Toledo (Coleoptera: Noteridae: Noterini), a new genus of burrowing water beetle from South America. The Coleopterists Bulletin 69: 477–488.

Baca, S.M., E.F.A. Toussaint, K.B. Miller & A.E.Z. Short. 2017. Molecular phylogeny of the aquatic beetle family Noteridae (Coleoptera: Adephaga) with an emphasis on data partitioning strategies. Molecular Phylogenetics and evolution 107: 282–292.

Benetti, C.J., J.A. Régil Cueto & G.L. Fiorentin. 2003. Gêneros de Hydradephaga (Coleoptera: Dytiscidae, Gyrinidae, Haliplidae, Noteridae) citados para o Brasil, com chaves para identificação. Biota Neotropica 3: 1–20.

Beutel, R.G. & R.E. Roughley. 2016. Chapter 7.1: Gyrinidae. Pages 80–88 *in*: R.G. Beutel & R.A.B. Leschen (eds.), Handbook of Zoology, Volume 4/38: Coleoptera, Beetles, Volume 1: Morphology and Systematics (Archostemata, Adephaga, Myxophaga, Polyphaga partim). De Gruyter, Berlin, Germany.

Brinck, Per. 1978. Derivation, taxonomy and history of distribution of the whirligig beetle genus Enhydrus (Coleoptera: Gyrinidae). Entomologica Germanica 4: 317–326.

Epler, J.H. 2010. The water beetles of Florida. An identification manual for the families Chrysomelidae, Curculionidae, Dryopidae, Dytiscidae, Elmidae, Gyrinidae, Haliplidae, Helophoridae, Hydraenidae, Hydrochidae, Hydrophilidae, Noteridae, Psephenidae, Ptilodactylidae and Scirtidae. Bureau of Laboratories, Florida Department of Environmental Protection, Tallahassee, FL. 414 pp.

Ferreira, N., Jr., B.H.L. Sampaio, A.S. Fernandes, B. Clarkson, R.B. Braga, M.I.S. Passos & A.D. Santos. 2014. Chapter 21: Ordem Coleoptera (coleos = invólucro, estojo; peteron = asa). Pages 349–375 *in*: N. Hamada, J.L. Nessimian & R.B. Querino (eds.), Insetos aquáticos na Amazônia brasileira: taxonomia, biologia e ecologia. Editora do Inpa, Manaus, Brazil.

Gomez, A. & K. Miller. 2013. Prionohydrus, a new genus of Noterini Thomson (Coleoptera: Noteridae) from South America with three new species and its phylogenetic considerations. Annals of the Entomological Society of America 106: 1–12.

Gustafson, G.T. & K.B. Miller. 2013. On the family- and genus-series nomina in Gyrinidae Latreille, 1810 (Coleoptera, Adephaga). Zootaxa 3731: 077–105.

Gustafson, G.T. & K.B. Miller. 2015. The New World whirligig beetles of the genus Dineutus Macleay, 1825 (Coleoptera, Gyrinidae, Gyrininae, Dineutini). ZooKeys 476: 1–135.

Gustafson, G.T. & K.B. Miller. 2017. Systematics and evolution of the whirligig beetle tribe Dineutini (Coleoptera: Gyrinidae: Gyrininae). Zoological Journal of the Linnean Society, 181: 118–150.

Holmen, M. 1987. The aquatic Adephaga (Coleoptera) of Fennoscandia and Denmark I: Gyrinidae, Haliplidae, Hygrobiidae and Noteridae. Fauna entomologica scandinavica 20: 1–168.

Jach, M.A. & M. Balke. 2008. Global diversity of water beetles (Coleoptera) in freshwater. Hydrobiologia 595: 419–442.

Michat, M., T.M. Alvarenga, M.S. Silva & Y. Alarie. 2016. First larval description and chaetotaxic analysis of the neotropical whirligig beetle genus Enhydrus Laporte (Coleoptera, Gyrinidae). Revista brasileira de Entomologia 60: 231–237.

Miller, K.B. 2009. On the systematics of Noteridae (Coleoptera: Adephaga: Hydradephaga): Phylogeny, description of a new tribe, genus and species, and survey of female genital morphology. Systematics and Biodiversity 7: 191–214.

Miller, K.B. & J. Bergsten. 2012. Phylogeny and classification of whirligig beetles (Coleoptera: Gyrinidae): relaxed-clock model outperforms parsimony and time-free Bayesian analyses. Systematic Entomology 37: 706–746.

Nilsson, A.N. 2011. A World Catalogue of the Family Noteridae, or the Burrowing Water Beetles (Coleoptera, Adephaga) Version 16.VIII.2011, 54 pp. URL: http://www2.emg.umu.se/projects/biginst/andersn/.

Pederzani, F. 1995. Keys to the identification of the genera and subgenera of adults Dytiscidae (sensu lato) of the world (Coleoptera Dytiscidae). Atti della Accademia Roveretana degli Agiati 244: 5–83..

Spangler, P.J. 1991. Chapter: Haliplidae (Adephaga). Pages 311–312 *in*: F.W. Stehr (ed.), Immature Insects. Vol. 2. Kendall/Hunt Publishing Company, Dubuque, Iowa.

Spangler, P.J. & G.W. Folkerts. 1973. Reassignment of Colpius inflatus and a description of its larva (Coleoptera: Noteridae). Proceedings of the Biological Society of Washington 86: 501–510.

Toledo, M. 2010. Chapter: Noteridae: Review of the species occurring east of the Wallace line (Coleoptera). Pages 195–236 *in*: M.A. Jäch & M. Balke (eds.), Water beetles of New Caledonia (Part 1), Monographs on Coleoptera 3.

Vondel, B.J. van 2016a. Chapter 7.2: Haliplidae. Pages 88–97 *in*: R.G. Beutel & R.A.B. Leschen (eds.), Handbook of Zoology, Volume 4/38: Coleoptera, Beetles, Volume 1: Morphology and Systematics (Archostemata, Adephaga, Myxophaga, Polyphaga partim). De Gruyter, Berlin, Germany.

Vondel, B.J. van 2016b. Corrections and additions to the "Revision of the Haliplidae of the Neotropical Region including Mexico" (Vondel & Spangler 2008) with description of a new species (Coleoptera: Haliplidae). Koleopterologische Rundschau 86: 51–59.

Vondel, B.J. van & P.J. Spangler. 2008. Revision of the Haliplidae of the Neotropical Region including Mexico (Coleoptera: Haliplidae). Koleopterologische Rundschau 78: 69–194.

White, D.S. & R.E. Roughley. 2008. Chapter 20: Aquatic Coleoptera. Pages 571–671 *in*: R.W. Merritt, K.W. Cummins & M.B. Berg (eds.), An Introduction to the Aquatic Insects of North America. Fourth Edition. Kendall/Hunt Publishing Co., Dubuque, IA. 1158 pp.

Chapter 15.3

Family Dytiscidae

Cesar J. Benetti
Coordenação de Biodiversidade, Instituto Nacional de Pesquisas da Amazônia, Manaus, Amazonas, Brazil

Mariano C. Michat
Laboratory of Entomology, IBBEA, CONICET-UBA., DBBE-FCEN, University of Buenos Aires, Buenos Aires, Argentina

Nelson Ferreira, Jr.
Laboratório de Entomologia, Instituto de Biologia, Universidade Federal do Rio de Janeiro, Rio de Janeiro, Brazil

Rafael Benzi Braga
Laboratório de Entomologia, Instituto de Biologia, Universidade Federal do Rio de Janeiro, Rio de Janeiro, Rio de Janeiro, Brazil

Yoandri S. Megna
Departamento de Biología, Facultad de Ciencias Naturales y Exactas, Universidad de Oriente, Santiago de Cuba, Cuba and Zoologische Staatssammlung München, Germany

Mario Toledo
Museo di Storia Naturale, Parma, Italy

FAMILY DYTISCIDAE

Chapter Outline

INTRODUCTION

Dytiscidae is the most speciose water beetle family, with over 4,400 species known worldwide (Nilsson & Hájek, 2017). This large family is distributed worldwide, although it is more common in tropical and subtropical areas. Dytiscids are commonly known as "predaceous diving beetles" or just "diving beetles" for their extraordinary ability to dive. They have a wide range of size (adults vary from ~1 to 50 mm) and are well adapted to aquatic life, being considered as "true water beetles" since both adults and larvae are aquatic (Jäch, 1998).

Diving beetles are aquatic in all stages, except the pupa. They occur in almost all types of aquatic habitats, although

Thorp and Covich's Freshwater Invertebrates, Fourth Edition. https://doi.org/10.1016/B978-0-12-804223-6.00024-X

they are much more common in lentic than lotic environments. Larvae and adults of several species can also be found in groundwater, seepages, and phytotelmata. A few species are terrestrial (humicolous) (Jäch & Balke, 2008; Garrido et al., 2011).

More than 4,400 species of Dytiscidae have been described in 181 genera (Nilsson & Hájek, 2017). A total of 11 subfamilies and 26 tribes are presently recognized (Miller & Bergsten, 2014). In the neotropical region, more than 700 species in 60 genera, 23 tribes, and 10 subfamilies are known: Agabinae (2 tribes, 4 genera), Colymbetinae (1 tribe, 3 genera), Copelatinae (1 tribe, 3 genera), Coptotominae (1 tribe, 1 genus), Cybistrinae (1 tribe, 2 genera), Dytiscinae (5 tribes, 5 genera), Hydrodytinae (1 tribe, 2 genera), Hydroporinae (9 tribes, 35 genera), Laccophilinae (1 tribe, 4 genera), and Lancetinae (1 tribe, 1 genus).

This manual follows the family taxonomy utilized in Miller & Bergsten (2014).

LIMITATIONS

Dytiscidae fauna in the Neotropical region is relatively well-known taxonomically, although the knowledge is uneven among the different groups. Several genera, including some of the richest ones, lack taxonomic revisions. In addition, some of the treatments are incomplete or outdated, making identification of the species difficult. In general, adults are much better known than larvae. This is especially notorious in some groups such as the tribe Bidessini, where less than half of the genera have their larvae described.

So far, no comprehensive keys to dytiscid genera covering the entire Neotropical Region are published. The most important contribution was provided by Archangelsky et al. (2009), including keys to the South American genera. Some works covering other biogeographic regions include keys that may be of some help, although they include only a fraction of the neotropical genera. Among them are those of White & Roughley (2008) and Epler (2010) which cover some of the North American fauna. In addition, there are some significant global studies that provide keys to the genera known to that date at a worldwide scale. The first worldwide study was published by Pederzani (1995). However, that study does not include figures, making the proper identification of some genera difficult. Recently, Miller & Bergsten (2016) published an interesting book about systematics and biology of the family which includes global keys to subfamilies, tribes, and genera. Michat et al. (2017) provide both traditional and interactive keys for the identification of larvae of the subfamilies and tribes of world Dytiscidae.

In recent years, a number of regional keys that include some of the Neotropical genera have been published, improving the knowledge on both adults (Benetti et al., 2003 to Brazil; Libonatti et al., 2011 to Argentina; Ferreira-Jr. et al., 2014 to Brazilian Amazon) and larvae (Michat et al., 2008 to Argentina). Several taxonomic revisions also present keys to genera and species, such as Young (1967) and Biström (1988, for Bidessini), Gueorguiev (1968, for Copelatinae), Gueorguiev (1971, for Agabinae), Spangler (1972, for Colymbetinae), Biström (1997, for Hydphydrini and Pachydrini), Miller (2005, for Vatellini), Miller et al. (2007, for Cybistrinae), and Toledo and Michat (2015, for Laccophilinae).

TERMINOLOGY AND MORPHOLOGY

Adults of Dytiscidae have a series of adaptations for swimming, such as a hydrodynamic body, and strong hind legs with swimming hairs. Either the hind tibiae, or both the hind and middle tibiae, and their corresponding tarsi are flattened to form swimming paddles. Fore legs (or sometimes middle legs) are frequently sexually dimorphic, with males bearing suction cups or dense groups of setae to grasp the females during copulation. Also, sexual dimorphism in pronotal and elytral sculpture is evident in many species, such as in the genus *Dytiscus*.

The three larval instars are morphologically rather similar except for the size. They are elongated, generally of light color although darker colors predominate in some groups. The head is subcircular or somewhat longer than wide, and the labrum is fused to the cephalic capsule. The mandibles are generally more or less falciform and bear a channel through which they suck the predigested juices of their prey. The legs are long and well developed, with six segments, the distal one bearing two claws. They lack tracheal gills (except in Coptotominae) but have well developed urogomphi (minute in Cybistrinae) and the last abdominal segment may or may not be projected backward in a more or less elongate siphon.

More detailed accounts on the morphology of adults and larvae of Dytiscidae can be found in Larson et al. (2000), Yee (2014), and Balke & Hendrich (2016).

Most of the terms used for identification of Dytiscidae are given in the introductory coverage of Coleoptera. Some additional terms are given here, as modified from Biström (1988), and Epler (2010):

Accessory stria: A stria of punctures between suture and basal stria on each elytron.
Cervical line: A suture on epicranium, apparently jointing the posterior corners of the eyes.
Metacoxal line: A line running parallel to the medial side of each metacoxa, ending caudally in one of the two lobes of metacoxal process.
Metasternal process: An odd, medial, bilobed process on fore side of metaventrite where the prosternal process fits.
Metasternal wings: Even lateral extentions on each side of metaventrite, triangular in shape.
Metepisternum: Anterior sclerite of metathoracic pleura, separated from metepimeron by the pleural suture.

Prosternal process: An odd medial elongate process on hind side of prosternum.

Sutural line: A line on each elytron, parallel to the elytral suture.

MATERIAL PREPARATION AND PRESERVATION

Methods used for the collection and preservation of dytiscids are similar to those used for other aquatic Coleoptera, which are explained in the general chapter. It may be important to concentrate efforts on collecting in as many different habitats as possible, since diving beetles can be found in a great variety of freshwater environments, and so to get the greatest diversity of a particular area. Collecting immature stages is also important, since the larvae and pupae of many genera and species have not been described. The proper association between larvae and adults can be carried out either by using molecular techniques or by rearing the larvae to obtain the adults.

KEYS TO NEOTROPICAL DYTISCIDAE

Dytiscidae: Subfamilies (adults)

1 Metatarsi without posterolateral lobes (Fig. 15.3.1 B), with one or two claws; body oblong, but wider posteriorly, rounded or with parallel sides (Fig. 15.3.1 D); scutellum fully exposed or concealed 2

1' Metatarsi with posterolateral lobes (Fig. 15.3.1 A), with a single claw; body oblong, wider anteriorly (Fig. 15.3.1 C); scutellum concealed .... **Laccophilinae [p. 541]**

2(1) Prosternal process in same plane as prosternum (Fig. 15.3.1 F); protarsi pentamerous; scutellum exposed, except in *Notaticus* (Aubehydrini), which have the prosternal process very broad with obtuse apex 3

2' Prosternal process declivous, not in same plane as prosternum (Fig. 15.3.1 E); protarsi pseudotetramerous, except in *Bidessonotus*, in which are pentamerous; scutellum concealed, except in *Celina*, which have elytral apices and last abdominal ventrite apices acuminated .... **Hydroporinae [p. 543]**

3(2) Eye anterolateral margin entire, not emarginate above antennal bases (Fig. 15.3.1 G); male protarsomeres I–III explanate, forming together rounded or elliptic adhesive disc (Figs. 15.3.1 I, J) 4

3' Eye anterolateral margin emarginate above antennal bases (Fig. 15.3.1 H); male protarsomeres I–III widened or not, but never forming disc (Fig. 15.3.1 K) 5

4(3) Anterior metatibial spur much wider than posterior (Fig. 15.3.1 L); male protarsomere disc, with filiform adhesive setae and adhesive setae apically with very elongate suckers (Fig. 15.3.1 N) .... **Cybistrinae [p. 545]**

4' Metatibial spurs similar in form and size (Fig. 15.3.1 M); male protarsomere disc, with adhesive setae apically with circular suckers (Fig. 15.3.1 O) .... **Dytiscinae [p. 545]**

5(3) Metafemur without setal series (Fig. 15.3.1 Q) 6

5' Metafemur anterior face near ventroapical angle with a linear series of setae (Fig. 15.3.1 P), except some species of *Platambus* .... **Agabinae [p. 556]**

6(5) Elytron with rounded apex (Fig. 15.3.1 S) 7

6' Elytron with truncate or sinuate apex (Fig. 15.3.1 R); example of habitus in Fig. 15.3.10 A .... Lancetinae, one genus: *Lancetes* [Argentina, Brazil, Chile, Peru]

7(6) Metacoxal lines very closely approximated posteriorly as almost to touch median line (Fig. 15.3.1 T) or absent 8

7' Metacoxal lines broadly separated, never almost touching median line (Fig. 15.3.1 U) or subparallel 9

8(7) Body length <4 mm; gonocoxae with apodema prominently forward (Fig. 15.3.1 V) .... **Hydrodytinae [p. 556]**

8' Body length usually >4 mm, except in *Agaporomorphus*, which have gonocoxae without apodema prominently forward .... **Copelatinae [p. 556]**

9(7) Terminal palpomeres (especially labial) apically notched or emarginate; prosternum anterior margin projecting beyond head margin (Fig. 15.3.1 W); metatarsal claws length equal or subequal .... Coptotominae, one genus: *Coptotomus* [Mexico. Nearctic]

9' Terminal palpomeres not apically emarginate; prosternum anterior margin near head margin (Fig. 15.3.1 X); metatarsal claws neatly unequal in length or not; if equal, body lateral margin between pronotum and elytron distinctly discontinuous .... **Colymbetinae [p. 556]**

Dytiscidae: Laccophilinae: Genera

1 Metatibial spurs acuminate (Fig. 15.3.6 M); body length ≤2.4 mm; elytra with or without a visible lateral border in dorsal view 2

1' Metatibial spurs apically bifid (Fig. 15.3.6 L); body length ≥2.3 mm; elytra without a visible lateral border in dorsal view; example of habitus in Fig. 15.3.11 G .... *Laccophilus* [Cosmopolitan]

2(1) Elytron, with a thin lateral border in dorsal view especially in humeral half; both mesotibial spurs short, as long as first or first two tarsomeres (Fig. 15.3.6 P); metacoxal lines almost straight, weakly narrowed before metacoxal lobes (Fig. 15.3.6 R) 3

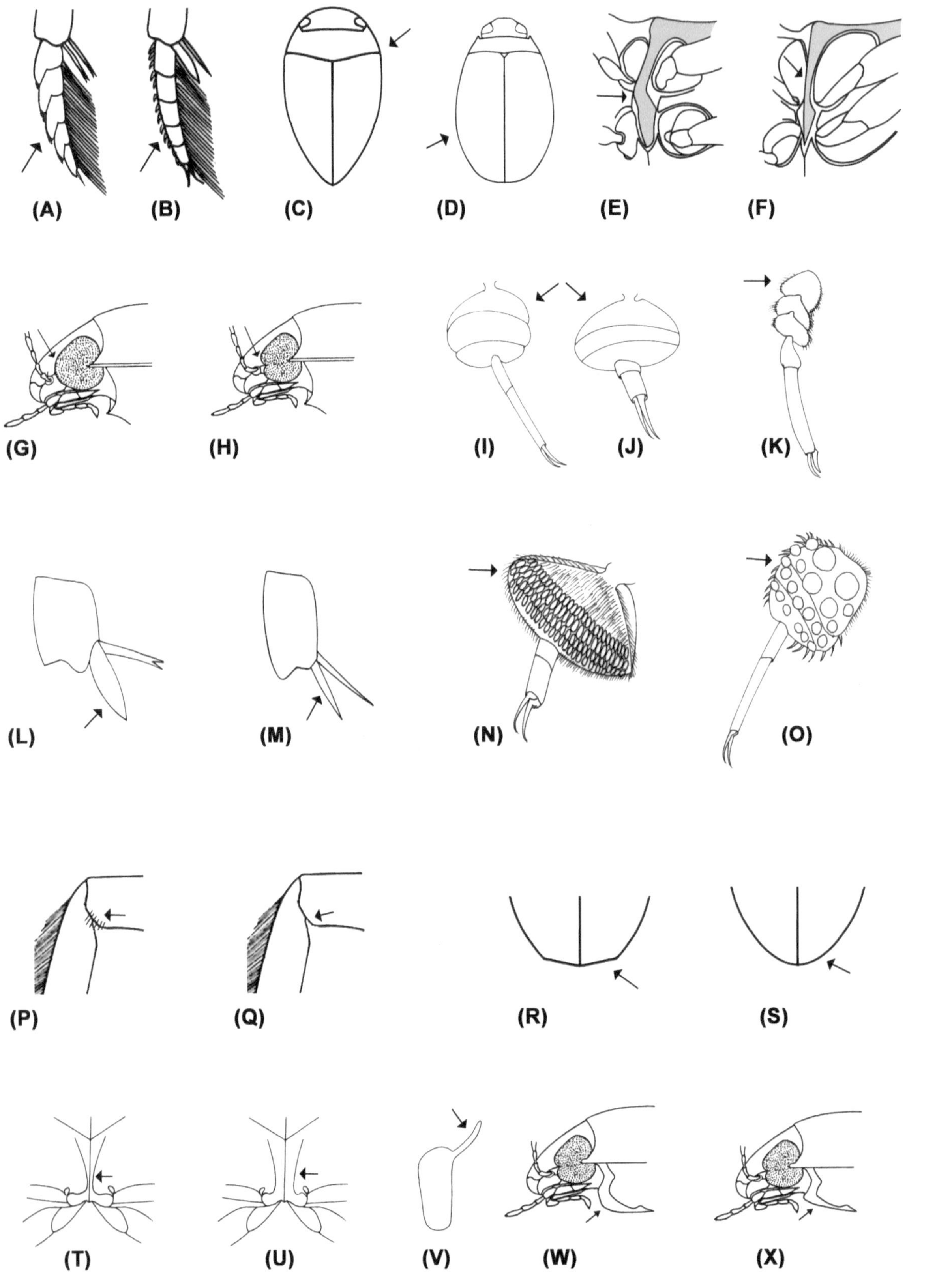

FIGURE 15.3.1 Dytiscidae, adult structures. Metatarsus, *arrow* indicates lobes: (A) *Laccophilus*, (B) *Megadytes*. Habitus, dorsal view, *arrow* indicates maximum width: (C) *Laccophilus*, (D) *Megadytes*. Ventral aspect, *arrow* indicates prosternal process: (E) Hydroporinae, (F) Copelatinae. Head, lateral view, *arrow* indicates eye: (G) Cybistrinae, (H) Colymbetinae. Male Protarsus, *arrow* indicates protarsomeres widened: (I) Dytiscinae, (J) Cybistrinae, (K) Copelatinae. Metatibia, *arrow* indicates anterior spur: (L) *Megadytes*, (M) *Dytiscus*. Male protarsus, *arrow* indicates protarsomere disc: (N) *Megadytes*, (O) *Thermonectus*. Metafemur, *arrow* indicates ventroapical angle: (P) *Platynectes*, (Q) *Rhantus*. Elytron, *arrow* indicates elytral apex: (R) *Lancetes*, (S) *Rhantus*. Metacoxae, *arrow* indicates metacoxal lines: (T) *Copelatus*, (U) *Rhantus*. Left gonocoxa of (V) *Hydrodytes*, *arrow* indicates apodema. Head, lateral view, *arrow* indicates prosternal process: (W) *Coptotomus*, (X) *Rhantus*.

2' Elytra without a visible lateral border in dorsal view; mesotibial spurs very long, longest one as long as mesotarsi (Fig. 15.3.6 O); metacoxal lines sinuate, strongly narrowed before metacoxal lobes (Fig. 15.3.6 Q) .... *Laccomimus*
[Neotropical]

3(2) Metatibiae with single apical spur (Fig. 15.3.6 N); male antennomeres enlarged and flattened .... *Napodytes*
[Ecuador]

3' Metatibiae with two apical spurs (Fig. 15.3.6 M); antennomeres never enlarged .... *Laccodytes*
[Cuba, Venezuela to Brazil. Nearctic]

Dytiscidae: Hydroporinae: Tribes

Modified from Benetti et al. (2003) and Miller & Bergsten (2016).

1 Metacoxal process not divided for posterior excavation (Fig. 15.3.3 D) 2

1' Metacoxal process divided for posterior excavation, forming a medium depressed area and lateral lobe that covers the base of trochanters (Fig. 15.3.3 C) .... **Hydrovatini [p. 543]**

2(1) Scutellum not visible; lateral body margin parallel or not; elytra and abdomen not extended in spiniform processes (Fig. 15.3.3 F)..... 3

2' Scutellum visible; lateral body margins parallel; elytra and abdomen extended in spiniform processes (Fig. 15.3.3 E); example of habitus in Fig. 15.3.10 I .... Methlini, one genus: *Celina*
[Neotropical. Nearctic]

3(2) Mesocoxal cavity closed, with metepisternum in contact with metasternum (Fig. 15.3.3 H); pro- and mesotarsi not lobed 4

3' Mesocoxal cavity open, with metepisternum separated from metasternum (Fig. 15.3.3 G); pro- and mesotarsi lobed .... **Vatellini [p. 543]**

4(3) Metafemora not extending to metacoxal process (Fig. 15.3.3 J); elytron apicolateral apex without any bead 5

4' Metafemora extend to metacoxal process (Fig. 15.3.3 I); elytron apicolateral apex with broad rounded bead; example of habitus in Fig. 15.3.11 A .... Laccornellini, one genus: *Laccornellus*
[Argentina, Brazil, Chile]

5(4) Metacoxae with broadly rounded lobes covering the metatrochanters bases (Fig. 15.3.3 K) 6

5' Metacoxae without broadly rounded lobes covering the metatrochanters bases Fig. 15.3.3 L) 7

6(5) Elytral epipleuron with transverse carina at humeral angle (Fig. 15.3.3 M) .... Hygrotini, one genus: *Hygrotus*
[Mexico. Holarctic]

6' Elytral epipleuron without transverse carina at humeral angle (Fig. 15.3.3 N) .... **Hydroporini [p. 543]**

7(6) Metatarsal claws unequal (Fig. 15.3.3 P) 8

7' Metatarsal claws equal (Fig. 15.3.3 O) .... **Bidessini [p. 544]**

8(7) Body length >3.4 mm; mesocoxae separated by the width of a coxa (Fig. 15.3.3 Q); example of habitus in Fig. 15.3.11 B .... Pachydrini, one genus: *Pachydrus*
[Neotropical. Nearctic]

8' Body length ≤3.5 mm; mesocoxae separated by half the width of a coxa (Fig. 15.3.3 R); example of habitus in Fig. 15.3.11 C .... Hyphidrini, one genus: *Desmopachria*
[Neotropical. Nearctic]

Dytiscidae: Hydroporinae: Hydrovatini: Genera

Modified from Benetti et al. (2003) and Miller & Bergsten (2016).

1 Metacoxal emargination deeper than wide (Fig. 15.3.6 H); labrum visible dorsally .... *Hydrovatus*
[Cosmopolitan]

1' Metacoxal emagination wider than deep (Fig. 15.3.6 I); labrum concealed dorsally; example of habitus in Fig. 15.3.11 E .... *Queda*
[Panama to Argentina]

Dytiscidae: Hydroporinae: Vatellini: Genera

Modified from Miller (2005).

1 Body length >5.0 mm; pronotum broadest at middle (Fig. 15.3.6 J); mesosternum visible ventrally; example of habitus in Fig. 15.3.11 F .... *Vatellus*
[Neotropical]

1 Body length <5.0 mm; pronotum broadest posteriorly (Fig. 15.3.6 K); mesosternum not visible ventrally .... *Derovatellus*
[Antilles, Guatemala to Argentina. Nearctic. Pantropical]

Dytiscidae: Hydroporinae: Hydroporini: Genera

1 Metacoxal process apically truncate or strongly sinuate, with medial line extending caudally as far as lateral lobes (Figs. 15.3.6 A, B); ventral surface smooth, microreticulated with punctae clearly distinguishable; Subtribe Hydroporina 2

1' Metacoxal process apically with lateral lobes divided by a medial notch (Fig. 15.3.6 C); ventral surface rough, microreticulation hardly or not visible and punctae often hardly distinguishable; Subtribe Deronectina 3
2(1) Metacoxal process apically truncate (Fig. 15.3.6 A); metatrochanter ratio of metafemur/metatrochanter length: 2.3–2.8 ...... *Hydroporus* [Mexico. Holarctic]
2' Metacoxal process apically strongly sinuate, with medial line extending caudally as far as lateral lobes (Fig. 15.3.6 B); metatrochanter ratio of metafemur/metatrochanter length: 1.9–2.2 .... *Sanfilippodytes* [Mexico. Nearctic]
3(1) Prosternum and head ventral surface and frons microreticulate, with a horizontal crease (Fig. 15.3.6 D); body elongate, not particularly convex, with angles between elytra and pronotum weak or absent (except two species, with pronotum lateral sides rounded), elytra typically with longitudinal vittae, rarely completely or nearly black .... *Boreonectes* [Mexico. Holarctic]
3' Prosternum and head ventral surface and frons rough and punctate, without visible crease (Fig. 15.3.6 E) .... ***Stictotarsus*** **[p. 544]**

Dytiscidae: Hydroporinae: Hydroporini: *Stictotarsus*: Species Groups

1 Metatibia ventral surface densely punctate, with a medial longitudinal row of larger, spiniferous punctae; prosternal process regularly lanceolate (Fig. 15.3.6 F); body elongate, with elytra vittate; visible angle between pronotum and elytra .... *Stictotarsus duodecimpustulatus*-group [Mexico. Holarctic]
1' Metatibia ventral surface smooth (or microreticulated) except for medial longitudinal row of spiniferous punctae; prosternal process usually broadly lanceolate (Fig. 15.3.6 G); body rather short and convex, elytra normally with transverse bands (rarely black or with longitudinal vittae) and angle between pronotum and elytra weak or absent .... *Stictotarsus roffi*-group [Mexico, Guatemala. Nearctic]

Dytiscidae: Hydroporinae: Bidessini: Genera

Modified from Biström (1988) and Miller & Bergsten (2016).

1 Cervical line absent (Fig. 15.3.4 A) 2
1' Cervical line present (Fig. 15.3.4 B) 14
2(1) Eyes not reduced (Fig. 15.3.4 D) 3
2' Eyes reduced (Fig. 15.3.4 C) .... *Trogloguignotus* [Venezuela]
3(2) Pronotal striae absent (Fig. 15.3.4 E) or faintly impressed, reduced (Fig. 15.3.4 G) 4
3' Pronotal striae present, clearly impressed (Fig. 15.3.4 F) 6
4(3) Pronotal striae absent or reduced (Fig. 15.3.4 G), if absent, then pronotum with two longitudinal, rather shallow depressions on each side (Fig. 15.3.4 H); humeral angle of elytron and metaventrite without carina; parameres two-segmented 5
4' Pronotal striae completely absent (Fig. 15.3.4 E); elytron with longitudinal lateral carina extending from humeral angle and along lateral surface (Fig. 15.3.4 I) or metaventrite lateral surface with longitudinal carinae (Fig. 15.3.4 J) or with both; parameres one-segmented .... *Hydrodessus* [Venezuela to Paraguay]
5(4) Pronotal striae reduced (Fig. 15.3.4 G); last abdominal ventrite posterior edge not margined (Fig 15.3.4 K) .... *Hypodessus* [French Guiana to Argentina]
5' Pronotal striae present; pronotum with two longitudinal, rather shallow depressions on each side (Fig. 15.3.4 H); last abdominal ventrite posterior edge strongly margined (Fig. 15.3.4 L) .... *Tepuidessus* [Venezuela]
6(3) Elytral striae present (Fig. 15.3.4 M) 7
6' Elytral striae absent (Fig. 15.3.4 N) 10
7(6) Clypeal margin not modified (Fig. 15.3.4 P) 8
7' Clypeal margin with a distinct marginal bead (Fig. 15.3.4 O) .... *Belladessus* [Venezuela, Suriname]
8(7) Metatrochanter variable, but not short and apically rounded (Fig. 15.3.4 R); lateral body outline continuous 9
8' Metatrochanter short, apically strongly rounded (Fig. 15.3.4 Q); lateral body outline discontinuous .... *Zimpherus* [Venezuela]
9(8) Body length <2.0 mm; median lobe terminating in four processes (Fig. 15.3.4 S) .... *Microdessus* [Brazil, Suriname]
9' Body length >2.0 mm; median lobe not terminating in four processes (Fig. 15.3.4 T) .... *Uvarus* [Mexico to Northern South America. Holarctic. Afrotropical. Oriental]
10(6) Parameres of two articles each (Fig. 15.3.4 V) 11
10' Parameres of one article each (Fig. 15.3.4 U) .... *Amarodytes* [South America]
11(10) Abdominal ventrites 3–5 posterior margin without denticles (Fig. 15.3.5 B); dorsum not iridescent; metatibiae without natatory setae 12

FAMILY DYTISCIDAE

11' Abdominal ventrites 3–5 posterior margin with series of minute denticles (Fig. 15.3.5 A); dorsum iridescent; metatibiae with natatory setae .... *Bidessodes*
[Guatemala to Argentina]

12(11) Body elongate, lateral outline continuous or discontinuous (Fig. 15.3.5 D); legs with natatory setae 13

12' Body robust, very broad, and broadly rounded laterally with a broad pronotal bead; lateral outline discontinuous (Fig. 15.3.5 C); legs without natatory setae .... *Spanglerodessus*
[Venezuela, Guyana]

13(12) Lateral outline continuous (Fig. 15.3.5 D); metatrochanters very large; male median lobe with ventral sclerite .... *Fontidessus*
[Venezuela, Guyana, Suriname]

13' Lateral outline discontinuous (Fig. 15.3.5 E); metatrochanters short; male median lobe without ventral sclerite .... *Novadessus*
[Peru]

14(1) Epipleura with basal cavity which is limited posteriorly by a transverse carina (Fig. 15.3.5 F) 15

14' Epipleura without transverse carina (Fig. 15.3.5 G) 16

15(14) Clypeus anterior margin with two prominences, one on each side (Fig. 15.3.5 H); pronotal striae present (Fig. 15.3.5 H) ... *Brachyvatus*
[Neotropical. Nearctic]

15' Clypeus anterior margin without prominences (Fig. 15.3.5 I); pronotal striae very short, nearly absent (Fig. 15.3.5 I) .... *Hemibidessus*
[South America]

16(14) Elytral striae present (Fig. 15.3.5 J) 17

16' Elytral striae absent (Fig. 15.3.5 K) .... *Incomptodessus*
[Venezuela]

17(16) Elytra without longitudinal carinae (Fig. 15.3.5 M) 18

17' Elytra with longitudinal carinae (Fig. 15.3.5 L) .... *Anodocheilus*
[Neotropical. Nearctic]

18(17) Anterior clypeal margin not modified (Fig. 15.3.5 O) 19

18' Anterior clypeal margin bordered anteriorly (Fig. 15.3.5 N) .... *Neoclypeodytes*
[Mexico to Panama, Antilles. Nearctic]

19(18) Pro and mesotarsi pseudotetramerous, with the fourth concealed in lobes of third (Fig. 15.3.5 Q) 20

19' Pro and mesotarsi pentamerous (Fig. 15.3.5 P) .... *Bidessonotus*
[Neotropical. Nearctic]

20(19) Parameres with the second segment not hooked (Fig. 15.3.5 S); median lobe variable 21

20' Parameres with the second segment hooked (Fig. 15.3.5 R); median lobe long, slender .... *Allodessus*
[Easter Island. Australian, Oriental, Palearctic]

21(20) Elytra with accessory stria between suture and basal stria (Fig. 15.3.5 T); last abdominal ventrite broad, not impressed at sides; example of habitus in Fig. 15.3.11 D .... *Neobidessus*
[Neotropical. Nearctic]

21' Elytra without accessory stria between suture and basal stria (Fig. 15.3.5 U); last abdominal ventrite narrow, impressed on either side .... *Liodessus*
[Neotropical. Nearctic. Afrotropical]

Dytiscidae: Cybistrinae: Genera

1 Both male and female with two metatarsal claws (Figs 15.3.2 N, O), females with posterior claw small (Fig. 15.3.2 O); posterior metatibial spur acute, bi-, or trifurcate; example of habitus in Fig. 15.3.10 E .... *Megadytes*
[Neotropical. Nearctic]

1' Both male and female with a single metatarsal claw (Fig. 15.3.2 P) or female with a second small claw; posterior metatibial spur acute .... *Cybister*
[Cosmopolitan]

Dytiscidae: Dytiscinae: Genera

1 Smaller beetles, body length 8–20 mm; metatarsomeres I–IV posterior margin with a dense fringe of golden setae (Fig. 15.3.2 R) 2

1' Large beetles, body length 20–40 mm; metatarsomeres I–IV posterior margin beardless, without a dense fringe of golden setae (Fig. 15.3.2 Q) .... Dytiscini, one genus: *Dytiscus*
[Mexico. Holarctic]

2(1) Scutellum exposed; prosternal process slender with acute apex (Fig. 15.3.2 T); elytral color pattern variable 3

2' Scutellum concealed; prosternal process very broad with lateral margin parallel and obtuse apex (Fig. 15.3.2 S); elytron dark with a transversal subbasal pale band; example of habitus in Fig. 15.3.10 F .... Aubehydrini, one genus: *Notaticus*
[South America]

3(2) Prosternal process with apex sharply rounded; meso- and metatarsomeres without appressed, flattened setae; elytron without posterolateral spines (Fig. 15.3.2 V) 4

3' Prosternal process with apex sharply pointed; metatarsomeres with appressed, flattened setae; elytron with short, curved spines in posterolateral margin (Fig. 15.3.2 U) .... Eretini, one genus: *Eretes*
[Cosmopolitan]

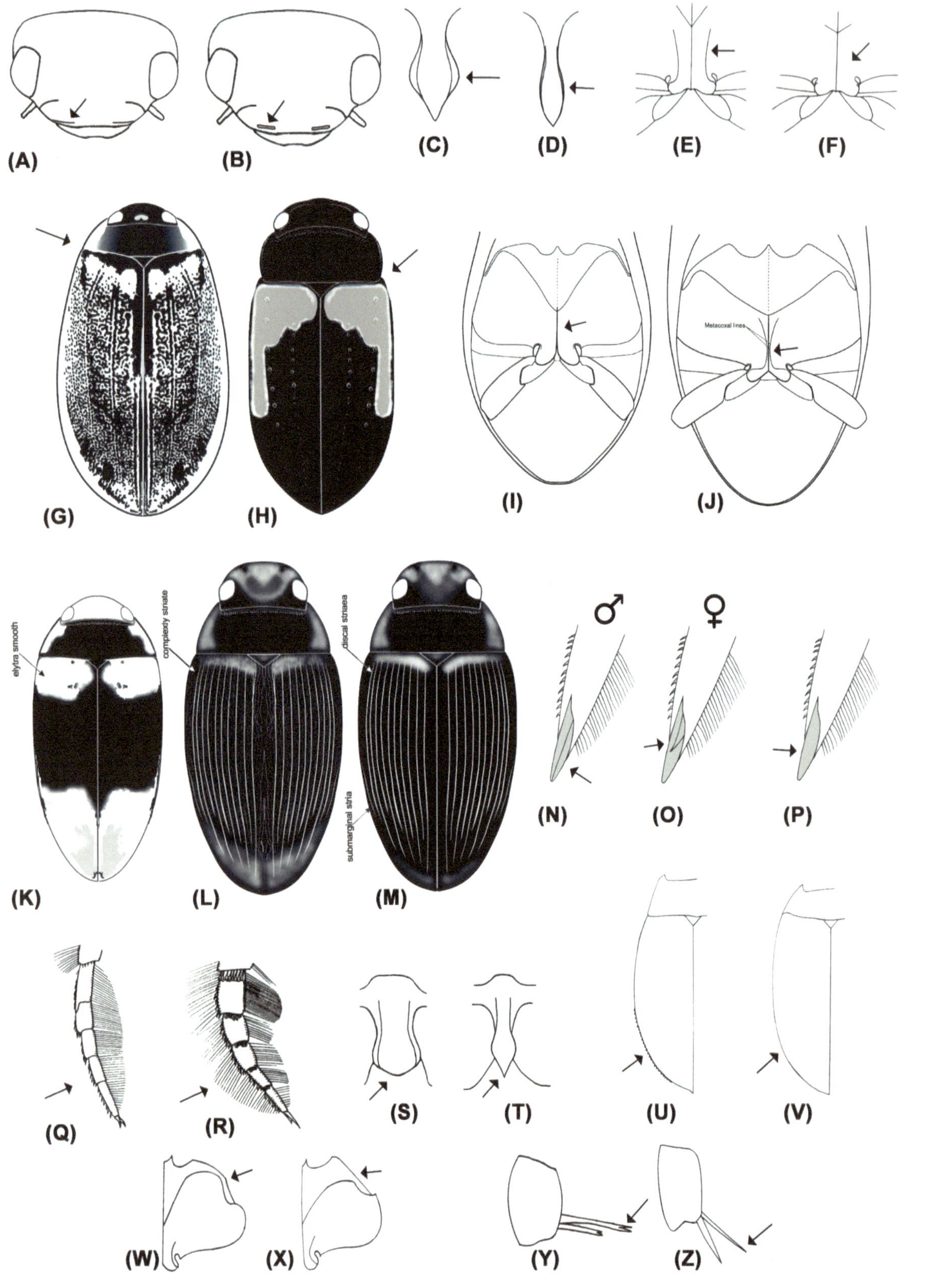

FIGURE 15.3.2 Dytiscidae, adult structures. Head, anterior view, *arrow* indicates clypeus: (A) *Ilybiosoma*, (B) *Platynectes*. Prosternal process (*arrows*): (C) *Platambus*, (D) *Ilybiosoma*. Metacoxae: (E) *Platynectes*, *arrow* indicates metacoxal lines, (F) *Andonectes*, *arrow* indicates absence of metacoxal lines. Habitus, dorsal view, *arrow* indicates lateral outline: (G) *Rhantus calidus* (Fabricius), (H) *Bunites distigma* (Brullé). Thorax, ventral view: (I) *Aglymbus pubis* (García), *arrow* indicates absence of metacoxal lines, (J) *Copelatus posticatus* (Fabricius), *arrow* indicates metacoxal lines. *Copelatus* species, habitus in dorsal view, *arrow* indicates elytral surface: (K) *Copelatus darlingtoni* Young; *Copelatus danyi* Megna & Epler; (L) female, (M) male. Last metatarsomere, *arrow* indicates metatarsal claw: (N–O) *Megadytes*, (P) *Cybister*. Metatarsus, *arrow* indicates posterior margin: (Q) *Dytiscus*, (R) *Thermonectus*. Prosternal process (*arrows*): (S) *Notaticus*, (T) *Thermonectus*. Dorsal aspect, *arrow* indicates elytral margin: (U) *Eretes*, (V) *Thermonectus*. Metacoxa and metaventrite, *arrow* indicates metasternal wing: (W) *Thermonectus*, (X) *Hydaticus*. Metatibia, *arrow* indicates metatibial spur: (Y) *Thermonectus*, (Z) *Hydaticus*.

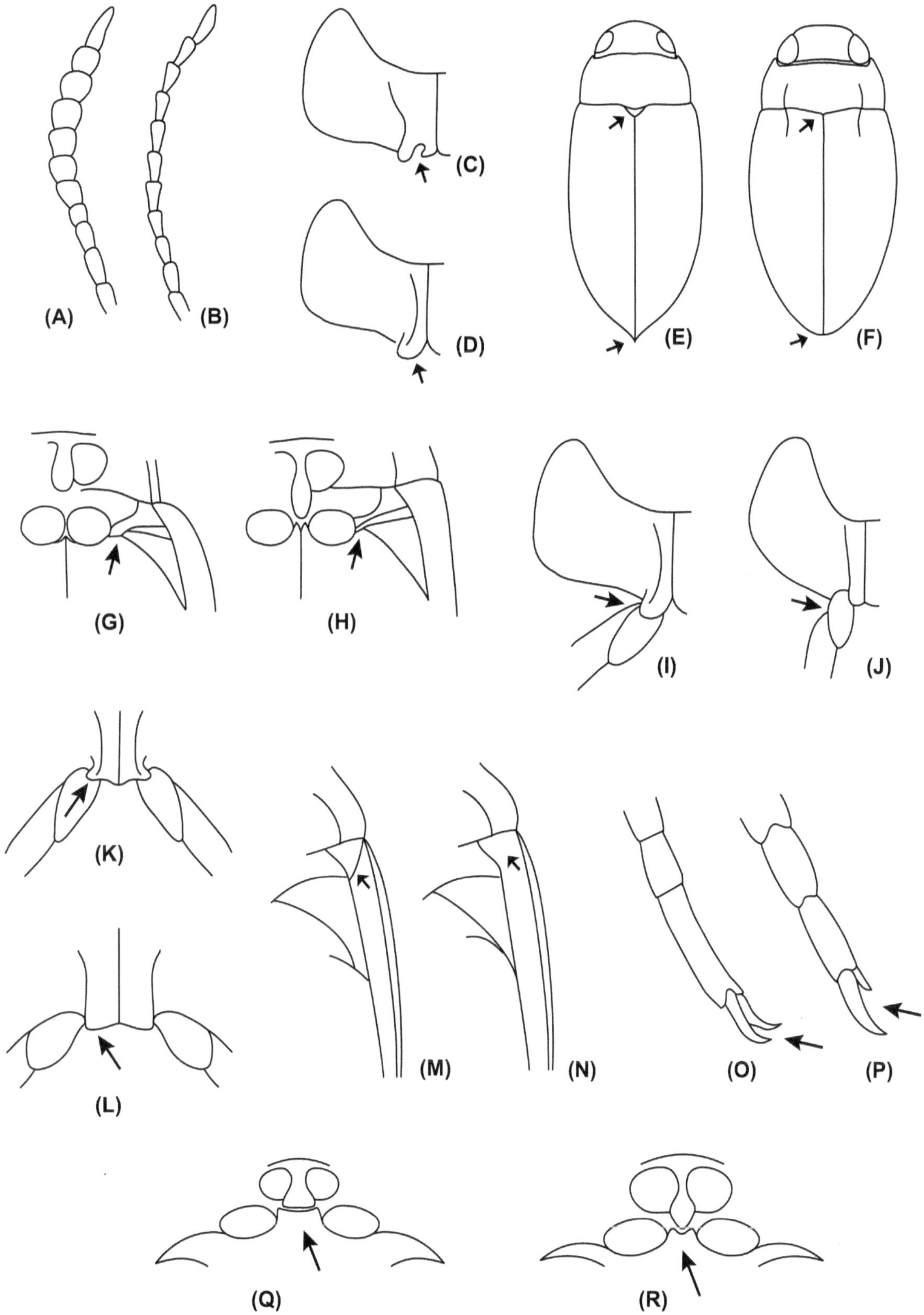

FIGURE 15.3.3 Dytiscidae, adult structures. Antenna: (A) *Microhydrodytes*, (B) *Hydrodytes*. Right metacoxa, *arrow* indicates metacoxal process: (C) *Hydrovatus*, (D) *Laccornellus*. Habitus, dorsal view: (E) *Celina*, *arrow* indicates scutellum visible and elytral apex, (F) *Liodessus*, *arrow* indicates scutellum not visible and elytral apex. Thorax, ventral view, *arrow* indicates mesocoxal cavity: (G) *Vatellus*, (H) *Laccornellus*. Right metacoxa and hindleg, *arrow* indicates metafemora: (I) *Laccornellus*, (J) *Liodessus*. Metacoxae and metatrochanters, *arrow* indicates metacoxae lobes: (K) *Stictotarsus*, (L) *Pachydrus*. Ventral surface showing epipleuron: (M) *Hygrotus*, *arrow* indicates carina, (N) *Stictotarsus*, *arrow* indicates absence of carina. Metatarsomere, *arrow* indicates metatarsal claws: (O) *Liodessus*, (P) *Desmopachria*. Thorax, ventral view, *arrow* indicates the distance between mesocoxae: (Q) *Pachydrus*, (R) *Desmopachria*.

FAMILY DYTISCIDAE

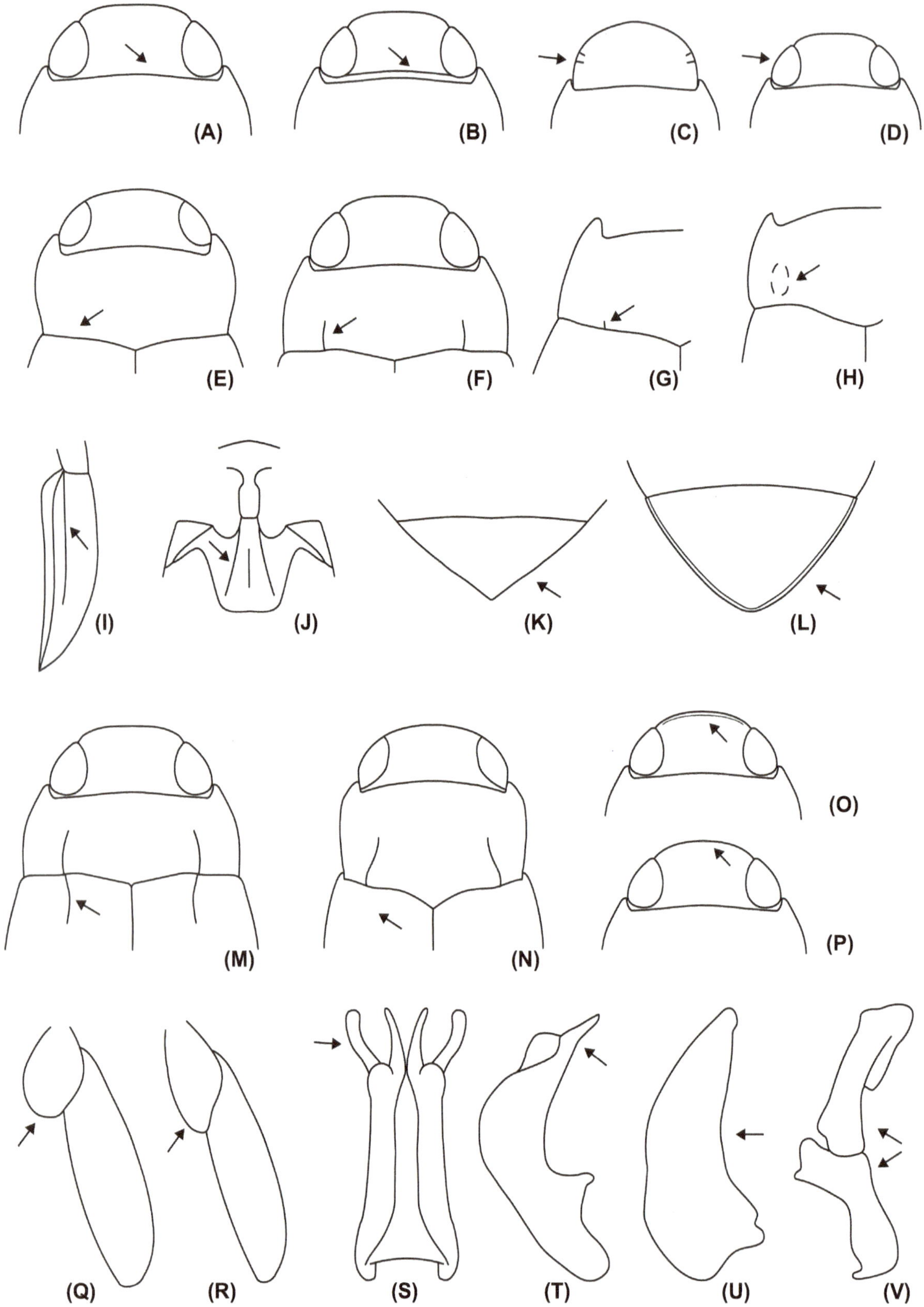

FIGURE 15.3.4 Dytiscidae, adult structures. Head, dorsal view: (A) *arrow* indicates absence of cervical line (e.g., in *Bidessodes*), (B) *arrow* indicates cervical line (e.g., in *Liodessus*). *Arrow* indicates eyes in (C) *Trogloguignotus*, D) *Bidessodes*. Head and pronotum, dorsal view: (E) *Hydrodessus*, *arrow* indicates absence of pronotal stria, (F) *Bidessodes*, *arrow* indicates pronotal stria. Pronotum, left side in dorsal view: G) *Hypodessus*, *arrow* indicates pronotal stria, H) *Tepuidessus*, *arrow* indicates depression. *Hydrodessus* adult: (I) elytron, lateral view, *arrow* indicates carina, (J) thorax, ventral view, *arrow* indicates carina. Last abdominal sternites (*arrow*): (K) *Hypodessus*, (L) *Tepuidessus*. Head, pronotum, and base of elytra, dorsal view: (M) *Uvarus*, *arrow* indicates elytral stria, (N) *Amarodytes*, *arrow* indicates absence of stria. Head, dorsal view, *arrow* indicates clypeal margin: (O) *Belladessus*, (P) *Microdessus*. Metatrochanter (*arrow*) and metafemur: (Q) *Zimpherus*, (R) *Uvarus*. Median lobe, (S) dorsal view of *Microdessus atomarius* (Sharp), *arrow* indicates processes Median lobe, (T) lateral view of *Uvarus lacustris* (Say), *arrow* indicates apex. Right paramere, lateral view: (U) *Amarodytes percosioides* Régimbart, *arrow* indicates single segment, (V) *Bidessodes semistriatus* Régimbart, *arrows* indicate the two segments. *(H, S–V) Modified from Biström (1988); (I–L, Q–R) Modified from Miller & Bergsten (2016); (O) Modified from Miller & Short (2015).*

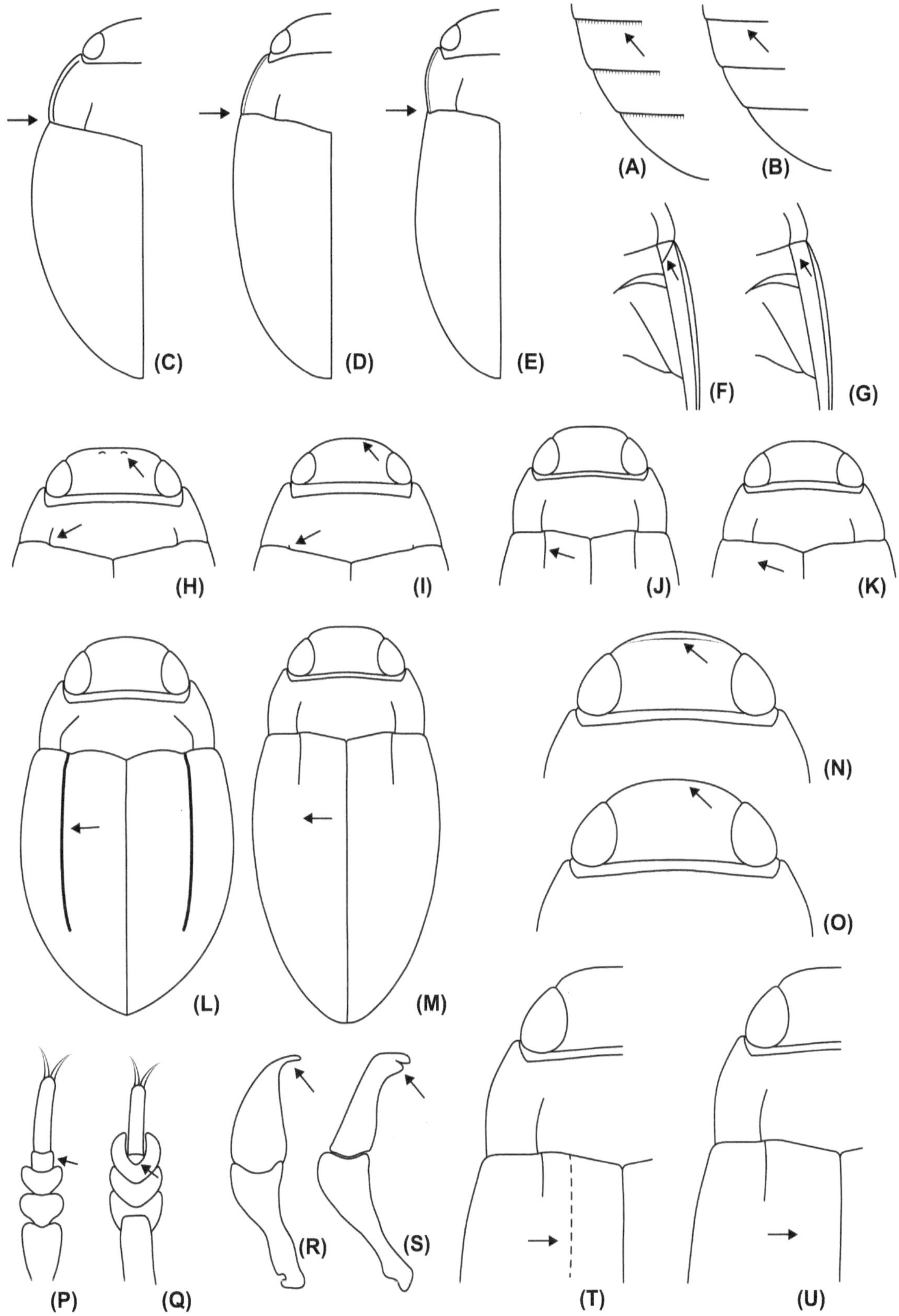

FIGURE 15.3.5 Dytiscidae, adult structures. Abdominal ventrites, *arrow* indicates posterior margin: (A) *Bidessodes*, (B) *Fontidessus*. Habitus, dorsal view, *arrow* indicates lateral outline: (C) *Spanglerodessus*, (D) *Fontidessus*, (E) *Novadessus*. Ventral surface showing epipleuron: (F) *Hemibidessus*, *arrow* indicates carina, (G) *Neobidessus*, *arrow* indicates absence of carina. Head and pronotum, *arrow* indicates pronotal stria and clypeal margin: (H) *Brachyvatus*, (I) *Hemibidessus*. Head, pronotum, and base of elytra: (J) *Liodessus*, *arrow* indicates elytral stria, (K) *Incomptodessus*, *arrow* indicates absence of elytral stria. Habitus, dorsal view: (L) *Anodocheilus*, *arrow* indicates longitudinal carina, (M) *Liodessus*, *arrow* indicates absence of longitudinal carina. Head, *arrow* indicates clypeal margin: (N) *Neoclypeodytes*, (O) *Bidessonotus*. Protarsus, *arrow* indicates fourth lobe: (P) *Bidessonotus*, (Q) *Neobidessus*. Parameres, lateral view, *arrow* indicates apex of second segment: (R) *Allodessus bistrigatus* Clark, (S) *Liodessus affinis* (Say). Head, pronotum and base of elytron: (T) *Neobidessus*, *arrow* indicates accessory stria, (U) *Liodessus*, *arrow* indicates absence of accessory stria. *(C, K) Modified from Miller & Garcia (2011); (E) modified from Miller (2016); (R–S) modified from Biström (1988).*

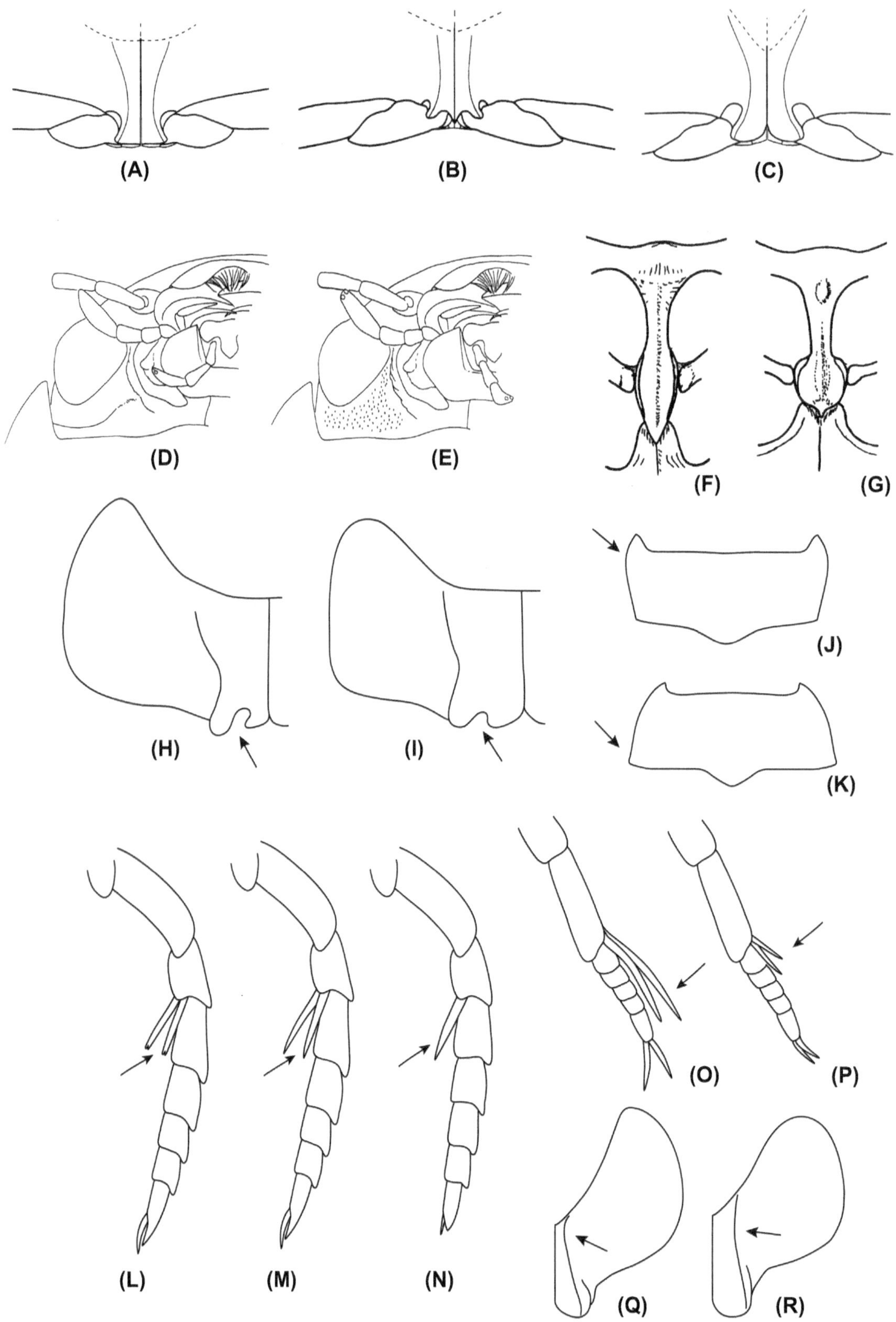

FIGURE 15.3.6 Dytiscidae, adult structures. Metacoxal process: (A) *Hydroporus*, (B) *Sanfilippodytes*, (C) *Boreonectes*. Ventral surface of head, right half: (D) *Boreonectes*, (E) *Stictotarsus*. Prosternal process: (F) *Stictotarsus grammicus* (Sharp), (G) *S. roffii* (Clark). Right metacoxa, *arrow* indicates emargination: (H) *Hydrovatus*, (I) *Queda*. Pronotum, *arrow* indicates maximum width: (J) *Vatellus*, (K) *Derovatellus*. Left hindleg, *arrow* indicates tibial spur: (L) *Laccophilus*, (M) *Laccodytes*, (N) *Napodytes*. Left midleg, *arrow* indicates tibial spur: (O) *Laccomimus*, (P) *Laccodytes*. Left metacoxa, *arrow* indicates metacoxal line: (Q) *Laccomimus*, (R) *Laccodytes*. *(G) From Zimmermann & Smith (1975); (N) Modified from Miller & Bergsten (2016).*

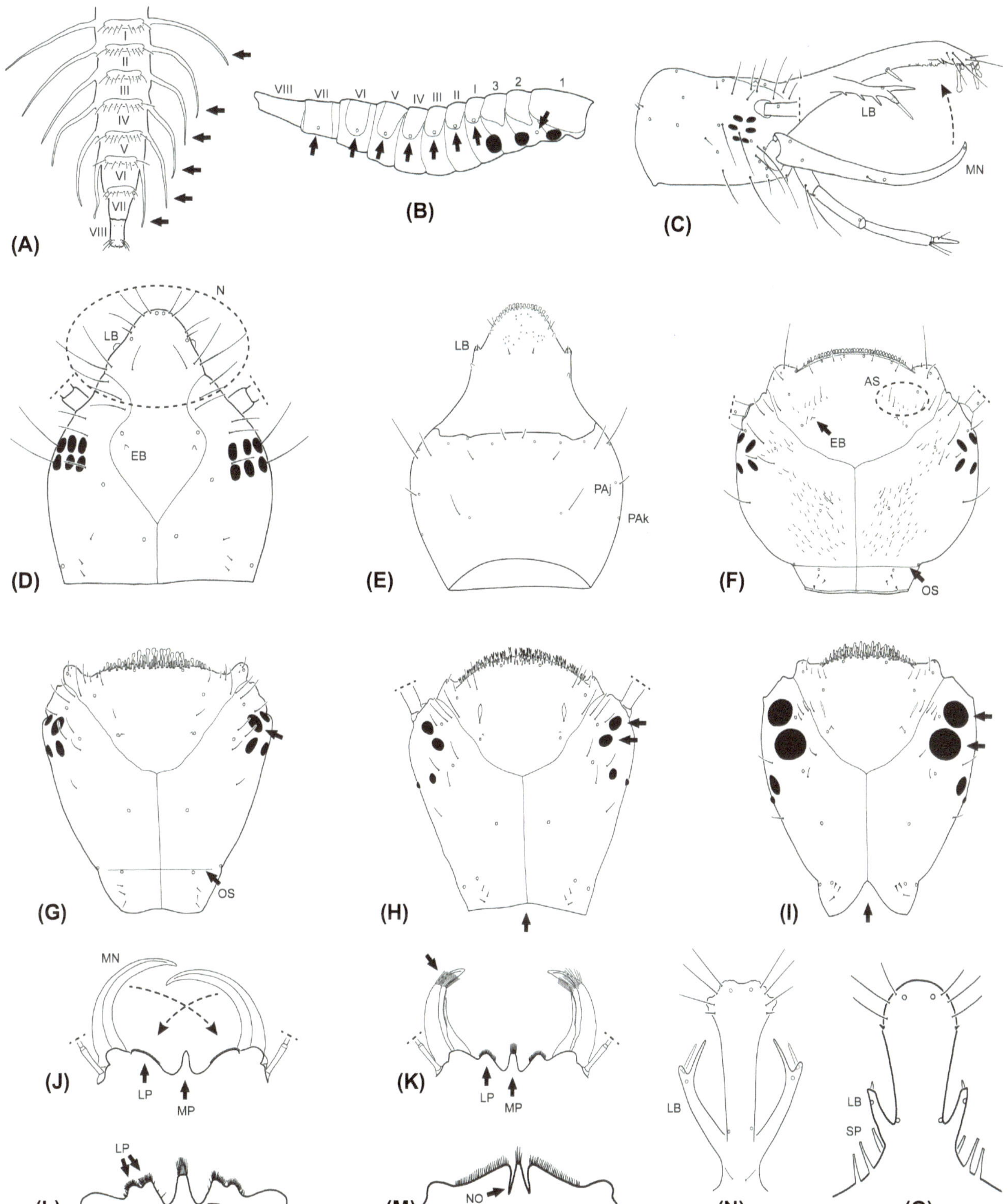

FIGURE 15.3.7 Dytiscidae, larval structures: (A) *Coptotomus*, instar I, abdomen, dorsal view, *arrow* indicates tracheal gills; (B) *Rhantus*, instar III, thorax, and abdomen, lateral view; *arrow* indicates spiracles; (C) *Vatellus*, instar I, head, lateral view, *arrow* indicates orientation of mandibles; (D) *Liodessus*, instar I, head, dorsal view; (E) Hydroporini, instar I, head, ventral view; (F) *Bunites*, instar I, head, dorsal view, *arrows* indicate egg burster and occipital suture; (G) *Hydaticus*, instar I, head, dorsal view, *arrow* indicates stemmata and occipital suture. *Arrow* indicates stemmata and occipital foramen: (H) *Eretes*, instar I, head, dorsal view; (I) *Thermonectus*, instar I, head, dorsal view. *Arrow* indicates mandible orientation and frontoclypeus: (J) *Megadytes* (*Trifurcitus*), instar III, anterior portion of head, dorsal view; (K) *Megadytes* (*Bifurcitus*), instar III, anterior portion of head, dorsal view; (L) *Megadytes* (*Paramegadytes*), instar III, anterior portion of frontoclypeus, dorsal view, *arrows* indicate lateral projections and notches; (M) *Megadytes* (s. str.), instar III, anterior portion of frontoclypeus, dorsal view, *arrow* indicates notches; (N) *Vatellus*, instar I, nasale, dorsal view; (O) *Pachydrus*, instar I, nasale, dorsal view. *AS*, additional setae; *EB*, egg burster; *LB*, lateral branch of nasale; *LP*, lateral projection of frontoclypeus; *MN*, mandible; *MP*, medial projection of frontoclypeus; *N*, nasale; *NO*, notch; *OS*, occipital suture; *SP*, spinula.

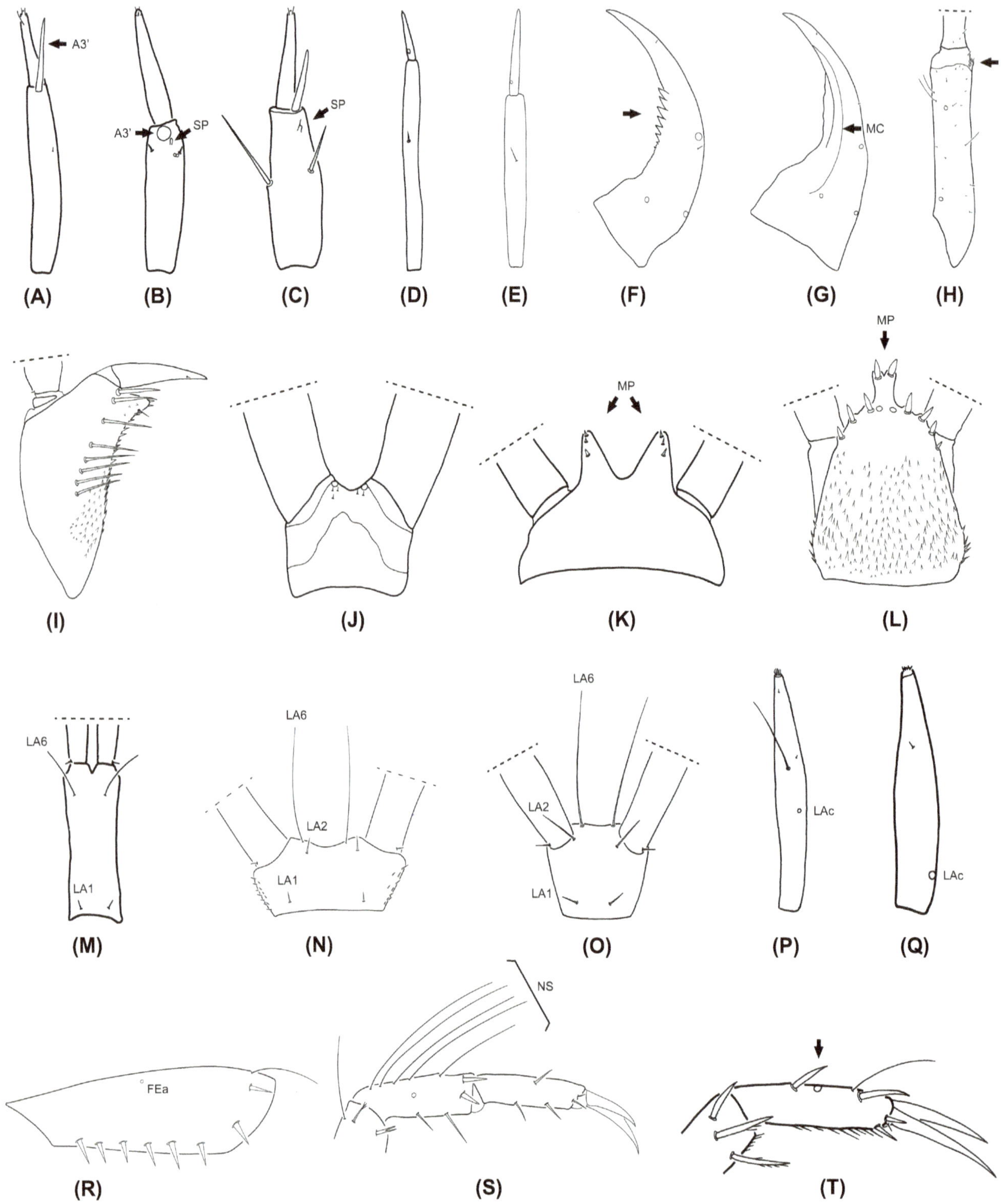

FIGURE 15.3.8 Dytiscidae, larval structures: (A) *Laccophilus*, instar I, ventral view, *arrow* indicates left antennomere 3; (B) *Leuronectes*, instar I, ventral view, *arrows* indicate left antennomere 3 and spinula; (C) *Hypodessus*, instar I, left antennomeres 3 and 4, ventral view, *arrow* indicates spinula; (D) *Vatellus*, instar I, left antennomeres 3 and 4, dorsal view; (E) *Derovatellus*, instar I, left antennomeres 3 and 4, dorsal view; (F) *Copelatus*, instar I, right mandible, dorsal view, *arrow* indicates denticle; (G) *Leuronectes*, instar I, right mandible, dorsal view, *arrow* indicates channel; (H) *Megadytes*, instar III, right stipes, ventral view, *arrow* indicates absence of galea; (I) *Thermonectus*, instar I, left stipes, dorsal view; (J) *Notaticus*, instar I, prementum, dorsal view; (K) *Hydaticus*, instar I, prementum, dorsal view, *arrow* indicates median process; (L) *Thermonectus*, instar I, prementum, dorsal view; (M) *Desmopachria*, instar I, prementum, ventral view; (N) Hydroporini, instar I, prementum, ventral view; (O) *Liodessus*, instar I, prementum, ventral view; (P) *Lancetes*, instar I, left labial palpomere 2, ventral view; (Q) *Leuronectes*, instar I, left labial palpomere 2, ventral view; (R) Hydroporini, instar III, right metafemur, posterior view; (S) *Brachyvatus*, instar III, right protibia, posterior view; (T) *Liodessus*, instar III, left protarsus, anterior view, *arrow* indicates secondary dorsal seta. *A3'*, ventroapical process of antennomere 3; *MC*, mandibular channel; *MP*, median process of prementum; *NS*, natatory setae; *SP*, spinula.

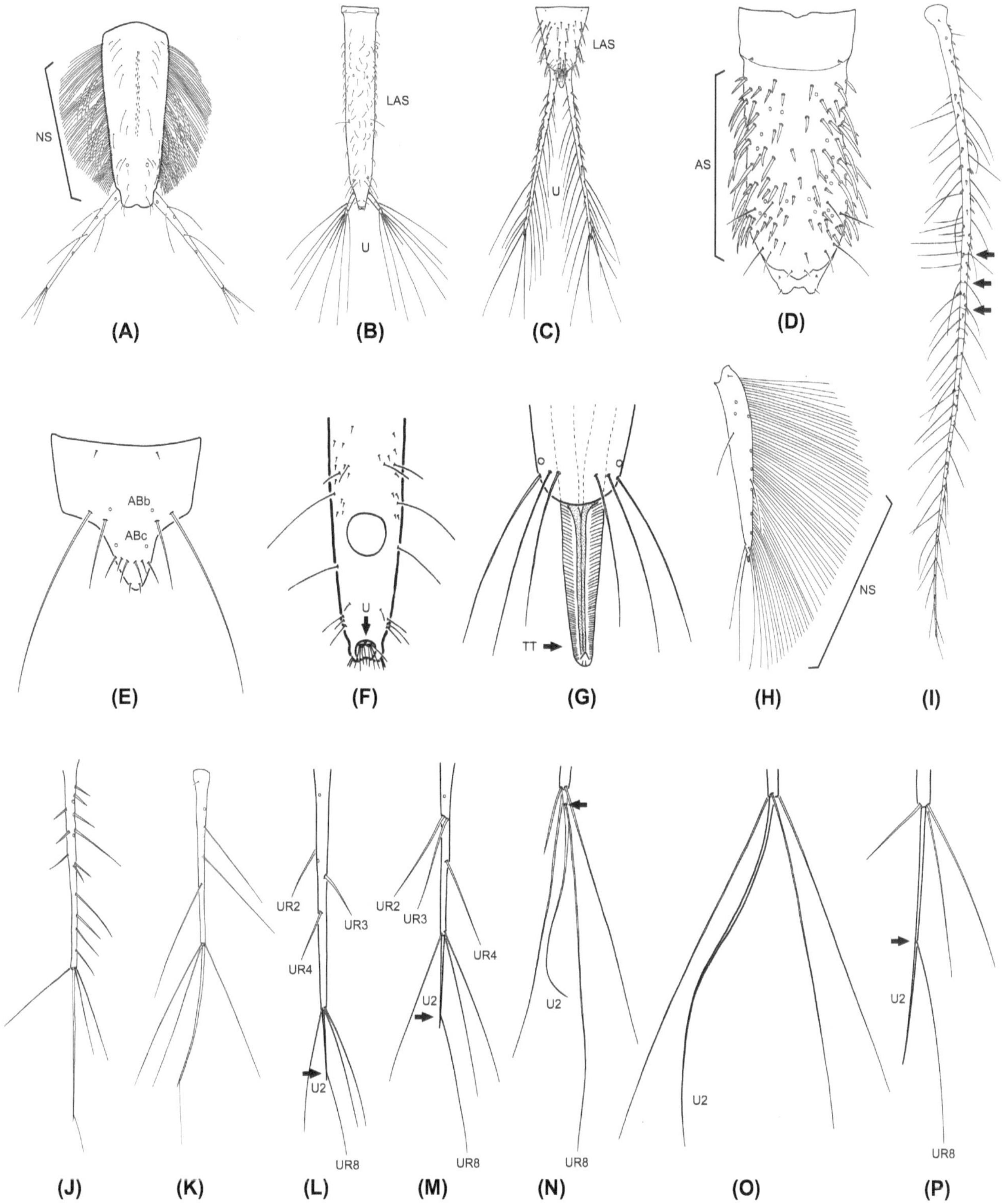

FIGURE 15.3.9 Dytiscidae, larval structures: (A) *Hydaticus*, instar III, last abdominal segment and urogomphi, dorsal view; (B) *Laccomimus*, instar III, last abdominal segment and urogomphi, dorsal view; (C) *Laccophilus*, instar III, last abdominal segment and urogomphi, dorsal view; (D) *Bunites*, instar I, last abdominal segment, dorsal view; (E) Hydroporini, instar I, last abdominal segment, dorsal view; (F) *Megadytes*, instar III, distal portion of last abdominal segment, ventral view, *arrow* indicates urogomphi; (G) *Celina*, instar I, distal portion of last abdominal segment, dorsal view, *arrow* indicates principal abdominal tracheal trunk; (H) *Dytiscus*, instar I, right urogomphus, dorsal view; (I) *Lancetes*, instar III, right urogomphus, dorsal view, *arrow* indicates fractures on urogomphus; (J) *Amarodytes*, instar III, right urogomphus, dorsal view; (K) *Liodessus*, instar III, right urogomphus, dorsal view; (L) *Boreonectes*, instar I, dorsal view, *arrow* indicates right urogomphus; (M) *Hydroporus*, instar I, dorsal view, *arrow* indicates right urogomphus; (N) *Desmopachria*, instar I, right urogomphus, dorsal view, *arrow* indicates insertion of seta UR8; (O) *Laccornellus*, instar I, right urogomphus, dorsal view; (P) *Hygrotus*, instar I, right urogomphus, dorsal view, *arrow* indicates insertion of seta UR8. *AS*, additional setae; *LAS*, last abdominal segment; *NS*, natatory setae; *TT*, tracheal trunks: *U*, urogomphus; *U2*, urogomphomere 2.

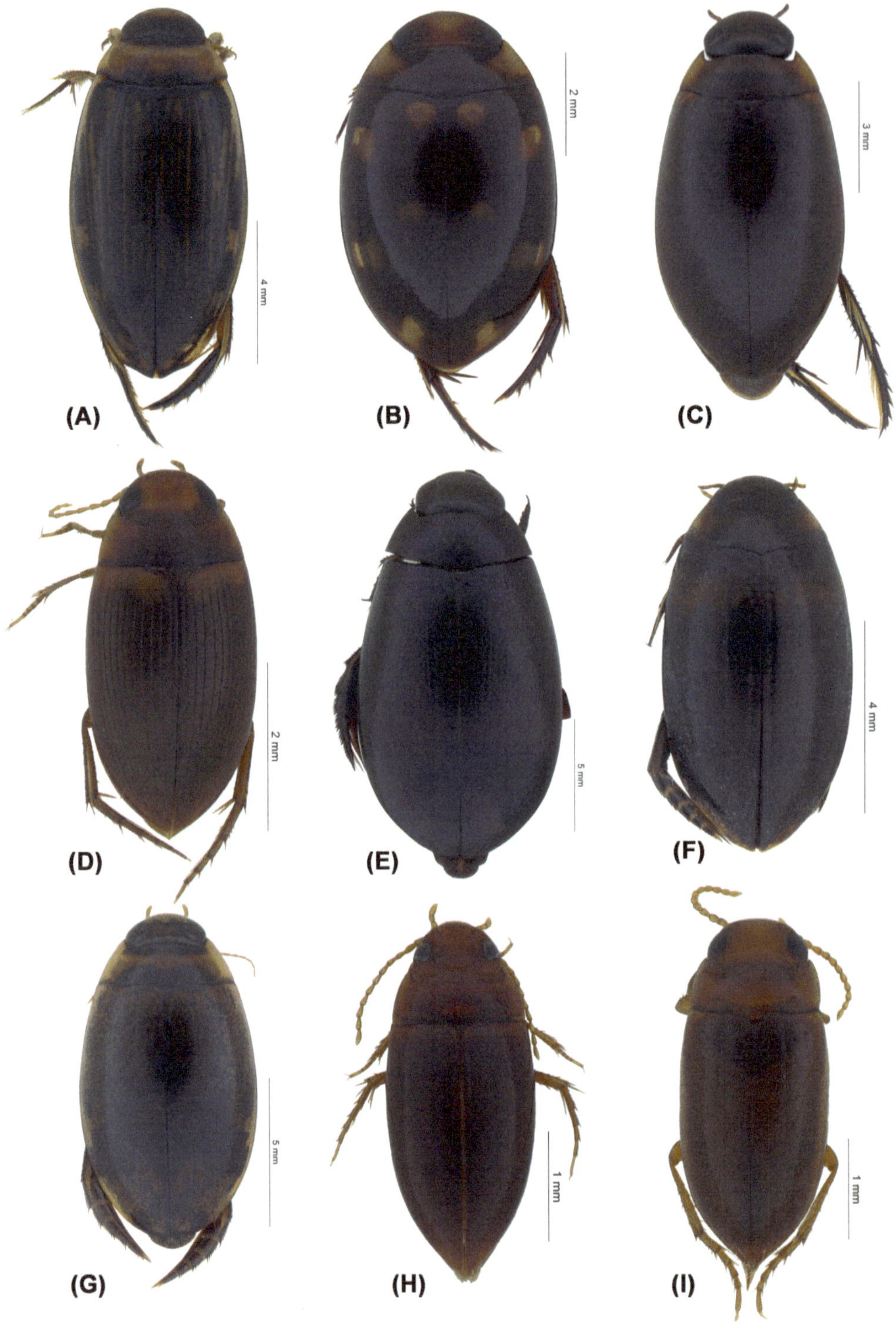

FIGURE 15.3.10 Dytiscidae adults, habitus, dorsal view: (A) *Lancetes*, (B) *Platynectes*, (C) *Rhantus*, (D) *Copelatus*, (E) *Megadytes*, (F) *Notaticus*, (G) *Thermonectus*, (H) *Hydrodytes*, (I) *Celina*.

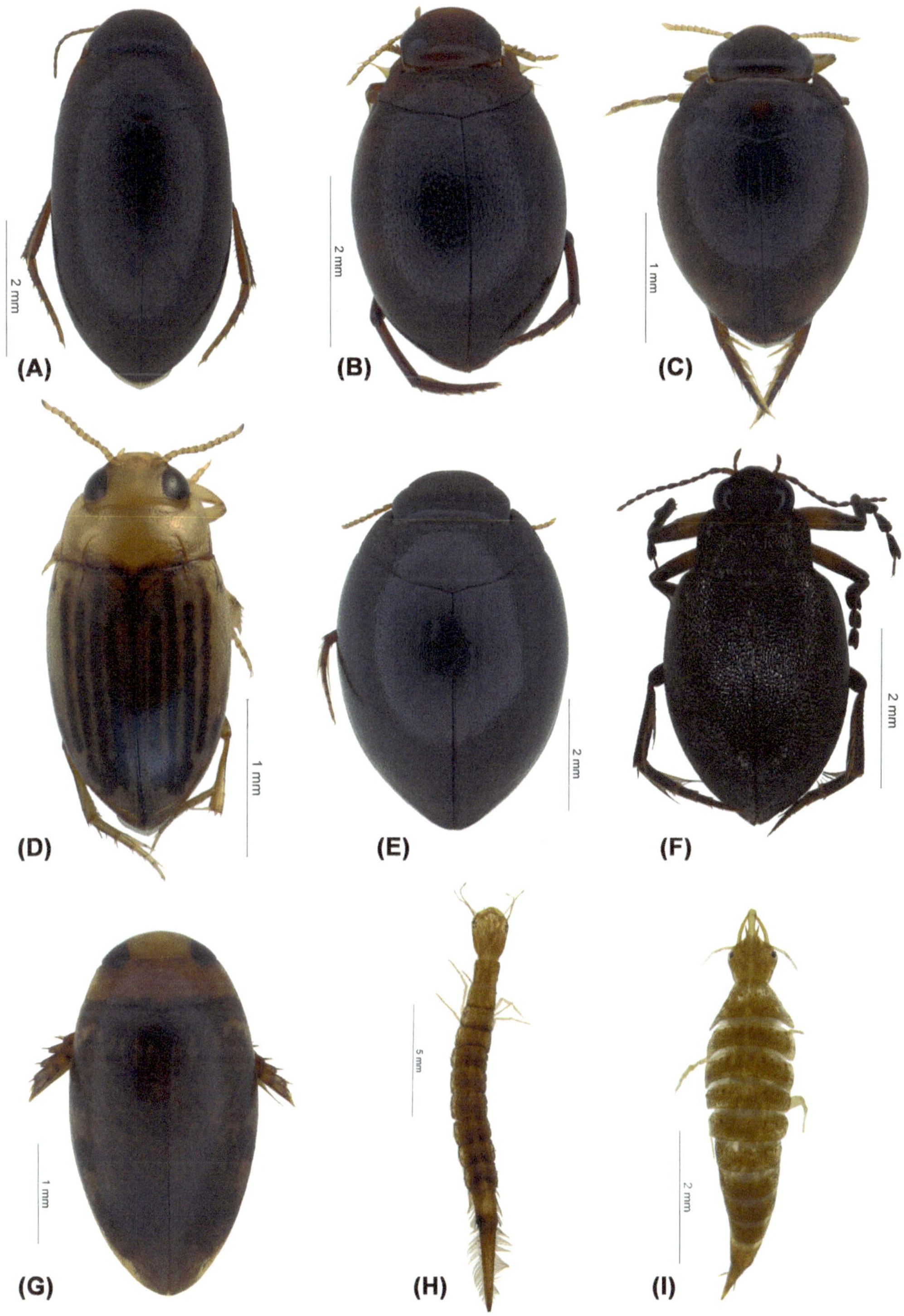

FIGURE 15.3.11 Dytiscidae habitus, dorsal view. Adults: (A) *Laccornellus*, (B) *Pachydrus*, (C) *Desmopachria*, (D) *Neobidessus*, (E) *Queda*, (F) *Vatellus*, (G) *Laccophilus*. Larvae: (H) *Megadytes*, (I) *Pachydrus*.

4(3) Metasternal wing lateral margin arcuate (Fig. 15.3.2 W); metatibial spurs with apex bifid (Fig. 15.3.2 Y); example of habitus in Fig. 15.3.10 G ...... Aciliini, one genus: *Thermonectus*
[Neotropical. Nearctic]

4' Metasternal wing lateral margin straight (Fig. 15.3.2 X); metatibial spurs with acute apex (Fig. 15.3.2 Z) ...... Hydaticini, one genus: *Hydaticus*
[Cosmopolitan]

Dytiscidae: Agabinae: Genera

1 Clypeus anterolateral angle with narrow, linear foveae (Fig. 15.3.2 A) or clypeal margin with marginal groove; Agabini 2

1' Clypeus anterolateral angle with elliptical, sublateral foveae (Fig. 15.3.2 B); Platynectini 3

2(1) Prosternal process lateral bead expanded posterior to procoxae (Fig. 15.3.2 C); mesocoxae broadly separated ...... *Platambus*
[Mexico. Holarctic. Oriental]

2' Prosternal process lateral bead unexpanded (Fig. 15.3.2 D); mesocoxae narrowly separated ...... *Ilybiosoma*
[Mexico. Holarctic. Afrotropical]

3(1) Metacoxal lines present (Fig. 15.3.2 E); example of habitus in Fig. 15.3.10 B ...... *Platynectes*
[Guatemala to Argentina and Chile. Australian. Oriental. Palearctic]

3' Metacoxal lines absent (Fig. 15.3.2 F) ...... *Andonectes*
[Brazil, Colombia, Ecuador and Venezuela]

Dytiscidae: Hydrodytinae: Genera

Modified from Miller (2002).

1 Elytral surface distinctly punctate, without opalescent sheen; body length <2.2 mm; antennomeres short and broad (Fig. 15.3.3 A) ...... *Mycrohydrodytes*
[Brazil]

1' Elytral surface impuctate with an opalescent sheen; body length >2.6 mm; antennomeres long and slender (Fig. 15.3.3 B); example of habitus in Fig. 15.3.10 H ...... *Hydrodytes*
[Honduras to Argentina]

Dytiscidae: Copelatinae: Genera

1 Metacoxal lines strongly convergent anterior to metacoxal lobes, almost touching anterior to metacoxal lobes, then diverging anteriorly (Fig. 15.3.2 J) 2

1' Metacoxal lines absent or faintly visible in ventral view (Fig. 15.3.2 I) ...... *Aglymbus*
[South America]

2(1) Size very small, body length <3.7 mm; metacoxae without oblique, fine strioles; elytra dorsal surfaces with very fine, short striae, evenly distibuted over surface; basal abdominal ventrites with curved, slender, longitudinal strioles ...... *Agaporomorphus*
[South America]

2' Moderate size, body length >4.0; metacoxae with oblique strioles; elytra dorsal surfaces smooth, densely striate or with a submarginal and discal striae (Figs. 15.3.2 K, L, M); abdominal ventrites with short curved strioles; example of habitus in Fig. 15.3.10 D ...... *Copelatus*
[Cosmopolitan]

Dytiscidae: Colymbetinae: Genera

Key modified from Balke et al. (2017).

1 Pronotum in dorsal view with posterior angle not produced, broadly rounded laterally and strongly discontinuous in outline with elytra in dorsal view (Fig. 15.3.2 H); Metatarsal claws length equal or unequal. 2

1' Pronotum in dorsal view with posterior angle produced and continuous in outline with elytra (Fig. 15.3.2 G); metatars lateral claw long and a medial claw rudimentary 3

2(1) Size large (length >13 mm) ...... *Bunites*
[high Andes in Argentina, Bolivia and Peru]

2' Size small (length <10 mm) ...... *Rhantus* (in part)
[Tristan da Cunha and Juan Fernandez]

3(1) Elytral reticulation simple, composed exclusively of fine isodiametric meshes ...... *Meridiorhantus*
[Mexico to Argentina and Antilles]

3' Elytral reticulation simple, composed of large polygonal meshes (meshes sometimes not closed, forming a net); or sculpture double, then large meshes filled with fine polygonal cells; example of habitus in Fig. 15.3.10 C ...... *Rhantus* (in part)
[Mexico to Argentina and Chile]

Dytiscidae: Subfamilies (Larvae)

For distribution of each taxon see Adult keys.

1	Nasale absent (Figs. 15.3.7 F, M); mandibles not oriented obliquely (Figs. 15.3.7 J, K) 2
1'	Nasale present (Figs. 15.3.7 C–E, N, O); mandibles oriented obliquely (Fig. 15.3.7 C) .... **Hydroporinae [p. 557]**
2(1)	Tracheal gills absent on abdominal segments I–VI 3
2'	Tracheal gills present on abdominal segments I–VI (Fig. 15.3.7 A) .... Coptotominae, Coptotomini, one genus: *Coptotomus*
3(3)	Abdominal segment VIII lateral margins with natatory setae (Fig. 15.3.9 A) 4
3'	Abdominal segment VIII lateral margins without natatory setae (Figs. 15.3.9 B–D) 5
4(3)	Frontoclypeus anterior margin sinuate (Figs. 15.3.7 J–M); galea absent (Fig. 15.3.8 H); urogomphi minute (Fig. 15.3.9 F) .... Cybistrinae, one genus: **Megadytes [p. 558]**
4'	Frontoclypeus anterior margin evenly curved (Figs. 15.3.7 G–I); galea (Fig. 15.3.8 I) and urogomphi (Fig. 15.3.9 A) well developed .... **Dytiscinae [p. 558]**
5(3)	Mandibles channeled or at least grooved on medial margin, lacking strong denticles (Fig. 15.3.8 G) 6
5'	Mandibles not channeled, not grooved, medial margin with strong denticles (Fig. 15.3.8 F) .... Copelatinae, one tribe: Copelatini
6(5)	Occipital suture absent 7
6'	Occipital suture present (Fig. 15.3.7 F) .... **Colymbetinae [p. 558]**
7(6)	Urogomphi of one or two articles, without fractures (Figs. 15.3.9 B, C); pore LAc inserted proximally on second labial palpomere (Fig. 15.3.8 Q) 8
7'	Urogomphi unarticulate, with at least three fractures along their length (Fig. 15.3.9 I); pore LAc inserted submedially on second labial palpomere (Fig. 15.3.8 P) .... Lancetinae, Lancetini, one genus: *Lancetes*
8(7)	Antennomere 3 ventroapical spinula absent (Fig. 15.3.8 A) .... **Laccophilinae [p. 558]**
8'	Antennomere 3 ventroapical spinula present (Fig. 15.3.8 B) .... **Agabinae [p. 558]**

Dytiscidae: Hydroporinae: Tribes

1	Principal abdominal tracheal trunks not protruding from siphon apex 2
1'	Principal abdominal tracheal trunks protruding from siphon apex (Fig. 15.3.9 G) .... Methlini, one genus: *Celina*
2(1)	Nasale lateral branches (if present) short (Fig. 15.3.7 O) to minute (Figs. 15.3.7 D, E), not bifid apically 3
2'	Nasale lateral branches strongly developed, bifid apically (Figs. 15.3.7 C, N) .... **Vatellini [p. 557]**
3(2)	Nasale base ventrolateral surface without elongate, robust spinulae 4
3'	Nasale base ventrolateral surface with a row of elongate, robust, spinulae (Fig. 15.3.7 O); example of habitus in Fig. 15.3.11 I .... Pachydrini, one genus: *Pachydrus*
4(3)	Prementum broader than long (Fig. 15.3.8 N) or subequal (Fig. 15.3.8 O); urogomphomere 2 with seta UR8 either absent (Fig. 15.3.9 O) or inserted submedially (Fig. 15.3.9 P) to distally (Figs. 15.3.9 L, M) 5
4'	Prementum clearly longer than broad (Fig. 15.3.8 M); urogomphomere 2 with seta UR8 inserted proximally (Fig. 15.3.9 N) .... Hyphydrini, one genus: *Desmopachria*
5(4)	Ventral surface of abdominal segment VI membranous 6
5'	Ventral surface of abdominal segment VI sclerotized .... Hydrovatini
6(5)	Urogomphal seta UR8 present (Figs. 15.3.9 L, M, P); premental seta LA2 present (Figs. 15.3.8 N, O) 7
6'	Urogomphal seta UR8 absent (Fig. 15.3.9 O); premental seta LA2 absent .... Laccornellini, one genus: *Laccornellus*
7(6)	Urogomphomere 2 with seta UR8 inserted distally (Figs. 15.3.9 L, M) 8
7'	Urogomphomere 2 with seta UR8 inserted submedially (Fig. 15.3.9 P) .... Hygrotini, one genus: *Hygrotus*
8(7)	Femoral pore FEa absent; last abdominal segment pore ABc absent; parietal pore PAj absent .... Bidessini
8'	Femoral pore FEa present (Fig. 15.3.8 R); last abdominal segment pore ABc present (Fig. 15.3.9 E); parietal pore PAj present (Fig. 15.3.7 E) .... **Hydroporini [p. 557]**

Dytiscidae: Hydroporinae: Vatellini: Genera

1	Antennomere 4 length $\leq 0.25\times$ antennomere 3 (Fig. 15.3.8 D) .... *Vatellus*
1'	Antennomere 4 length $\leq 0.40\times$ antennomere 3 (Fig. 15.3.8 E) .... *Derovatellus*

Dytiscidae: Hydroporinae: Hydroporini: Genera

1	Urogomphal setae UR2 and UR3 insertion points separate (Fig. 15.3.9 L) .... *Boreonectes*
1'	Urogomphal setae UR2 and UR3 inserted contiguously (Fig. 15.3.9 M) .... *Hydroporus*

Dytiscidae: Cybistrinae: *Megadytes:* Subgenera

Example of *Megadytes* habitus in Fig. 15.3.11 H, from Michat (2006).

1 Frontoclypeus medial projection truncate apically, with apical setae (Figs. 15.3.7 K–M); mandible distal third with crown of long setae (Fig. 15.3.7 K) 2
1' Frontoclypeus medial projection sharp apically, almost glabrous (Fig. 15.3.7 J); mandible distal third without crown of setae (Fig. 15.3.7 J) .... *Trifurcitus*
2(1) Frontoclypeus notches between medial and lateral projections wide (Figs. 15.3.7 K, L) 3
2' Frontoclypeus notches between medial and lateral projections very narrow (Fig. 15.3.7 M) .... *Megadytes* s. str.
3(2) Frontoclypeus lateral projections unifid (Fig. 15.3.7 K) .... *Bifurcitus*
3' Frontoclypeus lateral projections bifid (Fig. 15.3.7 L) .... *Paramegadytes*

Dytiscidae: Dytiscinae: Genera

1 Prementum median process absent (Fig. 15.3.8 J) 2
1' Prementum median process present (Figs. 15.3.8 K, L) 3
2(1) Occipital suture absent; urogomphus without natatory setae .... Aubehydrini, one genus: *Notaticus*
2' Occipital suture present (e.g., Fig. 15.3.7 G); urogomphus with natatory setae (Fig. 15.3.9 H) .... Dytiscini, one genus: *Dytiscus*
3(1) Stemmata dissimilar in size, anteriodorsal ones from slightly (Fig. 15.3.7 H) to considerably (Fig. 15.3.7 I) larger than the others; prementum median process single or bifid at apex (Fig. 15.3.8 L) 4
3' Stemmata subequal in size or anterodorsal one slightly larger (Fig. 15.3.7 G); prementum median process bifid from base, appearing as two separate lobes (Fig. 15.3.8 K) .... Hydaticini, *Hydaticus*
4(3) Occipital foramen dorsally slightly emarginate (Fig. 15.3.7 H); two anterodorsal stemmata somewhat larger than the others (Fig. 15.3.7 H) .... Eretini, *Eretes*
4' Occipital foramen dorsally deeply emarginate (Fig. 15.3.7 I); two anterodorsal stemmata considerably larger than the others (Fig. 15.3.7 I) .... Aciliini, *Thermonectus*

Dytiscidae: Colymbetinae: Genera

1 Spiracles absent on mesothorax and abdominal segments I–VII 2
1' Spiracles present on mesothorax and abdominal segments I–VII (Fig. 15.3.7 B) .... Instar III
2(1) Frontoclypeus with egg bursters (Fig. 15.3.7 F); Instar I 3
2' Frontoclypeus without egg bursters .... Instar II
3(2) Frontoclypeus and abdominal segment VIII with additional setae absent 4
3' Frontoclypeus and abdominal segment VIII with additional setae present (Figs. 15.3.7 F, 9 D) .... *Bunites*
4(3) Meso- and metafemoral seta FE5 short spiniform .... *Meridiorhantus*
4' Meso- and metafemoral seta FE5 elongate and setaform .... *Rhantus*

Dytiscidae: Laccophilinae: Genera

1 Urogomphus from much shorter to slightly longer than last abdominal segment (Fig. 15.3.9 B) .... *Laccomimus*
1' Urogomphus length >2× last abdominal segment (Fig. 15.3.9 C) .... *Laccophilus*

Dytiscidae: Agabinae: Tribes

1 Antennomere 3 apical lateroventral process protruding .... Agabini
1' Antennomere 3 apical lateroventral process not protruding (Fig. 15.3.8 B) .... Platynectini

ACKNOWLEDGMENTS

We are grateful to Kelly B. Miller for the loan of specimens. The work of Cesar J. Benetti was supported by CNPq and FAPEAM fellowships (processes 151461/2014-7 and 302031/2015-4). INPA/MCTI and CNPq provided financial support and laboratory infrastructure (PROTAX/CNPq - 440616/2015-8). The work of Mariano C. Michat was supported by CONICET under Grant PIP–2011–1087, by ANPCyT under Grant PICT–2014–0853, and by University of Buenos Aires under Grant UBACyT–20020150100170BA. The work of Nelson Ferreira-Jr was supported by FAPERJ and CAPES. The work of Rafael Benzi Braga was supported by PROTAX/CAPES. The work of Yoandri S. Megna was supported by Alexander von Humboldt Foundation through a Hermes fellowship and by the Department of Biology, Universidad de Oriente, Cuba.

FAMILY DYTISCIDAE

REFERENCES

Archangelsky, M., V. Manzo, M.C. Michat & P.L.M. Torres. 2009. Chapter 14: Coleoptera. Pages 411–468 in: E. Domínguez & H.R. Fernández (eds.), Macroinvertebrados bentónicos sudamericanos. Sistemática y biología. Fundación Miguel Lillo, Tucumán, Argentina.

Balke, M. & L. Hendrich. 2016. Chapter 7.6: Dytiscidae Leach, 1815. Pages 118–140 in: R.G. Beutel & R.A.B. Leschen (eds.), Handbook of Zoology, Volume 4/38: Coleoptera, Beetles, Volume 1: Morphology and Systematics (Archostemata, Adephaga, Myxophaga, Polyphaga partim). De Gruyter, Berlin, Germany.

Balke, M.; J. Hájek & L. Hendrich. 2017. Generic reclassification of species formerly included in Rhantus Dejean (Coleoptera, Dytiscidae, Colymbetinae). Zootaxa 4258: 091–100.

Benetti, C.J., J.A. Régil Cueto & G.L. Fiorentin. 2003. Gêneros de Hydradephaga (Coleoptera: Dytiscidae, Gyrinidae, Haliplidae, Noteridae) citados para o Brasil, com chaves para identificação. Biota Neotropica 3: 1–20.

Biström, O. 1988. Generic review of the Bidessini (Coleoptera,Dytiscidae). Acta Zoologica Fennica 184: 1–41.

Biström, O., A.N. Nilsson & G. Wewalka. 1997. A systematic review of the tribes Hyphydrini Sharp and Pachydrini n. trib. (Coleoptera, Dytiscidae). Entomologica Fennica 8: 57–82.

Epler, J.H. 2010. The water beetles of Florida. An identification manual for the families Chrysomelidae, Curculionidae, Dryopidae, Dytiscidae, Elmidae, Gyrinidae, Haliplidae, Helophoridae, Hydraenidae, Hydrochidae, Hydrophilidae, Noteridae, Psephenidae, Ptilodactylidae and Scirtidae. Bureau of Laboratories, Florida Department of Environmental Protection, Tallahassee, FL. 414 pp.

Ferreira-Jr., N., B.H.L. Sampaio, A.S. Fernandes, B. Clarkson, R.B. Braga, M.I.S. Passos & A.D. Santos. 2014. Chapter 21: Ordem Coleoptera. Pages 349–375 in: N. Hamada, J.L. Nessimian & R.B. Querino (eds.), Insetos aquáticos na Amazônia brasileira: taxonomia, biologia e ecologia. Editora do Inpa, Manaus, Brazil.

Garrido, J., C.J. Benetti & A. Pérez-Bilbao. 2011. Chapter: Chrysomelidae, Curculionidae, Dryopidae, Dytiscidae, Elmidae, Gyrinidae, Helophoridae, Haliplidae, Hydraenidae, Hydrochidae, Hydrophilidae, Hydroscaphidae, Hygrobiidae, Noteridae, Psephenidae and Scirtidae. Pages 102–111 in: J. Oscoz, D. Galicia & R. Miranda (eds.), Identification Guide of Freshwater Macroinvertebrates of Spain. Springer, The Netherlands.

Gueorguiev, V.B. 1968. Essai de classification des Coléoptères Dytiscidae. I. Tribus Copelatini (Colymbetinae). Izvestiya na Zoologicheskiya Institut s Muzey (Sofia) 28: 5–45.

Gueorguiev, V.B. 1971. Notes sur les Agabini (Coleoptera, Dytiscidae). I. Les genres Agametrus Sharp, Leuronectes Sharp et Andonectes gen. n. Izvestiya na Zoologicheskiya Institut s Muzey (Sofia) 33: 165–176.

Jäch, M. 1998. Annotated check list of aquatic and riparian/littoral beetle families of the world (Coleoptera). Water Beetles of China 2: 25–42.

Jach, M.A. & M. Balke. 2008. Global diversity of water beetles (Coleoptera) in freshwater. Hydrobiologia 595: 419–442.

Larson, D.J., Y. Alarie & R.E. Roughley. 2000. Predaceous diving beetles (Coleoptera: Dytiscidae) of the Nearctic region, with emphasis on the fauna of Canada and Alaska. NRC Research Press, Ottawa, Canada. 982 pp.

Libonatti, M.L., M.C. Michat & P.L.M. Torres. 2011. Key to the subfamilies, tribes and genera of adult Dytiscidae of Argentina (Coleoptera: Adephaga). Revista de la Sociedad Entomológica Argentina 70: 317–336.

Matta, J.F. 1983. Description of the larva of Uvarus granarius (Aubé) (Coleoptera: Dytiscidae) with a key to the Nearctic Hydroporinae larvae. The Coleopterists Bulletin 37: 203–207.

Michat, M.C. 2006. Descriptions of larvae of Megadytes (Coleoptera: Dytiscidae: Dytiscinae): The hypothesis of monophyletic origin revisited. European Journal of Entomology 103: 831–842.

Michat, M.C., Y. Alarie & K.B. Miller. 2017. Higher-level phylogeny of diving beetles (Coleoptera: Dytiscidae) based on larval characters. Systematic Entomology 42: 734–767.

Michat, M.C., M. Archangelsky & A.O. Bachmann. 2008. Generic keys for the identification of larval Dytiscidae from Argentina (Coleoptera: Adephaga). Revista de la Sociedad Entomológica Argentina 67: 17–36.

Miller, K.B. 2002. Revision of the subfamily Hydrodytinae Miller (Coleoptera: Dytiscidae) with description of a new genus. Insect Systematics and Evolution 33: 1–8.

Miller, K.B. 2005. Revision of the New World and south-east Asian Vatellini (Coleoptera: Dytiscidae: Hydroporinae) and phylogenetic analysis of the tribe. Zoological Journal of the Linnean Society 144: 415–510.

Miller, K.B. 2016. Novadessus viracocha, a new genus and species of Bidessini Sharp from Peru (Coleoptera, Adephaga, Dytiscidae, Hydroporinae). ZooKeys 623: 125–130.

Miller, K.B., J. Bergsten & M.F. Whiting. 2007. Phylogeny and classification of diving beetles in the tribe Cybistrini (Coleoptera, Dytiscidae, Dytiscinae). Zoologica Scripta 36: 41–59.

Miller, K.B. & J. Bergsten. 2014. Chapter 3: The phylogeny and classification of predaceous diving beetles. Pages 49–171 in: D.A. Yee (ed.), Ecology, Systematics, and the Natural History of Predaceous Diving Beetles (Coleoptera: Dytiscidae). Springer, The Netherlands.

Miller, K.B. & J. Bergsten. 2016. Diving Beetles of the World. Systematics and Biology of the Dytiscidae. Johns Hopkins University Press, Baltimore, MD. 336 pp.

Miller, K.B. & M. Garcia, 2011. Spanglerodessus shorti and Incomptodessus camachoi, new genera and species of Bidessini from Guyana and Venezuela (Coleoptera: Dytiscidae: Hydroporinae). Zootaxa 2996: 49–56.

Miller, K.B. & A.E.Z. Short. 2015. Belladessus Miller and Short (Coleoptera: Dytiscidae: Hydroporinae: Bidessini), New genus for two new species from northern South America: Parthenogenetic Diving Beetles? The Coleopterists Bulletin 69: 498–503.

Nilsson, A.N. & J. Hájek. 2017. A World Catalogue of the Family Dytiscidae, or the Diving Beetles (Coleoptera, Adephaga). Version 1.I.2017, 304 pp. URL: http://www.waterbeetles.eu/documents/W_CAT_Dytiscidae_2017.pdf.

Pederzani, F. 1995. Keys to the identification of the genera and subgenera of adults Dytiscidae (sensu lato) of the world (Coleoptera Dytiscidae). Atti della Accademia Roveretana degli Agiati 244, Ser. VII, 5–83.

Perkins, P.D. 1980. Larval and pupal stages of a predaceous diving beetle, Neoclypeodytes cinctellus (Leconte) (Dytiscidae: Hydroporinae: Bidessini). Proceedings of the Entomological Society of Washington 82: 474–481.

Spangler, P.J. 1972. A new genus and new species of water beetle from Bolivia with a key to the genera of the Western Hemisphere Colymbetini (Coleoptera: Dytiscidae). Proceedings of the Biological Society of Washington. 84: 427–434.

Toledo, M. & M.C. Michat. 2015. Description of Laccomimus gen. n. and eleven new species from the Neotropical region (Coleoptera, Dytiscidae, Laccophilinae). Zootaxa 3990: 301–354.

White, D.S. & R.E. Roughley. 2008. Chapter 20: Aquatic Coleoptera. Pages 571–671 in: R.W Merritt, K.W. Cummins & M.B. Berg (eds.), An Introduction to the Aquatic Insects of North America. Fourth Edition. Kendall/Hunt Publishing Co., Dubuque, IA. 1158 pp.

Yee, D.A. 2014. Ecology, systematics, and the natural history of predaceous diving beetles (Coleoptera: Dytiscidae). Springer, The Netherlands 484 pp.

Young, F.N. 1967. A key to the genera of American Bidessine water beetles, with descriptions of three new genera (Coleoptera: Dytiscidae, Hydroporinae). The Coleopterists Bulletin 21: 75–84.

Zimmerman, J.R. & A.H. Smith. 1975. A survey of the Deronectes (Coleoptera: Dytiscidae) of Canada, The United States and Northern Mexico. Transactions of the American Entomological Society 101: 651–726.

Family Hydrophilidae

Bruno Clarkson
Laboratório de Sistemática e Bioecologia de Coleoptera, Departamento de Zoologia, Universidade Federal do Paraná; Curitiba, Paraná, Brazil

Miguel Archangelsky
Laboratorio de Investigaciones en Ecología y Sistemática Animal (LIESA), Centro de Investigación Esquel de Montaña y Estepa Patagónica (CIEMEP, CONICET-UNPSJB), Esquel, Chubut, Argentina

Patrícia L.M. Torres
IBBEA, CONICET-UBA. Laboratory of Entomology, DBBE-FCEN, University of Buenos Aires, Buenos Aires, Buenos Aires, Argentina

Andrew E.Z. Short
Department of Ecology & Evolutionary Biology, Division of Entomology, Biology Institute, University of Kansas, Lawrence, United States

Chapter Outline

INTRODUCTION

Hydrophiloidea (*sensu* Hansen, 1991, 1999) is the most diverse group of aquatic Polyphaga. The monophyletic lineage currently comprises about 3340 species in 176 extant genera belonging to the families Georissidae, Helophoridae, Hydrochidae, Spercheidae (all monogeneric), Epimetopidae (three genera), and Hydrophilidae (169 genera). In this chapter, we will focus on the Hydrophilidae. They are worldwide in distribution, although most species are found in the Neotropical and Afrotropical regions (Hansen, 1995, 1999; Short & Hebauer, 2006; Short & Fikáček, 2011).

The ecologic habits of water scavenger beetles are extremely broad. The majority are found in aquatic freshwater habitats, such as ponds, pools, dams, wetlands, lake shores, and streams. Other aquatic and semiaquatic species are found in more specialized habitats such as brackish water, waterfalls, subterranean streams, and phytotelmata. Despite their common name, some taxa have terrestrial habitats (usually are generalists of decaying vegetation or dung, but some genera can be found in flowers or are myrmecophilous) (Crowson, 1981; Hansen, 1991; Jäch & Balke, 2008; Short & Fikáček, 2013). Few adults or larvae of the aquatic species are good swimmers, and most crawl over substrates or aquatic plants. Terrestrial species are also crawlers, though they can also penetrate soft substrates such as dung, decaying plant tissues, and fungi (Archangelsky et al., 2005).

Hydrophilids are commonly known as "water scavenger beetles" or "silver water beetles." These names are directly related to the biology of the group. Most adults feed on plant material and decaying organic matter, but they also may prey on snails, small fish, or tadpoles (Balfour-Browne, 1910; Balduf, 1935). Larvae are usually predacious and feed on other invertebrates such as crustaceans, adults and larvae of other insects, annelids, and snails, as well as vertebrates such as small fishes and tadpoles. Cannibalism has been recorded in many species (Balduf, 1935; Archangelsky, 1997). Adults get oxygen from atmospheric air kept in a cavity between the elytra and abdomen. The air under the elytra usually is connected with a ventral air bubble retained with dense hydrofuge pubescence (plastron), which is silver looking under water (Figs. 15.4.7 B–C).

Among the beetles, only a few representatives of Hydraenidae and all Hydrophiloidea lay their eggs wrapped in silk, forming a cocoon (silk case) (Archangelsky, 1997; Hansen, 1991, 1995, 2000). The shape, size, and number of eggs are variable but usually are constant for each species. Most hydrophilids have three larval instars, but some species of *Sphaeridium* may have only two (Wilson, 2000). Pupation usually takes place in moist soil. Prepupae dig a pupal chamber using their head and mouthparts. Some species build the pupal chamber with chewed plant tissues (Archangelsky, 1994, 1997).

The Hydrophilidae are the most diverse and frequently encountered group of Hydrophiloidea. The family contains approximately 2900 species in 169 genera worldwide distributed in six subfamilies (Acidocerinae, Chaetarthriinae [two tribes], Cylominae, Enochrinae, Hydrophilinae [five tribes], and Sphaeridiinae [six tribes]) (Short & Fikáček, 2013; Seidel et al., 2016). The Neotropics are likely the most species-rich, and more than 60 genera occur here (though, half of these are not aquatic).

LIMITATIONS

Hydrophilidae includes aquatic, riparian, and terrestrial species; in this book, only the aquatic and riparian biodiversity are treated, and most hydrophilid genera belonging to the subfamily Sphaeridiinae have been excluded. In general, the identification (both at adult and larval stages) can be a challenge at the genus level and often impossible at the species level for the Neotropical Region. The only comprehensive work including Hydrophilidae at the generic level is that of Archangelsky et al. (2009), but it is restricted to South American representatives. Several countries and regional specific keys to genus are available. For example, Arce-Pérez & Morón (2011) presented a key to adults of hydrophiloid genera occurring in México. Focusing on the Argentinian fauna, Oliva et al. (2002) treated only the aquatic Hydrophiloidea and provided keys for larvae and adults at the generic level, and keys for adults at the species level for selected groups (*Berosus, Chaetarthria, Derallus, Enochrus, Helobata, Helochares, Hemiosus, Hydrobiomorpha, Hydrophilus*, and *Tropisternus*).

As with other families of aquatic beetles, some groups or genera are taxonomically better known than others, and frequently, adults are far better known than larval stages. For larvae, the work of Archangelsky (1997) is profusely illustrated and covers in detail most of the genera that occur in the Neotropics. The identification of larval hydrophilids requires different techniques than those applied to the identification of adults. Larvae are small and delicate, and they have to be studied with greater detail than adults, making the use of compound microscopes essential for this task. The larval key presented in this chapter is meant to work with all larval instars; nevertheless, not all genera are known in all three instars (*Helobata, Notionotus*, and *Oocyclus* are only known from the first larval instar). It should also be considered that larvae of some genera remain undescribed: *Anacaena, Cylorygmus, Dieroxenus, Globulosis, Phaenostoma, Protistolophus, Quadriops, Radicitus, Tobochares*, and *Troglochares*. The larva of *Chasmogenus* has been described for the Australian Region; however, its description is poor and, therefore, has been omitted from the present key. Four sphaeridiine genera that contain at least some aquatic larvae have been included: *Dactylosternum, Pelosoma, Lachnodacnum*, and *Phaenonotum*; larvae of other terrestrial genera of sphaeridiines, such as *Cercyon, Cryptopleurum, Oosternum*, or *Sphaeridium* have been excluded. The present key is an update of Archangelsky (1997) and hence an important tool to identify Neotropical larvae. It includes all aquatic genera described until the date, which represent about two-thirds of the known adults.

TERMINOLOGY AND MORPHOLOGY

This section emphasizes those morphologic characters of both adults and larvae that are important in the identification of subfamilies, tribes, and genera. The discussion of these characters is arranged following the general body organization: head, thorax, and abdomen. Morphologic terminology follows mainly Hansen (1991), Archangelsky (1997), and Archangelsky et al. (2016).

Adult Morphology

Body

Strongly sclerotized, shape variable, from elongate to broad and from flat to strongly convex. Size ranging from 1 to 50 mm long. Outline not interrupted between pronotum and elytra (most Hydrophilidae, Fig. 15.4.7 A) or markedly interrupted (e.g., *Berosus* and *Hemiosus*). Dorsal surface usually glabrous and smooth (Figs. 15.4.7 A–C). Ventral surface usually with plastron (hydrofuge pubescence) (Figs. 15.4.7 B–C).

Head

Constricted behind the eyes, partially retracted within prothorax (Figs. 15.4.1 C–D). Compound eyes present (absent in *Troglochares*); sometimes divided (partly or completely) by a canthus (e.g., *Quadriops*, Fig. 15.4.1 L); ocelli absent. Clypeus large, covering antennal bases, with or without an anterolateral group of setiferous punctures. Frons with or without coarse setiferous punctures on inner margin of eye. Frontoclypeal line distinct, V- or Y-shaped. Gula present, variable in shape (Fig. 15.4.1 G, M), sometimes narrowed anteriorly. Labrum transverse, sclerotized, usually exposed (Figs. 15.4.1 E–F), but covered by clypeus in some taxa (e.g., *Helobata*, Fig. 15.4.1 C). Antenna short, clavate (Figs. 15.4.2 A–B), with seven to nine antennomeres; scapus longer than pedicellus; pedicellus followed by one to three small glabrous antennomeres and a cupule; club of three articles. Mandibles strong, covered by labrum, with large and asymmetric mola; subapical membranous lobe reduced, with antennomeres loosely united or forming a compact structure; mesal rows of articulated spines present or absent. Maxilla of six articles, composed of cardo, stipes, and a four-articled palpus; lacinia and galea well developed; maxillary palpi short (Fig. 15.4.1 D) or long (Fig. 15.4.1 C), with first palpomere short; some palpomeres may be curved outward (e.g., second palpomere in *Enochrus*, Fig. 15.4.2 C). Mentum large, lamellar, rectangular to trapezoidal in shape; prementum appearing as a paired structure due to a median incision, bearing a fringe or setae on anterior margin; labial palpi of three articles, basal palpomere never longer than second.

Thorax

Pronotum usually wider than long, convex. Lateral margins of pronotum with a sharp ridge, delimitating the dorsal portion from the ventral (hypomeron); lateral areas sometimes bearing coarse setiferous punctures. Hypomeron with mesal portions usually pubescent, lateral portions glabrous, shiny. Prosternum microsculptured, with fine and dense pubescence; when antennal grooves present, then glabrous; central portion of prosternum may be strongly raised (Fig. 15.4.1 N) or convex and smooth, forming a posteriorly directed process. Scutellar shield distinct, variable in size and length (e.g., elongate in Berosini) (Figs. 15.4.2 H–I). Mesoventrite narrow anteriorly, microsculptured, with fine and dense pubescence or partly glabrous; in many taxa raised medially or posteromedially, forming a longitudinal ridge, process or plate (Figs. 15.4.1 R, 3 A, C, G); raised parts usually glabrous and shiny. Metaventrite finely microsculptured, with fine and dense pubescence; in many taxa raised medially, raised parts usually glabrous and shiny; in some Hydrophilinae the metaventrite projects as a posteromedian spine (Figs. 15.4.3 A, 7 B). Elytra large, completely concealing abdomen in aquatic species (Fig. 15.4.7 A); sometimes with 10 rows of longitudinal striae of punctures (Figs. 15.4.2 F–G); setiferous punctures may appear in the alternate interstices 3, 5, 7, 9, and 11. Ventral side of the elytra may bear a small patch of microscopical spines, which serve as locking device; this patch of spines sometimes corresponds with a ribbed area on laterosternite III (plectrum), acting as a stridulatory organ. Epipleura subdivided into an inner pubescent part (true epipleuron) and a glabrous outer part (pseudepipleuron). Hind wings usually present, rarely absent or reduced. Procoxae narrowly separated or almost contiguous (separated in some terrestrial taxa), globular, conical, or transverse. Mesocoxae narrowly separated or almost contiguous, globular to transverse. Metacoxae narrowly separated, transverse, reaching epipleura. Trochanters well developed. Femora flattened, with or without tibial groove; base of femora with or without pubescence. Tibiae cylindrical to broadly flattened, with longitudinal series of spines and with apical spurs (Fig. 15.4.2 M); fringes of swimming hairs may be absent or present on both middle and hind tibiae (Fig. 15.4.2 J), or only on middle tibiae. Tarsi usually pentamerus, with fine spines or pubescence on ventral side; males of some genera with adhesive setae on basal tarsomeres (e.g., *Berosus*, Fig. 15.4.1 U). Pretarsi well developed, empodium usually present.

Abdomen

Tergites poorly sclerotized, tergite I divided in two plates; tergites I and II without corresponding sternites. Ventrites I to V (corresponding to sternites III to VII) complete and well sclerotized (Figs. 15.4.3 K–M); ventrite I with groove for the reception of metacoxae and a small process between them (Figs. 15.4.3 L, N); ventrites I and II may bear a large cavity on each side covered by a fringe of long setae (Chaetarthriinae, Fig. 15.4.3 N); ventrite V with or without emargination (Figs. 15.4.3 E, F, K–N); in some genera ventrite V may be truncate, leaving ventrite VI partially exposed. Remaining abdominal segments retracted while in repose. Abdominal spiracles on tergo-sternal membranes of segments I–VII, spiracle VIII nonfunctional. Stridulatory files (plectrum) on laterosternite III present or absent. Male aedeagus typically simple, trilobed, with basal piece, a pair of parameres and a median penis; aedeagus more complex in some genera (e.g., *Helochares* and *Hydrobiomorpha*). Female genitalia simple, tubular; sternite and tergite IX enclosed between sternite and tergite VIII; spermatheca small, membranous; terminal part of abdomen with strongly modified sclerites ending in a pair of styli inserted on a pair of gonocoxites (stylifers).

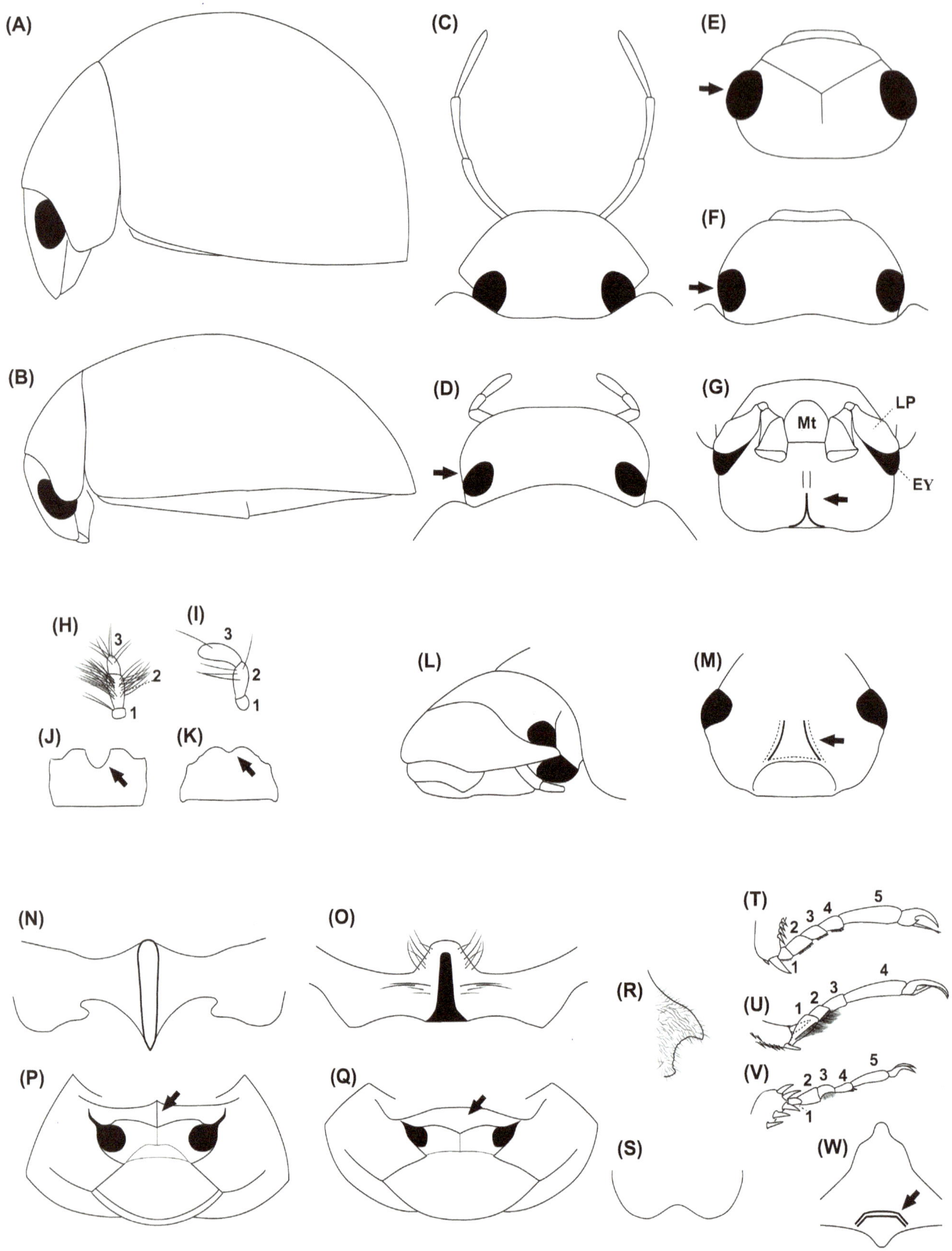

FIGURE 15.4.1 Hydrophilidae adult. Habitus in lateral view: (A) *Derallus* sp.; (B) *Berosus* sp. Head in dorsal view: (C) *Helobata* sp.; (D) *Guyanobius* sp., *black arrow* indicates the lateral projection of the frons (canthus); (E) *Crenitis* sp., *black arrow* indicates the slightly protruding eye; (F) *Anacaena* sp., *black arrow* indicates the eye not produced from lateral outlines of head. Head in ventral view: (G) *Pseudorygmodus* sp., *black arrow* indicates fused gular sutures, *EY*, eye; *LP*, labial palpus; *Mt*, mentum; (M) *Crenitis* sp., *black arrow* indicates developed gular sutures. Mouth parts: (H) labial palpus of *Phaenonotum* sp., numbers indicates palpomeres; (I) labial palpus of *Berosus* sp., numbers indicates palpomeres; (J) mentum in ventral view of *Chasmogenus* sp., *black arrow* indicates anteromedial notch; (K) mentum in ventral view of *Anacaena* sp., *black arrow* indicates anteromedial notch; (L) Head of *Quadriops* sp. in aterolateral view. Prosternum: (N) *Hydrobiomorpha* sp.; (O) *Hydrophilus* sp.; (P) *Paracymus* sp., *black arrow* indicates presence of medial carina; (Q) *Anacaena* sp., *black arrow* indicates absence of medial carina; (R) Mesoventral process in lateral view of *Berosus* sp. Clypeus in dorsal view of (S) *Anticura* sp. Male foretarsi: (T) *Hemiosus* sp.; (U) *Berosus* sp.; (V) *Laccobius* sp., numbers indicate tarsomeres. Mesoventrite of (W) *Cymbiodyta* sp. in ventral view, *black arrow* indicates transversal ridge. The following figures were modified from Short & Fikáček (2013): A and B; from Hansen (1991): C, E, F and M; from Archangelsky et al. (2009): L, P and Q; from Van Tassel (2000): W.

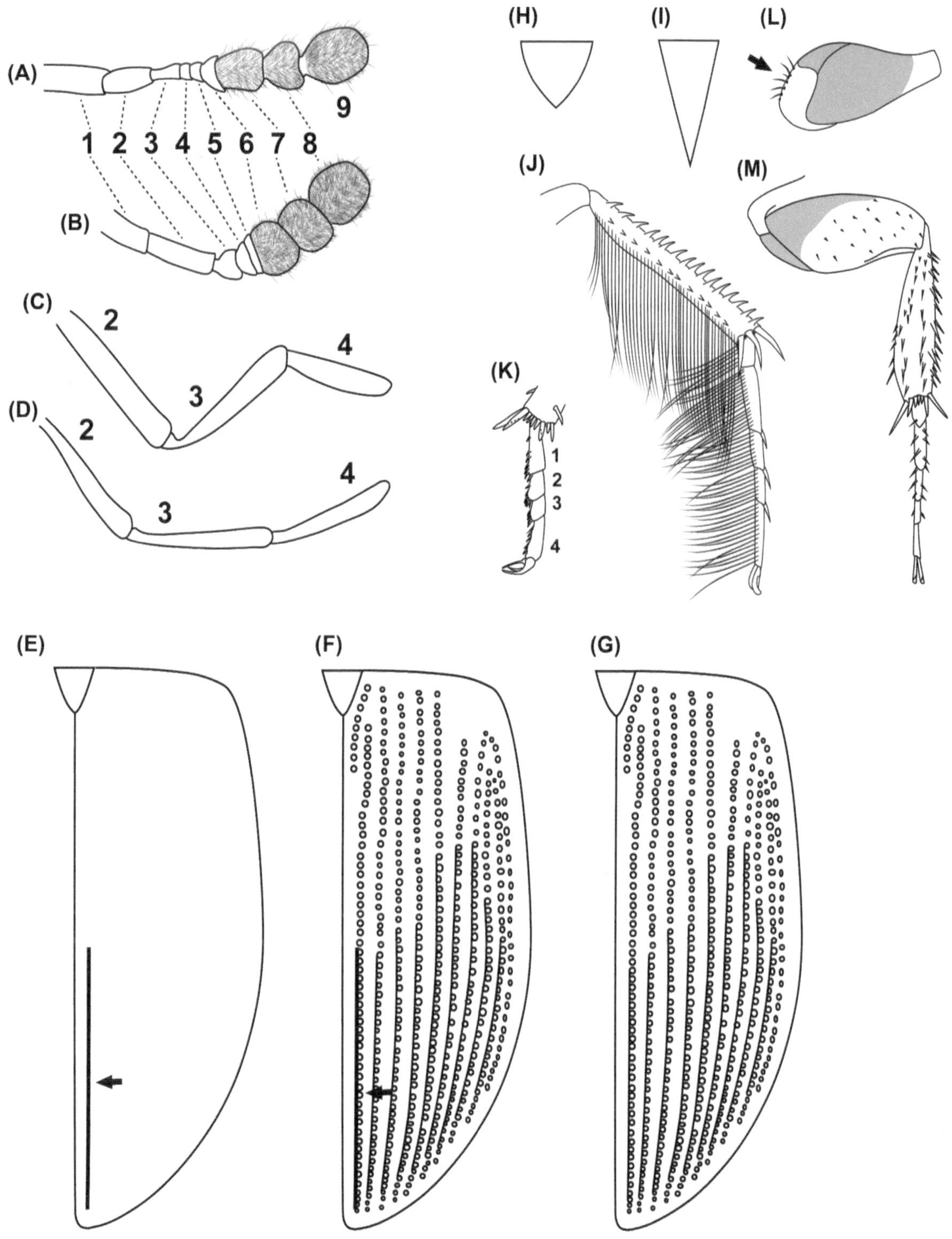

FIGURE 15.4.2 Hydrophilidae adult. Antennae: (A) *Radicitus* sp.; (B) *Tobochares* sp., numbers indicate antennomeres. Marxillary palps: (C) *Enochrus* sp.; (D) *Helochares* sp., numbers indicate antennomeres. Elytra in dorsal view: (E) *Chasmogenus* sp., *black arrow* indicates sutural stria; (F) *Tobochares* sp., *black arrow* indicates sutural stria; (G) *Helochares* sp. Scutellar shield: (H) *Anacaena* sp.; (I) *Berosus* sp. Legs: (J) hind leg of *Berosus* sp.; (K) hind tarsus of *Cymbiodyta* sp.; (L) Procoxae and profemur of *Crenitulus* sp., *black arrow* indicates presence of spiniform setae; (M) hind leg of *Crenitulus* sp. The following figures were modified from Short & Fikáček (2013): C and D; from Smetana (1988): K; from Komarek (2005): L and M.

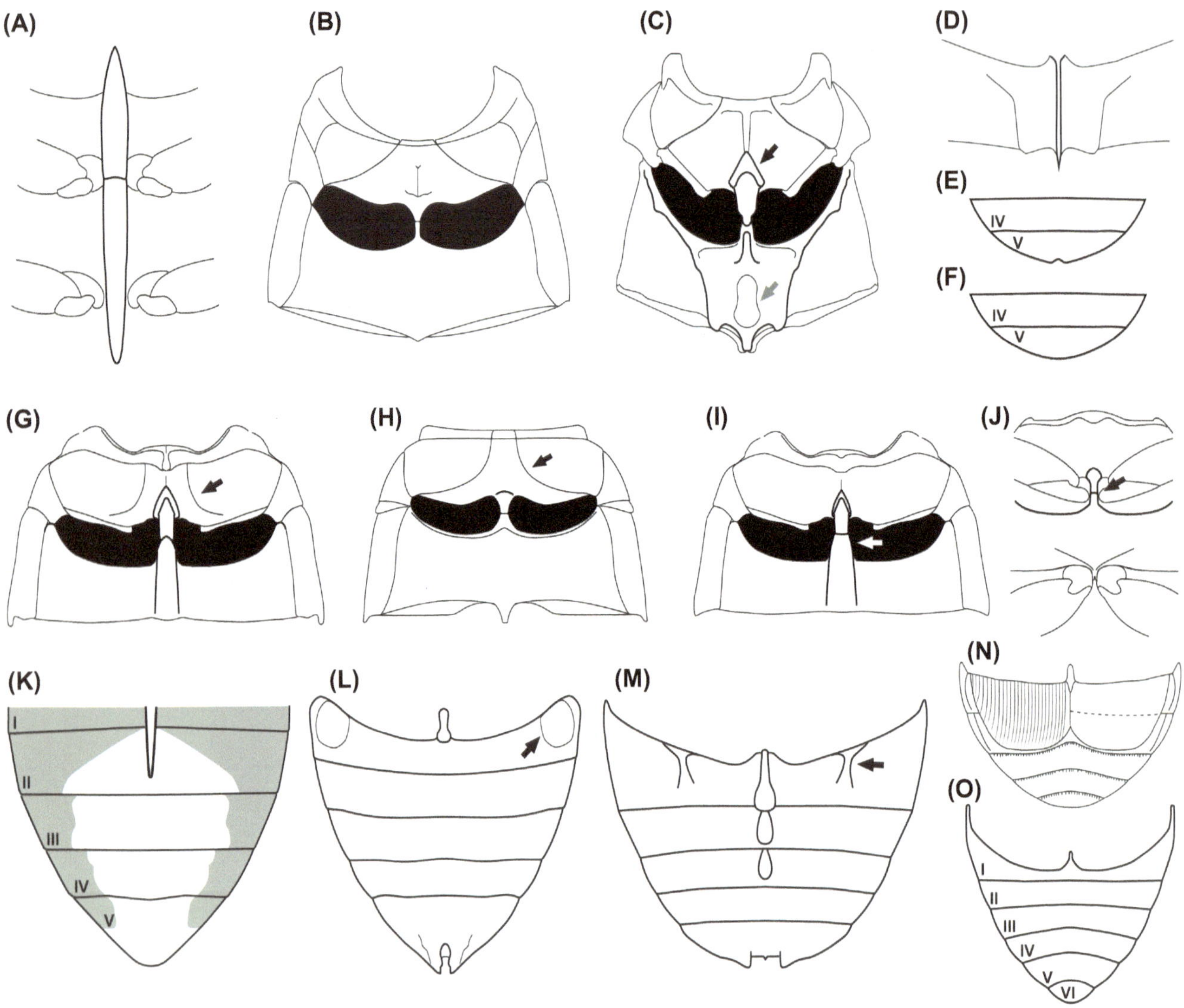

FIGURE 15.4.3 Hydrophilidae adult. (A) Meso and metaventral fused keels of *Hydrophilus*. Ventral plates of pterothorax: (B) *Paracymus* sp.; (C) *Hemiosus* sp., *black arrow* indicates mesoventral process, *grey arrow* indicates medial hollow of metaventrite. Metaventrite of (D) *Derallus* sp. Posterior portion of abdomen: (E) *Helochares* sp.; (F) *Paracymus* sp., numbers indicates abdominal ventrites. Ventral plates of pterothorax: (G) *Lachnodacnum* sp.; (H) *Cylorygmus* sp., *black arrow* indicates anapleural suture of mesoventrite; (I) *Phaenonotum* sp., *white arrow* indicates metaventral process projected between mesoventral cavities, making mesocoxae further apart. Meso and metaventrites of (J) *Hydroglobus* sp., *black arrow* indicates metaventrite without strong projection between mesoventral cavities. Abdominal ventrites: (K) *Hydrophilus* sp., numbers indicate ventrites; (L) *Berosus* sp., *black arrow* indicates lateral excavation; (M) *Hemiosus* sp., *black arrow* indicates side carina; (N) *Chaetarthria* sp.; (O) *Laccobius* sp., numbers indicate ventrites. The following figures were modified from Clarkson & Ferreira-Jr (2009): C and M; from Short & Torres (2006): D; from Short & McIntosh IV (2014): K; from Hansen (1991): N.

Larval Morphology

Body

Campodeiform except for *Chaetarthria* (Chaetarthriinae), and many terrestrial Sphaeridiinae, which show different degrees of leg reduction. Body elongate (Figs. 15.4.4 A–D), somewhat flattened dorsoventrally. Cuticle soft, lightly colored, except for sclerotized regions (head capsule, mouthparts, legs, thoracic, and abdominal plates). Tubercles, asperities, pubescence, tracheal gills, or protuberances on different parts of thorax or abdomen (Figs. 15.4.4 C–D, 6 I–K) may be present in different genera.

Head Capsule

Strongly sclerotized, prognathous (*Berosus*) or hyperprognathous (most Hydrophilidae). Ecdysial line distinct in first and second instar larvae (Figs. 15.4.6 A–H), formed by two frontal lines and a short coronal line (coronal line absent in many genera, Figs. 15.4.6 B–C, E–F, H); these lines divide the head capsule into an anterior frontoclypeolabrum and two epicranial halves. Gula reduced to

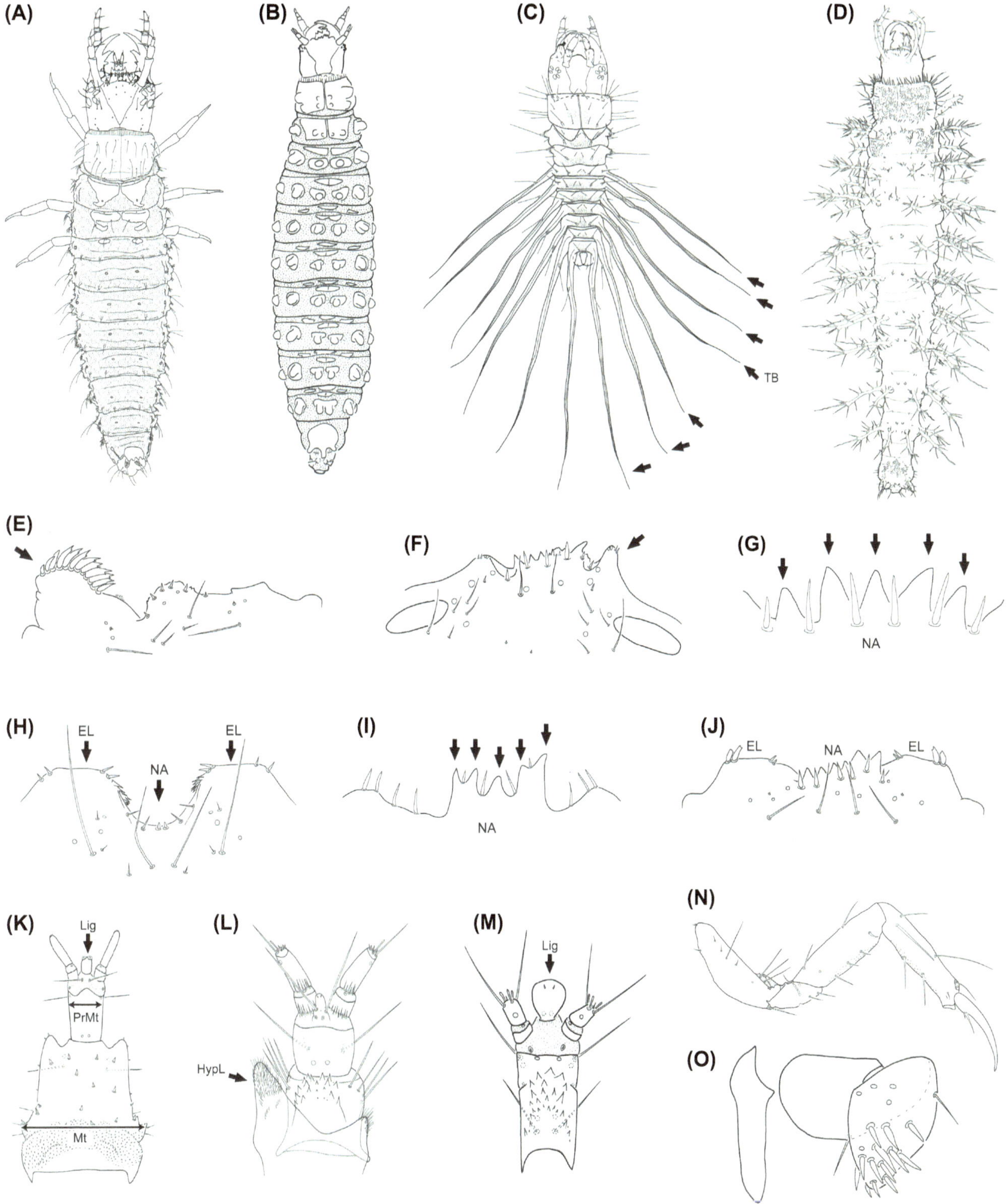

FIGURE 15.4.4 Hydrophilidae larva. Habitus in dorsal view: (A) *Tropisternus* sp.; (B) *Crenitulus* sp.; (C) *Berosus* sp., *black arrow* indicates the tracheobranchia, *TB*, tracheobranchia; (D) *Derallus* sp. Clypeolabrum: (E) *Hemiosus* sp., *black arrow* indicates left epistomal lobe strongly developed; (F) *Enochrus* (*Hugoscottia*) sp., *black arrow* indicates right epistomal lobe; (G) nasale of *Guyanobius* sp., *black arrows* indicate teeth of nasale; (H) *Helobata* sp., *black arrows* indicate epistomal lobes and nasale; (I) nasale of *Pseudorygmodus* sp., *black arrows* indicate teeth of nasale; (J) *Helochares* sp., *EL*, epistomal lobe; *NA*, nasale. Labium: (K) *Hydrobiomorpha* sp.; (L) *Dactylosternum* sp.; (M) *Chaetarthria* sp., *HypL*, hypopharingeal lobe; *Lig*, ligula; *Mt*, mentum; *PrMt*, prementum. Prothoracic leg: (N) *Berosus* sp.; (O) *Chaetarthria* sp. The following figures were modified from Torres et al. (2011): A; from Archangelsky & Fikáček (2004): B; from Rodriguez et al. (2015): C and N; from Archangelsky & Durand (1992): D; from Archangelsky (2000): E; from Bittebier & Torres (2009): F; from Archangelsky (1997): G, H, I, J and L; from Archangelsky et al. (2004): K; from Archangelsky (2002): M; from Archangelsky et al. (2009): O.

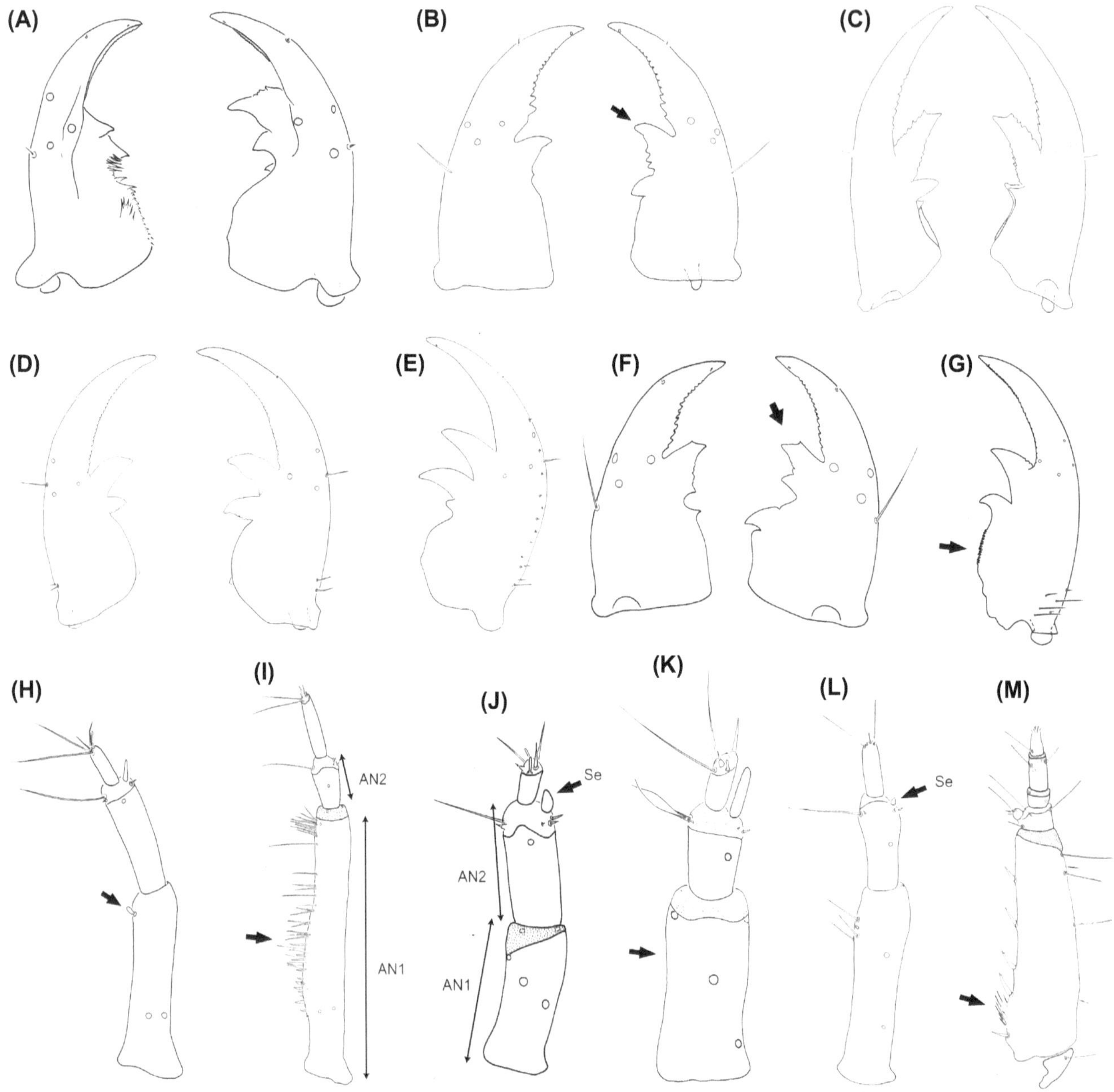

FIGURE 15.4.5 Hydrophilidae larva. Mandibles in dorsal view: (A) *Oocyclus* sp.; (B) *Enochrus* (*Methydrus*) sp., *black arrow* indicates inner tooth; (C) *Tropisternus* sp.; (D) *Helochares* sp.; (E) right mandible of *Hydramara* sp.; (F) *Notionotus* sp., *black arrow* indicates inner tooth; (G) right mandible of *Lachnodacnum* sp., *black arrow* indicates short cuticular spines directed backward. Right antennae in dorsal view: (H) *Hemiosus* sp., *black arrow* indicates a distal projection on inner margin; (I) *Hydrobiomorpha* sp., *black arrow* indicates numerous long setae on inner margin of first antennomere; (J) *Enochrus* sp.; (K) *Phaenonotum* sp., *black arrow* indicates bare inner margin of first antennomere; (L) *Hydramara* sp., *AN*, antennomere; *Se*, sensorium. Left maxilla of (M) *Helochares* in ventral view, *black arrow* indicates basal cuticular spines on stipes. The following figures were modified from Archangelsky (1997): A, D, E, F, K and L; from Torres et al. (2011): C; from Archangelsky (2000): G; from Archangelsky (2002): I; from Archangelsky et al. (2016): J; from Clarkson et al. (2014): M.

a gular suture. Clypeus and labrum fused, forming a clypeolabrum (Figs. 15.4.4 E–F, 6 A–H), which bears a median nasale and lateral lobes of epistome. Clypeolabrum symmetrical or asymmetrical (Figs. 15.4.4 E–F, 6 A–H); asymmetry may involve just the nasale, the lateral lobes of epistome, or both. Six pairs of stemmata (Figs. 15.4.6 A–H) or less, which may be fused into one or two ocular areas in some genera (many Sphaeridiinae and *Chaetarthria*). Occipital foramen large, wide; cervical sclerites present or absent (Figs. 15.4.6 A, D, F, H). Posterior tentorial grooves located in mid-ventral region of head capsule.

Head Appendages

Antennae inserted on anterolateral margins of head or on dorsal side of head (Figs. 15.4.6 A–H); three

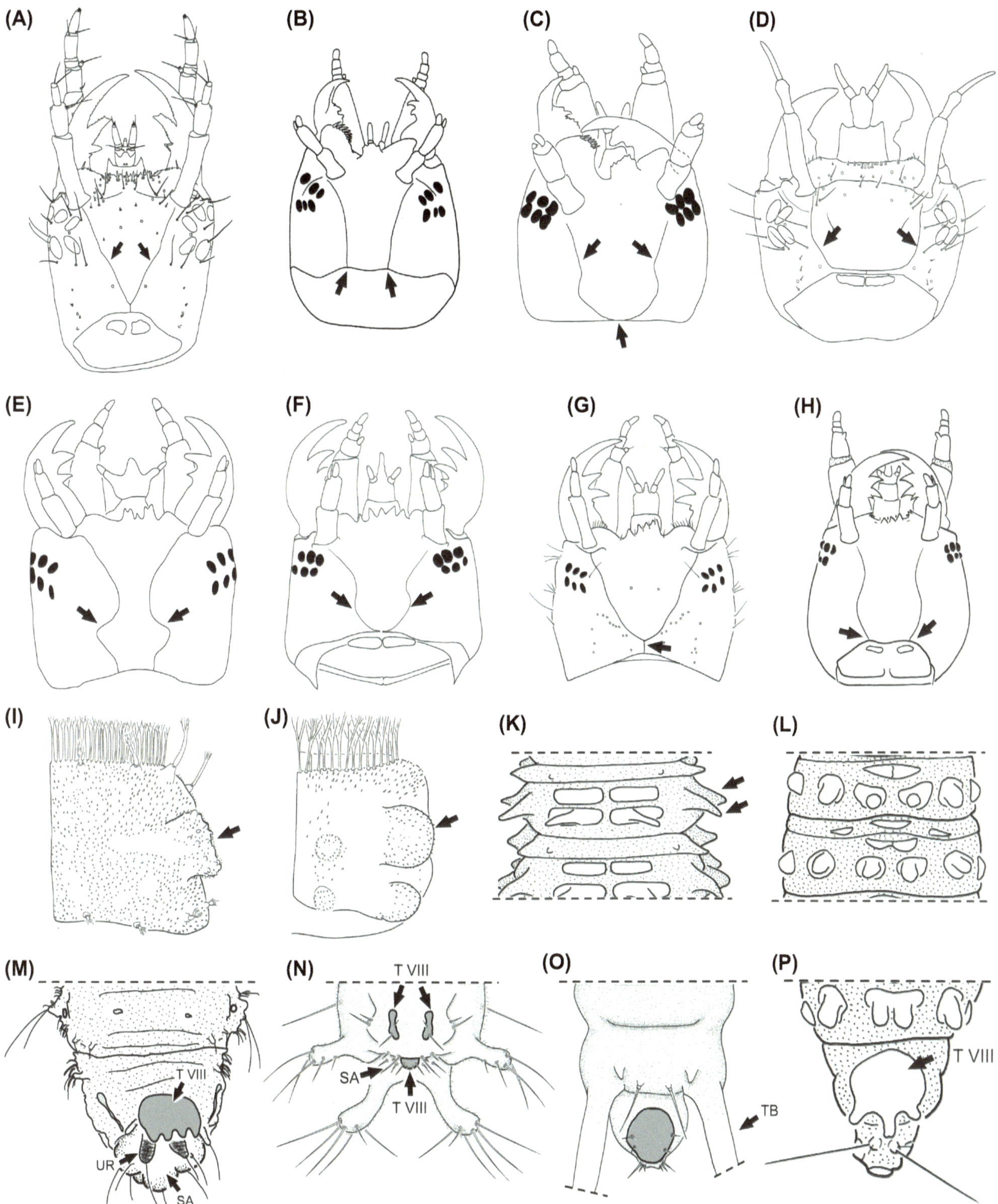

FIGURE 15.4.6 Hydrophilidae larva. Head capsules in dorsal view: (A) *Tropisternus* sp., *black arrow* indicates V-shaped frontal lines; (B) *Laccobius* sp., *black arrow* indicates subparallel frontal lines not merging before occipital foramen; (C) *Oocyclus* sp., *black arrow* indicates lyriform frontal lines merging at occipital foramen; (D) *Hydrophilus* sp., *black arrow* indicates U-shaped frontal lines; (E) *Pseudorygmodus* sp., *black arrow* indicates liriform frontal lines not converging toward occipital foramen; (F) *Guyanobius* sp., *black arrow* indicates inversely bell-shaped frontal lines converging toward occipital foramen; (G) *Hydramara* sp., *black arrow* indicates frontal lines united basally, coronal line; (H) *Paracymus* sp., *black arrow* indicates frontal lines not united basally. Lateral part of pronotum: (I) *Crenitis* sp., *black arrow* indicates lateral quadrangular lobes; (J) *Crenitulus* sp., *black arrow* indicates lateral rounded lobes. Abdominal segments: (K) *Crenitis* sp., *black arrow* indicates fleshy lateral projections; (L) *Crenitulus* sp. Abdominal apex: (M) *Tropisternus* sp.; (N) reduced spiracular atrium of *Hemiosus* sp.; (O) abdominal apex without spiracular atrium of *Berosus* sp.; (P) *Crenitulus* sp., *SA*, spiracular atrium; *TB*, tracheobranchia; *T VIII*, tergum of abdominal segment VIII; *UR*, urogomphus. The following figures were modified from Archangelsky (1997): B, C, E, F, G and O; from Archangelsky (1999): H; from Archangelsky & Fikáček (2004): I, J, L and P; from Archangelsky (1997): K; from Archangelsky (2002): N.

FIGURE 15.4.7 Hydrophilidae adult. *Tropisternus* spp.: (A) Habitus in dorsal view; (B, C): adults under water foraging on vegetation, showing the ventral air bubble retained with dense hydrofuge pubescence.

antennomeres (Figs. 15.4.5 H–L), apparently four antennomeres in some Hydrophilinae (e.g., *Hydrophilus*) due to a basal subdivision of the second antennomere; first antennomere usually the longest (Figs. 15.4.5 H–I); second antennomere bearing one sensory appendage (sensorium) (Figs. 15.4.5 J–L); sensory appendage reduced in several genera (Fig. 15.4.5 I). Mandibles falcate (Figs. 15.4.5 A–G, M), symmetrical to strongly asymmetrical (e.g., *Enochrus*, *Berosus*, many Sphaeridiinae); apex simple; inner margin with one to three retinacula (lost in some Sphaeridiinae); prostheca and mola absent; penicillus-like structure (tuft of hairs) absent. Maxilla of six articles (Fig. 15.4.5 M), cardo small, subdivided or represented by one sclerite; stipes tubular; lacinia missing; palp of four articles; first palpomere variable in length, bearing a small process on inner apical margin (reduced in many genera); remaining palpomeres simple, with variable relative lengths. Labium composed of submentum, mentum, and prementum, which bears one pair of palpi and usually a short ligula (Figs. 15.4.4 K–M); submentum subtriangular and attached to head capsule; mentum shifted to anterior margin of head capsule; prementum usually subrectangular; labial palpi biarticulate, basal palpomere the shortest; ligula present or absent, sclerotized or unsclerotized; larvae of some Sphaeridiinae bear a pubescent left hypopharyngeal lobe.

Thorax

Three-segmented. Prothoracic tergum large, strongly sclerotized, with sagittal line; meso- and metathoracic terga smaller, usually suboval or subtriangular, sometimes subdivided into smaller plates (e.g., *Crenitulus*, Fig. 15.4.4 B); prosternal plate well developed, with posterior projections articulating with procoxae. Meso- and metathorax may bear different types of lobes or spinose projections (e.g., *Derallus*, Fig. 15.4.4 D). Legs usually of five articles, composed of a large coxa, trochanter, femur, tibiotarsus, and a single pretarsus or claw (Fig. 15.4.4 N); reduction of leg articles occurs in several genera (e.g., *Chaetarthria* and several Sphaeridiinae, Fig. 15.4.4 O).

Abdomen

Ten-segmented, largely unsclerotized, terminal segments may be reduced or modified (Figs. 15.4.4 A–C). A well-developed spiracular atrium is present (Fig. 15.4.6 M), formed by segments VIII and IX, segment VIII usually with a large dorsal plate (Figs. 15.4.6 M, O–P) (subdivided in some genera, Fig. 15.4.6 N); urogomphi present on segment IX, unarticulated (Fig. 15.4.6 M). Abdominal segments I–VII may bear different types of lobes or spinose projections (e.g., *Derallus*, Fig. 15.4.4 D); *Berosus* larvae with tracheal gills present usually on segments I–VII (Fig. 15.4.4 C). Ventrally some genera may bear rows of spines or spinose lobes that function as prolegs (e.g., *Enochrus*, *Notionotus*, *Dactylosternum*), aiding in locomotion.

Spiracles

Present on mesothorax and abdominal segments I–VIII; first seven pairs nonfunctional, spiracles on segment VIII annular, enlarged and functional, within spiracular atrium (except in *Berosus*, in which all spiracles are nonfunctional).

MATERIAL PREPARATION AND PRESERVATION

Hydrophilidae can be preserved as either dry, pinned specimens or in ethanol. The following preservation methods are from various information sources (Martin, 1977; Dennis, 1974; Llorente et al., 1985; Steyskal et al., 1986; Morón & Lump, 1988; Borror et al., 1989; Imes, 1992; Merritt et al., 1996; Contreras-Ramos, 1999; Lawrence, 1991).

Hydrophilids are most easily collected directly into ethanol in the field (typically 70%–95%), and the alcohol should be changed at least once after collection (Merritt et al., 1996). To prevent excessive darkening, live or freshly collected larvae may be placed in hot water for 1–5 min (depending on specimen size), and then transferred to ethyl alcohol (Steyskal et al., 1986). Large adult specimens, such as those in *Hydrophilus*, *Hydrobiomorpha*, and *Tropisternus* (Hydrophilidae), are pinned directly through the right elytrum. Smaller specimen that may be damaged by pinning should be pointed (glued) on archival cardstock.

Examination of the aedeagus is usually required for species-level identifications. In these cases, it is best to dissect it from the abdomen. This can be done readily in ethanol-preserved specimens (80% or below); but dry specimens, or those in higher ethanol concentrations, may be placed in hot water (not boiling) for a few minutes to soften the tissues. The aedeagus may be pulled from the abdomen and left attached, or removed for clearing and subsequent preparation. The aedeagus can be cleared in 10% potassium hydroxide (KOH) and gradually heated for clearing (or kept in KOH at room temperature overnight). To make semipermanent preparations, the structure may be hydrated with water then placed on a slide with glycerin or glycerinated gelatin. As a temporary preparation, the structure may be placed on an excavated glass slide with glycerin. After examination, the dissected structures can be stored in a capsule or microvial with glycerin and pinned below the specimen or placed in its vial. Larvae are soft bodied, small, and delicate. Clearing techniques require the use of mild clearing agents to avoid damaging the specimens (e.g., warm lactic acid). Once cleared, larvae can be mounted on glass slides, and the identification process often requires the dissection of cephalic appendages.

KEY TO HYDROPHILIDAE

Hydrophilidae: Genera (Adults)

1 Labrum well sclerotized and exposed (Figs. 15.4.1 E, F), not concealed by clypeus (except *Helobata*); second labial palpomere with scattered setae (Fig. 15.4.1 I) 2

1' Labrum not well sclerotized, usually retracted under clypeus; second labial palpomere with subapical tufts of numerous setae (Fig. 15.4.1 H) 31

2(1) Meso- and metaventral keels fused to form a single structure, often ending posteriorly in a pronounced spine (Fig. 15.4.3 A); Body size 5–50 mm (subfamily Hydrophilinae: tribe Hydrophilini) 3

2' Sternal structure not as previous, never ending in large spine (Fig. 15.4.3 B); body size never exceeding 15 mm........ 6

3(2) Elytra without distinct sutural stria; body 5–50 mm........ 4

3' Elytra with distinct sutural stria (Fig. 15.4.2 E); body 8–10 mm........ *Protistolophus* [Venezuela]

4(3) Prosternum with large posterior emargination, sometimes completely divided to receive anterior portion sternal keel (Fig. 15.4.1 O) ... 5

4' Prosternum carinate centrally, never divided or emarginated (Fig. 15.4.1 N)........ *Hydrobiomorpha* [Widespread]

5(4) Body <20 mm; abdominal ventrites entirely pubescent........ *Tropisternus* [Neotropics. Nearctic]

5' Body >20 mm; abdominal ventrites at least partially glabrous medially (Fig. 15.4.3 K)........ *Hydrophilus* [Widespread]

6(2) Scutellar shield much longer than wide (Fig. 15.4.2 I); meso- and metatibia dorsally with fringe of natatory setae (Fig. 15.4.2 J) (subfamily Hydrophilinae: tribe Berosini)........ 7

6' Scutellar shield as long as wide, or rather longer than wide (Fig. 15.4.2 H); meso- and metatibia without natatory setal fringe (Fig. 15.4.2 M)........ 9

7(6) Dorsum at least partially (usually mostly) yellow or brown with scattered black markings; body slightly convex (Fig. 15.4.1 B), not laterally compressed; metaventrite with medial hollow more or less developed (Fig. 15.4.3 C); metanepistenum apparent........ 8

7' Dorsum entirely black; body subglobular (Fig. 15.4.1 A), laterally compressed; metaventrite with medial carina (Fig. 15.4.3 D); metanepistenum partially or completely concealed by the elytral margins........ *Derallus* [Neotropics. Nearctic]

8(7) Mesoventrite with elevated plate, usually medially hollowed (Fig. 15.4.3 C); abdominal ventrite I with one medial and two side carinae (Fig. 15.4.3 M); metafemur pubescence short, dense; male protarsus with five tarsomeres (Fig. 15.4.1 T); body <5 mm .... *Hemiosus* [Neotropics]

8'	Mesoventrite with laminar elevation (Fig. 15.4.1 R); abdominal ventrite I may have a median carina and/or lateral excavations but not side carinae (Fig. 15.4.3 L); metafemur pubescence long, less dense; male protarsus with four tarsomeres (Fig. 15.4.1 U); body 1.5–9 mm [Widespread]	*Berosus*
9(6)	Abdominal ventrites I and II length equal, glabrous, with large cavities (may be filled by hyaline mass), with a fringe of long setae; ventrites III to V shorter (Fig. 15.4.3 N); body form very round, <3.5 mm (subfamily Chaetarthriinae: tribe Chaetarthriini)	10
9'	Abdominal ventrites I–V length equal, without large cavities, more or less pilose; body form and size variable	11
10(9)	Eyes almost completely divided transversely by canthus (Fig. 15.4.1 D); elytra without sutural stria [Neotropics]	*Guyanobius*
10'	Canthus absent; elytra posterior half with sutural stria (e.g., Fig. 15.4.2 E) [Widespread]	*Chaetarthria*
11	Elytral margins finely serrate or denticulate, at least in anterior third. Mountain streams	12
11'	Elytral margins not serrate or denticulate	13
12(11)	Size <8 mm; pronotum with two size-pattern of punctuation irregularly distributed; both labrum and clypeus deeply emarginated (Fig. 15.4.1 S) (subfamily Cylominae in part) [Chile]	*Anticura*
12'	Size >12 mm; pronotum with evenly distributed punctuation; labrum and clypeus not deeply emarginated (subfamily Hydrophilinae tribe Hydrobiusini in part) [Argentina]	*Hydramara*
13(11)	Maxillary palps with second article bending laterally, giving it a "zig-zag" appearance (Fig. 15.4.2 C) (subfamily Enochrinae in part)...	14
13'	Maxillary palps with all articles bending the same direction (Figs. 15.4.1 C, 2 D)	15
14	Elytra without sutural stria; size <3 mm [Neotropics. Oriental]	*Notionotus*
14'	Elytra with sutural stria (e.g., Fig. 15.4.2 E); size 2–9 mm [Widespread]	*Enochrus*
15(13)	Eyes present	16
15'	Eyes absent (subfamily Acidocerinae in part) [Ecuador]	*Troglochares*
16(15)	Labrum not concealed by clypeus (Figs. 15.4.1 E, F), elytral margins not or at most weakly explanate; body form variable but rarely dorsoventrally compressed	17
16'	Labrum concealed by clypeus (Fig. 15.4.1 C), elytral margins broadly explanate; body extremely dorsoventrally compressed (subfamily Acidocerinae in part) [Neotropics. Nearctic]	*Helobata*
17(16)	Eyes not divided; size variable	18
17'	Eyes completely divided into dorsal and ventral sections by a canthus (Fig. 15.4.1 L); size <2.5 mm (subfamily Acidocerinae in part). [Costa Rica to Peru]	*Quadriops*
18(17)	Elytra without sutural stria or 10 rows of serial punctures (e.g., Fig. 15.4.2 G)	19
18'	Elytra with distinctly impressed sutural stria, or with 10 rows of distinct serial punctures (Figs. 15.4.2 E–F)	22
19(18)	Abdomen apex entire (Fig. 15.4.3 F); maxillary palpi length < head width; dorsum often with iridescent blue or green markings (subfamily Hydrophilinae: tribe Laccobiini in part)	20
19'	Abdomen apex with a small median emargination or truncation (Fig. 15.4.3 E); maxillary palpi length > head width; dorsum usually light brown to black, without iridescent markings (subfamily Acidocerinae in part)	21
20(19)	Abdomen with six ventrites (Fig. 15.4.3 O); male third protarsomere with modified plate (Fig. 15.4.1 V); size <3 mm [Mexico, Caribbean; widespread outside Neotropics]	*Laccobius*
20'	Abdomen with five ventrites; males protarsi unmodified; size variable but rarely <3 mm [Neotropics. Oriental]	*Oocyclus*
21(19)	Antennae with eight antennomeres (Fig. 15.4.2 B); size <3 mm [Northern South America]	*Globulosis*
21'	Antennae with nine antennomeres (Fig. 15.4.2 A); size rarely <4 mm [Widespread]	*Helochares*
22(18')	Elytra with 10 rows of serial punctures, often impressed in striae (Fig. 15.4.2 F) (subfamily Acidocerinae in part)	23
22'	Elytra with sutural stria only, lacking serial puncture rows (Fig. 15.4.2 E)	24
23(22)	Antennae with eight antennomeres (Fig. 15.4.2 B); maxillary palpi length < head width; size <3 mm [Venezuela, Suriname, Guyana]	*Tobochares*
23'	Antennae with nine antennomeres (Fig. 15.4.2 A); maxillary palpi length > head width; size >3.0 mm [Venezuela, Suriname, Guyana]	*Radicitus*
24(22)	Tarsi 5-5-5; mesoventrite with or without longitudinal elevation	25
24'	Tarsi 5-4-4 (Fig. 15.4.2 K); mesoventrite with transversal ridge (Fig. 15.4.1 W) (subfamily Enochrinae in part) [Mexico, Northern Central America. Nearctic, Palaearctic, Oriental]	*Cymbiodyta*
25(24)	Prosternum without distinct median carina (Fig. 15.4.1 Q); metafemora densely pubescent at least on basal third (Fig. 15.4.2 M); body without iridescent sheen; Body size variable	26

25' Prosternum with distinct median carina (Fig. 15.4.1 P); metafemora glabrous except for a few scattered individual setae; body often with distinct iridescent sheen; body size <3.5 mm (subfamily Hydrophilinae: tribe Laccobiini in part) .... *Paracymus* [Widespread]

26(25) Mentum deeply notched anteromedially (Fig. 15.4.1 J); body size typically >3 mm (subfamily Acidocerinae in part) 27

26' Mentum at most slightly notched anteromedially (Fig. 15.4.1 K); body size typically <3 mm (subfamily Chaetarthriinae: tribe Anacaenini) 28

27(26) Body elongate oval, sometimes subparallel; maxillary palpi length > head width; Widespread in the Neotropical region... *Chasmogenus* [Neotropics. Afrotropical, Australian, Oriental, Palaearctic]

27' Body broadly oval; maxillary palpi length < head with .... *Dieroxenus* [Ecuador]

28(26) Eyes slightly protruding from head lateral outlines (Fig. 15.4.1 E) 29

28' Eyes not protruding from head lateral outlines (Fig. 15.4.1 F) 30

29(28) Gula reduced, with gular sutures fused (Fig. 15.4.1 G); metafemora glabrous except for a few scattered individual setae .... *Pseudorygmodus* [Chile]

29' Gula well developed (Fig. 15.4.1 M); metafemora densely pubescent on basal half, sometimes pubescence restricted to extreme base (and a narrow stripe at anterior margin) .... *Crenitis* [Neotropics. Afrotropical, Nearctic, Palaearctic]

30(28) Body form more hemispherical and globular; procoxae never with spiniform setae; elytra glabrous, never set with scattered setae .... *Anacaena* [Widespread]

30' Body form more elongate, procoxae often with spiniform setae (Fig. 15.4.2 L); elytra glabrous or set with numerous short setae, especially in posterior half .... *Crenitulus* [Neotropics. Australia, Nearctic]

31(1) Mesoventrite partially or largely fused to anepisterna (Figs. 15.4.3 G, I), with mesoventral process strongly raised (Figs. 15.4.3 G, I); elytra without distinct serial puncture rows (Fig. 15.4.2 E). (subfamily Sphaeridiinae in part: tribe Coelostomatini) 32

31' Mesoventrite not fused to anepisterna; mesoventrite anapleural suture complete and distinct (Fig. 15.4.3 H); mesoventrite rather flat, without raised process, at most slightly bulging or with a small obtuse semicircular ridge posteromedially (Fig. 15.4.3 H); Elytra with 10 distinct rows of serial punctures (Fig. 14.5.2 G) (subfamily Cylominae in part) .... *Cylorygmus* [Chile]

32(31) Mesofemur glabrous, at most with very sparse setae; mesoventrite largely fused to anepisterna, anapleural sutures indistinct or totally absent (Fig. 15.4.3 I) 33

32' Mesofemur basal three-fourths with dense hydrofuge pubescence; mesoventrite anapleural suture very distinct, incomplete (Fig. 15.4.3 G) .... *Lachnodacnum* [Brazil, Trinidad]

33(32) Elytra without distinct sutural striae. 34

33' Elytra posterior half with distinct sutural stria (Fig. 15.4.2 E) .... *Phaenostoma* [Neotropical]

34(33) Meso- and metaventrite mediallyfused in a common keel more or less widely separating mesocoxal cavities (Fig. 15.4.3 I) .... *Phaenonotum* [Neotropical. Afrotropical, Nearctic]

34' Meso- and metaventrite well separated from each other, not forming a common ventral keel; mesocoxal cavities very narrowly separated from each other (Fig. 15.4.3 J) .... *Hydroglobus* [Argentina]

Hydrophilidae: Genera (Larvae)

Modified from Archangelsky (1997) and Archangelsky et al. (2009). Please see comments in the section on Limitations.

1 Spiracular atrium present (Fig. 15.4.6 M) (partially reduced in *Hemiosus*, Fig. 15.4.6 N); without tracheal gills on abdomen (fleshy lobes or projections may be present on the thoracic and abdominal segments, Figs. 15.4.4 A, D; but they never look like the gills of *Berosus*, Fig. 15.4.4 C) 2

1' Spiracular atrium absent (Fig. 15.4.6 O); at least some of the abdominal segments with long, lateral tracheal gills (usually on segments I to VII, 15.4.4 C) .... *Berosus*

2(1) Hypopharyngeal lobe absent (Fig. 15.4.4 K) 3

2' Hypopharyngeal lobe present, appearing as a pubescent tongue on the left side of labium (Fig. 15.4.4 L) 26

3(2) Legs not reduced, with five articles (Fig. 15.4.4 N); ligula present (Fig. 15.4.4 K) or absent, but never as follows 4

3' Legs reduced, with three articles (Fig. 15.4.4 O); labium with a large, rounded ligula (Fig. 15.4.4 M) .... *Chaetarthria*

4(3) Thorax and abdomen usually without setiferous projections, if projections present, then not branched (Figs. 15.4.4 A, B) 5

4' Thorax and abdomen with long and branched setiferous projections (Fig. 15.4.4 D) .... *Derallus*

5(4)

Mandibles clearly asymmetrical, with differences in shape and number of inner teeth (Figs. 15.4.5 A, B); in *Tropisternus*, (Figs. 15.4.5 C) the mandibles may appear as slightly asymmetrical but there are differences in the shape and position of the smaller basal teeth.......... 6

5' Mandibles symmetrical to slightly asymmetrical (difference in shape only, with equal number of inner teeth) (Figs. 15.4.5 D–F)...... 11

6(5) Clypeolabrum strongly asymmetrical with left epistomal lobe strongly developed, covering basal third of left mandible, bearing strong spines and projecting further than nasale (Fig. 15.4.4 E) 7

6' Clypeolabrum symmetrical or asymmetrical, if asymmetrical left epistomal lobe not as previous (Figs. 15.4.4 F, H, J) 9

7(6) Spiracular atrium well developed, tergum VIII entire (Fig. 15.4.6 M); first antennomere lacking distal projection on inner margin 8

7' Spiracular atrium reduced, tergum VIII subdivided (Fig. 15.4.6 N); first antennomere with a distal projection on inner margin (Fig. 15.4.5 H) .... *Hemiosus*

8(7) Frontal lines subparallel, not merging before occipital foramen (Fig. 15.4.6 B) .... *Laccobius*

8' Frontal lines lyriform, merging at occipital foramen (Fig. 15.4.6 C) .... *Oocyclus*

9(6) First antennomere much longer than second (Fig. 15.4.5 I); clypeolabrum symmetrical or asymmetrical 10

9' First antennomere slightly shorter or as long as second (Fig. 15.4.5 J); clypeolabrum asymmetrical, nasale with six to seven teeth, those on right side projecting further than those on left side .... *Enochrus* (*Methydrus*)

10(9) Head capsule suboval; frontal lines U-shaped (Fig. 15.4.6 D) .... *Hydrophilus*

10' Head capsule subquadrangular, frontal lines V-shaped (Fig. 15.4.6 A) .... *Tropisternus*

11(5) Posterior margin of tergite VIII trifid (Fig. 15.4.6 P) 12

11' Posterior margin of tergite VIII concave or convex, never trifid 13

12(11) Abdominal segments I to VII with fleshy lateral projections (Fig. 15.4.6 K); pronotum with lateral flat lobes of quadrangular shape (Fig. 15.4.6 I) .... *Crenitis*

12 Abdominal segments I to VII lobulated laterally (Fig. 15.4.4 B), but without fleshy projections (Fig. 15.4.6 L); pronotum with lateral rounded lobes (Fig. 15.4.6 J) .... *Crenitulus*

13(11) Mentum at most slightly wider than prementum; inner margin of first antennomere bare (Fig. 15.4.5 K) 14

13' Mentum twice as wide as prementum (Fig. 15.4.4 K); inner margin of first antennomere with numerous long setae (Fig. 15.4.5 I) .... *Hydrobiomorpha*

14(13) Tergum VIII subdivided 15

14' Tergum VIII entire 16

15(14) Head capsule strongly depressed; mandibles with inner basal part bearing short cuticular spines directed backward (Fig. 15.4.5 G) .... *Lachnodacnum*

15' Head capsule not strongly depressed; mandibles without inner basal spines .... *Phaenonotum*

16(14) Nasale with one or more teeth (Figs. 15.4.4 G, I, J) 17

16' Nasale concave, toothless (Fig. 15.4.4 H) .... *Helobata*

17(16) Nasale with two or more teeth 18

17' Nasale with one simple median tooth .... *Hydroglobus*

18(17) Nasale with three or more teeth 19

18' Nasale with two median teeth .... *Anticura*

19(18) Clypeolabrum symmetrical or nearly so (Fig. 15.4.4 G) (in *Hydramara*, it could exceptionally appear as symmetrical in first instar larvae); nasale with five teeth 20

19' Clypeolabrum clearly asymmetrical (Fig. 15.4.4 F); number of teeth on nasale variable 21

20(19) Lateral teeth of nasale closer to each other than to median tooth (Fig. 15.4.4 I); frontal lines liriform, not converging toward occipital foramen (Fig. 15.4.6 E); mandibles with two inner teeth .... *Pseudorygmodus*

20' Teeth of nasale equidistant (Fig. 15.4.4 G); frontal lines inversely bell-shaped, converging toward occipital foramen (Fig. 15.4.6 F); mandibles with three inner teeth .... *Guyanobius*

21(19) Frontal lines convergent, united basally (Fig. 15.4.6 G); first maxillary palpomere variable in length 22

21' Frontal lines subparallel, not united at base of head capsule (Fig. 15.4.6 H); first maxillary palpomere the longest (usually as long as the remaining three together) .... *Paracymus*

22(21) Mandibles with two inner teeth (in *Notionotus* the distal inner tooth is bifid and may look as if the mandible has three inner teeth, Fig. 15.4.5 F); sensorium on second antennomere long (Fig. 15.4.5 J); base of stipes with or without small cuticular spines 23

22' Mandibles with three inner teeth (Fig. 15.4.5 E); sensorium on second antennomere very short (Fig. 15.4.5 L); base of stipes with small cuticular spines .... *Hydramara*

23(22) Right epistomal lobe not projecting further than nasale (at most of same length) (Figs. 15.4.4 F, J); distal inner tooth of mandible simple (Fig. 15.4.5 B) 24

23' Right epistomal lobe projecting further than nasale; distal inner tooth of mandible bifid (Fig. 15.4.5 F) .... *Notionotus*

24(23) Right epistomal lobe rounded, similar to left one (Fig. 15.4.4 J) 25

24' Right epistomal lobe shaped as a tooth (Fig. 15.4.4 F), different from left one .... *Enochrus* (*Hugoscottia*)

25(24) Stipes with strong basal cuticular spines (Fig. 15.4.5 M) .... *Helochares*

25' Stipes lacking basal cuticular spines .... *Cymbiodyta*

26(2) Legs not reduced, with five articles .... *Dactylosternum*

26' Legs reduced, articles fused, undistinguishable .... *Pelosoma*

ACKNOWLEDGMENTS

We would like to thank Neusa Hamada and Cesar J. Benetti (INPA Manaus) for the opportunity to be a part of this work, as well as for all the support given to make this chapter possible. We are indebted to Martin Fikáček (MNPC Prague) for comments on the early version of the manuscript and for allowing us to use some of his unpublished drawings and to Maurício Garcia (MALUZ Venezuela) for his contribution to this chapter. BC would like to thank Inês C. Gonçalves for encouragement and aid during the development of this work. Financial support was provided in part by Fellowships from Coordenação de Aperfeiçoamento de Pessoal de Nível Superior (CAPES) for BC (UFRJ, PVE, Proc. 19/2013), by the grant PICT2014-0853 and from Agencia Nacional de Promoción Científica y Tecnológica (Argentina) for PT.

REFERENCES

Arce-Pérez, R. & M.A. Morón. 2011. Sinopsis de los Hydrophiloidea de México (Coleoptera: Hydrophilidae, Helophoridae, Epimetopidae, Georissidae e Hydrochidae), con una clave para la identificación de los géneros. Revista mexicana de biodiversidad 82: 491–514.

Archangelsky, M. 1994. Description of the preimaginal stages of *Dactylosternum cacti* (Coleoptera, Hydrophilidae: Hydrophilinae). Entomologica Scandinavica 25: 121–128.

Archangelsky, M. 1997. Studies on the biology, ecology, and systematics of the immature stages of New World Hydrophiloidea (Coleoptera: Staphyliniformia). Bulletin of the Ohio Biological Survey, New Series, 12: 1–206.

Archangelsky, M. 1999. Immature stages of *Paracymus rufocinctus* Bruch, 1915 and Enochrus (Methydrus) *vulgaris* (Steinheil, 1869) (Coleoptera: Hydrophilidae: Hydrophilinae). Elytron 13: 87–99.

Archangelsky, M. 2000. Immature stages of Neotropical Hydrophilidae (Coleoptera): *Hydramara argentina* (Knisch, 1925) and Hemiosus bruchi Knisch, 1924. Proceedings of the Entomological Society of Washington 102(2): 280–291.

Archangelsky, M. 2002. Nuevas larvas de Hydrophilidae (Coleoptera: Hydrophiloidea): *Hemiosus multimaculatus* y *Chaetarthria bruchi*. Revista de la Sociedad Entomológica Argentina 61(3–4): 89–97.

Archangelsky, M. 2002. Immature stages of Neotropical *Enochrus* (Coleoptera: Hydrophilidae): *E. (Methydrus) lampros* Knisch, 1924 and *E. (Hugoscottia) tremolerasi* Knisch, 1922. Aquatic Insects 24(1): 41–52.

Archangelsky, M. & M. E. Durand. 1992. Description of the preimaginal stages of *Derallus angustus* Sharp, 1882 (Coleoptera: Hydrophilidae, Hydrophilinae). Aquatic Insects 14(3): 169–178.

Archangelsky, M., R.G. Beutel & A. Komarek. 2005. Hydrophiloidea. Pages 157–183 *in*: R.G. Beutel & N.P. Kristensen (eds.), Handbook of Zoology, Arthropoda: Insecta Coleoptera, Beetles Volume 1: Morphology and Systematics, 1st edition. Walter de Gruyter GmbH & Co. KG, Berlin/Boston.

Archangelsky, M., R.G. Beutel & A. Komarek. 2016. Hydrophiloidea. Pages 231–272 *in*: R.G. Beutel & N.P. Kristensen (eds.), Handbook of Zoology, Arthropoda: Insecta Coleoptera, Beetles Volume 1: Morphology and Systematics, 2nd edition. Walter de Gruyter GmbH & Co. KG, Berlin/Boston.

Archangelsky, M. & M. Fikáček. 2004. Descriptions of the egg case and larva of *Anacaena* and a review of the knowledge and relationships between larvae of Anacaenini (Coleoptera: Hydrophilidae: Hydrophilinae). European Journal of Entomology 101(4): 629–636.

Archangelsky, M., V. Manzo, M.C. Michat & P.L. Torres. 2009. Coleoptera. Pages 411–468 *in*: Domínguez, E. & H.R. Fernández (eds.), Macroinvertebrados bentónicos sudamericanos. Sistemática y biología. Fundación Miguel Lillo, Tucumán, Argentina.

Archangelsky, M., P.L. Torres & M. C. Michat. 2004. Immature stages of *Hydrobiomorpha spinosa* (Orchymont, 1928) (Coleoptera: Hydrophilidae). Aquatic Insects 26(3–4): 253–264.

Archangelsky, M., G. Rodriguez & P.L. Torres. 2016. Primary chaetotaxy and larval morphometry of *Phaenonotum exstriatum* and *Dactylosternum cacti* (Coleoptera: Hydrophilidae). Acta Entomologica Musei Nationalis Pragae 56(1): 167–193.

Balduf, W.V. 1935. The bionomics of entomophagous Coleoptera. John S. Swift Co., St. Louis, Missouri. 220 pp.

Balfour-Browne, F. 1910. On the life-history of *Hydrobius fuscipes* L. Transactions of the Royal Society of Edinburgh 47: 317–340.

Bloom, D.D., M. Fikáček, & A.E.Z. Short. 2014. Clade age and diversification rate variation explain disparity in species richness among water scavenger beetle (Hydrophilidae) lineages. PLoS One 9: e98430.

Borror, D.J., C.A. Triplehorn & N.F. Johnson. 1989. An introduction to the study of insects. Saunders College Publishing, Philadelphia. 875 pp.

Clarkson, B. & N. Ferreira-Jr 2009. Three new species of *Hemiosus* Sharp (Coleoptera: Hydrophilidae) and new state records of *Hemiosus fittkaui* Oliva and *H. moreirai* d'Orchymont from Brazil. Zootaxa 2139: 61–68.

Contreras-Ramos, A. 1999. Métodos para estudios en sistemática de Megaloptera (Insecta: Neuropteridae) con base en morfología. Dugesiana 6: 1–15.

Crowson, R.A. 1981. The biology of the Coleoptera. Academic Press, London. 802pp.

Dennis, C.J. 1974. Laboratory manual for introductory entomology. W.C. Brown Company Publishers, Dubuque, Iowa. 166 pp.

Hansen, M. 1991. The hydrophiloid beetles. Phylogeny, classification and a revision of the genera (Coleoptera, Hydrophiloidea). Biologiske Skrifter 40: 1–367.

Hansen, M. 1995. Evolution and classification of the Hydrophiloidea - a systematic review. Pages 321–353 *in*: J. Pakaluk & S.A. Slipinki (eds.), Biology, Phylogeny, and Classification of Coleoptera; Papers Celebrating the 80th Birthday of Roy A. Crowson. Museum i Instytut Zoologii PAN, Warszawa.

Hansen, M. 1999. Hydrophiloidea (s. str.) (Coleoptera). World Catalogue of Insects 2: 1–416.

Imes, R. 1992. Practical Entomologist. Simon and Schuster, New York. 161 pp.

Jäch, M.A. & M. Balke. 2008. Global diversity of water beetles (Coleoptera) in freshwater. Hydrobiologia 595: 419–442.

Komarek, A. 2005. Taxonomic revision of *Anacaena* Thomson, 1859 II. Neotropical species. Koleopterologische Rundschau, 75: 253–301.

Lawrence, J.F. 1991. Order Coleoptera. Introduction. Pages 144–184 *in*: F.W. Stehr (ed.), Immature Insects, Volume 2. Kendall/Hunt, Dubuque.

Llorente J., A. Garcés, T. Pulido & I. Luna. 1985. Manual de Recolección y Preparación de Animales. Facultad de Ciencias, UNAM, México, D.F. 246 pp.

Martin, J.E. 1977. The insects and arachnids of Canada. Part 1. Collecting, preparing, and preserving insects, mites and spiders. Kromar Printing Ltd., Québec. 186 pp.

Merritt, R.W., V.H. Resh & K.W. Cummins. 1996. Design of aquatic insect studies: Collecting, sampling and rearing procedures. Pages 12–28 *in*: R.W. Merritt & K.W. Cummins (eds.), An introduction to the aquatic insects of North America. Kendall-Hunt Publishing Company, Iowa.

Morón, M.A. & R. Terrón 1988. Entomología Práctica. Instituto de Ecología A.C. México.

Newton, A.F. & M.K. Thayer. 1992. Current classification and family-group names in Staphyliniformia (Coleoptera). Fieldiana. Zoology 67: 1–92.

Oliva, A., L.A. Fernández & A.O. Bachmann. 2002. Sinopsis de los Hydrophiloidea acuáticos de la Argentina (Insecta, Coleoptera). Monografías del Museo Argentino de Ciencias Naturales 2: 1–67.

Seidel, M., E. Arriaga-Varela & M. Fikáček. 2016. Establishment of Cylominae Zaitzev, 1908 as a valid name for the subfamily Rygmodinae Orchymont, 1916 with an updated list of genera (Coleoptera: Hydrophilidae). Acta Entomologica Musei Nationalis Pragae 56: 159–165.

Short, A.E.Z. & M. Fikacek. 2013. Molecular phylogeny, evolution and classification of Hydrophilidae (Coleoptera). Systematic Entomology 38: 723–752.

Short, A.E.Z. & M. Fikacek 2011. World catalogue of the Hydrophiloidea (Coleoptera): additions and corrections II (2006–2010). Acta Entomologica Musei Nationalis Pragae 51: 83–122.

Short, A.E.Z. & F. Hebauer. 2006. World Catalogue of Hydrophiloidea (Coleoptera) – additions and corrections, I (1999–2005). Koleopterologische Rundschau 76: 315–359.

Short, A.E.Z. & C. E. McIntosh IV 2014. Review of the Giant Water Scavenger Beetle Genus *Hydrophilus* Geoffroy (Coleoptera: Hydrophilidae) of the United States and Canada. The Coleopterists Bulletin 68(2): 187–198.

Short, A.E.Z. & P.J. Torres 2006. Review of the *Hemiosus* Sharp and *Derallus* Sharp in North America, Central America and the Caribbean (Coleoptera: Hydrophilidae: Berosini). Zootaxa 1369: 1–17.

Smetana, A. 1988. Review of the family Hydrophilidae of Canada and Alaska (Coleoptera). The Memoirs of the Entomological Society of Canada 120: 3–316.

Steyskal, G.C., W.L. Murphy & E.M. Hoover. 1986. Insects and mites: Techniques for collection and preservation. U.S. Department of Agricultural, Miscellaneous Publication No. 1443.

Torres, P.L., M.C. Michat & M. Archangelsky 2011. The immature stages of *Tropisternus* (*Pleurhomus*) sahlbergi (Sharp) (Coleoptera: Hydrophilidae), with considerations on the primary chaetotaxy of *Tropisternus* Solier. Zootaxa 2777: 25–40.

Wilson, C. 2000. Eggs and egg cocoons of *Sphaeridium*. Latissimus 12: 22–23.

Van Tassell, E.R. 2000. Family 13. Hydrophilidae Latreille, 1802. Pages 272–418 in: Arnett R.H. Jr & M.C. Thomas (eds.), American Beetles. Volume 1. Archostemata, Myxophaga, Adephaga, Polyphaga. CRC Press, Boca Raton, Florida.

Family Hydraenidae

Juan A. Delgado
Departamento de Zoología y Antropología Física, Facultad de Biología, Universidad de Murcia, Murcia, Spain

Josefina Garrido
Departamento de Ecología y Biología Animal, Facultad de Biología, Universidad de Vigo, Vigo, Spain

Albert Deler-Hernández
Department of Zoology, Faculty of Science, Charles University, Prague, Czech Republic

Luis F. Valladares
Departamento de Biodiversidad y Gestión Ambiental, Facultad de Ciencias Biológicas y Ambientales, Universidad de León, León, Spain

Chapter Outline

INTRODUCTION

Hydraenidae is a cosmopolitan family of water beetles with around 1,500 described species and many others still awaiting description (Jäch & Balke, 2008; Jäch et al., 2016). Hydraenids occur in aquatic environment all over the world, although they are more common between temperate and tropical zones and are scarcely represented in cold regions. As its common name correctly suggests, most species of "minute moss beetles" are small, ranging in size from 0.8 to 2 mm, rarely reach more than 3 mm. However, only a small proportion is really associated with moss, and most species chiefly live among sand, mud, and vegetation in the aquatic habitats. Most species in this family can be considered "true water beetles," since adults and larvae are aquatic, but some members are semiaquatic or even terrestrial.

Although hydraenids are poor swimmers, they are nevertheless well adapted for aquatic existence. Adults can submerge for prolonged periods, breathing with the help of a ventral microplastron, which works as an effective physical gill. Helped by these adaptations, the members of this family occupy all types of aquatic habitats. Some genera or species groups prefer the margins of lentic environments and live among mud, decaying leaves, and aquatic vegetation. Others prefer running waters, where they are usually found in fast-flowing areas, associated to mats of aquatic mosses or plant debris. Some species are abundant in temporary habitats such as ephemeral ponds, temporary streams, and coastal rock-pools and lagoons. Hydraenids can also be found in hygropetric habitats, and a few genera are known to have humicolous species or live associated to coastal vegetation. A Chilean genus, *Ochtheosus* (not treated in this chapter) is terrestrial.

Preimaginal stages are also associated with aquatic or very humid environments. Eggs are laid under water or close to it, adhered to stones or submerged vegetation. Although some larvae are aquatic, in most species this stage is associated with moist terrestrial habitats. Pupation occurs outside the water, on humid soils, protected by an elliptical muddy cell built by the larvae. Adults rarely fly, although species associated with temporary waters must do regular

Thorp and Covich's Freshwater Invertebrates, Fourth Edition. https://doi.org/10.1016/B978-0-12-804223-6.00026-3

dispersive flights to reach new habitats. Larvae and adults are either phytophagous or detritivores.

Systematics

A total of four subfamilies (some of them probably ill-defined) are presently recognized in Hydraenidae, with about 1,500 described species in 42 genera (Jäch et al., 2016). Of these, two subfamilies with nine genera and about 150 species have been reported from the Neotropical region. Six genera are included in the subfamily Hydraeninae: *Adelphydraena* (two species, Fig. 15.5.1 B); *Haptaenida* (one species, Fig. 15.5.1 F); *Hydraena* (>80 species, Fig. 15.5.1 A); *Hydraenida* (five species, Fig. 15.5.1 E); *Limnebius* (seven species, Fig. 15.5.1 C); and *Parhydraenida* (nine species, Fig. 15.5.1 D). The subfamily Ochthebiinae comprises three genera: *Gymnochthebius* (20 species, Fig. 15.5.1 H); *Meropathus* (one species, Fig. 15.5.1 I); and *Ochthebius* (23 species, Fig. 15.5.1 G). We are following the family taxonomy utilized in Jäch et al. (2016).

LIMITATIONS

The Hydraenidae fauna of the Neotropical region is taxonomically quite well known, mainly thanks to the seminal contributions of Perkins (1980, 1989, 1997, 2011). Other contributions describing Neotropical species or with interesting information are Spangler (1980), Anderson (1983), Jäch (1998), Ribera (2000), Bameul & Jäch (2001), Perkins & Archangelsky (2001), Makhan (2008), Brojer & Jäch (2011), Valladares & Short (2011), and Deler-Hernández & Delgado (2012, 2017b). Adults are generally much better known than the preimaginal stages, where only a few species have been studied or described so far (Delgado et al., 1997a, 1997b; Delgado & Archangelsky, 2005; Deler-Hernández & Delgado, 2017a).

At the generic level, the first identification key was provided by Perkins (1980) for the Hydraenidae of the Western Hemisphere that were known at that time. Unfortunately, its keys are now outdated since the subfamilial classification has changed and some new genera have been described for the Neotropical fauna. Perkins (1997) proposed a new classification for the family and published keys to the new subfamilies and genera. These last keys are perfectly useful although, again, some genera have changed its status since 1997. Perkins's keys are adapted and updated in the present work to cover the genera of Hydraenidae that occur in the Neotropical region. A world catalog of Hydraenidae by Hansen (1998) contains much interesting information on the Neotropical genera and species.

TERMINOLOGY AND MORPHOLOGY

Hydraenidae adults exhibit a series of specific anatomical and behavioral adaptations for aquatic life. They are capable of breathing under water by means of a ventral plastron retained by a dense hydrofuge pubescence. The air is periodically renewed by the help of the antennae, which are used to break the air-water interface and hence are maxillary palps that take the role of antennae. Associated to these traits some genera have developed independently a remarkably exocrine secretion delivery system (Perkins, 1997), integrated by a set of glands, setae, grooves, and other cuticular structures, along with a complex secretion-grooming behavior. All this intended to maintain the ventral respiratory air bubble. Males and females do not show much sexual dimorphism in the Neotropical fauna of Hydraenidae. When it exists, it is usually reflected either in the tibiae (distinctly to slightly curved, enlarged, excavated, or with developed setal fringes in males) or in the shape of the labral margin (upturned, toothed, horned, or deeply emarginate in males).

Hydraenids are not only quite homogeneous in size, but also in shape, which vary only slightly among most of the Neotropical genera. With the exception of *Limnebius*, the body form is elongate and with an evident pronotoelytral angle. The color varies from dark to pale brown. Legs and maxillary palps are generally lighter, usually from reddish brown to testaceous. It is also frequent to find different dorsal sculptures, with more or less marked pronotal impressions and groves (Figs. 15.5.2 D, E), as well as elytral striae (Figs. 15.5.2 K, H).

Larvae are elongate, slightly tapering distally and generally lightly colored, from grayish to pale brown. The larval phase undergoes three instars. The head of larvae is ovate in dorsal view and globose when observed laterally. The labrum is free. The two mandibles are stout, slightly asymmetrical, and with a basal mola. Legs are well-developed and consist of four articles. The abdomen bears distally a pair of biarticulated urogomphi and is devoid of lateral tubercles or gills. They are clearly distinguished from other aquatic coleopteran larvae by the presence of two ventral recurved hooks in abdominal segment 10. More detailed accounts on the morphology of Hydraenidae can be found in Jäch et al. (2016).

Most of the terms used for identification of Hydraenidae are given in the general chapter on aquatic Coleoptera (Chapter 15). Some additional terms are given here:

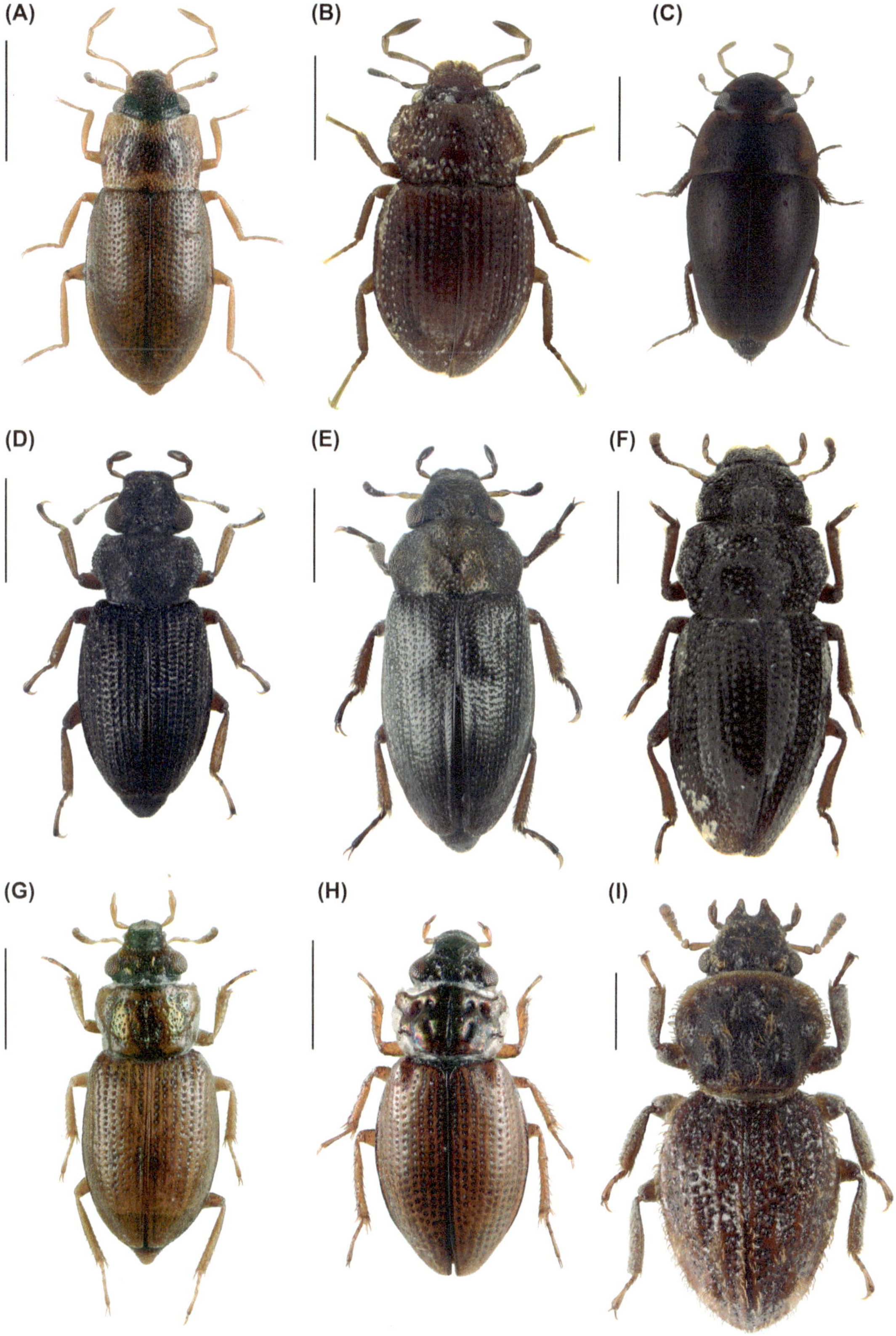

FIGURE 15.5.1 Hydraenidae (Coleoptera) adult habitus, in dorsal view: (A) *Hydraena*; (B) *Adelphydraena*; (C) *Limnebius*; (D) *Parhydraenida*; (E) *Hydraenida*; (F) *Haptaenida*; (G) *Ochthebius*; (H) *Gymnochthebius*; and (I) *Meropathus*. Scale bars = 500 μm.

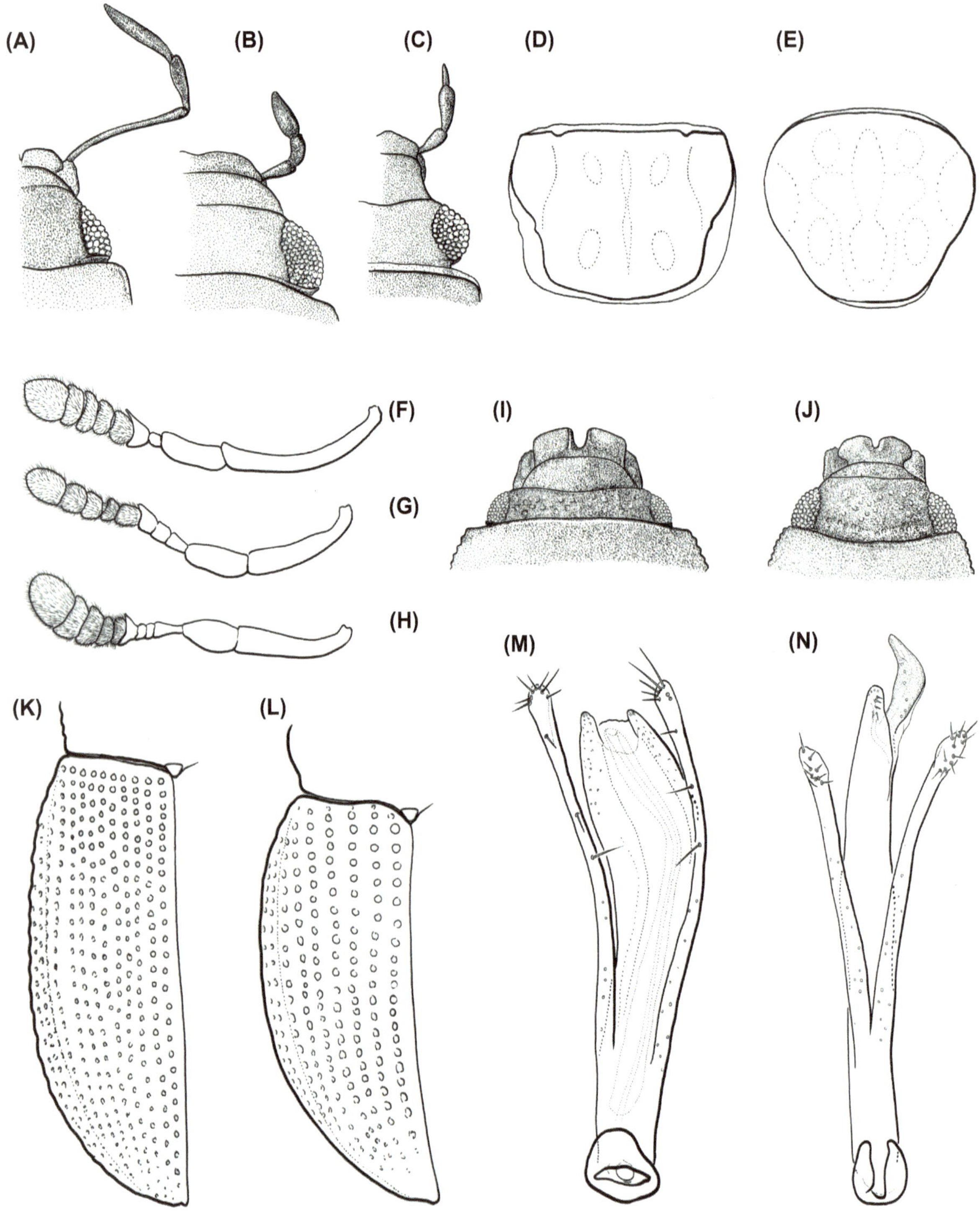

FIGURE 15.5.2 Hydraenidae (Coleoptera) adult, diagnostic features. Maxillary palpus: (A) *Hydraena*; (B) *Hydraenida*; (C) *Ochthebius*. Pronotum: (D) *Ochthebius*; (E) *Meropathus*. Antenna: (F) *Hydraena*; (G) *Adelphydraena*; and (H) *Hydraenida*. Labrum: (I) *Adelphydraena*; and (J) *Hydraena*. Elytral striae: (K) *Hydraena* and (L) *Adelphydraena*. Aedeagus in ventral view: (M) *Gymnochthebius* and (N) *Ochthebius*.

Aedeagal main piece: Sclerotized, elongate, tubular structure formed by the fusion of the basal piece and the median lobe of other Coleoptera. Generally, this structure bears two lateral parameres and a terminal distal lobe, which can be extremely complex (Figs. 15.5.2 M, N).

Hyaline membrane: Narrow fringe of hyaline membranous cuticle on the anterior, posterior, and lateral margins of pronotum (Figs. 15.5.2 D).

Prosternal intercoxal process: a median, intercoxal extension of the prosternum.

MATERIAL PREPARATION AND PRESERVATION

Methods used for the collection and preservation of Hydraenidae are similar to those used for other aquatic Coleoptera, which are explained in Chapter 15.

KEY TO HYDRAENIDAE

Hydraenidae: Genera

1	Maxillary apical palpomere shorter than penultimate (Fig. 15.5.2 C)	2
1'	Maxillary apical palpomere longer than penultimate (Fig. 15.5.2 B)	4
2(1)	All pronotal margins bordered by a hyaline membrane (Fig. 15.5.2 D); male aedeagus with parameres (Figs. 15.5.2 M, N)	3
2'	Pronotum with only anterior and posterior margins bordered by a very narrow hyaline membrane (Figs. 15.5.1 I, 2 E); male aedeagus without parameres	*Meropathus vectis* Perkins, 1980
3(2)	Pronotum anterolaterally lobate, frequently excised laterally and posterolaterally; anterior margin more or less deeply emarginated (Fig. 15.5.1 H); male aedeagus with main piece bifid apically (Fig. 15.5.2 M)	*Gymnochthebius*
3'	Pronotum not anterolaterally lobate; posterolateral margins and anterior pronotal corners sometimes excised but with lateral projections never markedly lobate (Figs. 15.5.1 G, 2 D), most frequently lateral margins gently sinuous; male aedeagus with main piece not bifid at apex (Fig. 15.5.2 N)	*Ochthebius*
4(1)	Body contour interrupted between pronotum and elytra; pronotum anterior and posterior margins subequal in width, broadest medially; pronotal surface with impressions, elytra punctuate	5
4'	Body contour evenly curved, without an evident angle between pronotum and elytra; pronotum broadest basally; pronotal surface without impressions, elytra not punctuate (Fig. 15.5.1 C)	*Limnebius*
5(4)	Maxillary palpi length > than head width; maxillary palpi second palpomere elongate and slender, at least as long as third and fourth palpomeres combined (Fig. 15.5.2 A); antennae of 9- or 10 antennomeres (Figs. 15.5.2 F, G); prosternal intercoxal process expanded laterally behind procoxae, closing procoxal cavities	6
5'	Maxillary palpi length < than head width; maxillary palpi second palpomere length < third and fourth palpomeres combined (Fig. 15.5.2 B); antennae with 11 antennomeres (Fig. 15.5.2 H); anterior coxal cavities not closed posteriorly	7
6(5)	Elytra, each with 10–15 series of punctures (Figs. 15.5.1 A, 2 K); labrum abruptly constricted posteriorly (Fig. 15.5.2 J); antenna 9-articulate (Fig. 15.5.2 F)	*Hydraena*
6'	Elytra, each with nine series of punctures (Figs. 15.5.1 B, 2 L); labrum not constricted posteriorly (Fig. 15.5.2 I); antenna 10-articulate (Fig. 15.5.2 G)	*Adelphydraena*
7(5)	Elytra at least 3× longer than thorax; elytral apices not markedly narrowed; male aedeagus with parameres	8
7'	Elytra short, <2.5 times thorax length; elytral apices markedly tapered (Fig. 15.5.1 F); male aedeagus without parameres	*Haptaenida huggerti* Perkins, 1997
8(7)	Pronotum subcordate, sides sinuate near posterior corners (Fig. 15.5.1 D); elytra serial punctures in sulcate impressions, interstices costate or subcostate	*Parhydraenida*
8'	Pronotum sides straight near posterior corners, not sinuate; elytra serial punctures not in sulcate impressions (Fig. 15.5.1 E); interstices flat or only slightly convex	*Hydraenida*

ACKNOWLEDGMENTS

We are grateful to Cesar Benetti who involved us in this project. Martin Fikáček, Ricardo Palma, and Philip Perkins kindly provided interesting exemplars and digital images for study.

REFERENCES

Anderson, J.J. 1983. *Ochthebius (Asiobates) otavalensis*, a new species of water beetle from the Ecuadorian Andes (Coleoptera: Hydraenidae). The Coleopterists Bulletin 37: 212–216.

Bameul, F. & M.A. Jäch. 2001. A new *Hydraena* Kugelann from Martinique Island, French West Indies (Coleoptera, Hydraenidae). Nouvelle Revue d'Entomologie (Nouvelle Série) 17: 273–285.

Brojer, M. & M.A. Jäch. 2011. Update on the hydraenid fauna of Costa Rica, with descriptions of six new species (Coleoptera: Hydraenidae). Koleopterologische Rundschau 81: 69–92.

Deler-Hernández, A. & J.A. Delgado. 2012. The Hydraenidae of Cuba (Insecta: Coleoptera) I: Contribution to the fauna of eastern Cuba. Zootaxa 3478: 213–238.

Deler-Hernández, A. & J.A. Delgado. 2017a. The Hydraenidae of Cuba (Insecta: Coleoptera) II: Morphology of preimaginal stages of six species and notes on their biology. Zootaxa 4238: 451–498.

Deler-Hernández, A. & J.A. Delgado. 2017b. The Hydraenidae of Cuba (Insecta: Coleoptera) III: Description of two new hygropetric species of Hydraena Kugelann. Zootaxa 4250: 434–446.

Delgado, J.A. & M. Archangelsky. 2005. Description of the larval stages of *Gymnochthebius jensenhaarupi* and phylogenetic analysis of the relationships with other species of the subfamily Ochthebiinae (Coleoptera: Hydraenidae). European Journal Entomology 102: 231–240.

Delgado, J.A. & F. Collantes. 1996. A new species of water beetle, *Hydraena delvasi,* from Colombia, South America (Coleoptera: Hydraenidae). Studies on Neotropical Fauna & Environment 31: 54–56.

Delgado, J.A., F. Collantes & A.G. Soler. 1997a. Description of the larval stages of the neotropical water beetle *Hydraena particeps* Perkins from Nicaragua. Studies on Neotropical Fauna and Environment 32: 47–51.

Delgado, J.A., F. Collantes & A.G. Soler 1997b. Distribución y ciclo vital de tres especies del género *Hydraena* en Nicaragua (Coleoptera: Hydraenidae). Revista de Biología Tropical 44/45 (1996/1997): 645–649.

Hansen, M. 1998. Hydraenidae (Coleoptera). Pages 1–169 *in*: M. Hansen (ed.), World catalogue of insects Vol. 1. Apollo Books, Copenhagen.

Jäch, M.A. 1998. Description of two new species of Hydraenida Germain (Coleoptera: Hydraenidae). Koleopterologische Rundschau 68: 167–169.

Jäch, M.A. & M. Balke. 2008 Global diversity of water beetles (Coleoptera) in freshwater. Hydrobiologia 595: 419–442.

Jäch, M.A., Beutel, R., Delgado, J.A. & Díaz, J.A. 2016. Hydraenidae. Pages 316–345 *in*: R. Beutel & R.A.B. Leschen (eds.), Handbook of Zoology Vol. 4, 2nd ed., Arthropoda Insecta. Part. 38. Coleoptera, Beetles Vol. 1: Morphology and Systematics (Archostemata, Adephaga, Myxophaga, Polyphaga partim). Berlin: De Gruiter, 684 pp.

Makhan, D. 2008. *Hydraena nelsonmandelai* sp. nov., a new water beetle from Suriname (Coleoptera: Hydraenidae). Calodema Supplementary Paper No. 78: 1–3.

Perkins, P.D. 1980. Aquatic beetles of the family Hydraenidae in the Western Hemisphere: classification, biogeography and inferred phylogeny (Insecta: Coleoptera). Quaestiones Entomologicae 16: 3–554.

Perkins, P.D. 1989. *Adelphydraena*, new genus, and two new species from Venezuela, and remarks on phylogenetic relationships within the subtribe Hydraenina (Coleoptera: Hydraenidae). Proceedings of the Biological Society of Washington 102: 447–457.

Perkins, P.D. 1997. Life on the effective bubble: Exocrine secretion delivery systems (ESDS) and the evolution and classification of beetles in the family Hydraenidae (Insecta: Coleoptera). Annals of Carnegie Museum 66: 89–207.

Perkins, P.D. & M. Archangelsky. 2002. A New Species of Water Beetle from Ischigualasto Park, Argentina (Coleoptera: Hydraenidae). The Coleopterists Bulletin 56: 237–240.

Perkins, P.D. 2011. New records and description of fifty-four new species of aquatic beetles in the genus *Hydraena* Kugelann from South America (Coleoptera: Hydraenidae). Zootaxa 3074: 1–198.

Ribera, I. 2000. Notes on the genus *Hydraenida*, with description of a new species from Chile (Hydraenidae). Koleopterologische Rundschau 70: 53–56.

Spangler, P.J. 1980. Two new species of aquatic beetles of the genus *Hydraena* from Cuba (Coleoptera: Hydraenidae). Proceeding of the Entomological Society of Washington 82: 329–333.

Valladares, LF. & A.E.Z. Short. 2011. Revision of the Ochthebiinae of Venezuela (Coleoptera: Hydraenidae). Zootaxa 2792: 63–67.

Chapter 15.6

Families Dryopidae, Elmidae, and Psephenidae

Maria Inês S. Passos

Laboratório de Insetos Aquáticos, Departamento de Zoologia/ Instituto de Biociências, Universidade Federal do Estado do Rio de Janeiro (Unirio), Rio de Janeiro, Rio de Janeiro, Brazil

Veronica Manzo

Instituto de Biodiversidad Neotropical (IBN), CONICET - UNT. Horco Molle, Yerba Buena, Tucumán, Argentina

Crystal A. Maier

Gantz Family Collection Center, Field Museum of Natural History, Chicago, IL, United States

Chapter Outline

INTRODUCTION

The suborder Polyphaga is the largest in the order Coleoptera, and it is also the most diverse, with over 300,000 described species in over 150 families. Ecologically, polyphagans run the gamut of life histories from xylophagous species to predators and specialist plant feeders; additionally, many species live a fully aquatic life. In the Neotropical Region, this suborder is represented by 13 families that are regarded as "predominantly aquatic" (Jäch & Balke, 2008). In this chapter, we will focus on the families Dryopidae, Elmidae, and Psephenidae, which belong to the superfamily Byrrhoidea.

Dryopidae is an ecologically diverse family, with some members being aquatic, while others are completely terrestrial or semiaquatic. Only five aquatic genera are recognized in the Neotropical Region. They are found on the sandy banks of lotic habitats in soft stream substrate, while others occur on emergent vegetation and sticks protruding from streams. They are often collected near lights at night; for these the habitat is unknown. In the terrestrial dryopids, adults and larvae can be found in forest leaf litter.

This family contains about 300 species in 33 genera worldwide. Dryopids occur in all biogeographic regions, but they are absent from the Australian continent.

Elmidae contains the taxa commonly called "riffle beetles," and it is the largest family of aquatic Byrrhoidea. It includes small- to medium-sized aquatic beetles that are common in running waters with high oxygen content. This family shows a worldwide distribution, though much of their diversity occurs in the Neotropical Region. Riffle beetles are always aquatic as larvae, whereas adults can be aquatic (Elminae) or riparian (Larainae).

Elmidae is a widespread family with two subfamilies, Elminae and Larainae, both present in the Neotropical Region. Riffle beetles are very common in lotic habitats and

Thorp and Covich's Freshwater Invertebrates, Fourth Edition. https://doi.org/10.1016/B978-0-12-804223-6.00027-5

are extraordinarily diverse: there are 47 genera and about 440 species described in this region (Manzo, 2013; Segura et al., 2013).

Psephenidae, or "water penny beetles," are reported for all continents except Antarctica, and they are quite diverse in the Neotropics. The adults are terrestrial, and larvae are strictly aquatic. The larvae are highly modified for a life in fast-flowing water: their very flattened body and the powerful claws permit them to explore food sources on the exposed surfaces of rocks while maintaining their position and resisting swift currents.

The family contains about 272 species in 35 genera. They are comprised of four subfamilies: Eubrianacinae, Eubriinae, Psepheninae, and Psephenoidinae. According to Fernandez et al. (2001), seven genera occur in the Neotropical Region.

LIMITATIONS

There have been few studies on aquatic Dryopidae in the Neotropical Region, perhaps due to the difficulty in collecting specimens in the flow of streams and rivers. The last work concerning the Neotropics is Shepard & Chaboo (2015).

Detailed studies of the Psephenidae are lacking, perhaps due the difficulty in associating their aquatic larvae with adults that are often found out of the water. Recently, work by Shepard & Barr (2014) resulted in the description of a new genus and new species, indicating there is still much work to be done on Neotropical Psephenidae.

Elmidae is the best studied family in Byrrhoidea; however, much of the diversity, particularly in the Neotropics, still awaits discovery. The lack of work on the group is likely due to their small size and external morphologic homogeneity. Also, the larvae of many genera are not known, or remain unassociated with their adult form; as a result, the key presented only applies to genera for which associated larvae are documented in the literature.

TERMINOLOGY AND MORPHOLOGY

Adults of Dryopidae are morphologically characterized by (1) anterior coxae transverse and with exposed trochantin; antennae usually short, with apical articles pectinate or lamellate and forming a club; and (2) female genitalia without styli, usually asymmetrical and resembling two knife blades, functioning as ovipositors; about 4–8 mm long. The larvae feature the following: body cylindrical, with abdominal sternites and pleurites greatly reduced, the tergites almost forming complete rings on first five segments and forming complete ones on segments 6–9; without retractile gills; abdominal spiracles lateral on segments 1–7 and dorsal on segment 8. Larvae of Dryopidae are unlikely to be found in streams (Brown, 1972).

Adults of the elmid subfamily Elminae are aquatic, body scarcely pubescent with plastron, procoxae globose, and trochantin not exposed; adults of the subfamily Larainae are occasionally aquatic, generally fly readily, body densely pubescent, procoxae transverse, and trochantin exposed. Elmid larvae are completely aquatic and are characterized by vermiform to onisciform body, abdomen nine-segmented with retractile tracheal gills.

Psephenidae adults can be distinguished by the following combination of characters: broadly ovoid in body shape and very much flattened, lateral margins of each segment greatly expanded, and the head completely concealed from a dorsal view by the expanded anterior pronotal margin (water pennies). The larvae are rather broad and depressed; the mandibles are typically concealed; and the labrum usually not visible from in front (Brown, 1972).

MATERIAL PREPARATION AND PRESERVATION

Methods used for the collection and preservation of Hydraenidae are similar to those used for other aquatic Coleoptera, which are explained in Chapter 15.

FAMILIES DRYOPIDAE, ELMIDAE, AND PSEPHENIDAE

KEYS TO DRYOPIDAE, ELMIDAE, AND PSEPHENIDAE

Keys to these families are in Chapter 15.0.

Dryopidae: Genera (Adults)

1	Dorsal surface matte or with a dense coating of long, erect setae (Figs. 15.6.1 B, D); antennae with 11 or 13 antennomeres (Fig. 15.6.2 B)	2
1'	Dorsal surface shining, with only sparse setae (Fig. 15.6.1 A); antennae with 9–10 antennomeres (Fig. 15.6.2 A)	*Elmoparnus*
2(1)	Pronotum without sublateral carinae or incomplete sulci (Fig. 15.6.3 B); pronotum and elytral pubescence variable	3

FIGURE 15.6.1 Dryopidae (Coleoptera) adults, habitus in dorsal view: (A) *Elmoparnus*, (B) *Helichus*, (C) *Dryops* sp., (D) *Pelonomus* sp.

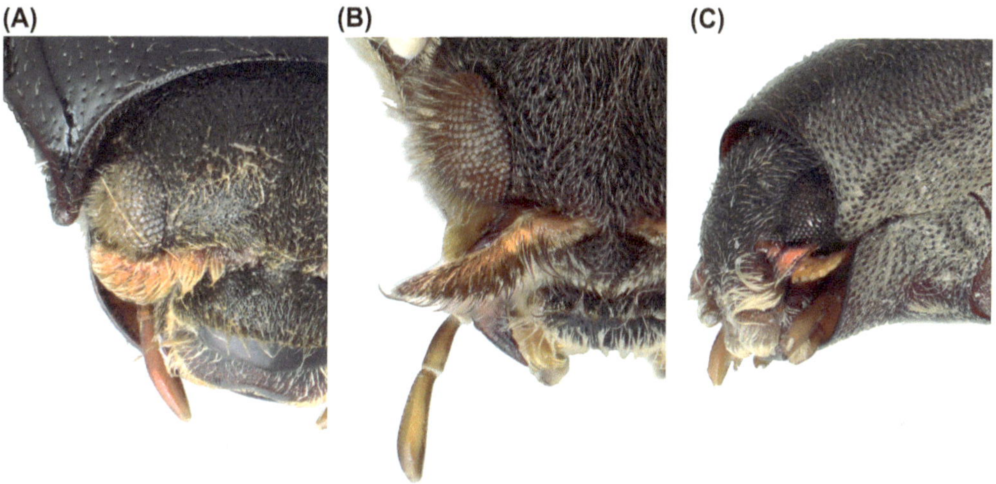

FIGURE 15.6.2 Dryopidae (Coleoptera) adults head, frontal view, showing antenna: (A) *Elmoparnus*, (B) *Pelonomus*. *Helichus* (C) head, oblique view, showing expanded antennomere.

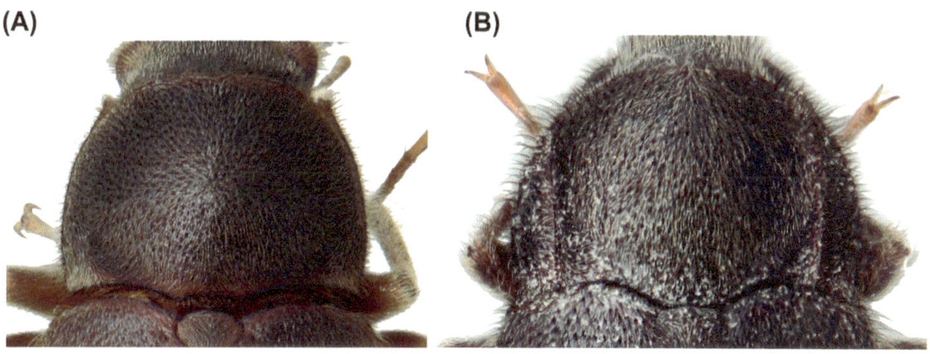

FIGURE 15.6.3 Dryopidae (Coleoptera) adults pronotum, dorsal view, showing sublateral carinae: (A) *Dryops*, (B) *Pelonomus*.

2' Pronotum usually with sublateral sulci (Fig. 15.6.3 A); elytron margin apical fourth without pubescent depression; pronotum and elytra densely and conspicuously pubescent (Fig. 15.6.1 C)........................ *Dryops*

3(2) Antenna article II not dilated (Fig. 15.6.2 B); dorsum typically densely pubescent, with conspicuous setae........................ 4

3' Antenna article II dilated, which may partially enclose apical antennomeres (Fig. 15.6.2 C); dorsum typically matte, lacking dense, conspicuous setae (Fig. 15.6.1 B)........................ *Helichus*

4(3) Antennae with 11 antennommeres (Figs. 15.6.1 D, 2 B) ... *Pelonomus*
4' Antennae with 13 antennommeres ... *Onopelmus*

Elmidae: Subfamilies (Adults)

1 Adults occasionally aquatic, generally fly; body densely pubescent, without plastron; procoxae transverse, trochantin exposed ... **Larainae [p. 586]**
1' Adults aquatic; body not or scarcely pubescent, with plastron; procoxae globose and trochantin not exposed ... **Elminae [p. 586]**

Elmidae: Larainae: Genera (Adults)

1 Body size 2.5–4.5 mm ... 2
1' Body size 5.1–10.1 mm ... 7
2(1) Elytron without accessory basal stria (Fig. 15.6.6 A) ... 3
2' Elytron with one accessory basal stria (Figs. 15.6.4 C, 6 B) ... *Hydora*
[Argentina, Chile]
3(2) Pronotum without a transverse groove, or with a shallow, wide, V-shaped groove across apical third (Figs. 15.6.5 D, F) ... 4
3' Pronotum with a deep transverse groove across apical third (Fig. 15.6.5 B) ... 6
4(3) Pronotum apical third with a shallow, wide, V-shaped groove (Fig. 15.6.5 D); southern Venezuela and Northern Brazil. ... 5
4' Pronotum without a transverse groove (Figs. 15.6.4 G, 5 F) ... *Phanocerus*
[Central America, Caribbean, Colombia, Peru. Nearctic]
5(4) Dorsal and ventral surfaces with similar pilosity, setae erect and light brownish; protibiae lacking tomentum fringe (Figs. 15.6.4 D, 6 C) ... *Hypsilara*
5' Dorsal and ventral surfaces with different pilosity, dorsal setae erect and light brownish. ventral setae recumbent, dense, and silvery (Figs. 15.6.4 F, 6 F); protibiae with tomentum fringe (Fig. 15.6.6 D) ... *Phanoceroides*
6(3) Pronotum with a median groove and without small prescutellar foveae; pronotum anterolateral angles rounded (Figs. 15.6.4 H, 5 E) ... *Pharceonus*
[Costa Rica, south to Venezuela]
6' Pronotum without a median groove and with two small prescutellar foveae; anterolateral angles of pronotum declivous (Figs. 15.6.4 B, 5 B) ... *Hexanchorus*
[Mexico, south to Peru and Caribbean]
7(1) Elytron without distinct longitudinal carinae (Fig. 15.6.4 B) ... 8
7' Elytron with distinct longitudinal carinae (Fig. 15.6.4 K) ... *Roraima*
[Venezuela]
8(7) Pronotum apical third with a distinct transverse groove (Fig. 15.6.5 H) ... 9
8' Pronotum apical third without a transverse groove (Figs. 15.6.4 A, 15.6.5 A) ... *Disersus*
[Costa Rica, south to Peru]
9(8) Pronotum basal third with a lateral longitudinal carina or arcuate-sinuate groove (Fig. 15.6.5 C) ... 10
9' Pronotum basal third without a carina or arcuate-sinuate groove (Fig. 15.6.5 H) ... 11
10(9) Pronotum basally with two prescutellar mammiform tubercles and each posterolateral angle with one similar tubercle, thus appearing bidentate (Figs. 15.6.4 I, 5 G). ... *Pseudodisersus*
[Panama, south to Ecuador]
10' Pronotum with two short, converging, prescutellar carinae, each with a deep pit laterally (Figs. 15.6.4 E, 5 C) ... *Neblinagena*
[Venezuela]
11(9) Body narrow; prosternal process sagittate, tapering apically (see Spangler & Santiago 1992: Fig. 215) ... *Hispaniolara*
[Hispaniola]
11' Body broad; prosternal process apically ligulate (Figs. 15.6.4 J, 6 E) ... *Potamophilops*
[Argentina, Brazil]

Elmidae: Elminae: Genera (Adults)

1 Eyes and wings greatly reduced or absent, Greater Antilles: Hispaniola ... 2
1' Eyes fully formed, wings variable ... 3
2(1) Eyes absent: pronotal sublateral carinae arcuate (see Spangler, 1981 Fig. 1) ... *Annomatelmis*
2' Eyes greatly reduced; pronotal sublateral carinae sinuate (see Spangler, 1981 Fig. 2) ... *Lemalelmis*
3(1) Head prognathous (Fig. 15.6.13 F); pronotum variable, but not as previous ... 4
3' Head opisthognathus (Fig. 15.6.7 L); pronotum with strong anterior and posterior depression adjacent to lateral margin, transverse groove at apical 1/3 and strong oblique groove in basal 1/3; without sublateral carinae (Fig. 15.6.9 F) ... *Luchoelmis*
4(3) Head, pronotum and elytra with large dorsal areas with plastron (Fig. 15.6.9 L), ventral areas of body with plastron; pronotum without carinae and depressions (Fig. 15.6.9 L); elytra without carinae, may have longitudinal rows of granules (Fig. 15.6.8 K) ... 5
4' Plastron mainly on ventral surface of body, if present dorsally, then only as small areas (Fig. 15.6.9 G); pronotum usually with carinae and depressions, seldom absent (Figs. 15.6.7 E, 9 G); elytra with or without longitudinal carinae (Figs. 15.6.7 M, 11 B) ... 6

FIGURE 15.6.4 Larainae (Elmidae: Coleoptera) adult, habitus in dorsal view: (A) *Disersus dasycolus* Spangler & Santiago-Fragoso; (B) *Hexanchorus dentitibialis* Maier, (C) *Hydora annectens* Spangler & Brown; (D) *Hypsilara royi* Maier & Spangler; (E) *Neblinagena doyeli* Kodada & Jäch; (F) *Phanoceroides* sp.; (G) *Phanocerus clavicornis* Sharp, (H) *Pharceonus volcanus* Spangler & Santiago-Fragoso; (I) *Pseudodisersus goudotii Guérin-Méneville*, (J) *Potamophilops bostrychophallus* Maier, (K) *Roraima carinata* Kodada & Jäch.

5(4) Tibiae without cleaning fringe (Fig. 15.6.12 A); prosternal process margins parallel and narrow, width ~1/3 length (Figs. 15.6.8 F, 10 C); male metatrochanters with apicoventral tubercle .... *Pagelmis*

5' Tibiae with apical cleaning fringes (Fig. 15.6.12 B); prosternal process large and wide (Fig. 15.6.10 D); male metatrochanters without apicoventral tubercle (Fig. 15.6.8 K) .... *Stenhelmoides*

FIGURE 15.6.5 Larainae (Elmidae: Coleoptera) adult pronotum, dorsal view, showing sculpture: (A) *Disersus dasycolus* Spangler & Santiago-Fragoso, (B) *Hexanchorus falconensis* Maier, (C) *Neblinagena prima* Spangler, (D) *Phanoceroides* sp., (E) *Pharceonus volcanus* Spangler & Santiago-Fragoso, (F) *Phanocerus clavicornis* Sharp, (G) *Pseudodisersus goudotii* Guérin-Méneville, (H) *Potamophilops bostrychophallus* Maier.

FIGURE 15.6.6 Larainae (Elmidae: Coleoptera) adults. (A) *Potamophilops bostrychophallus* Maier elytron, showing striae. (B) *Hydora annectens* Spangler & Brown elytron, showing accessory stria. (C) *Hypsilara royi* Maier & Spangler protibia, showing setation. (D) *Phanoceroides* sp. protibia, showing setation, with cleaning fringe of tomentum. (E) *Potamophilops bostrychophallus* Maier prosternum, ventral view. (F) *Phanoceroides* sp. habitus, lateral view, showing ventral setation of the plastron.

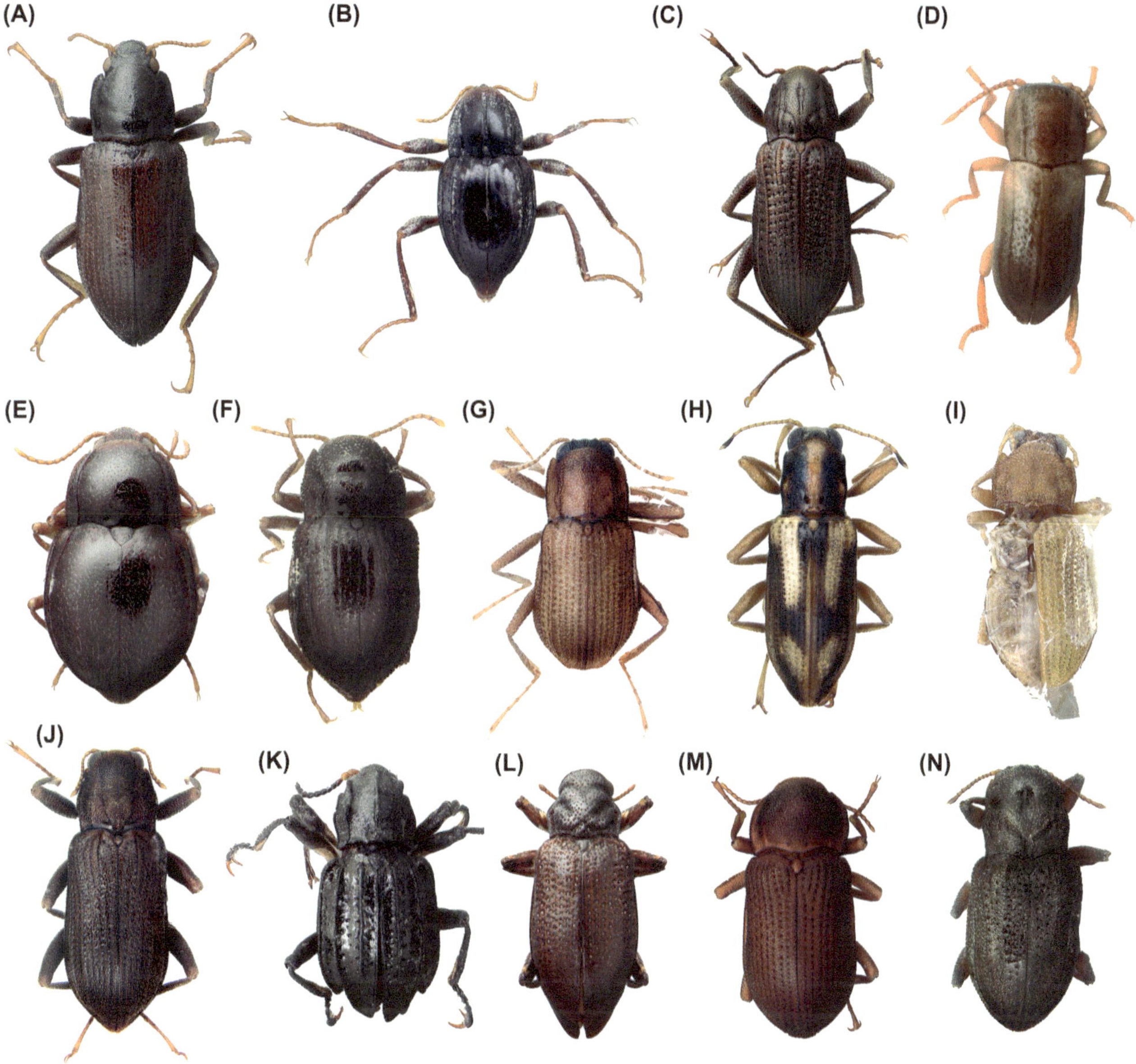

FIGURE 15.6.7 Elminae (Elmidae: Coleoptera) adults, habitus in dorsal view. (A) *Austrelmis consors* Brown; (B) *Austrolmnius* sp.; (C) *Cylloepus* sp.; (D) *Elachistelmis tetramera* Maier; (E) *Gyrelmis* sp.; (F) *Heterelmis* sp.; (G) *Hexacylloepus* sp.; (H) *Hintonelmis* sp.; (I) *Holcelmis woodruffi* Hinton; (J) *Huleechius* sp.; (K) *Jolyelmis* sp.; (L) *Luchoelmis penai* Spangler & Staines; (M) *Macrelmis* sp.; (N) *Microcylloepus similis* (Horn).

FAMILIES DRYOPIDAE, ELMIDAE, AND PSEPHENIDAE

6(4)	Epipleura without longitudinal line of granules (Fig. 15.6.13 A)	7
6'	Epipleura with one or two longitudinal lines of granules (Figs. 15.6.7 B, 13 B); elytra with prominent sublateral carinae on fifth and seventh intervals (Fig. 15.6.11 A)	*Austrolimnius*
7(6)	Elytron without accessory stria (Fig. 15.6.11 A)	8
7'	Elytron with short accessory stria (Figs. 15.6.7 M, 11 F)	*Macrelmis*
8(7)	Pronotum and elytra without rounded gibbosities or protuberances (Fig. 15.6.7 E); pronotum and elytra generally with sublateral carinae	9
8'	Pronotum basal 1/3 with rounded gibbosities or protuberances, usually four transversally, sometimes sublateral gibbosities absent, but medial ones always present (Fig. 15.6.9 O); pronotum with or without longitudinal depression and sublateral carinae; elytra with coarse punctate striae, humeral and subapical areas usually with gibbosities or protuberances (Fig. 15.6.8 J)	*Stegoelmis*
9(8)	Protonotum and elytra with one or two longitudinal carinae (Fig. 15.6.11 B), sometimes absent, metasternum variable, but not as follows	10
9'	Pronotum and elytra with three high, longitudinal carinae (Fig. 15.6.7 K); metasternum on each side of disc with an arcuate, lateral carina; each metacoxal cavity with an arcuate carina extending to posterior margin of first abdominal ventrite (Fig. 15.6.13 D)	*Jolyelmis*
10(9)	Elytra without sublateral carinae on intervals fifth and seventh (Fig. 15.6.11 G)	11

FIGURE 15.6.8 Elminae (Elmidae: Coleoptera) adults, habitus in dorsal view. (A) *Neocylloepus boeseli* Brown; *Neoelmis* sp.; (C) *Neolimnius palpalis* Hinton; (D) *Notelmis* sp.; (E) *Onychelmis* sp.; (F) *Pagelmis amazônica* Spangler; (G) *Pilielmis sara* Hinton; (H) *Portelmis nevermanni* Sanderson; (I) *Stethelmis kaszabi* Hinton; (J) *Stegoelmis* sp.; (K) *Stenhelmoides rufulus*-group; (L) *Tolmerelmis pubipes* (Hinton); (M) *Tyletelmis* sp.; (N) *Xenelmis* sp.

10'	Elytra with sublateral carinae on fifth and seventh intervals (Figs. 15.6.7 E, 11 B)	*Gyrelmis*
11(10)	Tarsal claws with one or two teeth (basal and subbasal) (Figs. 15.6.12 E, F)	12
11'	Tarsal claws without teeth (Fig. 15.6.12 D)	15
12(11)	Tarsal claws with only basal tooth (Fig. 15.6.12 F)	13
12'	Tarsal claws with basal and subbasal teeth (Figs. 15.6.8 E, 12 E)	*Onychelmis*
13(12)	Elytra usually without carina, but if present not on fourth interval; pronotum not as previous (Fig. 15.6.11 C)	14
13'	Elytra fourth interval basally with short granule row, with part of interval subcarinate; sixth interval with sublateral carina (Figs. 15.6.8 G, 11 G); pronotum with sublateral carinae, disc center with distinct oval depression, each side of disc in basal 1/5 with another distinct oval depression against sublateral carina (Fig. 15.6.9 M)	*Pilielmis*
14(13)	Hypomera and sides of mesosternum with tomentum; prosternum short, prosternal process long, moderately narrow with apex rounded; protibiae without cleaning fringes; elytra without carina (Fig. 15.6.8 I)	*Stethelmis*

FIGURE 15.6.9 Elminae (Elmidae: Coleoptera) adults head and pronotum. dorsal view. (A) *Austrelmis consors* (Hinton); (B) *Elachistelmis tetramera* Maier; (C) *Heterelmis* sp.; (D) *Hexanchorus* sp.; (E) *Holcelmis woodruffi* Hinton; (F) *Luchoelmis penai* Spangler & Staines; (G) *Macrelmis* sp.; (H) *Microcylloepus similis* (Horn); (I) *Neoelmis* sp.; (J) *Neolimnius palpalis* Hinton; (K) *Notelmis* sp.; (L) *Pagelmis amazônica* Spangler; (M) *Pilielmis sara* Hinton; (N) *Portelmis nevermanni* Sanderson; (O) *Stegoelmis selva* Spangler; (P) *Tolmerelmis pubipes* (Hinton); (Q) *Tyletelmis* sp.; (R) *Xenelmis* sp.

14' Hypomera anteriorly with tomentum, mesosternum without tomentum; prosternum long, prosternal process subquadrate, as broad between coxae as length of coxae, apex bisinuate; protibiae with cleaning fringes; elytra usually with carinae on sixth and eighth intervals, seldom absent (Figs. 15.6.7 H, 11 C) .... *Hintonelmis*

15(11) Lacinia and galea very long (Fig. 15.6.13 E) 16

FAMILIES DRYOPIDAE, ELMIDAE, AND PSEPHENIDAE

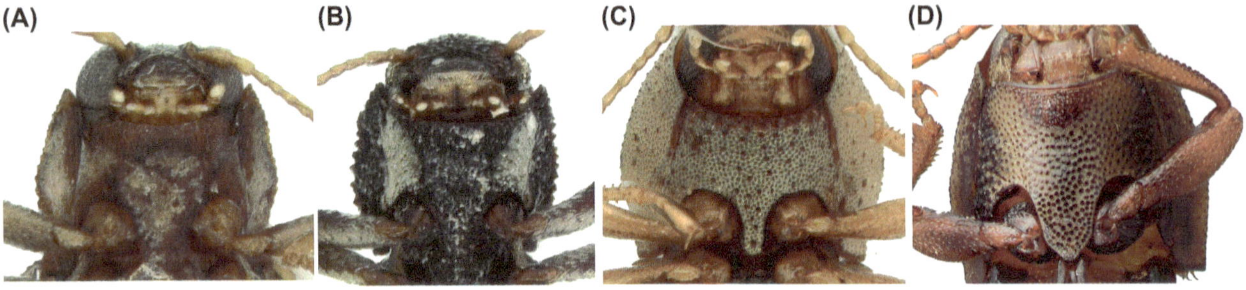

FIGURE 15.6.10 Elminae (Elmidae: Coleoptera) adults prosternum. (A) *Hexanchorus* sp.; (B) *Microcylloepus* sp.; (C) *Pagelmis amazônica* Spangler; (D) *Stenhelmoides variabilis* Spangler.

FIGURE 15.6.11 Elminae (Elmidae: Coleoptera) adults, left elytron. (A) *Austrolmnius* sp.; (B) *Gyrelmis* sp.; (C) *Hintonelmis* sp.; (D) *Neoelmis* sp.; (E) *Macrelmis* sp.; (F) *Macrelmis* sp.; (G) *Pilielmis sara* Hinton; (H) *Xenelmis* sp.

15' Lacinia and galea short and broad 17

16(15) Pronotum heavily sculptured, apical 2/5 with deep and complete transverse depression, each side of basal 1/4 of sublateral carinae with a nearly complete sublateral longitudinal carina and sublateral carina (Fig. 15.6.9 J); pronotum basal margin in front of scutellum not emarginate; tarsal formula 5-5-5 (Fig. 15.6.8 C)........ *Neolimnius*

16' Pronotum smooth, lacking transverse depression on apical 2/5, with complete sublateral and lateral carinae (Figs. 15.6.7 D, 9 B); tarsal formula 4-4-4 (Fig. 15.6.12 C) or 5-5-5 .... *Elachistelmis*

17(15) Disc of pronotum without longitudinal depression (Fig. 15.6.9 R) 18

17' Disc of pronotum with medial longitudinal depression (Fig. 15.6.9 D) 25

18(17) Body broad, oval, usually small (2.2 mm) (Fig. 15.6.8 N); pronotum without depressions, subtriangular (Fig. 15.6.9 R) 19

18' Body sides subparallel (Fig. 15.6.7 G); pronotum subrectangular, sublateral carinae and depressions present or absent (Fig. 15.6.9 D)........ 21

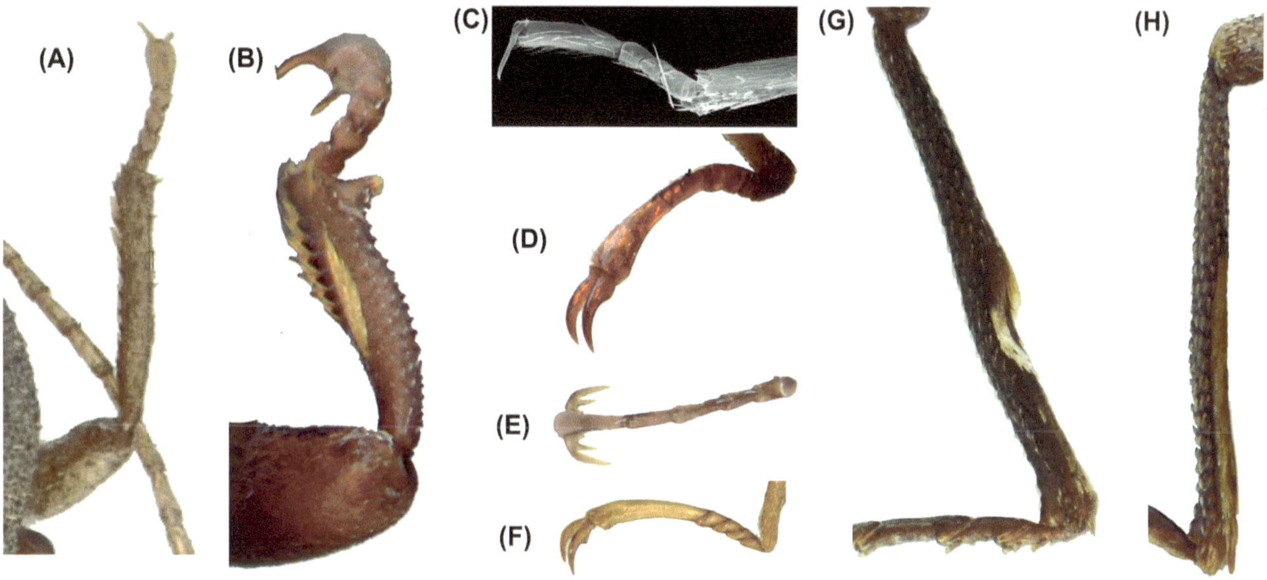

FIGURE 15.6.12 Elminae (Elmidae: Coleoptera) adults, leg. Protibia: (A) *Pagelmis amazônica* Spangler; (B) *Stenhelmoides variabilis* Spangler. Tarsus: (C) *Elachistelmis tetramera* Maier; (D) *Stegoelmis* sp.; (E) *Onychelmis* sp.; (F) *Hintonelmis* sp. tarsos. Metatibia: (G) *Huleechus* sp.; (H) *Cylloepus* sp.

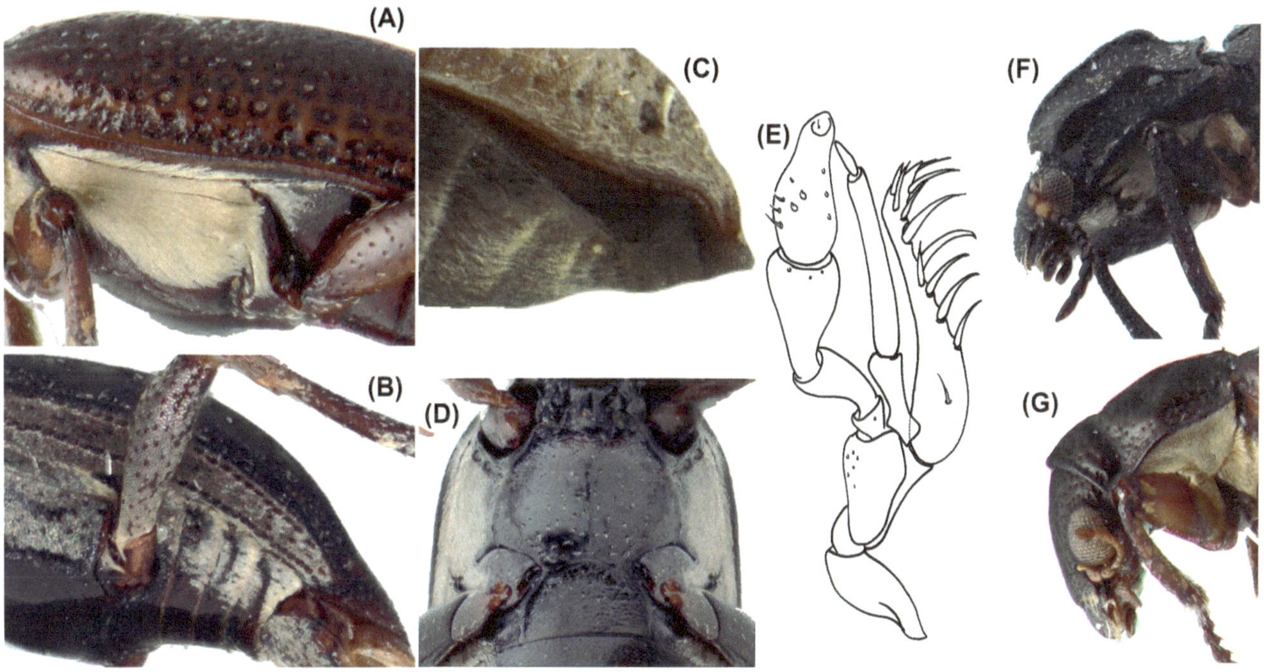

FIGURE 15.6.13 Elminae (Elmidae: Coleoptera) adults, thorax and abdomen, partial lateral view. (A) *Luchoelmis penai* Spangler & Staines epipleuron without granules; (B) *Austrolmnius* sp., epipleuron with granules. (C) *Portelmis nevermanni* Sanderson elytron, lateral view, showing posterior excavation. (D) *Jolyelmis* sp.metasternum and fist abdominal ventrite, showing carinae. (E) *Elachistelmis tetramera* Maier maxilla. Head, lateral view: (F) *Jolyelmis* sp., prognathous mouthparts; (G) *Luchoelmis penai* Spangler & Staines, opisthognathous mouthparts.

19(18) Body length <2 mm; pronotum tomentose, without sublateral carinae (only *X. granata* presents a sublateral carina extending from base to about apical 1/4) or depressions (Fig. 15.6.9 R) 20

19' Body length >2 mm; pronotum not tomentose, nearly complete sublateral carinae present (See Hinton, 1939, Fig. 40); elytra with sublateral carinae composed of low rows of granules .... *Oolimnius*

20(18) Elytra lacking sublateral carinae, body elongate in overall habitus .... *Xenhelmoides*

20 Elytra with two sublateral carinae (Fig. 15.6.8 N), body rounded in overall habitus (Fig. 15.6.8 N) .... *Xenelmis*

21(18) Pronotum and elytra generally carinate (Fig. 15.6.11 D). Elytra without marginal excavation near elytral apex ... 22

21' Pronotum usually without carinae (Fig. 15.6.9 N); if present each carina is short and basal; elytra without carinae, but with subapical deep, lateral, marginal excavation (Fig. 15.6.13 C), accompanied by prolonged lateral margins of the fifth sternite, which fits into this cavity (Fig. 15.6.8 H) ... *Portelmis*

22(21) Pronotum without transverse depression, if present never complete (Fig. 15.6.9 A); elytra with one or two sublateral carinae ... 23

22' Pronotum with deep and complete transverse depression on anterior 2/5 and with sublateral carinae (Fig. 15.6.9 I); elytra with sublateral carina on sixth interval (Fig. 15.6.8 B); body sometimes very small (Fig. 15.6.8 B) ... *Neoelmis*

23(22) Pronotum with carinae well-developed, transverse depression generally present, carinae never divided at base (Fig. 15.6.9 D). Elytra with one or two carinae ... 24

23' Pronotum with or without sublateral carinae; if present, carinae much higher and more prominent on apical half; if carinae extend to the base, they are completely divided at basal 2/5 by an oblique depression; transverse depression absent or very shallow (Fig. 15.6.9 A); elytra with two, one, or no sublateral carinae (if present, carinae are on intervals 6 and 8) (Fig. 15.6.7 A) ... *Austrelmis*

24(23) Body sides subparallel (Fig. 15.6.8 M); pronotum with well-developed carinae extending from base to about apical 1/6 (Fig. 15.6.9 Q); elytra with one carina (Fig. 15.6.8 M) ... *Tyletelmis*

24' Body subovate (Fig. 15.6.7 F); pronotum with sublateral carinae, which extend from base to anterior margin; pronotal disc generally with or rarely without a transverse depression at middle (Fig. 15.6.9 C); elytra with carinae on sixth and eighth intervals (Fig. 15.6.7 F) ... *Heterelmis*

25(17) Pronotum without transverse depression (Fig. 15.6.9 D); posterior cleaning fringe of middle tibia in males short ... 26

Pronotum with shallow transverse depression on basal 2/5 and sublateral carina well-developed; disc of pronotum with a broad, deep, median longitudinal depression (Fig. 15.6.9 P); posterior cleaning fringe of middle tibia, in male, exceptionally long (Fig. 15.6.8 L) ... *Tolmerelmis*

26(25) Hypomera completely tomentose or with complete transverse belt of tomentum (Fig. 15.6.10 A) ... 27

26' Hypomera without tomentum, shining (Fig. 15.6.10 B); occasionally a very narrow belt of tomentum adjacent to sternonotal suture or near procoxae present ... 29

27(26) Hypomera with a belt of tomentum on anterior part (Fig. 15.6.10 A) ... 28

27' Hypomera completely tomentose ... *Epodelmis*

28(26) Sides of pronotum strongly dilated (Fig. 15.6.9 E); prosternum and first abdominal sternite disc without sublateral carinae; cuticule without hard and rigid aspect (Fig. 15.6.7 I) ... *Holcelmis*

28' Sides of pronotum not dilated (Fig. 15.6.9 D); prosternum and first abdominal sternite disc with sublateral carinae; cuticle hard and rigid (Fig. 15.6.7 G) ... *Hexacylloepus*

29(26) Pronotum with oblique and transverse depressions (Fig. 15.6.9 H); longitudinal depressions sometimes present ... 30

29' Pronotum without oblique and transverse depressions; longitudinal depressions present ... 31

30(29) Pronotum with sublateral carina and a transverse depression, with a median longitudinal discal depression; basal half on each side with an oblique depression (Fig. 15.6.9 H); elytra with two sublateral carinae (sometimes with one or three) (Fig. 15.6.7 M); epipleura rarely tomentose (Fig. 15.6.7 M) ... *Microcylloepus*

30' Pronotum without sublateral carina; disc with deep and complete transverse depression and median longitudinal depression present; pronotum base with lateral, shallow, oblique depression (Fig. 15.6.9 K); elytra with two sublateral carinae; epipleura tomentose (Fig. 15.6.8 D) ... *Notelmis*

31(29) Labium with gula distinctly narrower than submentum and mentum; elytra with two sublateral carinae; hind wing without 3A2; male tibiae with pronounced secondary sexual characters (Figs. 15.6.7 J, 2 G) ... *Huleechius*

31' Labium with gula as wide as submentum and mentum; elytra with two sublateral carinae; hind wing with 3A2; male tibiae variable (Figs. 15.6.7 C, 12 H) ... *Cylloepus*

Elmidae: Elminae: Genera (Larvae)

See Limitations section.

1 Body strongly flattened dorsoventrally (platyform larvae); thorax and abdomen lateral margins falcate; abdominal segments I–VIII with pleural sclerites ... 2

1' Body cylindrical, subcylindrical or slightly flattened dorsoventrally (eruciform larvae); thorax and abdomen lateral margins never falcate; abdomen with pleural sclerites on segments I–IV, I–V, I–VI, I–VII, I–VIII, or pleural suture on abdominal segments I–IX ... 4

2(1) Prothorax ventrally with five sclerites, one anteriolateral pair, one lateral pair, and one posteromedian; thoracic and abdominal sclerite lateral margins strongly falcate ... 3

2' Prothorax ventrally with two large anteriolateral sclerites, one on each side, and a prosternal sclerite (Fig. 15.6.14 A); thoracic sclerite margins rounded; abdominal sclerite margins slightly falcate ... *Phanoceroides* [Brazil]

3(2) Last abdominal segment not abruptly narrowed at middle, apex slightly emarginated (Fig. 15.6.14 B) ... *Phanocerus*

3' Last abdominal segment abruptly narrowed at middle, apex deeply emarginate (Fig. 15.6.14 C) ... *Pharceonus*

4(1) Body shape variable; abdominal segment IX without pleural suture ... 5

4' Body cylindrical; pleural suture extending to middle of abdominal segment IX ... *Cylloepus*

5(4) Second antennomere sensory appendage short, < first antennomere; abdominal segment pleural sclerites number variable; thoracic and abdominal segment tubercles arrangement variable ... 6

FIGURE 15.6.14 Larainae (Elmidae: Coleoptera) larva, habitus. (A) *Phanoceroides* sp., ventral view; (B) *Phanocerus clavicornis* Sharp, ventral view; (C) *Pharceonus volcanus* Spangler & Santiago-Fragoso dorsal view.

5' Second antennomere sensory appendage very long, ≥ first antennomere; abdominal segments I–VII with pleural sclerites; thoracic and abdominal segment tubercles arranged in a variable number of longitudinal rows .... *Austrolimnius*

6(5) Abdominal segments I–V, I–VI, or I–VII with pleural sclerites 7

6' Abdominal segments I–IV with pleural sclerites (Fig. 15.6.15 B) .... *Luchoelmis*

7(6) Abdominal segments I–V with pleural sclerites 8

7' Abdominal segments I–VI or I–VII with pleural sclerites 9

8(7) Body elongate, subcylindrical; meso-, metathoracic, and abdominal segment terga lateral margins slightly expanded; abdominal segment VI with an incomplete pleural suture; mature larva length to 15.5 mm .... *Pseudodisersus*

8' Body elongate, narrowing behind abdominal segment I; meso-, metathoracic, and abdominal segment terga lateral margins not expanded; abdominal segment VI without a pleural suture; mature larva length to 6.0 mm .... *Stethelmis*

9(7) Abdominal segments I–VI with pleural sclerites, segment VII sometimes may have a partial pleural suture; procoxal cavities open .. 10

9' Abdominal segments I–VII with pleural sclerites; procoxal cavities open or closed 13

10(9) Prothorax ventrally with seven sclerites: one anterolateral pair, two lateral pairs, and one anteromedian sclerite; meso- and metathorax ventrally with five sclerites: one anterior sclerite and two lateral pairs; body shape variable. Mature larvae over 4.0 mm long 11

10' Prothorax ventrally with four sclerites: one anterior pair, one posterior pair; meso- and metathorax ventrally with seven sclerites: one anterior and three lateral pairs (Fig. 15.6.15 C); body convex, hemispherical in cross section; mature larva length to 3.0 mm .... *Xenelmis*

11(10) Body cylindrical or subcylindrical, sides subparallel; last abdominal segment apex emarginate; abdominal segment VIII posterior margin without large tubercles; abdominal terga I–VIII with tubercles arranged in longitudinal rows; length to 10.0 mm 12

11' Body fusiform, slightly dorsoventrally depressed; last abdominal segment apically rounded; abdominal segment VIII posterior margin with a large tubercle on each side; abdominal terga I–VIII without tubercles arranged in longitudinal rows; length to 5.0 mm (Fig. 15.6.16 A) .... *Hexanchorus*
[Widespread]

12(11) Body cylindrical, sides subparallel; prothorax and abdominal segments I–VIII with tergal tubercles arranged in four longitudinal rows, those of meso- and metathorax arranged in six rows; abdominal segment IX length < length of abdominal segments V–VIII combined; length to 16.0 mm .... *Disersus*

12' Body subcylindrical; thorax and abdominal segments I–VIII with tergal tubercles arranged in one longitudinal row on each side of median line (sometimes absent); abdominal segment IX length > length of abdominal segments V–VIII combined (Fig. 15.6.16 B); length to 10.0 mm .... *Stegoelmis*

13(9) Last abdominal segment variable in shape and length, but never with length 3× width 14

13' Last abdominal segment long and slender, length 3× width (Fig. 15.6.16 C) .... *Hexacylloepus*

14(13) Procoxal cavities closed; cuticle with tubercles and setae but not as follows 15

14' Procoxal cavities open; cuticle with modified setae and granules: prominent, round granules, each granule with rounded, adpressed scale with venous surface structures; papillate projections, each projection with an elongate scale with venous surface structures; small, flat, elliptical granules with short, fringed setae; clavate setae with setose surface and fine, moderately long, filiform setae .... *Roraima*

FIGURE 15.6.15 Elminae (Elmidae: Coleoptera) larva, habitus in dorsal view. (A) *Austrelmis* sp.; (B) *Luchoelmis* sp.; (C) *Xenelmis* sp.

FIGURE 15.6.16 Elminae (Elmidae: Coleoptera) larva, habitus. (A) *Hexanchorus* sp., lateral view; (B) *Stegoelmis* sp., dorsal view; (C) *Hexacylloepus* sp., dorsal view.

FIGURE 15.6.17 Elminae (Elmidae: Coleoptera) larva, habitus in dorsal view. (A) *Neoelmis* sp.; (B) *Macrelmis* sp.; (C) *Huleechius* sp.

FIGURE 15.6.18 Elminae (Elmidae: Coleoptera) larva, habitus in dorsal view. (A) *Microcylloepus* sp.; (B) *Heterelmis* sp.

15(14) Meso- and metathorax ventrally with five sclerites: one anterior subpentagonal, and two lateral pairs; thoracic and abdominal terga with setiferous tubercles, sometimes arranged in longitudinal rows 16

15' Meso- and metathorax ventrally with seven sclerites: one anterior subpentagonal, and three lateral pairs; thoracic and abdominal terga with setiferous tubercles arranged (although sometimes partially) in longitudinal rows 21

16(15) Meso- and metathoracic terga lateral margins not bent ventromedially; meso- and metathorax with five ventral sclerites: one anterior, subpentagonal, and two lateral pairs conspicuous, subrectangular 17

16' Meso- and metathoracic terga lateral margins terga strongly bent ventromedially; meso- and metathorax with five ventral sclerites: one anterior, subpentagonal, and two lateral pairs subtriangular (one pair inconspicuous, under the tergite margin) .... *Hydora*

17(16) Head capsule anterior border with a tooth on each side, between base of antennae and clypeus or on the sides of the clypeus 18

17' Head capsule anterior border without teeth (Fig. 15.6.17 A) .... *Neoelmis*

18(17) Frons anterior border with a large blunt tooth on each side, between base of antennae and clypeus; clypeus lateral margin without teeth; body shape variable 19

18' Frons anterior border without teeth between base of antennae and clypeus; clypeus lateral margin with a small tooth on each side; body subtriangular; cuticle usually with dark markings; abdominal segment IX tergum usually with a median keel (Fig. 15.6.15 A) .... *Austrelmis*

19(18) Body subcylindrical, not flattened; thoracic and first eight abdominal terga as wide as long 20

19' Body flattened dorsoventrally; thoracic and first eight abdominal terga wider than long; suture from procoxal cavity to lateral margin conspicuous; mature larva dark brown, almost black (Fig. 15.6.17 B) .... *Macrelmis*

20(19) Abdominal segment IX tergum with a median keel; suture from procoxal cavity to lateral margin conspicuous .... *Neocylloepus*

20' Abdominal segment IX tergum convex, without median keel; suture from procoxal cavity to lateral margin inconspicuous (Fig. 15.6.17 C) .... *Huleechius*

21(15) Pronotum and abdominal segment IX tergal tubercles randomly distributed, not forming longitudinal rows; remaining terga tubercles arranged in longitudinal rows (Fig. 15.6.18 A) .... *Microcylloepus*

21' Thorax and first eight abdominal segments with tergal tubercles arranged in 8 or 10 longitudinal rows; abdominal segment IX tergal tubercles arranged in two median and two lateral rows (Fig. 15.6.18 B) .... *Heterelmis*

Psephenidae: Genera (Larvae)

Modified from Archangelsky et al. (2009) following Artigas (1963), Spangler(1968), Fernández et al. (2001), and Jerez & Moroni (2006).

1	Abdominal segment IX without ventral operculum or gill chamber; abdominal segments with variable number of gills	2
1'	Abdominal segment IX with operculum or gill chamber; abdominal segments without gills	5
2(1)	Abdominal segment VIII without lateral expansions; abdomen with four or five pairs of gills	3
2'	Abdominal segment VIII with lateral expansions; abdomen with four pairs of gills	*Eubrianax*
3(2)	Abdomen segments II to VI with five pairs of gills	4
3'	Abdomen with four pairs of gills	*Pheneps*
4(3)	Clypeus emarginate; frons with long seta, close to frontoclypeal suture and antenna base; pleural sclerites with lower and anterior margin glabrous	*Psephenus*
4'	Clypeus truncate; frons without seta; pleural sclerites with lower and anterior margin with short setae densely arranged in a pecten pattern	*Psephenops*
5(1)	Abdominal segment IX gill chamber with three gill tufts; thoracic and abdominal pleurae expansions separate	*Ectopria*
5'	Abdominal segment IX gill chamber with two gill tufts; thoracic and abdominal pleurae expansions adjacent	*Tychepsephenus*

ACKNOWLEDGMENTS

The production of figures for this publication was supported by NSF #DEB-0816904 to Dr. Andrew Short (KU). The authors graciously thank the museums and curators that loaned specimens in support of this publication, including Louisiana State Arthropod Museum (Victoria Bayless), Naturhistorisches Museum Wien (Manfred Jäch), Smithsonian Institution (Charyn Micheli), and the University of Kansas (Zack Falin). Scanning electron micrographs were taken at the Microscopy and Analytical Imaging Laboratory at the University of Kansas.

REFERENCES

Artigas, J.N. 1963. Descripción de la larva de un psefenido chileno (Coleoptera - Psephenidae). Gayana 8: 3–8.

Archangelsky M., V. Manzo, M. Michat & P. Torres. 2009. Orden Coleoptera. Pages 411–468 *in*: E. Domínguez & H. Fernández (eds.), Macroinvertebrados bentonicos sudamericanos. Sistemática y Biología. Editorial Miguel Lillo Foundation, Tucumán, Argentina.

Brown, H.P. 1972. Aquatic dryopoid beetles (Coleoptera) of United States. Biota of Freshwater ecosystem, identification manual no 6. U.S. Environmental Protection Agency, Ohio, 82p.

Fernandez, M.F.S., C.R.V. Fonseca & P.J. Spangler. 2001. Estudo taxionômico dos Psephenidae (Coleoptera: Byrrhoidea) da Amazônia Brasileira. Acta Amazônica 31: 469–500.

Jäch, M.A. & M. Balke. 2008. Global diversity of water beetles (Coleoptera) in freshwater Hydrobiologia 595: 419–442.

Jerez, V. & J. Moroni. 2006. Diversidad de coleópteros acuáticos en Chile. Gayana 70: 72–81.

Manzo, V. 2013. Los élmidos de la región Neotropical (Coleoptera: Byrrhoidea: Elmidae): diversidad y distribución. Revista de la Sociedad Entomológica Argentina 72: 199–212.

Segura, M.O., M.I.S. Passos, A.A. Fonseca-Gessner & C.G. Froehlich. 2013. Elmidae Curtis, 1830 (Coleoptera, Polyphaga, Byrrhoidea) of the Neotropical region. Zootaxa 3731: 001–057.

Shepard, W.D. & C.B. Barr. 2014. Neoeubria inbionis Shepard & Barr, a new genus and new species of Neotropical water penny beetle (Coleoptera: Psephenidae: Eubriinae), with a key to the adult Eubriinae of the Neotropic Zone. Zootaxa 3811: 553–568.

Shepard, W. & C.S. Chaboo. 2015. Beetles (Coleoptera) of Peru: A Survey of the Families Dryopidae, Limnichidae, Luctrochidae, Psephenidae, and Ptilodactylidae. Journal of the Kansas Entomological Society 88: 163–166.

Spangler, J.P. 1968. A new Psephenus and its larva from Mexico (Coleoptera: Psephenidae) Entomological News 79: 91–97.

Family Scirtidae

María L. Libonatti
Laboratory of Entomology, IBBEA, CONICET-UBA, DBBE-FCEN, University of Buenos Aires, Buenos Aires, Argentina

Rafał Ruta
Department of Biodiversity and Evolutionary Taxonomy, Faculty of Biological Sciences, Wroclaw University, Poland

Chapter Outline

INTRODUCTION

Scirtidae comprises over 1800 species distributed worldwide, although the highest diversity of the family is in the tropics and the temperate parts of the Southern Hemisphere. This family is commonly known as "marsh beetles" since larvae often inhabit shallow bodies of stagnant water full of decaying plant material. Larvae, however, are found in other lotic and lentic habitats, including lowland ditches, ponds, microhabitats within rivers and streams, phytotelmata formed both in tree holes and leaf axils, and groundwater as deep as 10 m (Klausnitzer & Pospisil, 1991). Moreover, some larvae are terrestrial and live in wet soil or moist, rotten logs (Ruta et al., 2017). Preference of a particular kind of habitat is usually genus specific, and some geographic correlations can be observed as well. In tropical and subtropical lowland forests, a great diversity of species is especially common in phytotelmata formed in different families of plants (e.g., Agavaceae, Amaryllidaceae, Araceae, and Bromeliaceae, see Greeney, 2001). Cloud forests, tepuis, and high-altitude ecosystems, like paramo formation, are insufficiently studied and host some of the most interesting and the least studied taxa (Ruta, 2016).

Larvae go through 9–12 instars (Zwick & Hecht, 2008; Zwick & Zwick, 2008, 2010), and all are detritus feeders. Mature larvae usually leave the water and pupate in damp soil. Most genera pupate freely exposed or covered, but building pupal cells seems to be rare in Scirtidae (Zwick & Zwick, 2008). Pupal development is often much shorter than in many other families of Coleoptera and takes 1–11 days depending on the genus (Yoshitomi, 2005; Zwick & Zwick, 2008). *Scirtes hemisphaericus* (Linnaeus) (Palearctic), *S. tibialis* Guérin-Méneville (Nearctic), and all species of *Hydrocyphon* (Palearctic and Oriental), however, pupate underwater (Leech & Chandler, 1974; Zwick & Zwick, 2008; Klausnitzer, 2009). Adults are terrestrial, usually inhabit riparian vegetation, and are generally short lived. Scirtidae are commonly pollinophagous (Yoshitomi, 2005; Watts, 2007), although their diet also includes fungal spores and detritus (unpublished). Suggestions on predatory habits of Scirtidae (Klausnitzer, 1967; Lawrence, 2005) were based on morphology of mandibles only and are erroneous.

Scirtidae belong to the least known families of water beetles. Almost 10 years ago, Jäch & Balke (2008) estimated the total number of Scirtidae as 1700. Today, we already know 1800 species, and this is very likely less than a third of the actual number of existing species. It is even more evident in the case of South America, where 200 species are only a small fraction of the real number of species we expect.

We follow the classification of the family proposed by Lawrence & Yoshitomi (2007). About 1800 described species, 69 genera, and 3 subfamilies are presently recognized. In the Neotropical Region, ca. 200 species are known, classified in 10 genera and 2 subfamilies, as follows: Stenocyphoninae (one genus: *Stenocyphon*; one species) and Scirtinae (13 genera: *Anticyphon*, *Chilarboreus*,

Thorp and Covich's Freshwater Invertebrates, Fourth Edition. https://doi.org/10.1016/B978-0-12-804223-6.00028-7

Contacyphon, *Exagonthus*, *Ora*, *Prionocyphon*, *Prionoscirtes*, *Pseudomicrocara*, and *Scirtes* genera). One of the undescribed genera has terrestrial larva that inhabits wet logs (saproxylic) and lacks eyes (Fig. 15.7.1 C). An enigmatic monotypic Brazilian genus *Pentameria* is probably congeneric with one of the saltatorial genera, like *Ora* or *Scirtes* (Friedenreich, 1883).

Of the 10 described genera in the Neotropical Region, only five (*Contacyphon*, *Ora*, *Pseudomicrocara*, *Scirtes*, and *Prionocyphon*) have their mature larvae known; all of them are aquatic and have well-developed eyes (Figs. 15.7.1 A, B). *Ora*, *Pseudomicrocara*, *Scirtes*, and *Prionocyphon* have relatively long, reaching abdomen, antenna (Fig. 15.7.1 A), whereas *Contacyphon* has relatively short, not reaching mesothorax, antenna (Fig. 15.7.1 B). Note that the placement of Neotropical species in the genus *Contacyphon* is preliminary. The mandibles of *Ora* have an obtuse apex (Fig. 15.7.1 F) and maxillary palpomere 4 long, approximately as long as palpomere 3 (Fig. 15.7.1 I). *Pseudomicrocara*, *Scirtes*, and *Prionocyphon*, by contrast, have mandibles with acute apex (Figs. 15.7.1 D, E) and the maxillary palpomere 4 is short, less than 0.5x as long as palpomere 3 (Figs. 15.7.1 G, H). *Pseudomicrocara* has a mandible with bifid apex (Fig. 15.7.1 E), while *Scirtes* and *Prionocyphon* have a mandible with simple apex (Fig. 15.7.1 D). *Scirtes* has tergite 9 with apical setae well-separated (Fig. 15.7.1 L) and lobe of clypeolabrum with five inner setae (Fig. 15.7.1 J). On the other hand, *Prionocyphon* has tergite 9 with apical setae close together (Fig. 15.7.1 M) and lobe of clypeolabrum with nine inner setae (Fig. 15.7.1 K).

LIMITATIONS

Studies on South American Scirtidae date back to 1775 when *Pseudomicrocara livida* Fabricius was described (Ruta, 2013); but despite a long research history, the Neotropical fauna was never treated comprehensively. Except for a few recently described or redescribed species (Lawrence, 2001; Ruta, 2011, 2013; Libonatti & Ruta, 2013; Libonatti, 2014, 2015, 2017; Ruta, 2016; Ruta & Libonatti, 2016), most species are known from original descriptions that date back to the 19th century and the beginning of the 20th century. These are mainly brief, undetailed descriptions that are usually devoid of illustrations. Moreover, many species were misplaced in the Palearctic genera *Elodes* and *Microcara* (Ruta, 2011). Given that many of the species were described in 19th century, and were never redescribed, identification of Neotropical scirtids is difficult, and often impossible without examination of type specimens. In the last years, some local revisions of genera were published (Libonatti & Ruta, 2013; Libonatti, 2014, 2015, 2017; Ruta, 2016).

Adults and larvae may be identified to family level using the keys provided by Trémouilles et al. (1995), Benetti et al. (2006), and Archangelsky et al. (2009). Adults may be then identified to genus level using the valuable keys proposed by Trémouilles et al. (1995), for the genera found in Argentina and bordering countries, and by Ruta (2011) for genera of Chilean Scirtidae. So far, no key to all Neotropical genera, based either on adults or larvae, has ever been published. Since 50% of the genera in this region have described larvae, it is not possible to provide a safe identification key for this life stage.

Although they primarily focus on other biogeographic regions and include only some Neotropical genera, several contributions provide valuable keys to identify larvae of genera from the Nearctic (Leech & Chandler, 1974; White & Roughley, 2008), Florida (USA) (Epler, 2010), Japan (Yoshitomi, 2005), west Palearctic (Klausnitzer, 2009), and Australia (Watts, 2014).

TERMINOLOGY AND MORPHOLOGY

Adults are relatively soft-bodied, usually finely pubescent, slightly flattened to very convex, ovate to elongate, and range between 1−15 mm long, although Neotropical species rarely exceed 6 mm. It is noteworthy, however, that Chilean *"Elodes" russeli* Solier, 1849 is one of the largest representatives of the family, reaching length of 11 mm. Body coloration is often pallid, brown, black, and sometimes presents a distinctive elytral pattern or easily noticeable metallic hue. Many scirtids have enlarged metafemora and, therefore, can jump, like Chrysomelidae Halticinae, with whom they are often confused.

Larvae are elongated, slightly depressed, and sclerotized both in dorsal and ventral parts. Antennae are long and multiarticulate (a feature unique among the Holometabola). The head is wider than long, prognathous, with the labrum fused to clypeus. Mandibles, maxillae, the epipharynx, and the hypopharynx are highly specialized for detritus feeding. Legs are pentamerus, including a hamulate pretarsus. The abdomen has nine segments, and the tracheal system is metapneustic, with the eighth segment bearing the only pair of functional spiracles. Larvae have five digitiform anal papillae, which are involved in osmoregulation and measure oxygen content in water (Wichard & Komnick, 1974; Lawrence, 2005).

A more detailed description of the general morphology is found in Lawrence (2005), Yoshitomi (2005), and Klausnitzer (2009). Larval mouthparts were described in detail by Beier (1949), Hannappel (1991), and Hannappel & Paulus (1991).

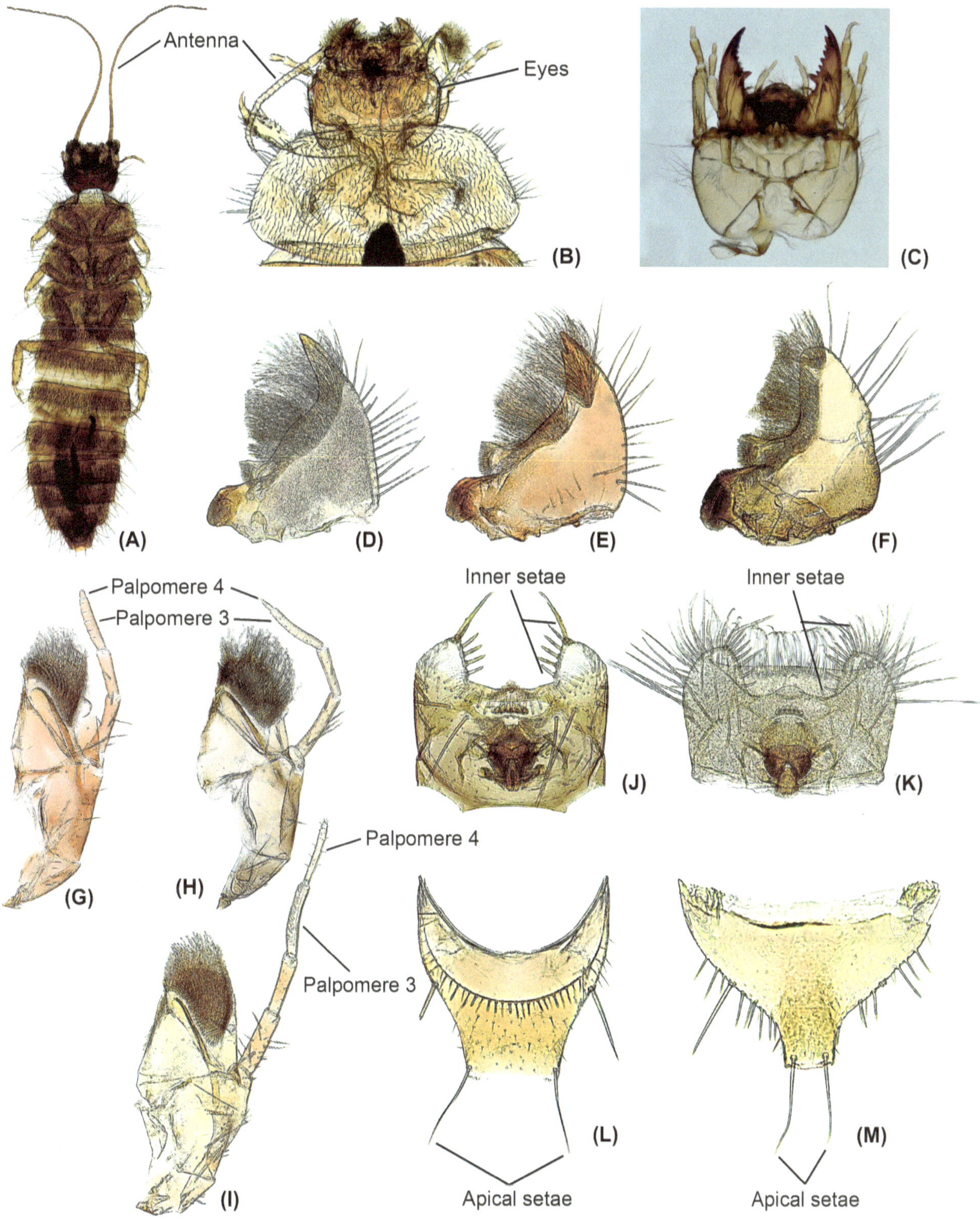

FIGURE 15.7.1 Scirtidae (Coleoptera), larval morphology: (A) *Scirtes* sp. (Argentina) habitus, dorsal view (antennae broken); (B) *Contacyphon cadornai* Pic, head and prothorax, dorsal view; and (C) saproxylic larva, head, dorsal view. Left mandible in ventral view: (D) *Prionocyphon* sp.; (E) *Pseudomicrocara antarctica* Libonatti & Ruta; and (F) *Ora depressa* (Fabricius). Left maxilla, ventral view: (G) *P. antarctica*; (H) *Scirtes* sp.; and (I) *O. depressa*. Clypeolabrum in ventral view: (J) *Scirtes* sp.; and (K) *Prionocyphon* sp. Tergite 9: (L) *Scirtes* sp.; and (M) *Prionocyphon* sp.

MATERIAL PREPARATION AND PRESERVATION

Larvae are collected using the same techniques as employed for other aquatic insects. A wide variety of collecting methods is detailed by Archangelsky et al. (2009) and Campos de Oliveira & Oliveira Pes (2014). Larvae inhabiting dendrotelms tend to float under the water surface when disturbed, and hence, their presence can be easily detected. Larvae inhabiting phytotelmata are collected either sucking out the water accumulated within the plant using a turkey baster or extracting the plant and examining the leaves thoroughly. Larvae living in running waters can be usually collected from underside of submerged stones or pieces of wood. Note that in many cases, even in the areas where adults are common, it is often difficult to locate larvae. Adults are captured using conventional techniques for collecting other terrestrial insects, such as sweeping and beating vegetation, and light and Malaise trapping. Larvae and adults are fixed and preserved in 75%–96% ethanol for examination purposes, and in 100% ethanol or propylene glycol for DNA sequencing. Pampel's fluid can also be recommended, as it preserves color pattern and will keep larvae flexible. Identification of larvae may often be impossible due to poor knowledge of Neotropical Scirtidae. In such cases, rearing is the easiest way to associate larvae and adults of a species. Several attempts have been undertaken, and it seems that rearing larvae developing in stagnant water as well as running water habitats is not difficult (see Zwick, 2007; Zwick & Zwick, 2010; and Watts, 2014 for details). Association of larvae and adults may be also achieved by molecular techniques (Watts, 2014; Cooper et al., 2014).

Identification of larvae to genus level requires close examination of clypeolabrum, mandibles, maxillae, and hypopharynx (abdominal segments 8th and 9th are also recommended to confirm determination) under a compound microscope. To soften and clear the larvae, place the specimens in 10% KOH or 10% NaOH overnight, or in lactid acid for a couple of days. The mouthparts and apical abdominal segments are then dissected and mounted on glass slides with commonly used mounting media, like polyvinyl-lacto-glycerol, gum-chloral medium or Euparal.

REFERENCES

Archangelsky, M., V. Manzo, M.C. Michat, & P.L.M. Torres. 2009. Capítulo 14: Coleoptera. Pages 411–468 *in*: E. Domínguez & H.R. Fernández (eds.), Macroinvertebrados bentónicos sudamericanos: sistemática y biología. Fundación Miguel Lillo, Tucumán.

Beier, M. 1949. Koerperbau und lebensweise der Larve von Helodes hausmanni Gredler (Col. Helodidae). Eos 25: 49–100.

Benetti, C.J., G.L. Fiorentin & J.A. Regil Cueto. 2006. Chaves de identificação para famílias de coleópteros aquáticos ocorrentes no Rio Grande do Sul, Brasil. Neotropical Biology and Conservation 1: 24–28.

Campos de Oliveira, V. & A.M. Oliveira Pes. 2014. Capítulo 10: Inventário da fauna de insetos aquáticos: coleta, preservação e criação. Pages 155–171 *in*: N. Hamada, J.L. Nessimian & R.B. Querino (eds.), Insetos aquáticos na Amazônia brasileira: taxonomia, biologia e ecología. Editora INPA, Manaus.

Cooper, S.J.B., C.H.S. Watts, K.M. Saint & R. Leijs. 2014. Phylogenetic relationships of Australian Scirtidae (Coleoptera) based on mitochondrial and nuclear sequences. Invertebrate Systematics 28: 628–642.

Epler, J.H. 2010. The water beetles of Florida, an identification manual for the families Chrysomelidae, Curculionidae, Dryopidae, Dytiscidae, Elmidae, Gyrinidae, Haliplidae, Helophoridae, Hydraenidae, Hydrochidae, Hydrophilidae, Noteridae, Psephenidae, Ptilodactylidae and Scirtidae. Division of Environmental Assessment and Restoration, Florida. 410 pp.

Friedenreich, C.W. 1883. *Pentameria bromeliarum*, eine pentamere Halticidae. Entomologische Zeitung XLIV: 140–144.

Greeney, H. 2001. The insects of plant-held waters: a review and bibliography. Journal of Tropical Ecology 17: 241–260.

Hannappel, U. 1991. Die Larven von *Helodes tournieri* Kiesw. und *Helodes gredleri* Kiesw. und ihre Stellung im phylogenetischen System (Coleoptera, Helodidae). Zeitschrift der Arbeitsgemeinschaft Oesterreichischer Entomologen 43: 89–100.

Hannappel, U. & H.F. Paulus.1991. Some undetermined Helodidae larvae from Australia and New Zealand: fine structure of mouthparts and phylogenetic position. Pages 89–128 *in*: M. Zunino, X. Bellés & M. Blas (eds.), Advances in Coleopterology. Asociacion Europea de Coleopterología, Barcelona.

Jäch, M.A. & M. Balke 2008. Global diversity of water beetles (Coleoptera) in freshwater. Hydrobiologia 595: 419–442.

Klausnitzer, B. 1967. Zur Bedeutung der Mandibein für die Imaginalsystematik der mitteleuropäischen *Helodidae*. Annotationes Zoologicae Botanicae 43: 1–2.

Klausnitzer, B. 2009. Insecta: Coleoptera: Scirtidae. Süßwasserfauna von Mitteleuropa. Bd. 20/17. Spektrum Akademischer Verlag, Heidelberg. 326 pp.

Klausnitzer, B. & P. Pospisil. 1991. Larvae of *Cyphon* sp. (Coleoptera, Helodidae) in ground water. Aquatic Insects 13: 161–165.

Lawrence, J. F. 2001. A new genus of Valdivian Scirtidae (Coleoptera) with comments on Scirtoidea and the beetles suborders. Special Publication of the Japan Coleopterological Society 1: 351–361.

Lawrence, J.F. 2005. 15.4. Scirtidae Fleming, 1821. Pages 443–450 *in*: R.G. Beutel & R.A.B. Leschen (eds.), Handbook of Zoology, Vol. IV (Part 38), Coleoptera, Beetles, Vol. 1: Morphology and Systematics (Archostemata, Adephaga, Mixophaga, Polyphaga partim). Walter de Gruyter, Berlin.

Lawrence, J.F. & H. Yoshitomi. 2007. *Nipponocyphon*, a new genus of Japanese Scirtidae (Coleoptera) and its phylogenetic significance. Elytra 35: 507–527.

Leech, H.B. & H.P. Chandler. 1974. Chapter 13: Aquatic Coleoptera. Pages 293–371 *in*: R.L. Usinger (ed.), Aquatic insects of California, with keys to North American genera and California species. University of California, California.

Libonatti, M.L. 2014. A revision of the genus *Ora* Clark, 1865 (Coleoptera: Scirtidae) in Argentina (part I) - descriptions of new species. Zootaxa 3884: 27–044.

Libonatti, M.L. 2015. A revision of the genus *Ora* Clark, 1865 (Coleoptera: Scirtidae) in Argentina (part II) - redescriptions, updated distributions and a key to species. Zootaxa 3985: 69–97.

Libonatti, M.L. 2017. Notes on some South American species of *Scirtes* Illiger, 1807 (Coleoptera: Scirtidae). Annales Zoologici 67(2): 349–368.

Libonatti, M.L. & R. Ruta. 2013. Review of the Argentinean species of *Pseudomicrocara* Armstrong (Coleoptera: Scirtidae). Zootaxa 3718: 137–157.

Ruta, R. 2011. *Chilarboreus* gen. nov., a new genus of Chilean Scirtidae (Coleoptera: Scirtoidea), with descriptions of three new species. Journal of Natural History 45: 1689–1713.

Ruta, R. 2013. Review of Scirtidae (Coleoptera: Scirtoidea) described by Johan Christian Fabricius (1745–1808). Zootaxa 3646: 51–67.

Ruta, R. 2016. *Anticyphon* gen. nov., a new genus of Scirtidae (Coleoptera: Scirtoidea) inhabiting high altitude Andean cloud forests and paramo formation. Zootaxa 4175: 301–318.

Ruta, R., B. Klausnitzer & A. Prokin. 2017. South American terrestrial larva of Scirtidae (Coleoptera: Scirtoidea): the adaptation of Scirtidae larvae to saproxylic habitat is more common than expected. Austral Entomology, doi: 10.1111/aen.12270.

Ruta, R. & M.L. Libonatti. 2016. Redescriptions of Scirtidae (Coleoptera: Scirtoidea) described by Carl Henrik Boheman (1796–1868) with notes on *Scirtes adustus diversenotatus* Pic, 1930. Zootaxa 4072: 203–216.

Trémouilles, E.R., A. Oliva, & A.O. Bachmann. 1995. Insecta Coleoptera. Pages 1152–1169 *in*: E.C. Lopretto & G. Tell (eds.), Ecosistemas de aguas continentales, Metodologías para su estudio, tomo III. Ediciones Sur, La Plata.

Watts, C.H.S. 2007. Revision of Australian *Pseudomicrocara* Armstrong (Coleoptera: Scirtidae). Transactions of the Royal Society of South Australia 131: 1–80.

Watts, C.H.S. 2014. The larvae of some Australian Scirtidae (Coleoptera) with a key to known genera. Transactions of the Royal Society of South Australia 138: 1–91.

White, D.S. & R.E. Roughley. 2008. Chapter 20: Aquatic Coleoptera. Pages 571–672 *in*: R.W. Merritt, K.W. Cummins & M.B. Berg (eds.), An Introduction to the Aquatic Insects of North America. Kendall/ Hunt Publishing Company, Iowa.

Wichard, N. & H. Komnick. 1974. Feinstruktur und Funktion der Analpapillen aquatischer Käferlarvaen (Coleoptera: Elodidae). International Journal of Insect Morphology and Embryology 3: 335–341.

Yoshitomi, H. 2005. Systematic revision of the family Scirtidae of Japan, with phylogeny, morphology and bionomics (Insecta: Coleoptera, Scirtoidea). Japanese Journal of Systematic Entomology Monographs Series 3: 1–212.

Zwick, P. 2007. *Elodes tricuspis*: Description of the larva and notes on biology (Coleoptera: Scirtidae). Lauterbornia 59: 85–93.

Zwick, P. & M. Hecht. 2008. Life history and instar number of *Elodes minuta*, with notes on additional species (Coleoptera: Scirtidae). Lauterbornia 62: 79–88.

Zwick, P. & H. Zwick. 2008. *Scirtes hemisphaericus* uses macrophyte snorkels to pupate under water. With notes on pupae of additional European genera of Scirtidae (Coleoptera). Aquatic Insects 30: 83–95.

Zwick, P. & H. Zwick. 2010. The number of larval instars in some Central European marsh beetles (Coleoptera, Scirtidae). Entomologische Blätter 106: 431–441.

Section II.4

Order Diptera

Chapter 16

Order Diptera

Lívia Maria Fusari

Depto. de Hidrobiologia, Universidade Federal de São Carlos, São Carlos, São Paulo, Brazil

Galileu P.S. Dantas

Coordenação de Biodiversidade, Instituto Nacional de Pesquisas da Amazônia, Manaus, Amazonas, Brazil

Luiz Carlos Pinho

Departamento de Ecologia e Zoologia, Universidade Federal de Santa Catarina, Florianópolis, Santa Catarina, Brazil

Chapter Outline

INTRODUCTION

Diptera representatives (flies, mosquitoes, and associated) are holometabolous insects, distinguished from others by the presence of a single pair of functional wings in adults. This characteristic gives the order its name: "two" (di-) and "wings" (ptera). Anterior wings are prominent, and the posterior wings, usually found in other insects, in dipterans are differentiated into small structures known as halteres. These halteres have balancing functions during flight.

This order, comprising over 153,000 species distributed in approximately 180 families (Brown et al., 2009), currently forms one of the most diverse group of insects. In the Neotropical region, there are over 31,000 species recognized in 118 families (Amorim, 2009b); however, these numbers are already obsolete. The diversity of Diptera is not restricted to the considerable number of species, but it also includes a wide variety of adaptations, which reflect the large morphologic and ecologic diversity. Groups with aquatic members are found in a wide range of environments, both natural and artificial (e.g., rivers, streams, ponds, phytotelmata, and temporary pools), as well as in marine habitats known for harboring few insect groups.

Adults of most mosquitoes, black flies, biting midges, deer flies, horse flies, and some other aquatic families feed on the blood of humans, livestock, and wild animals. These dipterans are vectors for several diseases, such as malaria, yellow fever, dengue fever, chikungunya fever, and leishmaniasis. Midges and other aquatic flies are of general importance to freshwater ecosystems, often being a major fish food and an important part of bottom-dwelling communities.

Thorp and Covich's Freshwater Invertebrates, Fourth Edition. https://doi.org/10.1016/B978-0-12-804223-6.00029-9

Among aquatic insect immatures, dipteran larvae are easily recognizable through the complete absence of articulated thoracic appendages (legs). Apod larvae (those lacking legs) also occur in some other orders, such as Siphonaptera, and some Hymenoptera and Coleoptera, but hardly any species of such groups inhabit aquatic or semiaquatic environments. Immature stages of aquatic flies can be found in virtually every type of aquatic habitat. Among Diptera families recorded for the Neotropical region, 31 show larvae inhabiting a wide variety of aquatic environments. They are found on or within the substrate, on or in aquatic plants, swimming or floating in open water, or associated with the air—water interface.

Diptera have been divided into "Nematocera" ("lower" Diptera) and Brachycera ("higher" Diptera). Brachycera is monophyletic, but the lower Diptera is paraphyletic, lacking any unifying characters (synapomorphies). This implies that Brachycera originated within lower Diptera. In this chapter, the lower Diptera larvae are considered in the following suborders, Culicomorpha, Blephariceromorpha, Tipulimorpha, Ptychopteromorpha, and Psychodomorpha, as suggested in Carvalho et al. (2012).

There are 31 aquatic or semiaquatic Diptera families in the Neotropical region, 14 of them are treated in detail in this book; see Chapters 16.1—16.14, listed subsequently:

Chapter 16.1 - Ceratopogonidae
Chapter 16.2 - Chironomidae
Chapter 16.3 - Simuliidae
Chapter 16.4 - Thaumaleidae
Chapter 16.5 - Culicidae
Chapter 16.6 - Dixidae
Chapter 16.7 - Corethrellidae
Chapter 16.8 - Chaoboridae
Chapter 16.9 - Blephariceridae
Chapter 16.10 - Psychodidae
Chapter 16.11 - Stratiomyidae
Chapter 16.12 - Syrphidae
Chapter 16.13 - Periscelididae
Chapter 16.14 - Ephydridae

Characteristics of the other 17 families, for which there is very scarce information on the aquatic or semiaquatic immature forms, are summarized next.

Tipulidae

It is cosmopolitan with 4,091 species described in 113 genera. In the Neotropical region, there are 838 species in 14 genera (Oosterbroek, 2017). Larvae are elongated, with a distinctive retractile head capsule. Tipulids inhabit habitats that range from strictly aquatic or accumulated water in plants (phytotelmata), to completely terrestrial, as in decomposing vegetable material, wood, or humid soil near streams (Alexander & Alexander, 1970). Aquatic species are shredders of fallen leaves and inhabit a large variety of environments, from fast current and very oxygenated streams, to temporary lakes and ponds, where they generally colonize the areas with vegetal detritus.

Limoniidae

It is the largest family in Tipulomorpha, comprising more than 10,000 species described worldwide. In the Neotropical region, 2,801 species are recognized in 65 genera (Oosterbroek, 2017). Larvae are hemicephalic, with two lateral and one dorsal, very well sclerotized sclerites. Mandibles of detritivore species are robust, with widened teeth, while predaceous species show slender and falciform mandibles. Larvae are metapneustic, rarely apneustic, and the last terminal segment bears disc-shaped spiracles. These are covered with hydrophobic secretions and are surrounded by five lobular projections. Species of this family are aquatic or semiaquatic and colonize habitats that go from deposit areas to vegetal detritus in the shores of rivers and lentic environments, and hygropetric surfaces.

Tanyderidae

Species in this family (primitive crane-flies) are mid-sized to large nematocerans, resembling Tipulidae. It is comprised of 38 extant species divided among 10 genera, and three species have a Neotropical distribution (Madriz & Courtney, 2016). Larvae of five genera are known: they occur in two types of habitats, the hyporheic zone of cobble and sand bottom streams, or in the outer layers of submerged rotting wood (Exner & Craig, 1976; Krzemiński & Judd, 1997). They are found in unpolluted mountainous streams.

Ptychopteridae

This family is composed of 78 species described worldwide, recorded in three genera: *Bittacomorpha*, *Bittacomorphella*, and *Ptychoptera* (Pape & Thompson, 2013); but in the Neotropical region, the family has been recorded by Hancock et al. (2006), after the discovery of adults of *Ptychoptera* from rain forest habitat in southeastern Mexico, and recently by Fasbender & Courtney (2017) with a record of *Bittacomorpha* from Costa Rica. The oviposition occurs in the limit between still water and sediment where their larvae and pupae live. Long respiratory tubes are used to breathe atmospheric oxygen. The specificity of their habitat is likely to be high, and habitats may be restricted (Wagner et al., 2008). Larvae are associated with saturated mud or soils near water (Brindle, 1962). They are considerably caudally elongated, presenting a superficial resemblance to the long-tailed larvae of the Eristalinae (Syrphidae). To cope with respiration after pupation, the right thoracic breathing horn is greatly elongated, up to twice the length of the pupa, whereas the other horn is atrophied.

Scatopsidae

This family is cosmopolitan, with 400 described species in 36 genera (Amorim, 2016). There are 67 named scatopsid species in the Neotropical region, but more than 130 undescribed species are known (Amorim, 2009c). Scatopsid larvae live in decaying plant or animal material, humid wood holes, and excrement (Cook, 1967; Amorim, 2009a). Few larvae are aquatic living under the surface of a thin water film among waterlogged dead tree leaves, while few others are dendrolimnobiontic (e.g., inhabit phytotelmata) (Wagner et al., 2008). Pupae are sometimes found involved in the exuvia of the last larval instar (Amorim, 2009a).

Athericidae

Worldwide, there are 123 recorded species in nine genera. The 27 Neotropical species are described in four genera: *Atherix*, *Dasyomma*, *Suragina*, and *Xeritha*. Neotropical species has no larvae described formally; however, immature stages of Athericidae from Brazil were recently collected in streams, and in other regions, immature stages, so far as known, are strictly aquatic. Larvae are usually found in rocky or gravelly riffle areas (Woodley, 2009) and are predaceous, feeding on larvae of other aquatic insects (Rozkosny & Nagatomi, 1997).

Rhagionidae

The family have about 720 species in 22 genera worldwide (Kerr, 2009). Most representatives of this family are terrestrial, with a single aquatic genus, *Chrysopilus* (Kerr, 2009; Oscoz et. al. 2011). In the Neotropical region, 70 species were reported in *Chrysopilus* (Santos, 2008). Larvae of Neotropical species are unknown, but larvae of other regions are recorded for lake and river shores, but also in areas where rock substrates are covered with a thin layer of water, or between aquatic bryophytes. They can also develop on humid soils and are predators that feed on aquatic invertebrates (Oscoz et. al., 2011).

Tabanidae

This family has global distribution, with approximately 4,500 species distributed in 230 genera; in the Neotropical region, there are over 1,200 species described (Carvalho et al., 2012). Most females are hematophagous and have mammals, reptiles, and even birds as hosts (Ferreira et al., 2002; Oliveira et al., 2002). Larvae are fusiform (Chrysopsinae, Tabaninae) or anteriorly and posteriorly widened (Pangoniinae). Their coloration varies from white, green, to dark brown. They have 11 segments, a completely retractile head capsule, and partially retractile respiratory siphon on the last abdominal segment. They are known to be predatory and to practice cannibalism (Ferreira & Rafael, 2006), but some species are detritivores. Many species, while in the larval phase, inhabit humid soil, and they are found in fresh water, saline swamps, and in the banks of rivers and ponds. Only a few species are restricted to sand and gravel (Pechuman & Teskey, 1981). Many larvae inhabit specialized niches, such as holes in trees, decomposing trunks, bromeliads, or submerged aquatic vegetation.

Pelecorhynchidae

This family is composed of 49 species distributed in two genera and has worldwide distribution. In the Neotropics, nine species of *Pelecorhynchus* are reported (Llanos et al., 2015), with distribution exclusive to Chile. Immature stages of Pelecorhynchidae from Chile were recently known, and only some immatures of the *Pelecorhynchus* from Australian species are known (Mackerras & Fuller, 1942; Llanos et al., 2015). Larvae were collected in humid margins of swampy areas, where they feed on earthworms and possibly other soft-bodied animals (Mackerras & Fuller, 1942). They move on to dryer levels to pupate. Larvae and pupae are similar to those of Tabanidae (Mackerras & Fuller, 1942).

Empididae

Empididae are cosmopolitan, with about 3,051 species worldwide in 68 genera (Pape & Thompson, 2013). In the Neotropical region there are approximately 500 species in 32 genera. Adults are mostly predaceous, and some species feed on pollen and nectar. Known larvae are predaceous, living in aquatic and terrestrial environments. Various species that have aquatic larvae prey on simuliid pupae (Hamada, 1993).

Dolichopodidae

This family is greatly diverse in tropical areas. There are 7,600 known species worldwide (Yang et al., 2006; Grichanov, 2014), of which 1,200 are Neotropical (Capellari, 2016). Adults prey on other smaller insects, and they are often observed on the dorsal face of foliage, trunks, stones, etc. Larvae are predominantly predaceous, and some are phytophagous. Larvae are found in mud, decomposing wood, under bark, and soil litter.

Lonchopteridae

About 58 species are known worldwide, all of them cited mostly in a single genus (Carvalho et al., 2012), *Lonchoptera*. An aquatic developmental stage is known for *Lonchoptera nigrociliata* Duda and *L. lutea* Panzer (Wagner et al., 2008; Vaillant & Rozkošný, 2002). Only *L. furcata* (Fallén) was recorded in the Neotropics (Amorim et al., 2002). The biology of species in this family is not well known. Larvae live in a wet microenvironment of decaying organic matter, among dead leaves, in plant debris, under stones on shores of water courses, or they are semiaquatic. They are saprophagous, microphagous, or presumably, mycetophagous.

Phoridae

Phoridae is a cosmopolitan family with about 4,087 species in 271 genera (Pape & Thompson, 2013); there are about 225 genera reported for the Neotropical region (Amorim et al., 2002). They are considered to be one of the groups with the most diverse biology among all insects, being found in several different types of environments, such as decomposing bodies, rodent and bird nests, flowers, fungi, wasp, ants, termite nests, and in caves; and with the greatest diversity of larval habits, but it is the least studied of the large families of flies due to identification difficulties. There are many genera that include species with larvae that are parasitoid of other insects, spiders, and myriapods, and others that are parasites or commensals in ant or termite nests. True aquatic larvae are found only in phytotelmata (Disney, 2004).

Muscidae

The family is cosmopolitan with about 5,190 species (Pape & Thompson, 2013) and 846 Neotropical species (Carvalho et al., 2005). The phylogeny of the family has been investigated by Carvalho (1989), Couri & Carvalho (2003), and Haseyama et al. (2015). Adults have extremely diverse habits, which include some predaceous species. Adults in most species feed on decomposing animal or vegetable matter (Carvalho & Couri, 2002). Muscid larvae are coprophagous, saprophagous, or carnivorous (Skidmore, 1985). Larvae of aquatic muscid species can be found in running and standing waters in streams and ponds. They are generally collected in decaying organic matter and algal mats, presumably where they find their prey. A few species of Muscidae are truly aquatic and prey primarily on other Diptera. The aquatic larvae of this group are not well studied.

Scathophagidae

The family has about 404 species worldwide in 57 genera (Pape & Thompson, 2013). About six species were recorded in Neotropical region, being divided in three genera, *Cordilura, Scathophaga*, and *Scatogera* (Vockeroth, 2010; Wolff et al., 2016). Of these, only *Scathophaga* include truly aquatic larvae, but their biology and ecology are still poorly described (Oscoz et. al., 2011). Females enter the waters of streams and ovoposit their eggs on the surface of the eggs masses of other invertebrates such as Trichoptera and other Diptera. The hatched larvae feed on the eggs and surrounding jelly of the host (Purcell et al., 2008).

Sarcophagidae

The family has more than 3,000 described species (Pape et al., 2011), distributed almost worldwide, but the greatest diversity is found in the tropical regions. About 750 species are known to occur in the Neotropical region, most found in tropical or warmer temperate areas. In aquatic environments, some larvae may be found mining bases of pitcher plants (Courtney & Merritt, 2008).

Sciomyzidae

It is cosmopolitan and has approximately 600 species described in 67 genera. In the Neotropical region, there are 74 species in 23 genera (Amorim et al., 2002). Larvae of most species are primarily parasitoid or predators of aquatic or terrestrial mollusks, including a few species of operculate mollusks. The feeding habits of larvae make it possible for them to be used in the biologic control of vectors of parasitic diseases, such as fascioliasis and schistosomiasis (Carvalho et al., 2012).

LIMITATIONS

Most systematic studies of Neotropical Diptera have focused on adult morphology, so the knowledge of aquatic stages of many families is fragmentary. Based on this fact, we included in the key to Neotropical families those aquatic larvae that are reported elsewhere, even though the larval habitats of the species in this region are not currently known (e.g., Pelecorhynchidae and Lonchopteridae).

TERMINOLOGY AND MORPHOLOGY

Larval dipterans are characterized by a highly diverse body shape that is fusiform (Tabanidae, Dolichopodidae), cylindrical (Chironomidae, Empididae), or sometimes anteriorly tapered (Muscidae, Phoridae) or strongly dorsoventrally flattened (Stratiomyidae). Others, such as Culicidae, Chaoboridae, and Corethrellidae, have an enlarged thorax, while Simuliidae have an enlarged abdomen (Teskey, 1981). The most common body segmentation pattern is that with 12 segments, 3 of which are thoracic, and 9 abdominal. There are numerous variations to this basic plan. In Brachycera, for example, there are frequently only 11 obvious segments. In Blephariceridae, in which the body is divided in six parts, the first one is correspondent to the entire thorax and the first abdominal segment. Similarly, pseudo-segmentation also occurs, as observed in Psychodidae, in which each segment has two or three subdivisions (Teskey, 1981; Courtney & Merritt, 2008). The cephalic head can be complete, sclerotized and exposed, with biting and chewing mouthparts, or reduced in many levels, partially or entirely inserted in the thorax, with mouthparts modified for stinging and scraping (Teskey, 1981; Guimarães & Amorim, 2006). Three states of head capsule reduction illustrate these patterns, as describe next.

Eucephalic larvae. These exhibit a totally exposed, well-developed, and sclerotized cephalic capsule (except for Tipulidae, in which the head is inserted in the thorax and sometimes reduced, e.g., Fig. 16.2 C; and Blephariceridae, that show reduction in the dorsal region of the cephalic capsule, e.g., Fig. 16.3 A), with mandibles operating in a horizontal or oblique plan (Figs. 16.2 A, 4 A, G, J).

Hemicephalic larvae. These have a relatively reduced head that is posteriorly incomplete and partially retracted in the thorax, with mandibles or mouth hooks in the shape of sickles, operating in a vertical plan (Figs. 16.6 A, I, 7 A, D).

Acephalic larvae. These show additional reduction and almost total retraction of cephalic capsule in the thorax. The cephalic thorax is modified in an internalized cephalopharyngean skeleton, of which only the vertical, sickle-shaped mandibles are exposed (Figs. 16.8 A, C, 9 E, M).

True thoracic appendages (articulated legs) are entirely absent in all Diptera larvae. Locomotion is conducted by pseudopods, various fleshy projections, typically bearing hairs or spines, present in anterior and posterior regions of the larva (Figs. 16.5 A, C, D, F), or through creeping welts, normally in greater number than pseudopods, and present in ventral intersegmental regions (Figs. 16.5 E, H). Such protrusions work as supporting points that, together with peristaltic motions of the body, allow for larval locomotion. Mandibles can aid this process, by becoming additional anchoring points (Teskey, 1981). Larvae show a variety of respiratory adaptations. Respiration may be directly from the atmosphere (e.g., Dixidae, most Culicidae), from plant tissues (e.g., some Culicidae), or from the surrounding water. According with the number and position of respiratory apertures (spiracles), aquatic dipteran larvae can be the following (Fig. 16.1):

Peripneustic. Loss of the posterior thoracic spiracles yields the peripneustic system, which is characteristic of the larvae (Fig. 16.1 A). An example is Scatopsidae.

Amphipneustic. These Diptera are characterized by spiracles on the first thoracic segment and also in the posterior body area (Fig. 16.1 B). Examples are Psychodidae, Syrphidae, and some Muscidae.

Metapneustic. Such dipterans have spiracles limited to the posterior body area (Fig. 16.1 C). Examples include Culicidae, Tabanidae, and certain Muscidae.

Apneustic. Functional spiracles are absent in these Diptera. In this case, gas exchanges occur throughout the entire tegument or in dedicated areas in the tegument, commonly named branchiae (Fig. 16.1 D). Examples are Chironomidae, Simuliidae, and Blephariceridae. In several groups the spiracles are located at the end of a retractile

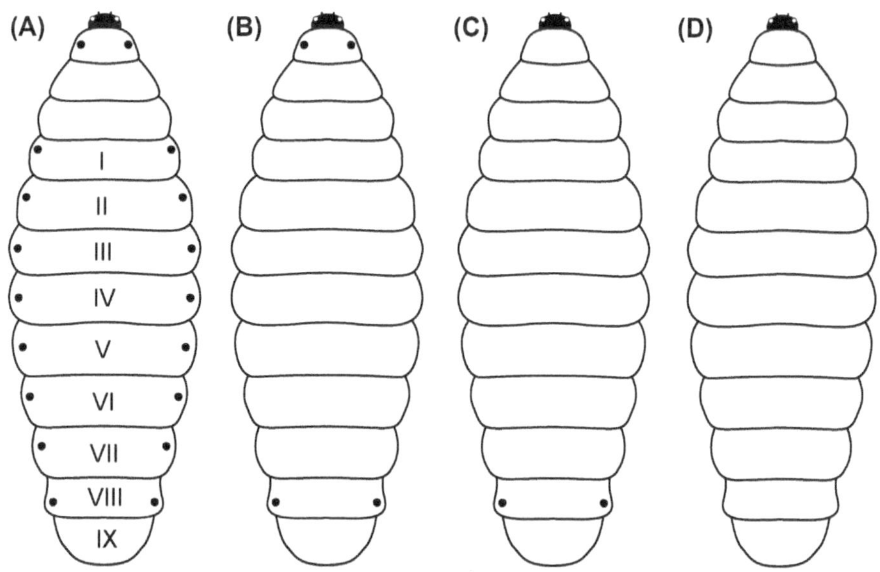

FIGURE 16.1 Position of respiratory apertures (spiracles) in Diptera larva: (A) peripneustic; (B) amphipneustic; (C) metapneustic; (D) apneustic.

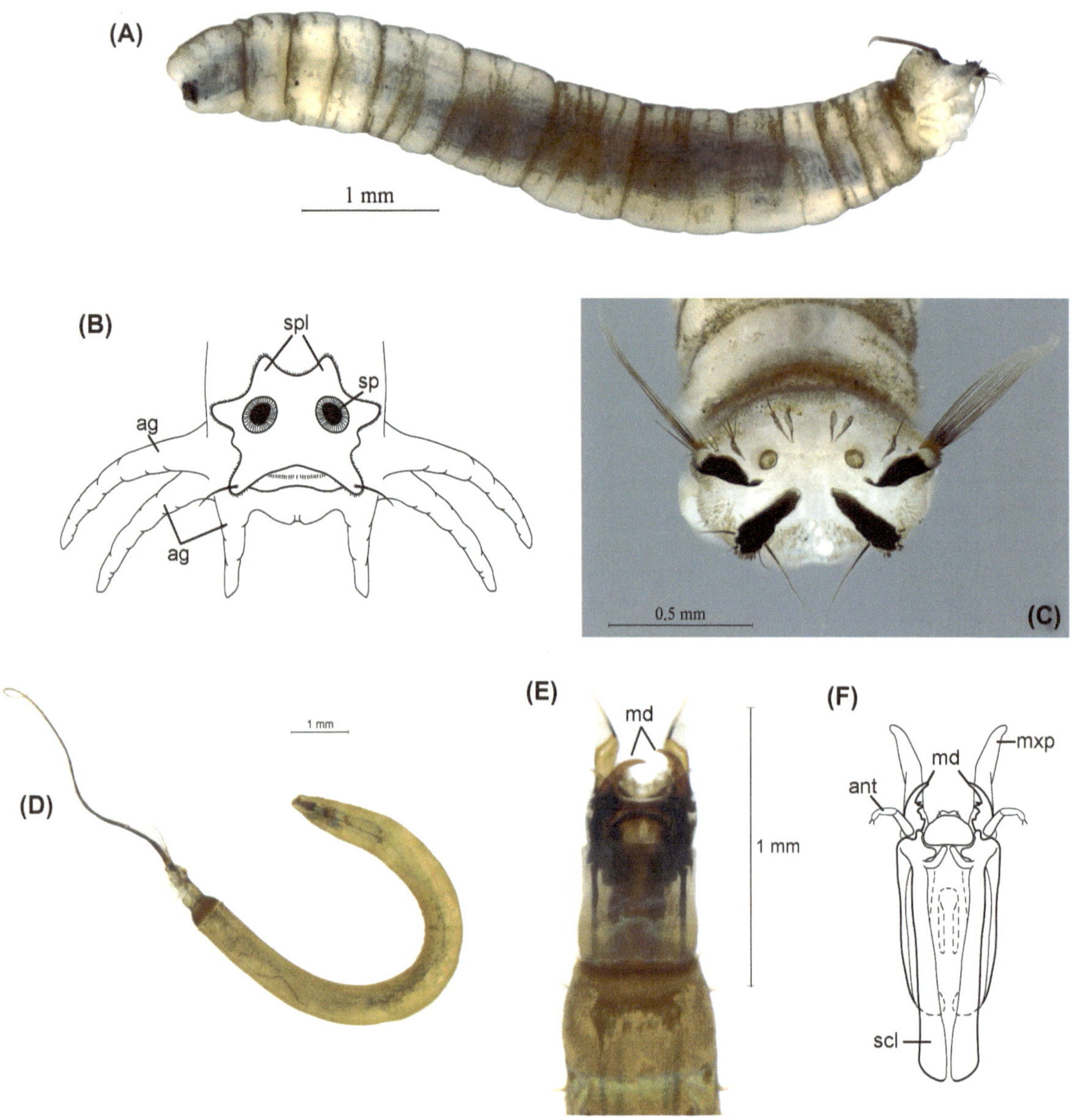

FIGURE 16.2 Tipulidae larva: (A) habitus, lateral view; head capsule retracted within thorax; (B) Posterior end of larva; (C) spiracular disk, in posterior view. Limoniidae larva (D) larval habitus; (E) head capsule fully exposed; (F) head capsule, dorsal view. Abbreviations: *ag*, anal gills; *ant*, antenna; *md*, mandible; *mxp*, maxillary palp; *scl*, sclerites; *sp*, spiracle; *spl*, spiracular lobes.

respiratory siphon (respiratory tube or breathing tube). Examples are Ptychopteridae, Ephydridae, and Syrphidae. Further information on the morphology of Diptera larvae is provided by Teskey (1981) and Guimarães & Amorim (2006).

MATERIAL PREPARATION AND PRESERVATION

Preservation

After sampling, immatures have to be fixed and preserved using the appropriate method for each taxonomic group and study goal. Alcohol-preserved immatures must be accommodated in tightly sealed jars, so that volatilization is reduced. It is frequently impossible to sort immatures while in the field; therefore biologic material can be fixed in 96% ethyl alcohol, along with the substrate in which it has been collected. After sorting, immatures should be preserved in 80% ethyl alcohol. Depending on the purpose of the study, other fixative solutions may be necessary. For example, to do chromosomal studies (e.g., on Chironomidae or Simuliidae), organisms must be fixed directly in Carnoy solution (3:1; absolute ethyl alcohol and glacial acetic acid). Molecular analyses require that organisms are fixed directly in ethylic alcohol. In both cases, fixative solutions

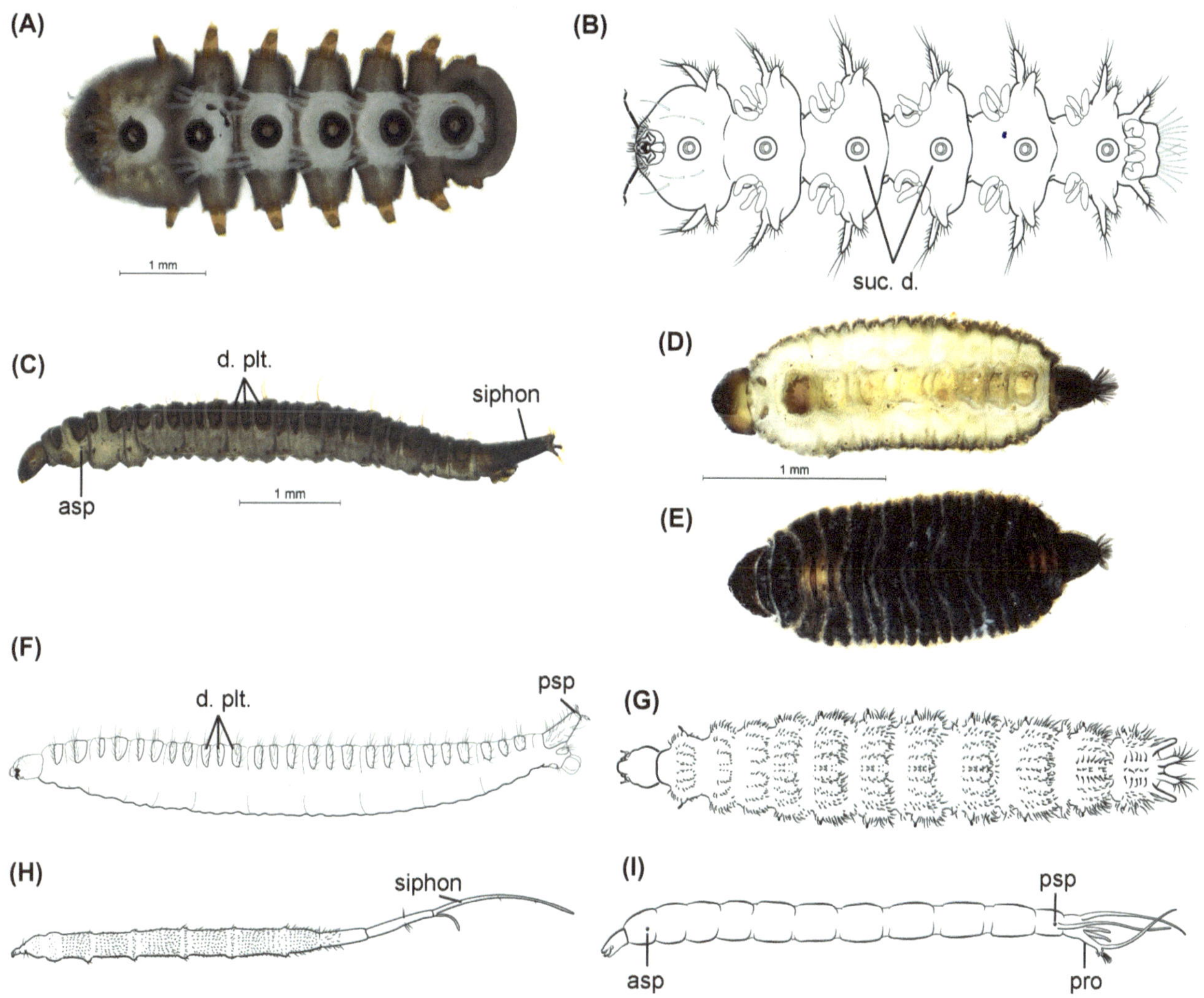

FIGURE 16.3 Blephariceridae larva: (A–B) ventral view. Psychodidae larva: (C) habitus, lateral view; habitus in (D) ventral view; (E) dorsal view; (F) habitus, lateral view. Scatopsidae larva: (G) habitus, dorsal view. Ptychopteridae larva: (H) habitus, lateral view. Tanyderidae larva: (I) habitus, lateral view. Abbreviations: *asp*, anterior spiracle; *d. plt*, dorsal plate; *psp*, posterior spiracle; *pro*, proleg; *suc. d.*, suctorial disc.

must be changed at least three times, and material must be stored in low-temperature (−20°C) and light-free environments.

Slide Preparation

The preparation of permanent or semipermanent slides requires that immature parts be separated (e.g., cephalic capsule and abdomen). For the observation on optical microscopes, muscles can be dissolved, which results in clarification. This process is used particularly for permanent slide mounting.

There are various methods for the clarification of specimens or selected structures before slide preparation. Three techniques are introduced as follows:

Lactic acid 85%. Heat insect or desired structure in lactic acid in a porcelain crucible, using a hot plate, for a period of 1–2 h (always following the process, ensuring that the medium does not evaporate completely). Heating time used for clarification will depend on the material's size and degree of sclerotization. Rinse the specimen or structure in water and transfer to an alcohol series (ethyl alcohol 70%, 80%, 90%, and absolute). The material should remain soaked in each alcohol concentration for 15–20 min. to ensure material dehydration.

Hot potassium hydroxide (KOH) 10% (or 5%) solution. Heat insect or desired structure in KOH on a hot plate, for a period of 1–2 h (always following the process, ensuring the structure is not overly transparent). Heating time used for clarification will depend on the material's size and degree of sclerotization. Rinse the specimen or structure in water, neutralize for 15 min in acetic acid 10% or 15%, and rinse in water again. Transfer the material to an alcohol series (ethyl alcohol 70%, 80%, 90%, and

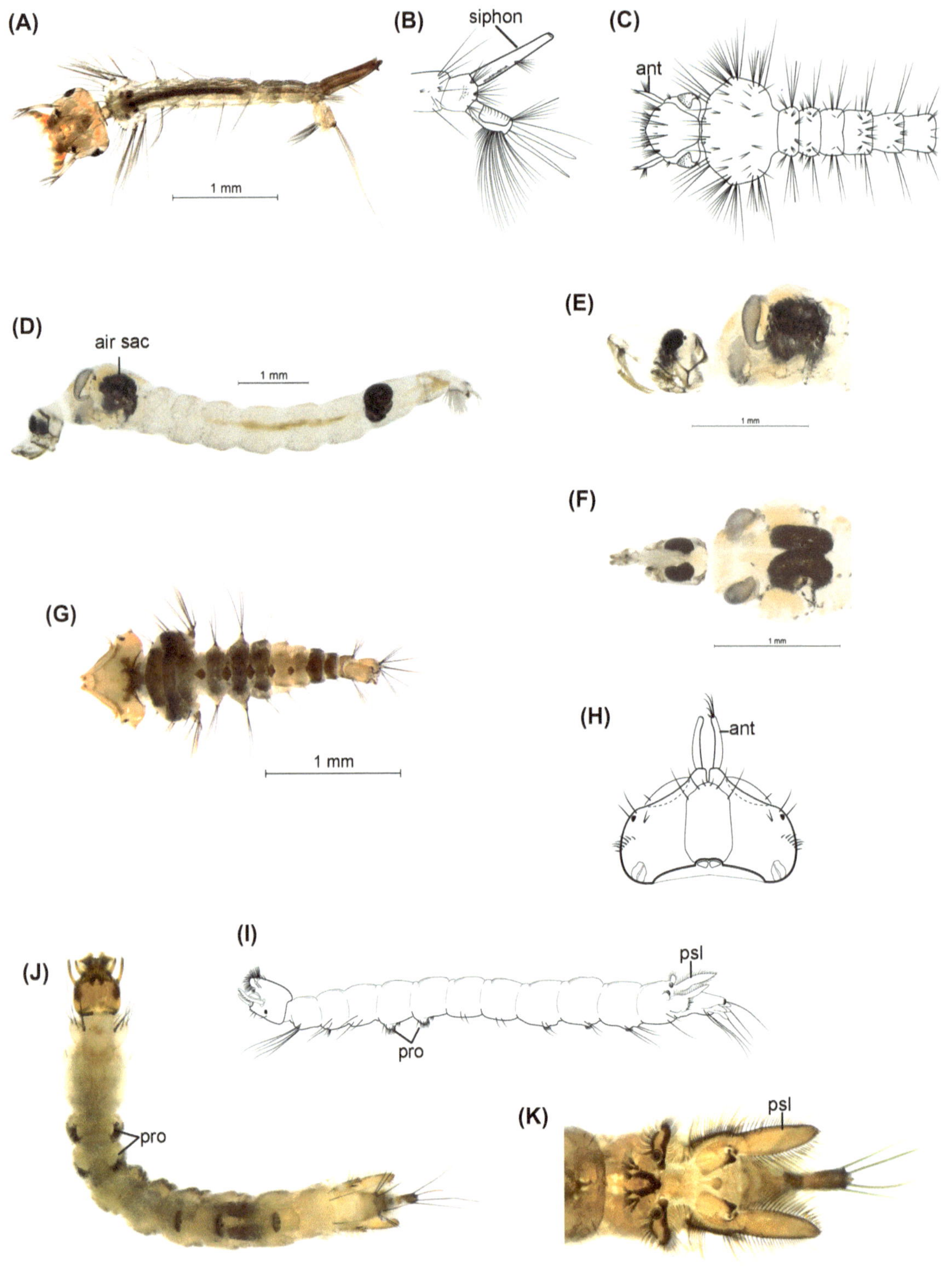

FIGURE 16.4 Culicidae larva: (A) habitus, dorsal view; (B) anal region, lateral view; (C) head, thorax, and six abdominal segments, in dorsal view. Chaoboridae larva: (D) habitus, lateral view; head and thorax in (E) lateral view and (F) dorsal view. Corethrellidae larva (G) habitus, dorsal view; (H) head, dorsal view. Dixidae larval habitus in (I) lateral view and (J) ventral view; (K) terminal segments, dorsal view. Abbreviations: *ant*, antenna; *pro*, proleg; *psl*, postspiracular lobe.

FIGURE 16.5 Ceratopogonidae larva: (A–C) habitus, lateral view. Chironomidae larva: (D–E) habitus, lateral view. Simuliidae larva habitus, (F) lateral view and (G) dorsal view. Thaumaleidae larva: head and thorax in (H) dorsal view and (I) in lateral view; (J) habitus, lateral view. Abbreviations: *asp*, anterior spiracle; *lf*, fans labral; *pro*, proleg; *psp*, posterior spiracle.

absolute). The material should remain soaked in each alcohol concentration for 15–20 min to ensure material dehydration.

Cold KOH 10% solution. Immerse specimen or desired structure in KOH for ~12–24 h (it is important to follow the process, to ensure the structure will not become overly transparent). Immersion time used for clarification will depend on the material's size and degree of sclerotization. Following this process, the same washing protocol used for heated 10% KOH should be used. Sometimes, it is necessary to remove with dissection needles the undissolved muscles from clarified specimens or separated structures.

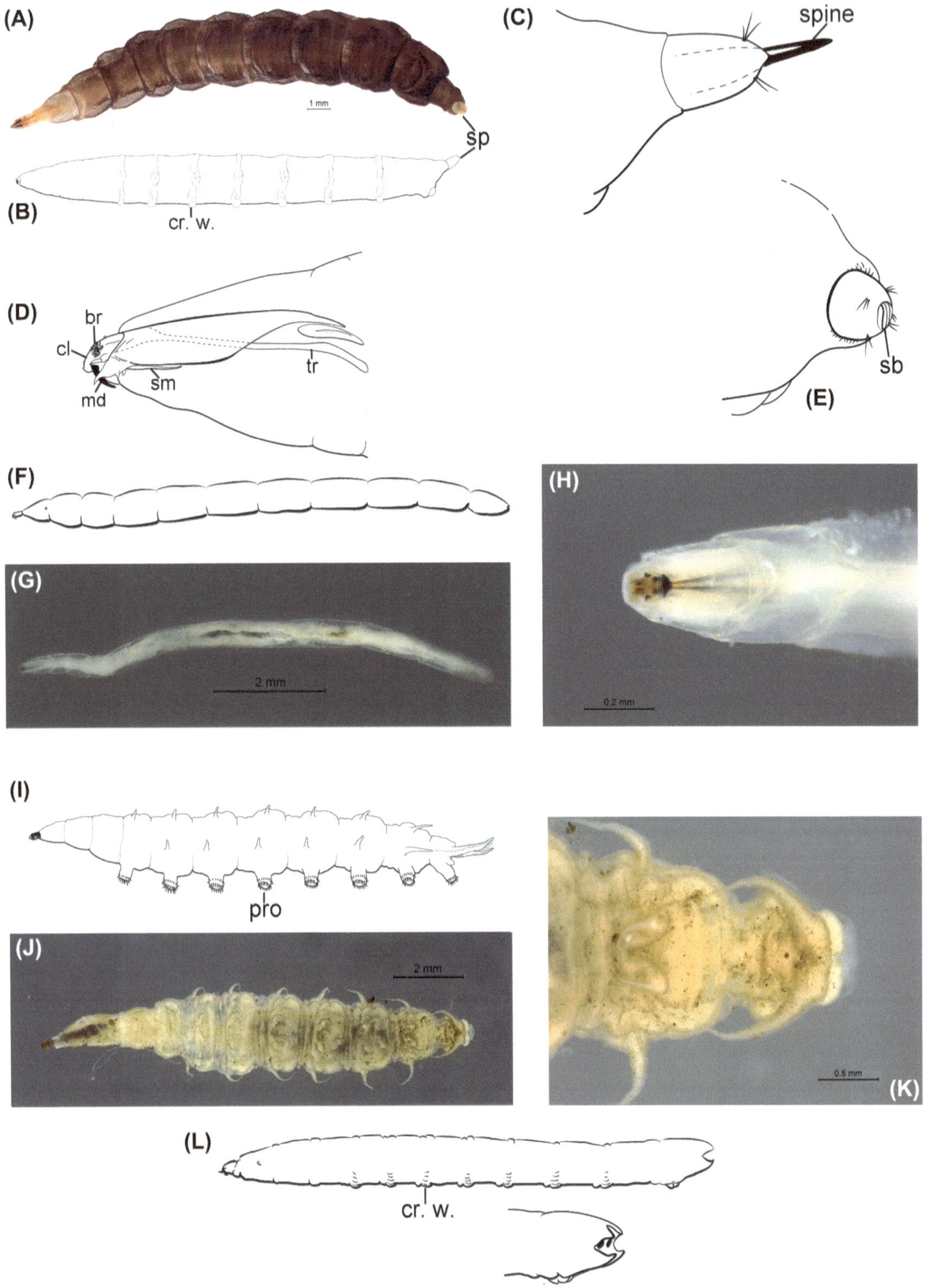

FIGURE 16.6 Tabanidae larva: habitus, (A) dorsal view, (B) lateral view; (C) respiratory siphon and spine; (D) lateral view of head capsule; (E) respiratory siphon. Pelecorhynchidae larva: (F) habitus, lateral view; (G) habitus; (H) head capsule, ventral view. Athericidae larva: (I) habitus, lateral view; (J) habitus, dorsal view; (K) posterior region of abdomen, dorsal view. Rhagionidae larva: (L) habitus, lateral view, with detailed posterior segments. Abbreviations: *br*, mandibular bristles; *cl*, clypeus; *cr. w.*, creeping welt; *md*, mandible; *pro*, proleg; *sb*, stigmatal bar; *sm*, submentum; *tr*, tentorial rod.

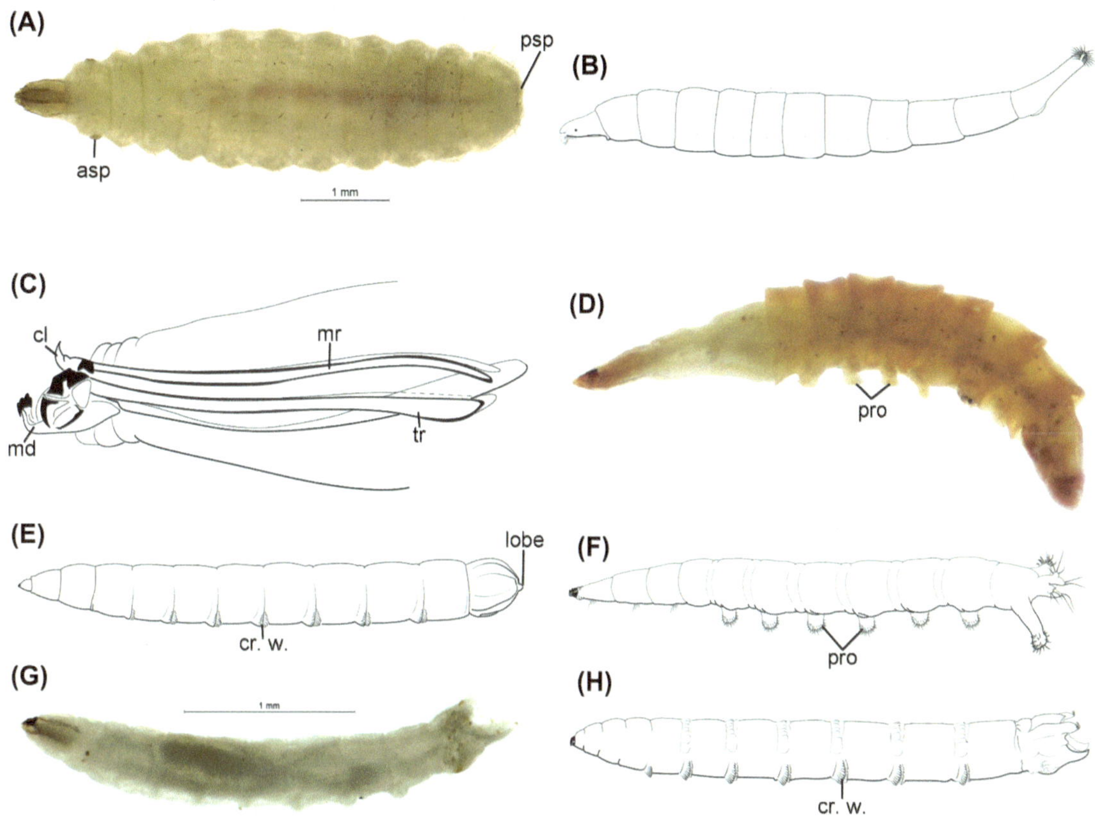

FIGURE 16.7 Stratiomyidae larva: habitus, (A) dorsal view; (B) lateral view. Empididae larva: (C) head capsule, ventral view; (D–F) habitus, lateral view. Dolichopodidae larva: (G–H) habitus, lateral view. Abbreviations: *asp*, anterior spiracle; *cl*, clypeus; *cr. w.*, creeping welt; *md*, mandible; *mr*, metacephalic rod; *pro*, proleg; *psp*, posterior spiracle; *tr*, tentorial rod.

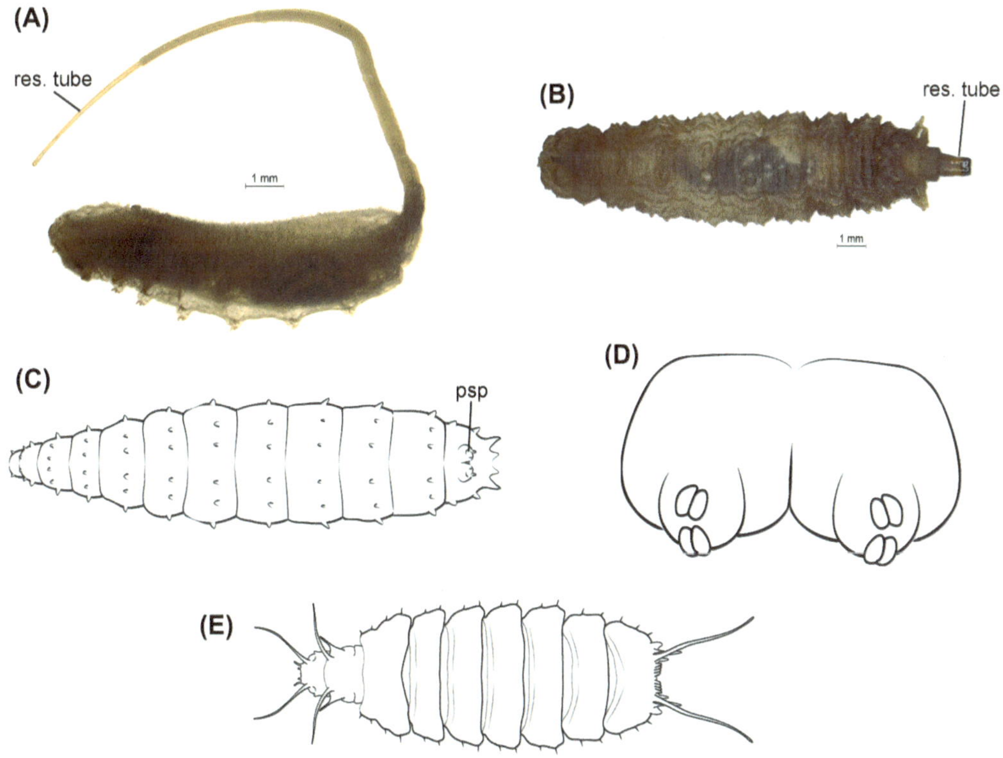

FIGURE 16.8 Syrphidae larva: habitus, (A) lateral view, (B) dorsal view. Phoridae larva: (C) habitus, dorsal view; (D) posterior spiracles. Lonchopteridae larva, habitus, in dorsal view. Abbreviations: *psp*, posterior spiracle; *res. tube*, respiratory tube.

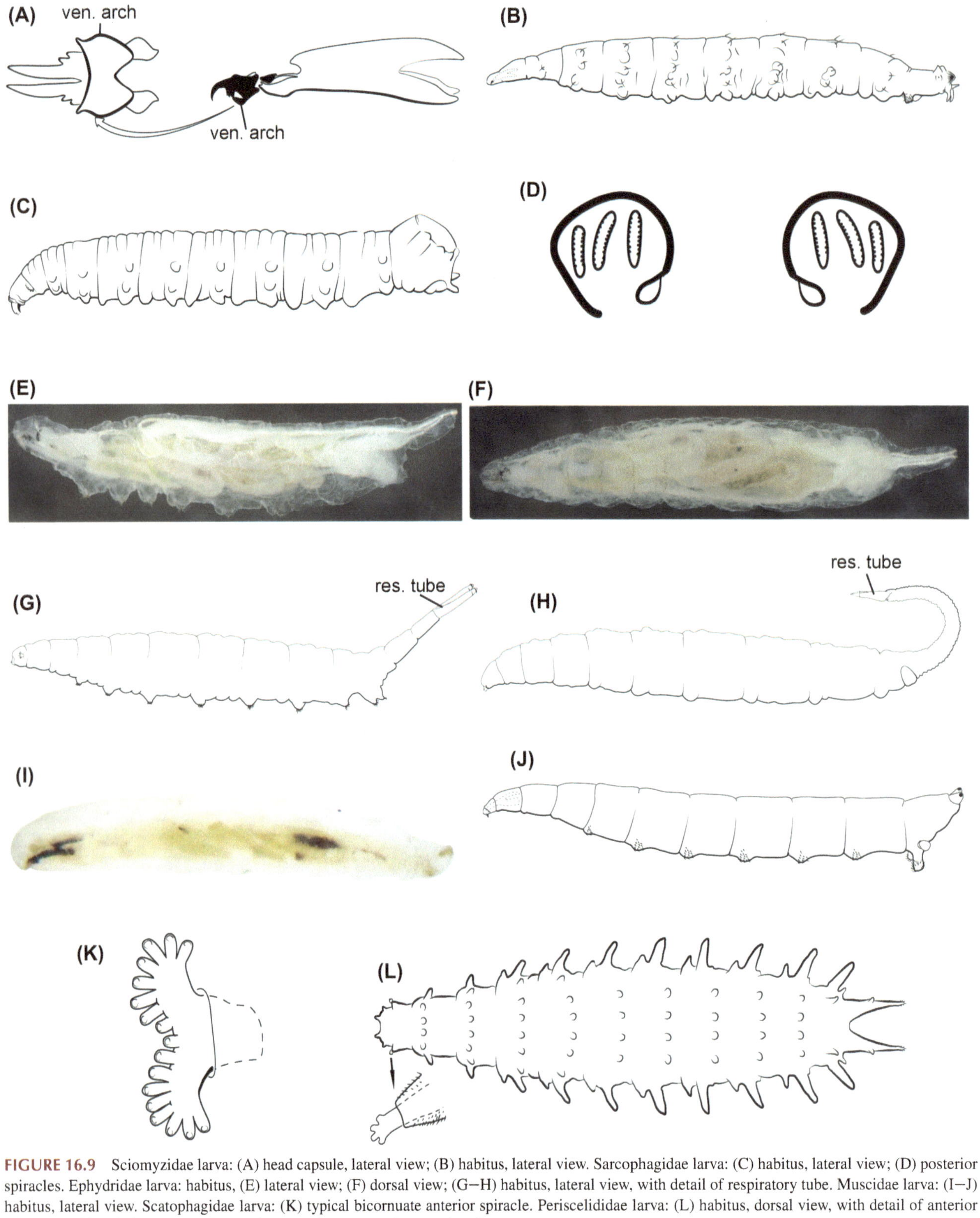

FIGURE 16.9 Sciomyzidae larva: (A) head capsule, lateral view; (B) habitus, lateral view. Sarcophagidae larva: (C) habitus, lateral view; (D) posterior spiracles. Ephydridae larva: habitus, (E) lateral view; (F) dorsal view; (G–H) habitus, lateral view, with detail of respiratory tube. Muscidae larva: (I–J) habitus, lateral view. Scatophagidae larva: (K) typical bicornuate anterior spiracle. Periscelididae larva: (L) habitus, dorsal view, with detail of anterior spiracles. Abbreviations: *res. tube*, respiratory tube; *ven. arch*, ventral arch.

For permanent slides, the most common preparation mediums are Euparal and Canada Balsam. Euparal is soluble in alcohol and, therefore, specimens can be transferred directly into this product after dehydration. For preparation in Canada Balsam, it is necessary that the dehydrated structure undergo creosote or eugenol treatment before specimens are mounted into slides. Following preparation, slides must be incubated at 50°C for drying. For semipermanent slides, the most frequently used medium for this type of preparation is Hoyer (solution of chloral hydrate, distilled water, glycerin, and gum arabic). Chironomidae larvae are normally mounted using this protocol (Trivinho-Strixino & Strixino, 2011; see Chapter 16.2 on Chironomidae).

Instant analysis with temporary slides in optical microscopes can be performed by mounting the specimen in an excavated slide with glycerin or a 1:1 mixture of glycerin and alcohol, and then covering this gel with a glass slide. Subsequently, specimen is stored in a microtube with 80% alcohol.

KEYS TO LARVAE OF AQUATIC AND SEMIAQUATIC DIPTERA

The following keys to suborders and families pertain only to taxa recorded in the Neotropical region. These keys are modified from Pinho (2008) and Pepinelli & Pinho (2014). Those keys were, in turn, based on Courtney & Merritt (2008) and have been adjusted according to groups found in the Neotropical region, both in the description of steps and in the associated illustrations.

Diptera: Suborders

1 Mandibles moving against one another in a horizontal or oblique plane (Figs. 16.2 E–F); head capsule usually complete and fully exposed, except retracted and reduced in Tipulidae .... **"Nematocera" [p. 619]**

1' Mandibles or mouth hooks moving parallel to one another in a vertical plane (Figs. 16.6 D, H, 7 C); head capsule variously reduced posteriorly, partially or almost completely retracted within thorax .... **Brachycera [p. 620]**

Diptera: "Nematocera": Families

1 Head capsule partially to fully retracted within thorax (Figs. 16.2 A, D) (when the specimen is observed alive, you can check the exposed head capsule), although sometimes head capsule consisting of elongated sclerites (Fig. 16.2 F), posterior spiracles usually bordered by 1–3 pairs of short lobes that are often fringed with short to very long hairs (Figs. 16.2 B–D) 2

1' Head capsule fully exposed, usually without longitudinal incisions dorsolaterally (Figs. 16.3 C, 4 A), except reduced dorsally in Blephariceridae (Fig. 16.3 A), posterior spiracles usually without bordering fringed lobes 3

2(1) Spiracular disk bordered by six (rarely eight) usually subconical lobes, usually two dorsal, two dorsolateral, two ventral spiracles (Figs. 16.2 B–C); lobes sometimes short and blunt .... Tipulidae

2' Spiracular disk, if present, bordered by five (rarely seven) or fewer lobes, usually one dorsomedial, two lateral, two ventral spiracles; lobes of variable shape .... Limoniidae

3(1) Head with distinct constrictive separation from thorax (Figs. 16.4 A, G, J); suctorial discs absent 4

3' Head not distinctly separated from thorax (Fig. 16.3 A); body divided into six major divisions, the first comprising fused head, thorax and first abdominal segment; each division with a ventromedial suctorial disc (Figs. 16.3 A–B) .... **Blephariceridae** (Chapter 16.9, p. 759)

4(3) Respiratory system not as follows 5

4' Respiratory system peripneustic (Fig. 16.1 A), with eight pairs of abdominal spiracles; posterior spiracles usually conspicuously larger than other abdominal spiracles (Fig. 16.3 G) .... Scatopsidae

5(4) Abdomen not as follows 6

5' Abdomen ending in a slender, respiratory siphon that when fully exerted is nearly as long as body (Fig. 16.3 H) .... Ptychopteridae

6(5) Thoracic segments fused and indistinctly differentiated, forming a single segment that is conspicuously wider than abdominal segments (Figs. 16.4 C, E, G); thoracic and abdominal segments with prominent, lateral flabelliform setal tufts, and/or terminal segment with a setal flabellum (Figs. 16.4 B, D) 7

6' Thoracic segments usually individually distinguishable, thorax about as wide as abdomen, rarely abdomen wider (Fig. 16.4 J); setae on thoracic and abdominal segments not tufted and terminal segment without setal flabellum 9

7(6) Antennae prehensile (raptorial), with long apical setae (Figs. 16.4 E, H); mouth brushes absent; respiratory siphon absent 8

7' Antennae not prehensile and with only short apical setae (Figs. 16.4 A, C); prominent oral brushes present on either side of labrum; respiratory siphon present (Figs. 16.4 A–B) or absent .... **Culicidae** (Chapter 16.5, p. 723)

8(7) Antennae inserted far apart; head laterally without a transverse setal row; terminal segment with longitudinal flabelliform setal row (Fig. 16.4 E); thoracic air sacs present (Fig. 16.4 D) or absent .... **Chaoboridae** (Chapter 16.8, p. 755)

8' Antennae inserted close together; head laterally with a transverse row of spiniform setae (Fig. 16.4 H); terminal segment ventrally with a tuft of long setae; thoracic air sacs absent (Fig. 16.4 G) .... **Corethrellidae** (Chapter 16.7, p. 751)

9(6) Abdominal segments I & II without prolegs; abdomen posteriorly lacking fringed postspiracular lobes 10

9' Prolegs present on abdominal segments I and usually II (Figs. 16.4 I–J); abdomen posteriorly with two flattened dorsolateral postspiracular lobes with setose margins (Fig. 16.4 k) .... **Dixidae** (Chapter 16.6, p. 747)

10(9) Prothorax with prolegs (Figs. 16.5 A, D, F) ... 11

10' Prothorax lacking prolegs (Figs. 16.3 F, I, 5 B) ... 14

11(10) Head capsule lacking labral fans (Fig. 16.5 D, H, I); posterior abdominal segments not markedly broad; terminal ring of hooks absent; anal prolegs bearing crochets may be present ... 12

11' Head capsule usually with a pair of conspicuous labral fans (Fig. 16.5 F–G); abdominal segments 5–8 broad, posterior segment ending in a ring of hooks ... **Simuliidae** (Chapter 16.3, p. 701)

2(11) Respiratory system apneustic (Fig. 16.1 D); prothoracic or anal prolegs usually paired, even if distinction is only a slight separation of the apical spines (Fig. 16.5 D) ... 13

12' Respiratory system amphipneustic (Fig. 16.1 B); prothorax with a pair of short respiratory tubes dorsolaterally (Fig. 16.5 I); eighth abdominal segment with a transverse respiratory opening near hind margin, flanked by a pair of sclerotized digitiform processes, bearing several longer dark setae apically (Fig. 16.5 J); prothoracic and anal prolegs unpaired ... **Thaumaleidae** (Chapter 16.4, p. 719)

13(12) All body segments dorsally with prominent tubercles (elevated fleshy processes) and/or setae (Figs. 16.5 A, C) ... **Ceratopogonidae** (in part) (Chapter 16.1, p. 621)

13' Body segments lacking prominent dorsal tubercles and setae (Fig. 16.5 D, E) ... **Chironomidae** (Chapter 16.2, p. 661)

14(10) Posterior abdominal segments without long filamentous processes; without prolegs (Figs. 16.3 F, 5 B), or at most, only a single anal proleg present ... 15

14' Posterior abdominal (8) segment laterally with long filamentous processes (Fig. 16.3 I), abdominal segment 9 with pairs of such processes arising dorsolaterally and from near apex of two elongate cylindrical posteroventrally projecting prolegs ... Tanyderidae

15(14) Thoracic and abdominal segments with three dorsal sclerotized plates (Figs. 16.3 C, E); remainder of integument with numerous dark spots that together with the dorsal plates impart a greyish brown coloration; respiratory system amphipneustic (Fig. 16.1 B); posterior spiracles usually at apex of a relatively short, conical respiratory siphon; sometimes with ventral suctorial discs (Fig. 16.3 D) ... **Psychodidae** (Chapter 16.10, p. 765)

15' Dorsal sclerotized plates absent; integument smooth, shiny, and creamy white, lacking all surface features except a few setae that may be noticeable at terminal segment tip (Fig. 16.5 B); respiratory siphon absent; ventral suctorial discs absent ... **Ceratopogonidae** (in part) (Chapter 16.1, p. 621)

Diptera: Brachycera: Families

1 Sclerotized portions of head capsule exposed externally (Figs. 16.6 A, 7A, D), if greatly reduced, with slender tentorial and metacephalic rods prominent internally (Figs. 16.6 H, 7 C) ... 2

1' External sclerotized portions of head capsule absent; head reduced to an internal cephalopharyngeal skeleton, with externally exposed bucal hooks (Figs. 16.9 A–B) ... 8

2(1) Larva usually not conspicuously depressed nor with a toughened integument bearing a network of facets; head capsule capable of extensive independent movement; without distinctive eye prominences ... 3

2' Larva tightly flattened, integument toughened and leathery from calcium deposits that are evident as numerous small reticulately arranged facets (Figs. 16.7 A–B); head capsule capable of only sight independent movement; usually with distinctive lateral eye prominences ... **Stratiomyidae** (Chapter 16.11, p. 771)

3(2) Head capsule well developed dorsally (Figs. 16.6 A, J), closed ventrally by a submental plate (Figs. 16.6 D, H); tentorial rods solidly fused with head capsule internally; each side of clypeus above and near base of each mandible usually a brush of backwardly curved bristles present ... 4

3' Head capsule reduced to a pair of slender metacephalic rods; these and tentorial rods flexibly articulate with anterior cephalic sclerites (Fig. 16.7 C); submental plate and brushes of bristles above mandibles absent ... 7

4(3) Posterior spiracles greatly reduced or situated within a small terminal cavity (Fig. 16.6 F); integument without striations or extensive covering of pubescence; prolegs, if present, limited to one to two ventral pairs on each abdominal segment ... 5

4' Posterior spiracles present, opening within slits on either side of a vertically linear stigmatal bar (Fig. 16.6 E) or a retractile, laterally compressed spine (Fig. 16.6 C); body integument with longitudinal striations, except in some species where integument is covered by short, velvety pubescence; first seven abdominal segments girdled by three or four pairs of fleshy creeping welts (setae-bearing swellings) or prolegs, these being the only projections from the segments (Figs. 16.6 A–B) ... Tabanidae

5(4) Larva cylindrical, lacking tubercles and prolegs, at most with ventral creeping welts (Figs. 16.6 F, L) ... 6

5' Larvae with slender lateral and dorsolateral tubercles of progressive size on abdominal segments 1–7; terminal segment with two longer tubercles, fringed with setae; all abdominal segments ventrally with pair of elongate prolegs, each proleg with semicircle of subapical and apical claws (Figs. 16.6 I–J) ... Athericidae

6(5) Labrum and apically on maxilla laterally with stout spines (Fig. 16.6 H); integument smooth, shiny, segmentation beadlike; respiratory system apneustic (Figs. 16.6 F–G) ... Pelecorhynchidae

6' Labrum and maxilla without stout spines; first seven abdominal segments with ventral creeping welts, segmentation not as previous; respiratory system metapneustic or amphipneustic (Fig. 16.6 L) ... Rhagionidae

7(3) Larva usually apneustic, terminal segment with one to four rounded lobes bearing apical setae, and abdominal segments bearing paired prolegs with apical crochets (Figs. 16.7 D, F); if metapneustic, then terminal segment with only a single lobe below spiracles, and abdominal segments with ventral creeping welts (Fig. 16.7 E); metacephalic rods slender posteriorly (Fig. 16.7 C) ... Empididae

7' Larva metapneustic; posterior spiracles situated at the base of upper two of four smooth primary lobes of terminal segment (Figs. 16.7 G–H); transverse ventral creeping welts present on abdominal segments; metacephalic rods expanded posteriorly .... Dolichopodidae

8(1) Body not distinctly flattened; all segments tergal plates smooth at margins; first two thoracic and terminal abdominal segments without slender processes 9

8' Body distinctly flattened; all segments tergal plates striated at margins; first two thoracic and terminal abdominal with slender processes (Fig. 16.8 E) .... Lonchopteridae

9(8) Posterior spiracular plates always distinctly separated whether mounted on a telescopic respiratory tube or not 10

9' Posterior spiracular plates fused or very closely approximated, usually on apex of a telescopic respiratory tube (Figs. 16.8 A–B) .... **Syrphidae** (Chapter 16.12, p. 779)

10(9) Body often dorsoventrally flattened and bearing a series of spicules or tubercles (Figs. 16.8 C, 9 L) 11

10' Body often cylindrical without spicules or tubercles 12

11(10) Anterior spiracles simple, each with one to several sensile spiracular openings arrayed peripherally at apex of short tubular or conical projection; tentoropharygeal and hypopharyngeal sclerites finely constructed and fused together; body often dorsoventrally flattened and bearing a series of spicules or tubercles (Fig. 16.8 C); posterior spiracles with openings arranged in two pairs placed one behind the other (Fig. 16.8 D) .... Phoridae

11' Anterior spiracles short, bulbiform with four to five small bulbous papillae; tentoropharygeal and hypopharyngeal sclerites often more strongly constructed than previous and distinctly separated (Fig. 16.9 L) .... **Periscelididae** (Chapter 16.13, p. 785)

12(10) Body segments not covered with short, fine setae; if present, a respiratory siphon present, or each spiracle situated on a short tubular posterior projection (Figs. 16.9 C, E); cephalopharyngeal skeleton lacking a ventral arch 13

12' Body segments often extensively covered with short, fine setae; abdomen often somewhat tapered posteriorly, its apex with tubercles surrounding posterior spiracles that are only slightly elevated (Fig. 16.9 B); cephalopharyngeal skeleton with a sclerotized ventral arch (Fig. 16.9 A) .... Sciomyzidae

13(12) Posterior spiracular often well exposed, not localized on cavity; one or more of the spiracular slits nearly horizontal 14

13' Posterior spiracular often on lowest point in depression on terminal segment (Fig. 16.9 D); spiracular slits inclined subvertically .... Sarcophagidae

14(13) Abdomen rather truncate posteriorly, with short respiratory tube or basally fused conical spiracular prominences, each bearing a dorsal tubercle; integument with setae only on intersegmental areas (Fig. 16.9 I); setaceous tubercles, if present, restricted to posterior end 15

14' Abdomen somewhat tapered posteriorly, usually ending in an elongate retractile respiratory tube that lacks dorsal tubercles (Figs. 16.9 E–H); integument of posterior abdominal segments covered with setae or spinules, or with setaceous tubercles on some segments (Fig. 16.9 F) .... **Ephydridae** (Chapter 16.14, p. 789)

15(14) Posterior spiracular plate without lobes, spines, or ridges along margins (Figs. 16.9 I–J) .... Muscidae

15' Posterior spiracular plate with one or more bordering lobes or spines, or with a sharp ridge along dorsal margin (Fig. 16.9 K) .... Scathophagidae

REFERENCES

Alexander, C.P. & M.M. Alexander. 1970. Family Tipulidae. Pages 1–259 *in*: N. Papavero (ed.), A catalogue of the Diptera of the Americas South of the United States. Universidade de São Paulo, São Paulo, Museu de Zoologia.

Amorim, D.S. 2009a. Catalogue of Neotropical Diptera. Scatopsidae. Neotropical Diptera 4: 1–18.

Amorim, D.S. 2009b. Neotropical Diptera diversity: richness, patterns, and perspectives. Pages 71–97 *in*: T. Pape, D. Bickel & R. Meier (eds.), Diptera diversity: status, challenges and tools. Koninklijke Brill NV, Leiden.

Amorim, D.S. 2009c. Scatopsidae (Minute black scavenger flies). Pages 347–355 *in*: B.V. Brown, A. Borkent, J.M. Cumming, D.M. Wood, N.E. Woodley & M.A. Zumbado (eds.), Manual of Central American Diptera, Vol. 1. NRC Research Press, Ottawa.

Amorim, D.S. 2016. Family Scatopsidae. Pages 239–245 *in*: M. Wolff, S.S. Nihei & C.J.B. Carvalho (eds.), Catalogue of Diptera of Colombia. Zootaxa 4122.

Amorim, D.S., V.C. Silva & M.I.P.A. Balbi. 2002. Estado do conhecimento dos Diptera neotropicais. Pages 29–36 *in*: C. Costa, S.A.Vanin, J.M. Lobo & A. Melic (eds.), Proyecto de Red Iberoamericana de Biogeografia y Entomologia Sistemática PriBES. M3m Monografias Tercer Milênio, vol. 2. Sociedad Entomológica Aragonesa (SEA) & Cyted, Zaragoza.

Brindle, A. 1962. Taxonomic notes on the larvae of British Diptera, 9: the family Ptychopteridae. Entomologist 95: 212–216.

Capellari, R.S. 2016. Family Dolichopodidae. Pages 404–413 *in*: M. Wolff, S.S. Nihei & C.J.B. de Carvalho (eds.), Catalogue of Diptera of Colombia. Zootaxa 4122.

Carvalho, C.J.B. De. 1989. Classificação de Muscidae (Diptera): uma proposta através da análise cladística. Revista Brasileira de Zoologia 6: 627–648.

Carvalho, C.J.B. De & M.S. Couri. 2002. Basal groups. Part I. Pages 17–132 *in*: C.J.B. de Carvalho (ed.), Muscidae (Diptera) of the Neotropical Region: taxonomy. Editora Universidade Federal do Paraná, Curitiba.

Carvalho, C.J.B. De, M.S. Couri, A.C. Pont, D. Pamplona & S.M. Lopes. 2005. A Catalogue of the Muscidae (Diptera) of the Neotropical Region. Zootaxa 860: 1–282.

Carvalho, C.J.B. De, J.A. Rafael, M.S. Couri & V.C. Silva. 2012. Diptera Linnaeus, 1758. Pages 702–743 *in*: J.A. Rafael, G.A.R Melo, C.J.B Carvalho, S.A. Casari & R. Constantino (eds.), Insetos do Brasil. Diversidade e Taxonomia. Ribeirão Preto, Holos Editora.

Cook, E.F. 1967. Family Scatopsidae. Pages 1–9 *in*: N. Papavero (ed.), A catalogue of the Diptera of the Americas South of the United States. Universidade de São Paulo, São Paulo, Museu de Zoologia.

Couri, M.S. & C.J.B. de Carvalho. 2003. Systematic relations among *Philornis* Meinert, *Passeromyia* Rodhain & Villeneuve and allied genera (Diptera, Muscidae). Brazilian Journal of Biology 63: 223–232.

Courtney, G.W. & R.W. Merritt. 2008. Aquatic Diptera: Part One. Larvae of Aquatic Diptera. Pages 687–722 *in*: R.W. Merritt, M.B. Berg & K.W. Cummins (eds.), An Introduction to the Aquatic Insects of North America (4th ed.). Kendall/Hunt Publ. Co., Dubuque, IA.

Courtney, G.W. 2009. Blephariceridae (Net-winged midges). Pages 237–243 *in*: B.V. Brown, A. Borkent, J.M. Cumming, D.M. Wood, N.E. Woodley & M.A. Zumbado (eds.), Manual of Central American Diptera, Vol. 1. NRC Research Press, Ottawa.

Crosskey, R.W. 1973. Simuliidae (black-flies, German: Kriebehnücken). Pages 109–153 *in*: K.G.V. Smith (ed.), Insects and other arthropods of medical importance. British Museum (Natural History), London.

Disney, R.H.L. 2004. Insecta: Diptera, Phoridae. Pages 818–824 *in*: C.M. Yule & H.S. Yong (eds.), Freshwater Invertebrates of the Malaysian Region. Academy of Sciences of Malaysia, Kuala Lumpur.

Exner, K. & D.A. Craig. 1976. Larvae of Alberta Tanyderidae (Diptera: Nematocera). Quaestiones Entomologicae 12: 219–237.

Fasbender, A. & G.W. Courtney. 2017. A revision of Bittacomorphinae with a review of the monophyly of the extant subfamilies of Ptychopteridae (Diptera). Zootaxa 4309: 1–69.

Ferreira, R.L.M & Rafael, J.A. 2006. Criação de imaturos de mutuca (Tabanidae: Diptera) utilizando briófitas e areia como substrato. Neotropical Entomology 35: 141–144.

Ferreira, R.L.M., A.L. Henriques & J.A. Rafael. 2002. Activity of tabanids (Insecta, Diptera, Tabanidae) attacking the reptiles *Caiman crocodilus* (Linn.) (Alligatoridae) and *Eunectes murinus* (Linn.) (Boidae), in Central Amazon, Brazil. Memórias do Instituto Oswaldo Cruz 97: 133–136.

Gil-Azevedo, L.H. & D.S. Santos. 2016. Family Blephariceridae. Pages 182–186 *in*: M. Wolff, S.S. Nihei & C.J.B. Carvalho (eds.), Catalogue of Diptera of Colombia. Zootaxa 4122.

Guimarães, J.H. & D.S. Amorim. 2006. Capítulo 14. Diptera. Pages 147–160 *in*: C. Costa, S. Ide & C.E. Simonka (eds.), Insetos Imaturos Metamorfose e Identificação. Holos Editora. Ribeirão Preto, SP.

Grichanov, I. 2014. A checklist of species of the family Dolichopodidae (Diptera) of the World arranged by alphabetic list of generic names. Accessed: jun/2016: http://dolicho.narod.ru/Genera3.htm.

Hamada, N. 1993. Association between *Hemerodromia* sp. (Diptera, Empididae) and *Simulium perflavum* (Diptera, Simuliidae) in Central Amazonia, Brazil. Memórias do Instituto Oswaldo Cruz 88: 169–170.

Hancock, G.E., M. Marcos-Garcis & G.E. Rotheray. 2006. Ptychopteridae - a family of flies (Diptera) new to the Neotropical Region and description of a new species. Zootaxa 1351: 61–68.

Haseyama, K.L.F., B.M. Wiegmann, E.A.B. Almeida & C.J.B. de Carvalho. 2015. Say goodbye to tribes in the new house fly classification: A new molecular phylogenetic analysis and an updated biogeographical narrative for the Muscidae (Diptera). Molecular Phylogenetics and Evolution 89: 1–12.

Kerr, P.H. 2009. Rhagionidae. Pages 485–489 *in*: B.V. Brown, A. Borkent, J.M. Cumming, D.M. Wood, N.E. Woodley & M. Zumbado (eds.), Manual of Central American Diptera, Vol. 1. National Research Council Press, Ottawa.

Krzemiński, W. & D.D. Judd. 1997. Family Tanyderidae. Pages 281–289 *in*: L. Papp & B. Darvas (eds), Contributions to a Manual of Palaearctic Diptera (with special reference to flies of economic importance), Vol. 2, Nematocera and Lower Brachycera. Science Herald, Budapest.

Llanos, L., C.R. Gonzalez & M. Saldarriaga-Córdoba. 2015. Revision of the New World species of the genus *Pelecorhynchus* Macquart, 1850 (Diptera: Pelecorhynchidae). Zootaxa 3955: 188–210.

Mackerras, I.M. & M.E. Fuller. 1942. The genus *Pelecorhynchus* (Diptera, Tabanoidea). Proceedings of the Linnean Society of New South Wales 67: 9–76.

Madriz, R.I. & G.W. Courtney. 2016. The Neotropical tanyderid *Araucoderus gloriosus* (Alexander) (Diptera, Tanyderidae), with description of the egg, larva and pupa, redescription of adults, and notes on ecology. Zootaxa 4158: 325–351.

Nagatomi, A. 1982. The genera of Rhagionidae (Diptera). Journal of Natural History, 16: 31–70.

Oliveira, F.L. De, J.A. Rafael & A.L. Henriques. 2002. *Phorcotabanus cinereus* (Wiedemann, 1821) (Diptera, Tabanidae), an omithophilic species of tabanid in Central Amazon, Brazil. Memórias do Instituto Oswaldo Cruz 97: 839–842.

Oosterbroek, P. 2017. Catalogue of the Craneflies of the World. Accessed: dez/2017. http://ccw.naturalis.nl/index.php.

Oscoz, J., D. Galicia, & R. Miranda. 2011. Taxa Description and Biology. Pages 47–148 *in*: J. Oscoz, D. Galicia & R. Miranda (eds.), Identification Guide of Freshwater Macroinvertebrates of Spain. Springer, Dordrecht.

Papavero, N. 1967. Family Lonchopteridae. Page 1–2 *in*: N. Papavero (ed.), A catalogue of the Diptera of the Americas South of the United States. Universidade de São Paulo, São Paulo, Museu de Zoologia.

Pape, T., V. Blagoderov, & M.B. Mostovski. 2011. Order Diptera Linnaeus, 1758. Pages 222–229 *in*: Z.Q. Zhang (ed.), Animal biodiversity: An outline of higher-level classification and survey of taxonomic richnes, Zootaxa 3148.

Pape, T. & Thompson, F. C. 2013. Systema Dipterorum, Version 1.5. Accessed: dez/2017. http://www.diptera.org.

Pechuman, L.L. & H.J. Teskey. 1981. Tabanidae. Pages 464–478 *in*: J.F. McAlpine, B.V. Peterson, G.E. Shewell, H.J. Teskey, J.R. Vockeroth & D.M. Wood. (eds.), Manual of Nearctic Diptera, Vol 1, Research Branch, Agriculture Canada, Canada.

Pepinelli, M. & Pinho, L.C. 2014. Ordem Diptera. Pages 447–456 *in*: N. Hamada, J.L. Nessimian & R.B. Querino. (eds.), Insetos aquáticos na Amazônia brasileira: taxonomia, biologia e ecologia, 1ed. Editora do INPA, Manaus.

Pinho, L.C. 2008. Diptera *in*: C.G. Froehlich (ed.), Guia *on-line*: Identificação de larvas de Insetos Aquáticos do estado de São Paulo. Accessed: jun/2016 http://sites.ffclrp.usp.br/aguadoce/guiaonline.

Robinson, H. 1970. Family Dolichopodidae. Pages 1–92 *in*: N. Papavero (ed.), A catalogue of the Diptera of the Americas South of the United States. Universidade de São Paulo, São Paulo, Museu de Zoologia.

Rozkosny, R. & A. Nagatomi 1997. Family Athericidae. Pages 439–446 *in*: L. Papp & B. Darvas (eds.), Manual of Palaearctic Diptera, Vol. 2. Nematocera and lower Brachycera, Budapest, Science.

Santos, C.M.D. dos. 2008. Geographical distribution of Tabanomorpha (Diptera, Brachycera): Athericidae, Austroleptidae, Oreoleptidae, Rhagionidae, and Vermileonidae. EntomoBrasilis 1: 43–50.

Sinclair, B.J. & J.M. Cumming. 2006. The morphology, higher-level phylogeny and classification of the Empidoidea (Diptera). Zootaxa 1180: 1–172.

Skidmore, P. 1985. The biology of the Muscidae of the world. Series Entomologica 29: 1–550.

Smith, K.G.V. 1989. Family Lonchopteridae *in*: N. Evenhuis (ed.), Catalog of the Diptera of the Australasian and Oceanian Regions. Accessed: jun/2016 http://hbs.bishopmuseum.org/aocat/aocathome.html.

Teskey, H.J. 1981. Morphology and terminology Larvae. Pages 65–68 *in*: J.F. McAlpine, B.V. Peterson, G.E. Shewell, H.J. Teskey, J.R. Vocheroth & D.M. Wood (eds.), Manual of Nearctic Diptera, Vol. 1, Res. Branch, Agric. Can. Monogr. 27, Ottawa.

Trivinho-Strixino, S. 2011. Larvas de Chironomidae: guia de identificação. Universidade Federal de São Carlos/UFSCar, São Carlos, 371p.

Vaillant, F. & R. Rozkósný. 2002. Insecta: Diptera: Lonchopteridae, Sciomyzidae. Süßwasserfauna von Mitteleuropa [Freshwater Fauna of Europe], Band 21/22,23. Springer Spektrum. 136pp.

Vockeroth, J.R. 2010. Scathophagidae. Pages 1267–1269 *in*: B.V. Brown, A. Borkent, J.M. Cumming, D.M. Wood, N.E. Woodley & M.A. Zumbado (eds.), Manual of Central American Diptera, Vol. 2. NCR Research Press, Ottawa.

Wagner, R., M. Barták, A. Borkent, G. Courtney, B. Goddeeris, J.P. Haenni, L. Knutson, A. Pont, G.E. Rotheray, R. Rozkošný, B. Sinclair, N. Woodley, T. Zatwarnicki & P. Zwick. 2008. Global diversity of dipteran families (Insecta Diptera) in freshwater (excluding Simulidae, Culicidae, Chironomidae, Tipulidae and Tabanidae). Pages 489–519 *in*: E. Balian, K. Martens, C. Lévêque & H. Segers (eds.), A Global Assessment of Animal Diversity in Freshwater. Hydrobiologia.

Wolff, M., D. Grisales & S.S. Nihei. 2016. Family Scathophagidae. Pages 854–855 *in*: M. Wolff, S.S. Nihei & C.J.B. Carvalho (eds.), Catalogue of Diptera of Colombia. Zootaxa 4122.

Woodley, N.E. 2009. Athericidae (Athericid flies). Pages 491–493 *in*: B.V. Brown, A. Borkent, J.M. Cumming, D.M. Wood, N.E. Woodley & M.A. Zumbado (eds.), Manual of Central American Diptera, Vol. 1. NRC Research Press, Ottawa.

Yang, D., Y. Zhu, M. Wang & L. Zhang. 2006. World Catalog of Dolichopodidae (Insecta: Diptera). China Agricultural University Press. Beijing, 740 p.

Chapter 16.1

Family Ceratopogonidae

Maria Marcela Ronderos
Centro de Estudios Parasitológicos y de Vectores (CEPAVE), Facultad de Ciencias Naturales y Museo (UNLP), Consejo Nacional de Investigaciones Científicas y Técnicas (CONICET), Buenos Aires, Argentina, UNLP, FCNYM, División Entomología, Museo de La Plata, La Plata, Buenos Aires, Argentina

Florentina Díaz
Centro de Estudios Parasitológicos y de Vectores (CEPAVE), Facultad de Ciencias Naturales y Museo (UNLP), Consejo Nacional de Investigaciones Científicas y Técnicas (CONICET), Buenos Aires, Argentina, UNLP, FCNYM, División Entomología, Museo de La Plata, La Plata, Buenos Aires, Argentina

Pablo Ignacio Marino
UNLP, FCNYM, División Entomología, Museo de La Plata, La Plata, Buenos Aires, Argentina

Ruth Leila Ferreira-Keppler
Coordenação de Biodiversidade, Instituto Nacional de Pesquisas da Amazônia, Manaus, Amazonas, Brazil

Chapter Outline

INTRODUCTION

Ceratopogonidae is placed in the infraorder Culicomorpha (Borkent, 2012). Their common names are "polvorines, manta blanca, chaquistes, or jejenes" (common name shared with Simuliidae) in Spanish-speaking countries; "mosquito pólvora or maruim" in Brazil; and "biting midges, no-see-ums, or punkies" in English-speaking countries (Spinelli & Ronderos, 2005).

It is an ancestral group, with numerous fossil records in amber from 17 to around 121 million years old (Borkent & Spinelli, 2007). It is one of the most diverse families of Diptera, occurring on all continents, except Antarctica, and is found in most habitats, including deserts. They are present from the tropics to the high Arctic and from lowlands to high mountains (4200 m on Mount Everest, Nepal) (Borkent, 2014). Despite their great diversity and their nearly ubiquitous presence, the immatures of Ceratopogonidae are still the least known among the biting Diptera.

The family includes six subfamilies, four of which are extant and worldwide in distribution, with 112 extant genera and 6268 species. In addition, the fossil record includes another 284 extinct species and 21 extinct genera (Borkent, 2016). Four subfamilies inhabit the Neotropical Region: Leptoconopinae, Forcipomyiinae, Dasyheleinae, and Ceratopogoninae (in seven tribes: Culicoidini Kieffer, Ceratopogonini, Heteromyiini Wirth, Johannsenomyiini Crampton, Sphaeromiini Newman, Palpomyiini Enderlein, and Stenoxenini Coquillett), including more than 1226 species in 53 genera, of which 16 are cosmopolitan (Borkent, 2016; Santarém & Felippe-Bauer, 2016).

Members of this family are holometabolous, with the immature stages inhabiting aquatic, semiaquatic, or

Thorp and Covich's Freshwater Invertebrates, Fourth Edition. https://doi.org/10.1016/B978-0-12-804223-6.00030-5

FIGURE 16.1.1 Environments inhabited by immatures of Ceratopogonidae: (A) river; (B) stream; (C) lakes, rich in organic matter; (D) lake with macrophytes; (E–H) phytotelmata.

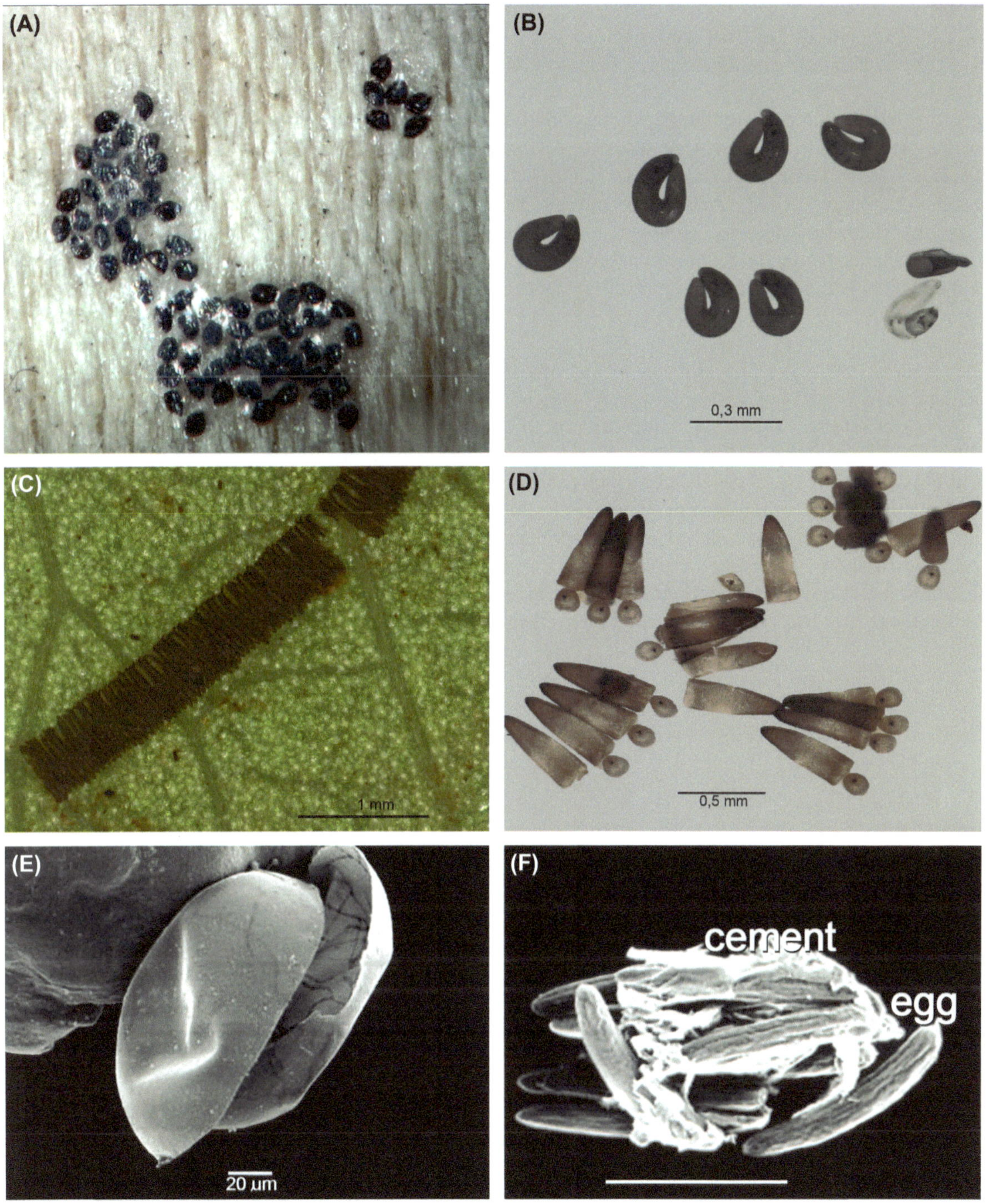

FIGURE 16.1.2 Ceratopogonidae eggs obtained from natural and artificial breeding sites: (A–B, E) subfamily Dasyheleinae; (C, F) subfamily Ceratopogoninae, tribe Ceratopogonini, and (D) tribe Palpomyiini.

terrestrial habitats. Females usually lay their eggs on or in organically rich aquatic or wet habitats, where larvae complete their development and pupate. Oviposition microhabitats are very diverse (Figs. 16.1.1 A–H). Egg production varies from 10 to 600, depending on the genera and species (Díaz et al., 2005). Egg clutches can be laid in strings or masses that are cemented with gelatinous substances (Figs. 16.1.2 A–F), depending on subfamily (Díaz et al., 2005; Ronderos et al., 2006). In Forcipomyiinae, eggs are oval to elongate, black or dark brown, and glossy; in Dasyheleinae (Figs. 16.1.2 A, B, E), they are horseshoe or C-shaped and enclosed in a gelatinous film. In Ceratopogoninae (Figs. 16.1.2 C, D, F), eggs are laid in loose groups on moist substrate and are slender, banana shaped, with a thin or often sculptured chorion, e.g., Culicoidini (Fig. 16.1.2 F), or in strings or masses with gelatinous coatings, e.g., Palpomyiini (Fig. 16.1.2 D).

Aquatic or semiaquatic larvae inhabit the superficial layer of the breeding site, usually at depths of 0–12 cm (Uslu & Dick, 2010; Gonzalez Gonzalez de Heredia & Goldarazena Lafuente, 2011). Active larvae can be recognized by their slow to very rapid serpentine. At the end of the final instar, larvae develop pupal respiratory organs and accumulate a large amount of body fat in the abdomen.

Ceratopogonidae larvae have diverse feeding habits—filter feeders, detritivores, or carnivores—and others feed on phytoplankton or small debris; however, many are omnivores (McCafferty, 1981). A summary of the morphology and food items of Ceratopogonidae is presented in Figs. 16.1.3 A–G, 5 A–H. From a morphobehavioral perspective, this variation is most evident in the size and shape of the head capsule, mandibles, epipharynx, and hypostoma.

Regarding larval locomotion, ceratopogonid larvae can be grouped in two categories based on their locomotory abilities: nonswimming (crawling) or swimming. Larvae of this family are unique among dipterous larvae in their adaptations for swimming. The morphological characteristics of the caudal segment are related to the environment in which they live (Figs. 16.1.8 A–G).

During the pupal stage, most aquatic species float on the surface of water where they rest and respire, but if disturbed, they may thrash about and/or sink. There are many questions about whether pupal character states are useful for differentiating closely related genera and species, as well as if they possess morphologic characters of group and intergroup relationships. Their respiratory organs and terminal processes of segment 9 are often unique to genus and species. Fox (1942) studied the correlation between some morphologic structures of the pupae of *Culicoides* and suggested that variation in the respiratory organ and segment 9 are related to environment.

Differences between genera and species could also be due to the different environments where they develop. Borkent & Craig (2001) noted associations between breeding sites and respiratory organs, length of body setae, and terminal processes of the pupa of

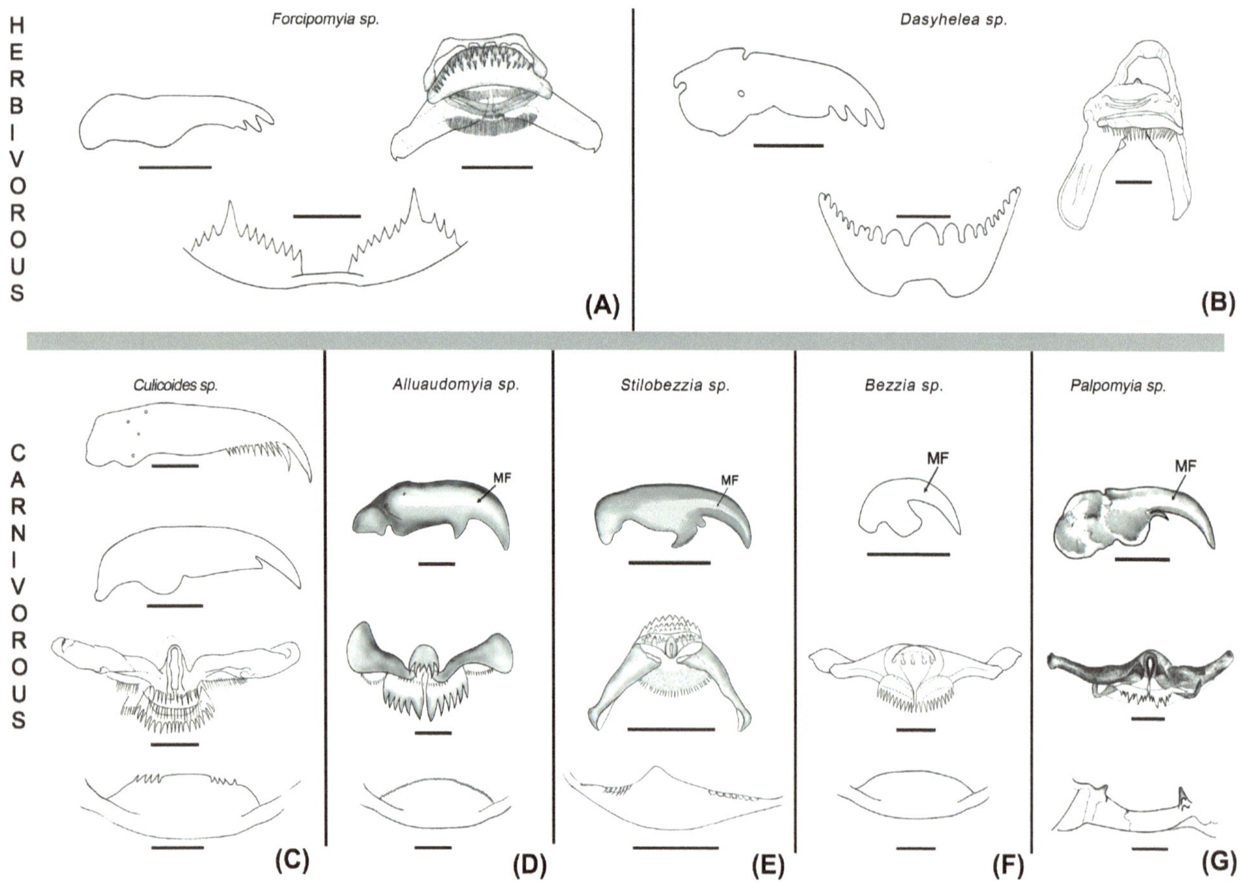

FIGURE 16.1.3 Relationships of Ceratopogonidae larval mouthparts and food (mandible, epipharynx, and hypostoma): (A–B) herbivorous; (C–G) carnivorous. Scale bar: 0.05 mm.

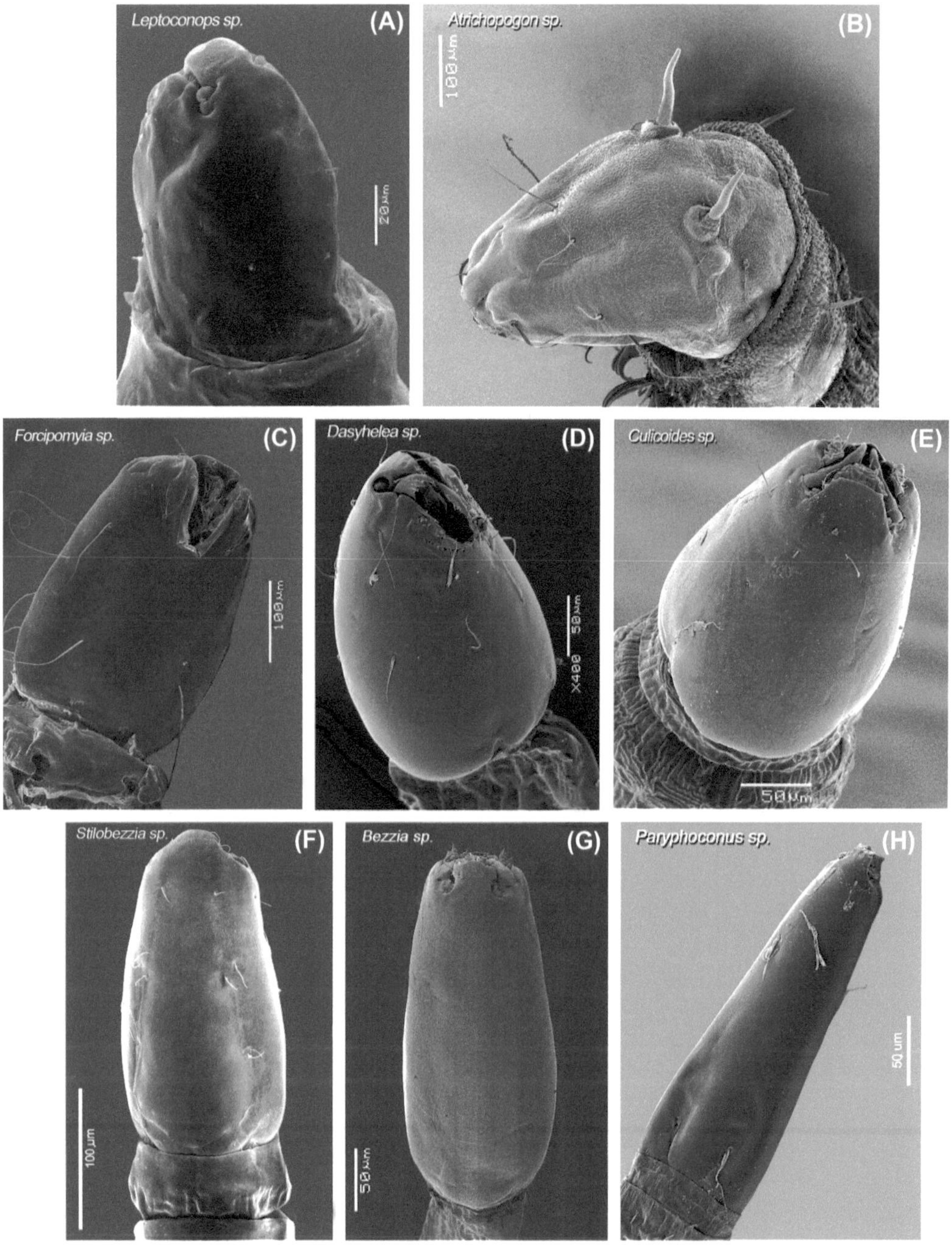

FIGURE 16.1.4 Head capsule of Ceratopogonidae larvae (scanning electron microscope): (A) subfamily Leptoconopinae; (B–C) subfamily Forcipomyiinae; (D) subfamily Dasyheleinae; (E–H) subfamily Ceratopogoninae; (E) tribe Culicoidini; (F) tribe Ceratopogonini; (G) tribe Palpomyiini; (H) tribe Stenoxenini.

Stilobezzia rabelloi Lane, which was later confirmed for several other ceratopogonids, including *S. punctulata* Lane, *Dasyhelea paulistana* Forattini & Rabello, *D. pseudopollinosa* Díaz & Ronderos, and *Culicoides lacustris* Ronderos. To obtain air, the pupae of these five species insert the pointed respiratory organ into the floating or dangling submerged leaves of aquatic plants (e.g., *Salvinia auriculata* Aubl., Salvinaceae, *Azolla filiculoides* Lam., Azollaceae).

After emergence, adults (teneral) are pale, poorly sclerotized, and require about 24 h to achieve full coloration. Males typically emerge earlier than females, and they are fertile about 24 h after emergence (Gonzalez Gonzalez de Heredia & Goldarazena Lafuente, 2011).

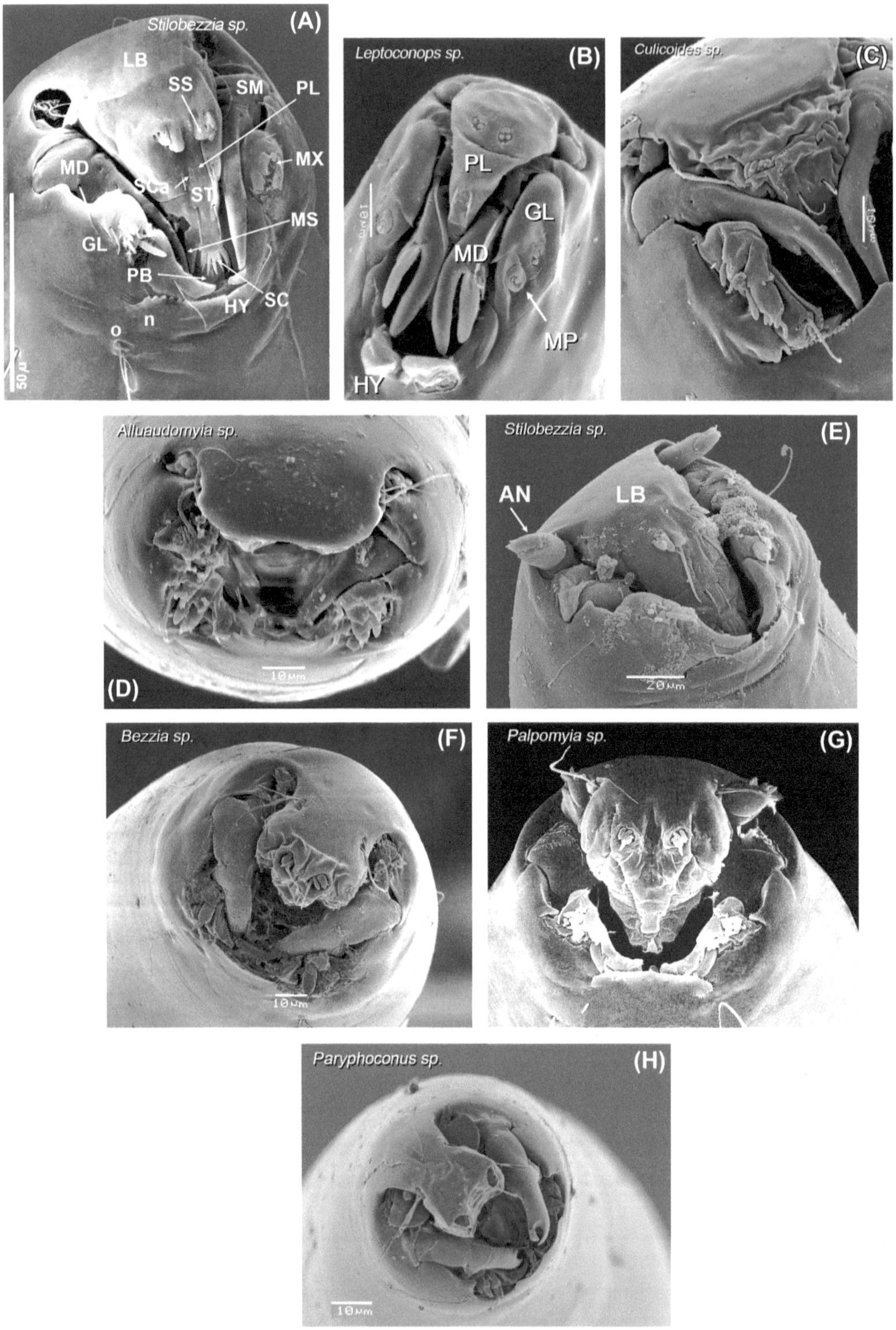

FIGURE 16.1.5 Details of head capsule of Ceratopogonidae larvae (frontal view) (scanning electron microscope): (A) *Stilobezzia* sp.; (B) *Leptoconops* sp.; (C) *Culicoides* sp.; (D) *Alluaudomyia* sp.; (E) *Stilobezzia* sp.; (F) *Bezzia* sp.; (G) *Palpomyia* sp.; (H) *Paryphoconus* sp.

FIGURE 16.1.6 Ceratopogonidae larvae habitus: (A) *Leptoconops* sp.; (B) *Forcipomyia* sp.; (C) *Atrichopogon* sp.; (D) *Dasyhelea* sp.; (E) *Culicoides* sp.; (F) *Bezzia* sp.; (G) *Palpomyia* sp.; (H) *Paryphoconus* sp.

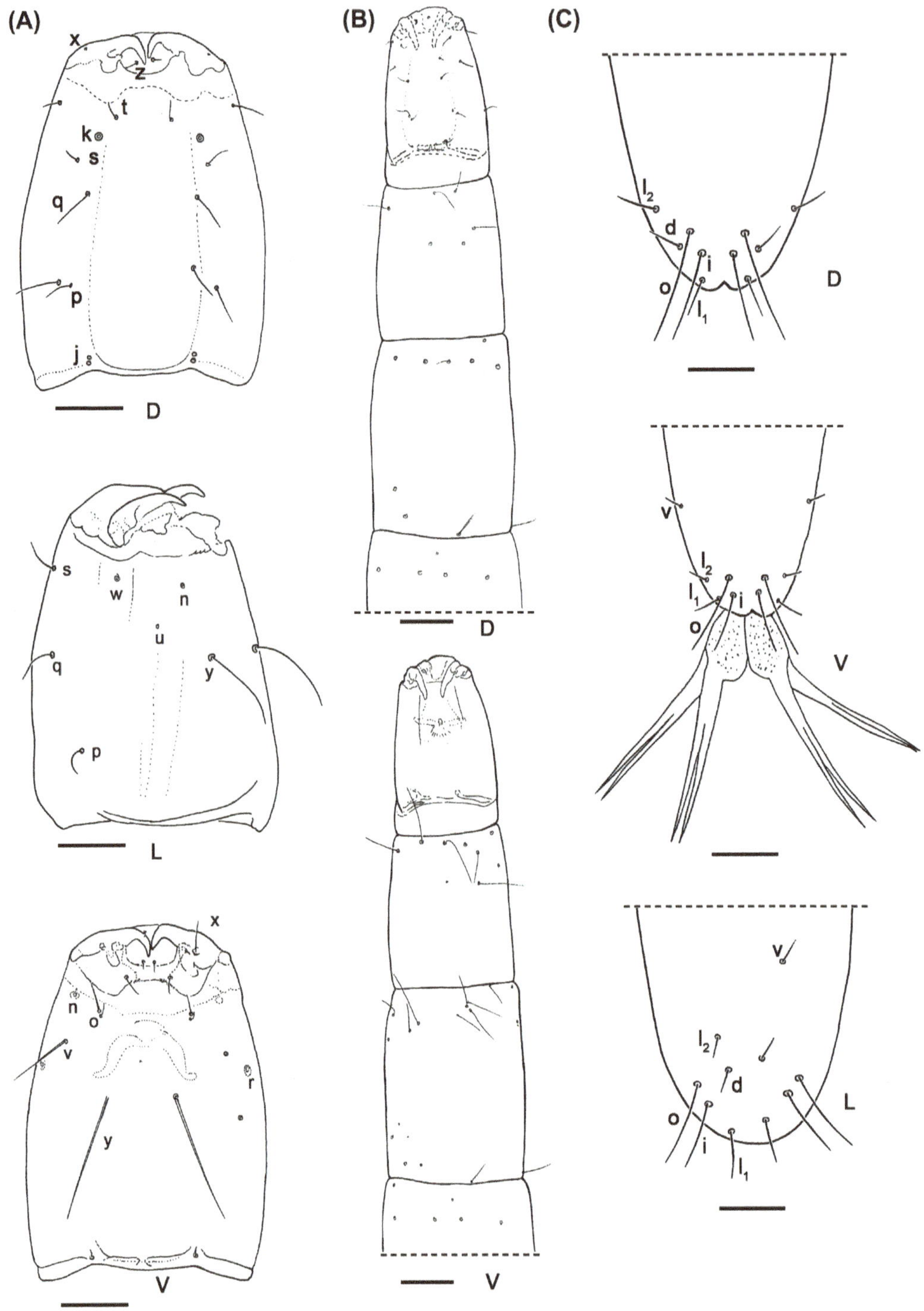

FIGURE 16.1.7 Ceratopogonidae larval chaetotaxy: (A) head capsule (dorsal, lateral, and ventral view); (B) thorax (dorsal and ventral view); (C) caudal segment. Scale bar: 0.05 mm.

LIMITATIONS

Our knowledge of Ceratopogonidae immatures of the Neotropical Region is incipient. Many species from this region are known only from adults, and most pupal descriptions are old and/or incomplete. The approximately 1225 Neotropical species are distributed in 52 genera, but only 185 species have eggs, larvae, or pupae described (Table 16.1.1). The numbers for the known genera are even worse: only eight genera have their eggs described, 11 have their larva identified, and 23 genera have their

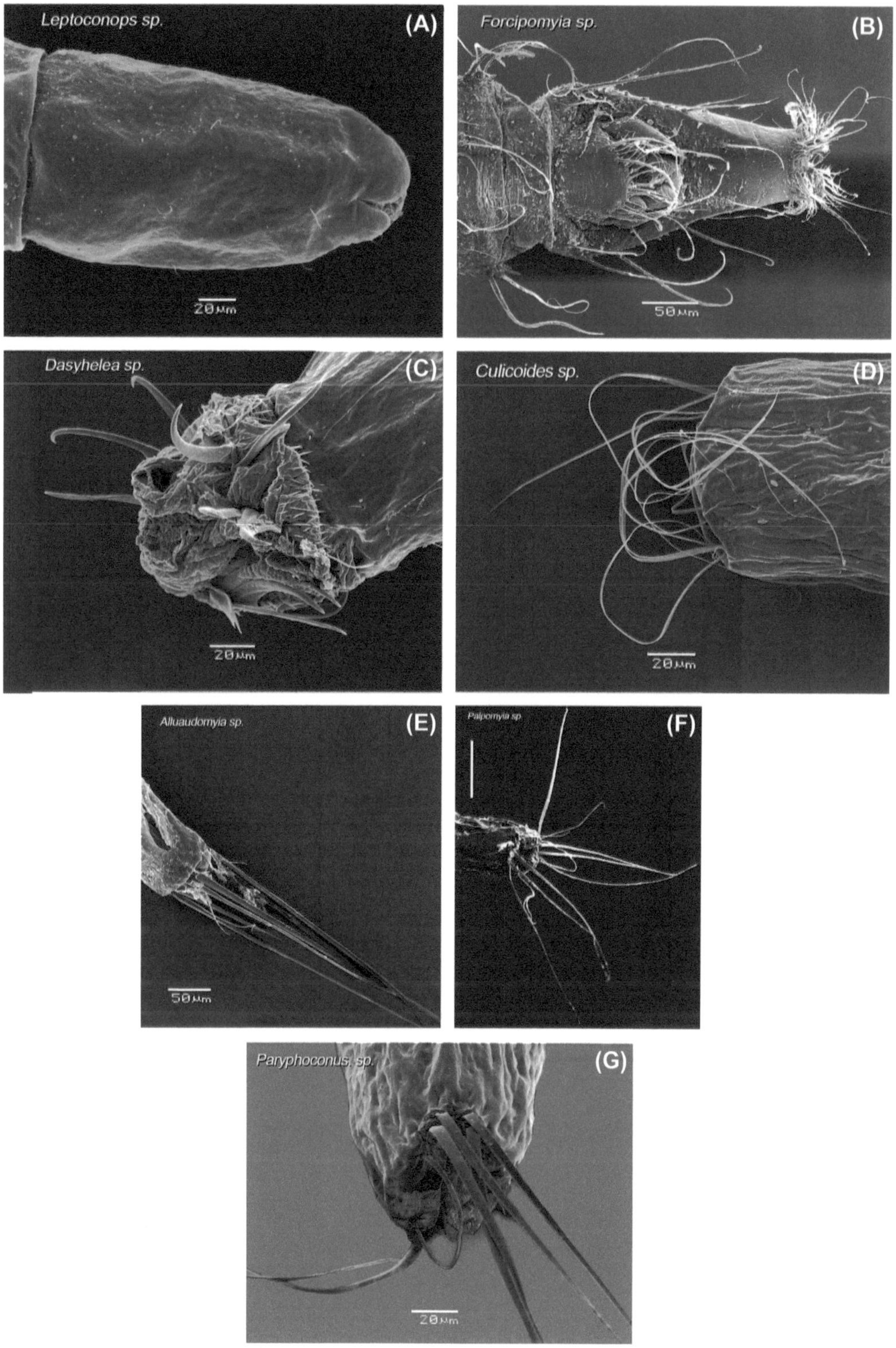

FIGURE 16.1.8 Detail of caudal segment of Ceratopogonidae larvae (scanning electron microscope): (A) subfamily Leptoconopinae; (B) subfamily Forcipomyiinae; (C) subfamily Dasyheleinae; subfamily Ceratopogoninae; (D) tribe Culicoidini; (E) tribe Ceratopogonini; (F) tribe Palpomyiini; (G) tribe Stenoxenini.

TABLE 16.1.1 Neotropical Ceratopogonidae Species Described as Eggs, Larvae, and/or Pupae.

Subfamily	Tribe	Genus	Subgenus	No	Species	Reference	Developmental Stage		
LEPTOCONOPINAE		*LEPTOCONOPS*	*HOLOCONOPS*		***bequaerti***	Painter, 1927 (as Leptoconops sp.)		larva	pupa
						Mayer, 1934			pupa
						Linley, 1965a	eggs		
						Linley, 1968		larva	
FORCIPOMYIINAE		*ATRICHOPOGON*	*ATRICHOPOGON*		***bifidus***	Ewen, in Ewen & Saunders, 1958		larva	pupa
					caribbeanus	Ewen, in Ewen & Saunders, 1958		larva	pupa
					incultus	Ewen, in Ewen & Saunders, 1958		larva	pupa
					longicornis	Ewen, in Ewen & Saunders, 1958		larva	
					obscurus	Ewen, in Ewen & Saunders, 1958		larva	pupa
					remigatus	Ewen, in Ewen & Saunders, 1958		larva	pupa
					saundersi	Ewen, in Ewen & Saunders, 1958		larva	pupa
					tuberculatus	Ewen, in Ewen & Saunders, 1958		larva	pupa
			LOPHOMYIDIUM		***delpontei***	Marino et al., 2017		larva	pupa
					fusculus	Malloch, 1915		larva	pupa
						Thomsen, 1937		larva	pupa
						Nielsen, 1951 (as polydactilus)		larva	pupa
						Ewen & Saunders, 1958 (as polydactilus)		larva	pupa
						Downes & Wirth, 1981 (as polydactilus)		larva	
						Wirth, 1994a		larva	pupa
		FORCIPOMYIA	*CALOFORCIPOMYIA*		***caerulea***	Saunders, 1957		larva	pupa

		glauca	Krivosheina, 1968 (as monilis)	larva	pupa
		varicolor	Saunders, 1957	larva	pupa
	EUPROJOANNISIA	***blantoni***	Soria & Bystrak, 1975	larva	pupa
			Bystrak & Wirth, 1978	larva	pupa
		bromeliae	Saunders, 1957	larva	pupa
		calcarata	Wirth, 1952a	larva	pupa
		falcifera	Saunders, 1959	larva	pupa
		longispina	Saunders, 1957	larva	pupa
		mortuifolii	Saunders, 1959	larva	pupa
			Bystrak & Wirth, 1978	larva	pupa
		setigera	Saunders, 1959	larva	pupa
		spatulifera	Saunders, 1957	larva	pupa
	FORCIPOMYIA	***argenteola***	Lane, 1947	larva	pupa
		calatheae	Wirth, 1982	larva	pupa
		genualis	Saunders, 1957 (as raleighi)	larva	pupa
			Gutsevich & Glukhova, 1970 (as raleighi)	larva	pupa
		harpegonata	Wirth & Soria, 1975	larva	pupa
		pictoni	Wirth, 1976	larva	pupa
		quatei	Wirth, 1982	larva	pupa
		rioplatensis	Spinelli et al., 2005	larva	pupa
		squamitibia	Wirth, 1991 (as soriai)	larva	pupa
		youngi	Wirth, 1982	larva	pupa
	LASIOHELEA	***attenuata***	Saunders, 1964	larva	pupa
		cornuta	Saunders, 1964	larva	pupa
			Krivosheina, 1968	larva	
		intermedia	Saunders, 1964	larva	pupa
		quasicornuta	Saunders, 1964	larva	pupa
		stylifer	Lane, 1947	larva	pupa
	LEPIDOHELEA	***brasiliensis***	Wirth & Spinelli, 1993a		pupa

(Continued)

TABLE 16.1.1 Neotropical Ceratopogonidae Species Described as Eggs, Larvae, and/or Pupae.—cont'd

Subfamily	Tribe	Genus	Subgenus	No	Species	Reference	Developmental Stage	
					edmistoni	Wirth & Spinelli, 1993b	larva	pupa
					herediae	Wirth & Spinelli, 1993a	larva	pupa
					luteigenua	Wirth & Spinelli, 1992	larva	pupa
					seminole	Wirth & Spinelli, 1992	larva	pupa
			METAFORCIPOMYIA		***cerifera***	Saunders, 1957	larva	pupa
			MICROHELEA		***fuliginosa***	Ingram & Macfie, 1924 (as inornatipennis)	larva	
						Lane, 1947 (as inornatipennis spp. ornaticrus)	larva	pupa
			PHYTOHELEA		***antiguensis***	Saunders, 1957	larva	pupa
					bilobata	Marino, in Marino et al., 2013a		pupa
					bromelicola	Saunders, 1925 (as an Apelma)	larva	pupa
						Marino et al., 2010	larva	pupa
					caribbeana	Saunders, 1957	larva	pupa
					dominicana	de Meillon & Wirth, 1979	larva	pupa
					edwardsi	(Saunders), 1925 (as an Apelma)	larva	pupa
					jocosa	Saunders, 1957	larva	pupa
					keilini	(Saunders), 1925 (as an Apelma)	larva	pupa
					magna	(Saunders), 1925 (as an Apelma)	larva	pupa
					musae	Clastrier & Delécolle, 1994		pupa
						Spinelli et al., 2007	larva	pupa
					oligarthra	Saunders, 1957	larva	pupa
						Tokunaga, 1961	larva	pupa
			THYRIDOMYIA		***nodosa***	Saunders, 1959	larva	pupa
			TRICHOHELEA		***trinidadensis***	Saunders, 1964		pupa
			WARMKEA		***aeria***	Saunders, 1957	larva	pupa
						Wirth & Soria, 1980	larva	pupa

Subfamily	Genus	Species	Reference			
		lesliei	Saunders, 1957 (as bicolor)		larva	pupa
			Wirth & Soria, 1980		larva	pupa
		spinosa	Saunders, 1957		larva	pupa
			Wirth & Soria, 1980		larva	pupa
		terrestris	Saunders, 1964		larva	pupa
			Wirth & Soria, 1980		larva	pupa
		tuberculata	Saunders, 1957		larva	pupa
			Wirth & Soria, 1980		larva	pupa
DASYHELEINAE	***DASYHELEA***	***allegrae***	Díaz, in Díaz et al., 2018a			pupa
		atlantis	Waugh & Wirth, 1976			pupa
			Borkent, 1991			pupa
		azteca	Díaz et al., 2018a		larva	pupa
		bahamensis	Ronderos et al., 2003a		larva	pupa
		calvescens	Williams, 1944	egg	larva	pupa
		cincta	Spinelli, 1983b (as penthesileae)			pupa
			Díaz et al., 2009			pupa
		correntina	Ronderos & Díaz, in Ronderos et al., 2004a			pupa
		eloyi	Díaz & Ronderos, in Díaz et al., 2013		larva	pupa
		filibranchia	(Lutz), 1914 (as a Ceratopogon)			pupa
		flavicauda	Spinelli & Ronderos, 1987			pupa
		mediomunda	Minaya, 1978			pupa
			Díaz et al., 2009			pupa
			Díaz et al., 2016		larva	pupa
		mutabilis	Thomsen, 1937		larva	pupa
			Wirth, 1952a			pupa
			Waugh & Wirth, 1976			pupa
		necrophila	Spinelli & Rodriguez, 1999		larva	pupa
			Ronderos et al., 2003a		larva	pupa
			Ronderos et al., 2006	egg	larva	

(Continued)

TABLE 16.1.1 Neotropical Ceratopogonidae Species Described as Eggs, Larvae, and/or Pupae.—cont'd

Subfamily	Tribe	Genus	Subgenus	No	Species	Reference	Developmental Stage		
						Ronderos et al., 2008b	egg		
						Martinez et al., 2010	egg		
					paracincta	Borkent, 1991		larva	pupa
					paulistana	Forattini & Rabello, 1957			pupa
						Díaz et al., 2014b			pupa
					pseudolacustris	Spinelli & Ronderos, 1987 (as lacustris)			pupa
						Díaz & Spinelli, in Díaz et al., 2014a			pupa
					pseudopollinosa	Díaz & Ronderos, in Díaz et al., 2014b			pupa
					pusilla	Forattini & Rabello, 1957			pupa
					serrana	Díaz & Spinelli, in Díaz et al., 2014a			pupa
					yunga	Díaz et al., 2018a			pupa
	CULICOIDINI	***CULICOIDES***	***AMOSSOVIA***		***oklahomensis***	Kann, 1980		larva	pupa
CERATOPOGONINAE						Murphree & Mullen, 1991		larva	
						Lamberson et al., 1992			pupa
			ANILOMYIA		***chaverrii***	Spinelli & Borkent, 2004		larva	pupa
					hayesi	Matta, 1967		larva	pupa
			BELTRANMYIA		***bermudensis***	Jamnback, 1965		larva y pupa	pupa
						Blanton & Wirth, 1979		larva	pupa
						Murphree & Mullen, 1991		larva	
					crepuscularis	Thomsen, 1937			pupa
						Fox, 1942			pupa
						Jones, 1955		larva	pupa
						Jamnback, 1965		larva	pupa
						Blanton & Wirth, 1979		larva pupa	pupa

			Murphree & Mullen, 1991	larva	
	COTOCRIPUS	***bambusicola***	Lutz, 1913	larva	
			Fox, 1942 (as sp. 2)		pupa
			Lane, 1947	larva	pupa
			Barbosa, 1952, 1953	larva	pupa
			Forattini, 1957		pupa
			Ronderos & Spinelli, 2000	larva	pupa
			Ronderos et al., 2000	larva	pupa
			Marino et al., 2013b	larva	pupa
	CULICOIDES	***luteovenus***	Wirth, 1952a	larva	pupa
		neopulicaris	Jones, 1961		pupa
	DIPHAOMYIA	***edeni***	Jones, 1961 (as possible haematopotus)		pupa
			Blanton & Wirth, 1979		pupa
		haematopotus	Thomsen, 1937		pupa
			Jones, 1955		pupa
			Jones, 1961		pupa
			Jamnback, 1965	larva	pupa
			Atchley & Wirth, 1979	larva	pupa
			Blanton & Wirth, 1979	larva	pupa
			Hribar & Mullen, 1991	larva	
			Murphree & Mullen, 1991	larva	
			Borkent, 2012		pupa
	DRYMODESMYIA	***borinqueni***	Linley, 1965b		pupa
	HAEMATOMYIDIUM	***annuliductus***	Wirth, in Vitale et al., 1981		pupa
		bayano	Wirth, in Vitale et al., 1981		pupa
		crucifer	Ronderos et al., 2013		pupa
		debilipalpis	Barbosa, 1952, 1953		pupa
			Forattini, 1957		pupa
			Ronderos et al., 2010	larva	pupa

(Continued)

TABLE 16.1.1 Neotropical Ceratopogonidae Species Described as Eggs, Larvae, and/or Pupae.—cont'd

Subfamily	Tribe	Genus	Subgenus	No	Species	Reference	Developmental Stage		
					filiductus	Wirth, in Vitale et al., 1981			pupa
					hoffmani	Linley & Kettle, 1964		larva	pupa
					lahillei	Lamberson et al., 1992			pupa
					paraensis	Murphree & Mullen, 1991		larva	
						Lamberson et al., 1992			pupa
			HOFFMANIA		***annettae***	Spinelli & Borkent, 2004			pupa
					charruus	Spinelli et al., 1993			pupa
						Ronderos et al., 2008a		larva	pupa
					heliconiae	Wirth & Blanton, 1968			pupa
					hylas	Fox, 1942 (as sp. 3)			pupa
						Forattini, 1957			pupa
					insignis	Lutz, 1913			pupa
						Forattini et al., 1956		larva	pupa
						Forattini, 1957		larva	pupa
						Linley, 1965b			pupa
						Blanton & Wirth, 1979			pupa
						Spinelli et al., 1993			pupa
						Marino et al., 2013b			pupa
					maruim	Lutz, 1913	egg		pupa
						Barbosa, 1952, 1953			pupa
						Forattini, 1957			pupa
						Spinelli et al., 1993			pupa
			MACFIELLA		***phlebotomus***	Painter, 1927		larva	pupa
						Fox, 1942			pupa
						Forattini, 1957			pupa
			MATAEMYIA		***dicrourus***	Wirth & Soria, 1981		larva	pupa
					felippebauerae	Spinelli, in Spinelli et al., 2007		larva	pupa

	macieli	Wirth & Soria, 1981		larva	pupa
OECACTA	***barbosai***	Linley, 1965b			pupa
		Linley & Davies, 1971	egg		
		Blanton & Wirth, 1979			pupa
		Murphree & Mullen, 1991		larva	
	cancer	Hogue & Wirth, 1968		larva	pupa
	furens	Painter, 1927		larva	pupa
		Dove et al., 1932 (as dovei)	egg	larva	pupa
		Fox, 1942			pupa
		Wirth, 1952b		larva	pupa
		Forattini, 1957		larva	pupa
		Jones, 1961			pupa
		Linley & Kettle, 1964		larva	pupa
		Jamnback, 1965		larva	pupa
		Linley & Davies, 1971	egg		
		Blanton & Wirth, 1979		larva	pupa
		Hribar & Mullen, 1991		larva	
		Murphree & Mullen, 1991		larva	
PSYCHOPHAENA	***lacustris***	Anjos-Santos et al., 2014		larva	pupa
	venezuelensis	Ronderos & Spinelli, 1989			pupa
		Díaz et al., 2005	egg	larva	pupa
		Ronderos et al. 2008b	egg		
unplaced, eublepharus species group	***guarani***	Ayala et al., 2015			pupa
unplaced, reticulatus species group	***guyanensis***	Forattini & Rabello, 1956		larva	pupa
		Forattini, 1957		larva	pupa
	reticulatus	Lutz, 1913		larva	
		Forattini et al., 1960		larva	pupa
unplaced, stonei species group	***melleus***	Wirth, 1952b		larva	pupa

(Continued)

TABLE 16.1.1 Neotropical Ceratopogonidae Species Described as Eggs, Larvae, and/or Pupae.—cont'd

Subfamily	Tribe	Genus	Subgenus	No	Species	Reference	Developmental Stage		
						Jamnback et al., 1958		larva	pupa
						Jones, 1961			pupa
						Jamback, 1965		larva	pupa
						Blanton & Wirth, 1979		larva	pupa
						Hribar & Mullen, 1991		larva	
						Murphree & Mullen, 1991		larva	
			miscellaneous species	**unplaced**	***arubae***	Fox, 1942 (as sp. 1)			pupa
						Forattini, 1957			pupa
						Murphree & Mullen, 1991		larva	
					floridensis	Linley, 1970			pupa
						Blanton & Wirth, 1979			pupa
					trilineatus	Díaz et al., 2018b		larva	pupa
	CERATOPOGONINI	***ALLUAUDOMYIA***			***amazonica***	Díaz et al., 2015		larva	pupa
					bella	Thomsen, 1937 (as splendida)		larva	pupa
						Wirth & Grogan, 1981		larva	pupa
					biestroi	Spinelli, 1988			pupa
					caribbeana	Spinelli & Wirth, 1984			pupa
					distispinulosa	Spinelli, 1997			pupa
					schnacki	Spinelli, 1983b		larva	pupa
						Díaz et al., 2015		larva	pupa
		BAEODASYMYIA			***christopheri***	Borkent, in Borkent & Craig, 1999	egg	larva	pupa
					michaeli	Borkent, in Borkent & Craig, 1999		larva	pupa
		ECHINOHELEA	***ECHINOHELEA***		***lanei***	Wirth, 1994b			pupa
		PARABEZZIA			***alexanderi***	Grogan & Wirth, 1977			pupa
						Wirth & Grogan, 1981			pupa

			Spinelli & Grogan, 1987			pupa
			Spinelli et al., 2012a			pupa
		balseiroi	Spinelli & Grogan, 1987			pupa
			Spinelli et al., 2012a			pupa
		brasiliensis	Ayala et al., 2015			pupa
SCHIZONYXHELEA		***forattinii***	Borkent, 2000		larva	pupa
STILOBEZZIA	***ACANTHOHELEA***	***megatheca***	Anjos-Santos et al., 2017		larva	pupa
	EUKRAIOHELEA	***dorsofasciata***	(Lutz), 1914 (as Palpomyia)	egg		
	STILOBEZZIA	***antennalis***	Thomsen, 1937		larva	pupa
			Wirth & Grogan, 1981		larva	pupa
			Hribar & Mullen, 1991		larva	
		bicolor	Forattini & Rabello, 1956			pupa
		chaconi	Lane & Forattini, 1961			pupa
		coquilletti	Lane & Forattini, 1961			pupa
			Wirth & Grogan, 1981			pupa
			Ronderos et al., 2008c		larva	pupa
		dubitans	Lane et al., 1955			pupa
		enigma	Ronderos et al., 2012a		larva	pupa
		fiebrigi	Forattini & Rabello, 1956		larva	pupa
			Cazorla et al., 2006			pupa
			Cazorla et al., 2013		larva	
		glauca	Lane et al., 1955			pupa
			Wirth & Grogan, 1981			pupa
		merceri	Cazorla & Spinelli, in Cazorla et al., 2005			pupa
		panamensis	Forattini et al., 1960			pupa
		pseudopunctulata	Cazorla & Ronderos, in Cazorla et al., 2012			pupa
		punctulata	Cazorla & Marino, 2004			pupa
			Cazorla et al., 2013		larva	
		rabelloi	Borkent & Craig, 2001			pupa

(*Continued*)

TABLE 16.1.1 Neotropical Ceratopogonidae Species Described as Eggs, Larvae, and/or Pupae.—cont'd

Subfamily	Tribe	Genus	Subgenus	No	Species	Reference	Developmental Stage	
					wygodzinskyi	Lane et al., 1955		pupa
	HETEROMYIINI	***CLINOHELEA***	***CLINOHELEA***		***horacioi***	Spinelli & Duret, 1993		pupa
					nigripes	Spinelli & Duret, 1993		pupa
		HETEROMYIA			***chaquensis***	Ronderos et al., 2017a		pupa
					clavata	Wirth & Grogan, 1977		pupa
						Ronderos et al., 2017a		pupa
					correntina	Ronderos et al., 2017a		pupa
					orellana	Ronderos et al., 2017a		pupa
					wokei	Wirth & Grogan, 1977		pupa
		PELLUCIDOMYIA			***oliveirai***	Ronderos et al., 2017b		pupa
	JOHANN SENOMYIINI	***MALLOCHOHELEA***			***termophila***	(Spinelli), 1984 (as Neobezzia)		pupa
						Ronderos & Spinelli, 2012b		pupa
		NEOBEZZIA			***amnicola***	Mayer, 1959 (as a Sphaeromias)		pupa
					fittkaui	Ronderos et al., 2011		pupa
		NILOBEZZIA			***neotropica***	Borkent, 2014		pupa
					schwarzii	Wirth, 1962		pupa
						Palchick, 1981		pupa
	PALPOMYIINI	***AMEROHELEA***			***sordidipes***	Lane et al., 1955		pupa
		BEZZIA			***bivittata***	Hribar & Mullen, 1991	larva	
						Borkent, 2014		pupa
					blantoni	Spinelli & Wirth, 1989		pupa
						Ronderos & Spinelli, 2009	larva	pupa
					brevicornis	Spinelli, 1983a		pupa
						Spinelli & Wirth, 1989		pupa
						Spinelli et al., 2012b		pupa
					bromeliae	Spinelli & Wirth, 1991		pupa

		galesa	Spinelli, in Spinelli et al., 2013		larva	pupa
		gibbera	Borkent, 2014			pupa
		glabra	Thomsem, 1937 (as a Probezzia)	egg	larva	pupa
			Wirth, 1952a		larva	pupa
			Chan & LeRoux, 1967	egg		
			Palchick, 1981			pupa
			Wirth, 1983a		larva	pupa
		nobilis	Johannsen, 1905 (as setulosa)		larva	pupa
			Malloch, 1915 (as setulosa)		larva	pupa
			Thienemann, 1928 (as chrysocoma)		larva	
			Mayer, 1934		larva	pupa
			Mayer, 1934 (as chrysocoma)			pupa
			Lenz, 1934 (as setulosa)			pupa
			Lenz, 1934 (as chrysocoma)			pupa
			Thomsen, 1937 (as setulosa)		larva	pupa
			Glukhova, 1979		larva	
			Palchick, 1981 (as setulosa)			pupa
			Wirth, 1983b		larva	pupa
			Spinelli, 1983a (as acanthodes)			pupa
			Knausenberger, 1987		larva	
			Hribar & Mullen, 1991		larva	
			Hribar, 1993		larva	
		pulchripes	Mayer, 1959			pupa
			Spinelli & Wirth, 1989			pupa
		roldani	Spinelli & Wirth, 1981		larva	pupa
			Spinelli & Wirth, 1989		larva	pupa
			Ronderos et al. 2007a		larva	pupa
		snowi	Spinelli & Wirth, 1991			pupa
		ventanensis	Spinelli, in Spinelli et al., 2012b			pupa

(Continued)

TABLE 16.1.1 Neotropical Ceratopogonidae Species Described as Eggs, Larvae, and/or Pupae.—cont'd

Subfamily	Tribe	Genus	Subgenus	No	Species	Reference	Developmental Stage		
		CLASTRIEROMYIA			***dycei***	Spinelli & Grogan, 1986			pupa
						Spinelli et al., 2016			pupa
		PACHYHELEA			***pachymera***	Spinelli, 1983a			pupa
						Spinelli et al., 2016			pupa
		PALPOMYIA			***guarani***	Lane, 1946			pupa
						Ronderos et al., 2004b		larva	pupa
					lacustris	Lane et al., 1955			pupa
					ryszardi	Spinelli & Ronderos, 2013			pupa
					subaspera	Wirth, 1952a (as essigi)			pupa
						Grogan & Wirth, 1979			pupa
					wirthi	Lane et al., 1955			pupa
	STENOXENINI	***PARYPHOCONUS***			***angustipennis***	Mayer, 1952	egg		
					flavidus	Mayer, 1959 (as lanei)			pupa
					grandis	Ronderos et al., 2012c			pupa
					mayeri	Mayer, 1959			pupa
					oliveirai	Wirth & Ratanaworabhan, 1972			pupa
						Ronderos et al., 2007b			pupa
						Ronderos et al., 2012b			pupa
		STENOXENUS			***johnsoni***	Wirth & Ratanaworabhan, 1972	egg		

pupa described. However, the immatures of 11 genera that also occur outside Neotropics are known from other regions. Accordingly, it is not possible to provide a key to eggs, larvae, and pupae at the generic level for this region.

TERMINOLOGY AND MORPHOLOGY

Larvae

Larvae, in general, have vermiform shape with a head capsule that is usually sclerotized. Mandibles are strong, toothed, and not opposed. The pharyngeal apparatus is conspicuous internally, with two strongly diverging arms and series of combs apparently serving to reduce and sort food. There is a short collar between the head and thorax. The head position and the relationship between the morphology of the mouthparts can be hypognathous or prognathous (Spinelli & Ronderos, 2005). Larvae have three thoracic and nine abdominal segments that are well defined. Functional spiracles are absent.

Larval Section: Head Capsule

The head capsule bears a characteristic number of setae and sensilla, excluding the sensilla associated with the mouthparts (chaetotaxy). Glukhova (1979) concluded that the larvae of all Ceratopogoninae, Dasyheleinae have 13 pairs of setae and five pairs of sensory pit, exclusive of sensory pit "j," which she considered a part of a collar (Figs. 16.1.4 A–H, 7 A).

Labrum (LB) forms the roof of the preoral cavity and can be membranous or sclerotized (Hribar & Mullen, 1991). The inner surface is the **palatum (PL)** (Cook, 1944). The shape varies from wide and short to elongated (projected) and bears different sensilla, which are probably used during feeding (Fig. 16.1.5 A–H).

Hypostoma (HY) is a sclerotized anterior projection of the ventral subgenal band which forms the ventral border of the preoral cavity. It can be serrate (Murphree & Mullen, 1991) (Figs. 16.1.5 A–H).

Mandibles (MD) are the major feeding appendages. They possess two articulation points and move independently of each other. They are strongly sclerotized, with variable shape: e.g., slightly curved (hooklike in *Culicoides* sp.), blade shape (in *Dasyhelea* sp.) among others (Hribar & Mullen, 1991) (Fig. 16.1.5 A–H).

Maxillae (MX) serve as accessory feeding structure to the mandibles. They are typically semicircular, membranous, structures that form the lateral edges of the preoral cavity. The different parts (cardo, stipes, galea and lacinia) are fused in one sclerite, the galeolacinia **(GL)** (Matsuda, 1965; Hribar, 1990) which has a maxillary palpus **(MP)** which is bisegmented and cylindrical (Murphree & Mullen, 1991) (Fig. 16.1.5 A–H).

Pharyngeal apparatus: It consists of the dorsal epipharynx and a ventral hypopharynx (Fig. 16.1.3 A–G).

Larval Section: Thorax (TS)

Thoracic segments are very uniform in their structure and do not differ dorsally and ventrally, have three segments with different small setae to the abdominal segments. In two subfamilies they are devoid of all protuberances (Dasyheleinae and Ceratopogoninae). The pigmentation is evident dorsal and laterally, when present is a useful character for distinguishing the larvae species (Fig. 16.1.7 B).

Prolegs are typically round or oval fleshy tubercles usually located in pairs ventrally on the prothorax and caudal segment of abdomen. The spinules often appear as relatively large, strongly hooked crochets arranged in one or more partially or completely encircling rows at the apex of the prolegs as in *Atrichopogon*. In *Dasyhelea* the terminal proleg comprises paired, sublateral clusters of a few curved spinules; the hooks and hooklets are sometime arranged as two prolegs (Figs. 16.1.4 B, 6 B).

Larval Section: Abdomen (Abd S)

The nine abdominal segments have distribution and number of setae as follows: 13 pairs (1–7); 9 pairs (8) and 10 pairs of caudal segment **(CS)** moderately developed, with different end (setae, papillae and/or hooked) (Murphree & Mullen, 1991) (Figs. 16.1.7 C, 8 A–G).

Pupae

Pupae of Ceratopogonidae have three major body sections: head, thorax and abdomen.

Borkent (2014) provided a general description of pupae of the genera of Ceratopogonidae of the World. In this contribution, we follow his methods and terminology. In general, the male and female pupae of lower lineages do not differ significantly, other than the obvious genital lobes of the males. In the higher lineages Heteromyiini + Sphaeromiini + Palpomyiini + Stenoxenini, males are somewhat smaller than females and differ somewhat in structure, e.g., in the shape of dorsal apotome. In general, pupae are diverse in size and overall body form (Figs. 16.1.9 A–F, 10 A–F).

FIGURE 16.1.9 Habitus of Ceratopogonidae pupae (scanning electron microscope): (A) *Leptoconops* sp.; (B) *Atrichopogon* sp.; (C) *Forcipomyia* sp.; (D) *Dasyhelea* sp.; (E) *Stilobezzia* sp.; (F) *Bezzia* sp.

Pupal Section: Head

The anterior portion of the pupal head is composed of three sclerites: the medial dorsal apotome and two dorsolateral cephalic sclerites, all important components of adult emergence.

Dorsal apotome (DA) is restricted dorsally and only partially separates the dorsolateral cephalic sclerites. The dorsal apotome is variable in shape, size, and presence or absence of different spicules, spinules, wrinkles, and tubercles (Fig. 16.1.11 A, C, E, G, I, K, M, Ñ, P, R).

Labium (Lb) forms the posteromedial margin of the mouthparts and is either entire, divided medially, or separated medially by the apex of the labrum (and possibly the hypopharynx).

FIGURE 16.1.10 Ceratopogonidae pupal exuviae: (A) *Forcipomyia* sp.; (B) *Dasyhelea* sp.; (C) *Culicoides* sp.; (D) *Neobezzia* sp.; (E) *Palpomyia* sp.; (F) *Paryphoconus* sp.

Labrum (**LB**) is a prominent medial section of the mouthparts and is rounded to narrowed posteriorly.

Mandible (**MD**) is an elongate lobe present in nearly all taxa (absent in *Dasyhelea*) along the posterolateral margin of the labrum.

Maxilla (**MX**) is composed of **lacinia**: present or absent as an elongate lobe.

Palpus is directed posteriorly or slightly posteromedially. Medially, it abuts the labium, posteriorly the foreleg, and laterally the foreleg, eye, and, when present, the prothoracic extension.

Sensilla vary in presence, variability in number, position, size, and form but are either setae or campaniform sensilla. **Dorsal apotome sensilla** (**DA-H**): usually one seta and one campaniform sensillum on dorsal surface. One to three **dorsolateral cephalic sclerite sensilla** (**DL-H**) are present, usually as one to two setae and one campaniform sensillum. The **clypeal/labrals** (**CL-H**) are zero to two, and absent in some genera (e.g., *Alluaudomyia* and some *Stilobezzia*). The **ocular sensilla** (**O−H**) are present laterally to the clypeus and labrum and are represented by zero to two setae and zero to two campaniform sensilla (Fig. 16.1.13 F−H).

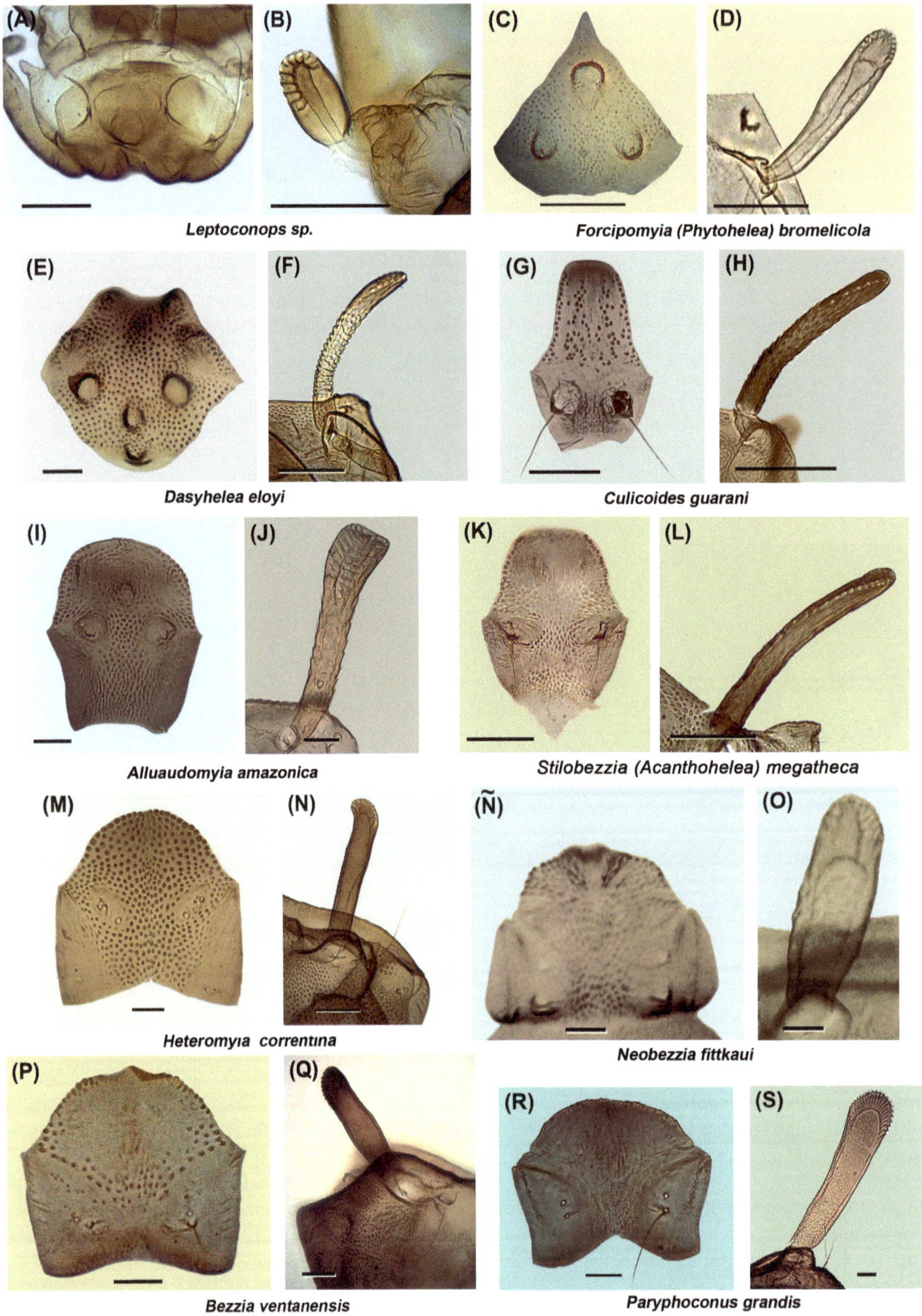

FIGURE 16.1.11 Ceratopogonidae pupae, dorsal apotome and respiratory organ: (A–B) *Leptoconops* sp.; (C–D) *Forcipomyia* (*Phytohelea*) *bromelicola* (Lutz); (E–F) *Dasyhelea eloyi* Díaz & Ronderos; (G–H) *Culicoides guarani* Ronderos & Spinelli; (I–J) *Alluaudomyia amazonica* Spinelli & Wirth. (K–L) *Stilobezzia* (*Acanthohelea*) *megatheca* Cazorla & Spinelli; (M–N) *Heteromyia correntina* Duret & Lane; (Ñ–O) *Neobezzia fittkaui* Wirth & Ratanaworabhan; (P–Q) *Bezzia ventanensis* Spinelli; (R–S) *Paryphoconus grandis* Macfie.

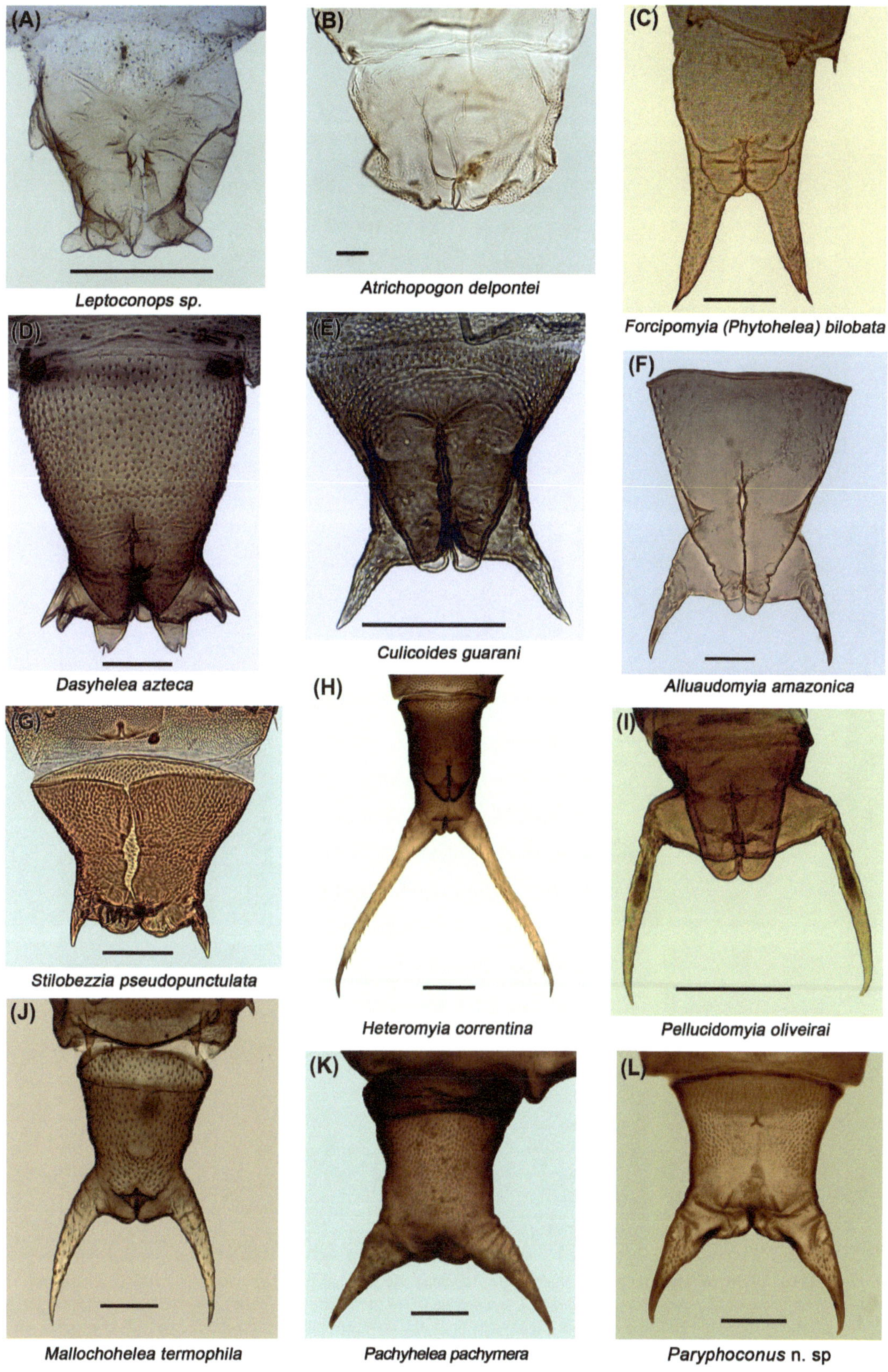

FIGURE 16.1.12 Ceratopogonidae pupae, segment 9: (A, C–F, H, I) Male pupa; (B, G, J–L) female pupa. (A) *Leptoconops* sp.; (B) *Atrichopogon delpontei* Cavalieri & Chiossone; (C) *Forcipomyia* (*Phytohelea*) *bilobata* Marino; (D) *Dasyhelea azteca* Huerta & Grogan; (E) *Culicoides guarani* Ronderos & Spinelli; (F) *Alluaudomyia amazonica* Spinelli & Wirth; (G) *Stilobezzia pseudopunctulata* Cazorla & Ronderos; (H) *Heteromyia correntina* Duret & Lane; (I) *Pellucidomyia oliveirai* (Lane); (J) *Mallochohelea termophila* (Spinelli); (K) *Pachyhelea pachymera* (Williston); (L) *Paryphoconus* sp.

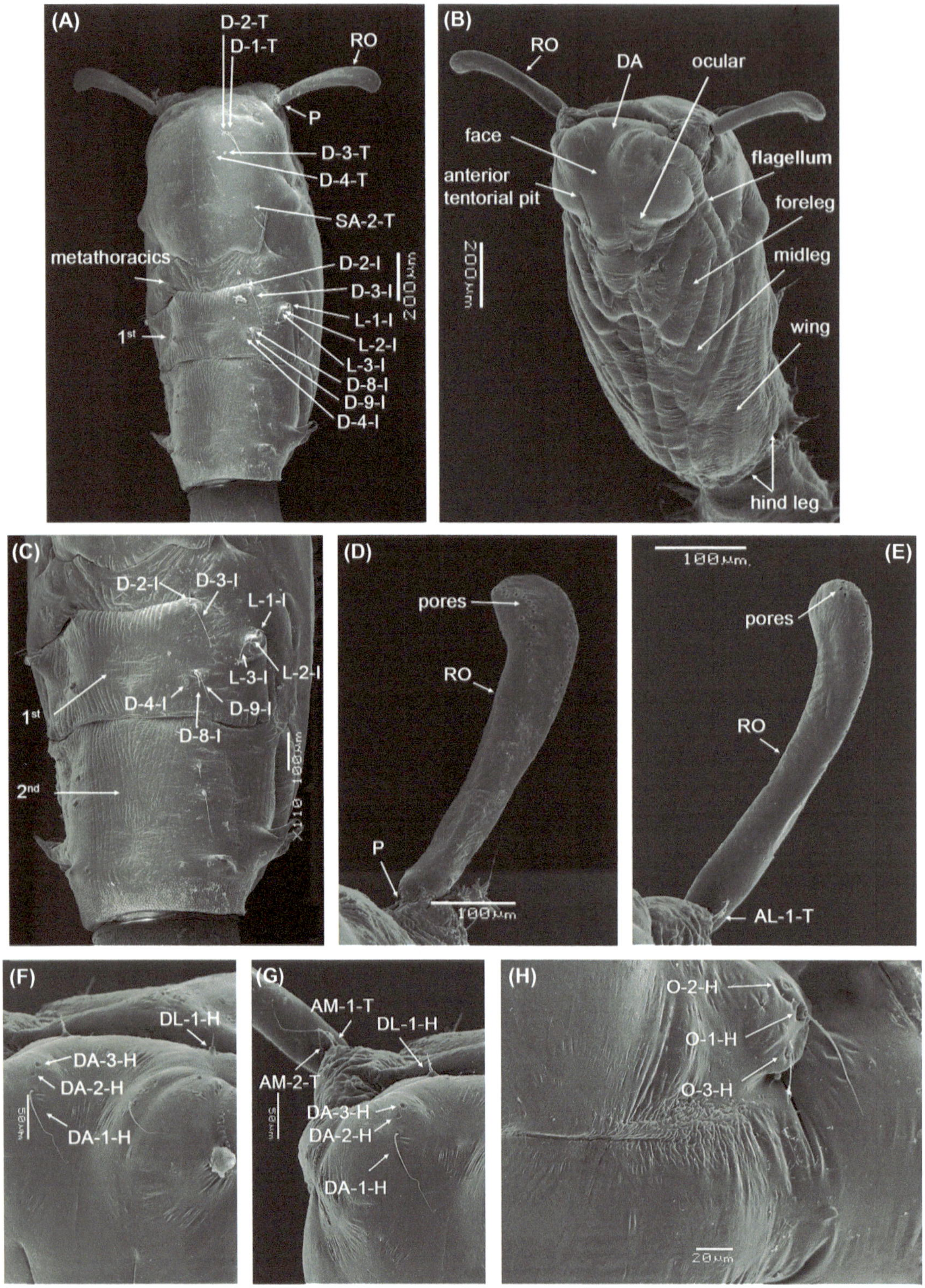

FIGURE 16.1.13 Detail of *Bezzia galesa* Spinelli, male pupa (scanning electron microscope): (A) cephalothorax chaetotaxy (dorsal view); (B) cephalothorax (ventral view); (C) chaetotaxy of segment 1 and segment 2 of cephalothorax, (dorsal view); (D) respiratory organ (dorsal view); (E) respiratory organ (lateral view); (F–G) Dorsal cephalic sclerite sensilla, dorsal apotomal and anteromedial sensilla; (H) Ocular sensilla. Anterolateral sensilla (AL-1-T, AL-2-T); anteromedial sensilla (AM-1-T, AM-2-T); dorsal apotomal sensilla (DA-1-H, DA-2-H, DA-3-H); dorsal apotome (DA); dorsolateral cephalic sclerite sensilla (DL-1-H); ocular sensilla (O-1-H, O-2-H, O-3-H); respiratory organ (RO); dorsal sensilla of first abdominal segment (D-2-I, D-3-I, D-4-I, D-8-I, D-9-I; lateral sensilla of first abdominal segment (L-1-I, L-2-I, L-3-I).

Pupal Section: Thorax

The thorax forms the widest and largest part of the pupa. Dorsally and anterolaterally, most of the thorax is a single unit that is composed of the combined pro- and mesothorax (mesonotum). The metathorax is present as a dorsal sclerite with posterolateral extensions (containing the developing halters) and lies between the mesonotum and the first abdominal segment and may be divided medially by a posteromedial projection of the mesonotum.

Prothoracic extension is separated from the dorsolateral portion of thorax by overlying antenna and externally forms a separate sclerite, extending from the antenna to the base of palpus. In few taxa the sclerite is short and abuts either only the antenna or the palpus.

Mesonotum forms an undivided unit occupying dorsal and the anterolateral region of the thorax (prothorax, scutum, scutellum, and postnotum).

Respiratory organs (RO): two located on the anterolateral portion of the mesonotum. They are usually somewhat compressed transversely, and each has at least some basal articulation. The respiratory organ is extremely variable in form and is either a single, undivided unit or with annulations at the base or at midlength in some. To obtain oxygen, each respiratory organ has at least a single apical row of pores, with various taxa having an additional apical row and/or scattered or groups of pores more basally along the respiratory organ (Figs. 16.1.11 B, D, F, H, J, L, N, O, Q, S, 16.1.13 A, B, D, E).

Legs are located ventrally in the body and, like the wings, are tightly fused with the rest of the thorax. The **foreleg** base lies posterior to the labium, where it overlays the developing adult coxa and/or trochanter and/or base of femur; most of the pupal **midleg** is only visible lateral to the anterolateral/posteromedial portion of the foreleg; and the bulk of the **hindleg** lies under the wing (Fig. 16.1.13 B).

Wings are broad structures separated ventrally by the forelegs in all but in female *Paryphoconus* Enderlein, where they meet medially (Fig. 16.1.13 B).

Sensilla: A variety of sensilla are present on the thorax of pupae, and they vary in number, position, size, and form. **Thoracic sensilla**: **anteromedial sensilla (AM-T)** are found from near the base of the respiratory organ to the anterior margin of the mesonotum; there are zero to six sensilla. **Anterolateral sensilla (AL-T)** are found on the anterolateral corner of the mesonotum and are generally in a tight group on a small lobe or projection; there are one to three setae. **Dorsal sensilla (D-T)** five on the area posteromedial to the base of the respiratory organs; one of these (D-3-T) is generally a campaniform sensillum. The single **supraalar** sensillum 2 (**SA-T**) is a campaniform sensillum in all Neotropical Ceratopogonidae subfamilies, and it is situated medial to the base of the wing and posterolateral to the dorsal sensilla (Fig. 16.1.13 A, E, G). There are zero to four metathoracic sensilla, with only some *Forcipomyia* lacking these (Fig. 16.1.13 A).

Pupal Section: Abdomen

The abdomen is composed of nine discrete segments with the last (ninth) segment bearing the genital lobes (**GL**) (Figs. 16.1.12 A, C–F, H, I, 16.1.14 C) that contain the developing adult genitalia and proctiger (elements of segment 10 such as the cercus). The shapes of the abdominal segments in a dorsoventral view vary between genera, with some comparatively wide and others nearly square. The width of abdominal segments 3-8 decreases successively anteriorly to posteriorly, and the shape of each segment subsequently changes from rectangular to square or elongate. In addition, there are differences between taxa in the relative shapes of their abdominal segments. Segment 9 is also variable in shape (short and wide to long and narrow) (Figs. 16.1.12 A–L, 14 C, D). In some Johannsenomyiini, sternites 6 and 7 (and sometimes 5) are mostly very thin membranous discs that adhere to dry substrate. Segments 2–7 have a spiracular scar that indicates the position of the underlying developing pharate adult spiracle (Clastrier, 1972).

The integument has a variety of ornamentation. Some taxa have elongated tubercles (only some Forcipomyiinae) or prominent rounded tubercles (e.g., some *Monohelea*, some *Parabezzia*), but generally the cuticle is more or less flat with a variety of sensilla (innervated) either arising directly from flat cuticles or on short to moderately elongate tubercles, which vary from apically rounded to pointed, from simple to bifid, or serrate.

Borkent (2012) provided a new system of numbering the abdominal sensilla of Culicomorpha, including those of Ceratopogonidae. Three basic groups are present as dorsal, lateral, and ventral sensilla on each of segments 1–8 (ventral sensilla are absent on segments 1 and 2).

Tergite 1 has two to six dorsal sensilla, with four to five setae and one to two campaniform sensilla in most taxa (reduced numbers in Forcipomyiinae, Dasyheleinae) (Figs. 16.1.13 A, C, 15 A–I). **Segment 4**: **Sensilla**: there are eight dorsal sensilla (never more than this), with the most anterior minute, either just anterior to the margin of the tergite or placed on it near its anterior margin. There are generally seven: with two setae anteriorly placed on the segment. The remaining five consist of three setae and

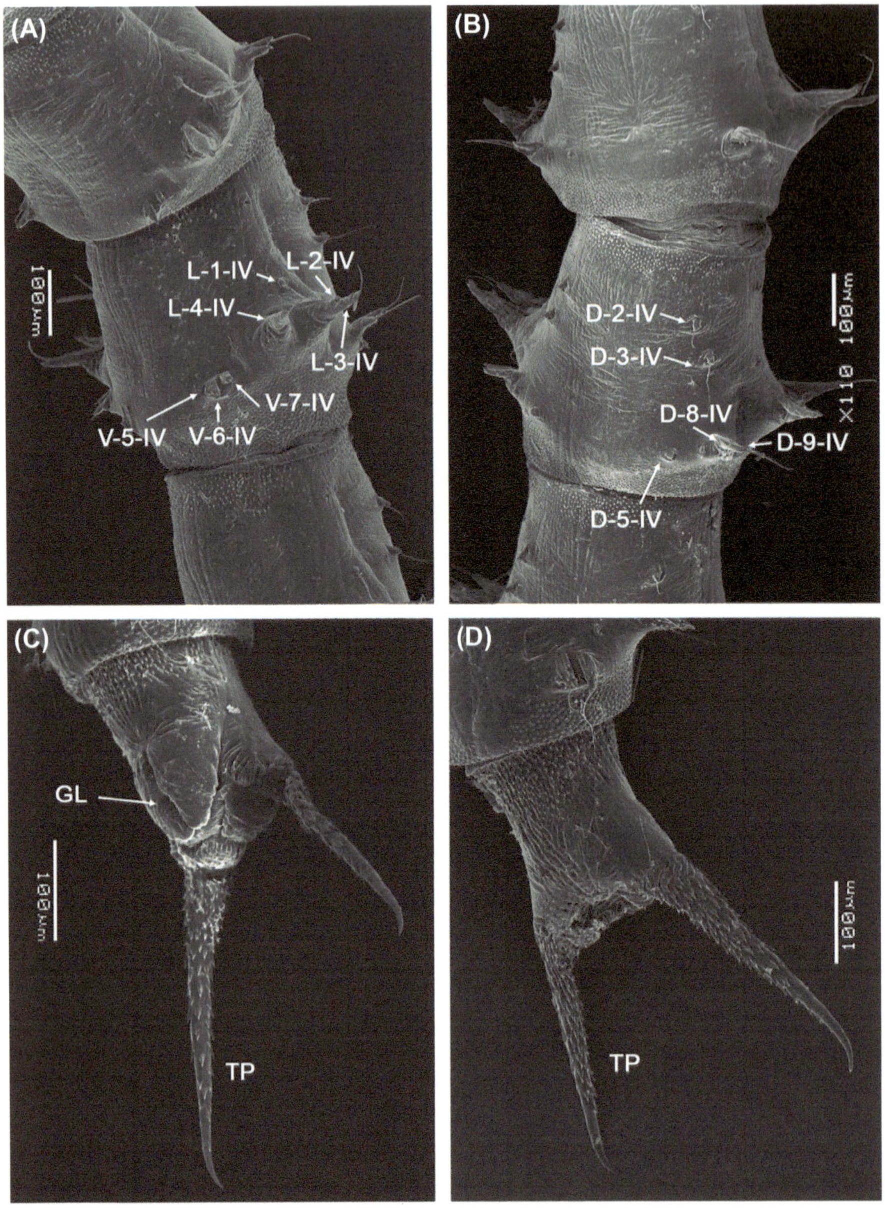

FIGURE 16.1.14 *Bezzia galesa* Spinelli, pupa (scanning electron microscope). (A–C) Male pupa. (D) Female pupa. (A) Segment 4 chaetotaxy (ventrolateral view); (B) Segment 4 chaetotaxy (dorsal view); (C, D) Segment 9 (ventral view). Dorsal sensilla (D-2-IV, D-3-IV, D-5-IV, D-7-IV, D-8-IV, D-9-IV); genital lobe (GL); lateral sensilla (L-1-IV, L-2-IV, L-3-IV, L-4-IV); terminal process (TP); ventral sensilla (V-5-IV, V-6-IV, V-7-IV).

two campaniform sensilla and are placed along the posterior margin of the tergite. There are generally four lateral sensilla, with one of them located more anteriorly, in some higher lineage of Ceratopogoninae, L-2-IV is on the most strongly developed tubercle. There are five ventral sensilla: two minute setae on or near the anterior margin of the sternite, which are difficult to observe. There are generally three (never more) ventral setae along the posterior margin of the sternite (Figs. 16.1.14 A, B, 16 A–E, 17 A–D).

Segment 9 is the last abdominal segment and, in males, bears the paired male genital lobes or, in the females of some taxa, a short medial genital lobe and cerci. In nearly all Ceratopogonidae, they are situated on the ventral surface of this terminal segment, but the male genital lobes are dorsal in some of subgenera of *Forcipomyia* (Figs. 16.1.12 A–L, 14 C–D).

Terminal processes (**TP**) are highly variable in shape and size and demonstrate their importance in pupal locomotion for pushing against substrate, as well as

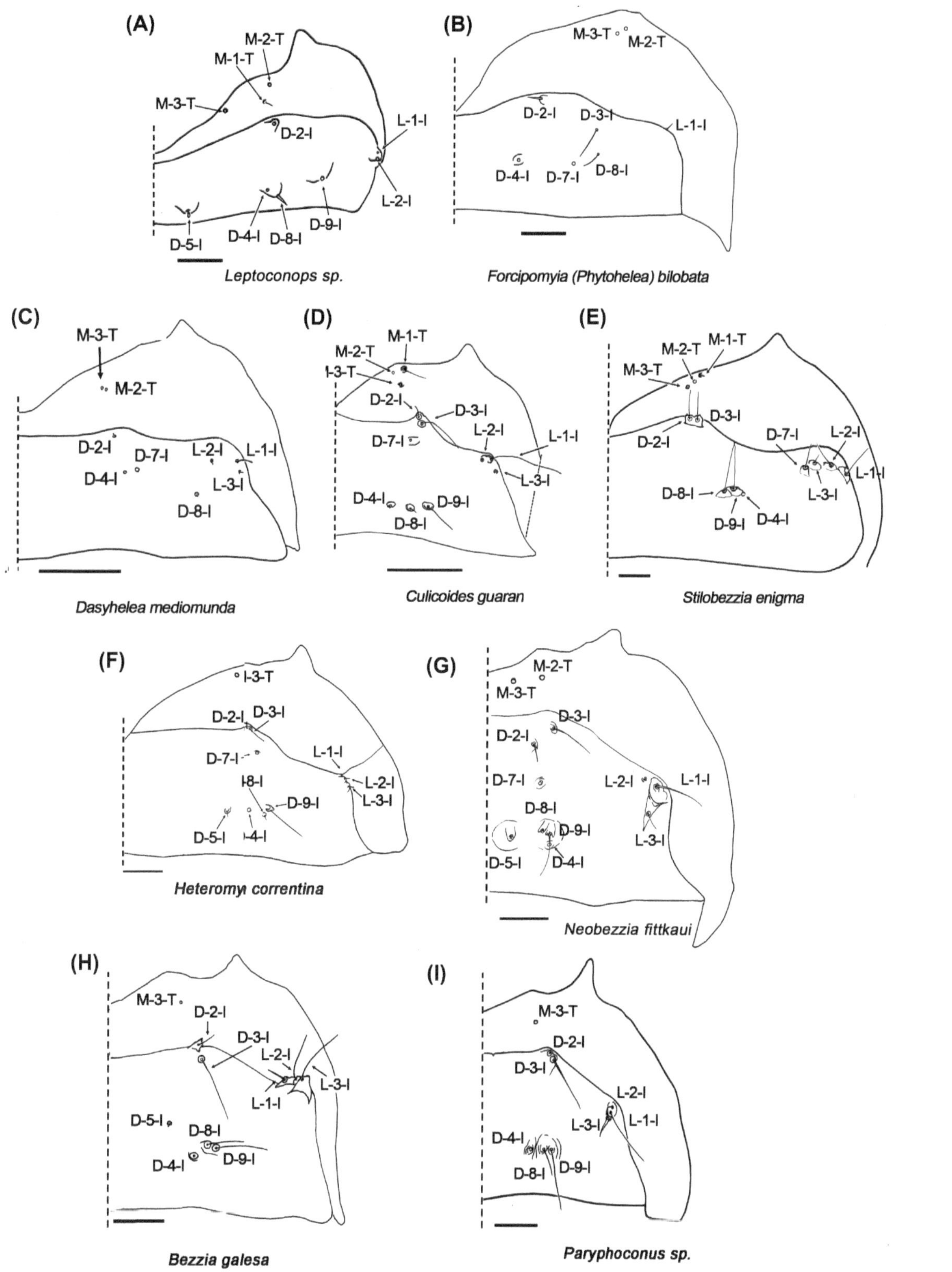

FIGURE 16.1.15 Metathorax and tergites 1 of Ceratopogonidae pupae, in dorsal view: (A) *Leptoconops* sp.; (B) *Forcipomyia (Phytohelea) bilobata* Marino; (C) *Dasyhelea mediomunda* Minaya; (D) *Culicoides guarani* Ronderos & Spinelli; (E) *Stilobezzia enigma* Ronderos, Spinelli & Borkent; (F) *Heteromyia correntina* Duret & Lane; (G) *Neobezzia fittkaui* Wirth & Ratanaworabhan; (H) *Bezzia galesa* Spinelli; (I) *Paryphoconus* sp. Scale bar: 0.05 mm.

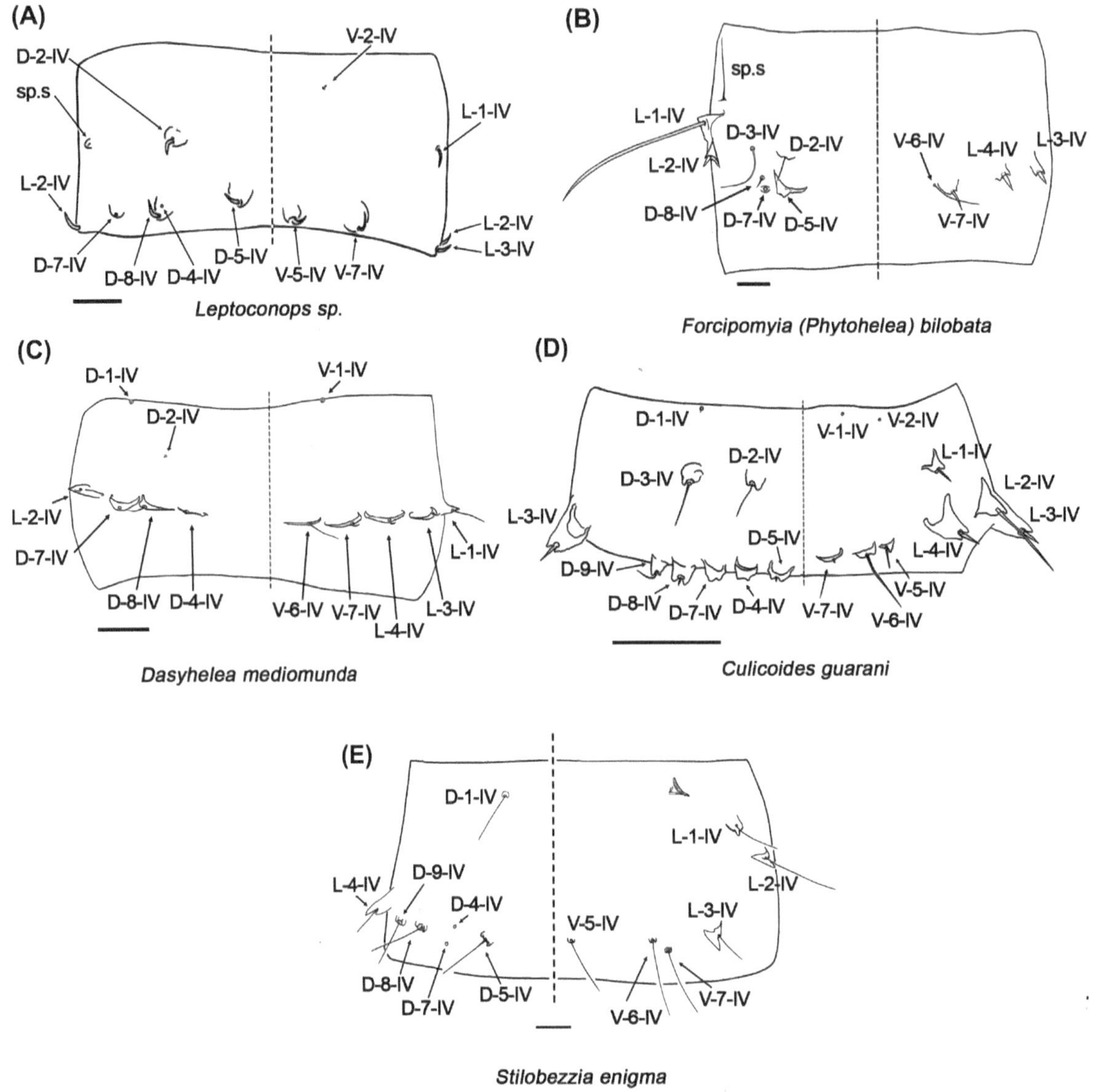

FIGURE 16.1.16 Segment 4 of Ceratopogonidae pupae, in dorsal and ventral view: (A) *Leptoconops* sp.; (B) *Forcipomyia* (*Phytohelea*) *bilobata* Marino; (C) *Dasyhelea mediomunda* Minaya; (D) *Culicoides guarani* Ronderos & Spinelli; (E) *Stilobezzia enigma* Ronderos, Spinelli & Borkent.

pushing out of the larval exuviae (abdomen free or partially encased by larval exuviae, as in Forcipomyiinae). Most taxa have two campaniform sensilla on segment 9, both of which are on the terminal process (Fig. 16.1.14C,D), except in *Dasyhelea*, which bears two additional setae.

MATERIAL PREPARATION AND PRESERVATION

Collected larvae and pupae should be examined to see live body shape, movements, and general coloration because after fixation, these characters are changed. Eggs can be collected from field substrates and later placed in the laboratory in plastic containers to record time of hatching and subsequent development of larval stages I to IV. Alternatively, they can be taken from gravid females. Larvae can be collected from standing water, mud, detritus, slit, leaves, and/or moist organic matter using pipettes and transported to the lab in plastic containers with water from their natural environment. They should be maintained individually in small boxes until pupation.

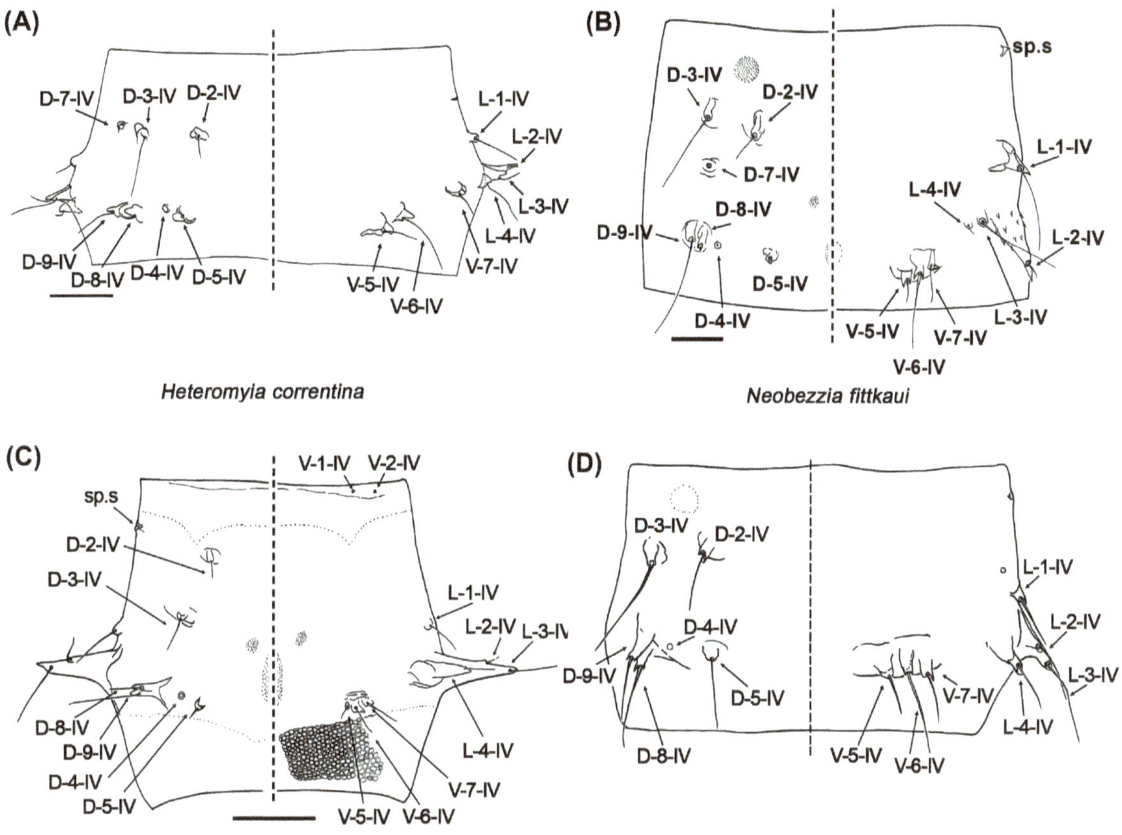

FIGURE 16.1.17 Ceratopogonidae pupae: (A) *Heteromyia correntina* Duret & Lane; (B) *Neobezzia fittkaui* Wirth & Ratanaworabhan; (C) *Bezzia galesa* Spinelli; (D) *Paryphoconus* sp. Scale bar: 0.05 mm.

Berlese funnels are a good method for extracting larvae from semiaquatic or moist material. Pupae are generally collected floating on the water surface. Detritus removed from the margins of ponds, streams, and lakes and placed in a white bucket allows for manual removal of pupae as they float. The pupae can be extracted from samples by pipette, and placed individually in small cotton-stoppered test tubes with 5 ml of water. They need to be checked daily for emergence of adults.

To prepare specimens for examination under a compound microscope, the eggs, larvae, and pupae should be cleared in hot 10% KOH for 5–10 min depending on the size, and then transferred to a saturated solution of phenol in ethanol. After this process, eggs, entire larvae, and pupae are mounted on microscope slides using Canada balsam as the mounting medium (Borkent & Spinelli, 2007). The body of the whole larvae and pupae should be oriented dorsoventrally and left in Canada balsam to dry overnight in an oven at 48°C, before placing coverslip over the specimen with an additional drop of Canada balsam to prevent specimens from drying or being crushed or distorted. These stages can also be examined on scanning electron microscopy using the technique of Ronderos et al. (2000, 2008). Emerged adults obtained from reared immatures in laboratory conditions should be stored individually, in vials of 70% ethanol, and clearly associated with their respective larval and/or pupal exuviae for later observation. For slide mounting, larval and pupal exuviae should be removed from 70% ethanol and placed over a saturated solution of phenol and then placed in Canada balsam on a microscope slide with the associated adult. In the case of pupal exuviae the dorsal apotome should be removed and placed in the same coverslip with its pupal exuvium. Associate larval and pupal exuviae by mounting them with the emerged adult on a single slide under a separated coverslip.

KEYS TO CERATOPOGONIDAE

Ceratopogonidae: Subfamilies (Larvae)

1 Prothorax without proleg; head capsule hypognathous to prognathous; head capsule and body without stout setae; mandible scooped or hooked; hypostoma crenulate or finely toothed 2

1' Prothorax with proleg (Figs. 16.1.4 B, 6 B); head capsule hypognathous; head capsule and body with stout setae (Figs. 16.1.4 B, 6 B, C); mandible scooped; hypostoma smooth .... Forcipomyiinae

2(1) Head capsule prognathous; caudal segment without hooks and hooklets, with or without elongate setae 3

2' Head capsule hypognathous (Figs. 16.1.4 D, 6 D); caudal segment with hooks and hooklets, sometime arranged as two prolegs (often retracted and difficult to see in material preserved in alcohol), never with elongate setae (Figs. 16.1.6 D, 8 C) .... Dasyheleinae

3(2) Head capsule with poorly developed pigmentation (Fig. 16.1.4 A); with well-developed apodemes, posteriorly directed apodemes extending into prothorax; abdomen with segments secondarily divided (Fig. 16.1.6 A) .... Leptoconopinae

3' Head capsule with well-developed pigmentation; without apodemes on the posterior margin; abdomen with segments not secondarily divided (Figs. 16.1.4 E–H, 6 E–H) .... Ceratopogoninae

Ceratopogonidae: Subfamilies (Pupae)

Modified from Borkent (2014).

1 Mouthparts with at least mandibles present, or if not evident, then the posterior margin of the labium forming a continuous (not medially divided) structure, or separated medially by posterior extension of labrum and/or hypopharynx; abdominal segments with posterior sensilla of variable sizes on short, bilobed tubercles and not all with same size and evenly spaced along posterior margin of abdominal segment; respiratory organ with pores grouped apically, with or without additional basal pores separate from apical group; terminal process simple, either short and apically rounded or tapering from base, with, at most, a single short to elongate seta (some *Forcipomyia*) 2

1' Mouthparts without mandibles or maxillae, two halves of the labium either broadly abutting medially or slightly separated (Fig. 16.1.9 D); abdominal segments with posterior sensilla on short, bilobed tubercles, all more or less the same size and evenly spaced along posterior margin of abdominal segment; respiratory organ with pores grouped apically to subapically, with or without additional separate basal pore; terminal process either tapering from narrow to broad base, or, with secondary divisions, with a pair of short to moderately elongate setae arising from or near its apex, some with elongate basal seta (Figs. 16.1.9 D, 10 B, 12 D) .... Dasyheleinae

2(1) Dorsolateral cephalic sclerites separated medially by dorsal apotome; hind leg visible at lateral margin of wing in lateral view; abdominal terminal process tapering and pointed apically 3

2' Dorsolateral cephalic sclerites abutting medially (Figs. 16.1.9 A, 11 A); hind leg not visible at lateral margin of wing in lateral view; abdominal terminal process rounded apically (Fig. 16.1.12 A) .... Leptoconopinae

3(2) Mesonotum smooth or with at least one pair of elongate tubercles of different shapes from short (about as long was wide) to very elongate (Figs. 16.1.9 B, 10 A); halter well separated from hind leg; prothoracic extension absent; mesonotum with, at most, one anterolateral seta on the mesonotum; mesonotum with two dorsal setae; male genital lobes dorsal or ventral to terminal processes (Figs. 16.1.9 C, 12 C); abdomen without or partially encased by larval exuviae .... Forcipomyiinae [Worldwide]

3' Mesonotum with bumps or with apically rounded tubercles at most about as long as wide (Fig. 16.1.10 C); halter and hind leg closely approximated or abutting; prothoracic extension well developed, although in some, it is restricted to abutting either the antenna or palpus (difficult to see in some slide preparations of *Culicoides* and *Paradasyhelea*, but these with two well-developed anterolateral setae and four dorsal setae on the mesonotum); mesonotum with at least three dorsal setae; male genital lobes ventral to terminal processes (Figs. 16.1.10 C, D, 12 E, 14 C); abdomen without attached larval exuviae .... **Ceratopogoninae** **[p. 658]**

Ceratopogonidae: Ceratopogoninae: Tribes (Pupae)

1 Prothoracic extension either extending from palpus to antenna, or only extending for part of this area, apex of halter broadly touching hind leg in lateral view, or if narrowly separated from or barely touching hind leg in lateral view then dorsal apotome with central dome bearing sensilla 2

1' Prothoracic extension extending from palpus to antenna, apex of halter narrowly separated from or barely touching hind leg in lateral view, and dorsal apotome lacking central dome bearing sensilla (Figs. 16.1.10 C, 11 G) .... Culicoidini

2(1) Abdominal segment 8 without D-3-VIII (*Macropeza* with anterior L-1-VIII); female with wings separated posteroventrally by forelegs. 3

2' Abdominal segment 8 with D-3-VIII; female with wings abutting posteroventrally (Fig. 16.1.10 F) .... Stenoxenini (*Paryphoconus*)

3(2) Abdominal sternites 6 and 7 without membranous discs 4

3' Abdominal sternites 6 and 7 with very thin membranous disc (Fig. 16.1.10 D) .... Johannsenomyiini

4(3) Mesonotum with two anteromedial setae of different lengths on anterior margin with one or both elongate, extending to respiratory organ; exuviae with dorsolateral cephalic sclerite fused to anterior margin of mesonotum (Fig. 16.1.9 F) 5

4' Mesonotum with zero to two very short anteromedial setae on anterior margin, not extending to respiratory organ; exuviae with dorsolateral cephalic sclerite either detached from anterior margin of mesonotum or fused with it (Fig. 16.1.9 E) .... Ceratopogonini

5(4) Mesonotum with one anteromedial setae elongate, the other 0.20× the length of longest seta, sometimes reduced to a small peg (Fig. 16.1.11 N); mesonotum with D-1-T, D-2-T, D-5-T either a small peg, or a short thick or an elongate seta .... Heteromyiini

5' Mesonotum with anteromedial setae very elongate, subequal in length, or with one shorter than the other, the shorter 0.75× the length of longest seta (Figs. 16.1.11 Q, 13 G); mesonotum with D-1-T, D-2-T, D-5-T elongate setae (Fig. 16.1.13 A) .... Palpomyiini

ACKNOWLEDGMENTS

We are grateful to William L. Grogan, Florida State Collection of Arthropods, Florida Department of Agriculture and Consumer Services, Gainesville, Florida, USA, for improving the English version and for his valuable comments and critical review of this chapter. We also thank Nélida Caligaris for technical assistance.

REFERENCES

Borkent, A. 2012. The pupae of Culicomorpha-morphology and a new phylogenetic tree. Zootaxa 3396: 1–98.

Borkent, A. 2014. The pupae of the biting midges of the World (Diptera: Ceratopogonidae), with a generic key and analysis of the phylogenetic relationships between genera. Monograph. Zootaxa 3879: 1–327. https://doi.org/10.11646/zootaxa.3879.1.1.

Borkent, A. 2016. World species of biting midges (Diptera: Ceratopogonidae). Web electronic publication http://www.inhs.uiuc.edu/research/FLYTREE/CeratopogonidaeCatalog.pdf, accessed (2017.05.17).

Borkent, A. & D.A. Craig. 2001. Submerged Stilobezzia rabelloi Lane (Diptera: Ceratopogonidae) pupae obtain oxygen from the aquatic fern Salvinia minima Baker. Proceedings of the Entomological Society of Washington 103: 655–665.

Borkent, A. & G.R. Spinelli. 2007. Neotropical Ceratopogonidae (Diptera: Insecta) Pages 1–198 *in*: J. Adis, J.R. Arias, G. Rueda-Delgado & K.M. Wantzen (eds.), Aquatic Biodiversity in Latin America (ABLA). Vol. 4. Pensoft, Sofia, Moscow.

Clastrier, J. 1972. Forcipomyia (F.) squamithorax n. sp. de la Guyane Française. Bulletin de la Société Entomologique de France 77: 170–176.

Cook, E.F. 1944. The morphology and musculature of the labrum and clypeus of insects. Microentomology 9: 1–35.

Díaz, F., M.M. Ronderos & G.R. Spinelli. 2005. The immatures of the Neotropical species Culicoides venezuelensis Ortiz and Mirsa (Diptera: Ceratopogonidae). Transactions of the American Entomological Society 131: 375–385.

Díaz, F. Spinelli, G.R. & Ronderos M.M. 2018a. Two new species of Dasyhelea Kieffer and the immature of D. azteca Huerta & Grogan from northwestern Argentina (Diptera: Ceratopogonidae). Zoologischer Anzeiger. 272: 6–19.

Díaz, F., Mangudo, C., Spinelli, G. R., Gleiser, R.M. & Ronderos M.M. 2018b. The Immatures of The Biting Midge Culicoides trilineatus (Diptera: Ceratopogonidae) Potential Vector of The Bluetongue Virus (BTV). Journal Medical Entomology. https://doi.org/10.1093/jme/tjy029.

Fox, I. 1942. The respiratory trumpet and anal segment of the pupa of some species of Culicoides (Diptera: Ceratopogonidae). Puerto Rico Journal of Public Health and Tropical Medicine 17: 412–425. [Spanish version, pages 426–434].

Glukhova, V.M. 1979. Larvae of the Biting Midges of the subfamilies Heteromyiini + Sphaeromiini + Palpomyiini + Stenoxenini and Ceratopogoninae of the fauna of the USSR (Diptera, Ceratopogonidae = Heleidae). AN SSSR, Opredeliteli po faune SSSR, 121, pp.1–230. [Nauk, SSSR, Leningrad, in Russian].

Gonzalez Gonzalez de Heredia, M. & A. Goldarazena Lafuente. 2011. El género Culicoides en el país Vasco. Guia práctica para su identificación y control. Departamento de Medio Ambiente Planificación territorial, Agricultura y Pesca, Eusko Jaurlaritzaren, pp. 1–250.

Hribar, L.J. 1990. Comparative morphology of the mouthparts and associated feeding structure of biting midge larvae. Mouthpart morphology and feeding behavior of biting midge larvae (Diptera: Ceratopogonidae). Contributions of the American Entomological Institute 26: 1–71.

Hribar, L.J. & G.R. Mullen. 1991. Comparative morphology of the mouthparts and associated feeding structure of biting midge larvae (Diptera: Ceratopogonidae). Contributions of the American Entomologist Institute 26: 1–71.

Matsuda, R. 1965. Morphology and evolution of the insect head. Memoirs of the American Entomological Institute 4: 1–334.

McCafferty, W.P. 1981. Aquatic Entomology. The fishermen's and Ecologists' Illustrated Guide to Insects and their Relatives. Science Book International, Boston, Massachusetts.

Murphree, C.S. & G.R. Mullen. 1991. Comparative larval morphology of the genus Culicoides Latreille (Diptera: Ceratopogonidae) in North America with a key to species. Bulletin of the Society for Vector Ecology 16: 269–399.

Ronderos, M.M., F. Díaz & P. Sarmiento. 2008. A new method using acid to clean and a technique for preparation of eggs of biting midges (Diptera: Ceratopogonidae) for Scanning Electron Microscope. Transactions of the American Entomological Society 134: 471–476.

Ronderos, M.M., G.R. Spinelli & F. Díaz. 2006. Observations on the immatures of Dasyhelea necrophila Spinelli & Rodríguez in laboratory (Diptera: Ceratopogonidae). Transactions of the American Entomological Society 132: 291–297.

Ronderos, M.M., G.R. Spinelli & P. Sarmiento. 2000. Preparation and mounting of biting midges of the genus Culicoides Latreille (Diptera: Ceratopogonidae) to be observed with a scanning electron microscope. Transactions of the American Entomological Society 126: 125–132.

Santarém, M.C.A. & M.L. Felippe-Bauer. 2016. Brazilian species of biting midges. (Diptera: Ceratopogonidae). Fiocruz, Rio de Janeiro, Brazil. Web publication http://www.fiocruz.br/ioc/media/especies_de_maruins_do_brasil.pdf, accessed (2017.03.20).

Spinelli, G.R. & M.M. Ronderos. 2005. Ceratopogonidae (Formas hematófagas). Pages 61–66 *in*: D. Salomón (ed.), Artrópodos de interés medico en la Argentina. Serie Enfermedades Transmisibles, Pub. Monogr. No 9. Fundación Mundo Sano, Buenos Aires, Argentina.

Uslu, U. & B. Dick. 2010. Chemical characteristics of breeding sites of Culicoides species (Diptera: Ceratopogonidae). Veterinary Parasitology 169: 178–184.

Chapter 16.2

Family Chironomidae

Fabio Laurindo da Silva
Department of Zoology, Institute of Biosciences, University of São Paulo, São Paulo, São Paulo, Brazil

Luiz Carlos Pinho
Departamento de Ecologia e Zoologia, Universidade Federal de Santa Catarina, Florianópolis, Santa Catarina, Brazil

Sofia Wiedenbrug
Zoologische Staatssammlung München, Munich, Germany

Galileu P.S. Dantas
Coordenação de Biodiversidade, Instituto Nacional de Pesquisas da Amazônia, Manaus, Amazonas, Brazil

Augusto Siri
Instituto de Limnología 'Dr. Raúl A. Ringuelet', (ILPLA, CONICET, UNLP), La Plata, Buenos Aires, Argentina

Trond Andersen
Department of Natural History, University Museum of Bergen, Bergen, Norway

Susana Trivinho-Strixino
Department of Hydrobiology, Federal University of São Carlos, São Carlos, Brazil

Chapter Outline

INTRODUCTION

Members of the nonbiting midge family Chironomidae (Diptera) are true flies, and they frequently dominate aquatic insect communities in both abundance and species richness (e.g., Cranston, 1982; Ferrington, 2008). Species of Chironomidae inhabit an enormous variety of aquatic ecosystems, from moist soils to pools in tree holes, and from low-oxygen lake sediments to fast-flowing mountain streams (Ferrington et al., 2008). The range of conditions under which chironomids are found is more extensive than that of any other family of aquatic insects. Several species lead semiterrestrial lives in moist habitats or are completely terrestrial. Moreover, Chironomidae is one of the very few insect families that has colonized the marine benthos. The

Thorp and Covich's Freshwater Invertebrates, Fourth Edition. https://doi.org/10.1016/B978-0-12-804223-6.00031-7

adult stage is short, with most of the lifespan spent as a larva (e.g., Thienemann, 1954; Tokeshi, 1995; Silva et al., 2015a). Oliver (1971), and Armitage et al. (1995) have provided an ample overview of chironomid biology and ecology. The great species and habitat diversity makes this family a valuable indicator species for lentic and lotic aquatic ecosystems.

The Chironomidae probably originated in the middle Triassic, approximately 248–210 million years ago (Cranston et al., 2010), and it has undergone extensive adaptive radiation to occupy a wider array of habitats at present than any other aquatic insect group (Jacobsen, 2008). The family comprises eleven subfamilies (Aphroteniinae, Buchonomyiinae, Chironominae, Chilenomyiinae, Diamesinae, Orthocladiinae, Podonominae, Prodiamesinae Tanypodinae, Telmatogetoninae and Usambaromyiinae), within which are at least 10,000 species in more than 400 genera (Cranston, 1995a; Sæther, 2000), with roughly 6200 of these are known to science (P. Ashe, personal communication). Nearly 900 species are recognized from the Neotropical Region (M. Spies, personal communication), of which 80% are assigned to one of the three subfamilies Chironominae, Orthocladiinae, and Tanypodinae (Spies et al., 2009). The remaining subfamilies are less species-rich: Aphroteniinae, Buchonomyiinae, Chilenomyiinae, Prodiamesinae, and Telmatogetoninae, for example, have one or a few species within one or a few genera.

Endemicity in the Antarctic and Neotropics is difficult to evaluate at present. Although many genera and species have been found only in some regions, their full geographic range is not known. While recent years have seen increased activity regarding Chironomidae in the Neotropical Region (e.g., Trivinho-Strixino et al., 2013, 2015; Andersen & Pinho, 2014; Andersen et al., 2015; Silva & Wiedenbrug, 2015; Siri & Donato, 2015; Dantas et al., 2016; Silva & Farrell 2017), the knowledge of the diversity and taxonomy as well as biogeography and phylogeny, particularly in the Central America and Caribbean regions, remains fragmentary. We provide updated information on the larval chironomid fauna in the Antarctic and Neotropical regions and will hopefully encourage further studies in the area.

Therefore, for this assessment, we recognize nine subfamilies, with known larvae, occurring in the Antarctic and Neotropical regions. In total, 144 genera were included in the key. The subfamily Chironominae is prevalent with 55 genera (38.2%), followed by the Orthocladiinae and Tanypodinae with 50 (34.7%) and 23 (16.0%) genera, respectively. The balance of 16 (11.1%) genera represent the six remaining subfamilies. The Antarctic fauna is represented by only three species: the orthoclads *Belgica antarctica* Jacobs and *Eretmoptera murphyi* Schaeffer and the podonomine *Parochlus steinenii* (Gerke) (Alegrucci et al., 2005).

LIMITATIONS

There has been no previous attempt to construct a comprehensive key to Chironomidae from the Antarctic and Neotropical regions. There are a few local keys including those by Paggi (2001, 2009) focusing on the Argentinean fauna, Spies et al. (2009) for Central America, and Trivinho-Strixino (2014), which covers mainly southeast Brazil. Several Neotropical chironomid larvae should key well in Epler (2001) or Ferrington et al. (2008), keeping in mind the divergences between regions in faunal composition (Spies et al., 2009). Andersen et al. (2013b) also included much useful information, but that should be employed with caution because of the geographic ranges. Spies & Reiss (1996) cataloged information and references for all taxa known from the Neotropics, including distribution records by country. Despite these works, the larval chironomid fauna remains poorly known ecologically and taxonomically in the aforementioned regions. Often, the incomplete knowledge of the full life history, associated with the lack of descriptions and keys to a local fauna prevent species identifications. Additionally, it is customary to overlook the Chironomidae in favor of other groups, such as Ephemeroptera, Plecoptera, and Trichoptera, which are more limited in number and diversity, and more easily identified (Spies et al., 2009).

A further limitation stems from species descriptions needing to be based on the adult (Cranston & Nolte, 1996). Association of chironomid immature stages can be achieved by individual rearing of larvae and collecting cast larval and pupal skins to establish the associations between each life stage (Ekrem et al., 2007). This procedure is time-consuming, however, and is not always successful for species with special environmental requirements (Silva et al., 2013). Regardless of these drawbacks, the Chironomidae has recently been studied taxonomically in the Antarctic and Neotropical regions. Several monographs on individual genera have been published (e.g., Mendes et al., 2004; Mendes & Andersen, 2008; Wiedenbrug & Trivinho-Strixino, 2011; Wiedenbrug et al., 2012; Trivinho-Strixino, 2012; Oliveira et al., 2013; Pinho et al., 2013; Silva et al., 2014; Trivinho-Strixino & Pepinelli, 2015), including species-level keys, as well as extensive reference collections. Finally, in this scenario where Chironomidae become an ideal group for investigating molecular/morphologic correspondence, it is important to point out that the emerging molecular data (e.g., Ekrem et al., 2010; Carew et al., 2011; Stur & Ekrem, 2011; Silva et al., 2013, 2015b; Silva & Wiedenbrug, 2014) support the concept of cryptic speciation in the family, and suggests that its diversity may be profoundly underestimated.

TERMINOLOGY AND MORPHOLOGY

Chironomidae are typical nematocerous dipterans, with four larval instars. Although morphologic and taxonomic examinations are made on the final instar, most structures seem to be present in earlier instars (Cranston, 1995c). The extensive ecologic amplitude displayed by species of Chironomidae is related to the very wide array of morphologic adaptations present among members of the group (Ferrington et al., 2008). The following diagnosis derives from Cranston (1995c; 2013) and Epler (2001). Mature (fourth instar) chironomid larvae possess an exposed, complete, sclerotized, nonretractable cephalic capsule, and an elongate, cylindrical body lacking jointed thoracic legs. The antenna is well developed and multi-articulated (retractile into the head in all Tanypodinae); it may be also mounted on a tubercle on some species. Larval chironomids bear mouthparts directed anteriorly (prognathous), with opposing mandibles, operating in oblique to the horizontal plane, and ventromedially with either transverse, usually toothed plate (the so-called mentum) or movable, toothed ligula (in Tanypodinae). There are usually no open spiracles, except in some members of the Podonominae. The body consists of three thoracic and nine abdominal segments and usually an anterior and posterior pair of parapods bearing spines or claws. The terminal segments usually bear paired procerci and usually four anal papillae, sometimes also with short lateral and/or longer ventral tubuli.

Several different terminologies have been used, with some being consistent with prevailing views of insect morphology, while others have been more idiosyncratic (Cranston, 1995c). Since the publication of Sæther (1980), a detailed assessment of morphologic homology, most discussions of morphologic structures have used the terms included in that glossary (Spies et al., 2009). Therefore, the subsequent list follows Sæther (1980), supplemented by Epler (2001), and it is not intended to be exhaustive. Figs. 16.2.1–5 illustrate most of the terminology used in the key and were modified from Oliver & Roussel (1983). In addition, some of the figures presented in this chapter were kindly provided by J.H. Epler, P.S. Cranston, and S. Trivinho-Strixino.

- accessory blade: smaller elongate appendage adjacent to antennal blade, usually partially fused with blade at base,
- accessory tooth: in Tanypodinae, small tooth between basal tooth and apical tooth of mandible,
- anal tubules: mostly oblong (but also ovoid, tapering, etc.) appendages of anal segment above and between posterior parapods,
- antennal blade: elongate structure adjacent to antennal flagellum, with a common base with accessory blade, arising from apex of first article,
- antennal ratio (AR): the ratio of length of basal antennal article to the combined lengths of remaining articles (the flagellum),
- antennal seta: seta of basal antennal article found in most Tanytarsini,
- antennal tubercle: base or platform from which the antenna arises,
- apical tooth: apical tooth at tip of mandible,
- cephalic index (IC): width of head capsule as a percentage of length of head capsule,
- cephalic setae: dorsal and lateral setae of head capsule, numbered with Arabic numerals from the anterior end to the posterior end of the head,
- clypeus: dorsal plate of the head immediately anterior to the frontal apotome that bears the S3 setae,
- dorsal accessory tooth: dorsal tooth or teeth of mandible in addition to the more apical and larger dorsal tooth,
- dorsal pore (DP): a sensory structure on the dorsum of the head capsule,
- dorsomentum: usually toothed, dorsal plate of double-walled mentum,
- flagellum: collective term for the apical articles of the antenna,
- frontal apotome: the most dorsal sclerite of the head carrying frontal setae S4 and S5, delimited anteriorly from the clypeus,
- frontoclypeal apotome: dorsal plate consisting of fusion of the clypeus and frontal apotome,
- labral lamella: one to several, smooth to pectinate lamellae near median anterior margin of labrum,
- labral sclerites: sclerite(s) on dorsal side of labrum carrying labral setae,
- labrum: the upper lip, consisting of the anterior dorsal portion of the head capsule,
- lateral tubules: small tubule laterally on each side of segment X in some Chironomini,
- Lauterborn organs: sensory organs, usually located on apex of second antennal article; usually digitiform but may be on pedicels and collectively may appear leaf-like (in Tanytarsini),
- ligula: in Tanypodinae, a sclerotized, toothed, tongue-like internal plate near center of head,
- M-appendage: in Tanypodinae, medioventral appendage of prementum, which may be reduced to fine teeth, very large and pronounced with several serrations, smaller with only median and paramedian lamellae, or consisting

of a triangular plate with lateral vesicles carrying a pseudoradula,

- mandible: the jaws, paired, usually toothed, structures inserted laterally, dorsal to the maxillae,
- maxilla: mouthpart located near base of mandible that bears the maxillary palp,
- mentum: usually toothed plate on anterior ventral margin of head capsule, consisting of a fused ventromentum and dorsomentum,
- paraligula: small sclerotized structure adjacent to ligula at ventral apex of prementum,
- parapods: fleshy, unjointed, ventral protuberances on prothorax (anterior parapods) and last abdominal segment (posterior parapods), nearly always carrying apical claws,
- pecten epipharyngis: usually three median scales, lamellae, spines, or rods directly behind labral margin,
- pecten hypopharyngis: in Tanypodinae, the comb-like structures on either side of the base of the ligula,
- pecten mandibularis: group of setae near ventral apex of mandible,
- premandible: a pair of ventral appendage-like structures beneath the labrum, taking the form of toothed plates or processes,
- premandibular brush: group of microtrichia sometimes present near premandible,
- premento-hypopharyngeal complex: double-lobed structure dorsal of, and to a large extent covered by, mentum in ventral view, and posterior and opposed to labrum,
- procercus: preanal tubercle, carrying 1–20 apical setae and usually two lateral setae; absent or vestigial in some taxa,
- pseudoradula: in Tanypodinae (except for *Tanypus*), longitudinal band of fine to coarse points present on middle of M-appendage,
- ring organ: campaniform sensillum present found on the basal article of the maxillary palp and the antenna,
- seta interna: seta on inner dorsomedial side of mandible, usually pectinate,
- seta subdentalis: seta on mandible proximal to inner teeth,
- setae submenti: pair of setae immediately posterior to mentum, anteriorly on submentum,
- S I, S II, S III, SIVA, S IVB: major setae of the anterodorsal surface of the labrum,
- ventral pore (VP): a sensory structure on the venter of the head capsule,
- ventral tubules: one to two pairs of tubules ventrally on abdominal segment 11,
- ventromental plate(s): lamellar ventral outgrowth of the head capsule adjacent to each side of the mentum.

MATERIAL PREPARATION AND PRESERVATION

Larval chironomids are best preserved in 70%–80% ethanol or isopropanol. Material intended for molecular studies should be preserved in ethanol or isopropanol 96%–100% (Andersen et al., 2013c). Formalin should be absolutely avoided, since it makes the material preserved in this way useless for molecular approaches. Although not also recommended, some workers add Rose Bengal stain to samples to facilitate to sort out the specimens from the debris, which nearly always excessively stains the mouthparts, making them too dark for appropriate light transmission (Epler, 2001). In this case, the specimens may be immersed in 2% acid alcohol, which acts as a clearing agent, for a few minutes before being transferred to ethanol (Trivinho-Strixino, 2014). The accurate larval identification relies on the examination of several structures in the cephalic capsule, usually requiring slide-mounting of the specimen, which may involve dissection and the removal of internal tissues. There are several techniques recommended by different authors (e.g., Epler, 2001; Andersen et al., 2013c) for preparation of permanent slides. Nearly all of them, however, involve the maceration of the larval head in a warm 5–10% solution of caustic potash (KOH) and posterior slide-mounting in Canada balsam or Euparal. Detailed descriptions of methods can be found in Epler (2001), Trivinho Strixino (2014), and Andersen et al. (2013c).

When dealing with reared series of an adult associated with larval and pupal exuviae, it is essential that all stages be mounted on the same slide to avoid the possibility of future misassociation (Andersen et al., 2013c). In the case of large amounts of larvae and whether routine identification is all that is required, the process of mounting permanent slides may be time consuming and needlessly complex. Numerous mounting media may be used to eliminate the need for dehydration in the preparation of permanent slides, and these can act as clearing agents. One such mounting medium, popularly used for mounting semipermanent slides, is the Hoyer's mounting medium. An advantage of this medium is its high refraction index and macerating properties of the small specimens. The main drawback of Hoyer's is its tendency to crystallize on drying. This may be overcome by ringing the cover slips with two layers of nail varnish or Euparal, after the initial drying period, which will prevent rehydration and medium deterioration (Andersen et al., 2013c). Hoyer's medium

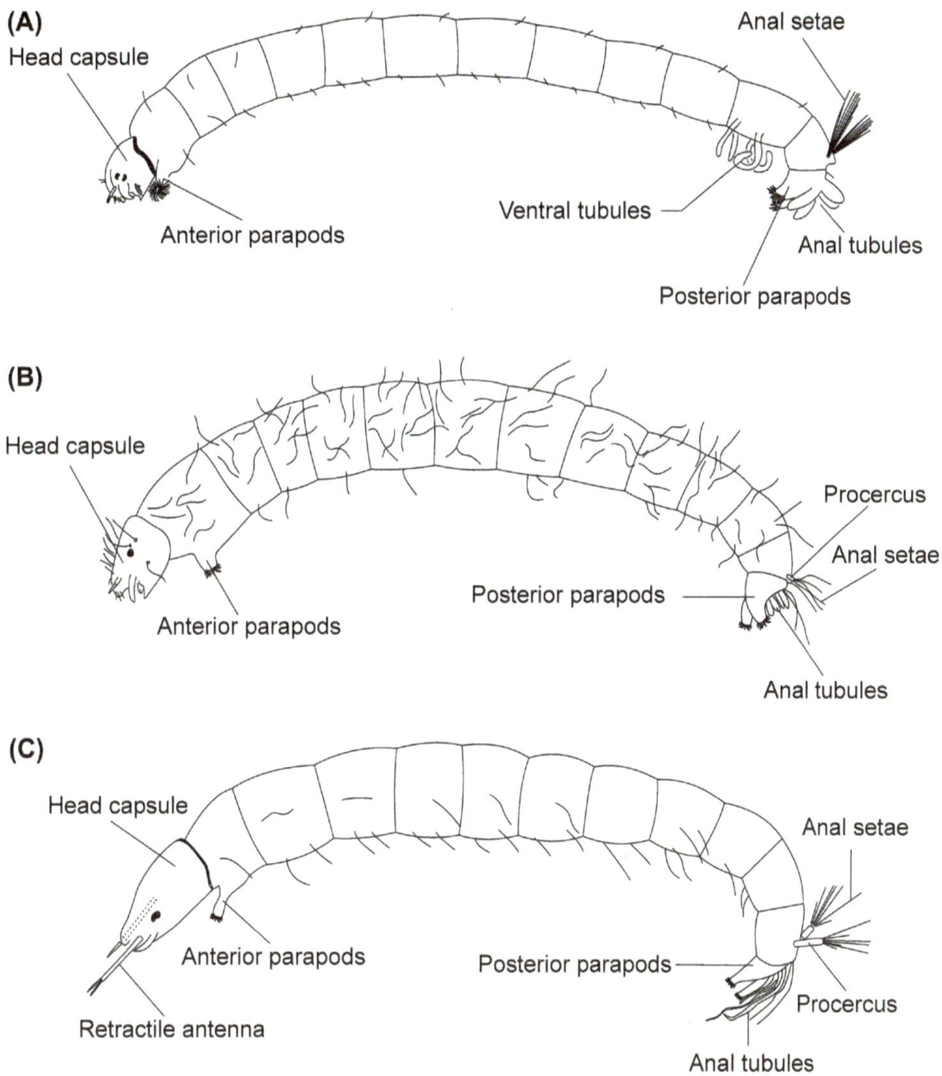

FIGURE 16.2.1 Chironomidae larva habitus, lateral view: (A) Chironominae; (B) Orthocladiinae; (C) Tanypodinae.

mounting may be prepared by dissolving gum arabic (30 g) in water (50ml), then adding in order chloral hydrate (200 g) and glycerol (16ml). This mixture is then filtered through glass wool before usage and/or storage (Epler, 2001). All slides should be labelled with information on where the specimen was collected, the date when it was collected, and the name of the collector (Figs. 16.2.1–16.2.29).

KEYS TO CHIRONOMIDAE

Chironomidae: Subfamilies

1 Antenna not retractile; prementum variously developed, ligula never developed as in Tanypodinae; mentum usually well developed and sclerotized 2

1' Antenna retractile into head; prementum with a distinctive ligula bearing four to eight teeth; mentum weakly developed, partially membranous with comb .... **Tanypodinae [p. 667]**

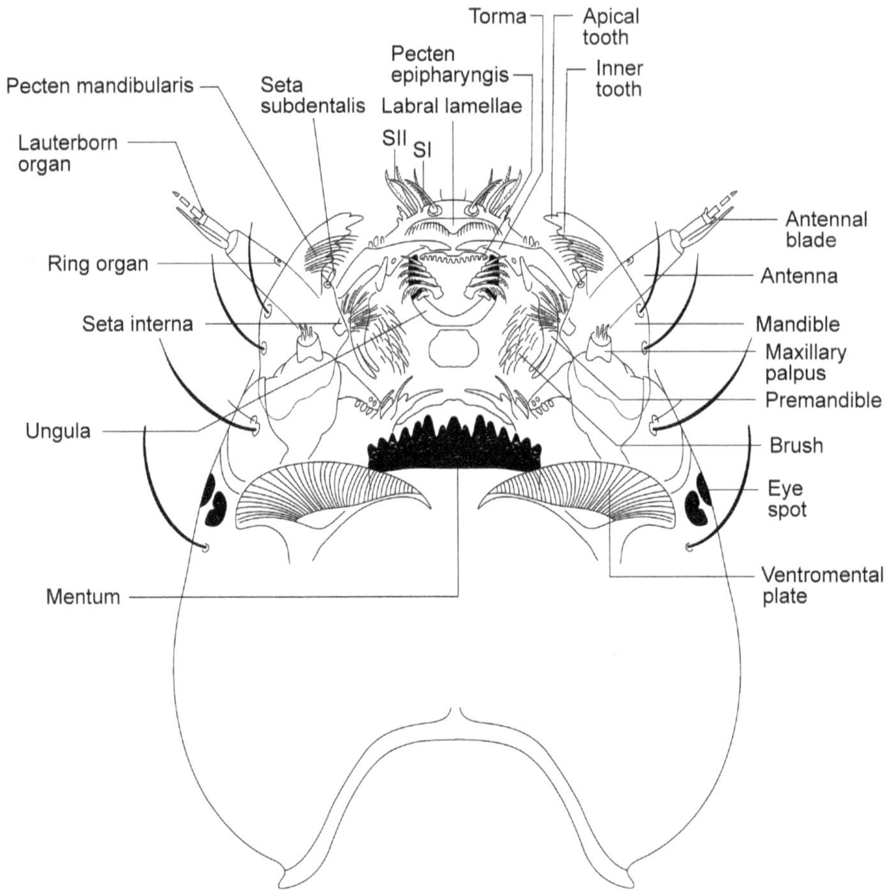

FIGURE 16.2.2 Chironominae (Chironomini) larva. Head capsule, ventral view.

2(1) Ventromental plates, if developed, never striated and usually with beard beneath 3
2' Ventral part of mentum laterally developed into ventromental plates, usually striate and never with beard beneath .... **Chironominae [p. 691]**
3(2) Mentum nearly always toothed; anterior labrum with short sensorial setae; body smooth, although may be strongly setose (Fig. 16.2.6 B) 4
3' Mentum untoothed; anterior labrum with elongated sensorial setae; body covered with sclerotized plates and strong setae (Fig. 16.2.6 A) (except for *Paraphrotenia*) .... **Aphroteniinae [p. 694]**
4(3) Premandibles absent 5
4' Premandibles present (Fig. 16.2.6 C) 6
5(4) Antenna weakly developed, third article not annulate (Fig. 16.2.6 D); procercus absent .... Buchonomyiinae, one genus: *Buchonomyia*
5' Antenna distinct, with four to five articles, with second/third article often annulate (Fig. 16.2.23 D); procercus well developed (Fig. 16.2.6 B) .... **Podonominae [p. 694]**
6(4) Third antennal article not annulate (Fig. 16.2.15 J); prementum with lamellae rather than brushes 7
6' Third antennal article annulate (except for Heptagiinii) (Fig. 16.2.15 I); prementum with three dense brushes of setaform processes .... **Diamesinae [p. 694]**
7(6) Ventromental plates, if developed, then either setal beard absent or antenna not with four articles (Fig. 16.2.15 J) 8
7' Ventromental plates well developed, with setal beard beneath; antenna with four articles (Fig. 16.2.23 J) .... **Prodiamesinae [p. 695]**
8(7) Antennae short, with four articles (Fig. 16.2.15 J); prementum with flabelliform appendage; premandible with strong brush .... **Telmatogetoninae [p. 695]**
8' Antennae variously developed (Fig. 16.2.18 E), if short, with four articles, then segmentation indistinct; prementum without brush-like appendage; premandible without brush, if present, weak .... **Orthocladiinae [p. 695]**

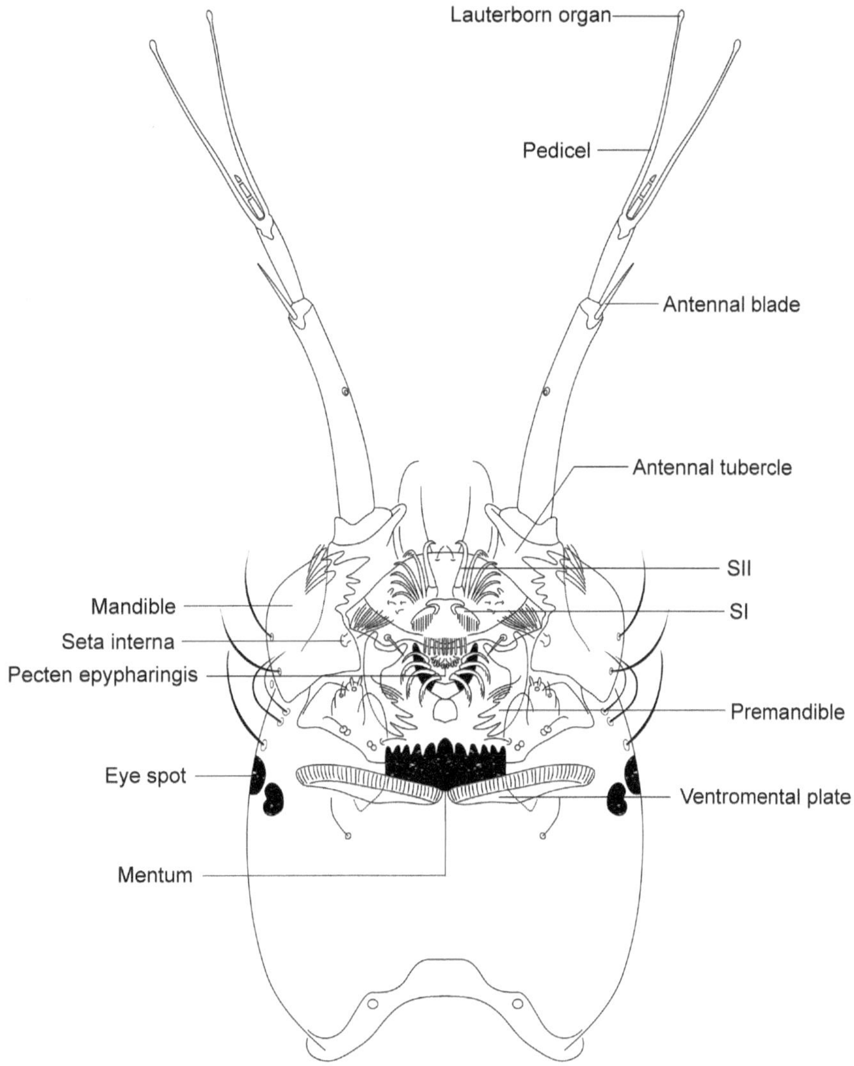

FIGURE 16.2.3 Chironominae (Tanytarsini) larva. Head capsule, ventral view.

Chironomidae: Tanypodinae: Tribes

This subfamily is the third most diverse, with species distributed widely across most of the globe, occupying numerous habits including small streams and ponds to lakes and bays (Silva et al., 2011). The larvae of the majority of species are free-living and none are known to produce larval or pupal cases (Ashe et al., 1987). Typically predators, some species feed on diatoms and detritus (Oliver, 1971). The monophyly of the subfamily is well supported (Cranston et al., 2012; Silva & Ekrem, 2016). Tanypodinae is subdivided into nine tribes (Silva & Ekrem, 2016). All, with exception of Anatopyniini, Coelopyniini, and Natarsiini, are found in the Neotropical Region. The *Thienemannimyia* group mature larvae may be easily identified. However, early instars should be regarded as *Thienemannimyia* group spp.

1	Head capsule rounded to oval; dorsomental teeth present in transverse plates or in longitudinal rows; body with lateral dense fringe of setae 2
1'	Head capsule more elongate; dorsomentum indistinct, teeth reduced or absent; lateral fringe of setae absent, although isolated hairs or groups of hairs may be present .... **Pentaneurini [p. 668]**
2(1)	Dorsomental teeth arranged at margin of transverse or diagonal plates (Fig. 16.2.24 B); ligula with four to five pale or dark teeth....... 3
2'	Dorsomental teeth aligned in longitudinal rows (Fig. 16.2.24 A), but not on distinct plates; ligula with six to seven pale teeth........ **Clinotanypodini [p. 691]**
3(2)	Mandible with base slenderer and apical tooth longer (Fig. 16.2.24 F); M-appendage with pseudoradula; pecten hypopharyngis present 4

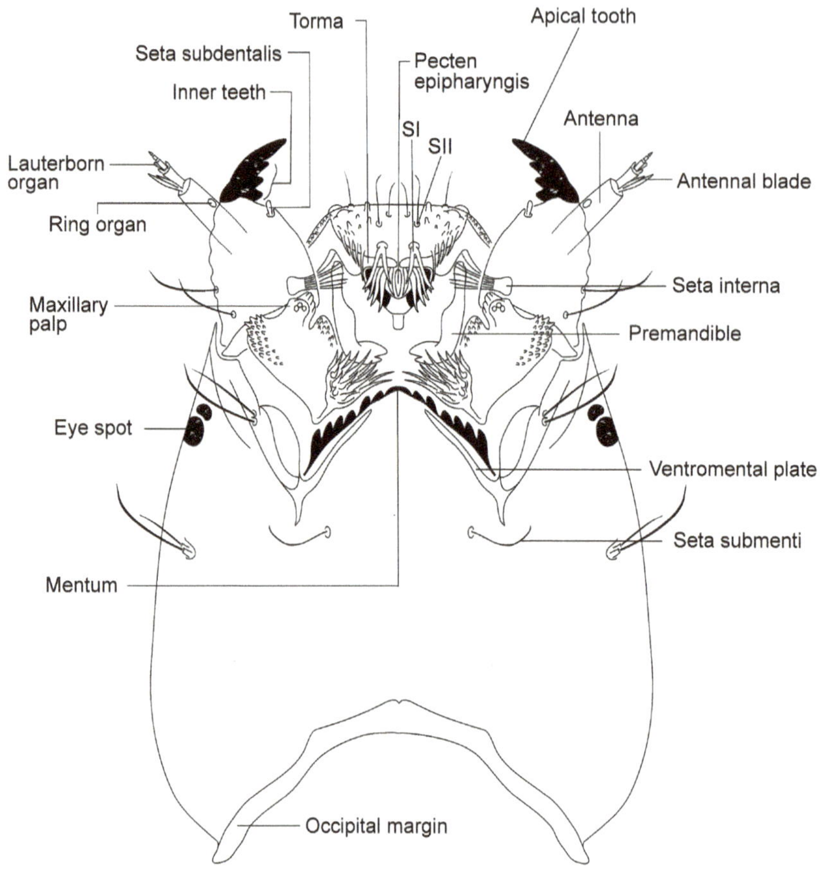

FIGURE 16.2.4 Orthocladiinae larva. Head capsule, ventral view.

3' Mandible with enlarged base and short apical tooth (Fig. 16.2.24 E); M-appendage without pseudoradula; pecten hypopharyngis absent .... Tanypodini, one genus: *Tanypus*

4(3) Dorsomentum not continuous, toothed plate distinctly subdivided into median and lateral sections; mandible without rows of additional teeth........ 5

4' Dorsomentum with continuous, concave-arched, toothed plate weakly subdivided into median and lateral sections; mandible with several rows of additional small dorsal and ventral teeth (Fig. 16.2.24 G) .... Fittkauimyiini, one genus: *Fittkauimyia*

5(4) Mandible with large blunt basal tooth; ligula with four or five dark teeth; paraligula pectinate (Fig. 16.2.24 H)...... **Procladiini [p. 691]**

5' Mandible with weak basal tooth; ligula with five teeth, dark or light; paraligula bifid (rarely trifid) (Fig. 16.2.24 I)........ **Macropelopiini [p. 691]**

Tanypodinae: Pentaneurini: Genera

1 Maxillary palp with two or more articles (Fig. 16.2.25 D)........ 2

1' Maxillary palp with single basal article (Fig. 16.2.25 E)........ 3

2(1) Maxillary palp with two unequal articles (Fig. 16.2.25 F); pseudoradula broadened posteriorly; posterior parapods with pale claws .... *Zavrelimyia* (*Paramerina*) (in part)

2' Maxillary palp with two to six articles; if only two articles, then articles subequal in length (Fig. 16.2.25 D); pseudoradula not broadened posteriorly; posterior parapods usually with one to three dark claws........ *Ablabesmyia*

3(2) Posterior parapod with bifid or pectinate claw (Fig. 16.2.25 G)........ 4

3' Posterior parapod with simple claws (Fig. 16.2.25 H)........ 7

4(3) Ligula with middle tooth distinctly longer than or subequal to inner tooth (Fig. 16.2.25 J); posterior parapod with one bifid claw; anal tubules shorter than parapods........ 5

4' Ligula with middle tooth longer than inner tooth; posterior parapod with one pectinate claw (Fig. 16.2.25 I) or claws with only inner serrations; anal tubules longer than parapods........ *Nilotanypus*

5(4) Ligula with subequal teeth (Fig. 16.2.25 K); bifid claws of posterior parapod with outer tooth longer than inner tooth........ 6

FAMILY CHIRONOMIDAE

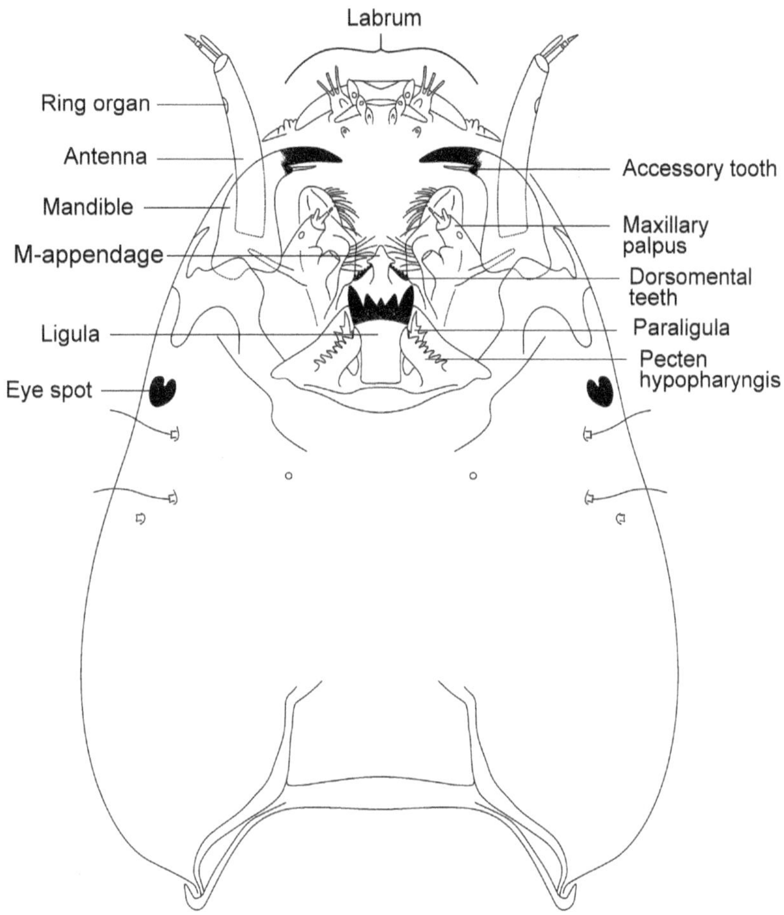

FIGURE 16.2.5 Tanypodinae larva. Head capsule, ventral view.

5' Ligula with middle tooth distinctly longer (Fig. 16.2.25 J); posterior parapod with a bifid claw with outer tooth shorter than inner tooth .... *Labrundinia*

6(5) Ligula with median and inner teeth lighter than outer teeth (Fig. 16.2.25 K); paraligula trifid .... *Denopelopia*

6' Ligula with dark teeth (Fig. 16.2.25 L); paraligula bifid .... *Zavrelimyia* (*Zavrelimyia*) (in part)

7(3) Posterior parapod with simple claws (Fig. 16.2.26 B) 8

7' Posterior parapod with one to two pectinate claws (Fig. 16.2.26 A), dark or light, with inner margin serrated .... *Monopelopia*

8(7) Ring organ of maxillary palp situated usually near middle of basal article (Fig. 16.2.26 D); mandible with weakly to moderately developed basal tooth 9

8' Ring organ of maxillary palp situated near apex of basal article (Fig. 16.2.26 C); mandible with no basal tooth *Thienemannimyia* Group 10

9(8) Length of antennal article 1 divided by mandible length >1.75; distance between ventrolateral setae 2 and 3 of mandible ½ as great as that between seta 2 and sensillum minusculum .... *Metapelopia*

9' Length of antennal article 1 divided by mandible length <1.90; distance between ventrolateral setae 2 and 3 of mandible smaller than ½ distance between seta 2 and sensillum minusculum .... *Thienemannimyia*

10(8) Ligula with subequal teeth (Fig. 16.2.26 F); supraanal seta arising from dark sclerotized area; anal tubules long and thin, surpassing length of posterior parapods 11

10' Ligula with median and inner teeth shorter than outer teeth (Fig. 16.2.26 E); supraanal not arising from dark sclerotized area; anal tubules shorter .... *Larsia*

11(10) Head pale yellow (Fig. 16.2.26 H); posterior parapods with light claws 12

11' Head usually dark brown (Fig. 16.2.26 G); posterior parapods with dark claws .... *Hudsonimyia*

12(11) Cephalic seta S10 between S9 and ventral pore (VP) forming an 80°–90° angle (Fig. 16.2.26 I) .... *Pentaneura*

12' Ventral pore (VP) between cephalic setae S9 and S10 forming a more or less straight line diagonal to longitudinal axis of head capsule (Fig. 16.2.26 J) .... *Parapentaneura*

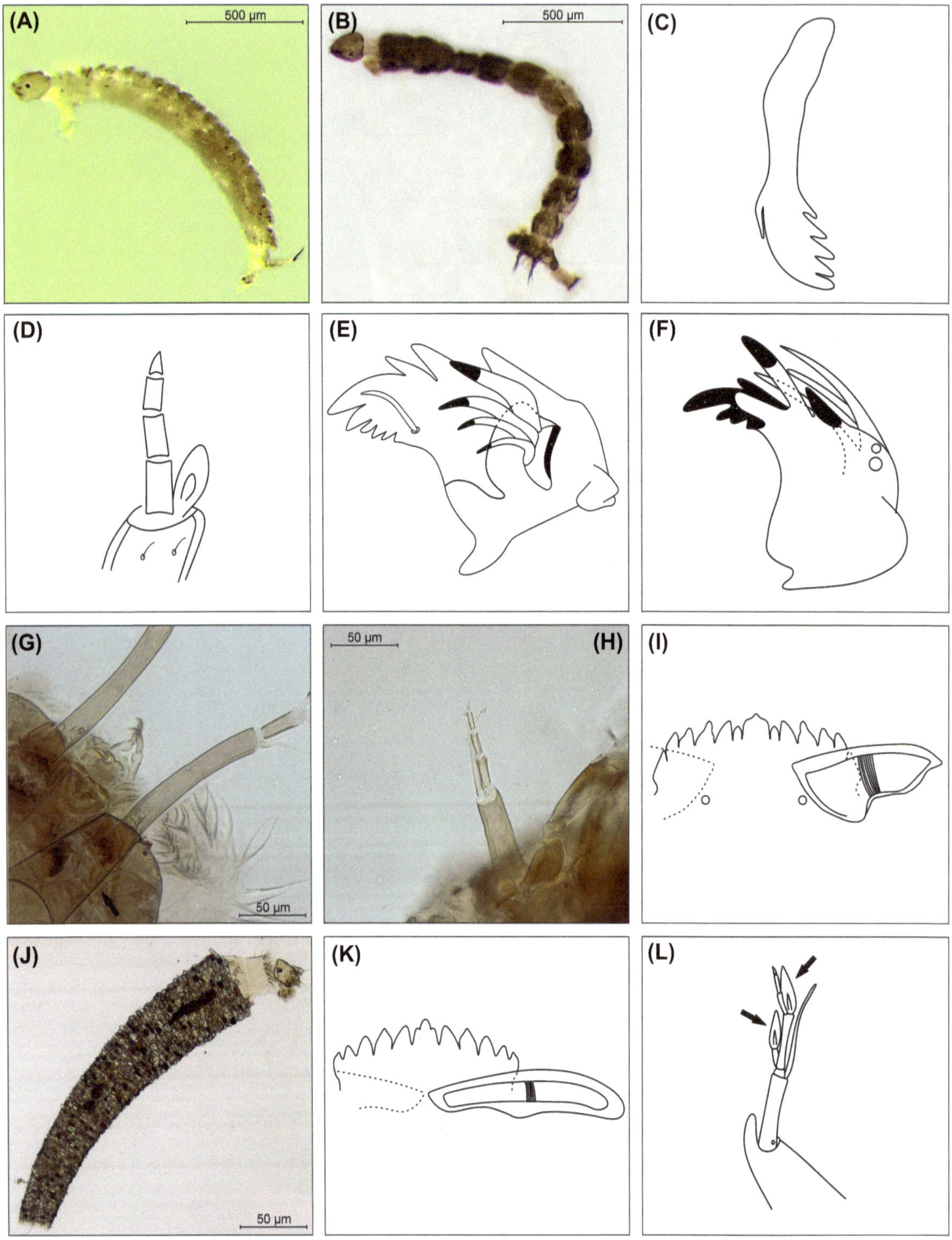

FIGURE 16.2.6 Chironomidae larva: (A) *Aphrotenia* sp.; (B) *Podonomus* sp.; (C) *Diamesa* sp.; (D) *Buchonomyia* sp.; (E) *Paraphrotenia* sp.; (F) *Aphroteniella* sp.; (G) *Caladomyia* sp.; (H) *Asheum* sp.; (I) *Stempellina* sp.; (J) *Constempellina* sp.; (K) *Tanytarsus* sp.; (L) *Stempellinella* sp.

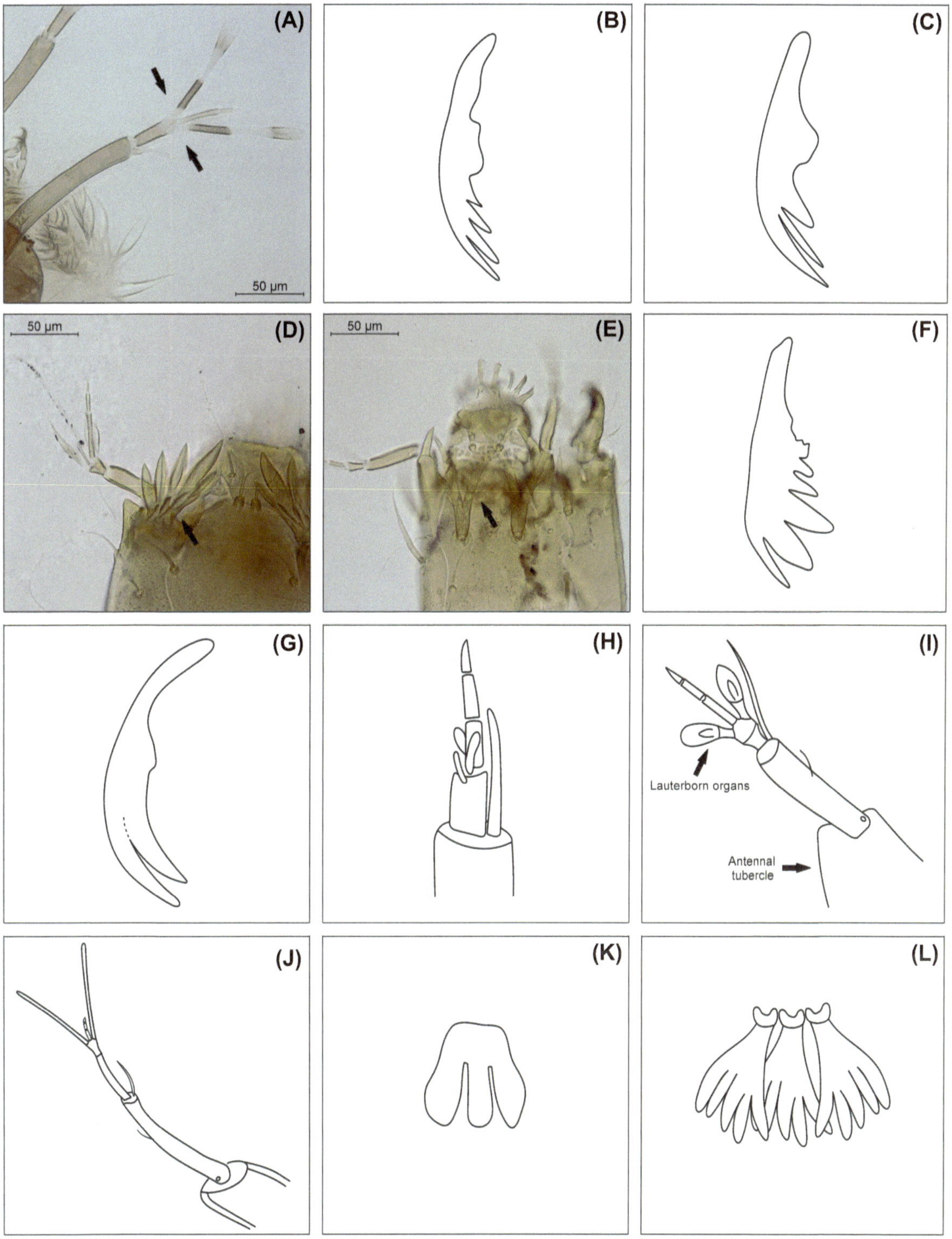

FIGURE 16.2.7 Chironomidae larva: (A) *Caladomyia* sp.; (B) *Zavrelia* sp.; (C) *Stempellinella* sp.; (D) *Stempellina* sp.; (E) *Constempellina* sp.; (F) *Pontomyia* sp.; (G). *Paratanytarsus* sp.; (H) *Pontomyia* sp.; (I) *Cladotanytarsus* sp.; (J) *Tanytarsus* sp.; (K). *Sublettea* sp.; (L) *Paratanytarsus* sp.

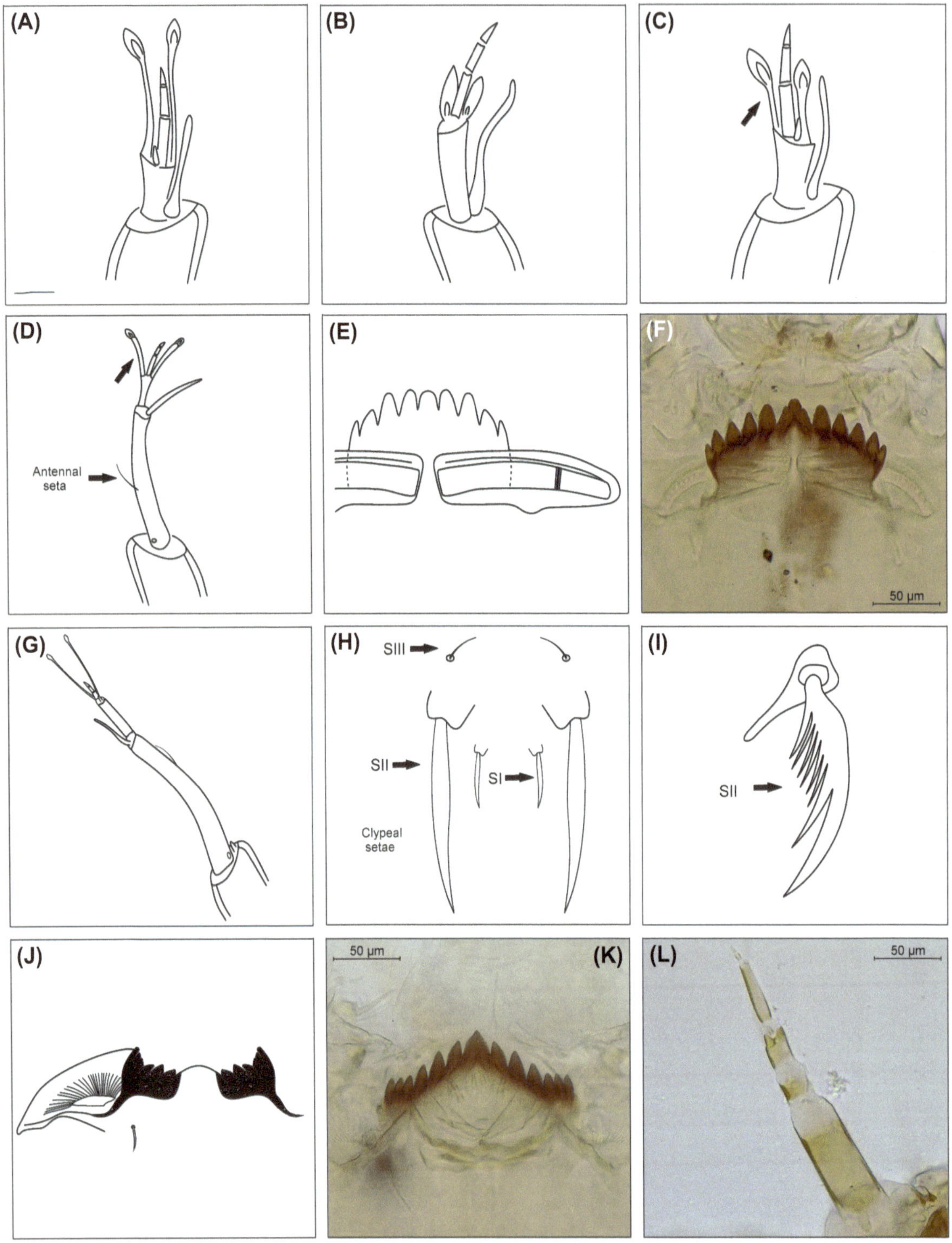

FIGURE 16.2.8 Chironomidae larva: (A) *Sublettea* sp.; (B, D) *Paratanytarsus* sp. (C, F) *Rheotanytarsus* sp.; (E) *Neozavrelia* sp.; (G) *Micropsectra* sp.; (H) *Cladopelma* sp.; (I) *Dicrotendipes* sp.; (J) *Cryptochironomus* sp.; (K) *Parachironomus* sp.; (L) *Demicryptochironomus* sp.

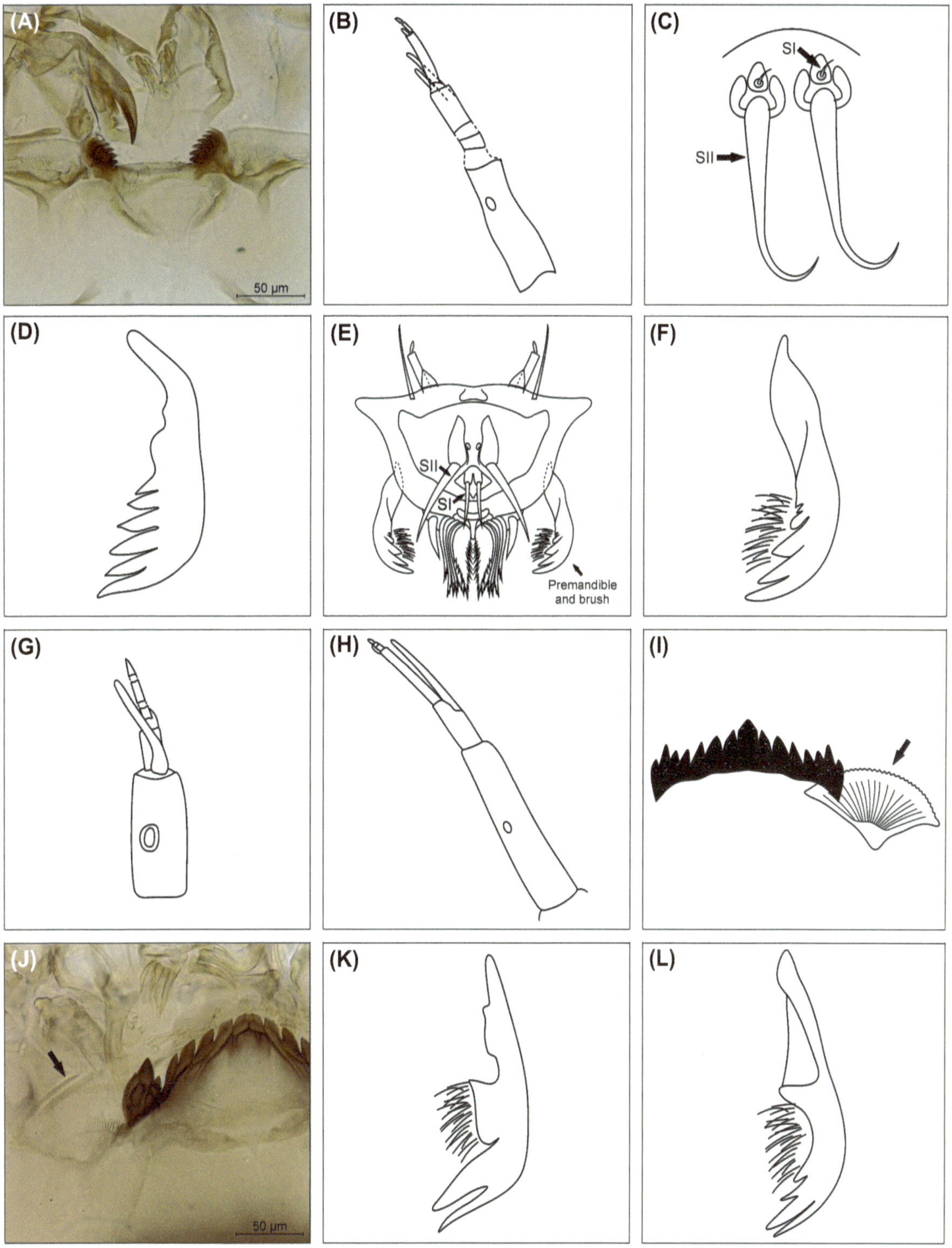

FIGURE 16.2.9 Chironomidae larva: (A) *Demicryptochironomus* sp.; (B) *Cryptochironomus* sp.; (C, D) *Gillotia* sp.; (E, F) *Cryptochironomus* sp.; (G) *Cryptotendipes* sp.; (H) *Pelomus* sp.; (I) *Parachironomus* sp.; (J) *Cladopelma* sp.; (K) *Microchironomus* sp.; (L) *Paracladopelma* sp.

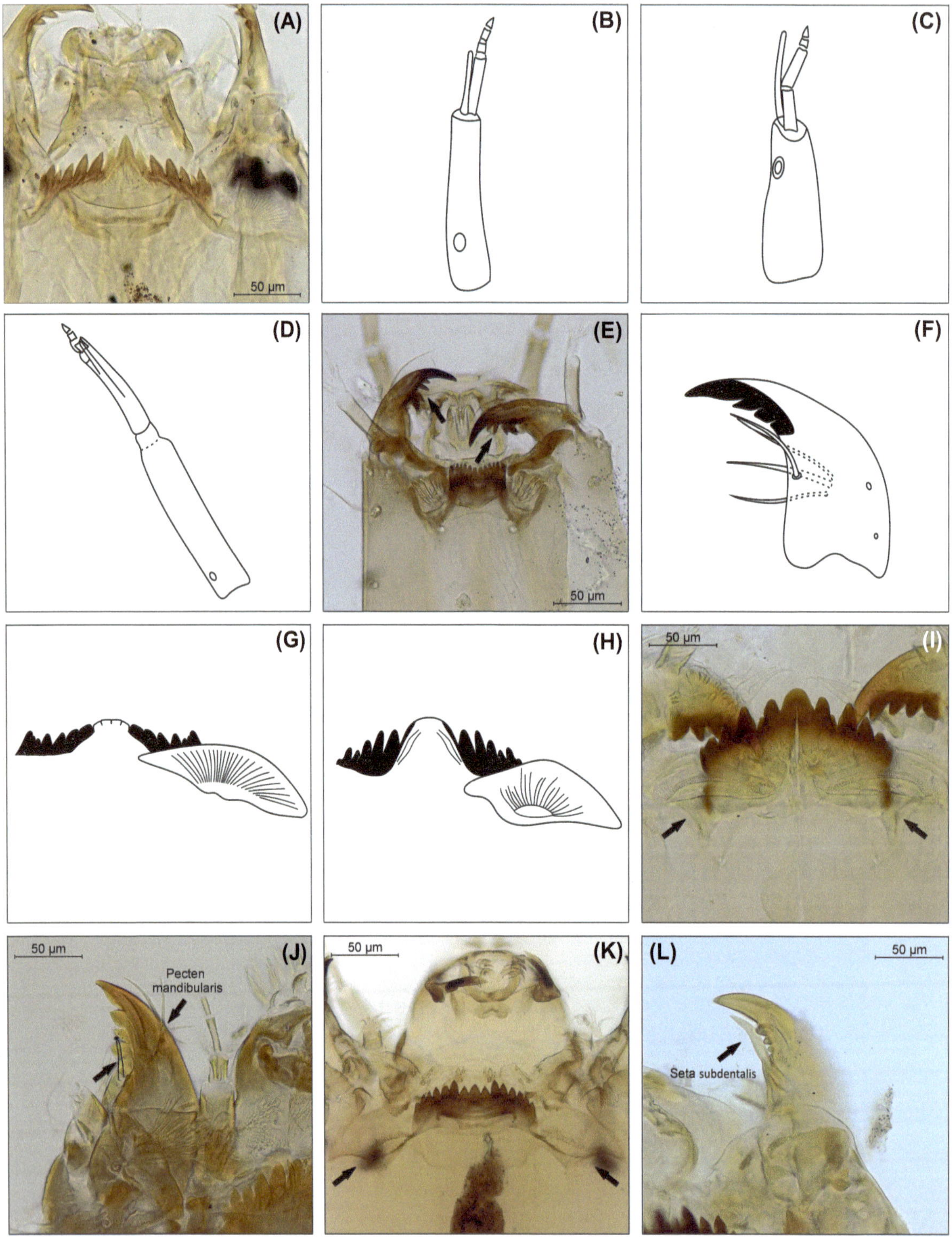

FIGURE 16.2.10 Chironomidae larva: (A) *Microchironomus* sp.; (B) *Cladopelma* sp.; (C) *Harnischia* sp.; (D) *Paracladopelma* sp.; (E) *Robackia* sp.; (F, G) *Pelomus* sp.; (H) tentatively identified as *Saetheria* sp.; (I) *Pseudochironomus* sp.; (J) *Aedokritus* sp.; (K) *Chironomus* sp.; (L) *Fissimentum* sp.

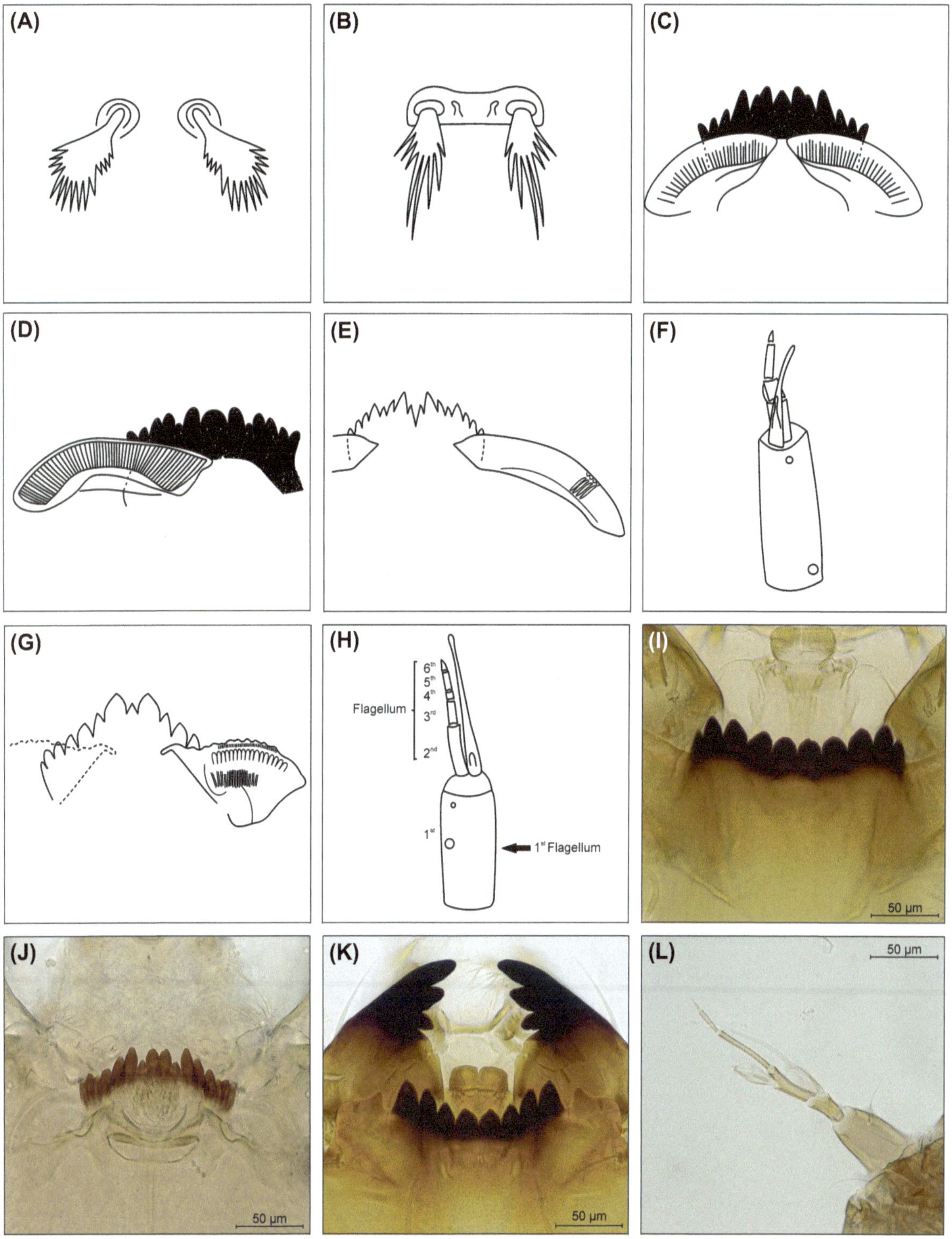

FIGURE 16.2.11 Chironomidae larva: (A) *Pseudochironomus* sp.; (B, C) *Manoa* sp.; (D) *Riethia* sp.; (E, H) *Fissimentum* sp.; (F, G) *Hyporhygma* sp.; (I) *Stenochironomus* sp.; (J) *Polypedilum* sp.; (K) *Xestochironomus* sp.; (L) *Zavreliella* sp.

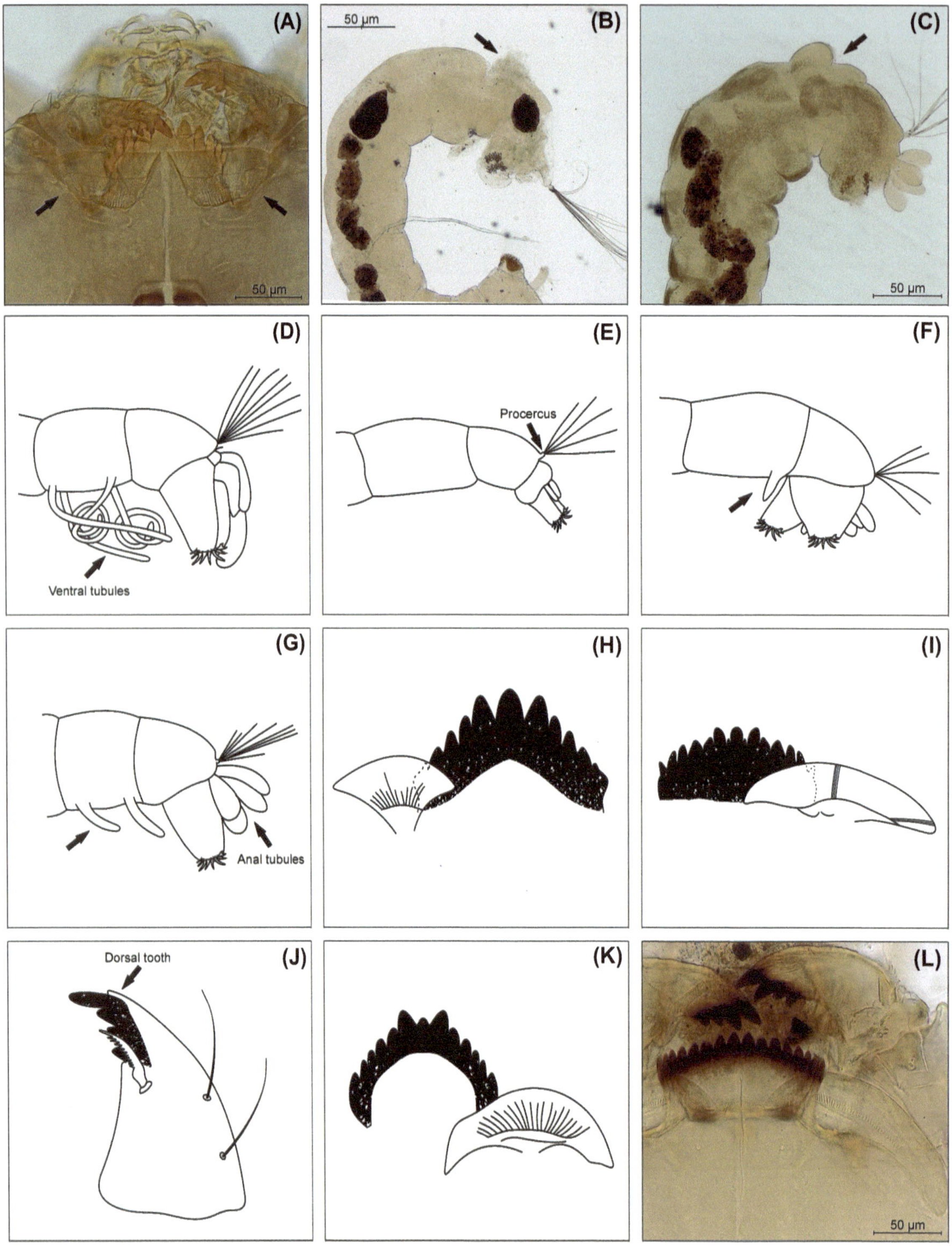

FIGURE 16.2.12 Chironomidae larva: (A, B) *Zavreliella* sp.; (C) *Lauterborniella* sp; (D) *Chironomus* sp.; (E) *Beardius* sp.; (F, H) *Dicrotendipes* sp.; (G, J, K) *Goeldichironomus* sp.; (I) *Kiefferulus* sp.; (L) *Polypedilum (Asheum)* sp.

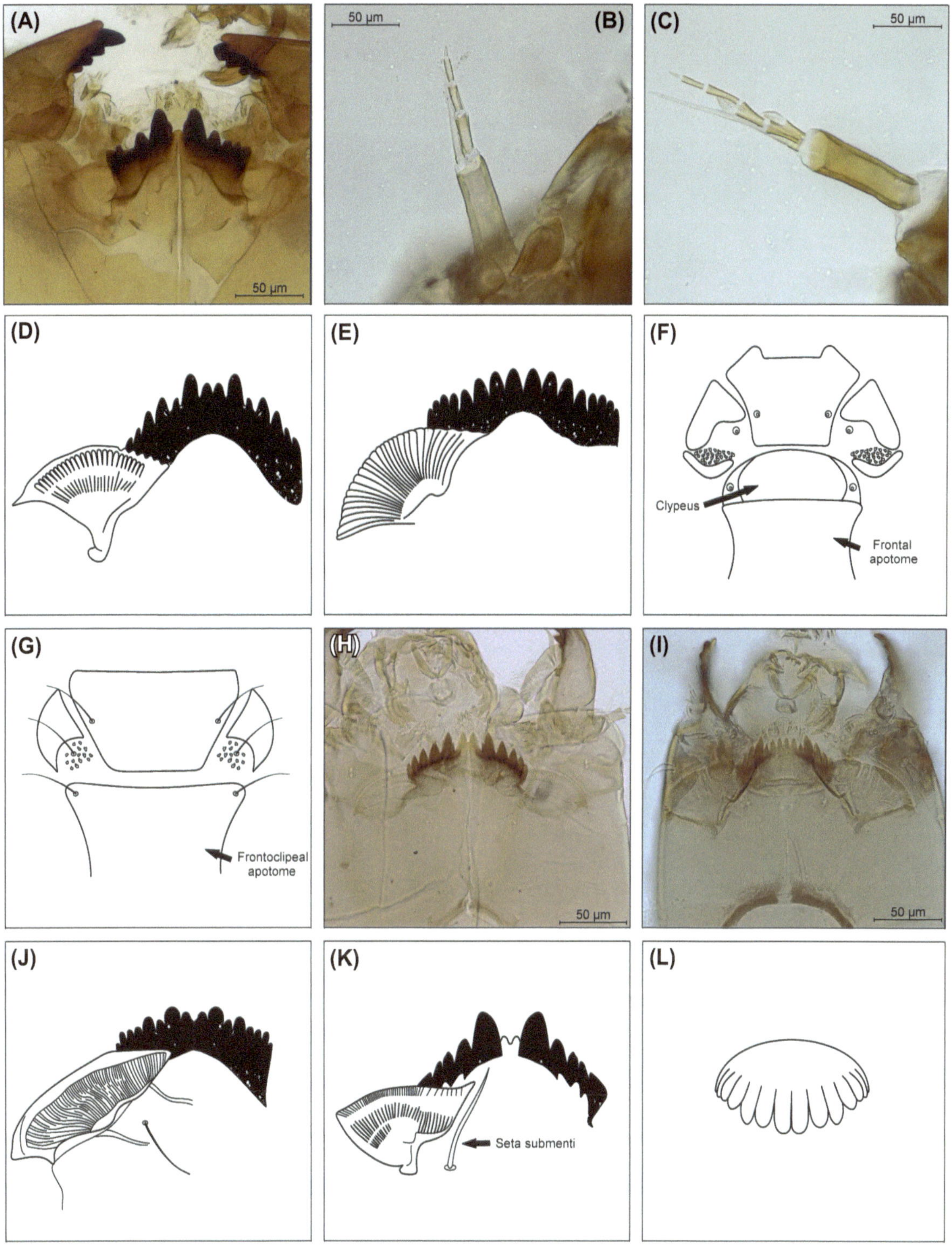

FIGURE 16.2.13 Chironomidae larva: (A, C, K) *Oukuriella* sp.; (B) *Polypedilum (Asheum)* sp.; (D, F) *Endotribelos* sp.; (E) *Polypedilum* sp.; (G) *Phaenopsectra* sp.; (H) *Apedilum* sp.; (I) *Paratendipes* sp.; (J) *Sigmoitendipes* sp.; (L) *Claudiotendipes* sp.

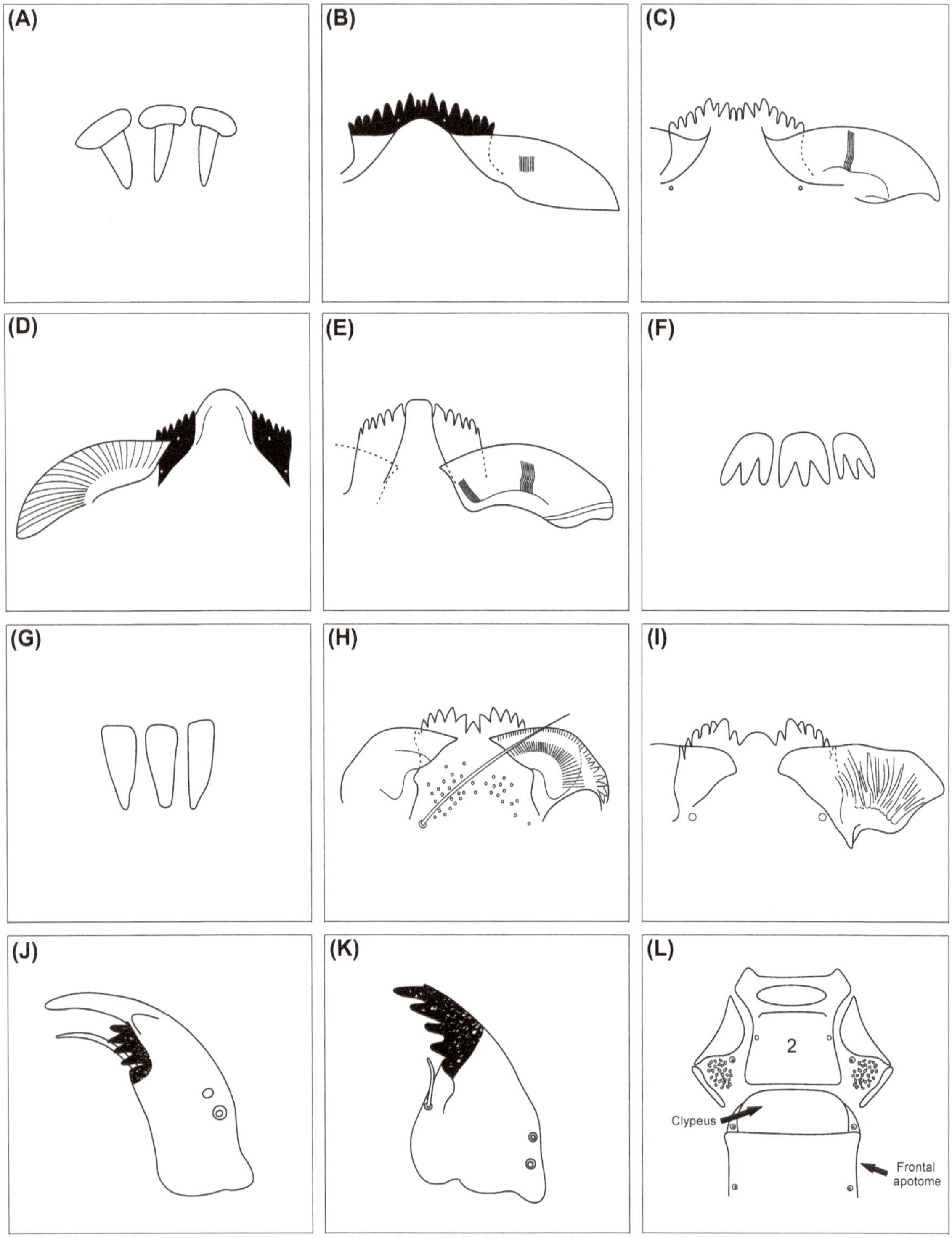

FIGURE 16.2.14 Chironomidae larva: (A) *Paratendipes* sp.; (B) *Stictochironomus* sp.; (C) *Omisus* sp.; (D) *Paralauterborniella* sp.; (E, J) *Nilothauma* sp.; (F) *Beardius* sp.; (G) *Oukuriella* sp.; (H) *Kribiodorum* sp.; (I) *Xenochironomus* sp; (K, L) *Endotribelos* sp.

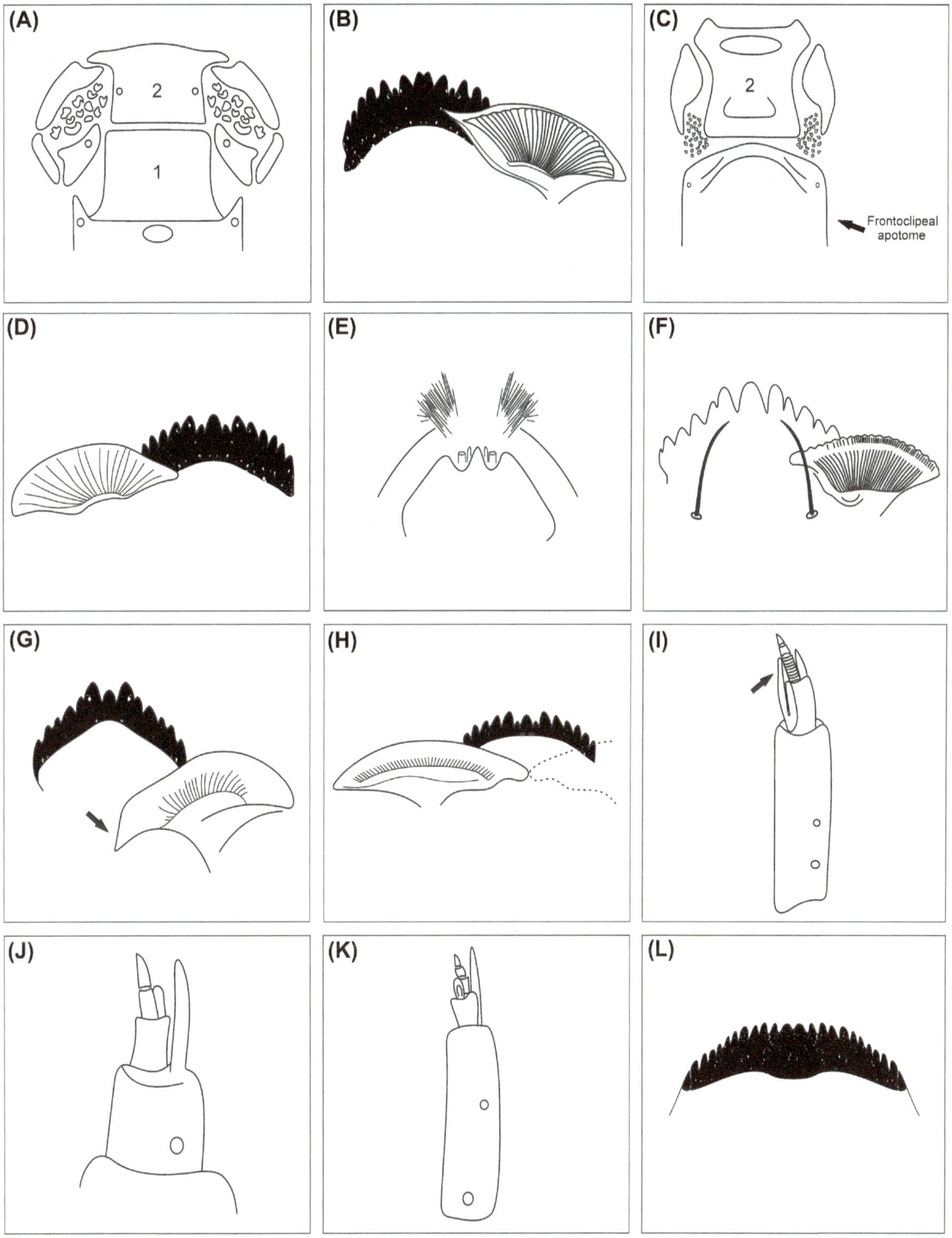

FIGURE 16.2.15 Chironomidae larva: (A, F) *Dicrotendipes* sp.; (B) *Endotribelos* sp.; (C, D) *Chironomus* sp.; (E) *Xenochironomus* sp.; (G) *Goeldichironomus* sp.; (H) *Axarus* sp.; (I) *Diamesa* sp.; (J) *Telmatogeton* sp.; (K, L) *Paraheptagyia* sp.

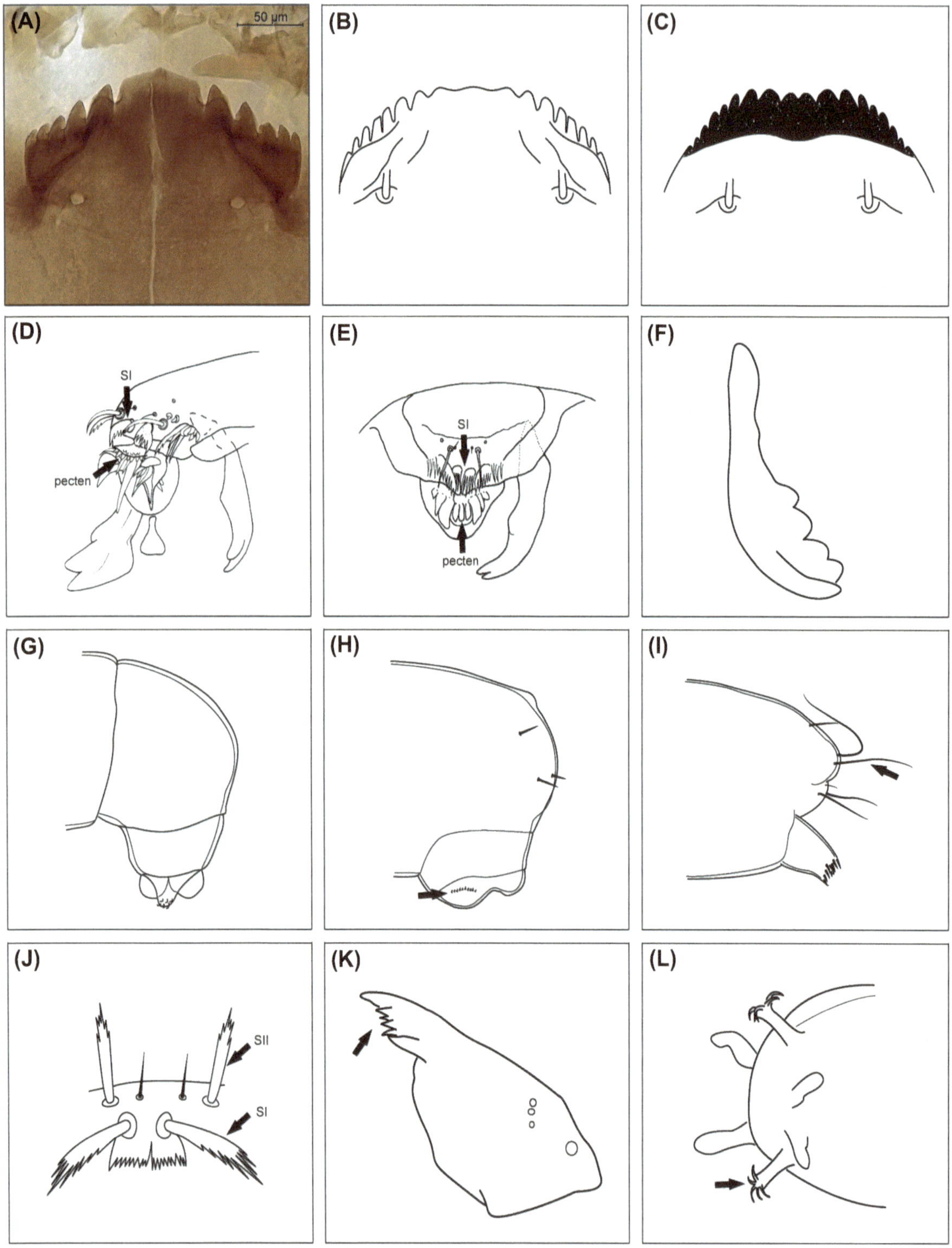

FIGURE 16.2.16 Chironomidae larva: (A) *Limayia* sp.; (B) *Reissmesa* sp.; (C) *Heptagyia* sp.; (D) *Antillocladius* sp.; (E) *Smittia* sp.; (F) *Eretmoptera* sp.; (G) *Bryophaenocladius* sp.; (H) *Gymnometriocnemus* sp.; (I, J) *Clunio* sp.; (K) *Symbiocladius* sp.; (L) *Pseudosmittia* sp.

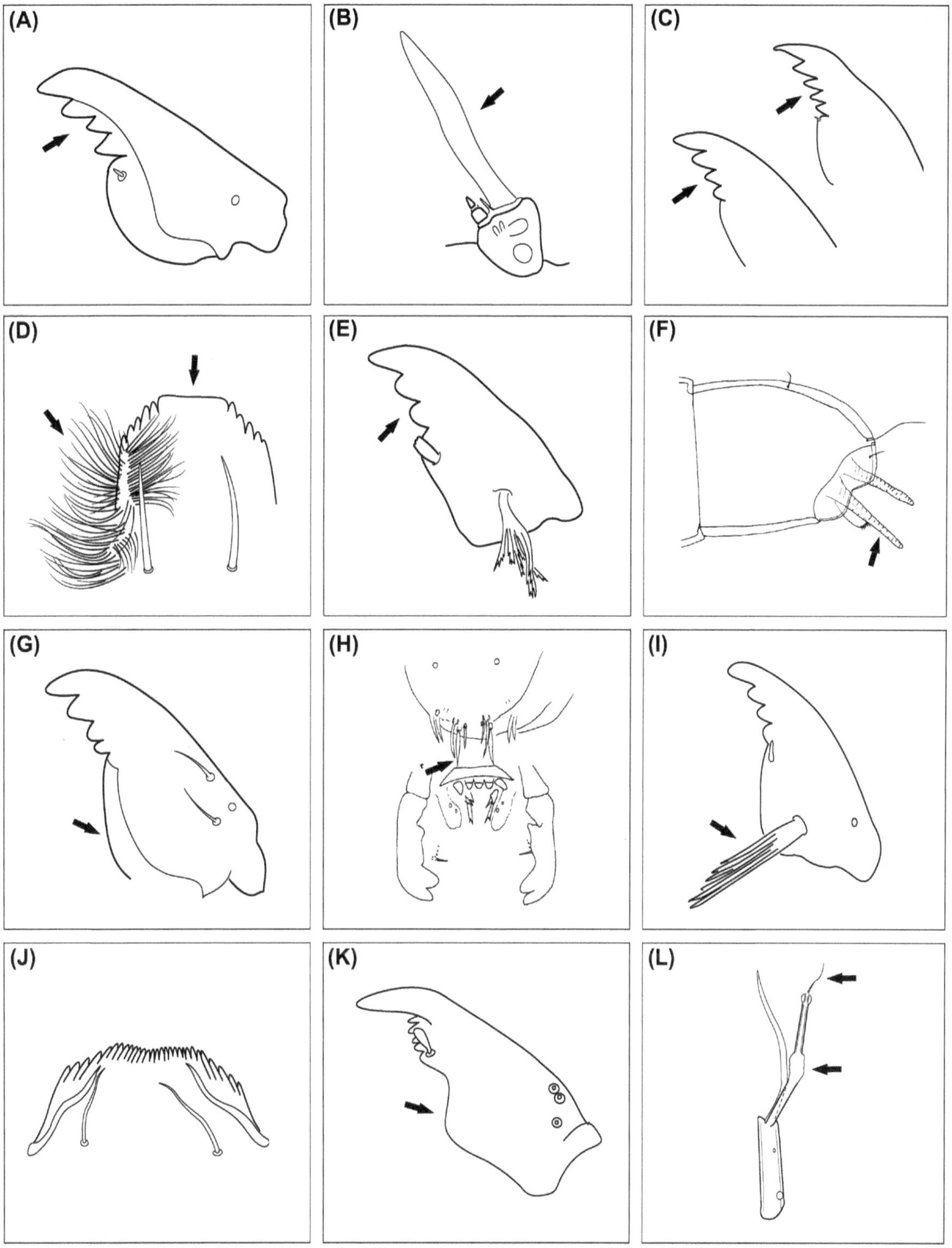

FIGURE 16.2.17 Chironomidae larva: (A, B) *Pseudosmittia* sp.; (C) *Allocladius* sp.; (D) *Barbadocladius* sp; (E, F) *Georthocladius* sp.; (G, H) *Mesosmittia* sp.; (I) *Smittia* sp.; (J, K) *Ichthyocladius* sp.; (L) *Stictocladius* sp.

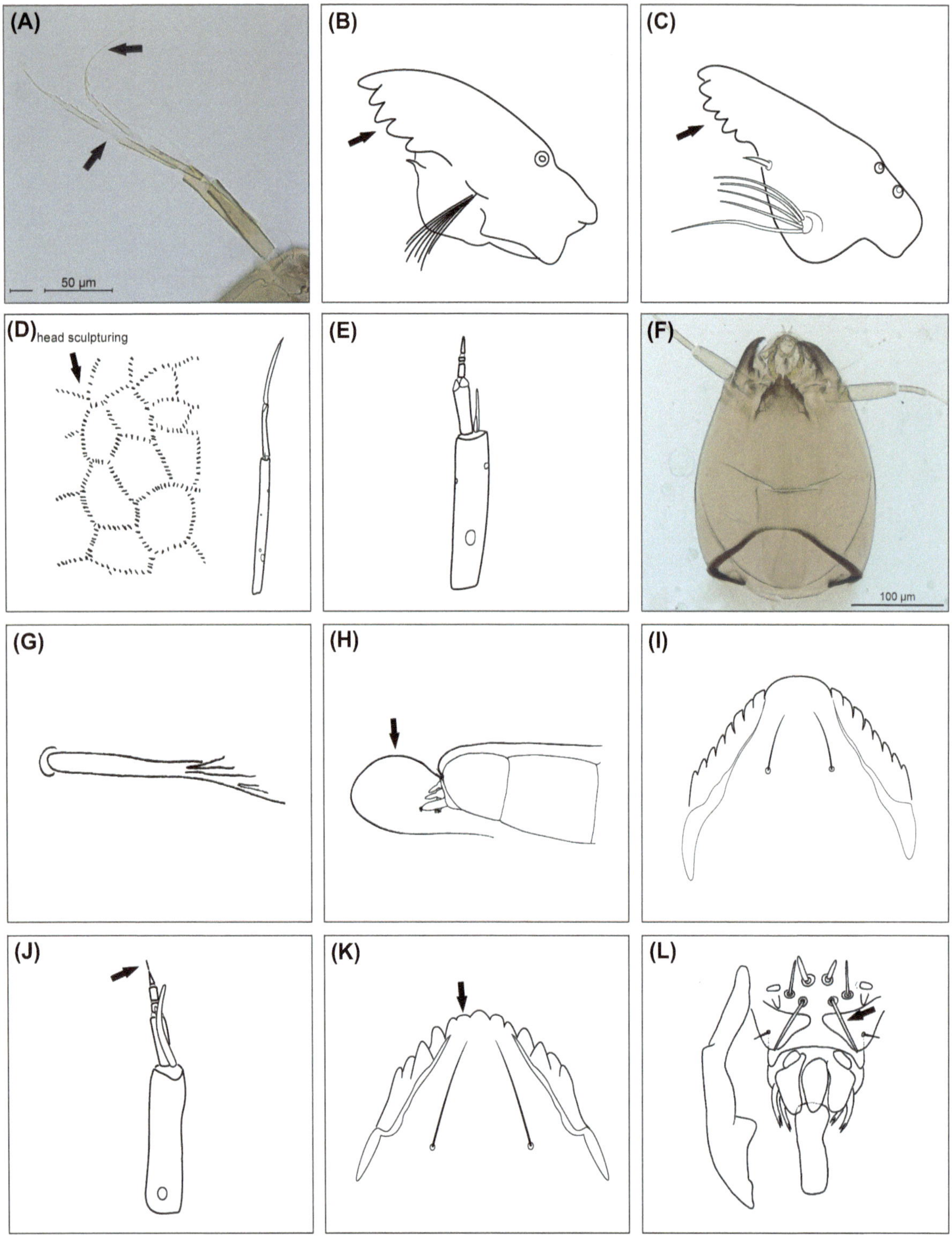

FIGURE 16.2.18 Chironomidae larva: (A, C) *Lopescladius* sp.; (B) *Stictocladius* sp.; (D) *Corynoneura* sp.; (E) *Ubatubaneura* sp.; (F) *Thienemanniella* sp.; (G) *Ubatubaneura* sp.; (H) *Pseudorthocladius* sp.; (I) *Orthocladius* sp.; (J) *Parakiefferiella* sp.; (K) *Orthocladius* sp.; (L) *Synorthocladius* sp.

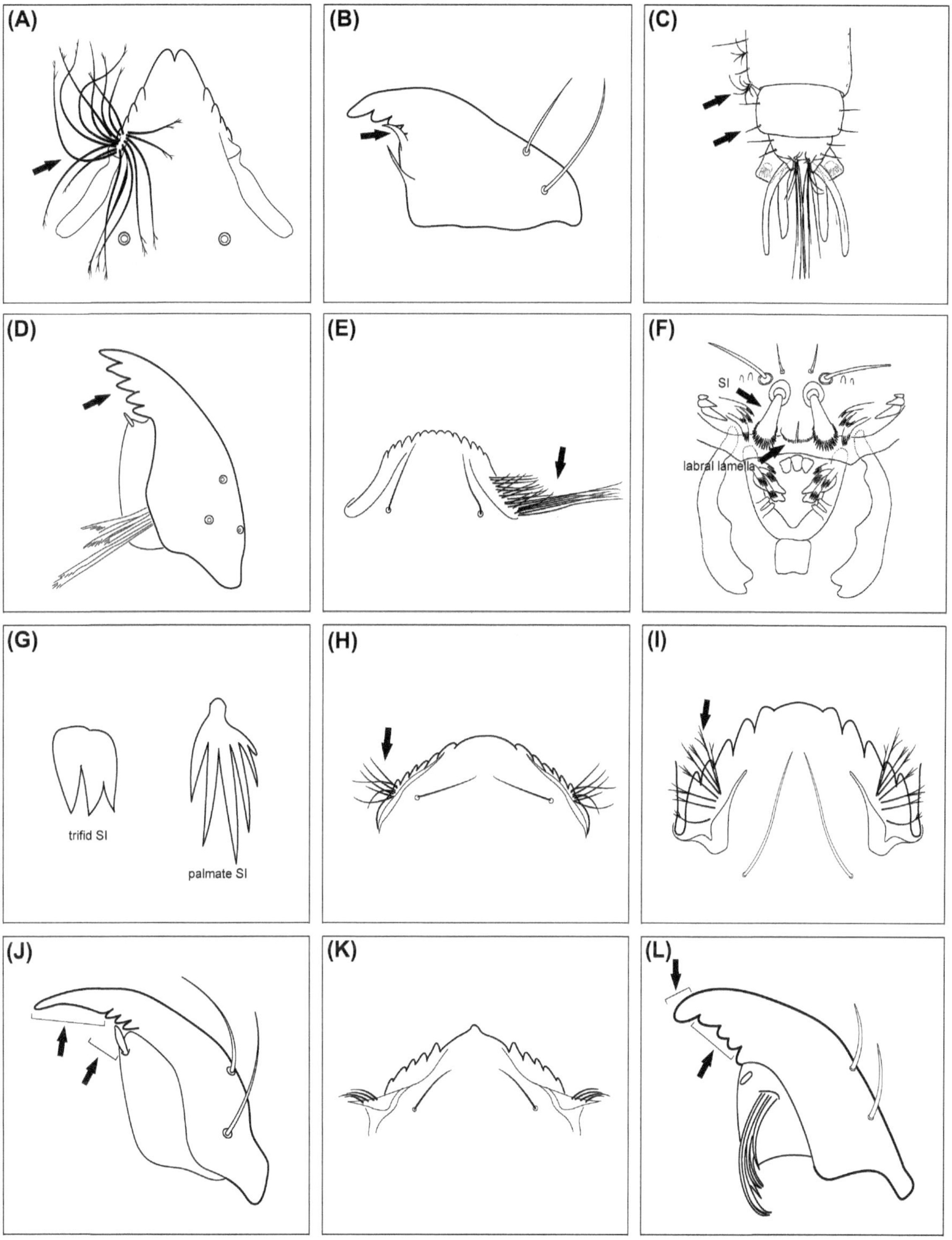

FIGURE 16.2.19 Chironomidae larva: (A–C) *Synorthocladius* sp; (D–F) *Diplocladius* sp.; (G, J, K) *Psectrocladius* sp.; (H) *Paracladius* sp.; (I) *Botryocladius* sp.; (L) *Rheocricotopus* sp.

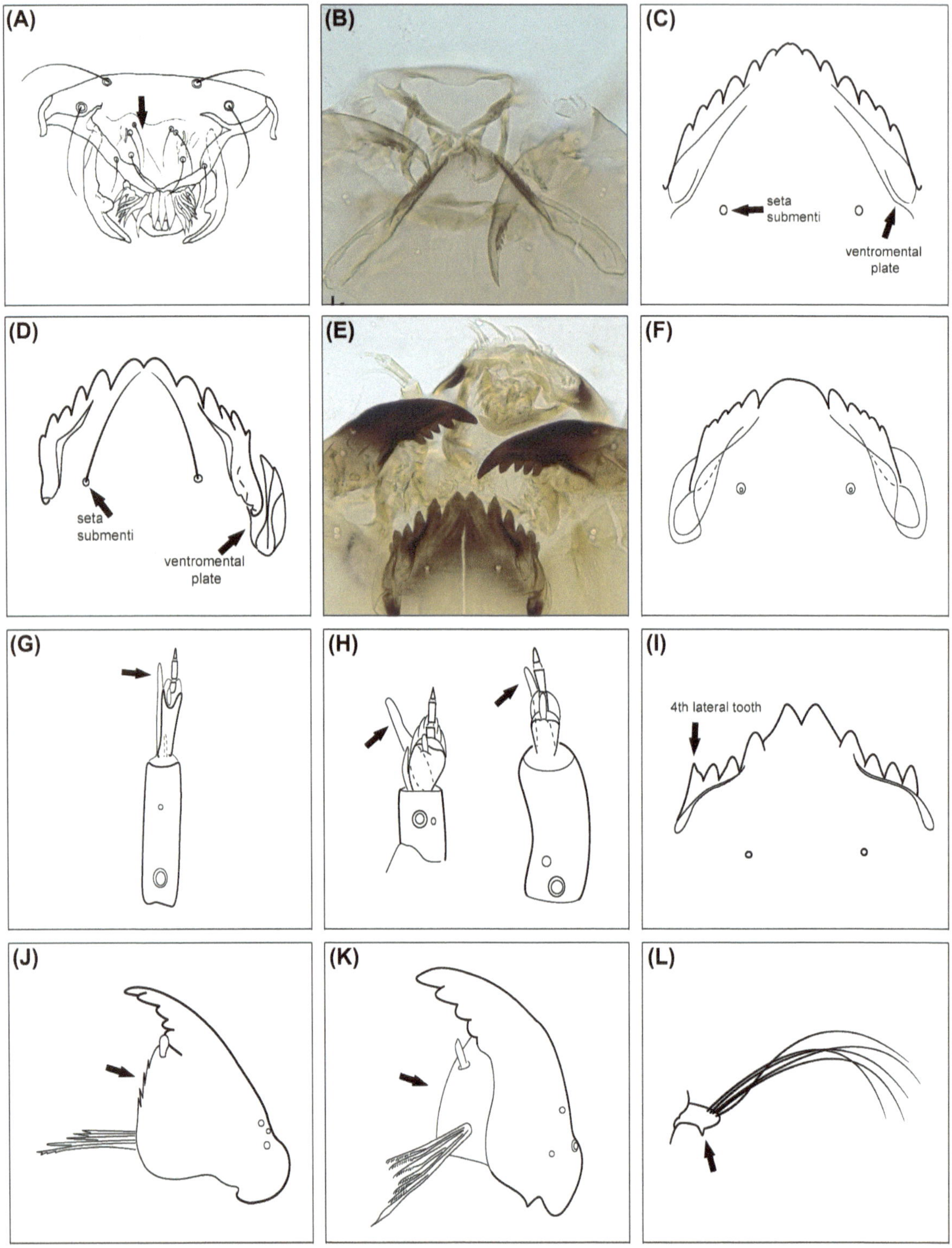

FIGURE 16.2.20 Chironomidae larva: (A, B) *Nanocladius* sp.; (C) *Parakiefferiella* sp.; (D, E, G) *Parametriocnemus* sp.; (F, H) *Paraphaenocladius* sp.; (I) *Paratrissocladius* sp.; (J) *Eukiefferiella* sp.; (K, L) *Paracricotopus* sp.

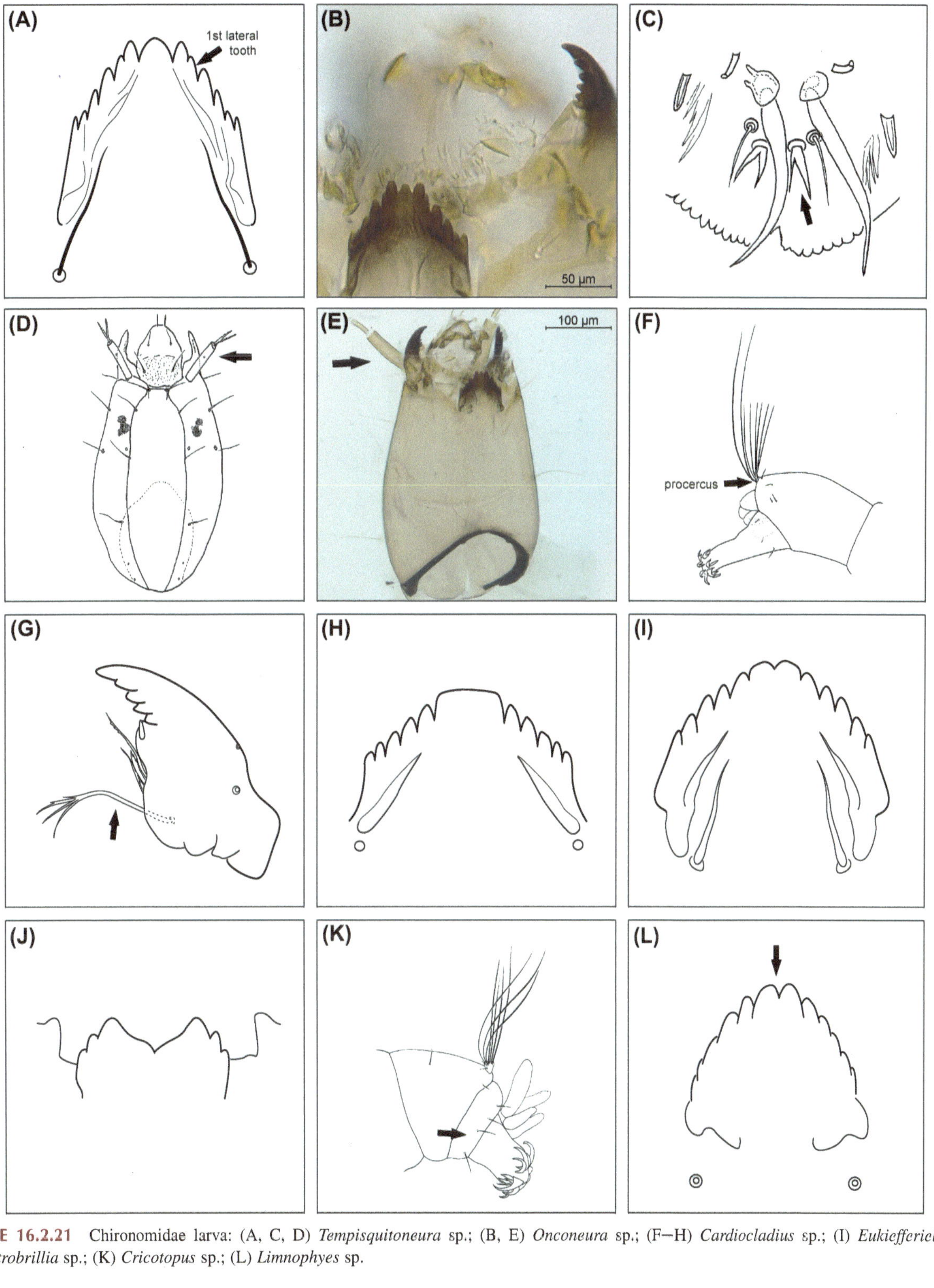

FIGURE 16.2.21 Chironomidae larva: (A, C, D) *Tempisquitoneura* sp.; (B, E) *Onconeura* sp.; (F–H) *Cardiocladius* sp.; (I) *Eukiefferiella* sp.; (J) *Austrobrillia* sp.; (K) *Cricotopus* sp.; (L) *Limnophyes* sp.

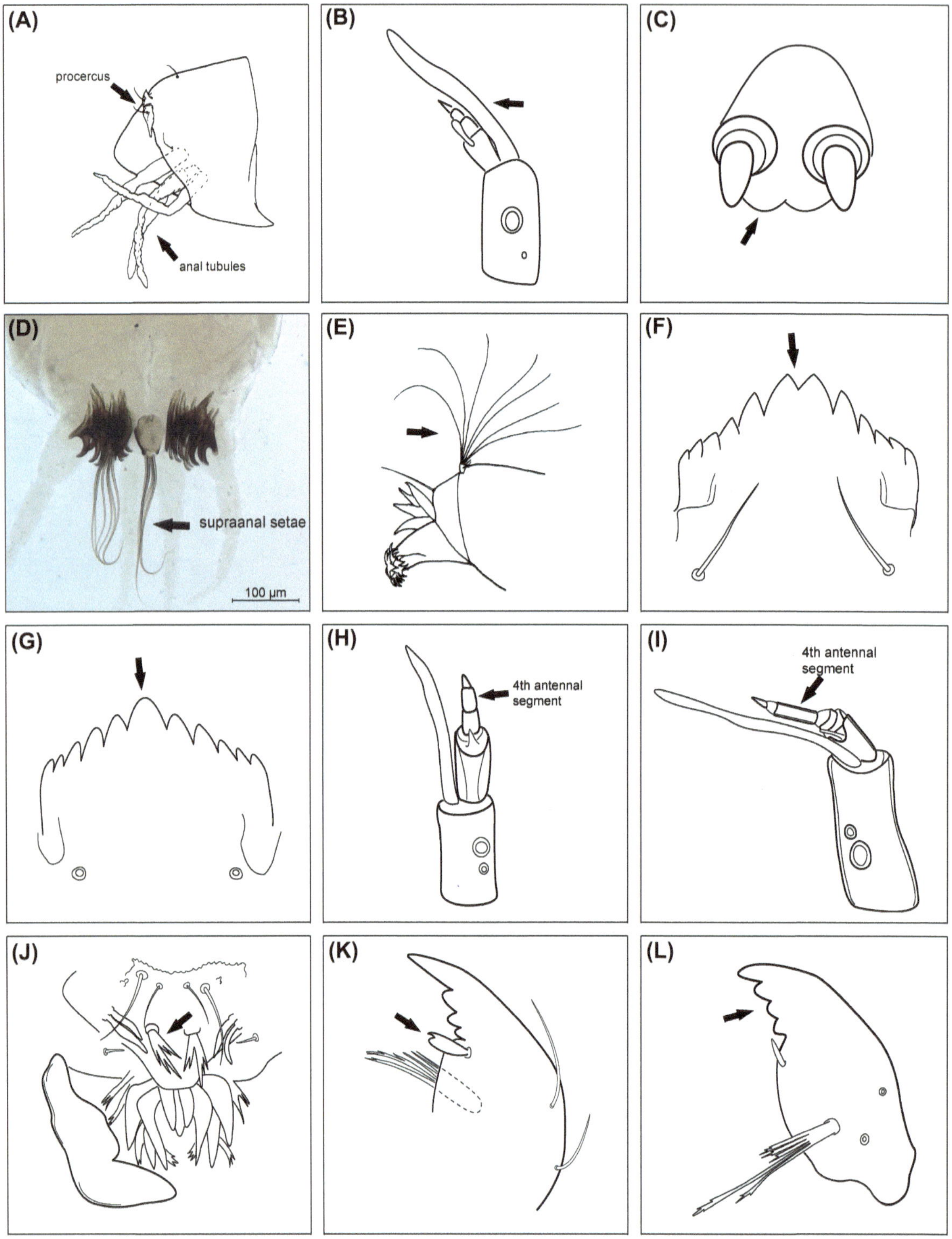

FIGURE 16.2.22 Chironomidae larva: (A, B) *Georthocladius* sp; (C, D) *Metriocnemus* sp.; (E) *Thienemannia* sp.; (F) *Phytotelmatocladius* sp.; (G, I) *Gynocladius* sp.; (H) *Gravatamberus* sp.; (J, K) *Parapsectrocladius* sp; (L) *Limnophyes* sp.

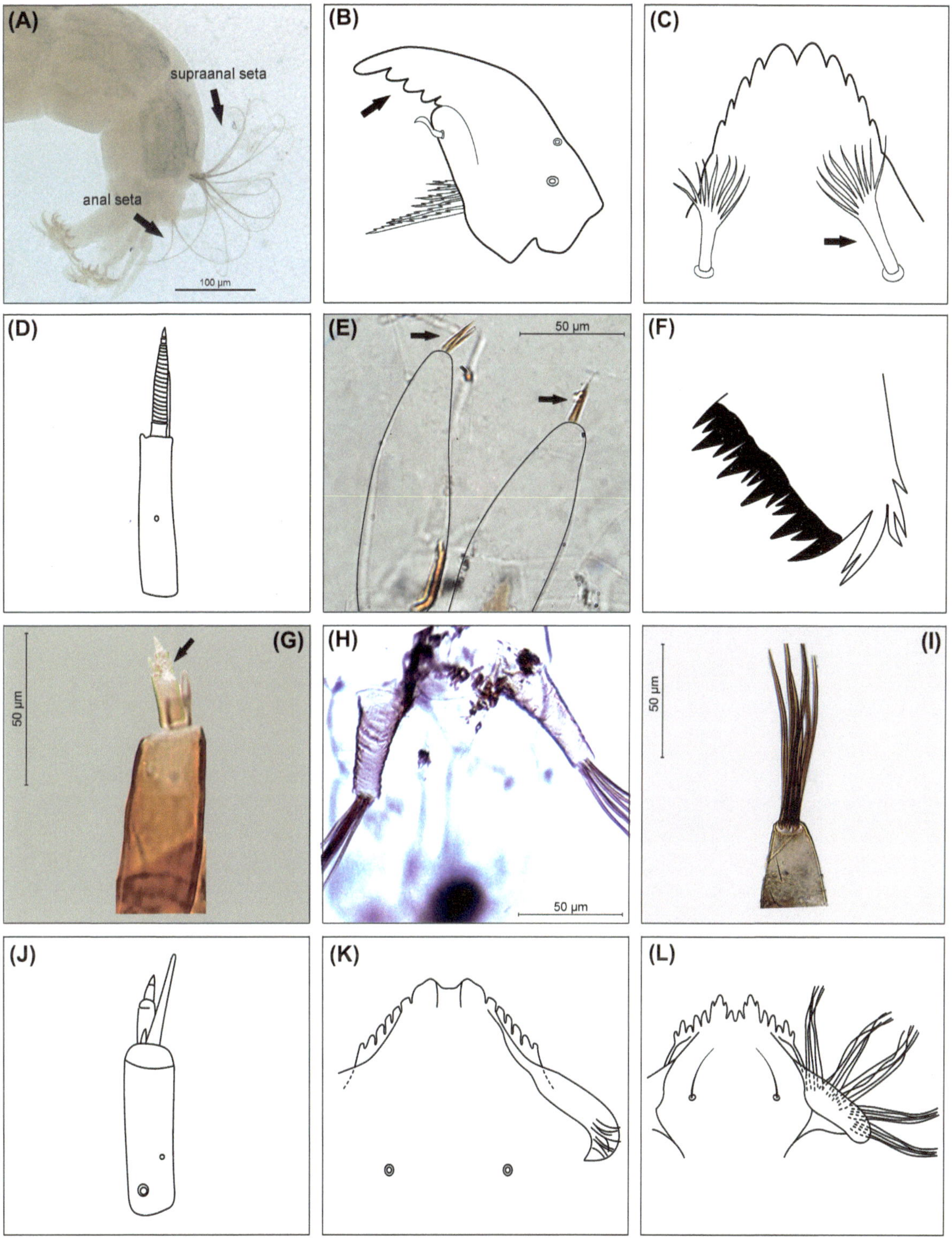

FIGURE 16.2.23 Chironomidae larva: (A) *Limnophyes* sp.; (B) *Compterosmittia* sp.; (C) *Pirara* sp.; (D, E) *Podonomopsis* sp.; (F) *Podochlus* sp.; (G, H) *Parochlus* sp.; (I) *Podonomus* sp.; (J, L) *Prodiamesa* sp.; (K) *Monodiamesa* sp.

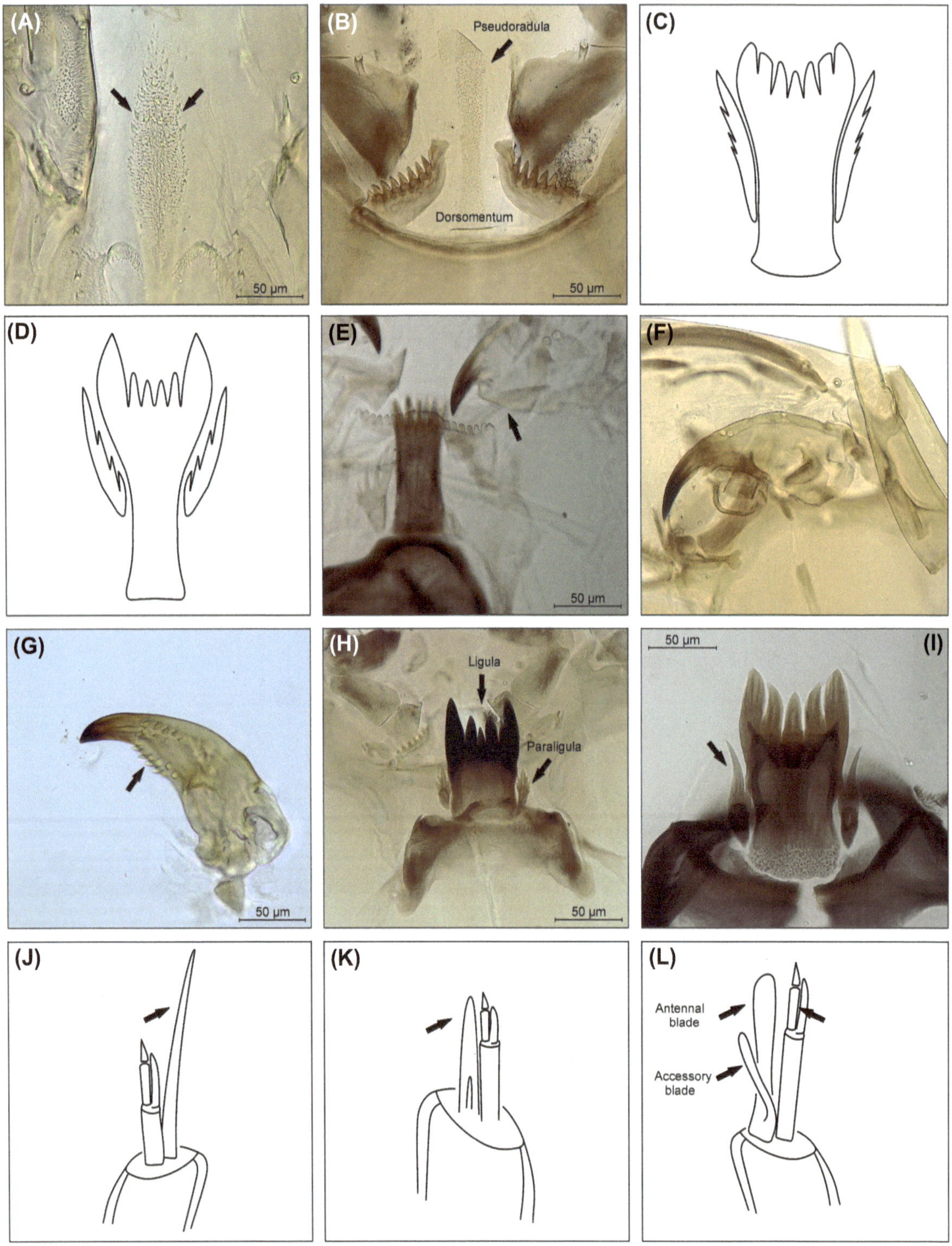

FIGURE 16.2.24 Chironomidae larva: (A, D) *Clinotanypus* sp.; (B, L) *Alotanypus* sp.; (C) *Coelotanypus* sp.; (E) *Tanypus* sp.; (F) *Brundiniella* sp.; (G) *Fittkauimyia* sp.; (H, J) *Djalmabatista* sp.; (I) *Apsectrotanypus* sp.; (K) *Procladius* sp.

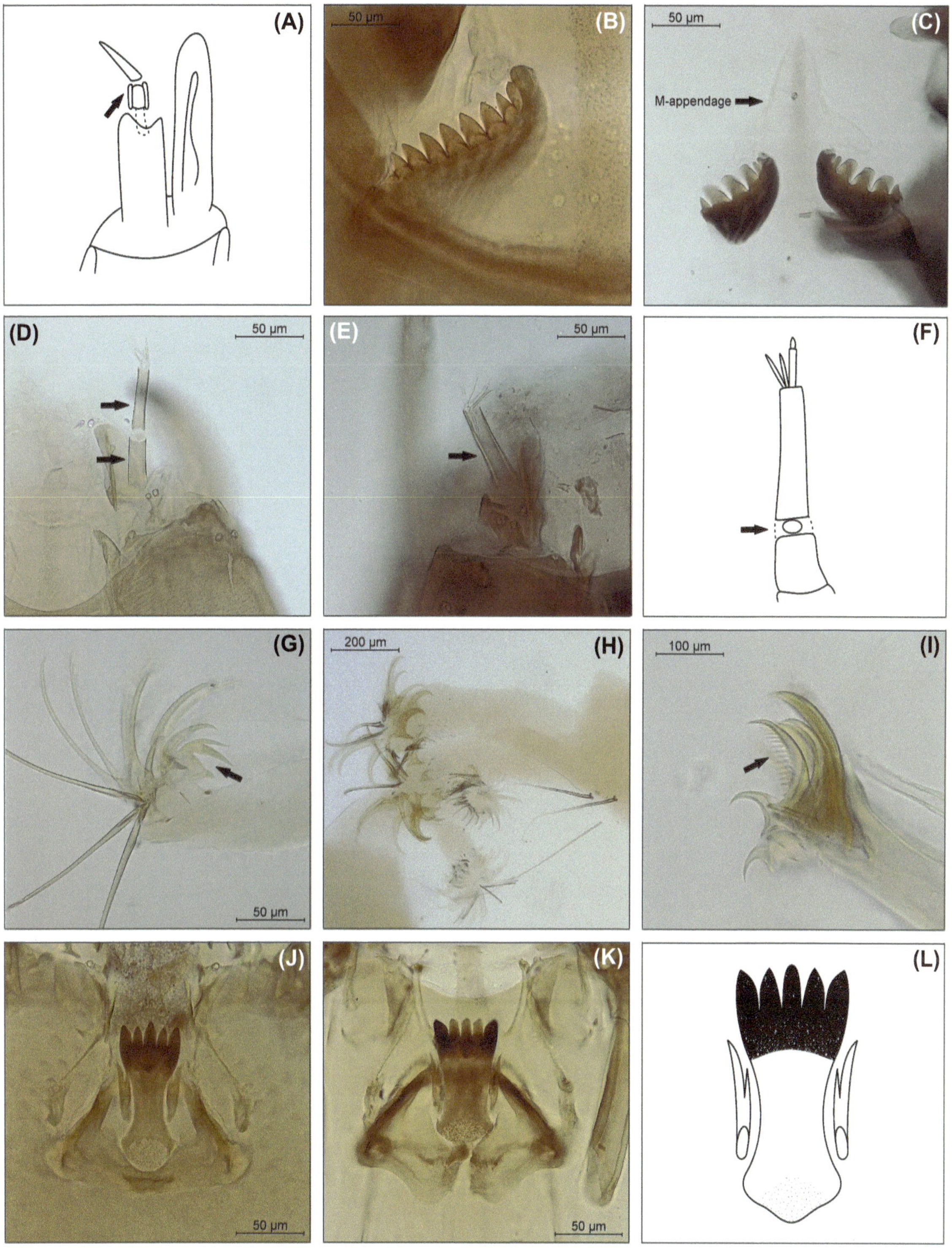

FIGURE 16.2.25 Chironomidae larva: (A) *Brundiniella* sp.; (B) *Macropelopia* sp.; (C) *Apsectrotanypus* sp.; (D) *Ablabesmyia* sp.; (E) *Metapelopia* sp.; (F) *Zavrelimyia* (*Paramerina*) sp.; (G, J) *Labrundinia* sp.; (H) *Pentaneura* sp.; (I) *Nilotanypus* sp.; (K) *Denopelopia* sp.; (l) *Zavrelimyia* (*Zavrelimyia*) sp.

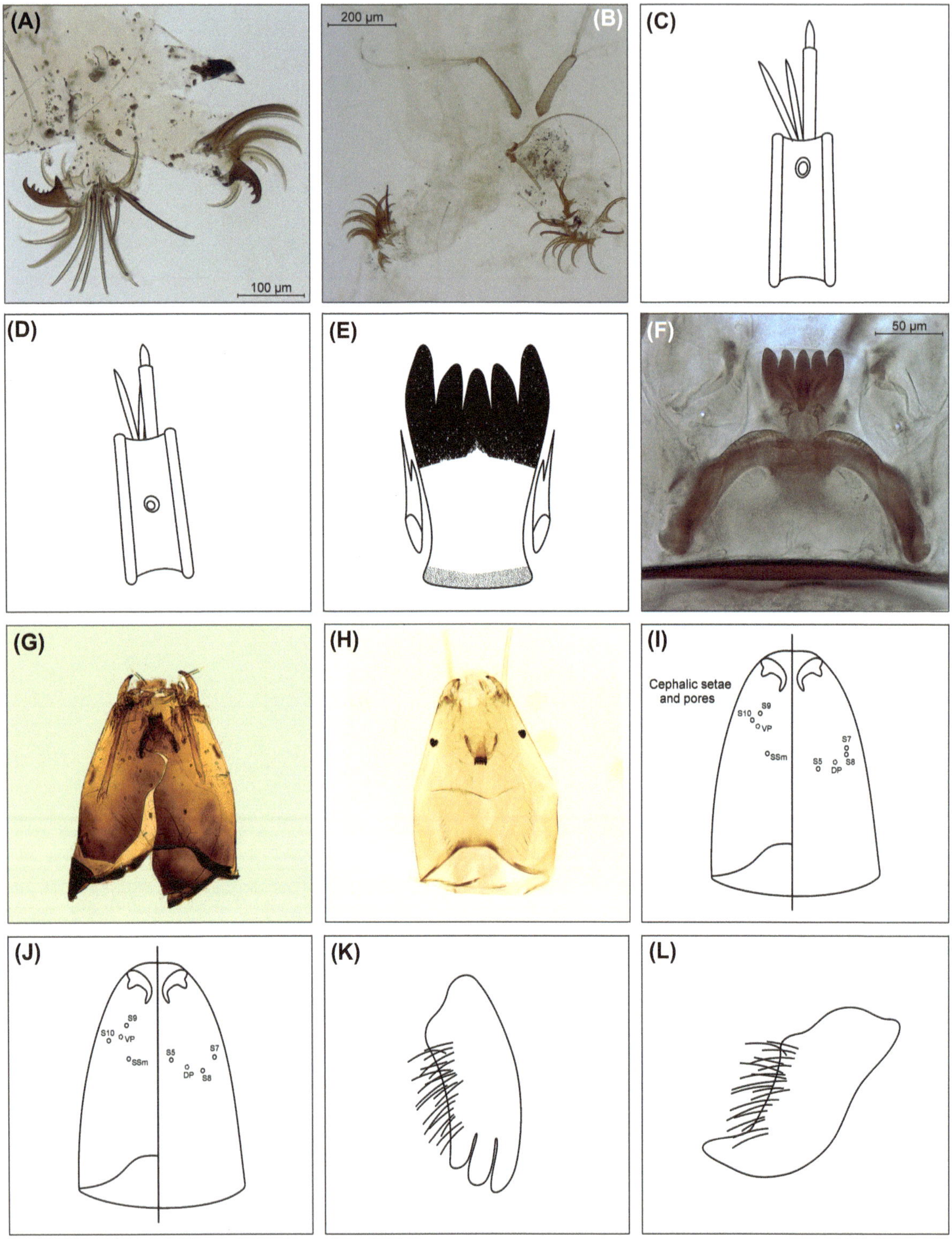

FIGURE 16.2.26 Chironomidae larva: (A) *Monopelopia* sp.; (B) *Hudsonimyia* sp.; (C) *Thienemannimyia* sp.; (D, E, H) *Larsia* sp.; (F, G) *Hudsonimyia* sp.; (I) *Pentaneura* sp.; (J) *Parapentaneura* sp.; (K) *Telmatogeton* sp.; (L) *Thalassomyia* sp.

Tanypodinae: Clinotanypodini: Genera

1 Ligula with seven teeth (rarely six or eight) (Fig. 16.2.24 C), first inner tooth strongly bent toward outer tooth; mandible slightly hooked; dorsal anterior margin of body segment 4 with a pair of sclerotized hooks .... *Coelotanypus*
1' Ligula with six teeth (rarely five or seven) (Fig. 16.2.24 D), first inner tooth slightly bent toward outer tooth; mandible strongly hooked; body segment 4 without sclerotized hooks .... *Clinotanypus*

Tanypodinae: Procladiini: Genera

1 Antennal blade extending well beyond apex of flagellum (Fig. 16.2.24 J); ligula with four or five teeth........ *Djalmabatista*
1' Antennal blade about as long as flagellum (Fig. 16.2.24 K); ligula with five teeth........ *Procladius*

Tanypodinae: Macropelopiini: Genera

1 Third antennal article shorter, about as long as wide (Fig. 16.2.25 A); ventrolateral mandibular setae 2 and 3 branched........ 2
1' Third antennal article at least twice as long as wide (Fig. 16.2.24 L); ventrolateral mandibular seta all simple........ *Alotanypus*
2(1) Apex of dorsomental plates not medially pointed, well separated pseudoradula........ 3
2' Apex of dorsomental plates with pointed medial extension, reaching almost to pseudoradula........ *Brundiniella*
3(2) Dorsomentum with four large teeth and one small tooth (Fig. 16.2.25 C); ligula with inner lateral teeth curve outward........ 4
3' Dorsomentum with six to eight large inner teeth and one small tooth (Fig. 16.2.25 B); ligula with inner lateral teeth directed forward... .... *Macropelopia*
4(3) Ventral cephalic setae S10 simple........ *Apsectrotanypus*
4' Ventral cephalic setae S10 branched........ *Paggipelopia*

Chironomidae: Chironominae: Tribes

This subfamily has the most described species. This group is dominant in tropical and subtropical lowlands, with immature stages adapted to a wide range of habitats, including brackish and marine (Ashe et al., 1987). Larvae utilize a wide array of foods, including aquatic plants, dead wood and leaves, and organic-enriched sediments. Several species, belonging to genera such as *Chironomus* Meigen, *Endotribelos* Grodhaus, and *Polypedilum* Kieffer, have hemoglobin, which may contribute to their dominance where oxygen concentrations are often low due to higher temperatures. The group can be distinguished based on the larval synapomorphy of ventromental plate associated with silk extrusion. The reduction or loss of this trait in the *Stenochironomus* Kieffer complex is confirmed as secondary (Cranston et al., 2012). Chironominae is subdivided into three tribes: Chironomini, Pseudochironomini, and Tanytarsini. It is important to mention that several morphotypes without generic status have been described in the *Harnischia* complex, and they were not included in the key.

1 Antennae not mounted on distinct elongated base (Fig. 16.2.6 H), Lauterborn organs not mounted on pedicels........ 2
1' Antennae mounted on distinct elongated base (Fig. 16.2.6 G), Lauterborn organs usually well developed and often mounted on short to long pedicels .... **Tanytarsini [p. 691]**
2(1) SI plumose or fringed (Fig. 16.2.8 I), SII never blade-like; pecten epipharyngis as a wide, multitoothed comb or as three separated small plates, toothed apically (occasionally smooth); mandible usually with dorsal tooth; maxillary palps usually reduced........ 3
2' SI and SII simple (Fig. 16.2.8 H), SII often large and blade-like; pecten epipharyngis rounded or subtriangular, consisting of a single plate or scale, which may be simple, serrated, notched, or toothed; mandible without dorsal tooth; maxillary palps often very long, as long as the half of the first antennal article .... **Chironomini [p. 692]** (in part)
3(2) Ventromental plates bar-like in near contact medially (Fig. 16.2.10 I); seta submentalis dorsal, on the same side of mandible as seta interna (Fig. 16.2.10 J)........ **Pseudochironomini [p. 694]**
3' Ventromental plates variable in shape (Fig. 16.2.10 K); seta submentalis ventral, on the opposite side of mandible from seta interna (rarely absent) (Fig. 16.2.10 L)........ **Chironomini [p. 692]** (in part)

Chironomidae: Chironominae: Tanytarsini: Genera

1 Ventromental plates narrow, separated medially by at least the width of the three median teeth (Fig. 16.2.6 I); Lauterborn organs not mounted on long pedicels; larvae build transportable cases (Fig. 16.2.6 J)........ 2
1' Ventromental plates wide, almost touching each other medially (Fig. 16.2.6 K); Lauterborn organs mounted on long pedicels; if larvae in cases, cases not transportable........ 5
2(1) One of the Lauterborn organs at the base of antennal article 2 and the other placed apically (Fig. 16.2.6 L)........ 3
2' Both Lauterborn organs at apex of antennal article 2 (Fig. 16.2.7 A)........ 4
3(2) Premandible with four teeth (Fig. 16.2.7 B)........ *Zavrelia*
3' Premandible with two to three teeth (Fig. 16.2.7 C)........ *Stempellinella*
4(2) Antennal base with palmate process (Fig. 16.2.7 D); frontoclypeal setae simple or bifid (Fig. 16.2.7 D)........ *Stempellina*
4' Antennal base with simple spur (Fig. 16.2.7 E); frontoclypeal setae branched (Fig. 16.2.7)........ *Constempellina*

5(1) Premandible with three to five main teeth (Fig. 16.2.7 F) 6
5' Premandible with two main teeth (Fig. 16.2.7 G) 9
6(5) Lauterborn organs on distinct pedicels (Fig. 16.2.7 A); larvae living in freshwater 7
6' Lauterborn organs small, sessile (Fig. 16.2.7 H); larvae living in marine habitats .... *Pontomyia*
7(6) Antennal article 2 longer than 3, Lauterborn organs usually small, on long pedicels (Fig. 16.2.7 J) 8
7' Antennal article 2 as long or shorter than antennal article 3, Lauterborn organs large, almost as large as their pedicels (Fig. 16.2.7 I) .... *Cladotanytarsus*
8(7) Pedicels of Lauterborn organs usually with basal half more sclerotized or annulated (Fig. 16.2.7 A) .... *Tanytarsus (Caladomyia)*
8' Pedicels of Lauterborn organs uniformly sclerotized (Fig. 16.2.7 J) .... *Tanytarsus* s. str.
9(5) Pecten epipharyngis as one plate with three to five lobes (Fig. 16.2.7 K) 10
9' Pecten epipharyngis as one or three plates with several apical teeth (Fig. 16.2.7 L) 11
10(9) Lauterborn organs on pedicels longer than flagellum (Fig. 16.2.8 A); mandible with distinct outer hump .... *Sublettea*
10' Lauterborn organs sessile or on pedicels shorter than flagellum (Fig. 16.2.8 B); mandible without outer hump... *Paratanytarsus* (in part)
11(9) Lauterborn organs on short pedicels, not surpassing the flagellum (Fig. 16.2.8 C) 12
11' Lauterborn organs on longer pedicels (Fig. 16.2.8 D) 13
12(11) Mentum with four or five pairs of lateral teeth (Fig. 16.2.8 E); Lauterborn organs on moderately long pedicels .... *Neozavrelia*
12' Mentum with five pairs of lateral teeth (Fig. 16.2.8 F); Lauterborn organs on short pedicels .... *Rheotanytarsus*
13(11) Pedicels 3−5× longer than combined length of antennal articles 3−5 (Fig. 16.2.8 G) .... *Micropsectra*
13' Pedicels slightly longer than combined length of antennal articles 3−5 (Fig. 16.2.8 D) .... *Paratanyarsus* (in part)

Chironomidae: Chironominae: Chironomini: Genera

1 SI and SII simple (Fig. 16.2.8 H), SII often large and laminate; pecten epipharyngis rounded or subtriangular, consisting of a single plate or scale, which may be simple, serrated, notched, or toothed; mandible without dorsal tooth; maxillary palps often very long, as long as the half of the first antennal article, *Harnischia* complex 2
1' SI plumose or fringed (Fig. 16.2.8 I), SII never laminate; pecten epipharyngis as a wide, multitoothed comb or as three separated small plates, toothed apically (occasionally smooth); mandible usually with dorsal tooth; maxillary palps usually reduced 13
2(1) Mentum concave with one large pale tooth (Fig. 16.2.8 J), or with a median large break between the dark lateral teeth (Fig. 16.2.9 A) 3
2' Mentum convex or linear (Fig. 16.2.8 K) 5
3(2) Antenna with five to six articles (Fig. 16.2.9 B); mentum with four to five pairs of lateral teeth (Fig. 16.2.8 J) 4
3' Antenna with seven articles (Fig. 16.2.8 L); mentum with seven pairs of lateral teeth (Fig. 16.2.9 A) .... *Demicryptochironomus*
4(3) S I minute, less than 1/5 length of S II (Fig. 16.2.9 C); pecten epipharyngis as a weakly trilobed scale; premandible without brush (Fig. 16.2.9 D); lateral teeth of mentum obliquely arranged .... *Gillotia*
4' S I at least ½ as long as S II (Fig. 16.2.9 E); pecten epipharyngis as a deeply trilobed scale; premandible with brush (Fig. 16.2.9 F); lateral teeth of mentum not obliquely arranged .... *Cryptochironomus*
5(2) Antenna with five articles (Fig. 16.2.9 G) 6
5' Antenna with six to seven articles (Fig. 16.2.9 H) 11
6(5) Pecten epipharyngis not scaliform undivided, or partially divided into two to three lobes or teeth, often greatly reduced; anterior margin of ventromental plate smooth (Fig. 16.2.9 J) 7
6' Pecten epipharyngis with numerous transparent teeth; anterior margin of ventromental plate serrated or sinuate (Fig. 16.2.9 I) .... *Parachironomus*
7(6) Premandible apically bifid (Fig. 16.2.9 K) 8
7' Premandible with three or more teeth (Fig. 16.2.9 L) 10
8(7) Antennal blade shorter than flagellum; mentum with single median tooth (occasionally notched, appearing trifid) (Fig. 16.2.9 J) 9
8' Antennal blade as long or longer than flagellum; mentum with median tooth trifid (Fig. 16.2.10 A) .... *Microchironomus*
9(8) First antennal article short, 2−2.5× longer than wide (Fig. 16.2.9 G); median tooth broadly rounded or laterally notched, extending far forward of lateral teeth .... *Cryptotendipes*
9' First antennal article longer than previous (Fig. 16.2.10 B); median tooth broad notched, may be double, not extending far forward of lateral teeth; mentum appearing rounded .... *Cladopelma*
10(7) Antennal articles 2 and 3 subequal (Fig. 16.2.10 C); ventromental plates weakly striated .... *Harnischia*
10' Antennal article 2 much longer than 3 (Fig. 16.2.10 D); ventromental plates coarsely striated .... *Paracladopelma*
11(5) Antenna with six articles; mandible without modified inner teeth (Fig. 16.2.10 F) 12
11' Antenna with seven articles; mandible with modified inner teeth (Fig. 16.2.10 E) .... *Robackia*
12(11) Mentum with median teeth paler than lateral teeth, with two or three notches (Fig. 16.2.10 G) .... *Pelomus*
12' Mentum with median tooth paler or not, not notched (Fig. 16.2.10 H) .... ? *Saetheria*
13(1) Mentum with median tooth or teeth deeply recessed (Fig. 16.2.11 E) 14
3' Mentum with median tooth or teeth not deeply recessed (Fig. 16.2.10 K) 315
14(13) Antennae with five articles (Fig. 16.2.11 F); mandible with slender seta subdentalis, seta interna and pecten mandibularis present; mentum with median tooth broad and extended (Fig. 16.2.11 G) .... *Hyporhygma*
14' Antennae with six articles (Fig. 16.2.11 H); mandible with enlarged seta subdentalis, seta interna and pecten mandibularis absent; mentum with median tooth deeply divided (Fig. 16.2.11 E) .... *Fissimentum*
15(13) Mentum concave with 8−10 teeth (Fig. 16.2.11 I); ventromental plates with obscure striae (Fig. 16.2.11 I); larvae mining in dead, submerged wood or in submerged leaves 16

15' Mentum variable; ventromental plates usually with numerous striae (Fig. 16.2.11 J); larvae in a variety of habitats ... 17

16(15) Antennal blade extends to apex of article 2; mentum with 8–10 teeth (Fig. 16.2.11 K) ... *Xestochironomus*

16' Antennal blade surpass apex of article 2; mentum with 10–12 teeth (Fig. 16.2.11 I) ... *Stenochironomus*

17(15) Antenna with six articles with alternate Lauterborn organs at apex of articles 2 and 3 (Fig. 16.2.11 L); ventromental plates triangular (Fig. 16.2.12 A); larvae with body segment 11 with dorsal hump (Fig. 16.2.12 B) ... 18

17' Antenna and ventromental plates with another character combination; larva with segment 11 without dorsal hump ... 19

18(17) Seta submenti simple; hump on segment 11 anteriorly directed (Fig. 16.2.12 B) ... *Zavreliella*

18' Seta submenti plumose; hump on segment 11 posteriorly directed (Fig. 16.2.12 C) ... *Lauterborniella*

19(17) Larva with one or two pairs of ventral tubules on abdominal segment 8 (Fig. 16.2.12 D) ... 20

19' Larva without ventral tubules on abdominal segment 8 (Fig. 16.2.12 E) ... 23

20(19) Larva with one pair of ventral tubules (Fig. 16.2.12 F) ... 21

20' Larva with two pairs of ventral tubules (Fig. 16.2.12 G) ... 22

21(20) Mentum with median tooth simple, robust (Fig. 16.2.12 H); ventromental plates narrow, curved, medially separated by at least $^{1}/_{3}$ of the mentum width ... *Dicrotendipes* (in part)

21' Mentum with median tooth trifid, crenulate (Fig. 16.2.12 I); ventromental plates wide as mentum, separated medially only by the width of median mental tooth (rarely meeting medially) ... *Kiefferulus*

22(20) Seta subdentalis with ventral margin fringed or toothed (Fig. 16.2.12 J); ventromental plates curved medially, almost touching (Fig. 16.2.12 K) ... *Goeldichironomus* (in part)

22' Seta subdentalis simple; ventromental plates medially not curved, well separated (Fig. 16.2.10 K) ... *Chironomus* (in part)

23(19) Mentum with an even number of teeth (Fig. 16.2.12 L) ... 24

23' Mentum with an odd number of teeth (Fig. 16.2.13 A) ... 35

24(23) Antenna with five articles (Fig. 16.2.13 B) ... 25

24' Antenna with six articles (Fig. 16.2.13 C) ... 29

25(24) Mentum with median teeth higher or subequal to the first lateral teeth (Fig. 16.2.11 J) ... 26

25' Mentum with median teeth lower than the first lateral teeth (Fig. 16.2.13 D) ... 28

26(25) Mentum with all teeth subequal (Fig. 16.2.12 L) ... 27

26' Mentum with median teeth higher than the first lateral teeth (Fig. 16.2.11 J) ... *Polypedilum* (in part)

27(26) Ventromental plates long with anterior margin smooth (Fig. 16.2.12 L) ... *Polypedilum* (*Asheum*)

27' Ventromental plates shorter with anterior margin sinuate (Fig. 16.2.13 E) ... *Polypedilum* (in part)

28(25) Frontal apotome (on anterior part of head, in dorsal view) separated from clypeus (Fig. 16.2.13 F) ... *Endotribelos* (in part)

28' Frontal apotome fused to clypeus (frontoclipeal apotome) (Fig. 16.2.13 G) ... *Phaenopsectra*

29(24) Mentum with two median teeth (Fig. 16.2.13 H) ... 30

29' Mentum with four median teeth (Fig. 16.2.13 I) ... 32

30(29) Mentum with median teeth pale, bearing six pairs of lateral teeth, third lateral tooth smaller than second lateral tooth (Fig. 16.2.13 H) ... 31

30' Mentum with all teeth even colored, bearing seven pairs of lateral teeth, third lateral tooth larger than second lateral tooth (Fig. 16.2.13 J) ... *Sigmoitendipes*

31(30) Mentum with two recessed, small median teeth, second lateral teeth usually much higher than median teeth, fused to first lateral tooth (Fig. 16.2.13 K); ventromental plates trapezoid ... *Oukuriella* (in part)

31' Mentum with two large median teeth, second lateral teeth subequal to median teeth, not fused to first lateral teeth; ventromental plates not trapezoid (Fig. 16.2.13 H) ... *Apedilum*

32(29) Pecten epipharyngis divided in three plates (Fig. 16.2.14 A), toothed or not ... 33

32' Pecten epipharyngis undivided and toothed (Fig. 16.2.13 L) ... *Claudiotendipes*

33(32) Mentum with median teeth pale or dark, lower than or subequal to second lateral teeth, second pair of lateral teeth higher than remaining lateral teeth (Fig. 16.2.13 I) ... 34

33' Mentum with four median teeth, with at least the outer pair of the median teeth higher than the remaining lateral teeth (Fig. 16.2.14 B) ... *Stictochironomus*

34(33) S I plumose with separate bases; mentum with median teeth darkened, central pair of the four median lower and more slender than outer median teeth (Fig. 16.2.14 C) ... *Omisus*

34' S I with bases fused; mentum with median teeth pale, central pair of the four median teeth equal to or higher than outer median teeth (Fig. 16.2.13 I) ... *Paratendipes*

35(23) Mentum with median tooth mostly pale (Fig. 16.2.14 D) ... 36

35' Mentum with all teeth even colored (Fig. 16.2.14 E) ... 38

36(35) Frontal apotome not narrowed anteriorly; antenna elongated, with five to seven articles; mandible with one to two dorsal teeth; mentum with median tooth shorter than first lateral teeth ... 37

36' Frontal apotome strongly narrowed anteriorly; antenna short with six articles; mandible without dorsal tooth; mentum with median tooth broad and higher than first lateral teeth (Fig. 16.2.14 D) ... *Paralauterborniella*

37(36) Antennae with five to seven articles; mandible with double dorsal tooth, seta subdentalis enlarged; pecten epipharyngis in three plates distally toothed (Fig. 16.2.14 F) ... *Beardius*

37' Antennae with six articles; mandible with dorsal tooth simple, seta subdentalis slender; pecten epipharyngis in three smooth plates (Fig. 16.2.14 G) ... *Oukuriella* (in part)

38(35) Mentum and mandible not as previous (Fig. 16.2.14 I) ... 39

38' Mentum with deeply recessed middle tooth (Fig. 16.2.14 H); mandible without seta interna and pecten mandibularis ... *Kribiodorum*

39(38) Mandible with apical tooth shorter, dorsal tooth present (Fig. 16.2.14 K); usually larger larva ... 40

39' Mandible with apical tooth very slender and long, dorsal tooth absent (Fig. 16.2.14 J); small larva .... *Nilothauma*
40(39) Labral sclerite 1 (on the head, dorsal view) absent (Fig. 16.2.14 L) 41
40' Labral sclerites 1(on the head, dorsal view) present (Fig. 16.2.15 A) 42
41(40) Frontal apotome separated from clypeus (Fig. 16.2.14 L); mentum with one median simple tooth, slightly notched or not (Fig. 16.2.15 B) .... *Endotribelos* (in part)
41' Frontal apotome fused to clypeus (frontoclipeal apotome) (Fig. 16.2.15 C); mentum with median tooth trifid. (Fig. 16.2.15 D) .... *Chironomus* (in part)
42(40) Labrum without setal brushes; larvae in a variety of habitats 43
42' Labrum with large setal brushes on each side (Fig. 16.2.15 E); larvae mining in sponges .... *Xenochironomus*
43(42) Ventromental plates almost touching each other medially (Fig. 16.2.15 G) 44
43' Ventromental plates median separated (Fig. 16.2.15 F) .... *Dicrotendipes* (in part)
44(43) Seta subdentalis on mandible with ventral margin fringed; ventromental plates strongly curved medially (Fig. 16.2.15 G) .... *Goeldichironomus* (in part)
44' Seta subdentalis simple; ventromental plates not curved (Fig. 16.2.15 H) .... *Axarus*

Chironomidae: Chironominae: Pseudochironomini: Genera

1 Pecten mandibularis weak or absent (Fig. 16.2.10 I); inner mandibular teeth and mentum darkened (Fig. 16.2.10 I) 2
1' Pecten mandibularis present (Fig. 16.2.10 J); inner mandibular teeth and mentum pale yellow (Fig. 16.2.10 J) .... *Aedokritus*
2(1) S I arising from a common base Fig. 16.2.11 B) 3
2' S I arising from separate bases (Fig. 16.2.11 A) .... *Pseudochironomus*
3(2) Mentum with second lateral tooth minute, fused to first lateral tooth (Fig. 16.2.11 C) .... *Manoa*
3' Mentum with second lateral tooth distinct, never fused to first lateral (Fig. 16.2.11 D) .... *Riethia*

Chironomidae: Aphroteniinae: Genera

This subfamily comprises three species-poor genera distributed exclusively in the Southern Hemisphere (Cranston, 1995b) in areas that were formerly part of Gondwana (Ashe & O'Connor, 2009). *Aphroteniella* and *Paraphrotenia* are present in the Neotropical Region. The first is only known by the pupae of *Aphroteniella* sp. "Peulla," while the larva of *Paraphrotenia excelens* Brundin is the only known larva of the genus in the region. Larvae are strictly rheobiontic and confined to mountain streams, living usually in the thin algal layer on the surface of stones and blocks fully exposed to the current (Brundin, 1966). Cranston & Edward (1992) have reported larvae associated with sandy substrates, predominantly where fine particulate organic matter was deposited.

1 Head relatively large (close to 20% of body length); antenna as long as mandible; mandible large, with five well-developed, acute teeth plus three long teeth located in the lateral margin (Fig. 16.2.6 E); body not tuberculate .... *Paraphrotenia*
1' Head close to 10% of body length; antenna clearly shorter than the mandible; mandible large, with up to seven teeth: one to three outside the apical tooth plus three to four inner teeth (Fig. 16.2.6 F); body tuberculate .... *Aphroteniella*

Chironomidae: Podonominae: Genera

This subfamily is distributed amphitropically. The group reaches its greatest diversity in the southern temperate belt, which comprises 84% of the world fauna (Sæther & Andersen, 2013a). Larvae are rheobiontic and cold-stenothermal, frequently associated with mosses in springs, brooks, and small streams (Epler, 2001); however, there have been reports of podonomines living in warm waters in Australia (Cranston & Edward, 1992). Podonominae is subdivided into two tribes: Boreochlini and Podonomini. So far, 90 valid species are known from the Neotropical and Antarctic regions (Ashe & O'Connor, 2009 and subsequent studies). Comments on the distribution of Podonominae in the Neotropical Region may be found in Siri & Donato (2012). Although immature stages for most Neotropical podonomines are known, larvae belonging to this subfamily are still understudied, with only a few associated at the species level. Within the genus *Podonomus*, usually two main larval morphotypes can be distinguished: (1) head slightly or not triangular, antenna half size of mandible; abdominal segments commonly with strong setae and (2) head broadly triangular, antenna as long as the mandible; abdominal segment without strong setae.

1 Procercus not bulbous nor hyaline, with six to seven well-developed setae 2
1' Procercus bulbous and hyaline, slightly darkened at the base, bearing very short setae (Fig. 16.2.23 E) .... *Podonomopsis*
2(1) Posterolateral border of segment 8 without comb of spines 3
2' Posterolateral border of segment 8 with strong comb of spines (Fig. 16.2.23 F) .... *Podochlus*
3(2) Antennal article 2 (or 2 and 3) annulated (Fig. 16.2.23 G); procercus commonly unicolor, more or less long and narrow (Fig. 16.2.23 H) .... *Parochlus*
3' Antennal article 2 and 3 not annulated; procercus commonly dark, variable in length and width (Fig. 16.2.23 I); some species with very strong setae in the body .... *Podonomus*

Chironomidae: Diamesinae: Tribes

This is the fourth largest subfamily. The group has worldwide distribution, occurring in all zoogeographic regions except the Antarctic and Oceania (Ashe & O'Connor, 2009). In general, larvae are cool-adapted, living in flowing water of circumpolar areas and mountain ranges, but some are found in springs and lakes (Epler, 2001). Three tribes, Harrisonini, Heptagyiini, and Lobodiamesini, occur in the in the Southern Hemisphere. The Heptagyiini and

Lobodiamesini are the only tribes in the Neotropical Region. Larval identification of Heptagyiini to generic level is considered relatively difficult; most species belonging to this tribe are known only from adults.

1 Antennal article 3 annulated (Fig. 16.2.15 I); procercus very reduced to absent, with at most four anal setae .... Diamesini, one genus *Diamesa*
1' Antennal article 3 nonannulate (Fig. 16.2.15 K); procercus very reduced or absent, with at least five anal setae...... **Heptagyiini [p. 695]**

Chironomidae: Diamesinae: Heptagyiini: Genera

1 Occipital margin strongly developed (in the fourth instar larva), mentum with at least seven lateral teeth (Fig. 16.2.15 L) 2
1' Occipital margin moderately developed (in the fourth instar larva), mentum with less than seven lateral teeth (Figs. 16.2.16 A, C)....... 3
2(1) Occipital margin close to 30% of the head; lateral incisure of the occipital margin weak or absent; mentum with two slightly smaller but nonrecessed median teeth plus more than eight lateral teeth (Fig. 16.2.15 L)........................ *Paraheptagyia*
2' Occipital margin 30% of head, lateral incisure of the occipital margin strong; mentum with seven to eight lateral teeth, the first lateral fused with the middle tooth like a "broad median tooth" (Fig. 16.2.16 B)........................ *Reissmesa*
3(2) Lateral incisure of the occipital margin weak or absent; mentum with two slightly recessed median teeth plus six lateral teeth diminishing in size (Fig. 16.2.16 C)........................ *Heptagyia*
3' Lateral incisure of the occipital margin well developed, mentum with a very broad middle tooth and five lateral teeth (Fig. 16.2.16 A). .... *Limaya*

Chironomidae: Prodiamesinae: Genera

This subfamily is small, comprising only four genera. Only *Monodiamesa* and *Prodiamesa* have been recorded in the Neotropical Region so far. Larvae seem to favor low to middle ranges of water temperature, and no species have been found in warm tropical waters (Ashe et al., 1987).

1 Ventromental plate narrow, with weak beard; mentum with a broad and concave median tooth (Fig. 16.2.23 K)........................ *Monodiamesa mariae* Andersen, 1996
1' Ventromental plate broad, with strong beard, mentum with two small median teeth (Fig. 16.2.23 L)........................ *Prodiamesa*

Chironomidae: Telmatogetoninae: Genera

This is a small subfamily, comprising two marine intertidal genera, with worldwide distribution (Cranston, 1995b). Larvae are nearly always associated with algae growing attached to rocks (Epler, 2001). According to Oliver (1971), larval *Telmatogetoninae* are euryhaline and often found in areas subject to direct freshwater runoff.

1 Area anterior to frontal apotome with two distinct sclerites; S3 placed on well-developed tubercle; premandible with three rounded apical teeth (Fig. 16.2.26 K)........................ *Telmatogeton*
1' Area anterior to frontal apotome uniform, without delimited sclerites; S3 not on tubercle; premandible simple (Fig. 16.2.26 L)........................ *Thalassomyia*

Chironomidae: Orthocladiinae: Genera

This subfamily is the second largest, with the greatest ecologic diversity (Ashe & Cranston, 1990). Several of the more primitive genera are aquatic, often rheophilous, while several genera are lentic. Some orthoclads exhibit hygropetric behavior in small standing waters, while several genera have developed semiterrestrial or fully terrestrial habits. Most larvae are scrapers, shredders, or collector-gatherers, while some taxa are predators (Epler, 2001). The monophyly of the subfamily remains uncertain, as no defining morphologic autapomorphies, in any life history stage, were found so far (Cranston et al., 2012). There is no accepted division of the subfamily into tribes. *Cricotopus* van der Wulp and *Orthocladius* van der Wulp have different larval types and are poorly studied in the Neotropical Region, and further studies are required to diagnose both genera. Additionally, the genus *Ferringtonia* Sæther & Andersen was not included in the key due to incomplete descriptions.

1 Anal end without procercus or, if present, without distinct anal setae and posterior parapods; anal setae present or absent; anterior parapods often partially fused........................ 2
1' Procercus present, but may be reduced, with variable number of anal setae; anterior parapods nearly always fully separated (e.g., *Belgica*, which anterior parapods are apparently fused)........................ 14
2(1) Pecten epipharyngis consisting of three scales; SI palmate only in *Smittia* (Fig. 16.2.16 E)........................ 3
2' Each scale of pecten epipharyngis divided into two to three teeth, forming continuous row of about eight teeth; SI nearly palmate (Fig. 16.2.16 D)........................ *Antillocladius*
3(2) Premandible with at most four teeth........................ 4
3' Premandible with five teeth (Fig. 16.2.16 F)........................ *Eretmoptera*
4(3') Preanal and anal segments and posterior parapods bent at right angles to axis of rest of body (Figs. 16.2.16 G, H)........................ 5

4' Preanal and anal segments in same axis as rest of body (Fig. 16.2.16 I) 6
5(4) Posterior parapods divided into two parts, anterior part with small claws, posterior part bare (Fig. 16.2.16 H); anal tubules small or absent. Anal setae present .... *Gymnometriocnemus*
5' Posterior parapods undivided, with or without claws (Fig. 16.2.16 G); anal tubules present; usually without anal setae .... *Bryophaenocladius*
6(4) Anal tubules absent; mostly marine 7
6' Anal tubules present; never marine 8
7(6) Single pair of anal seta present (Fig. 16.2.16 I); SI and SII broad and apically feathered (Fig. 16.2.16 J) .... *Clunio*
7' Two or three pairs, rarely 1, of anal setae present; SI feathered, SII always simple .... *Thalassosmittia*
8(6) Mandibular teeth rounded or subtriangular (Fig. 16.2.17 A) 9
8' Mandible with spiniform teeth (Fig. 16.2.16 K); ectoparasitic on Ephemeroptera .... *Symbiocladius*
9(8) SI and SII both bifid 10
9' SI bifid or not, SII never bifid 11
10(9) Posterior parapods with zero to five claws, claws 0–18 μm long (Fig. 16.2.16 L); mandible with three inner teeth (Fig. 16.2.17 A); antennal blade extending beyond flagellum (Fig. 16.2.17 B) .... *Pseudosmittia*
10' Posterior parapods with 7–12 claws, claws 15–80 μm long; mandible with three or four inner teeth (Fig. 16.2.17 C); antennal blade not or only slightly extending beyond flagellum .... *Allocladius*
11(9) Mentum different, beard absent 12
11' Mentum with broad flat median tooth and with five pairs of lateral teeth; large ventromental beard of many setae extending at least to margin of head (Fig. 16.2.17 D) .... *Barbadocladius*
12(11) Mandible with three inner teeth; anterior parapods fused; posterior parapods present; anal tubules short, without constrictions 13
12' Mandible with two inner teeth (Fig. 16.2.17 E) or, when with three, without posterior parapods; anterior parapods separate; anal tubules long, with many constrictions (Fig. 16.2.17 F) .... *Georthocladius*
13(12) Mandible without seta interna (Fig. 16.2.17 G); SI simple (Fig. 16.2.17 H) .... *Mesosmittia*
13' Mandible with seta interna (Fig. 16.2.17 I); SI plumose (Fig. 16.2.16 E) .... *Smittia*
14(1) Mentum not as previous; mandibular seta interna present or not 15
14' Mentum with 16–20 dark, equal-sized median and five pairs of slightly larger, lateral teeth (Fig. 16.2.17 J); mandibular seta interna absent (Fig. 16.2.17 K); larvae phoretic on catfish .... *Ichthyocladius*
15(14) Antenna length $\geq 0.5\times$ head length, or if shorter, then head capsule elongated, at least $1.5\times$ as long as wide 16
15' Antenna length $<0.5\times$ head length 20
16(15) Antenna terminal article flagelliform or filiform; second antennal article partly unsclerotized (Figs. 16.2.17 L, 18 A) 17
16' Antenna with normal, short terminal article, but may have short filiform extension; if second antennal article partly unsclerotized, Lauterborn organs alternate 18
17(16) SI simple or with a few apical teeth; mandible with three to four inner teeth (Fig. 16.2.18 B) .... *Stictocladius*
17' SI bifid; mandible with five inner teeth (Fig. 16.2.18 C) .... *Lopescladius*
18(16) Antenna with five distinct articles, at most as long as head, usually shorter; head capsule without surface sculpturing (Figs. 16.2.18 E–F) 19
18' Antenna with four articles, usually longer than head; head capsule sometimes with surface sculpturing (Fig. 16.2.18 D) .... *Corynoneura*
19(18) Posterior parapod subbasal seta apically split (Fig. 16.2.18 G) .... *Ubatubaneura*
19' Posterior parapod subbasal seta simple .... *Thienemanniella*
20(15) No anal seta as long as ¼ body length 21
20' Anal setae as long as ¼ body length (Fig. 16.2.18 H) .... *Pseudorthocladius*
21(20) Mentum with no more than 15 teeth 22
21' Mentum with at least 16 teeth (Fig. 16.2.18 I) .... *Orthocladius* (*Mesorthocladius*/*Euorthocladius*) (in part)
22(21) Beard present beneath or adjacent to ventromental plates (may be only a few setae and may require observation at $1000\times$ magnification) 23
22' Beard absent (one to four weak setae present in some *Orthocladius*) 31
23(22) Antenna with four or five distinct articles, if minute sixth visible (some *Botryocladius*), then premandibular brush moderately developed 24
23' Small, setaform sixth antennal article usually distinct (Fig. 16.2.18 J); premandibular brush absent .... *Parakiefferiella* (in part)
24(23) Antenna with five articles, Lauterborn organs variable; mentum rarely with second pair of lateral teeth smaller than first or third 25
24' Antenna with four articles and poorly developed Lauterborn organs; mentum with second pair of lateral teeth smaller than first or third (Fig. 16.2.18 K) .... *Orthocladius* (*Pogonocladius*) (in part)
25(24) SI simple, bifid, coarsely pectinate, palmate or plumose; mentum, beard and mandible not as previous; abdomen without alternating simple and plumose setae, but single pairs of tufts may be present posterolaterally 26
25' SI simple (Fig. 16.2.18 L); mentum with two median, equally sized, elongated teeth; beard well developed, radiating from common area (Fig. 16.2.19 A); seta interna absent; seta subdentalis large (Fig. 16.2.19 B); abdomen with alternating simple and plumose lateral setae (Fig. 16.2.19 C) .... *Synorthocladius*
26(25) Mandible with three or fewer inner teeth or, when with four, apical tooth longer than combined width of four inner teeth or anal tubules absent; mentum, beard, and SI variable; if labral lamellae present, weakly developed and never apically pectinate 27
26' Mandible with four inner teeth, apical tooth short (Fig. 16.2.19 D); ventromental plates with dense beard (Fig. 16.2.19 E); SI plumose; labral lamellae well developed, apically pectinate (Fig. 16.2.19 F) .... *Diplocladius*
27(26) SI usually bifid, occasionally simple, apically split into four or more short teeth, plumose or coarsely serrate; procercus with or without spurs 28

27' SI broadly trifid, palmate or with four long, narrow teeth (Fig. 16.2.19 G); procercus with small to large spurs .... *Psectrocladius* (in part)
28(27) Beard well developed; procercus usually with spurs, or anal tubules absent 29
28' Beard well developed to vestigial; procercus without spurs; anal tubules always present 30
29(28) Mentum with six lateral teeth; beard, if conspicuous, without broad and bifurcate setae (Fig. 16.2.19 H) .... *Paracladius*
29' Mentum with five lateral teeth; beard of sparse to strong setae that may be broad and apically bifurcate (Fig. 16.2.19 I) .... *Botryocladius*
30(28) Mandible apical tooth longer than combined width of three inner teeth (Fig. 16.2.19 J); ventromental plates large and triangular (Fig. 16.2.19 K); head capsule without ventral tubercle .... *Psectrocladius* (*Monopsectrocladius*) (in part)
30' Mandible apical tooth less than or equal to combined width of three inner teeth (Fig. 16.2.19 L); if ventromental plates large and triangular, then head capsule with pair of ventral tubercles .... *Rheocricotopus*
31(22) Ventromental plates well developed, extending beyond lateral margin of mentum 32
31' Ventromental plates absent or vestigial or, if present, not extending well beyond lateral margin of mentum or plates very thin 35
32(31) SI never simple; mentum not as previous; ventromental plate broader and less elongate; not phoretic on aquatic insects 33
32' All S setae simple (Fig. 16.2.20 A); mentum with small pair of median teeth, often well separated from the zero to six pairs of lateral teeth, which may be small and fused or closely adpressed to each other; ventromental plates long and elongate (Fig. 16.2.20 B); some species phoretic or parasitic on other aquatic insects .... *Nanocladius*
33(32) Ventromental plates appear double (Figs. 16.2.20 D–F); seta submenti located well anterior to posterior margin of ventromental plates 34
33' Ventromental plates single; seta submenti located near posterior margin of ventromental plates or more posteriad (Fig. 16.2.20 C) .... *Parakiefferiella* (in part)
34(33) Antenna with long basal article, antennal ratio >1.25; antennal blade shorter than flagellum (Fig. 16.2.20 G) .... *Parametriocnemus*
34' Antenna with shorter basal article, antennal ratio 0.5–1.0; antennal blade shorter or longer than flagellum (Fig. 16.2.20 H) .... *Paraphaenocladius*
35(31) Antenna with six to seven articles, terminal article vestigial, often setaform (Fig. 16.2.18 J); ventromental plates moderately developed 36
35' Antenna with five or fewer articles, last article not setaform; ventromental plates moderately developed to absent 37
36(35) Mentum with trifid median tooth, five pairs of lateral teeth, fifth lateral pair of teeth lower than fourth (Fig. 16.2.20 C) .... *Parakiefferiella* (in part)
36' Mentum with bifid median tooth, four pairs of lateral teeth, fourth as high as third (Fig. 16.2.20 I) .... *Paratrissocladius*
37(35) Abdomen with long simple setae, at least half as long as the segment bearing them 38
37' Abdomen without long simple setae, or if long setae present, they are arranged as pairs of single tufts, one on each side posterolaterally on body segments 40
38(37) SI simple or with indistinct apical dentations 39
38' SI bifid, with several apical teeth or plumose .... *Metriocnemus* (in part)
39(38) Mandible medial margin with several spines (Fig. 16.2.20 J); procerci without spurs .... *Eukiefferiella* (in part)
39' Mandible medial margin without spines (Fig. 16.2.20 K); procerci with spurs (Fig. 16.2.20 L) .... *Paracricotopus*
40(37) Mentum with three median teeth and five pairs of lateral teeth, first pair of lateral teeth closely adpressed to outer median teeth (Figs. 16.2.21 A–B); antenna ¼–⅓ as long as head; SII strong, situated on tubercle; SI bifid (Fig. 16.2.21 C) 41
40' Mentum with one to two median teeth and two to five pairs of lateral teeth; antenna at most ¼ as long as head; SII not on tubercle; SI simple, serrate, plumose, or bifid 42
41(40) Antenna length ~¼× head capsule (Fig. 16.2.21 D); anterior parapods with two well-sclerotized, large, hook-like claws .... *Tempisquitoneura*
41' Antenna length ~⅓ × head capsule (Fig. 16.2.21 E); anterior parapods without sclerotized claws .... *Onconeura*
42(40) SI simple 43
42' SI bifid, serrate, pectinate, or plumose 49
43(42) Inner margin of mandible with spines (these may be small) 44
43' Inner margin of mandible smooth 47
44(43) Procercus at least as long as wide, setae about equally thick; seta interna of mandible usually divided to near base, if divided near apex mentum with four pairs of lateral teeth 45
44' Procercus reduced, with two setae thicker than the others on each procercus (Fig. 16.2.21 F); seta interna of mandible with long stalk that branches near apex (Fig. 16.2.21 G); mentum with five pairs of lateral teeth (Fig. 16.2.21 H) .... *Cardiocladius* (in part)
45(44) Mentum with five pairs of lateral teeth 46
45' Mentum with four pairs of lateral teeth (Fig. 16.2.21 I) .... *Eukiefferiella* (in part)
46(45) Anal setae much shorter than posterior parapods .... *Cardiocladius* (in part)
46' Anal setae longer than posterior parapods .... *Eukiefferiella* (in part)
47(43) Mentum with single or moderately developed bifid median tooth and at least five pairs of lateral teeth 48
47' Mentum with large bifid median teeth and only two lateral teeth (Fig. 16.2.21 J) .... *Austrobrillia*
48(47) Body with posterolateral tufts (Fig. 16.2.21 K); mentum with single median tooth .... *Cricotopus* (in part)
48' Body without posterolateral tufts; mentum with bifid median tooth (Fig. 16.2.21 L) .... *Limnophyes* (in part)
49(42) SI serrate, apically fringed, pectinate, or plumose 50
49' SI bifid 59
50(49) Anal tubules short or normal, without numerous constrictions; procercus well developed to weak, but not vestigial; labral lamellae well developed, vestigial, or absent; blade shorter or longer than flagellum 51

50' Anal tubules very long, several times as long as posterior parapods, with many constrictions; procercus vestigial or consisting of two nearly contiguous, flat cones with basal, low plates, each with two to three setaform setae (Fig. 16.2.22 A); labral lamellae simple, rounded, weakly sclerotized; blade extending well beyond terminal antennal article (Fig. 16.2.22 B) .... *Georthocladius* (*Atelopodella*)
51(50) Well-developed labral lamellae present (Fig. 16.2.22 C) 52
51' Labral lamellae absent or vestigial 53
52(51) Procerci well developed, at least twice as long as wide; supraanal setae shorter than anal tubules (Fig. 16.2.22 D) .... *Metriocnemus* (in part)
52' Procerci weakly developed, about as long as wide; supraanal setae as long as or longer than anal tubules (Fig. 16.2.22 E) .... *Thienemannia*
53(51) Antennal blade longer than flagellum; SI apically pectinate or plumose; mandible with four inner teeth 54
53' Antennal blade subequal in length to flagellum; SI serrate or palmate; mandible with three or four inner teeth 56
54(53) Mentum with single median tooth (Fig. 16.2.22 G) 55
54' Mentum with double median tooth (Fig. 16.2.22 F) .... *Phytotelmatocladius*
55(54) Abdominal segments with one pair of setal tufts; fourth antennal article as long as the previous (Fig. 16.2.22 H) .... *Gravatamberus*
55' Abdominal segments without distinct setae; fourth antennal article about twice as long as the previous (Fig. 16.2.22 I) ...... *Gynocladius*
56(53) Seta subdentalis relatively long and narrow, extending beyond internal margin of mandible 57
56' Seta subdentalis very short, not extending beyond internal margin of mandible .... *Belgica*
57(56) SI serrate; Seta subdentalis narrow and simple 58
57' SI palmate, perhaps bifid, with each branch further divided into two to four lobes (Fig. 16.2.22 J); seta subdentalis subovoid, hooked (Fig. 16.2.22 K) .... *Parapsectrocladius*
58(57) Mandible with three inner teeth (Fig. 16.2.22 L); supraanal setae length subequal to anal setae (Fig. 16.2.23 A) ... *Limnophyes* (in part)
58' Mandible with four inner teeth (Fig. 16.2.23 B); supraanal setae length $\sim {}^{1}/_{3} \times$ anal setae .... *Compterosmittia*
59(49) Setae submenti palmate, with seven to eight branches (Fig. 16.2.23 C); similar setae in maxilla .... *Pirara*
59' Seta submenti not as previous .... *Cricotopus*/*Orthocladius*

REFERENCES

Allegrucci, G., G. Carchini, V. Todisco, P. Convey & V. Sbordoni. 2006. A molecular phylogeny of Antarctic Chironomidae and its implications for biogeographical history. Polar Biology 29: 320–326.

Andersen, T., H.F. Mendes & L.C. Pinho. 2015. *Mariambera*, a new genus of Orthocladiinae from Brazil (Insecta: Diptera, Chironomidae). Studies on Neotropical Fauna and Environment 50: 24–30.

Andersen, T. & L.C. Pinho. 2014. A new species of *Saetherocryptus* Andersen et Mendes, 2007 (Diptera: Chironomidae, Orthocladiinae) from the Amazon rainforest, Brazil. Norwegian Journal of Entomology 61: 160–164.

Andersen, T., O.A. Sæther, P.S. Cranston & J.H. Epler. 2013a. The larvae of Orthocladiinae (Diptera: Chironomidae) of the Holarctic Region - Keys and diagnoses. Pages 189–385 *in*: T. Andersen, P.S. Cranston & J.H. Epler (eds.), The larvae of Chironomidae (Diptera) of the Holarctic region - Keys and diagnoses. Insect Systematics & Evolution. Scandinavian Entomology, Lund, Sweden.

Andersen T., P.S. Cranston & J.H. Epler. 2013b. The larvae of Chironomidae (Diptera) of the Holarctic region - Keys and diagnoses. Insect Systematics & Evolution. Scandinavian Entomology, Lund, Sweden. 571pp.

Andersen, T., T. Ekrem & P.S. Cranston. 2013c. The larvae of the Holarctic Chironomidae - Introduction. Pages 7–12 *in*: T. Andersen, P.S. Cranston & J.H. Epler (eds.), The larvae of Chironomidae (Diptera) of the Holarctic region - Keys and diagnoses. Insect Systematics & Evolution. Scandinavian Entomology, Lund, Sweden.

Armitage, P.D., P.S. Cranston & L.C.V. Pinder. 1995. The Chironomidae. Biology and ecology of non-biting midges. Chapman & Hall, London, Glasgow, Weinheim, New York, Tokyo, Melbourne & Madras. xii + 572 pp.

Ashe, P., D.A. Murray & F. Reiss. 1987. The zoogeographical distribution of Chironomidae (Insecta: Diptera). Annales de Limnologie 23: 27–60.

Ashe, P. & J.P. O'Connor. 2009. A World Catalogue of Chironomidae (Diptera). Part 1. Buchonomyiinae, Chilenomyiinae, Podonominae, Aphroteniinae, Tanypodinae, Usambaromyiinae, Diamesinae, Prodiamesinae and Telmatogetoninae. Irish Biogeographical Society & National Museum of Ireland, Dublin. 445 pp.

Ashe, P. & P.S. Cranston. 1990. Family Chironomidae. Pages 113–335 *in*: Á. Soós & L. Papp (eds.), Catalogue of Palaearctic Diptera, Vol. 2. Psychodidae - Chironomidae. Akadémiai Kiadó, Budapest.

Brundin, L. 1966. Transantarctic relationships and their significance, as evidenced by chironomid midges. With a monograph of the subfamilies Podonominae and Aphroteniinae and the austral Heptagyiae. Kunglika Svenska Vetenskapsakademiens Handlingar 11: 1–472.

Brundin, L. 1983. *Chilenomyia paradoxa* gen. n., sp. n. and Chilenomyiinae, a new subfamily among the Chironomidae (Diptera). Entomologica Scandinavica 14: 33–45.

Carew, M.E., S.E. Marshall & A.A. Hoffmann. 2011. A combination of molecular and morphological approaches resolves species in the taxonomically difficult genus *Procladius* Skuse (Diptera: Chironomidae) despite high intra-specific morphological variation. Bulletin of Entomological Research 101: 505–519.

Cranston, P.S. 1982. A key to the larvae of the British Orthocladiinae (Chironomidae). Freshwater Biological Association, Scientific Publication 45. The Ferry Hause, Ambleside, Cumbria LA22 oLP. 152 pp.

Cranston, P.S. 1995a. Introduction. Pages 1–7 *in*: P.D. Armitage, P.S. Cranston & L.C.V. Pinder (eds.), The Chironomidae. Biology and ecology of non-biting midges. Chapman & Hall, London, Glasgow, Weinheim, New York, Tokyo, Melbourne & Madras. xii + 572 pp.

Cranston, P.S. 1995b. Systematics. Pages 31–61 *in*: P.D. Armitage, P.S. Cranston & L.C.V. Pinder (eds.), The Chironomidae. The biology and ecology of non-biting midges. Chapman & Hall, London, Glasgow, Weinheim, New York, Tokyo, Melbourne & Madras. xii + 572 pp.

Cranston, P.S. 1995c. Morphology. Pages 11–30 *in*: P.D. Armitage, P.S. Cranston & L.C.V. Pinder (eds.), The Chironomidae. The biology and ecology of non-biting midges. Chapman & Hall, London, Glasgow, Weinheim, New York, Tokyo, Melbourne & Madras. xii + 572 p.

Cranston, P.S. 2010. Chironomidae. *in*: http://chirokey.skullisland.info/.

Cranston P.S. 2013. The larvae of the Holarctic Chironomidae - Morphological terminology and keys to subfamilies. Pages 13–23 *in*: T. Andersen, P.S. Cranston & J.H. Epler (eds.), The larvae of Chironomidae (Diptera) of the Holarctic region - Keys and diagnoses. Insect Systematics & Evolution. Scandinavian Entomology, Lund, Sweden.

Cranston, P.S. & D. Edward. 1992. A systematic reappraisal of the Australian Aphroteniinae (Diptera: Chironomidae) with dating from vicariance biogeography. Systematic Entomology 17: 41–54.

Cranston, P.S., N.B. Hardy & G.E. Morse. 2012. A dated molecular phylogeny for the Chironomidae (Diptera). Systematic Entomology, 37: 172–188.

Cranston, P.S., N.B. Hardy, G.E. Morse, L. Puslednik & S.R. McCluen. 2010. When molecules and morphology concur: the 'Gondwanan' midges (Diptera: Chironomidae). Systematic Entomology 35: 636–648.

Cranston P.S. & P. Ashe. 2013. The larvae of the Holarctic Telmatogetoninae - Morphological terminology and keys to subfamilies. Pages 137–143 *in*: T. Andersen, P.S. Cranston & J.H. Epler (eds.), The larvae of Chironomidae (Diptera) of the Holarctic region - Keys and diagnoses. Insect Systematics & Evolution. Scandinavian Entomology, Lund, Sweden.

Cranston P.S. & U. Nolte. 1996. *Fissimentum*, a new genus of drought-tolerant Chironomini (Diptera: Chironomidae) from the Americas and Australia. Entomological News 107: 1–15.

Dantas, G.P.S., N. Hamada & H.F. Mendes. 2016. Denopelopia amicitia, a new Tanypodinae from Brazil (Diptera, Chironomidae). ZooKeys 553: 107–117.

Ekrem, T., E. Willassen & E. Stur. 2007. A comprehensive DNA library is essential for identification with DNA barcodes. Molecular Phylogenetics and Evolution 43: 530–542.

Ekrem, T., E. Willassen & E. Stur. 2010. Phylogenetic utility of five genes for dipteran phylogeny: a test case in the Chironomidae leads to generic synonymies. Molecular Phylogenetics and Evolution 57: 561–571.

Epler, J.H. 2001. Identification manual for the larval Chironomidae (Diptera) of North and South Carolina. A guide to the taxonomy of the midges of the southeastern United States, including Florida. Special Publication SJ2001–S2013. North Carolina Department of Environment and Natural Resources, Raleigh, North Carolina and St. Johns River Water Management District, Palatka, Florida. vi + 516 pp.

Ferrington, L.C. Jr. 2008. Global diversity of non-biting midges (Chironomidae; Insecta-Diptera) in freshwater. Hydrobiologia 595: 447–455.

Ferrington, L.C. Jr., M.B. Berg & W.P. Coffman. 2008. Chironomidae. Pages 847–989 *in*: R.W Merritt, K.W. Cummins & M.B. Berg (eds.), An introduction to the aquatic insects of North America. Kendall/ Hunt Publishing Co., Dubuque, Iowa.

Fittkau E.J. 1955. *Buchonomyia thienemanni* n. gen. n. sp. Chironomidenstudien IV (Diptera; Chironomidae). Beiträge zur Entomologie 5: 403–414.

Jacobsen, R.E. 2008. A key to the pupal exuviae of the midges (Diptera: Chironomidae) of Everglades National Park, Florida. U.S. Geological Survey Scientific Investigations Report 2008-5082. U.S. Department of the Interior and U.S. Geological Survey, Florida. vi + 516 p.

Mendes, H.F., T. Andersen & O.A. Sæther. 2004. A review of *Antillocladius* Sæther, 1981; *Compterosmittia* Sæther, 1981 and *Litocladius* new genus (Chironomidae, Orthocladiinae). Zootaxa 594: 1–82.

Mendes, H.F. & T. Andersen. 2008. A review of *Antillocladius* Sæther and *Litocladius* Mendes, Andersen et Sæther, with the description of two new Neotropical genera (Diptera, Chironomidae, Orthocladiinae). Zootaxa 1887: 1–75.

Oliveira, C.S.N., M.A.S. Navarro & A.A. Fonseca-Gessner. 2013. Neotropical *Ablabesmyia* Johannsen (Diptera: Chironomidae, Tanypodinae) - Part I. Zootaxa 3733(1): 1–123.

Oliver, D.R. 1971. Life history of the Chironomidae. Annual Review of Entomology 12: 211–230.

Oliver, D.R. & M.E. Roussel. 1983. The insects and arachnids of Canada, Part 11. The genera of larval midges of Canada. Diptera: Chironomidae. Agriculture Canada Publications, 1746: 1–263.

Paggi, A.C. 2001. Diptera: Chironomidae. Pages 167–193 *in*: H.R. Fernández & E. Domínguez (eds.), Guía para la determinación de los artrópodos bentónicos sudamericanos. Investigaciones de la Universidad Nacional de Tucumán. Subserie: Ciencias Exactas y Naturales, Tucumán, Argentina.

Paggi, A.C. 2009. Diptera Chironomidae. Pages 383–409 *in*: E. Domínguez & H.R. Fernández (eds.), Macroinvertebrados bentónicos sudamericanos. Sistemática y biología. Fundación Miguel Lillo, Tucumán, Argentina.

Pinho, L.C., H.F. Mendes & T. Andersen. 2013. Revision of *Beardius* Reiss et Sublette, 1985 (Diptera: Chironomidae), with the description of twenty new species. Zootaxa 3742: 1–78.

Sæther, O.A. 1980. Glossary of chironomid morphology terminology (Diptera: Chironomidae). Entomologica Scandinavica Supplement 14: 1–51.

Sæther, O.A. 2000. Phylogeny of the subfamilies of Chironomidae (Diptera). Systematic Entomology 25: 393–403.

Sæther, O.A. & T. Andersen. 2013a. The larvae of Podonominae (Diptera: Chironomidae) of the Holarctic Region - Keys and diagnoses. Pages 29–38 *in*: T. Andersen, P.S. Cranston & J.H. Epler (eds.), The larvae of Chironomidae (Diptera) of the Holarctic region - Keys and diagnoses. Insect Systematics & Evolution. Scandinavian Entomology, Lund, Sweden.

Sæther, O.A. & T. Andersen. 2013b. The larvae of Prodiamesinae (Diptera: Chironomidae) of the Holarctic Region - Keys and diagnoses. Pages 179–187 *in*: T. Andersen, P. S. Cranston & J.H. Epler (eds.), The larvae of Chironomidae (Diptera) of the Holarctic region - Keys and diagnoses. Insect Systematics & Evolution. Scandinavian Entomology, Lund, Sweden.

Silva, F.L., A.A. Fonseca-Gessner & T. Ekrem. 2011. Revision of *Labrundinia maculata* Roback, 1971, a new junior synonym of *L. longipalpis* (Goetghebuer, 1921) (Diptera: Chironomidae: Tanypodinae). Aquatic Insects 33: 293–303.

Silva, F.L., A.A. Fonseca-Gessner & T. Ekrem. 2014. A taxonomic revision of genus *Labrundinia* Fittkau, 1962 (Diptera: Chironomidae: Tanypodinae). Zootaxa 3769: 1–185.

Silva, F.L. & B.D. Farrell. 2017. Non-biting midges (Diptera: Chironomidae) research in South America: subsidizing biogeographic hypotheses. Annales de Limnologie - International Journal of Limnology 53: 111–128.

Silva, F.L. & S. Wiedenbrug. 2014. Integrating DNA barcodes and morphology for species delimitation in the *Corynoneura* group (Chironomidae: Orthocladiinae). Bulletin of Entomological Research 104: 65–78.

Silva, F.L. & S. Wiedenbrug. 2015. *Amazonimyia gigantea* gen. n., sp. n., a new Tanypodinae (Diptera: Chironomidae) from the Neotropical Region. Zootaxa 3947: 275–281.

Silva, F.L., S. Wiedenbrug & B.D. Farrell. 2015a. A preliminary survey of the non-biting midges (Diptera: Chironomidae) of the Dominican Republic. Chironomus Journal of Chironomidae Research 28: 12–19.

Silva, F.L. & T. Ekrem. 2016. Phylogenetic relationships of nonbiting midges in the subfamily Tanypodinae (Diptera: Chironomidae) inferred from morphology. Systematic Entomology 41: 73–92.

Silva, F.L., T. Ekrem & A.A. Fonseca-Gessner. 2013. DNA barcodes for species delimitation in Chironomidae (Diptera): a case study on genus *Labrundinia*. The Canadian Entomologist 145: 589–602.

Silva, F.L., T. Ekrem. & A.A. Fonseca-Gessner. 2015b. Out of South America: phylogeny of non-biting midges in the genus *Labrundinia* (Diptera: Chironomidae) suggests multiple dispersal events to Central and North America. Zoologica Scripta 44: 59–71.

Siri, A. & M. Donato. 2012. Two new species of *Podonomus* (Diptera: Chironomidae: Podonominae) of the Brundin's *decarthrus* group from Ventania system, Argentina. Zootaxa 3848: 39–54.

Siri, A. & M. Donato. 2015. Phylogenetic analysis of the tribe Macropelopiini (Chironomidae: Tanypodinae): adjusting homoplasies. Zoological Journal of the Linnean Society 174: 74–92.

Spies, M., T. Andersen, J.H. Epler. & C.N. Watson, Jr.. 2009. Chironomidae (non-biting midges). Pages 437–480 *in*: B.V. Brown, A. Borkent, J.M. Cumming, D.M. Wood, N.E. Woodley & M. Zumbado (eds.), Manual of Central American Diptera. NRC Research Press, Ottawa.

Spies M. & F. Reiss. 1996. Catalog and bibliography of Neotropical and Mexican Chironomidae (Insecta, Diptera). Spixiana Supplement 22: 61–119.

Stur, E. & T. Ekrem. 2011. Exploring unknown life stages of Arctic Tanytarsini (Diptera: Chironomidae) with DNA barcoding. Zootaxa 2743: 27–39.

Thienemann, A. 1954. Chironomus: Leben, Verbreitung und wirtschaftliche Bedeutung der Chironomiden. Die Binnengewässer 20. Stuttgart. 834 pp.

Tokeshi, M. 1995. Life cycles and population dynamics. Pages 225–268 *in*: P.D Armitage, P.S. Cranston & L.C.V. Pinder (eds.), The Chironomidae. Biology and ecology of non-biting midges. Chapman & Hall, London, Glasgow, Weinheim, New York, Tokyo, Melbourne & Madras. xii + 572 pp.

Trivinho-Strixino, S. 2012. A systematic review of Neotropical *Caladomyia* Säwedall (Diptera: Chironomidae). Zootaxa 3495: 1–41.

Trivinho-Strixino, S. 2014. Ordem Diptera. Família Chironomidae. Guia de identificação de larvas. Pages 457–660 *in*: N. Hamada, J.L. Nessimian & R.B. Querino (eds.), Insetos aquáticos na Amazônia brasileira: taxonomia, biologia e ecologia. Editora INPA, Manaus.

Trivinho-Strixino, S., F.L. Silva & C.S.N. Oliveira. 2013. *Tapajos cristinae* n. gen., n. sp. (Diptera: Chironomidae: Chironominae) from Neotropical Region. Zootaxa 3710: 395–399.

Trivinho-Strixino, S. & M. Pepinelli. 2015. A systematic study on *Endotribelos* Grodhaus (Diptera: Chironomidae) from Brazil including DNA barcoding to link males and females. Zootaxa 3936: 1–41.

Trivinho-Strixino, S., S. Wiedenbrug & F.L. Silva. 2015. New species of *Tanytarsus* van der Wulp (Diptera: Chironomidae: Tanytarsini) from Brazil. European Journal of Environmental Sciences 5(1): 92–100.

Wiedenbrug, S., C.J.E. Lamas & S. Trivinho-Strixino. 2012. A review of the genus *Corynoneura* Winnertz (Diptera: Chironomidae) from the Neotropical region. Zootaxa 3574: 1–61.

Wiedenbrug, S. & S. Trivinho-Strixino. 2011. New species of the genus *Corynoneura* Winnertz (Diptera, Chironomidae) from Brazil. Zootaxa 2822: 1–40.

Family Simuliidae

Neusa Hamada
Coordenação de Biodiversidade, Instituto Nacional de Pesquisas da Amazônia, Manaus, Amazonas, Brazil

Jeane Marcelle Cavalcante do Nascimento
Coordenação de Biodiversidade, Curso de Pós-Graduação em Entomologia, Instituto Nacional de Pesquisas da Amazônia, Manaus, Amazonas, Brazil

Chapter Outline

INTRODUCTION

Simuliidae has several vernacular names, among them "black fly" or "blackfly" (English), "jejen" (Spanish), and "borrachudo" or "pium" (Portuguese). The adults are terrestrial, while immatures inhabit a variety of lotic habitats, ranging from dripping water in hygropetric or swampy areas to highly turbulent water in large rapids or waterfalls. Some species can reach high larval and pupal densities (Santos-Neto et al., 2015). Simuliidae represent an important food resource for aquatic invertebrates and vertebrates (Crosskey, 1990). Simuliid larvae feed on suspended material in the water column, and thus have an important role in aquatic ecosystem function (Wotton et al., 1995; 1996; Malmqvist & Sckmann, 1996; Currie & Adler, 2008).

The family has several species of medical and veterinary importance worldwide; and, because of this, it is one of the best-known groups of aquatic insects (Currie & Adler, 2008). It belongs to the suborder Culicomorpha and has more than 2200 species in two subfamilies: Simuliinae (with two tribes, Prosimuliini and Simuliini) and Parasimuliinae (Adler & Crosskey, 2017). Only the first subfamily (and only the tribe Simuliini) is found in the Neotropical Region, where around 16% of the world's known simuliid fauna occurs (Currie & Adler, 2008; Adler & Crosskey, 2017). In this chapter, we follow the classification scheme presented in Adler & Crosskey (2017).

In the Neotropical Region are more than 350 species of Simuliidae in 10 genera (Adler & Crosskey, 2018). Four genera occur in southern South America (*Araucnephioides* Wygodzinsky & Coscarón, *Cnesia* Enderlein, *Cnesiamima* Wygodzinsky & Coscarón, and *Paraustrosimulium* Wygodzinsky & Coscarón) and one (*Tlalocomyia* Wygodzinsky & Díaz Nájera) occurs in Central America and Mexico. *Gigantodax* Enderlein is distributed from the Mesoamerican mountains through the Andes to Tierra del Fuego, while the distribution of *Pedrowygomyia* Coscarón & Miranda-Esquivel is more restricted, occurring in the Puna realm of the Andes (Coscarón & Miranda-Esquivel 1998). *Araucnephia* Wygodzinsky & Coscarón and *Lutzsimulium* d'Andretta & d'Andretta occur in the southern and northern regions of Brazil and in Argentina. *Simulium* Latreille is widely distributed in the Neotropics, and it is the most diverse genus, with several endemic subgenera (e.g., *Chirostilbia*, *Notolepria*, *Psaroniocompsa*, and *Pternaspatha*) (Adler & Crosskey, 2017).

LIMITATIONS

The high morphologic similarity among some black fly species makes the task of species identification difficult in some species groups in this family. The male and female genitalia are often very simple, providing few diagnostic characteristics, and a safe identification usually requires using associated characteristics of the pupae and sometimes

Thorp and Covich's Freshwater Invertebrates, Fourth Edition. https://doi.org/10.1016/B978-0-12-804223-6.00032-9

of the larvae. In recent decades, cryptic species have been detected using an integrated taxonomic approach, usually combining morphologic, cytotaxonomic, and molecular tools (e.g., Adler et al., 2004; Hamada et al., 2010).

The pupal stage is usually the one in which identification to species level is most feasible, especially when the regional fauna is not well known. Extra time in the field to associate the immatures (except eggs) with the adults is worthwhile, and this is relatively easy in Simuliidae. Last-instar larva can be associated with pupa by the shape and structure of the filaments and by the number of filaments on gills dissected from the histoblast as compared to the pupal gills. Pharate adult (completely formed adult still inside the pupal skin) can be reared under artificial conditions until adult emergence. Even if the adults do not emerge, completely formed genitalia can be dissected from the pharate adult, thereby facilitating species identification.

TERMINOLOGY AND MORPHOLOGY

Taxonomic terminology used in this chapter follows Adler et al. (2004) and, in part, Coscarón & Coscaron-Arias (2007). The terminology used in the key is shown in Fig. 16.3.1–3.

Pupa (Figs. 16.3.1, 44–55). The body is divided into three regions: head, thorax, and abdomen (Fig. 16.3.1 A), and depending on the genus, it can either have a defined shape (Figs. 16.3.1 B–F, 47, 48, 53, 54) or be a shapeless cocoon (Figs. 16.3.44–46, 49, 50–52, 55). Among the family, the cocoon is variable in shape and thickness of treads. Defined-shape cocoons can either have or lack the ventral anterior region elevated, conferring a shoe/boot shape (Figs. 16.3.1 B, C, E) or a slipper shape (Figs. 16.3.1 D, F) to the cocoon. Sexual dimorphism is usually present, the male having an elongated frontoclypeus and the female a shorter one. The length and shape of trichomes on the frontoclypeus (facial and frontal) and thorax also are variable, as are the shape and distribution of the tubercles on the frontoclypeus and thorax. The spiracular gills (Fig. 16.3.1 A) are inserted in the thorax, with the number, disposition, and shape of the filaments being variable. The abdominal tergites (Fig. 16.3.1 G) and sternites (Fig. 16.3.1 H) have hooks, spines, setae, and spine combs that are variable in size and distribution, and the posterior region may be either with or without a terminal spine, which can be variable in length (Fig. 16.3.1 G).

Last-Instar Larva (Figs. 16.3.2, 3, 56–66). The body is cylindrical and divided into head, thorax, and abdomen

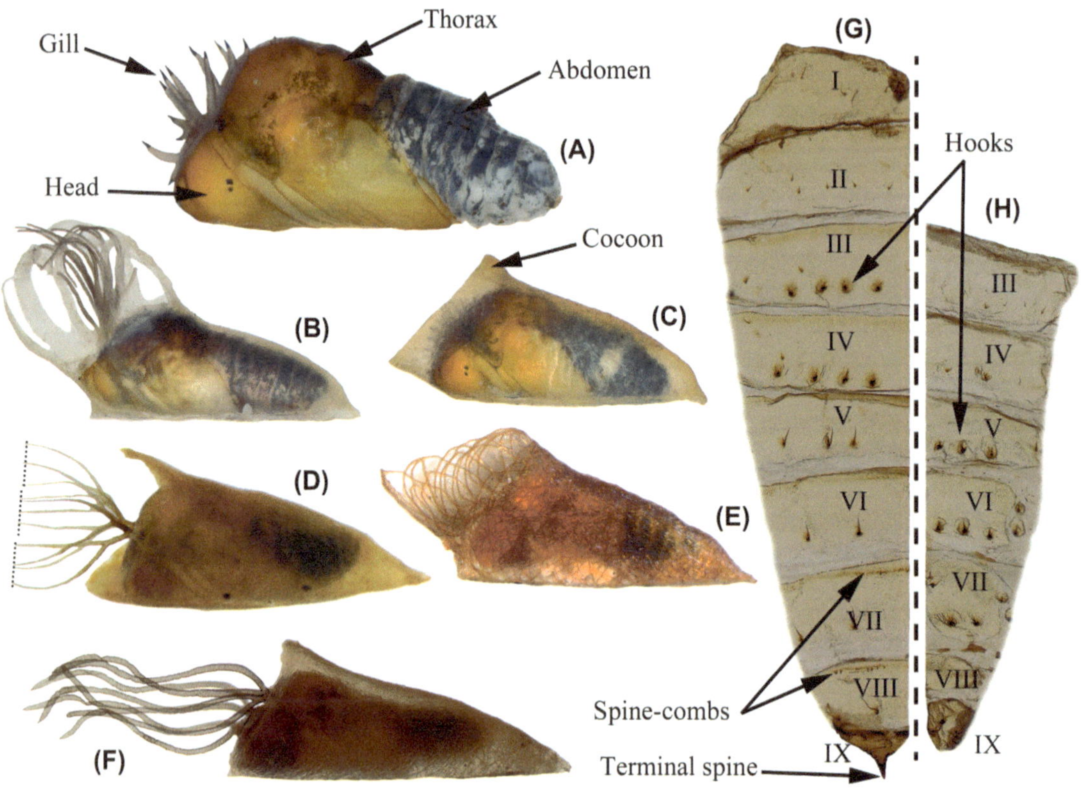

FIGURE 16.3.1 Simuliidae pupae: (A)–(F) habitus, lateral view: (A) without cocoon; (B–F) with cocoon; and (G–H) dissected abdomen in dorsal and ventral view, respectively.

(Fig. 16.3.2 A). The last instar can be characterized by having gill histoblast (Fig. 16.3.2 A) completely formed (whitish to black in color, depending on the larval age or gill characteristics), and the body cuticle can be bare or covered with setae of variable size and shape. The head (Figs. 16.3.2 B, C) is well structured and sclerotized, and usually has a pair of labral fans and well-developed antenna composed of three articles plus one small cone sensillum (Fig. 16.3.3 D); dorsally (Fig. 16.3.2 B) the frontoclypeal apotome has a characteristic pattern; ventrally (Fig. 16.3.2 C) the postgenal cleft is variable in shape and depth; the hypostoma has teeth variable in shape and distribution (Fig. 16.3.3 A). The mandibles (Fig. 16.3.3 B) have teeth and mandibular serrations that are variable in size and number. The thorax has one prothoracic proleg (Fig. 16.3.2 A) ventrally, and the histoblasts of gill, as well as legs, wing, and halter are well developed. The distal region of the abdomen has a posterior circlet (Figs. 16.3.2 A, 3 C), which is variable in both the number of rows of hooks and in the number of hooks per row. The anal sclerite (Fig. 16.3.3 C) is composed of an anterodorsal and a posteroventral arm.

MATERIAL PREPARATION AND PRESERVATION

Larvae and pupae can be fixed in 80% ethanol for general morphologic analysis. However, since cytological analysis can be useful to distinguish closely related species, last-instar larvae can be fixed in Carnoy solution (1:3, glacial acetic acid: absolute ethanol), which needs to be changed at least twice after collection. Carnoy also is an excellent fixative for morphologic analysis, since it provides good preservation of the tissues and color. Specimens fixed in Carnoy solution need to be kept in a refrigerator (4°C) or in a freezer (−20°C); if the material is not kept under refrigeration, the specimens need to be placed in 80% ethanol for better preservation. Preservation in absolute ethanol and maintenance at low temperature (−20°C) is also recommended because many taxonomic problems have been resolved in this family by using molecular tools. Specimens fixed in Carnoy or absolute ethanol need to have these solutions changed two or three times after collection since water and other liquids from the bodies of the specimens can dilute the fixative, resulting in poor tissue preservation.

To rear the pharate pupae under artificial conditions, a 2-mm plastic tube with a wet piece of filter paper can be used. If possible, cut the substrate from around the pharate adults using scissors and place the substrate on the humid filter paper inside the vial. If it is not possible to cut the substrate, for example when pupation occurs on rocks, then carefully remove the pupae from their substrate without damaging them.

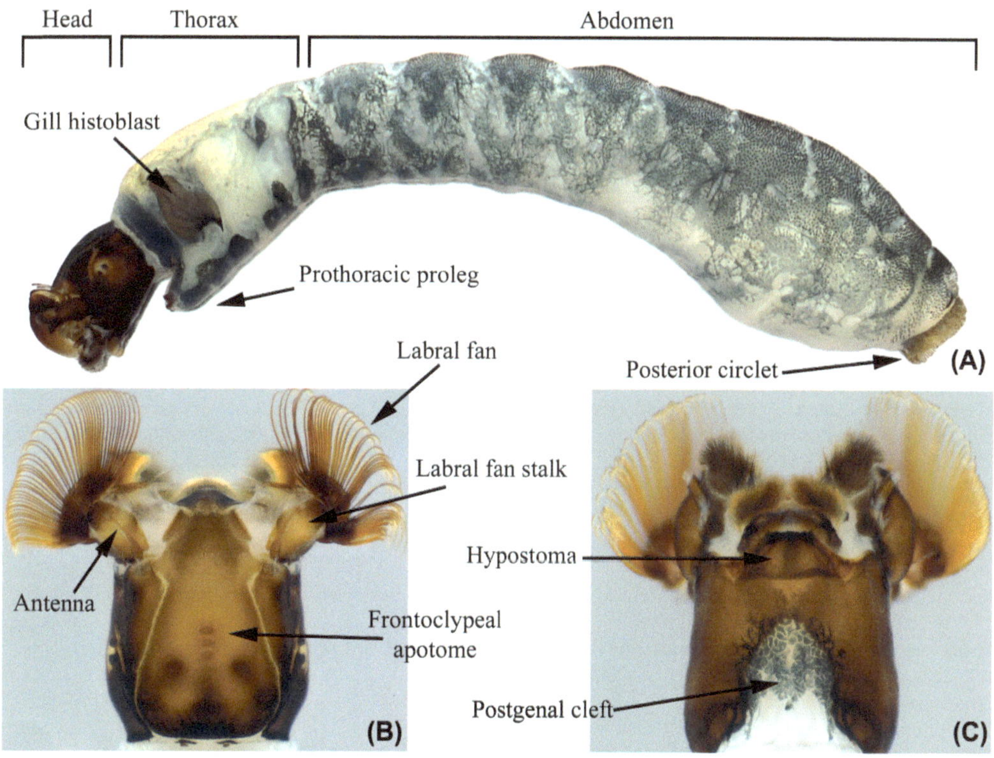

FIGURE 16.3.2 Simuliidae larva: (A) habitus, lateral view; (B) head, dorsal view; and (C) head, ventral view.

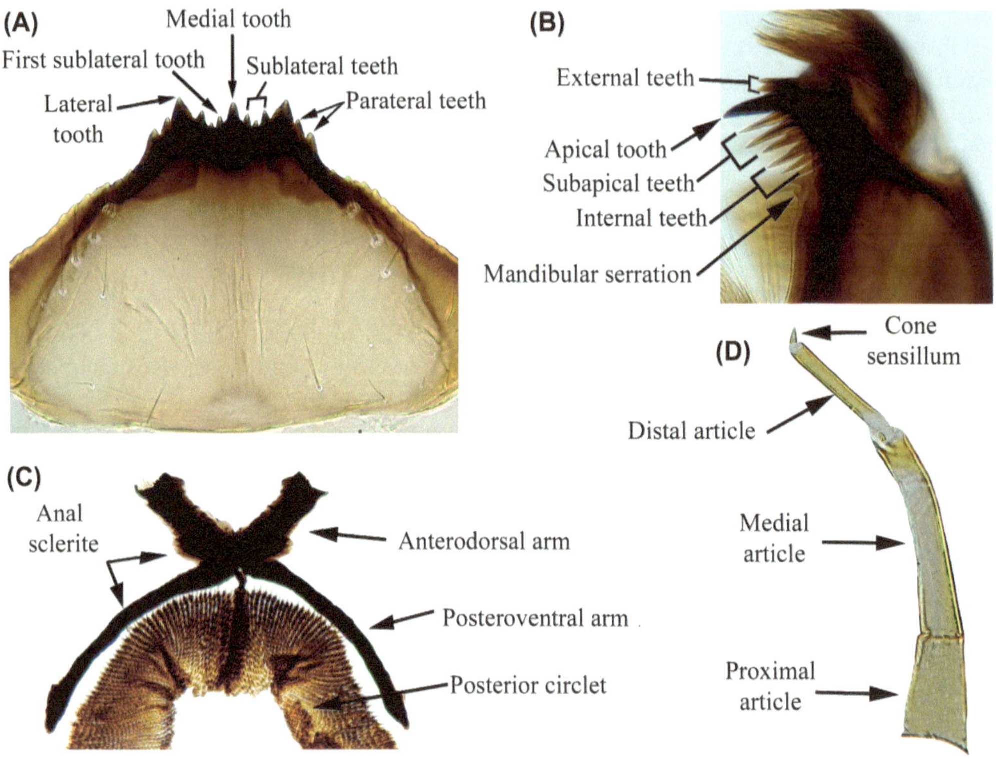

FIGURE 16.3.3 Simuliidae larva: (A) hypostoma, ventral view; (B) mandible, apical region; (C) anal sclerite; and (D) antenna.

KEYS TO SIMULIIDAE

Larval and pupal keys are based on Coscarón & Coscarón-Arias (2007).

Simuliidae: Genera: Pupae

1	Abdominal sterna VI and VII medially divided by membranous, striate area (Fig. 16.3.4 A); hooks of sterna VI–VIII always simple (Fig. 16.3.4 B); abdominal segments VIII and IX with strongly curved, twisted, looped, or unciform setae (Fig. 16.3.4 C) 2
1'	Abdominal sterna VI and VII not medially divided (Fig. 16.3.5 A); hooks on sterna VI–VIII simple, bifid, or trifid (Fig. 16.3.5 B); abdominal segments VIII and IX without aforementioned setae (Fig. 16.3.5 C) .... *Simulium* [Cosmopolitan]
2(1)	Abdomen terminal spines present (Fig. 16.3.4 D) 3
2'	Abdomen terminal spines absent .... *Tlalocomyia* (in part) [Central America. Nearctic]
3(2)	Abdomen terminal spines length ≤2× width (Fig. 16.3.6) 4
3'	Abdomen terminal spines length ≥3× width (Fig. 16.3.7) 5
4(3)	Gills tubular, perpendicularly inserted on short trunk and bearing delicate scattered filaments; gills without basal projections (Fig. 16.3.8) .... *Cnesiamima* [Argentina, Chile]
4'	Gills foliaceus, twisted, not perpendicularly inserted on short trunk and not bearing filaments; gills with two basal projections (Fig. 16.3.9) .... *Paraustrosimulium* [Argentina, Chile. Australia]
5(3)	Abdominal terga with spine combs anteriorly, at least on segments VII and VIII (Fig. 16.3.10 A); tergum VIII without four thick hooks (Fig. 16.3.10 B), or with four thin hooks posteriorly (Fig. 16.3.18) 6
5'	Abdominal terga without spine combs anteriorly (Fig. 16.3.11 A); tergum VIII with four thick hooks posteriorly (Fig. 16.3.11 B) .... *Cnesia* [Argentina, Chile]
6(5)	Gills filiform, arborescent, basally divided into two primary branches (Fig. 16.3.12) 7
6'	Gills of variable shape, never basally divided into two primary branches (Figs. 16.3.13, 19, 22) 8

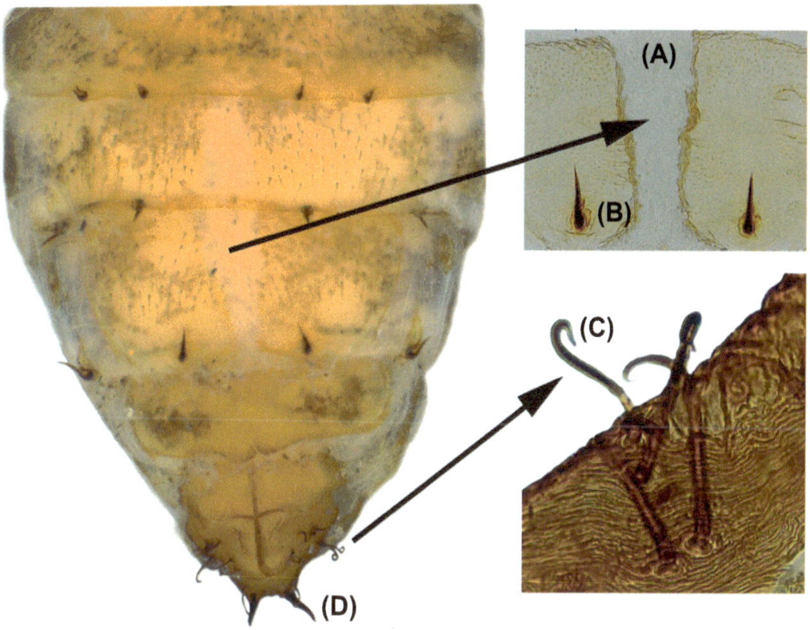

FIGURE 16.3.4 *Araucnephia* pupa: (A–D) abdomen, ventral view.

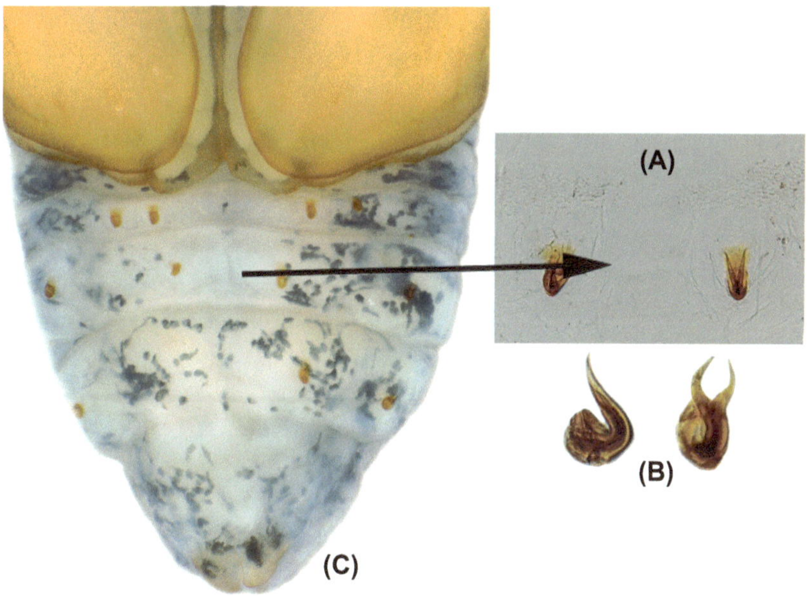

FIGURE 16.3.5 *Simulium* pupa: (A–C) abdomen, ventral view.

7(6)	Thorax with tubercles arranged in circles (Fig. 16.3.14); gills basal portion with minute lamellar cuticular structures (Fig. 16.3.15) .... *Araucnephioides* [Chile]
7'	Thorax without tubercles or with few tubercles randomly distributed, not arranged in circles (Fig. 16.3.16); gills basal portion without minute lamellar cuticular structures (Fig. 16.3.17) .... *Lutzsimulium* [Argentina, Brazil]
8(6)	Tergum VIII posteriorly with thin hooks (Fig. 16.3.18) 9
8'	Tergum VIII posteriorly without thin hooks (Fig. 16.3.10 B) 10

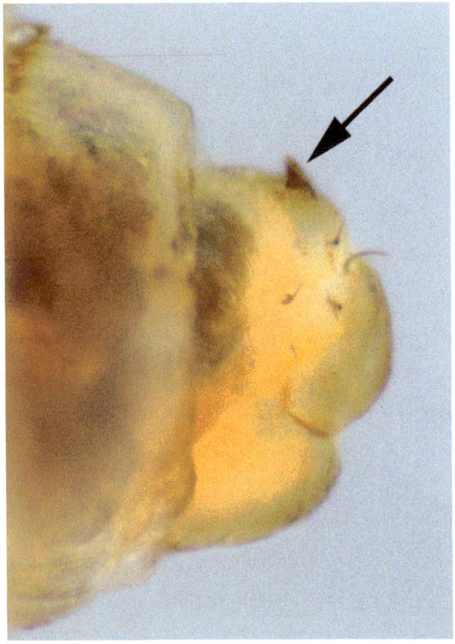

FIGURE 16.3.6 *Paraustrosimulium* pupa: terminal region of abdomen, lateral view.

FIGURE 16.3.8 *Cnesiamima* pupa: gill filament and partial region of cephalic plate and thorax, lateral view.

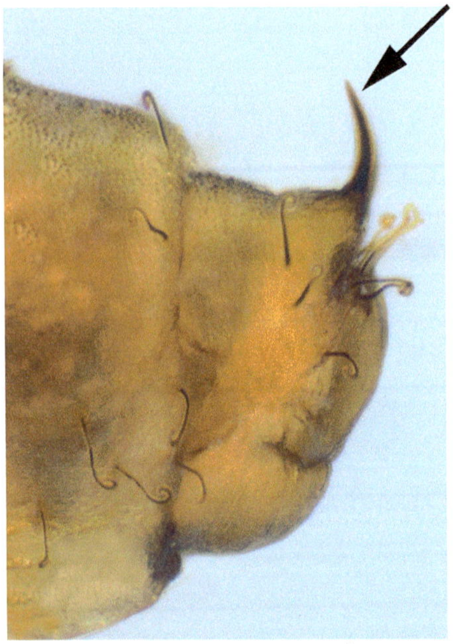

FIGURE 16.3.7 *Lutzsimulium* pupa: terminal region of abdomen, lateral view.

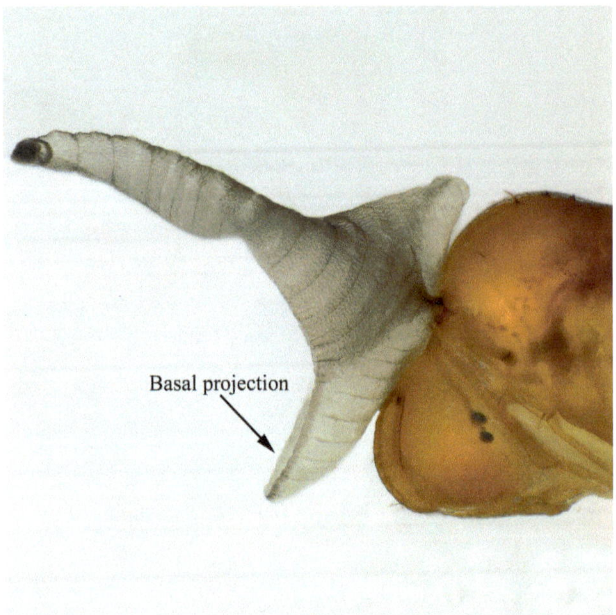

FIGURE 16.3.9 *Paraustrosimulium* pupa: gill filament and partial region of cephalic plate and thorax, lateral view.

9(8) Gills filiform, arborescent, with stout branches and abundant thin filamentous branches numbering more than 50; main branches of gills directed forward, distinctly longer than thorax (Fig. 16.3.19) .... *Pedrowygomyia* [Argentina, Chile, Peru]

9' Gills of variable shape from filamentous to arborescent, tubular or globular (Figs. 16.3.20, 21); rarely with stout branches and more than 50 thin filamentous branches (in this case the main branches of the gills are divergent and shorter than the thorax) .... *Gigantodax* [Neotropics. Nearctic]

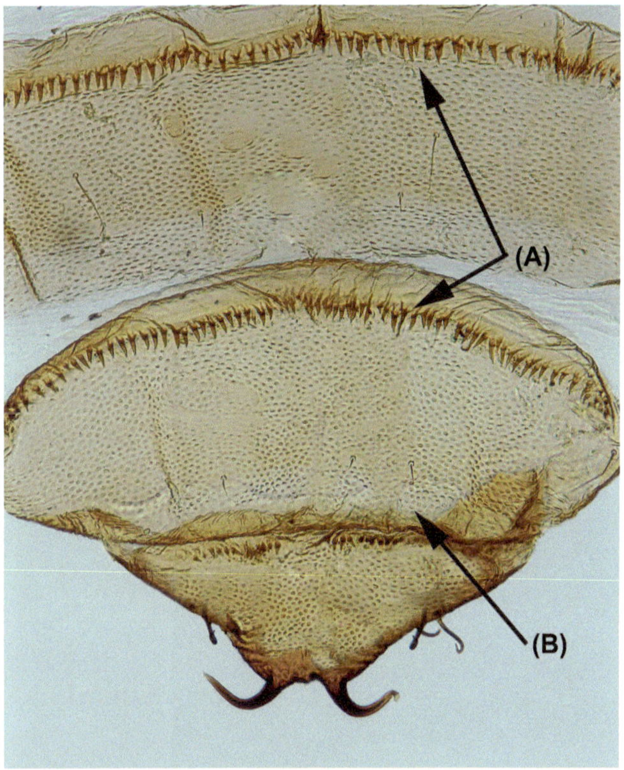

FIGURE 16.3.10 *Araucnephia* pupa: (A, B) abdomen, dorsal view.

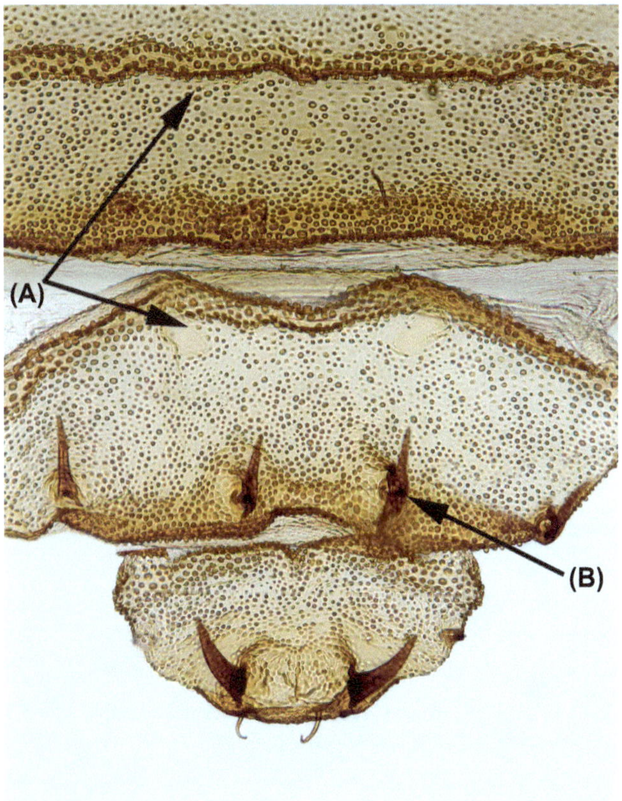

FIGURE 16.3.11 *Cnesia* pupa: (A, B): abdomen, dorsal view.

FIGURE 16.3.12 *Araucnephioides* pupa: gill filaments and partial region of cephalic plate and thorax, lateral view.

FIGURE 16.3.13 *Araucnephia* pupa: gill filaments and partial region of cephalic plate and thorax, lateral view.

FIGURE 16.3.14 *Araucnephioides* pupa: thorax, partial region.

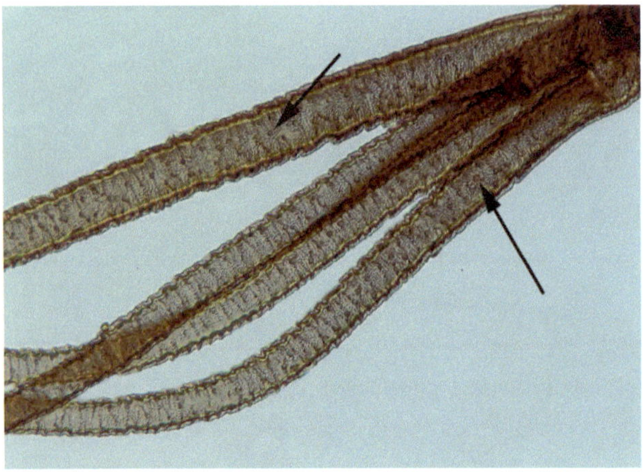

FIGURE 16.3.15 *Araucnephioides* pupa: gill filaments, detail.

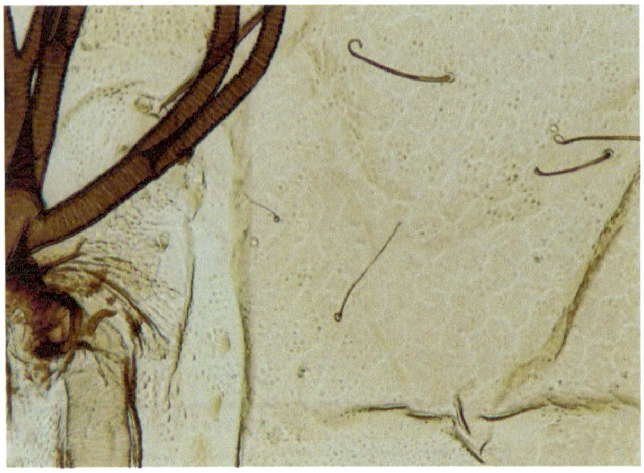

FIGURE 16.3.16 *Lutzsimulium* pupa: thorax, partial region.

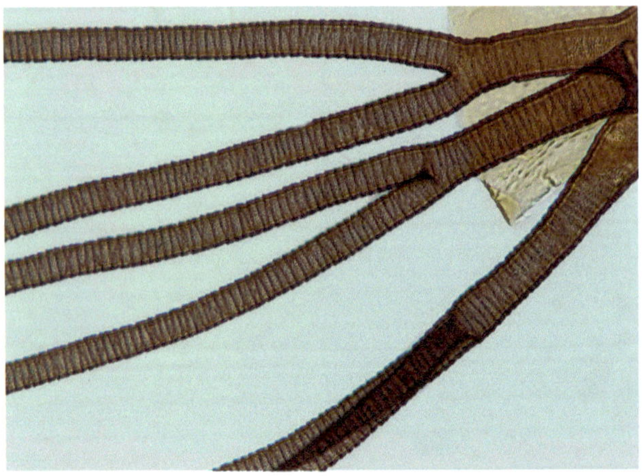

FIGURE 16.3.17 *Lutzsimulium* pupa: gill filaments, detail.

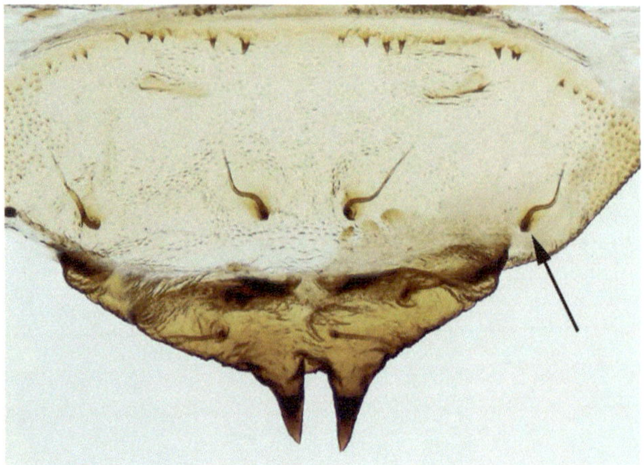

FIGURE 16.3.18 *Pedrowygomyia* pupa: terminal region of abdomen, dorsal view.

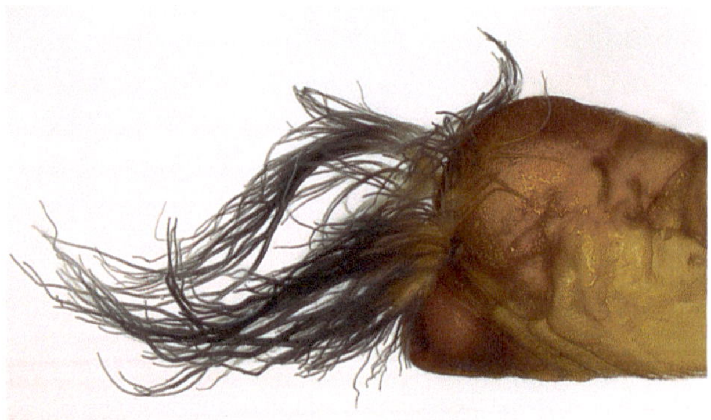

FIGURE 16.3.19 *Pedrowygomyia* pupa: gill filaments and partial region of cephalic plate and thorax, lateral view.

FIGURE 16.3.20 *Gigantodax* pupa: gill filaments and partial region of cephalic plate and thorax, lateral view.

FIGURE 16.3.21 *Gigantodax* pupa: gill filaments and partial region of cephalic plate and thorax, lateral view.

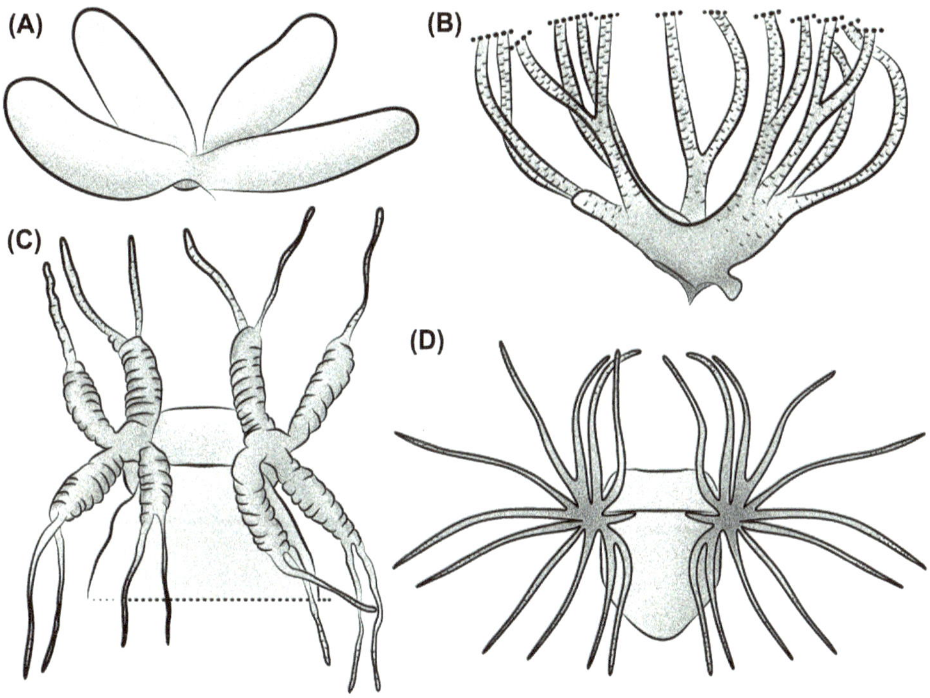

FIGURE 16.3.22 *Tlalocomyia* pupa. Line drawing of the gill filaments: (A) gill, dorsolateral view; (B) gill, lateral view, partial representation of the filaments; (C, D) gill filaments and partial region of cephalic plate and thorax, dorsal view. *Modified from Coscarón & Coscarón-Arias (2007).*

10(8) Gills filiform, slender, typically projected forward (Fig. 16.3.13) .... *Araucnephia*
[Argentina, Brazil, Chile]

10' Gills thickened, tubular, with branches typically radiated laterally from base (Figs 16.3.22 A, C); when filiform, branches radiating fanlike (Figs 16.3.22 B, D) .... *Tlalocomyia* (in part)
[Central America. Nearctic]

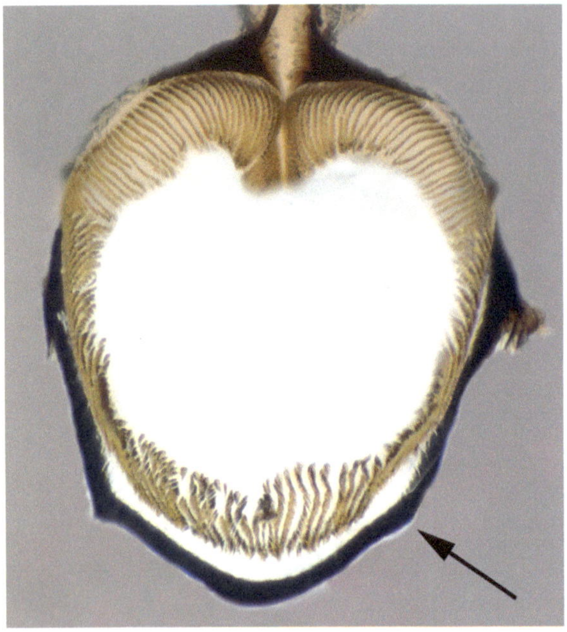

FIGURE 16.3.23 *Pedrowygomyia* larva: anal sclerite (anterodorsal arms not shown).

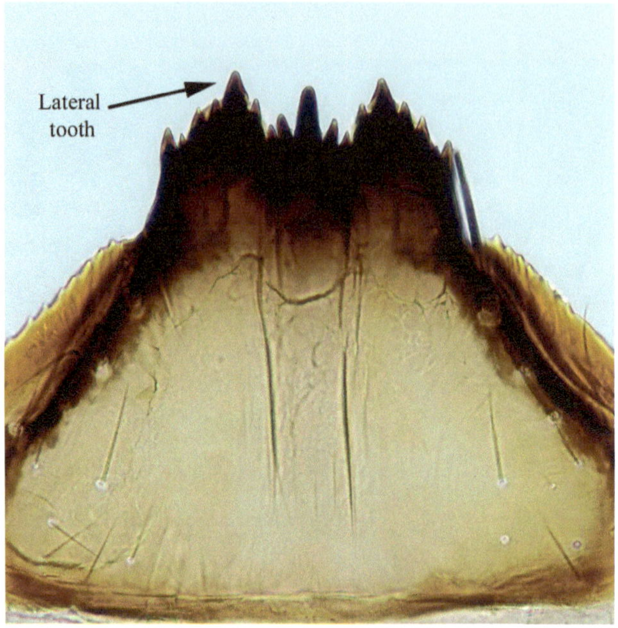

FIGURE 16.3.25 *Gigantodax* larva: hypostoma, ventral view.

FIGURE 16.3.24 *Simulium* larva: anal sclerite in situ (anterodorsal arms not shown).

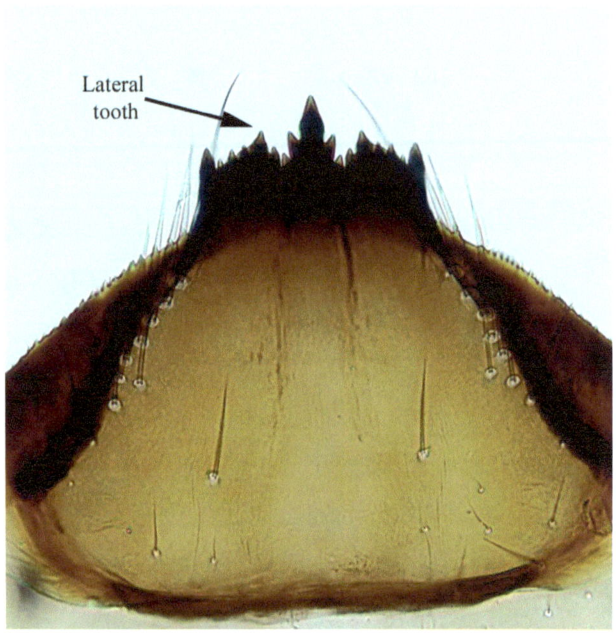

FIGURE 16.3.26 *Pedrowygomyia* larva: hypostoma, ventral view.

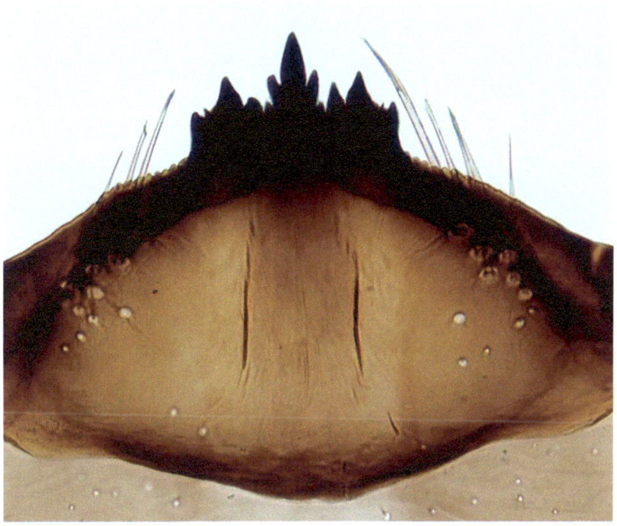

FIGURE 16.3.27 *Araucnephia* larva: hypostoma, ventral view.

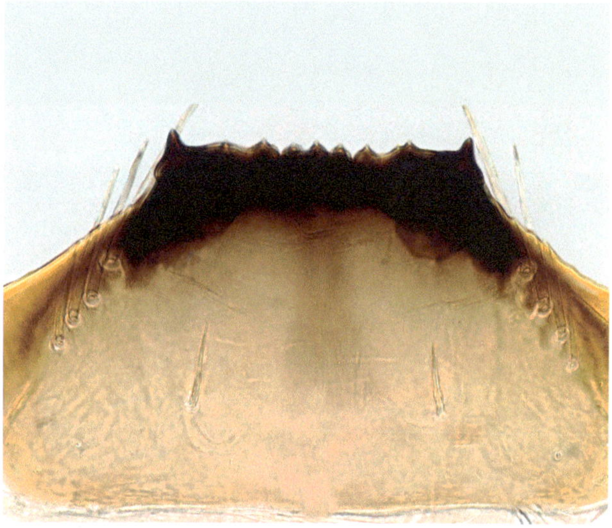

FIGURE 16.3.29 *Simulium* larva: hypostoma, ventral view.

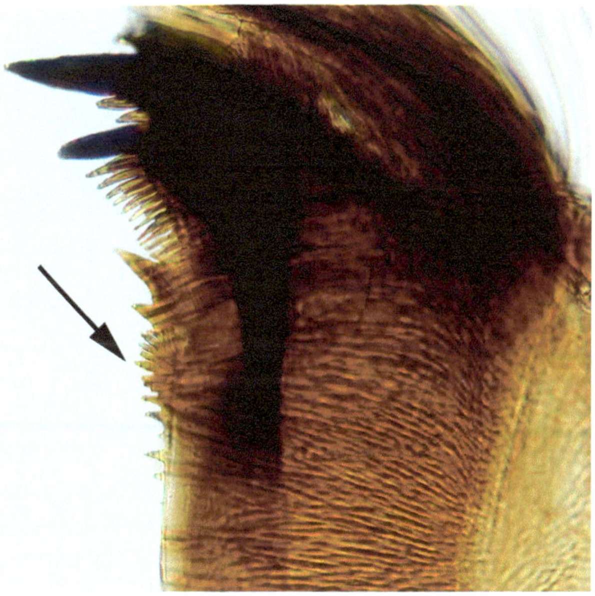

FIGURE 16.3.28 *Araucnephia* larva: mandible, apical region.

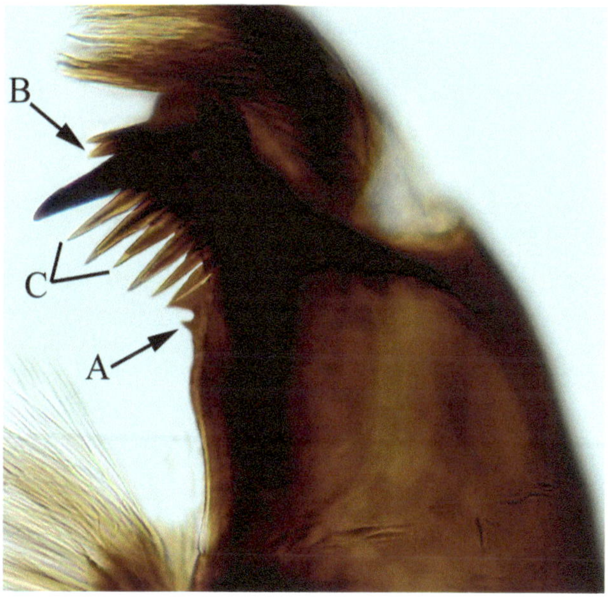

FIGURE 16.3.30 *Simulium* larva: mandible, apical region.

FIGURE 16.3.31 *Araucnephioides* larva: head, dorsal view.

FIGURE 16.3.32 *Araucnephia* larva: head, dorsal view.

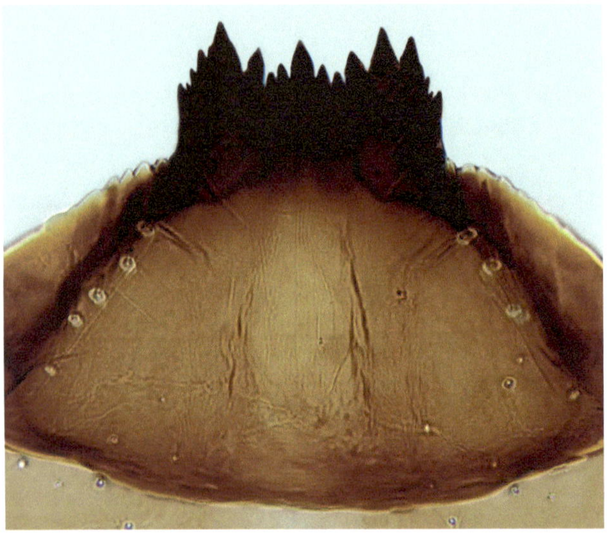

FIGURE 16.3.33 *Araucnephioides* larva: hypostoma, ventral view.

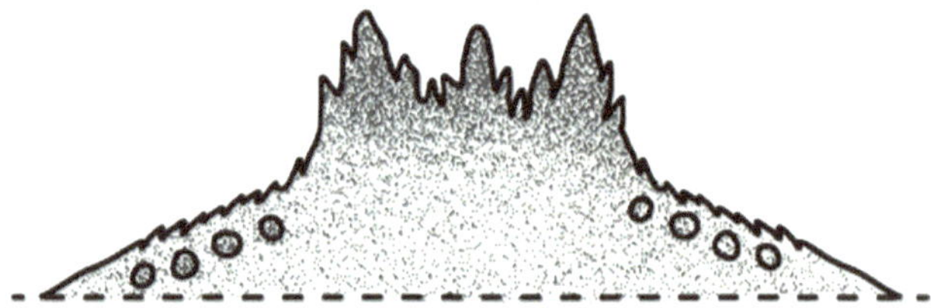

FIGURE 16.3.34 *Tlalocomyia* larva: hypostoma, distal region in ventral view. *Modified from Coscarón & Coscarón-Arias (2007).*

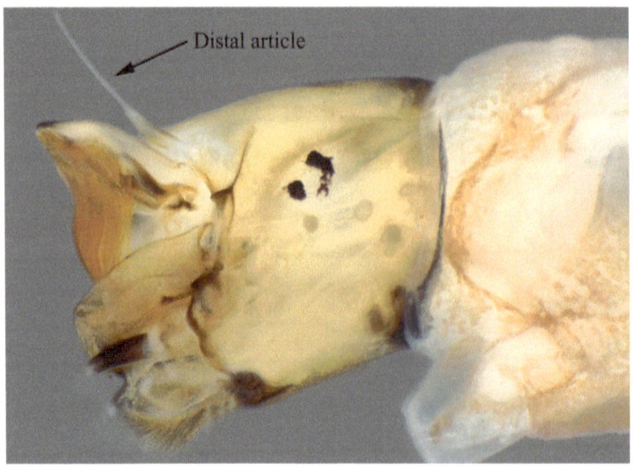

FIGURE 16.3.35 *Cnesiamima* larva: head and partial region of the thorax, lateral view.

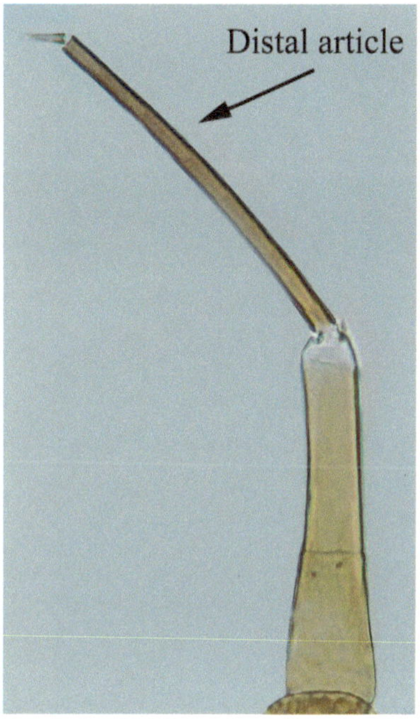

FIGURE 16.3.36 *Lutzsimulium* larva: antenna.

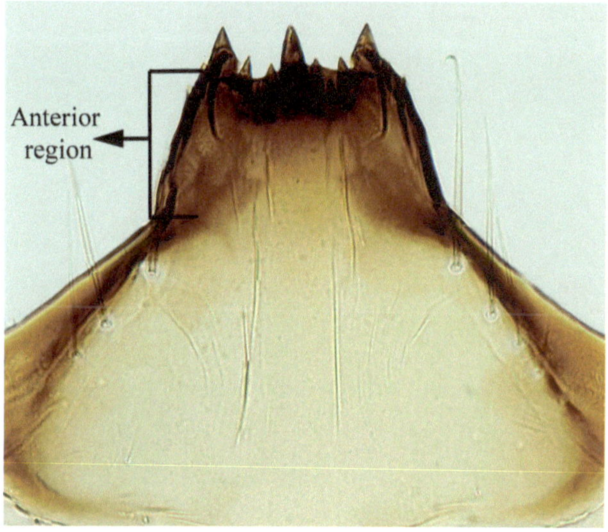

FIGURE 16.3.38 *Cnesiamima* larva: hypostoma, ventral view.

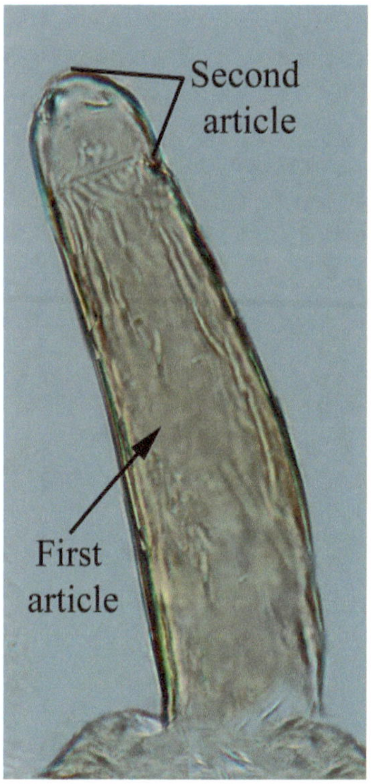

FIGURE 16.3.37 *Cnesiamima* larva: antenna, partial representation.

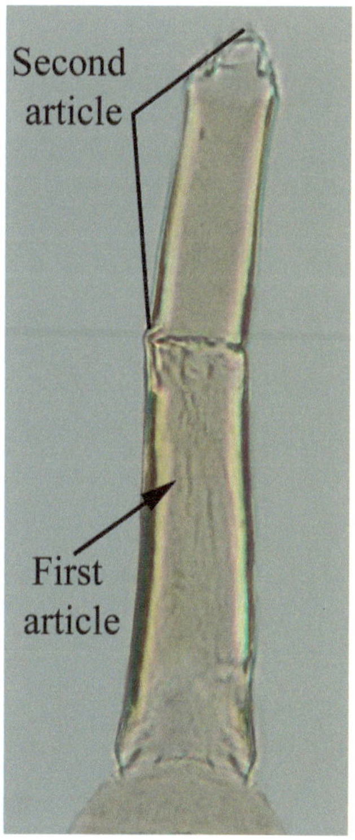

FIGURE 16.3.39 *Paraustrosimulium* larva: antenna, partial representation.

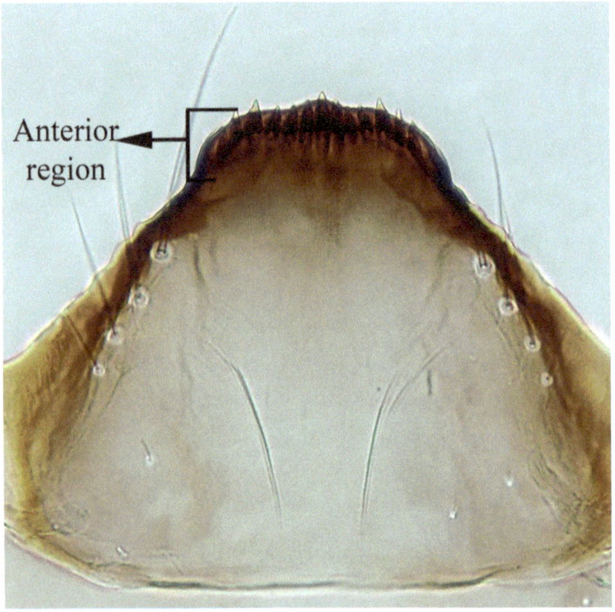

FIGURE 16.3.40 *Paraustrosimulium* larva: hypostoma, ventral view.

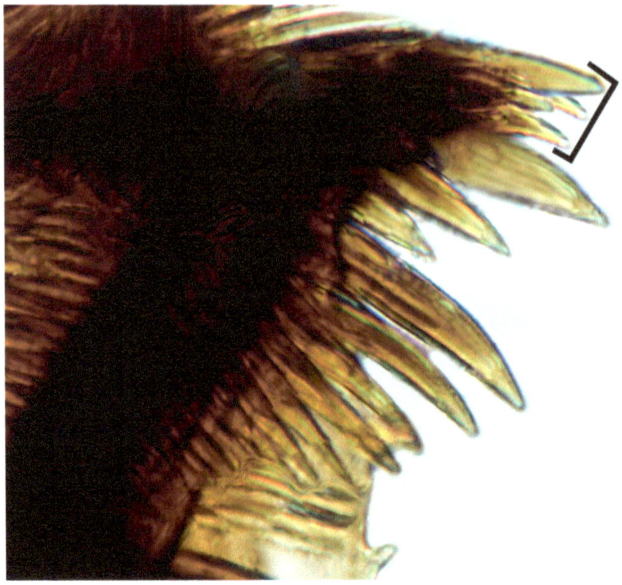

FIGURE 16.3.42 *Cnesia* larva: mandible, apical region.

FIGURE 16.3.41 *Cnesia* larva: hypostoma, distal region in ventral view.

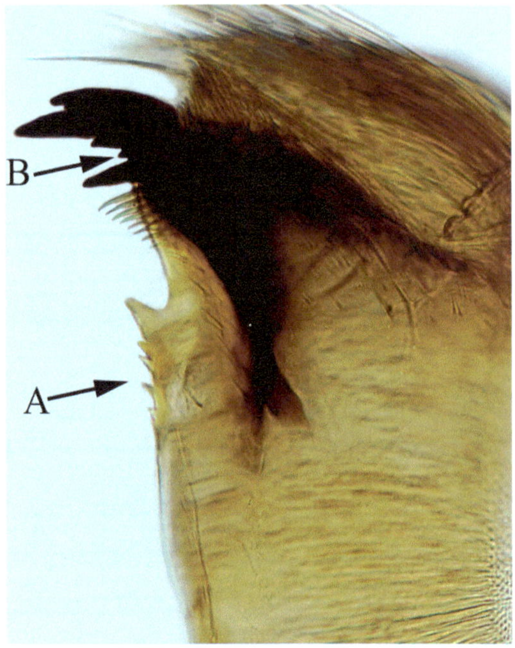

FIGURE 16.3.43 *Lutzsimulium* larva: mandible, apical region.

Simuliidae: Genera: Larvae

1	Anal sclerite posteroventral arms forming a complete ring around the ninth abdominal segment (Fig. 16.3.23)	2
1'	Anal sclerite posteroventral arms not forming a complete ring around the ninth abdominal segment (Fig. 16.3.24)	3
2(1)	Hypostoma lateral tooth projecting beyond level of first sublateral tooth (Fig. 16.3.25) [Neotropics. Nearctic]	*Gigantodax*
2'	Hypostoma lateral tooth not projecting beyond level of first sublateral tooth (Fig. 16.3.26) [Argentina, Chile, Peru]	*Pedrowygomyia*

3(1) Hypostoma teeth arranged in three conspicuous groups (Fig. 16.3.27); mandible with more than five mandibular serrations (Fig. 16.3.28) 4

3' Hypostoma teeth not arranged in three conspicuous groups (Fig. 16.3.29); mandible with five or fewer mandibular serrations (Figs 16.3.30 A, 43 A) 6

4(3) Antennae length subequal to labral fan stalk (Fig. 16.3.31) 5

4' Antennae length less than labral fan stalk (Fig. 16.3.32) .... *Araucnephia* [Argentina, Brazil, Chile]

5(4) Hypostoma with 17 teeth (Fig. 16.3.33) .... *Araucnephioides* [Chile]

5' Hypostoma with 13 teeth (Fig. 16.3.34) .... *Tlalocomyia* [Central America. Nearctic]

6(3) Antenna distal article length $\geq$1.5$\times$ proximal and medial articles combined (Fig. 16.3.35) 7

6' Antenna distal article length $\leq$ proximal and medial articles combined (Fig. 16.3.36) 8

7(6) Antenna first article length ~4$\times$ second article (Fig. 16.3.37); hypostoma anterior region width ~1.5$\times$ length (Fig. 16.3.38) .... *Cnesiamima* [Argentina, Chile]

7' Antenna first article length ~1.3$\times$ second article (Fig. 16.3.39); hypostoma anterior region width ~3$\times$ length (Fig. 16.3.40) .... *Parautrosimulium* [Argentina, Chile. Australia]

8(6) Hypostoma with at most 13 teeth (Fig. 16.3.29); mandible with two external teeth (Fig. 16.3.30 B) 9

8' Hypostoma with 17 teeth (Fig. 16.3.41); mandibles with four external teeth (Fig. 16.3.42) .... *Cnesia* [Argentina, Chile]

9(8) Mandible subapical teeth subequal in size or decreasing in size from first to third (Fig. 16.3.30 C) .... *Simulium* [Cosmopolitan]

9' Mandible subapical teeth with second smaller than either first or third (Fig. 16.3.43 B) .... *Lutzsimulium* [Argentina, Brazil]

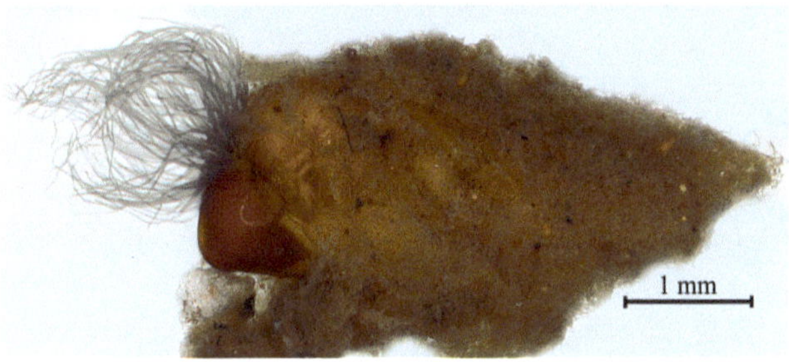

FIGURE 16.3.44 *Araucnephia* pupa: lateral view.

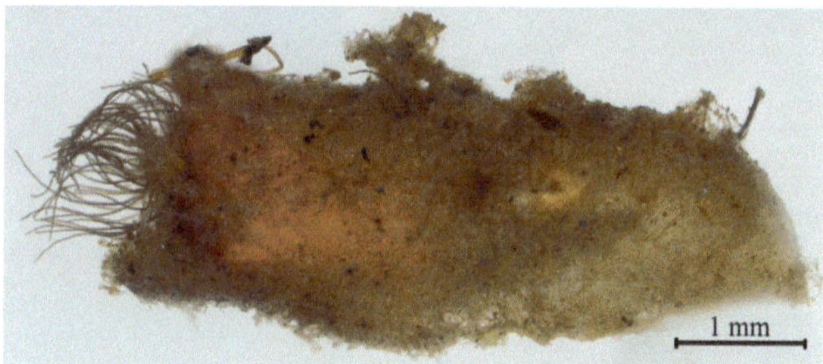

FIGURE 16.3.46 *Cnesia* pupa: lateral view.

FIGURE 16.3.45 *Araucnephioides* pupae.

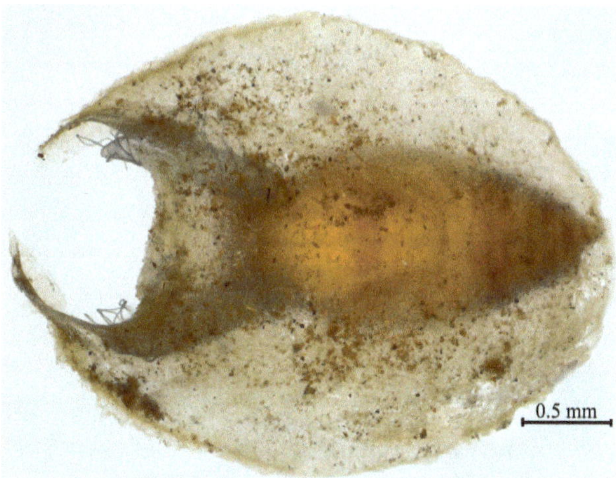

FIGURE 16.3.47 *Cnesiamima* pupa: dorsal view.

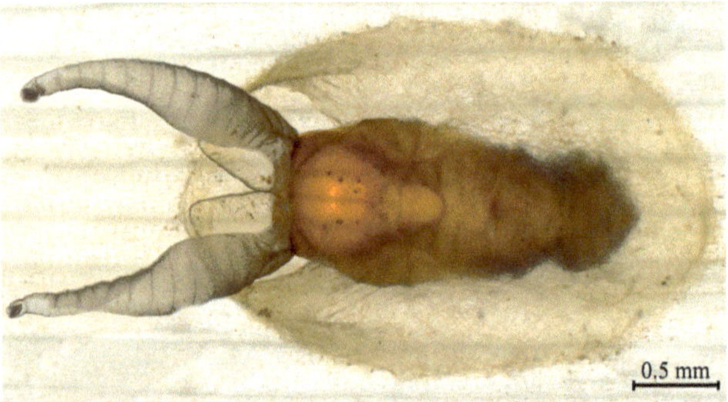

FIGURE 16.3.48 *Paraustrosimulium* pupa: dorsal view.

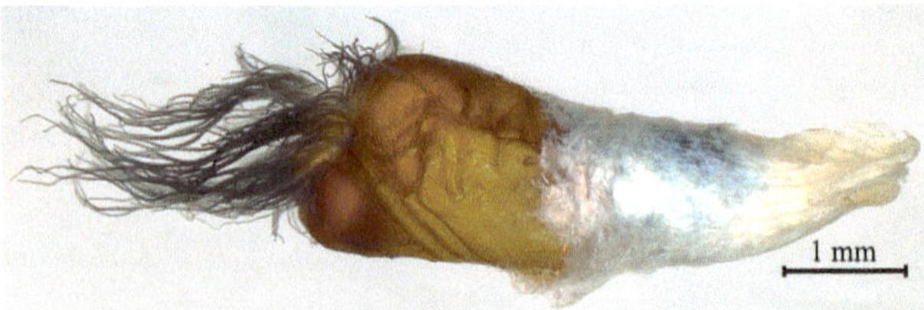

FIGURE 16.3.49 *Pedrowygomyia* pupa: lateral view.

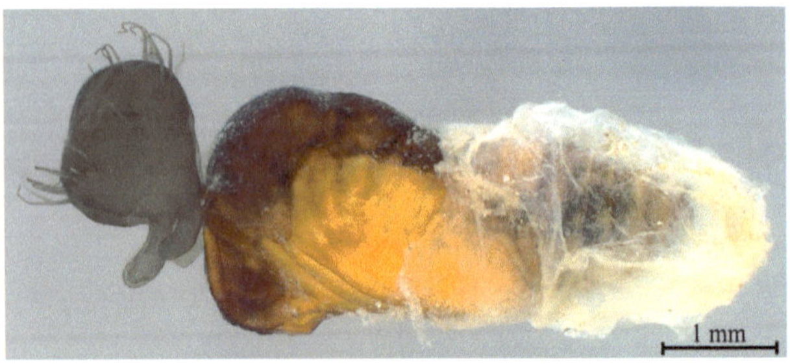

FIGURE 16.3.50 *Gigantodax* pupa: lateral view.

FIGURE 16.3.51 *Gigantodax* pupa: lateral view.

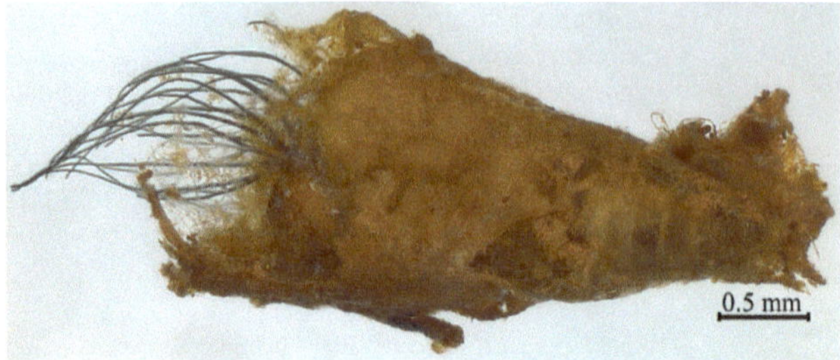

FIGURE 16.3.52 *Lutzsimulium* pupa: lateral view.

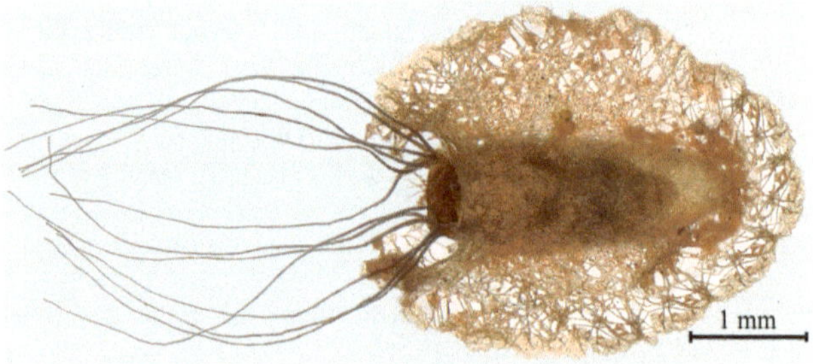

FIGURE 16.3.53 *Simulium* pupa: dorsal view.

FIGURE 16.3.54 *Simulium* pupa: lateral view.

FIGURE 16.3.55 *Tlalocomyia* pupa: lateral view.

FIGURE 16.3.56 *Araucnephia* larva: lateral view.

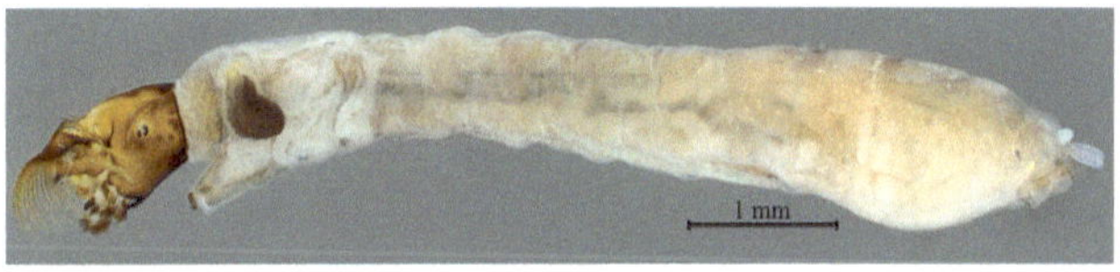

FIGURE 16.3.57 *Araucnephioides* larva: lateral view.

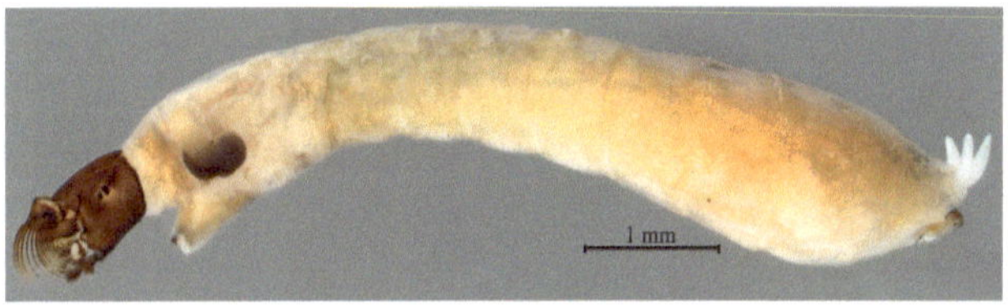

FIGURE 16.3.58 *Cnesia* larva: lateral view.

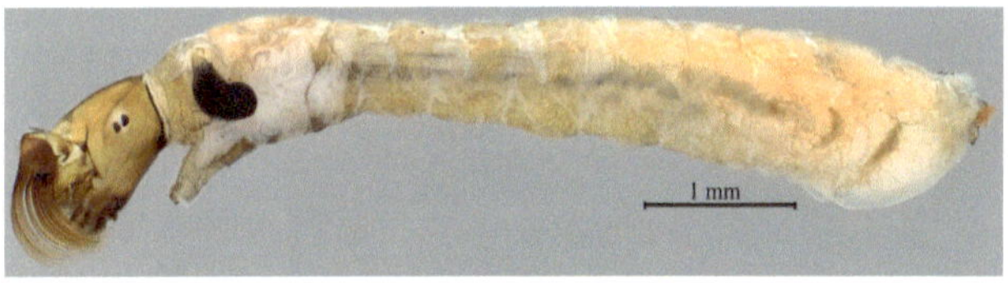

FIGURE 16.3.59 *Lutzsimulium* larva, lateral view.

FIGURE 16.3.60 *Cnesiamima* larva: lateral view.

FIGURE 16.3.61 *Paraustrosimulium* larva: lateral view.

FIGURE 16.3.62 *Gigantodax* larva, lateral view.

FIGURE 16.3.63 *Pedrowygomyia* larva: lateral view.

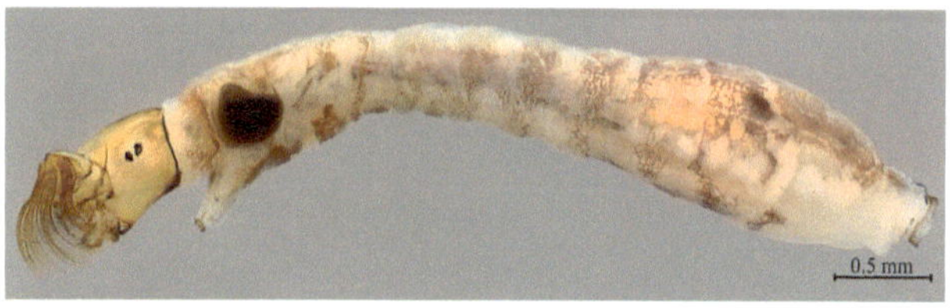

FIGURE 16.3.64 *Simulium* larva: lateral view.

FIGURE 16.3.65 *Simulium* larva: lateral view.

FIGURE 16.3.66 *Tlalocomyia* larva: lateral view.

ACKNOWLEDGMENTS

We thank Cristóbal Montedonico for guiding us safely in Chile; Dr. Pavel Stoev (Pensoft Editorial Director) for allowing the use of images 9F–I and 10N (Pages 58 and 64) of the book *Neotropical Simuliidae (Diptera: Insecta)* by Sixto Coscarón & Cecilia L. Coscarón Arias (In: Adis, J., Arias, J.R., Rueda-Delgado, G. & K.M. Wantzen (Eds.): *Aquatic Biodiversity in Latin America (ABLA)*. Vol. 3. Pensoft, Sofia-Moscow, 685 pp.); Heron Huerta for providing specimens from the INDRE collection; CNPq for constant financial support (e.g., PROTAX 440616/2015-8) and especially for fellowships 307849/2014-7 (provided to NH) and 141832/2014-2 (to JMCN), and MCTIC/INPA for internal grant support. CAPES Pro-Equipamentos provided laboratory equipment. Leo R. Trindade for the habitus photographs presented in this chapter. Philip M. Fearnside reviewed the English grammar.

REFERENCES

Adler, P.H. & R.W. Crosskey. 2018. World blackflies (Diptera: Simuliidae): a comprehensive revision of the taxonomic and geographical inventory. http://www.clemson.edu/cafls/biomia/pdfs/blackflyinventory.pdf, (accessed 2018.05.10).

Adler, P.H., D.C. Currie & D.M. Wood. 2004. The Black Flies (Simuliidae) of North America, p. 941. Cornell University Press, Ithaca, NY.

Coscarón, S. & C.L. Coscarón-Arias. 2007. Neotropical Simuliidae (Diptera: Insecta). Pages 1–694 *in*: J. Adis, J.R. Arias, G. Rueda-Delgado & K.M. Wantzen (eds.), Aquatic Biodiversity in Latin America (ABLA). Vol. 3. Pensoft, Sofia, Moscow.

Coscarón, S. & D.R. Miranda-Esquivel. 1998. *Pedrowygomyia* a new neotropical Prosimuliini genus (Diptera: Simuliidae): *Gigantodax* s. lat. split of into two genera. Entomologica Scandinavica 29: 161–168.

Crosskey, R.W. 1990. The natural history of blackflies, 711p. The Museum of Natural History, London.

Currie, D.C. & P.H. Adler. 2008. Global diversity of black flies (Diptera: Simuliidae) in freshwater. Hydrobiologia 595: 469–475.

Hamada, N., M. Pepinelli, A. Mattos-Glória & S.L. Luz. 2010. A new black fly species from Brazil, closely related to *Simulium guianense* Wise (Diptera, Simuliidae), revealed by morphological and DNA barcoding. Zootaxa 2428: 22–36.

Malmqvist, B. & B. Sackmann. 1996. Changing risk or predation for a filter-feeding insect along a current velocity gradient. Oecologia 108: 450–458.

Santos-Neto, C.R., N. Hamada & S.R.M. Couceiro. 2015. Bionomics of the black fly *Simulium guianense* (Diptera: Simuliidae) in northeast Brazil. Florida Entomologist 98: 446–450.

Wotton R.S., C.R. Joicey & B. Malmqvist. 1996. Spiralling of particles by suspension feeders in a small lake-outlet stream. Canadian Journal of Zoology 74: 758–761.

Wotton R.S., B. Malmqvist, K. Ashelford. 1995. The retention of particles intercepted by a dense aggregation of lake-outlet suspension feeders. Hydrobiologia 306: 125–129.

Chapter 16.4

Family Thaumaleidae

Luiz Carlos Pinho

Departamento de Ecologia e Zoologia, Universidade Federal de Santa Catarina, Florianópolis, Santa Catarina, Brazil

Heron Huerta

Instituto de Diagnóstico y Referencia Epidemiológicos, Laboratorio de Entomología, Ciudad de México, México

Chapter Outline

INTRODUCTION

Thaumaleidae is a small family of aquatic dipterans that belongs to the most basal lineage of the Chironomoidea (Wood & Borkent, 1989). Larvae are found in hygropetric habitats, such as wet rocks bordering or beneath waterfalls, seeps on roadcuts, cliff-face seepages, and the underfaces of large boulders in cascading mountain streams (Sinclair, 2000), where they feed on diatoms and debris (Leathers, 1922). Larvae can also glide rapidly over the water film on the rock surface in a characteristic sideways or jerking motion, presumably to avoid predators (Sinclair, 2000). Pupae are found on wet moss, leaves, or buried in mud (Stone & Peterson, 1981).

The family comprises about 174 species distributed in all biogeographic regions, except Antarctica (Wagner et al., 2008). Only seven species are found in the Neotropical Region: *Androprosopa zempoala* Sinclair & Huerta from Mexico, *Oterere setipennis* (Edwards) from Argentina and *Niphta halteris* (Edwards), *N. nudipennis* (Edwards), *Austrothaumalea chilensis* Edwards, *A. apicalis* Edwards, and *A. spatulata* Schmid from Chile. Other unknown species have been recorded to the mountains in Ecuador (Sinclair & Huerta, 2010) and Brazil (Pivar et al, in press).

LIMITATIONS

Little is known about the biology of Neotropical Thaumaleidae species, as no immature stages have been described. Exploring hygropetric habitats in the Neotropics, particularly in the Andes and Brazilian mountains, should reveal some more new taxa and insights on trans-Antarctic relationships of *Niphta* and *Austrothaumalea*.

TERMINOLOGY AND MORPHOLOGY

The larva (Figs. 16.4.1 A–C) of Thaumaleidae shows a unique combination of a complete head capsule, prothoracic spiracles on short tubes, and unpaired prolegs on the prothorax and at the end of the abdomen. The head capsule is complete, nonretractile, and may be smooth or ornamented with large spiny protuberances (Figs. 16.4.1 B–C). Thoracic and abdominal segments are sclerotized dorsally and bear stout dark setae laterally. Abdominal segment 8 shows a dorsal and transverse spiracular plate with two lateral, sclerotized, finger-like procerci (Fig. 16.4.1 A), bearing long, dark setae (Stone & Peterson, 1981).

The pupa (Fig. 16.4.1 D) is free and subcylindrical, with the cephalothorax being equal to or only slightly larger dorsoventrally than the abdomen; the apical portion of the abdomen is directed posteriorly in life. The thoracic respiratory organ is more or less tubular and truncated apically. The dorsal portions of abdominal segments are flat and distinctly angled laterally, with rounded tubercles on each corner, bearing one or two stout setae. Short, erect, tubular spiracles are present on most abdominal segments. Apex of abdomen with two long spinous, sometimes hooked, processes (Stone & Peterson, 1981; Borkent, 2012).

Thorp and Covich's Freshwater Invertebrates, Fourth Edition. https://doi.org/10.1016/B978-0-12-804223-6.00033-0

FIGURE 16.4.1 Thaumaleidae, new genus from Brazil (Pivar et al., in press): (A) larval habitus, lateral view; (B) larval head and thorax, lateral view; (C) larval head and thorax, dorsal view; and (D) pupal exuvia, dorsal view.

MATERIAL PREPARATION AND PRESERVATION

Immatures and adults can be preserved temporarily in 70%–80% ethanol inside a freezer to avoid bleaching. For permanent preservation, part of the adult material should be pinned and part slide mounted in Euparal or Canada Balsam.

For exuviae, the method we use for slide mounting consists in transferring the specimen from the 70% ethanol to absolute ethanol for about 20 min, putting a small drop of Euparal in a slide, and moving the exuvia to the center of the drop (dorsal side up for pupa, ventral side up for larva) with a small piece of glass shard on each side, to keep the cover slip from crushing the skin. It is common to separate the respiratory organs of pupae and place them by the side of the body, under the same coverslip. Whole specimens must be cleared in advance, using KOH (10%).

REFERENCES

Borkent, A. 2012. The pupae of Culicomorpha - morphology and a new phylogenetic tree. Zootaxa 3396: 1–98.

Leathers, A.L. 1922. Ecological study of aquatic midges and some related insects with special reference to feeding habits. Bulletin of the Bureau of Fisheries, Washington 38: 1–61.

Pivar, R.J., L.C. Pinho, W.E. Klingeman, J.K. Moulton & B.J. Sinclair. In press. A new genus and species, and first record of the family Thaumaleidae (Diptera) from Brazil. The Canadian Entomologist.

Sinclair, B.J. 2000. Immature stages of Australian *Austrothaumalea* Tonnoir and *Niphta* Theischinger (Diptera: Thaumaleidae). Australian Journal of Entomology 39: 171–176.

Sinclair, B.J. & H. Huerta. 2010. A new species of *Androprosopa* from Mexico (Diptera: Thaumaleidae). Canadian Entomologist 142: 443–447.

Stone, A. & B.V. Peterson. 1981. Thaumaleidae. Chapter 26. Pages 351–353 *in*: J.F. McAlpine, B.V. Peterson, G.E. Shewell, H.J. Teskey, J.R. Vockeroth & D.M. Wood (Coordinators). Manual of Nearctic Diptera Vol. 1. Research Branch, Agriculture Canada, Monograph 27.

Wagner, R., M. Barták, A. Borkent, G. Courtney, B. Goddeeris, J.-P. Haenni, L. Knutson, A. Pont, G.E. Rotheray, R. Rozkošný, B. Sinclair, N. Woodley, T. Zatarnicki & P. Zwick. 2008. Global diversity of dipteran families (Insecta: Diptera) in freshwater (excluding Simuliidae, Culicidae, Chironomidae, Tipulidae and Tabanidae). Hydrobiologia 595: 489–519.

Wood, D.M. & A. Borkent. 1989. Phylogeny and classification of the Nematocera. Chapter 114 *in*: J.F. McAlpine & D.M. Wood (eds.). Manual of Nearctic Diptera. Volume 3. Agriculture Canada Monograph 32: 1333–1370.

Family Culicidae

Jeane Marcelle Cavalcante do Nascimento
Coordenação de Biodiversidade, Curso de Pós-Graduação em Entomologia, Instituto Nacional de Pesquisas da Amazônia, Manaus, Amazonas, Brazil

Ruth Leila Ferreira Keppler
Coordenação de Biodiversidade, Instituto Nacional de Pesquisas da Amazônia, Manaus, Amazonas, Brazil

Neusa Hamada
Coordenação de Biodiversidade, Instituto Nacional de Pesquisas da Amazônia, Manaus, Amazonas, Brazil

Chapter Outline

INTRODUCTION

Culicidae (from the Latin *culex*: gnat) constitutes a large and abundant group of dipterans (Wood & Borkent, 1989; Wiegmann et al., 2011). They are commonly known as "mosquitoes," a word that comes from the Spanish "musketas," meaning "little fly." Native Hispanic-Americans called them "zancudos," a term that is still used in South and Central America, meaning "long-legged."

Mosquitoes are considered to be one of the most relevant groups of arthropods, mainly due to their medical importance. These insects act as global vectors of organisms causing many important human diseases, such as malaria, dengue, and yellow fever, which together are responsible for several million deaths and hundreds of millions of cases every year (WHO, 2016). Within Culicidae, *Aedes* Meigen, *Anopheles* Meigen, and *Culex* Linnaeus include the best-known disease-vector species. In addition to their impact on human health, mosquitoes play a key role in natural ecosystems. They are an important food source for small fish and aquatic insects, which, in turn, are food for medium-sized fish. Adults make up part of the diet of some spiders and insects, as well as some vertebrates such as amphibians, reptiles, birds, and mammals (Woodring & Davison, 1996; Chaverri, 2009).

This is a group with worldwide distribution, occurring in the humid tropics and subtropics, warm moist climates, temperate, and cool zones, everywhere except in deserts and in areas that are permanently frozen. Immature stages (larvae and pupae) of all species are exclusively water dependent. Depending on species, the eggs may either be deposited directly on the water or on moist soil or other substrates that will eventually be inundated with water (Wallace & Walker, 2008).

This family constitutes a group of extremely successful organisms due to their ability to adapt to a wide range of environments. The immature stages occur in both permanent (e.g., rivers, lakes, and marshes) or transient bodies of fresh water. Transient habitats can include any receptacle that accumulates water, such as phytotelmata (e.g., bamboo internodes, bromeliads, fallen fruit husks, leaf axils, and tree holes), crab holes, rock holes, and artificial containers (Harbach, 2007). Mosquito larvae also can be found in turbid or polluted water and even in brackish water, except in marine habitats with high salt concentrations (Rueda, 2008). Some species, and even some genera or subgenera, show specificity

Thorp and Covich's Freshwater Invertebrates, Fourth Edition. https://doi.org/10.1016/B978-0-12-804223-6.00034-2

in the use of these environments. In the Neotropical region it is estimated that approximately 22% of all mosquito species are known to inhabit bromeliad axils (Frank & Curtis, 1981).

The monophyly of the Culicidae within Diptera is recognized by all previous workers (e.g., Wood & Borkent, 1989; Harbach & Kitching, 1998; Wiegmann et al., 2011; Borkent, 2012). However, relationships within the family are still being discussed, principally as regards to the number of genera (Reinert et al., 2009; Wilkerson et al., 2015). In this chapter, the taxonomic classification of the tribe Aedini as given by Wilkerson et al. (2015) is followed.

Culicidae has approximately 3,500 described species and 41 genera around the world. The family is currently divided into two subfamilies: Anophelinae, with about 480 species, and Culicinae, with approximately 3,100 species distributed in 11 tribes. Aedini, with about 1,260 species, is the largest tribe (WRBU, 2016). The greatest mosquito diversity is found in the Neotropical and Oriental regions. The Nearctic region has the lowest diversity (Rueda, 2008). In the Neotropics there are nearly 1,030 species in 25 genera; 12 genera are predominantly confined to the Neotropical region. The greatest generic diversity in this realm is found in the tribe Sabethini, represented by nine genera, and the greatest species diversity is found in the genus *Culex* Linnaeus, represented by 360 valid species (WRBU, 2016).

FIGURE 16.5.1 Culicidae larvae, habitus, dorsal view: (A) *Anopheles*; and (B) *Aedeomyia*. Bars = 1 mm.

LIMITATIONS

Culicidae is relevant mainly because of its medical importance as vectors of dangerous diseases. Unfortunately, there have been few scientific studies aimed at increasing knowledge about the taxonomy of mosquito genera that are not important vectors or pests. In addition, most systematic studies have focused on adult morphology, so information about aquatic stages of many subgenera and species remains fragmented.

There are several published keys for identification of North and Central American Culicidae fauna (e.g., Wallace & Walker, 2008; Chaverri, 2009), but there has been no attempt to construct a comprehensive key to immature stages from the Neotropical region.

FIGURE 16.5.2 Culicidae larvae, habitus, dorsal view: (A) *Aedes*; (B) *Culex*; and (C) *Mansonia*. Bars = 1 mm.

TERMINOLOGY AND MORPHOLOGY

The terminology for anatomical structures and chaetotaxy used in this chapter follows Harbach & Knight (1980).

Larvae (Figs. 16.5.1–3), commonly called "wigglers" or "wrigglers," have a well-differentiated head, thorax, and abdomen (Fig. 16.5.6 A). Each of these regions has structures that are useful in identification, taxonomic study being done most reliably with mature (fourth-instar) larvae. The presence of well-developed mouth brushes, fusion, and enlargement of the thoracic segments, which are wider than the head and abdomen, and a tubular siphon on abdominal segment VIII (except in Anophelinae) can distinguish mosquito larvae from all other dipterans.

The larva is covered by soft, membranous tissue, but some parts consist of hardened sclerotized plates. The head and siphon are completely sclerotized, while the thorax and abdomen are mostly membranous. The characters commonly used for larval identification are found mainly on the head, terminal abdominal segments (segments VIII, X), and siphon; however, some thoracic and abdominal setae are also needed for identification.

FIGURE 16.5.3 Culicidae larvae, habitus, dorsal view: (A) *Limatus*; (B) *Wyeomya*; and (C) *Toxorhynchites*. Bars = 1 mm (A, B); 3 mm (C).

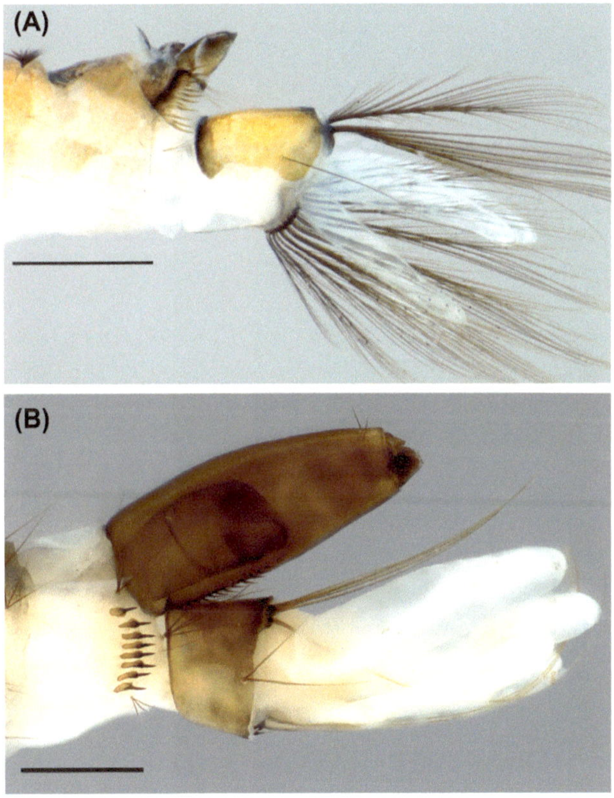

FIGURE 16.5.4 Culicidae larvae, terminal region of abdomen, lateral view: (A) *Anopheles*; and (B) *Aedes*. Bars = 0.3 mm.

The head is dorsoventrally flattened, prognathous, and consists of three large sclerites (Figs. 16.5.6 B, C). Posteriorly, there is an opening (the occipital foramen) to which the cervix is attached. This opening is usually surrounded (partially or completely) by a thickened rim-like collar. The mouthparts are ventral, but some portions extend anteriorly and are visible in dorsal view. Anterolaterally to the mouthparts are the lateral palatal brushes, or mouth brushes. These consist of a group of long, fine setae, but in certain predatory species, they are composed of a few, thick, curved filaments. The tubular antennae are located anterolateral to the mouth brushes and may vary in size, shape, and in the location of seta 1-A.

The thorax appears as a single undivided region of the body; however, the regions referring to the pro-, meso-, and metathorax can be distinguished by the setal group (three distinct setae) present on each segment (Fig. 16.5.7 A).

The abdomen contains 10 segments (Fig. 16.5.7 A). The first seven of the nine readily visible abdominal segments are somewhat similar and unmodified. The three terminal segments are modified for respiration and swimming (Figs. 16.5.4, 5). In most anophelines, at least some of abdominal segments I—VII have seta 1 palmate. Structurally, the eighth segment comprises embryonic abdominal segments VIII and IX. This segment bears the external respiratory organs, which provide useful taxonomic characters (Figs. 16.5.7 B, C). In anopheline larvae (Fig. 16.5.7 B), they consist of a sessile spiracular apparatus (SA) that has five lobes: anterior lobe, two anterolateral lobes, and two posterolateral lobes. Laterally, the SA is supported by pecten (Pe), which are heavily sclerotized. In culicines (Fig. 16.5.7 C), the SA, also five lobed, is borne on the end of a tube—the siphon (S). This structure varies greatly in length and form among the different groups in this family; and in most culicines, it has a row of specialized spines (the pecten). The pecten consists of a varying number of pecten spines that extend a variable distance from near the base of the siphon, although it is usually confined to the basal half. Laterally on segment VIII a patch of specialized spicules (the comb) is found; individually these spicules are called "comb scales." Segment X, or the anal segment, is generally smaller than the preceding segments. It bears a prominent sclerite—the saddle (Sa), which may completely encircle the segment; it also bears anal papillae (APP) and usually a row of setae called "ventral brush" (or seta 4-X).

Larval Chaetotaxy

In general, the larval body bears 223 pairs of symmetrically arranged setae and knowing their nomenclature, arrangement, numbers of branches, and location is essential for the taxonomic identification of this group (for further information on the larval chaetotaxy, see Harbach & Knight, 1980).

For each individual seta, a number is given followed by a hyphen and a capital letter or a Roman numeral (to indicate the body area involved). The abbreviations for

FIGURE 16.5.5 Culicidae larvae, terminal region terminal region of abdomen, lateral view: (A) *Aedeomyia*; (B) *Culex*; (C) *Mansonia*; (D) *Wyeomya*; (E) *Limatus*; and (F) *Toxorhynchites*. Bars = 0.5 mm (A, B, C, D); 0.3 mm (E); 1 mm (F).

the principal parts of the body are as follows: antenna (A); head (C= cranium); prothorax (P); mesothorax (M); metathorax (T); abdominal segments (I to X); siphon and spiracular apparatus (S). For example, "setae 2, 3-A" means seta 2 and seta 3 on the antenna; "seta 4-X" means seta 4 of segment X; "seta 6-IV-V" means seta 6 on abdominal segments IV and V.

MATERIAL PREPARATION AND PRESERVATION

The best way to preserve whole larvae is fixing them in 80% ethanol, which should be replaced at least twice to eliminate excess water. Specialists suggest first killing the larvae with very hot water (not boiling), because hot water

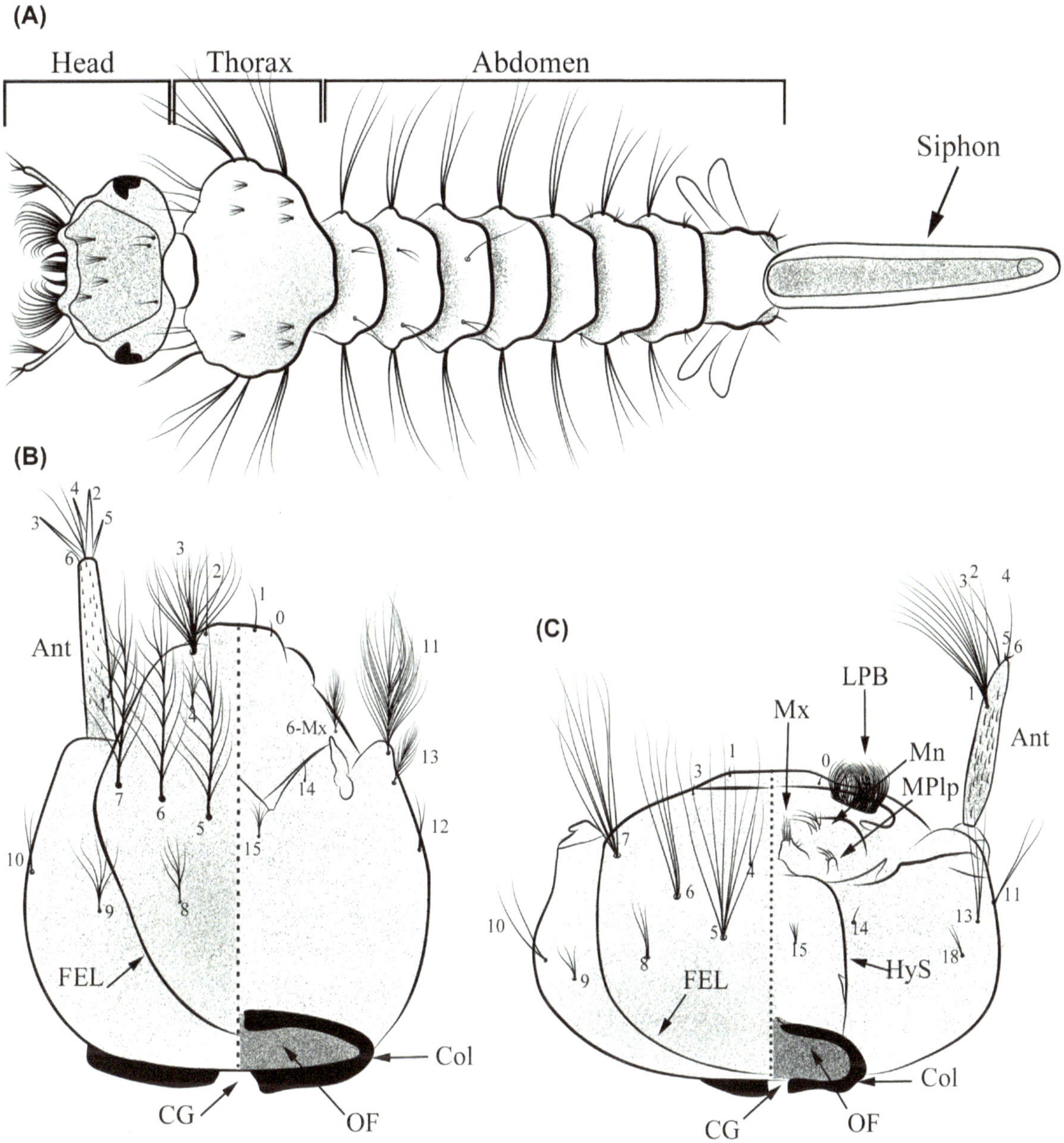

FIGURE 16.5.6 Schematic representation of a Culicidae larvae: (A) Body regions, dorsal view. (B) Anophelinae head (left: dorsal view; right: ventral view); and (C) Culicinae head (left: dorsal view; right: ventral view). Abbreviations as follows: *Ant*, antenna; *CG*, coronal gap; *Col*, collar; *FEL*, frontal ecdysial line; *HyS*, hypostomal suture; *LPB*, lateral palatal brush; *Mn*, mandible; *Mx*, maxilla; *MPlp*, maxillary palpus; *OF*, occipital foramen. Arabic numbers represent setae name.

fixes proteins, which prevents later darkening of the specimens. Molecular studies often require specimens preserved in 95% alcohol, followed by one or two changes of fresh alcohol, and storage in a refrigerator. Exuviae are preserved in 80% ethanol. For slide mounting, the following steps are recommended:

- Transfer specimens from alcohol to Cellosolve for 15 minutes or more (dark specimens can be stored in Cellosolve for 8 hours or overnight).
- Lift the specimen from the Cellosolve and place it in the center of glass microscope slide with the dorsal side up.
- Drop a small amount of Euparal on the specimen. Mount the specimen with the dorsal side up and with the head pointing down. Arrange the head, thorax, and abdomen in a natural position and then cut the abdomen between segments VI and VII (using a surgical blade). Place the terminal segments with the siphon to the left in culicine larvae or segment X to the right in anopheline larvae.
- Then, drop more Euparal on the specimen and check the arrangement of the setae and the position of the larva. Carefully cover the specimen with a glass coverslip.
- Dry in a drying oven at 45–55°C for two weeks or more.

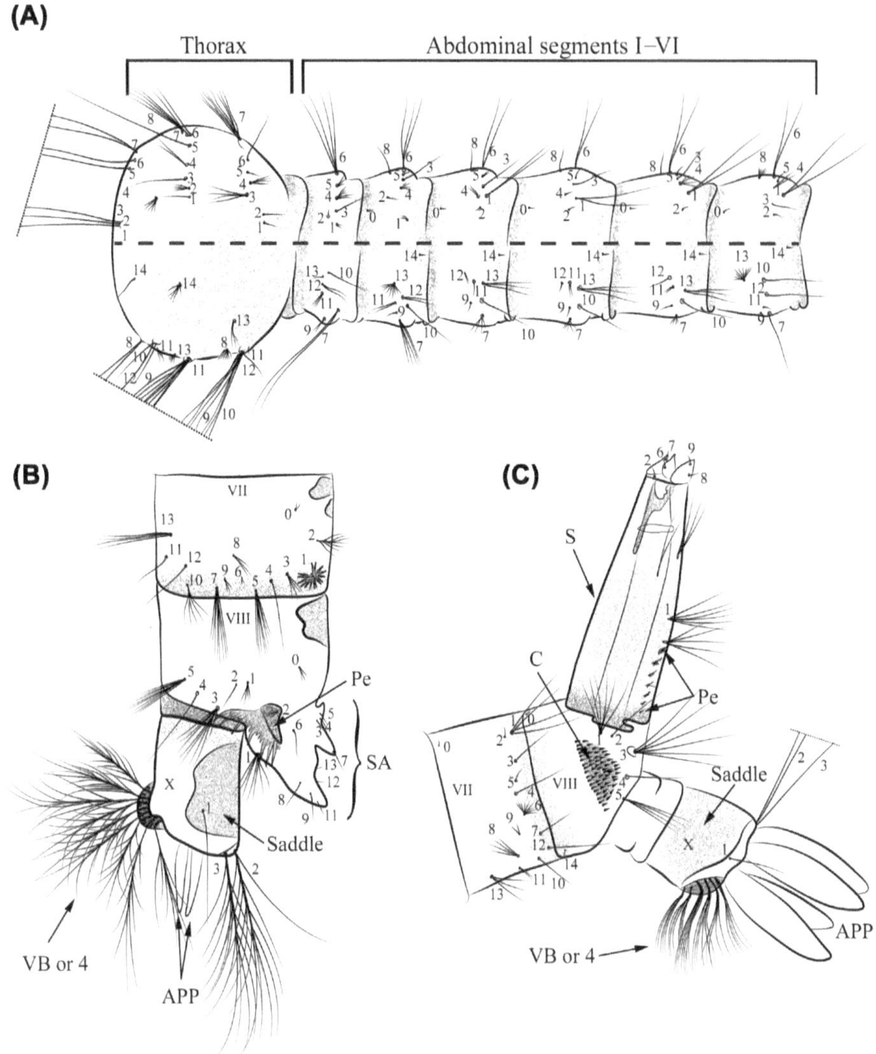

FIGURE 16.5.7 Schematic representation of a Culicidae larvae: (A) thorax and abdominal segments I–VI (portion above dashed line: dorsal view; portion below dashed line: ventral view); and (B–C): lateral aspects of abdominal segments VII–X (Anophelinae larvae and Culicinae larvae, respectively). Abbreviations as follows: *APP*, anal papillae; *C*, comb; *Pe*, pecten; *S*, siphon; *SA*, spiracular apparatus; *Sa*, saddle; *VB*, ventral brush. Arabic numbers represent setae name; roman numbers represent abdominal segments.

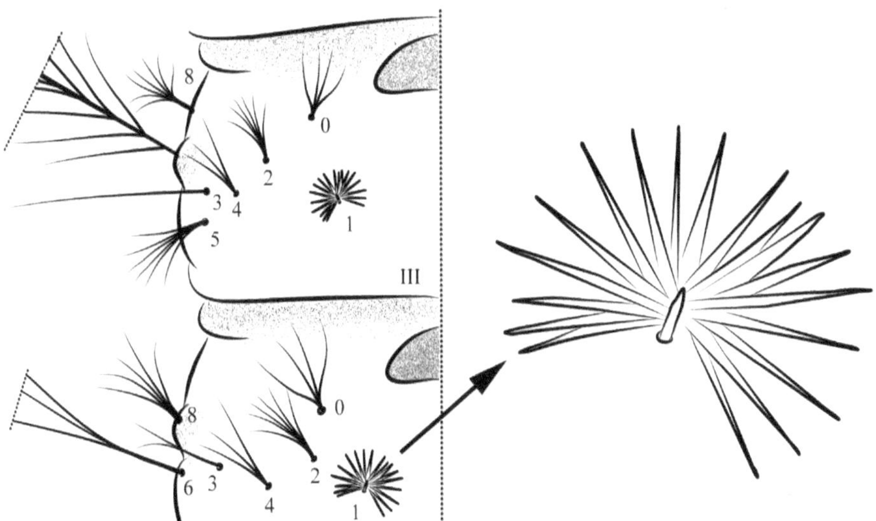

FIGURE 16.5.8 Anophelinae, abdominal segments III and IV.

KEYS TO CULICIDAE

This key is based on Clark-Gil & Darsie (1983).

Culicidae: Subfamilies

1 Respiratory siphon absent (Fig. 16.5.7 B); head length > width (Fig. 16.5.6 B); some or all abdominal segments with seta 1 palmate (I–VII) (Fig. 16.5.8)........ **Anophelinae [p. 729]**

1' Siphon present (Fig. 16.5.7 C); head width > length (Fig. 16.5.6 C); abdominal segment seta 1 not palmate (Fig. 16.5.9) .... **Culicinae [p. 729]**

Culicidae: Anophelinae: Genera

Anopheles are collected in a wide variety of natural breeding places, however, they occur predominantly in surface waters (Lane, 1953; Harbach, 2016). *Chagasia* is essentially forest dwelling, with their natural habitat being small shaded streams with sandy bottoms, especially streams that run down mountainsides (Arnet, 1950; Consoli & Oliveira, 1994).

1 Spiracular apparatus anterior lobe bearing a long median process ending in a filamentous seta (Fig. 16.5.10); abdominal segments III–V with palmate seta 1 racquet-shaped (Fig. 16.5.11)........ *Chagasia* [Neotropical]

1' Spiracular apparatus anterior lobe without a long median process (Fig. 16.5.7 B); abdominal segments III–V with palmate seta 1 lanceolate (Fig. 16.5.8) .... *Anopheles* [Cosmopolitan]

Culicidae: Culicinae: Tribes

Five of these tribes are monogeneric. Aedeomyiini (*Aedeomyia*) inhabit lakes and ponds with abundant aquatic vegetation, but they have also been collected from river banks (Lane, 1953). Culisetini (*Culiseta*) occur in temporary pools, streams with minimal flow, streams with polluted water, and some artificial containers (Clark-Gil & Darsie, 1983; Chaverri, 2009). Uranotaeniini (*Uranotaenia*) inhabit a wide variety of habitats, such as large open swamps with abundant growth of different types of aquatic vegetation, marshes, shaded side-pools of cool water along streams, reedy ditches, rocky pools in stream beds, phytotelmata, rock holes, and crab holes (Arnet, 1950; Galindo et al., 1954; Harbach, 2016). Orthopodomyiini (*Orthopodomyia*) are found mostly in large tree holes, internodes of bamboo, leaf axils of bromeliads, and, less frequently, in artificial containers and abandoned septic tanks (Arnet, 1950; Zavortink, 1968; Consoli & Oliveira, 1994). Toxorhynchitini (*Toxorhynchites*) species are mainly forest dwelling; immature stages are found in phytotelmata, rock holes, and artificial containers discarded in or near forested areas (Steffan & Evenhuis, 1981).

1 Siphon cylindrical or fusiform, without subapical strongly sclerotized serrate process (Fig. 16.5.16)........ 2

1' Siphon short and apically pointed, with subapical strongly sclerotized serrate process adapted for piercing and securing air from aquatic plants (Fig. 16.5.12)........ **Mansonini [p. 729]**

2(1) Segment X ventral brush (seta 4-X) with four or more seta pairs (Fig. 16.5.17) 3

2' Segment X, ventral brush (seta 4-X) reduced to one or rarely two seta pairs (Fig. 16.5.16)........ **Sabethini [p. 730]**

3(2) Siphon pecten absent (Fig. 16.5.38)........ 4

3' Siphon pecten present (Fig. 16.5.39) 6

4(3) Lateral palatal brush with at least 40 thin, simple or pectinate filaments (Fig. 16.5.42); abdominal setae arising separately, rarely borne on sclerotized plates (Fig. 16.5.43); segment VIII, comb present (Fig. 16.5.44)........ 5

4' Lateral palatal brush reduced to about 12 or fewer thick, curved filaments (Fig. 16.5.40); abdominal setae occurring in groups of 3–5 on common large sclerotized plates (Fig. 16.5.41); segment VIII, comb absent (Fig. 16.5.38).... Toxorhynchitini, one genus: *Toxorhynchites* [Circumtropical, Nearctic, Palaearctic]

5(4) Antennae greatly tumid, strongly curved, longer than head (Fig. 16.5.45); siphon pubescent, spiracular apparatus bearing a pair of large hamulate setae (seta 9-S) (Fig. 16.5 46) .... Aedeomyiini, one species: *Aedeomyia squamipennis* (Lynch Arribálzaga) [Neotropical]

5' Antennae simple, strait, shorter than head (Fig. 16.5.42); siphon glabrous, except for a single pair of seta 1-S, spiracular apparatus without a pair of large hamulate setae (Fig. 16.5.47)........ Orthopodomyiini, one genus: *Orthopodomyia* [Neotropics. Afrotropical, Nearctic, Oriental]

6(3) Head anterior margin seta 1-C not arising from a distinct prominence (Fig. 16.5.49); abdominal segment VIII without a sclerite (Fig. 16.5.50) 7

6' Head anterior margin seta 1-C arising from a distinct prominence (Fig. 16.5.48); abdominal segment VIII with a sclerite (comb plate) bearing a single row of comb scales on its posterior margin (Fig. 16.5.39) .... Uranotaeniini, one genus: *Uranotaenia* [Neotropical, Afrotropical, Oriental]

7(6) Siphon with three or more pairs of setae 1-S or with row of long filamentous spicules (Fig. 16.5.52)........ 8

7' Siphon with a single pair of seta 1-S inserted near or beyond the middle of the siphon; without a row of long filamentous spicules (Fig. 16.5.51) .... **Aedini [p. 730]**

8(7) Pecten spines followed by row of long filamentous spicules closely spaced (Fig. 16.5.52)........ Culisetini, one genus: *Culiseta* [Cosmopolitan]

8' Pecten spines not followed by row of long filamentous spicules closely spaced (Fig. 16.5.58) .... **Culicini [p. 743]**

Culicidae: Culicinae: Mansonini: Genera

Larvae and pupae inhabit temporary or permanent ponds with rooted or floating aquatic vegetation that is used as a source of oxygen. *Mansonia* larvae and pupae are normally found attached to floating aquatic plants, such as water lettuce and water hyacinth while *Coquillettidia* use plants attached to the substrate (Harbach, 2016).

1 Antenna distal part elongate (>basal part), setae 2, 3-A shorter than basal part (Fig. 16.5.13); segment X saddle without long, ventral setae (Fig. 16.5.14) .... *Coquillettidia*
[Cosmopolitan]

1' Antenna distal part short (<basal part), setae 2, 3-A as long as the basal part (Fig. 16.5.15); segment X saddle with 3–4 long, ventral setae (Fig. 16.5.12) .... *Mansonia*
[Cosmopolitan]

Culicidae: Culicinae: Sabethini: Genera

Immatures preferentially inhabit water accumulated in dead and living plant including those rich in debris and decomposing organic matter, such as leaf axils, bromeliads, bamboo, tree holes, flower bracts and spathes, and pitcher plants (Lane, 1953; Zavortink, 1979; Consoli & Oliveira, 1994; Harbach, 2016; Campos & Zavortink, 2010). Occasionally, some species of *Limatus*, *Trichoprosopon*, and *Wyeomyia* can inhabit artificial containers, snail shells and rock holes (Lane, 1953; Heinemann & Belkin, 1977; Chaverri, 2009).

1 Maxillary body with long dentiform apical process (Fig. 16.5.18) 2

1' Maxillary body without long dentiform apical process (Fig. 16.5.19) 8

2(1) Head capsule with a transverse slit-like occipital foramen (Fig. 16.5.20) 3

2' Head capsule with a normal, circular occipital foramen (Fig. 16.5.21) 6

3(2) Maxillary palpus not fused with maxillary body (Fig. 16.5.22) 4

3' Maxillary palpus fused with maxillary body (Fig. 16.5.18) .... *Sabethes*
[Neotropical]

4(3') Maxillary brush absent (Fig. 16.5.23) 5

4' Maxillary brush present (Fig. 16.5.22) .... *Wyeomyia* (in part)
[Neotropical, Nearctic]

5(4) Maxillary bundle absent, maxillary claw present (Fig. 16.5.23); segment X seta 3-X branched (with 2–8 branches) (Fig. 16.5.24); spiracular apparatus, anterolateral spiracular lobe, seta 6-S strongly developed, rigid and hooked at apex (Fig. 16.5.24) .... *Isostomyia*
[Neotropical]

5' Maxillary bundle present, maxillary claw absent (Fig. 16.5.25); segment X seta 3-X simple, unbranched (Fig. 16.5.26); spiracular apparatus, anterolateral spiracular lobe, seta 6-S weakly developed, flexible and not hooked at apex (Fig. 16.5.26) .... *Runchomyia*
[Neotropical]

6(2) Head hypostomal suture incomplete (Fig. 16.5.27) 7

6' Head hypostomal suture complete (Fig. 16.5.21) .... *Onirion*
[Neotropical]

7(6) Siphon midline region with a long row of branched setae (Fig. 16.5.28); maxillary bundle present, maxillary claw absent (Fig. 16.5.29); thorax seta 8-M well developed (Fig. 16.5.30) .... *Johnbelkinia*
[Neotropical]

7' Siphon midline region without a long row of branched setae (Fig. 16.5.31); maxillary bundle absent, maxillary claw present (Fig. 16.5.32); thorax seta 8-M weakly developed (Fig. 16.5.33) .... *Shannoniana*
[Neotropical]

8(1) Mandible not elongate and arising above level of base of antenna (Fig. 16.5.20); mesothorax seta 8-M present (Fig. 16.5.35) 9

8' Mandible elongate and conspicuous, arising approximately at level of base of antenna (Fig. 16.5.34); mesothorax seta 8-M absent .... *Trichoprosopon*
[Neotropical]

9(8) Segment X seta 4-X as long as seta 3X (Fig. 16.5.36); siphon setae branched (mostly with three or more branches) (Fig. 16.5.36); segment VIII, comb with 4–8 comb scales in row (Fig. 16.5.36) .... *Limatus*
[Neotropical]

9' Segment X seta 4-X shorter than seta 3X (Fig. 16.5.37); siphon setae mostly simple (sometimes with two or three branches) (Fig. 16.5.37); segment VIII, comb with more than eight comb scales in patch or united on plate (Fig. 16.5.37) .... *Wyeomyia* (in part)

Culicidae: Culicinae: Aedini: Genera

Members live in primary and secondary tropical rain forests, open deciduous forests, and mangroves. Immatures can be found in temporary freshwater ground pools, tree hollows or tree holes, cut or broken bamboo internodes, puddles, rock pools, phytotelmata, crab holes, animal footprints, holes in volcanic rocks, artificial containers, and brackish-water pools; some species can be found in disturbed habitats (Lane, 1953; Arnell, 1973; Consoli & Oliveira, 1994; Chaverri, 2009; Harbach, 2016).

1 Segment X completely encircled by saddle (Fig. 16.5.51) 2

1' Segment X not completely encircled by saddle (Fig. 16.5.53) 3

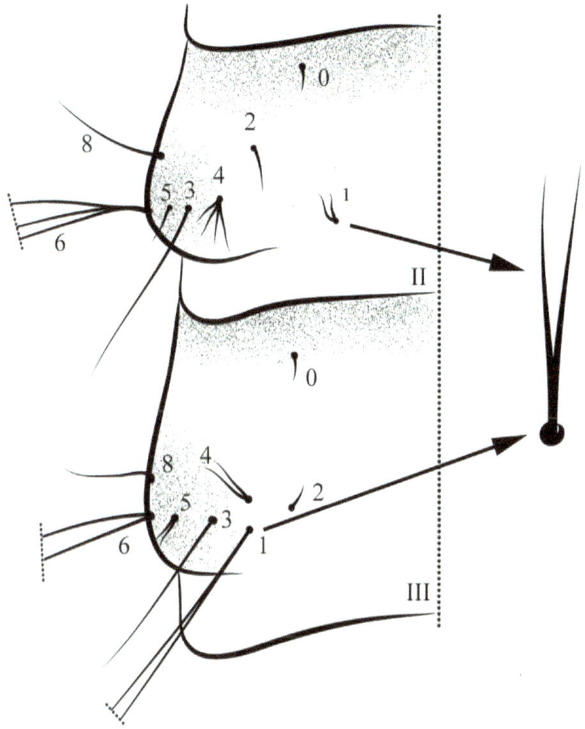

FIGURE 16.5.9 Culicinae, abdominal segments II and III.

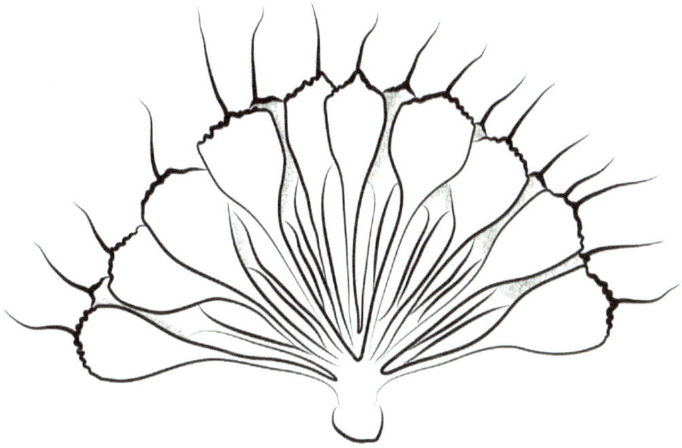

FIGURE 16.5.11 *Chagasia*, abdominal palmate setae 1.

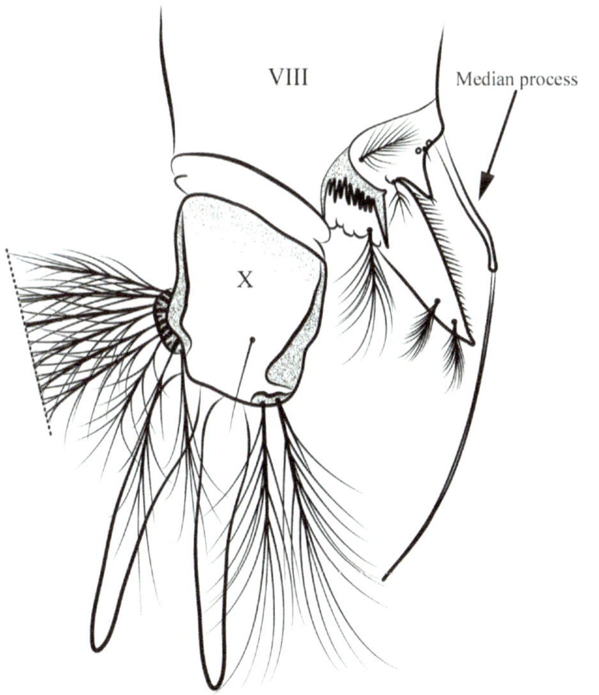

FIGURE 16.5.10 *Chagasia*, lateral aspects of abdominal segments VIII–X.

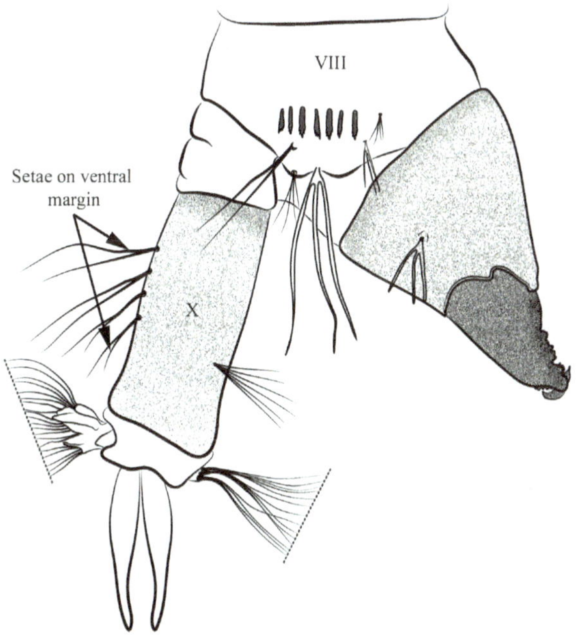

FIGURE 16.5.12 *Mansonia*, lateral aspect of abdominal segments VIII–X.

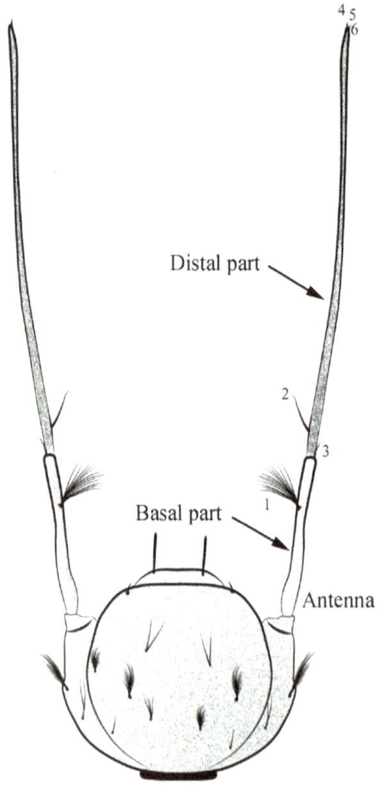

FIGURE 16.5.13 *Coquillettidia*, head, dorsal view.

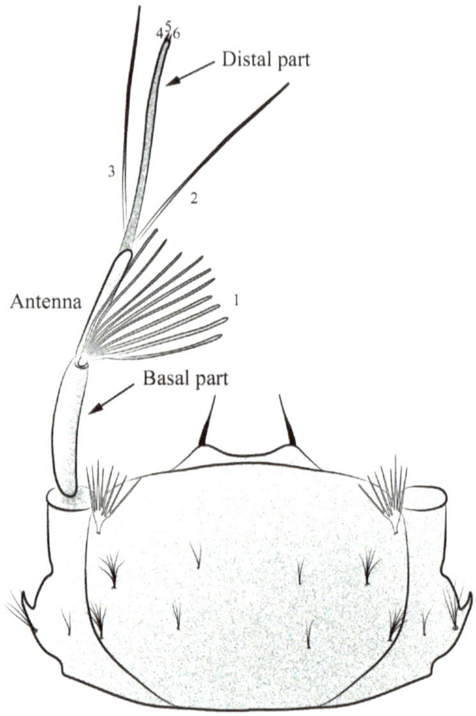

FIGURE 16.5.15 *Mansonia*, head, dorsal view.

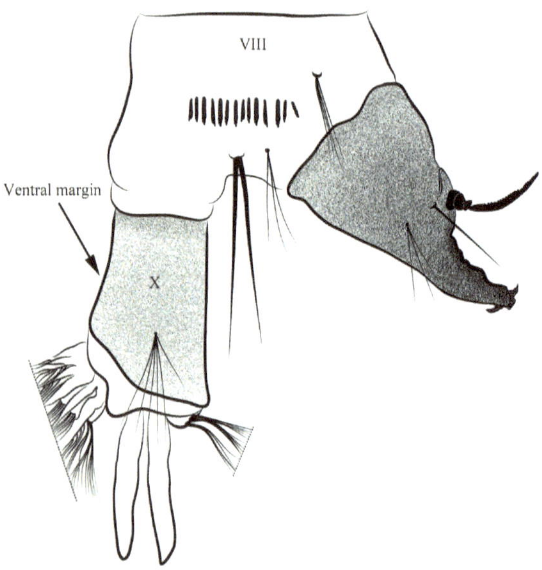

FIGURE 16.5.14 *Coquillettidia*, lateral aspect of abdominal segments VIII–X.

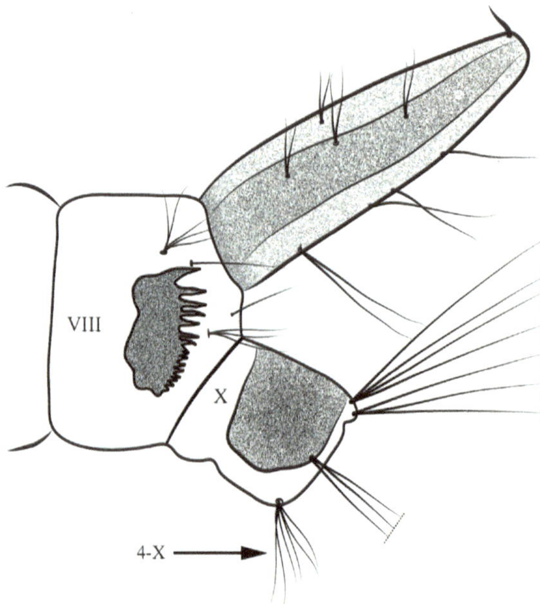

FIGURE 16.5.16 *Wyeomyia*, lateral aspect of abdominal segments VIII–X.

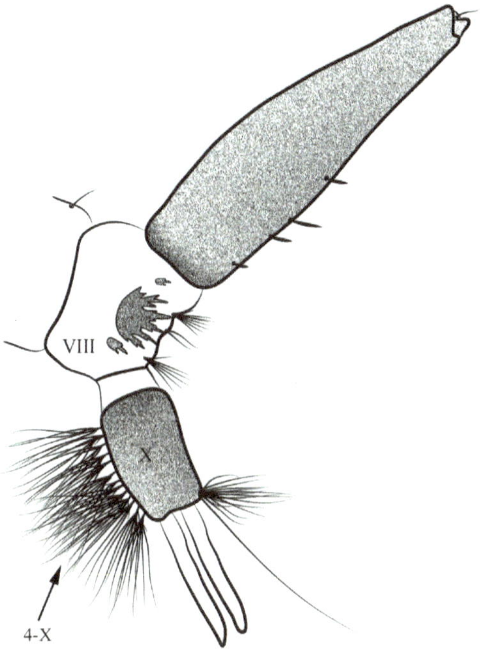

FIGURE 16.5.17 *Psorophora*, lateral aspect of abdominal segments VIII–X.

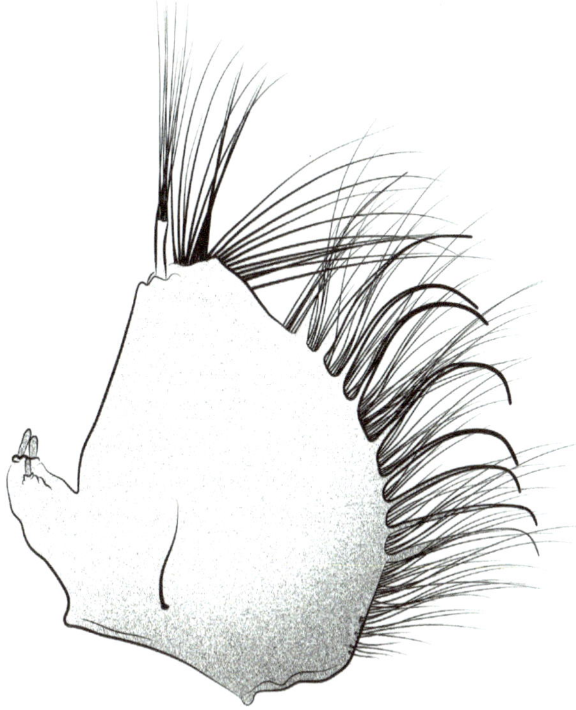

FIGURE 16.5.19 Sabethini, maxilla.

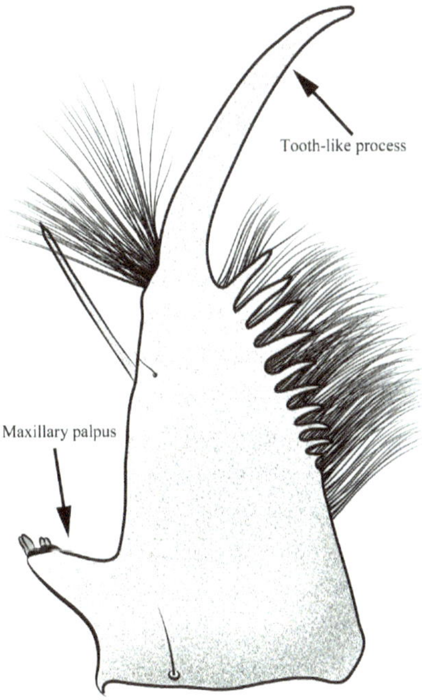

FIGURE 16.5.18 *Sabethes*, maxilla.

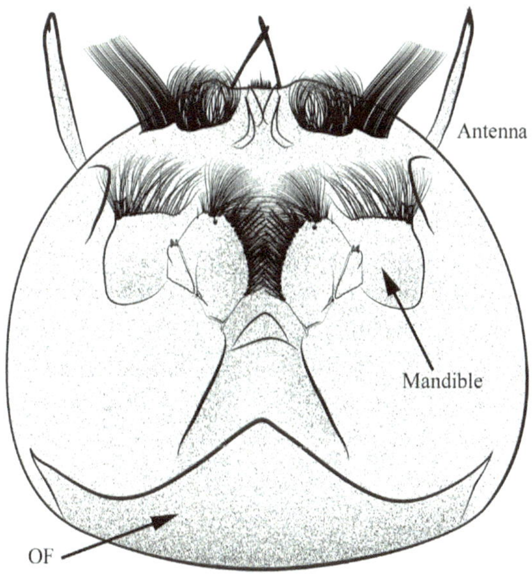

FIGURE 16.5.20 *Wyeomyia*, head, ventral view.

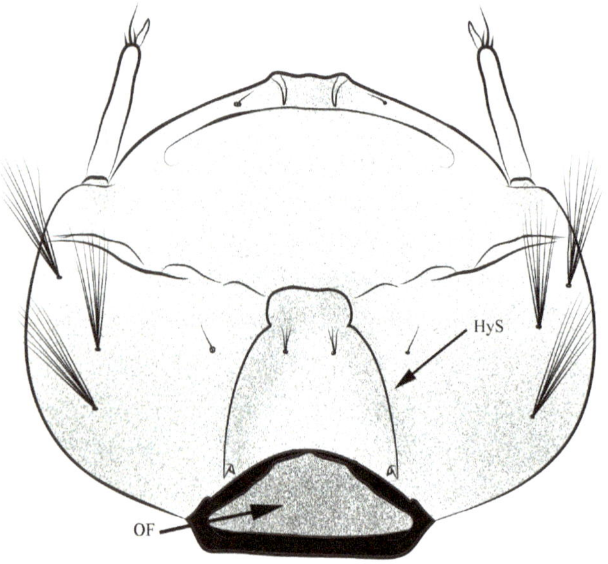

FIGURE 16.5.21 *Onirion*, head, ventral view.

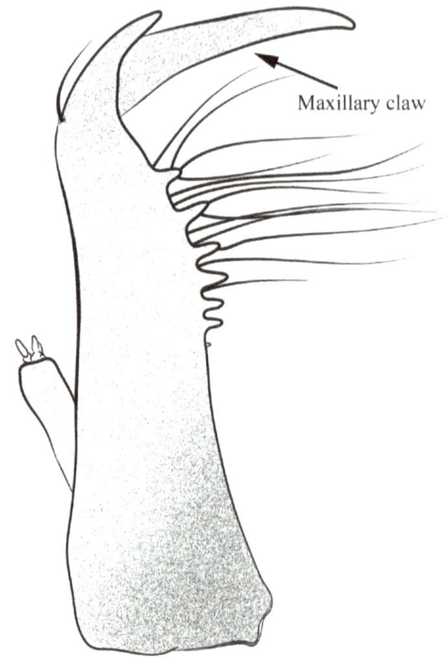

FIGURE 16.5.23 *Isostomyia*, maxilla.

FIGURE 16.5.22 *Wyeomyia*, maxilla.

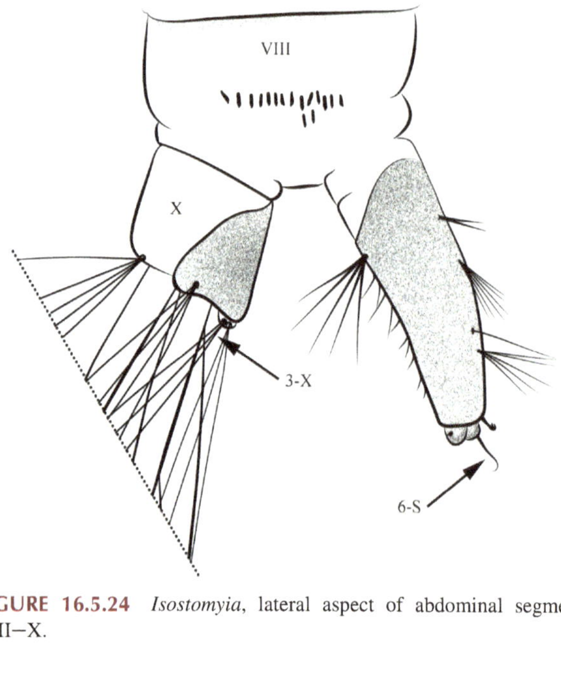

FIGURE 16.5.24 *Isostomyia*, lateral aspect of abdominal segments VIII–X.

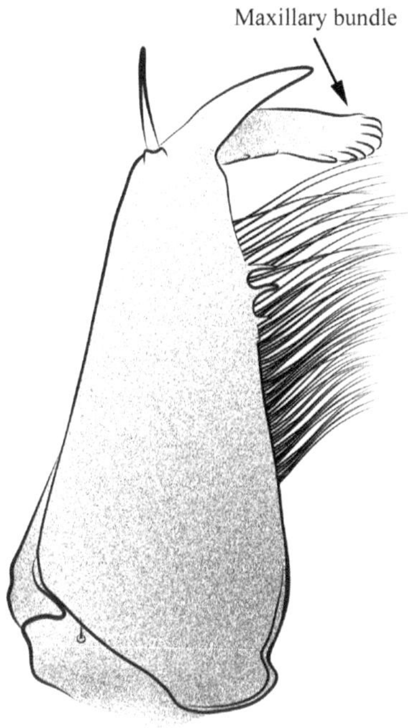

FIGURE 16.5.25 *Runchomyia*, maxilla.

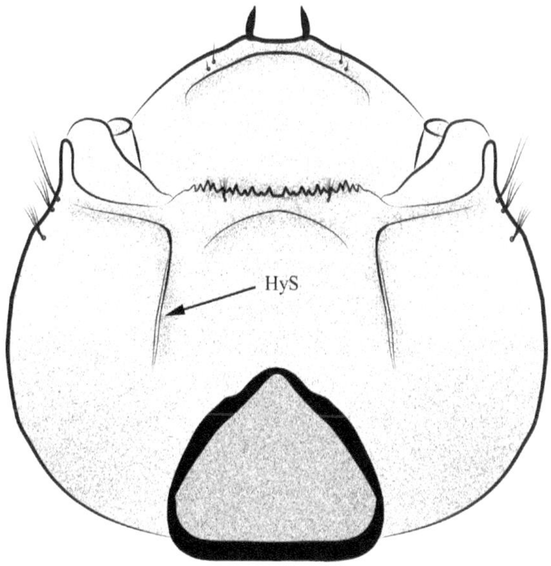

FIGURE 16.5.27 *Johnbelkinia*, head, ventral view.

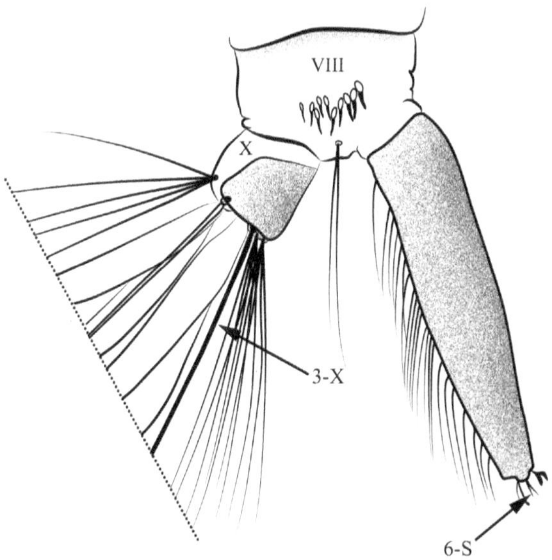

FIGURE 16.5.26 *Runchomyia*, lateral aspect of abdominal segments VIII–X.

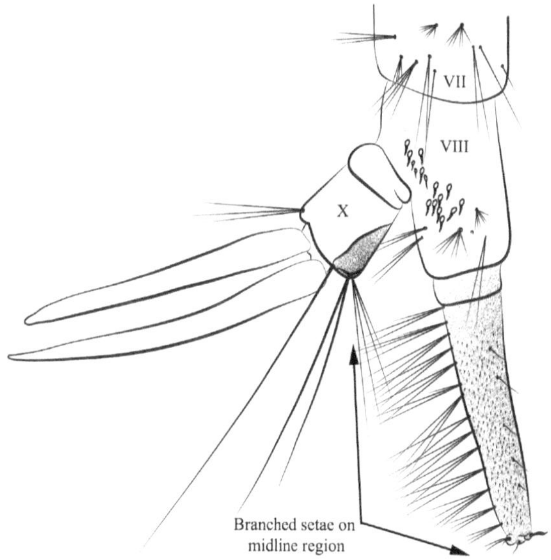

FIGURE 16.5.28 *Johnbelkinia*, lateral aspect of abdominal segments VII–X.

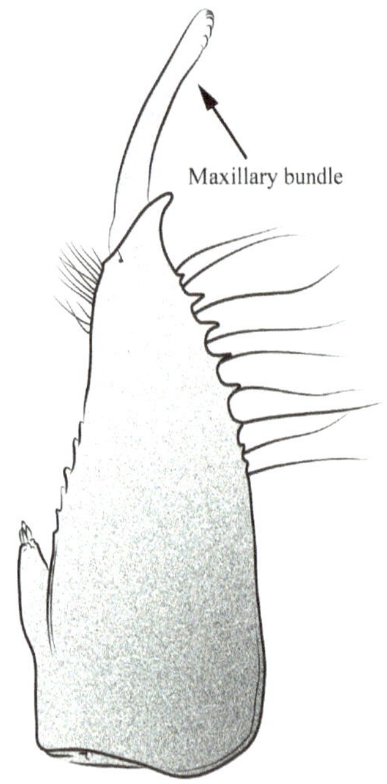

FIGURE 16.5.29 *Johnbelkinia*, maxilla.

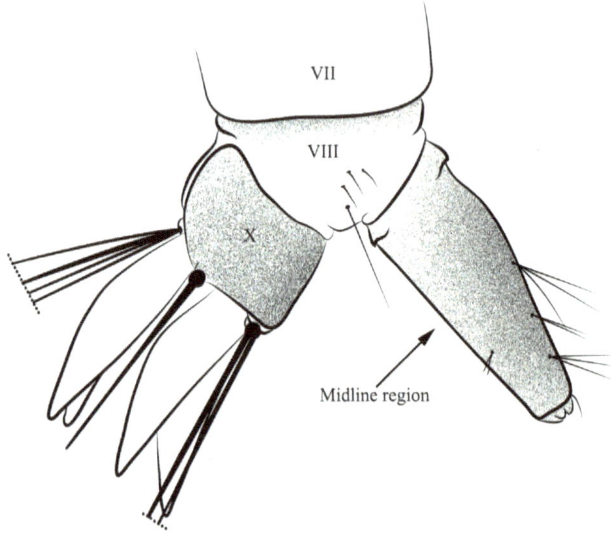

FIGURE 16.5.31 *Shannoniana*, lateral aspect of abdominal segments VII–X.

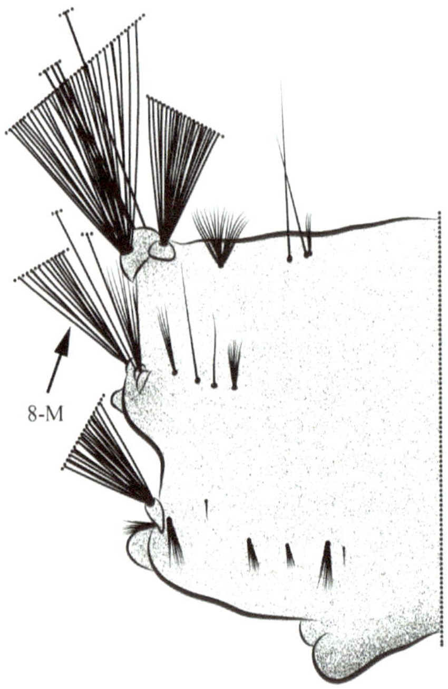

FIGURE 16.5.30 *Johnbelkinia*, left side of thorax, dorsal view.

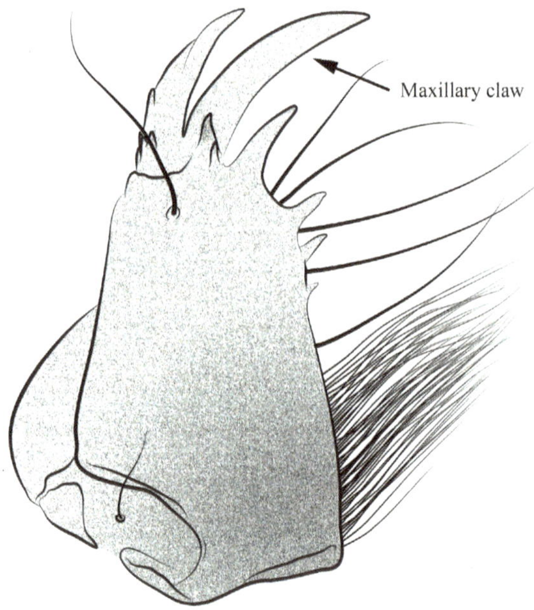

FIGURE 16.5.32 *Shannoniana*, maxilla.

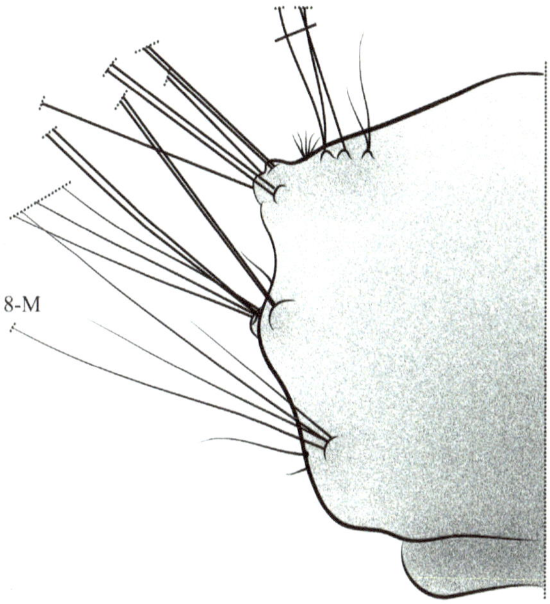

FIGURE 16.5.33 *Shannoniana*, left side of thorax, dorsal view.

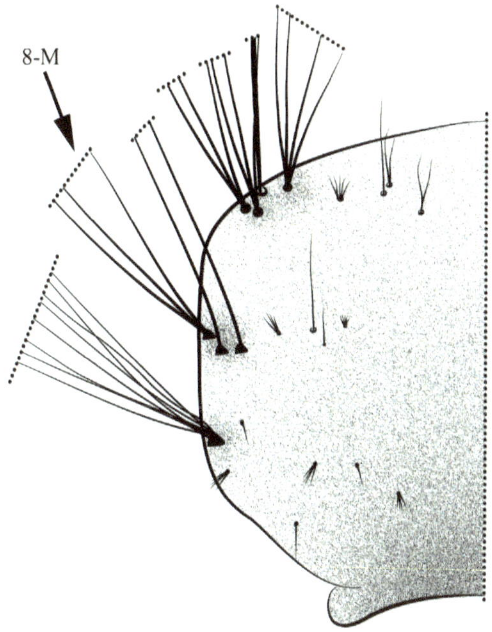

FIGURE 16.5.35 *Wyeomyia*, left side of thorax, dorsal view.

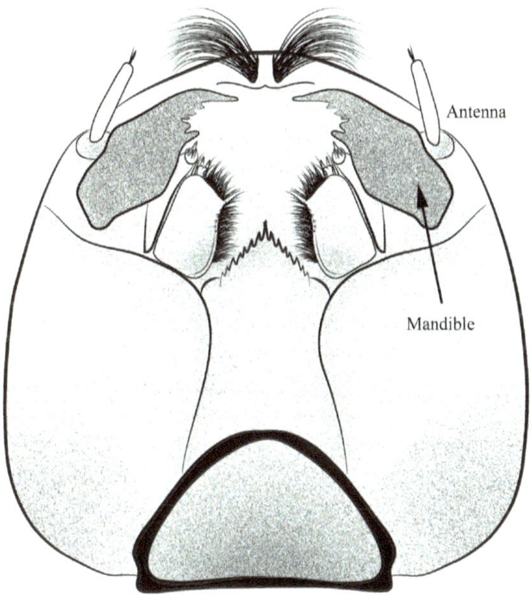

FIGURE 16.5.34 *Trichoprosopon*, head, ventral view.

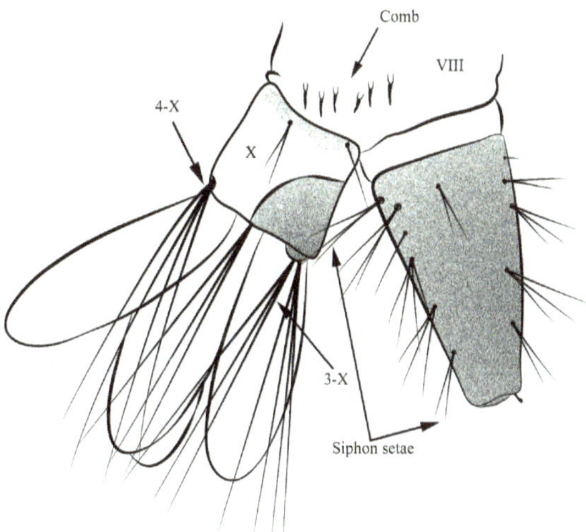

FIGURE 16.5.36 *Limatus*, lateral aspect of abdominal segments VIII–X.

2(1) Segment X saddle enclosing several setae within ventral margin (Fig. 16.5.54); siphon swollen in the middle (Fig. 16.5.54)...... *Psorophora* [Neotropical. Nearctic]

2' Segment X saddle lacking setae arising from within ventral margin (Fig. 16.5.51); siphon tapering from base to apex or cylindrical (Fig. 16.5.51) .... *Aedes* (in part) [Cosmopolitan]

3(1) Abdominal seta 3-VII at least as long as segment VII (Fig. 16.5.53) 4

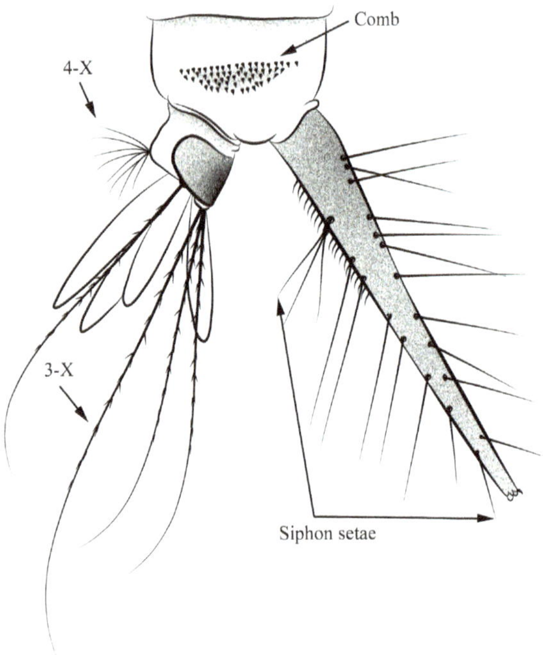

FIGURE 16.5.37 *Wyeomyia*, lateral aspect of abdominal segments VIII–X.

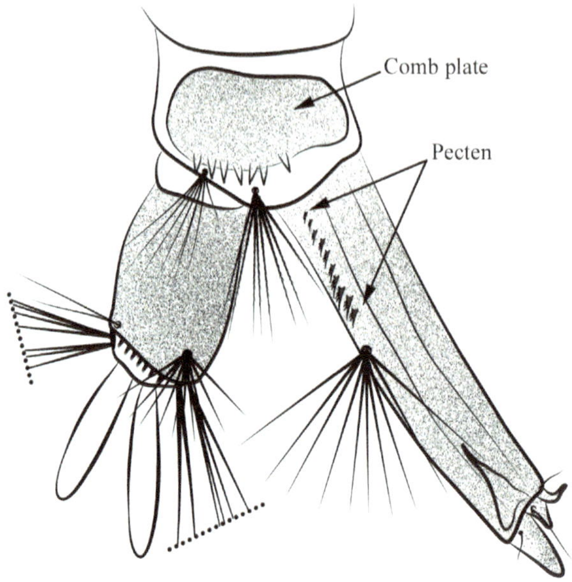

FIGURE 16.5.39 *Uranotaenia*, lateral aspect of abdominal segments VIII–X.

FIGURE 16.5.38 *Toxorhynchites*, lateral aspect of abdominal segments VIII–X.

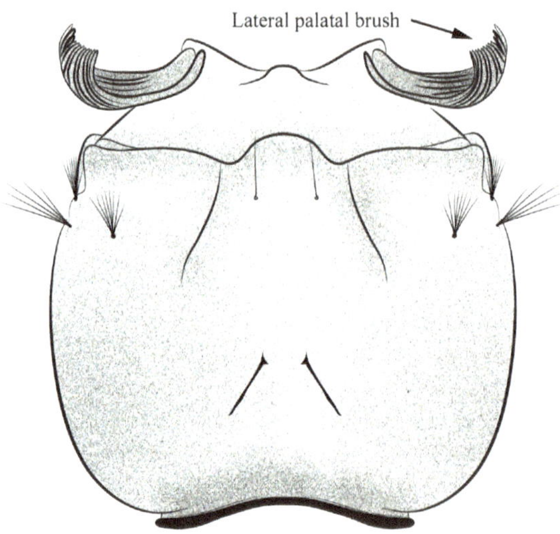

FIGURE 16.5.40 *Toxorhynchites*, head, ventral view.

3'	Abdominal seta 3-VII shorter than half length of the segment VII (Fig. 16.5.55) .... *Aedes* (in part)
4(3)	Abdominal segment I seta 12 usually present, abdominal segments III–V seta 9 shorter than seta 7 (Fig. 16.5.56)............ *Haemagogus* [Neotropical. Nearctic]
4'	Abdominal segment I seta 12 absent, abdominal segments III–V seta 9 equal to or longer than seta 7 (Fig. 16.5.57)...... *Aedes* (in part)

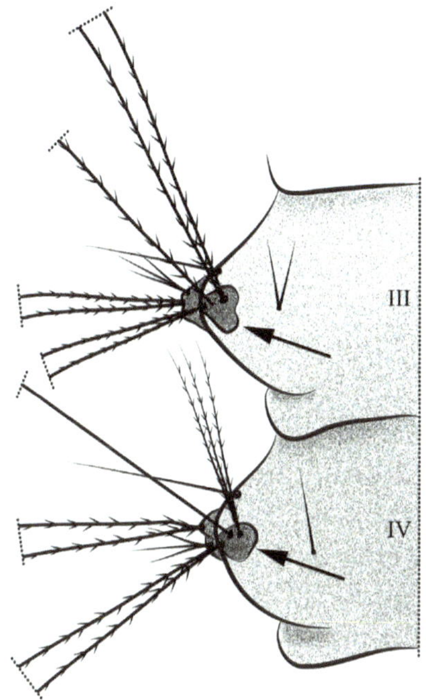

FIGURE 16.5.41 *Toxorhynchites*, abdominal segments III and IV, dorsal view.

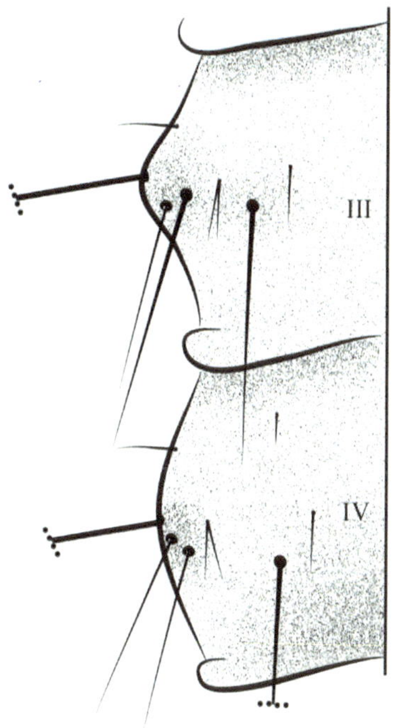

FIGURE 16.5.43 *Orthopodomyia*, abdominal segments III and IV.

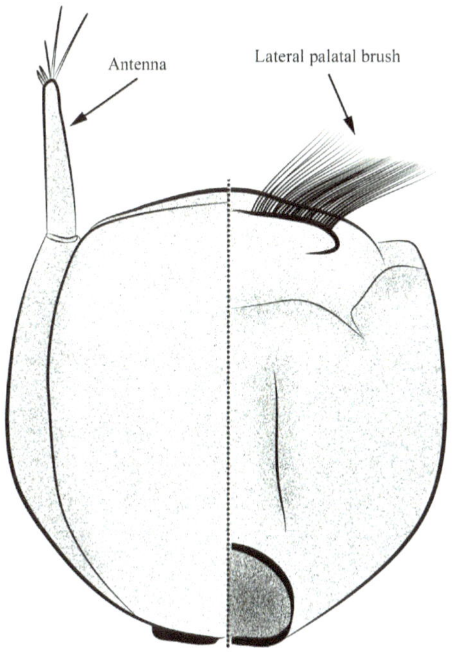

FIGURE 16.5.42 *Orthopodomyia*, head: left side = dorsal view, right side = ventral view.

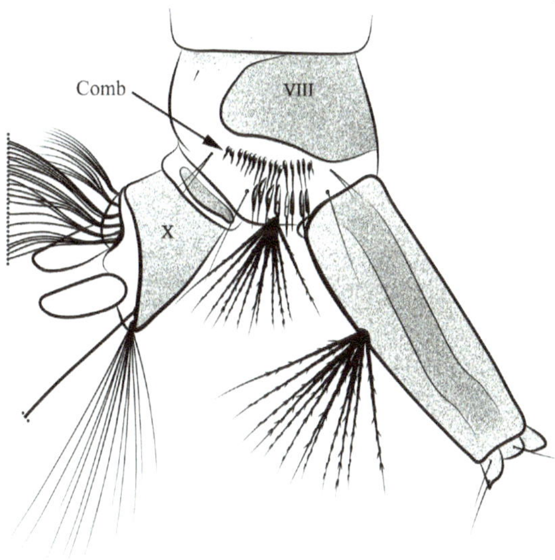

FIGURE 16.5.44 *Orthopodomyia*, lateral aspect of abdominal segments VIII–X.

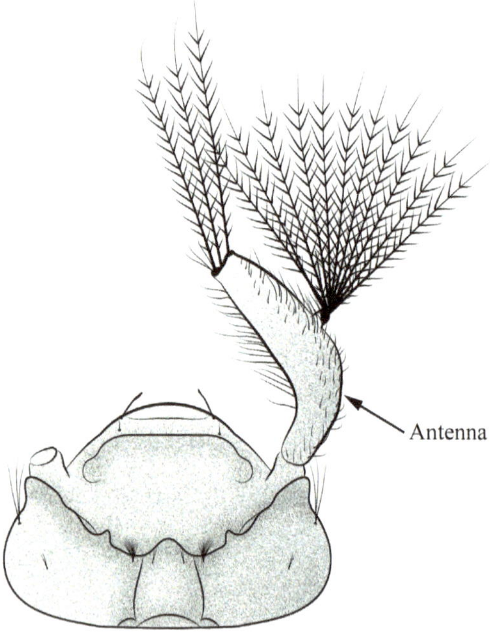

FIGURE 16.5.45 *Aedeomyia*, head, ventral view.

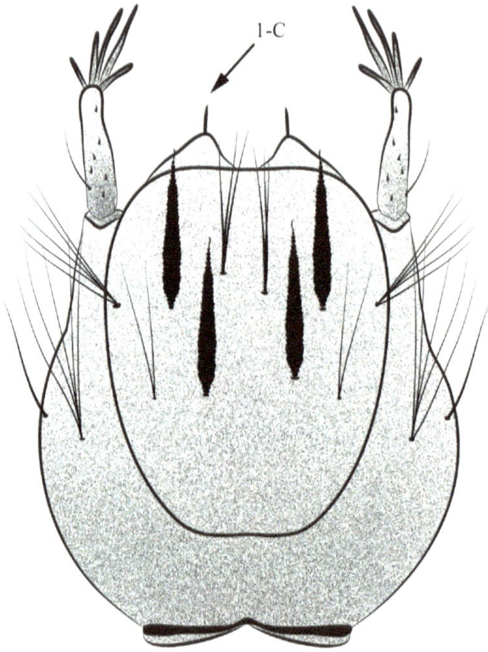

FIGURE 16.5.48 *Uranotaenia*, head, dorsal view.

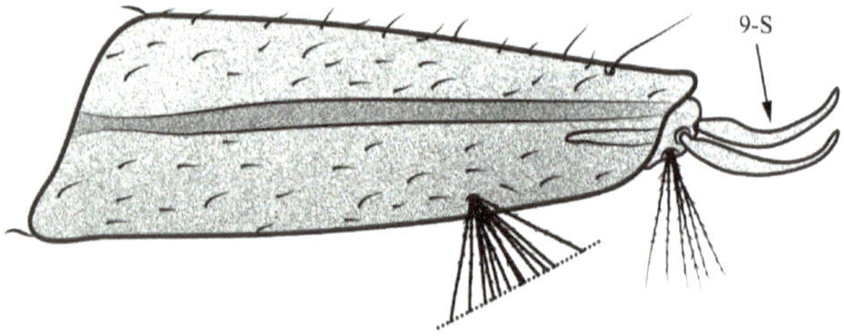

FIGURE 16.5.46 *Aedeomyia*, siphon, lateral view.

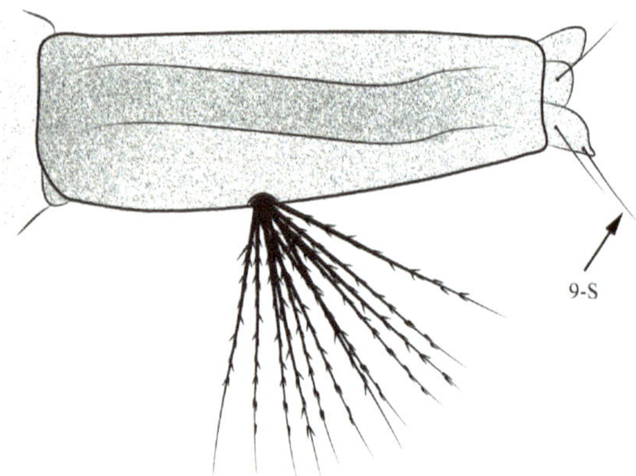

FIGURE 16.5.47 *Orthopodomyia*, siphon, lateral view.

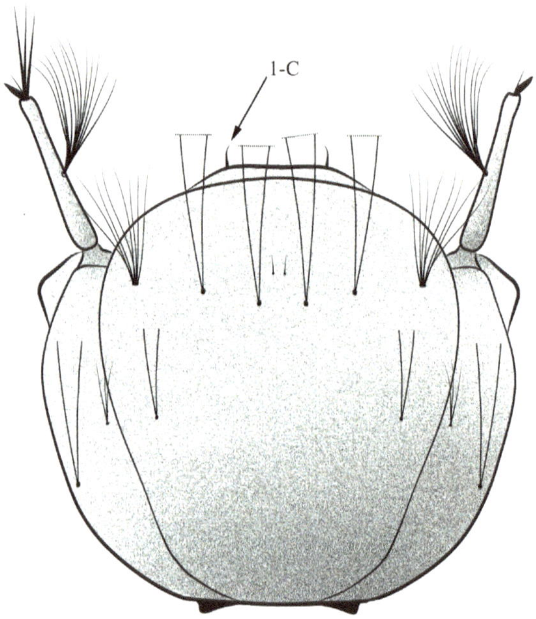

FIGURE 16.5.49 *Galindomyia*, head, dorsal view.

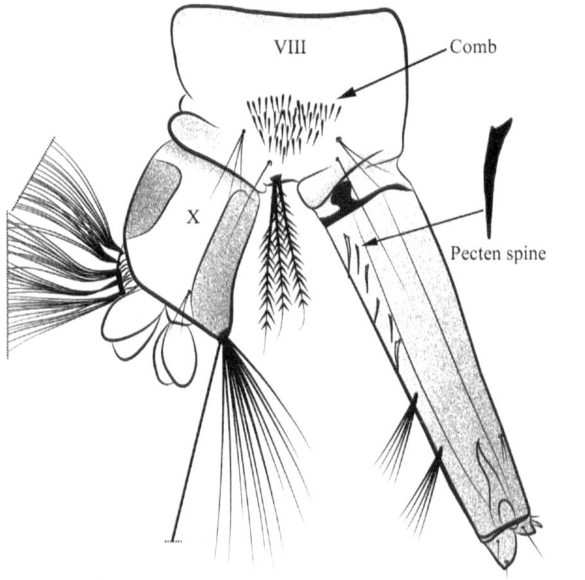

FIGURE 16.5.50 *Galindomyia*, lateral aspect of abdominal segments VIII–X.

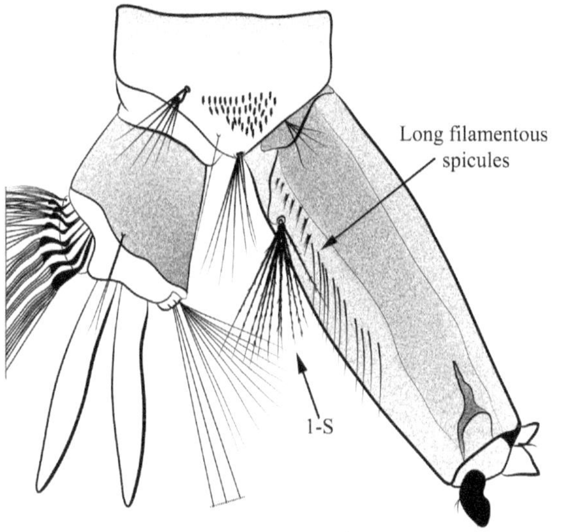

FIGURE 16.5.52 *Culiseta*, lateral aspect of abdominal segments VIII–X.

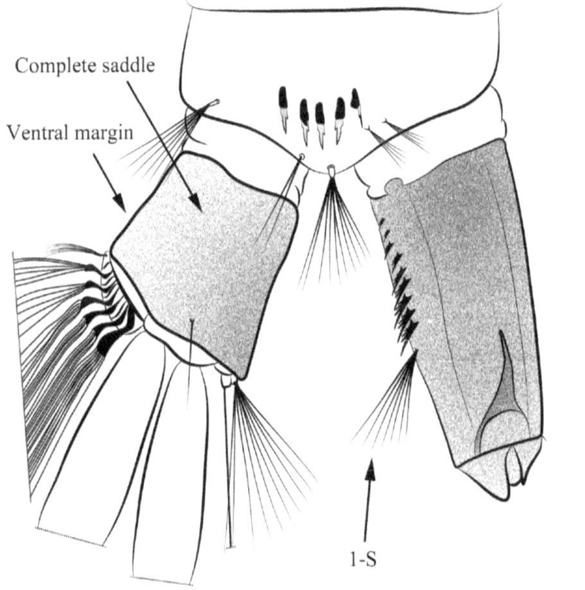

FIGURE 16.5.51 *Aedes*, lateral aspect of abdominal segments VIII–X.

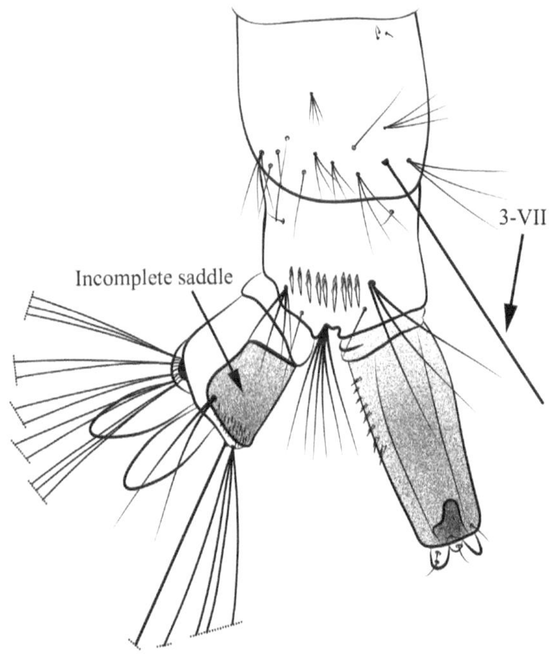

FIGURE 16.5.53 *Haemagogus*, lateral aspect of abdominal segments VII–X.

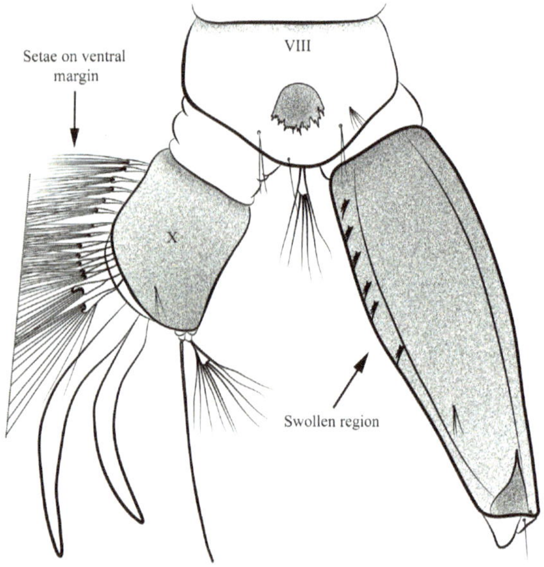

FIGURE 16.5.54 *Psorophora*, lateral aspect of abdominal segments VIII–X.

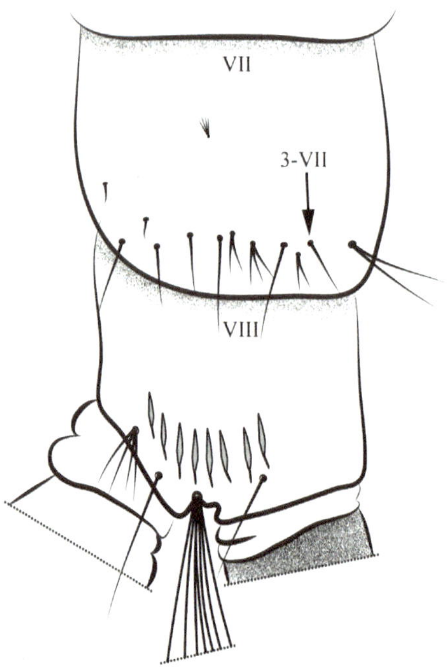

FIGURE 16.5.55 *Aedes*, lateral aspect of abdominal segments VII and VIII.

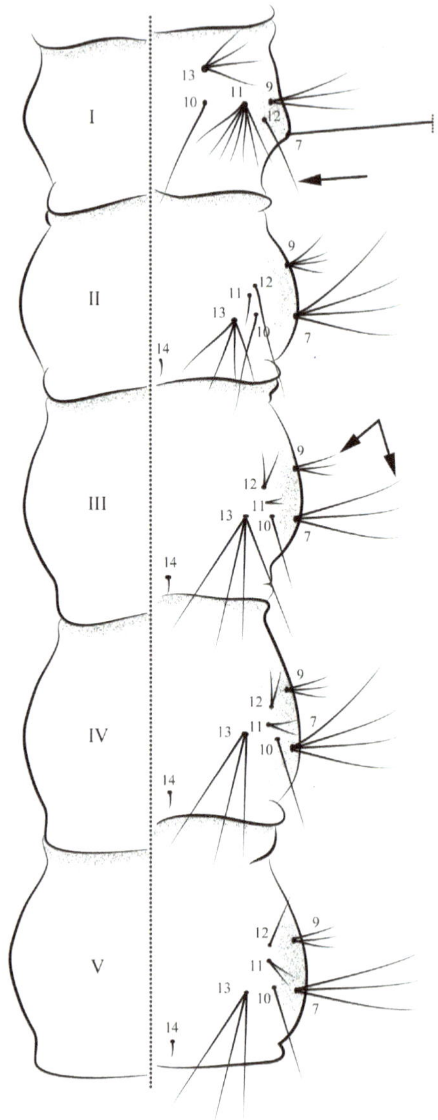

FIGURE 16.5.56 *Haemagogus*, abdominal segments I–V, right side = ventral view.

Culicidae: Culicinae: Culicini: Genera

Members have some specificity in relation to the habitats. *Deinocerites* and *Galindomyia* are crab hole specialists, being found in fresh or brackish water rarely are they found in other habitats, such as tree holes and rock holes (Lane, 1953; Adames & Galindo, 1975; Consoli & Oliveira, 1994). *Culex* breed in a wide range of habitats, such as temporary or permanent freshwater ground pools, marshes, leaf axils, tree holes, rock holes, crab holes, and stream pools; in addition, a few species are found in organically polluted waters (Consoli & Oliveira, 1994; Chaverri, 2009; Harbach, 2016). *Lutzia* occur predominantly in water habitats on the ground, but they are sometimes found in artificial containers and in phytotelm habitats (Kitching et al., 2015).

1	Segment X saddle divided into two plates widely separated, one dorsal and one ventral (Fig. 16.5.58)	2
1'	Segment X saddle not divided into two plates (Fig. 16.5.59)	3
2(1)	Mandibles base with a hairy lateral lobe (Fig. 16.5.60); siphon with bifid or trifid pecten spines (Fig. 16.5.58) [Neotropical. Nearctic]	*Deinocerites*
2'	Mandibles base without a hairy lateral lobe (Fig. 16.5.61); siphon with slender, tapered to a very tiny pointed tip pecten spines (Fig. 16.5.50) [Colombia, Ecuador]	*Galindomyia*

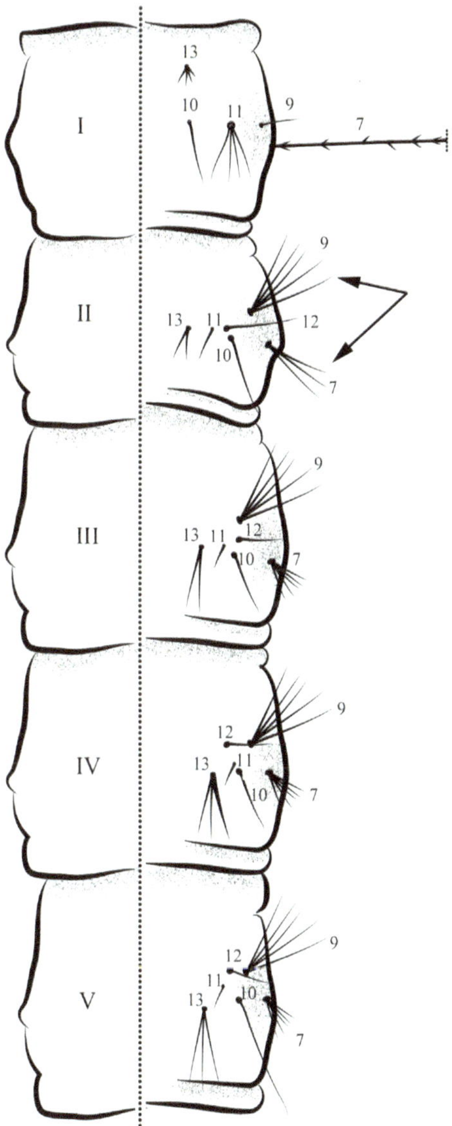

FIGURE 16.5.57 *Aedes*, abdominal segments I–V, right side = ventral view.

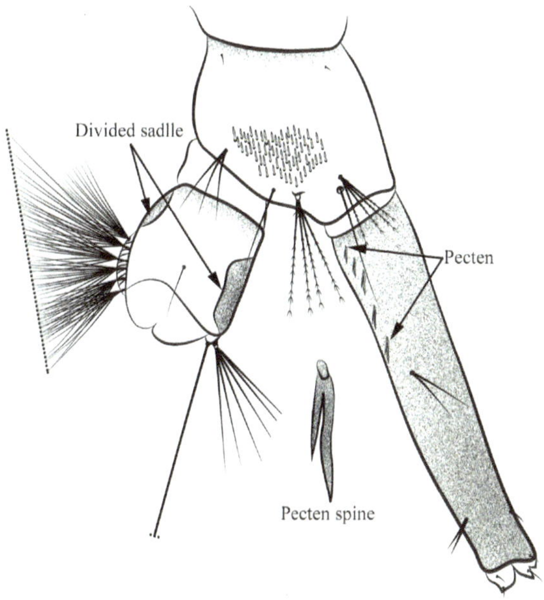

FIGURE 16.5.58 *Deinocerites*, lateral aspect of abdominal segments VIII–X.

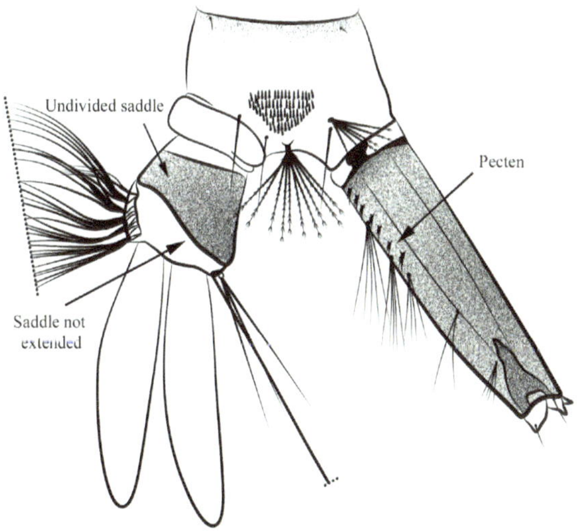

FIGURE 16.5.59 *Culex*, lateral aspect of abdominal segments VIII–X.

3(1) Segment X saddle with dorsal surface not strongly extended (Fig. 16.5.59); siphon long (about four times longer than wide), with pecten not extending to or near apex (Fig. 16.5.59) .... *Culex*
[Cosmopolitan]

3' Segment X saddle with dorsal surface strongly extended (Fig. 16.5.62); siphon short (about 1.5 times longer than wide), with pecten extending to or near apex (Fig. 16.5.62).. *Lutzia*
[Neotropical. Afrotropical, Australasian, Paleartic]

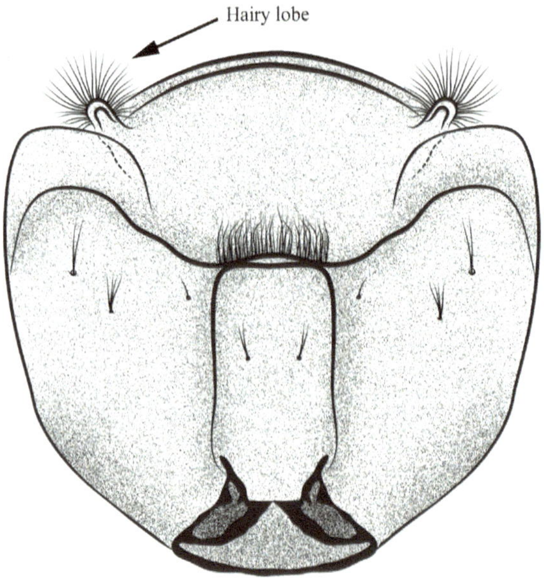

FIGURE 16.5.60 *Deinocerites*, head, ventral view.

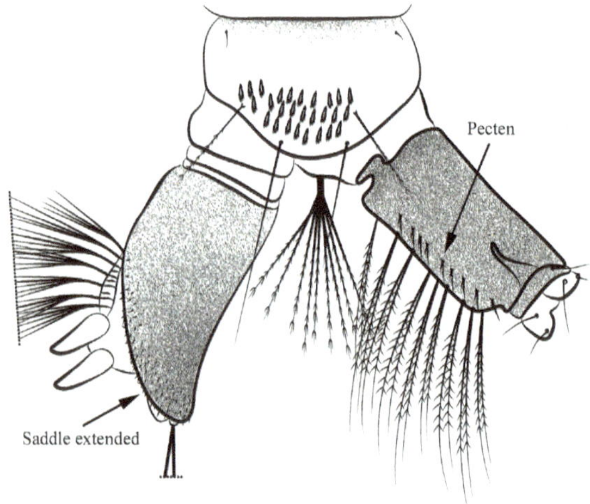

FIGURE 16.5.62 *Lutzia*, lateral aspect of abdominal segments VIII–X.

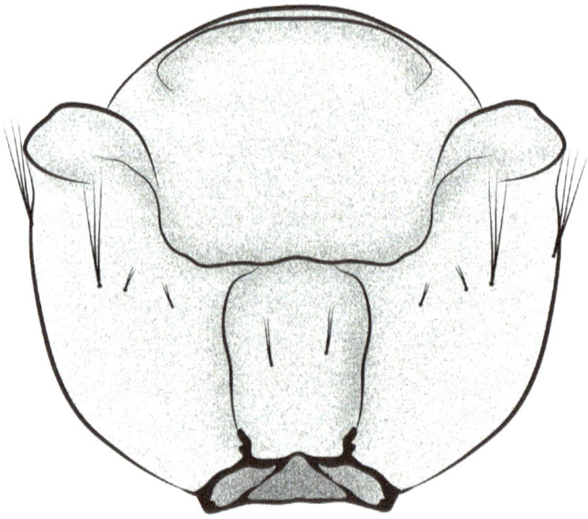

FIGURE 16.5.61 *Galindomyia*, head, ventral view.

ACKNOWLEDGMENTS

We are grateful to Leo R. Trindade for the habitus photographs presented in this chapter. Some line drawings used in the key were based on the photographs of the J. Stoffer, Walter Reed Biosystematics Unit. J.M.C do Nascimento, N. Hamada, and R.L.F. Keppler received CNPq fellowships (process numbers 141832/2014-2, 307849/2014-7 and 307039/2016-1, respectively). CAPES Pro-Equipamentos provided laboratory equipment. Philip M. Fearnside reviewed the English grammar.

REFERENCES

Adames, A.J. & P. Galindo. 1975. Description of the immature stages of Galindomyia leei Stone and Barreto, 1969. Mosquito Systematics 7: 132–136.

Arnell, H.H. 1973. Mosquitos studies (Diptera, Culicidae) XXXII. A revision of the genus Haemagogus. Contributions of the American Entomological Institute 10: 1–173.

Arnet, R.H.J. 1950. Notes on the distribution, habits, and habitats of some Panama Culicines (Diptera: Culicidae). Journal of the New York Entomological Society 58: 99–115.

Borkent, A. 2012. The pupae of Culicomorpha - Morphology and a new phylogenetic tree. Zootaxa 3396: 1–98.

Campos, R.E. & T.J. Zavortink. 2010. Description of the larva and pupa and redescription of the adults of Isostomyia paranensis (Brèthes) (Diptera: Culicidae). Zootaxa 2689: 27–36.

Clark-Gil, S. & R.F.J. Darsie. 1983. The mosquitoes of Guatemala, their identification, distribution and bionomics, with keys to adult females and larvae. Mosquito Systematics 15: 151–284.

Chaverri, L.G. 2009. Culicidae (Mosquitos, Zancudos). Pages 369–388 *in*: B.V. Brown, A. Borkent, J.M. Cumming, D.M. Wood, N.E. Woodley & M.A. Zumbado (eds.), Manual of Central American Diptera. Research Press, Ottawa, Ontario, Canada.

Consoli R.A.G.B. & R. Lourenço-de-Oliveira. 1994. Principais mosquitos de importância sanitária no Brasil. Fundação Instituto Oswaldo Cruz, Rio de Janeiro.

Frank, J.H. & G.A. Curtis. 1981. On the bionomics of bromeliad-inhabiting mosquitoes. VI. A review of the bromeliad-inhabiting species. Journal of the Florida Anti-mosquito Association 52: 4–23.

Galindo, P., F. Blanton & E.L. Peyton. 1954. A revision of the Uranotaenia of Panama with notes on other American species of the genus. Annals of the Entomological Society of America 47: 107–177.

Harbach, R.E., & K.L. Knight. 1980. Taxonomists' Glossary of Mosquito Anatomy. Plexus Publishing, Marlton, New Jersey.

Harbach, R.E. & I.J. Kitching. 1998. Phylogeny and classification of the Culicidae (Diptera). Systematic Entomology 23: 327–370.

Harbach, R.E. 2007. The Culicidae (Diptera) a review of taxonomy classification. Zootaxa, 1668: 591–638.

Harbach, R.E. 2016. Mosquito Taxonomic Inventory. http://mosquito-taxonomic-inventory.info/, accessed (2016.06.16).

Heinemann, S.J. & J.N. Belkin. 1977. Collection records of the project "Mosquitoes of Middle America". 8. Central America: Belize (BH), Guatemala (GUA), El Salvador (SAL), Honduras (HON), Nicaragua (NT, NIC). Mosquito Systematics 9: 403–453.

Kitching, I.J., C.L. Culverwell & R.E. Harbach. 2015. The phylogenetic conundrum of Lutzia (Diptera: Culicidae: Culicini): a cautionary account of conflict and support. Insect Systematics & Evolution 46: 269–290.

Lane, J. 1953. Neotropical Culicidae. 1112 p. Universidade de São Paulo, São Paulo.

Reinert J.F., R.E. Harbach & I.J. Kitching. 2009. Phylogeny and classification of tribe Aedini (Diptera: Culicidae). Zoological Journal of the Linnean Society 157: 700–794.

Rueda L.M. 2008. Global diversity of mosquitoes (Insecta: Diptera: Culicidae) in freshwater. Hydrobiologia 595: 477–487.

Steffan, W.A. & N.L. Evenhuis. 1981. Biology of Toxorhynchites. Annual Review of Entomology 26: 159–181.

Wallace, J.R. & E.D. Walker. 2008. Culicidae. Pages 801–823 *in*: R.W. Merritt, K.W. Cummins & M.B. Berg (eds.), An introduction to the aquatic insects of North America. Kendall/Hunt Publishing Company, Dubuque, IA.

WHO - World Health Organization. 2016. World health report. Executive summary: insect-borne diseases. WHO Press, World Health Organization, Geneva Switzerland http://www.who.int/whr/1996/media_centre/executive_summary1/en/index9.html, accessed (2016.06.10).

Wiegmann, B.M., M.D. Trautwein, I.S. Winkler, N.B. Barr, J. Kim, C. Lambkin, M.A. Bertone, B.K. Cassel, K.M. Bayless, A.M. Heimberg, B.M. Wheeler, K.J. Peterson, T. Pape, B.J. Sinclair, J.S. Skevington, V. Blagoderov, J. Caravas, S.N. Kutty, U. Schmidt-Ott, G.E. Kampmeier, F.C. Thompson, D.A. Grimaldi, A.T. Beckenbach, G.W. Courtney, M. Friedrich, R. Meier & D.K. Yeates. 2011. Episodic radiations in the fly tree of life. Proceedings of the National Academy of Sciences 108: 5690–5695.

Wilkerson, R.C., Y.M. Linton, D.M. Fonseca, T.R. Schultz, D.C. Price & D.A. Strickman. 2015. Making mosquito taxonomy useful: A stable classification of tribe Aedini that balances utility with current knowledge of evolutionary relationships. PLoS One 10: 1–26.

Wood, D.M. & A. Borkent. 1989. Phylogeny and classification of the Nematocera. Pages 1333–1370 *in*: J.F. McAlpine & D.M. Wood (eds.), Manual of Nearctic Diptera. Research Branch Agriculture Canada. Monograph n. 32, Canadian Government Publishing Centre, Hull, Quebec.

Woodring, J., & E.W. Davison. 1996. Biological control of mosquitoes. Pages 530–548 *in*: B.J. Beaty & C. Marquardt (eds.), The biology of disease vectors. University Press of Colorado, Niwot, Colorado.

WRBU - Walter Reed Bio-systematics Unit. Systematic catalog of Culicidae. 2016. Smithsonian Institution, Washington DC, USA. http://www.mosquitocatalog.org., accessed (2016.06.11).

Zavortink, T.J. 1968. Mosquito studies (Diptera, Culicidae). VIII. A prodrome of the genus Orthopodomyia. Contributions of the American Entomological Institute 3: 1–221.

Zavortink, T.J. 1979. A Reclassification of the Sabethine Genus Trichoprosopon. Mosquito Systematics 11: 255–257.

Family Dixidae

Caio C.D. Corrêa

Departamento de Entomologia, Museu Nacional, Universidade Federal do Rio de Janeiro, Rio de Janeiro, Rio de Janeiro, Brazil

Leonardo H. Gil-Azevedo

Departamento de Entomologia, Museu Nacional, Universidade Federal do Rio de Janeiro, Rio de Janeiro, Rio de Janeiro, Brazil

Chapter Outline

INTRODUCTION

Dixidae, known as meniscus midges, is found on all continents (except Antarctica), and comprises eight genera and about 180 species (Borkent, 2009). Adults are slender and frail, with long antennae and legs, similar to small Tipulidae. Their wings have an R vein with four branches that reach the wing margin, and an R_{2+3} that is strongly arched, but which never presents cross veins between R_1 and R_{2+3}. Adults do not feed and instead rest in vegetation near their breeding sites (Wagner et al., 2008). Female oviposition occurs in the water, and the resulting egg mass is composed of a gelatinous matrix with 10–100 eggs. The first-instar hatches in around four days (Nowell, 1951).

The immature stages occur in margins of water bodies, such as swamps, streams, rivers, lakes, and bromeliads (Lane, 1953). Larvae are cylindrical and assume a U-shape at rest position in the meniscus at the water surface. They breathe atmospheric air through their posterior spiracles, being able to assume a position above the waterline. All are filter-feeders, with a diet composed of microorganisms and organic material. Larvae swim in a rapid zigzag motion at the water surface (Borkent, 2009). Pupae respond slowly to external stimuli and are found on the waterline, usually near the margin of water body. Also, some larvae can leave the water and pupate in the moist edge.

Three dixid genera occur in the Neotropical Region: *Dixella* Dyar & Shannon, 1924, *Mesodixa* Belkin, Heinemann & Page, 1970, and *Nothodixa* Edwards, 1930. *Dixella* is found in all geographic regions and is the richest genus in Neotropics, with 26 species (Stone, 1966; Belkin et al., 1970; Peters, 1980; Chaverri & Borkent, 2007). *Nothodixa* is restricted in the Neotropics to southern South America (Chile and Argentina) and is comprised of four species (Stone, 1966). *Mesodixa* is monospecific and has been reported only from Jamaica (Belkin et al., 1970).

LIMITATIONS

The family is poorly known worldwide, but the situation is worse in the Neotropical region, were there are many undescribed species. Recently, Chaverri & Borkent (2007) described eight species from Costa Rica, increasing the number of species to 10 recorded for this country. The genera are poorly defined resulting in difficulties establishing their limits.

In view of current knowledge, it is not possible to distinguish the pupae, therefore, a key to pupae is not provided. We used data about *Nothodixa* of Australia (Tonnoir, 1923, 1924) because immature stages of this genus in the Neotropical region are unknown.

TERMINOLOGY AND MORPHOLOGY

Pupae of meniscus midges (Fig. 16.6.1 A) resemble Culicidae, with the abdomen curved under the larger cephalothorax, and with the respiratory organs corniform, the palpi are directed anteromedially, with the apices abutting or overlapping medially, and the metanotum is continuous medially, undivided. The abdomen has no

Thorp and Covich's Freshwater Invertebrates, Fourth Edition. https://doi.org/10.1016/B978-0-12-804223-6.00035-4

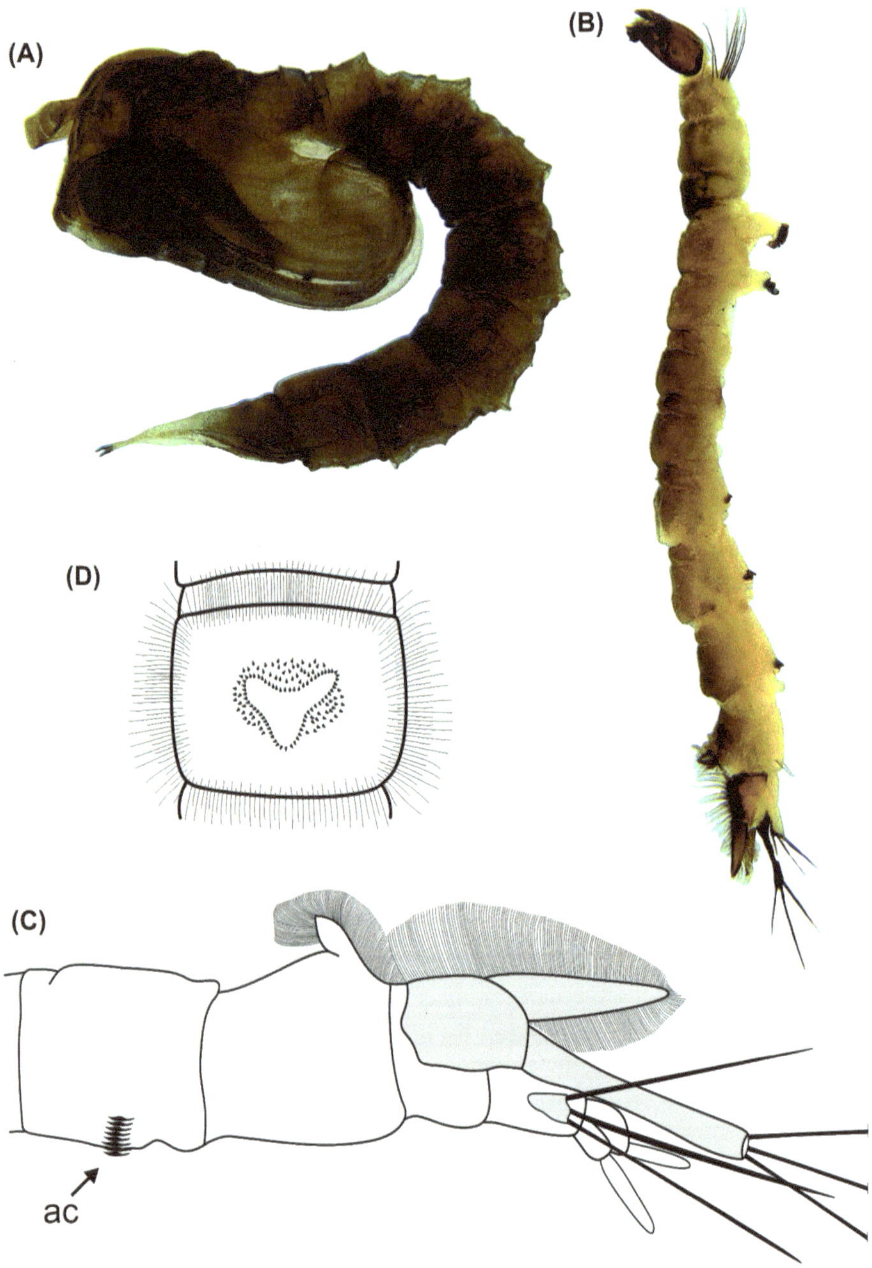

FIGURE 16.6.1 Dixidae (Diptera). *Dixella*: (A) pupa, in lateral view; (B) larva, in lateral view; and (C) larval abdomen, segments VII–X, in lateral view (ac = ventral ambulacral comb). *Nothodixa* larva from Australia: (D) abdominal segment VII, dorsal view.

tubercles, and the segments have no pleural membranes; the terminal processes are fused basally, triangular, and without paddles (Borkent, 2012).

Larvae (Fig. 16.6.1 B) have a cephalic capsule complete with lateral compound eyes. The antenna has one article, which is subtly curved; the labrum is subtriangular; the labral brushes have setae; the mandible is articulated and subquadrangular; the maxillary palpi have one article; and the labium is subtriangular. The thoracic segments are complete and distinct. Abdominal segments I–VII are subequal; abdominal segments dorsally with or without a setal crown (Fig. 16.6.1 D); segments I–II have a pair of prolegs with crochets and V–VII or only V–VI have ventral ambulacral combs (Fig. 16.6.1 C); segment VIII has a pair of spiracles, surrounded by paraspiracular setae, and a pair of postspiracular process distally; segment IX has one dorsal median plate, a pair of anterolateral plates with pectin, and a pair of postspiracular process elongate; segment X has a pair of ventrolateral plate with three setae, in the apex of segment is located the anus, anal papillae, and anal comb and the postanal process cylindrical, with two or three pairs of caudal setae (Smith, 1928; Peters, 1981; Peters & Adamski, 1982).

MATERIAL PREPARATION AND PRESERVATION

The immatures should be fixed, preserved and observed in ethanol (70–100%). In reared adults, pupal exuvia should be attached to the adult. When the adult is pinned, the pupal exuvia should be stored in a microvial with glycerin. Some small structures (as setae, mouth parts, etc.) can only be observed with a microscope, and the material should be mounted under slides and coverslips. To better observe the material, it should be macerated in unheated potassium hydroxide solution (KOH 10%) for about 4 h, or until soft parts are dissolved and clarified. After that, clean the material with destylated water, and stop the reaction by placing it in a 10% solution of acetic acid for about 20 min, and clean it in water again. Use an alcoholic series to dehydrate it to absolute ethanol, and mount with Euparal mounting medium.

KEYS TO GENERA OF DIXIDAE

Dixidae: Larvae

1	Abdominal segments without dorsal crown of setae	2
1'	Abdominal segments II–VII with dorsal crown of setae (Fig. 16.6.1 D)	*Nothodixa*
2(1)	Abdominal segment VII with 1 + 1 ventral ambulacral comb (Fig. 16.6.1 C)	*Dixella*
2'	Abdominal segment VII without ventral ambulacral comb	*Mesodixa*

REFERENCES

Belkin, J.N., S.J. Heinemann & W.A. Page. 1970. The Culicidae of Jamaica (Mosquito Studies, XXI). Contributions of the American Entomological Institute 6: 1–319.

Borkent, A. 2009. Dixidae (Meniscus Midges). Pages 359–360 *in*: B.V. Brown, A. Borkent, J.M. Cumming, D.M. Wood, N.E. Woodley & M.A. Zumbado (eds.), Manual of Central American Diptera. Volume 1. NRC Research Press, Ottawa.

Borkent, A. 2012. The pupae of Culicomorpha - morphology and a new phylogenetic tree. Zootaxa 3396: 1–98.

Chaverri, L.G. & A. Borkent. 2007. The meniscus midges of Costa Rica (Diptera: Dixidae). Zootaxa 1575: 1–34.

Lane, J. 1953. Neotropical Culicidae. Volume 1. University of São Paulo, Brazil.

Nowell, W. R. 1951. The dipterous family Dixidae in western North America (Insecta: Diptera). Microentomology 16: 187–270.

Peters, T.M. 1980. A new species of Dixella (Diptera: Dixidae) from Honduras, Central America. Proceedings of the Entomology Society of Washington 82: 681–684.

Peters, T.M. 1981. Chapter 23: Dixidae. Pages 329–333 *in*: J.F. McAlpine, B.V. Peterson, G.E. Shewell, H.J. Teskey, J.R. Vockeroth & D.M. Wood (eds.), Manual of Nearctic Diptera, Volume 1. Agriculture Canada Monograph 27, Ontario, Canada.

Peters, T.M. & D. Adamski. 1982. A description of the larva of Dixella nova (Walker) (Diptera: Dixidae). Proceedings of the Entomology Society of Washington 84: 521–528.

Smith, F.K. 1928. Larval characters of genus Dixa. Journal of the New York Entomological Society 36: 263–286.

Stone, A. 1966. Chapter 10: Family Dixidae. Pages 10.1–10.4 *in*: N. Papavero (ed.), Catalogue of the Diptera Americas South of the United States. Museu de Zoologia, Universidade de São Paulo, São Paulo.

Tonnoir, A.L. 1923. Australian Dixidae [Dipt.]. Papers and Proceedings of the Royal Society of Tasmania 1923: 58–71.

Tonnoir, A.L. 1924. New Zealand Dixidae (Dipt.). Records of the Canterbury Museum 2: 221–234.

Wagner, R., M. Barták, A. Borkent, G. Courtney, B. Goddeeris, J.P. Haenni, L. Knutson, A. Pont, G.E. Rotheray, R. Rozknosný, B. Sinclair, N. Woodley, T. Zatwarnicki & P. Zwick. 2008. Global diversity of dipteran families (Insecta Diptera) in freshwater (excluding Simulidae, Culicidae, Chironomidae, Tipulidae and Tabanidae). Hydrobiologia 595: 489–519.

Chapter 16.7

Family Corethrellidae

André Pereira Amaral
Departamento de Ecologia e Zoologia, Universidade Federal de Santa Catarina, Florianópolis, Santa Catarina, Brazil

Heron Huerta
Instituto de Diagnóstico y Referencia Epidemiológicos, Laboratorio de Entomología, Ciudad de México, México

Luiz Carlos Pinho
Departamento de Ecologia e Zoologia, Universidade Federal de Santa Catarina, Florianópolis, Santa Catarina, Brazil

Chapter Outline

INTRODUCTION

The family Corethrellidae is considered the sister taxon of Culicidae (mosquitoes) + Chaoboridae (phantom-midges), which together form the superfamily Culicoidea. Some synapomorphies supporting this group include larval thoracic segments fused in a single distinct region and larval spiracle elevated on a siphon structure. The early development of adult eye in larvae and movable paddles in pupae are characters not shared by Corethrellidae that support the relationships between the superfamily members (Wood & Borkent, 1989).

Corethrellidae is a monobasic family with only the genus *Corethrella* Coquillet. Its members are known as frog-biting midges, due to its female's habit of feeding on anuran blood after first locating the male hosts by eavesdropping on their mating calls. The group distribution ranges from 50°N to 50°S of latitude, although most of the species are found between 30°N and 30°S (Robert & Borkent, 2014). The distribution and diversity of habitats occupied by the aquatic stages of *Corethrella* is coincident to those of their anuran hosts, such as ponds, phytotelmata, swamps, and the lentic parts of streams (Borkent, 2008). A total of 108 extant and nine fossil species have been recorded, but the actual diversity of the family is expected to be much higher, as only a small part of its distribution has been sampled (Borkent, 2008; Yu et al., 2013; Amaral & Pinho, 2015; Baranov et al., 2016; Caldart et al., 2016).

Larvae of *Corethrella* are ambush predators and feed on a variety of aquatic invertebrates, including Culicidae and Chironomidae larvae, microcrustaceans, rotifers, and nematodes (Borkent, 2008). They are thought to alter the structure of communities, influencing the abundance and development of prey species (Alto et al., 2005; Griswold & Lounibos, 2005). Most of the time, they are found submerged, holding on the substrate by their siphon, but can also be found at the surface, lying horizontally on the meniscus (it is not well known how this behavior varies between species). Pupae have almost no motility and generally stay at the surface, although some may prefer hanging vertically or at about 45° angle (Borkent, 2008). According to Borkent (2008), the overall length of development stages, based on specimens reared in laboratory is 2–6 days for egg hatching, 11–37 days at larval stage and 3–8 days as pupa.

There are currently 73 valid species from the Neotropical region, 19 of which were described by Lane (1953) and Lane & Aitken (1956), who had been the main contributors to the taxonomy of the family in Neotropics until Borkent's publication in 2008. The latter summarized essentially all the knowledge about the taxonomy, phylogeny, biology, zoogeography, and evolution of the family at that time, while adding a vast amount of new information and analyses. That work encouraged novel research in the ecology and systematics

Thorp and Covich's Freshwater Invertebrates, Fourth Edition. https://doi.org/10.1016/B978-0-12-804223-6.00036-6

of Corethrellidae (Silva & Bernal, 2013; Amaral & Pinho, 2015; Baranov et al., 2016; Caldart et al., 2016); although, according to the author, there is still a lot to discover.

LIMITATIONS

The only available keys to immature stages of *Corethrella* were provided by McKeever & French (1991), comprising only larvae of four Nearctic species, and Amaral & Pinho (2015), for pupae of the Neotropical *peruviana* species group. Of the 73 valid extant Neotropical species, 28 were redescribed and 42 new ones were first described by Borkent (2008); however, those descriptions were focused mainly on morphology of adult females and some males, with a few incomplete descriptions of larvae and pupae that were not sufficient to create a comprehensive taxonomic key for the group. Descriptions from previous authors also do not give enough attention to immatures, resulting in 18 species that cannot be identified. Hence, a key will not be provided here, because the small number of taxa that would be included, compared to the diversity in the region, would probably lead to misidentifications.

TERMINOLOGY AND MORPHOLOGY

Larva (Figs. 16.7.1 A, B) shows unique posterolateral rows of spines on each side of head capsule (Borkent, 2008). The

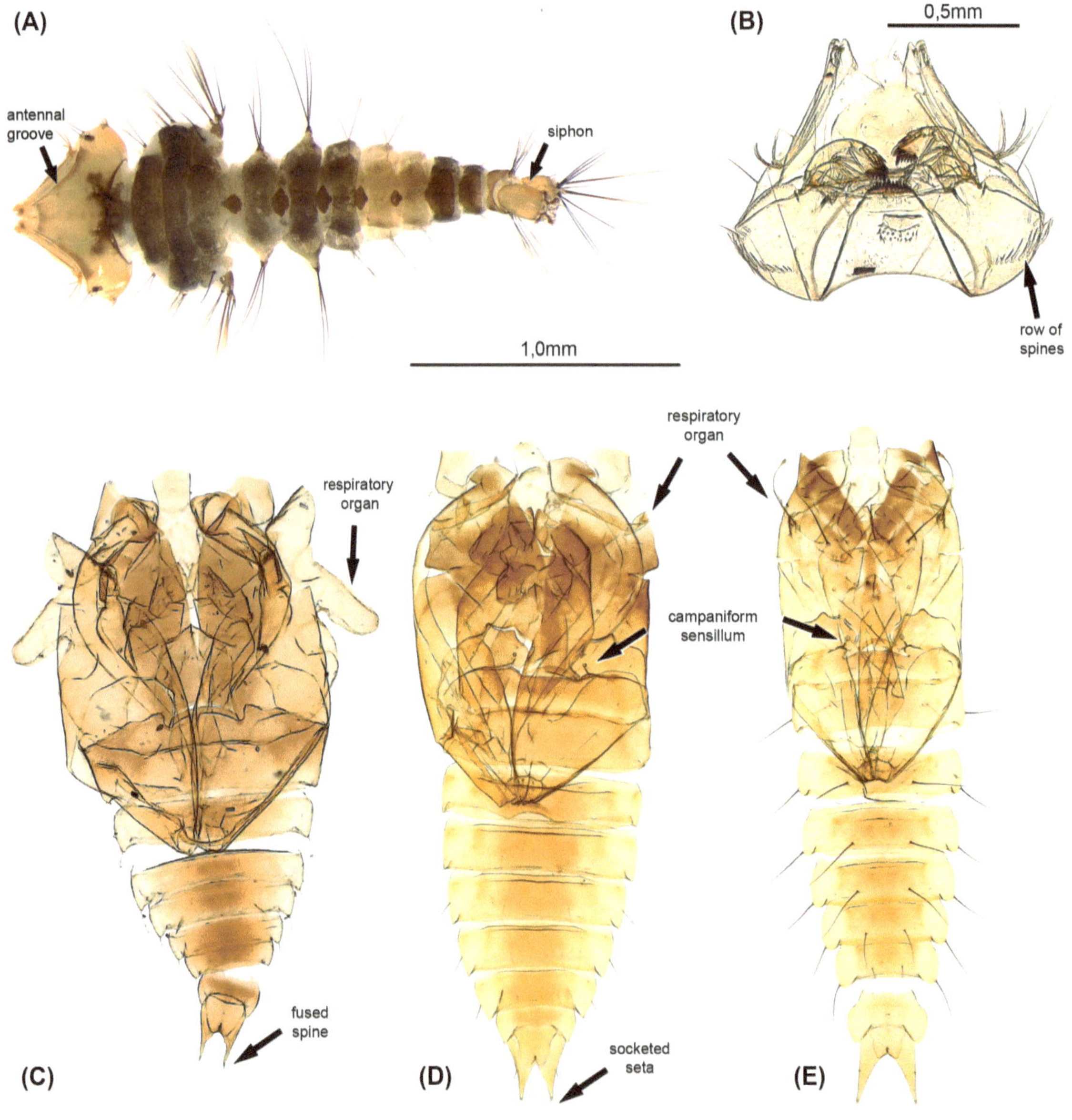

FIGURE 16.7.1 (A) *Corethrella* sp. larval habitus, dorsal view; (B) *Corethrella borkenti* Amaral & Pinho head capsule, ventral view. Dorsal view of pupal exuviae: (C) *Corethrella alticola* Lane; (D) *Corethrella infuscata* Lane; and (E) *C. borkenti*.

shape of larval head capsule ranges from somewhat round (Fig. 16.4. H) to anteriorly elongate (Fig. 16.7.1 A), with an anterolateral groove in which the antennae, apically inserted, is kept. Mandibles and mentum are well developed, with bold teeth. The thorax is large, with three fused segments and first instar have a ventral row of hooks. The abdomen is elongate, with 10 segments and a short but well developed siphon on eighth segment. Chaetotaxy is described in details in Belkin (1962).

The pupa body (Figs. 16.7.1 C, E) is oriented in one plane and has a broad cephalothorax, compared to the narrower and posteriorly tapered abdomen. Total length of pupa is about 2.5 mm. Cephalothorax bears dorsolateral respiratory organs, which may be funicular (Figs. 16.7.1 D, E) or expanded and flattened (Fig. 16.7.1 C). Most species possess campaniform sensilla on the metathorax (except those from *peruviana* species group), which are large spherical pits derived from a modified sensillum (Figs. 16.7.1 D, E). The abdomen is dorsoventrally compressed and slightly to strongly tapering distally, ending in a pair of tapering terminal processes, which may hold either a fused spine (Fig. 16.7.1 C) or a socketed seta (Fig. 16.7.1 E). The sensilla are variable and some species may have conspicuously long lateral and/or mesal setae, as well as other shorter ones distributed on dorsum and ventrum. Some species have multiple lateral, marginal setae on the abdomen (Borkent, 2012).

MATERIAL PREPARATION AND PRESERVATION

Preparation and preservation techniques are detailed in Borkent & Spinelli (2007) and Belkin (1962). Immatures and adults can be preserved temporarily in 70%–80% ethanol inside a freezer to avoid bleaching, and should be slide-mounted or pinned (adults) for permanent preservation.

For exuviae, the method we use for slide mounting consists in transferring the specimen from the 70% ethanol to absolute ethanol for about 30 min, transfer it to a small drop of Euparal (dorsal side up for pupa, ventral side up for larva) with a small piece of glass shard on each side, to keep the cover slip from crushing the skin. It is common to separate the respiratory organs of pupae and place them by the side of the body, under the same coverslip. Whole specimens must be cleared in advance, using KOH (10%).

REFERENCES

Alto, B.W., M.W. Griswold & L.P. Lounibos. 2005. Habitat complexity and sex-dependent predation of mosquito larvae in containers. Oecologia 146: 300–310.

Amaral, A.P. & L.C. Pinho. 2015. New species and records of frog-biting midges from southern Brazil (Diptera: Corethrellidae). Zootaxa 3946: 274–284.

Baranov, V.A., G.M. Kvifte & E.E. Perkovsky. 2016. Two new species of fossil *Corethrella* Coquillett from Late Eocene Rovno amber, with a species-level phylogeny for the family based on morphological traits (Diptera: Corethrellidae). Systematic Entomology 41: 1–11.

Belkin, J.N. 1962. The Mosquitoes of the South Pacific (Diptera, Culicidae). Volume 1. University of California Press, Berkeley and Los Angeles, CA.

Borkent, A. 2008. The frog-biting midges of the world (Corethrellidae: Diptera). Zootaxa 1804: 1–456.

Borkent, A. 2012. The pupae of Culicomorpha – Morphology and a New Phylogenetic Tree. Zootaxa 3396: 1–98.

Borkent, A. & G.R. Spinelli. 2007. Neotropical Ceratopogonidae (Diptera: Insecta). Pages 1–198 *in*: J. Adis, J.R. Arias, G. Rueda-Delgado and K.M. Wantzen (eds.), Aquatic Biodiversity in Latin America (ABLA). Volume 4. Pensoft, Sofia-Moscow.

Caldart, V.M., M.B. Santos, S. Iop, L.C. Pinho & S.Z. Cechin. 2016. Hematophagous flies attracted to frog calls in a preserved seasonal forest of the austral Neotropics, with the description of a new species of *Corethrella* (Diptera: Corethrellidae). Zoological Science 33: 527–536.

Griswold, M.W. & L.P. Lounibos. 2005. Competitive outcomes of aquatic container Diptera depend on predation andresource levels. Annals of the Entomological Society of America 98: 673–681.

Lane, J. 1953. Neotropical Culicidae. Volume 1. University of São Paulo.

Lane, J. & T.H.G. Aitken. 1956. Chaoborinae from Trinidad, B.W.I. (Diptera, Culicidae). Annals of the Entomological Society of America 49: 530–543.

McKeever, S. & F.E. French. 1991. *Corethrella* (Diptera: Corethrellidae) of North America north of Mexico: distribution and morphology of immature stages. Annals of the Entomological Society of America 84: 522–530.

Robert V. & A. Borkent. 2014. First record of Corethrellidae (Diptera), the frog-biting midges. Malagasy Nature 8: 49–53.

Silva, P.D. & X.E. Bernal. 2013. First report of the mating behavior of a species of frog-biting midge (Diptera: Corethrellidae). Florida Entomologist 96: 1522–1529.

Wood, D.M. & A. Borkent. 1989. Phylogeny and classification of the Nematocera. Chapter 114, Pages 1333–1370 *in*: J. F McAlpine and D. M. Wood (eds.). Manual of Nearctic Diptera. Volume 3. Agriculture Canada Monograph 32.

Yu, Y.-X., E.-J. Huang & J.-Q. Zhang. 2013. [A new species of Corethrellidae in China *Corethrella feipengi* sp. nov. (Diptera: Corethrelliidae).] Acta Parasitologica et Medica Entomologica Sinica 20: 261–264 (in Chinese, English summary).

Family Chaoboridae

Luiz Carlos Pinho
Departamento de Ecologia e Zoologia, Universidade Federal de Santa Catarina, Florianópolis, Santa Catarina, Brazil

André Pereira Amaral
Departamento de Ecologia e Zoologia, Universidade Federal de Santa Catarina, Florianópolis, Santa Catarina, Brazil

Chapter Outline

INTRODUCTION

Chaoboridae (phantom-midges) is a small family of Culicoidea, considered sister group of Culicidae (mosquitoes). Immature stages are found in standing freshwater habitats, from small pools to large lakes, with larvae preying small invertebrates and shaping freshwater communities by regulating abundance and distribution of prey populations, or even suppressing their presence (Hanazato & Yasuno, 1989; Yan et al., 1991; Bezerra-Neto & Pinto-Coelho, 2002).

Chaoborid larvae use inflatable air sacs to migrate on a diel basis vertically through the water column with changes in light and chemical cues (Dawidowicz et al., 1990; Persaud & Yan, 2001). This is thought to be a response to planktivorous fish, energetics (maximizing energy at different depths), and possibly avoidance of ultraviolet radiation. Mass emergence is recorded for some species and may possibly be triggered by lunar period (Fukuhara et al., 1992). Pupae are very motile and are found near the surface hanging vertically or slightly bent. Adults are very lightly colored and are attracted to light (Borkent, 2009).

Six extant genera are recognized, with 52 extant and 41 fossil species distributed worldwide (Borkent, 2014, Novais-Vicente et al., 2017). However, *Chaoborus* Lichtenstein is the only genus in Neotropical region and has 12 recorded species. *Chaoborus braziliensis* (Theobald) has a particularly large distribution, with several records from Jamaica to Argentina.

LIMITATIONS

Although phantom midges are considered important components of trophic nets in lenthic neotropical systems, the biology and systematics of neotropical *Chaoborus* is poorly understood. Of the 12 Neotropical species, only *C. borkenti* Novais-Vicente & Urso-Guimarães, *C. braziliensis*, *C. souzai* Lane, and *C. magnificus* Lane have their larvae and pupae described. For this reason, no taxonomic key is included in this chapter. Future biosystematic studies should focus on rearing and accurately describing immature stages.

TERMINOLOGY AND MORPHOLOGY

Larvae (Figs. 16.8.1 A, B) of *Chaoborus* have a head which is laterally compressed (Fig. 16.8.1 B) and bears a pair of medially set prehensile antennae. The frontoclipeal apotome is large, with 2 to 12 modified, often serrated, setae. Mandibles are well developed, bearing sharp teeth and an inner mandibular fan. Maxilla reduced to stipes and palpus, bearing stout modified setae. The thorax is enlarged with three segments fused into a single unit, each bearing numerous setae, as in Culicidae and Corethrellidae. The body is slender, almost transparent, except for the conspicuous pairs of dark pigmented air sacs in thorax and seventh segment of abdomen. The abdomen has nine developed segments and a reduced tenth segment which bears a flabelliform row of pectinate setae. Respiratory siphon is absent (Cook, 1981).

Thorp and Covich's Freshwater Invertebrates, Fourth Edition. https://doi.org/10.1016/B978-0-12-804223-6.00037-8

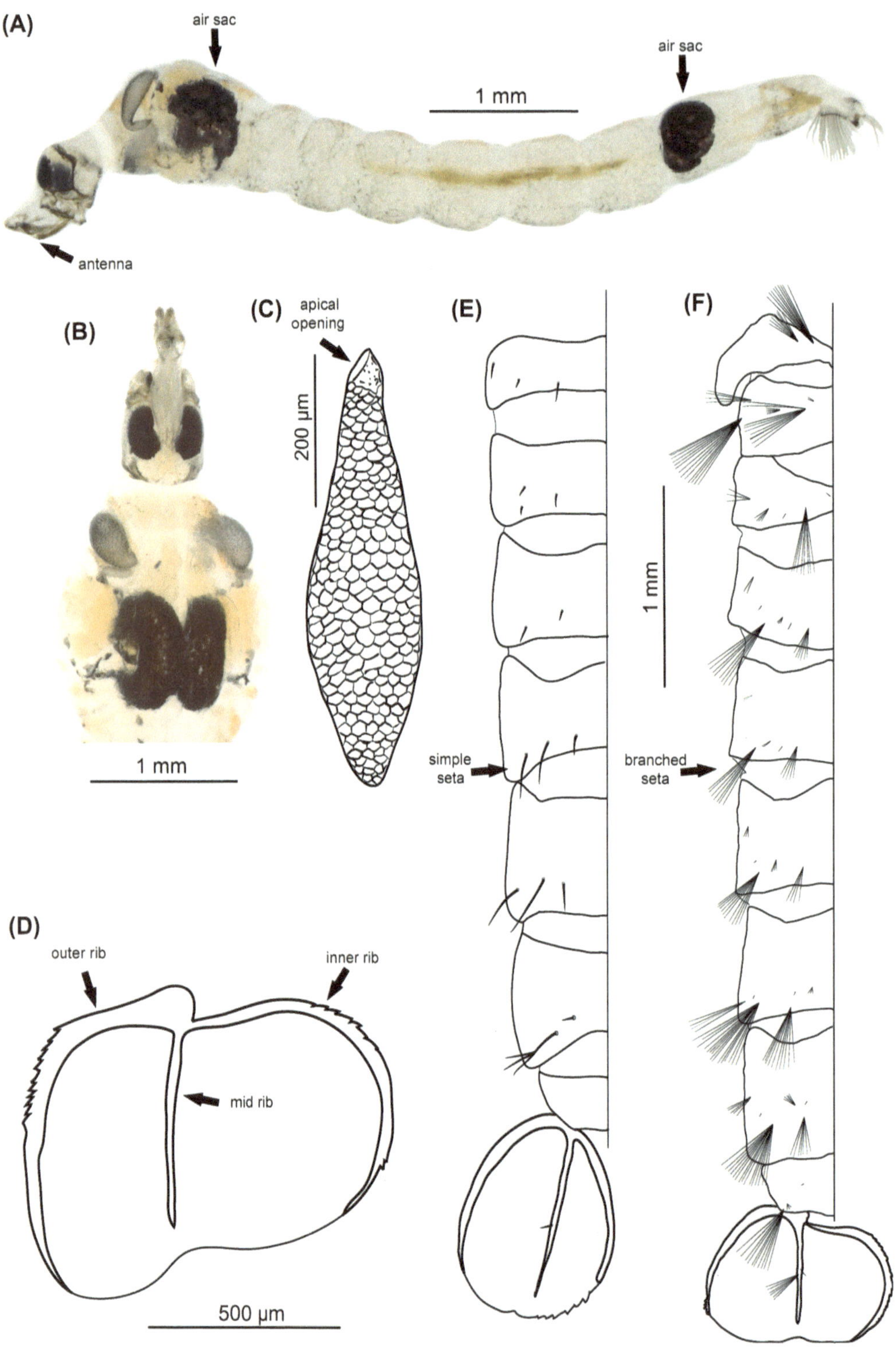

FIGURE 16.8.1 (A–C) *Chaoborus braziliensis* (Theobald): (A) larval habitus, lateral view; (B) head and thorax, dorsal view; (C) pupal respiratory organ; (D) *Chaoborus magnificus* Lane, pupal left terminal paddle; (E) *Chaoborus souzai* Lane, dorsal view of pupal abdomen, left side; and (F) *C. braziliensis*, dorsal view of pupal abdomen, left side.

Pupae (Figs. 16.8.1 C-F) are very similar to culicid pupae. Pupa of *Chaoborus* has a long thoracic respiratory organ (Fig. 16.8.1 C) and it is apically tapered, with a small apical opening or slit. The abdomen is long and cylindrical, covered with mostly simple (Fig. 16.8.1 E) or branched setae (Fig. 16.8.1 F). It bears a pair of articulated terminal paddles which are membranous, with a mid rib and thickened lateral margins (Fig. 16.8.1 D), smooth or minutely serrated, with erect or decumbent teeth (Lane & Aitken, 1956; Sæther, 1976; Belkin et al., 1970; Borkent, 2012).

MATERIAL PREPARATION AND PRESERVATION

Larvae and pupae can be temporarily preserved in 70%–80% ethanol but need to be slide mounted for taxonomic studies. Exuviae may be mounted in Euparal or Canada balsam (dorsal side up for pupa, laterally for larva) under coverslips. Whole specimens must have their musculature dissolved with KOH (10%) before mounting. Adults can also be preserved temporarily in 70–80% ethanol, but they should be pinned (along with slide-mounted male genitalia) or slide-mounted for permanent preservation. Detailed procedures are found in Belkin (1962).

REFERENCES

Belkin, J.N. 1962. The Mosquitoes of the South Pacific (Diptera, Culicidae). Volume 1. University of California Press, Berkeley and Los Angeles, CA.

Belkin, J.N., S.J. Heinemann & W.A. Page. 1970. Mosquito studies (Diptera: Culicidae). XXI. The Culicidae of Jamaica. Contributions of the American Entomological Institute 6: 1–458.

Bezerra-Neto, J.F. & R.M. Pinto-Coelho. 2002. A influência da larva de *Chaoborus brasiliensis* (Theobald, 1901) (Diptera, Chaoboridae) na distribuição vertical da comunidade zooplanctônica da lagoa do Nado, Belo Horizonte, Estado de Minas Gerais. Acta Scientiarum 24: 337–344.

Borkent, A. 2009. Chaoboridae (Phantom midges). Pages 365–368 *in*: B.V Brown, A. Borkent, J.M. Cumming, S.M. Wood, N.E. Woodley & Zumbado, M.A. (eds.), Manual of Central American Diptera. Volume 1. NRC Research Press, Ottawa.

Borkent, A. 2012. The pupae of Culicomorpha - Morphology and a New Phylogenetic Tree. Zootaxa 3396: 1–98.

Borkent, A. 2014. World catalog of extant and fossil Chaoboridae (Diptera). Zootaxa 3796: 469–493.

Cook, E.F. 1981. Chaoboridae. Chapter 24. Pages 1631–1637 *in*: J.F. McAlpine, B.V. Peterson, G.E. Shewell, H.J. Teskey, J.R. Vockeroth & D.M. Wood (Coordinators). Manual of Nearctic Diptera. Volume 1. Research Branch, Agriculture Canada, Monograph 27.

Dawidowicz, P., J. Pijanowska & K. Ciechomski. 1990. Vertical migration of *Chaoborus* larvae is induced by the presence of fish. Limnology and Oceanography 35: 1631–1637.

Fukuhara, H., G.E. Torres & S.M. Claro Monteiro. 1992. Initiation of mass emergence of a tropical chaoborid, *Chaoborus (Edwardsops) magnificus* (Insecta, Diptera), at new moon. Tropics 2: 29–34.

Hanazato, T. & M. Yasuno. 1989. Zooplankton community structure driven by vertebrate and invertebrate predators. Oecologia 81: 450–458.

Lane, J. & T.H.G. Aitken. 1956. Chaoborinae from Trinidad, B.W.I. (Diptera, Culicidae). Annals of the Entomological Society of America 49: 530–543.

Novais-Vicente, M.M., A.C.A. dos Santos, M.C. Calijuri & M.V. Urso-Guimarães. 2017. A new species of the genus *Chaoborus* Lichtenstein (Diptera: Chaoboridae) from the State of São Paulo, Brazil. Studia dipterologica 22: 201–207.

Persaud, A.D. & N.D. Yan. 2001. Accounting for spatial variability in the design of sampling programmes for *Chaoborus* larvae. Journal of Plankton Research 23: 279–285.

Sæther, O.A. 1976. Two species of *Chaoborus* (Diptera: Chaoboridae) from Venezuela. Acta Biológica Venezuélica 9: 195–212.

Yan, N.D., W. Keller, H.J. MacIsaac & L.J. McEachern. 1991. Regulation of zooplankton community structure of an acidified lake by *Chaoborus*. Ecological Applications 1: 52–65.

Chapter 16.9

Family Blephariceridae

Leonardo H. Gil-Azevedo

Departamento de Entomologia, Museu Nacional, Universidade Federal do Rio de Janeiro, Rio de Janeiro, Rio de Janeiro, Brazil

Chapter Outline

INTRODUCTION

Blephariceridae is found on all continents (except Antarctica), and comprises 27 genera and 320 species (Courtney, 2009). The immature stages of these net-winged midges inhabit a variety of streams, but in very specific habitats characterized by fast flowing, often torrential waters. Larvae are scrapers, feeding on thin films of periphyton, and attach to smooth rock substrata using six ventral suctorial discs (Wagner et al., 2008; Courtney, 2009). When disturbed, they exhibit a zig-zag escape response (Frutiger, 1998) without releasing the substrata, which explains their rarity in general collection samples. Pupae are oval, dorsoventrally flattened, and attached firmly to rocks with abdominal adhesive discs. Their immobile behavior, allied to dark color body, makes them hard to find during regular sampling procedures.

Adults are aerial and rarely collected because they are short-lived and hardly move away from their breeding sites. Adults may be recognized by the slender and small-to-medium body at 3.0–13.0 mm. The head is small and rounded with large eyes, three ocelli, and antenna with 11–13 flagellomeres, the thorax is convex, wings have netlike pattern of folds, and the legs are long and thin (Hogue, 1981; Courtney, 2009).

The classification adopted here is based on Zwick (1977), who recognize two subfamilies, Edwardsininae and Blepharicerinae, the latter of which is divided in three tribes: Blepharicerini, Apistomyiini, and Paltostomatini. Only Edwardsininae and Blepharicerinae (tribe Paltostomatini) occur in Neotropical region, with five genera and 76 species. The family is poorly known throughout the Neotropics, and the few published studies are limited to isolated descriptions and regional surveys.

Edwardsininae is restricted to temperate regions of the Southern Hemisphere (southern South America, Australia, and Madagascar). The three genera of the subfamily are probably of Gondwanian origin, but only *Edwardsina* Alexander, 1920 occurs in Neotropical region. *Edwardsina* comprises 32 species, 21 from eastern Australia (Zwick, 1977; 2006), and 11 from central to southern Chile and Argentina (Hogue, 1971). The most relevant work for South American species is the one of Edwards (1929).

Blepharicerinae is represented only by the Paltostomatini tribe in Neotropical region, and comprises four genera. *Aposonalco* Hogue, 1992 is monotypic (*A. amoyote* Hogue, 1992) and occurs in central Mexico. *Kelloggina* Williston, 1893 comprises 34 poorly defined species (Hogue, 1971), 32 of these species are reported from southeastern and southern regions of Brazil, mostly described in two regional surveys (Lutz, 1920; Lane & d'Andretta Jr., 1956). Also, one species each have been reported from Argentina (Missiones) and Chile (Santiago). *Limonicola* Lutz, 1928 comprises eight species,

Thorp and Covich's Freshwater Invertebrates, Fourth Edition. https://doi.org/10.1016/B978-0-12-804223-6.00038-X

from Venezuela, Colombia, Peru, and Argentina (Hogue, 1971; Hogue & Bedoya, 1989; Zwick, 2007). This is a well-defined genus, with diagnostic characters for larvae, pupae, and adults (see key; Hogue, 1982; and Hogue & Bedoya, 1989). *Paltostoma* Schiner, 1866 is a poorly defined genus, which include Neotropical species that do not fall readily into the other genera (Hogue & Bedoya, 1989), this genus comprises 22 species from Mexico to Argentina (Hogue, 1971; Hogue, 1979; Hogue & Bedoya, 1989; Hogue & Garcés, 1990).

LIMITATIONS

The Neotropical Blephariceridae are the least known of the family, with few fragmentary studies in a few poorly surveyed areas. The Neotropical genera are poorly defined, notably, *Kelloggina* and *Paltostoma* are potentially paraphyletic, which should be noted when using the key.

The current knowledge about the Neotropical fauna is limited by the lack of available material and systematic reviews.

TERMINOLOGY AND MORPHOLOGY

Pupae are attached immovably to the substrate, the body is ovoid, convex dorsally, and flattened ventrally (Fig. 16.9.1). The integument is rigid dorsally and thin and soft ventrally. The body is divided in cephalothorax and abdomen, with nine visible segments. The cephalothorax has a pair of anterodorsal gills, transverse in position, and composed of plates called lamellae (Fig. 16.9.2). In Blepharicerinae there are four, simple, parallel, flat, and erect lamellae (Figs. 16.9.2 B–E), while in Edwardsininae lamellae composition is variable (Fig. 16.9.2 A). The abdomen has three or four ventrolaterally adhesive pads (segments III–VI) (Figs. 16.9.1 B, D).

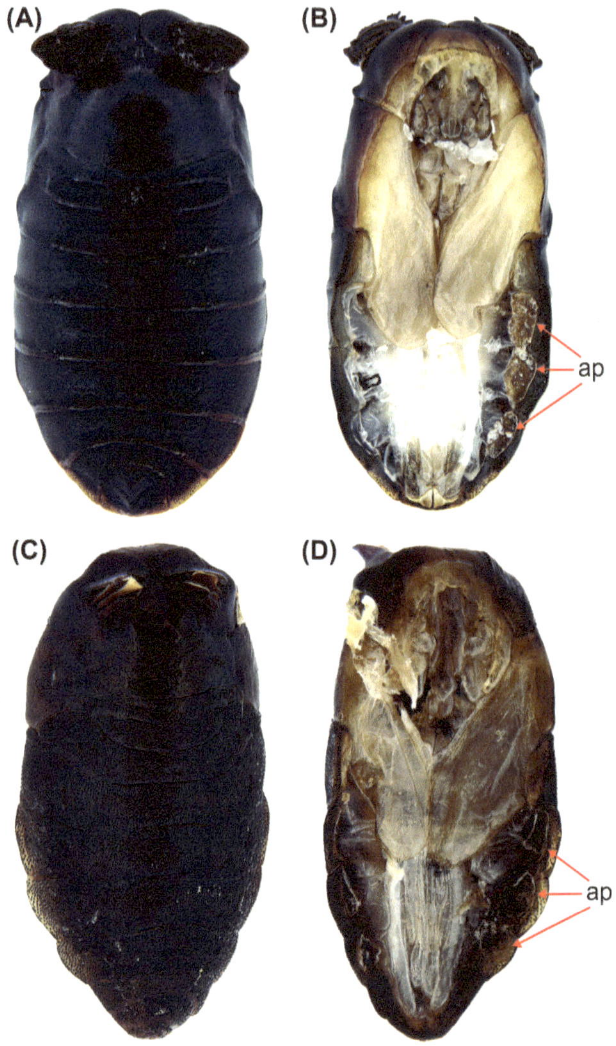

FIGURE 16.9.1 Blephariceridae pupa (ap = adhesive pads). *Edwardsina* habitus: (A) in dorsal; and (B) ventral view. *Kelloggina* habitus: (C) in dorsal, and (D) ventral views.

The larval body is cylindrical, ventrally flattened, with six major divisions (Fig. 16.9.3). The first division (= cephalic division) comprises fused head, thorax, and abdominal segment 1; divisions II–V represent one abdominal segment each, and the terminal division (= anal division) represent fused abdominal segments 7–10. In *Edwardsina* there is a narrow ring between each major division, termed the intercalary segment (Fig. 16.9.3 A). Usually, each major division has a single ventral suctorial disc and one pair of dorsolateral prolegs, which are rigid down-curved coniform structures, each bearing an apical dense patch of fine pubescence (Figs. 16.9.3 C–F), except in *Edwardsina* were prolegs are ill defined or absent (Figs. 16.9.3 A, B). Paltostomatini species usually have an additional pair of reduced prolegs (proleg VII) on the terminal body division (Figs. 16.9.3 E, F). Divisions II–VI have paired gill branches, each pair composed of a tuft of three to 10 digitiform filaments (Figs. 16.9.3 D–F); in *Edwardsina* they are on the intercalary segments (Fig. 16.9.3 B). The dorsal integument has variously shaped and distributed fine spiculae, setae, sclerotized plates, and projections; but some Paltostomatini species have conspicuous anterior and posterior rows of stout sensilla (= circumlateral sensilla) (Figs. 16.9.3 C, D).

MATERIAL PREPARATION AND PRESERVATION

The immatures should be fixed, preserved and observed in ethanol (70%–100%).

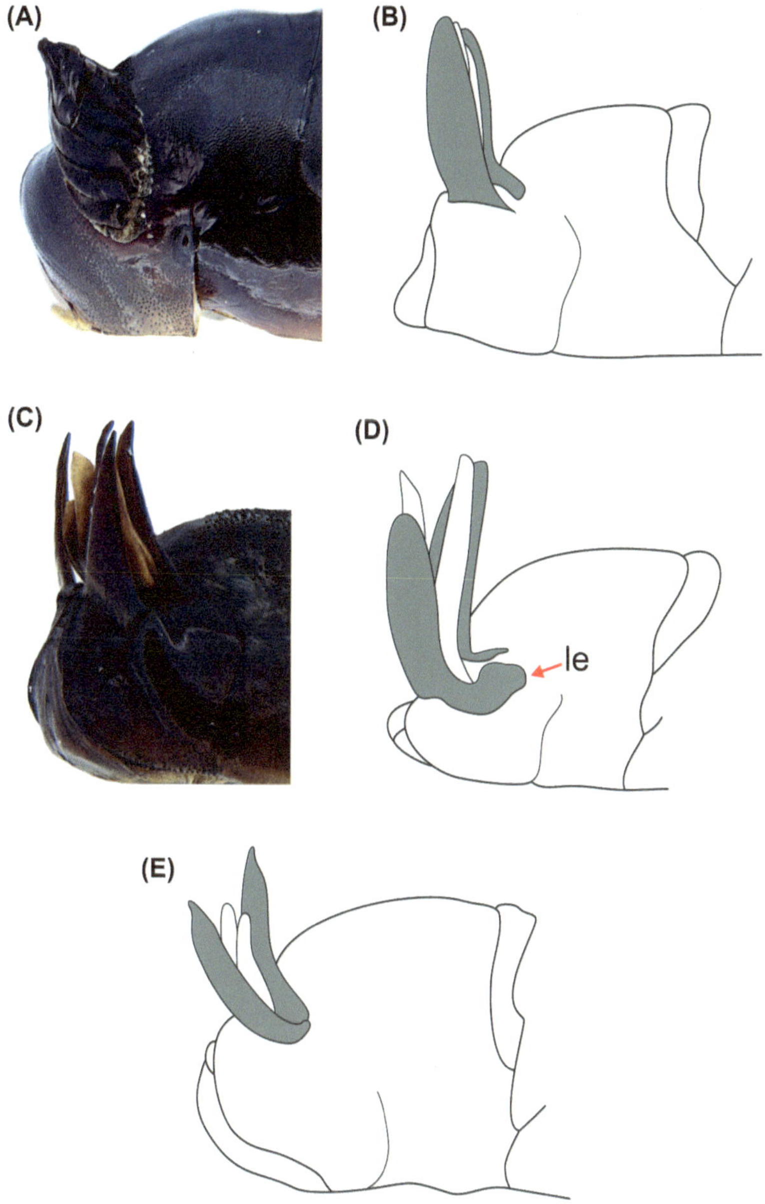

FIGURE 16.9.2 Blephariceridae pupa anterior region, in lateral view. Gill detail: (A) *Edwardsina*; (B) *Aposonalco*; (C) *Kelloggina*; (D) *Limonicola* (le = lobelike extension); and (E) *Paltostoma*.

KEYS TO BLEPHARICERIDAE

Blephariceridae: Subfamilies: Pupae

1 Gill composed of four flat, simple lamellae (Figs. 16.9.1 C, 2 B–E) .... **Blepharicerinae: Paltostomatini [p. 761]**

1' Gill morphology variable, but never composed of four flat, simple lamellae (Figs. 16.9.1 A, 2 A) .... Edwardsininae: *Edwardsina*

Blephariceridae: Blepharicerinae: Paltostomatini: Genera: Pupae

1 Abdomen with three adhesive discs (ventral view, segments IV–VI) (Figs. 16.9.1 B, D) 2

1' Abdomen with four adhesive discs (ventral view, segments III–VI) 3

2(1) Gill with respiratory lamellae compacted and slightly convergent, with apex blunt (Fig. 16.9.2 B) .... *Aposonalco*

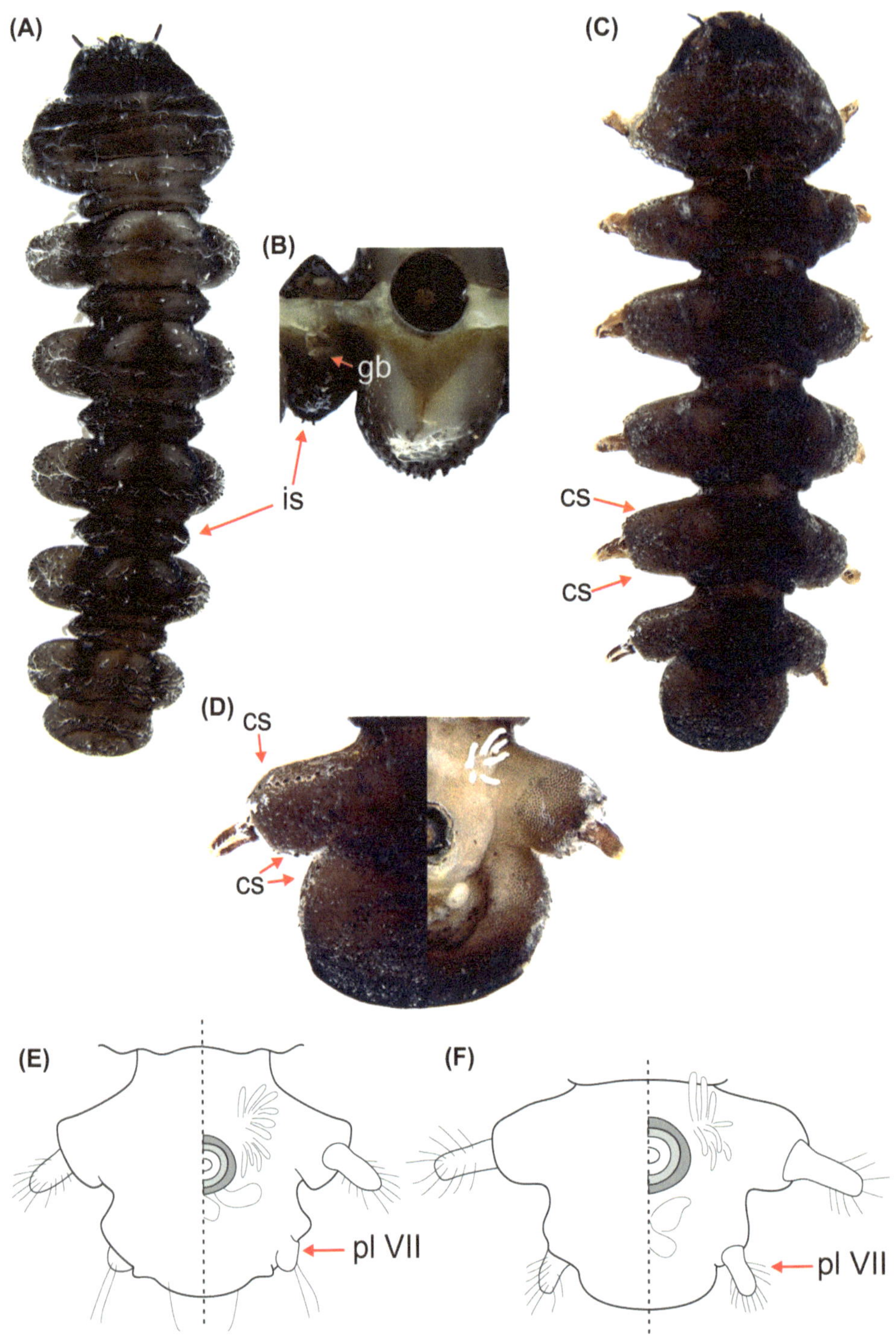

FIGURE 16.9.3 Blephariceridae larva (is = intercalary segment; gb = gill branches; cs = circumlateral sensilla; pl = proleg). *Edwardsina:* (A) habitus, dorsal view; and (B) abdominal segment in detail, ventral view. *Kelloggina*: (C) habitus, dorsal view; and (D) terminal body division with dorsal (left) and ventral views (right). *Paltostoma*: (E) terminal body division, dorsal view (left) and ventral view (right). *Limonicola*: (F) terminal body division, dorsal view (left) and ventral view (right).

2' Gill with respiratory lamellae distinctly separated and somehow divergent, with apex acute (Figs. 16.9.1 C, 2 C)................. *Kelloggina*

3(1) Respiratory lamellae with apex rounded; anteriormost lamella basally with lobiform extension (le) around base of other lamellae (Fig. 16.9.2 D)................. *Limonicola*

3' Respiratory lamellae with apex acute; anteriormost lamella base not extended around base of other lamellae (Fig. 16.9.2 E)................. *Paltostoma*

Blephariceridae: Subfamilies: Larvae

1	Intercalary segments absent, gill branches located on regular segments (Figs. 16.9.3 C–F); prolegs well definite and developed (Figs. 16.9.3 C–F)	**Blepharicerinae: Paltostomatini [p. 763]**
1'	Intercalary segments present, with gill branches (Figs. 16.9.3 A, B); prolegs ill-defined or absent (Figs. 16.9.3 A, B)	Edwardsininae: *Edwardsina*

Blephariceridae: Blepharicerinae: Paltostomatini: Genera: Larvae

1	Gills on body major divisions II–V with 8 + 8 to 10 + 10 branches each	2
1'	Gills on body major divisions II–V with 6 + 6 branches each	*Aposonalco*
2(1)	Abdominal segments dorsally with prominent anterior and posterior rows of stout sensilla (= circumlateral sensilla) (Figs. 16.9.3 C, D); proleg VII absent or very small and inserted on posterior or ventral margin (Figs. 16.9.3 D, E) (except *P. roldani* Hogue & Bedoya, 1989)	3
2'	Abdominal segments dorsally without rows of circumlateral sensilla; proleg VII developed, large, and projecting free of posterior margin (Fig. 16.9.3 F)	*Limonicola*
3(2)	Proleg VII usually absent or reduced, usually replaced by a tubercle with one or two setae (Figs. 16.9.3 C, D)	*Kelloggina*
3'	Proleg VII small, but developed (Fig. 16.9.3 E)	*Paltostoma*

REFERENCES

Courtney, G.W. 2009. Blephariceridae (Net-winged midges). Pages 237–243 *in*: B.V. Brown, A. Borkent, J.M. Cumming, D.M. Wood, N.E. Woodley & M.A. Zumbado (eds.), Manual of Central American Diptera. Vol. 1. NRC Research Press, Ottawa.

Edwards, F.W. 1929. Blepharoceridae. Pages 1–75 *in*: Diptera of Patagonia and South Chile. Part 2, Fascicle 2. British Museum (Natural History), London.

Frutiger, A. 1998. Walking on suckers - new insights into locomotory behavior of larval net-winged midges (Diptera: Blephariceridae) Journal of the North American Benthological Society 17: 104–120.

Hogue, C.L. 1971. Chapter 8: Family Blephariceridae. Pages 8.1–8.11 *in*: N. Papavero (ed.), Catalogue of the Diptera of the Americas South of the United States. Museu de Zoologia, Universidade de São Paulo, São Paulo.

Hogue, C.L. 1979. The family Blephariceridae in Costa Rica (Diptera). Contributions in Science 311: 1–22.

Hogue, C.L. 1981. Blephariceridae. Pages 191–197 *in*: J.F. McAlpine, B.V. Peterson, G.E. Shewell, H.J. Teskey, J.R. Vockeroth & D.M. Wood (eds.), Manual of Neartic Diptera. Vol. 1. Research Branch Monograph 27. Agriculture Canada, Ottawa.

Hogue, C.L. 1982. Description of a new species of net-winged midge (Diptera: Blephariceridae) from the Sierra Nevada de Santa Maria, Colombia. Contributions in Science, 340: 1–10.

Hogue, C.L. 1992. A new genus and species of net-winged midge (Diptera: Blephariceridae) from Mexico, with a redescription of *Paltostoma bellardii* Bezzi. Contributions in Science 435: 1–12.

Hogue, C.L. & I. Bedoya. 1989. The net-winged midge fauna (Diptera: Blephariceridae) of Antioquia Department, Colombia. Contributions in Science 413: 1–57.

Hogue, C.L. & G. Garcés. 1990. Discovery of the family Blephariceridae (Diptera) in Cuba, including the description of a new species. Contributions in Science 418: 1–9.

Lane, J. & C. d'Andretta, Jr. 1956. Brazilian Blepharoceridae (Diptera, Nematocera). Annals and Magazine of Natural History, Series 12, 9: 117–201.

Lutz, A. 1920. Dípteros da família Blepharoceridae, observados no Brazil. Memórias do Instituto Oswaldo Cruz 4: 75–83.

Wagner, R., M. Barták, A. Borkent, G. Courtney, B. Goddeeris, J.P. Haenni, L. Knutson, A. Pont, G.E. Rotheray, R. Rozknosný, B. Sinclair, N. Woodley, T. Zatwarnicki, & P. Zwick. 2008. Global diversity of dipteran families (Insecta Diptera) in freshwater (excluding Simulidae, Culicidae, Chironomidae, Tipulidae and Tabanidae). Hydrobiologia 595: 489–519.

Zwick, P. 1977. Autralian Blephariceridae (Diptera). Australian Journal of Zoology, Supplementary Series, 46: 1–121.

Zwick, P. 2006. Additions to the Australian net-winged midges (Diptera: Blephariceridae). Aquatic Insects 28: 139–157.

Zwick, P. 2007. Some net-winged midges from Argentina (Diptera: Blephariceridae). Studies on Neotropical Fauna and Environment 42: 109–119.

Family Psychodidae

Danilo P. Cordeiro
Coordenação de Biodiversidade, Instituto Nacional de Pesquisa da Amazônia, Manaus, Amazonas, Brazil

Rüdiger Wagner
Schlitz, Germany

Chapter Outline

INTRODUCTION

Psychodids are small hairy flies (except the rare and unusual Sycoracinae), which are commonly known as moth flies, drain flies (Psychodinae), or sand flies (Phlebotominae). The family consists of six extant subfamilies (all present in the Neotropics, except for Horaiellinae) and the extinct Datziinae. Although most of the known immatures are moisture loving (living on soil, decaying leaves, and rotten wood), many species of Horaiellinae, Sycoracinae, and Psychodinae are adapted to aquatic habitats like phytotelmata, mud pads, pond margins, and shallow rivers. Larvae of Phlebotominae and Bruchomyiinae appear to be edaphic, and the few reports on Trichomyiinae larvae have hypothesized that all are linked to decaying wood. Most of the available information on larval morphology and ecology refers to other parts of the world, mainly the Palearctic region (see studies by Vaillant, 1971–1983).

Although published reports (all from the Palearctic) of known larvae of Sycoracinae indicate that they are restricted to calcareous water in springs and moss pads, moist mosses and lichens are possible larval habitats to the Neotropical species. Adults of *Aposycorax* Duckhouse were commonly found resting on wet rock faces near aquatic habitats including springs, waterfalls and on a drainage culvert (Curler et al., 2015).

Among Psychodinae, fast flowing streams are home for the larvae of *Maruina* Müller, which are peculiar in having ventral suckers to attach the body to wet rocks. Immatures of *Lepidiella* Enderlein can be found in the edges of permanent and very small intermittent streams (Duckhouse, 1982), while the water retained between leaves of tank bromeliads and the water puddles in tree holes are the breeding sites for species of *Alepia* Enderlein, *Arisemus* Satchell, *Moruseodina* Bravo & Cordeiro, *Neurosystasis* Satchell, *Quatiella* Botosaneanu & Vaillant (Knab, 1914; Wagner & Hribar, 2010; Bravo et al., 2014) and *Tonnoira* Enderlein (unpublished). Dendrolimic moth flies are better known from the Paleartcic region (Oboňa & Ježek, 2012), with species also commonly associated with drainpipes of urban areas worldwide. *Psychodocha cinerea* (Banks), *Psychoda alternata* Say, and *Clogmia albipunctata* (Williston) are rather general in habitat choice, being also found at edge of streams and tank bromeliads (Cordeiro, unpublished data).

Among the Neotropical Psychodinae, descriptions of immatures are available for only five genera. Hogue (1973) described immatures of several *Maruina* species, and the larva of *Moruseodina cusucoensis* Bravo & Cordeiro, and some species of *Alepia* have been mentioned from water-filled plant containers (Wagner et al., 2008, 2010; Bravo et al., 2014). The immatures of the cosmopolitan species *C. albipunctata* and *P. alternata* were repeatedly described (Haseman, 1907; Vaillant, 1971–1983), although not from Neotropical specimens. One species of *Brunettia* Annandale was reared from oak tree holes in California, USA (Quate, 1955), but no immature stages of the Neotropical species of this genus have been described.

Thorp and Covich's Freshwater Invertebrates, Fourth Edition. https://doi.org/10.1016/B978-0-12-804223-6.00039-1

LIMITATIONS

Very few immature stages in the Neotropics have been described. As the knowledge on immature stages is incipient, mainly when we consider the large rate of endemism on tropical forests, identification even for genus level is generally only possible with the association of larvae and adults, which is a task for the future. Reports at the genus or species level for the aquatic stages are uncommon, due to the lack of descriptions and identifications keys. Descriptions of pupae are even more poorly detailed, commonly with only habitus and respiratory horn described or illustrated.

TERMINOLOGY AND MORPHOLOGY

Larvae of different subfamilies are very different morphologically, but in general they are elongated (dorsoventrally flattened with ventral suckers in *Maruina*), usually setose, with complete and well sclerotized head capsule and mouthparts. The trunk has three well marked thoracic and seven abdominal segments and a terminal complex (anal division) composed of abdomen segments 8 to 11, anus ventral, stigmata dorsal, or apical.

In Psychodinae, the thoracic and abdominal segments 1–7 are secondarily divided into 26 (or 27) body rings (pseudosegments or annuli), partitioned as follows: two annuli on the three thoracic and the first abdominal segment and three annuli on the abdomen segments 2 to 7 (in Brunettiini first abdomen segment has also three body annuli). The annuli may be covered with sclerites on all sides, often characteristic in shape and number; usually the dorsal sclerites (tergites) are the largest. Tergites of annuli 1–3 are medially divided (where the skin breaks during molt). All body segments have a characteristic pattern of true and accessorial setae (without basal ring) in terms of number, shape, and position. The shape of the terminal complex with posterior spiracles surrounded by two pairs of differently long seta-fringed siphonal processes is probably affected by ecological requirements of the larvae; the upper pair is usually shorter than the ventral pair, but all have more or less long rows of setae that can (in Palearctic *Berdeniella* and *Bazarella*) retain an air bubble when larvae are submerged.

The known Sycoracinae larvae are aselliform (isopod-like), and fourth instar larvae are <2 mm long. The segments of the trunk are not annulated, and one large tergite is present on each segment, except on the first thoracic. In this subfamily the eighth abdominal segment is partly separated from the terminal complex, bearing the posterior spiracles on the upper surface of the lateral plates of eighth abdominal segment. A series of 2-jointed lateral appendages are present on second and third thoracic and first to seventh abdominal segments.

FIGURE 16.10.1 *Alepia* sp. pupa: ventral view. Abbreviation: *rh* = respiratory horn.

Psychodidae pupae are fusiform (Fig. 16.10.1) or ventrally flattened (Fig. 16.10.2), with head, legs, and wings distinct, all close to the body. A pair of slender respiratory horns bearing rows of pits is present on the prothorax. The abdominal segments have several setae and spines posteriorly directed.

MATERIAL PREPARATION AND PRESERVATION

Larvae and pupae should be killed in hot water and subsequently fixed in 80% ethanol (Alencar et al., 2016). For clarification, the specimen should be maintained in 10% KOH solution at room temperature until sufficiently translucent, then transferred for 10 min to the following series of solutions: acetic acid (96%), a 1:1 mixture of acetic acid and clove oil, and finally pure clove oil. After 10 min in pure clove oil, the specimen is slide mounted in Canada balsam (Wagner et al., 2008).

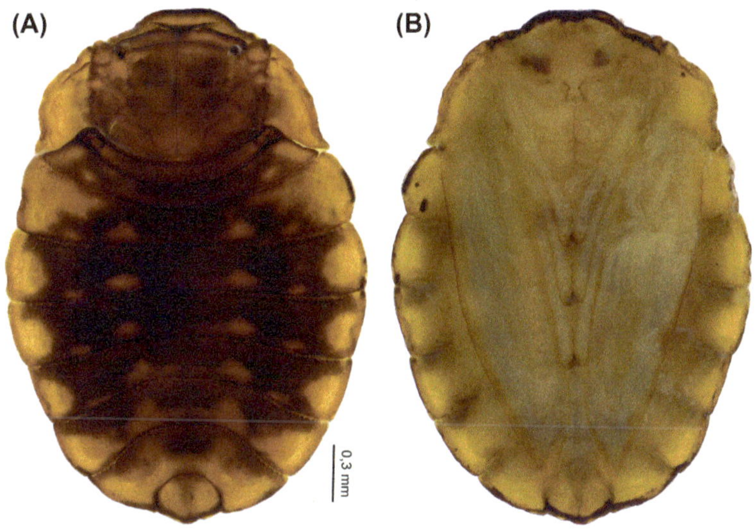

FIGURE 16.10.2 *Maruina* sp. pupa: (A) dorsal view; and (B) ventral view.

KEY TO LARVAE OF NEOTROPICAL PSYCHODIDAE SUBFAMILIES

Larval identification is mainly based on chaetotaxy, and morphology of the head, anterior spiracles, and tergal plates. Vaillant (1971) gave one of the most important contributions to the study of chaetotaxy and general morphology of immature Psychodidae. Sycoracinae larvae are known only in Paearctic region.

Psychodidae: Subfamilies

1 Larva short aselliform (isopod-like) (<2 mm), not pseudosegmented (Fig. 16.10.3); antennae longer than head capsule with four approximately filiform antennomeres; posterior spiracles separated from each other, situated dorsolaterally, on the lateral plates of eighth abdominal segment; head and prothorax retractable into body .... Sycoracinae
[All continents except North America and Antarctica]

1' Larva elongate (>2 mm) (Fig. 16.10.4), with 26 or 27 pseudosegments (annuli), mostly covered with sclerotized tergal plates (tergites); antennae very short, composed of several digitiform, setiform, or fungiform appendages arising from round base (Fig. 16.10.5); posterior spiracles close to one another, dorsally on a terminal complex (siphon) composed by segment 8 and following segments; head and prothorax not retractable into body .... **Psychodinae [p. 767]**

Psychodidae: Psychodinae: Genera

Hogue (1973, p. 3) mentioned *Maruina* larvae with only 25 annuli; however, there are only two lateral lobes of segment 7 but three tergites. This, in our opinion, indicates that two lateral lobes are superficially merged and that three tergites indicate a respective number of annuli.

1 Larvae with 26 pseudosegments (annuli), first abdominal segment with two annuli 2

1' Larvae with 27 pseudosegments, first abdominal segment with three annuli .... Brunettiini
[Cosmopolitan]

2(1) Larva usually dark to black, body well sclerotized, with 36 tergal plates on annuli usually large; antenna consists of two or more sensilla of different size and shape, hypostomium with one or more median teeth; preanal plate present 3

2' Larva generally whitish pale, weakly sclerotized; tergites on annuli often reduced in size or even completely absent (Fig. 16.10.6); antenna consisting of 1 or 2 elongate sensilla; hypostomium without median teeth, preanal plate generally absent...... Psychodini...... *Psychoda*
[Cosmopolitan]

3(2) Trunk without ventral suction discs, larvae elongate, cylindrical; accessorial setae present, larvae in other moist to wet habitats 4

3' Trunk with eight ventral suction discs (Fig. 16.10.7), dorsoventrally flattened, elongate to discoid (Fig. 16.10.8); no accessorial setae, larvae exclusively on smooth surfaces in running water .... Maruinini .... *Maruina*
[Argentina, Brazil, Costa Rica, Colombia, Mexico, Peru, Puerto Rico. Nearctic]

4(3) Mesothorax, mesotergite with only three unbranched true setae on each side; protergites of abdominal segments II–VII with one long setae on each side (Fig. 16.10.9) 5

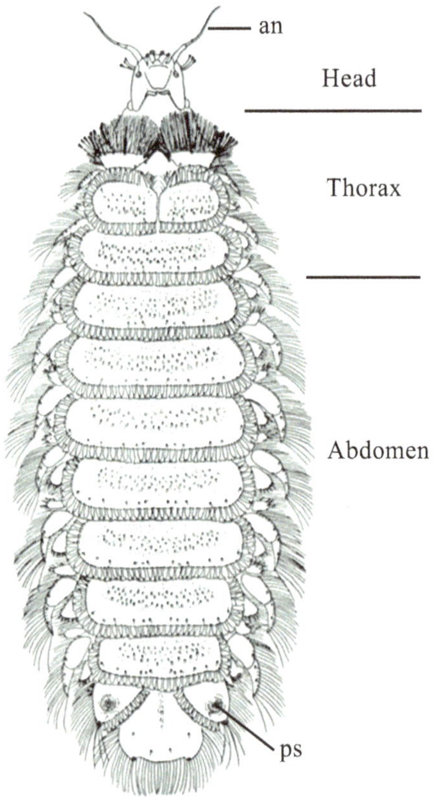

FIGURE 16.10.3 *Sycorax silacea* Curtis, larvae, dorsal view. Abbreviations: *an* = antennae, *ps* = posterior spiracle. *Modified from Jung (1958).*

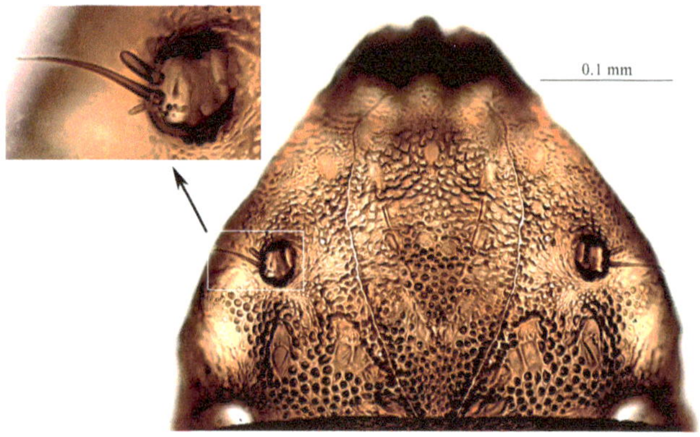

FIGURE 16.10.5 *Maruina garota* Hogue larva, head in dorsal view, with left antennae in detail.

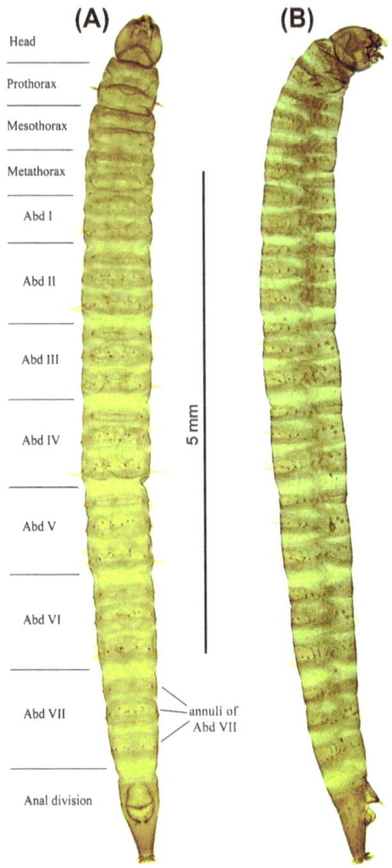

FIGURE 16.10.4 *Moruseodina cusucoensis* Bravo & Cordeiro, larva: (A) dorsal view; and (B) lateral view.

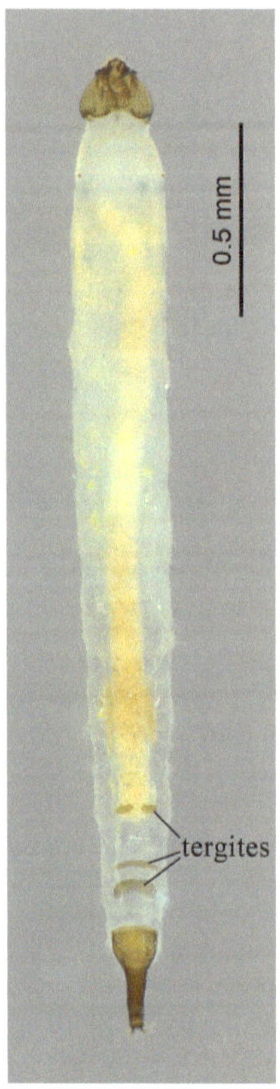

FIGURE 16.10.6 *Psychoda* sp. larva in dorsal view. *Photo by Dayana Andrade de Lima.*

FAMILY PSYCHODIDAE

FIGURE 16.10.7 *Maruina* sp. larva suctorial discs in ventral view. *Photo by Jéssica Luna Camico.*

FIGURE 16.10.9 *Moruseodina cusucoensis* Bravo & Cordeiro larva, abdominal segment VII in dorsal view, arrow indicating true setae on protergite.

FIGURE 16.10.8 *Maruina namorada* Hogue larva in dorsal view.

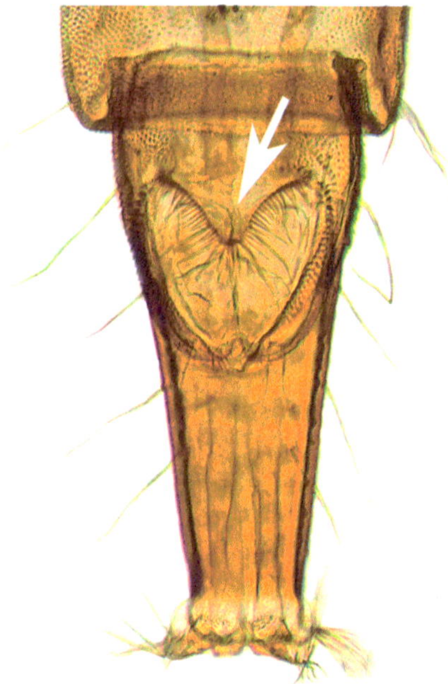

FIGURE 16.10.10 *Clogmia albipunctata* (Williston, 1893) larva, terminal complex, arrow indicating preanal plate.

4' Mesothorax, mesotergite generally with six true setae on each side, at least one true setae with two or more branches; if mesotergite with only three simple true setae, then protergites of abdominal segments II–VII without any setae .... Setomimini .... *Alepia* [Bolivia, Brazil, Costa Rica, French Guiana, Martinique, Nicaragua, Panama, Paraguay, Peru, Puerto Rico, Saba, Surinam, Trinidad and Tobago, Venezuela. Nearctic.]

5(4) Preanal plate wider than long, sagittiform (Fig. 16.10.10) .... Paramormiini.. *Clogmia* [Cosmopolitan]

5' Preanal plate longer than wide, subtrapezoidal (Fig. 16.10.11).................................. Paramormiini .... *Moruseodina* [Honduras. Nearctic]

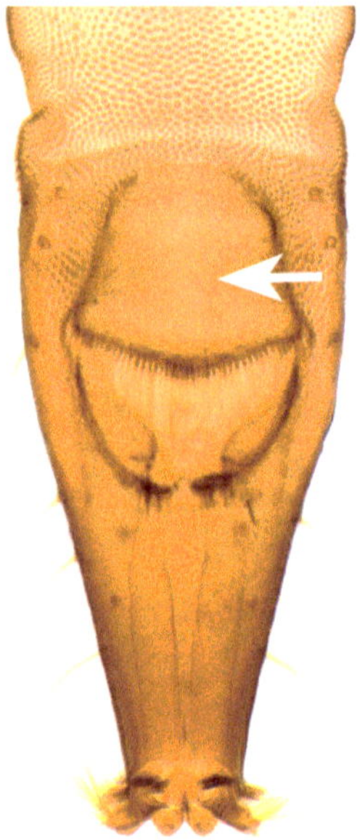

FIGURE 16.10.11 *Moruseodina cusucoensis* Bravo & Cordeiro larva, terminal complex, arrow indicating preanal plate.

REFERENCES

Alencar, R.B, V.T. Barrett & V.M. Scarpassa. 2016. Immature stages and larval chaetotaxy of Notofairchildia stenygros (Quate & Alexander) (Diptera: Psychodidae: Bruchomyiinae). Zootaxa 4169: 457–474.

Bravo, F., D. Cordeiro & M. Jocque. 2014. A new genus of Psychodinae (Diptera, Psychodidae) from phytotelmata in a Honduran cloud forest. Zootaxa 3841: 418–428.

Curler, G.R., J.K. Moulton & I. Madriz. 2015. Redescription of Aposycorax chilensis (Tonnoir) (Diptera, Psychodidae, Sycoracinae) with the first identification of blood meal host for the species. Zootaxa 4048: 114–126.

Duckhouse, D.A. 1982. Psychodidae. Pages 414–416 *in*: S.H. Hurlbert & A. Villalobos-Figueroa (eds.), Aquatic Biota of Mexico, Central America and the West Indies. San Diego State University, San Diego, CA.

Haseman, L. 1907. A monograph of the North American Psychodidae, including ten species and an aquatic psychodid from Florida. Transactions of the American Entomological Society, 33: 299–338.

Hogue, C.L. 1973. A taxonomic review of the genus Maruina (Diptera: Psychodidae). Science Bulletin of the Natural History Museum of Los Angeles County 17: 1–69.

Jung, H.F. 1958. 9b. Psychodidae – Trichomyiinae. Pages 1-16 in: E. Lindner (ed.), Die Fliegen der palaearktischen Region. E. Schweizerbart'sche Verlagsbuchhandlung (Nägele u. Obermüller), Stuttgart.

Knab, F. 1914. New moth-flies (Psychodidae) bred from Bromeliaceae and other plants. Proceedings of the United States National Museum 46: 103–106.

Oboňa, J. & J. Ježek. 2012. First records of dendrolimnetic moth flies (Diptera: Psychodidae) from Slovakia. Klapalekiana 48: 279–287.

Quate, L.W. 1955. A revision of the Psychodidae (Diptera) in America north of Mexico. University of California Publications in Entomology 10: 103–273.

Vaillant, F. 1971–1983. 9d. Psychodidae – Psychodinae. Pages 1–48 (1971), 49–78(1972), 79–108 (1972), 109–142 (1974), 143–182 (1975), 183–206 (1977), 207–238 (1978), 239–270 (1979), 271–310 (1981), 311–357 (1983) in: E. Lindner (ed.) Die Fliegen der palaearktischen Region. E. Schweizerbart'sche Verlagsbuchhandlung (Nägele und Obermiller), Stuttgart.

Wagner, R., B.A. Richardson & M.J. Richardson. 2008. A new psychodid species from Puerto Rican tank bromeliads. Studies on Neotropical Fauna and Environment 43: 209–216.

Wagner R. & L.J. Hribar. 2010. A new species of moth fly (Diptera, Psychodidae, Psychodinae) collected from Bromeliads in Florida. Zootaxa 2662: 66–68.

Wagner R., B.A. Richardson & M.J. Richardson. 2010. A new psychodid species from Saban tank bromeliads. Studies on Neotropical Fauna and Environment 45: 121–127.

Family Stratiomyidae

Fabio Siqueira Pitaluga de Godoi
Departamento de Biologia, Universidade Federal do Amazonas, Manaus, Amazonas, Brazil

José Roberto Pujol-Luz
Departamento de Zoologia, Universidade de Brası́lia, Brası́lia, Distrito Federal, Brazil

Chapter Outline

INTRODUCTION

Stratiomyidae, or soldier flies, as they are popularly known, are found on all continents except Antarctica but are especially diverse in tropical regions (James, 1973). The family consists of 12 subfamilies (Parhadestriinae, Chiromyzinae, Beridinae, Antissinae, Pachygastrinae, Clitellariinae, Sarginae, Chrysochlorininae, Hermetiinae, Raphiocerinae, Stratiomyinae, and Nemotelinae), with about 2,800 species distributed in 374 genera in the world (Woodley, 2001, 2011).

Adults exhibit a remarkable morphological diversity, including extraordinary mimetics of wasps and bees, and occupy a wide variety of habitats, normally associated with the larvae feeding sites (Woodley, 2001, 2009). Larvae, in general, are broad and somewhat flattened dorsoventrally. Their integument is impregnated with calcium carbonate, which is secreted by the Malpighian tubules, providing a very strong and flexible armor (McFadden, 1967; Rozkošný, 1982) and giving them a peculiar mosaic or honeycomb aspect by numerous small reticulate arranged facets (Brindle, 1959; Narchuk, 1988). Aquatic larvae have the anal segment (segment VIII) modified into a long or short respiratory tube with an apical spiracular cleft surrounded by a coronet of apical hydrofuge setae. With these setae, larvae can remain suspended on the surface of water for long periods while the spiracles are open to the atmosphere. When the larva submerges, these setae also aid respiration by enclosing an air bubble and keeping it in direct contact with the spiracular opening (McFadden, 1967; Rozkošný, 1982). While most Diptera larvae are predators or parasites, Stratiomyidae larvae are scavengers with highly specialized mouthparts to process granular food. In aquatic environments, the larvae feed on decaying leaves, organic detritus, and algae. In some aquatic larvae, the inner surface of the mandibular and maxillary parts of the mandibular-maxillary complex is armed with teeth to crush hard particles of food (Rozkošný, 1982). Eggs are usually placed in distinct groups on the overhanging vegetation. In general, the first instar is the shortest stage, whereas the last instar is the longest, often taking months to finish its development (Wesenberg-Lund, 1943; McFadden, 1967). The beginning of pupation is marked by the mature larva becoming more rigid, due to hardening of the integument. Puparium is formed upon the last larval skin that remains intact, forming a cover that protects the pupa inside. The pupa is often much smaller than the puparium, occupying only the anterior final one-third of the puparium; the empty space is filled with air, therefore, the puparia can float on the water surface. The adult emerges through a T-shaped slit along the anterodorsal part of the puparium (Brindle, 1959, 1962; McFadden, 1967; Rozkošný, 1982).

The Neotropical region has approximately 987 species, distributed in 148 genera, representing more than one-third of all known species in the world (Woodley, 2001); it is the only region where all subfamilies occur. Despite this diversity and species richness, little is known about the immature stages of Stratiomyidae in this region, with most

Thorp and Covich's Freshwater Invertebrates, Fourth Edition. https://doi.org/10.1016/B978-0-12-804223-6.00040-8

knowledge on stratiomyid larvae coming from Nearctic and Palearctic regions. According to Woodley (2009), studies on larvae of Neotropical fauna mainly are related to Brazilian species (Pujol-Luz, 1999; Pujol-Luz & Xerez, 1999; Pujol-Luz & Vieira, 2000; Pujol-Luz & Leite, 2001; Xerez & Pujol-Luz, 2001; Xerez et al., 2002, 2003a, b, 2004; Viana et al., 2003; Pujol-Luz et al., 2004, 2016a, b; Lopes et al., 2006; Pujol-Luz & Pujol-Luz, 2014). Virtually all publications on larvae in the Neotropics concern terrestrial forms, even those with known aquatic species, at least for Stratiomyinae. The exception was the discovery of the larva of *Raphiocera armata* (Wiedemann, 1830) by Pujol-Luz et al. (2004). Of all subfamilies, Parhadestriinae is the only one with unknown immature forms, and the subfamilies Raphiocerinae, Stratiomyinae and Nemotelinae have aquatic or subaquatic larvae (Rozkošný, 1982; Woodley, 2001; Pujol-Luz et al., 2004).

Raphiocerinae: Except for two species, one from Oriental region (*Rhaphioceroides pendleburyi* Brunetti, 1922: Thailand) and one from Australian region (*Raphiocera orientalis* Lindner, 1957: Indonesia, Irian Jaya), the distribution of this subfamily is essentially Neotropical (Woodley, 2001, Rozkošný & Woodley, 2010). Currently this region has 36 species distributed in 20 genera (Woodley, 2001, 2011; Pimentel & Pujol-Luz, 2002; Godoi & Rafael, 2013). The only known larva of the subfamily is *R. armata* (Pujol-Luz et al., 2004); it was found on land (in a decomposing tree trunk) a short distance from a stream with tidal influence. Despite being in a terrestrial habitat, this larva had posterior be hydrofuge setae as found in other aquatic larvae (Fig. 16.11.1). Thus, Raphiocerinae was included into the select group of larvae with aquatic or semiaquatic way of life, until then occupied only by Stratiomyinae and Nemotelinae. The larva differs essentially from the immatures of Stratiomyinae and Nemotelinae by the head with plumose ventral setae and tetramerous digitform setae; abdomen with lamellar and fringed ventral setae; and posterior spiracle of the abdominal segment VIII with ventral opening.

FIGURE 16.11.1 Larvae of *Raphiocera armata* (Wiedemann) (Diptera: Stratiomyidae). *Photo: J.R. Pujol-Luz.*

Stratiomyinae: With about 645 species in 46 genera, Stratiomyinae is the most diverse subfamily with 24% of all known species in the family. In the Neotropical region this subfamily is the second most diverse, both in number of recognized species (166) and in genera (22), comprising about 17% of the diversity of the family. The subfamily is divided into three tribes, Oxycerini, Prosopochrysini, and Stratiomyini, all found in Neotropics (Pujol-Luz & Papavero, 1998; Woodley, 2001, 2011). Unlike Raphiocerinae and Nemotelinae, larval forms of Stratiomyinae are relatively well known, especially in the Palearctic and Nearctic region. For the Neotropical region, the larvae of only five species are known, *Euparyphus* (*Euparyphus*) *cinctus* (Osten Sacken, 1886) (Oxycerini), *Hedriodiscus vertebratus* (Say, 1824), *Rhingiopsis enderleini* Lindner, 1928, *Stratiomicrodon convexa* (Wulp, 1857), *Zuerchermyia festiva* (Walker, 1854) (Stratiomyini). Prosopochrysini does not have any larva known for the region. Larvae of four undescribed Stratiomyini species, *Chloromelas barbata* Lindner, 1935, *Hedriodiscus punctifer* Bigot, 1879, *Hoplitimyia mutabilis* (Fabricius, 1787) and *Odontomyia vittata* Macquart, 1850, are known to the authors from material housed in Brazilian entomological collections. The aquatic or semiaquatic larvae of Stratiomyinae are found in virtually every type of aquatic environment, from ponds, lakes, and streams to hot springs and ocean shores. Larvae frequently remain hidden in the vegetation along the shore or at the bottom of the water, like *Stratiomys* Geoffroy, 1762 (McFadden, 1967; Rozkošný, 1982). An apical coronet of setae surrounding the posterior spiracular region on the anal segment characterizes the immature forms. Some larvae have a pair of strong ventral hooks on the segment VII or even segment VI (McFadden, 1967; Rozkošný, 1982; Woodley, 2009).

Nemotelinae: This subfamily is represented by 217 species distributed in four genera found in all biogeographical regions. With 48 species in two genera (*Brachycara* Thomson, 1869 and *Nemotelus* Geoffroy, 1762), the Neotropical region is the second-most diverse behind only the Palearctic region, which has 94 species in three genera (Woodley, 2001). Larval forms of Nemotelinae, except for some species of *Nemotelus* and a single species of *Lasiopa* Brullé, 1833, are unknown (Woodley, 2001). Larvae of *Nemotelus* are found in ponds and marshes.

LIMITATIONS

Very little or virtually nothing is known about the biology of the immature stages of the Neotropical species of aquatic Stratiomyidae. Despite the greater diversity of Diptera in the Neotropics, aquatic species are more abundant in other biogeographic regions, such as Nearctic and Palearctic. In these regions they are easier to locate and collect than terrestrial species. Stratiomyinae larvae, for example, are commonly used as representatives of the family in general collections of immature stages, as indicated by McFadden (1967). Larvae are generally found in the marginal vegetation of rivers, streams, lakes, or ponds, as well as buried in sand, mud, or fixed in stones, mosses, or plant roots in waterfalls and rapids. Despite this, they are not easily collected in these environments in the Neotropics, which are usually sporadic or accidental. These characteristics also make it difficult to collect specimens for rearing in the laboratory, making it virtually impossible to adequately study their biology, larval development, and life cycle, and to make the important association of immatures with adults.

The use of a scanning electron microscope is often necessary to better understand some very discrete morphological structures, particularly the integument, setae, and sternal patch of larvae and puparia as shown by Rozkošný (1982), and present in several articles (Rozkošný & Kovac, 1991, 1994a, b, c, 1997, 1998a, b, 2001; Pujol-Luz & Pujol-Luz, 2014; Pujol-Luz et al., 2016b). This is a valuable tool in the morphological study of the immature forms of Stratiomyidae, but in most research institutions this equipment is not available.

There are still many gaps on the knowledge of Stratiomyidae immatures in the Neotropical region, where there are few aquatic species with described immatures. Of the 250 species recorded in aquatic environments in this region (Woodley, 2001), only six have their larvae described, one in Raphiocerinae, and five in Stratiomyinae. No larvae of Nemotelinae is described for the Neotropical region. This scarcity of information represents a large obstacle for the study of the aquatic species of Stratiomyidae since it is not possible to identify their immatures and, consequently, identification keys for the aquatic stages are practically impossible to construct.

TERMINOLOGY AND MORPHOLOGY

The terminology follows Rozkošný (1982). The larval body consists of a distinct head followed by three thoracic segments (1–3) and eight abdominal segments (I–VIII) (Fig. 16.11.2 A). The main differences between the thoracic and abdominal segments lie in the chaetotaxy. Head capsule is longer than wide, but this is not readily noticed since the basal half is permanently retracted into the thorax (Fig. 16.11.2 B) (McFadden, 1967; Rozkošný, 1982), therefore, the head is capable of only slight independent movements. Eye prominences are situated on the dorsolateral margins and the antennae at the anterolateral margins of the head, near the apex of the ocular lobes (Figs. 16.11.2 B, C) (McFadden, 1967, 1972; Rozkošný, 1982). The thorax is broader than head, without legs or prolegs. Prothoracic spiracles are present and variable in shape and size (Fig. 16.11.2 B) (Brindle 1959; McFadden, 1967, 1972). In aquatic species, abdominal segments I–II usually are the broadest (Fig. 16.11.2 A), remaining segments can be either parallel sided, like in terrestrial species, with the last segment longer than wide (Figs. 16.11.3 A, 4 A) or tapering towards the terminal region of the body (Figs. 16.11.2 A, D). Also, often, aquatic larvae have the abdominal segment VIII, known as the anal segment, much longer than wide, modified into a long or short breathing tube with an apical spiracular cleft surrounded by a coronet of apical setae (Fig. 16.11.2 D). This segment has a transverse fissure that houses the spiracular chamber. Some aquatic species exhibit strongly curved sclerotized hooks in the posterior margin of the ventral surface of the anal segment (Fig. 16.11.2 D) (McFadden, 1967, 1972; Rozkošný, 1982).

Many subtle and discrete structures, such as the shape of the antennae, the mouthparts components, the sternal patch, the shape and position of spiracular opening, and the chaetotaxy are important taxonomic characters used in identification of larvae and puparia.

MATERIAL PREPARATION AND PRESERVATION

Once collected in field, larvae can be fixed in 70% alcohol, or in glycerinated alcohol to preserve delicate morphological structures like setae. Larvae or puparia do not require special or complex preparation to be studied. Usually immature forms are treated with a 10% hydrochloric acid solution to remove debris and a layer of calcium carbonate. For an accurate observation of morphological details, especially related to mouthparts, the larval head can be removed and placed in a hot solution of 10% sodium hydroxide for a few minutes and then transferred to a plate with water and acetic acid to neutralize the sodium hydroxide. After this preparation, the pieces should be placed in a solution of glycerin and water for observation or dissection. In some cases, additional preparation is required, such as the mandibular maxillary complexes that can be dissected and mounted on a polyvinyl alcohol medium in microscope slides. All morphological pieces, from the entire larva or puparium to the dissected structures, are conditioned in small plastic tubes containing glycerin, and for the puparium the tubes are attached to the adult specimen (McFadden, 1967; Rozkošný, 1982).

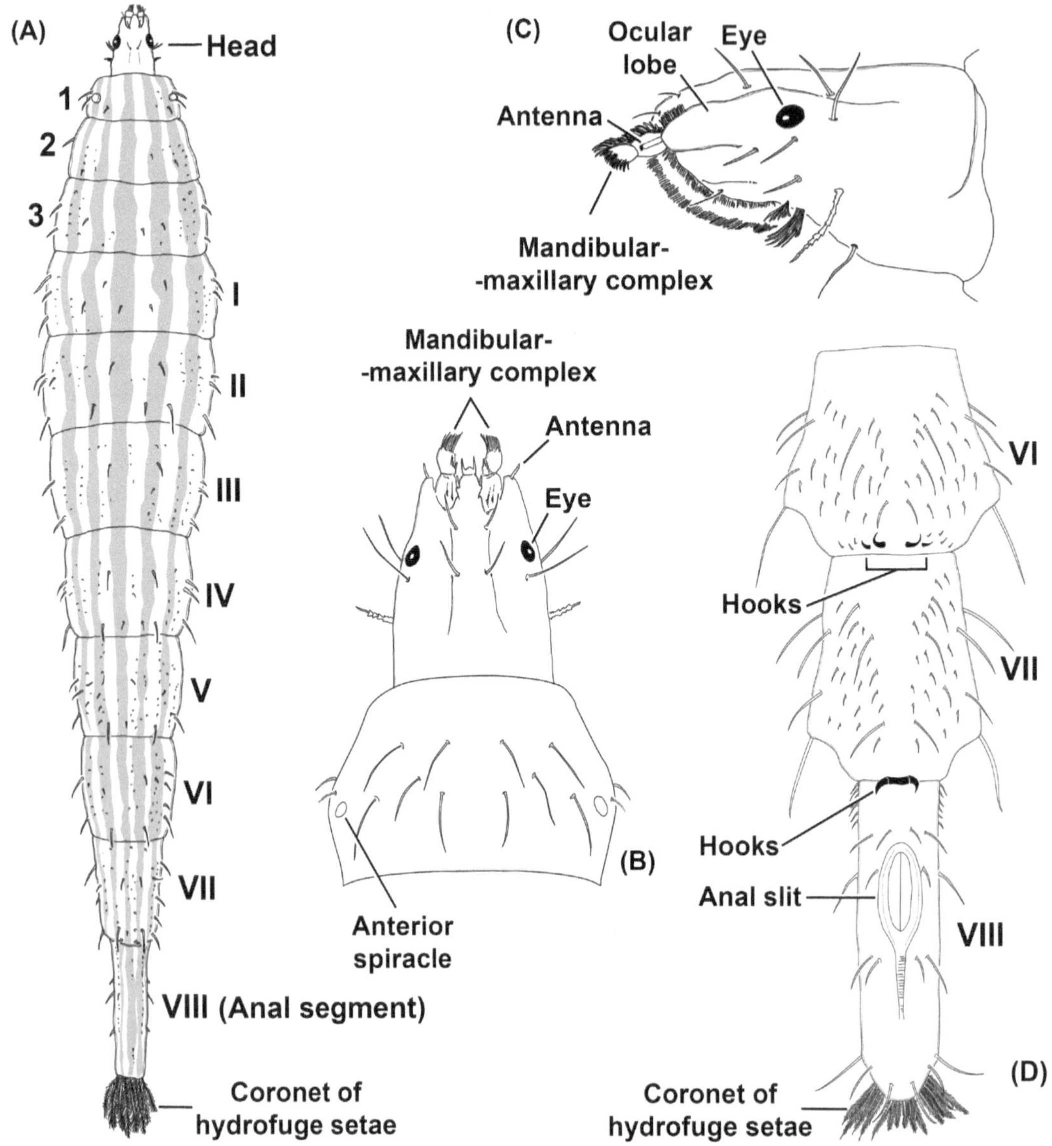

FIGURE 16.11.2 Larva of *Odontomyia* (Diptera: Stratiomyidae): (A) habitus, dorsal view; (B) head and thoracic segment, dorsal view; (C) head, lateral view; and (D) abdominal segments VI–VIII, ventral view.

KEY TO STRATIOMYIDAE

Larvae can be divided into two major ecological groups, terrestrial, and aquatic. Larval habitats of aquatic forms are considerably diverse, including littoral zones of ponds and lakes, springs and marshes, mud in marshy places, and the margins of slowly running and stagnant waters. The larvae can also be found in moist or wet moss alongside springs and streams, on aquatic plants on the water-surface, in hygropetric situations and even in salt water on the coast or inland and in thermal springs (Rozkošný, 1982).

Stratiomyidae: Subfamilies (Larvae)

1 Anal segment with apical (Figs. 16.11.2 A, B), ventral (Figs. 16.11.3 A–C) or dorsal (Figs. 16.11.4 A, C) coronet of hydrofuge setae Aquatic 2

1' Anal segment without a coronet of hydrofuge setae .... Terrestrial

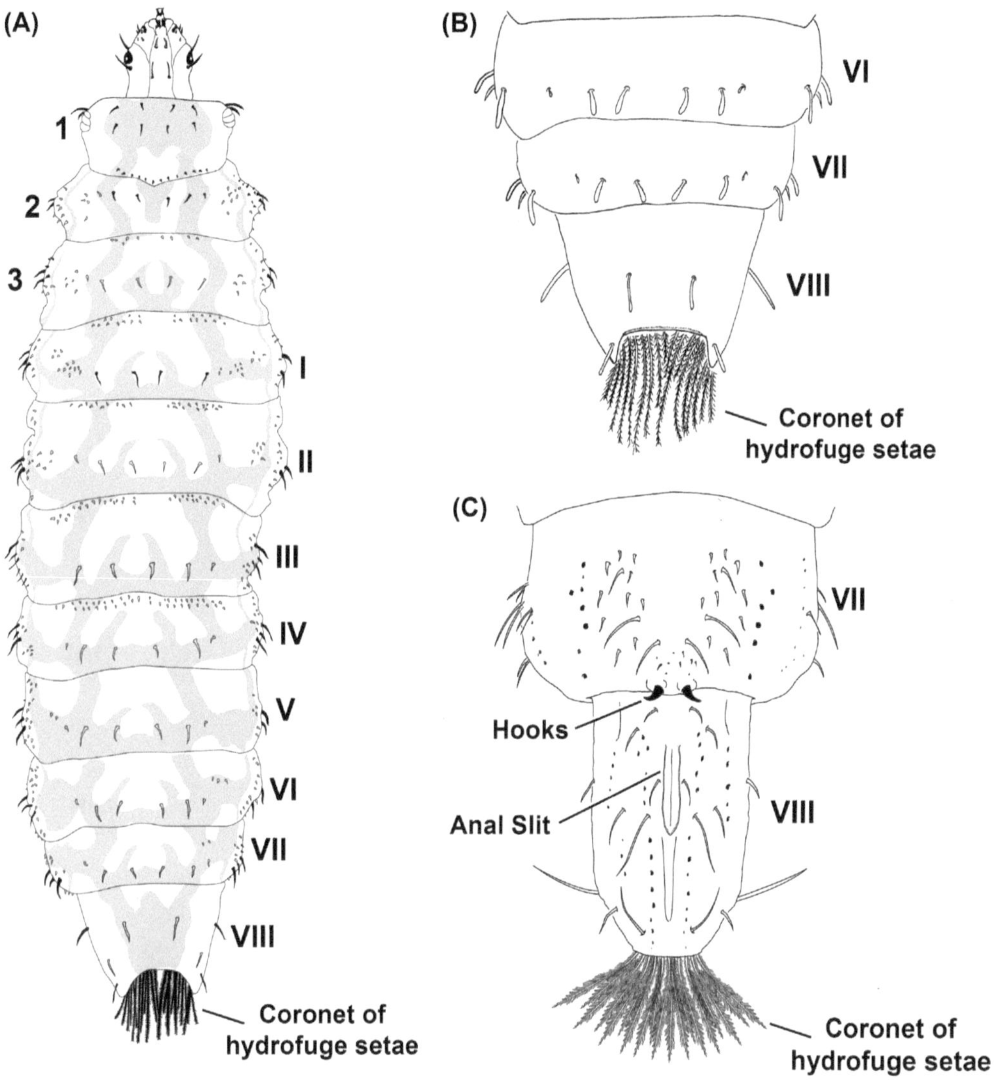

FIGURE 16.11.3 Larva of Stratiomyidae (Diptera). *Raphiocera armata* (Wiedemann): (A) habitus, dorsal view; and (B) abdominal segments VI–VIII, dorsal view. *Oplodontha*: (C) abdominal segments VII–VIII, ventral view.

2(1)	Abdomen with segments parallel sided and rounded posteriorly (Figs. 16.11.3 A, 4A, C)	3
2'	Abdomen with segments tapering toward the terminal region forming a tubular structure (Figs. 16.11.2a, d, 3c)	Stratiomyinae
3(2)	Anal segment ventral region with a coronet of hydrofuge setae; head and body with tufts of digitiform, branched, and fringed setae; dorsal chromatic pattern in characteristic zigzag (Fig. 16.11.3 A)	Raphiocerinae
3'	Anal segment dorsal region with a coronet of hydrofuge setae (Figs. 16.11.4 A, C); head and body with tufts of setae of varied shapes; dorsal chromatic pattern varied	Nemotelinae

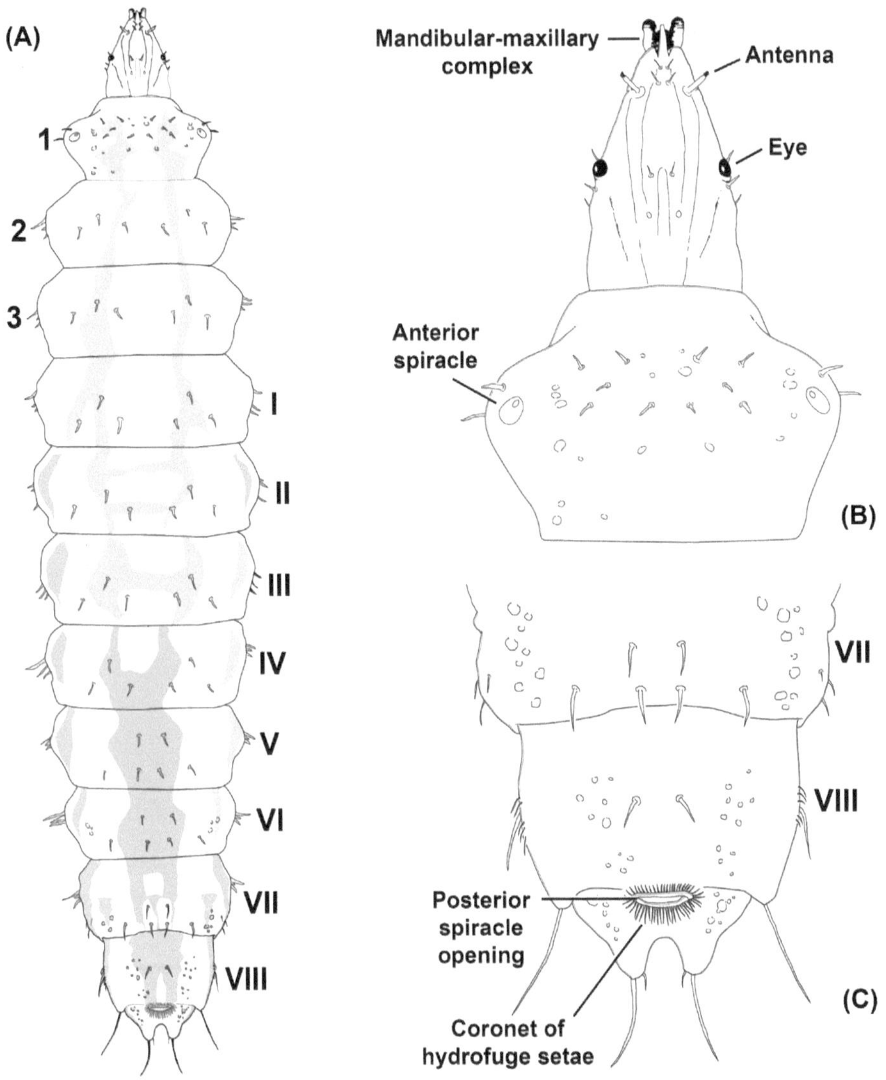

FIGURE 16.11.4 Larva of *Nemotelus* (Diptera: Stratiomyidae): (A) habitus, dorsal view; (B) head and first thoracic segment, dorsal view; and (C) abdominal segments VII–VIII, dorsal view.

ACKNOWLEDGMENTS

We are grateful to the editors for the invitation to write this chapter and for essential comments and suggestions, to Dr. Ciro Joko who kindly prepared the larvae illustrations, and for the support of Conselho Nacional de Desenvolvimento Científico e Tecnológico (CNPQ).

REFERENCES

Brindle, A. 1959. Notes on the larvae of the British Rhagionidae and Stratiomyidae with a key to the genera. The Entomologist's Record 71: 126–133.

Brindle, A. 1962. Taxonomic notes on the larvae of British Diptera - 7. The genus Pachygaster Meigen (Stratiomyidae). Entomologist 95: 77–82.

Godoi, F.S.P & J.A. Rafael. 2013. A new species of Dicranophora Macquart (Diptera, Stratiomyidae) from Bahia State, Brazil and a key to species of the genus. Zootaxa 3641: 83–91.

James, M.T. 1973. Chapter 26: Stratiomyidae. Pages 26.1–26.95 *in*: A catalogue of the Diptera of the Americas south of the United States. Universidade de São Paulo, Museu de Zoologia.

Lopes, A.L.F.; R. Xerez; R.R. Silva & J.R. Pujol-Luz. 2006. Descrição dos pupários de Manotes crassimanus James e Pedinocera longicornis Kertész (Diptera, Stratiomyidae, Pachygastrinae) da Ilha da Marambaia, Mangaratiba, Rio de Janeiro, Brasil. Revista Brasileira de Zoologia 23: 733–739.

McFadden, M.W. 1967. Soldier fly larvae in America north of Mexico. Proceedings of the United States National Museum 121: 1–72.

McFadden, M.W. 1972. The soldier flies of Canada and Alaska (Diptera: Stratiomyidae). - I. Beridinae, Sarginae and Clitellariinae. Canadian Entomologist 104: 531–562.

Narchuk, E.P. 1988. Chapter 36: Family Stratiomyidae. Pages 701–738 *in*: G.Ya. Bei-Bienko & G.C. Steykal (eds.), Keys to the insects of the European part of the USSR. Volume V. Diptera and Siphonaptera. Part I. Smithsonian Institution Libraries and The National Science Foundation, Washington, D.C.

Pimentel, T. & J.R. Pujol-Luz. 2002. A new species of the genus Raphiocera (Diptera, Stratiomyidae) from Santa Catarina, Brazil. Studia Dipterologica 9: 513–517.

Pujol-Luz, J.R. 1999. Notas sobre Chiromyza vittata Wiedemann ou mosca das raízes do café (Diptera, Stratiomyidae, Chiromyzinae). Revista Universidade Rural. Série Ciências da Vida 21: 161–163.

Pujol-Luz, J.R. & R. Xerez. 1999. The Larva of Chalcidomorphina aurata Enderlein, 1914 (Diptera, Stratiomyidae) from Ilha da Marambaia, Rio de Janeiro, Brazil. Proceedings of the Entomological Society of Washington 101: 295–299.

Pujol-Luz, J.R. & F.D. Vieira. 2000. A larva de Chiromyza vittata Wiedemann (Diptera: Stratiomyidae). Anais da Sociedade Entomológica do Brasil 29: 49–55.

Pujol-Luz, J.R. & F.M. Leite. 2001. Descrição do último ínstar larval e do pupário de Ptecticus testaceus (Fabr.) (Diptera: Stratiomyidae). Neotropical Entomology 30: 587–591.

Pujol-Luz, J.R. & N. Papavero. 1998. A new genus of Stratiomyini from Argentina (Diptera, Stratiomyidae). Annales de la Société Entomologique de France 34: 209–214.

Pujol-Luz, J.R., R. Xerez & G.G. Viana. 2004. Descrição do pupário de Raphiocera armata (Wiedemann) (Diptera, Stratiomyidae) da Ilha da Marambaia, Rio de Janeiro, Brasil. Revista Brasileira de Zoologia 21: 995–999.

Pujol-Luz, J.R. & C.V.A. Pujol-Luz. 2014. Chapter: Stratiomyidae. Pages 399–406 *in*: S. Roig-Juñent, L.E. Claps & J.J. Morrone (eds.), Biodiversidad de Artrópodos Argentinos. Volume 4. Editorial INSUE - UNT, San Miguel de Tucumán, Argentina.

Pujol-Luz, J.R., W.R. Lopes & G.G. Viana. 2016a. Description of the puparium and redescription of the imagoes of Chorophthalmyia brevicornis Lindner (Diptera: Stratiomyidae). Zootaxa 4121: 583–588.

Pujol-Luz, J.R., F.S.P. Godoi & K.B. Barros-Cordeiro. 2016b. Description of the puparium of Hermetia pulchra (Diptera: Stratiomyidae) from Brazil. Zootaxa 4205: 357–364.

Rozkošný, R. 1982. A biosystematics study of the European Stratiomyidae (Diptera). Introduction, Beridinae, Sarginae and Stratiomyinae. Volume 1. Dr. W. Junk, London.

Rozkošný, R. & D. Kovac. 1991. First description of the male and larva of Camptopteromyia fractipennis de Meijere from Malaysia (Diptera: Stratiomyidae). Entomology Scandinavian 22: 297–304.

Rozkošný, R. & D. Kovac. 1994a. Adults and larvae of two Ptecticus Loew from Peninsular Malaysia, (Diptera, Stratiomyidae). Tijdschrift voor Entomologie 137: 75–86.

Rozkošný, R. & D. Kovac. 1994b. A new species of Odontomyia Meigen (Insecta: Diptera: Stratiomyidae) from Sabah, Borneo. The Raffles Bulletin of Zoology 42: 859–867.

Rozkošný, R. & D. Kovac. 1994c. Adults and larvae of two Ptecticus Loew from Peninsular Malaysia, (Diptera, Stratiomyidae). Tijdschrift voor Entomologie 137: 75–86.

Rozkošný, R. & D. Kovac. 1997. Ptecticus minimus, a new species of Sarginae from West Malaysia including the description of its larva and puparium (Insecta: Diptera, Stratiomyidae). The Raffles Bulletin of Zoology 45: 39–51.

Rozkošný, R. & D. Kovac. 1998a. Descriptions of bamboo-inhabiting larvae and puparia of Oriental soldier flies Ptecticus brunettii and P. flavifemoratus (Diptera: Stratiomyidae: Sarginae) with observations on their biology. European Journal of Entomology 95: 65–86.

Rozkošný, R. & D. Kovac. 1998b. A new species of Pachygaster (Diptera: Stratiomyidae, Pachygasterinae) from West Malaysia and Thailand. Studia Dipterologica 5: 3–12.

Rozkošný, R. & D. Kovac. 2001. New male, larva and puparium of Odontomyia pulcherrima Brunetti (Insecta: Diptera: Stratiomyidae) from the Oriental Region. The Raffles Bulletin of Zoology 49: 101–108.

Rozkošný, R. & N. E. Woodley. 2010. A new genus and three new species of Oriental Oxycerini (Diptera: Stratiomyidae: Stratiomyinae) with notes on new generic synonyms in two other stratiomyine genera. Insect Systematics & Evolution 4: 275–294.

Viana, G.G., J.R. Pujol-Luz & R. Xerez. 2003. Descrição da larva e do pupário de Auloceromyia vespiformis Lindner (Diptera: Stratiomyidae, Clitellariinae). Neotropical Entomology 32: 69–74.

Wesenberg-Lund, C. 1943. Biologie der Siisswasserinsekten. Berlin-Wien.

Woodley, N.E. 2001. A World Catalog of the Stratiomyidae (Insecta: Diptera). Myia, Washington 11: 1–473.

Woodley, N.E. 2009. Chapter 38: Stratiomyidae (Soldier Flies). Pages 521–549 in: B.V. Brown, A. Borkent, J.M. Cumming, D.M. Wood, N.E. Woodley & M.A. Zumbado (eds.), Manual of Central American Diptera. NRC Research Press, Ottawa, Ontario, Canada.

Woodley, N.E. 2011. A World Catalog of the Stratiomyidae (Insecta: Diptera): A Supplement with Revisionary Notes and Errata. Myia 12: 379–415.

Xerez, R. & J.R. Pujol-Luz. 2001. Description of the larva of Vittiger schnusei Kertész, 1909 (Diptera: Stratiomyidae) from Ilha da Marambaia, Rio de Janeiro, Brazil. Studia Dipterologica 8: 337–341.

Xerez, R., J.R. Pujol-Luz & G.G. Viana 2002. Descrição da larva de Cosmariomyia argyrosticta Kertész e do pupário de Dactylodeictes lopesi Lindner (Diptera, Stratiomyidae). Revista Landa 19: 747–755.

Xerez, R., J.R. Pujol-Luz & G.G. Viana. 2003a. Descrição da larva de Popanomyia femoralis Kertész, 1909 e do pupário de Engicerus major Lindner, 1964 (Diptera, Stratiomyidae). Revista Brasileira de Entomologia 47: 403–408.

Xerez, R.; J.R. Pujol-Luz & G.G. Viana. 2003b. Description of the larva of Psephiocera modesta (Lindner, 1949) (Diptera: Stratiomyidae: Pachygastrinae). Studia Dipterologica 10: 189–193.

Xerez, R., G.G. Viana & J.R. Pujol-Luz. 2004. Description of the puparium of Cyphomyia aurifrons Wiedemann and Cyphomyia picta Schiner (Diptera, Stratiomyidae, Clitellariinae) from Ilha da Marambaia, Rio de Janeiro, Brazil. Revista Brasileira de Zoologia 21: 79–83.

Chapter 16.12

Family Syrphidae

Gil Felipe Gonçalves Miranda

Gerência da Região Administrativa Calha Norte III, Diretoria de Gestão e Monitoramento de Unidades de Conservação, Instituto de Desenvolvimento Florestal e da Biodiversidade, Monte Alegre, Pará, Brazil

Graham Rotheray

Department of Natural Sciences, National Museums, Scotland, Edinburgh, Great Britain

Chapter Outline

INTRODUCTION

The colloquial names for the Syrphidae, hoverflies, or flower flies, reflect two of their main attributes: hovering and flower visiting. In the former they demonstrate remarkable and unique flying abilities, and in the latter they play an important role in pollination (Rotheray & Gilbert, 2011). Another attribute are their colors that are sometimes cryptic in function (e.g., many *Copestylum* Macquart), but more frequently they are mimetic; indeed, included among syrphids are some of the most perfect mimics known of bees and wasps. A final attribute which sets syrphids apart is the adaptability of their immature stages. This is revealed by the exceptional range of form, habits, and behavior of their larvae and highly diversified lineages of phytophages, saprophages, and zoophages that exist in the family. Although focused on Palearctic genera, Table 3 in Rotheray & Gilbert (1999) gives an idea of the diversity of larval habitats.

Syrphid larvae can be distinguished from other fly larvae by the anus which is present on the anteroventral margin of the anal segment (Fig. 16.12.1), fused posterior breathing tubes (separate only in a few nonaquatic species) (Figs. 16.12.2, 3), and longitudinal grooves on the dorsum of the prothorax (Figs. 16.12.1, 4, 5; which are inconspicuous in some syrphines and microdontines) (Rotheray, 1993; Thompson et al., 2010). Pérez-Bañón et al. (2003) provided an identification key to Neotropical genera of "long-tailed" larvae. Thompson et al. (2010) provided a key to the genera of Central America with known immature stages.

The most diversified of aquatic syrphids are saprophages belonging to the Eristalinae, larvae of this group develop in water containing decaying organic matter and from which they filter feed microorganisms (Rotheray, 1993). The most well-known of these syrphids are the so-called "long-tailed" or "rat-tailed" larvae; these names refer to an extraordinarily long anal segment, which is drawn out to body length and more, and which acts as a protective sheath for the retractile breathing tube (Figs. 16.12.6, 7). With the apex anchored at the water surface, the breathing tube unravels from the body as the larva descends underwater, enabling it to breathe and feed at the same time. The tubular anterior respiratory organs of these larvae are also specialized, and in many taxa can be withdrawn into unique pockets formed by a fold in the outer integument, thus protecting them when the larva is burrowing. These larvae cannot swim; and when they are underwater, they hold on with crochets (Fig. 16.12.8) and prolegs. They can be found across a wide spectrum of water bodies from tree holes to water tanks of Bromeliaceae, in the bottom mud of rivers and lakes, among roots of water plants, within floating accumulations of decomposing plant matter, etc. They can also be found in wet, organic refuse from human activities; and where this material enters rivers and lakes, their presence has been used as an indicator of

Thorp and Covich's Freshwater Invertebrates, Fourth Edition. https://doi.org/10.1016/B978-0-12-804223-6.00041-X

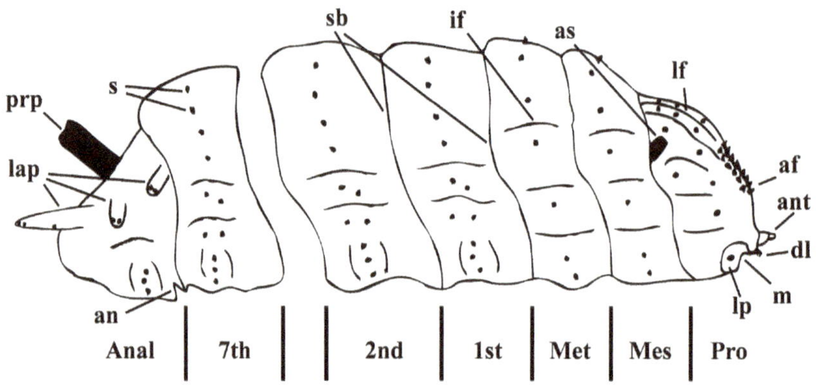

FIGURE 16.12.1 Syrphidae 3rd instar larva (abdominal segments 3 until 6 omitted), lateral view. Abbreviations: 1st, 2nd and 7th = first, second, and seventh abdominal segments; *af*, anterior fold; *an*, anus; *Anal*, anal segment; *ant*, antennomaxillary organ; *as*, anterior spiracle; *dl*, dorsal lip; *if*, integumental fold; *lap*, lappets; *lf*, longitudinal fold; *lp*, lateral lip; *m*, mouth; *Mes*, mesothorax; *Met*, metathorax; *Pro*, prothorax; *prp*, posterior respiratory process; *s*, sensillae; *sb*, segment boundary.

FIGURE 16.12.2 *Copestylum* sp. pupa, lateral view. Abbreviations: *prp*, posterior respiratory process; *ps*, pupal spiracles.

FIGURE 16.12.3 *Copestylum* sp. 3rd instar larva, dorsal view. Abbreviations: *lap*, lappets; *prp*, posterior respiratory process.

pollution (Pérez-Bañón et al. 2003, 2013). Occasionally they cause pseudomyiasis when a vertebrate accidentally ingests eggs or larvae (Perfetti et al., 2011 and references therein).

Although most aquatic syrphids are saprophages, aquatic predators are also known, such as certain species of *Ocyptamus sensu lato* that develop in the water tanks of bromeliads (Rotheray et al., 2000). Larvae move with the aid of ventral suckers and subdue their prey with venom before sucking out their contents. Prey are other aquatic insect larvae, including those of mosquitoes and larvae much larger than themselves. *Ocyptamus* has been revised recently, and distinctive species groups raised to generic rank. The aquatic species are part of a large and complex

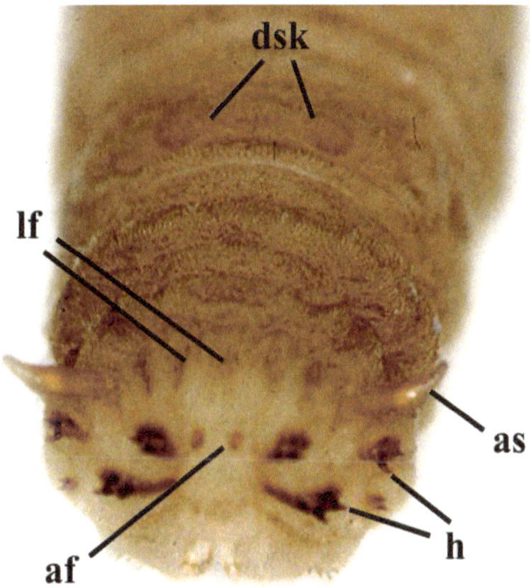

FIGURE 16.12.4 *Quichuana* sp. 3rd instar larva, anterior end, anterodorsal view. Abbreviations: *as*, anterior spiracle; *af*, anterior fold; *dsk*, differentiated disks (areas where pupal spiracles will emerge); *h*, hooks; *lf*, longitudinal fold.

clade, the *O. lepidus* group (Miranda et al., 2016), and another distinct lineage (*O. wulpianus* group), and since most of its species are unknown in their larval stages, it is likely that many will have larvae that are aquatic predators.

Traditionally, Syrphidae is divided into three subfamilies (Eristalinae, Microdontinae, and Syrphinae), but the Pipizinae have been added recently (Mengual et. al., 2015). Larvae of the three subfamilies with recognized or possible aquatic stages can be distinguished as follows (based on Thompson et al., 2010, and see Rotheray & Gilbert, 1999 for general habitus illustrations):

Eristalinae: Anal segment with up to three pairs of lappets (short, tapering, fleshy projections); mandibles hamulate or reduced and inconspicuous; thorax often as broad or broader than the abdomen; prolegs and crochets often present; anal segment often elongated with the posterior breathing tube at the apex.

Pipizinae and Syrphinae: Anal segment with no obvious lappets or only 1–2 pairs present; mandibles thin, inconspicuous, and without an apical hook or a serrated margin; thorax tapering and narrower than the abdomen; prolegs and crochets absent; posterior breathing tube not at the apex of the anal segment, but high up on this segment; colorful to whitish larvae. The Pipizinae are distinguished from the Syphinae by the presence of a pair of rounded, not tapering, projections on the apical margin of the anal segment.

Currently, aquatic larvae are only known in the Eristalinae and Syrphinae.

LIMITATIONS

Most immature stages of Neotropical Syrphidae are unknown, making it impossible to provide a key. This is because larvae being identified may belong to genera whose aquatic larvae have yet to be recognized or, they

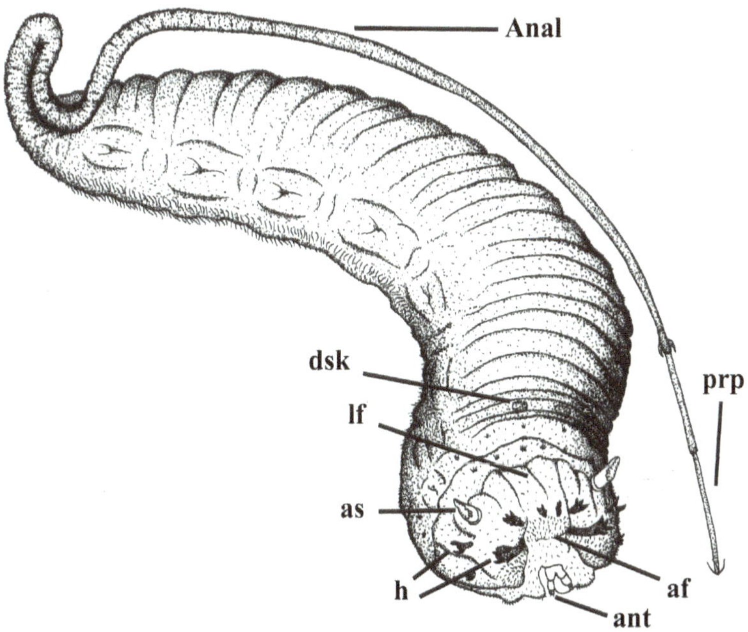

FIGURE 16.12.5 *Quichuana* sp. 3rd instar larva. Abbreviations: *af*, anterior fold; *Anal*, anal segment (extended); *ant*, antennomaxillary organ; *as*, anterior spiracle; *dsk*, differentiated disks (areas where pupal spiracles will emerge); *h*, hooks; *lf*, longitudinal fold; *prp*, posterior respiratory process.

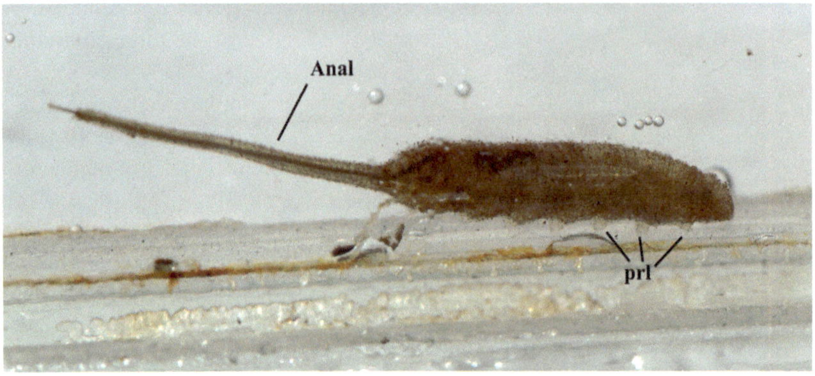

FIGURE 16.12.6 *Palpada* sp. 3rd instar larva, lateral view. Abbreviations: *Anal*, anal segment (extended); *prl*, prolegs.

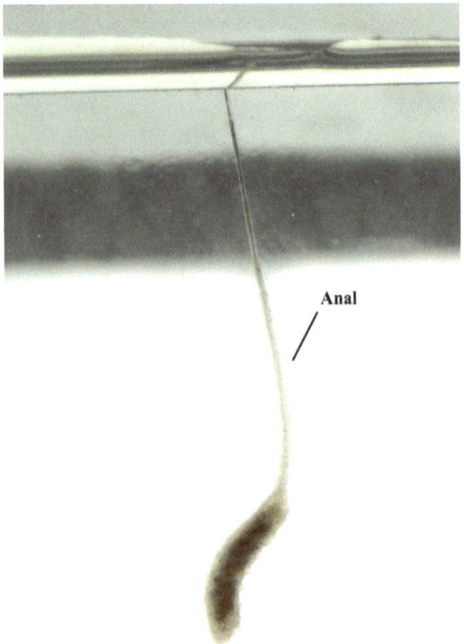

FIGURE 16.12.7 *Palpada* sp. 3rd instar larva. Abbreviations: *Anal*, anal segment (extended).

FIGURE 16.12.8 *Callicera* sp. 3rd instar larva, ventral view. Abbreviations: *crc*, crochets; *lp*, lateral lip; *m*, mouth.

belong to species within known aquatic genera, but have a different set of distinguishing characters.

TERMINOLOGY AND MORPHOLOGY

Terminology follows mostly Rotheray & Gilbert (1999), including sensilla numbering system, as presented in the "Results and Discussion" section, Figs. 2 and 3, and Table 2. Some terms follow Thompson et al. (2010).

In common with all Cyclorrhapha, larval Syrphidae have a head comprised of two sections: a fleshy external section, the **pseudocephalon**, and an inverted section comprising the sclerotized **cephalopharyngeal**/head skeleton, which houses a muscular pump for sucking in food. The **antennomaxillary** organs project forward at the apex of the **pseudocephalon**. The larval body is criss-crossed by a segmental pattern of grooves and folds, and the three thoracic and eight abdominal segments can be recognized by this pattern and the arrangement of sensilla on each segment, which are often accompanied by one or more setae (Fig. 16.12.1). The front of the larva varies in shape between the two aquatic groups, being tapered in Syrphinae and truncate in Eristalinae. In both groups, the body is covered in **vestiture** (integumental sculpturing and setae) which is variable in type, degree of sclerotization, size, shape, color, density, and distribution, and is a rich source of taxonomic characters (e.g., Ricarte et al., 2015). The anterior spiracles and the posterior respiratory process (Fig. 16.12.1), on the prothorax and anal segment respectively, and the pupal spiracles on the first abdominal segment of the **puparium** (Fig. 16.12.2), are each variable in size, shape, arrangement of the spiracular openings, etc., and are also major sources of taxonomic characters. The head skeleton (see Rotheray & Gilbert, 1999) also contains a useful set of taxonomic characters (Roberts, 1970; Rotheray & Gilbert, 1999; Rotheray et al., 2000).

There are three larval stages or instars that can be separated by their size. The larval stages can also be separated by the form of the posterior respiratory process and the head skeleton, which increase in relative size and complexity toward the third instar. Third instar larvae in the Eristalinae can additionally be recognized by a pair of differentiated **disks** on the dorsum of the first abdominal segment (Fig. 16.12.4). From these disks, the tubular, pupal spiracles emerge following pupariation. Most syrphines lack these disks and pupal spiracles, and third instar larvae are recognizable by their size and, at the end of feeding, by their empty hind guts, i.e., they defecate and lose the black material that has accumulated in the hind gut up to this point (Rotheray, 1993). The only known aquatic Syrphinae, *Ocyptamus luctuosus* (Bigot, 1883) and *O. wulpianus* (Lynch Arribálzaga, 1891), are maggots with ventral suckers on the posterior segments. Certain *Copestylum* larvae also develop in bromeliad water tanks and possess suckers (Rotheray et al., 2007), but they differ from aquatic syrphines in being truncate anteriorly, not tapering.

PREPARATION AND PRESERVATION

Usually the third instar larva and the puparium are preserved for study. Third instar larvae that have finished feeding are the best specimens for preservation (see "Terminology and Morphology"). Larvae are fixed in nearly boiling water for about 2–4 min, and preserved in 70% ethanol. Preserved larvae can be examined dry or wet; if dry they can be held by the fingers or placed on sand or glass beads in a solid watch glass. Puparia are best preserved dry in gelatin capsules, taking care to include any loosened sections caused by emergence of the adult, and the capsule pinned with the adult. Prior to examination, head skeletons need to be dissected from a preserved larva or a puparium. To dissect from a preserved larva, it is first cut across behind the head skeleton. The cut end or a whole puparium is then immersed in a hot solution of 10% KOH for about 5 min after which the specimen is placed in ethanol and the head skeleton removed using pins. The dissected head skeleton is immersed for a few minutes in acetic acid to halt further clearing, rinsed in distilled water and alcohol, and preserved in a microtube with glycerin (Rotheray, 1993; Rotheray & Gilbert, 1999; Rotheray et al., 2006).

ACKNOWLEDGMENTS

GFGM would like to thank for the fellowship PCI/INPA/CNPq (process #313083/2013-4), PRONEX 016/2006 (process #1437/2007) for the Leica stereomiscroscope, and FAPEAM/PRONEX (PRJ 12.39) and FAPEAM (062.03129/2012) for funding to study Diptera in the state of Amazonas, Brazil. GFGM also thanks PhD José Albertino Rafael for support.

REFERENCES

Mengual, X., G. Ståhls & S. Rojo. 2015. Phylogenetic relationships and taxonomic ranking of pipizine flower flies (Diptera: Syrphidae) with implications for the evolution of aphidophagy. Cladistics 31: 491–508.

Miranda, G.F.G., J.H. Skevington, S.A. Marshall & S. Kelso. 2016. The genus *Ocyptamus* Macquart (Diptera: Syrphidae): A molecular phylogenetic analysis. Arthropod Systematics and Phylogeny 74: 161–176.

Pérez-Bañón, C., G.E. Rotheray, G. Hancock, M.A. Marcos-García & M.A. Zumbado. 2003. Immature stages and breeding sites of some neotropical saprophagous syrphids (Diptera: Syrphidae). Anal of the Entomological Society of America 96: 458–471.

Pérez-Bañón, C., P. Hurtado, E. García-Gras & S. Rojo. 2013. SEM studies on immature stages of the drone flies (Diptera, Syrphidae): *Eristalis similis* (Fallen, 1817) and *Eristalis tenax* (Linnaeus, 1758). Microscopy Research and Technique 76: 853–861.

Perfetti, D.J.C., P.M. Moreno, M. Acosta & S. Bermúdez. 2011. First report of human intestinal pseudomiasis by *Eristalis tenax* (Diptera, Syrphidae) in urban semiarid area from Falcon state, Venezuela. Boletín de Malariología y Salud Ambiental 51: 225–228.

Ricarte, A., M.A. Marcos-García, E.G. Hancock & G.E. Rotheray. 2015. Neotropical *Copestylum* Macquart (Diptera: Syrphidae) breeding in fruits and flowers, including 7 new species. PLoS One 10: e0142441.

Roberts, M.J. 1970. The structure of the mouthparts of syrphid larvae (Diptera) in relation to feeding habits. Acta Zoologica 51: 43–65.

Rotheray, G.E. 1993. Colour guide to hoverfly larvae (Diptera, Syrphidae). Dipterists Digest 9: 1–156.

Rotheray, G.E., C. Dussaix, M.A. Marcos-García & C. Pérez-Bañón. 2006. The early stages of three palearctic species of saproxylic hoverflies (Syrphidae, Diptera). Micron 37: 73–80.

Rotheray, G.E., E.G. Hancock & M.A. Marcos-García. 2007. Neotropical *Copestylum* (Diptera, Syrphidae) breeding in bromeliads (Bromeliaceae) including 22 new species. Zoological Journal of the Linnean Society 150: 267–317.

Rotheray, G.E. & F. Gilbert. 1999. Phylogeny of Palaearctic Syrphidae (Diptera): evidence from larval stages. Zoological Journal of the Linnean Society 127: 1–112.

Rotheray, G.E. & F. Gilbert. 2011. The natural history of hoverflies. Forrest Text, Tresaith.

Rotheray, G.E., M. Zumbado, E.G. Hancock & F.C. Thompson. 2000. Remarkable aquatic predators in the genus *Ocyptamus* (Diptera, Syrphidae). Studia Dipterologica 7: 385–389.

Thompson, F.C., G.E. Rotheray & M. Zumbado. 2010. 53. Syrphidae (Flower flies). Pages 763–792 *in*: B.V. Brown, A. Borkent, J. M. Cumming, D. M. Wood, N. E. Woodley & M.A. Zumbado (eds.), Manual of Central American Diptera, Vol. 2. NRC Research Press, Ottawa, Canada.

Chapter 16.13

Family Periscelididae

Rosaly Ale-Rocha
Coordenação de Biodiversidade, Instituto Nacional de Pesquisas da Amazônia, Manaus, Amazonas, Brazil

Ruth Leila Ferreira-Keppler
Coordenação de Biodiversidade, Instituto Nacional de Pesquisas da Amazônia, Manaus, Amazonas, Brazil

Chapter Outline

INTRODUCTION

Periscelididae occur in all biogeographic regions, comprising about 120 species in 10 extant genera; however, the family is most diverse in the Neotropical Region. Knowledge on family diversity in this region has been greatly increased in the last 5 years, where the number of described species has almost doubled (Mathis & Rung, 2011; Ale-Rocha & Freitas, 2011; Ale-Rocha et al., 2014; Rung & Ale-Rocha, 2011; Freitas & Ale-Rocha, 2011; Mathis et al., 2012), with the Neotropical fauna currently comprising 55 species in nine genera.

Adults are small- to medium-sized flies (length: 2.5–5.0 mm), varying in color from brown or black to yellow, with convex and setose face and with hyaline or brown maculated wings. Adults can be recognized by having antenna with a cup-like pedicel that covers the postpedicel base and bears a lateral slit; the arista has dorsal and ventral branches; and they have the longitudinal fold in the wing cell *dm*. Additionally, the adults have the medial vertical seta usually present and the postocellar seta divergent or absent, with one to two frontoorbital setae present, one of them being reclinate; the thorax has one to two dorsocentral postsutural setae, and scutellum with one to two pairs of marginal setae; the wing has no true costal breaks, the cell *cup* usually is present, and *CuA2* is well developed or reduced.

Information about the immature stages is scarce, and most known larval forms are from the Nearctic and Palearctic. The eggs of three species of *Periscelis* Loew and one species each of *Cyamops* Melander and *Stenomicra* Coquillett have been briefly described (Duda, 1934; Sturtevant, 1954; Williams, 1939). Records of *Stenomicra* larvae in phytotelmata (Bromeliaceae, Gramineae, and Liliaceae) in Hawaii, including drawings and brief descriptions of the larva and pupa of *Stenomicra orientalis* Malloch, were provided by Williams (1939). *Stenomicra* larvae dwelling in phytotelmata native to Florida (USA) were reported feeding on *Wyeomyia* larvae (Fish, 1976). Teskey (1976) described the larva of *Periscelis flinti* (Malloch) (misidentified as *P. annulata* (Fallén) according to Papp, 1995), and Papp (1995) described the larva of *Periscelis annulata*. Campos et al. (2010) provided brief information on the life cycle of an unidentified *Stenomicra* and recorded four host plant species of the genus *Eryngium* L. (Apiaceae) for the immature, in Argentina. Papp (1998) provided data on larval development and on bionomics of adults of some Periscelidade species from Central Europe.

Several authors have discussed Periscelididae classification, whose monophyly is unresolved (Baptista & Mathis, 1994; Winkler et al., 2010), and this is reflected in different classification proposals (e.g., Hennig, 1969; Griffiths, 1972; McAlpine, 1978, 1983; Teskey, 1987; McAlpine, 1989). Previous authors suggested the very close relationship of Periscelididae and Aulacigastridae (Hennig, 1969; Baptista & Mathis, 1994), and some genera were allocated to one family or the other at different times.

Thorp and Covich's Freshwater Invertebrates, Fourth Edition. https://doi.org/10.1016/B978-0-12-804223-6.00042-1

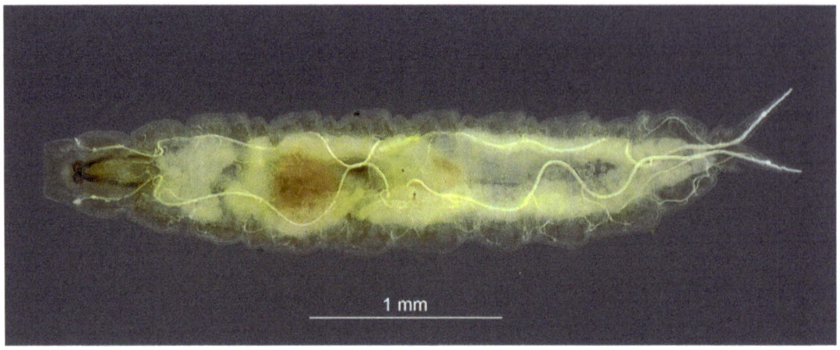

FIGURE 16.13.1 *Stenomicra* larva, dorsal view (Neotropical species).

We follow the classification of McAlpine (1978, 1983) that includes the genera *Cyamops, Planinasus*, and *Stenomicra* in Periscelididae; previously these genera were allocated in Aulacigastridae. Although Papp (1984) elevated Stenomicrinae to family status, the division of Periscelididae into two subfamilies, Periscelidinae and Stenomicrinae, has been widely accepted. Both subfamilies occur in the Neotropical region.

LIMITATIONS

Immature forms and their natural history are still poorly known, and the Neotropical taxa are mostly undescribed, so an identification key is not possible at this time. The larvae described to date belong to the genera *Periscelis* from the Nearctic Region (Papp, 1995) and *Stenomicra* from the Australasian/Oceanian Region (Williams, 1939).

TERMINOLOGY AND MORPHOLOGY

Eggs

The eggs are variable in form and coloration. In *Periscelis heegeri* Duda the eggs are white and smooth, roughly cylindrical, about 0.35 mm long, and half as wide as long (Duda, 1934). Eggs of *Periscelis annulata* and *P. wheeleri* are dark blackish-brown in color (Stutervant, 1926; 1954); eggs of *Cyamops nebulosus* Melander are white, with two short, thick anterodorsal filaments (Sturtevant, 1954); and eggs of *Stenomicra orientalis* are white, elongate, flattened, with finely reticulated surface and 0.65 mm long (Williams, 1939).

Larvae

The known larvae are very dissimilar, but in general, they have the posterior spiracles located on a bifurcated spiracular process with long and thin branches at the posterior region of the body. The third instar of *Periscelis* is strongly flattened dorsoventrally, body with numerous dorsal and lateral digitform or bulbous processes or spines and cuticle covered in fine spinules, subanal pads indistinct, anterior spiracles short, with 5(4) small lobes and cephalopharyngeal skeleton acalyptrate type with mouth hook very strong and extremely long. The third instar of *Stenomicra* has a fusiform, whitish colored to semitransparent body (Fig. 16.13.3), with the cephalic segment truncated anteriorly (Fig. 16.13.1); posterior spiracles are located on a bifurcated spiracular process with long and thin branches, and transverse rows of ventral spines on pseudopodia. The opening of the prothoracic spiracle is on a discrete, short, and digitiform process, with the spiracle ending in seven thin lobes; cephaloskeleton slender, black, and heavily sclerotized (Fig 16.13.2); and mouth hook robust and strongly inwardly curved (Fig. 16.13.4).

Puparium

The puparium of *Stenomicra* (Fig. 16.13.5) is yellow-orange colored, slightly robust, and widest on the middle, elliptical in profile, slightly tapering toward the apex but truncated apically, and ending in a bifid pointed process posteriorly.

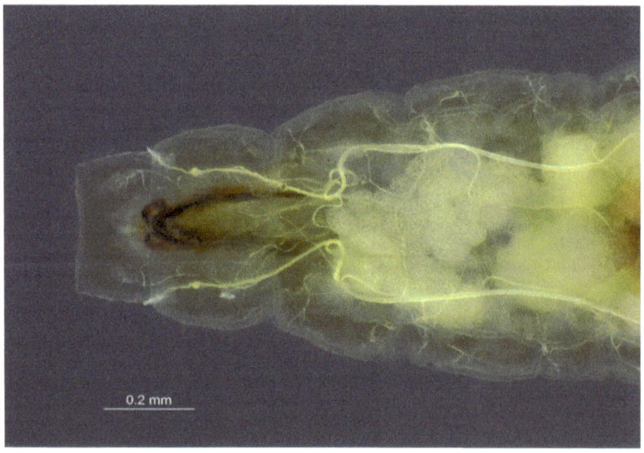

FIGURE 16.13.2 *Stenomicra* larva, dorsal view (Neotropical species).

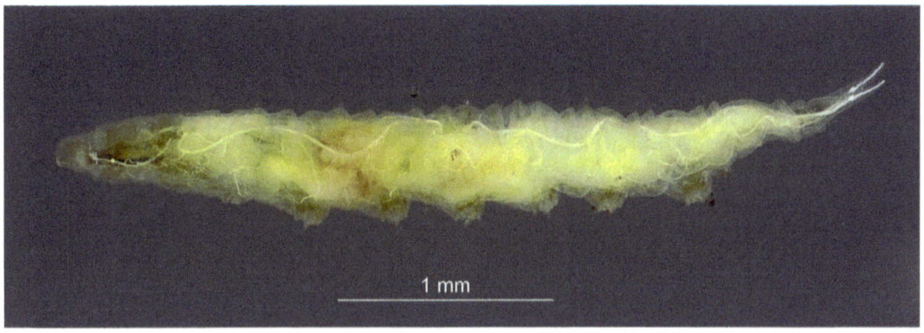

FIGURE 16.13.3 *Stenomicra* larva, lateral view (Neotropical species).

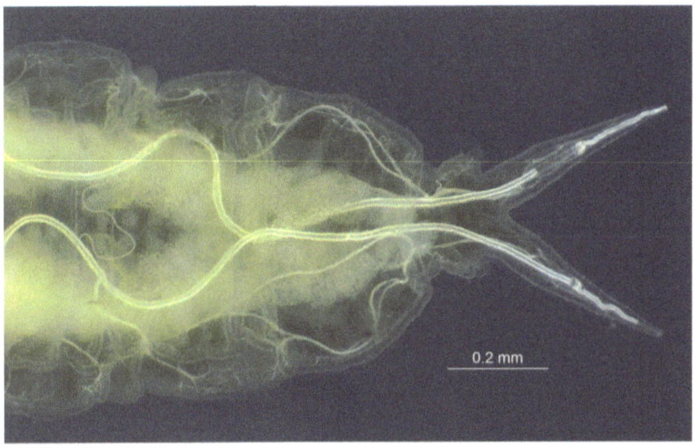

FIGURE 16.13.4 *Stenomicra* larva, dorsal view, posterior (Neotropical species).

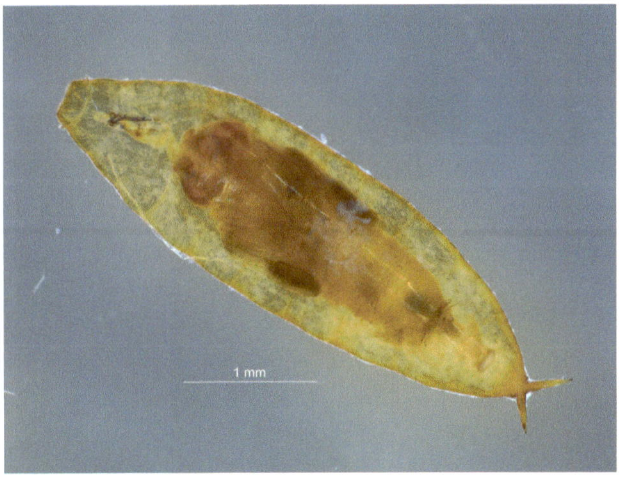

FIGURE 16.13.5 *Stenomicra* puparium, dorsal view, highlighting the adult in formation (Neotropical species).

MATERIAL PREPARATION AND PRESERVATION

Larvae are very delicate, with a thin cuticle and soft body. The common method for killing larvae using hot water has not proved to be very suitable, as verified by Papp (1995), because minute wood particles adhered to some of them. The use of cold seems more appropriate, and the larvae can be stored in a refrigerator at 5°C and then be preserved in 80% ethanol. To study the cephaloskeleton, larvae should be washed in water for 15 min, and placed in 6% potassium hydroxide (KOH) for maceration and clearing the sclerotized head structures. The maceration time varies from 15–30 min. Afterward, the specimen must be rinsed in acetic acid to wash out the KOH before be returned to 80% ethanol.

REFERENCES

Ale-Rocha, R. & G. Freitas. 2011. Revision of the Neotropical genus *Neoscutops* Malloch (Diptera: Periscelididae). Zootaxa 3016: 1–28.

Ale-Rocha, R., G. Freitas & W.N. Mathis. 2014. Revision of the Neotropical genus *Marbenia* Malloch (Diptera: Periscelididae). Zootaxa 3872: 365–375.

Baptista, A.R.P. & W.N. Mathis. 1994. A revision of new world *Cyamops* Melander (Diptera: Periscelididae). Smithsonian Contributions to Zoology 563: 1–25.

Campos, R.E., M.C. Gramajo & M. Lizarralde de Grosso. 2010. *Stenomicra* (Diptera: Opomyzoidea) in Argentina, with information on the biology of the genus. Revista de la Sociedad Entomológica Argentina 69: 281–285.

Courtney, G.W., B.J. Sinclair & R. Meier. 2000. Chapter 1.4: Morphology and terminology of Diptera larvae. Pages 85–161 *in*: L. Papp & B. Darvas (eds.), Contributions to a Manual of Palaearctic Diptera (With Special Reference to Flies Economic Importance). General and Applied Dipterology. Science Herald, Budapest.

Duda, O. 1934. Periscelidae. Pages 1–13 *in* E. Lindner (ed.), Die Fliegen Der Palaearktischen Region. Vol. 6. Stuttgart.

Fish, D. 1976. Structure and composition of the aquatic invertebrate community inhabiting epiphytic bromeliads in south Florida and the discovery of an insectivorous bromeliad. Ph.D. Thesis. University of Florida, Gainesville, 78 pp.

Freitas, G. & R. Ale-Rocha. 2011. Description of two new species of *Stenomicra* Coqquillett (Diptera, Periscelididae) from the Neotropics. Revista Brasileira de Entomologia 55: 348–354.

Griffiths, G.C.D. 1972. The phylogenetic classification of Diptera Cyclorrhapha, with special references to the structure of the male postabdomen. Series Entomologica 8: 1–340.

Hennig, W. 1969. Neue Gattungen und Arten der Acalyptratae. The Canadian Entomologist 101: 589–633.

Mathis, W.N. & A. Rung. 2011. World catalog and conspectus on the family Periscelididae (Diptera: Schizophora). Myia 12: 341–377.

Mathis, W.N., A. Rung & M. Kotrba. 2012. A revision of the genus *Planinasus* Cresson (Diptera, Periscelididae). Zookeys 255: 1–83.

McAlpine, D.K. 1978. Description and biology of a new genus of flies related to *Anthoclusia* and representing a new family (Diptera, Schizophora, Neurochaetidae). Annals of the Natal Museum 23: 273–295.

McAlpine, D.K. 1983. A new subfamily of Aulacigastridae (Diptera: Schizophora), with a discussion of Aulacigastridae classification. Australian Journal of Zoology 31: 55–78.

McAlpine, J.F. 1989. Chapter 116: Phylogeny and classification of the Muscomorpha. Pages 1397–1518 *in*: J.F. McAlpine & D.M. Wood (eds.), Manual of Neartic Diptera. Vol. 3. Research Branch, Agriculture Canada Monograph 32.

Papp, L. 1984. Chapter 10: Family Stenomicridae. Pages 61–62 *in*: Á. Soós & L. Papp (eds.), Catalogue of Palaearctic Diptera. Elsevier Science Publishers. Amsterdam.

Papp, L. 1995. Morphology of *Periscelis annulata* third instar larva and *Turanodinia tisciae* larva and puparium (Diptera: Periscelididae and Odiniidae). Acta Zoologica Academiae Scientiarum Hungaricae 41: 15–24.

Papp, L. 1998. Life-habits of the Central European species of Periscelididae (Diptera). Folia Entomologica Hungarica 59: 119–123.

Rung, A. & R. Ale-Rocha. 2011. New species of *Cyamops* (Diptera: Opomyzoidea: Periscelididae) from the old and new world tropics. Zoologia 28: 803–811.

Sturtevant, A.H. 1954. Nearctic flies of the family Periscelidae (Diptera) and certain Anthomyzidae referred to the family. Proceedings of the Unites States National Museum 103: 551–561.

Teskey, H.J. 1976. Diptera larvae associated with trees in North America. Memoirs of the Entomological Society of Canada 100: 1–53.

Teskey, H.J. 1987. Chapter 77: Family Aulacigastridae. Pages 891–894 *in*: J.F. McAlpine (ed.), Manual of Neartic Diptera. Vol. 2. Research Branch, Agriculture Canada Monograph 32.

Williams, F.X. 1939. Biological studies in Hawaiian water-loving insects. Part III. Diptera of flies. B. Asteiidae, Syrphidae, Dolichopodidae. Proceedings of the Hawaiian Entomological Society 10: 281–315.

Winkler, I.S., A. Rung & S.J. Scheffer. 2010. Hennig's orphangs revisited: Testing morphological hypotheses in the "Opomyzoidea". Molecular Phylogenetics and Evolution 54: 746–762.

Chapter 16.14

Family Ephydridae

Rosaly Ale-Rocha

Coordenação de Biodiversidade, Instituto Nacional de Pesquisas da Amazônia, Manaus, Amazonas, Brazil

Ruth Leila Ferreira-Keppler

Coordenação de Biodiversidade, Instituto Nacional de Pesquisas da Amazônia, Manaus, Amazonas, Brazil

Chapter Outline

INTRODUCTION

Adult Ephydridae are small- to medium-sized flies (body length 0.6−11.0 mm), brownish, whitish-grey, or shiny dark colored, with immature stages mainly aquatic or associated with aquatic habitats. This family is highly diverse, including around 1900 known species worldwide, distributed in 128 valid genera (Mathis & Zatwarnicki, 1995). Adults can be recognized by having large subcranial cavity, arista bare to pubescent or only dorsally pectinate, face flat to conspicuously protruding, scutum with one to five dorsocentral setae, two to three supraalar setae, 1 postalar seta and anepisternum setulose, *C* with humeral and subcostal breaks, *Sc* incomplete, cells *bm* and *dm* confluent, and crossvein *bm-cu* and cell cup absent. Larvae are variable in form and are similar in all three instars, except for the length and distribution of their spiracles (first instar metapneustic and second and third amphipneustic); they can measure up to 18 mm in the last instar.

This large and diverse family displays a considerable adaptive evolution and diverse feeding habits, although most species appear to be microphagous on autotrophic and heterotrophic organisms (Foote, 1983). Aquatic and semi-aquatic habits occur mainly in species whose larvae are collectors or filter feeders (Foote, 1995). Other species have terrestrial larvae, which can be leaf miners or feed on algae, detritus, decomposing carcasses, spider eggs, frog eggs, and midge larvae (Foote, 1983). Many species are adapted to inhospitable environments, such as highly alkaline hot springs, sulfur or saline ponds, or pools of crude oil (Foote, 1995). Some species are economically important because they are leaf miners of irrigated cereals such as rice and barley (Ferrar, 1987; Wirth et al., 1987). Except for a few species, the natural history of the family, especially the immature stages, is largely unknown.

The Neotropical Region contains about 69 genera and 600 species. At least 19 of these genera have aquatic immature stages or are associated with aquatic habitats, many of them from the subfamily Ephydrinae. Aquatic larvae of few Neotropical species belonging to 10 genera are known; however, few species have their immature stages described and bionomic data are rare for this region (Mathis, 2010). Examples of larval stages of this family as found in the Neotropical Region are shown in Figs. 16.14.1−5.

Zatwarnicki (1992) developed the most recent and widely accepted classification for Ephydridae, based on a phylogenetic analysis using adult characters, and recognized five subfamilies (Discomyzinae, Hydrelliinae, Gymnomyzinae, Ilytheinae, and Ephydrinae).

Discomyzinae adults have a flat or broadly convex face; pedicel with a long and conspicuous spine orbital setae proclinate and reclinate; and postsutural dorsocentral seta inconspicuous. Few species have known immature stages. The larvae are terrestrial or aquatic and have a cauliflower-shaped anterior spiracle (Zatwarnicki, 1992).

Hydrelliinae adults have a face that is usually flat or broadly convex, except Dryxini and some Atissini with a protruding face; other characteristics are a pedicel with a usually conspicuous spine except reduced in Dryxini;

Thorp and Covich's Freshwater Invertebrates, Fourth Edition. https://doi.org/10.1016/B978-0-12-804223-6.00043-3

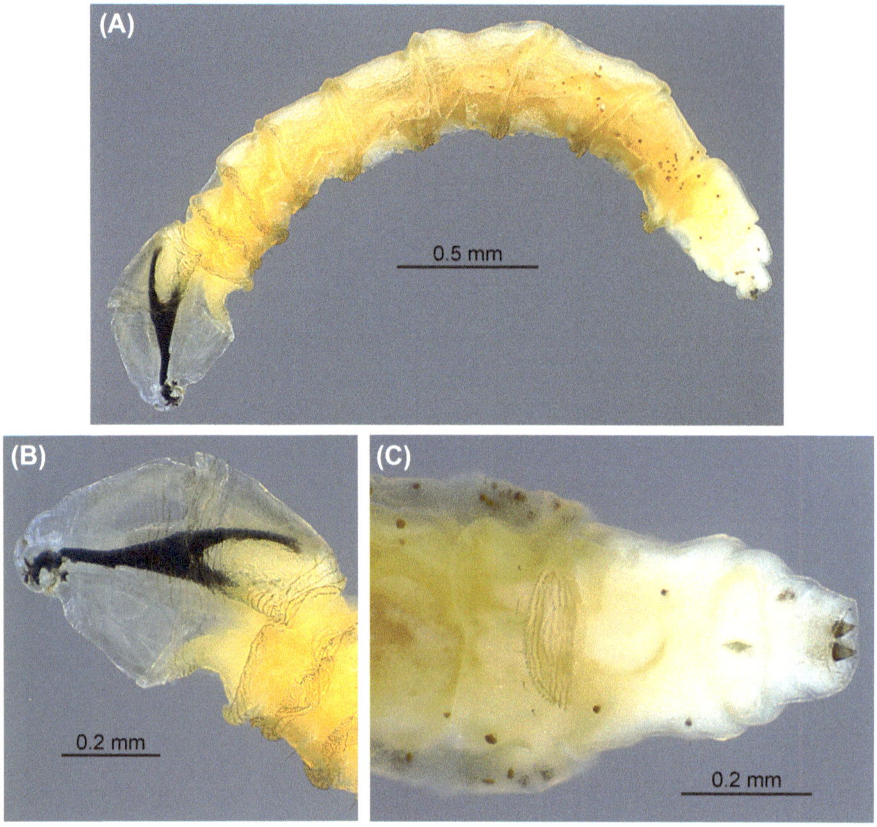

FIGURE 16.14.1 *Hydrellia* sp. larva: (A) habitus, lateral view; (B) cephaloskeleton, lateral view; and (C) abdomen apex, dorsal view.

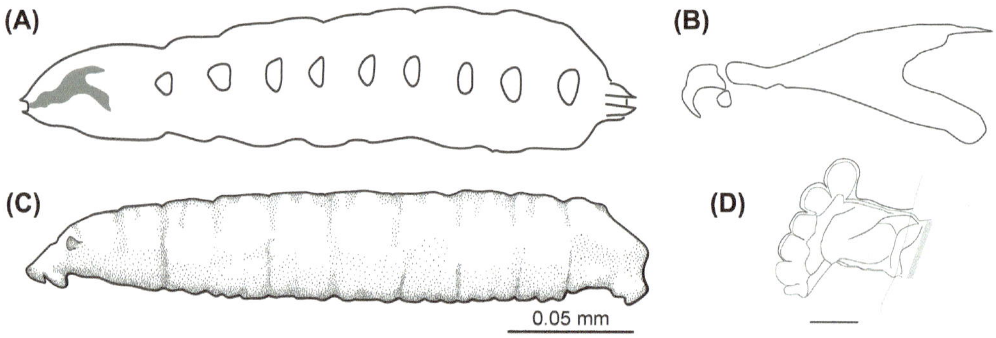

FIGURE 16.14.2 *Lemnaphila neotropica* Lizarralde de Grosso larva (from Lizarralde de Grosso, 1978) larva: (A) habitus, lateral view; (B) anterior region, cephaloskeleton, lateral view and *Discocerina obscurella* (Fallén) larva (from Foote & Eastin, 1974); (C) habitus, lateral view; and (D) anterior spiracle.

orbital setae proclinate and reclinate; and presutural and postsutural dorsocentral setae present. The larvae of all known species are phytophagous: most mine leaves and stems of aquatic plants, and some attack terrestrial plants too. The anterior spiracle of the larvae is usually reduced or absent, and posterior spiracle is inserted on a sharply pointed spine on the last abdominal segment (Figs. 16.14.1 A–C, 2 A, B).

Gymnomyzinae adults have a protruding face and usually carinate; pedicel usually with spine reduced or inconspicuous; orbital setae proclinate and reclinate; and postsutural dorsocentral seta reduced. The larvae have the anterior spiracle with several short branches, breathing tubes are undeveloped, and posterior spiracles are sessile (Figs 16.14.2 C, D, 3 F, G).

Ilytheinae adults have a protruding face; pedicel with an inconspicuous spine orbital setae proclinate and reclinate; and presutural dorsocentral setae are present, except in Hyadinini and some Phyligriini, where they are reduced. Larvae have anterior spiracle bifurcate at the base, with two

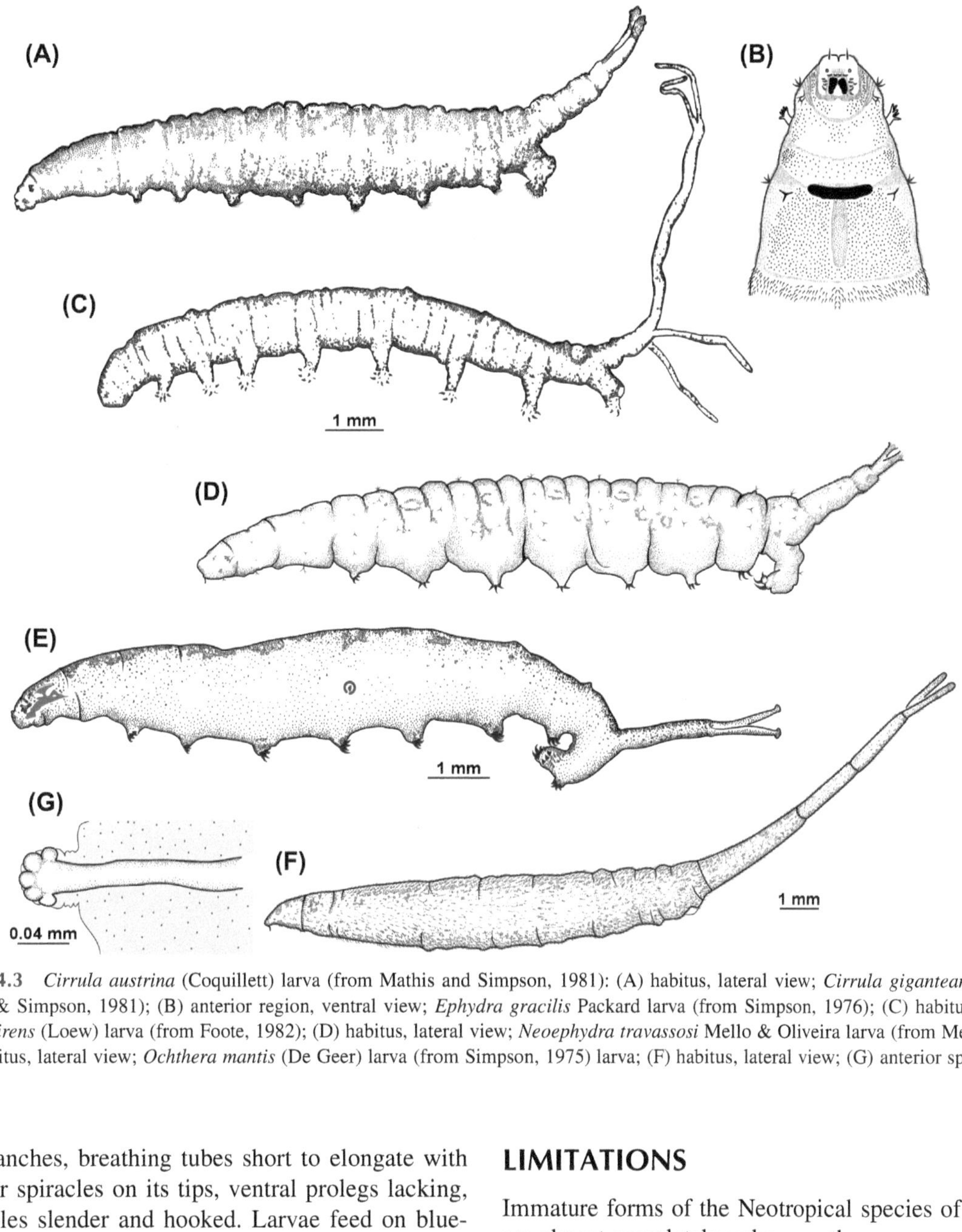

FIGURE 16.14.3 *Cirrula austrina* (Coquillett) larva (from Mathis and Simpson, 1981): (A) habitus, lateral view; *Cirrula gigantean* Cresson larva (from Mathis & Simpson, 1981); (B) anterior region, ventral view; *Ephydra gracilis* Packard larva (from Simpson, 1976); (C) habitus, lateral view; *Setacera atrovirens* (Loew) larva (from Foote, 1982); (D) habitus, lateral view; *Neoephydra travassosi* Mello & Oliveira larva (from Mello & Oliveira, 1992); (E) habitus, lateral view; *Ochthera mantis* (De Geer) larva (from Simpson, 1975) larva; (F) habitus, lateral view; (G) anterior spiracle.

elongate branches, breathing tubes short to elongate with the posterior spiracles on its tips, ventral prolegs lacking, and mandibles slender and hooked. Larvae feed on blue-green algae (Figs. 16.14.4 A–E).

Ephydrinae adults are characterized by the upper portion of the face projected anteriorly, central portion of the face setose in Scatellini and Ephydrini; pedicel usually with a spine; orbital setae lateroclinate; and pre- and postsutural dorsocentral setae usually present. Most of the known immature Ephydridae belong to this subfamily. Larvae have variable morphology, frequently have ventral prolegs, anterior spiracle clavate or with long, digitform branches, breathing tubes short to elongate with the posterior spiracles on its tips (Figs 16.14.3 A–E; 5 A–C).

LIMITATIONS

Immature forms of the Neotropical species of Ephydridae are almost completely unknown; thus, any current key will certainly not cover all the variation in size and morphology that is certainly found among species of each genus. Even a key to subfamilies is not useful since few genera of each subfamily have their larvae known and described. We recommend, therefore, that Neotropical members not be keyed beyond the family level until a series of comprehensive surveys of the larval stages are made. We recommend the works of Lizarralde de Grosso (1980) and Courtney & Merritt (2008) be considered a starting point for study of known aquatic larvae of Ephydridae genera occurring in the Neotropical Region. Since few Neotropical

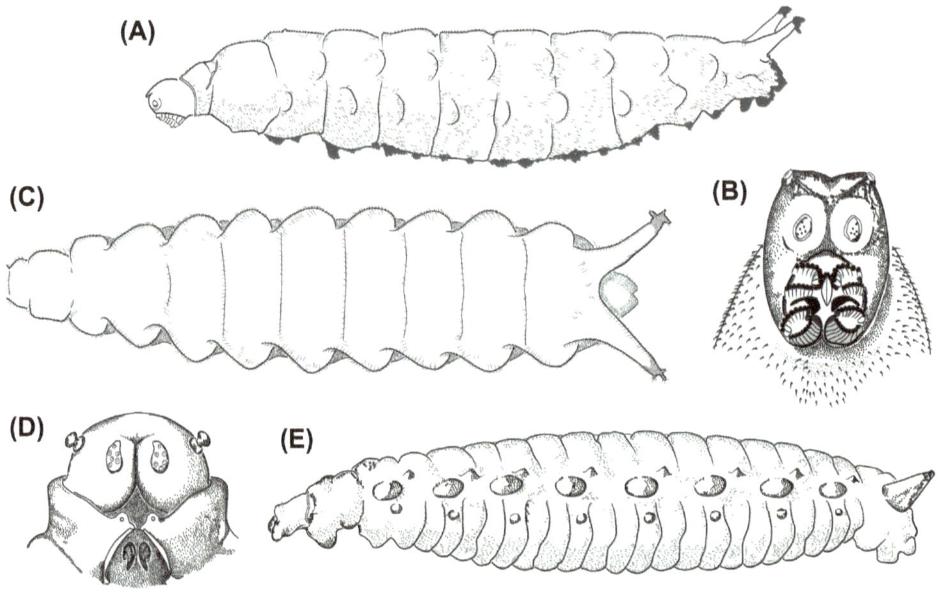

FIGURE 16.14.4 *Lytogaster excavata* (Sturtevant & Wheeler) larva, (A) habitus, lateral view (from Foote, 1981); (B) facial mask (from Foote, 1981); (C) habitus, dorsal view (from Foote, 1981); *Nostima approximata* Sturtevant and Wheeler larva, (D) facial mask (from Foote, 1983); and (E) habitus, lateral view (from Foote, 1983).

FIGURE 16.14.5 *Brachydeutera* sp. larva: (A) habitus, lateral view; (B) anterior region, cephaloskeleton, lateral view; and (C) posterior region, apex, dorsal view.

species have their larvae described, figures of non-Neotropical species were also used to illustrate some of the variation of larvae. A similar problem exists for the pupae.

TERMINOLOGY AND MORPHOLOGY

The terminology used follows Courtney et al. (2000). In general, third instar larvae of Ephydridae are characterized by having a cephaloskeleton without a ventral arch below the mandibles. The body is dorsoventrally flattened to cylindrical; integument usually is transparent (Figs. 16.14.1 A–C), with spicules or setulae of varying sizes and often extensively covering one or more segments, with the posterior abdominal segments covered with setae, spinules, or with setaceous (setae-bearing) tubercles on some segments (Figs. 16.14.3 D, 4 A, E, 5 A). Larvae can have up to eight pairs of abdominal prolegs present (Ephydrini)

(Fig. 16.14.3 A–E), each with two or three rows of hooks; last pair of prolegs larger, with opposable hooks; creeping welts sometimes present (*Notiphila caudata* Fállen) or lacking any apparent ambulatory processes. Anterior spiracle usually is present and positioned laterally (absent in Notiphilini); posterior spiracular plates are always distinctly separated; posterior spiracles usually are exposed on the apices of a bifurcate respiratory tube of variable length, at the end of the anal segment (Figs. 16.14.3 A–F, 4 A), rarely with posterior spiracles sessile on the anal division; posterior spiracle not spiniform (except in Hydrelliinae); and one or more of the spiracular slits nearly horizontal. Abdominal segments usually without lateral projections, but if present, these are slender (Wirth et al, 1987; Borkent & Rotheray, 2009; Mathis, 2010).

MATERIAL PREPARATION AND PRESERVATION

The most suitable method to examine relaxed specimens is to collect them straight into KAAD (a mixture of 1 vol. kerosene: 10 vol. 95% ethanol: 2 vol. glacial acetic acid: 1. vol. dioxane) or KAA (same mixture but without dioxane), leaving them in it for up to 24 h, and afterward placing them in 80% ethanol. An alternative is to place the larvae in hot but not boiling water, let it cool, and then transfer it to 80% ethanol. For the detailed examination of the cephaloskeleton and spiracles, the larva should be macerated in a cold 10% potassium hydroxide solution (KOH). The maceration time varies from 30–60 min, for small specimens, to several hours, for larger specimens. Afterward, the specimen must be rinsed in acetic acid to wash out the KOH. The specimen must then be dehydrated in absolute ethanol, cleared in clove oil, and then mounted on slides with Canada balsam, or transferred directly from the absolute ethanol to slides with Euparal.

ACKNOWLEDGMENTS

We are grateful to Rafael Freitas-Silva (National Institute of Amazonian Research) for the redrawing of the authorized original sources, to Leo R. R. Trindade and Laboratório de Insetos Aquáticos (INPA/MCTIC/COBIO) for the photographs, and to Conselho Nacional de Desenvolvimento Científico e Tecnológico (CNPq) from Brazil for support through the PQ fellowship.

REFERENCES

Borkent, A. & G. Rotheray. 2009. Chapter 7: Key to Diptera families - larvae. Pages 157–191 *in*: B.V. Brown, A. Borkent, J.M. Cumming, D.M. Wood, N.E. Woodley & M.A. Zumbado (eds.), Manual of Central American Diptera. Vol. 1. NRC Research Press, Ottawa, Ontario, Canada.

Courtney, G.W., B.J. Sinclair & R. Meier. 2000. Chapter 1.4: Morphology and terminology of Diptera larvae. Pages 85–161 *in*: L. Papp & B. Darvas (eds.), Contributions to a Manual of Palaearctic Diptera (with special reference to flies of economic importance). Science Herald Press, Budapest.

Courtney, G.W. & R.W. Merritt. 2008. Chapter 22: Aquatic Diptera. Part one. Larvae of aquatic Diptera. Pages 687–722 *in*: R.W. Merritt, K.W. Cummins & M.B. Berg (eds.), An Introduction to the Aquatic Insects of North America, Kendall/Hunt Publishing Co. Dubuque, Iowa.

Ferrar, P. 1987. A guide to the breeding habits and immature stages of Diptera Cyclorrhapha. E. J. Brill/Scandinavian Science Press, Leiden. 2 Vols. 907 pp.

Foote, B. A. 1981. Biology and immature stages of *Lytogaster excavata* a grazer of blue-green algae (Diptera: Ephydridae). Proceedings of the Entomological Society of Washington 83(2): 304–315.

Foote, B. A. 1982. Biology and immature stages of *Setacera atrovirens* a grazer of floating algal mats (Diptera: Ephydridae). Proceedings of the Entomological Society of Washington 84(4): 828–844.

Foote, B. A. 1983. Biology and immature stages of *Nostima approximata* (Diptera: Ephydridae), a grazer of blue-green algal genus *Oscillatoria*. Proceedings of the Entomological Society of Washington 85: 472–484.

Foote, B.A. 1995. Biology of shore flies. Pages 417–442 *in*: T.E. Mittler et al. (eds.), Annual Review of Entomology. Palo Alto, California.

Foote, B. A. & W. C. Eastin. 1974. Biology and immature stages of *Discocerina obscurella* (Diptera: Ephydridae). Proceedings of the Entomological Society of Washington 76(4): 401–408.

Lizarralde de Grosso, M. S. 1978. Nuevos aportes al conocimiento del género *Lemnaphila* Cresson (Diptera-Ephydridae). Neotropica 24(71): 13–20.

Lizarralde de Grosso, M. S. 1980. Estados preimaginales de Ephydridae (Diptera) Argentinos, con clave de Larvas. Physis 39: 55–60.

Mathis, W.N. 2010. Chapter 100: Ephydridae (shore flies). Pages 1211–1233 *in*: B.V. Brown, A. Borkent, J.M. Cumming, D.M. Wood, N.E. Woodley & M.A. Zumbado (eds.), Manual of Central American Diptera: Vol. 2. NRC Research Press, Ottawa, Ontario, Canada.

Mathis, W. N. & K. W. Simpson. 1981. Studies of Ephydrinae (Diptera: Ephydridae), V: The genera *Cirrula* Cresson and *Dimecoenia* Cresson in North America. Smithsonian Contributions to Zoology 329: 1–51.

Mathis, W. N. & T. Zatwarnicki. 1995. A world catalog of the shore flies (Diptera: Ephydridae). Memoirs on Entomology, International 4: 1–423.

Simpson, K. W. 1975. Biology and immature stages of three species of Nearctic *Ochthera* (Diptera: Ephydridae). Proceedings of the Entomological Society of Washington 77(1): 129–155.

Simpson, K. W. 1976. The mature larvae and puparia of *Ephydra (Halephydra) cinerea* Jones and *Ephydra (Hydropyrus) hians* Say (Diptera: Ephydridae). Proceedings of the Entomological Society of Washington 78(3): 263–269.

Wirth, W. W., W. N. Mathis & J.R. Vockeroth. 1987. Chapter 98: Ephydridae. Pages1027–1047 *in*: J.F. McAlpine (ed.), Manual of Nearctic Diptera Vol. 2. Monograph 28. Research Branch, Agriculture Canada, Ottawa.

Zatwarnicki, T. 1992. A new classification of Ephydridae based on phylogenetic reconstruction (Diptera: Cyclorrhapha). Genus 3: 65–119.

Taxonomic Index

'*Note*: Page numbers followed by "f" indicate figures and "t" indicate tables.'

B

H

M

O

P

T

U

V

W

X